STUDENT'S SOLUTIONS MANUAL

JEFFREY COLE

BEGINNING AND INTERMEDIATE ALGEBRA

FIFTH EDITION

Margaret L. Lial
American River College

John Hornsby
University of New Orleans

Terry McGinnis

Addison-Wesley
is an imprint of

PEARSON

Reproduced by Pearson Addison-Wesley from electronic files supplied by the author.

Publishing as Addison-Wesley, 75 Arlington Street, Boston, MA 02116.

ISBN-13: 978-0-321-71565-4
ISBN-10: 0-321-71565-9

1 2 3 4 5 6 BRR 14 13 12 11 10

Addison-Wesley
is an imprint of

PEARSON

www.pearsonhighered.com

Preface

This *Student's Solutions Manual* contains solutions to selected exercises in the text *Beginning and Intermediate Algebra, Fifth Edition* by Margaret L. Lial, John Hornsby, and Terry McGinnis. It contains solutions to the Now Try Exercises, the odd-numbered exercises in each section, all Relating Concepts exercises, as well as solutions to all the exercises in the review sections, the chapter tests, and the cumulative review sections.

This manual is a text supplement and should be read along *with* the text. You should read all exercise solutions in this manual because many concept explanations are given and then used in subsequent solutions. All concepts necessary to solve a particular problem are not reviewed for every exercise. If you are having difficulty with a previously covered concept, refer back to the section where it was covered for more complete help.

A significant number of today's students are involved in various outside activities, and find it difficult, if not impossible, to attend all class sessions; this manual should help meet the needs of these students. In addition, it is my hope that this manual's solutions will enhance the understanding of all readers of the material and provide insights to solving other exercises.

I appreciate feedback concerning errors, solution correctness or style, and manual style. Any comments may be sent directly to me at the address below, at jeff.cole@anokaramsey.edu, or in care of the publisher, Pearson Addison-Wesley.

I would like to thank Mary Johnson and Marv Riedesel, formerly of Inver Hills Community College, for their careful accuracy checking and valuable suggestions; Karen Hartpence, for creating the new art pieces; and the authors and Maureen O'Connor and Mary St. Thomas, of Pearson Addison-Wesley, for entrusting me with this project.

Jeffery A. Cole
Anoka-Ramsey Community College
11200 Mississippi Blvd. NW
Coon Rapids, MN 55433

Table of Contents

CHAPTER 1 THE REAL NUMBER SYSTEM

1.1 Fractions

1.1 Now Try Exercises

N1.

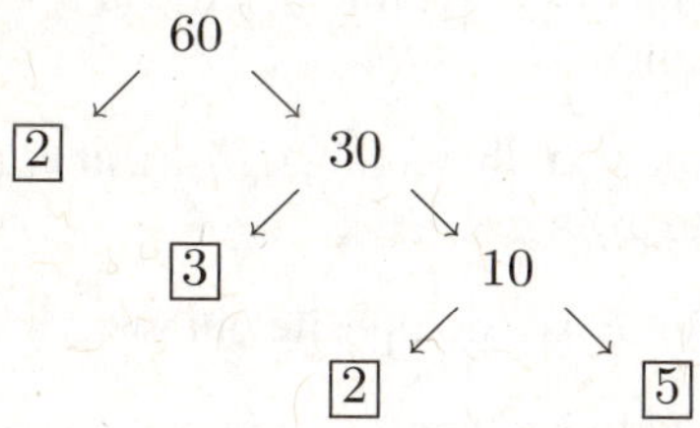

Writing 60 as the product of primes gives us

$$60 = 2 \cdot 2 \cdot 3 \cdot 5.$$

N2. $\dfrac{30}{42} = \dfrac{5 \cdot 6}{7 \cdot 6} = \dfrac{5 \cdot 1}{7 \cdot 1} = \dfrac{5}{7}$

N3. (a) $\dfrac{4}{7} \cdot \dfrac{5}{8} = \dfrac{4 \cdot 5}{7 \cdot 8}$ *Multiply numerators. Multiply denominators.*

$= \dfrac{4 \cdot 5}{7 \cdot 2 \cdot 4}$ *Factor.*

$= \dfrac{5}{14}$ *Write in lowest terms.*

(b)

$3\dfrac{2}{5} \cdot 6\dfrac{2}{3} = \dfrac{17}{5} \cdot \dfrac{20}{3}$ *Change both mixed numbers to improper fractions.*

$= \dfrac{17 \cdot 20}{5 \cdot 3}$ *Multiply numerators. Multiply denominators.*

$= \dfrac{17 \cdot 5 \cdot 4}{5 \cdot 3}$ *Factor.*

$= \dfrac{68}{3}$, or $22\dfrac{2}{3}$ *Write as a mixed number.*

N4. (a) $\dfrac{2}{7} \div \dfrac{8}{9} = \dfrac{2}{7} \cdot \dfrac{9}{8}$ *Multiply by the reciprocal of the second fraction.*

$= \dfrac{2 \cdot 3 \cdot 3}{7 \cdot 2 \cdot 4}$

$= \dfrac{9}{28}$

(b) $3\dfrac{3}{4} \div 4\dfrac{2}{7} = \dfrac{15}{4} \div \dfrac{30}{7}$ *Change both mixed numbers to improper fractions.*

$= \dfrac{15}{4} \cdot \dfrac{7}{30}$ *Multiply by the reciprocal of the second fraction.*

$= \dfrac{15 \cdot 7}{4 \cdot 2 \cdot 15}$

$= \dfrac{7}{8}$

N5. $\dfrac{1}{8} + \dfrac{3}{8} = \dfrac{1+3}{8}$ *Add numerators; denominator does not change.*

$= \dfrac{4}{8}$

$= \dfrac{1 \cdot 4}{2 \cdot 4}$ *Factor.*

$= \dfrac{1}{2}$

N6. (a) $\dfrac{5}{12} + \dfrac{3}{8}$

Since $12 = 2 \cdot 2 \cdot 3$ and $8 = 2 \cdot 2 \cdot 2$, the least common denominator must have three factors of 2 (from 8) and one factor of 3 (from 12), so it is $2 \cdot 2 \cdot 2 \cdot 3 = 24$.

Write each fraction with a denominator of 24.

$$\frac{5}{12} = \frac{5 \cdot 2}{12 \cdot 2} = \frac{10}{24} \quad \text{and} \quad \frac{3}{8} = \frac{3 \cdot 3}{8 \cdot 3} = \frac{9}{24}$$

Now add.

$$\frac{5}{12} + \frac{3}{8} = \frac{10}{24} + \frac{9}{24} = \frac{10+9}{24} = \frac{19}{24}$$

(b)

$3\dfrac{1}{4} + 5\dfrac{5}{8} = \dfrac{13}{4} + \dfrac{45}{8}$ *Change both mixed numbers to improper fractions.*

The least common denominator is 8, so write each fraction with a denominator of 8.

$$\frac{45}{8} \quad \text{and} \quad \frac{13}{4} = \frac{13 \cdot 2}{4 \cdot 2} = \frac{26}{8}$$

Now add.

$$\frac{13}{4} + \frac{45}{8} = \frac{26}{8} + \frac{45}{8} = \frac{26+45}{8}$$

$$= \frac{71}{8}, \text{ or } 8\frac{7}{8}$$

N7. (a) $\dfrac{5}{11} - \dfrac{2}{9}$

Since $11 = 11$ and $9 = 3 \cdot 3$, the least common denominator is $3 \cdot 3 \cdot 11 = 99$. Write each fraction with a denominator of 99.

$$\frac{5}{11} = \frac{5 \cdot 9}{11 \cdot 9} = \frac{45}{99} \quad \text{and} \quad \frac{2}{9} = \frac{2 \cdot 11}{9 \cdot 11} = \frac{22}{99}$$

Now subtract.

$$\frac{5}{11} - \frac{2}{9} = \frac{45}{99} - \frac{22}{99} = \frac{23}{99}$$

(b) $4\frac{1}{3} - 2\frac{5}{6} = \frac{13}{3} - \frac{17}{6}$ *Change each mixed number into an improper fraction.*

The least common denominator is 6. Write each fraction with a denominator of 6. $\frac{17}{6}$ remains unchanged, and

$$\frac{13}{3} = \frac{13 \cdot 2}{3 \cdot 2} = \frac{26}{6}.$$

Now subtract.

$$\frac{13}{3} - \frac{17}{6} = \frac{26}{6} - \frac{17}{6} = \frac{26 - 17}{6} = \frac{9}{6}$$

Now reduce.

$$\frac{9}{6} = \frac{3 \cdot 3}{2 \cdot 3} = \frac{3}{2}, \text{ or } 1\frac{1}{2}$$

N8. To find out how long each piece must be, divide the total length by the number of pieces.

$$10\frac{1}{2} \div 4 = \frac{21}{2} \div \frac{4}{1} = \frac{21}{2} \cdot \frac{1}{4} = \frac{21}{8}, \text{ or } 2\frac{5}{8}$$

Each piece should be $2\frac{5}{8}$ feet long.

N9. **(a)** In the circle graph, the sector for Other is the smallest, so Other had the least number of Internet users.

(b) As in Example 9(b) (using $\frac{1}{3}$ for $\frac{7}{20}$),

$$\tfrac{1}{3}(1000) \approx 333 \text{ million.}$$

(c) As in Example 9(c),

$$\frac{7}{20}(970) = \frac{7}{20} \cdot \frac{970}{1} = \frac{679}{2}$$
$$= 339\frac{1}{2} \text{ million, or } 339{,}500{,}000.$$

1.1 Section Exercises

1. True; the number above the fraction bar is called the numerator and the number below the fraction bar is called the denominator.

3. False; this is an improper fraction. Its value is 1.

5. False; the fraction $\frac{13}{39}$ can be written in lowest terms as $\frac{1}{3}$ since $\frac{13}{39} = \frac{13 \cdot 1}{13 \cdot 3} = \frac{1}{3}$.

7. False; *product* refers to multiplication, so the product of 10 and 2 is 20. The *sum* of 10 and 2 is 12.

9. Since 19 has only itself and 1 as factors, it is a prime number.

11. $30 = 2 \cdot 15$
$= 2 \cdot 3 \cdot 5$

Since 30 has factors other than itself and 1, it is a composite number.

13. $64 = 2 \cdot 32$
$= 2 \cdot 2 \cdot 16$
$= 2 \cdot 2 \cdot 2 \cdot 8$
$= 2 \cdot 2 \cdot 2 \cdot 2 \cdot 4$
$= 2 \cdot 2 \cdot 2 \cdot 2 \cdot 2 \cdot 2$

Since 64 has factors other than itself and 1, it is a composite number.

15. As stated in the text, the number 1 is neither prime nor composite, by agreement.

17. $57 = 3 \cdot 19$, so 57 is a composite number.

19. Since 79 has only itself and 1 as factors, it is a prime number.

21. $124 = 2 \cdot 62$
$= 2 \cdot 2 \cdot 31$,

so 124 is a composite number.

23. $500 = 2 \cdot 250$
$= 2 \cdot 2 \cdot 125$
$= 2 \cdot 2 \cdot 5 \cdot 25$
$= 2 \cdot 2 \cdot 5 \cdot 5 \cdot 5$,

so 500 is a composite number.

25. $3458 = 2 \cdot 1729$
$= 2 \cdot 7 \cdot 247$
$= 2 \cdot 7 \cdot 13 \cdot 19$

Since 3458 has factors other than itself and 1, it is a composite number.

27. $\frac{8}{16} = \frac{1 \cdot 8}{2 \cdot 8} = \frac{1}{2}$

29. $\frac{15}{18} = \frac{3 \cdot 5}{3 \cdot 6} = \frac{5}{6}$

31. $\frac{64}{100} = \frac{4 \cdot 16}{4 \cdot 25} = \frac{16}{25}$

33. $\frac{18}{90} = \frac{1 \cdot 18}{5 \cdot 18} = \frac{1}{5}$

35. $\frac{144}{120} = \frac{6 \cdot 24}{5 \cdot 24} = \frac{6}{5}$

37. $\frac{16}{24} = \frac{2 \cdot 8}{3 \cdot 8} = \frac{2}{3}$

Therefore, **C** is correct.

39. $\frac{4}{5} \cdot \frac{6}{7} = \frac{4 \cdot 6}{5 \cdot 7} = \frac{24}{35}$

41. $\frac{2}{3} \cdot \frac{15}{16} = \frac{2 \cdot 15}{3 \cdot 16} = \frac{2 \cdot 3 \cdot 5}{3 \cdot 2 \cdot 8} = \frac{5}{8}$

43. $\frac{1}{10} \cdot \frac{12}{5} = \frac{1 \cdot 12}{10 \cdot 5} = \frac{1 \cdot 2 \cdot 6}{2 \cdot 5 \cdot 5} = \frac{6}{25}$

45. $\frac{15}{4} \cdot \frac{8}{25} = \frac{15 \cdot 8}{4 \cdot 25}$

$= \frac{3 \cdot 5 \cdot 4 \cdot 2}{4 \cdot 5 \cdot 5}$

$= \frac{3 \cdot 2}{5}$

$= \frac{6}{5}$, or $1\frac{1}{5}$

47. $21 \cdot \frac{3}{7} = \frac{21 \cdot 3}{1 \cdot 7}$

$= \frac{3 \cdot 7 \cdot 3}{1 \cdot 7}$

$= \frac{3 \cdot 3}{1} = 9$

49. $3\frac{1}{4} \cdot 1\frac{2}{3}$

Change both mixed numbers to improper fractions.

$3\frac{1}{4} = 3 + \frac{1}{4} = \frac{12}{4} + \frac{1}{4} = \frac{13}{4}$

$1\frac{2}{3} = 1 + \frac{2}{3} = \frac{3}{3} + \frac{2}{3} = \frac{5}{3}$

$3\frac{1}{4} \cdot 1\frac{2}{3} = \frac{13}{4} \cdot \frac{5}{3}$

$= \frac{13 \cdot 5}{4 \cdot 3}$

$= \frac{65}{12}$, or $5\frac{5}{12}$

51. $2\frac{3}{8} \cdot 3\frac{1}{5}$

Change both mixed numbers to improper fractions.

$2\frac{3}{8} = 2 + \frac{3}{8} = \frac{16}{8} + \frac{3}{8} = \frac{19}{8}$

$3\frac{1}{5} = 3 + \frac{1}{5} = \frac{15}{5} + \frac{1}{5} = \frac{16}{5}$

$2\frac{3}{8} \cdot 3\frac{1}{5} = \frac{19}{8} \cdot \frac{16}{5}$

$= \frac{19 \cdot 16}{8 \cdot 5}$

$= \frac{19 \cdot 2 \cdot 8}{8 \cdot 5}$

$= \frac{38}{5}$, or $7\frac{3}{5}$

53. $\frac{5}{4} \div \frac{3}{8} = \frac{5}{4} \cdot \frac{8}{3}$ *Multiply by the reciprocal of the second fraction.*

$= \frac{5 \cdot 8}{4 \cdot 3}$

$= \frac{5 \cdot 4 \cdot 2}{4 \cdot 3}$

$= \frac{5 \cdot 2}{3}$

$= \frac{10}{3}$, or $3\frac{1}{3}$

55. $\frac{32}{5} \div \frac{8}{15} = \frac{32}{5} \cdot \frac{15}{8}$ *Multiply by the reciprocal of the second fraction.*

$= \frac{32 \cdot 15}{5 \cdot 8}$

$= \frac{8 \cdot 4 \cdot 3 \cdot 5}{1 \cdot 5 \cdot 8}$

$= \frac{4 \cdot 3}{1} = 12$

57. $\frac{3}{4} \div 12 = \frac{3}{4} \cdot \frac{1}{12}$ *Multiply by the reciprocal of 12.*

$= \frac{3 \cdot 1}{4 \cdot 12}$

$= \frac{3 \cdot 1}{4 \cdot 3 \cdot 4}$

$= \frac{1}{4 \cdot 4} = \frac{1}{16}$

59. $6 \div \frac{3}{5} = \frac{6}{1} \cdot \frac{5}{3}$ *Multiply by the reciprocal of the second fraction.*

$= \frac{6 \cdot 5}{1 \cdot 3}$

$= \frac{2 \cdot 3 \cdot 5}{1 \cdot 3}$

$= \frac{2 \cdot 5}{1} = 10$

61. $6\frac{3}{4} \div \frac{3}{8}$

Change the first number to an improper fraction.

$6\frac{3}{4} = 6 + \frac{3}{4} = \frac{24}{4} + \frac{3}{4} = \frac{27}{4}$

$6\frac{3}{4} \div \frac{3}{8} = \frac{27}{4} \cdot \frac{8}{3}$ *Multiply by the reciprocal of the second fraction.*

$= \frac{27 \cdot 8}{4 \cdot 3}$

$= \frac{3 \cdot 9 \cdot 2 \cdot 4}{4 \cdot 3}$

$= \frac{9 \cdot 2}{1} = 18$

63. $2\frac{1}{2} \div 1\frac{5}{7}$

Change both mixed numbers to improper fractions.

$$2\frac{1}{2} = 2 + \frac{1}{2} = \frac{4}{2} + \frac{1}{2} = \frac{5}{2}$$

$$1\frac{5}{7} = 1 + \frac{5}{7} = \frac{7}{7} + \frac{5}{7} = \frac{12}{7}$$

$$2\frac{1}{2} \div 1\frac{5}{7} = \frac{5}{2} \div \frac{12}{7}$$

$$= \frac{5}{2} \cdot \frac{7}{12}$$ *Multiply by the reciprocal of the second fraction.*

$$= \frac{5 \cdot 7}{2 \cdot 12}$$

$$= \frac{35}{24}, \text{ or } 1\frac{11}{24}$$

65. $2\frac{5}{8} \div 1\frac{15}{32}$

Change both mixed numbers to improper fractions.

$$2\frac{5}{8} = 2 + \frac{5}{8} = \frac{16}{8} + \frac{5}{8} = \frac{21}{8}$$

$$1\frac{15}{32} = 1 + \frac{15}{32} = \frac{32}{32} + \frac{15}{32} = \frac{47}{32}$$

$$2\frac{5}{8} \div 1\frac{15}{32} = \frac{21}{8} \div \frac{47}{32}$$

$$= \frac{21}{8} \cdot \frac{32}{47}$$

$$= \frac{21 \cdot 32}{8 \cdot 47}$$

$$= \frac{21 \cdot 8 \cdot 4}{8 \cdot 47}$$

$$= \frac{21 \cdot 4}{47}$$

$$= \frac{84}{47}, \text{ or } 1\frac{37}{47}$$

67. A common denominator for $\frac{p}{q}$ and $\frac{r}{s}$ must be a multiple of both denominators, q and s. Such a number is $q \cdot s$. Therefore, **A** is correct.

69. $\frac{7}{15} + \frac{4}{15} = \frac{7+4}{15} = \frac{11}{15}$

71. $\frac{7}{12} + \frac{1}{12} = \frac{7+1}{12}$

$$= \frac{8}{12}$$

$$= \frac{2 \cdot 4}{3 \cdot 4} = \frac{2}{3}$$

73. $\frac{5}{9} + \frac{1}{3}$

Since $9 = 3 \cdot 3$, and 3 is prime, the LCD (least common denominator) is $3 \cdot 3 = 9$.

$$\frac{1}{3} = \frac{1}{3} \cdot \frac{3}{3} = \frac{3}{9}$$

Now add the two fractions with the same denominator.

$$\frac{5}{9} + \frac{1}{3} = \frac{5}{9} + \frac{3}{9} = \frac{8}{9}$$

75. $\frac{3}{8} + \frac{5}{6}$

Since $8 = 2 \cdot 2 \cdot 2$ and $6 = 2 \cdot 3$, the LCD is $2 \cdot 2 \cdot 2 \cdot 3 = 24$.

$$\frac{3}{8} = \frac{3}{8} \cdot \frac{3}{3} = \frac{9}{24} \text{ and } \frac{5}{6} \cdot \frac{4}{4} = \frac{20}{24}$$

Now add fractions with the same denominator.

$$\frac{3}{8} + \frac{5}{6} = \frac{9}{24} + \frac{20}{24} = \frac{29}{24}, \text{ or } 1\frac{5}{24}$$

77. $3\frac{1}{8} + 2\frac{1}{4}$

$$3\frac{1}{8} = 3 + \frac{1}{8} = \frac{24}{8} + \frac{1}{8} = \frac{25}{8}$$

$$2\frac{1}{4} = 2 + \frac{1}{4} = \frac{8}{4} + \frac{1}{4} = \frac{9}{4}$$

$$3\frac{1}{8} + 2\frac{1}{4} = \frac{25}{8} + \frac{9}{4}$$

Since $8 = 2 \cdot 2 \cdot 2$ and $4 = 2 \cdot 2$, the LCD is $2 \cdot 2 \cdot 2$ or 8.

$$3\frac{1}{8} + 2\frac{1}{4} = \frac{25}{8} + \frac{9 \cdot 2}{4 \cdot 2}$$

$$= \frac{25}{8} + \frac{18}{8}$$

$$= \frac{43}{8}, \text{ or } 5\frac{3}{8}$$

79. $3\frac{1}{4} + 1\frac{4}{5}$

$$3\frac{1}{4} = 3 + \frac{1}{4} = \frac{12}{4} + \frac{1}{4} = \frac{13}{4}$$

$$1\frac{4}{5} = 1 + \frac{4}{5} = \frac{5}{5} + \frac{4}{5} = \frac{9}{5}$$

Since $4 = 2 \cdot 2$, and 5 is prime, the LCD is $2 \cdot 2 \cdot 5 = 20$.

$$3\frac{1}{4} + 1\frac{4}{5} = \frac{13 \cdot 5}{4 \cdot 5} + \frac{9 \cdot 4}{5 \cdot 4}$$

$$= \frac{65}{20} + \frac{36}{20}$$

$$= \frac{101}{20}, \text{ or } 5\frac{1}{20}$$

81. $\frac{7}{9}-\frac{2}{9}=\frac{7-2}{9}=\frac{5}{9}$

83. $\frac{13}{15}-\frac{3}{15}=\frac{13-3}{15}$

$=\frac{10}{15}$

$=\frac{2\cdot 5}{3\cdot 5}=\frac{2}{3}$

85. $\frac{7}{12}-\frac{1}{3}$

Since $12=4\cdot 3$ (12 is a multiple of 3), the LCD is 12.

$$\frac{1}{3}\cdot\frac{4}{4}=\frac{4}{12}$$

Now subtract fractions with the same denominator.

$$\frac{7}{12}-\frac{1}{3}=\frac{7}{12}-\frac{4}{12}=\frac{3}{12}=\frac{1\cdot 3}{4\cdot 3}=\frac{1}{4}$$

87. $\frac{7}{12}-\frac{1}{9}$

Since $12=2\cdot 2\cdot 3$ and $9=3\cdot 3$, the LCD is $2\cdot 2\cdot 3\cdot 3=36$.

$$\frac{7}{12}=\frac{7}{12}\cdot\frac{3}{3}=\frac{21}{36} \text{ and } \frac{1}{9}\cdot\frac{4}{4}=\frac{4}{36}$$

Now subtract fractions with the same denominator.

$$\frac{7}{12}-\frac{1}{9}=\frac{21}{36}-\frac{4}{36}=\frac{17}{36}$$

89. $4\frac{3}{4}-1\frac{2}{5}$

$4\frac{3}{4}=4+\frac{3}{4}=\frac{16}{4}+\frac{3}{4}=\frac{19}{4}$

$1\frac{2}{5}=1+\frac{2}{5}=\frac{5}{5}+\frac{2}{5}=\frac{7}{5}$

Since $4=2\cdot 2$, and 5 is prime, the LCD is $2\cdot 2\cdot 5=20$.

$$4\frac{3}{4}-1\frac{2}{5}=\frac{19\cdot 5}{4\cdot 5}-\frac{7\cdot 4}{5\cdot 4}$$
$$=\frac{95}{20}-\frac{28}{20}$$
$$=\frac{67}{20}, \text{ or } 3\frac{7}{20}$$

91. $6\frac{1}{4}-5\frac{1}{3}$

$6\frac{1}{4}=6+\frac{1}{4}=\frac{24}{4}+\frac{1}{4}=\frac{25}{4}$

$5\frac{1}{3}=5+\frac{1}{3}=\frac{15}{3}+\frac{1}{3}=\frac{16}{3}$

Since $4=2\cdot 2$, and 3 is prime, the LCD is $2\cdot 2\cdot 3=12$.

$$6\frac{1}{4}-5\frac{1}{3}=\frac{25}{4}-\frac{16}{3}$$
$$=\frac{25\cdot 3}{4\cdot 3}-\frac{16\cdot 4}{3\cdot 4}$$
$$=\frac{75}{12}-\frac{64}{12}$$
$$=\frac{11}{12}$$

93. Multiply the number of cups of water per serving by the number of servings.

$$\frac{3}{4}\cdot 8=\frac{3}{4}\cdot\frac{8}{1}$$
$$=\frac{3\cdot 8}{4\cdot 1}$$
$$=\frac{3\cdot 2\cdot 4}{4\cdot 1}$$
$$=\frac{3\cdot 2}{1}=6 \text{ cups}$$

For 8 microwave servings, 6 cups of water will be needed.

95. The difference in length is found by subtracting.

$$3\frac{1}{4}-2\frac{1}{8}=\frac{13}{4}-\frac{17}{8}$$
$$=\frac{13\cdot 2}{4\cdot 2}-\frac{17}{8} \quad LCD = 8$$
$$=\frac{26}{8}-\frac{17}{8}$$
$$=\frac{9}{8}, \text{ or } 1\frac{1}{8}$$

The difference is $1\frac{1}{8}$ inches.

97. The difference between the two measures is found by subtracting, using 16 as the LCD.

$$\frac{3}{4}-\frac{3}{16}=\frac{3\cdot 4}{4\cdot 4}-\frac{3}{16}$$
$$=\frac{12}{16}-\frac{3}{16}$$
$$=\frac{12-3}{16}=\frac{9}{16}$$

The difference is $\frac{9}{16}$ inch.

99. The perimeter is the sum of the measures of the 5 sides.

$$\begin{aligned}&196+98\frac{3}{4}+146\frac{1}{2}+100\frac{7}{8}+76\frac{5}{8}\\&=196+98\frac{6}{8}+146\frac{4}{8}+100\frac{7}{8}+76\frac{5}{8}\\&=196+98+146+100+76+\frac{6+4+7+5}{8}\\&=616+\frac{22}{8}\quad\left(\frac{22}{8}=2\frac{6}{8}=2\frac{3}{4}\right)\\&=618\frac{3}{4}\text{ feet}\end{aligned}$$

The perimeter is $618\frac{3}{4}$ feet.

101. Divide the total board length by 3.

$$\begin{aligned}15\frac{5}{8}\div 3&=\frac{125}{8}\div\frac{3}{1}\\&=\frac{125}{8}\cdot\frac{1}{3}\\&=\frac{125\cdot 1}{8\cdot 3}\\&=\frac{125}{24},\text{ or }5\frac{5}{24}\end{aligned}$$

The length of each of the three pieces must be $5\frac{5}{24}$ inches.

103. To find the number of cakes the caterer can make, divide $15\frac{1}{2}$ by $1\frac{3}{4}$.

$$\begin{aligned}15\frac{1}{2}\div 1\frac{3}{4}&=\frac{31}{2}\div\frac{7}{4}\\&=\frac{31}{2}\cdot\frac{4}{7}\\&=\frac{31\cdot 2\cdot 2}{2\cdot 7}\\&=\frac{62}{7},\text{ or }8\frac{6}{7}\end{aligned}$$

There is not quite enough sugar for 9 cakes. The caterer can make 8 cakes with some sugar left over.

105. Multiply the amount of fabric it takes to make one costume by the number of costumes.

$$\begin{aligned}2\frac{3}{8}\cdot 7&=\frac{19}{8}\cdot\frac{7}{1}\\&=\frac{19\cdot 7}{8\cdot 1}\\&=\frac{133}{8},\text{ or }16\frac{5}{8}\text{ yd}\end{aligned}$$

For 7 costumes, $16\frac{5}{8}$ yards of fabric would be needed.

107. Subtract the heights to find the difference.

$$\begin{aligned}10\frac{1}{2}-7\frac{1}{8}&=\frac{21}{2}-\frac{57}{8}\\&=\frac{21\cdot 4}{2\cdot 4}-\frac{57}{8}\qquad LCD=8\\&=\frac{84}{8}-\frac{57}{8}\\&=\frac{27}{8},\text{ or }3\frac{3}{8}\end{aligned}$$

The difference in heights is $3\frac{3}{8}$ inches.

109. The sum of the fractions representing the U.S. foreign-born population from Latin America, Asia, or Europe is

$$\begin{aligned}\frac{27}{50}+\frac{27}{100}+\frac{7}{50}&=\frac{27\cdot 2}{50\cdot 2}+\frac{27}{100}+\frac{7\cdot 2}{50\cdot 2}\\&=\frac{54+27+14}{100}\\&=\frac{95}{100}.\end{aligned}$$

So the fraction representing the U.S. foreign-born population from other regions is

$$\begin{aligned}1-\frac{95}{100}&=\frac{100}{100}-\frac{95}{100}\\&=\frac{5}{100}=\frac{1}{20}.\end{aligned}$$

111. Multiply the fraction representing the U.S. foreign-born population from Europe, $\frac{7}{50}$, by the total number of foreign-born people in the U.S., approximately 38 million.

$$\frac{7}{50}\cdot 38=\frac{7}{50}\cdot\frac{38}{1}=\frac{7\cdot 2\cdot 19}{2\cdot 25}=\frac{133}{25},\text{ or }5\frac{8}{25}$$

There were approximately $5\frac{8}{25}$ million (or 5,320,000) foreign-born people in the U.S. in 2006 who were born in Europe.

113. Observe that there are 24 dots in the entire figure, 6 dots in the triangle, 12 dots in the rectangle, and 2 dots in the overlapping region.

(a) $\frac{12}{24}=\frac{1}{2}$ of all the dots are in the rectangle.

(b) $\frac{6}{24}=\frac{1}{4}$ of all the dots are in the triangle.

(c) $\frac{2}{6}=\frac{1}{3}$ of the dots in the triangle are in the overlapping region.

(d) $\frac{2}{12}=\frac{1}{6}$ of the dots in the rectangle are in the overlapping region.

1.2 Exponents, Order of Operations, and Inequality

1.2 Now Try Exercises

N1. **(a)** $6^2 = 6 \cdot 6 = 36$

(b) $\left(\dfrac{4}{5}\right)^3 = \underbrace{\dfrac{4}{5} \cdot \dfrac{4}{5} \cdot \dfrac{4}{5}}_{\frac{4}{5} \text{ is used as a factor 3 times.}} = \dfrac{64}{125}$

N2. **(a)** $15 - 2 \cdot 6 = 15 - 12$ *Multiply.*
$= 3$ *Subtract.*

(b) $6(2 + 4) - 7 \cdot 5$
$= 6(6) - 7 \cdot 5$ *Add inside parentheses.*
$= 36 - 35$ *Multiply.*
$= 1$ *Subtract.*

(c) $8 \cdot 10 \div 4 - 2^3 + 3 \cdot 4^2$
$= 8 \cdot 10 \div 4 - 8 + 3 \cdot 16$ *Use exponents.*
$= 80 \div 4 - 8 + 48$ *Multiply.*
$= 20 - 8 + 48$ *Divide.*
$= 12 + 48$ *Subtract.*
$= 60$ *Add.*

N3. **(a)** $7[(3^2 - 1) + 4]$
$= 7[(9 - 1) + 4]$ *Use the exponent.*
$= 7[8 + 4]$ *Subtract inside parens.*
$= 7(12)$ *Add inside parentheses.*
$= 84$ *Multiply.*

(b) $\dfrac{9(14 - 4) - 2}{4 + 3 \cdot 6} = \dfrac{9(10) - 2}{4 + 3 \cdot 6}$ *Subt. inside parentheses.*
$= \dfrac{90 - 2}{4 + 18}$ *Multiply.*
$= \dfrac{88}{22}$ *Subtract and add.*
$= 4$ *Divide.*

N4. **(a)** The statement $12 \neq 10 - 2$ is *true* because 12 *is not equal to* 8.

(b) The statement $5 > 4 \cdot 2$ is *false* because 5 *is less than* 8.

(c) The statement $7 \leq 7$ is *true* since $7 = 7$.

(d) Write the fractions with a common denominator. The statement $\frac{5}{9} > \frac{7}{11}$ is equivalent to the statement $\frac{55}{99} > \frac{63}{99}$. Since 55 is *less* than 63, the original statement is *false*.

N5. **(a)** "Ten is not equal to eight minus two" is written $10 \neq 8 - 2$.

(b) "Fifty is greater than fifteen" is written $50 > 15$.

(c) "Eleven is less than or equal to twenty" is written $11 \leq 20$.

N6. $8 < 9$ may be written as $9 > 8$.

1.2 Section Exercises

1. False; 6^2 means that 6 is used as a factor 2 times, so $6^2 = 6 \cdot 6 = 36$.

3. False; 1 raised to *any* power is 1. Here, $1^3 = 1 \cdot 1 \cdot 1 = 1$.

5. False; $4 + 3(8 - 2) = 4 + 3 \cdot 6 = 4 + 18 = 22$. The common error leading to 42 is adding 4 to 3 and then multiplying by 6. One must follow the rules for order of operations.

7. $3^2 = 3 \cdot 3 = 9$

9. $7^2 = 7 \cdot 7 = 49$

11. $12^2 = 12 \cdot 12 = 144$

13. $4^3 = 4 \cdot 4 \cdot 4 = 64$

15. $10^3 = 10 \cdot 10 \cdot 10 = 1000$

17. $3^4 = 3 \cdot 3 \cdot 3 \cdot 3 = 81$

19. $4^5 = 4 \cdot 4 \cdot 4 \cdot 4 \cdot 4 = 1024$

21. $\left(\dfrac{1}{6}\right)^2 = \dfrac{1}{6} \cdot \dfrac{1}{6} = \dfrac{1}{36}$

23. $\left(\dfrac{2}{3}\right)^4 = \dfrac{2}{3} \cdot \dfrac{2}{3} \cdot \dfrac{2}{3} \cdot \dfrac{2}{3} = \dfrac{16}{81}$

25. $(0.4)^3 = (0.4)(0.4)(0.4) = 0.064$

27. $64 \div 4 \cdot 2 = (64 \div 4) \cdot 2$
$= 16 \cdot 2$
$= 32$

29. $13 + 9 \cdot 5 = 13 + 45$ *Multiply.*
$= 58$ *Add.*

31. $25.2 - 12.6 \div 4.2 = 25.2 - 3$ *Divide.*
$= 22.2$ *Subtract.*

33. $\dfrac{1}{4} \cdot \dfrac{2}{3} + \dfrac{2}{5} \cdot \dfrac{11}{3} = \dfrac{1}{6} + \dfrac{22}{15}$ *Multiply.*
$= \dfrac{5}{30} + \dfrac{44}{30}$ *LCD = 30*
$= \dfrac{49}{30}$, or $1\dfrac{19}{30}$ *Add.*

35. $9 \cdot 4 - 8 \cdot 3 = 36 - 24$ *Multiply.*
$= 12$ *Subtract.*

37. $20 - 4 \cdot 3 + 5 = 20 - 12 + 5$ *Multiply.*
$= 8 + 5$ *Subtract.*
$= 13$ *Add.*

39. $10 + 40 \div 5 \cdot 2 = 10 + 8 \cdot 2$ *Divide.*
$= 10 + 16$ *Multiply.*
$= 26$ *Add.*

41. $18 - 2(3+4) = 18 - 2(7)$ *Add inside parentheses.*
$= 18 - 14$ *Multiply.*
$= 4$ *Subtract.*

43. $3(4+2) + 8\cdot 3 = 3\cdot 6 + 8\cdot 3$ *Add.*
$= 18 + 24$ *Multiply.*
$= 42$ *Add.*

45. $18 - 4^2 + 3$
$= 18 - 16 + 3$ *Use the exponent.*
$= 2 + 3$ *Subtract.*
$= 5$ *Add.*

47. $2 + 3[5 + 4(2)]$
$= 2 + 3[5 + 8]$ *Multiply.*
$= 2 + 3[13]$ *Add.*
$= 2 + 39$ *Multiply.*
$= 41$ *Add.*

49. $5[3 + 4(2^2)]$
$= 5[3 + 4(4)]$ *Use the exponent.*
$= 5(3 + 16)$ *Multiply.*
$= 5(19)$ *Add.*
$= 95$ *Multiply.*

51. $3^2[(11+3) - 4]$
$= 3^2[14 - 4]$ *Add inside parentheses.*
$= 3^2[10]$ *Subtract.*
$= 9[10]$ *Use the exponent.*
$= 90$ *Multiply.*

53. Simplify the numerator and denominator separately; then divide.

$$\frac{6(3^2-1)+8}{8-2^2} = \frac{6(9-1)+8}{8-4} = \frac{6(8)+8}{4} = \frac{48+8}{4} = \frac{56}{4} = 14$$

55.
$$\frac{4(6+2)+8(8-3)}{6(4-2)-2^2} = \frac{4(8)+8(5)}{6(2)-2^2} = \frac{4(8)+8(5)}{6(2)-4} = \frac{32+40}{12-4} = \frac{72}{8} = 9$$

57. $9\cdot 3 - 11 \le 16$
$27 - 11 \le 16$
$16 \le 16$

The statement is true since $16 = 16$ is true.

59. $5\cdot 11 + 2\cdot 3 \le 60$
$55 + 6 \le 60$
$61 \le 60$

The statement is false since 61 *is greater than* 60.

61. $0 \ge 12\cdot 3 - 6\cdot 6$
$0 \ge 36 - 36$
$0 \ge 0$

The statement is true since $0 = 0$ is true.

63. $45 \ge 2[2 + 3(2+5)]$
$45 \ge 2[2 + 3(7)]$
$45 \ge 2[2 + 21]$
$45 \ge 2[23]$
$45 \ge 46$

The statement is false since 45 *is less than* 46.

65. $[3\cdot 4 + 5(2)]\cdot 3 > 72$
$[12 + 10]\cdot 3 > 72$
$[22]\cdot 3 > 72$
$66 > 72$

The statement is false since 66 *is less than* 72.

67.
$$\frac{3+5(4-1)}{2\cdot 4+1} \ge 3$$
$$\frac{3+5(3)}{8+1} \ge 3$$
$$\frac{3+15}{9} \ge 3$$
$$\frac{18}{9} \ge 3$$
$$2 \ge 3$$

The statement is false since 2 *is less than* 3.

69.
$$3 \ge \frac{2(5+1)-3(1+1)}{5(8-6)-4\cdot 2}$$
$$3 \ge \frac{2(6)-3(2)}{5(2)-8}$$
$$3 \ge \frac{12-6}{10-8}$$
$$3 \ge \frac{6}{2}$$
$$3 \ge 3$$

The statement is true since $3 = 3$ is true.

71. $3 \cdot 6 + 4 \cdot 2 = 60$

Listed below are some possibilities. We'll use trial and error until we get the desired result.

$$(3 \cdot 6) + 4 \cdot 2 = 18 + 8 = 26 \neq 60$$
$$(3 \cdot 6 + 4) \cdot 2 = 22 \cdot 2 = 44 \neq 60$$
$$3 \cdot (6 + 4 \cdot 2) = 3 \cdot 14 = 42 \neq 60$$
$$3 \cdot (6 + 4) \cdot 2 = 3 \cdot 10 \cdot 2 = 30 \cdot 2 = 60$$

73. $10 - 7 - 3 = 6$

$10 - (7 - 3) = 10 - 4 = 6$

75. $8 + 2^2 = 100$

$(8 + 2)^2 = 10^2 = 10 \cdot 10 = 100$

77. "$5 < 17$" means "five is less than seventeen." The statement is true.

79. "$5 \neq 8$" means "five is not equal to eight." The statement is true.

81. "$7 \geq 14$" means "seven is greater than or equal to fourteen." The statement is false.

83. "$15 \leq 15$" means "fifteen is less than or equal to fifteen." The statement is true.

85. "Fifteen is equal to five plus ten" is written

$$15 = 5 + 10.$$

87. "Nine is greater than five minus four" is written

$$9 > 5 - 4.$$

89. "Sixteen is not equal to nineteen" is written

$$16 \neq 19.$$

91. "One-half is less than or equal to two-fourths" is written

$$\tfrac{1}{2} \leq \tfrac{2}{4}.$$

93. $5 < 20$ becomes $20 > 5$ when the inequality symbol is reversed.

95. $2.5 \geq 1.3$ becomes $1.3 \leq 2.5$ when the inequality symbol is reversed.

97. **(a)** Substitute "40" for "age" in the expression for women.

$$14.7 - 40 \cdot 0.13$$

(b) $14.7 - 40 \cdot 0.13 = 14.7 - 5.2$ *Multiply.*
$= 9.5$ *Subtract.*

(c) 85% of $9.5 = 0.85(9.5) = 8.075$

Walking at 5 mph is associated with 8.0 METs, which is the table value closest to 8.075.

99. Answers will vary.

1.3 Variables, Expressions, and Equations

1.3 Now Try Exercises

N1. **(a)** $9k = 9 \cdot k$
$= 9 \cdot 6$ *Let k = 6.*
$= 54$ *Multiply.*

(b) $4k^2 = 4 \cdot k^2$
$= 4 \cdot 6^2$ *Let k = 6.*
$= 4 \cdot 36$ *Square 6.*
$= 144$ *Multiply.*

N2. Replace x with 4 and y with 7 in each expression.

(a) $3x + 4y = 3(4) + 4(7)$
$= 12 + 28$ *Multiply.*
$= 40$ *Add.*

(b) $\dfrac{6x - 2y}{2y - 9} = \dfrac{6(4) - 2(7)}{2(7) - 9}$
$= \dfrac{24 - 14}{14 - 9}$ *Multiply.*
$= \dfrac{10}{5} = 2$ *Subtract; reduce.*

(c) $4x^2 - y^2 = 4 \cdot 4^2 - 7^2$
$= 4 \cdot 16 - 49$ *Use exponents.*
$= 64 - 49$ *Multiply.*
$= 15$ *Subtract.*

N3. **(a)** Using x as the variable to represent the number, "the sum of a number and 10" translates as $x + 10$, or $10 + x$.

(b) "A number divided by 7" translates as $x \div 7$, or $\frac{x}{7}$.

(c) "The difference between 9 and a number" translates as $9 - x$. Thus, "the product of 3 and the difference between 9 and a number" translates as $3(9 - x)$.

N4. $8k + 5 = 61$
$8 \cdot 7 + 5 \stackrel{?}{=} 61$ *Replace k with 7.*
$56 + 5 \stackrel{?}{=} 61$ *Multiply.*
$61 = 61$ *True*

The number 7 is a solution of the equation.

N5. Using x as the variable to represent the number, "The sum of a number and nine is equal to the difference between 25 and the number" translates as

$x + 9 = 25 - x.$

Now try each number from the set $\{0, 2, 4, 6, 8, 10\}$.

$\boldsymbol{x = 4}$: $4 + 9 \stackrel{?}{=} 25 - 4$
$13 = 21$ *False*

$\boldsymbol{x = 6}$: $6 + 9 \stackrel{?}{=} 25 - 6$
$15 = 19$ *False*

$\boldsymbol{x = 8}$: $8 + 9 \stackrel{?}{=} 25 - 8$
$17 = 17$ *True*

Similarly, $x = 0, 2,$ or 10 result in false statements. Thus, 8 is the only solution.

N6. **(a)** $2x + 5 = 6$ has an equals symbol, so this is an *equation*.

(b) $2x + 5 - 6$ has no equals symbol, so this is an *expression*.

1.3 Section Exercises

1. The expression $8x^2$ means $8 \cdot x \cdot x$. The correct choice is **B**.

3. The sum of 15 and a number x is represented by the expression $15 + x$. The correct choice is **A**.

5. $2x^3 = 2 \cdot x \cdot x \cdot x$, while $2x \cdot 2x \cdot 2x = (2x)^3$. The last expression is equal to $8x^3$.

7. The exponent 2 applies only to its base, which is x. (The expression $(5x)^2$ would require multiplying 5 by $x = 4$ first.)

In part (a) of Exercises 9–22, replace x with 4. In part (b), replace x with 6. Then use the order of operations.

9. **(a)** $x + 7 = 4 + 7$
$= 11$

(b) $x + 7 = 6 + 7$
$= 13$

11. **(a)** $4x = 4(4) = 16$

(b) $4x = 4(6) = 24$

13. **(a)** $4x^2 = 4 \cdot 4^2$
$= 4 \cdot 16$
$= 64$

(b) $4x^2 = 4 \cdot 6^2$
$= 4 \cdot 36$
$= 144$

15. **(a)** $\dfrac{x+1}{3} = \dfrac{4+1}{3}$
$= \dfrac{5}{3}$

(b) $\dfrac{x+1}{3} = \dfrac{6+1}{3}$
$= \dfrac{7}{3}$

17. **(a)** $\dfrac{3x-5}{2x} = \dfrac{3 \cdot 4 - 5}{2 \cdot 4}$
$= \dfrac{12-5}{8}$
$= \dfrac{7}{8}$

(b) $\dfrac{3x-5}{2x} = \dfrac{3 \cdot 6 - 5}{2 \cdot 6}$
$= \dfrac{18-5}{12}$
$= \dfrac{13}{12}$

19. **(a)** $3x^2 + x = 3 \cdot 4^2 + 4$
$= 3 \cdot 16 + 4$
$= 48 + 4 = 52$

(b) $3x^2 + x = 3 \cdot 6^2 + 6$
$= 3 \cdot 36 + 6$
$= 108 + 6 = 114$

21. **(a)** $6.459x = 6.459 \cdot 4$
$= 25.836$

(b) $6.459x = 6.459 \cdot 6$
$= 38.754$

In part (a) of Exercises 23–38, replace x with 2 and y with 1. In part (b), replace x with 1 and y with 5.

23. **(a)** $8x + 3y + 5 = 8(2) + 3(1) + 5$
$= 16 + 3 + 5$
$= 19 + 5$
$= 24$

(b) $8x + 3y + 5 = 8(1) + 3(5) + 5$
$= 8 + 15 + 5$
$= 23 + 5$
$= 28$

25. **(a)** $3(x + 2y) = 3(2 + 2 \cdot 1)$
$= 3(2 + 2)$
$= 3(4)$
$= 12$

(b) $3(x + 2y) = 3(1 + 2 \cdot 5)$
$= 3(1 + 10)$
$= 3(11)$
$= 33$

27. **(a)** $x+\frac{4}{y}=2+\frac{4}{1}$
$=2+4$
$=6$

(b) $x+\frac{4}{y}=1+\frac{4}{5}$
$=\frac{5}{5}+\frac{4}{5}$
$=\frac{9}{5}$

29. **(a)** $\frac{x}{2}+\frac{y}{3}=\frac{2}{2}+\frac{1}{3}$
$=\frac{6}{6}+\frac{2}{6}$
$=\frac{8}{6}=\frac{4}{3}$

(b) $\frac{x}{2}+\frac{y}{3}=\frac{1}{2}+\frac{5}{3}$
$=\frac{3}{6}+\frac{10}{6}$
$=\frac{13}{6}$

31. **(a)** $\frac{2x+4y-6}{5y+2}=\frac{2(2)+4(1)-6}{5(1)+2}$
$=\frac{4+4-6}{5+2}$
$=\frac{8-6}{7}$
$=\frac{2}{7}$

(b) $\frac{2x+4y-6}{5y+2}=\frac{2(1)+4(5)-6}{5(5)+2}$
$=\frac{2+20-6}{25+2}$
$=\frac{22-6}{27}$
$=\frac{16}{27}$

33. **(a)** $2y^2+5x=2\cdot 1^2+5\cdot 2$
$=2\cdot 1+5\cdot 2$
$=2+10$
$=12$

(b) $2y^2+5x=2\cdot 5^2+5\cdot 1$
$=2\cdot 25+5\cdot 1$
$=50+5$
$=55$

35. **(a)** $\frac{3x+y^2}{2x+3y}=\frac{3(2)+1^2}{2(2)+3(1)}$
$=\frac{3(2)+1}{4+3}$
$=\frac{6+1}{7}$
$=\frac{7}{7}$
$=1$

(b) $\frac{3x+y^2}{2x+3y}=\frac{3(1)+5^2}{2(1)+3(5)}$
$=\frac{3(1)+25}{2+15}$
$=\frac{3+25}{17}$
$=\frac{28}{17}$

37. **(a)** $0.841x^2+0.32y^2$
$=0.841\cdot 2^2+0.32\cdot 1^2$
$=0.841\cdot 4+0.32\cdot 1$
$=3.364+0.32$
$=3.684$

(b) $0.841x^2+0.32y^2$
$=0.841\cdot 1^2+0.32\cdot 5^2$
$=0.841\cdot 1+0.32\cdot 25$
$=0.841+8$
$=8.841$

39. "Twelve times a number" translates as $12\cdot x$ or $12x$.

41. "Added to" indicates addition. "Nine added to a number" translates as $x+9$.

43. "Four subtracted from a number" translates as $x-4$.

45. "A number subtracted from seven" translates as $7-x$.

47. "The difference between a number and 8" translates as $x-8$.

49. "18 divided by a number" translates as $\frac{18}{x}$.

51. "The product of 6 and four less than a number" translates as $6(x-4)$.

53. An expression cannot by solved—it indicates a series of operations to perform. An expression is simplified. An equation is solved.

55. $4m + 2 = 6$; 1

$$4(1) + 2 \stackrel{?}{=} 6 \quad \textit{Let } m = 1.$$
$$4 + 2 \stackrel{?}{=} 6$$
$$6 = 6 \quad \textit{True}$$

Because substituting 1 for m results in a true statement, 1 is a solution of the equation.

57. $2y + 3(y - 2) = 14$; 3

$$2 \cdot 3 + 3(3 - 2) \stackrel{?}{=} 14 \quad \textit{Let } y = 3.$$
$$2 \cdot 3 + 3 \cdot 1 \stackrel{?}{=} 14$$
$$6 + 3 \stackrel{?}{=} 14$$
$$9 = 14 \quad \textit{False}$$

Because substituting 3 for y results in a false statement, 3 is not a solution of the equation.

59. $6p + 4p + 9 = 11$; $\frac{1}{5}$

$$6\left(\frac{1}{5}\right) + 4\left(\frac{1}{5}\right) + 9 \stackrel{?}{=} 11 \quad \textit{Let } p = \frac{1}{5}.$$
$$\frac{6}{5} + \frac{4}{5} + 9 \stackrel{?}{=} 11$$
$$\frac{10}{5} + 9 \stackrel{?}{=} 11$$
$$2 + 9 \stackrel{?}{=} 11$$
$$11 = 11 \quad \textit{True}$$

The true result shows that $\frac{1}{5}$ is a solution of the equation.

61. $3r^2 - 2 = 46$; 4

$$3(4)^2 - 2 \stackrel{?}{=} 46 \quad \textit{Let } r = 4.$$
$$3 \cdot 16 - 2 \stackrel{?}{=} 46$$
$$48 - 2 \stackrel{?}{=} 46$$
$$46 = 46 \quad \textit{True}$$

The true result shows that 4 is a solution of the equation.

63. $\frac{3}{8}x + \frac{1}{4} = 1$; 2

$$\frac{3}{8}(2) + \frac{1}{4} \stackrel{?}{=} 1 \quad \textit{Let } x = 2.$$
$$\frac{3}{4} + \frac{1}{4} \stackrel{?}{=} 1$$
$$1 = 1 \quad \textit{True}$$

The true result shows that 2 is a solution of the equation.

65. $0.5(x - 4) = 80$; 20

$$0.5(20 - 4) \stackrel{?}{=} 80 \quad \textit{Let } x = 20.$$
$$0.5(16) \stackrel{?}{=} 80$$
$$8 = 80 \quad \textit{False}$$

The false result shows that 20 is not a solution of the equation.

67. "The sum of a number and 8 is 18" translates as

$$x + 8 = 18.$$

Try each number from the given set, {2, 4, 6, 8, 10}, in turn.

$$x + 8 = 18 \quad \textit{Given equation}$$
$$2 + 8 = 18 \quad \textit{False}$$
$$4 + 8 = 18 \quad \textit{False}$$
$$6 + 8 = 18 \quad \textit{False}$$
$$8 + 8 = 18 \quad \textit{False}$$
$$10 + 8 = 18 \quad \textit{True}$$

The only solution is 10.

69. "Sixteen minus three-fourths of a number is 13" translates as

$$16 - \tfrac{3}{4}x = 13.$$

Try each number from the given set, {2, 4, 6, 8, 10}, in turn.

$$16 - \tfrac{3}{4}x = 13 \quad \textit{Given equation}$$
$$16 - \tfrac{3}{4}(2) = 13 \quad \textit{False}$$
$$16 - \tfrac{3}{4}(4) = 13 \quad \textit{True}$$
$$16 - \tfrac{3}{4}(6) = 13 \quad \textit{False}$$
$$16 - \tfrac{3}{4}(8) = 13 \quad \textit{False}$$
$$16 - \tfrac{3}{4}(10) = 13 \quad \textit{False}$$

The only solution is 4.

71. "One more than twice a number is 5" translates as

$$2x + 1 = 5.$$

Try each number from the given set. The only resulting true equation is

$$2 \cdot 2 + 1 = 5,$$

So the only solution is 2.

73. "Three times a number is equal to 8 more than twice the number" translates as

$$3x = 2x + 8.$$

Try each number from the given set.

$$3x = 2x + 8 \quad \textit{Given equation}$$
$$3(2) = 2(2) + 8 \quad \textit{False}$$
$$3(4) = 2(4) + 8 \quad \textit{False}$$
$$3(6) = 2(6) + 8 \quad \textit{False}$$
$$3(8) = 2(8) + 8 \quad \textit{True}$$
$$3(10) = 2(10) + 8 \quad \textit{False}$$

The only solution is 8.

75. There is no equals symbol, so $3x + 2(x - 4)$ is an expression.

77. There is an equals symbol, so $7t + 2(t + 1) = 4$ is an equation.

79. There is an equals symbol, so $x + y = 9$ is an equation.

81. $y = 0.212x - 347$
$= 0.212(1943) - 347$
$= 64.916 \approx 64.9$

The life expectancy of an American born in 1943 is about 64.9 years.

83. $y = 0.212x - 347$
$= 0.212(1985) - 347$
$= 73.82 \approx 73.8$

The life expectancy of an American born in 1985 is about 73.8 years.

85. Life expectancy has increased over 13 years during this time.

1.4 Real Numbers and the Number Line

1.4 Now Try Exercises

N1. Since the deepest point is *below* the water's surface, the depth is -136.

N2. $\left\{-7, -\frac{4}{5}, 0, \sqrt{3}, 2.7, \pi, 13\right\}$

(a) The whole numbers are 0, and 13.

(b) The integers are $-7, 0,$ and 13.

(c) The rational numbers are $-7, -\frac{4}{5}, 0, 2.7,$ and 13.

(d) The irrational numbers are $\sqrt{3}$ and π.

N3. Since -8 lies to the right of -9 on the number line, -8 is greater than -9. Therefore, the statement $-8 \leq -9$ is *false*.

N4. **(a)** $|4| = 4$

(b) $|-4| = -(-4) = 4$

(c) $-|-4| = -(4) = -4$

N5. The category *new cars* is negative in both years.

1.4 Section Exercises

1. Use the integer 2,866,000 since "increased by 2,866,000" indicates a positive number.

3. Use the integer $-52{,}000$ since "a decrease of 52,000" indicates a negative number.

5. Use the rational numbers -11.2 and 8.6 since "declined 11.2%" and "rose 8.6%" indicate a negative number and a positive number, respectively.

7. Use the rational number 82.60 since "closed up 82.60" indicates a positive number.

9. The only integer between 3.6 and 4.6 is 4.

11. There is only one whole number that is not positive and that is less than 1: the number 0.

13. An irrational number that is between $\sqrt{12}$ and $\sqrt{14}$ is $\sqrt{13}$. There are others.

15. True; every natural number is positive.

17. True; every integer is a rational number. For example, 5 can be written as $\frac{5}{1}$.

19. False; if a number is rational, it cannot be irrational, and vice versa.

21. Three examples of positive real numbers that are not integers are $\frac{1}{2}, \frac{5}{8},$ and $1\frac{3}{4}$. Other examples are $0.7, 4\frac{2}{3},$ and 5.1.

23. Three examples of real numbers that are not whole numbers are $-3\frac{1}{2}, -\frac{2}{3},$ and $\frac{3}{7}$. Other examples are $-4.3, -\sqrt{2},$ and $\sqrt{7}$.

25. Three examples of real numbers that are not rational numbers are $\sqrt{5}, \pi,$ and $-\sqrt{3}$. All irrational numbers are real numbers that are not rational.

27. $\left\{-9, -\sqrt{7}, -1\frac{1}{4}, -\frac{3}{5}, 0, 0.\overline{1}, \sqrt{5}, 3, 5.9, 7\right\}$

(a) The natural numbers in the given set are 3 and 7, since they are in the natural number set $\{1, 2, 3, \dots\}$.

(b) The set of whole numbers includes the natural numbers and 0. The whole numbers in the given set are 0, 3, and 7.

(c) The integers are the set of numbers $\{\dots, -3, -2, -1, 0, 1, 2, 3, \dots\}$. The integers in the given set are -9, 0, 3, and 7.

(d) Rational numbers are the numbers which can be expressed as the quotient of two integers, with denominators not equal to 0.

We can write numbers from the given set in this form as follows:

$$-9 = \frac{-9}{1}, -1\frac{1}{4} = \frac{-5}{4}, -\frac{3}{5} = \frac{-3}{5}, 0 = \frac{0}{1},$$

$$0.\overline{1} = \frac{1}{9}, 3 = \frac{3}{1}, 5.9 = \frac{59}{10}, \text{ and } 7 = \frac{7}{1}.$$

Thus, the rational numbers in the given set are $-9, -1\frac{1}{4}, -\frac{3}{5}, 0, 0.\overline{1}, 3, 5.9,$ and 7.

(e) Irrational numbers are real numbers that are not rational. $-\sqrt{7}$ and $\sqrt{5}$ can be represented by points on the number line but cannot be written as a quotient of integers. Thus, the irrational numbers in the given set are $-\sqrt{7}$ and $\sqrt{5}$.

(f) Real numbers are all numbers that can be represented on the number line. All the numbers in the given set are real.

29. Graph 0, 3, -5, and -6.

Place a dot on the number line at the point that corresponds to each number. The order of the numbers from smallest to largest is $-6, -5, 0, 3$.

31. Graph $-2, -6, -4, 3$, and 4.

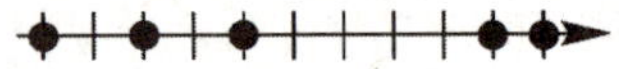

33. Graph $\frac{1}{4}, 2\frac{1}{2}, -3\frac{4}{5}, -4$, and $-1\frac{5}{8}$.

35. **(a)** $|-9| = 9$ **(A)**

The distance between -9 and 0 on the number line is 9 units.

(b) $-(-9) = 9$ **(A)**

The opposite of -9 is 9.

(c) $-|-9| = -(9) = -9$ **(B)**

(d) $-|-(-9)| = -|9|$ *Work inside absolute value symbols first*

$= -(9)$

$= -9$ **(B)**

37. **(a)** The opposite of -7 is found by changing the sign of -7. The opposite of -7 is 7.

(b) The absolute value of -7 is the distance between 0 and -7 on the number line.

$$|-7| = 7$$

The absolute value of -7 is 7.

39. **(a)** The opposite of 8 is -8.

(b) The distance between 0 and 8 on the number line is 8 units, so the absolute value of 8 is 8.

41. **(a)** The opposite of a number is found by changing the sign of a number, so the opposite of $-\frac{3}{4}$ is $\frac{3}{4}$.

(b) The distance between $-\frac{3}{4}$ and 0 on the number line is $\frac{3}{4}$ unit, so $\left|-\frac{3}{4}\right| = \frac{3}{4}$.

43. Since -6 is a negative number, its absolute value is the additive inverse of -6; that is,

$$|-6| = -(-6) = 6.$$

45. $-|12| = -(12) = -12$

47. $-\left|-\frac{2}{3}\right| = -\left[-\left(-\frac{2}{3}\right)\right] = -\left[\frac{2}{3}\right] = -\frac{2}{3}$

49. $|6-3| = |3| = 3$

51. The statement "Absolute value is always positive." is not true. The absolute value of 0 is 0, and 0 is not positive. We could say that *absolute value is never negative,* or *absolute value is always nonnegative.*

53. $-11, -3$
Since -11 is located to the left of -3 on the number line, -11 is the lesser number.

55. $-7, -6$
Since -7 is located to the left of -6 on the number line, -7 is the lesser number.

57. $4, |-5|$
Since $|-5| = 5$, 4 is the lesser of the two numbers.

59. $|-3.5|, |-4.5|$
Since $|-3.5| = 3.5$ and $|-4.5| = 4.5$, $|-3.5|$ or 3.5 is the lesser number.

61. $-|-6|, -|-4|$
Since $-|-6| = -6$ and $-|-4| = -4$, $-|-6|$ is to the left of $-|-4|$ on the number line, so $-|-6|$ or -6 is the lesser number.

63. $|5-3|, |6-2|$
Since $|5-3| = |2| = 2$ and $|6-2| = |4| = 4$, $|5-3|$ or 2 is the lesser number.

65. $-5 < -2$
Since -5 is to the *left* of -2 on the number line, -5 is *less than* -2, and the statement $-5 < -2$ is true.

67. $-4 \le -(-5)$
Since $-(-5) = 5$ and $-4 < 5$, $-4 \le -(-5)$ is true.

69. $|-6| < |-9|$
Since $|-6| = 6$ and $|-9| = 9$, and $6 < 9$, $|-6| < |-9|$ is true.

71. $-|8| > |-9|$
Since $-|8| = -8$ and $|-9| = -(-9) = 9$, $-|8| < |-9|$, so $-|8| > |-9|$ is false.

73. $-|-5| \ge -|-9|$
Since $-|-5| = -5$, $-|-9| = -9$, and $-5 > -9$, $-|-5| \ge -|-9|$ is true.

75. $|6-5| \geq |6-2|$
Since $|6-5| = |1| = 1$ and $|6-2| = |4| = 4$, $|6-5| < |6-2|$, so $|6-5| \geq |6-2|$ is false.

77. The number that represents the greatest percentage increase is 10.6, which corresponds to *fuel and other utilities* from 2004 to 2005.

79. The number with the smallest absolute value in the table is -0.4, so the least change corresponds to *apparel and upkeep* from 2006 to 2007.

1.5 Adding and Subtracting Real Numbers

1.5 Now Try Exercises

N1. **(a)** Start at 0 on a number line. Draw an arrow 3 units to the right to represent the positive number 3. From the right end of this arrow, draw a second arrow 5 units to the right to represent the addition of a positive number. The number below the end of this second arrow is 8, so $3 + 5 = 8$.

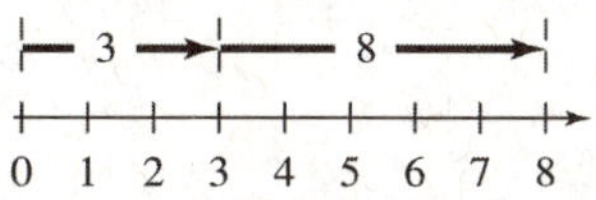

(b) Start at 0 on a number line. Draw an arrow 1 unit to the left to represent the negative number -1. From the left end of this arrow, draw a second arrow 3 units to the left. The number below the end of this second arrow is -4, so $-1 + (-3) = -4$.

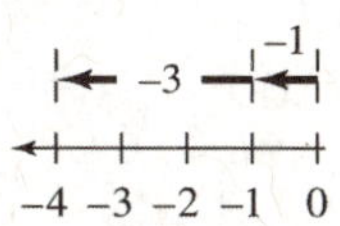

N2. $-6 + (-11) = -17$

The sum of two negative numbers is negative.

N3. Start at 0 on a number line. Draw an arrow 4 units to the right. From the right end of this arrow, draw a second arrow 8 units to the left. The number below the end of this second arrow is -4, so $4 + (-8) = -4$.

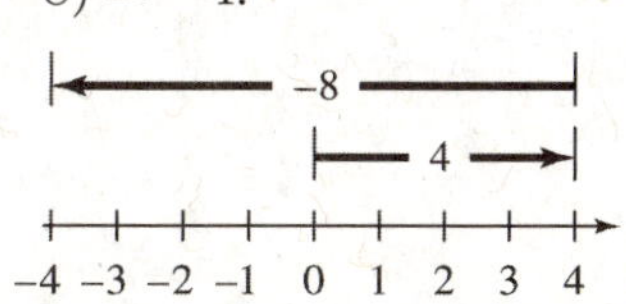

N4. $8 + (-17)$

Since the numbers have different signs, find the difference between their absolute values:

$$17 - 8 = 9.$$

Because -17 has the larger absolute value, the sum is negative:

$$8 + (-17) = -9.$$

N5. **(a)** $\frac{2}{3} + \left(-2\frac{1}{9}\right) = -1\frac{4}{9}$

The number with the greater absolute value is $-2\frac{1}{9}$, so the sum will be negative. The answer of $-1\frac{4}{9}$ is correct.

(b) $3.7 + (-5.7) = -2$

The number with the greater absolute value is -5.7, so the sum will be negative. The answer of -2 is correct.

N6. **(a)** $-5 - (-11) = -5 + (11)$ *Add the opposite.*
$= 6$

(b) $4 - 15 = 4 + (-15)$ *Add the opposite.*
$= -11$

(c) $-\frac{5}{7} - \frac{1}{3} = -\frac{5}{7} + \left(-\frac{1}{3}\right)$ *Add opposite.*

$$= -\frac{15}{21} + \left(-\frac{7}{21}\right)$$
$$= -\frac{22}{21}, \text{ or } -1\frac{1}{21}$$

N7. **(a)**
$$\begin{aligned} &8 - [(-3+7) - (3-9)] \\ &= 8 - [(4) - (3 + (-9))] \\ &= 8 - [4 - (-6)] \\ &= 8 - [4 + 6] \\ &= 8 - 10 \\ &= 8 + (-10) \\ &= -2 \end{aligned}$$

(b)
$$\begin{aligned} &3|6-9| - |4-12| \\ &= 3|6 + (-9)| - |4 + (-12)| \\ &= 3|-3| - |-8| \\ &= 3(3) - 8 \\ &= 9 - 8 \\ &= 1 \end{aligned}$$

N8. "The sum of -3 and 7, increased by 10" is written $(-3+7) + 10$.

$$(-3+7) + 10 = 4 + 10 = 14$$

N9. **(a)** "The difference between 5 and -8, decreased by 4" is written $[5 - (-8)] - 4$.

$$\begin{aligned} [5 - (-8)] - 4 &= [5 + 8] - 4 \\ &= 13 - 4 \\ &= 9 \end{aligned}$$

(b) "7 less than -2" is written $-2 - 7$.

$$\begin{aligned} -2 - 7 &= -2 + (-7) \\ &= -9 \end{aligned}$$

N10. The difference between a gain of 226 yards and a loss of 7 yards

$$\begin{aligned} 226 - (-7) &= 226 + 7 \\ &= 233. \end{aligned}$$

The difference is 233 yards.

N11. Subtract the CPI number for 2003 from the CPI number for 2004.

$$119.6 - 119.3 = 0.3$$

A positive result indicates an increase.

1.5 Section Exercises

1. The sum of two negative numbers will always be a *negative* number. In the illustration, we have $-2 + (-3) = -5$.

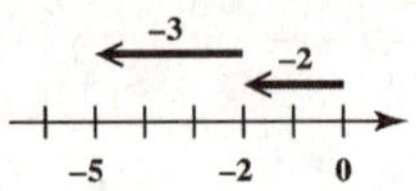

3. When adding a positive number and a negative number, where the negative number has the greater absolute value, the sum will be a *negative* number. In the illustration, the absolute value of -4 is larger than the absolute value of 2, so the sum is a negative number; that is, $-4 + 2 = -2$.

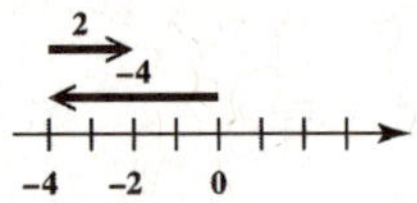

5. By the definition of subtraction, in order to perform the subtraction $-6 - (-8)$, we must add the opposite of $\underline{-8}$ to $\underline{-6}$ to get $\underline{2}$.

7. The expression $x - y$ would have to be *positive* since subtracting a negative number from a positive number is the same as adding a positive number to a positive number, which is a positive number.

9. $|x| = x$, since x is a positive number. Thus, $y - |x| = y - x$. Since x is a positive number and y is a negative number, $y - x$ will be the sum of two negative numbers, which is a *negative* number.

11. $-6 + (-2)$

The sum of two negative numbers is negative.

$$-6 + (-2) = -8$$

13. $-5 + (-7)$

Because the numbers have the same sign, add their absolute values:

$$5 + 7 = 12.$$

Because both numbers are negative, their sum is negative:

$$-5 + (-7) = -12.$$

15. $6 + (-4)$

To add $6 + (-4)$, find the difference between the absolute values of the numbers.

$$|6| = 6 \text{ and } |-4| = 4$$

$$6 - 4 = 2$$

Since $|6| > |-4|$, the sum will be positive:

$$6 + (-4) = 2.$$

17. $4 + (-6)$

Since the numbers have different signs, find the difference between their absolute values:

$$6 - 4 = 2.$$

Because -6 has the larger absolute value, the sum is negative:

$$4 + (-6) = -2.$$

19. $-3.5 + 12.4$

Since the numbers have different signs, find the difference between their absolute values:

$$12.4 - 3.5 = 8.9.$$

Since 12.4 has the larger absolute value, the answer is positive:

$$-3.5 + 12.4 = 8.9.$$

21. $4 + [13 + (-5)]$

Perform the operation inside the brackets first, then add.

$$4 + [13 + (-5)] = 4 + [8] = 12$$

23. $8 + [-2 + (-1)] = 8 + [-3] = 5$

25. $-2 + [5 + (-1)] = -2 + [4] = 2$

27. $-6 + [6 + (-9)] = -6 + [-3] = -9$

29. $[(-9) + (-3)] + 12 = [-12] + 12 = 0$

31. $-\frac{1}{6} + \frac{2}{3} = -\frac{1}{6} + \frac{4}{6} = \frac{3}{6} = \frac{1}{2}$

33. Since $8 = 2 \cdot 2 \cdot 2$ and $12 = 2 \cdot 2 \cdot 3$, the LCD is $2 \cdot 2 \cdot 2 \cdot 3 = 24$.

$$\begin{aligned} \frac{5}{8} + \left(-\frac{17}{12}\right) &= \frac{5 \cdot 3}{8 \cdot 3} + \left(-\frac{17 \cdot 2}{12 \cdot 2}\right) \\ &= \frac{15}{24} + \left(-\frac{34}{24}\right) \\ &= -\frac{19}{24} \end{aligned}$$

35. $$\begin{aligned} 2\tfrac{1}{2} + \left(-3\tfrac{1}{4}\right) &= \frac{5}{2} + \left(-\frac{13}{4}\right) \\ &= \frac{10}{4} + \left(-\frac{13}{4}\right) \\ &= -\frac{3}{4} \end{aligned}$$

37. $$\begin{aligned} -6.1 + [3.2 + (-4.8)] &= -6.1 + [-1.6] \\ &= -7.7 \end{aligned}$$

39. $[-3+(-4)]+[5+(-6)] = [-7]+[-1]$
$= -8$

41. $[-4+(-3)]+[8+(-1)] = [-7]+[7]$
$= 0$

43. $[-4+(-6)]+[(-3)+(-8)]+[12+(-11)]$
$= ([-10]+[-11])+[1]$
$= (-21)+1$
$= -20$

In Exercises 45–68, use the definition of subtraction to find the differences.

45. $4-7 = 4+(-7) = -3$

47. $5-9 = 5+(-9) = -4$

49. $-7-1 = -7+(-1) = -8$

51. $-8-6 = -8+(-6) = -14$

53. $7-(-2) = 7+(2) = 9$

55. $-6-(-2) = -6+(2) = -4$

57. $2-(3-5) = 2-[3+(-5)]$
$= 2-[-2]$
$= 2+(2)$
$= 4$

59. $\frac{1}{2}-\left(-\frac{1}{4}\right) = \frac{1}{2}+\frac{1}{4}$
$= \frac{2}{4}+\frac{1}{4} = \frac{3}{4}$

61. $-\frac{3}{4}-\frac{5}{8} = -\frac{3}{4}+\left(-\frac{5}{8}\right)$
$= -\frac{6}{8}+\left(-\frac{5}{8}\right)$
$= -\frac{11}{8}$, or $-1\frac{3}{8}$

63. $\frac{5}{8}-\left(-\frac{1}{2}-\frac{3}{4}\right)$
$= \frac{5}{8}-\left[-\frac{1}{2}+\left(-\frac{3}{4}\right)\right]$
$= \frac{5}{8}-\left[-\frac{2}{4}+\left(-\frac{3}{4}\right)\right]$
$= \frac{5}{8}-\left(-\frac{5}{4}\right)$
$= \frac{5}{8}+\frac{5}{4}$
$= \frac{5}{8}+\frac{10}{8}$
$= \frac{15}{8}$, or $1\frac{7}{8}$

65. $3.4-(-8.2) = 3.4+8.2$
$= 11.6$

67. $-6.4-3.5 = -6.4+(-3.5)$
$= -9.9$

69. $(4-6)+12 = [4+(-6)]+12$
$= [-2]+12$
$= 10$

71. $(8-1)-12 = [8+(-1)]+(-12)$
$= [7]+(-12)$
$= -5$

73. $6-(-8+3) = 6-(-5)$
$= 6+5$
$= 11$

75. $2+(-4-8) = 2+[-4+(-8)]$
$= 2+[-12]$
$= -10$

77. $|-5-6|+|9+2| = |-5+(-6)|+|11|$
$= |-11|+|11|$
$= -(-11)+11$
$= 11+11$
$= 22$

79. $|-8-2|-|-9-3|$
$= |-8+(-2)|-|-9+(-3)|$
$= |-10|-|-12|$
$= -(-10)-[-(-12)]$
$= 10-[12]$
$= -2$

81. $\left(-\frac{3}{4}-\frac{5}{2}\right)-\left(-\frac{1}{8}-1\right)$
$= \left(-\frac{3}{4}-\frac{10}{4}\right)-\left(-\frac{1}{8}-\frac{8}{8}\right)$
$= -\frac{13}{4}-\left(-\frac{9}{8}\right)$
$= -\frac{26}{8}+\frac{9}{8}$
$= -\frac{17}{8}$, or $-2\frac{1}{8}$

83. $\left(-\frac{1}{2}+0.25\right)-\left(-\frac{3}{4}+0.75\right)$
$= \left(-\frac{1}{2}+\frac{1}{4}\right)-\left(-\frac{3}{4}+\frac{3}{4}\right)$
$= \left(-\frac{2}{4}+\frac{1}{4}\right)-0$
$= -\frac{1}{4}$, or -0.25

85. $-9+[(3-2)-(-4+2)]$
$=-9+[1-(-2)]$
$=-9+[1+2]$
$=-9+3$
$=-6$

87. $-3+[(-5-8)-(-6+2)]$
$=-3+[(-5+(-8))-(-4)]$
$=-3+[-13+4]$
$=-3+[-9]$
$=-12$

89. $-9.1237+[(-4.8099-3.2516)+11.27903]$
$=-9.1237+[(-4.8099+(-3.2516))+11.27903]$
$=-9.1237+[-8.0615+11.27903]$
$=-9.1237+3.21753$
$=-5.90617$

91. "The sum of -5 and 12 and 6" is written $-5+12+6$.

$$-5+12+6=[-5+12]+6$$
$$=7+6=13$$

93. "14 added to the sum of -19 and -4" is written $[-19+(-4)]+14$.

$$[-19+(-4)]+14=(-23)+14$$
$$=-9$$

95. "The sum of -4 and -10, increased by 12" is written$[-4+(-10)]+12$.

$$[-4+(-10)]+12=-14+12$$
$$=-2$$

97. "$\frac{2}{7}$ more than the sum of $\frac{5}{7}$ and $-\frac{9}{7}$" is written $[\frac{5}{7}+(-\frac{9}{7})]+\frac{2}{7}$.

$$\left[\frac{5}{7}+\left(-\frac{9}{7}\right)\right]+\frac{2}{7}=-\frac{4}{7}+\frac{2}{7}$$
$$=-\frac{2}{7}$$

99. "The difference between 4 and -8" is written $4-(-8)$.

$$4-(-8)=4+8=12$$

101. "8 less than -2" is written $-2-8$.

$$-2-8=-2+(-8)=-10$$

103. "The sum of 9 and -4, decreased by 7" is written $[9+(-4)]-7$.

$$[9+(-4)]-7=5+(-7)=-2$$

105. "12 less than the difference between 8 and -5" is written $[8-(-5)]-12$.

$$[8-(-5)]-12=[8+(5)]-12$$
$$=13-12$$
$$=13+(-12)$$
$$=1$$

107. $[-5+(-4)]+(-3)=-9+(-3)$
$=-12$

The total number of seats that New York, Pennsylvania, and Ohio are projected to lose is twelve, which can be represented by the signed number -12.

109. To find the low temperature, start with 44 and subtract 100.

$$44-100=44+(-100)$$
$$=-56$$

The low temperature was -56°F.

111. 33°F lower than -36°F can be represented as

$$-36-33=-36+(-33)$$
$$=-69.$$

The record low in Utah is -69°F.

113. $0+(-130)+(-54)=-130+(-54)$
$=-184$

Their new altitude is 184 meters below the surface, which can be represented by the signed number -184.

115. **(a)** $6.9-(-0.5)=6.9+0.5$
$=7.4$

The difference is 7.4%.

(b) Americans spent more money than they earned, which means they had to dip into savings or increase borrowing.

117. $1879+869-579+1004$
$=2748-579+1004$ *Add.*
$=2169+1004$ *Subtract.*
$=3173$ *Add.*

The average was \$3173.

119. Add the scores of the four turns to get the final score.

$$-19+28+(-5)+13=9+(-5)+13$$
$$=4+13$$
$$=17$$

Her final score for the four turns was 17.

121. Sum of checks:

$$\$35.84 + \$26.14 + \$3.12 = \$61.98 + \$3.12 = \$65.10$$

Sum of deposits:

$$\$85.00 + \$120.76 = \$205.76$$

$$\begin{aligned}\text{Final balance} &= \text{Beginning balance} - \text{checks} + \text{deposits}\\ &= \$904.89 - \$65.10 + \$205.76\\ &= \$839.79 + \$205.76\\ &= \$1045.55\end{aligned}$$

Her account balance at the end of August was \$1045.55.

123.

−870.00	*amount owed*
+ 185.90	*2 return credits* (\$35.90 + \$150.00)
−684.10	
−102.50	*3 purchases* (\$82.50 + \$10.00 + \$10.00)
−786.60	
+ 500.00	*payment*
−286.60	
−37.23	*finance charge*
−323.83	

She still owes \$323.83.

125. The outlay for 2005 is \$38.7 billion and the outlay for 2006 is \$69.1 billion. Thus, the *change in outlay* is

$$\begin{aligned}69.1 - 38.7 &= 69.1 + (-38.7)\\ &= 30.4\end{aligned}$$

billion dollars (an increase).

127. The outlay for 2007 is \$39.2 billion and the outlay for 2008 is \$42.3 billion. Thus, the *change in outlay* is

$$\begin{aligned}42.3 - 39.2 &= 42.3 + (-39.2)\\ &= 3.1\end{aligned}$$

billion dollars (an increase).

129. $17{,}400 - (-32{,}995) = 17{,}400 + 32{,}995 = 50{,}395$

The difference between the height of Mt. Foraker and the depth of the Philippine Trench is 50,395 feet.

131. $-23{,}376 - (-24{,}721) = -23{,}376 + 24{,}721 = 1345$

The Cayman Trench is 1345 feet deeper than the Java Trench.

133. $14{,}246 - 14{,}110 = 14{,}246 + (-14{,}110) = 136$

Mt. Wilson is 136 feet higher than Pikes Peak.

1.6 Multiplying and Dividing Real Numbers

1.6 Now Try Exercises

N1. **(a)** $-11(9) = -(11 \cdot 9) = -99$

(b) $3.1(-2.5) = -(3.1 \cdot 2.5) = -7.75$

N2. $-\frac{1}{7}\left(-\frac{5}{2}\right) = \frac{1}{7} \cdot \frac{5}{2} = \frac{1 \cdot 5}{7 \cdot 2} = \frac{5}{14}$

N3. **(a)** $\frac{15}{-3} = 15 \cdot \left(-\frac{1}{3}\right) = -5$

(b) $\frac{9.81}{-0.9} = 9.81\left(-\frac{1}{0.9}\right) = -10.9$

(c) $-\frac{5}{6} \div \frac{17}{9} = -\frac{5}{\cancel{6}_2} \cdot \frac{\cancel{9}^3}{17} = -\frac{15}{34}$

N4. **(a)** $\frac{-10}{5} = -2$

(b) $\frac{-1.44}{-0.12} = 12$

(c) $-\frac{3}{8} \div \frac{7}{10} = -\frac{3}{\cancel{8}_4} \cdot \frac{\cancel{10}^5}{7} = -\frac{15}{28}$

N5. **(a)**

$$\begin{aligned}-4(6) - (-5)5 &= -24 - (-25)\\ &= -24 + 25\\ &= 1\end{aligned}$$

(b)

$$\begin{aligned}\frac{12(-4) - 6(-3)}{-4(7 - 16)} &= \frac{-48 - (-18)}{-4(-9)}\\ &= \frac{-48 + 18}{36}\\ &= \frac{-30}{36} = -\frac{5}{6}\end{aligned}$$

N6. Replace x with -4 and y with -3.

$$\begin{aligned}\frac{3x^2 - 12}{y} &= \frac{3(-4)^2 - 12}{-3}\\ &= \frac{3(16) - 12}{-3}\\ &= \frac{48 - 12}{-3}\\ &= \frac{36}{-3} = -12\end{aligned}$$

N7. **(a)** "Twice the sum of -10 and 7" is written $2(-10 + 7)$.

$$2(-10 + 7) = 2(-3) = -6$$

(b) "40% of the difference between 45 and 15" is written $0.40(45 - 15)$.

$$0.40(45 - 15) = 0.40(30) = 12$$

N8. "The quotient of 21 and the sum of 10 and -7" is written $\frac{21}{10+(-7)}$.

$$\frac{21}{10+(-7)} = \frac{21}{3} = 7$$

N9. **(a)** "The sum of a number and -4 is 7" is written

$$x + (-4) = 7.$$

Here, x must be 4 more than 7, so the solution is 11.

$$11 + (-4) = 7, \ldots$$

(b) "The difference between -8 and a number is -11" is written

$$-8 - x = -11.$$

If we start at -8 on a number line, we must move 3 units to the left to get to -11, so the solution is 3.

1.6 Section Exercises

1. The product or the quotient of two numbers with the same sign is <u>greater than 0</u>, since the product or quotient of two positive numbers is positive and the product or quotient of two negative numbers is positive.

3. If three negative numbers are multiplied, the product is <u>less than 0</u>, since a negative number times a negative number is a positive number, and that positive number times a negative number is a negative number.

5. If a negative number is squared and the result is added to a positive number, the result is <u>greater than 0</u>, since a negative number squared is a positive number, and a positive number added to another positive number is a positive number.

7. If three positive numbers, five negative numbers, and zero are multiplied, the product is <u>equal to 0</u>. Since one of the numbers is zero, the product is zero (regardless of what the other numbers are).

9. The quotient formed by any nonzero number divided by 0 is <u>undefined</u>, and the quotient formed by 0 divided by any nonzero number is <u>0</u>. Examples include $\frac{1}{0}$, which is undefined, and $\frac{0}{1}$, which equals 0.

11. $5(-6) = -(5 \cdot 6) = -30$

Note that the product of a positive number and a negative number is negative.

13. $-5(-6) = 5 \cdot 6 = 30$

Note that the product of two negative numbers is positive.

15. $-10(-12) = 10 \cdot 12 = 120$

17. $3(-11) = -(3 \cdot 11) = -33$

19. $-0.5(0) = 0$

21. $-6.8(0.35) = -(6.8 \cdot 0.35) = -2.38$

23.
$$\begin{aligned} -\frac{3}{8} \cdot \left(-\frac{10}{9}\right) &= \frac{3}{8}\left(\frac{10}{9}\right) \\ &= \frac{3 \cdot 10}{8 \cdot 9} \\ &= \frac{3 \cdot (2 \cdot 5)}{(4 \cdot 2) \cdot (3 \cdot 3)} \\ &= \frac{3 \cdot 2 \cdot 5}{4 \cdot 2 \cdot 3 \cdot 3} \\ &= \frac{5}{4 \cdot 3} = \frac{5}{12} \end{aligned}$$

25.
$$\begin{aligned} \frac{2}{15}\left(-1\frac{1}{4}\right) &= \frac{2}{15}\left(-\frac{5}{4}\right) \\ &= -\frac{2 \cdot 5}{15 \cdot 4} \\ &= -\frac{2 \cdot 5}{3 \cdot 5 \cdot 2 \cdot 2} \\ &= -\frac{1}{3 \cdot 2} = -\frac{1}{6} \end{aligned}$$

27. $-8\left(-\frac{3}{4}\right) = 8\left(\frac{3}{4}\right) = \frac{24}{4} = 6$

29. Using only positive integer factors, 32 can be written as $1 \cdot 32$, $2 \cdot 16$, or $4 \cdot 8$. Including the negative integer factors, we see that the integer factors of 32 are $-32, -16, -8, -4, -2, -1, 1, 2, 4, 8, 16$, and 32.

31. The integer factors of 40 are $-40, -20, -10, -8, -5, -4, -2, -1, 1, 2, 4, 5, 8, 10, 20$, and 40.

33. The integer factors of 31 are $-31, -1, 1$, and 31.

35. $\frac{15}{5} = \frac{5 \cdot 3}{5} = \frac{3}{1} = 3$

37. $\frac{-42}{6} = -\frac{2 \cdot 3 \cdot 7}{2 \cdot 3} = -7$

Note that the quotient of two numbers having different signs is negative.

39. $\frac{-32}{-4} = \frac{4 \cdot 8}{4} = 8$

Note that the quotient of two numbers having the same sign is positive.

41. $\frac{96}{-16} = -\frac{6 \cdot 16}{16} = -6$

43. Dividing by a fraction (in this case, $-\frac{1}{8}$) is the same as multiplying by the reciprocal of the fraction (in this case, $-\frac{8}{1}$).

$$\left(-\frac{4}{3}\right) \div \left(-\frac{1}{8}\right) = \left(-\frac{4}{3}\right)\cdot\left(-\frac{8}{1}\right)$$
$$= \frac{4\cdot 8}{3\cdot 1}$$
$$= \frac{32}{3}, \text{ or } 10\frac{2}{3}$$

45. $\dfrac{-8.8}{2.2} = -\dfrac{4(2.2)}{2.2} = -4$

47. $\dfrac{0}{-5} = 0$, because 0 divided by any nonzero number is 0.

49. $\dfrac{11.5}{0}$ is *undefined* because we cannot divide by 0.

In Exercises 51–68, use the order of operations.

51. $7 - 3\cdot 6 = 7 - 18$
$= -11$

53. $-10 - (-4)(2) = -10 - (-8)$
$= -10 + 8$
$= -2$

55. $-7(3 - 8) = -7[3 + (-8)]$
$= -7(-5) = 35$

57. $7 + 2(4 - 1) = 7 + 2(3)$
$= 7 + 6$
$= 13$

59. $-4 + 3(2 - 8) = -4 + 3(-6)$
$= -4 - 18$
$= -22$

61. $(12 - 14)(1 - 4) = (-2)(-3)$
$= 6$

63. $(7 - 10)(10 - 4) = (-3)(6)$
$= -18$

65. $(-2 - 8)(-6) + 7 = (-10)(-6) + 7$
$= 60 + 7$
$= 67$

67. $3(-5) + |3 - 10| = -15 + |-7|$
$= -15 + 7$
$= -8$

69. $\dfrac{-5(-6)}{9 - (-1)} = \dfrac{30}{10}$
$= \dfrac{3\cdot 10}{10} = 3$

71. $\dfrac{-21(3)}{-3 - 6} = \dfrac{-63}{-3 + (-6)}$
$= \dfrac{-63}{-9} = 7$

73. $\dfrac{-10(2) + 6(2)}{-3 - (-1)} = \dfrac{-20 + 12}{-3 + 1}$
$= \dfrac{-8}{-2} = 4$

75. $\dfrac{3^2 - 4^2}{7(-8 + 9)} = \dfrac{9 - 16}{7(1)} = \dfrac{-7}{7} = -1$

77. $\dfrac{8(-1) - |(-4)(-3)|}{-6 - (-1)} = \dfrac{-8 - |12|}{-6 + 1}$
$= \dfrac{-8 - 12}{-5}$
$= \dfrac{-20}{-5} = 4$

79. $\dfrac{-13(-4) - (-8)(-2)}{(-10)(2) - 4(-2)}$
$= \dfrac{52 - 16}{-20 - (-8)}$
$= \dfrac{36}{-20 + 8}$
$= \dfrac{36}{-12} = -3$

In Exercises 81–92, replace x with 6, y with -4, and a with 3. Then use the order of operations to evaluate the expression.

81. $5x - 2y + 3a = 5(6) - 2(-4) + 3(3)$
$= 30 - (-8) + 9$
$= 30 + 8 + 9$
$= 38 + 9$
$= 47$

83. $(2x + y)(3a) = [2(6) + (-4)][3(3)]$
$= [12 + (-4)](9)$
$= (8)(9)$
$= 72$

85. $\left(\frac{1}{3}x-\frac{4}{5}y\right)\left(-\frac{1}{5}a\right)$

$$=\left[\frac{1}{3}(6)-\frac{4}{5}(-4)\right]\left[-\frac{1}{5}(3)\right]$$
$$=\left[2-\left(-\frac{16}{5}\right)\right]\left(-\frac{3}{5}\right)$$
$$=\left(2+\frac{16}{5}\right)\left(-\frac{3}{5}\right)$$
$$=\left(\frac{10}{5}+\frac{16}{5}\right)\left(-\frac{3}{5}\right)$$
$$=\left(\frac{26}{5}\right)\left(-\frac{3}{5}\right)$$
$$=-\frac{78}{25}$$

87. $(-5+x)(-3+y)(3-a)$
$$=(-5+6)[-3+(-4)][3-3]$$
$$=(1)(-7)(0)$$
$$=0$$

89. $-2y^2+3a=-2(-4)^2+3(3)$
$$=-2(16)+9$$
$$=-32+9$$
$$=-23$$

91. $\dfrac{2y^2-x}{a+10}=\dfrac{2(-4)^2-(6)}{3+10}$
$$=\frac{2(16)-6}{13}$$
$$=\frac{32-6}{13}$$
$$=\frac{26}{13}$$
$$=2$$

93. "The product of -9 and 2, added to 9" is written $9+(-9)(2)$.

$$9+(-9)(2)=9+(-18)$$
$$=-9$$

95. "Twice the product of -1 and 6, subtracted from -4" is written $-4-2[(-1)(6)]$.

$$-4-2[(-1)(6)]=-4-2(-6)$$
$$=-4-(-12)$$
$$=-4+12=8$$

97. "Nine subtracted from the product of 1.5 and -3.2 is written $(1.5)(-3.2)-9$.

$$(1.5)(-3.2)-9=-4.8-9$$
$$=-4.8+(-9)$$
$$=-13.8$$

99. "The product of 12 and the difference between 9 and -8" is written $12[9-(-8)]$.

$$12[9-(-8)]=12[9+8]$$
$$=12(17)=204$$

101. "The quotient of -12 and the sum of -5 and -1" is written

$$\frac{-12}{-5+(-1)},$$

and

$$\frac{-12}{-5+(-1)}=\frac{-12}{-6}=2.$$

103. "The sum of 15 and -3, divided by the product of 4 and -3" is written

$$\frac{15+(-3)}{4(-3)},$$

and

$$\frac{15+(-3)}{4(-3)}=\frac{12}{-12}=-1.$$

105. "Two-thirds of the difference between 8 and -1" is written

$$\tfrac{2}{3}[8-(-1)],$$

and

$$\tfrac{2}{3}[8-(-1)]=\tfrac{2}{3}[8+(1)]=\tfrac{2}{3}[9]=6.$$

107. "20% of the product of -5 and 6" is written

$$0.20(-5\cdot 6),$$

and

$$0.20(-5\cdot 6)=0.20(-30)=-6.$$

109. "The sum of $\frac{1}{2}$ and $\frac{5}{8}$, times the difference between $\frac{3}{5}$ and $\frac{1}{3}$" is written

$$\left(\tfrac{1}{2}+\tfrac{5}{8}\right)\left(\tfrac{3}{5}-\tfrac{1}{3}\right),$$

and

$$\left(\tfrac{1}{2}+\tfrac{5}{8}\right)\left(\tfrac{3}{5}-\tfrac{1}{3}\right)=\left(\tfrac{4}{8}+\tfrac{5}{8}\right)\left(\tfrac{9}{15}-\tfrac{5}{15}\right)$$
$$=\tfrac{9}{8}\left(\tfrac{4}{15}\right)$$
$$=\frac{3\cdot 3\cdot 4}{2\cdot 4\cdot 3\cdot 5}=\frac{3}{10}.$$

111. "The product of $-\frac{1}{2}$ and $\frac{3}{4}$, divided by $-\frac{2}{3}$" is written $\dfrac{-\frac{1}{2}\left(\frac{3}{4}\right)}{-\frac{2}{3}}$. Simplifying gives us:

$$\frac{-\frac{1}{2}\left(\frac{3}{4}\right)}{-\frac{2}{3}}=\frac{-\frac{3}{8}}{-\frac{2}{3}}$$
$$=-\frac{3}{8}\cdot\left(-\frac{3}{2}\right)$$
$$=\frac{9}{16}$$

113. "The quotient of a number and 3 is -3" is written

$$\frac{x}{3} = -3.$$

The solution is -9, since

$$\frac{-9}{3} = -3.$$

115. "6 less than a number is 4" is written

$$x - 6 = 4.$$

The solution is 10, since

$$10 - 6 = 4.$$

117. "When 5 is added to a number, the result is -5" is written

$$x + 5 = -5.$$

The solution is -10, since

$$-10 + 5 = -5.$$

119. Add the numbers and divide by 5.

$$\frac{(23 + 18 + 13) + [(-4) + (-8)]}{5}$$
$$= \frac{54 - 12}{5}$$
$$= \frac{42}{5}, \text{ or } 8\frac{2}{5}$$

121. Add the numbers and divide by 4.

$$\frac{(29 + 8) + [(-15) + (-6)]}{4}$$
$$= \frac{37 - 21}{4}$$
$$= \frac{16}{4} = 4$$

123. Add the integers from -10 to 14.

$$(-10) + (-9) + \cdots + 14 = 50$$

[the 3 dots indicate that the pattern continues]

There are 25 integers from -10 to 14 (10 negative, zero, and 14 positive). Thus, the average is

$$\frac{50}{25} = 2.$$

125. **(a)** 3,473,986 is divisible by 2 because its last digit, 6, is divisible by 2.

(b) 4,336,879 is not divisible by 2 because its last digit, 9, is not divisible by 2.

127. **(a)** 6,221,464 is divisible by 4 because the number formed by its last two digits, 64, is divisible by 4.

(b) 2,876,335 is not divisible by 4 because the number formed by its last two digits, 35, is not divisible by 4.

129. **(a)** 1,524,822 is divisible by 2 because its last digit, 2, is divisible by 2. It is also divisible by 3 because the sum of its digits,

$$1 + 5 + 2 + 4 + 8 + 2 + 2 = 24,$$

is divisible by 3.

Because 1,524,822 is divisible by *both* 2 and 3, it is divisible by 6.

(b) 2,873,590 is divisible by 2 because its last digit, 0, is divisible by 2. However, it is not divisible by 3 because the sum of its digits,

$$2 + 8 + 7 + 3 + 5 + 9 + 0 = 34,$$

is not divisible by 3.

Because 2,873,590 is not divisible by *both* 2 and 3, it is not divisible by 6.

131. **(a)** 4,114,107 is divisible by 9 because the sum of its digits,

$$4 + 1 + 1 + 4 + 1 + 0 + 7 = 18,$$

is divisible by 9.

(b) 2,287,321 is not divisible by 9 because the sum of its digits,

$$2 + 2 + 8 + 7 + 3 + 2 + 1 = 25,$$

is not divisible by 9.

Summary Exercises on Operations with Real Numbers

1.
$$\begin{aligned} 14 - 3 \cdot 10 &= 14 - 30 \\ &= 14 + (-30) \\ &= -16 \end{aligned}$$

3.
$$\begin{aligned} (3 - 8)(-2) - 10 &= (-5)(-2) - 10 \\ &= 10 - 10 \\ &= 0 \end{aligned}$$

5.
$$\begin{aligned} 7 + 3(2 - 10) &= 7 + 3(-8) \\ &= 7 - 24 \\ &= -17 \end{aligned}$$

7.
$$\begin{aligned} (-4)(7) - (-5)(2) &= (-28) - (-10) \\ &= -28 + (10) \\ &= -18 \end{aligned}$$

9.
$$\begin{aligned} 40 - (-2)[8 - 9] &= 40 - (-2)[-1] \\ &= 40 - (2) \\ &= 38 \end{aligned}$$

11. $$\frac{-3-(-9+1)}{-7-(-6)} = \frac{-3-(-8)}{-7+6} = \frac{-3+8}{-1} = \frac{5}{-1} = -5$$

13. $$\frac{6^2-8}{-2(2)+4(-1)} = \frac{36-8}{-4+(-4)} = \frac{28}{-8} = -\frac{4\cdot 7}{2\cdot 4} = -\frac{7}{2}, \text{ or } -3\frac{1}{2}$$

15. $$\frac{9(-6)-3(8)}{4(-7)+(-2)(-11)} = \frac{-54-24}{-28+22} = \frac{-78}{-6} = 13$$

17. $$\frac{(2+4)^2}{(5-3)^2} = \frac{(6)^2}{(2)^2} = \frac{36}{4} = 9$$

19. $$\frac{-9(-6)+(-2)(27)}{3(8-9)} = \frac{(54)+(-54)}{3(-1)} = \frac{0}{-3} = 0$$

21. $$\frac{6(-10+3)}{15(-2)-3(-9)} = \frac{6(-7)}{(-30)-(-27)} = \frac{-42}{-30+27} = \frac{-42}{-3} = 14$$

23. $$\frac{(-10)^2+10^2}{-10(5)} = \frac{100+100}{-50} = \frac{200}{-50} = -4$$

25. $$\frac{1}{2} \div \left(-\frac{1}{2}\right) = \frac{1}{2}\cdot\left(-\frac{2}{1}\right) = -\frac{2}{2} = -1$$

27. $$\left[\frac{5}{8}-\left(-\frac{1}{16}\right)\right]+\frac{3}{8} = \left[\frac{10}{16}+\frac{1}{16}\right]+\frac{6}{16} = \left[\frac{11}{16}\right]+\frac{6}{16} = \frac{17}{16}, \text{ or } 1\frac{1}{16}$$

29. $$-0.9(-3.7) = 0.9(3.7) = 3.33$$

31. $$-3^2-2^2 = -(3^2)-(2^2) = -9-4 = -13$$

33. $$40+2[-5-3] = 40+2[-8] = 40-16 = 24$$

In Exercises 34–42, replace x with -2, y with 3, and a with 4. Then use the order of operations to evaluate the expression.

35. $$(x+6)^3-y^3 = (-2+6)^3-3^3 = (4)^3-27 = 64-27 = 37$$

37. $$\left(\frac{1}{2}x+\frac{2}{3}y\right)\left(-\frac{1}{4}a\right) = \left(\frac{1}{2}(-2)+\frac{2}{3}(3)\right)\left(-\frac{1}{4}(4)\right) = (-1+2)(-1) = (1)(-1) = -1$$

39. $$\frac{x^2-y^2}{x^2+y^2} = \frac{(-2)^2-3^2}{(-2)^2+3^2} = \frac{4-9}{4+9} = \frac{-5}{13} = -\frac{5}{13}$$

41. $$\left(\frac{x}{y}\right)^3 = \left(\frac{-2}{3}\right)^3 = \left(-\frac{2}{3}\right)\left(-\frac{2}{3}\right)\left(-\frac{2}{3}\right) = -\frac{8}{27}$$

1.7 Properties of Real Numbers

1.7 Now Try Exercises

N1. **(a)** $7+(-3) = -3+\underline{7}$

(b) $(-5)4 = 4\cdot\underline{(-5)}$

N2. **(a)** $-9+(3+7) = \underline{(-9+3)}+7$

(b) $5[(-4)\cdot 9] = \underline{[5\cdot(-4)]\cdot 9}$

N3. $5+(7+6) = 5+(6+7)$

While the same numbers are grouped inside the two pairs of parentheses, the order of the numbers has been changed. This illustrates a *commutative* property.

N4. **(a)** $8+54+7+6+32$
$=(8+32)+(54+6)+7$
$=40+60+7$
$=100+7$
$=107$

(b) $5(37)(20)=5(20)(37)=100(37)=3700$

N5. **(a)** $\frac{2}{5}\cdot \underline{1}=\frac{2}{5}$ *Multiplicative identity*

(b) $8+\underline{0}=8$ *Additive identity*

N6. **(a)** $\frac{16}{20}=\frac{4\cdot 4}{5\cdot 4}$ *Factor.*

$=\frac{4}{5}\cdot\frac{4}{4}$ *Write as a product.*

$=\frac{4}{5}\cdot 1$ *Divide.*

$=\frac{4}{5}$ *Identity property*

(b) $\frac{2}{5}+\frac{3}{20}=\frac{2}{5}\cdot 1+\frac{3}{20}$ *Identity property*

$=\frac{2}{5}\cdot\frac{4}{4}+\frac{3}{20}$ *Use $1=\frac{4}{4}$ to get a common denominator.*

$=\frac{8}{20}+\frac{3}{20}$ *Multiply.*

$=\frac{11}{20}$ *Add.*

N7. **(a)** $10+\underline{-10}=0$ *Inverse property*

(b) $-9\cdot\underline{(-\frac{1}{9})}=1$ *Inverse property*

N8. $-\frac{1}{3}x+7+\frac{1}{3}x$

$=\left(-\frac{1}{3}x+7\right)+\frac{1}{3}x$ *Order of operations*

$=\left[7+\left(-\frac{1}{3}x\right)\right]+\frac{1}{3}x$ *Commutative property*

$=7+\left[\left(-\frac{1}{3}x\right)+\frac{1}{3}x\right]$ *Associative property*

$=7+0$ *Inverse property*

$=7$ *Identity property*

N9. **(a)** $-5(4x+1)=-5\cdot 4x+(-5\cdot 1)$ *Dist. prop.*
$=-20x-5$ *Multiply.*

(b) $6(2r+t-5z)=6(2r)+6t+6(-5z)$
$=12r+6t-30z$

(c) $5x-5y=5(x-y)$

N10. **(a)** $-(2-r)=-1(2-r)$
$=-2+r$

(b) $-(2x-5y-7)=-1(2x-5y-7)$
$=-2x+5y+7$

1.7 Section Exercises

1. **(a)** **B**, since 0 is the identity element for addition.

(b) **F**, since 1 is the identity element for multiplication.

(c) **C**, since $-a$ is the additive inverse of a.

(d) **I**, since $\frac{1}{a}$ is the multiplicative inverse, or reciprocal, of any nonzero number a.

(e) **B**, since 0 is the only number that is equal to its negative; that is, $0=-0$.

(f) **D** and **F**, since -1 has reciprocal $\frac{1}{(-1)}=-1$ and 1 has a reciprocal $\frac{1}{(1)}=1$; that is, -1 and 1 are their own multiplicative inverses.

(g) **B**, since the multiplicative inverse of a number a is $\frac{1}{a}$ and the only number that we *cannot* divide by is 0.

(h) **A**

(i) **G**, since we can consider $(5\cdot 4)$ to be one number, $(5\cdot 4)\cdot 3$ is the same as $3\cdot(5\cdot 4)$ by the commutative property.

(j) **H**

3. "Washing your face" and "brushing your teeth" *are* commutative.

5. "Preparing a meal" and "eating a meal" *are not* commutative.

7. "Putting on your socks" and "putting on your shoes" *are not* commutative.

9. "(Foreign sales) clerk" is a clerk dealing with foreign sales, whereas "foreign (sales clerk)" is a sales clerk who is foreign.

11. $-15+9=9+\underline{(-15)}$
by the *commutative property of addition.*

13. $-8\cdot 3=\underline{3}\cdot(-8)$
by the *commutative property of multiplication.*

15. $(3+6)+7=3+(\underline{6}+7)$
by the *associative property of addition.*

17. $7\cdot(2\cdot 5)=(\underline{7}\cdot 2)\cdot 5$
by the *associative property of multiplication.*

19. $25-(6-2)=25-(4)$
$=21$
$(25-6)-2=19-2$
$=17$

Since $21\neq 17$, this example shows that subtraction is not associative.

21.

Number	Additive inverse	Multiplicative inverse
5	-5	$\frac{1}{5}$
-10	10	$-\frac{1}{10}$
$-\frac{1}{2}$	$\frac{1}{2}$	-2
$\frac{3}{8}$	$-\frac{3}{8}$	$\frac{8}{3}$
$x\ (x \neq 0)$	$-x$	$\frac{1}{x}$
$-y\ (y \neq 0)$	y	$-\frac{1}{y}$

In general, a number and its additive inverse have opposite signs. A number and its multiplicative inverse have the same signs.

23. $4 + 15 = 15 + 4$

The order of the two numbers has been changed, so this is an example of the commutative property of addition: $a + b = b + a$.

25. $5 \cdot (13 \cdot 7) = (5 \cdot 13) \cdot 7$

The numbers are in the same order but grouped differently, so this is an example of the associative property of multiplication: $(ab)c = a(bc)$.

27. $-6 + (12 + 7) = (-6 + 12) + 7$

The numbers are in the same order but grouped differently, so this is an example of the associative property of addition: $(a + b) + c = a + (b + c)$.

29. $-9 + 9 = 0$

The sum of the two numbers is 0, so they are additive inverses (or opposites) of each other. This is an example of the additive inverse property: $a + (-a) = 0$.

31. $\frac{2}{3}\left(\frac{3}{2}\right) = 1$

The product of the two numbers is 1, so they are multiplicative inverses (or reciprocals) of each other. This is an example of the multiplicative inverse property: $a \cdot \frac{1}{a} = 1\ (a \neq 0)$.

33. $1.75 + 0 = 1.75$

The sum of a number and 0 is the original number. This is an example of the identity property of addition: $a + 0 = a$.

35. $(4 + 17) + 3 = 3 + (4 + 17)$

The order of the numbers has been changed, but not the grouping, so this is an example of the commutative property of addition: $a + b = b + a$.

37. $2(x + y) = 2x + 2y$

The number 2 outside the parentheses is "distributed" over the x and y. This is an example of the distributive property.

39. $-\frac{5}{9} = -\frac{5}{9} \cdot \frac{3}{3} = -\frac{15}{27}$

$\frac{3}{3}$ is a form of the number 1. We use it to rewrite $-\frac{5}{9}$ as $-\frac{15}{27}$. This is an example of the identity property of multiplication.

41. $4(2x) + 4(3y) = 4(2x + 3y)$

This is an example of the distributive property. The number 4 is "distributed " over $2x$ and $3y$.

43.
$$\begin{aligned} 97 + 13 + 3 + 37 &= (97 + 3) + (13 + 37) \\ &= 100 + 50 \\ &= 150 \end{aligned}$$

45.
$$\begin{aligned} 1999 + 2 + 1 + 8 &= (1999 + 1) + (2 + 8) \\ &= 2000 + 10 \\ &= 2010 \end{aligned}$$

47.
$$\begin{aligned} 159 + 12 + 141 + 88 &= (159 + 141) + (12 + 88) \\ &= 300 + 100 \\ &= 400 \end{aligned}$$

49.
$$\begin{aligned} &843 + 627 + (-43) + (-27) \\ &= [843 + (-43)] + [627 + (-27)] \\ &= 800 + 600 \\ &= 1400 \end{aligned}$$

51. $5(47)(2) = 5(2)(47) = 10(47) = 470$

53.
$$\begin{aligned} -4 \cdot 5 \cdot 93 \cdot 5 &= -4 \cdot 5 \cdot 5 \cdot 93 \\ &= -20 \cdot 5 \cdot 93 \\ &= -100 \cdot 93 \\ &= -9300 \end{aligned}$$

55. $6t + 8 - 6t + 3$

$= 6t + 8 + (-6t) + 3$ *Definition of subtraction*

$= (6t + 8) + (-6t) + 3$ *Order of operations*

$= (8 + 6t) + (-6t) + 3$ *Commutative property*

$= 8 + [6t + (-6t)] + 3$ *Associative property*

$= 8 + 0 + 3$ *Inverse property*

$= (8 + 0) + 3$ *Order of operations*

$= 8 + 3$ *Identity property*

$= 11$ *Add.*

57. $\frac{2}{3}x - 11 + 11 - \frac{2}{3}x$

$= \frac{2}{3}x + (-11) + 11 + \left(-\frac{2}{3}x\right)$

Definition of subtraction

$= \left[\frac{2}{3}x + (-11)\right] + 11 + \left(-\frac{2}{3}x\right)$

Order of operations

$= \frac{2}{3}x + (-11 + 11) + \left(-\frac{2}{3}x\right)$

Associative property

$= \frac{2}{3}x + 0 + \left(-\frac{2}{3}x\right)$ *Inverse property*

$= \left(\frac{2}{3}x + 0\right) + \left(-\frac{2}{3}x\right)$

Order of operations

$= \frac{2}{3}x + \left(-\frac{2}{3}x\right)$ *Identity property*

$= 0$ *Inverse property*

59. $\left(\frac{9}{7}\right)(-0.38)\left(\frac{7}{9}\right)$

$= \left[\left(\frac{9}{7}\right)(-0.38)\right]\left(\frac{7}{9}\right)$ *Order of operations*

$= \left[(-0.38)\left(\frac{9}{7}\right)\right]\left(\frac{7}{9}\right)$ *Commutative property*

$= (-0.38)\left[\left(\frac{9}{7}\right)\left(\frac{7}{9}\right)\right]$ *Associative property*

$= (-0.38)(1)$ *Inverse property*

$= -0.38$ *Identity property*

61. $t + (-t) + \frac{1}{2}(2)$

$= t + (-t) + 1$ *Inverse property*

$= [t + (-t)] + 1$ *Order of operations*

$= 0 + 1$ *Inverse property*

$= 1$ *Identity property*

63. $-3(4 - 6)$

When distributing a negative number over a quantity, be careful not to "lose" a negative sign. The problem should be worked in the following way.

$$\begin{aligned} -3(4-6) &= -3(4) - 3(-6) \\ &= -12 + 18 \\ &= 6 \end{aligned}$$

65. $\begin{aligned} 5(9+8) &= 5 \cdot 9 + 5 \cdot 8 \\ &= 45 + 40 \\ &= 85 \end{aligned}$

67. $\begin{aligned} 4(t+3) &= 4 \cdot t + 4 \cdot 3 \\ &= 4t + 12 \end{aligned}$

69. $\begin{aligned} 7(z-8) &= 7[z + (-8)] \\ &= 7z + 7(-8) \\ &= 7z - 56 \end{aligned}$

71. $\begin{aligned} -8(r+3) &= -8(r) + (-8)(3) \\ &= -8r + (-24) \\ &= -8r - 24 \end{aligned}$

73. $-\frac{1}{4}(8x + 3)$

$= -\frac{1}{4}(8x) + (-\frac{1}{4})(3)$

$= [(-\frac{1}{4}) \cdot 8]x - \frac{3}{4}$

$= -2x - \frac{3}{4}$

75. $\begin{aligned} -5(y-4) &= -5(y) + (-5)(-4) \\ &= -5y + 20 \end{aligned}$

77. $-\frac{4}{3}(12y + 15z)$

$= -\frac{4}{3}(12y) + (-\frac{4}{3})(15z)$

$= [(-\frac{4}{3}) \cdot 12]y + [(-\frac{4}{3}) \cdot 15]z$

$= -16y + (-20)z$

$= -16y - 20z$

79. $8z + 8w = 8(z + w)$

81. $7(2v) + 7(5r) = 7(2v + 5r)$

83. $8(3r + 4s - 5y)$

$= 8(3r) + 8(4s) + 8(-5y)$

Distributive property

$= (8 \cdot 3)r + (8 \cdot 4)s + [8(-5)]y$

Associative property

$= 24r + 32s - 40y$ *Multiply.*

85. $-3(8x + 3y + 4z)$

$= -3(8x) + (-3)(3y) + (-3)(4z)$

Distributive property

$= (-3 \cdot 8)x + (-3 \cdot 3)y + (-3 \cdot 4)z$

Associative property

$= -24x - 9y - 12z$ *Multiply.*

87. $\begin{aligned} 5x + 15 &= 5x + 5 \cdot 3 \\ &= 5(x + 3) \end{aligned}$

89. $-(4t + 3m)$

$= -1(4t + 3m)$ *Identity property*

$= -1(4t) + (-1)(3m)$ *Distributive property*

$= (-1 \cdot 4)t + (-1 \cdot 3)m$ *Associative property*

$= -4t - 3m$ *Multiply.*

91. $-(-5c-4d)$

$= -1(-5c-4d)$	*Identity property*
$= -1(-5c)+(-1)(-4d)$	*Distributive property*
$= (-1\cdot -5)c+(-1\cdot -4)d$	*Associative property*
$= 5c+4d$	*Multiply.*

93. $-(-q+5r-8s)$
$= -1(-q+5r-8s)$
$= -1(-q)+(-1)(5r)+(-1)(-8s)$
$= (-1\cdot -1)q+(-1\cdot 5)r+(-1\cdot -8)s$
$= q-5r+8s$

1.8 Simplifying Expressions

1.8 Now Try Exercises

N1. **(a)** $3(2x-4y) = 3(2x)-3(4y)$
$= (3\cdot 2)x-(3\cdot 4)y$
$= 6x-12y$

(b) $-4-(-3y+5) = -4-1(-3y+5)$
$= -4-1(-3y)-1(5)$
$= -4+3y+(-5)$
$= -4+(-5)+3y$
$= -9+3y$, or $3y-9$

N2. **(a)** $4x+6x-7x=(4+6-7)x=3x$

(b) $z+z=1z+1z=(1+1)z=2z$

(c) $4p^2-3p^2=(4-3)p^2=1p^2$, or p^2

N3. **(a)** $5k-6-(3-4k)$
$= 5k-6-1(3-4k)$
$= 5k-6-1(3)-1(-4k)$
$= 5k-6-3+4k$
$= 9k-9$

(b) $\frac{1}{4}x-\frac{2}{3}(x-9) = \frac{1}{4}x-\frac{2}{3}(x)-\frac{2}{3}(-9)$
$= \frac{3}{12}x-\frac{8}{12}x+6$
$= -\frac{5}{12}x+6$

N4. "Twice a number, subtracted from the sum of the number and 5" is written $(x+5)-2x$.

$(x+5)-2x = x+5-2x$
$= -x+5$, or $5-x$

1.8 Section Exercises

1. $-(6x-3) = -1(6x-3)$
$= -1(6x)-1(-3)$
$= -6x+3$
The correct response is **B**.

3. Examples **A**, **B**, and **D** are pairs of *unlike* terms since either the variables or their powers are different. Example **C** is a pair of *like* terms, since both terms have the same variables (r and y) and the same exponents (both variables are to the first power). Note that we can use the commutative property to rewrite $6yr$ as $6ry$.

5. $4r+19-8=4r+11$

7. $5+2(x-3y) = 5+2(x)+2(-3y)$
$= 5+2x-6y$

9. $-2-(5-3p) = -2-1(5-3p)$
$= -2-1(5)-1(-3p)$
$= -2-5+3p$
$= -7+3p$

11. $6+(4-3x)-8 = 6+4-3x-8$
$= 10-3x-8$
$= 10-8-3x$
$= 2-3x$

13. The numerical coefficient of the term $-12k$ is -12.

15. The numerical coefficient of the term $3m^2$ is 3.

17. Because xw can be written as $1\cdot xw$, the numerical coefficient of the term xw is 1.

19. Since $-x=-1x$, the numerical coefficient of the term $-x$ is -1.

21. Since $\frac{x}{2}=\frac{1}{2}x$, the numerical coefficient of the term $\frac{x}{2}$ is $\frac{1}{2}$.

23. Since $\frac{2x}{5}=\frac{2}{5}x$, the numerical coefficient of the term $\frac{2x}{5}$ is $\frac{2}{5}$.

25. The numerical coefficient of the term 10 is 10.

27. $8r$ and $-13r$ are *like* terms since they have the same variable with the same exponent (which is understood to be 1).

29. $5z^4$ and $9z^3$ are *unlike* terms. Although both have the variable z, the exponents are not the same.

31. All numerical terms (constants) are considered like terms, so 4, 9, and -24 are *like* terms.

33. x and y are *unlike* terms because they do not have the same variable.

35. The student made a sign error when applying the distributive property.

$7x-2(3-2x) = 7x-2(3)-2(-2x)$
$= 7x-6+4x$
$= 11x-6$

The correct answer is $11x-6$.

37. $7y + 6y = (7+6)y$
$= 13y$

39. $-6x - 3x = (-6-3)x$
$= -9x$

41. $12b + b = 12b + 1b$
$= (12+1)b$
$= 13b$

43. $3k + 8 + 4k + 7 = 3k + 4k + 8 + 7$
$= (3+4)k + 15$
$= 7k + 15$

45. $-5y + 3 - 1 + 5 + y - 7$
$= (-5y + 1y) + (3+5) + (-1-7)$
$= (-5+1)y + (8) + (-8)$
$= -4y + 8 - 8$
$= -4y$

47. $-2x + 3 + 4x - 17 + 20$
$= (-2x + 4x) + (3 - 17 + 20)$
$= (-2+4)x + 6$
$= 2x + 6$

49. $16 - 5m - 4m - 2 + 2m$
$= (16 - 2) + (-5m - 4m + 2m)$
$= 14 + (-5 - 4 + 2)m$
$= 14 - 7m$

51. $-10 + x + 4x - 7 - 4x$
$= (-10 - 7) + (1x + 4x - 4x)$
$= -17 + (1 + 4 - 4)x$
$= -17 + 1x$
$= -17 + x$

53. $1 + 7x + 11x - 1 + 5x$
$= (1-1) + (7x + 11x + 5x)$
$= 0 + (7 + 11 + 5)x$
$= 23x$

55. $-\frac{4}{3} + 2t + \frac{1}{3}t - 8 - \frac{8}{3}t$
$= \left(2t + \frac{1}{3}t - \frac{8}{3}t\right) + \left(-\frac{4}{3} - 8\right)$
$= \left(2 + \frac{1}{3} - \frac{8}{3}\right)t + \left(-\frac{4}{3} - 8\right)$
$= \left(\frac{6}{3} + \frac{1}{3} - \frac{8}{3}\right)t + \left(-\frac{4}{3} - \frac{24}{3}\right)$
$= -\frac{1}{3}t - \frac{28}{3}$

57. $6y^2 + 11y^2 - 8y^2 = (6 + 11 - 8)y^2$
$= 9y^2$

59. $2p^2 + 3p^2 - 8p^3 - 6p^3$
$= (2p^2 + 3p^2) + (-8p^3 - 6p^3)$
$= (2+3)p^2 + (-8-6)p^3$
$= 5p^2 - 14p^3$ or $-14p^3 + 5p^2$

61. $2(4x + 6) + 3 = 2(4x) + 2(6) + 3$
$= 8x + 12 + 3$
$= 8x + 15$

63. $100[0.05(x+3)]$
$= [100(0.05)](x+3)$ *Associative property*
$= 5(x+3)$
$= 5(x) + 5(3)$ *Distributive property*
$= 5x + 15$

65. $-6 - 4(y - 7)$
$= -6 - 4(y) + (-4)(-7)$ *Distributive property*
$= -6 - 4y + 28$
$= -4y + 22$

67. $-\frac{4}{3}(y - 12) - \frac{1}{6}y$
$= -\frac{4}{3}y - \frac{4}{3}(-12) - \frac{1}{6}y$
$= -\frac{4}{3}y + 16 - \frac{1}{6}y$
$= -\frac{4}{3}y - \frac{1}{6}y + 16$
$= \left(-\frac{8}{6} - \frac{1}{6}\right)y + 16$
$= -\frac{3}{2}y + 16$ $\left[-\frac{9}{6} = -\frac{3}{2}\right]$

69. $-5(5y - 9) + 3(3y + 6)$
$= -5(5y) + (-5)(-9) + 3(3y) + 3(6)$
Distributive property
$= -25y + 45 + 9y + 18$
$= (-25y + 9y) + (45 + 18)$
$= (-25 + 9)y + 63$
$= -16y + 63$

71. $-3(2r - 3) + 2(5r + 3)$
$= -3(2r) + (-3)(-3) + 2(5r) + 2(3)$
Distributive property
$= -6r + 9 + 10r + 6$
$= (-6r + 10r) + (9 + 6)$
$= (-6 + 10)r + 15$
$= 4r + 15$

73. $8(2k-1)-(4k-3)$
$=8(2k-1)-1(4k-3)$ *Replace − with −1.*
$=8(2k)+8(-1)+(-1)(4k)+(-1)(-3)$
$=16k-8-4k+3$
$=12k-5$

75. $-2(-3k+2)-(5k-6)-3k-5$
$=-2(-3k)+(-2)(2)-1(5k-6)-3k-5$
$=6k-4+(-1)(5k)+(-1)(-6)-3k-5$
$=6k-4-5k+6-3k-5$
$=-2k-3$

77. $-4(-3x+3)-(6x-4)-2x+1$
$=-4(-3x+3)-1(6x-4)-2x+1$
$=12x-12-6x+4-2x+1$ *Distributive property*
$=(12x-6x-2x)+(-12+4+1)$ *Group like terms.*
$=4x-7$ *Combine like terms.*

79. $-7.5(2y+4)-2.9(3y-6)$
$=-7.5(2y)-7.5(4)-2.9(3y)-2.9(-6)$ *Distributive property*
$=-15y-30-8.7y+17.4$ *Multiply.*
$=-23.7y-12.6$ *Combine like terms.*

81. "Five times a number, added to the sum of the number and three" is written $(x+3)+5x$.

$$\begin{aligned}(x+3)+5x &= x+3+5x\\ &= (x+5x)+3\\ &= 6x+3\end{aligned}$$

83. "A number multiplied by −7, subtracted from the sum of 13 and six times the number" is written $(13+6x)-(-7x)$.

$$\begin{aligned}(13+6x)-(-7x) &= 13+6x+7x\\ &= 13+13x\end{aligned}$$

85. "Six times a number added to −4, subtracted from twice the sum of three times the number and 4" is written $2(3x+4)-(-4+6x)$.

$$\begin{aligned}&2(3x+4)-(-4+6x)\\ &= 2(3x+4)-1(-4+6x)\\ &= 6x+8+4-6x\\ &= 6x+(-6x)+8+4\\ &= 0+12=12\end{aligned}$$

♦♦♦ Relating Concepts 87–90 ♦♦♦

87. For widgets, the fixed cost is \$1000 and the variable cost is \$5 per widget, so the cost to produce x widgets is

$$1000+5x \text{ (dollars).}$$

88. For gadgets, the fixed cost is \$750 and the variable cost is \$3 per gadget, so the cost to produce y gadgets is

$$750+3y \text{ (dollars).}$$

89. The total cost to make x widgets and y gadgets is

$$1000+5x+750+3y \text{ (dollars).}$$

90. $1000+5x+750+3y$
$=(1000+750)+5x+3y$
$=1750+5x+3y,$

so the total cost to make x widgets and y gadgets is

$$1750+5x+3y \text{ (dollars).}$$

Chapter 1 Review Exercises

1. $$\begin{aligned}\frac{8}{5}\div\frac{32}{15} &= \frac{8}{5}\cdot\frac{15}{32}\\ &= \frac{8\cdot(3\cdot5)}{5\cdot(8\cdot4)}\\ &= \frac{8\cdot3\cdot5}{5\cdot8\cdot4}\\ &= \frac{3}{4}\end{aligned}$$

2. $$\begin{aligned}2\frac{4}{5}\cdot1\frac{1}{4} &= \frac{14}{5}\cdot\frac{5}{4}\\ &= \frac{2\cdot7\cdot5}{5\cdot2\cdot2}\\ &= \frac{7}{2},\ \text{or}\ 3\frac{1}{2}\end{aligned}$$

3. $$\begin{aligned}\frac{5}{8}-\frac{1}{6} &= \frac{5\cdot3}{8\cdot3}-\frac{1\cdot4}{6\cdot4} \quad \textit{LCD} = 24\\ &= \frac{15}{24}-\frac{4}{24}\\ &= \frac{11}{24}\end{aligned}$$

4. $$\begin{aligned}\frac{3}{8}+3\frac{1}{2}-\frac{3}{16} &= \frac{3}{8}+\frac{7}{2}-\frac{3}{16}\\ &= \frac{3\cdot2}{8\cdot2}+\frac{7\cdot8}{2\cdot8}-\frac{3}{16} \quad \textit{LCD} = 16\\ &= \frac{6}{16}+\frac{56}{16}-\frac{3}{16}\\ &= \frac{62}{16}-\frac{3}{16}\\ &= \frac{59}{16},\ \text{or}\ 3\frac{11}{16}\end{aligned}$$

5. $\frac{1}{6} \cdot 7618 = \frac{1}{6} \cdot \frac{7618}{1}$
$= \frac{2 \cdot 3809}{2 \cdot 3}$
$= \frac{3809}{3} \approx 1269.7$

About 1270 thousand luxury cars were sold in the United States in 2007.

6. Since the entire pie chart represents $\frac{3}{3}$, this leaves $\frac{3}{3} - \frac{1}{3} = \frac{2}{3}$ of the cars that are *not* small.

$\frac{2}{3} \cdot 7618 = \frac{2}{3} \cdot \frac{7618}{1}$
$= \frac{2 \cdot 7618}{3 \cdot 1}$
$= \frac{15{,}236}{3} \approx 5078.7$

About 5079 thousand cars sold in the United States in 2007 were *not* small.

7. $5^4 = 5 \cdot 5 \cdot 5 \cdot 5 = 625$

8. $\left(\frac{3}{5}\right)^3 = \frac{3}{5} \cdot \frac{3}{5} \cdot \frac{3}{5} = \frac{27}{125}$

9. $(0.02)^2 = (0.02)(0.02)$
$= 0.0004$

10. $(0.1)^3 = (0.1)(0.1)(0.1)$
$= 0.001$

11. $8 \cdot 5 - 13 = 40 - 13 = 27$

12. $16 + 12 \div 4 - 2 = 16 + (12 \div 4) - 2$
$= 16 + 3 - 2$
$= 19 - 2$
$= 17$

13. $20 - 2(5+3) = 20 - 2(8)$
$= 20 - 16$
$= 4$

14. $7[3 + 6(3^2)] = 7[3 + 6(9)]$
$= 7(3 + 54)$
$= 7(57)$
$= 399$

15. $\frac{9(4^2 - 3)}{4 \cdot 5 - 17} = \frac{9(16 - 3)}{20 - 17}$
$= \frac{9(13)}{3}$
$= \frac{3 \cdot 3 \cdot 13}{3} = 39$

16. $\frac{6(5-4) + 2(4-2)}{3^2 - (4+3)} = \frac{6(1) + 2(2)}{9 - (4+3)}$
$= \frac{6+4}{9-7}$
$= \frac{10}{2} = 5$

17. $12 \cdot 3 - 6 \cdot 6 = 36 - 36 = 0$

Since $0 = 0$ is true, so is $0 \leq 0$, and therefore, the statement "$12 \cdot 3 - 6 \cdot 6 \leq 0$" is true.

18. $3[5(2) - 3] = 3(10 - 3) = 3(7) = 21$

Therefore, the statement "$3[5(2) - 3] > 20$" is true.

19. $4^2 - 8 = 16 - 8 = 8$

Since $9 \leq 8$ is false, the statement "$9 \leq 4^2 - 8$" is false.

20. "Thirteen is less than seventeen" is written $13 < 17$.

21. "Five plus two is not equal to ten" is written $5 + 2 \neq 10$.

22. "Two-thirds is greater than or equal to four-sixths" is written $\frac{2}{3} \geq \frac{4}{6}$.

In Exercises 23–26, replace x with 6 and y with 3.

23. $2x + 6y = 2(6) + 6(3)$
$= 12 + 18 = 30$

24. $4(3x - y) = 4[3(6) - 3]$
$= 4(18 - 3)$
$= 4(15) = 60$

25. $\frac{x}{3} + 4y = \frac{6}{3} + 4(3)$
$= 2 + 12 = 14$

26. $\frac{x^2 + 3}{3y - x} = \frac{6^2 + 3}{3(3) - 6}$
$= \frac{36 + 3}{9 - 6}$
$= \frac{39}{3} = 13$

27. "Six added to a number" translates as $x + 6$.

28. "A number subtracted from eight" translates as $8 - x$.

29. "Nine subtracted from six times a number" translates as $6x - 9$.

30. "Three-fifths of a number added to 12" translates as $12 + \frac{3}{5}x$.

31. $5x + 3(x + 2) = 22;\ 2$

$$\begin{aligned} 5x + 3(x+2) &= 5(2) + 3(2+2) \quad \textit{Let } x = 2. \\ &= 5(2) + 3(4) \\ &= 10 + 12 = 22 \end{aligned}$$

Since the left side and the right side are equal, 2 is a solution of the given equation.

32. $\dfrac{t+5}{3t} = 1;\ 6$

$$\begin{aligned} \frac{t+5}{3t} &= \frac{6+5}{3(6)} \quad \textit{Let } t = 6. \\ &= \frac{11}{18} \end{aligned}$$

Since the left side, $\frac{11}{18}$, is not equal to the right side, 1, 6 is not a solution of the equation.

33. "Six less than twice a number is 10" is written

$$2x - 6 = 10.$$

Letting x equal 0, 2, 4, 6, and 10 results in a false statement, so those values are not solutions.

Since $2(8) - 6 = 16 - 6 = 10$, the solution is 8.

34. "The product of a number and 4 is 8" is written

$$4x = 8.$$

Since $4(2) = 8$, the solution is 2.

35. $-4, -\frac{1}{2}, 0, 2.5, 5$

Graph these numbers on a number line. They are already arranged in order from smallest to largest.

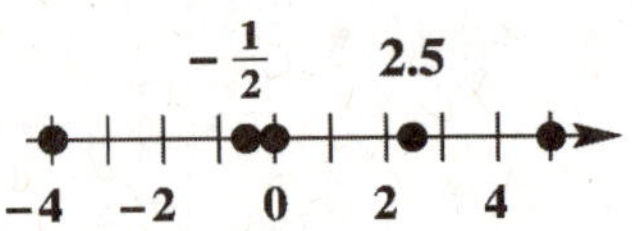

36. $-2, |-3|, -3, |-1|$

Recall that $|-3| = 3$ and $|-1| = 1$. From smallest to largest, the numbers are $-3, -2, |-1|, |-3|$.

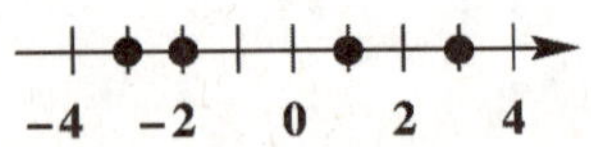

37. Since $\frac{4}{3}$ is the quotient of two integers, it is a *rational number*. Since all rational numbers are also real numbers, $\frac{4}{3}$ is a *real number*.

38. Since the decimal representation of $0.\overline{63}$ repeats, it is a *rational number*. Since all rational numbers are also real numbers, $0.\overline{63}$ is a *real number*.

39. Since 19 is a *natural number*, it is also a *whole number* and an *integer*. We can write it as $\frac{19}{1}$, so it is a *rational number* and hence, a *real number*.

40. Since the decimal representation of $\sqrt{6}$ does not terminate or repeat, it is an *irrational number*. Since all irrational numbers are also real numbers, $\sqrt{6}$ is a *real number*.

41. $-10, 5$
Since any negative number is less than any positive number, -10 is the lesser number.

42. $-8, -9$
Since -9 is to the left of -8 on the number line, -9 is the lesser number.

43. $-\dfrac{2}{3}, -\dfrac{3}{4}$

To compare these fractions, use a common denominator.

$$-\frac{2}{3} = -\frac{8}{12}, \quad -\frac{3}{4} = -\frac{9}{12}$$

Since $-\frac{9}{12}$ is to the left of $-\frac{8}{12}$ on the number line, $-\frac{3}{4}$ is the lesser number.

44. $0, -|23|$
Since $-|23| = -23$ and $-23 < 0$, $-|23|$ is the lesser number.

45. $12 > -13$
This statement is true since 12 is to the right of -13 on the number line.

46. $0 > -5$
This statement is true since 0 is to the right of -5 on the number line.

47. $-9 < -7$
This statement is true since -9 is to the left of -7 on the number line.

48. $-13 \geq -13$
This is a true statement since $-13 = -13$.

49. **(a)** The opposite of the number -9 is its negative; that is, $-(-9) = 9$.

(b) Since $-9 < 0$, the absolute value of the number -9 is $|-9| = -(-9) = 9$.

50. 0 **(a)** $-0 = 0$ **(b)** $|0| = 0$

51. 6 **(a)** $-(6) = -6$ **(b)** $|6| = 6$

52. $-\frac{5}{7}$ **(a)** $-(-\frac{5}{7}) = \frac{5}{7}$

(b) $|-\frac{5}{7}| = -(-\frac{5}{7}) = \frac{5}{7}$

53. $|-12| = -(-12) = 12$

54. $-|3| = -3$

55. $-|-19| = -[-(-19)] = -19$

56. $-|9-2| = -|7| = -7$

57. $-10+4=-6$

58. $14+(-18)=-4$

59. $-8+(-9)=-17$

60. $$\begin{aligned}\frac{4}{9}+\left(-\frac{5}{4}\right) &= \frac{4\cdot 4}{9\cdot 4}+\left(-\frac{5\cdot 9}{4\cdot 9}\right) \quad \textit{LCD = 36}\\ &= \frac{16}{36}+\left(-\frac{45}{36}\right)\\ &= -\frac{29}{36}\end{aligned}$$

61. $-13.5+(-8.3)=-21.8$

62. $$\begin{aligned}(-10+7)+(-11) &= (-3)+(-11)\\ &= -14\end{aligned}$$

63. $$\begin{aligned}&[-6+(-8)+8]+[9+(-13)]\\ &= \{[-6+(-8)]+8\}+(-4)\\ &= [(-14)+8]+(-4)\\ &= (-6)+(-4) = -10\end{aligned}$$

64. $$\begin{aligned}&(-4+7)+(-11+3)+(-15+1)\\ &= (3)+(-8)+(-14)\\ &= [3+(-8)]+(-14)\\ &= (-5)+(-14) = -19\end{aligned}$$

65. $-7-4=-7+(-4)=-11$

66. $-12-(-11)=-12+(11)=-1$

67. $5-(-2)=5+(2)=7$

68. $$\begin{aligned}-\frac{3}{7}-\frac{4}{5} &= -\frac{3\cdot 5}{7\cdot 5}-\frac{4\cdot 7}{5\cdot 7}\\ &= -\frac{15}{35}-\frac{28}{35} \quad \textit{LCD = 35}\\ &= -\frac{15}{35}+\left(-\frac{28}{35}\right)\\ &= -\frac{43}{35}, \text{ or } -1\frac{8}{35}\end{aligned}$$

69. $$\begin{aligned}2.56-(-7.75) &= 2.56+(7.75)\\ &= 10.31\end{aligned}$$

70. $$\begin{aligned}(-10-4)-(-2) &= [-10+(-4)]+2\\ &= (-14)+(2)\\ &= -12\end{aligned}$$

71. $$\begin{aligned}(-3+4)-(-1) &= (-3+4)+1\\ &= 1+1\\ &= 2\end{aligned}$$

72. $$\begin{aligned}-(-5+6)-2 &= -(1)+(-2)\\ &= -1+(-2)\\ &= -3\end{aligned}$$

73. "19 added to the sum of -31 and 12" is written

$$\begin{aligned}(-31+12)+19 &= (-19)+19\\ &= 0.\end{aligned}$$

74. "13 more than the sum of -4 and -8" is written

$$\begin{aligned}[-4+(-8)]+13 &= -12+13\\ &= 1.\end{aligned}$$

75. "The difference between -4 and -6" is written

$$\begin{aligned}-4-(-6) &= -4+6\\ &= 2.\end{aligned}$$

76. "Five less than the sum of 4 and -8" is written

$$\begin{aligned}[4+(-8)]-5 &= (-4)+(-5)\\ &= -9.\end{aligned}$$

77. $x+(-2)=-4$

Because $(-2)+(-2)=-4$,
the solution is -2.

78. $12+x=11$

Because $12+(-1)=1$,
the solution is -1.

79. $-23.75+50.00=26.25$

He now has a positive balance of \$26.25.

80. $-26+16=-10$

The high temperature was $-10°$F.

81. $$\begin{aligned}-28+13-14 &= (-28+13)-14\\ &= (-28+13)+(-14)\\ &= -15+(-14)\\ &= -29\end{aligned}$$

His present financial status is $-\$29$.

82. $$\begin{aligned}-3-7 &= -3+(-7)\\ &= -10\end{aligned}$$

The new temperature is $-10°$.

83. $$\begin{aligned}8-12+42 &= [8+(-12)]+42\\ &= -4+42\\ &= 38\end{aligned}$$

The total net yardage is 38.

84. To get the closing value for the previous day, we can add the amount it was down to the amount at which it closed.

$$47.92+9496.28=9544.20$$

85. $(-12)(-3)=36$

86. $$\begin{aligned}15(-7) &= -(15\cdot 7)\\ &= -105\end{aligned}$$

87. $-\frac{4}{3}\left(-\frac{3}{8}\right) = \frac{4}{3}\cdot\frac{3}{8}$
$= \frac{4\cdot 3}{3\cdot 4\cdot 2}$
$= \frac{1}{2}$

88. $(-4.8)(-2.1) = 10.08$

89. $5(8-12) = 5[8+(-12)]$
$= 5(-4) = -20$

90. $(5-7)(8-3) = [5+(-7)][8+(-3)]$
$= (-2)(5) = -10$

91. $2(-6)-(-4)(-3) = -12-(12)$
$= -12+(-12)$
$= -24$

92. $3(-10)-5 = -30+(-5) = -35$

93. $\frac{-36}{-9} = \frac{4\cdot 9}{9} = 4$

94. $\frac{220}{-11} = -\frac{20\cdot 11}{11} = -20$

95. $-\frac{1}{2}\div\frac{2}{3} = -\frac{1}{2}\cdot\frac{3}{2} = -\frac{3}{4}$

96. $-33.9\div(-3) = \frac{-33.9}{-3} = 11.3$

97. $\frac{-5(3)-1}{8-4(-2)} = \frac{-15+(-1)}{8-(-8)}$
$= \frac{-16}{8+8}$
$= \frac{-16}{16} = -1$

98. $\frac{5(-2)-3(4)}{-2[3-(-2)]-1} = \frac{-10-12}{-2(3+2)-1}$
$= \frac{-10+(-12)}{-2(5)-1}$
$= \frac{-22}{-10+(-1)}$
$= \frac{-22}{-11} = 2$

99. $\frac{10^2-5^2}{8^2+3^2-(-2)} = \frac{100-25}{64+9+2}$
$= \frac{75}{75} = 1$

100. $\frac{(0.6)^2+(0.8)^2}{(-1.2)^2-(-0.56)} = \frac{0.36+0.64}{1.44+0.56}$
$= \frac{1.00}{2.00} = 0.5$

In Exercises 101–104, replace x with -5, y with 4, and z with -3.

101. $6x-4z = 6(-5)-4(-3)$
$= -30-(-12)$
$= -30+12 = -18$

102. $5x+y-z = 5(-5)+(4)-(-3)$
$= (-25+4)+3$
$= -21+3 = -18$

103. $5x^2 = 5(-5)^2$
$= 5(25)$
$= 125$

104. $z^2(3x-8y) = (-3)^2[3(-5)-8(4)]$
$= 9(-15-32)$
$= 9[-15+(-32)]$
$= 9(-47) = -423$

105. "Nine less than the product of -4 and 5" is written

$$-4(5)-9 = -20+(-9) = -29.$$

106. "Five-sixths of the sum of 12 and -6" is written

$$\tfrac{5}{6}[12+(-6)] = \tfrac{5}{6}(6) = 5.$$

107. "The quotient of 12 and the sum of 8 and -4" is written

$$\frac{12}{8+(-4)} = \frac{12}{4} = 3.$$

108. "The product of -20 and 12, divided by the difference between 15 and -15" is written

$$\frac{-20(12)}{15-(-15)} = \frac{-240}{15+15} = \frac{-240}{30} = -8.$$

109. "8 times a number is -24" is written

$$8x = -24.$$

If $x = -3$,

$$8x = 8(-3) = -24.$$

The solution is -3.

110. "The quotient of a number and 3 is -2" is written

$$\frac{x}{3} = -2.$$

If $x = -6$,

$$\frac{x}{3} = \frac{-6}{3} = -2.$$

The solution is -6.

111. Find the average of the eight numbers.

$$\frac{26 + 38 + 40 + 20 + 4 + 14 + 96 + 18}{8}$$

$$= \frac{256}{8} = \frac{8 \cdot 32}{8} = 32$$

112. Find the average of the six numbers.

$$\frac{-12 + 28 + (-36) + 0 + 12 + (-10)}{6}$$

$$= \frac{-18}{6} = -3$$

113. $6 + 0 = 6$

This is an example of an identity property.

114. $5 \cdot 1 = 5$

This is an example of an identity property.

115. $-\frac{2}{3}\left(-\frac{3}{2}\right) = 1$

This is an example of an inverse property.

116. $17 + (-17) = 0$

This is an example of an inverse property.

117. $5 + (-9 + 2) = [5 + (-9)] + 2$

This is an example of an associative property.

118. $w(xy) = (wx)y$

This is an example of an associative property.

119. $3x + 3y = 3(x + y)$

This is an example of the distributive property.

120. $(1 + 2) + 3 = 3 + (1 + 2)$

This is an example of a commutative property.

121. $7y + 14 = 7y + 7 \cdot 2$
$= 7(y + 2)$

122. $-12(4 - t) = -12(4) - (-12)(t)$
$= -48 + 12t$

123. $3(2s) + 3(5y) = 3(2s + 5y)$

124. $-(-4r + 5s) = -1(-4r + 5s)$
$= (-1)(-4r) + (-1)(5s)$
$= 4r - 5s$

125. $2m + 9m = (2 + 9)m$ *Distributive property*
$= 11m$

126. $15p^2 - 7p^2 + 8p^2$
$= (15 - 7 + 8)p^2$ *Distributive property*
$= 16p^2$

127. $5p^2 - 4p + 6p + 11p^2$
$= (5 + 11)p^2 + (-4 + 6)p$
Distributive property
$= 16p^2 + 2p$

128. $-2(3k - 5) + 2(k + 1)$
$= -6k + 10 + 2k + 2$
Distributive property
$= -4k + 12$

129. $7(2m + 3) - 2(8m - 4)$
$= 14m + 21 - 16m + 8$
Distributive property
$= (14 - 16)m + 29$
$= -2m + 29$

130. $-(2k + 8) - (3k - 7)$
$= -1(2k + 8) - 1(3k - 7)$
Replace $-$ with -1.
$= -2k - 8 - 3k + 7$
Distributive property
$= -5k - 1$

131. "Seven times a number, subtracted from the product of -2 and three times the number" is written

$$-2(3x) - 7x = -6x - 7x = -13x.$$

132. "A number multiplied by 8, added to the sum of 5 and four times the number" is written

$$(5 + 4x) + 8x = 5 + (4x + 8x) = 5 + 12x.$$

133. [1.6] $\frac{6(-4) + 2(-12)}{5(-3) + (-3)} = \frac{-24 + (-24)}{-15 + (-3)}$
$= \frac{-48}{-18} = \frac{8 \cdot 6}{3 \cdot 6}$
$= \frac{8}{3}$, or $2\frac{2}{3}$

134. [1.5] $\frac{3}{8} - \frac{5}{12} = \frac{3 \cdot 3}{8 \cdot 3} - \frac{5 \cdot 2}{12 \cdot 2}$
$= \frac{9}{24} - \frac{10}{24}$
$= \frac{9}{24} + \left(-\frac{10}{24}\right)$
$= -\frac{1}{24}$

135. [1.6] $\frac{8^2+6^2}{7^2+1^2} = \frac{64+36}{49+1}$
$= \frac{100}{50} = 2$

136. [1.6] $-\frac{12}{5} \div \frac{9}{7} = -\frac{12}{5} \cdot \frac{7}{9}$
$= -\frac{12 \cdot 7}{5 \cdot 9}$
$= -\frac{3 \cdot 4 \cdot 7}{5 \cdot 3 \cdot 3}$
$= -\frac{28}{15}$, or $-1\frac{13}{15}$

137. [1.5] $2\frac{5}{6} - 4\frac{1}{3} = \frac{17}{6} - \frac{13}{3}$
$= \frac{17}{6} - \frac{13 \cdot 2}{3 \cdot 2}$
$= \frac{17}{6} - \frac{26}{6}$
$= \frac{17}{6} + \left(-\frac{26}{6}\right)$
$= -\frac{9}{6} = -\frac{3}{2}$, or $-1\frac{1}{2}$

138. [1.6] $\left(-\frac{5}{6}\right)^2 = \left(-\frac{5}{6}\right)\left(-\frac{5}{6}\right)$
$= \frac{25}{36}$

139. [1.5] $[(-2)+7-(-5)]+[-4-(-10)]$
$= \{[(-2)+7]-(-5)\}+(-4+10)$
$= (5+5)+6$
$= 10+6 = 16$

140. [1.6] $-16(-3.5)-7.2(-3)$
$= 56-[(7.2)(-3)]$
$= 56-(-21.6)$
$= 56+21.6$
$= 77.6$

141. [1.5] $-8+[(-4+17)-(-3-3)]$
$= -8+\{(13)-[-3+(-3)]\}$
$= -8+[13-(-6)]$
$= -8+(13+6)$
$= -8+19 = 11$

142. [1.8] $-4(2t+1)-8(-3t+4)$
$= -4(2t)-4(1)-8(-3t)-8(4)$
$= -8t-4+24t-32$
$= 16t-36$

143. [1.8] $5x^2-12y^2+3x^2-9y^2$
$= (5x^2+3x^2)+(-12y^2-9y^2)$
$= (5+3)x^2+(-12-9)y^2$
$= 8x^2-21y^2$

144. [1.6] $(-8-3)-5(2-9)$
$= [-8+(-3)]-5[2+(-9)]$
$= -11-5(-7)$
$= -11-(-35)$
$= -11+35 = 24$

145. [1.6] Dividing 0 *by* a nonzero number gives a quotient of 0. However, dividing a number *by* 0 is undefined.

146. [1.5] $118-165 = 118+(-165)$
$= -47$

The lowest temperature ever recorded in Iowa was $-47°$F.

147. [1.5] The change in enrollment from 1980 to 1985 was $12.39-13.23 = -0.84$ million students. Expressed as an integer, this number is $-840{,}000$.

148. [1.5] The change in enrollment from 1985 to 1990 was $11.34-12.39 = -1.05$ million students. Expressed as an integer, this number is $-1{,}050{,}000$.

149. [1.5] The change in enrollment from 1995 to 2000 was $13.52-12.50 = 1.02$ million students. Expressed as an integer, this number is 1,020,000.

150. [1.5] The change in enrollment from 2000 to 2005 was $14.91-13.52 = 1.39$ million students. Expressed as an integer, this number is 1,390,000.

Chapter 1 Test

1. $\frac{63}{99} = \frac{7 \cdot 9}{11 \cdot 9} = \frac{7}{11}$

2. The denominators are 8, 12, and 15; or equivalently, $2^3, 2^2 \cdot 3,$ and $3 \cdot 5$. So the LCD is $2^3 \cdot 3 \cdot 5 = 120$.

$\frac{5}{8}+\frac{11}{12}+\frac{7}{15}$
$= \frac{5 \cdot 15}{8 \cdot 15}+\frac{11 \cdot 10}{12 \cdot 10}+\frac{7 \cdot 8}{15 \cdot 8}$
$= \frac{75}{120}+\frac{110}{120}+\frac{56}{120}$
$= \frac{241}{120}$, or $2\frac{1}{120}$

3. $\frac{19}{15} \div \frac{6}{5} = \frac{19}{15} \cdot \frac{5}{6} = \frac{19 \cdot 5}{3 \cdot 5 \cdot 6} = \frac{19}{18}$, or $1\frac{1}{18}$

4. $4[-20+7(-2)] = 4[-20+(-14)]$
$= 4(-34) = -136$

Since $-136 \leq 135$, the statement "$4[-20+7(-2)] \leq 135$" is true.

5. $-1, -3, |-4|, |-1|$

Recall that $|-4| = 4$ and $|-1| = 1$. From smallest to largest, the numbers are $-3, -1, |-1|, |-4|$.

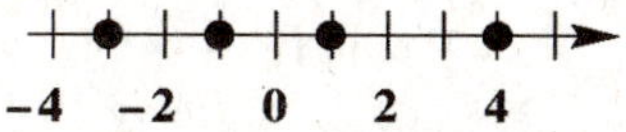

6. The number $-\frac{2}{3}$ can be written as a quotient of two integers with denominator not 0, so it is a *rational number*. Since all rational numbers are real numbers, it is also a *real number*.

7. If -8 and -1 are both graphed on a number line, we see that the point for -8 is to the *left* of the point for -1. This indicates that -8 is *less than* -1.

8. "The quotient of -6 and the sum of 2 and -8" is written $\frac{-6}{2+(-8)}$,

$$\text{and } \frac{-6}{2+(-8)} = \frac{-6}{-6} = 1.$$

9.
$$\begin{aligned} &-2-(5-17)+(-6) \\ &= -2-[5+(-17)]+(-6) \\ &= -2-(-12)+(-6) \\ &= (-2+12)+(-6) \\ &= 10+(-6) = 4 \end{aligned}$$

10.
$$\begin{aligned} -5\frac{1}{2}+2\frac{2}{3} &= -\frac{11}{2}+\frac{8}{3} \\ &= -\frac{11\cdot 3}{2\cdot 3}+\frac{8\cdot 2}{3\cdot 2} \\ &= -\frac{33}{6}+\frac{16}{6} \\ &= -\frac{17}{6}, \text{ or } -2\frac{5}{6} \end{aligned}$$

11.
$$\begin{aligned} &-6-[-7+(2-3)] \\ &= -6-[-7+(-1)] \\ &= -6-(-8) \\ &= -6+8 = 2 \end{aligned}$$

12.
$$\begin{aligned} &4^2+(-8)-(2^3-6) \\ &= 16+(-8)-(8-6) \\ &= [16+(-8)]-2 \\ &= 8-2 = 6 \end{aligned}$$

13.
$$\begin{aligned} &(-5)(-12)+4(-4)+(-8)^2 \\ &= (-5)(-12)+4(-4)+64 \\ &= [60+(-16)]+64 \\ &= 44+64 = 108 \end{aligned}$$

14.
$$\begin{aligned} &\frac{30(-1-2)}{-9[3-(-2)]-12(-2)} \\ &= \frac{30(-3)}{-9(5)-(-24)} \\ &= \frac{-90}{-45+24} \\ &= \frac{-90}{-21} \\ &= \frac{30\cdot 3}{7\cdot 3} = \frac{30}{7}, \text{ or } 4\frac{2}{7} \end{aligned}$$

15. $-x+3=-3$

If $x=6$,

$$-6+3=-3.$$

Therefore, the solution is 6.

16. $-3x=-12$

If $x=4$,

$$-3x=-3(4)=-12.$$

Therefore, the solution is 4.

17.
$$\begin{aligned} &3x-4y^2 \\ &= 3(-2)-4(4^2) \qquad \textit{Let } x=-2,\ y=4. \\ &= 3(-2)-4(16) \\ &= -6-64 = -70 \end{aligned}$$

18.
$$\begin{aligned} &\frac{5x+7y}{3(x+y)} \\ &= \frac{5(-2)+7(4)}{3(-2+4)} \qquad \textit{Let } x=-2,\ y=4. \\ &= \frac{-10+28}{3(2)} \\ &= \frac{18}{6} = 3 \end{aligned}$$

19. The difference between the highest and lowest elevations is

$6960-(-40)=6960+40=7000$ meters.

20. 4 saves (3 points per save)

+ 3 wins (3 points per win)

+ 2 losses (-2 points per loss)

+ 1 blown save (-2 points per blown save)

$$\begin{aligned} &= 4(3)+3(3)+2(-2)+1(-2) \\ &= 12+9-4-2 \\ &= 15 \text{ points} \end{aligned}$$

He has a total of 15 points.

21. $2.10 - 3.52 = 2.10 + (-3.52) = -1.42$

As a signed number, the federal budget deficit is $-\$1.42$ trillion.

22. Commutative property

$$(5+2)+8 = 8+(5+2)$$

illustrates a commutative property because the order of the numbers is changed, but not the grouping. The correct response is **B**.

23. Associative property

$$-5+(3+2) = (-5+3)+2$$

illustrates an associative property because the grouping of the numbers is changed, but not the order. The correct response is **D**.

24. Inverse property

$$-\frac{5}{3}\left(-\frac{3}{5}\right) = 1$$

illustrates an inverse property. The correct response is **E**.

25. Identity property

$$3x + 0 = 3x$$

illustrates an identity property. The correct response is **A**.

26. Distributive property

$$-3(x+y) = -3x + (-3y)$$

illustrates the distributive property. The correct response is **C**.

27. $3(x+1) = 3\cdot x + 3\cdot 1$
$= 3x + 3$

The distributive property is used to rewrite $3(x+1)$ as $3x+3$.

28. **(a)** $-6[5+(-2)] = -6(3) = -18$

(b) $-6[5+(-2)] = -6(5) + (-6)(-2)$
$= -30 + 12 = -18$

(c) The distributive property assures us that the answers must be the same, because $a(b+c) = ab + ac$ for all a, b, c.

29. $8x + 4x - 6x + x + 14x$
$= (8+4-6+1+14)x$
$= 21x$

30. $5(2x-1) - (x-12) + 2(3x-5)$
$= 5(2x-1) - 1(x-12) + 2(3x-5)$
$= 10x - 5 - x + 12 + 6x - 10$
$= (10-1+6)x + (-5+12-10)$
$= 15x - 3$

CHAPTER 2 LINEAR EQUATIONS AND INEQUALITIES IN ONE VARIABLE

2.1 The Addition Property of Equality

2.1 Now Try Exercises

N1. Note: When solving equations we will write "Add 5" as a shorthand notation for "Add 5 to each side" and "Subtract 5" as a notation for "Subtract 5 from each side."

$$\begin{aligned} x - 13 &= 4 && \textit{Given} \\ x - 13 + 13 &= 4 + 13 && \textit{Add 13.} \\ x &= 17 && \textit{Combine like terms.} \end{aligned}$$

We check by substituting 17 for x in the *original* equation.

Check: $x - 13 = 4$ *Original equation*

$17 - 13 \stackrel{?}{=} 4$ *Let x = 17.*

$4 = 4$ *True*

Since a true statement results, $\{17\}$ is the solution set.

N2.
$$\begin{aligned} t - 5.7 &= -7.2 \\ t - 5.7 + 5.7 &= -7.2 + 5.7 && \textit{Add 5.7.} \\ t &= -1.5 \end{aligned}$$

Check $t = -1.5$: $-7.2 = -7.2$ *True*

The solution set is $\{-1.5\}$.

N3.
$$\begin{aligned} -15 &= x + 12 \\ -15 - 12 &= x + 12 - 12 && \textit{Subtract 12.} \\ -27 &= x \end{aligned}$$

Check $x = -27$: $-15 = -15$ *True*

The solution set is $\{-27\}$.

N4.
$$\begin{aligned} \frac{2}{3}x - 4 &= \frac{5}{3}x \\ \frac{2}{3}x - 4 - \frac{2}{3}x &= \frac{5}{3}x - \frac{2}{3}x && \textit{Subtract } \tfrac{2}{3}x. \\ -4 &= \frac{3}{3}x && \textit{Combine terms.} \\ -4 &= x \end{aligned}$$

Check $x = -4$: $-\frac{20}{3} = -\frac{20}{3}$ *True*

The solution set is $\{-4\}$.

N5.
$$\begin{aligned} 6x - 8 &= 12 + 5x && \textit{Given.} \\ 6x - 8 - 5x &= 12 + 5x - 5x && \textit{Subtract 5x.} \\ x - 8 &= 12 && \textit{Combine terms.} \\ x - 8 + 8 &= 12 + 8 && \textit{Add 8.} \\ x &= 20 && \textit{Combine terms.} \end{aligned}$$

Check $x = 20$: $120 - 8 = 12 + 100$ *True*

The solution set is $\{20\}$.

N6.
$$\begin{aligned} 5x - 10 - 12x &= 4 - 8x - 9 && \textit{Given} \\ -7x - 10 &= -8x - 5 && \textit{Combine terms.} \\ -7x - 10 + 8x &= -8x - 5 + 8x && \textit{Add 8x.} \\ x - 10 &= -5 && \textit{Combine terms.} \\ x - 10 + 10 &= -5 + 10 && \textit{Add 10.} \\ x &= 5 && \textit{Combine terms.} \end{aligned}$$

Check $x = 5$: $-45 = -45$ *True*

The solution set is $\{5\}$.

N7.
$$\begin{aligned} 4(3x - 2) - (11x - 4) &= 3 && \textit{Given} \\ 4(3x - 2) - 1(11x - 4) &= 3 && -a = -1a \\ 12x - 8 - 11x + 4 &= 3 && \textit{Distributive property} \\ x - 4 &= 3 && \textit{Combine terms.} \\ x - 4 + 4 &= 3 + 4 && \textit{Add 4.} \\ x &= 7 \end{aligned}$$

Check $x = 7$: $3 = 3$ *True*

The solution set is $\{7\}$.

2.1 Section Exercises

1. **(a)** $5x + 8 - 4x + 7$

This is an expression, not an equation, since there is no equals symbol. It can be simplified by rearranging terms and then combining like terms.

$$\begin{aligned} 5x + 8 - 4x + 7 &= 5x - 4x + 8 + 7 \\ &= x + 15 \end{aligned}$$

(b) $-6y + 12 + 7y - 5$

This is an expression, not an equation, since there is no equals symbol. It can be simplified by rearranging terms and then combining like terms.

$$\begin{aligned} -6y + 12 + 7y - 5 &= -6y + 7y + 12 - 5 \\ &= y + 7 \end{aligned}$$

(c) $5x + 8 - 4x = 7$

This is an equation because of the equals symbol.

$$\begin{aligned} 5x + 8 - 4x &= 7 \\ x + 8 &= 7 \\ x &= -1 \end{aligned}$$

The solution set is $\{-1\}$.

(d) $-6y + 12 + 7y = -5$

This is an equation because of the equals symbol.

$$\begin{aligned} -6y + 12 + 7y &= -5 \\ y + 12 &= -5 \\ y &= -17 \end{aligned}$$

The solution set is $\{-17\}$.

3. Equations **A** $x^2 - 5x + 6 = 0$ and **B** $x^3 = x$ are *not* linear equations in one variable because they cannot be written in the form $Ax + B = C$. Note that in a linear equation the exponent on the variable must be 1.

For Exercises 5–52, all solutions should be checked by substituting into the original equation. Checks will be shown here for only a few of the exercises.

5.
$$\begin{aligned} x - 3 &= 9 \\ x - 3 + 3 &= 9 + 3 \\ x &= 12 \end{aligned}$$

Check this solution by replacing x with 12 in the original equation.

$$\begin{aligned} x - 3 &= 9 \\ 12 - 3 &\stackrel{?}{=} 9 \quad \textit{Let } x = 12. \\ 9 &= 9 \quad \textit{True} \end{aligned}$$

Because the final statement is true, $\{12\}$ is the solution set.

7.
$$\begin{aligned} x - 12 &= 19 \\ x - 12 + 12 &= 19 + 12 \\ x &= 31 \end{aligned}$$

Check $x = 31$:

$$\begin{aligned} 31 - 12 &\stackrel{?}{=} 19 \quad \textit{Let } x = 31. \\ 19 &= 19 \quad \textit{True} \end{aligned}$$

Thus, $\{31\}$ is the solution set.

9.
$$\begin{aligned} x - 6 &= -9 \\ x - 6 + 6 &= -9 + 6 \\ x &= -3 \end{aligned}$$

Checking yields a true statement, so $\{-3\}$ is the solution set.

11.
$$\begin{aligned} r + 8 &= 12 \\ r + 8 - 8 &= 12 - 8 \\ r &= 4 \end{aligned}$$

Checking yields a true statement, so $\{4\}$ is the solution set.

13.
$$\begin{aligned} x + 28 &= 19 \\ x + 28 - 28 &= 19 - 28 \\ x &= -9 \end{aligned}$$

Checking yields a true statement, so $\{-9\}$ is the solution set.

15.
$$\begin{aligned} x + \frac{1}{4} &= -\frac{1}{2} \\ x + \frac{1}{4} - \frac{1}{4} &= -\frac{1}{2} - \frac{1}{4} \\ x &= -\frac{2}{4} - \frac{1}{4} \\ x &= -\frac{3}{4} \end{aligned}$$

Check $x = -\frac{3}{4}$: $-\frac{1}{2} = -\frac{1}{2}$ *True*

The solution set is $\{-\frac{3}{4}\}$.

17.
$$\begin{aligned} 7 + r &= -3 \\ r + 7 &= -3 \\ r + 7 - 7 &= -3 - 7 \\ r &= -10 \end{aligned}$$

The solution set is $\{-10\}$.

19.
$$\begin{aligned} 2 &= p + 15 \\ 2 - 15 &= p + 15 - 15 \\ -13 &= p \end{aligned}$$

The solution set is $\{-13\}$.

21.
$$\begin{aligned} -4 &= x - 14 \\ -4 + 14 &= x - 14 + 14 \\ 10 &= x \end{aligned}$$

The solution set is $\{10\}$.

23.
$$\begin{aligned} -\frac{1}{3} &= x - \frac{3}{5} \\ -\frac{1}{3} + \frac{3}{5} &= x - \frac{3}{5} + \frac{3}{5} \\ -\frac{5}{15} + \frac{9}{15} &= x \\ \frac{4}{15} &= x \end{aligned}$$

Check $x = \frac{4}{15}$: $-\frac{5}{15} = \frac{4}{15} - \frac{9}{15}$ *True*

The solution set is $\{\frac{4}{15}\}$.

25.
$$\begin{aligned} x - 8.4 &= -2.1 \\ x - 8.4 + 8.4 &= -2.1 + 8.4 \\ x &= 6.3 \end{aligned}$$

The solution set is $\{6.3\}$.

27.
$$\begin{aligned} t + 12.3 &= -4.6 \\ t + 12.3 - 12.3 &= -4.6 - 12.3 \\ t &= -16.9 \end{aligned}$$

The solution set is $\{-16.9\}$.

29.
$$\begin{aligned} 3x &= 2x + 7 \\ 3x - 2x &= 2x + 7 - 2x \quad \textit{Subtract } 2x. \\ 1x &= 7 \quad \text{or} \quad x = 7 \end{aligned}$$

Check $x = 7$: $21 = 21$ *True*

The solution set is $\{7\}$.

31.
$$\begin{aligned} 10x+4&=9x \\ 10x+4-9x&=9x-9x \quad \textit{Subtract 9x.} \\ 1x+4&=0 \\ x+4-4&=0-4 \quad \textit{Subtract 4.} \\ x&=-4 \end{aligned}$$

Check $x=-4$: $-36=-36$ *True*

The solution set is $\{-4\}$.

33.
$$\begin{aligned} 3x+7&=2x+4 \\ 3x+7-2x&=2x+4-2x \\ x+7&=4 \\ x+7-7&=4-7 \\ x&=-3 \end{aligned}$$

The solution set is $\{-3\}$.

35.
$$\begin{aligned} 8t+6&=7t+6 \\ 8t+6-7t&=7t+6-7t \\ t+6&=6 \\ t+6-6&=6-6 \\ t&=0 \end{aligned}$$

The solution set is $\{0\}$.

37.
$$\begin{aligned} -4x+7&=-5x+9 \\ -4x+7+5x&=-5x+9+5x \\ x+7&=9 \\ x+7-7&=9-7 \\ x&=2 \end{aligned}$$

The solution set is $\{2\}$.

39.
$$\begin{aligned} \frac{2}{5}w-6&=\frac{7}{5}w \\ \frac{2}{5}w-6-\frac{2}{5}w&=\frac{7}{5}w-\frac{2}{5}w \quad \textit{Subtract } \frac{2}{5}w. \\ -6&=\frac{5}{5}w \\ -6&=w \end{aligned}$$

The solution set is $\{-6\}$.

41.
$$\begin{aligned} 5.6x+2&=4.6x \\ 5.6x+2-4.6x&=4.6x-4.6x \\ 1.0x+2&=0 \\ x+2-2&=0-2 \\ x&=-2 \end{aligned}$$

The solution set is $\{-2\}$.

43.
$$\begin{aligned} 1.4x-3&=0.4x \\ 1.4x-3-0.4x&=0.4x-0.4x \\ 1.0x-3&=0 \\ 1.0x-3+3&=0+3 \\ x&=3 \end{aligned}$$

The solution set is $\{3\}$.

45.
$$\begin{aligned} 5p&=4p \\ 5p-4p&=4p-4p \\ p&=0 \end{aligned}$$

The solution set is $\{0\}$.

47.
$$\begin{aligned} 1.2y-4&=0.2y-4 \\ 1.2y-4-0.2y&=0.2y-4-0.2y \\ 1.0y-4&=-4 \\ y-4+4&=-4+4 \\ y&=0 \end{aligned}$$

The solution set is $\{0\}$.

49.
$$\begin{aligned} \frac{1}{2}x+5&=-\frac{1}{2}x \\ \frac{1}{2}x+\frac{1}{2}x+5&=-\frac{1}{2}x+\frac{1}{2}x \\ x+5&=0 \\ x+5-5&=0-5 \\ x&=-5 \end{aligned}$$

The solution set is $\{-5\}$.

51.
$$\begin{aligned} 3x+7-2x&=0 \\ x+7&=0 \\ x+7-7&=0-7 \\ x&=-7 \end{aligned}$$

The solution set is $\{-7\}$.

53.
$$\begin{aligned} 5t+3+2t-6t&=4+12 \\ (5+2-6)t+3&=16 \\ t+3-3&=16-3 \\ t&=13 \end{aligned}$$

Check $t=13$: $16=16$ *True*

The solution set is $\{13\}$.

55.
$$\begin{aligned} 6x+5+7x+3&=12x+4 \\ 13x+8&=12x+4 \\ 13x+8-12x&=12x+4-12x \\ x+8&=4 \\ x+8-8&=4-8 \\ x&=-4 \end{aligned}$$

Check $x=-4$: $-44=-44$ *True*

The solution set is $\{-4\}$.

57.
$$\begin{aligned} 5.2q-4.6-7.1q&=-0.9q-4.6 \\ -1.9q-4.6&=-0.9q-4.6 \\ -1.9q-4.6+0.9q&=-0.9q-4.6+0.9q \\ -1.0q-4.6&=-4.6 \\ -1.0q-4.6+4.6&=-4.6+4.6 \\ -q&=0 \\ q&=0 \end{aligned}$$

Check $q=0$: $-4.6=-4.6$ *True*

The solution set is $\{0\}$.

59.
$$\frac{5}{7}x+\frac{1}{3}=\frac{2}{5}-\frac{2}{7}x+\frac{2}{5}$$
$$\frac{5}{7}x+\frac{1}{3}=\frac{4}{5}-\frac{2}{7}x$$
$$\frac{5}{7}x+\frac{2}{7}x+\frac{1}{3}=\frac{4}{5}-\frac{2}{7}x+\frac{2}{7}x \quad \textit{Add } \tfrac{2}{7}x.$$
$$\frac{7}{7}x+\frac{1}{3}=\frac{4}{5} \quad \textit{Combine like terms.}$$
$$1x+\frac{1}{3}-\frac{1}{3}=\frac{4}{5}-\frac{1}{3} \quad \textit{Subtract } \tfrac{1}{3}.$$
$$x=\frac{12}{15}-\frac{5}{15} \quad \textit{LCD = 15}$$
$$x=\tfrac{7}{15}$$

Check $x=\frac{7}{15}$: $\frac{2}{3}=\frac{2}{3}$ *True*
The solution set is $\{\frac{7}{15}\}$.

61. $(5y+6)-(3+4y)=10$
$$5y+6-3-4y=10 \quad \textit{Distributive property}$$
$$y+3=10 \quad \textit{Combine terms.}$$
$$y+3-3=10-3 \quad \textit{Subtract 3.}$$
$$y=7$$

Check $y=7$: $10=10$ *True*

The solution set is $\{7\}$.

63. $2(p+5)-(9+p)=-3$
$$2p+10-9-p=-3$$
$$p+1=-3$$
$$p+1-1=-3-1$$
$$p=-4$$

Check $p=-4$: $-3=-3$ *True*

The solution set is $\{-4\}$.

65. $-6(2b+1)+(13b-7)=0$
$$-12b-6+13b-7=0$$
$$b-13=0$$
$$b-13+13=0+13$$
$$b=13$$

Check $b=13$: $0=0$ *True*

The solution set is $\{13\}$.

67.
$$10(-2x+1)=-19(x+1)$$
$$-20x+10=-19x-19$$
$$-20x+10+19x=-19x-19+19x$$
$$-x+10=-19$$
$$-x+10-10=-19-10$$
$$-x=-29$$
$$x=29$$

Check $x=29$: $-570=-570$ *True*

The solution set is $\{29\}$.

69. $-2(8p+2)-3(2-7p)-2(4+2p)=0$
$$-16p-4-6+21p-8-4p=0$$
$$p-18=0$$
$$p-18+18=0+18$$
$$p=18$$

Check $p=18$: $0=0$ *True*

The solution set is $\{18\}$.

71. $4(7x-1)+3(2-5x)-4(3x+5)=-6$
$$28x-4+6-15x-12x-20=-6$$
$$x-18=-6$$
$$x-18+18=-6+18$$
$$x=12$$

Check $x=12$: $-6=-6$ *True*

The solution set is $\{12\}$.

73. Answers will vary. One example is $x-6=-8$.

75. "Three times a number is 17 more than twice the number."
$$3x=2x+17$$
$$3x-2x=2x+17-2x$$
$$x=17$$

The number is 17 and $\{17\}$ is the solution set.

77. "If six times a number is subtracted from seven times the number, the result is -9."
$$7x-6x=-9$$
$$x=-9$$

The number is -9 and $\{-9\}$ is the solution set.

79. $\frac{2}{3}\left(\frac{3}{2}\right)=\frac{2\cdot 3}{3\cdot 2}=1$

81.
$$-\frac{5}{4}\left(-\frac{4}{5}x\right)=-\frac{5}{4}\left(-\frac{4}{5}\right)x$$
$$=\frac{5\cdot 4}{4\cdot 5}x$$
$$=1x=x$$

83.
$$9\left(\frac{r}{9}\right)=9\left(\frac{1}{9}r\right)$$
$$=9\left(\frac{1}{9}\right)r$$
$$=1r=r$$

2.2 The Multiplication Property of Equality

2.2 Now Try Exercises

N1. $8x=80$
$$\frac{8x}{8}=\frac{80}{8} \quad \textit{Divide by 8.}$$
$$x=10$$

Check $x=10$: $80=80$ *True*

The solution set is $\{10\}$.

N2. $10x = -24$

$$\frac{10x}{10} = -\frac{24}{10} \quad \textit{Divide by 10.}$$

$$x = -\tfrac{24}{10} = -\tfrac{12}{5} \quad \textit{Write in lowest terms.}$$

Check $x = -\frac{12}{5}$: $-24 = -24$ *True*

The solution set is $\{-\frac{12}{5}\}$.

N3. $-1.3x = 7.02$

$$\frac{-1.3x}{-1.3} = \frac{7.02}{-1.3} \quad \textit{Divide by } -1.3.$$

$$x = -5.4$$

Check $x = -5.4$: $7.02 = 7.02$ *True*

The solution set is $\{-5.4\}$.

N4. $\frac{x}{5} = -7$

$$\frac{1}{5}x = -7$$

$$5 \cdot \frac{1}{5}x = 5(-7) \quad \textit{Multiply by 5, the reciprocal of } \tfrac{1}{5}.$$

$$p = -35$$

Check $p = -35$: $-7 = -7$ *True*

The solution set is $\{-35\}$.

N5. $\frac{4}{7}z = -16$

$$\frac{7}{4}\left(\frac{4}{7}z\right) = \frac{7}{4}(-16) \quad \textit{Multiply by } \frac{7}{4}.$$

$$1 \cdot t = \frac{7}{4} \cdot \frac{-16}{1} \quad \textit{Multiplicative inverse property}$$

$$t = -28 \quad \textit{Multiplicative identity property; multiply fractions.}$$

Check $t = -28$: $-16 = -16$ *True*

The solution set is $\{-28\}$.

N6. $-x = 9$

$$-1 \cdot x = 9 \quad -x = -1 \cdot x$$

$$(-1)(-1) \cdot x = (-1)(9) \quad \textit{Multiply by } -1.$$

$$1 \cdot x = -9$$

$$x = -9$$

Check $x = -9$: $9 = 9$ *True*

The solution set is $\{-9\}$.

N7. $9n - 6n = 21$

$$3n = 21 \quad \textit{Combine terms.}$$

$$\frac{3n}{3} = \frac{21}{3} \quad \textit{Divide by 3.}$$

$$n = 7$$

Check $n = 7$: $21 = 21$ *True*

The solution set is $\{7\}$.

2.2 Section Exercises

1. **(a)** multiplication property of equality; to get x alone on the left side of the equation, multiply each side by $\frac{1}{3}$ (or divide each side by 3).

(b) addition property of equality; to get x alone on the left side of the equation, add -3 (or subtract 3) on each side.

(c) multiplication property of equality; to get x alone on the left side of the equation, multiply each side by -1 (or divide each side by -1).

(d) addition property of equality; to get x alone on the right side of the equation, add -6 (or subtract 6) on each side.

3. To find the solution of $-x = 5$, multiply (or divide) each side by -1, or use the rule "If $-x = a$, then $x = -a$."

5. $\frac{4}{5}x = 8$

To get just x on the left side, multiply both sides of the equation by the reciprocal of $\frac{4}{5}$, which is $\frac{5}{4}$.

7. $\frac{x}{10} = 5$

This equation is equivalent to $\frac{1}{10}x = 5$. To get just x on the left side, multiply both sides of the equation by the reciprocal of $\frac{1}{10}$, which is 10.

9. $-\frac{9}{2}x = -4$

To get just x on the left side, multiply both sides of the equation by the reciprocal of $-\frac{9}{2}$, which is $-\frac{2}{9}$.

11. $-x = 0.75$

This equation is equivalent to $-1x = 0.75$. To get just x on the left side, multiply both sides of the equation by the reciprocal of -1, which is -1.

13. $6x = 5$

To get just x on the left side, divide both sides of the equation by the coefficient of x, which is 6.

15. $-4x = 16$

To get just x on the left side, divide both sides of the equation by the coefficient of x, which is -4.

17. $0.12x = 48$

To get just x on the left side, divide both sides of the equation by the coefficient of x, which is 0.12.

19. $-x = 25$

This equation is equivalent to $-1x = 25$. To get just x on the left side, divide both sides of the equation by the coefficient of x, which is -1.

21.
$$6x = 36$$
$$\frac{6x}{6} = \frac{36}{6} \quad \textit{Divide by 6.}$$
$$1x = 6$$
$$x = 6$$

Check $x = 6$: $36 = 36$ *True*

The solution set is $\{6\}$.

23.
$$2m = 15$$
$$\frac{2m}{2} = \frac{15}{2} \quad \textit{Divide by 2.}$$
$$m = \tfrac{15}{2}$$

Check $m = \frac{15}{2}$: $15 = 15$ *True*

The solution set is $\{\frac{15}{2}\}$.

25.
$$4x = -20$$
$$\frac{4x}{4} = \frac{-20}{4} \quad \textit{Divide by 4.}$$
$$x = -5$$

Check $x = -5$: $-20 = -20$ *True*

The solution set is $\{-5\}$.

27.
$$-7x = 28$$
$$\frac{-7x}{-7} = \frac{28}{-7} \quad \textit{Divide by } -7.$$
$$x = -4$$

Check $x = -4$: $28 = 28$ *True*

The solution set is $\{-4\}$.

29.
$$10t = -36$$
$$\frac{10t}{10} = \frac{-36}{10} \quad \textit{Divide by 10.}$$
$$t = -\tfrac{36}{10} = -\tfrac{18}{5} \quad \textit{Lowest terms}$$

Check $t = -\frac{18}{5}$: $-36 = -36$ *True*

The solution set is $\{-\frac{18}{5}\}$, or $\{-3.6\}$.

31.
$$-6x = -72$$
$$\frac{-6x}{-6} = \frac{-72}{-6} \quad \textit{Divide by } -6.$$
$$x = 12$$

Check $x = 12$: $-72 = -72$ *True*

The solution set is $\{12\}$.

33.
$$4r = 0$$
$$\frac{4r}{4} = \frac{0}{4} \quad \textit{Divide by 4.}$$
$$r = 0$$

Check $r = 0$: $0 = 0$ *True*

The solution set is $\{0\}$.

35.
$$-x = 12$$
$$-1 \cdot (-x) = -1 \cdot 12 \quad \textit{Multiply by } -1.$$
$$x = -12$$

Check $x = -12$: $12 = 12$ *True*

The solution set is $\{-12\}$.

37.
$$-x = -\tfrac{3}{4}$$
$$-1 \cdot (-x) = -1 \cdot (-\tfrac{3}{4})$$
$$x = \tfrac{3}{4}$$

Check $x = \frac{3}{4}$: $-\frac{3}{4} = -\frac{3}{4}$ *True*

The solution set is $\{\frac{3}{4}\}$.

39.
$$0.2t = 8$$
$$\frac{0.2t}{0.2} = \frac{8}{0.2}$$
$$t = 40$$

Check $t = 40$: $8 = 8$ *True*

The solution set is $\{40\}$.

41.
$$-2.1m = 25.62$$
$$\frac{-2.1m}{-2.1} = \frac{25.62}{-2.1}$$
$$m = -12.2$$

Check $m = -12.2$: $25.62 = 25.62$ *True*

The solution set is $\{-12.2\}$.

43.
$$\tfrac{1}{4}x = -12$$
$$4 \cdot \tfrac{1}{4}x = 4(-12) \quad \textit{Multiply by 4.}$$
$$1x = -48$$
$$x = -48$$

Check $x = -48$: $-12 = -12$ *True*

The solution set is $\{-48\}$.

45.
$$\frac{z}{6} = 12$$
$$\tfrac{1}{6}z = 12$$
$$6 \cdot \tfrac{1}{6}z = 6 \cdot 12$$
$$z = 72$$

Check $z = 72$: $12 = 12$ *True*

The solution set is $\{72\}$.

47.
$$\frac{x}{7} = -5$$
$$\tfrac{1}{7}x = -5$$
$$7(\tfrac{1}{7}x) = 7(-5)$$
$$x = -35$$

Check $x = -35$: $-5 = -5$ *True*

The solution set is $\{-35\}$.

49. $$\frac{2}{7}p = 4$$

$$\frac{7}{2}\left(\frac{2}{7}p\right) = \frac{7}{2}(4) \quad \textit{Multiply by the reciprocal of } \tfrac{2}{7}.$$

$$p = 14$$

Check $p = 14$: $4 = 4$ *True*

The solution set is $\{14\}$.

51. $$-\frac{5}{6}t = -15$$

$$-\frac{6}{5}\left(-\frac{5}{6}t\right) = -\frac{6}{5}(-15) \quad \textit{Multiply by the reciprocal of } -\tfrac{5}{6}.$$

$$t = 18$$

Check $t = 18$: $-15 = -15$ *True*

The solution set is $\{18\}$.

53. $$-\frac{7}{9}x = \frac{3}{5}$$

$$-\frac{9}{7}\left(-\frac{7}{9}x\right) = -\frac{9}{7} \cdot \frac{3}{5} \quad \textit{Multiply by the reciprocal of } -\tfrac{7}{9}.$$

$$x = -\tfrac{27}{35}$$

Check $x = -\frac{27}{35}$: $\frac{3}{5} = \frac{3}{5}$ *True*

The solution set is $\left\{-\frac{27}{35}\right\}$.

55. $$-0.3x = 9$$

$$\frac{-0.3x}{-0.3} = \frac{9}{-0.3} \quad \textit{Divide by } -0.3.$$

$$x = -30$$

Check $x = -30$: $9 = 9$ *True*

The solution set is $\{-30\}$.

57. $$4x + 3x = 21$$
$$7x = 21$$
$$\frac{7x}{7} = \frac{21}{7}$$
$$x = 3$$

Check $x = 3$: $21 = 21$ *True*

The solution set is $\{3\}$.

59. $$6r - 8r = 10$$
$$-2r = 10$$
$$\frac{-2r}{-2} = \frac{10}{-2}$$
$$r = -5$$

Check $r = -5$: $10 = 10$ *True*

The solution set is $\{-5\}$.

61. $$\frac{2}{5}x - \frac{3}{10}x = 2$$
$$\frac{4}{10}x - \frac{3}{10}x = 2$$
$$\frac{1}{10}x = 2$$
$$10 \cdot \frac{1}{10}x = 10 \cdot 2$$
$$x = 20$$

Check $x = 20$: $8 - 6 = 2$ *True*

The solution set is $\{20\}$.

63. $$7m + 6m - 4m = 63$$
$$9m = 63$$
$$\frac{9m}{9} = \frac{63}{9}$$
$$m = 7$$

Check $m = 7$: $63 = 63$ *True*

The solution set is $\{7\}$.

65. $$-6x + 4x - 7x = 0$$
$$-9x = 0$$
$$\frac{-9x}{-9} = \frac{0}{-9}$$
$$x = 0$$

Check $x = 0$: $0 = 0$ *True*
The solution set is $\{0\}$.

67. $$8w - 4w + w = -3$$
$$5w = -3$$
$$\frac{5w}{5} = \frac{-3}{5}$$
$$w = -\tfrac{3}{5}$$

Check $w = -\frac{3}{5}$: $-3 = -3$ *True*
The solution set is $\left\{-\frac{3}{5}\right\}$.

69. $$\frac{1}{3}x - \frac{1}{4}x + \frac{1}{12}x = 3$$
$$\left(\frac{1}{3} - \frac{1}{4} + \frac{1}{12}\right)x = 3 \quad \textit{Distributive property}$$
$$\left(\frac{4}{12} - \frac{3}{12} + \frac{1}{12}\right)x = 3 \quad \textit{LCD = 12}$$
$$\tfrac{1}{6}x = 3 \quad \textit{Lowest terms}$$
$$6\left(\tfrac{1}{6}x\right) = 6(3) \quad \textit{Multiply by 6.}$$
$$x = 18$$

Check $x = 18$: $6 - 4.5 + 1.5 = 3$ *True*

The solution set is $\{18\}$.

71. Answers will vary. One example is

$$\tfrac{3}{2}x = -6.$$

73. "When a number is multiplied by 4, the result is 6."

$$4x = 6$$
$$\frac{4x}{4} = \frac{6}{4}$$
$$x = \tfrac{3}{2}$$

The number is $\frac{3}{2}$ and $\left\{\frac{3}{2}\right\}$ is the solution set.

75. "When a number is divided by -5, the result is 2."

$$\frac{x}{-5} = 2$$
$$(-5)(-\tfrac{1}{5}x) = (-5)(2)$$
$$x = -10$$

The number is -10 and $\{-10\}$ is the solution set.

77. $-(3m+5)$
$= -1(3m+5)$
$= -1(3m) - 1(5)$
$= -3m - 5$

79. $4(-5+2p) - 3(p-4)$
$= 4(-5) + 4(2p) - 3(p) - 3(-4)$
$= -20 + 8p - 3p + 12$
$= -20 + 12 + 8p - 3p$
$= -8 + 5p$

81. $4x + 5 + 2x = 7x$
$6x + 5 = 7x$ *Combine terms.*
$5 = x$ *Subtract 6x.*

Check $x = 5$: $20 + 5 + 10 = 35$ *True*

The solution set is $\{5\}$.

2.3 More on Solving Linear Equations

2.3 Now Try Exercises

N1. *Step 1* (not necessary)

Step 2
$7 + 2m = -3$
$7 + 2m - 7 = -3 - 7$ *Subtract 7.*
$2m = -10$ *Combine terms.*

Step 3
$\frac{2m}{2} = \frac{-10}{2}$ *Divide by 2.*
$m = -5$

Step 4
Check $m = -5$: $7 - 10 = -3$ *True*

The solution set is $\{-5\}$.

N2. *Step 1* (not necessary)

Step 2
$2q + 3 = 4q - 9$
$2q + 3 - 2q = 4q - 9 - 2q$ *Subtract 2q.*
$3 = 2q - 9$ *Combine terms.*
$3 + 9 = 2q - 9 + 9$ *Add 9.*
$12 = 2q$ *Combine terms.*

Step 3
$\frac{12}{2} = \frac{2q}{2}$ *Divide by 2.*
$6 = q$

Step 4
Check $q = 6$: $12 + 3 = 24 - 9$ *True*

The solution set is $\{6\}$.

N3. *Step 1*
$3(z-6) - 5z = -7z + 7$
$3z - 18 - 5z = -7z + 7$ *Distributive property*
$-2z - 18 = -7z + 7$ *Combine terms.*

Step 2
$-2z - 18 + 18 = -7z + 7 + 18$ *Add 18.*
$-2z = -7z + 25$
$-2z + 7z = -7z + 25 + 7z$ *Add 7z.*
$5z = 25$

Step 3
$\frac{5z}{5} = \frac{25}{5}$ *Divide by 5.*
$z = 5$

Step 4
Check $z = 5$: $-28 = -28$ *True*

The solution set is $\{5\}$.

N4. *Step 1*
$5x - (x+9) = x - 4$
$5x - x - 9 = x - 4$ *Distributive property*
$4x - 9 = x - 4$

Step 2
$4x - 9 + 9 = x - 4 + 9$ *Add 9.*
$4x = x + 5$
$4x - x = x + 5 - x$ *Subtract x.*
$3x = 5$

Step 3
$\frac{3x}{3} = \frac{5}{3}$ *Divide by 3.*
$x = \frac{5}{3}$

Step 4
Check $x = \frac{5}{3}$: $-\frac{7}{3} = -\frac{7}{3}$ *True*

The solution set is $\left\{\frac{5}{3}\right\}$.

N5. *Step 1*
$24 - 4(7 - 2t) = 4(t - 1)$
$24 - 28 + 8t = 4t - 4$ *Dist. prop.*
$-4 + 8t = 4t - 4$

Step 2
$-4 + 8t + 4 = 4t - 4 + 4$ *Add 4.*
$8t = 4t$
$8t - 4t = 4t - 4t$ *Subtract 4t.*
$4t = 0$

Step 3

$$\frac{4t}{4}=\frac{0}{4} \quad \textit{Divide by 4.}$$
$$t=0$$

Step 4

Check $t=0$: $24-4(7)\stackrel{?}{=}4(-1)$

$$-4=-4 \quad \textit{True}$$

The solution set is $\{0\}$.

N6. *Step 1*

$$\frac{1}{2}x+\frac{5}{8}x=\frac{3}{4}x-6$$

The LCD of all the fractions in the equation is 8, so multiply each side by 8 to clear the fractions.

$$8\left(\frac{1}{2}x+\frac{5}{8}x\right)=8\left(\frac{3}{4}x-6\right)$$
$$8\left(\frac{1}{2}x\right)+8\left(\frac{5}{8}x\right)=8\left(\frac{3}{4}x\right)-8(6)$$

Distributive property

$$4x+5x=6x-48$$
$$9x=6x-48$$

Step 2

$9x-6x=6x-48-6x$ *Subtract 6x.*

$$3x=-48$$

Step 3

$$\frac{3x}{3}=\frac{-48}{3} \quad \textit{Divide by 3.}$$
$$x=-16$$

Step 4

Check $x=-16$: $-8-10\stackrel{?}{=}-12-6$

$$-18=-18 \quad \textit{True}$$

The solution set is $\{-16\}$.

N7. *Step 1*

$$\tfrac{2}{3}(x+2)-\tfrac{1}{2}(3x+4)=-4$$
$$6\left[\tfrac{2}{3}(x+2)-\tfrac{1}{2}(3x+4)\right]=6(-4)$$

Multiply by 6.

$$6\left[\tfrac{2}{3}(x+2)\right]-6\left[\tfrac{1}{2}(3x+4)\right]=-24$$

Distributive property

$$4(x+2)-3(3x+4)=-24$$

Multiply.

$$4x+8-9x-12=-24$$

Distributive property

$$-5x-4=-24$$

Combine terms.

Step 2

$-5x-4+4=-24+4$ *Add 4.*

$-5x=-20$ *Combine terms.*

Step 3

$$\frac{-5x}{-5}=\frac{-20}{-5} \quad \textit{Divide by } -5.$$
$$x=4$$

Step 4

Check $x=4$: $4-8=-4$ *True*

The solution set is $\{4\}$.

N8. *Step 1*

$$0.05(13-t)-0.2t=0.08(30)$$

To clear decimals, multiply both sides by 100.

$$100[0.05(13-t)-0.2t]=100[0.08(30)]$$
$$5(13-t)-20t=8(30)$$
$$65-5t-20t=240$$
$$65-25t=240$$

Step 2

$$65-25t-65=240-65$$
$$-25t=175$$

Step 3

$$\frac{-25t}{-25}=\frac{175}{-25}$$
$$t=-7$$

Step 4

Check $t=-7$: $1+1.4\stackrel{?}{=}2.4$

$$2.4=2.4 \quad \textit{True}$$

The solution set is $\{-7\}$.

N9.

$$-3(x-7)=2x-5x+21$$
$$-3x+21=-3x+21$$
$$-3x+21-21=-3x+21-21 \quad \textit{Subtract 21.}$$
$$-3x=-3x$$
$$-3x+3x=-3x+3x \quad \textit{Add 3x.}$$
$$0=0 \quad \textit{True}$$

The variable x has "disappeared," and a *true* statement has resulted. The original equation is an identity. This means that for every real number value of x, the equation is true. Thus, the solution set is {all real numbers}.

N10.

$$-4x+12=3-4(x-3)$$
$$-4x+12=3-4x+12 \quad \textit{Distributive property}$$
$$-4x+12=-4x+15 \quad \textit{Combine terms.}$$
$$-4x+12+4x=-4x+15+4x \quad \textit{Add 4x.}$$
$$12=15 \quad \textit{False}$$

The variable x has "disappeared," and a *false* statement has resulted. This means that for every real number value of x, the equation is false. Thus, the equation has ***no solution*** and its solution set is the **empty set**, or **null set**, symbolized $\emptyset$.

N11. First, suppose that the sum of two numbers is 18, and one of the numbers is 10. How would you find the other number? You would subtract 10 from 18. Instead of using 10 as one of the numbers, use m. This gives us the expression $18-m$ for the other number.

2.3 Section Exercises

1. $7x + 8 = 1$

Use the addition property of equality to subtract 8 from each side.

3. $3(2t - 4) = 20 - 2t$

Clear the parentheses by using the distributive property.

5. $\frac{2}{3}x - \frac{1}{6} = \frac{3}{2}x + 1$

Clear fractions by multiplying by the LCD, 6.

7. Equations **A**, **B**, and **C** each have {all real numbers} for their solution set. However, equation **D** gives

$$\begin{aligned} 3x &= 2x \\ 3x - 2x &= 2x - 2x \\ x &= 0. \end{aligned}$$

The only solution of this equation is 0, so the correct choice is **D**.

Use the four-step method for solving linear equations as given in the text. The details of these steps will only be shown for a few of the exercises.

9.

$$\begin{aligned} 3x + 2 &= 14 \\ 3x + 2 - 2 &= 14 - 2 && \textit{Subtract 2.} \\ 3x &= 12 && \textit{Combine terms.} \\ \frac{3x}{3} &= \frac{12}{3} && \textit{Divide by 3.} \\ x &= 4 \end{aligned}$$

Check $x = 4$: $12 + 2 = 14$ *True*

The solution set is $\{4\}$.

11.

$$\begin{aligned} -5z - 4 &= 21 \\ -5z - 4 + 4 &= 21 + 4 && \textit{Add 4.} \\ -5z &= 25 && \textit{Combine terms.} \\ \frac{-5z}{-5} &= \frac{25}{-5} && \textit{Divide by } -5. \\ z &= -5 \end{aligned}$$

Check $z = -5$: $25 - 4 = 21$ *True*

The solution set is $\{-5\}$.

13.

$$\begin{aligned} 4p - 5 &= 2p \\ 4p - 5 - 4p &= 2p - 4p && \textit{Subtract 4p.} \\ -5 &= -2p && \textit{Combine terms.} \\ \frac{-5}{-2} &= \frac{-2p}{-2} && \textit{Divide by } -2. \\ \frac{5}{2} &= p \end{aligned}$$

Check $p = \frac{5}{2}$: $10 - 5 = 5$ *True*

The solution set is $\{\frac{5}{2}\}$.

15.

$$\begin{aligned} 2x + 9 &= 4x + 11 \\ -2x + 9 &= 11 && \textit{Subtract 4x.} \\ -2x &= 2 && \textit{Subtract 9.} \\ x &= -1 && \textit{Divide by } -2. \end{aligned}$$

Check $x = -1$: $7 = 7$ *True*

The solution set is $\{-1\}$.

17. $5m + 8 = 7 + 3m$

For this equation, step 1 is not needed.

Step 2

$$\begin{aligned} 5m + 8 - 8 &= 7 + 3m - 8 && \textit{Subtract 8.} \\ 5m &= 3m - 1 \\ 5m - 3m &= 3m - 1 - 3m && \textit{Subtract 3m.} \\ 2m &= -1 \end{aligned}$$

Step 3

$$\begin{aligned} \frac{2m}{2} &= \frac{-1}{2} && \textit{Divide by 2.} \\ m &= -\tfrac{1}{2} \end{aligned}$$

Step 4

Substitute $-\frac{1}{2}$ for m in the original equation.

$$\begin{aligned} 5m + 8 &= 7 + 3m \\ 5(-\tfrac{1}{2}) + 8 &\stackrel{?}{=} 7 + 3(-\tfrac{1}{2}) && \textit{Let } m = -\tfrac{1}{2}. \\ -\tfrac{5}{2} + 8 &\stackrel{?}{=} 7 + (-\tfrac{3}{2}) \\ \tfrac{11}{2} &= \tfrac{11}{2} && \textit{True} \end{aligned}$$

The solution set is $\{-\frac{1}{2}\}$.

19.

$$\begin{aligned} -12x - 5 &= 10 - 7x \\ -12x - 5 + 7x &= 10 - 7x + 7x && \textit{Add 7x.} \\ -5x - 5 &= 10 \\ -5x - 5 + 5 &= 10 + 5 && \textit{Add 5.} \\ -5x &= 15 \\ \frac{-5x}{-5} &= \frac{15}{-5} && \textit{Divide by } -5. \\ x &= -3 \end{aligned}$$

Check $x = -3$: $36 - 5 = 10 + 21$ *True*

The solution set is $\{-3\}$.

21.

$$\begin{aligned} 12h - 5 &= 11h + 5 - h \\ 12h - 5 &= 10h + 5 && \textit{Combine terms.} \\ 2h - 5 &= 5 && \textit{Subtract 10h.} \\ 2h &= 10 && \textit{Add 5.} \\ h &= 5 && \textit{Divide by 2.} \end{aligned}$$

Check $h = 5$: $55 = 55$ *True*

The solution set is $\{5\}$.

23. $7r - 5r + 2 = 5r + 2 - r$

$2r + 2 = 4r + 2$ *Combine terms.*

$2 = 2r + 2$ *Subtract 2r.*

$0 = 2r$ *Subtract 2.*

$0 = r$ *Divide by 2.*

Check $r = 0$: $2 = 2$ *True*

The solution set is $\{0\}$.

25. $3(4x + 2) + 5x = 30 - x$

$12x + 6 + 5x = 30 - x$ *Distributive property*

$17x + 6 = 30 - x$ *Combine terms.*

$18x + 6 = 30$ *Add 1x.*

$18x = 24$ *Subtract 6.*

$x = \frac{24}{18} = \frac{4}{3}$ *Divide by 18.*

Check $x = \frac{4}{3}$: $\frac{86}{3} = \frac{86}{3}$ *True*

The solution set is $\{\frac{4}{3}\}$.

27. $-2p + 7 = 3 - (5p + 1)$

$-2p + 7 = 3 - 5p - 1$ *Distributive property*

$-2p + 7 = -5p + 2$ *Combine terms.*

$3p + 7 = 2$ *Add 5p.*

$3p = -5$ *Subtract 7.*

$p = -\frac{5}{3}$

Check $p = -\frac{5}{3}$: $\frac{31}{3} = \frac{31}{3}$ *True*

The solution set is $\{-\frac{5}{3}\}$.

29. $6(3w + 5) = 2(10w + 10)$

$18w + 30 = 20w + 20$

$18w = 20w - 10$ *Subtract 30.*

$-2w = -10$ *Subtract 20w.*

$w = 5$ *Divide by −2.*

Check $w = 5$: $120 = 120$ *True*

The solution set is $\{5\}$.

31. $-(4x + 2) - (-3x - 5) = 3$

$-1(4x + 2) - 1(-3x - 5) = 3$

$-4x - 2 + 3x + 5 = 3$

$-x + 3 = 3$

$-x = 0$

$x = 0$

Check $x = 0$: $3 = 3$ *True*

The solution set is $\{0\}$.

33. $6(4x - 1) = 12(2x + 3)$

$24x - 6 = 24x + 36$

$-6 = 36$ *Subtract 24x.*

The variable has "disappeared," and the resulting equation is false. Therefore, the equation has no solution set, symbolized by $\emptyset$.

35. $3(2x - 4) = 6(x - 2)$

$6x - 12 = 6x - 12$

$-12 = -12$ *Subtract 6x.*

$0 = 0$ *Add 12.*

The variable has "disappeared." Since the resulting statement is a *true* one, *any* real number is a solution. We indicate the solution set as {all real numbers}.

37. $11x - 5(x + 2) = 6x + 5$

$11x - 5x - 10 = 6x + 5$

$6x - 10 = 6x + 5$

$-10 = 5$ *Subtract 6x.*

The variable has "disappeared," and the resulting equation is false. Therefore, the equation has no solution set, symbolized by $\emptyset$.

39. $\frac{3}{5}t - \frac{1}{10}t = t - \frac{5}{2}$

The least common denominator of all the fractions in the equation is 10.

$10(\frac{3}{5}t - \frac{1}{10}t) = 10(t - \frac{5}{2})$

Multiply both sides by 10.

$10(\frac{3}{5}t) + 10(-\frac{1}{10}t) = 10t + 10(-\frac{5}{2})$

Distributive property

$6t - t = 10t - 25$

$5t = 10t - 25$

$-5t = -25$ *Subtract 10t.*

$\frac{-5t}{-5} = \frac{-25}{-5}$ *Divide by −5.*

$t = 5$

Check $t = 5$: $\frac{5}{2} = \frac{5}{2}$ *True*

The solution set is $\{5\}$.

41. $\frac{3}{4}x - \frac{1}{3}x + 5 = \frac{5}{6}x$

The least common denominator of all the fractions in the equation is 12, so multiply both sides by 12 and solve for x.

$$12\left(\frac{3}{4}x - \frac{1}{3}x + 5\right) = 12\left(\frac{5}{6}x\right)$$

$9x - 4x + 60 = 10x$

$5x + 60 = 10x$

$60 = 5x$

$\frac{60}{5} = \frac{5x}{5}$

$12 = x$

Check $x = 12$: $9 - 4 + 5 = 10$ *True*

The solution set is $\{12\}$.

43. $\frac{1}{7}(3x+2)-\frac{1}{5}(x+4)=2$

The least common denominator of all the fractions in the equation is 35, so multiply both sides by 35 and solve for x.

$$\begin{aligned} 35\left[\frac{1}{7}(3x+2)-\frac{1}{5}(x+4)\right] &= 35(2)\\ 5(3x+2)-7(x+4) &= 70\\ 15x+10-7x-28 &= 70\\ 8x-18 &= 70\\ 8x &= 88\\ \frac{8x}{8} &= \frac{88}{8}\\ x &= 11 \end{aligned}$$

Check $x=11$: $5-3=2$ *True*

The solution set is $\{11\}$.

45. $-\frac{1}{4}(x-12)+\frac{1}{2}(x+2)=x+4$

The LCD of all the fractions is 4.

$$\begin{aligned} 4\left[-\frac{1}{4}(x-12)+\frac{1}{2}(x+2)\right] &= 4(x+4) && \textit{Multiply by 4.}\\ 4\left(-\frac{1}{4}\right)(x-12)+4\left(\frac{1}{2}\right)(x+2) &= 4x+16 && \textit{Distributive property}\\ (-1)(x-12)+2(x+2) &= 4x+16 && \textit{Multiply.}\\ -x+12+2x+4 &= 4x+16 && \textit{Distributive property}\\ x+16 &= 4x+16\\ -3x+16 &= 16\\ -3x &= 0\\ \frac{-3x}{-3} &= \frac{0}{-3} && \textit{Divide by } -3.\\ x &= 0 \end{aligned}$$

Check $x=0$: $4=4$ *True*

The solution set is $\{0\}$.

47. $\frac{2}{3}k-\left(k-\frac{1}{2}\right)=\frac{1}{6}(k-51)$

The least common denominator of all the fractions in the equation is 6, so multiply both sides by 6 and solve for k.

$$\begin{aligned} 6\left[\tfrac{2}{3}k-\left(k-\tfrac{1}{2}\right)\right] &= 6\left[\tfrac{1}{6}(k-51)\right]\\ 6\left(\tfrac{2}{3}k\right)-6\left(k-\tfrac{1}{2}\right) &= 6\left[\tfrac{1}{6}(k-51)\right] && \textit{Distributive property}\\ 4k-6k+3 &= 1(k-51)\\ -2k+3 &= k-51\\ -3k+3 &= -51\\ -3k &= -54\\ k &= 18 \end{aligned}$$

Check $k=18$: $-\frac{11}{2}=-\frac{11}{2}$ *True*

The solution set is $\{18\}$.

49. $0.2(60)+0.05x=0.1(60+x)$

To eliminate the decimal in 0.2 and 0.1, we need to multiply both sides by 10. But to eliminate the decimal in 0.05, we need to multiply by 100, so we choose 100.

$$\begin{aligned} 100[0.2(60)+0.05x] &= 100[0.1(60+x)] && \textit{Multiply by 100.}\\ 100[0.2(60)]+100(0.05x) &= 100[0.1(60+x)] && \textit{Distributive property}\\ 20(60)+5x &= 10(60+x) && \textit{Multiply.}\\ 1200+5x &= 600+10x\\ 1200-5x &= 600\\ -5x &= -600\\ x &= \tfrac{-600}{-5} = 120 \end{aligned}$$

Check $x=120$: $18=18$ *True*

The solution set is $\{120\}$.

51. $1.00x+0.05(12-x)=0.10(63)$

To clear the equation of decimals, we multiply both sides by 100.

$$\begin{aligned} 100[1.00x+0.05(12-x)] &= 100[0.10(63)]\\ 100(1.00x)+100[0.05(12-x)] &= (100)(0.10)(63)\\ 100x+5(12-x) &= 10(63)\\ 100x+60-5x &= 630\\ 95x+60 &= 630\\ 95x &= 570\\ x &= \tfrac{570}{95} = 6 \end{aligned}$$

Check $x=6$: $6.3=6.3$ *True*

The solution set is $\{6\}$.

53. $0.6(10{,}000)+0.8x=0.72(10{,}000+x)$

$$\begin{aligned} 100[0.6(10{,}000)]+100(0.8x) &= 100[0.72(10{,}000+x)] && \textit{Multiply by both sides by 100.}\\ 60(10{,}000)+80x &= 72(10{,}000+x)\\ 600{,}000+80x &= 720{,}000+72x\\ 600{,}000+8x &= 720{,}000\\ 8x &= 120{,}000\\ x &= \tfrac{120{,}000}{8} = 15{,}000 \end{aligned}$$

Check $x=15{,}000$: $18{,}000=18{,}000$ *True*

The solution set is $\{15{,}000\}$.

55. $10(2x-1)=8(2x+1)+14$
$20x-10=16x+8+14$
$20x-10=16x+22$
$4x-10=22$
$4x=32$
$x=8$

Check $x=8$: $150=150$ *True*

The solution set is $\{8\}$.

57. $\frac{1}{2}(x+2)+\frac{3}{4}(x+4)=x+5$

To clear fractions, multiply both sides by the LCD, which is 4.

$$4\left[\tfrac{1}{2}(x+2)+\tfrac{3}{4}(x+4)\right]=4(x+5)$$
$$4\left(\tfrac{1}{2}\right)(x+2)+4\left(\tfrac{3}{4}\right)(x+4)=4x+20$$
$$2(x+2)+3(x+4)=4x+20$$
$$2x+4+3x+12=4x+20$$
$$5x+16=4x+20$$
$$x+16=20$$
$$x=4$$

Check $x=4$: $9=9$ *True*

The solution set is $\{4\}$.

59. $0.1(x+80)+0.2x=14$

To eliminate the decimals, multiply both sides by 10.

$$10[0.1(x+80)+0.2x]=10(14)$$
$$1(x+80)+2x=140$$
$$x+80+2x=140$$
$$3x+80=140$$
$$3x=60$$
$$x=20$$

Check $x=20$: $14=14$ *True*

The solution set is $\{20\}$.

61. $4(x+8)=2(2x+6)+20$
$4x+32=4x+12+20$
$4x+32=4x+32$
$4x=4x$
$0=0$

Since $0=0$ is a *true* statement, the solution set is {all real numbers}.

63. $9(v+1)-3v=2(3v+1)-8$
$9v+9-3v=6v+2-8$
$6v+9=6v-6$
$9=-6$

Because $9=-6$ is a *false* statement, the equation has no solution set, symbolized by $\emptyset$.

65. The sum of q and the other number is 11. To find the other number, you would subtract q from 11, so an expression for the other number is $11-q$.

67. The product of x and the other number is 9. To find the other number, you would divide 9 by x, so an expression for the other number is $\frac{9}{x}$.

69. An expression for the total number of yards is $x+9$.

71. If a baseball player gets 65 hits in one season, and h of the hits are in one game, then $65-h$ of the hits came in the rest of the games.

73. If Monica is x years old now, then 15 years from now, she will be $x+15$ years old. Five years ago, she was $x-5$ years old.

75. Since the value of each quarter is 25 cents, the value of r quarters is $25r$ cents.

77. Since each bill is worth 5 dollars, the number of bills is $\frac{t}{5}$.

79. Since each adult ticket costs x dollars, the cost of 3 adult tickets is $3x$. Since each child's ticket costs y dollars, the cost of 2 children's tickets is $2y$. Therefore, the total cost is $3x+2y$ (dollars).

81. "A number added to -6" is written

$$-6+x.$$

83. "The difference between -5 and a number" is written

$$-5-x.$$

85. "The product of 12 and the difference between a number and 9" is written

$$12(x-9).$$

Summary Exercises on Solving Linear Equations

1. $x+2=-3$
$x=-5$ *Subtract 2.*

Check $x=-5$: $-3=-3$ *True*

The solution set is $\{-5\}$.

3. $12.5x=-63.75$
$x=\frac{-63.75}{12.5}$ *Divide by 12.5.*
$=-5.1$

Check $x=-5.1$: $-63.75=-63.75$ *True*

The solution set is $\{-5.1\}$.

5. $\frac{4}{5}x=-20$
$x=\left(\frac{5}{4}\right)(-20)$ *Multiply by $\frac{5}{4}$.*
$=-25$

Check $x=-25$: $-20=-20$ *True*

The solution set is $\{-25\}$.

7. $5x - 9 = 3(x - 3)$

$5x - 9 = 3x - 9$ *Distributive property*

$2x - 9 = -9$ *Subtract 3x.*

$2x = 0$ *Add 9.*

$x = 0$ *Divide by 2.*

Check $x = 0$: $-9 = -9$ *True*

The solution set is $\{0\}$.

9. $-x = 6$

$x = -6$ *Multiply by −1.*

Check $x = -6$: $6 = 6$ *True*

The solution set is $\{-6\}$.

11. $4x + 2(3 - 2x) = 6$

$4x + 6 - 4x = 6$

$6 = 6$

Since $6 = 6$ is a *true* statement, the solution set is {all real numbers}.

13. $-3(m - 4) + 2(5 + 2m) = 29$

$-3m + 12 + 10 + 4m = 29$

$m + 22 = 29$

$m = 7$

Check $m = 7$: $29 = 29$ *True*

The solution set is $\{7\}$.

15. $0.08x + 0.06(x + 9) = 1.24$

To eliminate the decimals, multiply both sides by 100.

$100[0.08x + 0.06(x + 9)] = 100(1.24)$

$8x + 6(x + 9) = 124$

$8x + 6x + 54 = 124$

$14x + 54 = 124$

$14x = 70$

$x = 5$

Check $x = 5$: $0.4 + 0.84 = 1.24$ *True*

The solution set is $\{5\}$.

17. $7m - (2m - 9) = 39$

$7m - 2m + 9 = 39$

$5m + 9 = 39$

$5m = 30$

$m = 6$

Check $m = 6$: $39 = 39$ *True*

The solution set is $\{6\}$.

19. $-2t + 5t - 9 = 3(t - 4) - 5$

$-2t + 5t - 9 = 3t - 12 - 5$

$3t - 9 = 3t - 17$

$-9 = -17$

Because $-9 = -17$ is a *false* statement, the equation has no solution set, symbolized by $\emptyset$.

21. $0.2(50) + 0.8r = 0.4(50 + r)$

To eliminate the decimals, multiply both sides by 10.

$10[0.2(50) + 0.8r] = 10[0.4(50 + r)]$

$2(50) + 8r = 4(50 + r)$

$100 + 8r = 200 + 4r$

$100 + 4r = 200$

$4r = 100$

$r = 25$

Check $r = 25$: $10 + 20 = 30$ *True*

The solution set is $\{25\}$.

23. $2(3 + 7x) - (1 + 15x) = 2$

$6 + 14x - 1 - 15x = 2$

$-x + 5 = 2$

$-x = -3$

$x = 3$

Check $x = 3$: $48 - 46 = 2$ *True*

The solution set is $\{3\}$.

25. $2(4 + 3r) = 3(r + 1) + 11$

$8 + 6r = 3r + 3 + 11$

$8 + 6r = 3r + 14$

$8 + 3r = 14$

$3r = 6$

$r = 2$

Check $r = 2$: $20 = 20$ *True*

The solution set is $\{2\}$.

27. $\frac{1}{4}x - 4 = \frac{3}{2}x + \frac{3}{4}x$

To clear fractions, multiply both sides by the LCD, which is 4.

$4\left(\frac{1}{4}x - 4\right) = 4\left(\frac{3}{2}x + \frac{3}{4}x\right)$

$x - 16 = 6x + 3x$

$x - 16 = 9x$

$-16 = 8x$

$x = -2$

Check $x = -2$: $-4.5 = -3 - 1.5$ *True*

The solution set is $\{-2\}$.

29. $\frac{3}{4}(z - 2) - \frac{1}{3}(5 - 2z) = -2$

To clear fractions, multiply both sides by the LCD, which is 12.

$12\left[\frac{3}{4}(z - 2) - \frac{1}{3}(5 - 2z)\right] = 12(-2)$

$9(z - 2) - 4(5 - 2z) = -24$

$9z - 18 - 20 + 8z = -24$

$17z - 38 = -24$

$17z = 14$

$z = \frac{14}{17}$

Check $z = \frac{14}{17}$: $-\frac{15}{17} - \frac{19}{17} = -2$ *True*

The solution set is $\left\{\frac{14}{17}\right\}$.

2.4 An Introduction to Applications of Linear Equations

2.4 Now Try Exercises

N1. *Step 2*
Let x = the number.

Step 3

If 5 is	added to	a number,	the result is	7 less than 3 times the number.
↓	↓	↓	↓	↓
5	+	x	=	$3x - 7$

Step 4
Solve the equation.

$$\begin{aligned} 5 + x &= 3x - 7 \\ 5 + x - 5 &= 3x - 7 - 5 && \textit{Subtract 5.} \\ x &= 3x - 12 \\ x - 3x &= 3x - 12 - 3x && \textit{Subtract } 3x. \\ -2x &= -12 \\ \frac{-2x}{-2} &= \frac{-12}{-2} && \textit{Divide by } -2. \\ x &= 6 \end{aligned}$$

Step 5
The number is 6.

Step 6
5 added to 6 is 11. 3 times 6 is 18, and 7 less than 18 is 11, so 6 is the number.

N2. *Step 2*
Let x = the number of medals Germany won.
Let $x - 7$ = the number of medals Russia won.

Step 3

The total	is	the number of medals Germany won	plus	the number of medals Russia won.
↓	↓	↓	↓	↓
51	=	x	+	$(x - 7)$

Step 4
Solve this equation.

$$\begin{aligned} 51 &= 2x - 7 \\ 51 + 7 &= 2x - 7 + 7 \\ 58 &= 2x \\ \frac{58}{2} &= \frac{2x}{2} \\ 29 &= x \end{aligned}$$

Step 5
Germany won 29 medals and Russia won $29 - 7 = 22$ medals.

Step 6
22 is 7 fewer than 29 and the sum of 22 and 29 is 51.

N3. *Step 2*
Let x = the number of orders for chocolate scones.
Then $\frac{2}{3}x$ = the number of orders for bagels.

Step 3

The total	is	orders for choc. scones	plus	orders for bagels.
↓	↓	↓	↓	↓
525	=	x	+	$\frac{2}{3}x$

Step 4
Solve this equation.

$$\begin{aligned} 525 &= 1x + \tfrac{2}{3}x && x = 1x \\ 525 &= \tfrac{3}{3}x + \tfrac{2}{3}x && \textit{LCD} = 3 \\ 525 &= \tfrac{5}{3}x && \textit{Combine like terms.} \\ \tfrac{3}{5}(525) &= \tfrac{3}{5}\left(\tfrac{5}{3}x\right) && \textit{Multiply by } \tfrac{3}{5}. \\ 315 &= x \end{aligned}$$

Step 5
The number of orders for chocolate scones was 315, so the number of orders for bagels was $\frac{2}{3}(315) = 210$.

Step 6
Two-thirds of 315 is 210 and the sum of 315 and 210 is 525.

N4. *Step 2*
Let x = the number of residents.
Then $4x$ = the number of guests.
(If each resident brought four guests, there would be four times as many guests as residents.)

Step 3

Number of residents	plus	number of guests	is	the total in attendance
↓	↓	↓	↓	↓
x	+	$4x$	=	175

Step 4
Solve this equation.

$$\begin{aligned} x + 4x &= 175 \\ 5x &= 175 \\ \frac{5x}{5} &= \frac{175}{5} \\ x &= 35 \end{aligned}$$

Step 5
There were 35 residents and $4 \cdot 35 = 140$ guests.

Step 6
140 is four times as much as 35 and the sum of 35 and 140 is 175.

N5. *Step 2*
Let $x =$ the time spent practicing free throws.
Then $2x =$ the time spent lifting weights
and $x + 2 =$ the time spent watching game films.

Step 3

Time spent practicing free throws	plus	time spent lifting weights
↓	↓	↓
x	$+$	$2x$

plus	time spent watching game films	is	total time.
↓	↓	↓	↓
$+$	$x + 2$	$=$	6

Step 4
Solve this equation.

$$x + 2x + (x + 2) = 6$$
$$4x + 2 = 6$$
$$4x = 4$$
$$\frac{4x}{4} = \frac{4}{4}$$
$$x = 1$$

Step 5
The time spent practicing free throws is 1 hour, the time spent lifting weights is $2(1) = 2$ hours, and the time spent watching game films is $1 + 2 = 3$ hours.

Step 6
Since 2 hours is twice as much time as 1 hour, and 3 hours is 2 more hours than 1 hour, and the sum of the times is $1 + 2 + 3 = 6$ hours (the total time spent), the answers are correct.

N6. *Step 2*
Let $x =$ the lesser page number.
Then $x + 1 =$ the greater page number.

Step 3
Because the sum of the page numbers is 593, an equation is

$$x + (x + 1) = 593$$

Step 4

$2x + 1 = 593$ *Combine like terms.*
$2x = 592$ *Subtract 1.*
$x = 296$ *Divide by 2.*

Step 5
The lesser page number is 296, and the greater page number is $296 + 1 = 297$.

Step 6
297 is one more than 296 and the sum of 296 and 297 is 593.

N7. Let $x =$ the lesser odd integer.
Then $x + 2 =$ the greater odd integer.

From the given information, we have

$$2 \cdot x + 3(x + 2) = 191.$$

Solve this equation.

$$2x + 3x + 6 = 191$$
$$5x + 6 = 191$$
$$5x = 185$$
$$x = 37$$

The lesser odd integer is 37 and the greater consecutive odd integer is $37 + 2 = 39$. Two times 37 is 74, three times 39 is 117, and 74 plus 117 is 191.

N8. Let $x =$ the degree measure of the angle.
Then $90 - x =$ the degree measure of its complement.

The complement	is	twice the angle.
↓	↓	↓
$90 - x$	$=$	$2x$

Solve this equation.

$$90 - x = 2x$$
$$90 = 3x$$
$$30 = x$$

The measure of the angle is 30°.

N9. *Step 2*
Let $x =$ the degree measure of the angle.
Then $90 - x =$ the degree measure of its complement, and $180 - x =$ the degree measure of its supplement.

Step 3

supplement	equals	46° less than 3 times its complement
↓	↓	↓
$180 - x$	$=$	$3(90 - x) - 46$

Step 4

$180 - x = 3(90 - x) - 46$
$180 - x = 270 - 3x - 46$ *Dist. property*
$180 - x = 224 - 3x$ *Combine terms.*
$180 + 2x = 224$ *Add 3x.*
$2x = 44$ *Subtract 180.*
$x = 22$ *Divide by 2.*

Step 5
The measure of the angle is 22°.

Step 6
The complement of 22° is $90° - 22° = 68°$ and 46° less than 3 times 68° is
$3(68°) - 46° = 204° - 46° = 158°$.
The supplement is $180° - 22° = 158°$.

2.4 Section Exercises

1. Choice **D**, $6\frac{1}{2}$, is *not* a reasonable answer in an applied problem that requires finding the number of cars on a dealer's lot, since you cannot have $\frac{1}{2}$ of a car. The number of cars must be a whole number.

3. Choice **A**, -10, is *not* a reasonable answer since distance cannot be negative.

The applied problems in this section should be solved by using the six-step method shown in the text. These steps will only be listed in a few of the solutions, but all of the solutions are based on this method.

5. *Step 2*
Let $x =$ the number.

Step 3

The product of 8,		and a number increased by 6,	is	104.
↓		↓	↓	↓
8	$\cdot$	$(x+6)$	$=$	104

Step 4
Solve this equation.

$$\begin{aligned} 8(x+6) &= 104 \\ 8x+48 &= 104 \\ 8x &= 56 \\ x &= 7 \end{aligned}$$

Step 5
The number is 7.

Step 6
7 increased by 6 is 13. The product of 8 and 13 is 104, so 7 is the number.

7. *Step 2*
Let $x =$ the unknown number. Then $5x+2$ represents "2 is added to five times a number," and $4x+5$ represents "5 more than four times the number."

Step 3 $\quad 5x+2 = 4x+5$

Step 4

$$\begin{aligned} 5x+2 &= 4x+5 \\ x+2 &= 5 \\ x &= 3 \end{aligned}$$

Step 5
The number is 3.

Step 6
Check that 3 is the correct answer by substituting this result into the words of the original problem. 2 added to five times a number is $2+5(3) = 17$ and 5 more than four times the number is $5+4(3) = 17$. The values are equal, so the number 3 is the correct answer.

9. *Step 2*
Let $x =$ the unknown number. Then $x-2$ is two subtracted from the number, $3(x-2)$ is triple the difference, and $x+6$ is six more than the number.

Step 3 $\quad 3(x-2) = x+6$

Step 4

$$\begin{aligned} 3x-6 &= x+6 \\ 2x-6 &= 6 \\ 2x &= 12 \\ x &= 6 \end{aligned}$$

Step 5
The number is 6.

Step 6
Check that 6 is the correct answer by substituting this result into the words of the original problem. Two subtracted from the number is $6-2=4$. Triple this difference is $3(4)=12$, which is equal to 6 more than the number, since $6+6=12$.

11. *Step 2*
Let $x =$ the unknown number. Then $3x$ is three times the number, $x+7$ is 7 more than the number, $2x$ is twice the number, and $-11-2x$ is the difference between -11 and twice the number.

Step 3 $\quad 3x+(x+7) = -11-2x$

Step 4

$$\begin{aligned} 4x+7 &= -11-2x \\ 6x+7 &= -11 \\ 6x &= -18 \\ x &= -3 \end{aligned}$$

Step 5
The number is -3.

Step 6
Check that -3 is the correct answer by substituting this result into the words of the original problem. The sum of three times a number and 7 more than the number is $3(-3)+(-3+7) = -5$ and the difference between -11 and twice the number is $-11-2(-3) = -5$. The values are equal, so the number -3 is the correct answer.

13. Let $x =$ the number of drive-in movie screens in Ohio.
Then $x+2 =$ the number of drive-in movie screens in Pennsylvania.
Since the total number of screens was 68, we can write the equation

$$x+(x+2) = 68.$$

Solve this equation.

$$\begin{aligned} 2x+2 &= 68 \\ 2x &= 66 \\ x &= 33 \end{aligned}$$

Since $x = 33$, $x+2 = 35$.

There were 33 drive-in movie screens in Ohio and 35 in Pennsylvania. Since 35 is 2 more than 33 and $33+35 = 68$, this answer checks.

15. Let $x =$ the number of Republicans.
Then $x + 18 =$ the number of Democrats.

The number of Republicans	plus	the number of Democrats
↓	↓	↓
x	$+$	$(x+18)$

equals	the total number of Republican and Democrat members in the Senate.
↓	↓
$=$	98

Solve the equation.

$$\begin{aligned} x + (x + 18) &= 98 \\ 2x + 18 &= 98 \\ 2x &= 80 \\ x &= 40 \end{aligned}$$

There were 40 Republicans and $40 + 18 = 58$ Democrats.

17. Let $x =$ revenue from ticket sales for Bon Jovi.
Then $x - 6.1 =$ revenue from ticket sales for Bruce Springsteen.

Since the total revenue from ticket sales was \$415.3 (all numbers in millions), we can write the equation

$$x + (x - 6.1) = 415.3.$$

Solve this equation.

$$\begin{aligned} 2x - 6.1 &= 415.3 \\ 2x &= 421.4 \\ x &= 210.7 \end{aligned}$$

Since $x = 210.7$, $x - 6.1 = 204.6$.

Bon Jovi took in \$210.7 million and Bruce Springsteen took in \$204.6 million. Since 204.6 is 6.1 less than 210.7 and $210.7 + 204.6 = 415.3$, this answer checks.

19. Let $x =$ the number of games the Celtics lost.
Then $3x + 2 =$ the number of games the Celtics won.

Since the total number of games played was 82, we can write the equation

$$x + (3x + 2) = 82.$$

Solve this equation.

$$\begin{aligned} 4x + 2 &= 82 \\ 4x &= 80 \\ x &= 20 \end{aligned}$$

Since $x = 20$, $3x + 2 = 62$.

The Celtics won 62 games and lost 20 games. Since $62 + 20 = 82$, this answer checks.

21. Let $x =$ the number of mg of vitamin C in a one-cup serving of pineapple juice.
Then $4x - 3 =$ the number of mg of vitamin C in a one-cup serving of orange juice.

Since the total amount of vitamin C in a serving of the two juices is 122 mg, we can write

$$x + (4x - 3) = 122.$$

Solve this equation.

$$\begin{aligned} 5x - 3 &= 122 \\ 5x &= 125 \\ x &= 25 \end{aligned}$$

Since $x = 25$, $4x - 3 = 97$.

A one-cup serving of pineapple juice has 25 mg of vitamin C and a one-cup serving of orange juice has 97 mg of vitamin C. Since 97 is 3 less than four times 25 and $25 + 97 = 122$, this answer checks.

23. Let $x =$ the number of CDs sold.
Then $\frac{8}{5}x =$ the number of DVDs sold.

The total number of CDs and DVDs sold was 273, so

$$x + \tfrac{8}{5}x = 273.$$

Solve this equation.

$$\begin{aligned} 1x + \tfrac{8}{5}x &= 273 \\ \tfrac{13}{5}x &= 273 \\ \tfrac{5}{13}\left(\tfrac{13}{5}x\right) &= \tfrac{5}{13}(273) \\ x &= \frac{5}{\cancel{13}_{\,1}} \cdot \frac{\cancel{273}^{\,21}}{1} = 105 \end{aligned}$$

Since $x = 105$, $\frac{8}{5}x = \frac{8}{5}(105) = 168$.

There were 168 DVDs sold.

25. Let $x =$ the number of kg of onions.
Then $6.6x =$ the number of kg of grilled steak.

The total weight of these two ingredients was 617.6 kg, so

$$x + 6.6x = 617.6.$$

Solve this equation.

$$\begin{aligned} 1x + 6.6x &= 617.6 \\ 7.6x &= 617.6 \\ x &= \tfrac{617.6}{7.6} \approx 81.3 \end{aligned}$$

Since $x = \frac{617.6}{7.6}$, $6.6x = 6.6\left(\frac{617.6}{7.6}\right) \approx 536.3$.

To the nearest tenth of a kilogram, 81.3 kg of onions and 536.3 kg of grilled steak were used to make the taco.

27. Let $x =$ the value of the 1945 nickel.
Then $2x =$ the value of the 1950 nickel.
The total value of the two coins is \$24.00, so

$$x + 2x = 24.$$

Solve this equation.

$$3x = 24$$
$$x = 8 \quad \textit{Divide by 8.}$$

Since $x = 8$, $2x = 2(8) = 16$.

The value of the 1945 Philadelphia nickel is \$8.00 and the value of the 1950 Denver nickel is \$16.00.

29. Let $x =$ the number of ounces of rye flour.
Then $4x =$ the number of ounces of whole wheat flour.

The total number of ounces would be 32, so

$$x + 4x = 32.$$

Solve this equation.

$$5x = 32$$
$$x = \tfrac{32}{5} = 6.4$$

Since $x = 6.4$, $4x = 4(6.4) = 25.6$. To make a loaf of bread weighing 32 oz, use 6.4 oz of rye flour and 25.6 oz of whole wheat flour.

31. Let $x =$ the number of tickets booked on United Airlines.
Then $x + 7 =$ the number of tickets booked on American Airlines, and
$2x + 4 =$ the number of tickets booked on Southwest Airlines.

The total number of tickets booked was 55, so

$$x + (x + 7) + (2x + 4) = 55.$$

Solve this equation.

$$4x + 11 = 55$$
$$4x = 44$$
$$x = \tfrac{44}{4} = 11$$

Since $x = 11$, $x + 7 = 11 + 7 = 18$, and $2x + 4 = 2(11) + 4 = 26$. He booked 18 tickets on American, 11 tickets on United, and 26 tickets on Southwest.

33. Let $x =$ the length of the shortest piece.
Then $x + 5 =$ the length of the middle piece, and
$x + 9 =$ the length of the longest piece.

The total length is 59 inches, so

$$x + (x + 5) + (x + 9) = 59.$$

Solve this equation.

$$3x + 14 = 59$$
$$3x = 45$$
$$x = 15$$

Since $x = 15$, $x + 5 = 20$, and $x + 9 = 24$.

The shortest piece should be 15 inches, the middle piece should be 20 inches, and the longest piece should be 24 inches. The answer checks since

$$15 + 20 + 24 = 59.$$

35. Let $x =$ the distance of Mercury from the sun.
Then $x + 31.2 =$ the distance of Venus from the sun, and
$x + 57 =$ the distance of Earth from the sun.

Since the total of the distances from these three planets is 196.2 (all distances in millions of miles), we can write the equation

$$x + (x + 31.2) + (x + 57) = 196.2.$$

Solve this equation.

$$3x + 88.2 = 196.2$$
$$3x = 108$$
$$x = 36$$

Mercury is 36 million miles from the sun, Venus is $36 + 31.2 = 67.2$ million miles from the sun, and Earth is $36 + 57 = 93$ million miles from the sun. The answer checks since

$$36 + 67.2 + 93 = 196.2.$$

37. Let $x =$ the measure of angles A and B.
Then $x + 60 =$ the measure of angle C.

The sum of the measures of the angles of any triangle is 180°, so

$$x + x + (x + 60) = 180.$$

Solve this equation.

$$3x + 60 = 180$$
$$3x = 120$$
$$x = 40$$

Angles A and B have measures of 40 degrees, and angle C has a measure of $40 + 60 = 100$ degrees. The answer checks since

$$40 + 40 + 100 = 180.$$

39. Let $x =$ the number on the first locker.
Then $x + 1 =$ the number on the next locker.

Since the numbers have a sum of 137, we can write the equation

$$x + (x + 1) = 137.$$

Solve the equation.

$$2x + 1 = 137$$
$$2x = 136$$
$$x = \tfrac{136}{2} = 68$$

Since $x = 68$, $x + 1 = 69$.

The lockers have numbers 68 and 69. Since $68 + 69 = 137$, this answer checks.

41. Because the two pages are back-to-back, they must have page numbers that are consecutive integers.

Let $x =$ the lesser page number.
Then $x + 1 =$ the greater page number.

$$\begin{aligned} x + (x + 1) &= 203 \\ 2x + 1 &= 203 \\ 2x &= 202 \\ x &= 101 \end{aligned}$$

Since $x = 101$, $x + 1 = 102$.

The page numbers are 101 and 102. This answer checks since the sum is 203.

43. Let $x =$ the lesser even integer.
Then $x + 2 =$ the greater even integer.

The lesser added to three times the greater gives a sum of 46 can be written as

$$\begin{aligned} x + 3(x + 2) &= 46. \\ x + 3x + 6 &= 46 \\ 4x + 6 &= 46 \\ 4x &= 40 \\ x &= 10 \end{aligned}$$

Since $x = 10$, $x + 2 = 12$.

The integers are 10 and 12. This answer checks since $10 + 3(12) = 46$.

45. Let $x =$ the lesser integer.
Then $x + 1 =$ the greater integer.

$$\begin{aligned} x + 3(x + 1) &= 43 \\ x + 3x + 3 &= 43 \\ 4x + 3 &= 43 \\ 4x &= 40 \\ x &= 10 \end{aligned}$$

Since $x = 10$, $x + 1 = 11$.

The integers are 10 and 11. This answer checks since $10 + 3(11) = 43$.

47. Let $x =$ the first even integer.
Then $x + 2 =$ the second even integer, and
$x + 4 =$ the third even integer.

$$\begin{aligned} x + (x + 2) + (x + 4) &= 60 \\ 3x + 6 &= 60 \\ 3x &= 54 \\ x &= 18 \end{aligned}$$

Since $x = 18$, $x + 2 = 20$, and $x + 4 = 22$.

The first even integer is 18. This answer checks since $18 + 20 + 22 = 60$.

49. Let $x =$ the first odd integer.
Then $x + 2 =$ the second odd integer, and
$x + 4 =$ the third odd integer.

$$\begin{aligned} 2[(x + 4) - 6] &= [x + 2(x + 2)] - 23 \\ 2(x - 2) &= x + 2x + 4 - 23 \\ 2x - 4 &= 3x - 19 \\ -4 &= x - 19 \\ 15 &= x \end{aligned}$$

Since $x = 15$, $x + 2 = 17$, and $x + 4 = 19$.

The integers are 15, 17, and 19.

51. Let $x =$ the measure of the angle.
Then $90 - x =$ the measure of its complement.

The "complement is four times its measure" can be written as

$$90 - x = 4x.$$

Solve this equation.

$$\begin{aligned} 90 &= 5x \\ x &= \tfrac{90}{5} = 18 \end{aligned}$$

The measure of the angle is 18°. The complement is $90° - 18° = 72°$, which is four times 18°.

53. Let $x =$ the measure of the angle.
Then $180 - x =$ the measure of its supplement.

The "supplement is eight times its measure" can be written as

$$180 - x = 8x.$$

Solve this equation.

$$\begin{aligned} 180 &= 9x \\ x &= \frac{180}{9} = 20 \end{aligned}$$

The measure of the angle is 20°. The supplement is $180° - 20° = 160°$, which is eight times 20°.

55. Let $x =$ the measure of the angle. Then
$90 - x =$ the measure of its complement, and
$180 - x =$ the measure of its supplement.

Its supplement	measures	39°
↓	↓	↓
$180 - x$	$=$	39

more than	twice its complement.
↓	↓
$+$	$2(90 - x)$

Solve the equation.

$$\begin{aligned} 180 - x &= 39 + 2(90 - x) \\ 180 - x &= 39 + 180 - 2x \\ 180 - x &= 219 - 2x \\ x + 180 &= 219 \\ x &= 39 \end{aligned}$$

The measure of the angle is 39°. The complement is $90° - 39° = 51°$. Now 39° more than twice its complement is $39° + 2(51°) = 141°$, which is the supplement of $39°$ since $180° - 39° = 141°$.

57. Let $x =$ the measure of the angle. Then
$180 - x =$ the measure of its supplement, and
$90 - x =$ the measure of its complement.

between the measure of its supplement and	The difference	three times the measure of its complement		
↓	↓	↓		
$(180 - x)$	$-$	$3(90 - x)$		
			is	10°.
			↓	↓
			$=$	10

Solve the equation.

$$\begin{aligned}(180 - x) - 3(90 - x) &= 10\\ 180 - x - 270 + 3x &= 10\\ 2x - 90 &= 10\\ 2x &= 100\\ x &= 50\end{aligned}$$

The measure of the angle is $50°$. The supplement is $180° - 50° = 130°$ and the complement is $90° - 50° = 40°$. The answer checks since $130° - 3(40°) = 10°$.

59. $L = 6$ and $W = 4$, so $LW = 6 \cdot 4 = 24$.

61. $L = 8$ and $W = 2$, so
$2L + 2W = 2(8) + 2(2) = 16 + 4 = 20$.

2.5 Formulas and Additional Applications from Geometry

2.5 Now Try Exercises

N1.
$$\begin{aligned}P &= 2a + 2b\\ 78 &= 2(12) + 2b \quad \textit{Let } P = 78 \textit{ and } a = 12.\\ 78 &= 24 + 2b\\ 54 &= 2b \quad \textit{Subtract 24.}\\ 27 &= b \quad \textit{Divide by 2.}\end{aligned}$$

N2. The fence will enclose the perimeter of the rectangular garden, so use the formula for the perimeter of a rectangle. Find the width of the garden by substituting $P = 160$ and $L = 2W - 10$ into the formula and solving for W.

$$\begin{aligned}P &= 2L + 2W\\ 160 &= 2(2W - 10) + 2W\\ 160 &= 4W - 20 + 2W\\ 180 &= 6W\\ 30 &= W\end{aligned}$$

Since $W = 30$, $L = 2(30) - 10 = 50$. The dimensions of the garden are 50 ft by 30 ft.

N3. Let $s =$ the length of the medium side, in feet;
$s + 1 =$ the length of the longest side, and,
$s - 7 =$ the length of the shortest side.

The perimeter is 30 feet, so

$$\begin{aligned}s + (s + 1) + (s - 7) &= 30.\\ 3s - 6 &= 30\\ 3s &= 36\\ s &= 12\end{aligned}$$

Since $s = 12$, $s + 1 = 13$, and $s - 7 = 5$. The lengths of the sides are 5, 12, and 13 feet. The perimeter is $5 + 12 + 13 = 30$, as required.

N4. Use the formula for the area of a triangle.

$$\begin{aligned}A &= \tfrac{1}{2}bh\\ 77 &= \tfrac{1}{2}(14)h \quad \textit{Let } A = 77,\ h = 14.\\ 77 &= 7h\\ 11 &= h\end{aligned}$$

The length of the height is 11 centimeters.

N5. Since the marked angles are vertical angles, they have equal measures.

$$\begin{aligned}6x + 2 &= 8x - 8\\ 2 &= 2x - 8\\ 10 &= 2x\\ 5 &= x\end{aligned}$$

If $x = 5$, $6x + 2 = 6(5) + 2 = 32$ and $8x - 8 = 8(5) - 8 = 32$.

The measure of the angles is $32°$.

N6. Solve $W = Fd$ for F.

$$\frac{W}{d} = \frac{Fd}{d} \quad \textit{Divide by d.}$$
$$\frac{W}{d} = F, \text{ or } F = \frac{W}{d}$$

N7. Solve $Ax + By = C$ for A.

$$Ax = C - By \quad \textit{Subtract By.}$$
$$\frac{Ax}{x} = \frac{C - By}{x} \quad \textit{Divide by x.}$$
$$A = \frac{C - By}{x}$$

N8. Solve $x = u + zs$ for z.

$$x - u = u + zs - u \quad \textit{Subtract u.}$$
$$x - u = zs$$
$$\frac{x - u}{s} = \frac{zs}{s} \quad \textit{Divide by s.}$$
$$\frac{x - u}{s} = z, \text{ or } z = \frac{x - u}{s}$$

N9. Solve $S = \frac{1}{2}(a + b + c)$ for a.

$$\begin{aligned}2S &= a + b + c \quad \textit{Multiply by 2.}\\ 2S - b - c &= a \quad \textit{Subtract b and c.}\end{aligned}$$

2.5 Section Exercises

1. **(a)** The perimeter of a plane geometric figure is the distance around the figure. It can be found by adding up the lengths of all the sides. Perimeter is a one-dimensional (linear) measurement, so it is given in linear units (inches, centimeters, feet, etc.).

(b) The area of a plane geometric figure is the measure of the surface covered or enclosed by the figure. Area is a two-dimensional measurement, so it is given in square units (square centimeters, square feet, etc.).

3. **(a)** The measure of a straight angle is $\underline{180°}$.

(b) Vertical angles have <u>the same</u> measures.

5. Carpeting for a bedroom covers the surface of the bedroom floor, so *area* would be used.

7. To measure fencing for a yard, use *perimeter* since you would need to measure the lengths of the sides of the yard.

9. Tile for a bathroom covers the surface of the bathroom floor, so *area* would be used.

11. To determine the cost of replacing a linoleum floor with a wood floor, use *area* since you need to know the measure of the surface covered by the wood.

In Exercises 13–32, substitute the given values into the formula and then solve for the remaining variable.

13. $P = 2L + 2W;\ L = 8,\ W = 5$

$$\begin{aligned} P &= 2L + 2W \\ &= 2(8) + 2(5) \\ &= 16 + 10 \\ P &= 26 \end{aligned}$$

15. $\mathcal{A} = \frac{1}{2}bh;\ b = 8,\ h = 16$

$$\begin{aligned} \mathcal{A} &= \tfrac{1}{2}bh \\ &= \tfrac{1}{2}(8)(16) \\ \mathcal{A} &= 64 \end{aligned}$$

17. $P = a + b + c;\ P = 12,\ a = 3,\ c = 5$

$$\begin{aligned} P &= a + b + c \\ 12 &= 3 + b + 5 \\ 12 &= b + 8 \\ 4 &= b \end{aligned}$$

19. $d = rt;\ d = 252,\ r = 45$

$$\begin{aligned} d &= rt \\ 252 &= 45t \\ \frac{252}{45} &= \frac{45t}{45} \\ 5.6 &= t \end{aligned}$$

21. $I = prt;\ p = 7500,\ r = 0.035,\ t = 6$

$$\begin{aligned} I &= prt \\ &= (7500)(0.035)(6) \\ I &= 1575 \end{aligned}$$

23. $\mathcal{A} = \frac{1}{2}h(b + B);\ \mathcal{A} = 91,\ h = 7,\ b = 12$

$$\begin{aligned} \mathcal{A} &= \tfrac{1}{2}h(b + B) \\ 91 &= \tfrac{1}{2}(7)(12 + B) \\ 182 &= (7)(12 + B) \\ 12 + B &= \tfrac{1}{7}(182) \\ B &= 26 - 12 = 14 \end{aligned}$$

25. $C = 2\pi r;\ C = 16.328,\ \pi = 3.14$

$$\begin{aligned} C &= 2\pi r \\ 16.328 &= 2(3.14)r \\ 16.328 &= 6.28r \\ 2.6 &= r \end{aligned}$$

27. $C = 2\pi r;\ C = 20\pi$

$$\begin{aligned} C &= 2\pi r \\ 20\pi &= 2\pi r \\ 10 &= r \qquad \textit{Divide by } 2\pi. \end{aligned}$$

29. $\mathcal{A} = \pi r^2;\ r = 4,\ \pi = 3.14$

$$\begin{aligned} \mathcal{A} &= \pi r^2 \\ &= 3.14(4)^2 \\ &= 3.14(16) \\ \mathcal{A} &= 50.24 \end{aligned}$$

31. $S = 2\pi rh;\ S = 120\pi,\ h = 10$

$$\begin{aligned} S &= 2\pi rh \\ 120\pi &= 2\pi r(10) \\ 120\pi &= 20\pi r \\ 6 &= r \qquad \textit{Divide by } 20\pi. \end{aligned}$$

In Exercises 33–38, substitute the given values into the formula and then evaluate V.

33. $V = LWH;\ L = 10,\ W = 5,\ H = 3$

$$\begin{aligned} V &= LWH \\ &= (10)(5)(3) \\ V &= 150 \end{aligned}$$

35. $V = \frac{1}{3}Bh;\ B = 12,\ h = 13$

$$\begin{aligned} V &= \tfrac{1}{3}Bh \\ &= \tfrac{1}{3}(12)(13) \\ V &= 52 \end{aligned}$$

37. $V = \frac{4}{3}\pi r^3;\ r = 12,\ \pi = 3.14$

$$\begin{aligned} V &= \tfrac{4}{3}\pi r^3 \\ &= \tfrac{4}{3}(3.14)(12)^3 \\ &= \tfrac{4}{3}(3.14)(1728) \\ V &= 7234.56 \end{aligned}$$

39.
$$\begin{aligned} P &= 2L + 2W \\ 54 &= 2(W+9) + 2W \quad \text{Let } L = W + 9. \\ 54 &= 2W + 18 + 2W \\ 54 &= 4W + 18 \\ 36 &= 4W \\ 9 &= W \end{aligned}$$

The width is 9 inches and the length is $9 + 9 = 18$ inches.

41.
$$\begin{aligned} P &= 2l + 2w \\ 36 &= 2(3w+2) + 2w \quad \text{Let } l = 3w + 2. \\ 36 &= 6w + 4 + 2w \\ 36 &= 8w + 4 \\ 32 &= 8w \\ 4 &= w \end{aligned}$$

The width is 4 meters and the length is $3(4) + 2 = 14$ meters.

43. Let $s =$ the length of the shortest side, in inches;
$s + 2 =$ the length of the medium side, and,
$s + 3 =$ the length of the longest side.

The perimeter is 20 inches, so

$$\begin{aligned} s + (s+2) + (s+3) &= 20. \\ 3s + 5 &= 20 \\ 3s &= 15 \\ s &= 5 \end{aligned}$$

Since $s = 5$, $s + 2 = 7$, and $s + 3 = 8$. The lengths of the sides are 5, 7, and 8 inches. The perimeter is $5 + 7 + 8 = 20$, as required.

45. Let $s =$ the length of the two sides that have equal length, in meters; and $2s - 4 =$ the length of the third side.

The perimeter is 24 meters, so

$$\begin{aligned} s + s + (2s - 4) &= 24. \\ 4s - 4 &= 24 \\ 4s &= 28 \\ s &= 7 \end{aligned}$$

Since $s = 7$, $2s - 4 = 10$. The lengths of the sides are 7, 7, and 10 meters. The perimeter is $7 + 7 + 10 = 24$, as required.

47. The diameter of the circle is 443 feet, so its radius is $\frac{443}{2} = 221.5$ ft. Use the area of a circle formula to find the enclosed area.

$$\begin{aligned} \mathcal{A} &= \pi r^2 \\ &= \pi(221.5)^2 \\ &\approx 154{,}133.6 \text{ ft}^2, \end{aligned}$$

or about 154,000 ft^2. (If 3.14 is used for π, the value is 154,055.465.)

49. The page is a rectangle with length 1.5 m and width 1.2 m, so use the formulas for the perimeter and area of a rectangle.

$$\begin{aligned} P &= 2L + 2W \\ &= 2(1.5) + 2(1.2) \\ &= 3 + 2.4 \\ P &= 5.4 \text{ meters} \end{aligned}$$

$$\begin{aligned} \mathcal{A} &= LW \\ &= (1.5)(1.2) \\ \mathcal{A} &= 1.8 \text{ square meters} \end{aligned}$$

51. Use the formula for the area of a triangle with $\mathcal{A} = 70$ and $b = 14$.

$$\begin{aligned} \mathcal{A} &= \tfrac{1}{2}bh \\ 70 &= \tfrac{1}{2}(14)h \\ 70 &= 7h \\ 10 &= h \end{aligned}$$

The height of the sign is 10 feet.

53. To find the area of the drum face, use the formula for the area of a circle, $\mathcal{A} = \pi r^2$. Since the diameter of the circle is 15.74 feet, the radius is $(\frac{1}{2})(15.74) = 7.87$ feet.

$$\begin{aligned} \mathcal{A} &= \pi r^2 \\ &\approx (3.14)(7.87)^2 \\ &= (3.14)(61.9369) \\ \mathcal{A} &\approx 194.48 \end{aligned}$$

The area of the drum face is about 194.48 square feet.

Use the circumference of a circle formula.

$$\begin{aligned} C &\approx 2(3.14)(7.87) \\ &= 49.4236 \end{aligned}$$

The circumference is about 49.42 feet.

55. Use the formula for the area of a trapezoid with $B = 115.80$, $b = 171.00$, and $h = 165.97$.

$$\begin{aligned} \mathcal{A} &= \tfrac{1}{2}(B + b)h \\ &= \tfrac{1}{2}(115.80 + 171.00)(165.97) \\ &= \tfrac{1}{2}(286.80)(165.97) \\ &= 23{,}800.098 \end{aligned}$$

To the nearest hundredth of a square foot, the combined area of the two lots is 23,800.10 square feet.

57. The girth is $4 \cdot 18 = 72$ inches. Since the length plus the girth is 108, we have

$$\begin{aligned} L + G &= 108 \\ L + 72 &= 108 \\ L &= 36 \text{ in.} \end{aligned}$$

The volume of the box is

$$\begin{aligned} V &= LWH \\ &= (36)(18)(18) \\ &= 11{,}664 \text{ in.}^3 \end{aligned}$$

59. The two angles are supplementary, so the sum of their measures is 180°.

$$\begin{aligned} (x+1)+(4x-56) &= 180 \\ 5x-55 &= 180 \\ 5x &= 235 \\ x &= 47 \end{aligned}$$

Since $x = 47$, $x + 1 = 47 + 1 = 48$, and $4x - 56 = 4(47) - 56 = 132$.

The measures of the angles are 48° and 132°.

61. In the figure, the two angles are complementary, so their sum is 90°.

$$\begin{aligned} (8x-1)+5x &= 90 \\ 13x-1 &= 90 \\ 13x &= 91 \\ x &= 7 \end{aligned}$$

Since $x = 7$, $8x - 1 = 8(7) - 1 = 55$, and $5x = 5(7) = 35$.

The two angle measures are 55° and 35°.

63. The two angles are vertical angles, which have equal measures. Set their measures equal to each other and solve for x.

$$\begin{aligned} 5x-129 &= 2x-21 \\ 3x-129 &= -21 \\ 3x &= 108 \\ x &= 36 \end{aligned}$$

Since $x = 36$, $5x - 129 = 5(36) - 129 = 51$, and $2x - 21 = 2(36) - 21 = 51$.

The measure of each angle is 51°.

65. The angles are vertical angles, so their measures are equal.

$$\begin{aligned} 12x-3 &= 10x+15 \\ 2x-3 &= 15 \\ 2x &= 18 \\ x &= 9 \end{aligned}$$

Since $x = 9$, $12x - 3 = 12(9) - 3 = 105$, and $10x + 15 = 10(9) + 15 = 105$.

The measure of each angle is 105°.

67. $d = rt$ for t

$$\frac{d}{r} = \frac{rt}{r} \quad \textit{Divide by r.}$$
$$\frac{d}{r} = t \quad \text{or} \quad t = \frac{d}{r}$$

69. $A = bh$ for b

$$\frac{A}{h} = \frac{bh}{h} \quad \textit{Divide by h.}$$
$$\frac{A}{h} = b \quad \text{or} \quad b = \frac{A}{h}$$

71. $C = \pi d$ for d

$$\frac{C}{\pi} = \frac{\pi d}{\pi} \quad \textit{Divide by } \pi.$$
$$\frac{C}{\pi} = d \quad \text{or} \quad d = \frac{C}{\pi}$$

73. $V = LWH$ for H

$$\frac{V}{LW} = \frac{LWH}{LW} \quad \textit{Divide by LW.}$$
$$\frac{V}{LW} = H \quad \text{or} \quad H = \frac{V}{LW}$$

75. $I = prt$ for r

$$\frac{I}{pt} = \frac{prt}{pt} \quad \textit{Divide by pt.}$$
$$\frac{I}{pt} = r \quad \text{or} \quad r = \frac{I}{pt}$$

77. $A = \frac{1}{2}bh$ for h

$$2A = 2\left(\tfrac{1}{2}bh\right) \quad \textit{Multiply by 2.}$$
$$2A = bh$$
$$\frac{2A}{b} = \frac{bh}{b} \quad \textit{Divide by b.}$$
$$\frac{2A}{b} = h \quad \text{or} \quad h = \frac{2A}{b}$$

79. $V = \frac{1}{3}\pi r^2 h$ for h

$$3V = 3\left(\tfrac{1}{3}\right)\pi r^2 h \quad \textit{Multiply by 3.}$$
$$3V = \pi r^2 h$$
$$\frac{3V}{\pi r^2} = \frac{\pi r^2 h}{\pi r^2} \quad \textit{Divide by } \pi r^2.$$
$$\frac{3V}{\pi r^2} = h \quad \text{or} \quad h = \frac{3V}{\pi r^2}$$

81. $P = a + b + c$ for b

$$P - a - c = a + b + c - a - c$$
Subtract a and c.
$$P - a - c = b \quad \text{or} \quad b = P - a - c$$

83. $P = 2L + 2W$ for W

$$P - 2L = 2L + 2W - 2L \quad \textit{Subtract 2L.}$$
$$P - 2L = 2W$$
$$\frac{P-2L}{2} = \frac{2W}{2} \quad \textit{Divide by 2.}$$
$$\frac{P-2L}{2} = W \quad \text{or} \quad W = \frac{P-2L}{2}$$

85. $y = mx + b$ for m

$$\begin{aligned} y - b &= mx + b - b && \textit{Subtract b.} \\ y - b &= mx \\ \frac{y-b}{x} &= \frac{mx}{x} && \textit{Divide by x.} \\ \frac{y-b}{x} &= m \quad \text{or} \quad m = \frac{y-b}{x} \end{aligned}$$

87. $Ax + By = C$ for y

$$\begin{aligned} By &= C - Ax && \textit{Subtract Ax.} \\ \frac{By}{B} &= \frac{C - Ax}{B} && \textit{Divide by B.} \\ y &= \frac{C - Ax}{B} \end{aligned}$$

89. $M = C(1 + r)$ for r

$$\begin{aligned} M &= C + Cr && \textit{Distributive Property} \\ M - C &= Cr && \textit{Subtract C.} \\ \frac{M-C}{C} &= \frac{Cr}{C} && \textit{Divide by C.} \\ \frac{M-C}{C} &= r \quad \text{or} \quad r = \frac{M-C}{C} \end{aligned}$$

Alternative solution:

$$\begin{aligned} M &= C(1+r) \\ \frac{M}{C} &= 1 + r && \textit{Divide by C.} \\ \frac{M}{C} - 1 &= r && \textit{Subtract 1.} \end{aligned}$$

91. $P = 2(a + b)$ for a

$$\begin{aligned} P &= 2a + 2b && \textit{Distributive property} \\ P - 2b &= 2a && \textit{Subtract 2b.} \\ \frac{P-2b}{2} &= \frac{2a}{2} && \textit{Divide by 2.} \\ \frac{P-2b}{2} &= a \end{aligned}$$

93.

$$\begin{aligned} 0.06x &= 300 \\ x &= \tfrac{300}{0.06} = 5000 \end{aligned}$$

The solution set is $\{5000\}$.

95.

$$\begin{aligned} \tfrac{3}{4}x &= 21 \\ \tfrac{4}{3}\left(\tfrac{3}{4}x\right) &= \tfrac{4}{3}(21) \\ x &= 28 \end{aligned}$$

The solution set is $\{28\}$.

97.

$$\begin{aligned} -3x &= \tfrac{1}{4} \\ \frac{-3x}{-3} &= \frac{\frac{1}{4}}{-3} \\ x &= \tfrac{1}{4}\left(-\tfrac{1}{3}\right) = -\tfrac{1}{12} \end{aligned}$$

The solution set is $\left\{-\tfrac{1}{12}\right\}$.

2.6 Ratio, Proportion, and Percent

2.6 Now Try Exercises

N1. **(a)** The ratio of 7 inches to 4 inches is

$$\frac{7 \text{ inches}}{4 \text{ inches}} = \frac{7}{4}.$$

(b) 45 seconds $= \frac{45}{60} = \frac{3}{4}$ minute

The ratio of 45 seconds to 2 minutes is then

$$\frac{45 \text{ seconds}}{2 \text{ minutes}} = \frac{3}{4} \div 2 = \frac{3}{4} \cdot \frac{1}{2} = \frac{3}{8}.$$

N2. The results in the following table are rounded to the nearest thousandth.

Size	**Unit Cost** (dollars per oz)
150 oz	$\frac{\$19.97}{150} = \$0.133$
100 oz	$\frac{\$13.97}{100} = \$0.140$
75 oz	$\frac{\$8.94}{75} = \$0.119$ (*)

Because the 75 oz size produces the lowest unit cost, it is the best buy. The unit cost, to the nearest thousandth, is \$0.119 per oz.

N3. **(a)** $\frac{1}{3} = \frac{33}{100}$

Compare the cross products.

$$\begin{aligned} 1 \cdot 100 &= 100 \\ 3 \cdot 33 &= 99 \end{aligned}$$

The cross products are *different*, so the proportion is *false*.

(b) $\frac{4}{13} = \frac{16}{52}$

Check to see whether the cross products are equal.

$$\begin{aligned} 4 \cdot 52 &= 208 \\ 13 \cdot 16 &= 208 \end{aligned}$$

The cross products are *equal*, so the proportion is *true*.

N4. $\frac{9}{7} = \frac{x}{56}$

$$\begin{aligned} 7x &= 9 \cdot 56 && \textit{Cross products} \\ x &= \frac{9 \cdot 56}{7} && \textit{Divide by 7.} \\ &= \frac{9 \cdot 7 \cdot 8}{7} && \textit{Factor.} \\ &= 72 && \textit{Cancel.} \end{aligned}$$

The solution set is $\{72\}$.

Note: We could have multiplied $9 \cdot 56$ to get 504 and then divided 504 by 7 to get 72. This may be the best approach if you are doing these calculations on a calculator. The factor and cancel method is preferable if you're not using a calculator.

N5. $\frac{k-3}{6} = \frac{3k+2}{4}$

$4(k-3) = 6(3k+2)$ *Cross products*
$4k - 12 = 18k + 12$ *Distributive property*
$-14k - 12 = 12$ *Subtract 18k.*
$-14k = 24$ *Add 12.*
$a = -\frac{24}{14} = -\frac{12}{7}$ *Divide by −14.*

The solution set is $\{-\frac{12}{7}\}$.

N6. Let x = the cost for 27 gallons.

$\frac{\$49.80}{20} = \frac{x}{27}$
$20x = 27(49.80)$ *Cross products*
$20x = 1344.60$ *Multiply.*
$x = 67.23$ *Divide by 20.*

It would cost $67.23

N7. **(a)** $16\% = 16 \cdot 1\% = 16 \cdot 0.01 = 0.16$

(b) $1.5 = 150 \cdot 0.01 = 150 \cdot 1\% = 150\%$

N8. **(a)** What is 20% of 70?

$20\% \cdot 70 = 0.20 \cdot 70 = 14$

(b) 40% of what number is 130?

As in Example 8(b), let n denote the number.

$0.40 \cdot n = 130$
$n = \frac{130}{0.40}$ *Divide by 0.40.*
$= 325$ *Simplify.*

40% of 325 is 130.

(c) 121 is what percent of 484?

As in Example 8(c), let p denote the percent.

$121 = p \cdot 484$
$p = \frac{121}{484} = 0.25 = 25\%$

121 is 25% of 484.

N9. We can think of this problem as

48 is what percent of 120?

Let p denote the percent.

$48 = p \cdot 120$
$p = \frac{48}{120} = 0.40 = 40\%$

The coat cost 40% of the regular price, so the savings is 100% − 40% = 60% of the regular price.

2.6 Section Exercises

1. **(a)** 75 to 100 is $\frac{75}{100} = \frac{3}{4}$ or 3 to 4.
The answer is **C**.

(b) 5 to 4 or $\frac{5}{4} = \frac{5 \cdot 3}{4 \cdot 3} = \frac{15}{12}$ or 15 to 12.
The answer is **D**.

(c) $\frac{1}{2} = \frac{1 \cdot 50}{2 \cdot 50} = \frac{50}{100}$ or 50 to 100
The answer is **B**.

(d) 4 to 5 or $\frac{4}{5} = \frac{4 \cdot 20}{5 \cdot 20} = \frac{80}{100}$ or 80 to 100.
The answer is **A**.

3. The ratio of 40 miles to 30 miles is

$$\frac{40 \text{ miles}}{30 \text{ miles}} = \frac{40}{30} = \frac{4}{3}.$$

5. The ratio of 120 people to 90 people is

$$\frac{120 \text{ people}}{90 \text{ people}} = \frac{4 \cdot 30}{3 \cdot 30} = \frac{4}{3}.$$

7. To find the ratio of 20 yards to 8 feet, first convert 20 yards to feet.

$$20 \text{ yards} = 20 \text{ yards} \cdot \frac{3 \text{ feet}}{1 \text{ yard}} = 60 \text{ feet}$$

The ratio of 20 yards to 8 feet is then

$$\frac{60 \text{ feet}}{8 \text{ feet}} = \frac{60}{8} = \frac{15 \cdot 4}{2 \cdot 4} = \frac{15}{2}.$$

9. Convert 2 hours to minutes.

$$2 \text{ hours} = 2 \text{ hours} \cdot \frac{60 \text{ minutes}}{1 \text{ hour}} = 120 \text{ minutes}$$

The ratio of 24 minutes to 2 hours is then

$$\frac{24 \text{ minutes}}{120 \text{ minutes}} = \frac{24}{120} = \frac{1 \cdot 24}{5 \cdot 24} = \frac{1}{5}.$$

11. 2 yards $= 2 \cdot 3 = 6$ feet
6 feet $= 6 \cdot 12 = 72$ inches
The ratio of 60 inches to 2 yards is then

$$\frac{60 \text{ inches}}{72 \text{ inches}} = \frac{5 \cdot 12}{6 \cdot 12} = \frac{5}{6}.$$

In Exercises 13–22, to find the best buy, divide the price by the number of units to get the unit cost. Each result was found by using a calculator and rounding the answer to three decimal places. The *best buy* (based on price per unit) is the smallest unit cost.

13.

Size	**Unit Cost** (dollars per lb)
4 lb	$\frac{\$1.78}{4} = \$0.445$
10 lb	$\frac{\$4.29}{10} = \$0.429$ (*)

The 10 lb size is the best buy.

15.

Size	Unit Cost (dollars per oz)
16 oz	$\frac{\$2.44}{16} = \$0.153$
32 oz	$\frac{\$2.98}{32} = \$0.093$ (*)
48 oz	$\frac{\$4.95}{48} = \$0.103$

The 32 oz size is the best buy.

17.

Size	Unit Cost (dollars per oz)
16 oz	$\frac{\$1.66}{16} = \$0.104$
32 oz	$\frac{\$2.59}{32} = \$0.081$
64 oz	$\frac{\$4.29}{64} = \$0.067$
128 oz	$\frac{\$6.49}{128} = \$0.051$ (*)

The 128-oz size is the best buy.

19.

Size	Unit Cost (dollars per oz)
14 oz	$\frac{\$1.39}{14} = \$0.099$
24 oz	$\frac{\$1.55}{24} = \$0.065$
36 oz	$\frac{\$1.78}{36} = \$0.049$ (*)
64 oz	$\frac{\$3.99}{64} = \$0.062$

The 36 oz size is the best buy.

21.

Size	Unit Cost (dollars per oz)
87 oz	$\frac{\$7.88}{87} = \$0.091$
131 oz	$\frac{\$10.98}{131} = \$0.084$
263 oz	$\frac{\$19.96}{263} = \$0.076$ (*)

The 263 oz size is the best buy.

23. $\frac{5}{35} = \frac{8}{56}$

Check to see whether the cross products are equal.

$$5 \cdot 56 = 280$$
$$35 \cdot 8 = 280$$

The cross products are *equal*, so the proportion is *true*.

25. $\frac{120}{82} = \frac{7}{10}$

Compare the cross products.

$$120 \cdot 10 = 1200$$
$$82 \cdot 7 = 574$$

The cross products are *different*, so the proportion is *false*.

27. $\frac{\frac{1}{2}}{5} = \frac{1}{10}$

Compare the cross products.

$$\tfrac{1}{2} \cdot 10 = 5$$
$$5 \cdot 1 = 5$$

The cross products are *equal*, so the proportion is *true*.

29. $\frac{k}{4} = \frac{175}{20}$

$20k = 4(175)$ *Cross products are equal.*

$20k = 700$

$\frac{20k}{20} = \frac{700}{20}$ *Divide by 20.*

$k = 35$

The solution set is $\{35\}$.

31. $\frac{49}{56} = \frac{z}{8}$

$56z = 49(8)$ *Cross products are equal.*

$56z = 392$

$\frac{56z}{56} = \frac{392}{56}$ *Divide by 56.*

$z = 7$

The solution set is $\{7\}$.

33. $\frac{x}{24} = \frac{15}{16}$

$16x = 24(15)$ *Cross products are equal.*

$16x = 360$

$\frac{16x}{16} = \frac{360}{16}$ *Divide by 16.*

$x = \frac{45 \cdot 8}{2 \cdot 8} = \frac{45}{2}$

The solution set is $\{\frac{45}{2}\}$.

35. $\frac{z}{2} = \frac{z+1}{3}$

$3z = 2(z+1)$ *Cross products are equal.*

$3z = 2z + 2$ *Distributive property*

$z = 2$ *Subtract 2z.*

The solution set is $\{2\}$.

37. $\frac{3y-2}{5} = \frac{6y-5}{11}$

$11(3y-2) = 5(6y-5)$ *Cross products are equal.*

$33y - 22 = 30y - 25$ *Dist. prop.*

$3y - 22 = -25$ *Subtract 30y.*

$3y = -3$ *Add 22.*

$y = -1$ *Divide by 3.*

The solution set is $\{-1\}$.

39. $\frac{5k+1}{6} = \frac{3k-2}{3}$

$3(5k+1) = 6(3k-2)$ *Cross products*

$15k + 3 = 18k - 12$ *Dist. prop.*

$-3k + 3 = -12$ *Subtract 18k.*

$-3k = -15$ *Subtract 3.*

$k = 5$ *Divide by −3.*

The solution set is $\{5\}$.

41.
$$\frac{2p+7}{3} = \frac{p-1}{4}$$

$4(2p+7) = 3(p-1)$ *Cross products*

$8p+28 = 3p-3$ *Dist. prop.*

$5p+28 = -3$ *Subtract 3p.*

$5p = -31$ *Subtract 28.*

$p = -\frac{31}{5}$ *Divide by 5.*

The solution set is $\{-\frac{31}{5}\}$.

43. Let $x =$ the cost of 24 candy bars.
Set up a proportion.

$$\frac{x}{24} = \frac{\$20.00}{16}$$
$$16x = 24(20)$$
$$16x = 480$$
$$x = 30$$

The cost of 24 candy bars is \$30.00.

45. Let $x =$ the cost of 5 quarts of oil.
Set up a proportion.

$$\frac{x}{5} = \frac{\$14.00}{8}$$
$$8x = 5(14)$$
$$8x = 70$$
$$x = 8.75$$

The cost of five quarts of oil is \$8.75.

47. Let $x =$ the cost of five pairs of jeans.

$$\frac{9 \text{ pairs}}{\$121.50} = \frac{5 \text{ pairs}}{x}$$
$$9x = 5(121.50)$$
$$9x = 607.5$$
$$\frac{9x}{9} = \frac{607.5}{9}$$
$$x = 67.5$$

The cost of five pairs is \$67.50.

49. Let $x =$ the cost for filling a 15-gallon tank.
Set up a proportion.

$$\frac{x \text{ dollars}}{\$19.56} = \frac{15 \text{ gallons}}{6 \text{ gallons}}$$
$$6x = 15(19.56)$$
$$6x = 293.4$$
$$x = 48.90$$

It would cost \$48.90 to completely fill a 15-gallon tank.

51. Let $x =$ the distance between Memphis and Philadelphia on the map (in feet).

Set up a proportion with one ratio involving map distances and the other involving actual distances.

$$\frac{x \text{ feet}}{2.4 \text{ feet}} = \frac{1000 \text{ miles}}{600 \text{ miles}}$$
$$\frac{x}{2.4} = \frac{1000}{600}$$
$$600x = (2.4)(1000)$$
$$600x = 2400$$
$$x = 4$$

The distance on the map between Memphis and Philadelphia would be 4 feet.

53. Let $x =$ the number of inches between St. Louis and Des Moines on the map.
Set up a proportion.

$$\frac{8.5 \text{ inches}}{x \text{ inches}} = \frac{1040 \text{ miles}}{333 \text{ miles}}$$
$$1040x = 8.5(333)$$
$$1040x = 2830.5$$
$$x \approx 2.72$$

St. Louis and Des Moines are about 2.7 inches apart on the map.

55. Let $x =$ the number of inches between Moscow and Berlin on the globe.
Set up a proportion.

$$\frac{12.4 \text{ inches}}{x \text{ inches}} = \frac{10{,}080 \text{ km}}{1610 \text{ km}}$$
$$10{,}080x = 12.4(1610)$$
$$10{,}080x = 19{,}964$$
$$x \approx 1.98$$

Moscow and Berlin are about 2.0 inches apart on the globe.

57. Let $x =$ the number of cups of cleaner.
Set up a proportion with one ratio involving the number of cups of cleaner and the other involving the number of gallons of water.

$$\frac{x \text{ cups}}{\frac{1}{4} \text{ cup}} = \frac{10\frac{1}{2} \text{ gallons}}{1 \text{ gallon}}$$
$$x \cdot 1 = \tfrac{1}{4}(10\tfrac{1}{2})$$
$$x = \tfrac{1}{4}(\tfrac{21}{2}) = \tfrac{21}{8}$$

The amount of cleaner needed is $2\frac{5}{8}$ cups.

59. Let $x =$ the number of U.S. dollars Ashley exchanged.

Set up a proportion.

$$\frac{\$1.4294}{x \text{ dollars}} = \frac{1 \text{ euro}}{300 \text{ euros}}$$
$$x \cdot 1 = 1.4294(300)$$
$$x = 428.82$$

She exchanged \$428.82.

61. Let $x =$ the number of fish in North Bay.

Set up a proportion with one ratio involving the sample and the other involving the total number of fish.

$$\frac{7 \text{ fish}}{700 \text{ fish}} = \frac{500 \text{ fish}}{x \text{ fish}}$$
$$7x = (700)(500)$$
$$7x = 350{,}000$$
$$x = 50{,}000$$

We estimate that there are 50,000 fish in North Bay.

63. $\frac{x}{12} = \frac{3}{9}$
$9x = 12 \cdot 3 = 36$
$x = 4$

Other possibilities for the proportion are:

$$\frac{12}{x} = \frac{9}{3}, \quad \frac{x}{12} = \frac{5}{15}, \quad \frac{12}{x} = \frac{15}{5}$$

65. Ratios of corresponding sides are equal, so $\frac{2}{3} = \frac{x}{3}$, and we must have $x = 2$.

67. $\frac{x}{3} = \frac{2}{6}$ $\qquad$ $\frac{y}{\frac{4}{3}} = \frac{6}{2}$

$6x = 3 \cdot 2 = 6$ $\qquad$ $2y = 6\left(\frac{4}{3}\right) = 8$

$x = 1$ $\qquad$ $y = 4$

69. **(a)**

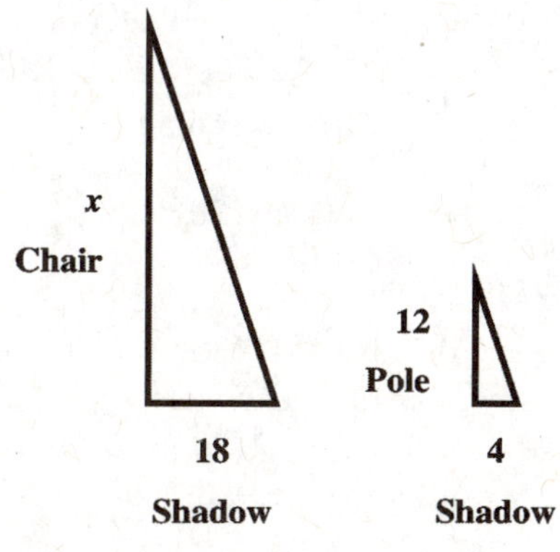

(b) These two triangles are similar, so their sides are proportional.

$$\frac{x}{12} = \frac{18}{4}$$
$$4x = 18(12)$$
$$4x = 216$$
$$x = 54$$

The chair is 54 feet tall.

71. Let $x =$ the 1997 price of electricity.

$$\frac{1995 \text{ price}}{1995 \text{ index}} = \frac{1997 \text{ price}}{1997 \text{ index}}$$
$$\frac{225}{152.4} = \frac{x}{160.5}$$
$$152.4x = 225(160.5)$$
$$x = \frac{225(160.5)}{152.4} \approx 236.96$$

The 1997 price would be about \$237.

73. Let $x =$ the 2003 price of electricity.

$$\frac{1995 \text{ price}}{1995 \text{ index}} = \frac{2003 \text{ price}}{2003 \text{ index}}$$
$$\frac{225}{152.4} = \frac{x}{184.0}$$
$$152.4x = 225(184.0)$$
$$x = \frac{225(184.0)}{152.4} \approx 271.65$$

The 2003 price would be about \$272.

75. $53\% = 53 \cdot 1\% = 53(0.01) = 0.53$

An alternative method is

$53 = 53 \cdot 1\% = 53\left(\frac{1}{100}\right) = \frac{53}{100} = 0.53.$

77. $96\% = 96 \cdot 1\% = 96(0.01) = 0.96$

79. $9\% = 9 \cdot 1\% = 9(0.01) = 0.09$

81. $129\% = 129 \cdot 1\% = 129(0.01) = 1.29$

83. $0.80 = 80(0.01) = 80 \cdot 1\% = 80\%$

85. $0.02 = 2(0.01) = 2 \cdot 1\% = 2\%$

87. $0.125 = 12.5(0.01) = 12.5 \cdot 1\% = 12.5\%$

89. $2.2 = 220(0.01) = 220 \cdot 1\% = 220\%$

91. What is 14% of 780?

$$14\% \cdot 780 = 0.14 \cdot 780 = 109.2$$

93. 42% of what number is 294?

As in Example 8(b), let n denote the number.

$$0.42 \cdot n = 294$$
$$n = \frac{294}{0.42} \quad \textit{Divide by 0.42.}$$
$$= 700 \quad \textit{Simplify.}$$

42% of 700 is 294.

95. 120% of what number is 510?

As in Example 8(b), let n denote the number.

$$1.20 \cdot n = 510$$
$$n = \frac{510}{1.20} \quad \textit{Divide by 1.20.}$$
$$= 425 \quad \textit{Simplify.}$$

120% of 425 is 510.

97. 4 is what percent of 50?

As in Example 8(c), let p denote the percent.

$$4 = p \cdot 50$$
$$p = \tfrac{4}{50} = 0.08 = 8\%$$

4 is 8% of 50.

99. What percent of 30 is 36?

As in Example 8(c), let p denote the percent.

$$36 = p \cdot 30$$
$$p = \tfrac{36}{30} = 1.2 = 120\%$$

36 is 120% of 30.

101. The discount is 15% of \$795.

$$15\% \cdot 795 = 0.15 \cdot 795 = 119.25$$

The discount is \$119.25.

The sale price is \$795 − \$119.25 = \$675.75.

103. 48 is what percent of 60?

As in Example 8(c), let p denote the percent.

$$48 = p \cdot 60$$
$$p = \tfrac{48}{60} = 0.8 = 80\%$$

He earned 80% of the total points.

105. 65% of what number is 1950?

As in Example 8(b), let n denote the number.

$$\begin{aligned} 0.65 \cdot n &= 1950 \\ n &= \frac{1950}{0.65} && \textit{Divide by 0.65.} \\ &= 3000 && \textit{Simplify.} \end{aligned}$$

She needs \$3000 for the car.

107.
$$\begin{aligned} 0.15x + 0.30(3) &= 0.20(3 + x) \\ 100[0.15x + 0.30(3)] &= 100[0.20(3 + x)] \\ 15x + 30(3) &= 20(3 + x) \\ 15x + 90 &= 60 + 20x \\ 90 &= 60 + 5x \\ 30 &= 5x \\ 6 &= x \end{aligned}$$

The solution set is $\{6\}$.

109.
$$\begin{aligned} 0.92x + 0.98(12 - x) &= 0.96(12) \\ 100[0.92x + 0.98(12 - x)] &= 100[0.96(12)] \\ 92x + 98(12 - x) &= 96(12) \\ 92x + 1176 - 98x &= 1152 \\ 1176 - 6x &= 1152 \\ -6x &= -24 \\ x &= 4 \end{aligned}$$

The solution set is $\{4\}$.

2.7 Further Applications of Linear Equations

2.7 Now Try Exercises

N1. **(a)** The amount of pure alcohol in 70 L of a 20% alcohol solution is

70 L	×	0.20	= 14 L.
↑		↑	↑
Amount of solution		Rate of concentration	Amount of pure alcohol

(b) If \$3200 is invested for one year at 2% simple interest, the amount of interest earned is

\$3200	×	0.02	= \$64.
↑		↑	↑
Principal		Interest Rate	Interest earned

N2. Let $x =$ the number of ounces of seasoning that is 70% salt.

Then $x + 30 =$ the number of ounces of seasoning that is 50% salt.

Number of ounces of seasoning	x	+	30	=	$x + 30$
Percent of salt in seasoning	0.70		0.10		0.50

Salt in 70% seasoning	plus	salt in 10% seasoning	is	salt in 50% seasoning.
↓	↓	↓	↓	↓
$0.70x$	+	$0.10(30)$	=	$0.50(x + 30)$

Solve the equation.

$$0.70x + 0.10(30) = 0.50x + 15$$

Multiply by 10 to clear decimals.

$$\begin{aligned} 7x + 1(30) &= 5x + 150 \\ 7x + 30 &= 5x + 150 \\ 2x + 30 &= 150 \\ 2x &= 120 \\ x &= 60 \end{aligned}$$

60 ounces of seasoning that is 70% salt are needed.

Check $x = 60$:

LS and RS refer to the left side and right side of the original equation.

LS: $0.70(60) + 0.10(30) = 42 + 3 = 45$
RS: $0.50(60 + 30) = 0.50(90) = 45$

N3. Let $x =$ the amount invested at 3%.
Then $x + 5000 =$ the amount invested at 4%.

Amount invested (in dollars)	Rate of interest	Interest for one year
x	0.03	$0.03x$
$x + 5000$	0.04	$0.04(x + 5000)$

Since the total annual interest is \$410, the equation is

$$0.03x + 0.04(x + 5000) = 410.$$

Multiply both sides by 100 to eliminate the decimals.

$$\begin{aligned} 3x + 4(x + 5000) &= 100(410) \\ 3x + 4x + 20{,}000 &= 41{,}000 \\ 7x + 20{,}000 &= 41{,}000 \\ 7x &= 21{,}000 \\ x &= 3000 \end{aligned}$$

The financial advisor should invest \$3000 at 3% and \$8000 at 4%.

N4. Let $x =$ the number of dimes.
Then $x + 10 =$ the number of quarters.

The value of dimes plus the value of quarters is \$5.65.

$$0.10x + 0.25(x + 10) = 5.65$$

Multiply by 100.

$$\begin{aligned} 10x + 25(x + 10) &= 565 \\ 10x + 25x + 250 &= 565 \\ 35x + 250 &= 565 \\ 35x &= 315 \\ x &= 9 \end{aligned}$$

Since $x = 9$, $x + 10 = 19$.
Clayton has 19 quarters and 9 dimes.

N5. $r = \dfrac{d}{t} = \dfrac{400 \text{ miles}}{6 \text{ hours}} \approx 66.6667$

The rate was about 66.67 miles per hour.

N6. Let $t =$ the time it takes for the bicyclists to be 5 miles apart. Use the formula $d = rt$.

$$\begin{aligned} d_{\text{faster}} - d_{\text{slower}} &= d_{\text{total}} \\ 20t - 18t &= 5 \\ 2t &= 5 \\ t &= \tfrac{5}{2} = 2.5 \end{aligned}$$

It will take 2.5 hours for the bicyclists to be 5 miles apart.

N7. Let $x =$ the rate of the slower car.
Then $x + 6 =$ the rate of the faster car.

Use the formula $d = rt$ and the fact that each car travels for $\frac{1}{4}$ hour.

$$\begin{aligned} d_{\text{slower}} + d_{\text{faster}} &= d_{\text{total}} \\ (x + 6)(\tfrac{1}{4}) + x(\tfrac{1}{4}) &= 35 \\ \tfrac{1}{4}x + \tfrac{3}{2} + \tfrac{1}{4}x &= \tfrac{70}{2} \\ \tfrac{2}{4}x + \tfrac{3}{2} &= \tfrac{70}{2} \\ \tfrac{1}{2}x &= \tfrac{67}{2} \\ \tfrac{2}{1}(\tfrac{1}{2}x) &= \tfrac{2}{1}(\tfrac{67}{2}) \\ x &= 67 \end{aligned}$$

Since $x = 67$, $x + 6 = 73$. The slower car had a rate of 67 mph and the faster car had a rate of 73 mph.

Check: The slower car traveled $67(\frac{1}{4}) = \frac{67}{4}$ miles and the faster car traveled $73(\frac{1}{4}) = \frac{73}{4}$ miles. The total miles traveled is $\frac{67}{4} + \frac{73}{4} = \frac{140}{4} = 35$, as required.

2.7 Section Exercises

1. The amount of pure alcohol in 150 liters of a 30% alcohol solution is

150	×	0.30	= 45 liters.
↑		↑	↑
Amount of solution		Rate of concentration	Amount of pure alcohol

3. If \$25,000 is invested at 3% simple interest for one year, the amount of interest earned is

\$25,000	×	0.03	×	1	= \$750.
↑		↑		↑	↑
Principal		Interest Rate		Time	Interest earned

5. The monetary value of 35 half-dollars is

35	×	\$0.50	= \$17.50.
↑		↑	↑
Number of coins		Denomination	Monetary value

7. 44.4% is about 45%. The population of 1,917,000 is about 1.9 million, and 45% of 1.9 million is 855,000. So choice **A** is the correct answer.

9. **(a)** 14% of 3.8 million $= 0.14(3{,}800{,}000)$
$= 532{,}000$ white cars.

(b) 21% of 3.8 million $= 0.21(3{,}800{,}000)$
$= 798{,}000$ silver cars.

(c) 13% of 3.8 million $= 0.13(3{,}800{,}000)$
$= 494{,}000$ red cars.

11. The concentration of the new solution could not be more than the strength of the stronger of the original solutions, so the correct answer is **D**, since 32% is stronger than both 20% and 30%.

13. *Step 2*
Let $x =$ the number of liters of 25% acid solution to be used.

Step 3
Use the box diagram in the textbook to write the equation.

Pure acid in 25% solution		Pure acid in 40% solution		Pure acid in 30% solution
↓		↓		↓
$0.25x$	$+$	$0.40(80)$	$=$	$0.30(x+80)$

Step 4
Multiply by 100 to clear decimals.

$$\begin{aligned} 25x + 40(80) &= 30(x+80) \\ 25x + 3200 &= 30x + 2400 \\ 25x + 800 &= 30x \\ 800 &= 5x \\ 160 &= x \end{aligned}$$

Step 5
160 liters of 25% acid solution must be added.

Step 6
25% of 160 liters plus 40% of 80 liters is 40 liters plus 32 liters, or 72 liters, of pure acid; which is equal to 30% of (160 + 80) liters.
[0.30(240) = 72]

15. Let $x =$ the number of liters of 5% drug solution.

Number of liters of solution	20	+	x	=	$x+20$
Strength of solution	0.10		0.05		0.08

Pure drug solution in 10% solution	plus	pure drug solution in 5% solution
↓	↓	↓
$0.10(20)$	$+$	$0.05x$

is	pure drug solution in 8% solution.
↓	↓
$=$	$0.08(x+20)$

Solve the equation.

$$\begin{aligned} 0.10(20) + 0.05x &= 0.08(x+20) \\ 10(20) + 5x &= 8(x+20) \\ 200 + 5x &= 8x + 160 \\ 200 &= 3x + 160 \\ 40 &= 3x \\ x &= \tfrac{40}{3} = 13\tfrac{1}{3} \end{aligned}$$

The pharmacist needs $13\frac{1}{3}$ liters of 5% drug solution.

Check $x = 13\frac{1}{3}$:

LS and RS refer to the left side and right side of the original equation.

LS: $0.10(20) + 0.05(13\frac{1}{3}) = 2\frac{2}{3}$
RS: $0.08(13\frac{1}{3} + 20) = 2\frac{2}{3}$

17. Let $x =$ the number of liters of the 20% alcohol solution.
Complete the table.

Strength	Liters of solution	Liters of pure alcohol
12%	12	$0.12(12) = 1.44$
20%	x	$0.20x$
14%	$x + 12$	$0.14(x+12)$

From the last column, we can formulate an equation that compares the number of liters of pure alcohol.

Alcohol in 12% solution	+	alcohol in 20% solution	=	alcohol in 14% solution.

$$\begin{aligned} 1.44 + 0.20x &= 0.14(x+12) \\ 1.44 + 0.20x &= 0.14x + 1.68 \\ 0.06x &= 0.24 \\ x &= 4 \end{aligned}$$

4 L of the 20% alcohol solution are needed.

19. Let $x =$ the amount of water to be added.
Then $20 + x =$ the amount of 2% solution.

There is no minoxidil in water.

Number of milliliters of solution	x	+	20	=	$20+x$
Concentration of solution	0		0.04		0.02

Pure minoxidil in x milliliters of water	plus	pure minoxidil in 4% solution	is	pure minoxidil in 2% solution
↓	↓	↓	↓	↓
$0(x)$	$+$	$0.04(20)$	$=$	$0.02(20+x)$

Solve the equation.

$$\begin{aligned} 0x + 0.04(20) &= 0.02(20+x) \\ 4(20) &= 2(20+x) \\ 80 &= 40 + 2x \\ 40 &= 2x \\ 20 &= x \end{aligned}$$

20 milliliters of water should be used.

Check $x = 20$:
LS: $0(20) + 0.04(20) = 0.8$
RS: $0.02(20 + 20) = 0.8$

This answer should make common sense; that is, equal amounts of 0% and 4% solutions should produce a 2% solution.

21. Let $x =$ the number of liters of 60% acid solution.
Then $20 - x =$ the number of liters of 75% acid solution.

Number of liters of solution	x	+	$20 - x$	=	20
Concentration of acid	0.60		0.75		0.72

Pure acid in 60% solution plus pure acid in 75% solution
↓ ↓ ↓
$0.60x$ + $0.75(20 - x)$

is pure acid in 72% solution.
↓ ↓
= $0.72(20)$

Solve the equation.

$$\begin{aligned} 0.60x + 0.75(20 - x) &= 0.72(20) \\ 60x + 75(20 - x) &= 72(20) \\ 60x + 1500 - 75x &= 1440 \\ 1500 - 15x &= 1440 \\ -15x &= -60 \\ x &= 4 \end{aligned}$$

4 liters of 60% acid solution must used.

Check $x = 4$:
LS: $0.60(4) + 0.75(20 - 4) = 14.4$
RS: $0.72(20) = 14.4$

23. Let $x =$ the amount invested at 5% (in dollars).
Then $x - 1200 =$ the amount invested at 4% (in dollars).

Amount invested (in dollars)	Rate of interest	Interest for one year
x	0.05	$0.05x$
$x - 1200$	0.04	$0.04(x - 1200)$

Since the total annual interest was \$141, the equation is

$$\begin{aligned} 0.05x + 0.04(x - 1200) &= 141. \\ 5x + 4(x - 1200) &= 100(141) \\ 5x + 4x - 4800 &= 14{,}100 \\ 9x - 4800 &= 14{,}100 \\ 9x &= 18{,}900 \\ x &= 2100 \end{aligned}$$

Since $x = 2100$, $x - 1200 = 900$.
Arlene invested \$2100 at 5% and \$900 at 4%.

25. Let $x =$ the amount invested at 6%.
Then $3x + 6000 =$ the amount invested at 5%.

$$\begin{aligned} 0.06x + 0.05(3x + 6000) &= 825 \\ 6x + 5(3x + 6000) &= 100(825) \\ 6x + 15x + 30{,}000 &= 82{,}500 \\ 21x + 30{,}000 &= 82{,}500 \\ 21x &= 52{,}500 \\ x &= 2500 \end{aligned}$$

Since $x = 2500$, $3x + 6000 = 13{,}500$.
The artist invested \$2500 at 6% and \$13,500 at 5%.

27. Let $x =$ the number of nickels.
Then $x + 2 =$ the number of dimes.

The value of nickels plus the value of dimes is \$1.70
↓ ↓ ↓ ↓ ↓
$0.05x$ + $0.10(x + 2)$ = 1.70

$$\begin{aligned} 5x + 10(x + 2) &= 100(1.70) \\ 5x + 10x + 20 &= 170 \\ 15x + 20 &= 170 \\ 15x &= 150 \\ x &= 10 \end{aligned}$$

The collector has 10 nickels.

29. Let $x =$ the number of 44-cent stamps.
Then $45 - x =$ the number of 17-cent stamps.

The value of the 44-cent stamps is $0.44x$ and the value of the 17-cent stamps is $0.17(45 - x)$. The total value is \$14.40, so

$$\begin{aligned} 0.44x + 0.17(45 - x) &= 14.40. \\ 44x + 17(45 - x) &= 1440 \\ 44x + 765 - 17x &= 1440 \\ 27x + 765 &= 1440 \\ 27x &= 675 \\ x &= 25 \end{aligned}$$

Since $x = 25$, $45 - x = 20$. She bought 25 44-cent stamps valued at \$11.00 and 20 17-cent stamps valued at \$3.40, for a total value of \$14.40.

31. Let $x =$ the number of pounds of Colombian Decaf beans.
Then $2x =$ the number of pounds of Arabian Mocha beans.

Number of Pounds	Cost per Pound	Total Value (in \$)
x	\$8.00	$8x$
$2x$	\$8.50	$8.5(2x)$

The total value is \$87.50, so

$$\begin{aligned} 8x + 8.5(2x) &= 87.50. \\ 8x + 17x &= 87.50 \\ 25x &= 87.50 \\ x &= 3.5 \end{aligned}$$

Since $x = 3.5$, $2x = 7$. She can buy 3.5 pounds of Colombian Decaf and 7 pounds of Arabian Mocha.

33. To estimate the average rate of the trip, round 405 to 400 and 8.2 to 8.

Use $r = \frac{d}{t}$ with $d = 400$ and $t = 8$.

$$r = \frac{d}{t} = \frac{400}{8} = 50$$

The best estimate is **A**, 50 miles per hour.

35. Use the formula $d = rt$ with $r = 53$ and $t = 10$.

$$\begin{aligned} d &= rt \\ &= (53)(10) \\ &= 530 \end{aligned}$$

The distance between Memphis and Chicago is 530 miles.

37. Use $d = rt$ with $d = 500$ and $r = 143.567$.

$$\begin{aligned} d &= rt \\ 500 &= 143.567t \\ t &= \tfrac{500}{143.567} \approx 3.483 \end{aligned}$$

His time was about 3.483 hours.

39. $r = \dfrac{d}{t} = \dfrac{100 \text{ meters}}{12.54 \text{ seconds}} \approx 7.974$

Her rate was about 7.97 meters per second.

41. $r = \dfrac{d}{t} = \dfrac{400 \text{ meters}}{47.25 \text{ seconds}} \approx 8.466$

His rate was about 8.47 meters per second.

43. Let t = the number of hours until John and Pat meet.
The distance John travels and the distance Pat travels total 440 miles.

John's distance	and	Pat's distance	equal	total distance.
↓	↓	↓	↓	↓
$60t$	$+$	$28t$	$=$	440

$$\begin{aligned} 88t &= 440 \\ t &= 5 \end{aligned}$$

It will take 5 hours for them to meet.

45. Let t = the number of hours until the trains are 315 kilometers apart.

Distance of northbound train	plus	distance of southbound train	is	total distance.
↓	↓	↓	↓	
$85t$	$+$	$95t$	$=$	315

$$\begin{aligned} 180t &= 315 \\ t &= \tfrac{315}{180} = \tfrac{7}{4} \end{aligned}$$

It will take $1\frac{3}{4}$ hours for the trains to be 315 kilometers apart.

47. Let t = the number of hours Marco and Celeste traveled.

Make a chart using the formula $d = rt$.

	r	t	d
Marco	10	t	$10t$
Celeste	12	t	$12t$

Marco's distance	minus	Celeste's distance	is	15.
↓	↓	↓	↓	↓
$12t$	$-$	$10t$	$=$	15

$$\begin{aligned} 12t - 10t &= 15 \\ 2t &= 15 \\ t &= \tfrac{15}{2} \text{ or } 7\tfrac{1}{2} \end{aligned}$$

They will be 15 miles apart in $7\frac{1}{2}$ hours.

49. Let x = the rate of the westbound plane.
Then $x - 150$ = the rate of the eastbound plane.

Using the formula $d = rt$ and the chart in the text, we see that

$$\begin{aligned} d_{\text{west}} + d_{\text{east}} &= d_{\text{total}} \\ x(3) + (x - 150)(3) &= 2250 \\ 3x + 3x - 450 &= 2250 \\ 6x &= 2700 \\ x &= 450 \end{aligned}$$

Since $x = 450$, $x - 150 = 300$.

The rate of the westbound plane is 450 mph and the rate of the eastbound plane is 300 mph.

51. Let x = the rate of the slower car.
Then $x + 20$ = the rate of the faster car.

Use the formula $d = rt$ and the fact that each car travels for 4 hours.

$$\begin{aligned} d_{\text{faster}} + d_{\text{slower}} &= d_{\text{total}} \\ (x + 20)(4) + (x)(4) &= 400 \\ 4x + 80 + 4x &= 400 \\ 8x &= 320 \\ x &= 40 \end{aligned}$$

The rate of the slower car is 40 mph and the rate of the faster car is 60 mph.

53. Let x = Bob's current age.
Then $3x$ = Kevin's current age.

Three years ago, Bob's age was $x - 3$ and Kevin's age was $3x - 3$, and this sum was 22.

$$\begin{aligned} (x - 3) + (3x - 3) &= 22 \\ 4x - 6 &= 22 \\ 4x &= 28 \\ x &= 7 \end{aligned}$$

Bob is 7 years old and Kevin is $3(7) = 21$ years old.

55. Let $w =$ the width of the table.
Then $3w =$ the length of the table.

If we subtract 3 feet from the length $(3w - 3)$ and add 3 feet to the width $(w + 3)$, then the length and the width would be equal.

$$\begin{aligned} 3w - 3 &= w + 3 \\ 3w &= w + 6 \\ 2w &= 6 \\ w &= 3 \end{aligned}$$

The width is 3 feet and the length is $3(3) = 9$ feet.

57. Let $x =$ her gross pay (pay before deductions).

$$\begin{aligned} \text{gross pay} - \text{deductions} &= \text{take-home pay} \\ x - 0.10(x) &= 585 \\ 0.90x &= 585 \\ x &= \tfrac{585}{0.90} = 650 \end{aligned}$$

She was paid \$650 before deductions.

59. Since $6 = 6$, the statement $6 > 6$ is *false*.

61. Since -4 is to the left of -3 on the real number line, the statement $-4 \le -3$ is *true*.

63. Since 0 is to the right of $-\frac{1}{2}$ on the real number line, the statement $0 > -\frac{1}{2}$ is *true*.

2.8 Solving Linear Inequalities

2.8 Now Try Exercises

N1. **(a)** The statement $x < -1$ says that x can represent any number less than -1. The interval is written as $(-\infty, -1)$. Graph this inequality by placing a parenthesis at -1 on a number line and drawing an arrow to the left.

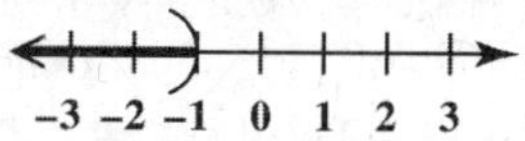

(b) The statement $-2 \le x$ is the same as $x \ge -2$. The interval is written as $[-2, \infty)$. Graph this inequality by placing a bracket at -2 on a number line and drawing an arrow to the right.

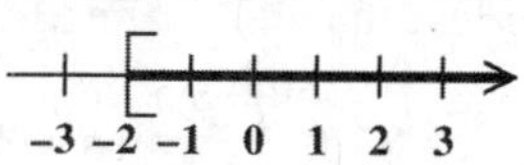

N2.

$$\begin{aligned} 5 + 5x &\ge 4x + 3 & \\ 5 + 5x - 5 &\ge 4x + 3 - 5 & \textit{Subtract 5.} \\ 5x &\ge 4x - 2 & \\ 5x - 4x &\ge 4x - 2 - 4x & \textit{Subtract 4x.} \\ x &\ge -2 & \end{aligned}$$

Graph the solution set $[-2, \infty)$.

To graph this inequality, place a bracket at -2 on a number line and draw an arrow to the right.

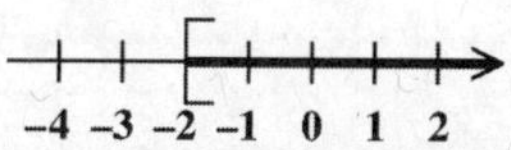

N3.

$$\begin{aligned} -5k &\ge 15 & \\ \frac{-5k}{-5} &\le \frac{15}{-5} & \textit{Divide by } -5\textit{; reverse the symbol.} \\ k &\le -3 & \end{aligned}$$

Graph the solution set $(-\infty, -3]$.

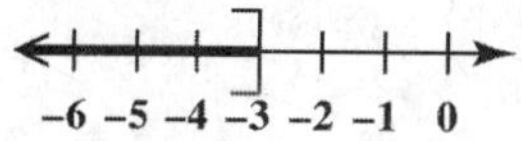

N4.

$$\begin{aligned} 6 - 2t + 5t &\le 8t - 4 & \\ 3t + 6 &\le 8t - 4 & \\ 3t + 6 - 8t &\le 8t - 4 - 8t & \textit{Subtract 8t.} \\ -5t + 6 &\le -4 & \\ -5t + 6 - 6 &\le -4 - 6 & \textit{Subtract 6.} \\ -5t &\le -10 & \\ \frac{-5t}{-5} &\ge \frac{-10}{-5} & \textit{Divide by } -5\textit{; reverse the symbol.} \\ t &\ge 2 & \end{aligned}$$

Graph the solution set $[2, \infty)$.

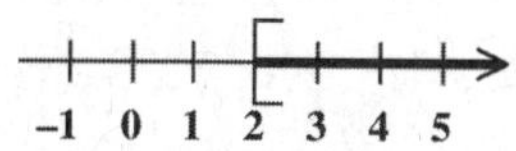

N5.

$$\begin{aligned} 2x - 3(x - 6) &< 4(x + 7) & \\ 2x - 3x + 18 &< 4x + 28 & \textit{Distributive property} \\ -x + 18 &< 4x + 28 & \\ -x + 18 + x &< 4x + 28 + x & \textit{Add x.} \\ 18 &< 5x + 28 & \\ 18 - 28 &< 5x + 28 - 28 & \textit{Subtract 28.} \\ -10 &< 5x & \\ \frac{-10}{5} &< \frac{5x}{5} & \textit{Divide by 5.} \\ -2 &< x & \\ x &> -2 & \end{aligned}$$

Graph the solution set $(-2, \infty)$.

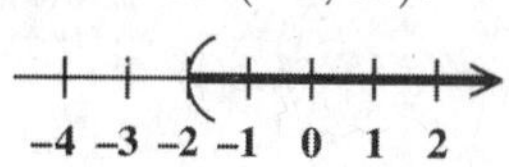

N6.

$$\begin{aligned} \tfrac{3}{4}(x - 2) + \tfrac{1}{2} &> \tfrac{1}{5}(x - 8) & \\ 20\left[\tfrac{3}{4}(x - 2) + \tfrac{1}{2}\right] &> 20\left[\tfrac{1}{5}(x - 8)\right] & \textit{Mult. by 20.} \\ 15(x - 2) + 10 &> 4(x - 8) & \\ 15x - 30 + 10 &> 4x - 32 & \\ 15x - 20 &> 4x - 32 & \\ 11x - 20 &> -32 & \textit{Subtract 4x.} \\ 11x &> -12 & \textit{Add 20.} \\ x &> -\tfrac{12}{11} & \textit{Divide by 11.} \end{aligned}$$

Check that the solution set is the interval $(-\frac{12}{11}, \infty)$.

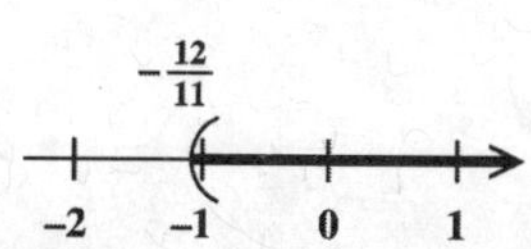

N7. *Step 2*
Let $x =$ the number of months she belongs to the health club.

Step 3
She must pay \$40, plus \$$35x$, to belong to the health club for x months, and this amount must be *no more than* \$355.

Cost of belonging	is no more than	355 dollars.
↓	↓	↓
$40 + 35x$	$\leq$	355

Step 4

$$35x \leq 315 \quad \textit{Subtract 40.}$$
$$x \leq 9 \quad \textit{Divide by 35.}$$

Step 5
She can belong to the health club for a maximum of 9 months. (She may belong for less time, as indicated by the inequality $x \leq 9$.)

Step 6
If Sara belongs for 9 months, she will spend $40 + 35(9) = 355$, the maximum amount.

N8. Let $x =$ Kristine's score on the third test.

The average	is at least	90.
↓	↓	↓
$\frac{98 + 85 + x}{3}$	$\geq$	90

Solve the inequality.

$$3\left(\frac{183 + x}{3}\right) \geq 3(90) \quad \textit{Add in the numerator; multiply by 3.}$$
$$183 + x \geq 270$$
$$183 + x - 183 \geq 270 - 183 \quad \textit{Subtract 183.}$$
$$x \geq 87 \quad \textit{Combine terms.}$$

She must score 87 or more on the third test to have an average of *at least* 90.

N9. The statement $0 \leq x < 2$ says that x can represent any number between 0 and 2, including 0 and excluding 2. To graph the inequality, place a bracket at 0 and a parenthesis at 2 and draw a line segment between them. The interval is written as $[0, 2)$.

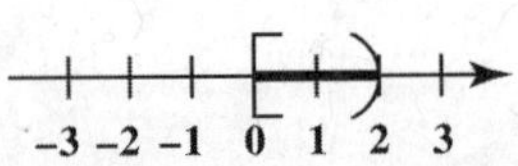

N10.

$$-4 \leq \frac{3}{2}x - 1 \leq 0$$
$$-4 + 1 \leq \frac{3}{2}x - 1 + 1 \leq 0 + 1 \quad \textit{Add 1 to each part.}$$
$$-3 \leq \frac{3}{2}x \leq 1$$
$$\frac{2}{3}(-3) \leq \frac{2}{3}\left(\frac{3}{2}x\right) \leq \frac{2}{3}(1) \quad \textit{Mult. each part by } \tfrac{2}{3}.$$
$$-2 \leq x \leq \frac{2}{3}$$

Graph the solution set $[-2, \frac{2}{3}]$.

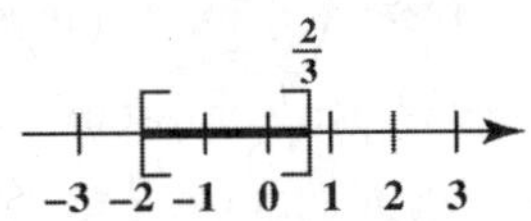

2.8 Section Exercises

1. When graphing an inequality, use a parenthesis if the inequality symbol is $\underline{>}$ or $\underline{<}$. Use a square bracket if the inequality symbol is $\underline{\geq}$ or $\underline{\leq}$.
Examples:

A parenthesis would be used for the inequalities $x < 2$ and $x > 3$. A square bracket would be used for the inequalities $x \leq 2$ and $x \geq 3$.

3. In interval notation, the set $\{x|x > 0\}$ is written $\underline{(0, \infty)}$.

5. The set of numbers graphed corresponds to the inequality $x > -4$.

7. The set of numbers graphed corresponds to the inequality $x \leq 4$.

9. The statement $z \leq 4$ says that z can represent any number less than or equal to 4. The interval is written as $(-\infty, 4]$. To graph the inequality, place a square bracket at 4 (to show that 4 is part of the graph) and draw an arrow extending to the left.

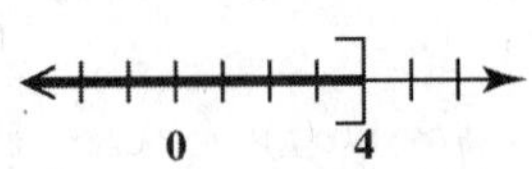

11. The statement $x < -3$ says that x can represent any number less than -3. The interval is written as $(-\infty, -3)$. To graph the inequality, place a parenthesis at -3 (to show that -3 is *not* part of the graph) and draw an arrow extending to the left.

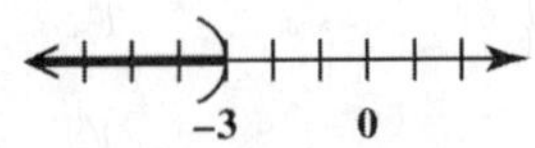

13. The statement $t > 4$ says that t can represent any number greater than 4. The interval is written $(4, \infty)$. To graph the inequality, place a parenthesis at 4 (to show that 4 is *not* part of the graph) and draw an arrow extending to the right.

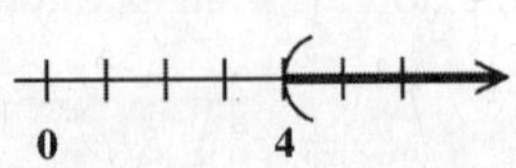

15. The statement $0 \geq x$ (or $x \leq 0$) says that x can represent any number less than or equal to 0. The interval is written as $(-\infty, 0]$. To graph the inequality, place a bracket at 0 (to show that 0 is part of the graph) and draw an arrow extending to the left.

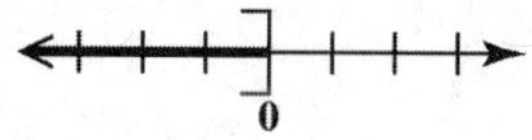

17. The statement $-\frac{1}{2} \leq x$ (or $x \geq -\frac{1}{2}$) says that x can represent any number greater than or equal to $-\frac{1}{2}$. The interval is written $[-\frac{1}{2}, \infty)$. To graph the inequality, place a bracket at $-\frac{1}{2}$ (to show that $-\frac{1}{2}$ is part of the graph) and draw an arrow extending to the right.

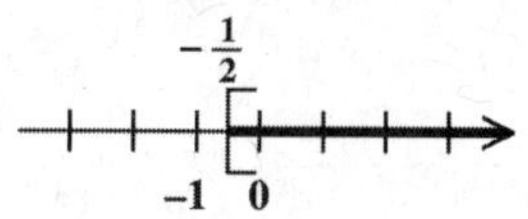

19.

$$\begin{aligned} z - 8 &\geq -7 \\ z - 8 + 8 &\geq -7 + 8 \quad \textit{Add 8.} \\ z &\geq 1 \end{aligned}$$

Graph the solution set $[1, \infty)$.

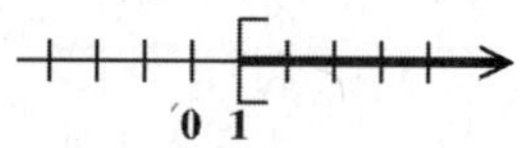

21.

$$\begin{aligned} 2k + 3 &\geq k + 8 \\ 2k + 3 - k &\geq k + 8 - k \quad \textit{Subtract k.} \\ k + 3 &\geq 8 \\ k + 3 - 3 &\geq 8 - 3 \quad \textit{Subtract 3.} \\ k &\geq 5 \end{aligned}$$

Graph the solution set $[5, \infty)$.

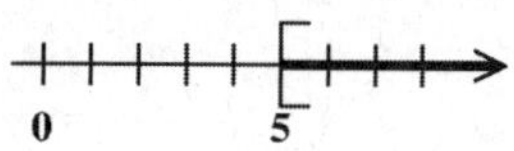

23.

$$\begin{aligned} 3n + 5 &< 2n - 6 \\ 3n - 2n + 5 &< 2n - 2n - 6 \quad \textit{Subtract 2n.} \\ n + 5 &< -6 \\ n + 5 - 5 &< -6 - 5 \quad \textit{Subtract 5.} \\ n &< -11 \end{aligned}$$

Graph the solution set $(-\infty, -11)$.

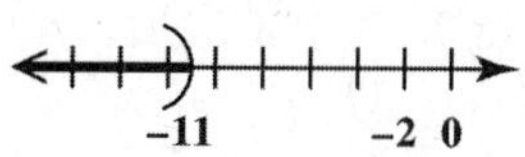

25.

$$\begin{aligned} 3x &< 18 \\ \frac{3x}{3} &< \frac{18}{3} \quad \textit{Divide by 3.} \\ x &< 6 \end{aligned}$$

Graph the solution set $(-\infty, 6)$.

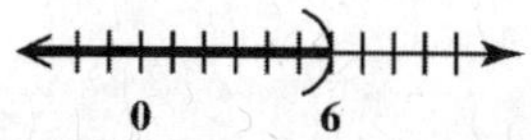

27.

$$\begin{aligned} 2y &\geq -20 \\ \frac{2y}{2} &\geq \frac{-20}{2} \quad \textit{Divide by 2.} \\ y &\geq -10 \end{aligned}$$

Graph the solution set $[-10, \infty)$.

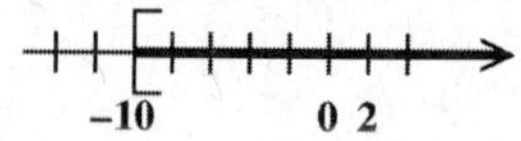

29.

$$\begin{aligned} -8t &> 24 \\ \frac{-8t}{-8} &< \frac{24}{-8} \quad \textit{Divide by } -8\textit{; reverse the symbol from} > \textit{to} < . \\ t &< -3 \end{aligned}$$

Graph the solution set $(-\infty, -3)$.

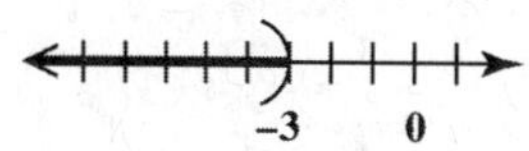

31.

$$\begin{aligned} -x &\geq 0 \\ -1x &\geq 0 \\ \frac{-1x}{-1} &\leq \frac{0}{-1} \quad \textit{Divide by } -1\textit{; reverse the symbol from} \geq \textit{to} \leq . \\ x &\leq 0 \end{aligned}$$

Graph the solution set $(-\infty, 0]$.

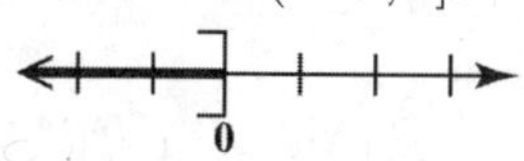

33.

$$-\tfrac{3}{4}r < -15$$

$$(-\tfrac{4}{3})(-\tfrac{3}{4}r) > (-\tfrac{4}{3})(-15)$$

Multiply by $-\frac{4}{3}$ *(the reciprocal of* $-\frac{3}{4}$*); reverse the symbol from* $<$ *to* $>$.

$$r > 20$$

Graph the solution set $(20, \infty)$.

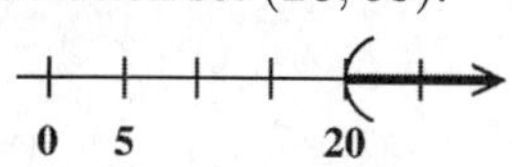

35.

$$\begin{aligned} -0.02x &\leq 0.06 \\ \frac{-0.02x}{-0.02} &\geq \frac{0.06}{-0.02} \quad \textit{Divide by } -0.02\textit{; reverse the symbol from} \leq \textit{to} \geq . \\ x &\geq -3 \end{aligned}$$

Graph the solution set $[-3, \infty)$.

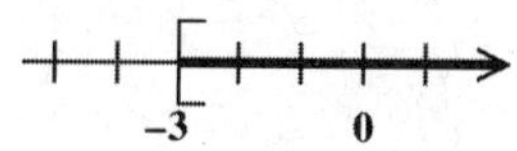

37.

$$\begin{aligned} 8x + 9 &\leq -15 \\ 8x &\leq -24 \quad \textit{Subtract 9.} \\ \frac{8x}{8} &\leq \frac{-24}{8} \quad \textit{Divide by 8.} \\ x &\leq -3 \end{aligned}$$

Graph the solution set $(-\infty, -3]$.

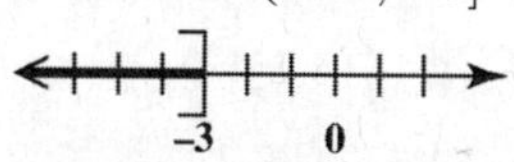

39.
$$\begin{aligned} -4x - 3 &< 1 \\ -4x &< 4 && \textit{Add 3.} \\ \frac{-4x}{-4} &> \frac{4}{-4} && \textit{Divide by } -4\textit{; reverse the symbol from < to >.} \\ x &> -1 \end{aligned}$$

Graph the solution set $(-1, \infty)$.

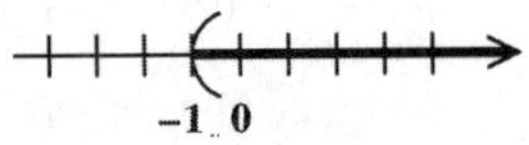

41.
$$\begin{aligned} 5r + 1 &\geq 3r - 9 \\ 2r + 1 &\geq -9 && \textit{Subtract 3r.} \\ 2r &\geq -10 && \textit{Subtract 1.} \\ r &\geq -5 && \textit{Divide by 2.} \end{aligned}$$

Graph the solution set $[-5, \infty)$.

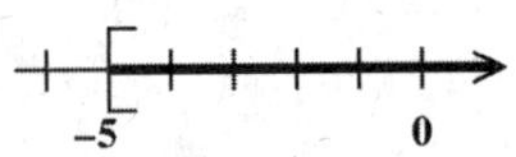

43.
$$\begin{aligned} 6x + 3 + x &< 2 + 4x + 4 \\ 7x + 3 &< 4x + 6 && \textit{Combine like terms.} \\ 3x + 3 &< 6 && \textit{Subtract 4x.} \\ 3x &< 3 && \textit{Subtract 3.} \\ x &< 1 && \textit{Divide by 3.} \end{aligned}$$

Graph the solution set $(-\infty, 1)$.

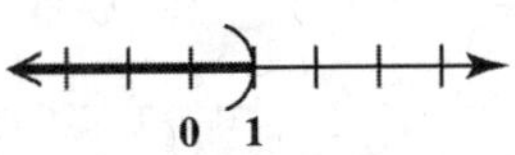

45.
$$\begin{aligned} -x + 4 + 7x &\leq -2 + 3x + 6 \\ 6x + 4 &\leq 4 + 3x \\ 3x + 4 &\leq 4 \\ 3x &\leq 0 \\ x &\leq 0 \end{aligned}$$

Graph the solution set $(-\infty, 0]$.

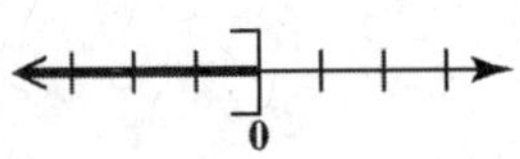

47.
$$\begin{aligned} 5(t-1) &> 3(t-2) \\ 5t - 5 &> 3t - 6 \\ 2t - 5 &> -6 \\ 2t &> -1 \\ t &> -\tfrac{1}{2} \end{aligned}$$

Graph the solution set $(-\frac{1}{2}, \infty)$.

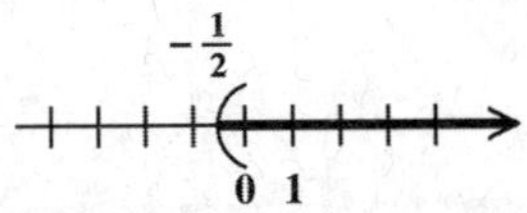

49.
$$\begin{aligned} 5(x+3) - 6x &\leq 3(2x+1) - 4x \\ 5x + 15 - 6x &\leq 6x + 3 - 4x \\ -x + 15 &\leq 2x + 3 \\ -3x + 15 &\leq 3 \\ -3x &\leq -12 \\ \frac{-3x}{-3} &\geq \frac{-12}{-3} && \textit{Divide by } -3\textit{; reverse the symbol.} \\ x &\geq 4 \end{aligned}$$

Graph the solution set $[4, \infty)$.

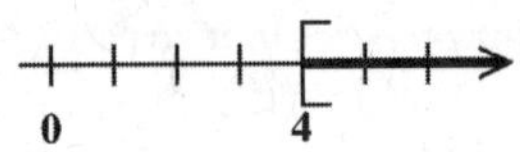

51.
$$\begin{aligned} 4x - (6x+1) &\leq 8x + 2(x-3) \\ 4x - 6x - 1 &\leq 8x + 2x - 6 \\ -2x - 1 &\leq 10x - 6 \\ -12x - 1 &\leq -6 \\ -12x &\leq -5 \\ \frac{-12x}{-12} &\geq \frac{-5}{-12} && \textit{Divide by } -12\textit{; reverse the symbol.} \\ x &\geq \tfrac{5}{12} \end{aligned}$$

Graph the solution set $[\frac{5}{12}, \infty)$.

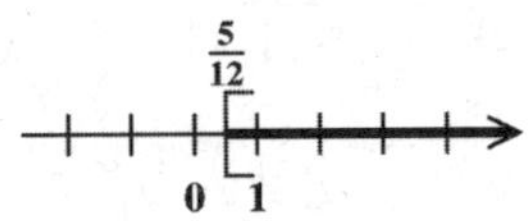

53.
$$\begin{aligned} \tfrac{2}{3}(p+3) &> \tfrac{5}{6}(p-4) \\ 6\left(\tfrac{2}{3}\right)(p+3) &> 6\left(\tfrac{5}{6}\right)(p-4) \end{aligned}$$

Multiply by 6, the LCD.

$$\begin{aligned} 4(p+3) &> 5(p-4) \\ 4p + 12 &> 5p - 20 \\ -p + 12 &> -20 \\ -p &> -32 \\ \frac{-p}{-1} &< \frac{-32}{-1} && \textit{Divide by } -1\textit{; reverse the symbol.} \\ p &< 32 \end{aligned}$$

Graph the solution set $(-\infty, 32)$.

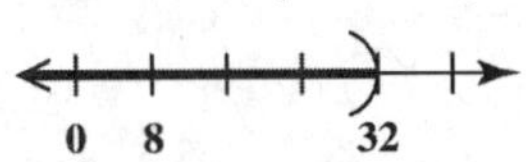

55.
$$\tfrac{2}{3}(3x-1) \geq \tfrac{3}{2}(2x-3)$$

Multiply both sides by 6 to clear the fractions.

$$\begin{aligned} 6 \cdot \tfrac{2}{3}(3x-1) &\geq 6 \cdot \tfrac{3}{2}(2x-3) \\ 4(3x-1) &\geq 9(2x-3) \\ 12x - 4 &\geq 18x - 27 && \textit{Distributive property} \\ -6x &\geq -23 && \textit{Subtract 18x; add 4.} \end{aligned}$$

Divide by -6, and reverse the inequality symbol..

$$x \leq \frac{23}{6}$$

Check that the solution set is the interval $(-\infty, \frac{23}{6}]$.

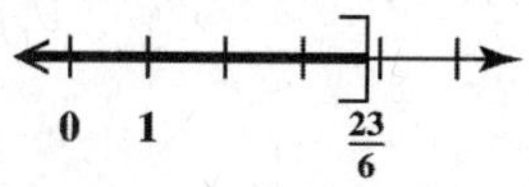

57. $-\frac{1}{4}(p+6)+\frac{3}{2}(2p-5)<10$

Multiply each term by 4 to clear the fractions.

$$\begin{aligned} -1(p+6)+6(2p-5) &< 40 \\ -p-6+12p-30 &< 40 \\ 11p-36 &< 40 \\ 11p &< 76 \\ p &< \tfrac{76}{11} \end{aligned}$$

Check that the solution set is the interval $(-\infty, \frac{76}{11})$.

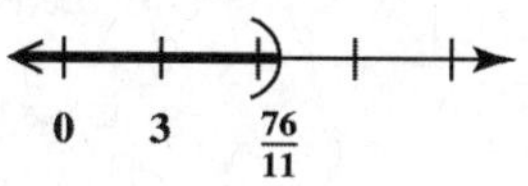

♦♦♦ Relating Concepts 59–62 ♦♦♦

59. $3x+2=14$

$$\begin{aligned} 3x &= 12 \\ x &= 4 \end{aligned}$$

Solution set: $\{4\}$

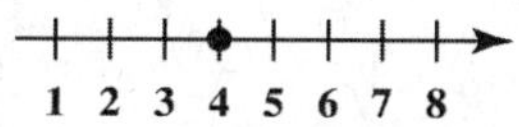

60. $3x+2>14$

$$\begin{aligned} 3x &> 12 \\ x &> 4 \end{aligned}$$

Solution set: $(4, \infty)$

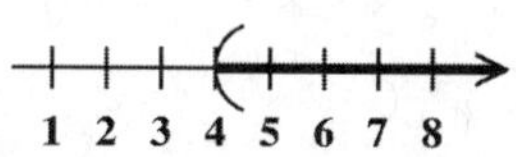

61. $3x+2<14$

$$\begin{aligned} 3x &< 12 \\ x &< 4 \end{aligned}$$

Solution set: $(-\infty, 4)$

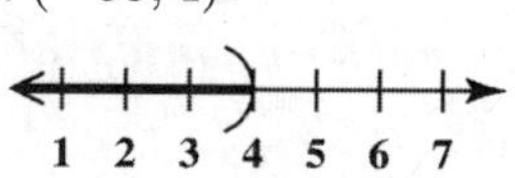

62. If you were to graph all the solutions from Exercises 59–61 on the same number line, the graph would be the complete number line, that is, all real numbers.

63. The statement "You must be at least 18 yr old to vote" translates as $x \geq 18$.

65. The statement "Chicago received more than 5 in. of snow" translates as $x > 5$.

67. The statement "Tracy could spend at most $20 on a gift" translates as $x \leq 20$.

69. Let $x =$ the score on the third test.

The average of the three tests	is at least	80.
↓	↓	↓
$\frac{76+81+x}{3}$	$\geq$	80

$$\begin{aligned} \frac{157+x}{3} &\geq 80 \\ 3\left(\frac{157+x}{3}\right) &\geq 3(80) \\ 157+x &\geq 240 \\ x &\geq 83 \end{aligned}$$

In order to average at least 80, Christy's score on her third test must be 83 or greater.

71. Let $n =$ the number.

"When 2 is added to the difference between six times a number and 5, the result is greater than 13 added to five times the number" translates to

$$(6n-5)+2>5n+13.$$

Solve the inequality.

$$\begin{aligned} 6n-5+2 &> 5n+13 \\ 6n-3 &> 5n+13 \\ n-3 &> 13 \quad \textit{Subtract } 5n. \\ n &> 16 \quad \textit{Add } 3. \end{aligned}$$

All numbers greater than 16 satisfy the given condition.

73. The Fahrenheit temperature must correspond to a Celsius temperature that is greater than or equal to -25 degrees.

$$\begin{aligned} C = \tfrac{5}{9}(F-32) &\geq -25 \\ \tfrac{9}{5}\left[\tfrac{5}{9}(F-32)\right] &\geq \tfrac{9}{5}(-25) \\ F-32 &\geq -45 \\ F &\geq -13 \end{aligned}$$

The temperature in Minneapolis on a certain winter day is never less than $-13°$ Fahrenheit.

75. $P=2L+2W$; $P \geq 400$

From the figure, we have $L=4x+3$ and $W=x+37$. Thus, we have the inequality

$$2(4x+3)+2(x+37) \geq 400.$$

Solve this inequality.

$$\begin{aligned} 8x+6+2x+74 &\geq 400 \\ 10x+80 &\geq 400 \\ 10x &\geq 320 \\ x &\geq 32 \end{aligned}$$

The rectangle will have a perimeter of at least 400 if the value of x is 32 or greater.

77.
$$\begin{aligned} 2+0.30x &\le 5.60 \\ 10(2+0.30x) &\le 10(5.60) \\ 20+3x &\le 56 \\ 3x &\le 36 \\ x &\le 12 \end{aligned}$$

Alan can use the phone for a maximum of 12 minutes after the first three minutes. This means that the maximum *total* time he can use the phone is 15 minutes.

79. "Revenue from the sales of the DVDs is \$5 per DVD less sales costs of \$100" translates to

$$R = 5x - 100,$$

where x represents the number of DVDs to be produced.

81.
$$\begin{aligned} P &= R - C \\ &= (5x-100)-(125+4x) \\ &= 5x - 100 - 125 - 4x \\ &= x - 225 \end{aligned}$$

We can use this expression for P to solve the inequality.

$$\begin{aligned} P &> 0 \\ x - 225 &> 0 \\ x &> 225 \end{aligned}$$

83. The graph corresponds to the inequality $-1 < x < 2$, excluding both -1 and 2.

85. The graph corresponds to the inequality $-1 < x \le 2$, excluding -1 but including 2.

87. The statement $8 \le x \le 10$ says that x can represent any number between 8 and 10, including 8 and 10. To graph the inequality, place brackets at 8 and 10 (to show that 8 and 10 are part of the graph) and draw a line segment between the brackets. The interval is written as $[8, 10]$.

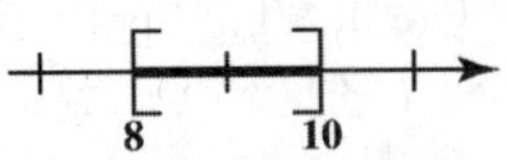

89. The statement $0 < y \le 10$ says that y can represent any number between 0 and 10, excluding 0 and including 10. To graph the inequality, place a parenthesis at 0 and a bracket at 10 and draw a line segment between them. The interval is written as $(0, 10]$.

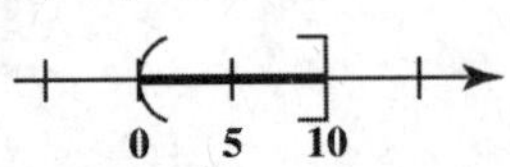

91. The statement $4 > x > -3$ can be written as $-3 < x < 4$. Graph the solution set $(-3, 4)$.

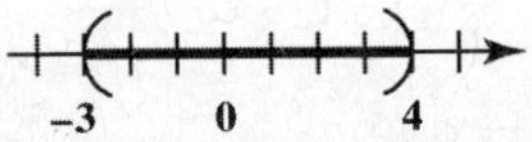

93.
$$-5 \le 2x - 3 \le 9$$
$$-5+3 \le 2x-3+3 \le 9+3 \quad \textit{Add 3 to each part.}$$
$$-2 \le 2x \le 12$$
$$\frac{-2}{2} \le \frac{2x}{2} \le \frac{12}{2} \quad \textit{Divide each part by 2.}$$
$$-1 \le x \le 6$$

Graph the solution set $[-1, 6]$.

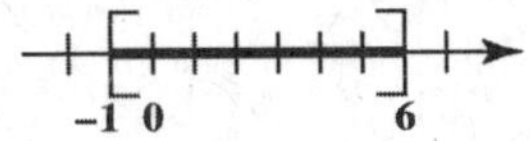

95.
$$5 < 1 - 6m < 12$$
$$5-1 < 1-6m-1 < 12-1 \quad \textit{Subtract 1 from each part.}$$
$$4 < -6m < 11$$
$$\frac{4}{-6} > \frac{-6m}{-6} > \frac{11}{-6} \quad \textit{Divide each part by –6; reverse both symbols.}$$
$$-\tfrac{2}{3} > m > -\tfrac{11}{6}$$
$$\text{or } -\tfrac{11}{6} < m < -\tfrac{2}{3} \quad \textit{Equivalent inequality}$$

Graph the solution set $\left(-\frac{11}{6}, -\frac{2}{3}\right)$.

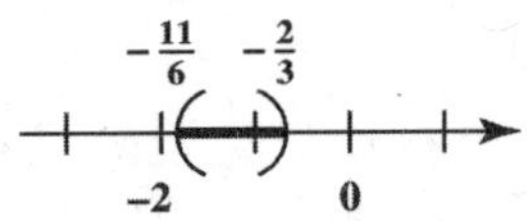

97.
$$\begin{aligned} 10 < 7p + 3 &< 24 \\ 7 < 7p &< 21 \quad \textit{Subtract 3.} \\ 1 < p &< 3 \quad \textit{Divide by 7.} \end{aligned}$$

Graph the solution set $(1, 3)$.

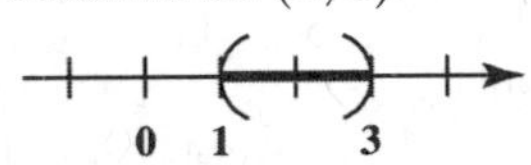

99.
$$-12 \le \tfrac{1}{2}z + 1 \le 4$$
$$2(-12) \le 2\left(\tfrac{1}{2}z+1\right) \le 2(4) \quad \textit{Multiply each part by 2.}$$
$$-24 \le z + 2 \le 8$$
$$-26 \le z \le 6 \quad \textit{Subtract 2.}$$

Note: We could have started this solution by subtracting 1 from each part.

Graph the solution set $[-26, 6]$.

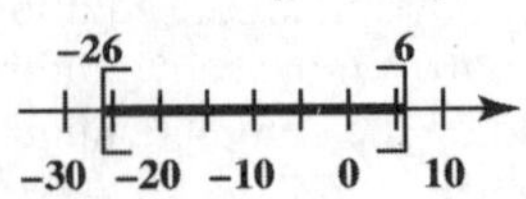

101.
$$1 \le 3 + \tfrac{2}{3}p \le 7$$
$$3(1) \le 3\left(3+\tfrac{2}{3}p\right) \le 3(7) \quad \textit{Multiply by 3.}$$
$$3 \le 9 + 2p \le 21$$
$$-6 \le 2p \le 12 \quad \textit{Subtract 9.}$$
$$-3 \le p \le 6 \quad \textit{Divide by 2.}$$

Graph the solution set $[-3, 6]$.

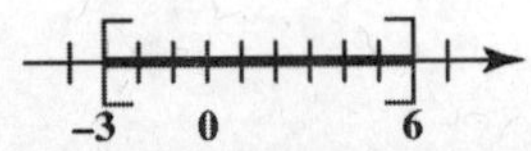

103. $-7 \le \frac{5}{4}r - 1 \le -1$

$-6 \le \frac{5}{4}r \le 0$ *Add 1.*

$\frac{4}{5}(-6) \le \frac{4}{5}(\frac{5}{4}r) \le \frac{4}{5}(0)$ *Multiply by $\frac{4}{5}$.*

$-\frac{24}{5} \le r \le 0$

Graph the solution set $\left[-\frac{24}{5}, 0\right]$.

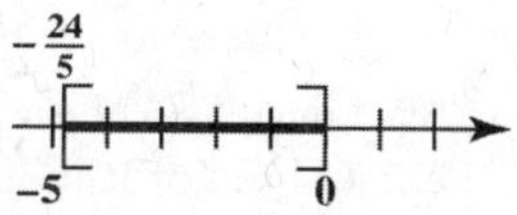

In part (a) of Exercises 105–110, replace x with -2. In part (b), replace x with 4. Then use the order of operations.

105. (a) $y = 5x + 3$
$y = 5(-2) + 3$
$y = -10 + 3$
$y = -7$

(b) $y = 5x + 3$
$y = 5(4) + 3$
$y = 20 + 3$
$y = 23$

107. (a) $6x - 2 = y$
$6(-2) - 2 = y$
$-12 - 2 = y$
$-14 = y$

(b) $6x - 2 = y$
$6(4) - 2 = y$
$24 - 2 = y$
$22 = y$

109. (a) $2x - 5y = 10$
$2(-2) - 5y = 10$
$-4 - 5y = 10$
$-5y = 14$
$y = -\frac{14}{5}$

(b) $2x - 5y = 10$
$2(4) - 5y = 10$
$8 - 5y = 10$
$-5y = 2$
$y = -\frac{2}{5}$

Chapter 2 Review Exercises

1. $x - 5 = 1$
$x = 6$ *Add 5.*

The solution set is $\{6\}$.

2. $x + 8 = -4$
$x = -12$ *Subtract 8.*

The solution set is $\{-12\}$.

3. $3t + 1 = 2t + 8$
$t + 1 = 8$ *Subtract 2t.*
$t = 7$ *Subtract 1.*

The solution set is $\{7\}$.

4. $5z = 4z + \frac{2}{3}$
$z = \frac{2}{3}$ *Subtract 4z.*

The solution set is $\left\{\frac{2}{3}\right\}$.

5. $(4r - 2) - (3r + 1) = 8$
$(4r - 2) - 1(3r + 1) = 8$ *Replace – with –1.*
$4r - 2 - 3r - 1 = 8$ *Dist. prop.*
$r - 3 = 8$
$r = 11$ *Add 3.*

The solution set is $\{11\}$.

6. $3(2x - 5) = 2 + 5x$
$6x - 15 = 2 + 5x$ *Dist. prop.*
$x - 15 = 2$ *Subtract 5x.*
$x = 17$ *Add 15.*

The solution set is $\{17\}$.

7. $7x = 35$
$x = 5$ *Divide by 7.*

The solution set is $\{5\}$.

8. $12r = -48$
$r = -4$ *Divide by 12.*

The solution set is $\{-4\}$.

9. $2p - 7p + 8p = 15$
$3p = 15$
$p = 5$ *Divide by 3.*

The solution set is $\{5\}$.

10. $\frac{x}{12} = -1$
$x = -12$ *Multiply by 12.*

The solution set is $\{-12\}$.

11. $\frac{5}{8}q = 8$
$\frac{8}{5}(\frac{5}{8}q) = \frac{8}{5}(8)$ *Multiply by $\frac{8}{5}$.*
$q = \frac{64}{5}$

The solution set is $\left\{\frac{64}{5}\right\}$.

12. $12m + 11 = 59$
$12m = 48$ *Subtract 11.*
$m = 4$ *Divide by 12.*

The solution set is $\{4\}$.

13. $3(2x + 6) - 5(x + 8) = x - 22$
$6x + 18 - 5x - 40 = x - 22$
$x - 22 = x - 22$

This is a true statement, so the solution set is {all real numbers}.

14. $5x + 9 - (2x - 3) = 2x - 7$
$5x + 9 - 2x + 3 = 2x - 7$
$3x + 12 = 2x - 7$
$x + 12 = -7$
$x = -19$

The solution set is $\{-19\}$.

15. $\frac{1}{2}r - \frac{r}{3} = \frac{r}{6}$
$6\left(\frac{1}{2}r\right) - 6\left(\frac{r}{3}\right) = 6\left(\frac{r}{6}\right)$ *Multiply by 6.*
$3r - 2r = r$
$r = r$

This is a true statement, so the solution set is {all real numbers}.

16.
$$0.1(x+80)+0.2x = 14$$
$$10[0.1(x+80)+0.2x] = 10(14) \quad \textit{Multiply by 10.}$$
$$(x+80)+2x = 140 \quad \textit{Dist. prop.}$$
$$3x+80 = 140$$
$$3x = 60$$
$$x = 20$$

The solution set is $\{20\}$.

17.
$$3x-(-2x+6) = 4(x-4)+x$$
$$3x+2x-6 = 4x-16+x$$
$$5x-6 = 5x-16$$
$$-6 = -16$$

This statement is false, so there is no solution set, symbolized by $\emptyset$.

18.
$$\tfrac{1}{2}(x+3)-\tfrac{2}{3}(x-2) = 3$$

Multiply both sides by 6 to eliminate fractions.
$$6\left[\tfrac{1}{2}(x+3)-\tfrac{2}{3}(x-2)\right] = 6(3)$$
$$6\left[\tfrac{1}{2}(x+3)\right]-6\left[\tfrac{2}{3}(x-2)\right] = 18$$
$$3(x+3)-4(x-2) = 18$$
$$3x+9-4x+8 = 18$$
$$-x+17 = 18$$
$$-x = 1$$
$$x = -1$$

The solution set is $\{-1\}$.

19. Let x represent the number.
$$5x+7 = 3x$$
$$7 = -2x \quad \textit{Subtract 5x.}$$
$$-\frac{7}{2} = x \quad \textit{Divide by } -2.$$

The number is $-\frac{7}{2}$.

20. *Step 2*
Let $x =$ the number of Republicans.
Then $x+22 =$ the number of Democrats.

Step 3 $\quad x+(x+22) = 118$

Step 4
$$2x+22 = 118$$
$$2x = 96$$
$$x = 48$$

Step 5
Since $x = 48$, $x+22 = 70$.
There were 70 Democrats and 48 Republicans.

Step 6
There are 22 more Democrats than Republicans and the total is 118.

21. *Step 2*
Let $x =$ the land area of Rhode Island.
Then $x+5213 =$ land area of Hawaii.

Step 3
The areas total 7637 square miles, so
$$x+(x+5213) = 7637.$$

Step 4
$$2x+5213 = 7637$$
$$2x = 2424$$
$$x = 1212$$

Step 5
Since $x = 1212$, $x+5213 = 6425$. The land area of Rhode Island is 1212 square miles and that of Hawaii is 6425 square miles.

Step 6
The land area of Hawaii is 5213 square miles greater than the land area of Rhode Island and the total is 7637 square miles.

22. *Step 2*
Let $x =$ the height of Twin Falls.
Then $\frac{5}{2}x =$ the height of Seven Falls.

Step 3
The sum of the heights is 420 feet, so
$$x+\tfrac{5}{2}x = 420.$$

Step 4
$$2(x+\tfrac{5}{2}x) = 2(420)$$
$$2x+5x = 840$$
$$7x = 840$$
$$x = 120$$

Step 5
Since $x = 120$, $\frac{5}{2}x = \frac{5}{2}(120) = 300$. The height of Twin Falls is 120 feet and that of Seven Falls is 300 feet.

Step 6
The height of Seven Falls is $\frac{5}{2}$ the height of Twin Falls and the sum is 420.

23. *Step 2*
Let $x =$ the measure of the angle.
Then $90-x =$ the measure of its complement and $180-x =$ the measure of its supplement.

Step 3 $\quad 180-x = 10(90-x)$

Step 4
$$180-x = 900-10x$$
$$9x+180 = 900$$
$$9x = 720$$
$$x = 80$$

Step 5
The measure of the angle is $80°$.
Its complement measures $90°-80° = 10°$, and its supplement measures $180°-80° = 100°$.

Step 6
The measure of the supplement is 10 times the measure of the complement.

24. *Step 2*
Let $x =$ lesser odd integer.
Then $x+2 =$ greater odd integer.

Step 3 $\quad x+2(x+2) = (x+2)+24$

Step 4

$$\begin{aligned} x + 2x + 4 &= x + 26 \\ 3x + 4 &= x + 26 \\ 2x + 4 &= 26 \\ 2x &= 22 \\ x &= 11 \end{aligned}$$

Step 5
Since $x = 11$, $x + 2 = 13$. The consecutive odd numbers are 11 and 13.

Step 6
The lesser plus twice the greater is $11 + 2(13) = 37$, which is 24 more than the greater.

In Exercises 25–28, substitute the given values into the given formula and then solve for the remaining variable.

25. $\mathcal{A} = \frac{1}{2}bh$; $\mathcal{A} = 44$, $b = 8$

$$\begin{aligned} \mathcal{A} &= \tfrac{1}{2}bh \\ 44 &= \tfrac{1}{2}(8)h \\ 44 &= 4h \\ 11 &= h \end{aligned}$$

26. $\mathcal{A} = \frac{1}{2}h(b + B)$; $h = 8$, $b = 3$, $B = 4$

$$\begin{aligned} \mathcal{A} &= \tfrac{1}{2}h(b + B) \\ \mathcal{A} &= \tfrac{1}{2}(8)(3 + 4) \\ &= \tfrac{1}{2}(8)(7) \\ &= (4)(7) \\ \mathcal{A} &= 28 \end{aligned}$$

27. $C = 2\pi r$; $C = 29.83$, $\pi = 3.14$

$$\begin{aligned} C &= 2\pi r \\ 29.83 &= 2(3.14)r \\ 29.83 &= 6.28r \\ \frac{29.83}{6.28} &= \frac{6.28r}{6.28} \\ 4.75 &= r \end{aligned}$$

28. $V = \frac{4}{3}\pi r^3$; $r = 6$, $\pi = 3.14$

$$\begin{aligned} V &= \tfrac{4}{3}\pi r^3 \\ &= \tfrac{4}{3}(3.14)(6)^3 \\ &= \tfrac{4}{3}(3.14)(216) \\ &= \tfrac{4}{3}(678.24) \\ V &= 904.32 \end{aligned}$$

29. $\mathcal{A} = bh$ for h

$$\frac{\mathcal{A}}{b} = \frac{bh}{b} \quad \textit{Divide by } b.$$

$$\frac{\mathcal{A}}{b} = h, \text{ or } h = \frac{\mathcal{A}}{b}$$

30. $\mathcal{A} = \frac{1}{2}h(b + B)$ for h

$$2\mathcal{A} = 2\left[\tfrac{1}{2}h(b + B)\right] \quad \textit{Multiply by 2.}$$

$$2\mathcal{A} = h(b + B)$$

$$\frac{2\mathcal{A}}{(b + B)} = \frac{h(b + B)}{(b + B)} \quad \textit{Divide by } b + B.$$

$$\frac{2\mathcal{A}}{b + B} = h, \text{ or } h = \frac{2\mathcal{A}}{b + B}$$

31. Because the two angles are supplementary,

$$\begin{aligned} (8x - 1) + (3x - 6) &= 180. \\ 11x - 7 &= 180 \\ 11x &= 187 \\ x &= 17 \end{aligned}$$

Since $x = 17$, $8x - 1 = 135$, and $3x - 6 = 45$. The measures of the two angles are 135° and 45°.

32. The angles are vertical angles, so their measures are equal.

$$\begin{aligned} 3x + 10 &= 4x - 20 \\ 10 &= x - 20 \\ 30 &= x \end{aligned}$$

Since $x = 30$, $3x + 10 = 100$, and $4x - 20 = 100$. Each angle has a measure of 100°.

33. Let W = the width of the rectangle.
Then $W + 12$ = the length of the rectangle.

The perimeter of the rectangle is 16 times the width can be written as

$$2L + 2W = 16W$$

since the perimeter is $2L + 2W$.

Because $L = W + 12$, we have

$$\begin{aligned} 2(W + 12) + 2W &= 16W. \\ 2W + 24 + 2W &= 16W \\ 4W + 24 &= 16W \\ -12W + 24 &= 0 \\ -12W &= -24 \\ W &= 2 \end{aligned}$$

The width is 2 cm and the length is $2 + 12 = 14$ cm.

34. First, use the formula for the circumference of a circle to find the value of r.

$$C = 2\pi r$$

$$62.5 \approx 2(3.14)(r) \quad \textit{Let } C = 62.5, \ \pi = 3.14.$$

$$\begin{aligned} 62.5 &= 6.28r \\ \frac{62.5}{6.28} &= \frac{6.28r}{6.28} \\ 9.95 &\approx r \end{aligned}$$

The radius of the turntable is approximately 9.95 feet. The diameter is twice the radius, so the diameter is approximately 19.9 feet.

Now use the formula for the area of a circle.

$$\begin{aligned} \mathcal{A} &= \pi r^2 \\ &\approx (3.14)(9.95)^2 \quad \textit{Let } \pi = 3.14,\ r = 9.95. \\ &= (3.14)(99.0025) \\ \mathcal{A} &\approx 311 \end{aligned}$$

The area of the turntable is approximately 311 square feet.

35. The sum of the three marked angles in the triangle is 180°.

$$\begin{aligned} 45° + (x + 12.2)° + (3x + 2.8)° &= 180° \\ 4x + 60 &= 180 \\ 4x &= 120 \\ x &= 30 \end{aligned}$$

Since $x = 30$, $(x + 12.2)° = 42.2°$, and $(3x + 2.8)° = 92.8°$.

36. The ratio of 60 centimeters to 40 centimeters is

$$\frac{60 \text{ cm}}{40 \text{ cm}} = \frac{3 \cdot 20}{2 \cdot 20} = \frac{3}{2}.$$

37. To find the ratio of 5 days to 2 weeks, first convert 2 weeks to days.

$$2 \text{ weeks} = 2 \cdot 7 = 14 \text{ days}$$

Thus, the ratio of 5 days to 2 weeks is $\frac{5}{14}$.

38. To find the ratio of 90 inches to 10 feet, first convert 10 feet to inches.

$$10 \text{ feet} = 10 \cdot 12 = 120 \text{ inches}$$

Thus, the ratio of 90 inches to 10 feet is

$$\frac{90}{120} = \frac{3 \cdot 30}{4 \cdot 30} = \frac{3}{4}.$$

39. $\dfrac{p}{21} = \dfrac{5}{30}$

$$30p = 105 \quad \textit{Cross products are equal.}$$

$$\frac{30p}{30} = \frac{105}{30} \quad \textit{Divide by 30.}$$

$$p = \frac{105}{30} = \frac{7 \cdot 15}{2 \cdot 15} = \frac{7}{2}$$

The solution set is $\left\{\frac{7}{2}\right\}$.

40. $\dfrac{5 + x}{3} = \dfrac{2 - x}{6}$

$$\begin{aligned} 6(5 + x) &= 3(2 - x) && \textit{Cross products are equal.} \\ 30 + 6x &= 6 - 3x && \textit{Distributive property} \\ 30 + 9x &= 6 && \textit{Add 3x.} \\ 9x &= -24 && \textit{Subtract 30.} \\ x &= \tfrac{-24}{9} = -\tfrac{8}{3} \end{aligned}$$

The solution set is $\left\{-\frac{8}{3}\right\}$.

41. Let $x =$ the tax on a \$36.00 item.
Set up a proportion with one ratio involving sales tax and the other involving the costs of the items.

$$\begin{aligned} \frac{x \text{ dollars}}{\$2.04} &= \frac{\$36}{\$24} \\ 24x &= (2.04)(36) = 73.44 \\ x &= \tfrac{73.44}{24} = 3.06 \end{aligned}$$

The sales tax on a \$36.00 item is \$3.06.

42. Let $x =$ the actual distance between the second pair of cities (in kilometers).

Set up a proportion with one ratio involving map distances and the other involving actual distances.

$$\begin{aligned} \frac{x \text{ kilometers}}{150 \text{ kilometers}} &= \frac{80 \text{ centimeters}}{32 \text{ centimeters}} \\ 32x &= (150)(80) = 12{,}000 \\ x &= \tfrac{12{,}000}{32} = 375 \end{aligned}$$

The cities are 375 kilometers apart.

43. Let $x =$ the number of bronze medals earned by Japan.

$$\begin{aligned} \frac{x \text{ bronze medals}}{25 \text{ medals}} &= \frac{2 \text{ bronze medals}}{5 \text{ medals}} \\ 5x &= 2(25) = 50 \\ x &= 10 \end{aligned}$$

At the 2008 Olympics, 10 bronze medals were earned by Japan.

In Exercise 44, to find the best buy, divide the price by the number of units to get the unit cost. Each result was found by using a calculator and rounding the answer to three decimal places. The *best buy* (based on price per unit) is the smallest unit cost.

44.

Size	**Unit Cost** (dollars per oz)
15 oz	$\frac{\$2.69}{15} = \$0.179$
20 oz	$\frac{\$3.29}{20} = \$0.165$
25.5 oz	$\frac{\$3.49}{25.5} = \$0.137$ (*)

The 25.5 oz size is the best buy.

45. What is 8% of 75?

$$8\% \cdot 75 = 0.08 \cdot 75 = 6$$

46. What percent of 12 is 21?

Let p denote the percent.

$$\begin{aligned} 21 &= p \cdot 12 \\ p &= \tfrac{21}{12} = 1.75 = 175\% \end{aligned}$$

Thus, 175% of 12 is 21.

47. 36% of what number is 900?

Let n denote the number.

$$\begin{aligned} 0.36 \cdot n &= 900 \\ n &= \frac{900}{0.36} \quad \textit{Divide by 0.36.} \\ &= 2500 \quad \textit{Simplify.} \end{aligned}$$

Thus, 36% of 2500 is 900.

48. Let $x =$ the number of liters of the 60% solution to be used.
Then $x + 15 =$ the number of liters of the 20% solution.

Liters of solution	15	+	x	=	$x + 15$
Strength of solution	0.10		0.60		0.20

Drug amount in 10% solution	plus	drug amount in 60% solution	is	drug amount in 20% solution.
↓	↓	↓	↓	↓
$0.10(15)$	$+$	$0.60(x)$	$=$	$0.20(x + 15)$

Multiply by 10 to clear decimals.

$$\begin{aligned} 1(15) + 6x &= 2(x + 15) \\ 15 + 6x &= 2x + 30 \\ 15 + 4x &= 30 \\ 4x &= 15 \\ x &= \tfrac{15}{4} = 3.75 \end{aligned}$$

3.75 liters of 60% solution are needed.

49. Let $x =$ the amount invested at 5%.
Then $10{,}000 - x =$ the amount invested at 6%.

Interest at 5%	plus	interest at 6%	equals	\$550.
↓	↓	↓	↓	↓
$0.05x$	$+$	$0.06(10{,}000 - x)$	$=$	550

$$\begin{aligned} 5x + 6(10{,}000 - x) &= 100(550) \\ 5x + 60{,}000 - 6x &= 55{,}000 \\ -x &= -5000 \\ x &= 5000 \end{aligned}$$

Robert invested \$5000 at 5% and $10{,}000 - 5000 = \$5000$ at 6%.

50. Use the formula $d = rt$ or $r = \frac{d}{t}$.

$$r = \tfrac{d}{t} = \tfrac{3150}{384} \approx 8.203$$

Rounded to the nearest tenth, the *Yorkshire's* average rate was 8.2 mph.

51. Use the formula $d = rt$ or $t = \frac{d}{r}$.

$$t = \tfrac{d}{r} = \tfrac{819}{63} = 13$$

Janet drove for 13 hours.

52. Let $t =$ the number of hours until the planes are 1925 miles apart.
Use $d = rt$.

The distance one plane flies north	plus	the distance the other plane flies south	is	the distance between the planes.
↓	↓	↓	↓	↓
$350t$	$+$	$420t$	$=$	1925

$$\begin{aligned} 770t &= 1925 \\ t &= \tfrac{1925}{770} = \tfrac{5}{2} = 2\tfrac{1}{2} \end{aligned}$$

The planes will be 1925 miles apart in $2\frac{1}{2}$ hours.

53. The statement $x \geq -4$ can be written as $[-4, \infty)$.

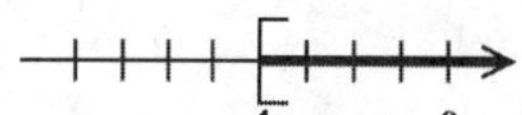

54. The statement $x < 7$ can be written as $(-\infty, 7)$.

55. The statement $-5 \leq x < 6$ can be written as $[-5, 6)$.

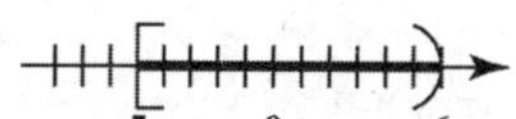

56. By examining the choices, we see that $-4x \leq 36$ is the only inequality that has a negative coefficient of x. Thus, **B** is the only inequality that requires a reversal of the inequality symbol when it is solved.

57.

$$\begin{aligned} x + 6 &\geq 3 \\ x &\geq -3 \quad \textit{Subtract 6.} \end{aligned}$$

Graph the solution set $[-3, \infty)$.

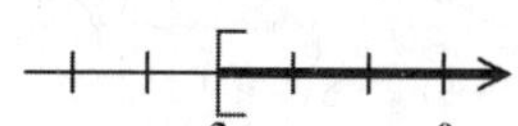

58.

$$\begin{aligned} 5x &< 4x + 2 \\ x &< 2 \qquad \textit{Subtract 4x.} \end{aligned}$$

Graph the solution set $(-\infty, 2)$.

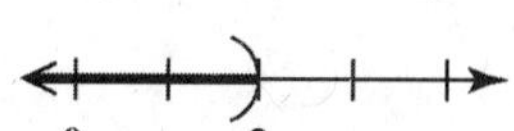

59.

$$\begin{aligned} -6x &\leq -18 \\ \frac{-6x}{-6} &\geq \frac{-18}{-6} \quad \textit{Divide by –6; reverse the symbol.} \\ x &\geq 3 \end{aligned}$$

Graph the solution set $[3, \infty)$.

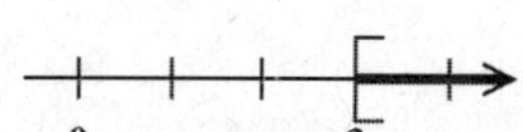

60. $8(x-5)-(2+7x) \geq 4$
$$8x-40-2-7x \geq 4$$
$$x-42 \geq 4$$
$$x \geq 46$$

Graph the solution set $[46, \infty)$.

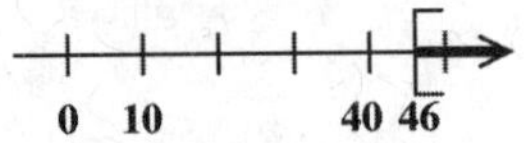

61. $4x-3x > 10-4x+7x$
$$x > 10+3x$$
$$-2x > 10$$
$$\frac{-2x}{-2} < \frac{10}{-2} \quad \textit{Divide by } -2\textit{; reverse the symbol.}$$
$$x < -5$$

Graph the solution set $(-\infty, -5)$.

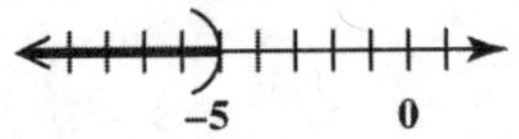

62. $3(2x+5)+4(8+3x) < 5(3x+7)$
$$6x+15+32+12x < 15x+35$$
$$18x+47 < 15x+35$$
$$3x+47 < 35$$
$$3x < -12$$
$$x < -4$$

Graph the solution set $(-\infty, -4)$.

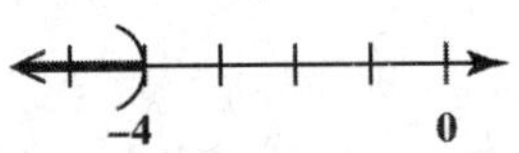

63. $-3 \leq 2x+1 \leq 4$
$$-4 \leq 2x \leq 3 \quad \textit{Subtract 1.}$$
$$-2 \leq x \leq \frac{3}{2} \quad \textit{Divide by 2.}$$

Graph the solution set $[-2, \frac{3}{2}]$.

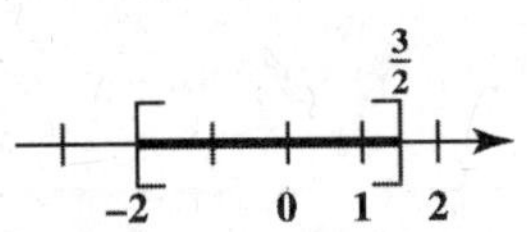

64. $9 < 3x+5 \leq 20$
$$4 < 3x \leq 15 \quad \textit{Subtract 5.}$$
$$\tfrac{4}{3} < x \leq 5 \quad \textit{Divide by 3.}$$

Graph the solution set $\left(\frac{4}{3}, 5\right]$.

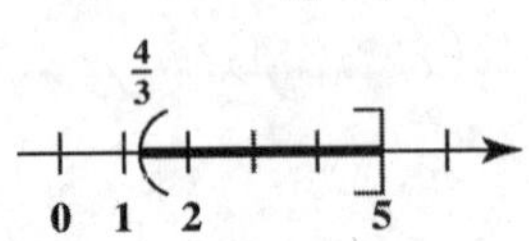

65. Let $x =$ the score on the third test.

The average of the three tests	is at least	90.
↓	↓	↓
$\frac{94+88+x}{3}$	$\geq$	90

$$\frac{182+x}{3} \geq 90$$
$$3\left(\frac{182+x}{3}\right) \geq 3(90)$$
$$182+x \geq 270$$
$$x \geq 88$$

In order to average at least 90, Awilda's score on her third test must be 88 or more.

66. Let $n =$ the number.
"If nine times a number is added to 6, the result is at most 3" can be written as
$$9n+6 \leq 3.$$

Solve the inequality.
$$9n \leq -3 \quad \textit{Subtract 6.}$$
$$n \leq \tfrac{-3}{9} \quad \textit{Divide by 9.}$$

All numbers less or equal to $-\frac{1}{3}$ satisfy the given condition.

67. **[2.6]** $\frac{x}{7} = \frac{x-5}{2}$
$$2x = 7(x-5) \quad \textit{Cross products are equal.}$$
$$2x = 7x-35$$
$$-5x = -35$$
$$x = 7$$

The solution set is $\{7\}$.

68. **[2.5]** $I = prt$ for r
$$\frac{I}{pt} = \frac{prt}{pt} \quad \textit{Divide by pt.}$$
$$\frac{I}{pt} = r \quad \text{or} \quad r = \frac{I}{pt}$$

69. **[2.8]** $-2x > -4$
$$\frac{-2x}{-2} < \frac{-4}{-2} \quad \textit{Divide by } -2\textit{; reverse the symbol.}$$
$$x < 2$$

The solution set is $(-\infty, 2)$.

70. **[2.3]** $2k-5 = 4k+13$
$$-2k-5 = 13 \quad \textit{Subtract 4k.}$$
$$-2k = 18 \quad \textit{Add 5.}$$
$$k = -9 \quad \textit{Divide by } -2.$$

The solution set is $\{-9\}$.

71. **[2.2]** $0.05x+0.02x = 4.9$
To clear decimals, multiply both sides by 100.
$$100(0.05x+0.02x) = 100(4.9)$$
$$5x+2x = 490$$
$$7x = 490$$
$$x = 70$$

The solution set is $\{70\}$.

72. **[2.3]** $2 - 3(x - 5) = 4 + x$
$$\begin{aligned} 2 - 3x + 15 &= 4 + x \\ 17 - 3x &= 4 + x \\ 17 - 4x &= 4 \\ -4x &= -13 \\ x &= \tfrac{-13}{-4} = \tfrac{13}{4} \end{aligned}$$

The solution set is $\{\frac{13}{4}\}$.

73. **[2.3]** $9x - (7x + 2) = 3x + (2 - x)$
$$\begin{aligned} 9x - 7x - 2 &= 3x + 2 - x \\ 2x - 2 &= 2x + 2 \\ -2 &= 2 \end{aligned}$$

Because $-2 = 2$ is a false statement, the given equation has no solution, symbolized by $\emptyset$.

74. **[2.3]** $\frac{1}{3}s + \frac{1}{2}s + 7 = \frac{5}{6}s + 5 + 2$
$$\tfrac{1}{3}s + \tfrac{1}{2}s = \tfrac{5}{6}s \qquad \textit{Subtract 7.}$$

The least common denominator is 6.
$$\begin{aligned} 6\left(\tfrac{1}{3}s + \tfrac{1}{2}s\right) &= 6\left(\tfrac{5}{6}s\right) \\ 2s + 3s &= 5s \\ 5s &= 5s \end{aligned}$$

Because $5s = 5s$ is a true statement, the solution set is {all real numbers}.

75. **[2.6]** The problem may be stated as follows:

"What is 8% of $3800?"
$$8\% \cdot 3800 = 0.08 \cdot 3800 = 304$$

The family plans to spend $304 on entertainment.

76. **[2.6]** Let $x =$ the number of calories a 175-pound athlete can consume.
Set up a proportion with one ratio involving calories and the other involving pounds.
$$\begin{aligned} \frac{x \text{ calories}}{50 \text{ calories}} &= \frac{175 \text{ pounds}}{2.2 \text{ pounds}} \\ 2.2x &= 50(175) \\ x &= \tfrac{8750}{2.2} \approx 3977.3 \end{aligned}$$

To the nearest hundred calories, a 175-pound athlete in a vigorous training program can consume 4000 calories per day.

77. **[2.4]** Let $x =$ the length of the Brooklyn Bridge.
Then $x + 2604 =$ the length of the Golden Gate Bridge.
$$\begin{aligned} x + (x + 2604) &= 5796 \\ 2x + 2604 &= 5796 \\ 2x &= 3192 \\ x &= 1596 \end{aligned}$$

Since $x = 1596$, $x + 2604 = 4200$.

The length of the Brooklyn Bridge is 1596 feet and that of the Golden Gate Bridge is 4200 feet.

78. **[2.6]**

Size	**Unit Cost** (dollars per oz)
50 oz	$\frac{\$4.69}{50} = \$0.094$
100 oz	$\frac{\$5.98}{100} = \$0.060$ (*)
200 oz	$\frac{\$13.68}{200} = \$0.068$

The 100-oz size is the best buy.

79. **[2.6]** Let $x =$ the number of quarts of oil needed for 192 quarts of gasoline.
Set up a proportion with one ratio involving oil and the other involving gasoline.
$$\begin{aligned} \frac{x \text{ quarts}}{1 \text{ quart}} &= \frac{192 \text{ quarts}}{24 \text{ quarts}} \\ x \cdot 24 &= 1 \cdot 192 \qquad \textit{Cross products} \\ x &= 8 \qquad \textit{Divide by 24.} \end{aligned}$$

The amount of oil needed is 8 quarts.

80. **[2.7]** Let $x =$ the rate of the slower train.
Then $x + 30 =$ the rate of the faster train.

	r	t	d
Slower train	x	3	$3x$
Faster train	$x + 30$	3	$3(x + 30)$

The sum of the distances traveled by the two trains is 390 miles, so
$$\begin{aligned} 3x + 3(x + 30) &= 390. \\ 3x + 3x + 90 &= 390 \\ 6x + 90 &= 390 \\ 6x &= 300 \\ x &= 50 \end{aligned}$$

Since $x = 50$, $x + 30 = 80$.

The rate of the slower train is 50 miles per hour and the rate of the faster train is 80 miles per hour.

81. **[2.5]** Let $x =$ the length of the first side.
Then $2x =$ the length of the second side.

Use the formula for the perimeter of a triangle, $P = a + b + c$, with perimeter 96 and third side 30.
$$\begin{aligned} x + 2x + 30 &= 96 \\ 3x + 30 &= 96 \\ 3x &= 66 \\ x &= 22 \end{aligned}$$

The sides have lengths 22 meters, 44 meters, and 30 meters. The length of the longest side is 44 meters.

82. **[2.8]** Let $s =$ the length of a side of the square.
The formula for the perimeter of a square is $P = 4s$.

The perimeter	cannot be greater than	200.
↓	↓	↓
$4s$	$\leq$	200

$$\begin{aligned} 4s &\leq 200 \\ s &\leq 50 \end{aligned}$$

The length of a side is 50 meters or less.

Chapter 2 Test

1.
$$\begin{aligned} 5x+9&=7x+21 \\ -2x+9&=21 && \textit{Subtract } 7x. \\ -2x&=12 && \textit{Subtract } 9. \\ x&=-6 && \textit{Divide by } -2. \end{aligned}$$

The solution set is $\{-6\}$.

2.
$$\begin{aligned} -\tfrac{4}{7}x&=-12 \\ \left(-\tfrac{7}{4}\right)\left(-\tfrac{4}{7}x\right)&=\left(-\tfrac{7}{4}\right)(-12) \\ x&=21 \end{aligned}$$

The solution set is $\{21\}$.

3.
$$\begin{aligned} 7-(x-4)&=-3x+2(x+1) \\ 7-x+4&=-3x+2x+2 \\ -x+11&=-x+2 \\ 11&=2 \end{aligned}$$

Because the last statement is false, the equation has no solution set, symbolized by $\emptyset$.

4. $0.6(x+20)+0.8(x-10)=46$
To clear decimals, multiply both sides by 10.

$$\begin{aligned} 10[0.6(x+20)+0.8(x-10)]&=10(46) \\ 6(x+20)+8(x-10)&=460 \\ 6x+120+8x-80&=460 \\ 14x+40&=460 \\ 14x&=420 \\ x&=30 \end{aligned}$$

The solution set is $\{30\}$.

5.
$$\begin{aligned} -8(2x+4)&=-4(4x+8) \\ -16x-32&=-16x-32 \end{aligned}$$

Because the last statement is true, the solution set is {all real numbers}.

6. Let $x =$ the number of games the Angels lost.
Then $2x-24 =$ the number of games the Angels won.

The total number of games played was 162.

$$\begin{aligned} x+(2x-24)&=162 \\ 3x-24&=162 \\ 3x&=186 \\ x&=62 \end{aligned}$$

Since $x=62$, $2x-24=100$.
The Angels won 100 games and lost 62 games.

7. Let $x =$ the area of Kauai (in square miles).
Then $x+177 =$ the area of Maui (in square miles), and
$(x+177)+3293=x+3470 =$ the area of Hawaii.

$$\begin{aligned} x+(x+177)+(x+3470)&=5300 \\ 3x+3647&=5300 \\ 3x&=1653 \\ x&=551 \end{aligned}$$

Since $x=551$, $x+177=728$, and $x+3470=4021$.

The area of Hawaii is 4021 square miles, the area of Maui is 728 square miles, and the area of Kauai is 551 square miles.

8. Let $x =$ the measure of the angle.
Then $90-x =$ the measure of its complement, and $180-x =$ the measure of its supplement.

$$\begin{aligned} 180-x&=3(90-x)+10 \\ 180-x&=270-3x+10 \\ 180-x&=280-3x \\ 180+2x&=280 \\ 2x&=100 \\ x&=50 \end{aligned}$$

The measure of the angle is 50°. The measure of its supplement, 130°, is 10° more than three times its complement, 40°.

9. **(a)** Solve $P=2L+2W$ for W.

$$\begin{aligned} P-2L&=2W \\ \frac{P-2L}{2}&=W \quad \text{or} \quad W=\frac{P-2L}{2} \end{aligned}$$

(b) Substitute 116 for P and 40 for L in the formula obtained in part (a).

$$\begin{aligned} W&=\frac{P-2L}{2} \\ &=\frac{116-2(40)}{2} \\ &=\frac{116-80}{2}=\frac{36}{2}=18 \end{aligned}$$

10. The angles are vertical angles, so their measures are equal.

$$\begin{aligned} 3x+15&=4x-5 \\ 15&=x-5 \\ 20&=x \end{aligned}$$

Since $x=20$, $3x+15=75$ and $4x-5=75$.

Both angles have measure 75°.

11.

$$\frac{z}{8}=\frac{12}{16}$$

$$16z=8(12) \quad \textit{Cross products are equal.}$$

$$16z=96$$

$$\frac{16z}{16}=\frac{96}{16} \quad \textit{Divide by 16.}$$

$$z=6$$

The solution set is $\{6\}$.

12.

$$\frac{x+5}{3}=\frac{x-3}{4}$$

$$4(x+5)=3(x-3)$$

$$4x+20=3x-9$$

$$x+20=-9$$

$$x=-29$$

The solution set is $\{-29\}$.

13. The results in the following table are rounded to the nearest thousandth.

Size	**Unit Cost** (dollars per oz)
8 oz	$\frac{\$2.79}{8}=\$0.349$
16 oz	$\frac{\$4.99}{16}=\$0.312$
32 oz	$\frac{\$7.99}{32}=\$0.250$ (*)

The best buy is the 32-oz size.

14. Let $x=$ the actual distance between Seattle and Cincinnati.

$$\frac{x \text{ miles}}{1050 \text{ miles}}=\frac{92 \text{ inches}}{42 \text{ inches}}$$

$$42x=92(1050)=96{,}600$$

$$x=\frac{96{,}600}{42}=2300$$

The actual distance between Seattle and Cincinnati is 2300 miles.

15. Let $x=$ the amount invested at 3%.
Then $x+6000=$ the amount invested at 4.5%.

Amount invested (in dollars)	**Rate of interest**	**Interest for one year**
x	0.03	$0.03x$
$x+6000$	0.045	$0.045(x+6000)$

$$0.03x+0.045(x+6000)=870$$

$$1000(0.03x)+1000[0.045(x+6000)]=1000(870)$$

$$30x+45(x+6000)=870{,}000$$

$$30x+45x+270{,}000=870{,}000$$

$$75x+270{,}000=870{,}000$$

$$75x=600{,}000$$

$$x=8000$$

Since $x=8000$, $x+6000=14{,}000$.

Carlos invested \$8000 at 3% and \$14,000 at 4.5%.

16. Use the formula $d=rt$ and let t be the number of hours they traveled.

	r	t	d
First car	50	t	$50t$
Second car	65	t	$65t$

First car's distance ↓ $50t$ | and ↓ $+$ | second car's distance ↓ $65t$ | is ↓ $=$ | total distance. ↓ 460

$$50t+65t=460$$

$$115t=460$$

$$t=4$$

The two cars will be 460 miles apart in 4 hours.

17.

$$-4x+2(x-3)\geq 4x-(3+5x)-7$$

$$-4x+2x-6\geq 4x-3-5x-7$$

$$-2x-6\geq -x-10$$

$$-x-6\geq -10$$

$$-x\geq -4$$

$$\frac{-1x}{-1}\leq\frac{-4}{-1} \quad \textit{Divide by –1; reverse the symbol}$$

$$x\leq 4$$

Graph the solution set $(-\infty,4]$.

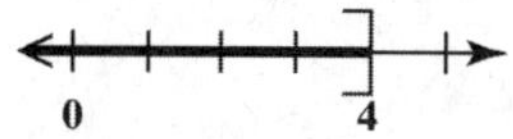

18.

$$-10<3x-4\leq 14$$

$$-6<3x\leq 18 \quad \textit{Add 4.}$$

$$-2<x\leq 6 \quad \textit{Divide by 3.}$$

Graph the solution set $(-2,6]$.

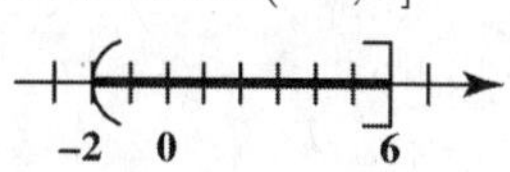

19. Let $x=$ the score on the third test.

The average of the three tests ↓ $\frac{76+81+x}{3}$ | is at least ↓ $\geq$ | 80. ↓ 80

$$\frac{157+x}{3}\geq 80$$

$$3\left(\frac{157+x}{3}\right)\geq 3(80)$$

$$157+x\geq 240$$

$$x\geq 83$$

In order to average at least 80, Susan's score on her third test must be 83 or more.

20. When an inequality is multiplied or divided by a negative number, the direction of the inequality symbol must be reversed.

Cumulative Review Exercises (Chapters 1–2)

1. $$\frac{5}{6}+\frac{1}{4}-\frac{7}{15}=\frac{50}{60}+\frac{15}{60}-\frac{28}{60}$$
$$=\frac{65-28}{60}$$
$$=\frac{37}{60}$$

2. $$\frac{9}{8}\cdot\frac{16}{3}\div\frac{5}{8}=\frac{9}{8}\cdot\frac{16}{3}\cdot\frac{8}{5}$$
$$=\frac{3\cdot 3\cdot 16\cdot 8}{8\cdot 3\cdot 5}$$
$$=\frac{48}{5}$$

3. "The difference between half a number and 18" is written

$$\tfrac{1}{2}x-18.$$

4. "The quotient of 6 and 12 more than a number is 2" is written

$$\frac{6}{x+12}=2.$$

5. $$\frac{8(7)-5(6+2)}{3\cdot 5+1}\geq 1$$
$$\frac{8(7)-5(8)}{3\cdot 5+1}\geq 1$$
$$\frac{56-40}{15+1}\geq 1$$
$$\frac{16}{16}\geq 1$$
$$1\geq 1$$

The statement is *true*.

6. $$\frac{-4(9)(-2)}{-3^2}=\frac{-36(-2)}{-1\cdot 3^2}$$
$$=\frac{72}{-9}$$
$$=-8$$

7. $$(-7-1)(-4)+(-4)=(-8)(-4)+(-4)$$
$$=32+(-4)$$
$$=28$$

8. $$\frac{3x^2-y^3}{-4z}=\frac{3(-2)^2-(-4)^3}{-4(3)}$$ *Let $x=-2$, $y=-4, z=3$.*
$$=\frac{3(4)-(-64)}{-12}$$
$$=\frac{12+64}{-12}$$
$$=\frac{76}{-12}$$
$$=-\frac{19}{3}$$

9. $7(p+q)=7p+7q$

The multiplication of 7 is distributed over the sum, which illustrates the distributive property.

10. $3+(5+2)=3+(2+5)$

The order of the numbers added in the parentheses is changed, which illustrates the commutative property.

11. $$2r-6=8r$$
$$-6=6r$$
$$-1=r$$

Check $r=-1$: $-8=-8$ *True*

The solution set is $\{-1\}$.

12. $$4-5(s+2)=3(s+1)-1$$
$$4-5s-10=3s+3-1$$
$$-5s-6=3s+2$$
$$-8s-6=2$$
$$-8s=8$$
$$s=-1$$

Check $s=-1$: $-1=-1$ *True*

The solution set is $\{-1\}$.

13. $$\tfrac{2}{3}x+\tfrac{3}{4}x=-17$$
$$12\left(\tfrac{2}{3}x+\tfrac{3}{4}x\right)=12(-17)$$ *LCD = 12*
$$8x+9x=-204$$
$$17x=-204$$
$$x=-12$$

Check $x=-12$: $-17=-17$ *True*

The solution set is $\{-12\}$.

14.
$$\begin{aligned}\frac{2x+3}{5} &= \frac{x-4}{2}\\ (2x+3)(2) &= (5)(x-4)\\ 4x+6 &= 5x-20\\ 6 &= x-20\\ 26 &= x\end{aligned}$$

Check $x = 26$: $11 = 11$ *True*

The solution set is $\{26\}$.

15. Solve $3x + 4y = 24$ for y.
$$\begin{aligned}4y &= 24 - 3x\\ y &= \frac{24-3x}{4}\end{aligned}$$

16.
$$\begin{aligned}6(r-1)+2(3r-5) &\le -4\\ 6r-6+6r-10 &\le -4\\ 12r-16 &\le -4\\ 12r &\le 12\\ r &\le 1\end{aligned}$$

Graph the solution set $(-\infty, 1]$.

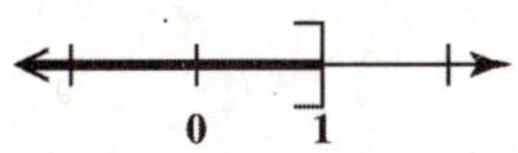

17. $-18 \le -9z < 9$

$2 \ge z \quad > -1$ *Divide by –9; reverse the symbols.*

$-1 < z \quad \le 2$ *Equivalent inequality*

Graph the solution set $(-1, 2]$.

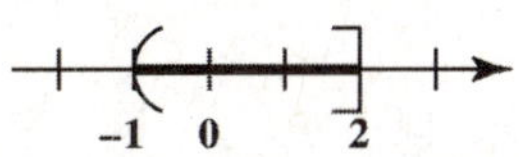

18. Let $x =$ the length of the middle-sized piece. Then $3x =$ the length of the longest piece, and $x - 5 =$ the length of the shortest piece.
$$\begin{aligned}x+3x+(x-5) &= 40\\ 5x-5 &= 40\\ 5x &= 45\\ x &= 9\end{aligned}$$

The length of the middle-sized piece is 9 centimeters, of the longest piece is 27 centimeters, and of the shortest piece is 4 centimeters.

19. Let $r =$ the radius and use 3.14 for π. Using the formula for circumference, $C = 2\pi r$, and $C = 78$, we have
$$\begin{aligned}2\pi r &= 78.\\ r &= \frac{78}{2\pi} \approx 12.4204\end{aligned}$$

To the nearest hundredth, the radius is 12.42 cm.

20. Let $x =$ rate of slower car. Then $x + 20 =$ rate of faster car.

Use the formula $d = rt$.
$$\begin{aligned}d_{\text{slower}} + d_{\text{faster}} &= d_{\text{total}}\\ (x)(4)+(x+20)(4) &= 400\\ 4x+4x+80 &= 400\\ 8x+80 &= 400\\ 8x &= 320\\ x &= 40\end{aligned}$$

The rates are 40 mph and 60 mph.

CHAPTER 3 LINEAR EQUATIONS IN TWO VARIABLES

3.1 Linear Equations in Two Variables; The Rectangular Coordinate System

3.1 Now Try Exercises

N1. **(a)** Move up from 2006 on the horizontal scale to the point plotted for 2006. Looking across at the vertical scale, this point is close to the line on the vertical scale for \$2.60. So, it cost about \$2.60 for a gallon of gasoline in 2006.

(b) As in part (a), we see that the point for 2008 is about one-fourth of the way between the lines on the vertical scale for \$3.20 and \$3.40. Halfway between the lines for \$3.20 and \$3.40 would be \$3.30. So, it cost about \$3.25 for a gallon of gasoline in 2008. The increase is \$3.25 − \$2.60 = \$0.65.

N2. **(a)** $(3,4)$

$$\begin{aligned} 3x - 7y &= 19 \\ 3(3) - 7(4) &\stackrel{?}{=} 19 \quad \textit{Let } x=3,\ y=4. \\ 9 - 28 &\stackrel{?}{=} 19 \\ -19 &= 19 \quad \textit{False} \end{aligned}$$

No, $(3,4)$ is not a solution.

(b) $(-3,-4)$

$$\begin{aligned} 3x - 7y &= 19 \\ 3(-3) - 7(-4) &\stackrel{?}{=} 19 \quad \textit{Let } x=-3,\ y=-4. \\ -9 + 28 &\stackrel{?}{=} 19 \\ 19 &= 19 \quad \textit{True} \end{aligned}$$

Yes, $(-3,-4)$ is a solution.

N3. **(a)** In the ordered pair $(4, \_\_)$, $x = 4$.

$$\begin{aligned} y &= 3x - 12 \\ y &= 3(4) - 12 \quad \textit{Let } x=4. \\ y &= 12 - 12 \\ y &= 0 \end{aligned}$$

The ordered pair is $(4,0)$.

(b) In the ordered pair $(\_\_, 3)$, $y = 3$. Find the corresponding value of x by replacing y with 3 in the given equation.

$$\begin{aligned} y &= 3x - 12 \\ 3 &= 3x - 12 \quad \textit{Let } y=3. \\ 15 &= 3x \quad \textit{Add 12.} \\ 5 &= x \quad \textit{Divide by 3.} \end{aligned}$$

The ordered pair is $(5,3)$.

N4. To complete the first ordered pair, let $x = 0$.

$$\begin{aligned} 5x - 4y &= 20 \\ 5(0) - 4y &= 20 \quad \textit{Let } x=0. \\ -4y &= 20 \\ y &= -5 \end{aligned}$$

The first ordered pair is $(0,-5)$.

To complete the second ordered pair, let $y = 0$.

$$\begin{aligned} 5x - 4y &= 20 \\ 5x - 4(0) &= 20 \quad \textit{Let } y=0. \\ 5x &= 20 \\ x &= 4 \end{aligned}$$

The second ordered pair is $(4,0)$.

To complete the third ordered pair, let $x = 2$.

$$\begin{aligned} 5x - 4y &= 20 \\ 5(2) - 4y &= 20 \quad \textit{Let } x=2. \\ 10 - 4y &= 20 \\ -4y &= 10 \\ y &= -\tfrac{5}{2} \end{aligned}$$

The third ordered pair is $(2, -\frac{5}{2})$.

The completed table of ordered pairs follows.

x	y
0	-5
4	0
2	$-\frac{5}{2}$

N5. To plot the ordered pairs, start at the origin in each case.

(a) $(-3,1)$

Go 3 units to the left along the x-axis; then go up 1 unit, parallel to the y-axis.

graph on next page

(b) $(2,-4)$

Go 2 units to the right along the x-axis; then go down 4 units.

(c) $(0,-1)$

Go down 1 unit. This point is on the y-axis since the x-coordinate is 0.

(d) $(\frac{5}{2}, 3)$

Go 2.5 units to the right; then go up 3 units.

(e) $(-4,-3)$

Go 4 units to the left; then go down 3 units.

(f) $(-4,0)$

Go 4 units to the left. This point is on the x-axis since the y-coordinate is 0.

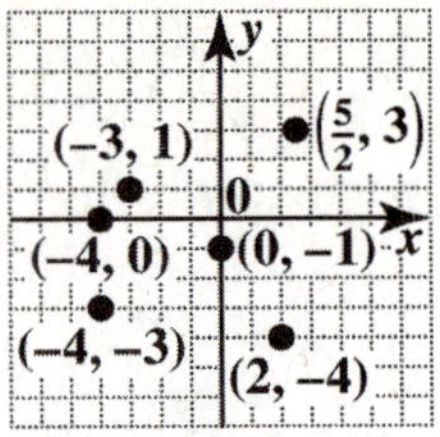

N6. To find y when $x = 2004$, substitute 2004 for x into the equation and use a calculator.

$$\begin{aligned} y &= 3.049x - 5979.0 \\ y &= 3.049(2004) - 5979.0 \\ y &\approx 131 \end{aligned}$$

There were about 131,000 twin births in the United States in 2004.

3.1 Section Exercises

1. The line segments between 2003 and 2004, 2004 and 2005, and 2005 and 2006 fall, so these are the pairs of consecutive years in which the unemployment rate decreased.

3. For 2003, the unemployment rate was about 6%. For 2004, the unemployment rate was about 5.5%. The decline is $6\% - 5.5\% = 0.5\%$.

5. The symbol (x, y) *does* represent an ordered pair, while the symbols $[x, y]$ and $\{x, y\}$ *do not* represent ordered pairs. (Note that only parentheses are used to write ordered pairs.)

7. All points having coordinates in the form

(negative, positive)

are in quadrant II, so the point whose graph has coordinates $(-4, 2)$ is in quadrant II.

9. All ordered pairs that are solutions of the equation $y = 3$ have y-coordinates equal to 3, so the ordered pair $(4, 3)$ is a solution of the equation $y = 3$.

11. $x + y = 8$; $(0, 8)$

To determine whether $(0, 8)$ is a solution of the given equation, substitute 0 for x and 8 for y.

$$\begin{aligned} x + y &= 8 \\ 0 + 8 &\stackrel{?}{=} 8 \quad \textit{Let } x = 0,\ y = 8. \\ 8 &= 8 \quad \textit{True} \end{aligned}$$

The result is true, so $(0, 8)$ is a solution of the given equation $x + y = 8$.

13. $2x + y = 5$; $(3, -1)$

Substitute 3 for x and -1 for y.

$$\begin{aligned} 2x + y &= 5 \\ 2(3) + (-1) &\stackrel{?}{=} 5 \quad \textit{Let } x = 3,\ y = -1. \\ 6 - 1 &\stackrel{?}{=} 5 \\ 5 &= 5 \quad \textit{True} \end{aligned}$$

The result is true, so $(3, -1)$ is a solution of $2x + y = 5$.

15. $5x - 3y = 15$; $(5, 2)$

Substitute 5 for x and 2 for y.

$$\begin{aligned} 5x - 3y &= 15 \\ 5(5) - 3(2) &\stackrel{?}{=} 15 \quad \textit{Let } x = 5,\ y = 2. \\ 25 - 6 &\stackrel{?}{=} 15 \\ 19 &= 15 \quad \textit{False} \end{aligned}$$

The result is false, so $(5, 2)$ is not a solution of $5x - 3y = 15$.

17. $x = -4y$; $(-8, 2)$

Substitute -8 for x and 2 for y.

$$\begin{aligned} x &= -4y \\ -8 &\stackrel{?}{=} -4(2) \quad \textit{Let } x = -8,\ y = 2. \\ -8 &= -8 \quad \textit{True} \end{aligned}$$

The result is true, so $(-8, 2)$ is a solution of $x = -4y$.

19. $y = 2$; $(4, 2)$

Since x does not appear in the equation, we just substitute 2 for y.

$$\begin{aligned} y &= 2 \\ 2 &= 2 \quad \textit{Let } y = 2\textit{; true} \end{aligned}$$

The result is true, so $(4, 2)$ is a solution of $y = 2$.

21. $x - 6 = 0$; $(4, 2)$

Since y does not appear in the equation, we just substitute 4 for x.

$$\begin{aligned} x - 6 &= 0 \\ 4 - 6 &\stackrel{?}{=} 0 \quad \textit{Let } x = 4. \\ -2 &= 0 \quad \textit{False} \end{aligned}$$

The result is false, so $(4, 2)$ is not a solution of $x - 6 = 0$.

23. $y = 2x + 7$; $(5, \_\_)$

In this ordered pair, $x = 5$. Find the corresponding value of y by replacing x with 5 in the given equation.

$$\begin{aligned} y &= 2x + 7 \\ y &= 2(5) + 7 \quad \textit{Let } x = 5. \\ y &= 10 + 7 \\ y &= 17 \end{aligned}$$

The ordered pair is $(5, 17)$.

25. $y = 2x + 7$; $(\_\_, -3)$

In this ordered pair, $y = -3$. Find the corresponding value of x by replacing y with -3 in the given equation.

$$\begin{aligned} y &= 2x + 7 \\ -3 &= 2x + 7 \quad \textit{Let } y = -3. \\ -10 &= 2x \\ -5 &= x \end{aligned}$$

The ordered pair is $(-5, -3)$.

27. $y = -4x - 4$; $(\_\_, 0)$

$$\begin{aligned} y &= -4x - 4 \\ 0 &= -4x - 4 \quad \textit{Let } y = 0. \\ 4 &= -4x \\ -1 &= x \end{aligned}$$

The ordered pair is $(-1, 0)$.

29. $y = -4x - 4$; $(\_\_, 24)$

$$\begin{aligned} y &= -4x - 4 \\ 24 &= -4x - 4 \quad \textit{Let } y = 24. \\ 28 &= -4x \\ -7 &= x \end{aligned}$$

The ordered pair is $(-7, 24)$.

31. $4x + 3y = 24$

If $x = 0$,

$$\begin{aligned} 4(0) + 3y &= 24 \\ 0 + 3y &= 24 \\ y &= 8. \end{aligned}$$

If $y = 0$,

$$\begin{aligned} 4x + 3(0) &= 24 \\ 4x + 0 &= 24 \\ x &= 6. \end{aligned}$$

If $y = 4$,

$$\begin{aligned} 4x + 3(4) &= 24 \\ 4x + 12 &= 24 \\ 4x &= 12 \\ x &= 3. \end{aligned}$$

The completed table of values is shown below.

x	y	ordered pair
0	8	$(0, 8)$
6	0	$(6, 0)$
3	4	$(3, 4)$

33. $4x - 9y = -36$

If $y = 0$,

$$\begin{aligned} 4x - 9(0) &= -36 \\ 4x - 0 &= -36 \\ 4x &= -36 \\ x &= -9. \end{aligned}$$

If $x = 0$,

$$\begin{aligned} 4(0) - 9y &= -36 \\ 0 - 9y &= -36 \\ -9y &= -36 \\ y &= 4. \end{aligned}$$

If $y = 8$,

$$\begin{aligned} 4x - 9(8) &= -36 \\ 4x - 72 &= -36 \\ 4x &= 36 \\ x &= 9. \end{aligned}$$

The completed table of values is shown below.

x	y	ordered pair
-9	0	$(-9, 0)$
0	4	$(0, 4)$
9	8	$(9, 8)$

35. $x = 12$

No matter which value of y is chosen, the value of x will always be 12. Each ordered pair can be completed by placing 12 in the first position.

x	y	ordered pair
12	3	$(12, 3)$
12	8	$(12, 8)$
12	0	$(12, 0)$

37. $y = -10$

No matter which value of x is chosen, the value of y will always be -10. Each ordered pair can be completed by placing -10 in the second position.

x	y	ordered pair
4	-10	$(4, -10)$
0	-10	$(0, -10)$
-4	-10	$(-4, -10)$

39. The given equation, $y + 2 = 0$, may be written $y = -2$. For any value of x, the value of y will always be -2.

x	y	ordered pair
9	-2	$(9, -2)$
2	-2	$(2, -2)$
0	-2	$(0, -2)$

41. The given equation, $x - 4 = 0$, may be written $x = 4$. For any value of y, the value of x will always be 4.

x	y	ordered pair
4	4	$(4, 4)$
4	0	$(4, 0)$
4	-4	$(4, -4)$

43. No, the ordered pair $(3, 4)$ represents the point 3 units to the right of the origin and 4 units up from the x-axis. The ordered pair $(4, 3)$ represents the point 4 units to the right of the origin and 3 units up from the x-axis.

45. Point A was plotted by starting at the origin, going 2 units to the right along the x-axis, and then going up 4 units on a line parallel to the y-axis. The ordered pair for this point is $(2, 4)$, which is in quadrant I.

Point B was plotted by starting at the origin, going 3 units to the left along the x-axis, and then going up 2 units on a line parallel to the y-axis. The ordered pair for this point is $(-3, 2)$, which is in quadrant II.

Point C was plotted by starting at the origin, going 5 units to the left along the x-axis, and then going up 4 units on a line parallel to the y-axis. The ordered pair for this point is $(-5, 4)$, which is in quadrant II.

Point D was plotted by starting at the origin, going 5 units to the left along the x-axis, and then going down 2 units on a line parallel to the y-axis. The ordered pair for this point is $(-5, -2)$, which is in quadrant III.

Point E was plotted by starting at the origin, going 3 units to the right along the x-axis. The ordered pair for this point is $(3, 0)$, which is not in any quadrant.

Point F is located on the y-axis. It was plotted by starting at the origin and going down 2 units along the y-axis. The ordered pair for this point is $(0, -2)$, which is not in any quadrant.

For Exercises 47–58, the ordered pairs are plotted on the graph following the solution for Exercise 57.

47. To plot $(6, 2)$, start at the origin, go 6 units to the right, and then go up 2 units.

49. To plot $(-4, 2)$, start at the origin, go 4 units to the left, and then go up 2 units.

51. To plot $\left(-\frac{4}{5}, -1\right)$, start at the origin, go $\frac{4}{5}$ unit to the left, and then go down 1 unit.

53. To plot $(3, -1.75)$, start at the origin, go 3 units to the right, and then go down 1.75 (or $1\frac{3}{4}$) units.

55. To plot $(0, 4)$, start at the origin and go up 4 units. The point lies on the y-axis.

57. To plot $(4, 0)$, start at the origin and go right 4 units. The point lies on the x-axis.

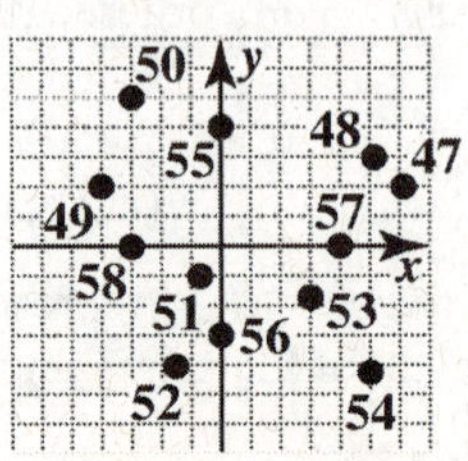

59. The point with coordinates (x, y) is in quadrant III if x is <u>negative</u> and y is <u>negative</u>.

61. The point with coordinates (x, y) is in quadrant IV if x is <u>positive</u> and y is <u>negative</u>.

63. If $xy < 0$, then either $x < 0$ and $y > 0$ or $x > 0$ and $y < 0$. If $x < 0$ and $y > 0$, then the point lies in quadrant II. If $x > 0$ and $y < 0$, then the point lies in quadrant IV.

65. $x - 2y = 6$

x	y
0	
	0
2	
	-1

Substitute the given values to complete the ordered pairs.

$$\begin{aligned} x - 2y &= 6 \\ 0 - 2y &= 6 \quad \textit{Let } x = 0. \\ -2y &= 6 \\ y &= -3 \end{aligned}$$

$$\begin{aligned} x - 2y &= 6 \\ x - 2(0) &= 6 \quad \textit{Let } y = 0. \\ x - 0 &= 6 \\ x &= 6 \end{aligned}$$

$$\begin{aligned} x - 2y &= 6 \\ 2 - 2y &= 6 \quad \textit{Let } x = 2. \\ -2y &= 4 \\ y &= -2 \end{aligned}$$

$$\begin{aligned} x - 2y &= 6 \\ x - 2(-1) &= 6 \quad \textit{Let } y = -1. \\ x + 2 &= 6 \\ x &= 4 \end{aligned}$$

The completed table of values follows.

x	y
0	-3
6	0
2	-2
4	-1

Plot the points $(0, -3)$, $(6, 0)$, $(2, -2)$, and $(4, -1)$ on a coordinate system.

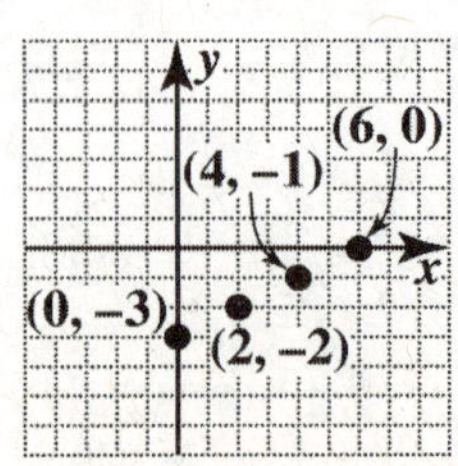

67. $3x - 4y = 12$

x	y
0	
	0
-4	
	-4

Substitute the given values to complete the ordered pairs.

$$
\begin{aligned}
3x - 4y &= 12 \\
3(0) - 4y &= 12 \quad \textit{Let } x = 0. \\
0 - 4y &= 12 \\
-4y &= 12 \\
y &= -3
\end{aligned}
$$

$$
\begin{aligned}
3x - 4y &= 12 \\
3x - 4(0) &= 12 \quad \textit{Let } y = 0. \\
3x - 0 &= 12 \\
3x &= 12 \\
x &= 4
\end{aligned}
$$

$$
\begin{aligned}
3x - 4y &= 12 \\
3(-4) - 4y &= 12 \quad \textit{Let } x = -4. \\
-12 - 4y &= 12 \\
-4y &= 24 \\
y &= -6
\end{aligned}
$$

$$
\begin{aligned}
3x - 4y &= 12 \\
3x - 4(-4) &= 12 \quad \textit{Let } y = -4. \\
3x + 16 &= 12 \\
3x &= -4 \\
x &= -\tfrac{4}{3}
\end{aligned}
$$

The completed table is as follows.

x	y
0	-3
4	0
-4	-6
$-\frac{4}{3}$	-4

Plot the points $(0, -3)$, $(4, 0)$, $(-4, -6)$, and $(-\frac{4}{3}, -4)$ on a coordinate system.

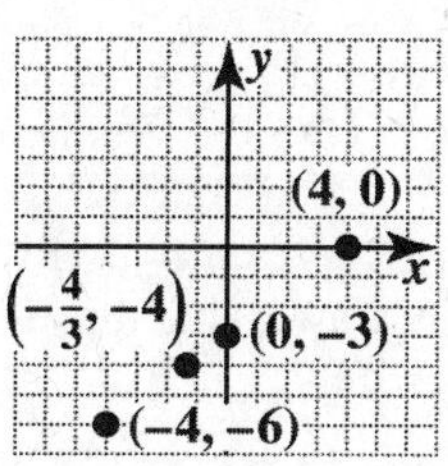

69. The given equation, $y + 4 = 0$, can be written as $y = -4$. So regardless of the value of x, the value of y is -4.

x	y
0	-4
5	-4
-2	-4
-3	-4

Plot the points $(0, -4)$, $(5, -4)$, $(-2, -4)$, and $(-3, -4)$ on a coordinate system.

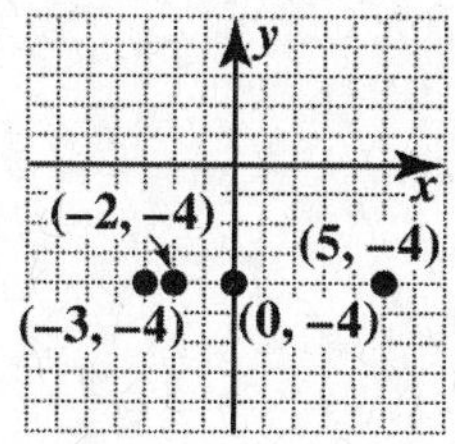

71. The points in each graph appear to lie on a straight line.

73. **(a)** When $x = 5$, $y = 45$. The ordered pair is $(5, 45)$.

(b) When $y = 50$, $x = 6$. The ordered pair is $(6, 50)$.

75. **(a)** We can write the results from the table as ordered pairs (x, y).

$(2002, 31.6)$, $(2003, 30.1)$, $(2004, 29.0)$, $(2005, 27.5)$, $(2006, 26.6)$, $(2007, 26.9)$

(b) $(2007, 26.9)$ means that 26.9 percent of 2-year college students in 2007 received a degree within 3 years.

(c)

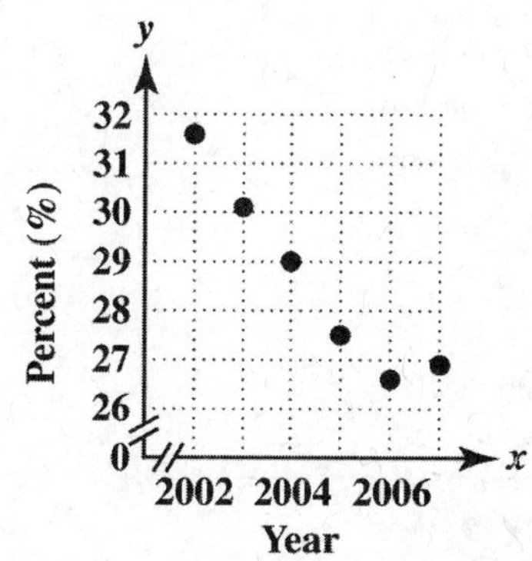

(d) With the exception of the point for 2007, the points lie approximately on a straight line. Rates at which 2-year college students complete a degree within 3 years were generally decreasing.

77. **(a)** Substitute $x = 20, 40, 60, 80$ in the equation $y = -0.65x + 143$.

$$
\begin{aligned}
y &= -0.65(20) + 143 = -13 + 143 = 130 \\
y &= -0.65(40) + 143 = -26 + 143 = 117 \\
y &= -0.65(60) + 143 = -39 + 143 = 104 \\
y &= -0.65(80) + 143 = -52 + 143 = 91
\end{aligned}
$$

The completed table follows.

Age	Heartbeats (per minute)
20	130
40	117
60	104
80	91

(b) In the ordered pairs (x, y), x represents age and y represents heartbeats. $(20, 130)$, $(40, 117)$, $(60, 104)$, $(80, 91)$

(c)

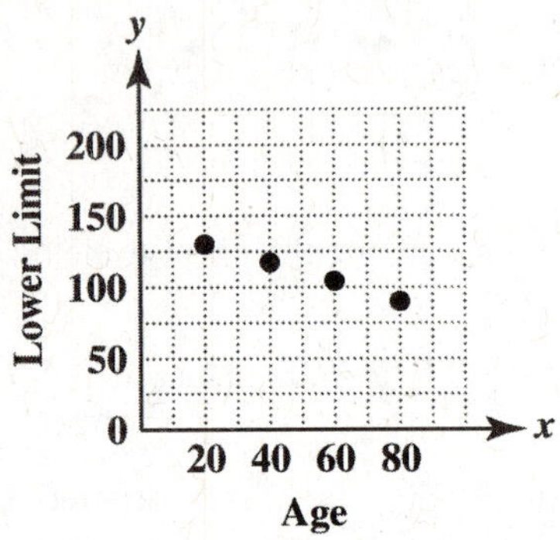

Yes, the points lie in a linear pattern.

79. Refer to the tables and locate age 20 and age 40 in each table. For age 20, the target heart rate is between 130 and 170. For age 40, the target heart rate is between 117 and 153.

81. $3x + 6 = 0$

$$3x = -6$$
$$x = \frac{-6}{3} = -2$$

Check $x = -2$: $-6 + 6 = 0$

The solution set is $\{-2\}$.

83. $9 - x = -4$

$$-x = -13$$
$$x = 13$$

Check $x = 13$: $9 - 13 = -4$
The solution set is $\{13\}$.

3.2 Graphing Linear Equations in Two Variables

3.2 Now Try Exercises

N1. $2x - 4y = 8$

Find the x-intercept by letting $y = 0$.

$$\begin{aligned} 2x - 4y &= 8 \\ 2x - 4(0) &= 8 \quad \textit{Let } y = 0. \\ 2x - 0 &= 8 \\ 2x &= 8 \\ x &= 4 \end{aligned}$$

The x-intercept is $(4, 0)$.

Find the y-intercept by letting $x = 0$.

$$\begin{aligned} 2x - 4y &= 8 \\ 2(0) - 4y &= 8 \quad \textit{Let } x = 0. \\ 0 - 4y &= 8 \\ -4y &= 8 \\ y &= -2 \end{aligned}$$

The y-intercept is $(0, -2)$.

Get a third point as a check. For example, choosing $x = -2$ gives $y = -3$. Plot $(4, 0)$, $(0, -2)$, and $(-2, -3)$ and draw a line through them.

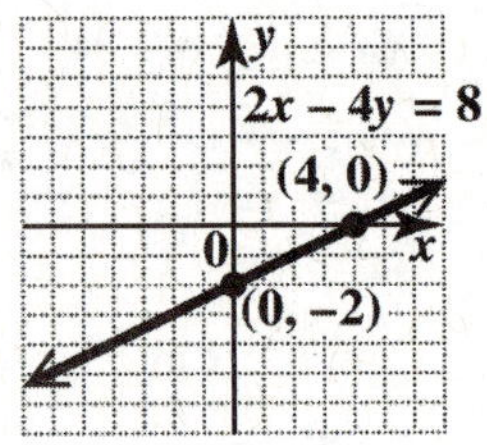

N2. $y = \frac{1}{3}x + 1$

Find the x-intercept by letting $y = 0$.

$$\begin{aligned} y &= \tfrac{1}{3}x + 1 \\ 0 &= \tfrac{1}{3}x + 1 \quad \textit{Let } y = 0. \\ -1 &= \tfrac{1}{3}x \\ 3(-1) &= 3\left(\tfrac{1}{3}x\right) \quad \textit{Multiply by 3.} \\ -3 &= x \end{aligned}$$

The x-intercept is $(-3, 0)$.

Find the y-intercept by letting $x = 0$.

$$\begin{aligned} y &= \tfrac{1}{3}x + 1 \\ y &= \tfrac{1}{3}(0) + 1 \quad \textit{Let } x = 0. \\ y &= 0 + 1 \\ y &= 1 \end{aligned}$$

The y-intercept is $(0, 1)$.

Get a third point as a check. For example, choosing $x = 3$ gives $y = 2$. Plot $(-3, 0)$, $(0, 1)$, and $(3, 2)$ and draw a line through them.

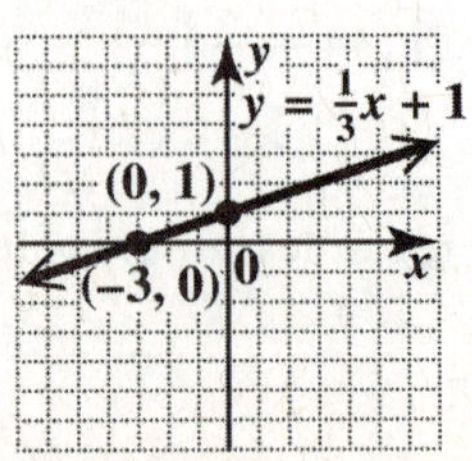

N3. $x + 2y = 2$

Find the x-intercept by letting $y = 0$.

$$\begin{aligned} x + 2y &= 2 \\ x + 2(0) &= 2 \quad \textit{Let } y = 0. \\ x + 0 &= 2 \\ x &= 2 \end{aligned}$$

The x-intercept is $(2, 0)$.

Find the y-intercept by letting $x = 0$.

$$\begin{aligned} x + 2y &= 2 \\ 0 + 2y &= 2 \quad \textit{Let } x = 0. \\ 2y &= 2 \\ y &= 1 \end{aligned}$$

The y-intercept is $(0, 1)$.

Get a third point as a check. For example, choosing $x = 4$ gives $y = -1$. Plot $(2, 0)$, $(0, 1)$, and $(4, -1)$ and draw a line through them.

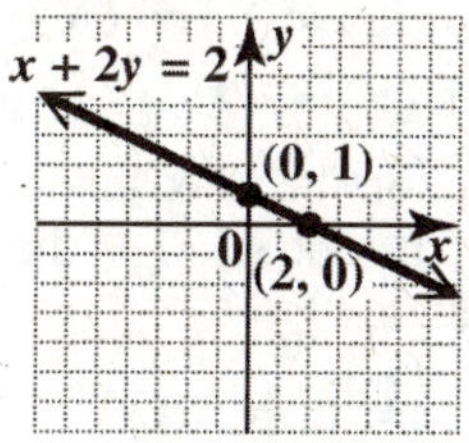

N4. $2x + y = 0$

The x-intercept and the y-intercept are the *same* point, $(0, 0)$. Select two other values for x or y to find two other points on the graph, as in the following table.

x	y
0	0
1	-2
-1	2

Graph the equation by plotting these points and drawing a line through them.

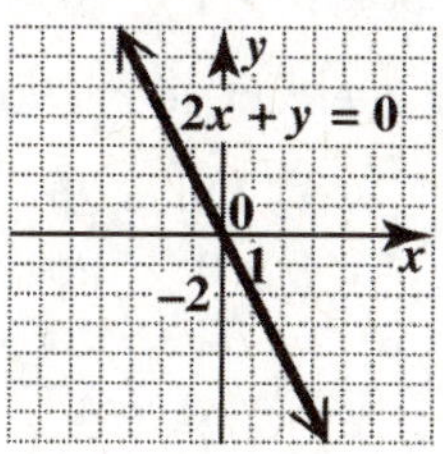

N5. $y = 2$

No matter what value we choose for x, y is always 2, as shown in the table of ordered pairs.

x	y
0	2
-5	2
3	2

The graph is a horizontal line through $(0, 2)$.

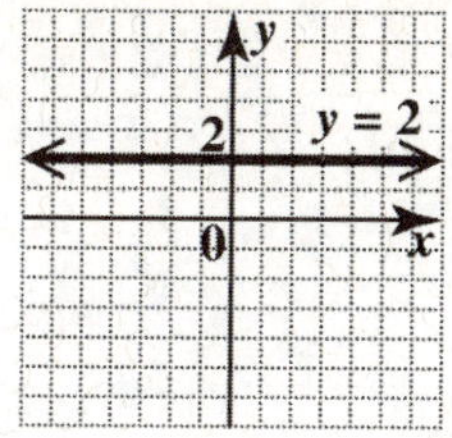

N6. $x + 4 = 0$ or $x = -4$

x is always -4 regardless of the value of y, as shown in the table of ordered pairs.

x	y
-4	0
-4	-5
-4	2

The graph is a vertical line through $(-4, 0)$.

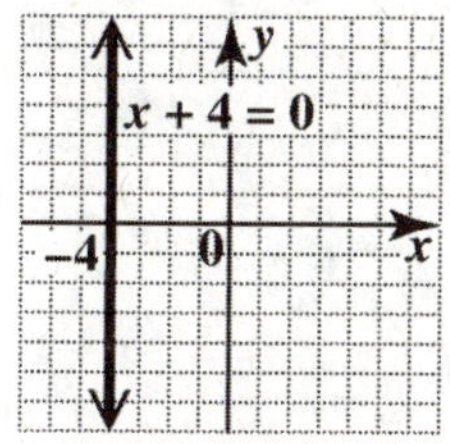

N7. **(a)** $x = 0$ represents 2000, so $x = 6$ represents 2006. On the graph, find 6 on the horizontal axis, move up to the graphed line, then across to the vertical axis. It appears that credit card debt in 2006 was about 875 billion dollars.

(b) To use the equation, substitute 6 for x.

$$\begin{aligned} y &= 32.0x + 684 \\ y &= 32.0(6) + 684 \qquad \textit{Let } x = 6. \\ y &= 876 \text{ billion dollars} \end{aligned}$$

3.2 Section Exercises

1. $x + y = 5$

$(0, \_\_)$, $(\_\_, 0)$, $(2, \_\_)$

If $x = 0$,
$0 + y = 5$
$y = 5$.

If $y = 0$,
$x + 0 = 5$
$x = 5$.

If $x = 2$,
$2 + y = 5$
$y = 3$.

The ordered pairs are $(0, 5)$, $(5, 0)$, and $(2, 3)$. Plot the corresponding points and draw a line through them.

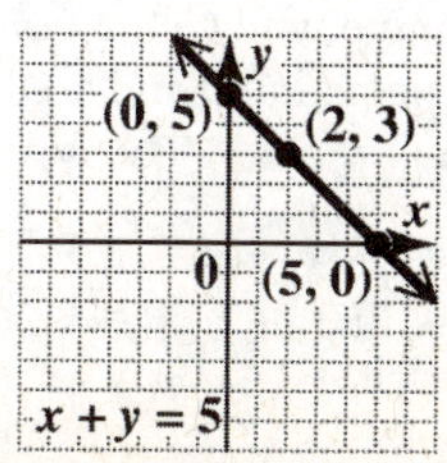

3. $y = \frac{2}{3}x + 1$

$(0, \_\_), (3, \_\_), (-3, \_\_)$

If $x = 0$,
$$y = \tfrac{2}{3}(0) + 1$$
$$y = 0 + 1$$
$$y = 1.$$

If $x = 3$,
$$y = \tfrac{2}{3}(3) + 1$$
$$y = 2 + 1$$
$$y = 3.$$

If $x = -3$,
$$y = \tfrac{2}{3}(-3) + 1$$
$$y = -2 + 1$$
$$y = -1.$$

The ordered pairs are $(0, 1)$, $(3, 3)$, and $(-3, -1)$. Plot the corresponding points and draw a line through them.

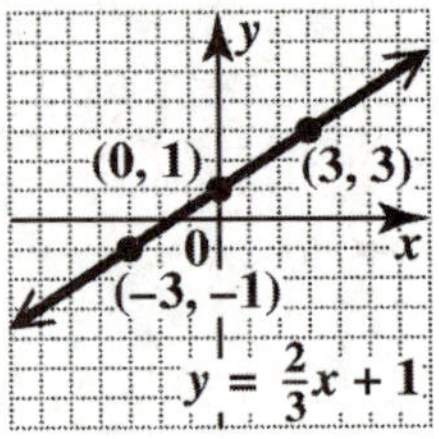

5. $3x = -y - 6$

$(0, \_\_), (\_\_, 0), (-\frac{1}{3}, \_\_)$

If $x = 0$,
$$3(0) = -y - 6$$
$$0 = -y - 6$$
$$y = -6.$$

If $y = 0$,
$$3x = -0 - 6$$
$$3x = -6$$
$$x = -2.$$

If $x = -\frac{1}{3}$,
$$3\left(-\tfrac{1}{3}\right) = -y - 6$$
$$-1 = -y - 6$$
$$y - 1 = -6$$
$$y = -5.$$

The ordered pairs are $(0, -6)$, $(-2, 0)$, and $(-\frac{1}{3}, -5)$. Plot the corresponding points and draw a line through them.

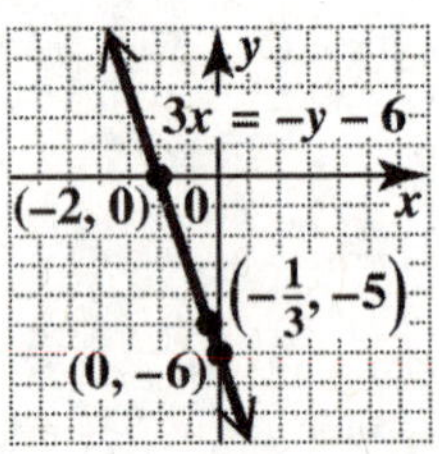

7. **(a)** To determine which equation has y-intercept $(0, -4)$, set x equal to 0 and see which equation is equivalent to $y = -4$. Choice **A** is correct since

$$3x + y = -4$$
$$3(0) + y = -4$$
$$y = -4$$

(b) If the graph of the equation goes through the origin, then substituting 0 for x and 0 for y will result in a true statement. Choice **C** is correct since

$$y = 4x$$
$$0 = 4(0)$$
$$0 = 0$$

is a true statement.

(c) Choice **D** is correct since the graph of $y = 4$ is a horizontal line.

(d) To determine which equation has x-intercept $(4, 0)$, set y equal to 0 and see which equation is equivalent to $x = 4$. Choice **B** is correct since

$$x - 4 = 0$$
$$x = 4$$

9. The dot at 4 on the x-axis is the x-intercept, that is, $(4, 0)$. The dot at -4 on the y-axis is the y-intercept, that is, $(0, -4)$.

11. The dot at -2 on the x-axis is the x-intercept, that is, $(-2, 0)$. The dot at -3 on the y-axis is the y-intercept, that is, $(0, -3)$.

13. To find the x-intercept, let $y = 0$.

$$x - y = 8$$
$$x - 0 = 8$$
$$x = 8$$

The x-intercept is $(8, 0)$.

To find the y-intercept, let $x = 0$.

$$x - y = 8$$
$$0 - y = 8$$
$$-y = 8$$
$$y = -8$$

The y-intercept is $(0, -8)$.

15. To find the x-intercept, let $y = 0$.

$$5x - 2y = 20$$
$$5x - 2(0) = 20$$
$$5x = 20$$
$$x = 4$$

The x-intercept is $(4, 0)$.

To find the y-intercept, let $x = 0$.

$$5x - 2y = 20$$
$$5(0) - 2y = 20$$
$$-2y = 20$$
$$y = -10$$

The y-intercept is $(0, -10)$.

17. To find the x-intercept, let $y = 0$.

$$\begin{aligned} x + 6y &= 0 \\ x + 6(0) &= 0 \\ x + 0 &= 0 \\ x &= 0 \end{aligned}$$

The x-intercept is $(0, 0)$. Since we have found the point with x equal to 0, this is also the y-intercept.

19. To find the x-intercept, let $y = 0$.

$$\begin{aligned} y &= -2x + 4 \\ 0 &= -2x + 4 \\ 2x &= 4 \\ x &= 2 \end{aligned}$$

The x-intercept is $(2, 0)$.

To find the y-intercept, let $x = 0$.

$$\begin{aligned} y &= -2x + 4 \\ y &= -2(0) + 4 \\ y &= 0 + 4 \\ y &= 4 \end{aligned}$$

The y-intercept is $(0, 4)$.

21. To find the x-intercept, let $y = 0$.

$$\begin{aligned} y &= \tfrac{1}{3}x - 2 \\ 0 &= \tfrac{1}{3}x - 2 \\ 2 &= \tfrac{1}{3}x \\ 6 &= x \end{aligned}$$

The x-intercept is $(6, 0)$.

To find the y-intercept, let $x = 0$.

$$\begin{aligned} y &= \tfrac{1}{3}x - 2 \\ y &= \tfrac{1}{3}(0) - 2 \\ y &= 0 - 2 \\ y &= -2 \end{aligned}$$

The y-intercept is $(0, -2)$.

23. To find the x-intercept, let $y = 0$.

$$\begin{aligned} 2x - 3y &= 0 \\ 2x - 3(0) &= 0 \\ 2x - 0 &= 0 \\ 2x &= 0 \\ x &= 0 \end{aligned}$$

The x-intercept is $(0, 0)$. Since we have found the point with x equal to 0, this is also the y-intercept.

25. $x - 4 = 0$ is equivalent to $x = 4$. This is an equation of a vertical line. Its x-intercept is $(4, 0)$ and there is no y-intercept.

27. $y = 2.5$ is an equation of a horizontal line. Its y-intercept is $(0, 2.5)$ and there is no x-intercept.

29. **(a)** The graph of $x = -2$ is a vertical line with x-intercept $(-2, 0)$, which is choice **D**.

(b) The graph of $y = -2$ is a horizontal line with y-intercept $(0, -2)$, which is choice **C**.

(c) The graph of $x = 2$ is a vertical line with x-intercept $(2, 0)$, which is choice **B**.

(d) The graph of $y = 2$ is a horizontal line with y-intercept $(0, 2)$, which is choice **A**.

31. Begin by finding the intercepts.

$$\begin{aligned} x &= y + 2 \\ x &= 0 + 2 \quad \textit{Let } y = 0. \\ x &= 2 \end{aligned} \qquad \begin{aligned} x &= y + 2 \\ 0 &= y + 2 \quad \textit{Let } x = 0. \\ -2 &= y \end{aligned}$$

The x-intercept is $(2, 0)$ and the y-intercept is $(0, -2)$. To find a third point, choose $y = 1$.

$$\begin{aligned} x &= y + 2 \\ x &= 1 + 2 \quad \textit{Let } y = 1. \\ x &= 3 \end{aligned}$$

This gives the ordered pair $(3, 1)$. Plot $(2, 0)$, $(0, -2)$, and $(3, 1)$ and draw a line through them.

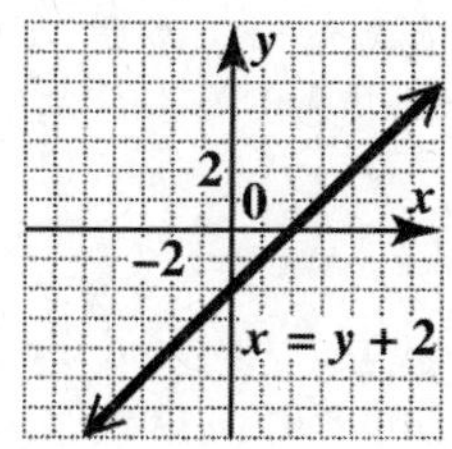

33. Find the intercepts.

$$\begin{aligned} x - y &= 4 \\ x - 0 &= 4 \quad \textit{Let } y = 0. \\ x &= 4 \end{aligned} \qquad \begin{aligned} x - y &= 4 \\ 0 - y &= 4 \quad \textit{Let } x = 0. \\ y &= -4 \end{aligned}$$

The x-intercept is $(4, 0)$ and the y-intercept is $(0, -4)$. To find a third point, choose $y = 1$.

$$\begin{aligned} x - y &= 4 \\ x - 1 &= 4 \quad \textit{Let } y = 1. \\ x &= 5 \end{aligned}$$

This gives the ordered pair $(5, 1)$. Plot $(4, 0)$, $(0, -4)$, and $(5, 1)$ and draw a line through them.

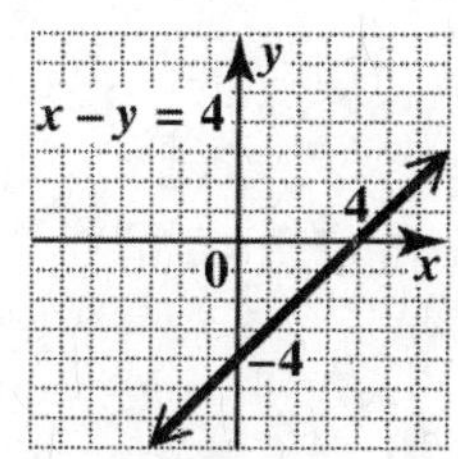

35. Find the intercepts.

$$\begin{aligned} 2x + y &= 6 \\ 2x + 0 &= 6 \quad \textit{Let } y = 0. \\ 2x &= 6 \\ x &= 3 \end{aligned} \qquad \begin{aligned} 2x + y &= 6 \\ 2(0) + y &= 6 \quad \textit{Let } x = 0. \\ 0 + y &= 6 \\ y &= 6 \end{aligned}$$

The x-intercept is $(3, 0)$ and the y-intercept is $(0, 6)$. To find a third point, choose $x = 1$.

$$\begin{aligned} 2x + y &= 6 \\ 2(1) + y &= 6 \quad \textit{Let } x = 1. \\ 2 + y &= 6 \\ y &= 4 \end{aligned}$$

This gives the ordered pair $(1, 4)$. Plot $(3, 0)$, $(0, 6)$, and $(1, 4)$ and draw a line through them.

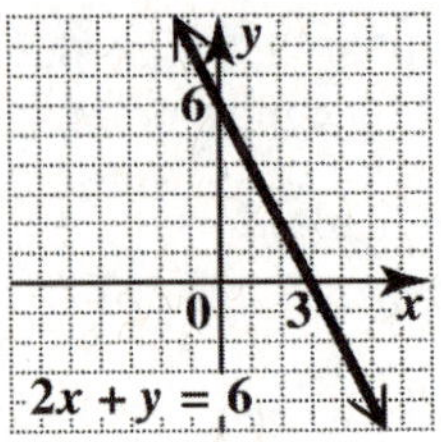

37. Find the intercepts.

$$\begin{aligned} y &= 2x - 5 \\ 0 &= 2x - 5 \quad \textit{Let } y = 0. \\ 5 &= 2x \\ \tfrac{5}{2} &= x \end{aligned} \qquad \begin{aligned} y &= 2x - 5 \\ y &= 2(0) - 5 \quad \textit{Let } x = 0. \\ y &= 0 - 5 \\ y &= -5 \end{aligned}$$

The x-intercept is $\left(\frac{5}{2}, 0\right)$ and the y-intercept is $(0, -5)$. To find a third point, choose $x = 1$.

$$\begin{aligned} y &= 2x - 5 \\ y &= 2(1) - 5 \quad \textit{Let } x = 1. \\ y &= 2 - 5 \\ y &= -3 \end{aligned}$$

This gives the ordered pair $(1, -3)$. Plot $\left(\frac{5}{2}, 0\right)$, $(0, -5)$, and $(1, -3)$ and draw a line through them.

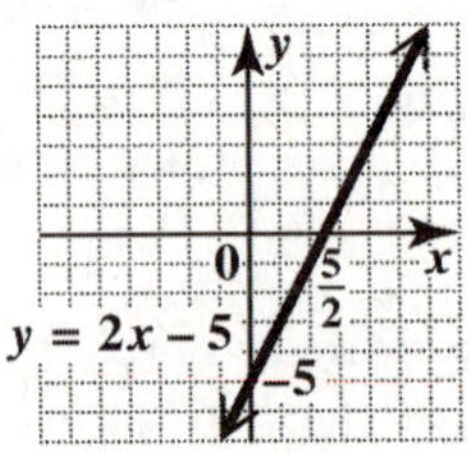

39. Find the intercepts.

$$\begin{aligned} 3x + 7y &= 14 \\ 3x + 7(0) &= 14 \quad \textit{Let } y = 0. \\ 3x + 0 &= 14 \\ 3x &= 14 \\ x &= \tfrac{14}{3} \end{aligned}$$

$$\begin{aligned} 3x + 7y &= 14 \\ 3(0) + 7y &= 14 \quad \textit{Let } x = 0. \\ 0 + 7y &= 14 \\ 7y &= 14 \\ y &= 2 \end{aligned}$$

The x-intercept is $\left(\frac{14}{3}, 0\right)$ and the y-intercept is $(0, 2)$. To find a third point, choose $x = 2$.

$$\begin{aligned} 3x + 7y &= 14 \\ 3(2) + 7y &= 14 \quad \textit{Let } x = 2. \\ 6 + 7y &= 14 \\ 7y &= 8 \\ y &= \tfrac{8}{7} \end{aligned}$$

This gives the ordered pair $\left(2, \frac{8}{7}\right)$. Plot $\left(\frac{14}{3}, 0\right)$, $(0, 2)$, and $\left(2, \frac{8}{7}\right)$. Writing $\frac{14}{3}$ as the mixed number $4\frac{2}{3}$ and $\frac{8}{7}$ as $1\frac{1}{7}$ will be helpful for plotting. Draw a line through these three points.

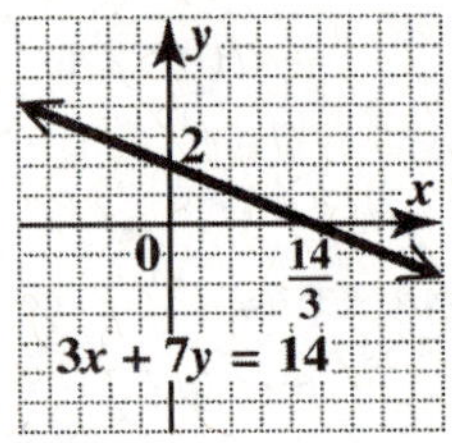

41. $y = -\frac{3}{4}x + 3$

x-intercept ($y = 0$)

$0 = -\frac{3}{4}x + 3$

$\frac{3}{4}x = 3$

$x = \frac{4}{3}(3) = 4$ **(4, 0)**

y-intercept ($x = 0$)

$y = -\frac{3}{4}(0) + 3$

$y = 0 + 3$

$y = 3$ **(0, 3)**

third point ($x = 2$)

$y = -\frac{3}{4}(2) + 3$

$y = -\frac{3}{2} + 3$

$y = \frac{3}{2}$ $\left(\mathbf{2, \frac{3}{2}}\right)$

Plot $(4, 0)$, $(0, 3)$, and $\left(2, \frac{3}{2}\right)$ and draw a line through these points.

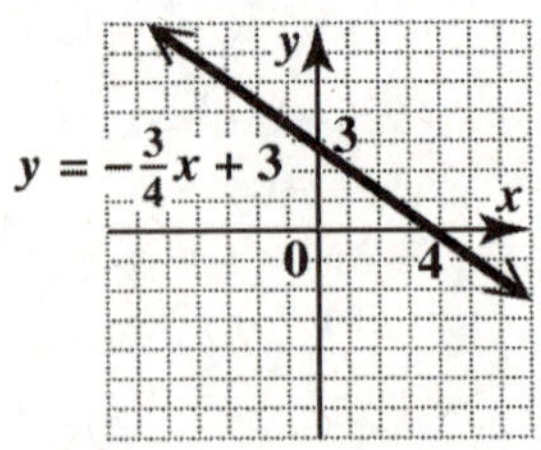

43. $y - 2x = 0$

If $y = 0$, $x = 0$. Both intercepts are the origin, $(0, 0)$. Find two additional points.

$$\begin{aligned} y - 2x &= 0 \\ y - 2(1) &= 0 \quad \textit{Let } x = 1. \\ y - 2 &= 0 \\ y &= 2 \end{aligned}$$

$$y - 2x = 0$$
$$y - 2(-3) = 0 \quad \textit{Let } x = -3.$$
$$y + 6 = 0$$
$$y = -6$$

Plot $(0, 0)$, $(1, 2)$, and $(-3, -6)$ and draw a line through them.

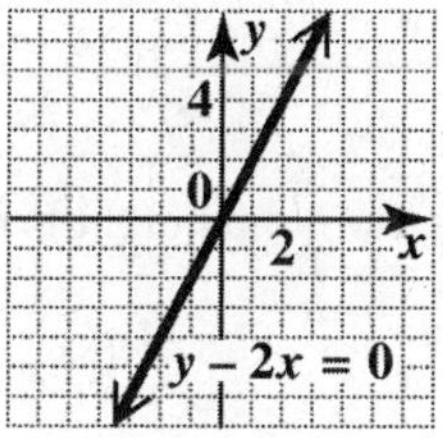

45. $y = -6x$

Find three points on the line.

If $x = 0$, $y = -6(0) = 0$.

If $x = 1$, $y = -6(1) = -6$.

If $x = -1$, $y = -6(-1) = 6$.

Plot $(0, 0)$, $(1, -6)$, and $(-1, 6)$ and draw a line through these points.

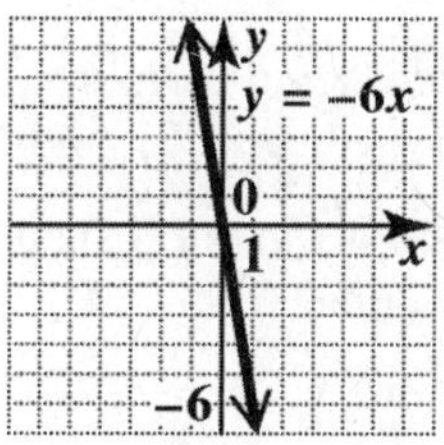

47. $y = -1$

For any value of x, the value of y is -1. Three ordered pairs are $(-4, -1)$, $(0, -1)$, and $(3, -1)$. Plot these points and draw a line through them. The graph is a horizontal line.

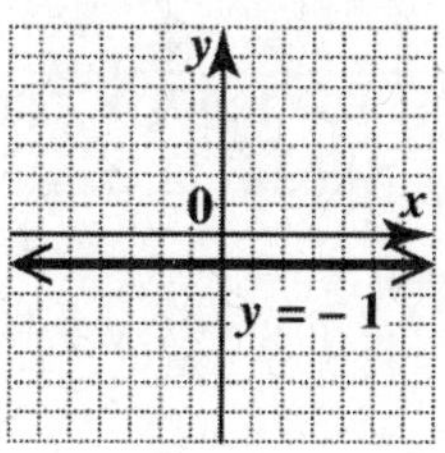

49. $x + 2 = 0$

$$x = -2$$

For any value of y, the value of x is -2. Three ordered pairs are $(-2, 3)$, $(-2, 0)$, and $(-2, -4)$. Plot these points and draw a line through them. The graph is a vertical line.

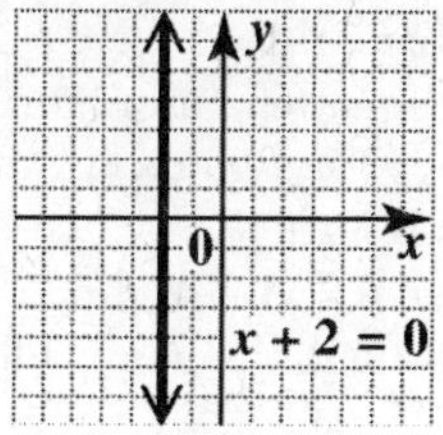

51. $-3y = 15$

$$y = \frac{15}{-3} = -5$$

For any value of x, the value of y is -5. Three ordered pairs are $(-2, -5)$, $(0, -5)$, and $(1, -5)$. Plot these points and draw a line through them. The graph is a horizontal line.

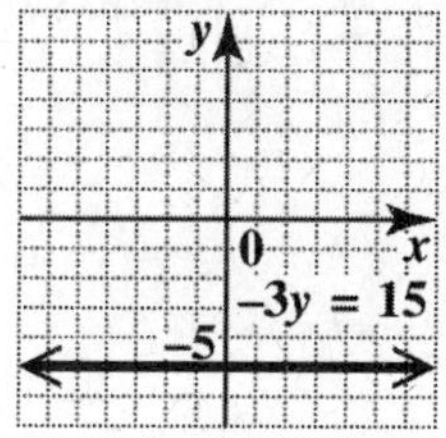

53. $x + 2 = 8$

$$x = 6$$

For any value of y, the value of x is 6. Three ordered pairs are $(6, -2)$, $(6, 0)$, and $(6, 1)$. Plot these points and draw a line through them. The graph is a vertical line.

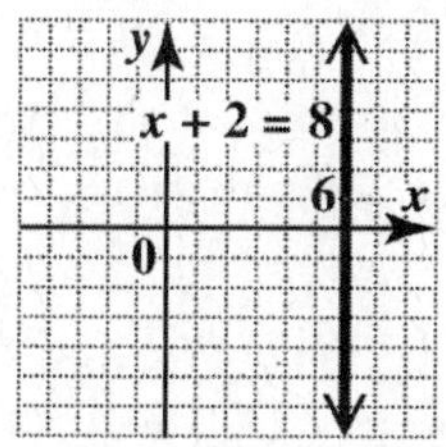

55. $3x = y - 9$

If $x = 0$,	If $y = 0$,
$3(0) = y - 9$	$3x = 0 - 9$
$0 = y - 9$	$3x = -9$
$9 = y$.	$x = -3$.

The graph of this equation is a line with x-intercept $(-3, 0)$ and y-intercept $(0, 9)$.

57. $x - 10 = 1$

$$x = 11 \quad \textit{Add 10.}$$

The graph of this equation is a vertical line with x-intercept $(11, 0)$.

59. $3y = -6$

$$y = -2 \quad \textit{Divide by 3.}$$

The graph of this equation is a horizontal line with y-intercept $(0, -2)$.

61. $2x = 4y$ can be written as $2x - 4y = 0$, so the graph of this equation passes through the origin, $(0, 0)$, which is the x-intercept and the y-intercept. If we let $x = 2$, then $4 = 4y$, so $y = 1$; and if we let $x = 4$, then $8 = 4y$, so $y = 2$. Thus, the graph also passes through the points $(2, 1)$ and $(4, 2)$.

63. The points $(3, 5)$, $(3, 0)$, and $(3, -3)$ all have x-coordinate 3, so $x = 3$ is the equation of the line.

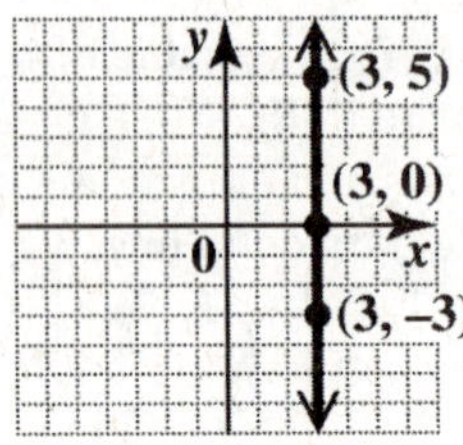

65. The points $(-3, -3)$, $(0, -3)$, and $(4, -3)$ all have y-coordinate -3, so $y = -3$ is the equation of the line.

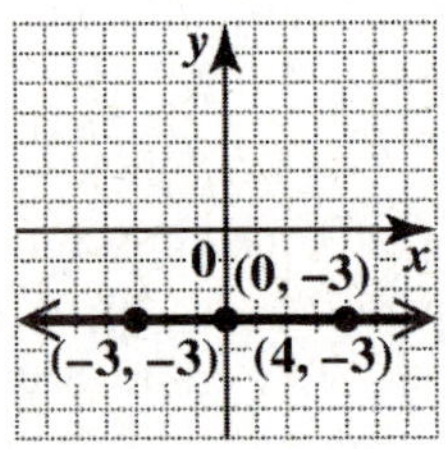

67. **(a)** $y = 5.5x - 220$

$y = 5.5(62) - 220 = 121$ *Let x = 62.*
$y = 5.5(66) - 220 = 143$ *Let x = 66.*
$y = 5.5(72) - 220 = 176$ *Let x = 72.*

The approximate weights of men whose heights are 62 in., 66 in., and 72 in. are 121 lb, 143 lb, and 176 lb, respectively.

(b) The three ordered pairs from part (a) are $(62, 121)$, $(66, 143)$, and $(72, 176)$.

(c) Plot the points $(62, 121)$, $(66, 143)$, and $(72, 176)$ and connect them with a smooth line.

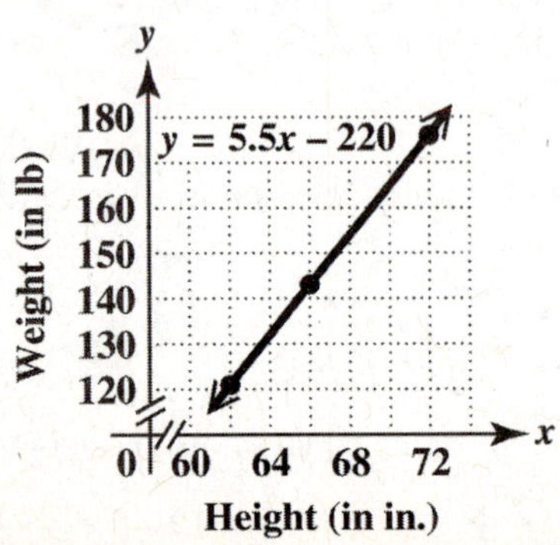

(d) Locate 155 on the vertical scale, then move across to the line, then down to the horizontal scale. From the graph, the height of a man who weighs 155 lb is about 68 in. Now substitute 155 for y in the equation.

$$\begin{aligned} y &= 5.5x - 220 \\ 155 &= 5.5x - 220 \\ 375 &= 5.5x \\ x &= \tfrac{375}{5.5} \approx 68.18 \end{aligned}$$

From the equation, the height is 68 in. to the nearest inch.

69. **(a)** Let $x = 50$. Let $x = 100$.

$y = 0.75x + 25$ $y = 0.75x + 25$
$y = 0.75(50) + 25$ $y = 0.75(100) + 25$
$y = 62.50$ $y = 100$

Thus, 50 posters cost \$62.50 and 100 posters cost \$100.

(b) Let $y = 175$.

$$\begin{aligned} 175 &= 0.75x + 25 \\ 150 &= 0.75x \\ x &= \frac{150}{0.75} = 200 \end{aligned}$$

Thus, 200 posters cost \$175.

(c) The three ordered pairs from parts (a) and (b) are $(50, 62.50)$, $(100, 100)$, and $(200, 175)$.

(d) Plot the points in part (c) and connect them with a smooth line.

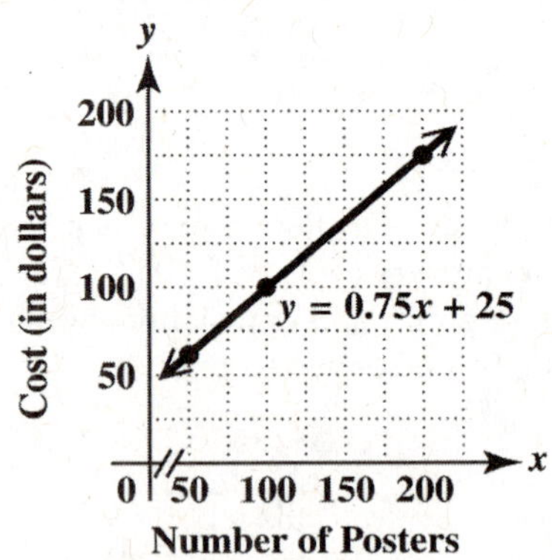

71. **(a)** At $x = 0$, $y = 30{,}000$.
The initial value of the SUV is \$30,000.

(b) At $x = 3$, $y = 15{,}000$.
$30{,}000 - 15{,}000 = 15{,}000$
The depreciation after the first 3 years is \$15,000.

(c) The line connecting consecutive years from 0 to 1, 1 to 2, 2 to 3, and so on, drops 5000 for each segment. Therefore, the annual depreciation in each of the first 5 years is \$5000.

(d) $(5, 5000)$ means after 5 years the SUV has a value of \$5000.

73. **(a)** $y = 3.018x + 72.52$

For 2000, let $x = 0$.
$y = 3.018(0) + 72.52 = 72.52$

For 2004, let $x = 4$.
$y = 3.018(4) + 72.52 = 84.592$

For 2006, let $x = 6$.
$y = 3.018(6) + 72.52 = 90.628$

The approximate sales for 2000, 2004, and 2006 are \$73 billion, \$85 billion, and \$91 billion, respectively.

(b) Locate 0, 4, and 6 on the horizontal scale, then find the corresponding value on the vertical scale.

The approximate sales for 2000, 2004, and 2006 are \$74 billion, \$86 billion, and \$91 billion, respectively.

(c) The corresponding values are quite close.

75. $\dfrac{4-2}{8-5} = \dfrac{2}{3}$

77. $\dfrac{-2-(-4)}{3-(-1)} = \dfrac{-2+4}{3+1} = \dfrac{2}{4} = \dfrac{1}{2}$

3.3 The Slope of a Line

3.3 Now Try Exercises

N1. The given points are $(-2,-2)$ and $(-1,1)$.

$$\text{slope} = \frac{\text{change in } y \text{ (rise)}}{\text{change in } x \text{ (run)}} = \frac{1-(-2)}{-1-(-2)} = \frac{1+2}{-1+2} = \frac{3}{1} = 3$$

N2. Use the slope formula with $(4,-5)$ and $(-2,-4)$.

$$\text{slope } m = \frac{\text{change in } y}{\text{change in } x} = \frac{y_2 - y_1}{x_2 - x_1} = \frac{-4-(-5)}{-2-4} = \frac{1}{-6} = -\frac{1}{6}$$

N3. Use the slope formula with $(1,-3)$ and $(4,-3)$.

$$\text{slope } m = \frac{\text{change in } y}{\text{change in } x} = \frac{-3-(-3)}{4-1} = \frac{0}{3} = 0$$

N4. Use the slope formula with $(-2,1)$ and $(-2,-4)$.

$$\text{slope } m = \frac{\text{change in } y}{\text{change in } x} = \frac{-4-1}{-2-(-2)} = \frac{-5}{0},$$

which is *undefined*.

N5. Solve the equation for y.

$$\begin{aligned} 3x + 5y &= -1 \\ 5y &= -3x - 1 \quad \textit{Subtract } 3x. \\ y &= -\tfrac{3}{5}x - \tfrac{1}{5} \quad \textit{Divide by } 5. \end{aligned}$$

The slope is given by the coefficient of x, so the slope is $-\frac{3}{5}$.

N6. Find the slope of each line by first solving each equation for y.

$$\begin{aligned} 2x - 3y &= 1 \\ -3y &= -2x + 1 \\ y &= \tfrac{2}{3}x - \tfrac{1}{3} \end{aligned} \qquad \begin{aligned} 4x + 6y &= 5 \\ 6y &= -4x + 5 \\ y &= -\tfrac{2}{3}x + \tfrac{5}{6} \end{aligned}$$

Slope is $\frac{2}{3}$. Slope is $-\frac{2}{3}$.

The slopes are not equal, so the lines are not parallel. The product of the slopes is not -1, so the lines are not perpendicular. Thus, the lines are *neither* parallel nor perpendicular.

3.3 Section Exercises

1. The indicated points have coordinates $(-1,-4)$ and $(1,4)$.

$$\text{slope} = \frac{\text{change in } y \text{ (rise)}}{\text{change in } x \text{ (run)}} = \frac{4-(-4)}{1-(-1)} = \frac{8}{2} = 4$$

3. The indicated points have coordinates $(-3,2)$ and $(5,-2)$.

$$\text{slope} = \frac{\text{change in } y \text{ (rise)}}{\text{change in } x \text{ (run)}} = \frac{-2-2}{5-(-3)} = \frac{-4}{8} = -\frac{1}{2}$$

5. The indicated points have coordinates $(-2,-4)$ and $(4,-4)$.

$$\text{slope} = \frac{\text{change in } y \text{ (rise)}}{\text{change in } x \text{ (run)}} = \frac{-4-(-4)}{4-(-2)} = \frac{-4+4}{4+2} = \frac{0}{6} = 0$$

7. Rise is the vertical change between two different points on a line.

Run is the horizontal change between two different points on a line.

9. **(a)** The indicated points have coordinates $(-1,2)$ and $(2,0)$.

$$\text{slope} = \frac{\text{change in } y \text{ (rise)}}{\text{change in } x \text{ (run)}} = \frac{0-2}{2-(-1)} = \frac{-2}{3} = -\frac{2}{3} \quad \text{(choice } \mathbf{C}\text{)}$$

(b) The indicated points have coordinates $(-1,2)$ and $(-4,0)$.

$$\text{slope} = \frac{\text{change in } y \text{ (rise)}}{\text{change in } x \text{ (run)}} = \frac{0-2}{-4-(-1)} = \frac{-2}{-4+1} = \frac{-2}{-3} = \frac{2}{3} \quad \text{(choice } \mathbf{A}\text{)}$$

(c) The indicated points have coordinates $(-1, 2)$ and $(1, -1)$.

$$\text{slope} = \frac{\text{change in } y \text{ (rise)}}{\text{change in } x \text{ (run)}}$$

$$= \frac{-1-2}{1-(-1)} = \frac{-3}{2} = -\frac{3}{2} \quad \text{(choice } \mathbf{D}\text{)}$$

(d) The indicated points have coordinates $(-1, 2)$ and $(-3, -1)$.

$$\text{slope} = \frac{\text{change in } y \text{ (rise)}}{\text{change in } x \text{ (run)}}$$

$$= \frac{-1-2}{-3-(-1)} = \frac{-3}{-2} = \frac{3}{2} \quad \text{(choice } \mathbf{B}\text{)}$$

11. Negative slope

Sketches will vary. The line must fall from left to right. One such line is shown in the following graph.

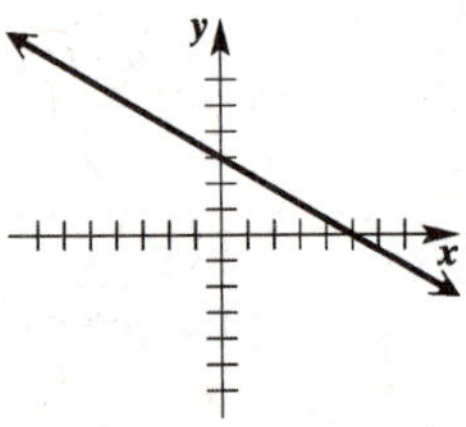

13. Undefined slope

Sketches will vary. The line must be vertical. One such line is shown in the following graph.

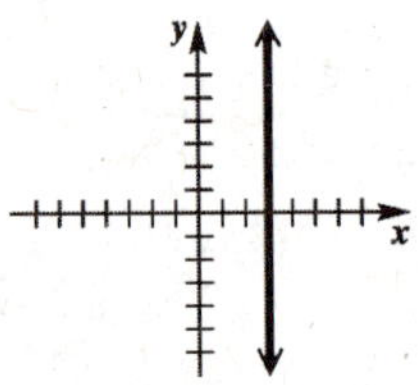

15. **(a)** Because the line *falls* from left to right, its slope is *negative.*

(b) Because the line intersects the y-axis *at* the origin, the y-value of its y-intercept is *zero.*

17. **(a)** Because the line *rises* from left to right, its slope is *positive.*

(b) Because the line intersects the y-axis *below* the origin, the y-value of its y-intercept is *negative.*

19. **(a)** The line is *horizontal*, so its slope is *zero.*

(b) The line intersects the y-axis *below* the origin, so the y-value of its y-intercept is *negative.*

21. The slope (or grade) of the hill is the ratio of the rise to the run, or the ratio of the vertical change to the horizontal change. Since the rise is 32 and the run is 108, the slope is

$$\frac{32}{108} = \frac{8 \cdot 4}{27 \cdot 4} = \frac{8}{27}.$$

23. $$\text{slope} = \frac{\text{vertical change (rise)}}{\text{horizontal change (run)}}$$

$$= \tfrac{-8}{12} \quad \text{("drops" indicates negative)}$$

$$= -\tfrac{2}{3}$$

25. Because he found the difference $3 - 5 = -2$ in the numerator, he should have subtracted in the same order in the denominator to get $-1 - 2 = -3$. The correct slope is $\frac{-2}{-3} = \frac{2}{3}$. Note that the student's slope and the correct slope are opposites of one another.

27. Use the slope formula with $(1, -2) = (x_1, y_1)$ and $(-3, -7) = (x_2, y_2)$.

$$\text{slope } m = \frac{\text{change in } y}{\text{change in } x} = \frac{y_2 - y_1}{x_2 - x_1}$$

$$= \frac{-7-(-2)}{-3-1} = \frac{-5}{-4} = \frac{5}{4}$$

29. Use the slope formula with $(0, 3) = (x_1, y_1)$ and $(-2, 0) = (x_2, y_2)$.

$$\text{slope } m = \frac{\text{change in } y}{\text{change in } x} = \frac{y_2 - y_1}{x_2 - x_1}$$

$$= \frac{0-3}{-2-0} = \frac{-3}{-2} = \frac{3}{2}$$

31. Use the slope formula with $(4, 3) = (x_1, y_1)$ and $(-6, 3) = (x_2, y_2)$.

$$\text{slope } m = \frac{\text{change in } y}{\text{change in } x} = \frac{y_2 - y_1}{x_2 - x_1}$$

$$= \frac{3-3}{-6-4} = \frac{0}{-10} = 0$$

33. Use the slope formula with $(-2, 4) = (x_1, y_1)$ and $(-3, 7) = (x_2, y_2)$.

$$\text{slope } m = \frac{\text{change in } y}{\text{change in } x} = \frac{y_2 - y_1}{x_2 - x_1}$$

$$= \frac{7-4}{-3-(-2)} = \frac{3}{-1} = -3$$

35. Use the slope formula with $(-12, 3) = (x_1, y_1)$ and $(-12, -7) = (x_2, y_2)$.

$$\text{slope } m = \frac{\text{change in } y}{\text{change in } x} = \frac{y_2 - y_1}{x_2 - x_1}$$

$$= \frac{-7-3}{-12-(-12)} = \frac{-10}{0},$$

which is *undefined.*

37. Use the slope formula with $(4.8, 2.5) = (x_1, y_1)$ and $(3.6, 2.2) = (x_2, y_2)$.

$$\text{slope } m = \frac{\text{change in } y}{\text{change in } x} = \frac{y_2 - y_1}{x_2 - x_1}$$

$$= \frac{2.2-2.5}{3.6-4.8} = \frac{-0.3}{-1.2} = \frac{1}{4}$$

39. Use the slope formula with $\left(-\frac{7}{5}, \frac{3}{10}\right) = (x_1, y_1)$ and $\left(\frac{1}{5}, -\frac{1}{2}\right) = (x_2, y_2)$.

$$\text{slope } m = \frac{\text{change in } y}{\text{change in } x} = \frac{y_2 - y_1}{x_2 - x_1}$$

$$= \frac{-\frac{1}{2} - \frac{3}{10}}{\frac{1}{5} - \left(-\frac{7}{5}\right)} = \frac{-\frac{5}{10} - \frac{3}{10}}{\frac{1}{5} + \frac{7}{5}} = \frac{-\frac{8}{10}}{\frac{8}{5}}$$

$$= \left(-\frac{8}{10}\right)\left(\frac{5}{8}\right) = -\frac{5}{10} = -\frac{1}{2}$$

41. $y = 5x + 12$

Since the equation is already solved for y, the slope is given by the coefficient of x, which is 5. Thus, the slope of the line is 5.

43. Solve the equation for y.

$$4y = x + 1$$
$$y = \frac{1}{4}x + \frac{1}{4} \quad \textit{Divide by 4.}$$

The slope of the line is given by the coefficient of x, so the slope is $\frac{1}{4}$.

45. Solve the equation for y.

$$3x - 2y = 3$$
$$-2y = -3x + 3 \quad \textit{Subtract 3x.}$$
$$y = \frac{3}{2}x - \frac{3}{2} \quad \textit{Divide by –2.}$$

The slope of the line is given by the coefficient of x, so the slope is $\frac{3}{2}$.

47. Solve the equation for y.

$$-3x + 2y = 5$$
$$2y = 3x + 5 \quad \textit{Add 3x.}$$
$$y = \frac{3}{2}x + \frac{5}{2} \quad \textit{Divide by 2.}$$

The slope of the line is given by the coefficient of x, so the slope is $\frac{3}{2}$.

49. $y = -5$

This is an equation of a horizontal line. Its slope is 0. (This equation may be rewritten in the form $y = 0x - 5$, where the coefficient of x gives the slope.)

51. $x = 6$

This is an equation of a vertical line. Its slope is *undefined.*

53. Solve the equation for y.

$$3x + y = 7$$
$$y = -3x + 7 \quad \textit{Subtract 3x.}$$

The slope of the given line is -3, so the slope of a line whose graph is parallel to the graph of the given line is also -3.

The slope of a line whose graph is perpendicular to the graph of the given line is the negative reciprocal of -3, that is, $\frac{1}{3}$.

55. If two lines are both vertical or both horizontal, they are *parallel.* Choice **A** is correct.

57. Find the slope of each line by solving the equations for y.

$$2x + 5y = 4$$
$$5y = -2x + 4 \quad \textit{Subtract 2x.}$$
$$y = -\frac{2}{5}x + \frac{4}{5} \quad \textit{Divide by 5.}$$

The slope of the first line is $-\frac{2}{5}$.

$$4x + 10y = 1$$
$$10y = -4x + 1 \quad \textit{Subtract 4x.}$$
$$y = -\frac{4}{10}x + \frac{1}{10} \quad \textit{Divide by 10.}$$
$$y = -\frac{2}{5}x + \frac{1}{10} \quad \textit{Lowest terms}$$

The slope of the second line is $-\frac{2}{5}$.

The slopes are equal, so the lines are *parallel.*

59. Find the slope of each line by solving the equations for y.

$$8x - 9y = 6$$
$$-9y = -8x + 6 \quad \textit{Subtract 8x.}$$
$$y = \frac{8}{9}x - \frac{2}{3} \quad \textit{Divide by –9.}$$

The slope of the first line is $\frac{8}{9}$.

$$8x + 6y = -5$$
$$6y = -8x - 5 \quad \textit{Subtract 8x.}$$
$$y = -\frac{4}{3}x - \frac{5}{6} \quad \textit{Divide by 6.}$$

The slope of the second line is $-\frac{4}{3}$.

The slopes are not equal, so the lines are not parallel. The slopes are not negative reciprocals (the negative reciprocal of $\frac{8}{9}$ is $-\frac{9}{8}$), so the lines are not perpendicular. Thus, the lines are *neither* parallel nor perpendicular.

61. Find the slope of each line by solving the equations for y.

$$3x - 2y = 6$$
$$-2y = -3x + 6 \quad \textit{Subtract 3x.}$$
$$y = \frac{3}{2}x - 3 \quad \textit{Divide by –2.}$$

The slope of the first line is $\frac{3}{2}$.

$$2x + 3y = 3$$
$$3y = -2x + 3 \quad \textit{Subtract 2x.}$$
$$y = -\frac{2}{3}x + 1 \quad \textit{Divide by 3.}$$

The slope of the second line is $-\frac{2}{3}$.

The product of the slopes is

$$\frac{3}{2}\left(-\frac{2}{3}\right) = -1,$$

so the lines are *perpendicular*.

63. Find the slope of each line by solving the equations for y.

$$\begin{aligned} 5x - y &= 1 \\ -y &= -5x + 1 && \textit{Subtract } 5x. \\ y &= 5x - 1 && \textit{Divide by } -1. \end{aligned}$$

The slope of the first line is 5.

$$\begin{aligned} x - 5y &= -10 \\ -5y &= -x - 10 && \textit{Subtract } x. \\ y &= \tfrac{1}{5}x + 2 && \textit{Divide by } -5. \end{aligned}$$

The slope of the second line is $\frac{1}{5}$.

The slopes are not equal, so the lines are not parallel. The slopes are not negative reciprocals (the negative reciprocal of 5 is $-\frac{1}{5}$), so the lines are not perpendicular. Thus, the lines are *neither* parallel nor perpendicular.

♦♦♦ Relating Concepts 65–70 ♦♦♦

65. We use the points with coordinates $(1990, 11{,}338)$ and $(2005, 14{,}818)$.

$$\begin{aligned} m &= \frac{14{,}818 \text{ thousand} - 11{,}338 \text{ thousand}}{2005 - 1990} \\ &= \frac{3480 \text{ thousand}}{15} \\ &= 232 \text{ thousand} \quad \text{or} \quad 232{,}000. \end{aligned}$$

66. The slope of the line in Figure A is *positive*. This means that during the period represented, enrollment *increased* in grades 9–12.

67. The increase is approximately 232,000 students per year.

68. We use the points with coordinates $(1990, 20)$ and $(2007, 3.8)$.

$$\begin{aligned} m &= \frac{3.8 - 20}{2007 - 1990} \\ &= \frac{-16.2}{17} \\ &\approx -0.95 \text{ students per computer} \end{aligned}$$

69. The slope of the line in Figure B is *negative*. This means that the number of students per computer *decreased* during the period shown.

70. The decrease is 0.95 students per computer per year.

71. **(a)** The ordered pairs that represent music purchases are $(2004, 817)$ and $(2008, 1513)$.

(b) $m = \dfrac{1513 - 817}{2008 - 2004} = \dfrac{696}{4} = 174$

(c) Music purchases increased by 696 million units in 4 years, or 174 million units per year.

73. Two of the ordered pairs have coordinates $(-12, -0.8)$ and $(0, 4)$. Therefore, the slope is

$$m = \frac{4 - (-0.8)}{0 - (-12)} = \frac{4.8}{12} = 0.4.$$

75. From the table, when $X = 0$, $Y_1 = 4$. Therefore, the y-intercept is $(0, 4)$.

77.
$$\begin{aligned} 2x + 5y &= 15 \\ 5y &= -2x + 15 \\ y &= -\tfrac{2}{5}x + 3 \end{aligned}$$

79.
$$\begin{aligned} 10x &= 30 + 3y \\ 10x - 30 &= 3y \\ \tfrac{10}{3}x - 10 &= y \end{aligned}$$

81.
$$\begin{aligned} y - (-8) &= 2(x - 4) \\ y + 8 &= 2x - 8 \\ y &= 2x - 16 \end{aligned}$$

3.4 Writing and Graphing Equations of Lines

3.4 Now Try Exercises

N1. **(a)** $y = -\frac{3}{5}x - 9$

The slope is the coefficient of x, that is, $-\frac{3}{5}$. The y-intercept has x-coordinate 0 and y-coordinate -9, that is, $(0, -9)$.

(b) $y = -\frac{x}{3} + \frac{7}{3}$ or $y = -\frac{1}{3}x + \frac{7}{3}$

The slope is $-\frac{1}{3}$ and the y-intercept is $(0, \frac{7}{3})$.

N2. The slope-intercept form of the equation of a line with slope -4 and y-intercept $(0, 2)$ is

$$\begin{aligned} y &= mx + b \\ y &= -4x + 2. \end{aligned}$$

N3. *Step 1*
Begin by solving for y.

$$\begin{aligned} 3x + 2y &= 8 && \textit{Given equation} \\ 2y &= -3x + 8 && \textit{Subtract } 3x. \\ y &= -\tfrac{3}{2}x + 4 && \textit{Divide by } 2. \end{aligned}$$

Step 2
The y-intercept is $(0, 4)$. Graph this point.

Step 3
The slope is $-\dfrac{3}{2} = \dfrac{\text{change in } y \text{ (rise)}}{\text{change in } x \text{ (run)}}$.

Starting at the y-intercept, we count 3 units down and 2 units right to obtain another point on the graph, $(2, 1)$.

Step 4
Draw the line through the points $(0, 4)$ and $(2, 1)$ to obtain the graph of $3x + 2y = 8$.

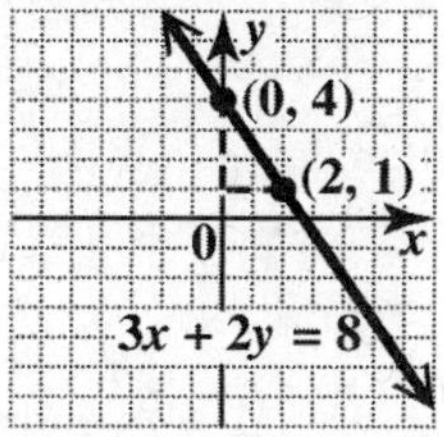

N4. Graph the line through $(-3,-4)$ with slope $\frac{5}{2}$.

To graph the line, remember the slope is

$$m=\frac{\text{change in } y \text{ (rise)}}{\text{change in } x \text{ (run)}}=\frac{5}{2}.$$

Locate $(-3,-4)$. Count 5 units up and 2 units to the right. Draw a line through this point, $(-1,1)$, and $(-3,-4)$.

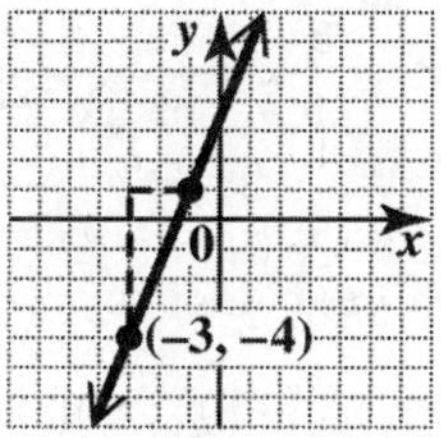

N5.

$$\begin{aligned} y &= mx+b && \textit{slope-intercept form}\\ 1 &= 3(-2)+b && \textit{Let } x=-2,\ y=1, \textit{ and } m=3.\\ 1 &= -6+b && \textit{Multiply.}\\ 7 &= b && \textit{Add 6.} \end{aligned}$$

The y-intercept is $(0,7)$ and the slope-intercept form of the equation of the line is

$$y=3x+7.$$

N6. Write an equation of the line through $(3,-1)$ with slope $-\frac{2}{5}$.

Use the point-slope form of the equation of a line with $x_1=3$, $y_1=-1$, and $m=-\frac{2}{5}$.

$$\begin{aligned} y-y_1 &= m(x-x_1)\\ y-(-1) &= -\tfrac{2}{5}(x-3)\\ y+1 &= -\tfrac{2}{5}x+\tfrac{6}{5}\\ y &= -\tfrac{2}{5}x+\tfrac{6}{5}-\tfrac{5}{5}\\ y &= -\tfrac{2}{5}x+\tfrac{1}{5} \end{aligned}$$

N7. Write an equation of the line through the points $(4,1)$ and $(6,-2)$.

(a) First, find the slope of the line.

$$m=\frac{-2-1}{6-4}=\frac{-3}{2}=-\frac{3}{2}$$

Now use the point $(4,1)$ for (x_1,y_1) and $m=-\frac{3}{2}$ in the point-slope form.

$$\begin{aligned} y-y_1 &= m(x-x_1)\\ y-1 &= -\tfrac{3}{2}(x-4)\\ y-1 &= -\tfrac{3}{2}x+6\\ y &= -\tfrac{3}{2}x+7 \end{aligned}$$

(b)

$$\begin{aligned} y &= -\tfrac{3}{2}x+7 &&\\ 2y &= -3x+14 && \textit{Multiply by 2.}\\ 3x+2y &= 14 && \textit{Add } x. \end{aligned}$$

N8. Let $(x_1,y_1)=(3,4645)$ and $(x_2,y_2)=(5,5491)$.

$$\begin{aligned} m &= \frac{y_2-y_1}{x_2-x_1}\\ &= \frac{5491-4645}{5-3}\\ &= \frac{846}{2}\\ &= 423 \end{aligned}$$

Now use this slope and the point $(3,4645)$ in the point-slope form to find an equation of the line.

$$\begin{aligned} y-y_1 &= m(x-x_1)\\ y-4645 &= 423(x-3)\\ y-4645 &= 423x-1269\\ y &= 423x+3376 \end{aligned}$$

For 2007, let $x=7$.

$$\begin{aligned} y &= 423(7)+3376\\ y &= 6337 \end{aligned}$$

The equation gives $y=6337$ when $x=7$, which approximates the cost given in the table, 6185, reasonable well.

3.4 Section Exercises

1. The point-slope form of the equation of a line with slope -2 passing through the point $(4,1)$ is

$$y-1=-2(x-4).$$

Choice **E** is correct.

3. A line that passes through the points $(0,0)$ [its y-intercept] and $(4,1)$ has slope

$$m=\frac{\text{rise}}{\text{run}}=\frac{1-0}{4-0}=\frac{1}{4}.$$

Its slope-intercept form is

$$y=\tfrac{1}{4}x+0 \quad\text{or}\quad y=\tfrac{1}{4}x.$$

Choice **B** is correct.

5. $y=x+3=1x+3$

The graph of this equation is a line with slope 1 and y-intercept $(0,3)$. The only graph which has a *positive* slope and intersects the y-axis *above* the origin is **C**.

7. $y=x-3=1x-3$

The graph of this equation is a line with slope 1 and y-intercept $(0,-3)$. The only graph which has a *positive* slope and intersects the y-axis *below* the origin is **A**.

9. $y=\frac{5}{2}x-4$

The slope is $\frac{5}{2}$ and the y-intercept is $(0,-4)$.

11. $y = -x + 9$ or $y = -1x + 9$

The slope is -1 and the y-intercept is $(0, 9)$.

13. $y = \frac{x}{5} - \frac{3}{10}$ or $y = \frac{1}{5}x - \frac{3}{10}$

The slope is $\frac{1}{5}$ and the y-intercept is $(0, -\frac{3}{10})$.

15. The rise is 3 and the run is 1, so the slope is given by

$$m = \frac{\text{rise}}{\text{run}} = \frac{3}{1} = 3.$$

The y-intercept is $(0, -3)$, so $b = -3$. The equation of the line, written in slope-intercept form, is

$$y = 3x - 3.$$

17. Since the line falls from left to right, the "rise" is negative. For this line, the rise is -3 and the run is 3, so the slope is

$$m = \frac{\text{rise}}{\text{run}} = \frac{-3}{3} = -1.$$

The y-intercept is $(0, 3)$, so $b = 3$. The slope-intercept form of the equation of the line is

$$y = -1x + 3$$
$$y = -x + 3.$$

19. Since the line falls from left to right, the "rise" is negative. For this line, the rise is -2 and the run is 4, so the slope is

$$m = \frac{\text{rise}}{\text{run}} = \frac{-2}{4} = -\frac{1}{2}.$$

The y-intercept is $(0, 2)$, so $b = 2$. The slope-intercept form of the equation of the line is

$$y = -\tfrac{1}{2}x + 2.$$

21. $m = 4$, $(0, -3)$

Since the y-intercept is $(0, -3)$, we have $b = -3$. Use the slope-intercept form.

$$y = mx + b$$
$$y = 4x + (-3)$$
$$y = 4x - 3$$

23. $m = -1$, $(0, -7)$

Since the y-intercept is $(0, -7)$, we have $b = -7$. Use the slope-intercept form.

$$y = mx + b$$
$$y = -1x - 7$$
$$y = -x - 7$$

25. $m = 0$, $(0, 3)$

Since the y-intercept is $(0, 3)$, we have $b = 3$. Use the slope-intercept form.

$$y = mx + b$$
$$y = 0x + 3$$
$$y = 3$$

27. Undefined slope, $(0, -2)$

Since the slope is undefined, the line is vertical and has equation $x = k$. Because the line goes through a point $(x, y) = (0, -2)$, we must have $x = 0$.

29. $y = 3x + 2$

The slope is $3 = \frac{3}{1} = \frac{\text{change in } y}{\text{change in } x}$, and the y-intercept is $(0, 2)$. Graph that point and count up 3 units and right 1 unit to get to the point $(1, 5)$. Draw a line through the points.

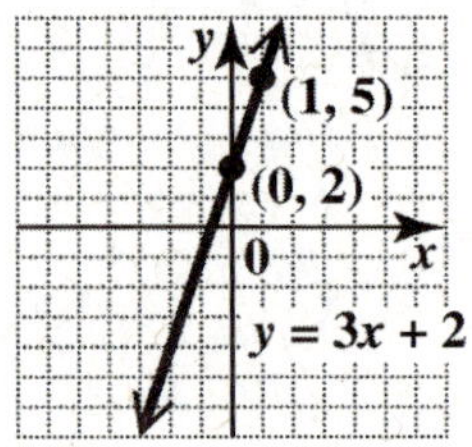

31. $y = -\frac{1}{3}x + 4$

The slope is $-\frac{1}{3} = \frac{-1}{3} = \frac{\text{change in } y}{\text{change in } x}$, and the y-intercept is $(0, 4)$. Graph that point and count down 1 unit and right 3 units to get to the point $(3, 3)$. Draw a line through the points.

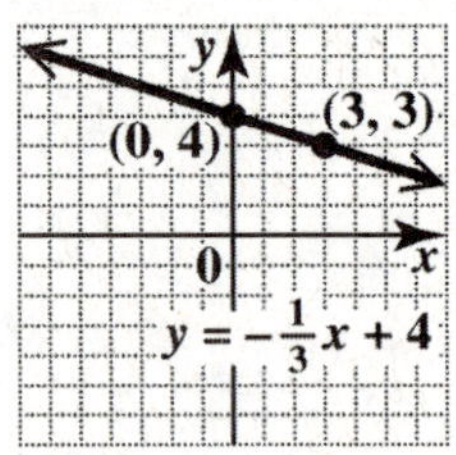

33. $2x + y = -5$, so $y = -2x - 5$

The slope is $-2 = \frac{-2}{1} = \frac{\text{change in } y}{\text{change in } x}$, and the y-intercept is $(0, -5)$. Graph that point and count down 2 units and right 1 unit to get to the point $(1, -7)$. Draw a line through the points.

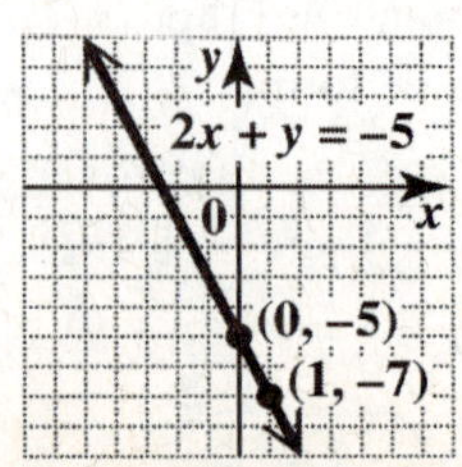

35. Solve the equation for y.

$$\begin{aligned} 4x - 5y &= 20 && \textit{Given equation} \\ -5y &= -4x + 20 && \textit{Subtract 4x.} \\ y &= \tfrac{4}{5}x - 4 && \textit{Divide by } -5. \end{aligned}$$

The slope is the coefficient of x, $\frac{4}{5}$.

The slope is $\frac{4}{5} = \frac{\text{change in } y}{\text{change in } x}$, and the y-intercept is $(0, -4)$. Graph that point and count up 4 units and right 5 units to get to the point $(5, 0)$. Draw a line through the points.

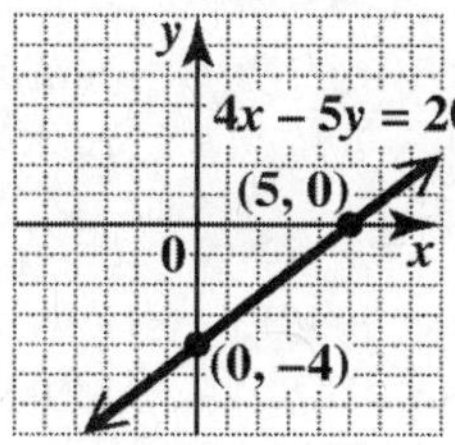

37. $(0, 1)$, $m = 4$

First, locate the point $(0, 1)$, which is the y-intercept of the line to be graphed. Write the slope as

$$m = \frac{\text{rise}}{\text{run}} = \frac{4}{1}.$$

Locate another point by counting 4 units up and then 1 unit to the right. Draw a line through this new point, $(1, 5)$, and the given point, $(0, 1)$.

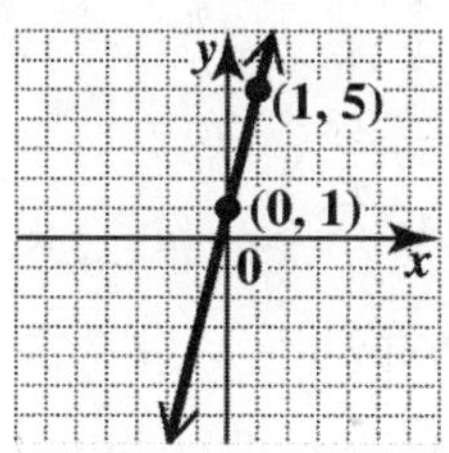

39. $(1, -5)$, $m = -\frac{2}{5}$

First, locate the point $(1, -5)$. Write the slope as

$$m = \frac{\text{rise}}{\text{run}} = \frac{-2}{5}.$$

Locate another point by counting 2 units down (because of the negative sign) and then 5 units to the right. Draw a line through this new point, $(6, -7)$, and the given point, $(1, -5)$.

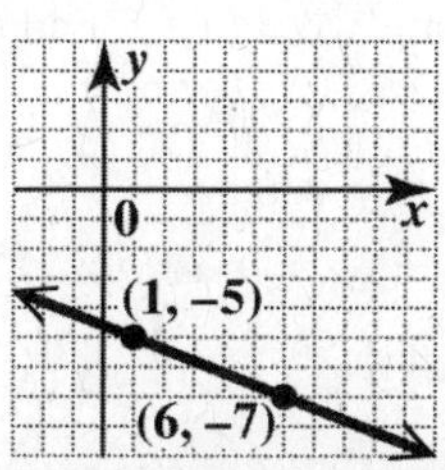

41. $(-1, 4)$, $m = \frac{2}{5}$

First, locate the point $(-1, 4)$. The slope is

$$m = \frac{\text{rise}}{\text{run}} = \frac{2}{5}.$$

Locate another point by counting 2 units up and then 5 units to the right. Draw a line through this new point, $(4, 6)$, and the given point, $(-1, 4)$.

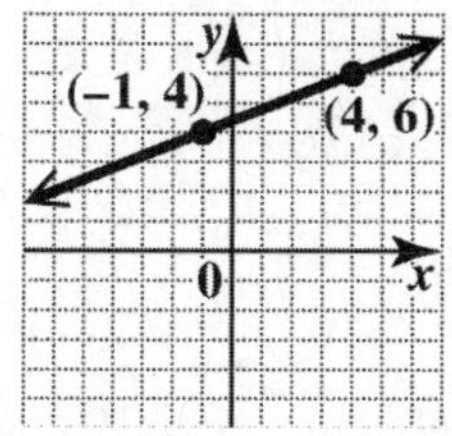

43. $(0, 0)$, $m = -2$

First, locate the point $(0, 0)$. The slope is

$$m = \frac{\text{rise}}{\text{run}} = \frac{-2}{1}.$$

Locate another point by counting 2 units down and then 1 unit to the right. Draw a line through this new point, $(1, -2)$, and the given point, $(0, 0)$.

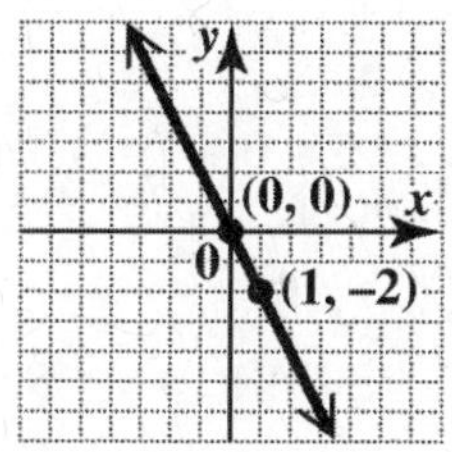

45. $(-2, 3)$, $m = 0$

First, locate the point $(-2, 3)$. Since the slope is 0, the line will be horizontal. Draw a horizontal line through the point $(-2, 3)$.

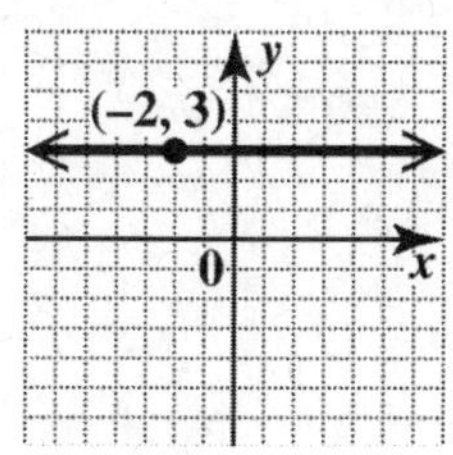

47. $(2, 4)$, undefined slope

First, locate the point $(2, 4)$. Since the slope is undefined, the line will be vertical. Draw the vertical line through the point $(2, 4)$.

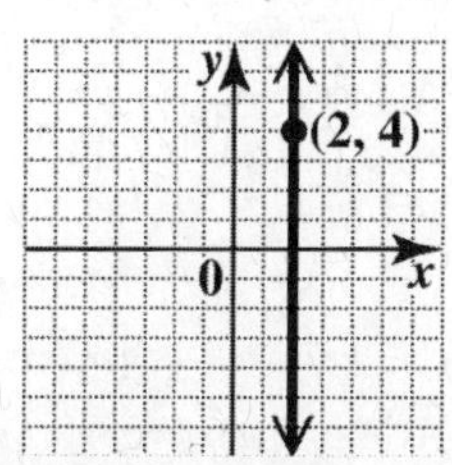

49. The common name given to a vertical line whose x-intercept is the origin is the y-axis.

51. $(4, 1), m = 2$

The given point is $(4, 1)$, so $x_1 = 4$ and $y_1 = 1$. Also, $m = 2$. Substitute these values into the point-slope form. Then solve for y to obtain the slope-intercept form.

$$\begin{aligned} y - y_1 &= m(x - x_1) \\ y - 1 &= 2(x - 4) \\ y - 1 &= 2x - 8 && \textit{Dist. property} \\ y &= 2x - 7 && \textit{Add 1.} \end{aligned}$$

53. $(-1, 3), m = -4$

The given point is $(-1, 3)$, so $x_1 = -1$ and $y_1 = 3$. Also, $m = -4$. Substitute these values into the point-slope form. Then solve for y to obtain the slope-intercept form.

$$\begin{aligned} y - y_1 &= m(x - x_1) \\ y - 3 &= -4[x - (-1)] \\ y - 3 &= -4(x + 1) \\ y - 3 &= -4x - 4 && \textit{Dist. property} \\ y &= -4x - 1 && \textit{Add 3.} \end{aligned}$$

55. $(9, 3), m = 1$

Use the values $x_1 = 9$, $y_1 = 3$, and $m = 1$ in the point-slope form.

$$\begin{aligned} y - y_1 &= m(x - x_1) \\ y - 3 &= 1(x - 9) \\ y - 3 &= 1x - 9 && \textit{Dist. property} \\ y &= x - 6 && \textit{Add 3.} \end{aligned}$$

57. $(-4, 1), m = \frac{3}{4}$

Use the values $x_1 = -4$, $y_1 = 1$, and $m = \frac{3}{4}$ in the point-slope form.

$$\begin{aligned} y - y_1 &= m(x - x_1) \\ y - 1 &= \tfrac{3}{4}[x - (-4)] \\ y - 1 &= \tfrac{3}{4}(x + 4) \\ y - 1 &= \tfrac{3}{4}x + 3 \\ y &= \tfrac{3}{4}x + 4 \end{aligned}$$

59. $(-2, 5), m = \frac{2}{3}$

Use the values $x_1 = -2$, $y_1 = 5$, and $m = \frac{2}{3}$ in the point-slope form.

$$\begin{aligned} y - y_1 &= m(x - x_1) \\ y - 5 &= \tfrac{2}{3}[x - (-2)] \\ y - 5 &= \tfrac{2}{3}(x + 2) \\ y - 5 &= \tfrac{2}{3}x + \tfrac{4}{3} \\ y &= \tfrac{2}{3}x + \tfrac{19}{3} && \textit{Add } 5 = \tfrac{15}{3}. \end{aligned}$$

61. $(6, -3), m = -\frac{4}{5}$

Use the values $x_1 = 6$, $y_1 = -3$, and $m = -\frac{4}{5}$ in the point-slope form.

$$\begin{aligned} y - y_1 &= m(x - x_1) \\ y - (-3) &= -\tfrac{4}{5}(x - 6) \\ y + 3 &= -\tfrac{4}{5}x + \tfrac{24}{5} \\ y &= -\tfrac{4}{5}x + \tfrac{9}{5} && \textit{Subtract } 3 = \tfrac{15}{5}. \end{aligned}$$

63. **A.**

$$\begin{aligned} y &= \tfrac{2}{3}x - 2 && \textit{Given} \\ 3(y) &= 3(\tfrac{2}{3}x - 2) && \textit{Multiply by 3.} \\ 3y &= 3(\tfrac{2}{3}x) - 3(2) && \textit{Distributive property} \\ 3y &= 2x - 6 && \textit{Multiply.} \\ 6 &= 2x - 3y && \textit{Standard form} \end{aligned}$$

B.

$$\begin{aligned} -2x + 3y &= -6 && \textit{Given} \\ -1(-2x + 3y) &= -1(-6) && \textit{Mult. by } -1. \\ -1(-2x) - 1(3y) &= -1(-6) && \textit{Dist. property} \\ 2x - 3y &= 6 && \textit{Multiply.} \end{aligned}$$

C.

$$\begin{aligned} y &= -\tfrac{3}{2}x + 3 && \textit{Given} \\ 2(y) &= 2(-\tfrac{3}{2}x + 3) && \textit{Multiply by 2.} \\ 2y &= 2(-\tfrac{3}{2}x) + 2(3) && \textit{Dist. property} \\ 2y &= -3x + 6 && \textit{Multiply.} \\ 3x + 2y &= 6 && \textit{Standard form} \end{aligned}$$

D.

$$\begin{aligned} y - 2 &= \tfrac{2}{3}(x - 6) && \textit{Given} \\ 3(y - 2) &= 3[\tfrac{2}{3}(x - 6)] && \textit{Multiply by 3.} \\ 3y - 6 &= 2(x - 6) && \textit{Distributive prop. Associative prop.} \\ 3y - 6 &= 2x - 12 && \textit{Distributive prop.} \\ 6 &= 2x - 3y && \textit{Standard form} \end{aligned}$$

So **A**, **B**, and **D** are equivalent to $2x - 3y = 6$.

65. **(a)** $(4, 10)$ and $(6, 12)$

First, find the slope of the line.

$$m = \frac{12 - 10}{6 - 4} = \frac{2}{2} = 1$$

Now use the point $(4, 10)$ for (x_1, y_1) and $m = 1$ in the point-slope form.

$$\begin{aligned} y - y_1 &= m(x - x_1) \\ y - 10 &= 1(x - 4) \\ y - 10 &= x - 4 \\ y &= x + 6 \end{aligned}$$

The same result would be found by using $(6, 12)$ for (x_1, y_1).

(b)

$$\begin{aligned} y &= x + 6 \\ -x + y &= 6 && \textit{Subtract } x. \\ x - y &= -6 && \textit{Multiply by } -1. \end{aligned}$$

67. **(a)** $(-4,0)$ and $(0,2)$

$$m=\frac{2-0}{0-(-4)}=\frac{2}{4}=\frac{1}{2}$$

Use the point-slope form with $(x_1,y_1)=(-4,0)$ and $m=\frac{1}{2}$.

$$\begin{aligned} y-y_1&=m(x-x_1)\\ y-0&=\tfrac{1}{2}[x-(-4)]\\ y&=\tfrac{1}{2}(x+4)\\ y&=\tfrac{1}{2}x+2 \end{aligned}$$

(b)

$$\begin{aligned} y&=\tfrac{1}{2}x+2 &&\\ 2y&=x+4 && \textit{Multiply by 2.}\\ -x+2y&=4 && \textit{Subtract x.}\\ x-2y&=-4 && \textit{Multiply by } -1. \end{aligned}$$

69. **(a)** $(-2,-1)$ and $(3,-4)$

$$m=\frac{-4-(-1)}{3-(-2)}=\frac{-3}{5}=-\frac{3}{5}$$

Use the point-slope form with $(x_1,y_1)=(3,-4)$ and $m=-\frac{3}{5}$.

$$\begin{aligned} y-y_1&=m(x-x_1)\\ y-(-4)&=-\tfrac{3}{5}(x-3)\\ y+4&=-\tfrac{3}{5}(x-3)\\ y+4&=-\tfrac{3}{5}x+\tfrac{9}{5}\\ y&=-\tfrac{3}{5}x-\tfrac{11}{5} \end{aligned}$$

(b)

$$\begin{aligned} y&=-\tfrac{3}{5}x-\tfrac{11}{5} &&\\ 5y&=-3x-11 && \textit{Multiply by 5.}\\ 3x+5y&=-11 && \textit{Add 3x.} \end{aligned}$$

71. **(a)** $\left(-\frac{2}{3},\frac{8}{3}\right)$ and $\left(\frac{1}{3},\frac{7}{3}\right)$

$$m=\frac{\frac{7}{3}-\frac{8}{3}}{\frac{1}{3}-\left(-\frac{2}{3}\right)}=\frac{-\frac{1}{3}}{\frac{3}{3}}=\frac{-\frac{1}{3}}{1}=-\frac{1}{3}$$

Use the point-slope form with $(x_1,y_1)=\left(\frac{1}{3},\frac{7}{3}\right)$ and $m=-\frac{1}{3}$.

$$\begin{aligned} y-y_1&=m(x-x_1)\\ y-\tfrac{7}{3}&=-\tfrac{1}{3}\left(x-\tfrac{1}{3}\right)\\ y-\tfrac{7}{3}&=-\tfrac{1}{3}x+\tfrac{1}{9}\\ y&=-\tfrac{1}{3}x+\tfrac{22}{9} \end{aligned}$$

(b)

$$\begin{aligned} y&=-\tfrac{1}{3}x+\tfrac{22}{9} &&\\ 9y&=-3x+22 && \textit{Multiply by 9.}\\ 3x+9y&=22 && \textit{Add 3x.} \end{aligned}$$

73. Solve the equation for y.

$$\begin{aligned} x-2y&=7\\ -2y&=-x+7\\ y&=\tfrac{1}{2}x-\tfrac{7}{2} \end{aligned}$$

The slope is $\frac{1}{2}$. A line perpendicular to this line has slope -2 (the negative reciprocal of $\frac{1}{2}$). Now use the slope-intercept form with $m=-2$ and y-intercept $(0,-3)$.

$$\begin{aligned} y&=mx+b\\ y&=-2x-3 \end{aligned}$$

75. Solve the equation for y.

$$\begin{aligned} 4x-y&=-2\\ 4x+2&=y \end{aligned}$$

The slope is 4. A line parallel to this line has the same slope. Now use the point-slope form with $m=4$ and $(x_1,y_1)=(2,3)$.

$$\begin{aligned} y-y_1&=m(x-x_1)\\ y-3&=4(x-2)\\ y-3&=4x-8\\ y&=4x-5 \end{aligned}$$

77. Solve the equation for y.

$$\begin{aligned} 3x&=4y+5\\ -4y&=-3x+5\\ y&=\tfrac{3}{4}x-\tfrac{5}{4} \end{aligned}$$

The slope is $\frac{3}{4}$. A line parallel to this line has the same slope. Now use the point-slope form with $m=\frac{3}{4}$ and $(x_1,y_1)=(2,-3)$.

$$\begin{aligned} y-y_1&=m(x-x_1)\\ y-(-3)&=\tfrac{3}{4}(x-2)\\ y+3&=\tfrac{3}{4}x-\tfrac{3}{2}\\ y&=\tfrac{3}{4}x-\tfrac{3}{2}-\tfrac{6}{2}\\ y&=\tfrac{3}{4}x-\tfrac{9}{2} \end{aligned}$$

79. **(a)** The fixed cost is \$400.

(b) The variable cost is \$0.25.

(c) Substitute $m=0.25$ and $b=400$ into $y=mx+b$ to get the cost equation

$$y=0.25x+400.$$

(d) Let $x=100$ in the cost equation.

$$\begin{aligned} y&=0.25(100)+400\\ y&=25+400\\ y&=425 \end{aligned}$$

The cost to produce 100 snow cones will be \$425.

(e) Let $y=775$ in the cost equation.

$$\begin{aligned} 775&=0.25x+400 &&\\ 375&=0.25x && \textit{Subtract 400.}\\ x&=\tfrac{375}{0.25}=1500 && \textit{Divide by 0.25.} \end{aligned}$$

If the total cost is \$775, 1500 snow cones will be produced.

81. **(a)** x represents the year and y represents the cost in the ordered pairs (x, y). The ordered pairs are $(1, 2079)$, $(2, 2182)$, $(3, 2272)$, $(4, 2361)$, and $(5, 2402)$.

(b)

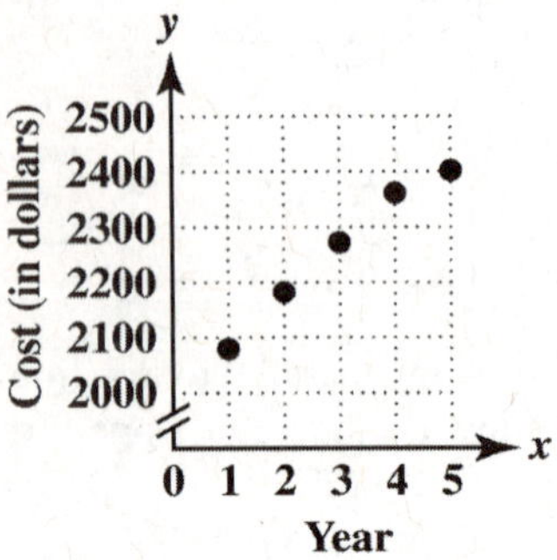

Yes, the points lie approximately in a straight line.

(c) Find the slope using

$(x_1, y_1) = (1, 2079)$ and $(x_2, y_2) = (4, 2361)$.

$$m = \frac{y_2 - y_1}{x_2 - x_1} = \frac{2361 - 2079}{4 - 1} = \frac{282}{3} = 94$$

Now use the point-slope form with $m = 94$ and $(x_1, y_1) = (1, 2079)$.

$$\begin{aligned} y - y_1 &= m(x - x_1) \\ y - 2079 &= 94(x - 1) \\ y - 2079 &= 94x - 94 \\ y &= 94x + 1985 \end{aligned}$$

(d) Since year 1 represents 2004, year 0 represents 2003.

For 2009, $x = 2009 - 2003 = 6$.

$$\begin{aligned} y &= 94x + 1985 \\ y &= 94(6) + 1985 \\ y &= 2549 \end{aligned}$$

In 2009, the estimate of the average annual cost at 2-year colleges is \$2549.

83. Find the slope using $(x_1, y_1) = (2002, 52)$ and $(x_2, y_2) = (2007, 127)$.

$$m = \frac{y_2 - y_1}{x_2 - x_1} = \frac{127 - 52}{2007 - 2002} = \frac{75}{5} = 15$$

Now use the point-slope form with $m = 15$ and $(x_2, y_2) = (2007, 127)$.

$$\begin{aligned} y - y_1 &= m(x - x_1) \\ y - 127 &= 15(x - 2007) \\ y - 127 &= 15x - 30{,}105 \\ y &= 15x - 29{,}978 \end{aligned}$$

85. $2 \cdot 2 \cdot 2 \cdot 2 \cdot 2 \cdot 2 = 64$

87. $5 \cdot 5 \cdot 5 \cdot 5 = 625$

89. $\frac{2}{3} \cdot \frac{2}{3} \cdot \frac{2}{3} = \frac{8}{27}$

Summary Exercises on Linear Equations and Graphs

1. Solve the equation for y.

$$\begin{aligned} x - 2y &= -4 && \textit{Given equation} \\ -2y &= -x - 4 && \textit{Subtract } x. \\ y &= \tfrac{1}{2}x + 2 && \textit{Divide by } -2. \end{aligned}$$

The slope is the coefficient of x, $\frac{1}{2}$.

The slope is $\frac{1}{2} = \frac{\text{change in } y}{\text{change in } x}$, and the y-intercept is $(0, 2)$. Graph that point and count up 1 unit and right 2 units to get to the point $(2, 3)$. Draw a line through the points.

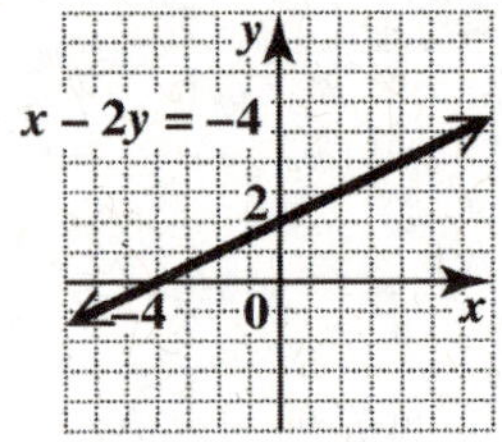

3. $m = 1$, y-intercept $(0, -2)$

The slope is $1 = \frac{1}{1} = \frac{\text{change in } y}{\text{change in } x}$, and the y-intercept is $(0, -2)$. Graph that point and count up 1 unit and right 1 unit to get to the point $(1, -1)$. Draw a line through the points.

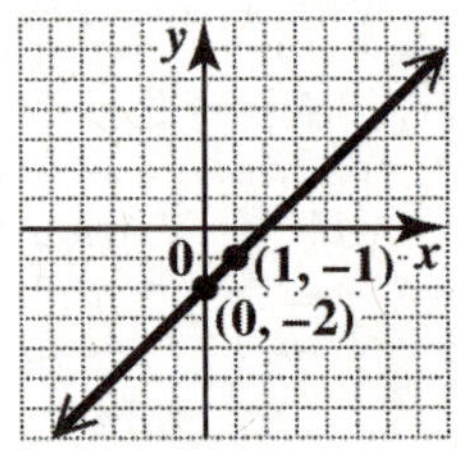

5. $m = -\frac{2}{3}$, passes through $(3, -4)$

The slope is $-\frac{2}{3} = \frac{2}{-3} = \frac{\text{change in } y}{\text{change in } x}$. Graph the point $(3, -4)$ and count up 2 units and left 3 units to get to the point $(0, -2)$.

Draw a line through the points.

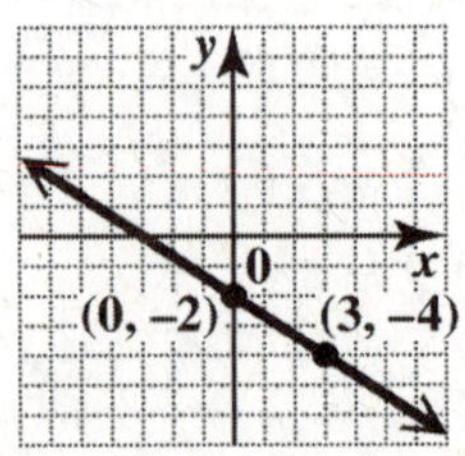

7. Solve the equation for y.

$$\begin{aligned} x - 4y &= 0 && \textit{Given equation} \\ -4y &= -x && \textit{Subtract } x. \\ y &= \tfrac{1}{4}x && \textit{Divide by } -4. \end{aligned}$$

The slope is the coefficient of x, $\frac{1}{4}$.

The slope is $\frac{1}{4} = \frac{\text{change in } y}{\text{change in } x}$, and the y-intercept is $(0, 0)$. Graph that point and count up 1 unit and right 4 units to get to the point $(4, 1)$. Draw a line through the points.

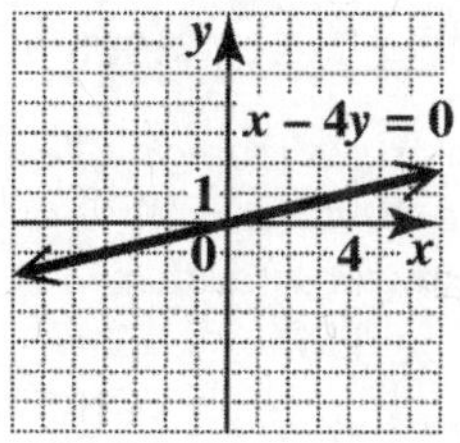

9. $8x = 6y + 24$

***x*-intercept ($y = 0$)**
$8x = 6(0) + 24$
$8x = 24$
$x = 3$ **(3, 0)**

***y*-intercept ($x = 0$)**
$8(0) = 6y + 24$
$-24 = 6y$
$-4 = y$ **(0, −4)**

third point ($x = 2$)
$8(2) = 6y + 24$
$16 = 6y + 24$
$-8 = 6y$
$-\frac{4}{3} = y$ **$(2, -\frac{4}{3})$**

Plot $(3, 0)$, $(0, -4)$, and $(2, -\frac{4}{3})$ and draw a line through these points.

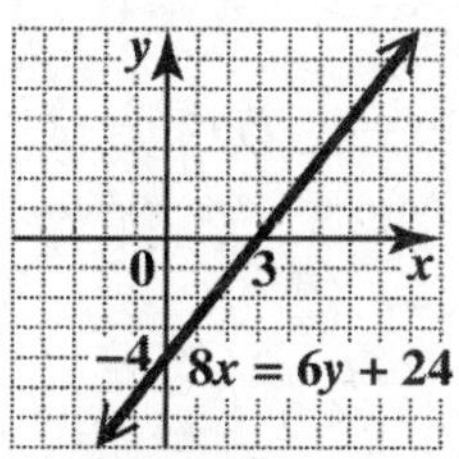

11. Solve the equation for y.

$5x + 2y = 10$	*Given equation*
$2y = -5x + 10$	*Subtract 5x.*
$y = -\frac{5}{2}x + 5$	*Divide by 2.*

The slope is the coefficient of x, $-\frac{5}{2}$.

The slope is $-\frac{5}{2} = \frac{-5}{2} = \frac{\text{change in } y}{\text{change in } x}$, and the y-intercept is $(0, 5)$. Graph that point and count down 5 units and right 2 units to get to the point $(2, 0)$. Draw a line through the points.

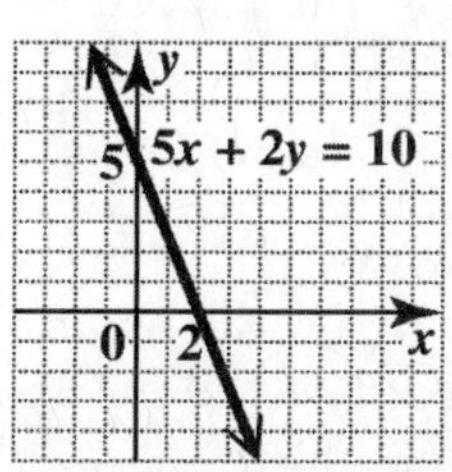

13. $m = 0$, passes through $(0, \frac{3}{2})$

The slope is 0, so the graph is a horizontal line. Graph the y-intercept, $(0, \frac{3}{2})$, and draw a horizontal line through it.

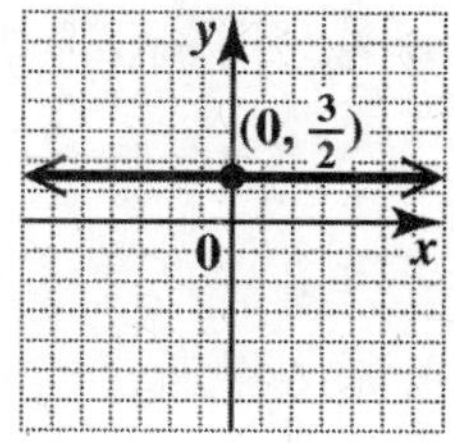

15. $y = -x + 6$

The slope is $-1 = \frac{-1}{1} = \frac{\text{change in } y}{\text{change in } x}$, and the y-intercept is $(0, 6)$. Graph that point and count down 1 unit and right 1 unit to get to the point $(1, 5)$. Draw a line through the points.

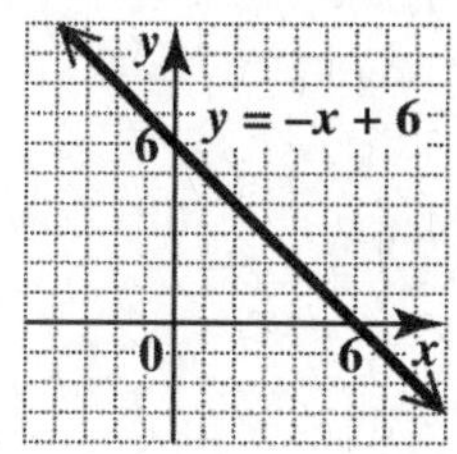

17. $x + 4 = 0$

$x = -4$ *Subtract 4.*

This is an equation of the vertical line with x-intercept $(-4, 0)$. The slope of thc line is undefined.

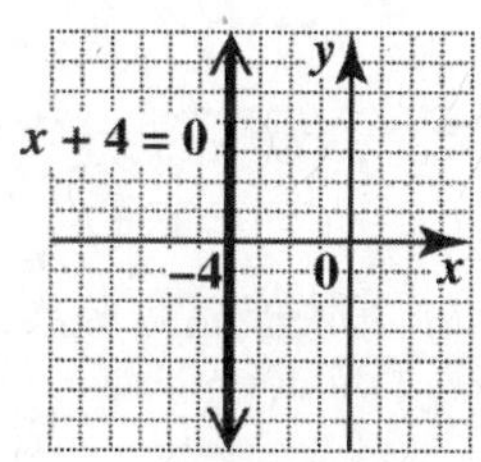

19. **(a)** "Slope −0.5" indicates that $m = -\frac{1}{2}$ and "$b = -2$" indicates that the y-intercept is $(0, -2)$. Given the slope and y-intercept, we know that one form of the equation of the line is $y = -\frac{1}{2}x - 2$, choice **B**.

(b) The slope of the line passing through the x-intercept of $(4, 0)$ and the y-intercept of $(0, 2)$ is

$$m = \frac{2 - 0}{0 - 4} = \frac{2}{-4} = -\frac{1}{2}.$$

The slope-intercept form is $y = -\frac{1}{2}x + 2$. None of the choices has this equation, so change the form of $y = -\frac{1}{2}x + 2$ by multiplying each side by 2 to clear fractions.

$$2(y) = 2(-\tfrac{1}{2}x + 2)$$
$$2y = -x + 4$$

Add x to both sides to get $x + 2y = 4$, choice **D**.

(c) The slope of the line passing through $(4, -2)$ and $(0, 0)$ is

$$m = \frac{0-(-2)}{0-4} = \frac{2}{-4} = -\frac{1}{2}.$$

The slope-intercept form is $y = -\frac{1}{2}x$, choice **A**.

(d) $m = \frac{1}{2}$; passes through $(-2, -2)$

$y = mx + b$	*Slope-intercept form*
$-2 = \frac{1}{2}(-2) + b$	*Let $x = -2$, $y = -2$, $m = \frac{1}{2}$.*
$-2 = -1 + b$	*Multiply.*
$-1 = b$	*Add 1.*

The line has equation $y = \frac{1}{2}x - 1$. Multiply each side by 2 to get $2y = x - 2$, which is the same as $2 = x - 2y$, choice **C**.

21. $m = -3$, $b = -6$

$$y = mx + b$$
$$y = -3x - 6$$

23. Through $(1, -7)$ and $(-2, 5)$

$$m = \frac{5-(-7)}{-2-1} = \frac{12}{-3} = -4$$

$$y - y_1 = m(x - x_1)$$
$$y - (-7) = -4(x - 1)$$
$$y + 7 = -4x + 4$$
$$y = -4x - 3$$

25. Through $(0, 0)$, undefined slope

A line with undefined slope has equation $x = x$-intercept value; in this case, $x = 0$.

27. Through $(0, 0)$ and $(3, 2)$

$$m = \frac{2-0}{3-0} = \frac{2}{3}$$

$$y = mx + b$$
$$y = \tfrac{2}{3}x + 0$$
$$y = \tfrac{2}{3}x$$

29. Through $(5, 0)$ and $(0, -5)$

$$m = \frac{-5-0}{0-5} = \frac{-5}{-5} = 1$$

$$y = mx + b$$
$$y = 1x - 5$$
$$y = x - 5$$

31. $m = \frac{5}{3}$, through $(-3, 0)$

$$y - y_1 = m(x - x_1)$$
$$y - 0 = \tfrac{5}{3}[x - (-3)]$$
$$y = \tfrac{5}{3}(x + 3)$$
$$y = \tfrac{5}{3}x + 5$$

Chapter 3 Review Exercises

1. **(a)** The graph rises from 2002 to 2003, 2003 to 2004, 2004 to 2005, and 2006 to 2007, so the number of real trees purchased increased from 2002 to 2005 and 2006 to 2007.

(b) The graph falls from 2005 to 2006, so the number of real trees purchased decreased from 2005 to 2006.

(c) Locate 2005 on the horizontal scale and follow the line up to the line graph. Then move across to read the value on the vertical scale. The number of real trees purchased in 2005 is about 33 million. Similarly, we get about 29 million for 2006.

(d) The change from 2005 to 2006 was

29 million − 33 million = −4 million.

This represents a decrease of about 4 million.

2. $y = 3x + 2$; $(-1, \_\_)$ $(0, \_\_)$ $(\_\_, 5)$

$y = 3x + 2$	
$y = 3(-1) + 2$	*Let $x = -1$.*
$y = -3 + 2$	
$y = -1$	

$y = 3x + 2$	
$y = 3(0) + 2$	*Let $x = 0$.*
$y = 0 + 2$	
$y = 2$	

$y = 3x + 2$	
$5 = 3x + 2$	*Let $y = 5$.*
$3 = 3x$	
$1 = x$	

The ordered pairs are $(-1, -1)$, $(0, 2)$, and $(1, 5)$.

3. $4x + 3y = 6$; $(0, \_\_)$ $(\_\_, 0)$ $(-2, \_\_)$

$4x + 3y = 6$	
$4(0) + 3y = 6$	*Let $x = 0$.*
$3y = 6$	
$y = 2$	

$4x + 3y = 6$	
$4x + 3(0) = 6$	*Let $y = 0$.*
$4x = 6$	
$x = \frac{6}{4} = \frac{3}{2}$	

$$4x + 3y = 6$$
$$4(-2) + 3y = 6 \quad \textit{Let } x = -2.$$
$$-8 + 3y = 6$$
$$3y = 14$$
$$y = \tfrac{14}{3}$$

The ordered pairs are $(0, 2)$, $(\frac{3}{2}, 0)$, and $(-2, \frac{14}{3})$.

4. $x = 3y$; $(0, \_\_)$ $(8, \_\_)$ $(\_\_, -3)$

$$x = 3y$$
$$0 = 3y \quad \textit{Let } x = 0.$$
$$0 = y$$

$$x = 3y$$
$$8 = 3y \quad \textit{Let } x = 8.$$
$$\tfrac{8}{3} = y$$

$$x = 3y$$
$$x = 3(-3) \quad \textit{Let } y = -3.$$
$$x = -9$$

The ordered pairs are $(0, 0)$, $(8, \frac{8}{3})$, and $(-9, -3)$.

5. $x - 7 = 0$; $(\_\_, -3)$ $(\_\_, 0)$ $(\_\_, 5)$
The given equation may be written $x = 7$. For any value of y, the value of x will always be 7. The ordered pairs are $(7, -3)$, $(7, 0)$, and $(7, 5)$.

6. $x + y = 7$; $(2, 5)$
Substitute 2 for x and 5 for y in the given equation.

$$x + y = 7$$
$$2 + 5 \stackrel{?}{=} 7$$
$$7 = 7 \quad \textit{True}$$

Yes, $(2, 5)$ is a solution of the given equation.

7. $2x + y = 5$; $(-1, 3)$
Substitute -1 for x and 3 for y in the given equation.

$$2x + y = 5$$
$$2(-1) + 3 \stackrel{?}{=} 5$$
$$-2 + 3 \stackrel{?}{=} 5$$
$$1 = 5 \quad \textit{False}$$

No, $(-1, 3)$ is not a solution of the given equation.

8. $3x - y = 4$; $(\frac{1}{3}, -3)$
Substitute $\frac{1}{3}$ for x and -3 for y in the given equation.

$$3x - y = 4$$
$$3(\tfrac{1}{3}) - (-3) \stackrel{?}{=} 4$$
$$1 + 3 \stackrel{?}{=} 4$$
$$4 = 4 \quad \textit{True}$$

Yes, $(\frac{1}{3}, -3)$ is a solution of the given equation.

9. To plot $(2, 3)$, start at the origin, go 2 units to the right, and then go up 3 units. The point lies in quadrant I. (Graph follows Exercise 12.)

10. To plot $(-4, 2)$, start at the origin, go 4 units to the left, and then go up 2 units. The point lies in quadrant II.

11. To plot $(3, 0)$, start at the origin, go 3 units to the right. The point lies on the x-axis (not in any quadrant).

12. To plot $(0, -6)$, start at the origin, go down 6 units. The point lies on the y-axis (not in any quadrant).

Graph for Exercises 9–12

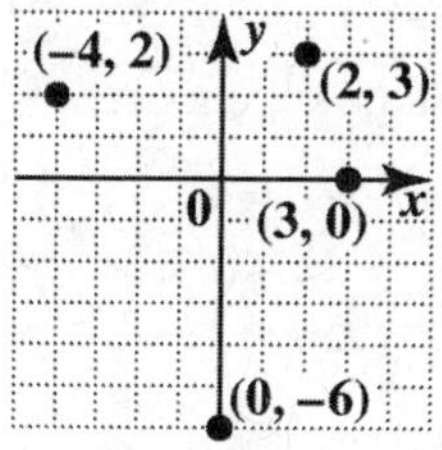

13. The product of two numbers is positive whenever the two numbers have the same sign. If $xy > 0$, either $x > 0$ and $y > 0$, so that (x, y) lies in quadrant I, or $x < 0$ and $y < 0$, so that (x, y) lies in quadrant III.

14. To find the x-intercept, let $y = 0$.

$$y = 2x + 5$$
$$0 = 2x + 5 \quad \textit{Let } y = 0.$$
$$-2x = 5$$
$$x = -\tfrac{5}{2}$$

The x-intercept is $(-\frac{5}{2}, 0)$.

To find the y-intercept, let $x = 0$.

$$y = 2x + 5$$
$$y = 2(0) + 5 \quad \textit{Let } x = 0.$$
$$y = 5$$

The y-intercept is $(0, 5)$.

To find a third point, choose $x = -1$.

$$y = 2x + 5$$
$$y = 2(-1) + 5 \quad \textit{Let } x = -1.$$
$$y = 3$$

This gives the ordered pair $(-1, 3)$. Plot $(-\frac{5}{2}, 0)$, $(0, 5)$, and $(-1, 3)$ and draw a line through them.

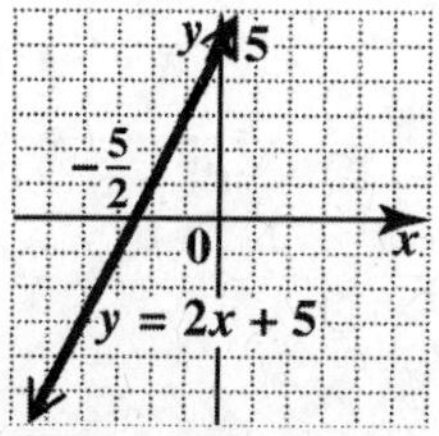

15. To find the x-intercept, let $y = 0$.

$$\begin{aligned} 3x + 2y &= 8 \\ 3x + 2(0) &= 8 \quad \textit{Let } y = 0. \\ 3x &= 8 \\ x &= \tfrac{8}{3} \end{aligned}$$

The x-intercept is $(\frac{8}{3}, 0)$.

To find the y-intercept, let $x = 0$.

$$\begin{aligned} 3x + 2y &= 8 \\ 3(0) + 2y &= 8 \quad \textit{Let } x = 0. \\ 2y &= 8 \\ y &= 4 \end{aligned}$$

The y-intercept is $(0, 4)$.

To find a third point, choose $x = 1$.

$$\begin{aligned} 3x + 2y &= 8 \\ 3(1) + 2y &= 8 \quad \textit{Let } x = 1. \\ 2y &= 5 \\ y &= \tfrac{5}{2} \end{aligned}$$

This gives the ordered pair $(1, \frac{5}{2})$.
Plot $(\frac{8}{3}, 0)$, $(0, 4)$, and $(1, \frac{5}{2})$ and draw a line through them.

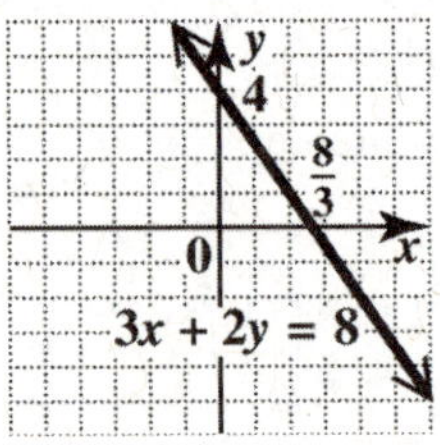

16. $x + 2y = -4$

Find the intercepts.

If $y = 0$, $x = -4$, so the x-intercept is $(-4, 0)$.

If $x = 0$, $y = -2$, so the y-intercept is $(0, -2)$.

To find a third point, choose $x = 2$.

$$\begin{aligned} x + 2y &= -4 \\ 2 + 2y &= -4 \quad \textit{Let } x = 2. \\ 2y &= -6 \\ y &= -3 \end{aligned}$$

This gives the ordered pair $(2, -3)$. Plot $(-4, 0)$, $(0, -2)$, and $(2, -3)$ and draw a line through them.

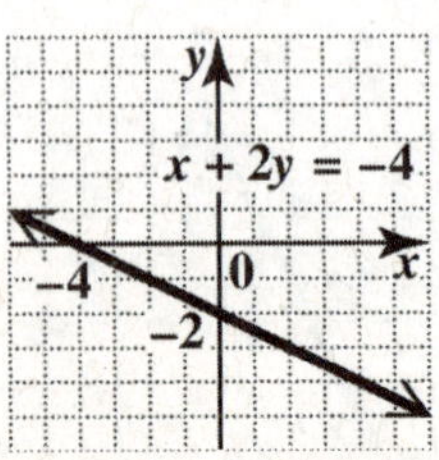

17. Let $(2, 3) = (x_1, y_1)$ and $(-4, 6) = (x_2, y_2)$.

$$\text{slope } m = \frac{\text{change in } y}{\text{change in } x} = \frac{y_2 - y_1}{x_2 - x_1} = \frac{6 - 3}{-4 - 2} = \frac{3}{-6} = -\frac{1}{2}$$

18. Let $(2, 5) = (x_1, y_1)$ and $(2, 8) = (x_2, y_2)$.

$$\text{slope } m = \frac{\text{change in } y}{\text{change in } x} = \frac{y_2 - y_1}{x_2 - x_1} = \frac{8 - 5}{2 - 2} = \frac{3}{0},$$

which is *undefined.*

19. $y = 3x - 4$

The equation is already solved for y, so the slope of the line is given by the coefficient of x. Thus, the slope is 3.

20. $y = 5$ is an equation of a horizontal line. Its slope is 0.

21. The indicated points have coordinates $(0, -2)$ and $(2, 1)$. Use the definition of slope with $(0, -2) = (x_1, y_1)$ and $(2, 1) = (x_2, y_2)$.

$$m = \frac{\text{change in } y}{\text{change in } x} = \frac{y_2 - y_1}{x_2 - x_1} = \frac{1 - (-2)}{2 - 0} = \frac{3}{2}$$

Note: We could also simply count grid marks to get $m = \frac{\text{rise}}{\text{run}} = \frac{+3}{+2} = \frac{3}{2}$.

22. The indicated points have coordinates $(0, 1)$ and $(3, 0)$. Use the definition of slope with $(0, 1) = (x_1, y_1)$ and $(3, 0) = (x_2, y_2)$.

$$m = \frac{\text{change in } y}{\text{change in } x} = \frac{y_2 - y_1}{x_2 - x_1} = \frac{0 - 1}{3 - 0} = \frac{-1}{3} = -\frac{1}{3}$$

Note: We could also simply count grid marks to get $m = \frac{\text{rise}}{\text{run}} = \frac{-1}{+3} = -\frac{1}{3}$.

23. From the table, we choose the two points $(0, 1)$ and $(2, 4)$. Therefore,

$$\text{slope } m = \frac{\text{change in } y}{\text{change in } x} = \frac{y_2 - y_1}{x_2 - x_1} = \frac{4 - 1}{2 - 0} = \frac{3}{2}.$$

24. **(a)** Because parallel lines have equal slopes and the slope of the graph of $y = 2x + 3$ is 2, the slope of a line parallel to it will also be 2.

(b) Because perpendicular lines have slopes which are negative reciprocals of each other and the slope of the graph of $y = -3x + 3$ is -3, the slope of a line perpendicular to it will be

$$-\frac{1}{-3} = \frac{1}{3}.$$

25. Find the slope of each line by solving the equations for y.

$$\begin{aligned} 3x + 2y &= 6 \\ 2y &= -3x + 6 \quad \textit{Subtract 3x.} \\ y &= -\tfrac{3}{2}x + 3 \quad \textit{Divide by 2.} \end{aligned}$$

The slope of the first line is $-\frac{3}{2}$.

$$\begin{aligned} 6x + 4y &= 8 \\ 4y &= -6x + 8 \quad \textit{Subtract 6x.} \\ y &= -\tfrac{6}{4}x + 2 \quad \textit{Divide by 4.} \\ y &= -\tfrac{3}{2}x + 2 \quad \textit{Lowest terms} \end{aligned}$$

The slope of the second line is $-\frac{3}{2}$. The slopes are equal so the lines are *parallel*.

26. Find the slope of each line by solving the equations for y.

$$\begin{aligned} x - 3y &= 1 \\ -3y &= -x + 1 \quad \textit{Subtract x.} \\ y &= \tfrac{1}{3}x - \tfrac{1}{3} \quad \textit{Divide by −3.} \end{aligned}$$

The slope of the first line is $\frac{1}{3}$.

$$\begin{aligned} 3x + y &= 4 \\ y &= -3x + 4 \quad \textit{Subtract 3x.} \end{aligned}$$

The slope of the second line is -3.

The product of the slopes is

$$\tfrac{1}{3}(-3) = -1,$$

so the lines are *perpendicular*.

27. Find the slope of each line by solving the equations for y.

$$\begin{aligned} x - 2y &= 8 \\ -2y &= -x + 8 \quad \textit{Subtract x.} \\ y &= \tfrac{1}{2}x - 4 \quad \textit{Divide by −2.} \end{aligned}$$

The slope of the first line is $\frac{1}{2}$.

$$\begin{aligned} x + 2y &= 8 \\ 2y &= -x + 8 \quad \textit{Subtract x.} \\ y &= -\tfrac{1}{2}x + 4 \quad \textit{Divide by 2.} \end{aligned}$$

The slope of the second line is $-\frac{1}{2}$.

The slopes are not equal and their product is

$$\left(\tfrac{1}{2}\right)\left(-\tfrac{1}{2}\right) = -\tfrac{1}{4} \neq -1,$$

so the lines are *neither* parallel nor perpendicular.

28. $m = -1, b = \frac{2}{3}$

Use the slope-intercept form, $y = mx + b$.

$$y = -1 \cdot x + \tfrac{2}{3} \quad \text{or} \quad y = -x + \tfrac{2}{3}$$

29. Through $(2, 3)$ and $(-4, 6)$

$$m = \frac{6-3}{-4-2} = \frac{3}{-6} = -\frac{1}{2}$$

Use $(2, 3)$ and $m = -\frac{1}{2}$ in the point-slope form.

$$\begin{aligned} y - y_1 &= m(x - x_1) \\ y - 3 &= -\tfrac{1}{2}(x - 2) \\ y - 3 &= -\tfrac{1}{2}x + 1 \\ y &= -\tfrac{1}{2}x + 4 \end{aligned}$$

30. Through $(4, -3)$, $m = 1$

Use the point-slope form.

$$\begin{aligned} y - y_1 &= m(x - x_1) \\ y - (-3) &= 1(x - 4) \\ y + 3 &= x - 4 \\ y &= x - 7 \end{aligned}$$

31. Through $(-1, 4)$, $m = \frac{2}{3}$

Use the point-slope form.

$$\begin{aligned} y - y_1 &= m(x - x_1) \\ y - 4 &= \tfrac{2}{3}[x - (-1)] \\ y - 4 &= \tfrac{2}{3}(x + 1) \\ y - 4 &= \tfrac{2}{3}x + \tfrac{2}{3} \\ y &= \tfrac{2}{3}x + \tfrac{14}{3} \end{aligned}$$

32. Through $(1, -1)$, $m = -\frac{3}{4}$

Use the point-slope form.

$$\begin{aligned} y - (-1) &= -\tfrac{3}{4}(x - 1) \\ y + 1 &= -\tfrac{3}{4}x + \tfrac{3}{4} \\ y &= -\tfrac{3}{4}x - \tfrac{1}{4} \end{aligned}$$

33. $m = -\frac{1}{4}, b = \frac{3}{2}$

Use the slope-intercept form, $y = mx + b$.

$$y = -\tfrac{1}{4}x + \tfrac{3}{2}$$

34. Slope 0, through $(-4, 1)$

Horizontal lines have 0 slope and equations of the form $y = k$. In this case, k must equal 1 since the line goes through $(-4, 1)$, so the equation is $y = 1$.

35. Through $\left(\frac{1}{3}, -\frac{5}{4}\right)$ with undefined slope

Vertical lines have undefined slope and equations of the form $x = k$. In this case, k must equal $\frac{1}{3}$ since the line goes through $\left(\frac{1}{3}, -\frac{5}{4}\right)$, so the equation is $x = \frac{1}{3}$.

It is not possible to express this equation as $y = mx + b$.

36. **[3.3]** Vertical lines have undefined slopes. The answer is **A**.

37. **[3.2]** Two graphs pass through $(0, -3)$. **C** and **D** are the answers.

38. **[3.2]** Three graphs pass through the point $(-3, 0)$. **A**, **B**, and **D** are the answers.

39. **[3.3]** Lines that fall from left to right have negative slope. The answer is **D**.

40. **[3.2]** $y = -3$ is a horizontal line passing through $(0, -3)$. **C** is the answer.

41. **[3.3]** **B** is the only graph that has a positive slope, so it is the only one we need to investigate. **B** passes through the points $(0, 3)$ and $(-3, 0)$. Find the slope.

$$m = \frac{0 - 3}{-3 - 0} = \frac{-3}{-3} = 1$$

B is the answer.

42. **[3.3]** $y = -2x - 5$

The equation is in the slope-intercept form, so the slope is -2 and the y-intercept is $(0, -5)$. To find the x-intercept, let $y = 0$.

$$\begin{aligned} 0 &= -2x - 5 \quad \textit{Let } y = 0. \\ 2x &= -5 \\ x &= -\tfrac{5}{2} \end{aligned}$$

The x-intercept is $(-\frac{5}{2}, 0)$.

Graph the line using the intercepts.

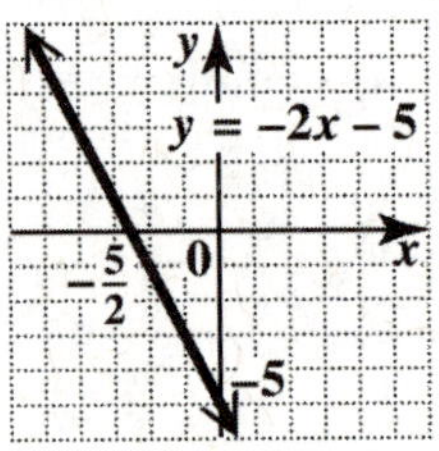

43. **[3.3]** $x + 3y = 0$

Solve the equation for y.

$$\begin{aligned} x + 3y &= 0 \\ 3y &= -x \quad \textit{Subtract } x. \\ y &= -\tfrac{1}{3}x \quad \textit{Divide by 3.} \end{aligned}$$

From this slope-intercept form, we see that the slope is $-\frac{1}{3}$ and the y-intercept is $(0, 0)$, which is also the x-intercept. To find another point, let $y = 1$.

$$\begin{aligned} x + 3(1) &= 0 \quad \textit{Let } y = 1. \\ x &= -3 \end{aligned}$$

So the point $(-3, 1)$ is on the graph. Graph the line through $(0, 0)$ and $(-3, 1)$.

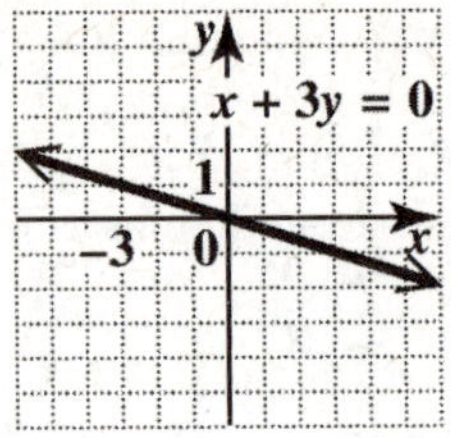

44. **[3.3]** $y - 5 = 0$ or $y = 5$

This is a horizontal line passing through the point $(0, 5)$, which is the y-intercept.

There is no x-intercept.
Horizontal lines have slopes of 0.

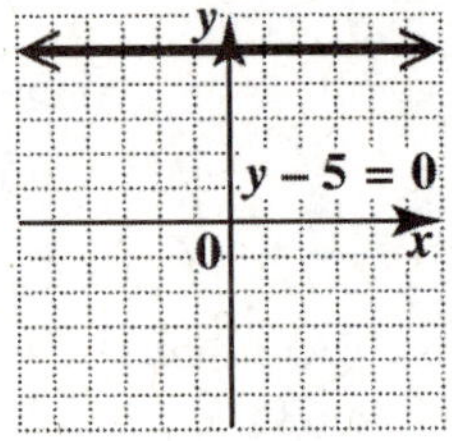

45. **[3.4]** $m = -\frac{1}{4}, b = -\frac{5}{4}$

Substitute the values $m = -\frac{1}{4}$ and $b = -\frac{5}{4}$ into the slope-intercept form.

$$\begin{aligned} y &= mx + b \\ y &= -\tfrac{1}{4}x - \tfrac{5}{4} \end{aligned}$$

46. **[3.4]** Through $(8, 6)$, $m = -3$

Use the point-slope form with $(x_1, y_1) = (8, 6)$ and $m = -3$.

$$\begin{aligned} y - y_1 &= m(x - x_1) \\ y - 6 &= -3(x - 8) \\ y - 6 &= -3x + 24 \\ y &= -3x + 30 \end{aligned}$$

47. **[3.4]** Through $(3, -5)$ and $(-4, -1)$

First, find the slope of the line.

$$m = \frac{-1 - (-5)}{-4 - 3} = \frac{4}{-7} = -\frac{4}{7}$$

Now use either point and the slope in the point-slope form. If we use $(3, -5)$, we get the following.

$$\begin{aligned} y - y_1 &= m(x - x_1) \\ y - (-5) &= -\tfrac{4}{7}(x - 3) \\ y + 5 &= -\tfrac{4}{7}x + \tfrac{12}{7} \\ y &= -\tfrac{4}{7}x - \tfrac{23}{7} \quad \textit{Subtract } 5 = \tfrac{35}{7}. \end{aligned}$$

48. **[3.4]** Slope 0, through $(5, -5)$

Since the slope is 0, we have a horizontal line. Knowing that one y-value is -5, tells us that all y-values are -5, so the equation is $y = -5$.

Chapter 3 Test

1. $3x + 5y = -30$; $(0, \_\_)$, $(\_\_, 0)$, $(\_\_, -3)$

$$\begin{aligned} 3x + 5y &= -30 \\ 3(0) + 5y &= -30 \quad \textit{Let } x = 0. \\ 5y &= -30 \\ y &= -6 \end{aligned}$$

$$\begin{aligned} 3x + 5y &= -30 \\ 3x + 5(0) &= -30 \quad \textit{Let } y = 0. \\ 3x &= -30 \\ x &= -10 \end{aligned}$$

$$\begin{aligned} 3x + 5y &= -30 \\ 3x + 5(-3) &= -30 \quad \textit{Let } y = -3. \\ 3x - 15 &= -30 \\ 3x &= -15 \\ x &= -5 \end{aligned}$$

The ordered pairs are $(0, -6)$, $(-10, 0)$, and $(-5, -3)$.

2.

$$\begin{aligned} 4x - 7y &= 9 \\ 4(4) - 7(-1) &\stackrel{?}{=} 9 \quad \textit{Let } x = 4,\ y = -1. \\ 16 + 7 &\stackrel{?}{=} 9 \\ 23 &\stackrel{?}{=} 9 \quad \textit{No} \end{aligned}$$

So $(4, -1)$ is not a solution of $4x - 7y = 9$.

3. To find the x-intercept, let $y = 0$ and solve for x. To find the y-intercept, let $x = 0$ and solve for y.

4. $3x + y = 6$

If $y = 0$, $x = 2$, so the x-intercept is $(2, 0)$.
If $x = 0$, $y = 6$, so the y-intercept is $(0, 6)$.

A third point, such as $(1, 3)$, can be used as a check. Draw a line through $(0, 6)$, $(1, 3)$, and $(2, 0)$.

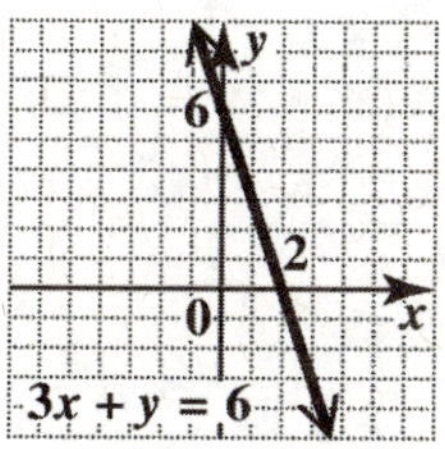

5. $y - 2x = 0$
Solving for y gives us the slope-intercept form of the line, $y = 2x$. We see that the y-intercept is $(0, 0)$ and so the x-intercept is also $(0, 0)$. The slope is 2 and can be written as

$$m = \frac{\text{rise}}{\text{run}} = \frac{2}{1}.$$

Starting at the origin and moving to the right 1 unit and then up 2 units gives us the point $(1, 2)$. Draw a line through $(0, 0)$ and $(1, 2)$.

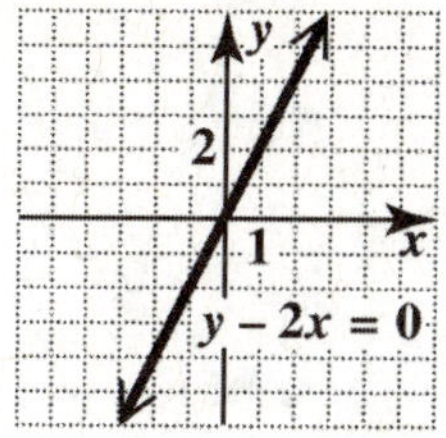

6. $x + 3 = 0$ can also be written as $x = -3$. Its graph is a vertical line with x-intercept $(-3, 0)$. There is no y-intercept.

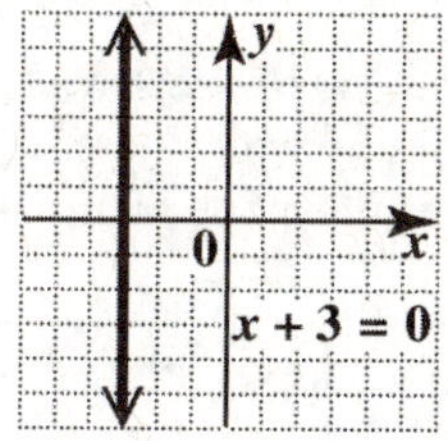

7. $y = 1$ is the graph of a horizontal line with y-intercept $(0, 1)$. There is no x-intercept.

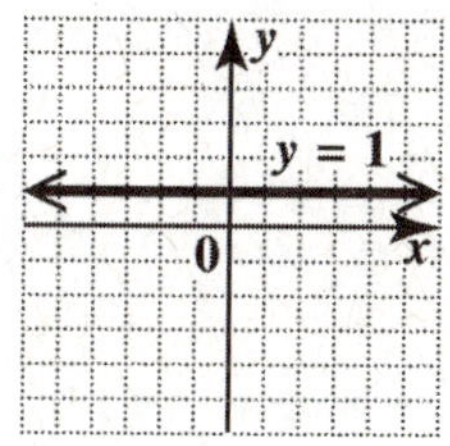

8. $x - y = 4$

If $y = 0$, $x = 4$, so the x-intercept is $(4, 0)$.
If $x = 0$, $y = -4$, so the y-intercept is $(0, -4)$.

A third point, such as $(2, -2)$, can be used as a check. Draw a line through $(0, -4)$, $(2, -2)$, and $(4, 0)$.

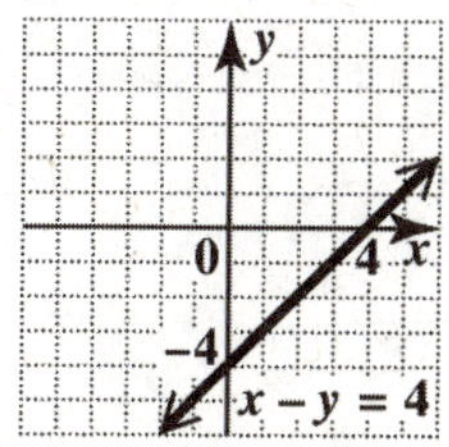

9. Through $(-4, 6)$ and $(-1, -2)$

Use the definition of slope with $(x_1, y_1) = (-4, 6)$ and $(x_2, y_2) = (-1, -2)$.

$$\text{slope } m = \frac{\text{change in } y}{\text{change in } x} = \frac{y_2 - y_1}{x_2 - x_1}$$

$$= \frac{-2 - 6}{-1 - (-4)} = \frac{-8}{3} = -\frac{8}{3}$$

10. $2x + y = 10$

To find the slope, solve the given equation for y.

$$\begin{aligned} 2x + y &= 10 \\ y &= -2x + 10 \end{aligned}$$

The equation is now written in $y = mx + b$ form, so the slope is given by the coefficient of x, which is -2.

11. $x + 12 = 0$ can also be written as $x = -12$. Its graph is a vertical line with x-intercept $(-12, 0)$. The slope is undefined.

12. $y - 4 = 6$ can also be written as $y = 10$. Its graph is a horizontal line with y-intercept $(0, 10)$. Its slope is 0, as is the slope of any line parallel to it.

13. The indicated points are $(0, -4)$ and $(2, 1)$. Use the definition of slope with $(x_1, y_1) = (0, -4)$ and $(x_2, y_2) = (2, 1)$.

$$\begin{aligned} \text{slope } m &= \frac{\text{change in } y}{\text{change in } x} = \frac{y_2 - y_1}{x_2 - x_1} \\ &= \frac{1 - (-4)}{2 - 0} = \frac{5}{2} \end{aligned}$$

14. Through $(-1, 4)$; $m = 2$

Let $x_1 = -1$, $y_1 = 4$, and $m = 2$ in the point-slope form.

$$\begin{aligned} y - y_1 &= m(x - x_1) \\ y - 4 &= 2[x - (-1)] \\ y - 4 &= 2(x + 1) \\ y - 4 &= 2x + 2 \\ y &= 2x + 6 \end{aligned}$$

15. The indicated points are $(0, -4)$ and $(2, 1)$. The slope of the line through those points is

$$m = \frac{1 - (-4)}{2 - 0} = \frac{5}{2}.$$

The y-intercept is $(0, -4)$, so the slope-intercept form is

$$y = \tfrac{5}{2}x - 4.$$

16. Through $(2, -6)$ and $(1, 3)$

The slope of the line through these points is

$$m = \frac{3 - (-6)}{1 - 2} = \frac{9}{-1} = -9.$$

Use the point-slope form of a line with $(1, 3) = (x_1, y_1)$ and $m = -9$.

$$\begin{aligned} y - y_1 &= m(x - x_1) \\ y - 3 &= -9(x - 1) \\ y - 3 &= -9x + 9 \\ y &= -9x + 12 \end{aligned}$$

17. The slope is negative since worldwide snowmobile sales are decreasing, indicated by the line which falls from left to right.

18. Two ordered pairs are $(0, 209)$ and $(7, 160)$. Use these points to find the slope.

$$m = \frac{y_2 - y_1}{x_2 - x_1} = \frac{160 - 209}{7 - 0} = \frac{-49}{7} = -7$$

The slope is -7. Using the ordered pair $(0, 209)$, which is the y-intercept, and the slope of -7, we get the slope-intercept form

$$y = -7x + 209.$$

19. For 2005, $x = 2005 - 2000 = 5$.

$$\begin{aligned} y &= -7x + 209 \\ y &= -7(5) + 209 \\ y &= 174 \end{aligned}$$

This represents worldwide snowmobile sales of 174 thousand in 2005. The equation gives a good approximation of the actual sales, which are 173.7 thousand.

20. The ordered pair $(7, 160)$ means that in the year 2007, worldwide snowmobile sales were 160 thousand.

Cumulative Review Exercises (Chapters 1–3)

1. $$\begin{aligned} 10\tfrac{5}{8} - 3\tfrac{1}{10} &= \tfrac{85}{8} - \tfrac{31}{10} \\ &= \tfrac{425}{40} - \tfrac{124}{40} \\ &= \tfrac{301}{40}, \text{ or } 7\tfrac{21}{40} \end{aligned}$$

2. $$\frac{3}{4} \div \frac{1}{8} = \frac{3}{4} \cdot \frac{8}{1} = \frac{3 \cdot 2 \cdot 4}{4 \cdot 1} = 3 \cdot 2 = 6$$

3. $5 - (-4) + (-2) = 9 + (-2) = 7$

4. $$\begin{aligned} &\frac{(-3)^2 - (-4)(2^4)}{5(2) - (-2)^3} \\ &= \frac{9 - (-4)(16)}{10 - (-8)} \quad \textit{Do exponents first.} \\ &= \frac{9 - (-64)}{10 - (-8)} \quad \textit{Multiply.} \\ &= \frac{9 + 64}{10 + 8} = \frac{73}{18}, \text{ or } 4\frac{1}{18} \end{aligned}$$

5. $$\begin{aligned} \frac{4(3 - 9)}{2 - 6} &\overset{?}{\geq} 6 \\ \frac{4(-6)}{-4} &\overset{?}{\geq} 6 \\ \frac{-24}{-4} &\overset{?}{\geq} 6 \\ 6 &\geq 6 \end{aligned}$$

The statement is *true* since $6 = 6$.

6. $xz^3 - 5y^2 = (-2)(-1)^3 - 5(-3)^2$ *Let x = −2, y = −3, z = −1.*

$= (-2)(-1) + (-5)(9)$

$= 2 + (-45)$

$= -43$

7. $3(-2 + x) = 3 \cdot (-2) + 3(x)$

$= -6 + 3x$

illustrates the *distributive property.*

8. $-4p - 6 + 3p + 8 = (-4p + 3p) + (-6 + 8)$

$= -p + 2$

9. $V = \frac{1}{3}\pi r^2 h$ for h

$3V = \pi r^2 h$ *Multiply by 3.*

$\dfrac{3V}{\pi r^2} = h$ *Divide by πr^2.*

10. $6 - 3(1 + x) = 2(x + 5) - 2$

$6 - 3 - 3x = 2x + 10 - 2$

$3 - 3x = 2x + 8$

$-5x = 5$

$x = -1$

The solution set is $\{-1\}$.

11. $-(m - 3) = 5 - 2m$

$-m + 3 = 5 - 2m$ *Distributive property*

$m + 3 = 5$ *Add 2m.*

$m = 2$ *Subtract 3.*

The solution set is $\{2\}$.

12. $\dfrac{x - 2}{3} = \dfrac{2x + 1}{5}$

$(x - 2)(5) = (3)(2x + 1)$ *Cross products*

$5x - 10 = 6x + 3$

$-10 = x + 3$

$-13 = x$

The solution set is $\{-13\}$.

13. $-2.5x < 6.5$

$\dfrac{-2.5x}{-2.5} > \dfrac{6.5}{-2.5}$

Divide by −2.5; reverse the symbol.

$x > -2.6$

Thus, the solution set is $(-2.6, \infty)$.

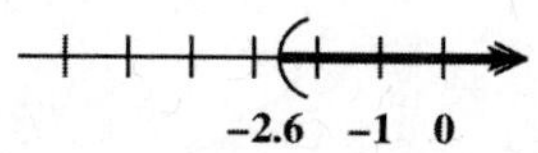

14. $4(x + 3) - 5x < 12$

$4x + 12 - 5x < 12$ *Distributive property*

$-x + 12 < 12$ *Combine like terms.*

$-x < 0$ *Subtract 12.*

$x > 0$ *Divide by −1; reverse the symbol.*

Thus, the solution set is $(0, \infty)$.

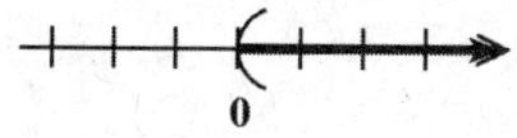

15. $\frac{2}{3}x - \frac{1}{6}x \le -2$

$6\left(\frac{2}{3}x - \frac{1}{6}x\right) \le 6(-2)$ *Multiply by 6.*

$6\left(\frac{2}{3}x\right) - 6\left(\frac{1}{6}x\right) \le 6(-2)$ *Dist. property*

$4x - x \le -12$

$3x \le -12$

$x \le -4$

Thus, the solution set is $(-\infty, -4]$.

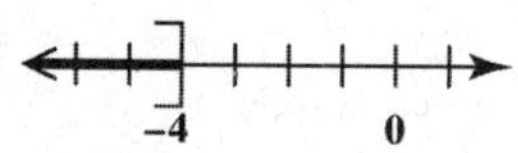

16. Let x = average annual earnings for a person with a high school diploma.
Then $x + 20{,}124$ = average annual earnings for a person with a bachelor's degree.

$x + (x + 20{,}124) = 81{,}588$

$2x + 20{,}124 = 81{,}588$

$2x = 61{,}464$

$x = 30{,}732$

A person with a high school diploma can expect to earn \$30,732 per year while a person with a bachelor's degree can expect to earn

\$30,732 + \$20,124 = \$50,856 per year.

17. **(a)** Multiply 14% (or 0.14) by the total of \$50,000.

$0.14(50{,}000) = 7000$

\$7000 is expected to go toward home purchase.

(b) Multiply 20% (or 0.20) by the total of \$50,000.

$0.20(50{,}000) = 10{,}000$

\$10,000 is expected to go toward retirement.

18. Since the sector for paying off debt or funding children's education is about three times larger than the sector for retirement, 3($10,000) or about $30,000 is expected to go toward paying off debt or funding children's education.

19. To find the x-intercept, let $y = 0$.

$$\begin{aligned} -3x + 4y &= 12 \\ -3x + 4(0) &= 12 \\ -3x &= 12 \\ x &= -4 \end{aligned}$$

The x-intercept is $(-4, 0)$.

To find the y-intercept, let $x = 0$.

$$\begin{aligned} -3x + 4y &= 12 \\ -3(0) + 4y &= 12 \\ 4y &= 12 \\ y &= 3 \end{aligned}$$

The y-intercept is $(0, 3)$.

20. To find the slope of the line, solve the equation for y.

$$\begin{aligned} -3x + 4y &= 12 \\ 4y &= 3x + 12 \\ y &= \tfrac{3}{4}x + 3 \end{aligned}$$

The slope is the coefficient of x, $\frac{3}{4}$.

21. To find a third point, let $x = 4$.

$$\begin{aligned} -3x + 4y &= 12 \\ -3(4) + 4y &= 12 \\ 4y &= 12 + 12 = 24 \\ y &= 6 \end{aligned}$$

Plot the points $(-4, 0)$, $(0, 3)$, and $(4, 6)$ and draw a line through them.

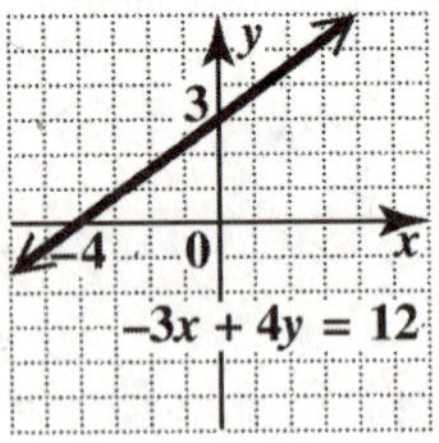

22.

$$\begin{aligned} x + 5y &= -6 \\ 5y &= -x - 6 \\ y &= -\tfrac{1}{5}x - \tfrac{6}{5} \end{aligned}$$

The slope of the first line is $-\frac{1}{5}$.

The slope of the second line, $y = 5x - 8$, is 5. Since $-\frac{1}{5}$ is the negative reciprocal of 5, the lines are *perpendicular*.

23. Through $(2, -5)$ with slope 3.

Use the point-slope form of a line.

$$\begin{aligned} y - y_1 &= m(x - x_1) \\ y - (-5) &= 3(x - 2) \\ y + 5 &= 3x - 6 \\ y &= 3x - 11 \end{aligned}$$

24. Through $(0, 4)$ and $(2, 4)$

$$\text{slope } m = \frac{\text{change in } y}{\text{change in } x} = \frac{4-4}{2-0} = \frac{0}{2} = 0$$

Since the slope is 0, the line is horizontal. Horizontal lines have equations of the form $y = k$. An equation of the line is $y = 4$.

CHAPTER 4 EXPONENTS AND POLYNOMIALS

4.1 The Product Rule and Power Rules for Exponents

4.1 Now Try Exercises

N1. $4 \cdot 4 \cdot 4 = 4^3$ 4 occurs as a factor 3 times.
$= 64$

N2. **(a)** $(-3)^4 = (-3)(-3)(-3)(-3)$
$= 81$

Base: -3; exponent: 4

(b) $-3^4 = -1 \cdot 3^4$
$= -1 \cdot (3 \cdot 3 \cdot 3 \cdot 3)$
$= -81$

Base: 3; exponent: 4

N3. **(a)** $(-5)^2(-5)^4 = (-5)^{2+4}$ *Product rule*
$= (-5)^6$

(b) $y^2 \cdot y \cdot y^5 = y^2 \cdot y^1 \cdot y^5 = y^{2+1+5} = y^8$

(c) $(2x^3)(4x^7) = (2 \cdot 4) \cdot (x^3 \cdot x^7)$
$= 8(x^{3+7})$
$= 8x^{10}$

(d) $2^4 \cdot 5^3$: the product rule does not apply because the bases are different.

$$2^4 \cdot 5^3 = 16 \cdot 125 = 2000$$

(e) $3^2 + 3^3$: the product rule does not apply because it is a sum, not a product.

$$3^2 + 3^3 = 9 + 27 = 36$$

N4. **(a)** $(4^7)^5 = 4^{7 \cdot 5}$ *Power rule (a)*
$= 4^{35}$

(b) $(y^4)^7 = y^{4 \cdot 7}$ *Power rule (a)*
$= y^{28}$

N5. **(a)** $(-5ab)^3$
$= (-1 \cdot 5 \cdot a \cdot b)^3$ $\quad -x = -1 \cdot x$
$= (-1)^3 \cdot 5^3 \cdot a^3 \cdot b^3$ *Power rule (b)*
$= -5^3a^3b^3$ or $-125a^3b^3$

(b) $3(4t^3p^5)^2 = 3 \cdot 4^2(t^3)^2(p^5)^2$ *Power rule (b)*
$= 48t^6p^{10}$ *Power rule (a)*

N6. **(a)** $\left(\frac{p}{q}\right)^5 = \frac{p^5}{q^5}$ *Power rule (c)*, $q \neq 0$

(b) $\left(\frac{1}{4}\right)^3 = \frac{1^3}{4^3}$ *Power rule (c)*
$= \frac{1}{64}$

N7. **(a)** $\left(\frac{3}{5}\right)^3 \cdot 3^2 = \frac{3^3}{5^3} \cdot \frac{3^2}{1}$ *Power rule (c)*
$= \frac{3^{3+2}}{5^3}$ *Power rule (b)*
$= \frac{3^5}{5^3}$ or $\frac{243}{125}$

(b) $(8k)^5(8k)^4 = (8k)^{5+4}$ *Product rule*
$= (8k)^9$
$= 8^9k^9$ *Power rule (b)*

(c) $(x^4y)^5(-2x^2y^5)^3$
$= (x^4)^5y^5(-1)^3 2^3(x^2)^3(y^5)^3$ *Power rule (b)*
$= x^{4 \cdot 5}y^5(-1)2^3x^{2 \cdot 3}y^{5 \cdot 3}$ *Power rule (a)*
$= -2^3x^{20}y^5x^6y^{15}$
$= -2^3x^{20+6}y^{5+15}$ *Product rule*
$= -2^3x^{26}y^{20}$ or $-8x^{26}y^{20}$

N8. Use the formula for the area of a triangle, $\mathcal{A} = \frac{1}{2}bh$, with $b = 10x^8$ and $h = 3x^7$.

$\mathcal{A} = \frac{1}{2}(10x^8)(3x^7)$ *Area formula*
$= \frac{1}{2} \cdot 10 \cdot 3 \cdot x^{8+7}$ *Product rule*
$= 15x^{15}$

4.1 Section Exercises

1. $3^3 = 3 \cdot 3 \cdot 3 = 27$, so the statement $3^3 = 9$ is *false*.

3. $(x^2)^3 = x^{2(3)} = x^6$, so the statement $(x^2)^3 = x^5$ is *false*.

5. $\underbrace{w \cdot w \cdot w \cdot w \cdot w \cdot w}_{6w\text{'s}} = w^6$

7. $\left(\frac{1}{2}\right)\left(\frac{1}{2}\right)\left(\frac{1}{2}\right)\left(\frac{1}{2}\right)\left(\frac{1}{2}\right)\left(\frac{1}{2}\right) = \left(\frac{1}{2}\right)^6$

9. $(-4)(-4)(-4)(-4) = (-4)^4$

11. $(-7y)(-7y)(-7y)(-7y) = (-7y)^4$

13. In $(-3)^4$, -3 is the base.

$$(-3)^4 = (-3)(-3)(-3)(-3) = 81$$

In -3^4, 3 is the base.

$$-3^4 = -(3 \cdot 3 \cdot 3 \cdot 3) = -81$$

15. In the exponential expression 3^5, the base is 3 and the exponent is 5.

$$3^5 = 3 \cdot 3 \cdot 3 \cdot 3 \cdot 3 = 243$$

17. In the expression $(-3)^5$, the base is -3 and the exponent is 5.

$$(-3)^5 = (-3)(-3)(-3)(-3)(-3) = -243$$

19. In the expression $(-6x)^4$, the base is $-6x$ and the exponent is 4.

21. In the expression $-6x^4$, -6 is not part of the base. The base is x and the exponent is 4.

23. The product rule does not apply to $5^2 + 5^3$ because the expression is a sum, not a product. The product rule would apply if we had $5^2 \cdot 5^3$.

$$5^2 + 5^3 = 25 + 125 = 150$$

25. $5^2 \cdot 5^6 = 5^{2+6} = 5^8$

27. $4^2 \cdot 4^7 \cdot 4^3 = 4^{2+7+3} = 4^{12}$

29. $(-7)^3(-7)^6 = (-7)^{3+6} = (-7)^9$

31. $t^3 \cdot t^8 \cdot t^{13} = t^{3+8+13} = t^{24}$

33. $$\begin{aligned}(-8r^4)(7r^3) &= -8 \cdot 7 \cdot r^4 \cdot r^3 \\ &= -56r^{4+3} \\ &= -56r^7\end{aligned}$$

35. $$\begin{aligned}(-6p^5)(-7p^5) &= (-6)(-7)p^5 \cdot p^5 \\ &= 42p^{5+5} \\ &= 42p^{10}\end{aligned}$$

37. $$\begin{aligned}&(5x^2)(-2x^3)(3x^4) \\ &= (5)(-2)(3)x^2 \cdot x^3 \cdot x^4 \\ &= (-10)(3)x^{2+3+4} \\ &= -30x^9\end{aligned}$$

39. $3^8 + 3^9$ is a sum, so the product rule does not apply.

41. $5^8 \cdot 3^9$ is a product with different bases, so the product rule does not apply.

43. $(4^3)^2 = 4^{3 \cdot 2}$ *Power rule (a)*
$= 4^6$

45. $(t^4)^5 = t^{4 \cdot 5} = t^{20}$ *Power rule (a)*

47. $(7r)^3 = 7^3r^3$ *Power rule (b)*

49. $(5xy)^5 = 5^5x^5y^5$ *Power rule (b)*

51. $(-5^2)^6 = (-1 \cdot 5^2)^6$
$= (-1)^6 \cdot (5^2)^6$ *Power rule (b)*
$= 1 \cdot 5^{2 \cdot 6}$ *Power rule (a)*
$= 1 \cdot 5^{12} = 5^{12}$

53. $(-8^3)^5 = (-1 \cdot 8^3)^5$
$= (-1)^5 \cdot (8^3)^5$ *Power rule (b)*
$= -1 \cdot 8^{3 \cdot 5}$ *Power rule (a)*
$= -8^{15}$

55. $8(qr)^3 = 8q^3r^3$ *Power rule (b)*

57. $\left(\frac{9}{5}\right)^8 = \frac{9^8}{5^8}$ *Power rule (c)*

59. $\left(\frac{1}{2}\right)^3 = \frac{1^3}{2^3} = \frac{1}{2^3}$ *Power rule (c)*

61. $\left(\frac{a}{b}\right)^3 = \frac{a^3}{b^3}$ *Power rule (c), ($b \neq 0$)*

63. $\left(\frac{x}{2}\right)^3 = \frac{x^3}{2^3}$ *Power rule (c)*

65. $\left(\frac{5}{2}\right)^3 \cdot \left(\frac{5}{2}\right)^2 = \left(\frac{5}{2}\right)^{3+2}$ *Product rule*
$= \left(\frac{5}{2}\right)^5$
$= \frac{5^5}{2^5}$ *Power rule (c)*

67. $\left(\frac{9}{8}\right)^3 \cdot 9^2 = \frac{9^3}{8^3} \cdot \frac{9^2}{1}$ *Power rule (c)*
$= \frac{9^3 \cdot 9^2}{8^3 \cdot 1}$ *Multiply fractions*
$= \frac{9^{3+2}}{8^3}$ *Product rule*
$= \frac{9^5}{8^3}$

69. $(2x)^9(2x)^3 = (2x)^{9+3}$ *Product rule*
$= (2x)^{12}$
$= 2^{12}x^{12}$ *Product rule (b)*

71. $(-6p)^4(-6p)$
$= (-6p)^4(-6p)^1$
$= (-6p)^5$ *Product rule*
$= (-1)^5 6^5 p^5$ *Power rule (b)*
$= -6^5p^5$

73. $(6x^2y^3)^5 = 6^5(x^2)^5(y^3)^5$ *Power rule (b)*
$= 6^5x^{2 \cdot 5}y^{3 \cdot 5}$ *Power rule (a)*
$= 6^5x^{10}y^{15}$

75. $(x^2)^3(x^3)^5 = x^6 \cdot x^{15}$ *Power rule (a)*
$= x^{21}$ *Product rule*

77. $(2w^2x^3y)^2(x^4y)^5$
$= [2^2(w^2)^2(x^3)^2y^2][(x^4)^5y^5]$ *Power rule (b)*
$= (2^2w^4x^6y^2)(x^{20}y^5)$ *Power rule (a)*
$= 2^2w^4(x^6x^{20})(y^2y^5)$
Commutative and associative properties
$= 4w^4x^{26}y^7$

79. $(-r^4s)^2(-r^2s^3)^5$
$= [(-1)r^4s]^2[(-1)r^2s^3]^5$
$= [(-1)^2(r^4)^2s^2][(-1)^5(r^2)^5(s^3)^5]$
Power rule (b)
$= [(-1)^2r^8s^2][(-1)^5r^{10}s^{15}]$ *Power rule (a)*
$= (-1)^7r^{18}s^{17}$ *Product rule*
$= -r^{18}s^{17}$

81. $\left(\dfrac{5a^2b^5}{c^6}\right)^3 \quad (c \neq 0)$

$$= \frac{(5a^2b^5)^3}{(c^6)^3} \qquad \textit{Power rule (c)}$$
$$= \frac{5^3(a^2)^3(b^5)^3}{(c^6)^3} \qquad \textit{Power rule (b)}$$
$$= \frac{125a^6b^{15}}{c^{18}} \qquad \textit{Power rule (a)}$$

83. To simplify $(10^2)^3$ as 1000^6 is not correct. Using power rule (a) to simplify $(10^2)^3$, we obtain

$$\begin{aligned}(10^2)^3 &= 10^{2\cdot 3}\\ &= 10^6\\ &= 10\cdot 10\cdot 10\cdot 10\cdot 10\cdot 10\\ &= 1{,}000{,}000.\end{aligned}$$

85. Use the formula for the area of a rectangle, $\mathcal{A} = LW$, with $L = 4x^3$ and $W = 3x^2$.

$$\begin{aligned}\mathcal{A} &= (4x^3)(3x^2)\\ &= 4\cdot 3\cdot x^3\cdot x^2\\ &= 12x^5\end{aligned}$$

87. Use the formula for the area of a parallelogram, $\mathcal{A} = bh$, with $b = 2p^5$ and $h = 3p^2$.

$$\begin{aligned}\mathcal{A} &= (2p^5)(3p^2)\\ &= 2\cdot 3\cdot p^5\cdot p^2\\ &= 6p^7\end{aligned}$$

89. Use the formula for the volume of a cube, $V = e^3$, with $e = 5x^2$.

$$\begin{aligned}V &= (5x^2)^3\\ &= 5^3(x^2)^3\\ &= 125x^6\end{aligned}$$

91. If a is a number greater than 1, then:

(1) $-(-a)^3 = -1(-a^3) = a^3$ is positive
(2) $-a^3 = -1(a^3)$ is negative
(3) $(-a)^4 = a^4$ is positive
(4) $-a^4 = -1(a^4)$ is negative

Since $a^4 > a^3$, $-a^4 < -a^3$, so in order from least to greatest, we have

$$-a^4,\ -a^3,\ -(-a)^3,\ (-a)^4.$$

Another way to determine the order is to choose a number greater than 1 and substitute it for a in each expression, and then arrange the terms from least to greatest.

93. Use the formula $A = P(1+r)^n$ with $P = \$250$, $r = 0.04$, and $n = 5$.

$$\begin{aligned}A &= 250(1+0.04)^5\\ &= 250(1.04)^5\\ &\approx 304.16\end{aligned}$$

The amount of money in the account will be \$304.16.

95. Use the formula $A = P(1+r)^n$ with $P = \$1500$, $r = 0.035$, and $n = 6$.

$$\begin{aligned}A &= 1500(1+0.035)^6\\ &= 1500(1.035)^6\\ &\approx 1843.88\end{aligned}$$

The amount of money in the account will be \$1843.88.

97. The reciprocal of 9 is $\frac{1}{9}$.

99. The reciprocal of $-\frac{1}{8}$ is $\frac{1}{-\frac{1}{8}} = 1\left(-\frac{8}{1}\right) = -8$.

101. $8 - (-4) = 8 + 4 = 12$

103. "Subtract -6 from -3" translates and simplifies as:

$$\begin{aligned}-3-(-6) &= -3+6\\ &= 3\end{aligned}$$

4.2 Integer Exponents and the Quotient Rule

4.2 Now Try Exercises

N1. **(a)** $6^0 = 1$ *Definition of zero exponent*

(b) $-12^0 = -(12^0) = -1$

(c) $(-12x)^0 = 1\ (x \neq 0)$

(d) $14^0 - 12^0 = 1 - 1 = 0$

N2. **(a)** $2^{-3} = \dfrac{1}{2^3}$ *Definition of negative exponent*

$$= \frac{1}{8}$$

(b) $\left(\frac{1}{7}\right)^{-2} = \left(\frac{7}{1}\right)^2$ $\quad\frac{1}{7}$ *and 7 are reciprocals.*

$$= 49$$

(c) $\left(\frac{3}{2}\right)^{-4} = \left(\frac{2}{3}\right)^4 = \frac{16}{81}$

(d) $3^{-2} + 4^{-2} = \dfrac{1}{3^2} + \dfrac{1}{4^2}$

$$\begin{aligned}&= \tfrac{1}{9} + \tfrac{1}{16}\\ &= \tfrac{16}{144} + \tfrac{9}{144} = \tfrac{25}{144}\end{aligned}$$

(e) $p^{-4} = \dfrac{1}{p^4},\ p \neq 0$

N3. **(a)** $\dfrac{5^{-3}}{6^{-2}} = \dfrac{6^2}{5^3}$, or $\dfrac{36}{125}$

(b) $m^2n^{-4} = \dfrac{m^2}{1}\cdot\dfrac{1}{n^4} = \dfrac{m^2}{n^4}$

(c) $\dfrac{x^2y^{-3}}{5z^{-4}} = \dfrac{x^2}{5}\cdot\dfrac{y^{-3}}{z^{-4}} = \dfrac{x^2}{5}\cdot\dfrac{z^4}{y^3} = \dfrac{x^2z^4}{5y^3}$

N4. (a) $\frac{6^3}{6^4} = 6^{3-4} = 6^{-1} = \frac{1}{6^1} = \frac{1}{6}$

(b) $\frac{t^4}{t^{-5}} = t^{4-(-5)} = t^{4+5} = t^9$

(c) $\frac{(p+q)^{-3}}{(p+q)^{-7}} = (p+q)^{-3-(-7)}$
$= (p+q)^{-3+7}$
$= (p+q)^4$

(d) $\frac{5^2xy^{-3}}{3^{-1}x^{-2}y^2} = \frac{5^2}{3^{-1}} \cdot \frac{x}{x^{-2}} \cdot \frac{y^{-3}}{y^2}$
$= 5^2 3^1 \cdot x^{1-(-2)} \cdot y^{-3-2}$
$= 25 \cdot 3 \cdot x^3 \cdot y^{-5}$
$= \frac{75}{1} \cdot \frac{x^3}{1} \cdot \frac{1}{y^5}$
$= \frac{75x^3}{y^5}$

N5. (a) $\frac{3^{15}}{(3^3)^4} = \frac{3^{15}}{3^{12}} = 3^{15-12} = 3^3$, or 27

(b) $(4t)^5(4t)^{-3} = (4t)^{5+(-3)}$
$= (4t)^2$
$= 4^2t^2$, or $16t^2$

(c) $\left(\frac{7y^4}{10}\right)^{-3} = \left(\frac{10}{7y^4}\right)^3$
$= \frac{10^3}{(7y^4)^3}$
$= \frac{10^3}{7^3(y^4)^3}$
$= \frac{10^3}{7^3y^{12}}$, or $\frac{1000}{343y^{12}}$

(d) $\frac{(a^2b^{-2}c)^{-3}}{(2ab^3c^{-4})^5} = \frac{(a^2)^{-3}(b^{-2})^{-3}c^{-3}}{2^5a^5(b^3)^5(c^{-4})^5}$
$= \frac{a^{-6}b^6c^{-3}}{2^5a^5b^{15}c^{-20}}$
$= \frac{b^6c^{20}}{2^5a^5a^6b^{15}c^3}$
$= \frac{c^{20-3}}{2^5a^{5+6}b^{15-6}}$
$= \frac{c^{17}}{2^5a^{11}b^9}$, or $\frac{c^{17}}{32a^{11}b^9}$

4.2 Section Exercises

1. By definition, $a^0 = 1$ $(a \neq 0)$, so $9^0 = 1$.

3. $(-2)^0 = 1$ *Definition of zero exponent*

5. $-8^0 = -(8^0) = -(1) = -1$

7. $-(-6)^0 = -1 \cdot (-6)^0 = -1 \cdot 1 = -1$

9. $(-4)^0 - 4^0 = 1 - 1 = 0$

11. $\frac{0^{10}}{12^0} = \frac{0}{1} = 0$

13. $8^0 - 12^0 = 1 - 1 = 0$

15. $\frac{0^2}{2^0 + 0^2} = \frac{0}{1+0} = \frac{0}{1} = 0$

17. **(a)** $x^0 = 1$ (Choice **B**)

(b) $-x^0 = -1 \cdot x^0 = -1 \cdot 1 = -1$ (Choice **C**)

(c) $7x^0 = 7 \cdot x^0 = 7 \cdot 1 = 7$ (Choice **D**)

(d) $(7x)^0 = 1$ (Choice **B**)

(e) $-7x^0 = -7 \cdot x^0 = -7 \cdot 1 = -7$ (Choice **E**)

(f) $(-7x)^0 = 1$ (Choice **B**)

19. $6^0 + 8^0 = 1 + 1 = 2$

21. $4^{-3} = \frac{1}{4^3}$ *Definition of negative exponent*
$= \frac{1}{64}$

23. When we evaluate a fraction raised to a negative exponent, we can use a shortcut. Note that

$$\left(\frac{a}{b}\right)^{-n} = \frac{1}{\left(\frac{a}{b}\right)^n} = \frac{1}{\frac{a^n}{b^n}} = \frac{b^n}{a^n} = \left(\frac{b}{a}\right)^n.$$

In words, a fraction raised to the negative of a number is equal to its reciprocal raised to the number. We will use the simple phrase "$\frac{a}{b}$ and $\frac{b}{a}$ are reciprocals" to indicate our use of this evaluation shortcut.

$\left(\frac{1}{2}\right)^{-4} = 2^4 = 16$ *$\frac{1}{2}$ and 2 are reciprocals.*

25. $\left(\frac{6}{7}\right)^{-2} = \left(\frac{7}{6}\right)^2$ *$\frac{6}{7}$ and $\frac{7}{6}$ are reciprocals.*
$= \frac{7^2}{6^2}$ *Power rule (c)*
$= \frac{49}{36}$

27. $(-3)^{-4} = \frac{1}{(-3)^4}$
$= \frac{1}{81}$

29. $5^{-1} + 3^{-1} = \frac{1}{5} + \frac{1}{3}$
$= \frac{3}{15} + \frac{5}{15} = \frac{8}{15}$

31. $3^{-2} - 2^{-1} = \frac{1}{3^2} - \frac{1}{2^1}$
$= \frac{1}{9} - \frac{1}{2}$
$= \frac{2}{18} - \frac{9}{18} = -\frac{7}{18}$

33. $(\frac{1}{2})^{-1} + (\frac{2}{3})^{-1} = (\frac{2}{1})^1 + (\frac{3}{2})^1$
$= \frac{4}{2} + \frac{3}{2}$
$= \frac{7}{2}$

35. $\dfrac{5^8}{5^5} = 5^{8-5} = 5^3$, or 125 *Quotient rule*

37. $\dfrac{3^{-2}}{5^{-3}} = \dfrac{5^3}{3^2}$, or $\frac{125}{9}$ *Changing from negative to positive exponents*

39. $\dfrac{5}{5^{-1}} = \dfrac{5^1}{5^{-1}} = 5^1 \cdot 5^1$ *Changing from negative to positive exponents*
$= 5^{1+1} = 5^2$, or 25

41. $\dfrac{x^{12}}{x^{-3}} = x^{12} \cdot x^3$ *Changing from negative to positive exponents*
$= x^{12+3} = x^{15}$

43. $\dfrac{1}{6^{-3}} = 6^3$, or 216 *Changing from negative to positive exponents*

45. $\dfrac{2}{r^{-4}} = 2r^4$ *Changing from negative to positive exponents*

47. $\dfrac{4^{-3}}{5^{-2}} = \dfrac{5^2}{4^3}$, or $\dfrac{25}{64}$ *Changing from negative to positive exponents*

49. $p^5q^{-8} = \dfrac{p^5}{q^8}$ *Changing from negative to positive exponents*

51. $\dfrac{r^5}{r^{-4}} = r^5 \cdot r^4 = r^{5+4} = r^9$

Or we can use the quotient rule:

$$\frac{r^5}{r^{-4}} = r^{5-(-4)} = r^{5+4} = r^9$$

53. $\dfrac{x^{-3}y}{4z^{-2}} = \dfrac{yz^2}{4x^3}$

55. Treat the expression in parentheses as a single variable; that is, treat $(a+b)$ as you would treat x.

$$\frac{(a+b)^{-3}}{(a+b)^{-4}} = (a+b)^{-3-(-4)}$$
$$= (a+b)^{-3+4}$$
$$= (a+b)^1 = a+b$$

Another Method:

$$\frac{(a+b)^{-3}}{(a+b)^{-4}} = \frac{(a+b)^4}{(a+b)^3}$$
$$= (a+b)^{4-3}$$
$$= (a+b)^1 = a+b$$

57. $\dfrac{(x+2y)^{-3}}{(x+2y)^{-5}} = (x+2y)^{-3-(-5)}$
$= (x+2y)^{-3+5}$
$= (x+2y)^2$

♦♦♦ Relating Concepts 59–62 ♦♦♦

59. In simplest form, $\frac{25}{25} = 1$.

60. $\dfrac{25}{25} = \dfrac{5^2}{5^2}$

61. $\dfrac{5^2}{5^2} = 5^{2-2} = 5^0$

62. $1 = 5^0$; this supports the definition of 0 as an exponent.

63. $\dfrac{(7^4)^3}{7^9} = \dfrac{7^{4\cdot 3}}{7^9}$ *Power rule (a)*
$= \dfrac{7^{12}}{7^9}$
$= 7^{12-9}$ *Quotient rule*
$= 7^3$, or 343

65. $x^{-3} \cdot x^5 \cdot x^{-4}$
$= x^{-3+5+(-4)}$ *Product rule*
$= x^{-2}$
$= \dfrac{1}{x^2}$ *Definition of negative exponent*

67. $\dfrac{(3x)^{-2}}{(4x)^{-3}} = \dfrac{(4x)^3}{(3x)^2}$ *Changing from negative to positive exponents*
$= \dfrac{4^3x^3}{3^2x^2}$ *Power rule (b)*
$= \dfrac{4^3x^{3-2}}{3^2}$ *Quotient rule*
$= \dfrac{4^3x}{3^2}$
$= \dfrac{64x}{9}$

69. $\left(\dfrac{x^{-1}y}{z^2}\right)^{-2} = \dfrac{(x^{-1}y)^{-2}}{(z^2)^{-2}}$ *Power rule (c)*
$= \dfrac{(x^{-1})^{-2}y^{-2}}{(z^2)^{-2}}$ *Power rule (b)*
$= \dfrac{x^2y^{-2}}{z^{-4}}$ *Power rule (a)*
$= \dfrac{x^2z^4}{y^2}$ *Definition of negative exponent*

71. $(6x)^4(6x)^{-3} = (6x)^{4+(-3)}$ *Product rule*
$= (6x)^1 = 6x$

73. $\dfrac{(m^7n)^{-2}}{m^{-4}n^3} = \dfrac{(m^7)^{-2}n^{-2}}{m^{-4}n^3}$
$= \dfrac{m^{7(-2)}n^{-2}}{m^{-4}n^3}$
$= \dfrac{m^{-14}n^{-2}}{m^{-4}n^3}$
$= m^{-14-(-4)}n^{-2-3}$
$= m^{-10}n^{-5}$
$= \dfrac{1}{m^{10}n^5}$

75. $\dfrac{(x^{-1}y^2z)^{-2}}{(x^{-3}y^3z)^{-1}} = \dfrac{(x^{-1})^{-2}(y^2)^{-2}z^{-2}}{(x^{-3})^{-1}(y^3)^{-1}z^{-1}}$
$= \dfrac{x^2y^{-4}z^{-2}}{x^3y^{-3}z^{-1}}$
$= \dfrac{x^2y^3z^1}{x^3y^4z^2}$
$= \dfrac{1}{xyz}$

77. $\left(\dfrac{xy^{-2}}{x^2y}\right)^{-3} = \dfrac{x^{-3}(y^{-2})^{-3}}{(x^2)^{-3}y^{-3}}$
$= \dfrac{x^{-3}y^6}{x^{-6}y^{-3}}$
$= \dfrac{x^6y^6y^3}{x^3}$
$= x^3y^9$

79. $\dfrac{(4a^2b^3)^{-2}(2ab^{-1})^3}{(a^3b)^{-4}}$
$= \dfrac{(a^3b)^4(2ab^{-1})^3}{(4a^2b^3)^2}$
$= \dfrac{(a^3)^4b^42^3a^3(b^{-1})^3}{4^2(a^2)^2(b^3)^2}$
$= \dfrac{a^{12}b^48a^3b^{-3}}{16a^4b^6}$
$= \dfrac{8a^{15}b^1}{16a^4b^6}$
$= \dfrac{a^{11}}{2b^5}$

81. $\dfrac{(2y^{-1}z^2)^2(3y^{-2}z^{-3})^3}{(y^3z^2)^{-1}}$
$= \dfrac{2^2y^{-2}z^43^3y^{-6}z^{-9}}{y^{-3}z^{-2}}$
$= \dfrac{4 \cdot 27y^{-8}z^{-5}}{y^{-3}z^{-2}}$
$= 108y^{-5}z^{-3} = \dfrac{108}{y^5z^3}$

83. $\dfrac{(9^{-1}z^{-2}x)^{-1}(4z^2x^4)^{-2}}{(5z^{-2}x^{-3})^2}$
$= \dfrac{9^1z^2x^{-1}4^{-2}z^{-4}x^{-8}}{5^2z^{-4}x^{-6}}$
$= \dfrac{9z^{-2}x^{-9}}{25 \cdot 4^2z^{-4}x^{-6}}$
$= \dfrac{9z^2x^{-3}}{400}$
$= \dfrac{9z^2}{400x^3}$

85. The student attempted to use the quotient rule with unequal bases. The correct way to simplify this expression is

$$\frac{16^3}{2^2} = \frac{(2^4)^3}{2^2} = \frac{2^{12}}{2^2} = 2^{10} = 1024.$$

87. $10 = 10^1$; move the decimal point to the *right* 1 place when *multiplying*.

$$10(6428) = 64{,}280$$

89. $1000(1.53) = 1530 \quad (1000 = 10^3)$

91. $10 = 10^1$; move the decimal point to the *left* 1 place when *dividing*.

$$38 \div 10 = 3.8$$

93. $277 \div 1000 = 0.277 \quad (1000 = 10^3)$

Summary Exercises on the Rules for Exponents

1. $(10x^2y^4)^2(10xy^2)^3$
$= 10^2(x^2)^2(y^4)^2 \cdot 10^3x^3(y^2)^3$
$= 10^2x^4y^810^3x^3y^6$
$= 10^5x^7y^{14}$

3. $\left(\dfrac{9wx^3}{y^4}\right)^3 = \dfrac{(9wx^3)^3}{(y^4)^3}$
$= \dfrac{9^3w^3(x^3)^3}{y^{12}}$
$= \dfrac{729w^3x^9}{y^{12}}$

5. $\dfrac{c^{11}(c^2)^4}{(c^3)^3(c^2)^{-6}} = \dfrac{c^{11}c^8}{c^9c^{-12}}$
$= \dfrac{c^{19}}{c^{-3}}$
$= c^{19-(-3)} = c^{22}$

7. $5^{-1} + 6^{-1} = \dfrac{1}{5^1} + \dfrac{1}{6^1}$
$= \frac{6}{30} + \frac{5}{30}$
$= \frac{11}{30}$

9. $\dfrac{(2xy^{-1})^3}{2^3x^{-3}y^2} = \dfrac{2^3x^3y^{-3}}{2^3x^{-3}y^2}$
$= x^6y^{-5}$
$= \dfrac{x^6}{y^5}$

11. $(z^4)^{-3}(z^{-2})^{-5}$
$= z^{-12}z^{10} = z^{-2} = \dfrac{1}{z^2}$

13. $\dfrac{(3^{-1}x^{-3}y)^{-1}(2x^2y^{-3})^2}{(5x^{-2}y^2)^{-2}}$
$= \dfrac{(5x^{-2}y^2)^2(2x^2y^{-3})^2}{(3^{-1}x^{-3}y)^1}$
$= \dfrac{5^2x^{-4}y^4 2^2x^4y^{-6}}{3^{-1}x^{-3}y}$
$= \dfrac{25x^0y^{-2}\cdot 4}{3^{-1}x^{-3}y}$
$= 100x^3y^{-3}\cdot 3$
$= \dfrac{300x^3}{y^3}$

15. $\left(\dfrac{-9x^{-2}}{9x^2}\right)^{-2} = \left(\dfrac{9x^2}{-9x^{-2}}\right)^2$
$= \left(\dfrac{x^4}{-1}\right)^2 = \dfrac{(x^4)^2}{(-1)^2}$
$= \dfrac{x^8}{1} = x^8$

17. $\dfrac{(a^{-2}b^3)^{-4}}{(a^{-3}b^2)^{-2}(ab)^{-4}}$
$= \dfrac{a^8b^{-12}}{a^6b^{-4}a^{-4}b^{-4}}$
$= \dfrac{a^8b^{-12}}{a^2b^{-8}}$
$= a^6b^{-4}$
$= \dfrac{a^6}{b^4}$

19. $5^{-2} + 6^{-2} = \dfrac{1}{5^2} + \dfrac{1}{6^2}$
$= \dfrac{1}{25} + \dfrac{1}{36}$
$= \dfrac{36}{25\cdot 36} + \dfrac{25}{25\cdot 36}$
$= \dfrac{36+25}{900} = \dfrac{61}{900}$

21. $\left(\dfrac{7a^2b^3}{2}\right)^3 = \dfrac{(7a^2b^3)^3}{2^3}$
$= \dfrac{7^3a^6b^9}{8} = \dfrac{343a^6b^9}{8}$

23. $-(-13)^0 = -1$

25. $\dfrac{(2xy^{-3})^{-2}}{(3x^{-2}y^4)^{-3}}$
$= \dfrac{(3x^{-2}y^4)^3}{(2xy^{-3})^2}$
$= \dfrac{3^3x^{-6}y^{12}}{2^2x^2y^{-6}}$
$= \dfrac{27x^{-8}y^{18}}{4}$
$= \dfrac{27y^{18}}{4x^8}$

27. $(6x^{-5}z^3)^{-3}$
$= 6^{-3}x^{15}z^{-9}$
$= \dfrac{x^{15}}{6^3z^9}$
$= \dfrac{x^{15}}{216z^9}$

29. $\dfrac{(xy)^{-3}(xy)^5}{(xy)^{-4}} = (xy)^{-3+5-(-4)}$
$= (xy)^6 = x^6y^6$

31. $\dfrac{(7^{-1}x^{-3})^{-2}(x^4)^{-6}}{7^{-1}x^{-3}}$
$= \dfrac{7^2x^6x^{-24}}{7^{-1}x^{-3}}$
$= 7^{2-(-1)}x^{6-24-(-3)}$
$= 7^3x^{-15} = \dfrac{343}{x^{15}}$

33. $(5p^{-2}q)^{-3}(5pq^3)^4$
$= 5^{-3}p^6q^{-3}5^4p^4q^{12}$
$= 5p^{10}q^9$

35. $\left[\dfrac{4r^{-6}s^{-2}t}{2r^8s^{-4}t^2}\right]^{-1}$
$= \left[\dfrac{2s^2}{r^{14}t}\right]^{-1}$
$= \dfrac{r^{14}t}{2s^2}$

37. $\dfrac{(8pq^{-2})^4}{(8p^{-2}q^{-3})^3}$
$= \dfrac{8^4p^4q^{-8}}{8^3p^{-6}q^{-9}}$
$= 8^{4-3}p^{4-(-6)}q^{-8-(-9)}$
$= 8p^{10}q$

39. $-(-8^0)^0 = -1$

4.3 An Application of Exponents: Scientific Notation

4.3 Now Try Exercises

N1. **(a)** $12{,}600{,}000 = 1.26 \times 10^7$

The decimal point has been moved 7 places to put it after the first nonzero digit. Since 1.26 is less than the original number, it must be multiplied by a number greater than 1 to get 12,600,000. Thus, the exponent on 10 must be positive.

(b) $0.00027 = 2.7 \times 10^{-4}$

The decimal point has been moved 4 places to put it after the first nonzero digit. Since 2.7 is greater than the original number, it must be multiplied by a number less than 1 to get 0.00027. Thus, the exponent on 10 must be negative.

(c) $-0.0000341 = -3.41 \times 10^{-5}$

Work with the absolute value of the number, then apply the negative sign to the answer.

N2. **(a)** $5.71 \times 10^4 = 57{,}100$

Since the exponent is positive, move the decimal point 4 places to the *right*.

(b) $2.72 \times 10^{-5} = 0.000\,027\,2$

Since the exponent is negative, move the decimal point 5 places to the *left*.

N3. **(a)**
$$\begin{aligned}&(6 \times 10^7)(7 \times 10^{-4})\\ &= (6 \times 7)(10^7 \times 10^{-4})\\ &= 42 \times 10^3\\ &= (4.2 \times 10^1) \times 10^3\\ &= 4.2 \times (10^1 \times 10^3)\\ &= 4.2 \times 10^4, \text{ or } 42{,}000\end{aligned}$$

(b)
$$\begin{aligned}\frac{18 \times 10^{-3}}{6 \times 10^4} &= \frac{18}{6} \times \frac{10^{-3}}{10^4}\\ &= 3 \times 10^{-3-4}\\ &= 3 \times 10^{-7}, \text{ or } 0.000\,000\,3\end{aligned}$$

N4.
$$\begin{aligned}&(8{,}000{,}000)(0.000\,000\,039\,37)\\ &= (8 \times 10^6) \times (3.937 \times 10^{-8})\\ &= (8 \times 3.937) \times (10^6 \times 10^{-8})\\ &= 31.496 \times 10^{6+(-8)}\\ &= 31.496 \times 10^{-2}\\ &= 3.1496 \times 10^{-1}, \text{ or } 0.31496\end{aligned}$$

Thus, 8,000,000 nanometers would measure 3.1496×10^{-1} in., or 0.31496 in.

N5. Divide the land area by the population.

$$\begin{aligned}\frac{1.6 \times 10^5}{4 \times 10^7} &= \frac{1.6}{4} \times \frac{10^5}{10^7}\\ &= 0.4 \times 10^{5-7}\\ &= 0.4 \times 10^{-2}\\ &= 4 \times 10^{-3}\\ &= 0.004\end{aligned}$$

In 2008, the number of square miles per person living in California was about 4×10^{-3}, or 0.004.

4.3 Section Exercises

1. **(a)** Move the decimal point to the left 4 places due to the exponent, −4.

$$4.6 \times 10^{-4} = 0.000\,46$$

Choice **C** is correct.

(b) $4.6 \times 10^4 = 46{,}000$

Choice **A** is correct.

(c) Move the decimal point to the right 5 places due to the exponent, 5.

$$4.6 \times 10^5 = 460{,}000$$

Choice **B** is correct.

(d) $4.6 \times 10^{-5} = 0.000\,046$

Choice **D** is correct.

3. 4.56×10^4 is written in scientific notation because 4.56 is between 1 and 10, and 10^4 is a power of 10.

5. 5,600,000 is not in scientific notation. It can be written in scientific notation as 5.6×10^6.

7. 0.8×10^2 is not in scientific notation because $|0.8| = 0.8$ is not greater than or equal to 1 and less than 10. It can be written in scientific notation as 8×10^1.

9. 0.004 is not in scientific notation because $|0.004| = 0.004$ is not between 1 and 10. It can be written in scientific notation as 4×10^{-3}.

11. A number is written in scientific notation if it is written as a product of two numbers, the first of which has absolute value less than 10 and greater than or equal to 1 and the second of which is a power of 10. Some examples are 2.3×10^{-4} and 6.02×10^{23}.

13. 5,876,000,000

Move the decimal point to the right of the first nonzero digit and count the number of places the decimal point was moved.

5,876,000,000 *9 places*

Because moving the decimal point to the *left* made the number *smaller*, we must multiply by a *positive* power of 10 so that the product 5.876×10^n will equal the larger number. Thus, $n = 9$, and

$$5{,}876{,}000{,}000 = 5.876 \times 10^9.$$

15. 82,350

Move the decimal point left 4 places so it is to the right of the first nonzero digit.

$$8.2350 \quad \textit{4 places}$$

Since the number got smaller, multiply by a positive power of 10.

$$82{,}350 = 8.2350 \times 10^4 = 8.235 \times 10^4$$

(Note that the final zero need not be written.)

17. 0.000 007

Move the decimal point to the right of the first nonzero digit.

$$0.000007. \quad \textit{6 places}$$

Since moving the decimal point to the *right* made the number *larger*, we must multiply by a *negative* power of 10 so that the product 7×10^n will equal the smaller number. Thus, $n = -6$, and

$$0.000\,007 = 7 \times 10^{-6}.$$

19. 0.002 03

To move the decimal point to the right of the first nonzero digit, we move it 3 places. Since 2.03 is larger than 0.00203, the exponent on 10 must be negative.

$$0.002\,03 = 2.03 \times 10^{-3}$$

21. −13,000,000

Move the decimal point to the right of the first nonzero digit and count the number of places the decimal point was moved.

$$-13{,}000{,}000 \quad \textit{7 places}$$

Because moving the decimal point to the *left* made the number *smaller*, we must multiply by a *positive* power of 10 so that the product -1.3×10^n will equal the larger number. Thus, $n = 7$, and

$$-13{,}000{,}000 = -1.3 \times 10^7.$$

23. −0.006

To move the decimal point to the right of the first nonzero digit, we move it 3 places. Since 6 is larger than 0.006, the exponent on 10 must be negative. Remember to include the negative sign.

$$-0.006 = -6 \times 10^{-3}$$

25. 7.5×10^5

Since the exponent is positive, make 7.5 larger by moving the decimal point 5 places to the right.

$$7.5 \times 10^5 = 750{,}000$$

27. 5.677×10^{12}

Since the exponent is positive, make 5.677 larger by moving the decimal point 12 places to the right. We need to add 9 zeros.

$$5.677 \times 10^{12} = 5{,}677{,}000{,}000{,}000$$

29. 1×10^{12}

Since the exponent is positive, make 1 larger by moving the decimal point 12 places to the right. We need to add 12 zeros.

$$1 \times 10^{12} = 1{,}000{,}000{,}000{,}000$$

31. 6.21×10^0

Because the exponent is 0, the decimal point should not be moved.

$$6.21 \times 10^0 = 6.21$$

We know this result is correct because $10^0 = 1$.

33. 7.8×10^{-4}

Since the exponent is negative, make 7.8 smaller by moving the decimal point 4 places to the left.

$$7.8 \times 10^{-4} = 0.000\,78$$

35. 5.134×10^{-9}

Since the exponent is negative, make 5.134 smaller by moving the decimal point 9 places to the left.

$$5.134 \times 10^{-9} = 0.000\,000\,005\,134$$

37. -4×10^{-3}

Move the decimal point 3 places to the left.

$$-4 \times 10^{-3} = -0.004$$

39. -8.1×10^5

Move the decimal point 5 places to the right.

$$-8.1 \times 10^5 = -810{,}000$$

41. **(a)** $(2 \times 10^8)(3 \times 10^3)$

$= (2 \times 3)(10^8 \times 10^3)$ *Commutative and associative properties*

$= 6 \times 10^{11}$ *Product rule for exponents*

(b) $6 \times 10^{11} = 600{,}000{,}000{,}000$

43. **(a)** $(5 \times 10^4)(3 \times 10^2)$
$= (5 \times 3)(10^4 \times 10^2)$
$= 15 \times 10^6$
$= 1.5 \times 10^7$

(b) $15 \times 10^6 = 15{,}000{,}000$

45. **(a)** $(3 \times 10^{-4})(-2 \times 10^8)$
$= (3 \times (-2))(10^{-4} \times 10^8)$
$= -6 \times 10^4$

(b) $-6 \times 10^4 = -60{,}000$

47. **(a)** $(6 \times 10^3)(4 \times 10^{-2})$
$= (6 \times 4)(10^3 \times 10^{-2})$
$= 24 \times 10^1$
$= 2.4 \times 10^2$

(b) $2.4 \times 10^2 = 240$

49. **(a)** $(9 \times 10^4)(7 \times 10^{-7})$
$= (9 \times 7)(10^4 \times 10^{-7})$
$= 63 \times 10^{-3}$
$= 6.3 \times 10^{-2}$

(b) $6.3 \times 10^{-2} = 0.063$

51. **(a)** $\dfrac{9 \times 10^{-5}}{3 \times 10^{-1}} = \dfrac{9}{3} \times \dfrac{10^{-5}}{10^{-1}}$
$= 3 \times 10^{-5-(-1)}$
$= 3 \times 10^{-4}$

(b) $3 \times 10^{-4} = 0.0003$

53. **(a)** $\dfrac{8 \times 10^3}{-2 \times 10^2} = \dfrac{8}{-2} \times \dfrac{10^3}{10^2}$
$= -4 \times 10^1 = -4 \times 10$

(b) $-4 \times 10^1 = -40$

55. **(a)** $\dfrac{2.6 \times 10^{-3}}{2 \times 10^2} = \dfrac{2.6}{2} \times \dfrac{10^{-3}}{10^2}$
$= 1.3 \times 10^{-5}$

(b) $1.3 \times 10^{-5} = 0.000\,013$

57. **(a)** $\dfrac{4 \times 10^5}{8 \times 10^2} = \dfrac{4}{8} \times \dfrac{10^5}{10^2}$
$= 0.5 \times 10^{5-2}$
$= 0.5 \times 10^3$
$= 5 \times 10^2$

(b) $5 \times 10^2 = 500$

59. **(a)** $\dfrac{-4.5 \times 10^4}{1.5 \times 10^{-2}} = \dfrac{-4.5}{1.5} \times \dfrac{10^4}{10^{-2}}$
$= -3 \times 10^{4-(-2)}$
$= -3 \times 10^6$

(b) $-3 \times 10^6 = -3{,}000{,}000$

61. **(a)** $\dfrac{-8 \times 10^{-4}}{-4 \times 10^3} = \dfrac{-8}{-4} \times \dfrac{10^{-4}}{10^3}$
$= 2 \times 10^{-4-3}$
$= 2 \times 10^{-7}$

(b) $2 \times 10^{-7} = 0.000\,000\,2$

63. To work in scientific mode on the TI-83/84 Plus, press MODE and then change from "Normal" to "Sci."

$$0.000\,000\,47 = 4.7 \times 10^{-7}$$

Prediction: 4.7 E ⁻7

65. $(8\text{E}5)/(4\text{E}^{-}2) = \dfrac{8 \times 10^5}{4 \times 10^{-2}}$
$= \frac{8}{4} \times 10^{5-(-2)}$
$= 2 \times 10^7$

Prediction: 2 E7

67. $(2\text{E}6) * (2\text{E}^{-}3)/(4\text{E}2)$
$= \dfrac{(2 \times 10^6)(2 \times 10^{-3})}{4 \times 10^2}$
$= \dfrac{(2 \times 2) \times (10^6 \times 10^{-3})}{4 \times 10^2}$
$= \dfrac{4 \times 10^3}{4 \times 10^2}$
$= \frac{4}{4} \times 10^{3-2}$
$= 1 \times 10^1$

Prediction: 1 E1

69. $\dfrac{650{,}000{,}000(0.000\,003\,2)}{0.000\,02}$
$= \dfrac{(6.5 \times 10^8)(3.2 \times 10^{-6})}{2 \times 10^{-5}}$
$= \dfrac{20.8 \times 10^2}{2 \times 10^{-5}}$
$= \dfrac{20.8}{2} \times \dfrac{10^2}{10^{-5}}$
$= 10.4 \times 10^7$
$= 1.04 \times 10^8$

71. $\dfrac{0.000\,000\,72(0.000\,23)}{0.000\,000\,018}$
$= \dfrac{(7.2 \times 10^{-7})(2.3 \times 10^{-4})}{1.8 \times 10^{-8}}$
$= \dfrac{16.56 \times 10^{-11}}{1.8 \times 10^{-8}}$
$= 9.2 \times 10^{-3}$

73. $\dfrac{0.000\,001\,6(240{,}000{,}000)}{0.000\,02(0.0032)}$
$= \dfrac{(1.6 \times 10^{-6})(2.4 \times 10^8)}{(2 \times 10^{-5})(3.2 \times 10^{-3})}$
$= \dfrac{3.84 \times 10^2}{6.4 \times 10^{-8}}$
$= 0.6 \times 10^{10}$
$= 6 \times 10^9$

75. 10 billion $= 10{,}000{,}000{,}000$
$= 1 \times 10^{10}$

77. $2 \times 10^9 = 2{,}000{,}000{,}000$

79. $3{,}305{,}000{,}000 = 3.305 \times 10^9$

81. $\$28{,}100{,}000{,}000 = \$2.81 \times 10^{10}$

83.
$$\begin{aligned} 1200(2.3 \times 10^{-4}) &= (1.2 \times 10^3)(2.3 \times 10^{-4}) \\ &= (1.2 \times 2.3) \times (10^3 \times 10^{-4}) \\ &= 2.76 \times 10^{3+(-4)} \\ &= 2.76 \times 10^{-1} \\ &= 0.276 \end{aligned}$$

There is about 2.76×10^{-1}, or 0.276, lb of copper in 1200 such people.

85. Use the formula $d = rt$.

$$\begin{aligned} \text{distance} &= \text{rate} \times \text{time} \\ 6.68 \times 10^7 &= (1.86 \times 10^5) \times \text{time} \\ \frac{6.68 \times 10^7}{1.86 \times 10^5} &= \text{time} \\ \text{time} &= \frac{6.68}{1.86} \times \frac{10^7}{10^5} \\ &\approx 3.59 \times 10^2 \end{aligned}$$

It takes about 359 seconds for light to travel from the sun to Venus.

87. To get the average ticket price, divide the total gross by the attendance.

$$\begin{aligned} \frac{\$9.38 \times 10^8}{1.23 \times 10^7} &= \frac{\$9.38}{1.23} \times \frac{10^8}{10^7} \\ &\approx \$7.626 \times 10^1 \end{aligned}$$

The average ticket price was about \$76.26.

89. To find the debt per person, divide the debt by the population.

$$\begin{aligned} \frac{\$1.2 \times 10^{13}}{300 \text{ million}} &= \frac{\$1.2 \times 10^{13}}{300{,}000{,}000} \\ &= \frac{\$1.2 \times 10^{13}}{3 \times 10^8} \\ &= \$0.4 \times 10^5 \\ &= \$4 \times 10^4 \end{aligned}$$

The debt will be about \$40,000 for every person.

91. To find the number of miles, multiply the number of light-years by the number of miles per light-year.

$$\begin{aligned} &(25{,}000)(6{,}000{,}000{,}000{,}000) \\ &= (2.5 \times 10^4)(6 \times 10^{12}) \\ &= (2.5 \times 6)(10^4 \times 10^{12}) \\ &= 15 \times 10^{16} \\ &= 1.5 \times 10^{17} \end{aligned}$$

The nebula is about 1.5×10^{17} miles from Earth.

93.
$$\begin{aligned} &\$4013(304{,}000{,}000) \\ &= (\$4.013 \times 10^3)(3.04 \times 10^8) \\ &= (\$4.013 \times 3.04)(10^3 \times 10^8) \\ &\approx \$12.2 \times 10^{11} \\ &= \$1.22 \times 10^{12} \end{aligned}$$

In 2008, the U.S. government collected about \$1,220,000,000,000 in taxes.

95. To find the amount each person would have to contribute, divide the amount by the population.

$$\begin{aligned} \frac{\$1{,}000{,}000{,}000{,}000}{307.5 \text{ million}} &= \frac{\$1 \times 10^{12}}{307{,}500{,}000} \\ &= \frac{\$1 \times 10^{12}}{3.075 \times 10^8} \\ &= \frac{\$1}{3.075} \times \frac{10^{12}}{10^8} \\ &\approx \$0.3252 \times 10^4 \\ &= \$3.252 \times 10^3 \end{aligned}$$

Each person would have to contribute about \$3252.

97.
$$\begin{aligned} &-3(2x + 4) + 4(2x - 6) \\ &= -6x - 12 + 8x - 24 \\ &= 2x - 36 \end{aligned}$$

99.
$$\begin{aligned} &2x^2 - 3x + 10 \\ &= 2(3)^2 - 3(3) + 10 \quad \textit{Let } x = 3. \\ &= 2(9) - 9 + 10 \\ &= 19 \end{aligned}$$

101.
$$\begin{aligned} &4x^3 - 5x^2 + 2x - 5 \\ &= 4(3)^3 - 5(3)^2 + 2(3) - 5 \quad \textit{Let } x = 3. \\ &= 4(27) - 5(9) + 6 - 5 \\ &= 108 - 45 + 6 - 5 \\ &= 63 + 6 - 5 \\ &= 69 - 5 \\ &= 64 \end{aligned}$$

4.4 Adding and Subtracting Polynomials; Graphing Simple Polynomials

4.4 Now Try Exercises

N1. The coefficient of t, or $1t$, is 1 and the coefficient of $-10t^2$, is -10.

N2.
$$\begin{aligned} &3x^2 - x^2 + 2x \\ &= (3x^2 - 1x^2) + 2x \quad \textit{Associative property} \\ &= (3 - 1)x^2 + 2x \quad \textit{Distributive property} \\ &= 2x^2 + 2x \end{aligned}$$

N3. $x^2 + 4x - 2x - 8 = x^2 + (4 - 2)x - 8$ simplifies to $x^2 + 2x - 8$. This polynomial has three terms, and the largest exponent on the variable x is 2. So the polynomial has degree 2 and is a trinomial.

N4. $4t^3 - t^2 - t$ *Replace t with −3.*

$= 4(-3)^3 - (-3)^2 - (-3)$
$= 4(-27) - (9) + 3$
$= -108 - 9 + 3$
$= -117 + 3$
$= -114$

N5. Add, column by column.

$4y^3 - 2y^2 + y - 1$	*First polynomial*
$y^3 + 0y^2 - y - 7$	*Second polynomial*
$5y^3 - 2y^2 \quad - 8$	*Sum*

N6. Combine like terms.

$(10x^4 - 3x^2 - x) + (x^4 - 3x^2 + 5x)$
$= (10x^4 + x^4) + (-3x^2 - 3x^2) + (-x + 5x)$
$= 11x^4 - 6x^2 + 4x$

N7. Change subtraction to addition.

$(4t^4 - t^2 + 7) - (5t^4 - 3t^2 + 1)$
$= (4t^4 - t^2 + 7) + (-5t^4 + 3t^2 - 1)$

Combine like terms.

$= (4t^4 - 5t^4) + (-t^2 + 3t^2) + (7 - 1)$
$= -t^4 + 2t^2 + 6$

Check by adding.

$5t^4 - 3t^2 + 1$	*Second polynomial*
$-t^4 + 2t^2 + 6$	*Answer*
$4t^4 - \ t^2 + 7$	*First polynomial*

N8. Arrange like terms in columns. Insert zeros for missing terms.

$$\begin{array}{r} 12x^2 - 9x + 4 \\ \underline{-10x^2 - 3x + 7} \end{array}$$

Change all signs in the second row, and then add.

$$\begin{array}{r} 12x^2 - 9x + 4 \\ \underline{10x^2 + 3x - 7} \\ 22x^2 - 6x - 3 \end{array}$$

N9. $(4x^2 - 2xy + y^2) - (6x^2 - 7xy + 2y^2)$
$= 4x^2 - 2xy + y^2 - 6x^2 + 7xy - 2y^2$
$= (4x^2 - 6x^2) + (-2xy + 7xy) + (y^2 - 2y^2)$
$= -2x^2 + 5xy - y^2$

N10. Graph $y = -x^2 - 1$.

Make a table of ordered pairs whose x-values are on either side of the vertex's x-value of $x = 0$.

x	$y = -x^2 - 1$
0	$-(0)^2 - 1 = 0 - 1 = -1$
± 1	$-(\pm 1)^2 - 1 = -1 - 1 = -2$
± 2	$-(\pm 2)^2 - 1 = -4 - 1 = -5$

Plot these seven ordered pairs and connect them with a smooth curve.

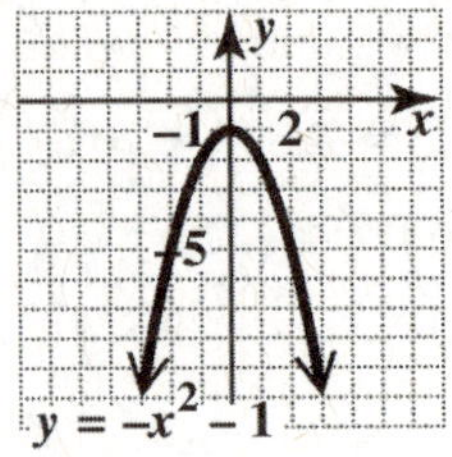

4.4 Section Exercises

1. In the term $4x^6$, the coefficient is $\underline{4}$ and the exponent is $\underline{6}$.

3. The degree of the term $-3x^9$ is $\underline{9}$, the exponent.

5. When $x^2 + 10$ is evaluated for $x = 3$, the result is

$$3^2 + 10 = 9 + 10 = \underline{19}.$$

7. $-3xy - 2xy + 5xy = (-3 - 2 + 5)xy$
$= (0)xy$
$= \underline{0}$

9. The polynomial $6x^4$ has one term. The coefficient of this term is 6.

11. The polynomial t^4 has one term. Since $t^4 = 1 \cdot t^4$, the coefficient of this term is 1.

13. The polynomial $-19r^2 - r$ has two terms. The coefficient of r^2 is -19 and the coefficient of r is -1.

15. The polynomial $x + 8x^2 + 5x^3$ has three terms. The coefficient of x is 1, the coefficient of x^2 is 8, and the coefficient of x^3 is 5.

In Exercises 17–28, use the distributive property to add like terms.

17. $-3m^5 + 5m^5 = (-3 + 5)m^5$
$= 2m^5$

19. $2r^5 + (-3r^5) = [2 + (-3)]r^5$
$= -1r^5 = -r^5$

21. The polynomial $0.2m^5 - 0.5m^2$ cannot be simplified. The two terms are unlike because the exponents on the variables are different, so they cannot be combined.

23. $-3x^5 + 3x^5 - 5x^5 = (-3 + 3 - 5)x^5$
$= -5x^5$

25. $-4p^7 + 8p^7 + 5p^9 = (-4 + 8)p^7 + 5p^9$
$= 4p^7 + 5p^9$

In descending powers of the variable, this polynomial is written $5p^9 + 4p^7$.

27. $-4xy^2 + 3xy^2 - 2xy^2 + xy^2$
$= (-4 + 3 - 2 + 1)xy^2$
$= -2xy^2$

29. $6x^4 - 9x$

This polynomial has no like terms, so it is already simplified. It is already written in descending powers of the variable x. The highest degree of any nonzero term is 4, so the degree of the polynomial is 4. There are two terms, so this is a *binomial*.

31. $5m^4 - 3m^2 + 6m^4 - 7m^3$
$= (5m^4 + 6m^4) + (-7m^3) + (-3m^2)$
$= 11m^4 - 7m^3 - 3m^2$

The resulting polynomial is a *trinomial* of degree 4.

33. $\frac{5}{3}x^4 - \frac{2}{3}x^4 = (\frac{5}{3} - \frac{2}{3})x^4$
$= \frac{3}{3}x^4 = x^4$

The resulting polynomial is a *monomial* of degree 4.

35. $0.8x^4 - 0.3x^4 - 0.5x^4 + 7$
$= (0.8 - 0.3 - 0.5)x^4 + 7$
$= 0x^4 + 7 = 7$

Since 7 can be written as $7x^0$, the degree of the polynomial is 0. The simplified polynomial has one term, so it is a *monomial*.

37. **(a)** $2x^2 - 3x - 5$
$= 2(2)^2 - 3(2) - 5$ *Let x = 2.*
$= 2(4) - 6 - 5$
$= 8 - 6 - 5$
$= 2 - 5$
$= -3$

(b) $2x^2 - 3x - 5$
$= 2(-1)^2 - 3(-1) - 5$ *Let x = –1.*
$= 2(1) + 3 - 5$
$= 2 + 3 - 5$
$= 5 - 5$
$= 0$

39. **(a)** $-3x^2 + 14x - 2$
$= -3(2)^2 + 14(2) - 2$ *Let x = 2.*
$= -3(4) + 28 - 2$
$= -12 + 28 - 2$
$= 16 - 2$
$= 14$

(b) $-3x^2 + 14x - 2$
$= -3(-1)^2 + 14(-1) - 2$ *Let x = –1.*
$= -3(1) - 14 - 2$
$= -3 - 14 - 2$
$= -17 - 2$
$= -19$

41. **(a)** $2x^5 - 4x^4 + 5x^3 - x^2$
$= 2(2)^5 - 4(2)^4 + 5(2)^3 - (2)^2$ *Let x = 2.*
$= 2(32) - 4(16) + 5(8) - 4$
$= 64 - 64 + 40 - 4$
$= 36$

(b) $2x^5 - 4x^4 + 5x^3 - x^2$
$= 2(-1)^5 - 4(-1)^4 + 5(-1)^3 - (-1)^2$ *Let x = –1.*
$= 2(-1) - 4(1) + 5(-1) - 1$
$= -2 - 4 - 5 - 1$
$= -12$

43. Add, column by column.

$$\begin{array}{rcr} 2x^2 & - & 4x \\ 3x^2 & + & 2x \\ \hline 5x^2 & - & 2x \end{array}$$

45. Add, column by column.

$$\begin{array}{rcrcr} 3m^2 & + & 5m & + & 6 \\ 2m^2 & - & 2m & - & 4 \\ \hline 5m^2 & + & 3m & + & 2 \end{array}$$

47. Add, column by column.

$$\begin{array}{rcrcr} \frac{2}{3}x^2 & + & \frac{1}{5}x & + & \frac{1}{6} \\ \frac{1}{2}x^2 & - & \frac{1}{3}x & + & \frac{2}{3} \\ \hline \end{array}$$

Rewrite the fractions so that the fractions in each column have a common denominator; then add column by column.

$$\begin{array}{rcrcr} \frac{4}{6}x^2 & + & \frac{3}{15}x & + & \frac{1}{6} \\ \frac{3}{6}x^2 & - & \frac{5}{15}x & + & \frac{4}{6} \\ \hline \frac{7}{6}x^2 & - & \frac{2}{15}x & + & \frac{5}{6} \end{array}$$

49. $(9m^3 - 5m^2 + 4m - 8) + (-3m^3 + 6m^2 - 6)$
$= (9m^3 - 3m^3) + (-5m^2 + 6m^2) + 4m$
$+ (-8 - 6)$
$= 6m^3 + m^2 + 4m - 14$

51. Subtract, column by column.

$$\begin{array}{rcr} 5y^3 & - & 3y^2 \\ 2y^3 & + & 8y^2 \\ \hline \end{array}$$

Change all signs in the second row, and then add.

$$\begin{array}{rcr} 5y^3 & - & 3y^2 \\ -2y^3 & - & 8y^2 \\ \hline 3y^3 & - & 11y^2 \end{array}$$

53. Subtract, column by column.

$$\begin{array}{rrr} 12x^4 & -\ x^2 & +\ x \\ \underline{8x^4} & \underline{+\ 3x^2} & \underline{-\ 3x} \end{array}$$

Change all signs in the second row, and then add.

$$\begin{array}{rrr} 12x^4 & -\ x^2 & +\ x \\ -8x^4 & -\ 3x^2 & +\ 3x \\ \hline 4x^4 & -\ 4x^2 & +\ 4x \end{array}$$

55. Subtract, column by column.

$$\begin{array}{rrrr} 12m^3 & -\ 8m^2 & +\ 6m & +\ 7 \\ -3m^3 & +\ 5m^2 & -\ 2m & -\ 4 \\ \hline \end{array}$$

Change all signs in the second row, and then add.

$$\begin{array}{rrrr} 12m^3 & -\ 8m^2 & +\ 6m & +\ 7 \\ 3m^3 & -\ 5m^2 & +\ 2m & +\ 4 \\ \hline 15m^3 & -\ 13m^2 & +\ 8m & +\ 11 \end{array}$$

57. Vertical addition and subtraction of polynomials is preferable because like terms are arranged in columns.

59. $(8m^2 - 7m) - (3m^2 + 7m - 6)$
$= (8m^2 - 7m) + (-3m^2 - 7m + 6)$
$= (8 - 3)m^2 + (-7 - 7)m + 6$
$= 5m^2 - 14m + 6$

61. $(16x^3 - x^2 + 3x) + (-12x^3 + 3x^2 + 2x)$
$= 16x^3 - x^2 + 3x - 12x^3 + 3x^2 + 2x$
$= (16 - 12)x^3 + (-1 + 3)x^2 + (3 + 2)x$
$= 4x^3 + 2x^2 + 5x$

63. $(7y^4 + 3y^2 + 2y) - (18y^4 - 5y^2 + y)$
$= (7y^4 + 3y^2 + 2y) + (-18y^4 + 5y^2 - y)$
$= (7 - 18)y^4 + (3 + 5)y^2 + (2 - 1)y$
$= -11y^4 + 8y^2 + y$

65. $(9a^4 - 3a^2 + 2) + (4a^4 - 4a^2 + 2)$
$+ (-12a^4 + 6a^2 - 3)$
$= (9a^4 + 4a^4 - 12a^4)$
$+ (-3a^2 - 4a^2 + 6a^2) + (2 + 2 - 3)$
$= a^4 - a^2 + 1$

67. $[(8m^2 + 4m - 7) - (2m^2 - 5m + 2)]$
$- (m^2 + m + 1)$
$= (8m^2 + 4m - 7) + (-2m^2 + 5m - 2)$
$+ (-m^2 - m - 1)$
$= (8 - 2 - 1)m^2 + (4 + 5 - 1)m$
$+ (-7 - 2 - 1)$
$= 5m^2 + 8m - 10$

69. $[(3x^2 - 2x + 7) - (4x^2 + 2x - 3)]$
$- [(9x^2 + 4x - 6) + (-4x^2 + 4x + 4)]$
$= [(3 - 4)x^2 + (-2 - 2)x + (7 + 3)]$
$- [(9 - 4)x^2 + (4 + 4)x + (-6 + 4)]$
$= (-x^2 - 4x + 10) - (5x^2 + 8x - 2)$
$= -x^2 - 4x + 10 - 5x^2 - 8x + 2$
$= -6x^2 - 12x + 12$

71. The coefficients of the x^2 terms are -4, $-(-2)$, and -8. The sum of these numbers is

$$-4 + 2 - 8 = -10.$$

73. $(6b + 3c) + (-2b - 8c)$
$= (6b - 2b) + (3c - 8c)$
$= 4b - 5c$

75. $(4x + 2xy - 3) - (-2x + 3xy + 4)$
$= (4x + 2xy - 3) + (2x - 3xy - 4)$
$= (4x + 2x) + (2xy - 3xy) + (-3 - 4)$
$= 6x - xy - 7$

77. $(5x^2y - 2xy + 9xy^2)$
$- (8x^2y + 13xy + 12xy^2)$
$= (5x^2y - 2xy + 9xy^2)$
$+ (-8x^2y - 13xy - 12xy^2)$
$= (5x^2y - 8x^2y) + (-2xy - 13xy)$
$+ (9xy^2 - 12xy^2)$
$= -3x^2y - 15xy - 3xy^2$

79. Use the formula for the perimeter of a rectangle, $P = 2L + 2W$, with length $L = 4x^2 + 3x + 1$ and width $W = x + 2$.

$$\begin{aligned} P &= 2L + 2W \\ &= 2(4x^2 + 3x + 1) + 2(x + 2) \\ &= 8x^2 + 6x + 2 + 2x + 4 \\ &= 8x^2 + 8x + 6 \end{aligned}$$

A polynomial that represents the perimeter of the rectangle is $8x^2 + 8x + 6$.

81. Use the formula for the perimeter of a square, $P = 4s$, with $s = \frac{1}{2}x^2 + 2x$.

$$\begin{aligned} P &= 4s \\ &= 4(\tfrac{1}{2}x^2 + 2x) \\ &= 4(\tfrac{1}{2}x^2) + 4(2x) \\ &= 2x^2 + 8x \end{aligned}$$

A polynomial that represents the perimeter of the square is $2x^2 + 8x$.

83. Use the formula for the perimeter of a triangle, $P = a + b + c$, with $a = 3t^2 + 2t + 7$, $b = 5t^2 + 2$, and $c = 6t + 4$.

$$\begin{aligned}P &= (3t^2+2t+7)+(5t^2+2)+(6t+4)\\ &= (3t^2+5t^2)+(2t+6t)+(7+2+4)\\ &= 8t^2+8t+13\end{aligned}$$

A polynomial that represents the perimeter of the triangle is $8t^2+8t+13$.

85. **(a)** Use the formula for the perimeter of a triangle, $P=a+b+c$, with $a=2y-3t$, $b=5y+3t$, and $c=16y+5t$.

$$\begin{aligned}P &= (2y-3t)+(5y+3t)+(16y+5t)\\ &= (2y+5y+16y)+(-3t+3t+5t)\\ &= 23y+5t\end{aligned}$$

The perimeter of the triangle is $23y+5t$.

(b) Use the fact that the sum of the angles of any triangle is 180°.

$$\begin{aligned}(7x-3)^\circ+(5x+2)^\circ+(2x-1)^\circ &= 180^\circ\\ (7x+5x+2x)+(-3+2-1) &= 180\\ 14x-2 &= 180\\ 14x &= 182\\ x &= \tfrac{182}{14}=13\end{aligned}$$

If $x=13$,

$$\begin{aligned}7x-3 &= 7(13)-3=88,\\ 5x+2 &= 5(13)+2=67,\\ \text{and}\quad 2x-1 &= 2(13)-1=25.\end{aligned}$$

The measures of the angles are 25°, 67°, and 88°.

87. First find the two sums and then find their difference.

$$\begin{aligned}&[(5x^2+2x-3)+(x^2-8x+2)]\\ &\quad-[(7x^2-3x+6)+(-x^2+4x-6)]\\ &=(6x^2-6x-1)-(6x^2+x)\\ &=(6x^2-6x-1)+(-6x^2-x)\\ &=(6x^2-6x^2)+(-6x-x)+(-1)\\ &=-7x-1\end{aligned}$$

89. $y=x^2-4$

x	$y=x^2-4$
-2	$(-2)^2-4=4-4=0$
-1	$(-1)^2-4=1-4=-3$
0	$(0)^2-4=0-4=-4$
1	$(1)^2-4=1-4=-3$
2	$(2)^2-4=4-4=0$

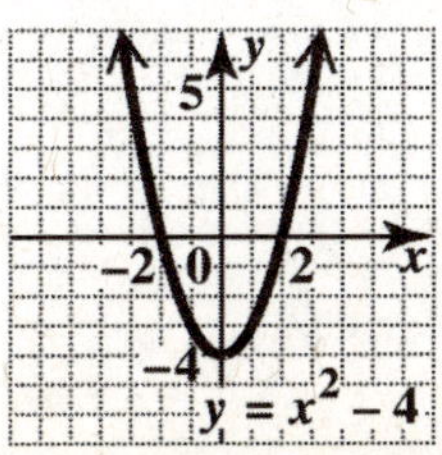

91. $y=2x^2-1$

x	$y=2x^2-1$
-2	$2(-2)^2-1=2\cdot4-1=7$
-1	$2(-1)^2-1=2\cdot1-1=1$
0	$2(0)^2-1=2\cdot0-1=-1$
1	$2(1)^2-1=2\cdot1-1=1$
2	$2(2)^2-1=2\cdot4-1=7$

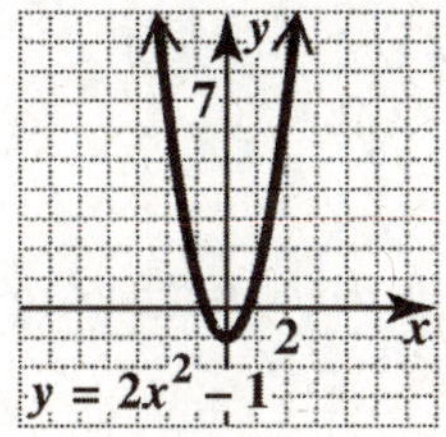

93. $y=-x^2+4$

x	$y=-x^2+4$
-2	$-(-2)^2+4=-4+4=0$
-1	$-(-1)^2+4=-1+4=3$
0	$-(0)^2+4=-0+4=4$
1	$-(1)^2+4=-1+4=3$
2	$-(2)^2+4=-4+4=0$

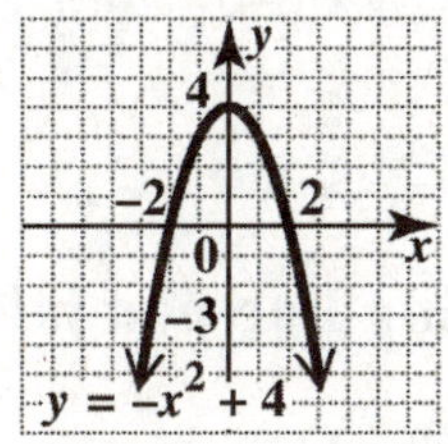

95. $y=(x+3)^2$

x	$y=(x+3)^2$
-5	$(-5+3)^2=(-2)^2=4$
-4	$(-4+3)^2=(-1)^2=1$
-3	$(-3+3)^2=(0)^2=0$
-2	$(-2+3)^2=(1)^2=1$
-1	$(-1+3)^2=(2)^2=4$

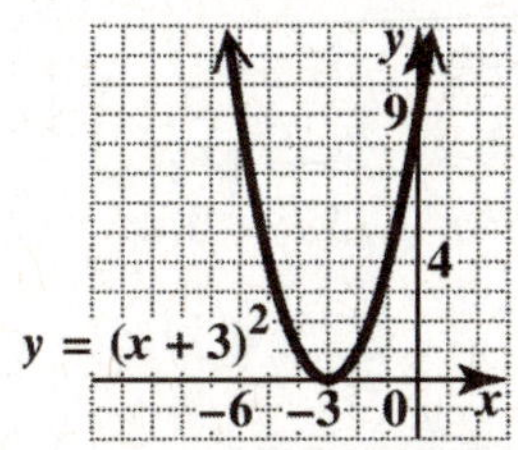

◆◆◆ Relating Concepts 97–100 ◆◆◆

97. If $x=9$, then $y=7x=7(9)=63$. If a dog is 9 in dog years, then it is 63 in human years.

98. $-0.0545x^2 + 5.047x + 11.78$
$= -0.0545(3)^2 + 5.047(3) + 11.78$ *Let x = 3.*
$= -0.4905 + 15.141 + 11.78$
$= 26.4305$

If a dog is 3 in dog years, then it is about 26 in human years.

99. $-16x^2 + 60x + 80$
$= -16(2.5)^2 + 60(2.5) + 80$ *Let x = 2.5.*
$= -16(6.25) + 150 + 80$
$= -100 + 150 + 80$
$= 130$

If 2.5 seconds have elapsed, the height of the object is 130 feet.

100. If $x = 6$,

$$2x + 15 = 2(6) + 15 = 12 + 15 = 27.$$

If the saw is rented for 6 days, the cost is $27.

101. $5(x + 4) = 5x + 5(4) = 5x + 20$

103. $4(2a + 6b) = 4(2a) + 4(6b)$
$= 8a + 24b$

105. $(2a)(-5ab) = (2)(-5)(a \cdot a)b$
$= -10a^2b$

107. $(-m^2)(m^5) = -m^{2+5} = -m^7$

4.5 Multiplying Polynomials

4.5 Now Try Exercises

N1. $-3x^5(2x^3 - 5x^2 + 10)$
$= -3x^5(2x^3) + (-3x^5)(-5x^2) + (-3x^5)(10)$ *Distributive property*
$= -6x^8 + 15x^7 + (-30x^5)$ *Multiply monomials.*
$= -6x^8 + 15x^7 - 30x^5$

N2. $(x^2 - 4)(2x^2 - 5x + 3)$
$= x^2(2x^2) + x^2(-5x) + x^2(3)$
$- 4(2x^2) - 4(-5x) - 4(3)$
$= 2x^4 - 5x^3 + 3x^2 - 8x^2 + 20x - 12$
$= 2x^4 - 5x^3 - 5x^2 + 20x - 12$

N3.

$$\begin{array}{rl} 5t^2 - 7t + 4 & \\ 2t - 6 & \\ \hline -30t^2 + 42t - 24 & \leftarrow -6(5t^2 - 7t + 4) \\ 10t^3 - 14t^2 + 8t \quad & \leftarrow 2t(5t^2 - 7t + 4) \\ \hline 10t^3 - 44t^2 + 50t - 24 & \leftarrow \textit{Add like terms.} \end{array}$$

N4.

$$\begin{array}{rl} 9x^3 - 12x^2 + 3 & \\ \frac{1}{3}x^2 - \frac{2}{3} & \\ \hline -6x^3 + 8x^2 - 2 & (\textit{times } -\tfrac{2}{3}) \\ 3x^5 - 4x^4 \qquad + x^2 \qquad & (\textit{times } \tfrac{1}{3}x^2) \\ \hline 3x^5 - 4x^4 - 6x^3 + 9x^2 - 2 & \textit{Add.} \end{array}$$

The product is $3x^5 - 4x^4 - 6x^3 + 9x^2 - 2$.

N5. $(t - 6)(t + 5)$

F: $t(t) = t^2$
O: $t(5) = 5t$
I: $-6(t) = -6t$
L: $-6(5) = -30$

$(t - 6)(t + 5) = t^2 + 5t - 6t - 30$
$= t^2 - t - 30$

N6. $(7y - 3)(2x + 5)$

F: $7y(2x) = 14yx$
O: $7y(5) = 35y$
I: $-3(2x) = -6x$
L: $-3(5) = -15$

The product $(7y - 3)(2x + 5)$ is equal to $14yx + 35y - 6x - 15$.

N7. **(a)** $(3p - 5q)(4p - q)$

F: $3p(4p) = 12p^2$
O: $3p(-q) = -3pq$
I: $-5q(4p) = -20pq$
L: $-5q(-q) = 5q^2$

$(3p - 5q)(4p - q)$
$= 12p^2 - 3pq - 20pq + 5q^2$
$= 12p^2 - 23pq + 5q^2$

(b) $5x^2(3x + 1)(x - 5)$

First multiply the binomials and then multiply that result by $5x^2$.

F: $3x(x) = 3x^2$
O: $3x(-5) = -15x$
I: $1(x) = x$
L: $1(-5) = -5$

$5x^2(3x + 1)(x - 5) = 5x^2(3x^2 - 15x + x - 5)$
$= 5x^2(3x^2 - 14x - 5)$
$= 15x^4 - 70x^3 - 25x^2$

4.5 Section Exercises

1. **(a)** $5x^3(6x^7)$
$= 5 \cdot 6x^{3+7}$
$= 30x^{10}$

Choice **B** is correct.

(b) $-5x^7(6x^3)$
$= -5 \cdot 6x^{7+3}$
$= -30x^{10}$

Choice **D** is correct.

(c) $(5x^7)^3 = (5)^3(x^7)^3$
$= 125x^{7\cdot 3}$
$= 125x^{21}$

Choice **A** is correct.

(d) $(-6x^3)^3 = (-6)^3(x^3)^3$
$= -216x^{3\cdot 3}$
$= -216x^9$

Choice **C** is correct.

3. $5y^4(3y^7) = 5(3)y^{4+7}$
$= 15y^{11}$

5. $-15a^4(-2a^5) = -15(-2)a^{4+5}$
$= 30a^9$

7. $5p(3q^2) = 5(3)pq^2$
$= 15pq^2$

9. $-6m^3(3n^2) = -6(3)m^3n^2$
$= -18m^3n^2$

11. $y^5 \cdot 9y \cdot y^4 = 9y^{5+1+4}$
$= 9y^{10}$

13. $(4x^3)(2x^2)(-x^5) = (-4\cdot 2)x^{3+2+5}$
$= -8x^{10}$

15. $2m(3m+2) = 2m(3m) + 2m(2)$
$= 6m^2 + 4m$

17. $3p(-2p^3 + 4p^2) = 3p(-2p^3) + 3p(4p^2)$
$= -6p^4 + 12p^3$

19. $-8z(2z + 3z^2 + 3z^3)$
$= -8z(2z) + (-8z)(3z^2) + (-8z)(3z^3)$
$= -16z^2 - 24z^3 - 24z^4$

21. $2y^3(3 + 2y + 5y^4)$
$= 2y^3(3) + 2y^3(2y) + 2y^3(5y^4)$
$= 6y^3 + 4y^4 + 10y^7$

23. $-4r^3(-7r^2 + 8r - 9)$
$= -4r^3(-7r^2) + (-4r^3)(8r) + (-4r^3)(-9)$
$= 28r^5 - 32r^4 + 36r^3$

25. $3a^2(2a^2 - 4ab + 5b^2)$
$= 3a^2(2a^2) + 3a^2(-4ab) + 3a^2(5b^2)$
$= 6a^4 - 12a^3b + 15a^2b^2$

27. $7m^3n^2(3m^2 + 2mn - n^3)$
$= 7m^3n^2(3m^2) + 7m^3n^2(2mn) + 7m^3n^2(-n^3)$
$= 21m^5n^2 + 14m^4n^3 - 7m^3n^5$

In Exercises 29–40, we can multiply the polynomials horizontally or vertically. The following solutions illustrate these two methods.

29. $(6x+1)(2x^2 + 4x + 1)$
$= 6x(2x^2) + 6x(4x) + 6x(1)$
$\quad + 1(2x^2) + 1(4x) + 1(1)$
$= 12x^3 + 24x^2 + 6x + 2x^2 + 4x + 1$
$= 12x^3 + 26x^2 + 10x + 1$

31. $(9y-2)(8y^2 - 6y + 1)$

Multiply vertically.

$$\begin{array}{rrrr}
 & 8y^2 & -\ 6y & +\ 1 \\
 & & 9y & -\ 2 \\
\hline
 & -\ 16y^2 & +\ 12y & -\ 2 \\
72y^3 & -\ 54y^2 & +\ 9y & \\
\hline
72y^3 & -\ 70y^2 & +\ 21y & -\ 2
\end{array}$$

33. $(4m+3)(5m^3 - 4m^2 + m - 5)$

Multiply vertically.

$$\begin{array}{rrrrr}
 & 5m^3 & -\ 4m^2 & +\ m & -\ 5 \\
 & & & 4m & +\ 3 \\
\hline
 & 15m^3 & -\ 12m^2 & +\ 3m & -\ 15 \\
20m^4 & -\ 16m^3 & +\ 4m^2 & -\ 20m & \\
\hline
20m^4 & -\ m^3 & -\ 8m^2 & -\ 17m & -\ 15
\end{array}$$

35. $(2x-1)(3x^5 - 2x^3 + x^2 - 2x + 3)$

Multiply vertically.

$$\begin{array}{rrrrrrr}
 & 3x^5 & & -\ 2x^3 & +\ x^2 & -\ 2x & +3 \\
 & & & & & 2x & -\ 1 \\
\hline
 & -\ 3x^5 & & +\ 2x^3 & -\ x^2 & +\ 2x & -\ 3 \\
6x^6 & & -\ 4x^4 & +\ 2x^3 & -\ 4x^2 & +\ 6x & \\
\hline
6x^6 & -\ 3x^5 & -\ 4x^4 & +\ 4x^3 & -\ 5x^2 & +\ 8x & -\ 3
\end{array}$$

37. $(5x^2 + 2x + 1)(x^2 - 3x + 5)$

Multiply vertically.

$$\begin{array}{rrrrr}
 & & 5x^2 & +\ 2x & +1 \\
 & & x^2 & -\ 3x & +5 \\
\hline
 & & 25x^2 & +10x & +5 \\
 & -\ 15x^3 & -\ 6x^2 & -\ 3x & \\
5x^4 & +\ 2x^3 & +\ x^2 & & \\
\hline
5x^4 & -\ 13x^3 & +20x^2 & +\ 7x & +5
\end{array}$$

39. $(6x^4 - 4x^2 + 8x)(\frac{1}{2}x + 3)$
$= 6x^4(\frac{1}{2}x) + -4x^2(\frac{1}{2}x) + 8x(\frac{1}{2}x)$
$+ 6x^4(3) + -4x^2(3) + 8x(3)$
$= 3x^5 - 2x^3 + 4x^2 + 18x^4 - 12x^2 + 24x$
$= 3x^5 + 18x^4 - 2x^3 - 8x^2 + 24x$

41. $(m + 7)(m + 5)$

F O I L
$= m(m) + m(5) + 7(m) + 7(5)$
$= m^2 + 5m + 7m + 35$
$= m^2 + 12m + 35$

43. $(n - 1)(n + 4)$

F O I L
$= n(n) + n(4) + (-1)(n) + (-1)(4)$
$= n^2 + 4n + (-1n) + (-4)$
$= n^2 + 3n - 4$

45. $(x + 5)(x - 5)$

F O I L
$= x(x) + x(-5) + 5(x) + 5(-5)$
$= x^2 - 5x + 5x - 25$
$= x^2 - 25$

47. $(2x + 3)(6x - 4)$

F O I L
$= 2x(6x) + 2x(-4) + 3(6x) + 3(-4)$
$= 12x^2 - 8x + 18x - 12$
$= 12x^2 + 10x - 12$

49. $(9 + t)(9 - t)$

F O I L
$= 9(9) + 9(-t) + t(9) + t(-t)$
$= 81 - 9t + 9t - t^2$
$= 81 - t^2$

51. $(3x - 2)(3x - 2)$

F O I L
$= 3x(3x) + 3x(-2) + (-2)(3x) + (-2)(-2)$
$= 9x^2 - 6x - 6x + 4$
$= 9x^2 - 12x + 4$

53. $(5a + 1)(2a + 7)$

F O I L
$= 5a(2a) + 5a(7) + 1(2a) + 1(7)$
$= 10a^2 + 35a + 2a + 7$
$= 10a^2 + 37a + 7$

55. $(6 - 5m)(2 + 3m)$

F O I L
$= 6(2) + 6(3m) + (-5m)(2) + (-5m)(3m)$
$= 12 + 18m - 10m - 15m^2$
$= 12 + 8m - 15m^2$

57. $(5 - 3x)(4 + x)$

F O I L
$= 5(4) + 5(x) + (-3x)(4) + (-3x)(x)$
$= 20 + 5x - 12x - 3x^2$
$= 20 - 7x - 3x^2$

59. $(3t - 4s)(t + 3s)$

F O I L
$= 3t(t) + 3t(3s) + (-4s)(t) + (-4s)(3s)$
$= 3t^2 + 9st - 4st - 12s^2$
$= 3t^2 + 5st - 12s^2$

61. $(4x + 3)(2y - 1)$

F O I L
$= 4x(2y) + 4x(-1) + 3(2y) + 3(-1)$
$= 8xy - 4x + 6y - 3$

63. $(3x + 2y)(5x - 3y)$

F O I L
$= 3x(5x) + 3x(-3y) + 2y(5x) + 2y(-3y)$
$= 15x^2 - 9xy + 10xy - 6y^2$
$= 15x^2 + xy - 6y^2$

65. $3y^3(2y + 3)(y - 5)$

$(2y + 3)(y - 5)$
F O I L
$= 2y(y) + 2y(-5) + 3(y) + 3(-5)$
$= 2y^2 - 10y + 3y - 15$
$= 2y^2 - 7y - 15$

Now multiply this result by $3y^3$.

$3y^3(2y^2 - 7y - 15)$
$= 3y^3(2y^2) + 3y^3(-7y) + 3y^3(-15)$
$= 6y^5 - 21y^4 - 45y^3$

67. $-8r^3(5r^2 + 2)(5r^2 - 2)$

$(5r^2 + 2)(5r^2 - 2)$
F O I L
$= 5r^2(5r^2) + 5r^2(-2) + 2(5r^2) + 2(-2)$
$= 25r^4 - 10r^2 + 10r^2 - 4$
$= 25r^4 - 4$

Now multiply this result by $-8r^3$.

$-8r^3(25r^4 - 4)$
$= -8r^3(25r^4) + (-8r^3)(-4)$
$= -200r^7 + 32r^3$

69. **(a)** Use the formula for the area of a rectangle, $\mathcal{A} = LW$, with $L = 3y + 7$ and $W = y + 1$.

$$\begin{aligned}\mathcal{A} &= (3y+7)(y+1)\\ &= (3y+7)(y) + (3y+7)(1)\\ &= 3y^2 + 7y + 3y + 7\\ &= 3y^2 + 10y + 7\end{aligned}$$

(b) Use the formula for the perimeter of a rectangle, $P = 2L + 2W$, with $L = 3y + 7$ and $W = y + 1$.

$$\begin{aligned}P &= 2(3y+7) + 2(y+1)\\ &= 2(3y) + 2(7) + 2(y) + 2(1)\\ &= 6y + 14 + 2y + 2\\ &= 8y + 16\end{aligned}$$

71. $(3p + \frac{5}{4}q)(2p - \frac{5}{3}q)$ *Use FOIL.*

$$\begin{aligned}&= 3p(2p) + 3p(-\tfrac{5}{3}q) + \tfrac{5}{4}q(2p) + \tfrac{5}{4}q(-\tfrac{5}{3}q)\\ &= 6p^2 - 5pq + \tfrac{5}{2}pq - \tfrac{25}{12}q^2\\ &= 6p^2 + (-\tfrac{10}{2} + \tfrac{5}{2})pq - \tfrac{25}{12}q^2\\ &= 6p^2 - \tfrac{5}{2}pq - \tfrac{25}{12}q^2\end{aligned}$$

73. $(x+7)^2 = (x+7)(x+7)$ *Use FOIL.*

$$\begin{aligned}&= x^2 + 7x + 7x + 49\\ &= x^2 + 14x + 49\end{aligned}$$

75. $(a-4)(a+4)$ *Use FOIL.*

$$\begin{aligned}&= a^2 + 4a - 4a - 16\\ &= a^2 - 16\end{aligned}$$

77. $(2p-5)^2 = (2p-5)(2p-5)$ *Use FOIL.*

$$\begin{aligned}&= 4p^2 - 10p - 10p + 25\\ &= 4p^2 - 20p + 25\end{aligned}$$

79. $(5k+3q)^2 = (5k+3q)(5k+3q)$ *Use FOIL.*

$$\begin{aligned}&= 25k^2 + 15kq + 15kq + 9q^2\\ &= 25k^2 + 30kq + 9q^2\end{aligned}$$

81. Recall that a^3 means $(a)(a)(a)$, so $(m-5)^3 = (m-5)(m-5)(m-5)$. We'll start by finding

$$\begin{aligned}(m-5)(m-5) &= m^2 - 5m - 5m + 25\\ &= m^2 - 10m + 25.\end{aligned}$$

Now multiply that result by $m - 5$.

$$\begin{array}{rrrr} & m^2 & -\ 10m & +\ 25\\ & & m & -\ 5\\ \hline & -\ 5m^2 & +\ 50m & -\ 125\\ m^3 & -\ 10m^2 & +\ 25m & \\ \hline m^3 & -\ 15m^2 & +\ 75m & -\ 125\end{array}$$

83. $(2a+1)^3 = (2a+1)(2a+1)(2a+1)$

$$\begin{aligned}(2a+1)(2a+1) &= 4a^2 + 2a + 2a + 1\\ &= 4a^2 + 4a + 1\end{aligned}$$

Now multiply vertically.

$$\begin{array}{rrrr} & 4a^2 & +\ 4a & +\ 1\\ & & 2a & +\ 1\\ \hline & 4a^2 & +\ 4a & +\ 1\\ 8a^3 & +\ 8a^2 & +\ 2a & \\ \hline 8a^3 & +\ 12a^2 & +\ 6a & +\ 1\end{array}$$

85. $-3a(3a+1)(a-4)$

$$\begin{aligned}&= -3a(3a^2 - 12a + a - 4) \quad \textit{FOIL}\\ &= -3a(3a^2 - 11a - 4)\\ &= -9a^3 + 33a^2 + 12a\end{aligned}$$

87. $7(4m-3)(2m+1)$

$$\begin{aligned}&= 7(8m^2 + 4m - 6m - 3) \quad \textit{FOIL}\\ &= 7(8m^2 - 2m - 3)\\ &= 56m^2 - 14m - 21\end{aligned}$$

89. $(3r-2s)^4 = (3r-2s)^2(3r-2s)^2$

First we find $(3r-2s)^2$.

$$\begin{aligned}(3r-2s)^2 &= (3r-2s)(3r-2s)\\ &= 9r^2 - 6rs - 6rs + 4s^2\\ &= 9r^2 - 12rs + 4s^2\end{aligned}$$

Now multiply this result by itself.

$$\begin{array}{rrrrr} & & 9r^2 & -\ 12rs & +\ 4s^2\\ & & 9r^2 & -\ 12rs & +\ 4s^2\\ \hline & & 36r^2s^2 & -\ 48rs^3 & +\ 16s^4\\ & -\ 108r^3s & +\ 144r^2s^2 & -\ 48rs^3 & \\ 81r^4 & -\ 108r^3s & +\ 36r^2s^2 & & \\ \hline 81r^4 & -\ 216r^3s & +\ 216r^2s^2 & -\ 96rs^3 & +\ 16s^4\end{array}$$

91. $3p^3(2p^2 + 5p)(p^3 + 2p + 1)$

$$\begin{aligned}&= [3p^3(2p^2) + 3p^3(5p)](p^3 + 2p + 1)\\ &\qquad \textit{Distributive property}\\ &= (6p^5 + 15p^4)(p^3 + 2p + 1)\end{aligned}$$

Now multiply vertically.

$$\begin{array}{rrrrr} & & p^3 & +\ 2p & +1\\ & & & 6p^5 & +15p^4\\ \hline & 15p^7 & & +30p^5 & +15p^4\\ 6p^8 & & +12p^6 & +\ 6p^5 & \\ \hline 6p^8 & +15p^7 & +12p^6 & +36p^5 & +15p^4\end{array}$$

93. $-2x^5(3x^2+2x-5)(4x+2)$
$= [-2x^5(3x^2)+(-2x^5)(2x)$
$+(-2x^5)(-5)](4x+2)$
Distributive property
$= (-6x^7-4x^6+10x^5)(4x+2)$

Now multiply vertically.

$$\begin{array}{rrrrr} & -6x^7 & -\ 4x^6 & +10x^5 & \\ & & 4x & +\ 2 & \\ \hline & -\ 12x^7 & -\ 8x^6 & +20x^5 & \\ -24x^8 & -\ 16x^7 & +40x^6 & & \\ \hline -24x^8 & -\ 28x^7 & +32x^6 & +20x^5 & \end{array}$$

95. The area $\mathcal{A}$ of the shaded region is the difference between the area of the larger square, which has sides of length $x+7$, and the area of the smaller square, which has sides of length x.

$$\begin{aligned} \mathcal{A} &= (x+7)^2-(x)^2 \\ &= (x+7)(x+7)-x^2 \\ &= (x^2+7x+7x+49)-x^2 \\ &= x^2+14x+49-x^2 \\ &= 14x+49 \end{aligned}$$

97. The area $\mathcal{A}$ of the shaded region is the difference between the area of the circle, which has radius x, and the area of the square, which has sides of length 3.

$$\begin{aligned} \mathcal{A} &= \pi r^2-s^2 \\ &= \pi(x)^2-(3)^2 \\ &= \pi x^2-9 \end{aligned}$$

99. $(3m)^2 = 3^2m^2 = 9m^2$

101. $(-2r)^2 = (-2)^2r^2 = 4r^2$

103. $(4x^2)^2 = 4^2(x^2)^2 = 16x^4$

4.6 Special Products

4.6 Now Try Exercises

N1. $(x+5)^2 = x^2+2(x)(5)+5^2$
$= x^2+10x+25$

N2. (a) $(3x-1)^2 = (3x)^2-2(3x)(1)+1^2$
$= 9x^2-6x+1$

(b) $(4p-5q)^2 = (4p)^2-2(4p)(5q)+(5q)^2$
$= 16p^2-40pq+25q^2$

(c) $(6t-\frac{1}{3})^2 = (6t)^2-2(6t)(\frac{1}{3})+(\frac{1}{3})^2$
$= 36t^2-4t+\frac{1}{9}$

(d) $m(2m+3)^2 = m(4m^2+12m+9)$
$= 4m^3+12m^2+9m$

N3. $(t+10)(t-10) = t^2-10^2$
$= t^2-100$

N4. (a) $(4x-6)(4x+6) = (4x)^2-6^2$
$= 16x^2-36$

(b) $(5r-\frac{4}{5})(5r+\frac{4}{5}) = (5r)^2-(\frac{4}{5})^2$
$= 25r^2-\frac{16}{25}$

(c) $y(3y+1)(3y-1)$
First use the rule for the product of the sum and difference of two terms.

$$\begin{aligned} (3y+1)(3y-1) &= (3y)^2-1^2 \\ &= 9y^2-1 \end{aligned}$$

Now multiply this result by y.

$$\begin{aligned} y(9y^2-1) &= y(9y^2)+y(-1) \\ &= 9y^3-y \end{aligned}$$

N5. $(2m-1)^3$

Since $(2m-1)^3 = (2m-1)^2(2m-1)$, the first step is to find the product $(2m-1)^2$.

$$\begin{aligned} (2m-1)^2 &= (2m)^2-2(2m)(1)+1^2 \\ &= 4m^2-4m+1 \end{aligned}$$

Now multiply this result by $2m-1$.

$$\begin{aligned} &(2m-1)^3 \\ &= (2m-1)(4m^2-4m+1) \\ &= 8m^3-8m^2+2m-4m^2+4m-1 \\ &= 8m^3-12m^2+6m-1 \end{aligned}$$

4.6 Section Exercises

1. $(4x+3)^2$

(a) The square of the first term is

$$(4x)^2 = (4x)(4x) = 16x^2.$$

(b) Twice the product of the two terms is

$$2(4x)(3) = 24x.$$

(c) The square of the last term is

$$3^2 = 9.$$

(d) The final product is the trinomial

$$16x^2+24x+9.$$

In Exercises 3–20, use one of the following formulas for the square of a binomial:

$$\begin{aligned} (x+y)^2 &= x^2+2xy+y^2 \\ (x-y)^2 &= x^2-2xy+y^2 \end{aligned}$$

3. $(m+2)^2 = m^2+2(m)(2)+2^2$
$= m^2+4m+4$

5. $(r-3)^2 = r^2-2(r)(3)+3^2$
$= r^2-6r+9$

7. $(x+2y)^2 = x^2+2(x)(2y)+(2y)^2$
$= x^2+4xy+4y^2$

9. $(5p+2q)^2 = (5p)^2 + 2(5p)(2q) + (2q)^2$
$= 25p^2 + 20pq + 4q^2$

11. $(4a+5b)^2 = (4a)^2 + 2(4a)(5b) + (5b)^2$
$= 16a^2 + 40ab + 25b^2$

13. $(6m - \frac{4}{5}n)^2 = (6m)^2 - 2(6m)(\frac{4}{5}n) + (\frac{4}{5}n)^2$
$= 36m^2 - \frac{48}{5}mn + \frac{16}{25}n^2$

15. $t(3t-1)^2$

$(3t-1)^2 = (3t)^2 - 2(3t)(1) + 1^2$
$= 9t^2 - 6t + 1$

Now multiply by t.

$$t(9t^2 - 6t + 1) = 9t^3 - 6t^2 + t$$

17. $3t(4t+1)^2$

$(4t+1)^2 = (4t)^2 + 2(4t)(1) + 1^2$
$= 16t^2 + 8t + 1$

Now multiply by $3t$.

$3t(16t^2 + 8t + 1) = 48t^3 + 24t^2 + 3t$

19. $-(4r-2)^2$

$(4r-2)^2 = (4r)^2 - 2(4r)(2) + 2^2$
$= 16r^2 - 16r + 4$

Now multiply by -1.

$-1(16r^2 - 16r + 4) = -16r^2 + 16r - 4$

21. **(a)** $7x(7x) = 49x^2$

(b) $7x(-3y) + 3y(7x)$
$= -21xy + 21xy = 0$

(c) $3y(-3y) = -9y^2$

(d) $49x^2 - 9y^2$

The sum found in part (b) is omitted because it is 0. Adding 0, the identity element for addition, would not change the answer.

In Exercises 23–40, use the formula for the product of the sum and difference of two terms.

$$(x+y)(x-y) = x^2 - y^2$$

23. $(k+5)(k-5) = k^2 - 5^2$
$= k^2 - 25$

25. $(4-3t)(4+3t) = 4^2 - (3t)^2$
$= 16 - 9t^2$

27. $(5x+2)(5x-2) = (5x)^2 - 2^2$
$= 25x^2 - 4$

29. $(5y+3x)(5y-3x) = (5y)^2 - (3x)^2$
$= 25y^2 - 9x^2$

31. $(10x+3y)(10x-3y) = (10x)^2 - (3y)^2$
$= 100x^2 - 9y^2$

33. $(2x^2-5)(2x^2+5) = (2x^2)^2 - 5^2$
$= 4x^4 - 25$

35. $(\frac{3}{4} - x)(\frac{3}{4} + x) = (\frac{3}{4})^2 - x^2$
$= \frac{9}{16} - x^2$

37. $(9y + \frac{2}{3})(9y - \frac{2}{3}) = (9y)^2 - (\frac{2}{3})^2$
$= 81y^2 - \frac{4}{9}$

39. $q(5q-1)(5q+1)$

$(5q-1)(5q+1) = (5q)^2 - 1^2$
$= 25q^2 - 1$

Now multiply by q.

$$q(25q^2 - 1) = 25q^3 - q$$

41. No. In general, $(a+b)^2$ equals $a^2 + 2ab + b^2$, which is not equivalent to $a^2 + b^2$.

43. $(x+1)^3$

$= (x+1)^2(x+1)$ *$a^3 = a^2 \cdot a$*

$= (x^2 + 2x + 1)(x+1)$ *Square the binomial.*

$= x^3 + 2x^2 + x + x^2 + 2x + 1$ *Multiply polynomials.*

$= x^3 + 3x^2 + 3x + 1$ *Combine like terms.*

45. $(t-3)^3$

$= (t-3)^2(t-3)$ *$a^3 = a^2 \cdot a$*

$= (t^2 - 6t + 9)(t-3)$ *Square the binomial.*

$= t^3 - 6t^2 + 9t - 3t^2 + 18t - 27$ *Multiply polynomials.*

$= t^3 - 9t^2 + 27t - 27$ *Combine like terms.*

47. $(r+5)^3$

$= (r+5)^2(r+5)$ *$a^3 = a^2 \cdot a$*

$= (r^2 + 10r + 25)(r+5)$ *Square the binomial.*

$= r^3 + 10r^2 + 25r + 5r^2 + 50r + 125$ *Multiply polynomials.*

$= r^3 + 15r^2 + 75r + 125$ *Combine like terms.*

49. $(2a+1)^3$

$= (2a+1)^2(2a+1)$ *$a^3 = a^2 \cdot a$*

$= (4a^2 + 4a + 1)(2a+1)$ *Square the binomial.*

$= 8a^3 + 8a^2 + 2a + 4a^2 + 4a + 1$ *Multiply polynomials.*

$= 8a^3 + 12a^2 + 6a + 1$ *Combine like terms.*

51. $(4x-1)^4$
$= (4x-1)^2(4x-1)^2$ $\quad a^4 = a^2 \cdot a^2$
$= (16x^2-8x+1)(16x^2-8x+1)$
Square each binomial.
$= 256x^4 - 128x^3 + 16x^2 - 128x^3 + 64x^2$
$\quad - 8x + 16x^2 - 8x + 1$
Multiply polynomials.
$= 256x^4 - 256x^3 + 96x^2 - 16x + 1$
Combine like terms.

53. $(3r-2t)^4$
$= (3r-2t)^2(3r-2t)^2$ $\quad a^4 = a^2 \cdot a^2$
$= (9r^2-12rt+4t^2)(9r^2-12rt+4t^2)$
Square each binomial.
$= 81r^4 - 108r^3t + 36r^2t^2 - 108r^3t$
$\quad + 144r^2t^2 - 48rt^3 + 36r^2t^2 - 48rt^3 + 16t^4$
Multiply polynomials.
$= 81r^4 - 216r^3t + 216r^2t^2 - 96rt^3 + 16t^4$
Combine like terms.

55. $2x(x+1)^3$
$= 2x(x+1)(x+1)^2$
$= 2x(x+1)(x^2+2x+1)$
$= 2x(x^3+2x^2+x+x^2+2x+1)$
$= 2x(x^3+3x^2+3x+1)$
$= 2x^4+6x^3+6x^2+2x$

57. $-4t(t+3)^3$
$= -4t(t+3)(t+3)^2$
$= -4t(t+3)(t^2+6t+9)$
$= -4t(t^3+6t^2+9t+3t^2+18t+27)$
$= -4t(t^3+9t^2+27t+27)$
$= -4t^4-36t^3-108t^2-108t$

59. $(x+y)^2(x-y)^2$
$= [(x+y)(x-y)]^2$ $\quad a^2b^2 = (ab)^2$
$= (x^2-y^2)^2$
$= (x^2)^2 - 2x^2y^2 + (y^2)^2$
$= x^4 - 2x^2y^2 + y^4$

Another method is to expand both binomials first and then multiply those trinomials.

◆◆◆ Relating Concepts 61–70 ◆◆◆

61. The large square has sides of length $a+b$, so its area is $(a+b)^2$.

62. The red square has sides of length a, so its area is a^2.

63. Each blue rectangle has length a and width b, so each has an area of ab. Thus, the sum of the areas of the blue rectangles is

$$ab + ab = 2ab.$$

64. The yellow square has sides of length b, so its area is b^2.

65. Sum $= a^2 + 2ab + b^2$

66. The area of the largest square equals the sum of the areas of the two smaller squares and the two rectangles. Therefore, $(a+b)^2$ must equal $a^2+2ab+b^2$.

67. $35^2 = (35)(35)$

$$\begin{array}{r} 35 \\ \underline{35} \\ 175 \\ \underline{105} \\ 1225 \end{array}$$

68. $(a+b)^2 = a^2+2ab+b^2$
$(30+5)^2 = 30^2 + 2(30)(5) + 5^2$

69. $30^2 + 2(30)(5) + 5^2$
$= 900 + 60(5) + 25$
$= 900 + 300 + 25$
$= 1225$

70. The answers are equal.

71. $101 \times 99 = (100+1)(100-1)$
$= 100^2 - 1^2$
$= 10{,}000 - 1$
$= 9999$

73. $201 \times 199 = (200+1)(200-1)$
$= 200^2 - 1^2$
$= 40{,}000 - 1$
$= 39{,}999$

75. $20\frac{1}{2} \times 19\frac{1}{2} = (20+\frac{1}{2})(20-\frac{1}{2})$
$= 20^2 - (\frac{1}{2})^2$
$= 400 - \frac{1}{4}$
$= 399\frac{3}{4}$

77. Use the formula for area the of a triangle, $\mathcal{A} = \frac{1}{2}bh$, with $b = m+2n$ and $h = m-2n$.

$$\mathcal{A} = \tfrac{1}{2}(m+2n)(m-2n)$$
$$= \tfrac{1}{2}[m^2-(2n)^2]$$
$$= \tfrac{1}{2}(m^2-4n^2)$$
$$= \tfrac{1}{2}m^2 - 2n^2$$

79. Use the formula for the area of a parallelogram, $\mathcal{A} = bh$, with $b = 3a+2$ and $h = 3a-2$.

$$\mathcal{A} = (3a+2)(3a-2)$$
$$= (3a)^2 - 2^2$$
$$= 9a^2 - 4$$

81. Use the formula for the area of a circle, $\mathcal{A} = \pi r^2$, with $r = x + 2$.

$$\begin{aligned}\mathcal{A} &= \pi(x+2)^2\\ &= \pi(x^2+4x+4)\\ &= \pi x^2 + 4\pi x + 4\pi\end{aligned}$$

83. Use the formula for the volume of a cube, $V = e^3$, with $e = x + 2$.

$$\begin{aligned}V &= (x+2)^3\\ &= (x+2)^2(x+2)\\ &= (x^2+4x+4)(x+2)\\ &= x^3+4x^2+4x+2x^2+8x+8\\ &= x^3+6x^2+12x+8\end{aligned}$$

85. $\frac{1}{2p}(4p^2+2p+8)$

$$\begin{aligned}&= \frac{1}{2p}(4p^2) + \frac{1}{2p}(2p) + \frac{1}{2p}(8)\\ &= 2p + 1 + \frac{4}{p}\end{aligned}$$

87. $\frac{1}{3m}(m^3+9m^2-6m)$

$$\begin{aligned}&= \frac{1}{3m}(m^3) + \frac{1}{3m}(9m^2) + \frac{1}{3m}(-6m)\\ &= \frac{m^2}{3} + 3m - 2\end{aligned}$$

89. $-3k(8k^2-12k+2)$

$$\begin{aligned}&= -3k(8k^2) + (-3k)(-12k) + (-3k)(2)\\ &= -24k^3 + 36k^2 - 6k\end{aligned}$$

91. $(-2k+1)(8k^2+9k+3)$

$$\begin{aligned}&= -2k(8k^2) + (-2k)(9k) + (-2k)(3)\\ &\quad + 1(8k^2) + 1(9k) + 1(3)\\ &= -16k^3 - 18k^2 - 6k + 8k^2 + 9k + 3\\ &= -16k^3 - 10k^2 + 3k + 3\end{aligned}$$

93. Subtract.

$$\begin{array}{rrr} 5t^2 & +\ 2t & -\ 6\\ \underline{5t^2} & \underline{-\ 3t} & \underline{-\ 9}\end{array}$$

Change all signs in the second row, and then add.

$$\begin{array}{rrr} 5t^2 & +\ 2t & -\ 6\\ \underline{-5t^2} & \underline{+\ 3t} & \underline{+\ 9}\\ & 5t & +\ 3\end{array}$$

4.7 Dividing Polynomials

4.7 Now Try Exercises

N1. $\frac{16a^6 - 12a^4}{4a^2}$

$$\begin{aligned}&= \frac{16a^6}{4a^2} - \frac{12a^4}{4a^2}\\ &= 4a^4 - 3a^2\end{aligned}$$

N2. $\frac{36x^5 + 24x^4 - 12x^3}{6x^4}$

$$\begin{aligned}&= \frac{36x^5}{6x^4} + \frac{24x^4}{6x^4} - \frac{12x^3}{6x^4}\\ &= 6x + 4 - \frac{2}{x}\end{aligned}$$

N3. Divide $7y^4 - 40y^5 + 100y^2$ by $-5y^2$.

$$\frac{7y^4 - 40y^5 + 100y^2}{-5y^2}$$

$$\begin{aligned}&= \frac{7y^4}{-5y^2} - \frac{40y^5}{-5y^2} + \frac{100y^2}{-5y^2}\\ &= -\frac{7y^2}{5} + 8y^3 - 20 \quad \text{or} \quad 8y^3 - \frac{7y^2}{5} - 20\end{aligned}$$

N4. Divide $35m^5n^4 - 49m^2n^3 + 12mn$ by $7m^2n$.

$$\frac{35m^5n^4 - 49m^2n^3 + 12mn}{7m^2n}$$

$$\begin{aligned}&= \frac{35m^5n^4}{7m^2n} - \frac{49m^2n^3}{7m^2n} + \frac{12mn}{7m^2n}\\ &= 5m^3n^3 - 7n^2 + \frac{12}{7m}\end{aligned}$$

N5. Divide. $\frac{4x^2 + x - 18}{x - 2}$

Step 1 (see the division following Step 4)
$4x^2$ divided by x is $4x$;
$4x(x-2) = 4x^2 - 8x$.

Step 2
Subtract $4x^2 - 8x$ from $4x^2 + x$.
Bring down -18.

Step 3
$9x$ divided by x is 9;
$9(x-2) = 9x - 18$.

Step 4
Subtract $9x - 18$ from $9x - 18$.
The remainder is 0.
The answer is the quotient, $4x + 9$.

$$\begin{array}{r} 4x + 9\\ x-2\,\overline{)\,4x^2 + x - 18}\\ \underline{4x^2 - 8x}\\ 9x - 18\\ \underline{9x - 18}\\ 0\end{array}$$

Check: Multiply the divisor, $x - 2$, by the quotient, $4x + 9$. The product must be the original dividend, $4x^2 + x - 18$.

$$\begin{aligned}&(x-2)(4x+9)\\ &= 4x^2 - 8x + 9x - 18\\ &= 4x^2 + x - 18\end{aligned}$$

N6. Divide $6k^3 - 20k - k^2 + 1$ by $2k - 3$.

Write the numerator in descending powers and divide.

$$\begin{array}{r|rrrrl}
 & 3k^2 & + 4k & - 4 & & \\
\hline
2k - 3 & 6k^3 & - k^2 & - 20k & + 1 & \\
 & 6k^3 & - 9k^2 & & & \frac{6k^3}{2k} = 3k^2 \\
\hline
 & & 8k^2 & - 20k & & \frac{8k^2}{2k} = 4k \\
 & & 8k^2 & - 12k & & \\
\hline
 & & & - 8k & + 1 & \frac{-8k}{2k} = -4 \\
 & & & - 8k & + 12 & \\
\hline
 & & & & - 11 &
\end{array}$$

The remainder is -11.

The answer is $3k^2 + 4k - 4 + \dfrac{-11}{2k - 3}$.

N7. Divide $m^3 - 1000$ by $m - 10$.

Use 0 as the coefficient of the missing m^2- and m-terms.

$$\begin{array}{r|rrrr}
 & m^2 & + 10m & + 100 & \\
\hline
m - 10 & m^3 & + 0m^2 & + 0m & - 1000 \\
 & m^3 & - 10m^2 & & \\
\hline
 & & 10m^2 & + 0m & \\
 & & 10m^2 & - 100m & \\
\hline
 & & & 100m & - 1000 \\
 & & & 100m & - 1000 \\
\hline
 & & & & 0
\end{array}$$

$(m^3 - 1000)$ divided by $(m - 10)$ is

$$m^2 + 10m + 100.$$

N8. Divide $y^4 - 5y^3 + 6y^2 + y - 4$ by $y^2 + 2$.

$$\begin{array}{r|rrrrr}
 & & & y^2 & - 5y & + 4 \\
\hline
y^2 + 0y + 2 & y^4 & - 5y^3 & + 6y^2 & + y & - 4 \\
 & y^4 & + 0y^3 & + 2y^2 & & \\
\hline
 & & - 5y^3 & + 4y^2 & + y & \\
 & & - 5y^3 & - 0y^2 & - 10y & \\
\hline
 & & & 4y^2 & + 11y & - 4 \\
 & & & 4y^2 & + 0y & + 8 \\
\hline
 & & & & 11y & - 12
\end{array}$$

$(y^4 - 5y^3 + 6y^2 + y - 4)$ divided by $(y^2 + 2)$ is
$y^2 - 5y + 4 + \dfrac{11y - 12}{y^2 + 3}$.

N9. Divide $10x^3 + 21x^2 + 5x - 8$ by $2x + 4$.

$$\begin{array}{r|rrrrl}
 & 5x^2 & + \frac{1}{2}x & + \frac{3}{2} & & \\
\hline
2x + 4 & 10x^3 & + 21x^2 & + 5x & - 8 & \\
 & 10x^3 & + 20x^2 & & & \\
\hline
 & & x^2 & + 5x & & \frac{x^2}{2x} = \frac{1}{2}x \\
 & & x^2 & + 2x & & \\
\hline
 & & & 3x & - 8 & \frac{3x}{2x} = \frac{3}{2} \\
 & & & 3x & + 6 & \\
\hline
 & & & & - 14 &
\end{array}$$

$(10x^3 + 21x^2 + 5x - 8)$ divided by $(2x + 4)$ is

$$5x^2 + \frac{1}{2}x + \frac{3}{2} + \frac{-14}{2x + 4}.$$

4.7 Section Exercises

1. In the statement $\dfrac{10x^2 + 8}{2} = 5x^2 + 4$, $\underline{10x^2 + 8}$ is the dividend, $\underline{2}$ is the divisor, and $\underline{5x^2 + 4}$ is the quotient.

3. To check the division shown in Exercise 1, multiply $\underline{5x^2 + 4}$ by $\underline{2}$ (or $\underline{2}$ by $\underline{5x^2 + 4}$) and show that the product is $\underline{10x^2 + 8}$.

5. In this section, we are dividing a polynomial by a monomial. The problem

$$\frac{16m^3 - 12m^2}{4m}$$

is an example of such a division. However, in the problem

$$\frac{4m}{16m^3 - 12m^2},$$

we are dividing a monomial by a binomial. Therefore, the methods of this section do not apply.

7. $\dfrac{60x^4 - 20x^2 + 10x}{2x}$

$$= \frac{60x^4}{2x} - \frac{20x^2}{2x} + \frac{10x}{2x}$$

$$= \tfrac{60}{2}x^{4-1} - \tfrac{20}{2}x^{2-1} + \tfrac{10}{2}$$

$$= 30x^3 - 10x + 5$$

9. $\dfrac{20m^5 - 10m^4 + 5m^2}{5m^2}$

$$= \frac{20m^5}{5m^2} - \frac{10m^4}{5m^2} + \frac{5m^2}{5m^2}$$

$$= \tfrac{20}{5}m^{5-2} - \tfrac{10}{5}m^{4-2} + \tfrac{5}{5}$$

$$= 4m^3 - 2m^2 + 1$$

11. $\dfrac{8t^5 - 4t^3 + 4t^2}{2t}$

$= \dfrac{8t^5}{2t} - \dfrac{4t^3}{2t} + \dfrac{4t^2}{2t}$

$= 4t^4 - 2t^2 + 2t$

13. $\dfrac{4a^5 - 4a^2 + 8}{4a}$

$= \dfrac{4a^5}{4a} - \dfrac{4a^2}{4a} + \dfrac{8}{4a}$

$= a^4 - a + \dfrac{2}{a}$

15. $\dfrac{18p^5 + 12p^3 - 6p^2}{-6p^3}$

$= \dfrac{18p^5}{-6p^3} + \dfrac{12p^3}{-6p^3} - \dfrac{6p^2}{-6p^3}$

$= -3p^2 - 2 + \dfrac{1}{p}$

17. $\dfrac{-7r^7 + 6r^5 - r^4}{-r^5}$

$= \dfrac{-7r^7}{-r^5} + \dfrac{6r^5}{-r^5} - \dfrac{r^4}{-r^5}$

$= 7r^2 - 6 + \dfrac{1}{r}$

19. $\dfrac{12x^5 - 9x^4 + 6x^3}{3x^2}$

$= \dfrac{12x^5}{3x^2} - \dfrac{9x^4}{3x^2} + \dfrac{6x^3}{3x^2}$

$= 4x^3 - 3x^2 + 2x$

21. $\dfrac{3x^2 + 15x^3 - 27x^4}{3x^2}$

$= \dfrac{3x^2}{3x^2} + \dfrac{15x^3}{3x^2} - \dfrac{27x^4}{3x^2}$

$= 1 + 5x - 9x^2$ or $-9x^2 + 5x + 1$

23. $\dfrac{36x + 24x^2 + 6x^3}{3x^2}$

$= \dfrac{36x}{3x^2} + \dfrac{24x^2}{3x^2} + \dfrac{6x^3}{3x^2}$

$= \dfrac{12}{x} + 8 + 2x$ or $2x + 8 + \dfrac{12}{x}$

25. $\dfrac{4x^4 + 3x^3 + 2x}{3x^2}$

$= \dfrac{4x^4}{3x^2} + \dfrac{3x^3}{3x^2} + \dfrac{2x}{3x^2}$

$= \dfrac{4x^2}{3} + x + \dfrac{2}{3x}$

27. $\dfrac{-81x^5 + 30x^4 + 12x^2}{3x^2}$

$= \dfrac{-81x^5}{3x^2} + \dfrac{30x^4}{3x^2} + \dfrac{12x^2}{3x^2}$

$= -27x^3 + 10x^2 + 4$

29. $\dfrac{-27r^4 + 36r^3 - 6r^2 - 26r + 2}{-3r}$

$= \dfrac{-27r^4}{-3r} + \dfrac{36r^3}{-3r} - \dfrac{6r^2}{-3r} - \dfrac{26r}{-3r} + \dfrac{2}{-3r}$

$= 9r^3 - 12r^2 + 2r + \dfrac{26}{3} - \dfrac{2}{3r}$

31. $\dfrac{2m^5 - 6m^4 + 8m^2}{-2m^3}$

$= \dfrac{2m^5}{-2m^3} - \dfrac{6m^4}{-2m^3} + \dfrac{8m^2}{-2m^3}$

$= -m^2 + 3m - \dfrac{4}{m}$

33. $(20a^4 - 15a^5 + 25a^3) \div (5a^4)$

$= \dfrac{20a^4 - 15a^5 + 25a^3}{5a^4}$

$= \dfrac{20a^4}{5a^4} - \dfrac{15a^5}{5a^4} + \dfrac{25a^3}{5a^4}$

$= 4 - 3a + \dfrac{5}{a}$ or $-3a + 4 + \dfrac{5}{a}$

35. $(120x^{11} - 60x^{10} + 140x^9 - 100x^8) \div (10x^{12})$

$= \dfrac{120x^{11} - 60x^{10} + 140x^9 - 100x^8}{10x^{12}}$

$= \dfrac{120x^{11}}{10x^{12}} - \dfrac{60x^{10}}{10x^{12}} + \dfrac{140x^9}{10x^{12}} - \dfrac{100x^8}{10x^{12}}$

$= \dfrac{12}{x} - \dfrac{6}{x^2} + \dfrac{14}{x^3} - \dfrac{10}{x^4}$

37. $(120x^5y^4 - 80x^2y^3 + 40x^2y^4 - 20x^5y^3)$
$\div (20xy^2)$

$= \dfrac{120x^5y^4 - 80x^2y^3 + 40x^2y^4 - 20x^5y^3}{20xy^2}$

$= \dfrac{120x^5y^4}{20xy^2} - \dfrac{80x^2y^3}{20xy^2} + \dfrac{40x^2y^4}{20xy^2} - \dfrac{20x^5y^3}{20xy^2}$

$= 6x^4y^2 - 4xy + 2xy^2 - x^4y$

♦♦♦ Relating Concepts 39–42 ♦♦♦

39. $2\overline{)2846}$ with quotient 1423

40. $1423 = (1 \times 10^3) + (4 \times 10^2)$
$+ (2 \times 10^1) + (3 \times 10^0)$

41. $\dfrac{2x^3+8x^2+4x+6}{2}$

$= \dfrac{2x^3}{2}+\dfrac{8x^2}{2}+\dfrac{4x}{2}+\dfrac{6}{2}$

$= x^3+4x^2+2x+3$

42. They are similar in that the coefficients of powers of 10 are equal to the coefficients of the powers of x. They are different in that one is a constant while the other is a polynomial. They are equal if $x = 10$ (the base of our decimal system).

43. $\dfrac{x^2-x-6}{x-3}$

$$\begin{array}{r}
x+2\\
x-3\overline{)x^2-x-6}\\
\underline{x^2-3x}\\
2x-6\\
\underline{2x-6}\\
0
\end{array}$$

The remainder is 0. The answer is the quotient, $x+2$.

45. $\dfrac{2y^2+9y-35}{y+7}$

$$\begin{array}{r}
2y-5\\
y+7\overline{)2y^2+9y-35}\\
\underline{2y^2+14y}\\
-5y-35\\
\underline{-5y-35}\\
0
\end{array}$$

The remainder is 0. The answer is the quotient, $2y-5$.

47. $\dfrac{p^2+2p+20}{p+6}$

$$\begin{array}{r}
p-4\\
p+6\overline{)p^2+2p+20}\\
\underline{p^2+6p}\\
-4p+20\\
\underline{-4p-24}\\
44
\end{array}$$

The remainder is 44. Write the remainder as the numerator of a fraction that has the divisor $p+6$ as its denominator. The answer is

$$p-4+\frac{44}{p+6}.$$

49. $\dfrac{12m^2-20m+3}{2m-3}$

$$\begin{array}{r}
6m-1\\
2m-3\overline{)12m^2-20m+3}\\
\underline{12m^2-18m}\\
-2m+3\\
\underline{-2m+3}\\
0
\end{array}$$

The remainder is 0. The answer is the quotient, $6m-1$.

51. $\dfrac{4a^2-22a+32}{2a+3}$

$$\begin{array}{r}
2a-14\\
2a+3\overline{)4a^2-22a+32}\\
\underline{4a^2+6a}\\
-28a+32\\
\underline{-28a-42}\\
74
\end{array}$$

The remainder is 74. The answer is

$$2a-14+\frac{74}{2a+3}.$$

53. $\dfrac{8x^3-10x^2-x+3}{2x+1}$

$$\begin{array}{r}
4x^2-7x+3\\
2x+1\overline{)8x^3-10x^2-x+3}\\
\underline{8x^3+4x^2}\\
-14x^2-x\\
\underline{-14x^2-7x}\\
6x+3\\
\underline{6x+3}\\
0
\end{array}$$

The remainder is 0. The answer is the quotient,

$$4x^2-7x+3.$$

55. $\dfrac{8k^4-12k^3-2k^2+7k-6}{2k-3}$

$$\begin{array}{r}
4k^3-k+2\\
2k-3\overline{)8k^4-12k^3-2k^2+7k-6}\\
\underline{8k^4-12k^3}\\
-2k^2+7k\\
\underline{-2k^2+3k}\\
+4k-6\\
\underline{+4k-6}\\
0
\end{array}$$

The remainder is 0. The answer is the quotient,

$$4k^3-k+2.$$

57. $\dfrac{5y^4 + 5y^3 + 2y^2 - y - 8}{y+1}$

$$\begin{array}{r} 5y^3 \qquad\quad + 2y - 3 \\ y+1\,\overline{)\,5y^4 + 5y^3 + 2y^2 - y - 8} \\ \underline{5y^4 + 5y^3} \qquad\qquad\qquad \\ 2y^2 - y \qquad \\ \underline{2y^2 + 2y} \qquad \\ -3y - 8 \\ \underline{-3y - 3} \\ -5 \end{array}$$

The remainder is -5.

The quotient is $5y^3 + 2y - 3$.

The answer is $5y^3 + 2y - 3 + \dfrac{-5}{y+1}$.

59. $\dfrac{3k^3 - 4k^2 - 6k + 10}{k-2}$

$$\begin{array}{r} 3k^2 + 2k - 2 \\ k-2\,\overline{)\,3k^3 - 4k^2 - 6k + 10} \\ \underline{3k^3 - 6k^2} \qquad\qquad \\ 2k^2 - 6k \qquad \\ \underline{2k^2 - 4k} \qquad \\ -2k + 10 \\ \underline{-2k + 4} \\ 6 \end{array}$$

The remainder is 6.

The quotient is $3k^2 + 2k - 2$.

The answer is $3k^2 + 2k - 2 + \dfrac{6}{k-2}$.

61. $\dfrac{6p^4 - 16p^3 + 15p^2 - 5p + 10}{3p+1}$

$$\begin{array}{r} 2p^3 - 6p^2 + 7p - 4 \\ 3p+1\,\overline{)\,6p^4 - 16p^3 + 15p^2 - 5p + 10} \\ \underline{6p^4 + 2p^3} \qquad\qquad\qquad\quad \\ -18p^3 + 15p^2 \qquad\qquad \\ \underline{-18p^3 - 6p^2} \qquad\qquad \\ 21p^2 - 5p \qquad \\ \underline{21p^2 + 7p} \qquad \\ -12p + 10 \\ \underline{-12p - 4} \\ 14 \end{array}$$

The remainder is 14.

The quotient is $2p^3 - 6p^2 + 7p - 4$.

The answer is $2p^3 - 6p^2 + 7p - 4 + \dfrac{14}{3p+1}$.

63. $(x^3 + 2x^2 - 3) \div (x - 1)$

Use 0 as the coefficient of the missing x-term.

$$\begin{array}{r} x^2 + 3x + 3 \\ x-1\,\overline{)\,x^3 + 2x^2 + 0x - 3} \\ \underline{x^3 - x^2} \qquad\qquad\quad \\ 3x^2 + 0x \qquad \\ \underline{3x^2 - 3x} \qquad \\ 3x - 3 \\ \underline{3x - 3} \\ 0 \end{array}$$

$(x^3 + 2x^2 - 3) \div (x - 1) = x^2 + 3x + 3$

65. $(2x^3 + x + 2) \div (x + 1)$

Use 0 as the coefficient of the missing x^2-term.

$$\begin{array}{r} 2x^2 - 2x + 3 \\ x+1\,\overline{)\,2x^3 + 0x^2 + x + 2} \\ \underline{2x^3 + 2x^2} \qquad\qquad \\ -2x^2 + x \qquad \\ \underline{-2x^2 - 2x} \qquad \\ 3x + 2 \\ \underline{3x + 3} \\ -1 \end{array}$$

$(2x^3 + x + 2) \div (x + 1) = 2x^2 - 2x + 3 + \dfrac{-1}{x+1}$

67. $\dfrac{5 - 2r^2 + r^4}{r^2 - 1}$

Use 0 as the coefficient for the missing terms. Rearrange terms of the dividend in descending powers.

$$\begin{array}{r} r^2 \qquad\qquad - 1 \\ r^2 + 0r - 1\,\overline{)\,r^4 + 0r^3 - 2r^2 + 0r + 5} \\ \underline{r^4 + 0r^3 - r^2} \qquad\qquad \\ -r^2 + 0r + 5 \\ \underline{-r^2 + 0r + 1} \\ 4 \end{array}$$

The remainder is 4.

The quotient is $r^2 - 1$.

The answer is $r^2 - 1 + \dfrac{4}{r^2 - 1}$.

69. $\dfrac{-4x+3x^3+2}{x-1}$

Use 0 as the coefficient for the missing terms. Rearrange terms of the dividend in descending powers.

$$\begin{array}{r l}
 & \quad 3x^2 + 3x - 1 \\
x-1 \big) & \overline{3x^3 + 0x^2 - 4x + 2} \\
 & \underline{3x^3 - 3x^2} \\
 & \qquad 3x^2 - 4x \\
 & \qquad \underline{3x^2 - 3x} \\
 & \qquad\qquad -x + 2 \\
 & \qquad\qquad \underline{-x + 1} \\
 & \qquad\qquad\qquad 1
\end{array}$$

$$\frac{-4x+3x^3+2}{x-1} = 3x^2+3x-1+\frac{1}{x-1}$$

71. $\dfrac{y^3+1}{y+1}$

$$\begin{array}{r l}
 & \quad y^2 - y + 1 \\
y+1 \big) & \overline{y^3 + 0y^2 + 0y + 1} \\
 & \underline{y^3 + y^2} \\
 & \quad -y^2 + 0y \\
 & \quad \underline{-y^2 - y} \\
 & \qquad\qquad y + 1 \\
 & \qquad\qquad \underline{y + 1} \\
 & \qquad\qquad\quad 0
\end{array}$$

The remainder is 0. The answer is the quotient,

$$y^2 - y + 1.$$

73. $\dfrac{a^4-1}{a^2-1}$

$$\begin{array}{r l}
 & \quad a^2 \qquad\qquad + 1 \\
a^2+0a-1 \big) & \overline{a^4 + 0a^3 + 0a^2 + 0a - 1} \\
 & \underline{a^4 + 0a^3 - a^2} \\
 & \qquad\qquad a^2 + 0a - 1 \\
 & \qquad\qquad \underline{a^2 + 0a - 1} \\
 & \qquad\qquad\qquad\quad 0
\end{array}$$

The remainder is 0. The answer is the quotient,

$$a^2 + 1.$$

75. $\dfrac{x^4-4x^3+5x^2-3x+2}{x^2+3}$

$$\begin{array}{r l}
 & \quad x^2 - 4x + 2 \\
x^2+0x+3 \big) & \overline{x^4 - 4x^3 + 5x^2 - 3x + 2} \\
 & \underline{x^4 + 0x^3 + 3x^2} \\
 & \quad -4x^3 + 2x^2 - 3x \\
 & \quad \underline{-4x^3 + 0x^2 - 12x} \\
 & \qquad\qquad 2x^2 + 9x + 2 \\
 & \qquad\qquad \underline{2x^2 + 0x + 6} \\
 & \qquad\qquad\qquad 9x - 4
\end{array}$$

$$\frac{x^4-4x^3+5x^2-3x+2}{x^2+3} = x^2-4x+2+\frac{9x-4}{x^2+3}$$

77. $\dfrac{2x^5+9x^4+8x^3+10x^2+14x+5}{2x^2+3x+1}$

$$\begin{array}{r l}
 & \quad x^3 + 3x^2 - x + 5 \\
2x^2+3x+1 \big) & \overline{2x^5 + 9x^4 + 8x^3 + 10x^2 + 14x + 5} \\
 & \underline{2x^5 + 3x^4 + x^3} \\
 & \qquad 6x^4 + 7x^3 + 10x^2 \\
 & \qquad \underline{6x^4 + 9x^3 + 3x^2} \\
 & \qquad\qquad -2x^3 + 7x^2 + 14x \\
 & \qquad\qquad \underline{-2x^3 - 3x^2 - x} \\
 & \qquad\qquad\qquad 10x^2 + 15x + 5 \\
 & \qquad\qquad\qquad \underline{10x^2 + 15x + 5} \\
 & \qquad\qquad\qquad\qquad\qquad 0
\end{array}$$

The remainder is 0. The answer is the quotient,

$$x^3 + 3x^2 - x + 5.$$

79. $(3a^2 - 11a + 17) \div (2a + 6)$

$$\begin{array}{r l}
 & \quad \tfrac{3}{2}a - 10 \\
2a+6 \big) & \overline{3a^2 - 11a + 17} \\
 & \underline{3a^2 + 9a} \\
 & \quad -20a + 17 \\
 & \quad \underline{-20a - 60} \\
 & \qquad\qquad 77
\end{array}$$

The remainder is 77.

The quotient is $\frac{3}{2}a - 10$.

The answer is $\dfrac{3}{2}a - 10 + \dfrac{77}{2a+6}$.

81. $\dfrac{3x^3+5x^2-9x+5}{3x-3}$

$$\begin{array}{r l l}
 & \quad x^2 + \tfrac{8}{3}x - \tfrac{1}{3} & \\
3x-3 \big) & \overline{3x^3 + 5x^2 - 9x + 5} & \\
 & \underline{3x^3 - 3x^2} & \\
 & \qquad 8x^2 - 9x & \tfrac{8x^2}{3x} = \tfrac{8}{3}x \\
 & \qquad \underline{8x^2 - 8x} & \\
 & \qquad\qquad -x + 5 & \tfrac{-x}{3x} = -\tfrac{1}{3} \\
 & \qquad\qquad \underline{-x + 1} & \\
 & \qquad\qquad\qquad 4 &
\end{array}$$

$(3x^3+5x^2-9x+5)$ divided by $(3x-3)$ is

$$x^2 + \frac{8}{3}x - \frac{1}{3} + \frac{4}{3x-3}.$$

83. Use $A = LW$ with

$$A = 5x^3 + 7x^2 - 13x - 6$$

and $W = 5x + 2$.

$$\begin{aligned} A &= LW \\ 5x^3+7x^2-13x-6 &= L(5x+2) \\ \frac{5x^3+7x^2-13x-6}{5x+2} &= L \end{aligned}$$

$$\begin{array}{r|rrrr} & x^2 & +\ x & -\ 3 & \\ \hline 5x+2 & 5x^3 & +7x^2 & -13x & -6 \\ & 5x^3 & +2x^2 & & \\ \hline & & 5x^2 & -13x & \\ & & 5x^2 & +\ 2x & \\ \hline & & & -15x & -6 \\ & & & -15x & -6 \\ \hline & & & & 0 \end{array}$$

The length L is (x^2+x-3) units.

85.
$$\begin{aligned} A &= \tfrac{1}{2}bh \\ 24m^3+48m^2+12m &= \tfrac{1}{2}(b)m \\ 48m^3+96m^2+24m &= bm \qquad \textit{Multiply by 2.} \\ \frac{48m^3+96m^2+24m}{m} &= b \qquad \textit{Divide by m.} \\ \frac{48m^3}{m}+\frac{96m^2}{m}+\frac{24m}{m} &= b \\ 48m^2+96m+24 &= b \end{aligned}$$

87. Use the distance formula, $d = rt$, with $d = (5x^3-6x^2+3x+14)$ miles and $r = (x+1)$ miles per hour.

$$\begin{aligned} (5x^3-6x^2+3x+14) &= (x+1)t \\ \frac{(5x^3-6x^2+3x+14)}{(x+1)} &= t \end{aligned}$$

$$\begin{array}{r|rrrr} & 5x^2 & -11x & +14 & \\ \hline x+1 & 5x^3 & -\ 6x^2 & +\ 3x & +14 \\ & 5x^3 & +\ 5x^2 & & \\ \hline & & -11x^2 & +\ 3x & \\ & & -11x^2 & -11x & \\ \hline & & & 14x & +14 \\ & & & 14x & +14 \\ \hline & & & & 0 \end{array}$$

The time t is $(5x^2-11x+14)$ hours.

89. We can write 18 as

$$1\cdot 18,\quad 2\cdot 9,\quad \text{or}\quad 3\cdot 6,$$

so the positive integer factors of 18 are

$$1, 2, 3, 6, 9, \quad \text{and} \quad 18.$$

91. We can write 48 as

$$1\cdot 48,\ 2\cdot 24,\ 3\cdot 16,\ 4\cdot 12, \quad \text{or} \quad 6\cdot 8,$$

so the positive integer factors of 48 are

$$1, 2, 3, 4, 6, 8, 12, 16, 24, \quad \text{and} \quad 48.$$

Chapter 4 Review Exercises

1. $4^3\cdot 4^8 = 4^{3+8} = 4^{11}$

2. $(-5)^6(-5)^5 = (-5)^{6+5} = (-5)^{11}$

3. $\begin{aligned}[t] (-8x^4)(9x^3) &= -8(9)(x^4)(x^3) \\ &= -72x^{4+3} = -72x^7 \end{aligned}$

4. $\begin{aligned}[t] (2x^2)(5x^3)(x^9) &= 2(5)(x^2)(x^3)(x^9) \\ &= 10x^{2+3+9} = 10x^{14} \end{aligned}$

5. $(19x)^5 = 19^5x^5$

6. $(-4y)^7 = (-4)^7y^7$

7. $5(pt)^4 = 5p^4t^4$

8. $\left(\dfrac{7}{5}\right)^6 = \dfrac{7^6}{5^6}$

9. $\begin{aligned}[t] (3x^2y^3)^3 &= 3^3(x^2)^3(y^3)^3 \\ &= 3^3x^{2\cdot 3}y^{3\cdot 3} \\ &= 3^3x^6y^9 \end{aligned}$

10. $\begin{aligned}[t] (t^4)^8(t^2)^5 &= t^{4\cdot 8}t^{2\cdot 5} \\ &= t^{32}t^{10} \\ &= t^{32+10} \\ &= t^{42} \end{aligned}$

11. $\begin{aligned}[t] &(6x^2z^4)^2(x^3yz^2)^4 \\ &\quad = 6^2(x^2)^2(z^4)^2(x^3)^4(y)^4(z^2)^4 \\ &\quad = 6^2x^4z^8x^{12}y^4z^8 \\ &\quad = 6^2x^{4+12}y^4z^{8+8} \\ &\quad = 6^2x^{16}y^4z^{16} \end{aligned}$

12. $\begin{aligned}[t] \left(\frac{2m^3n}{p^2}\right)^3 &= \frac{2^3(m^3)^3n^3}{(p^2)^3} \\ &= \frac{2^3m^9n^3}{p^6} \end{aligned}$

13. The product rule does not apply to 7^2+7^4 because you are adding powers of 7, not multiplying them.

14. $6^0+(-6)^0 = 1+1 = 2$

15. $-(-23)^0 = -1\cdot(-23)^0 = -1\cdot 1 = -1$

16. $-10^0 = -(10^0) = -(1) = -1$

17. $-7^{-2} = -\dfrac{1}{7^2} = -\dfrac{1}{49}$

18. $\left(\dfrac{5}{8}\right)^{-2} = \left(\dfrac{8}{5}\right)^2 = \dfrac{64}{25}$

19. $\begin{aligned}[t] (5^{-2})^{-4} &= 5^{(-2)(-4)} \qquad \textit{Power rule} \\ &= 5^8 \end{aligned}$

20. $9^3\cdot 9^{-5} = 9^{3+(-5)} = 9^{-2} = \dfrac{1}{9^2} = \dfrac{1}{81}$

21. $2^{-1}+4^{-1}=\frac{1}{2^1}+\frac{1}{4^1}$
$=\frac{1}{2}+\frac{1}{4}$
$=\frac{2}{4}+\frac{1}{4}=\frac{3}{4}$

22. $\frac{6^{-5}}{6^{-3}}=\frac{6^3}{6^5}=\frac{1}{6^2}=\frac{1}{36}$

23. $\frac{x^{-7}}{x^{-9}}=\frac{x^9}{x^7}=x^{9-7}=x^2$ *Quotient rule*

24. $\frac{y^4\cdot y^{-2}}{y^{-5}}=\frac{y^4\cdot y^5}{y^2}=\frac{y^9}{y^2}=y^7$

25. $(3r^{-2})^{-4}=(3)^{-4}(r^{-2})^{-4}$
$=\left(3^{-4}\right)\left(r^{-2(-4)}\right)$
$=\frac{1}{3^4}r^8$
$=\frac{r^8}{81}$

26. $(3p)^4(3p^{-7})=(3^4p^4)(3p^{-7})$
$=3^{4+1}\cdot p^{4+(-7)}$
$=3^5p^{-3}$
$=\frac{3^5}{p^3}$

27. $\frac{ab^{-3}}{a^4b^2}=\frac{a}{a^4b^2b^3}=\frac{1}{a^3b^5}$

28. $\frac{(6r^{-1})^2(2r^{-4})}{r^{-5}(r^2)^{-3}}=\frac{(6^2r^{-2})(2r^{-4})}{r^{-5}r^{-6}}$
$=\frac{72r^{-6}}{r^{-11}}$
$=\frac{72r^{11}}{r^6}$
$=72r^5$

29. $48{,}000{,}000=4.8\times10^7$

Move the decimal point left 7 places so it is to the right of the first nonzero digit. 48,000,000 is *greater* than 4.8, so the power is *positive.*

30. $28{,}988{,}000{,}000=2.8988\times10^{10}$

Move the decimal point left 10 places so it is to the right of the first nonzero digit. 28,988,000,000 is *greater* than 2.8988, so the power is *positive.*

31. $0.000\,000\,082\,4=8.24\times10^{-8}$

Move the decimal point right 8 places so it is to the right of the first nonzero digit. 0.000 000 082 4 is *less* than 8.24, so the power is *negative.*

32. $2.4\times10^4=24{,}000$

Move the decimal point 4 places to the right.

33. $7.83\times10^7=78{,}300{,}000$

Move the decimal point 7 places to the right.

34. $8.97\times10^{-7}=0.000\,000\,897$

Move the decimal point 7 places to the left.

35. $(2\times10^{-3})\times(4\times10^5)$
$=(2\times4)(10^{-3}\times10^5)$
$=8\times10^{-3+5}=8\times10^2=800$

36. $\frac{8\times10^4}{2\times10^{-2}}=\frac{8}{2}\times\frac{10^4}{10^{-2}}=4\times10^{4-(-2)}$
$=4\times10^6=4{,}000{,}000$

37. $\frac{12\times10^{-5}\times5\times10^4}{4\times10^3\times6\times10^{-2}}$
$=\frac{12\times5}{4\times6}\times\frac{10^{-5}\times10^4}{10^3\times10^{-2}}$
$=\frac{60}{24}\times\frac{10^{-1}}{10^1}$
$=\frac{5}{2}\times10^{-1-1}$
$=2.5\times10^{-2}$
$=0.025$

38. $2\times10^{-6}=0.000\,002$

39. $1.6\times10^{-12}=0.000\,000\,000\,001\,6$

40. There are 41 zeros, so the number is 4.2×10^{42}.

41. $97{,}000=9.7\times10^4$; $5000=5\times10^3$

42. There are 100 zeros, so the number is 1×10^{100}.

43. $1000=1\times10^3$; $2000=2\times10^3$;
$50{,}000=5\times10^4$; $100{,}000=1\times10^5$

44. $9m^2+11m^2+2m^2=(9+11+2)m^2$
$=22m^2$

The degree is 2.

To determine if the polynomial is a monomial, binomial, or trinomial, count the number of terms in the final expression.

There is one term, so this is a *monomial.*

45. $-4p+p^3-p^2+8p+2$
$=p^3-p^2+(-4+8)p+2$
$=p^3-p^2+4p+2$

The degree is 3.

To determine if the polynomial is a monomial, binomial, or trinomial, count the number of terms in the final expression. Since there are four terms, it is none of these.

46. $12a^5 - 9a^4 + 8a^3 + 2a^2 - a + 3$ cannot be simplified further and is already written in descending powers of the variable.

The degree is 5.

This polynomial has 6 terms, so it is none of the names listed.

47.
$$\begin{aligned} &-7y^5 - 8y^4 - y^5 + y^4 + 9y \\ &= -7y^5 - 1y^5 - 8y^4 + 1y^4 + 9y \\ &= (-7-1)y^5 + (-8+1)y^4 + 9y \\ &= -8y^5 - 7y^4 + 9y \end{aligned}$$

The degree is 5.

There are three terms, so the polynomial is a *trinomial.*

48.
$$\begin{aligned} &(12r^4 - 7r^3 + 2r^2) - (5r^4 - 3r^3 + 2r^2 - 1) \\ &= (12r^4 - 7r^3 + 2r^2) \\ &\quad + (-5r^4 + 3r^3 - 2r^2 + 1) \end{aligned}$$
Change signs in the second polynomial and add.
$$\begin{aligned} &= 12r^4 - 7r^3 + 2r^2 - 5r^4 + 3r^3 - 2r^2 + 1 \\ &= 7r^4 - 4r^3 + 1 \end{aligned}$$

The degree is 4.

The polynomial is a *trinomial.*

49.
$$\begin{aligned} &(5x^3y^2 - 3xy^5 + 12x^2) \\ &\quad - (-9x^2 - 8x^3y^2 + 2xy^5) \\ &= (5x^3y^2 - 3xy^5 + 12x^2) \\ &\quad + (9x^2 + 8x^3y^2 - 2xy^5) \\ &= (5x^3y^2 + 8x^3y^2) + (-3xy^5 - 2xy^5) \\ &\quad + (12x^2 + 9x^2) \\ &= 13x^3y^2 - 5xy^5 + 21x^2 \end{aligned}$$

50. Add.

$$\begin{array}{r} -2a^3 + 5a^2 \\ \underline{3a^3 - a^2} \\ a^3 + 4a^2 \end{array}$$

51. Subtract.

$$\begin{array}{r} 6y^2 - 8y + 2 \\ \underline{5y^2 + 2y - 7} \end{array}$$

Change all signs in the second row and then add.

$$\begin{array}{r} 6y^2 - 8y + 2 \\ \underline{-5y^2 - 2y + 7} \\ y^2 - 10y + 9 \end{array}$$

52. Subtract.

$$\begin{array}{r} -12k^4 - 8k^2 + 7k \\ \underline{k^4 + 7k^2 - 11k} \end{array}$$

Change all signs in the second row and then add.

$$\begin{array}{r} -12k^4 - 8k^2 + 7k \\ \underline{-k^4 - 7k^2 + 11k} \\ -13k^4 - 15k^2 + 18k \end{array}$$

53. $y = -x^2 + 5$

$$\begin{aligned} x = -2 &: y = -(-2)^2 + 5 = -4 + 5 = 1 \\ x = -1 &: y = -(-1)^2 + 5 = -1 + 5 = 4 \\ x = 0 &: y = -(0)^2 + 5 = 0 + 5 = 5 \\ x = 1 &: y = -(1)^2 + 5 = -1 + 5 = 4 \\ x = 2 &: y = -(2)^2 + 5 = -4 + 5 = 1 \end{aligned}$$

x	-2	-1	0	1	2
y	1	4	5	4	1

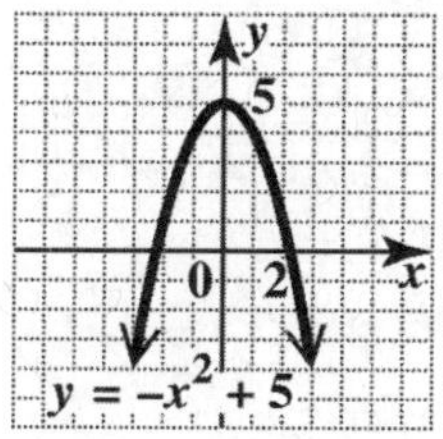

54. $y = 3x^2 - 2$

$$\begin{aligned} x = -2 &: y = 3(-2)^2 - 2 = 3 \cdot 4 - 2 = 10 \\ x = -1 &: y = 3(-1)^2 - 2 = 3 \cdot 1 - 2 = 1 \\ x = 0 &: y = 3(0)^2 - 2 = 3 \cdot 0 - 2 = -2 \\ x = 1 &: y = 3(1)^2 - 2 = 3 \cdot 1 - 2 = 1 \\ x = 2 &: y = 3(2)^2 - 2 = 3 \cdot 4 - 2 = 10 \end{aligned}$$

x	-2	-1	0	1	2
y	10	1	-2	1	10

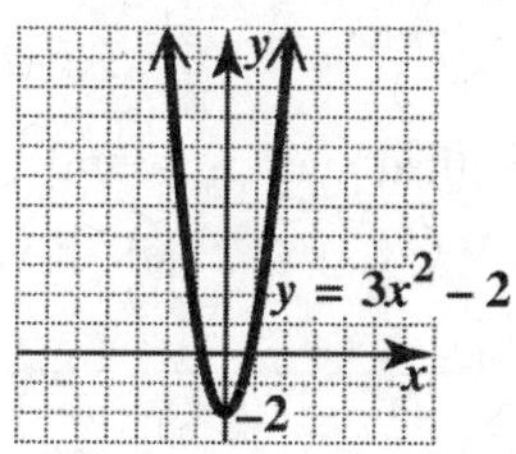

55. $(a + 2)(a^2 - 4a + 1)$

Multiply vertically.

$$\begin{array}{r} a^2 - 4a + 1 \\ \underline{a + 2} \\ 2a^2 - 8a + 2 \\ \underline{a^3 - 4a^2 + a } \\ a^3 - 2a^2 - 7a + 2 \end{array}$$

56. $(3r-2)(2r^2+4r-3)$

Multiply vertically.

$$\begin{array}{rrrr} & 2r^2 & +\ 4r & -\ 3 \\ & & 3r & -\ 2 \\ \hline & -\ 4r^2 & -\ 8r & +\ 6 \\ 6r^3 & +\ 12r^2 & -\ 9r & \\ \hline 6r^3 & +\ 8r^2 & -\ 17r & +\ 6 \end{array}$$

57. $(5p^2+3p)(p^3-p^2+5)$
$= 5p^2(p^3)+5p^2(-p^2)+5p^2(5)$
$\quad +3p(p^3)+3p(-p^2)+3p(5)$
$= 5p^5-5p^4+25p^2+3p^4-3p^3+15p$
$= 5p^5-2p^4-3p^3+25p^2+15p$

58. $(m-9)(m+2)$
F O I L
$= m(m)+m(2)+(-9)(m)+(-9)(2)$
$= m^2+2m-9m-18$
$= m^2-7m-18$

59. $(3k-6)(2k+1)$
F O I L
$= 3k(2k)+3k(1)+(-6)(2k)+(-6)(1)$
$= 6k^2+3k-12k-6$
$= 6k^2-9k-6$

60. $(a+3b)(2a-b)$
F O I L
$= a(2a)+a(-b)+3b(2a)+3b(-b)$
$= 2a^2-ab+6ab-3b^2$
$= 2a^2+5ab-3b^2$

61. $(6k+5q)(2k-7q)$
F O I L
$= 6k(2k)+6k(-7q)+5q(2k)+5q(-7q)$
$= 12k^2-42kq+10kq-35q^2$
$= 12k^2-32kq-35q^2$

62. $(s-1)^3 = (s-1)^2(s-1)$
$\quad = (s^2-2s+1)(s-1)$

Now, use vertical multiplication.

$$\begin{array}{rrrr} & s^2 & -\ 2s & +\ 1 \\ & & s & -\ 1 \\ \hline & -\ s^2 & +\ 2s & -\ 1 \\ s^3 & -\ 2s^2 & +\ s & \\ \hline s^3 & -\ 3s^2 & +\ 3s & -\ 1 \end{array}$$

$(s-1)^3 = s^3-3s^2+3s-1$

63. $(a+4)^2 = a^2+2(a)(4)+4^2$
$\quad = a^2+8a+16$

64. $(2r+5t)^2 = (2r)^2+2(2r)(5t)+(5t)^2$
$\quad = 4r^2+20rt+25t^2$

65. $(6m-5)(6m+5) = (6m)^2-5^2$
$\quad = 36m^2-25$

66. $(5a+6b)(5a-6b) = (5a)^2-(6b)^2$
$\quad = 25a^2-36b^2$

67. $(r+2)^3 = (r+2)^2(r+2)$
$\quad = (r^2+4r+4)(r+2)$
$\quad = r^3+4r^2+4r+2r^2+8r+8$
$\quad = r^3+6r^2+12r+8$

68. $t(5t-3)^2 = t(25t^2-30t+9)$
$\quad = 25t^3-30t^2+9t$

69. Answers will vary. One example is given here.

(a) $(x+y)^2 \neq x^2+y^2$

Let $x=2$ and $y=3$.

$$(x+y)^2 = (2+3)^2 = 5^2 = 25$$
$$x^2+y^2 = 2^2+3^2 = 4+9 = 13$$

Since $25 \neq 13$,

$$(x+y)^2 \neq x^2+y^2.$$

(b) $(x+y)^3 \neq x^3+y^3$

Let $x=2$ and $y=3$.

$$(x+y)^3 = (2+3)^3 = 5^3 = 125$$
$$x^3+y^3 = 2^3+3^3 = 8+27 = 35$$

Since $125 \neq 35$,

$$(x+y)^3 \neq x^3+y^3.$$

70. To find the third power of a binomial, such as $(a+b)^3$, first square the binomial and then multiply that result by the binomial:

$$\begin{aligned}(a+b)^3 &= (a+b)^2(a+b) \\ &= (a^2+2ab+b^2)(a+b) \\ &= (a^3+2a^2b+ab^2) \\ &\quad + (a^2b+2ab^2+b^3) \\ &= a^3+3a^2b+3ab^2+b^3\end{aligned}$$

71. If we chose to let $x=0$ and $y=1$, we would get the true equation $1=1$ for both (a) and (b). These results would not be sufficient to illustrate the truth, in general, of the inequalities. The next step in working the exercise would be to use two other values instead of $x=0$ and $y=1$.

72. Use the formula for the volume of a cube, $V=e^3$, with $e=x^2+2$ centimeters.

$$\begin{aligned}V &= (x^2+2)^3 \\ &= (x^2+2)^2(x^2+2) \\ &= (x^4+4x^2+4)(x^2+2)\end{aligned}$$

Now use vertical multiplication.

$$\begin{array}{r} x^4 + 4x^2 + 4 \\ x^2 + 2 \\ \hline 2x^4 + 8x^2 + 8 \\ x^6 + 4x^4 + 4x^2 \quad\quad \\ \hline x^6 + 6x^4 + 12x^2 + 8 \end{array}$$

The volume of the cube is

$$x^6 + 6x^4 + 12x^2 + 8$$

cubic centimeters.

73. Use the formula for the volume of a sphere, $V = \frac{4}{3}\pi r^3$, with $r = x + 1$ inches.

$$\begin{aligned} V &= \tfrac{4}{3}\pi(x+1)^3 \\ &= \tfrac{4}{3}\pi(x+1)^2(x+1) \\ &= \tfrac{4}{3}\pi(x^2+2x+1)(x+1) \end{aligned}$$

Now use vertical multiplication.

$$\begin{array}{r} x^2 + 2x + 1 \\ x + 1 \\ \hline x^2 + 2x + 1 \\ x^3 + 2x^2 + x \quad\quad \\ \hline x^3 + 3x^2 + 3x + 1 \end{array}$$

$$\begin{aligned} V &= \tfrac{4}{3}\pi(x^3 + 3x^2 + 3x + 1) \\ &= \tfrac{4}{3}\pi x^3 + 4\pi x^2 + 4\pi x + \tfrac{4}{3}\pi \end{aligned}$$

The volume of the sphere is

$$\tfrac{4}{3}\pi x^3 + 4\pi x^2 + 4\pi x + \tfrac{4}{3}\pi$$

cubic inches.

74. $\dfrac{-15y^4}{9y^2} = \dfrac{-15y^{4-2}}{9} = \dfrac{-5y^2}{3}$

75. $\dfrac{6y^4 - 12y^2 + 18y}{6y} = \dfrac{6y^4}{6y} - \dfrac{12y^2}{6y} + \dfrac{18y}{6y}$

$= y^3 - 2y + 3$

76. $(-10m^4n^2 + 5m^3n^2 + 6m^2n^4) \div (5m^2n)$

$$\begin{aligned} &= \frac{-10m^4n^2 + 5m^3n^2 + 6m^2n^4}{5m^2n} \\ &= \frac{-10m^4n^2}{5m^2n} + \frac{5m^3n^2}{5m^2n} + \frac{6m^2n^4}{5m^2n} \\ &= -2m^2n + mn + \frac{6n^3}{5} \end{aligned}$$

77. Let P be the polynomial that when multiplied by $6m^2n$ gives the product

$$12m^3n^2 + 18m^6n^3 - 24m^2n^2.$$

$$\begin{aligned} (P)(6m^2n) &= 12m^3n^2 + 18m^6n^3 - 24m^2n^2 \\ P &= \frac{12m^3n^2 + 18m^6n^3 - 24m^2n^2}{6m^2n} \\ &= \frac{12m^3n^2}{6m^2n} + \frac{18m^6n^3}{6m^2n} - \frac{24m^2n^2}{6m^2n} \\ &= 2mn + 3m^4n^2 - 4n \end{aligned}$$

78. The error made was not dividing both terms in the numerator by 6. The correct method is as follows:

$$\frac{6x^2 - 12x}{6} = \frac{6x^2}{6} - \frac{12x}{6} = x^2 - 2x.$$

79. $\dfrac{2r^2 + 3r - 14}{r - 2}$

$$\begin{array}{r} 2r + 7 \\ r - 2 \overline{\smash{)}\, 2r^2 + 3r - 14} \\ 2r^2 - 4r \quad\quad \\ \hline 7r - 14 \\ 7r - 14 \\ \hline 0 \end{array}$$

The remainder is 0.

The answer is the quotient, $2r + 7$.

80. $\dfrac{10a^3 + 9a^2 - 14a + 9}{5a - 3}$

$$\begin{array}{r} 2a^2 + 3a - 1 \\ 5a - 3 \overline{\smash{)}\, 10a^3 + 9a^2 - 14a + 9} \\ 10a^3 - 6a^2 \quad\quad\quad\quad \\ \hline 15a^2 - 14a \quad\quad \\ 15a^2 - 9a \quad\quad \\ \hline -5a + 9 \\ -5a + 3 \\ \hline 6 \end{array}$$

The answer is

$$2a^2 + 3a - 1 + \frac{6}{5a - 3}.$$

81. $\dfrac{x^4 - 5x^2 + 3x^3 - 3x + 4}{x^2 - 1}$

Write the dividend in descending powers and use 0 as the coefficient of the missing term.

$$\begin{array}{r} x^2 + 3x - 4 \\ x^2 + 0x - 1 \overline{\smash{)}\, x^4 + 3x^3 - 5x^2 - 3x + 4} \\ x^4 + 0x^3 - x^2 \quad\quad\quad\quad \\ \hline 3x^3 - 4x^2 - 3x \quad\quad \\ 3x^3 + 0x^2 - 3x \quad\quad \\ \hline -4x^2 \quad\quad + 4 \\ -4x^2 \quad\quad + 4 \\ \hline 0 \end{array}$$

The remainder is 0. The answer is the quotient,

$$x^2 + 3x - 4.$$

82. $\dfrac{m^4+4m^3-12m-5m^2+6}{m^2-3}$

Write the dividend in descending powers and use 0 as the coefficient of the missing term.

$$\begin{array}{r|l} & m^2 + 4m - 2 \\ \hline m^2+0m-3 & m^4 + 4m^3 - 5m^2 - 12m + 6 \\ & \underline{m^4 + 0m^3 - 3m^2} \\ & \quad 4m^3 - 2m^2 - 12m \\ & \quad \underline{4m^3 + 0m^2 - 12m} \\ & \qquad -2m^2 + 0m + 6 \\ & \qquad \underline{-2m^2 + 0m + 6} \\ & \qquad\qquad 0 \end{array}$$

The remainder is 0. The answer is the quotient,

$m^2+4m-2.$

83. $\dfrac{16x^2-25}{4x+5}$

$$\begin{array}{r|l} & 4x - 5 \\ \hline 4x+5 & 16x^2 + 0x - 25 \\ & \underline{16x^2 + 20x} \\ & \quad -20x - 25 \\ & \quad \underline{-20x - 25} \\ & \qquad 0 \end{array}$$

The remainder is 0.
The answer is the quotient, $4x-5$.

84. $\dfrac{25y^2-100}{5y+10}$

$$\begin{array}{r|l} & 5y - 10 \\ \hline 5y+10 & 25y^2 + 0y - 100 \\ & \underline{25y^2 + 50y} \\ & \quad -50y - 100 \\ & \quad \underline{-50y - 100} \\ & \qquad 0 \end{array}$$

The remainder is 0.
The answer is the quotient, $5y-10$.

85. $\dfrac{y^3-8}{y-2}$

$$\begin{array}{r|l} & y^2 + 2y + 4 \\ \hline y-2 & y^3 + 0y^2 + 0y - 8 \\ & \underline{y^3 - 2y^2} \\ & \quad 2y^2 + 0y \\ & \quad \underline{2y^2 - 4y} \\ & \qquad 4y - 8 \\ & \qquad \underline{4y - 8} \\ & \qquad\quad 0 \end{array}$$

The remainder is 0.
The answer is the quotient, y^2+2y+4.

86. $\dfrac{1000x^6+1}{10x^2+1}$

$$\begin{array}{r|l} & 100x^4 - 10x^2 + 1 \\ \hline 10x^2+1 & 1000x^6 + 0x^4 + 0x^2 + 1 \\ & \underline{1000x^6 + 100x^4} \\ & \quad -100x^4 + 0x^2 \\ & \quad \underline{-100x^4 - 10x^2} \\ & \qquad 10x^2 + 1 \\ & \qquad \underline{10x^2 + 1} \\ & \qquad\quad 0 \end{array}$$

The remainder is 0.
The answer is the quotient, $100x^4-10x^2+1$.

87. $\dfrac{6y^4-15y^3+14y^2-5y-1}{3y^2+1}$

$$\begin{array}{r|l} & 2y^2 - 5y + 4 \\ \hline 3y^2+1 & 6y^4 - 15y^3 + 14y^2 - 5y - 1 \\ & \underline{6y^4 \qquad\quad + 2y^2} \\ & \quad -15y^3 + 12y^2 - 5y \\ & \quad \underline{-15y^3 \qquad\quad - 5y} \\ & \qquad 12y^2 \qquad - 1 \\ & \qquad \underline{12y^2 \qquad + 4} \\ & \qquad\qquad\quad -5 \end{array}$$

The answer is $2y^2-5y+4+\dfrac{-5}{3y^2+1}$.

88. $\dfrac{4x^5-8x^4-3x^3+22x^2-15}{4x^2-3}$

$$\begin{array}{r|l} & x^3 - 2x^2 \qquad + 4 \\ \hline 4x^2-3 & 4x^5 - 8x^4 - 3x^3 + 22x^2 + 0x - 15 \\ & \underline{4x^5 \qquad - 3x^3} \\ & \quad -8x^4 \qquad + 22x^2 \\ & \quad \underline{-8x^4 \qquad + 6x^2} \\ & \qquad 16x^2 \qquad - 15 \\ & \qquad \underline{16x^2 \qquad - 12} \\ & \qquad\qquad\quad -3 \end{array}$$

The answer is $x^3-2x^2+4+\dfrac{-3}{4x^2-3}$.

89. **[4.2]** $5^0+7^0=1+1=2$

90. **[4.1]** $\left(\dfrac{6r^2p}{5}\right)^3=\dfrac{6^3r^{2\cdot 3}p^3}{5^3}$

$$=\frac{6^3r^6p^3}{5^3}$$

91. **[4.6]** $(12a+1)(12a-1)=(12a)^2-1^2$

$$=144a^2-1$$

92. **[4.2]** $2^{-4}=\dfrac{1}{2^4}=\dfrac{1}{16}$

93. [4.2] $(8^{-3})^4 = 8^{(-3)(4)}$
$$= 8^{-12}$$
$$= \frac{1}{8^{12}}$$

94. [4.7] $\dfrac{2p^3 - 6p^2 + 5p}{2p^2}$
$$= \frac{2p^3}{2p^2} - \frac{6p^2}{2p^2} + \frac{5p}{2p^2}$$
$$= p - 3 + \frac{5}{2p}$$

95. [4.2] $\dfrac{(2m^{-5})(3m^2)^{-1}}{m^{-2}(m^{-1})^2}$
$$= \frac{(2m^{-5})(3^{-1}m^{-2})}{m^{-2}(m^{-2})}$$
$$= \frac{2}{3} \cdot \frac{m^{-5+(-2)}}{m^{-2+(-2)}}$$
$$= \frac{2}{3} \cdot \frac{m^{-7}}{m^{-4}}$$
$$= \frac{2}{3} \cdot \frac{m^4}{m^7}$$
$$= \frac{2}{3m^3}$$

96. [4.5] $(3k - 6)(2k^2 + 4k + 1)$

Multiply vertically.

$$\begin{array}{rrrrrrr} & & 2k^2 & + & 4k & + & 1 \\ & & & & 3k & - & 6 \\ \hline & - & 12k^2 & - & 24k & - & 6 \\ 6k^3 & + & 12k^2 & + & 3k & & \\ \hline 6k^3 & & & - & 21k & - & 6 \end{array}$$

97. [4.2] $\dfrac{r^9 \cdot r^{-5}}{r^{-2} \cdot r^{-7}} = \dfrac{r^9 r^2 r^7}{r^5}$
$$= \frac{r^{9+2+7}}{r^5}$$
$$= \frac{r^{18}}{r^5} = r^{13}$$

98. [4.6] $(2r + 5s)^2$
$$= (2r)^2 + 2(2r)(5s) + (5s)^2$$
$$= 4r^2 + 20rs + 25s^2$$

99. [4.4] $(-5y^2 + 3y - 11) + (4y^2 - 7y + 15)$
$$= -5y^2 + 4y^2 + 3y - 7y - 11 + 15$$
$$= -y^2 - 4y + 4$$

100. [4.5] $(2r + 5)(5r - 2)$

F O I L
$$= 2r(5r) + 2r(-2) + 5(5r) + 5(-2)$$
$$= 10r^2 - 4r + 25r - 10$$
$$= 10r^2 + 21r - 10$$

101. [4.7] $\dfrac{2y^3 + 17y^2 + 37y + 7}{2y + 7}$

$$\begin{array}{r|l} & y^2 + 5y + 1 \\ \hline 2y + 7 & 2y^3 + 17y^2 + 37y + 7 \\ & \underline{2y^3 + 7y^2} \\ & \quad 10y^2 + 37y \\ & \quad \underline{10y^2 + 35y} \\ & \qquad 2y + 7 \\ & \qquad \underline{2y + 7} \\ & \qquad\quad 0 \end{array}$$

The remainder is 0.
The answer is the quotient, $y^2 + 5y + 1$.

102. [4.7] $(25x^2y^3 - 8xy^2 + 15x^3y) \div (10x^2y^3)$
$$= \frac{25x^2y^3 - 8xy^2 + 15x^3y}{10x^2y^3}$$
$$= \frac{25x^2y^3}{10x^2y^3} - \frac{8xy^2}{10x^2y^3} + \frac{15x^3y}{10x^2y^3}$$
$$= \frac{5}{2} - \frac{4}{5xy} + \frac{3x}{2y^2}$$

103. [4.4] $(6p^2 - p - 8) - (-4p^2 + 2p - 3)$
$$= (6p^2 - p - 8) + (4p^2 - 2p + 3)$$
$$= 10p^2 - 3p - 5$$

104. [4.7] $\dfrac{3x^3 - 2x + 5}{x - 3}$

Use 0 as the coefficient for the missing term.

$$\begin{array}{r|l} & 3x^2 + 9x + 25 \\ \hline x - 3 & 3x^3 + 0x^2 - 2x + 5 \\ & \underline{3x^3 - 9x^2} \\ & \quad 9x^2 - 2x \\ & \quad \underline{9x^2 - 27x} \\ & \qquad 25x + 5 \\ & \qquad \underline{25x - 75} \\ & \qquad\quad 80 \end{array}$$

$$\frac{3x^3 - 2x + 5}{x - 3} = 3x^2 + 9x + 25 + \frac{80}{x - 3}$$

105. [4.6] $(-7 + 2k)^2$
$$= (-7)^2 + 2(-7)(2k) + (2k)^2$$
$$= 49 - 28k + 4k^2$$

106. [4.2] $\left(\dfrac{x}{y^{-3}}\right)^{-4} = \dfrac{x^{-4}}{(y^{-3})^{-4}}$
$$= \frac{x^{-4}}{y^{12}}$$
$$= \frac{1}{x^4 y^{12}}$$

107. **[4.5]** **(a)** Use the formula for the perimeter of a rectangle, $P = 2L + 2W$, with $L = 2x - 3$ and $W = x + 2$.

$$\begin{aligned} P &= 2(2x-3) + 2(x+2) \\ &= 4x - 6 + 2x + 4 \\ &= 6x - 2 \end{aligned}$$

The perimeter of the rectangle is

$(6x - 2)$ units.

(b) Use the formula for the area of a rectangle, $\mathcal{A} = LW$, with $L = 2x - 3$ and $W = x + 2$.

$$\begin{aligned} \mathcal{A} &= (2x-3)(x+2) \\ &= (2x)(x) + (2x)(2) + (-3)(x) + (-3)(2) \\ &= 2x^2 + 4x - 3x - 6 \\ &= 2x^2 + x - 6 \end{aligned}$$

The area of the rectangle is

$(2x^2 + x - 6)$ square units.

108. **[4.6]** **(a)** Use the formula for the perimeter of a square, $P = 4s$, with $s = 5x^4 + 2x^2$.

$$\begin{aligned} P &= 4(5x^4 + 2x^2) \\ &= 20x^4 + 8x^2 \end{aligned}$$

The perimeter of the square is

$(20x^4 + 8x^2)$ units.

(b) Use the formula for the area of a square, $\mathcal{A} = s^2$, with $s = 5x^4 + 2x^2$.

$$\begin{aligned} \mathcal{A} &= (5x^4 + 2x^2)^2 \\ &= (5x^4)^2 + 2(5x^4)(2x^2) + (2x^2)^2 \\ &= 25x^8 + 20x^6 + 4x^4 \end{aligned}$$

The area of the square is

$(25x^8 + 20x^6 + 4x^4)$ square units.

Chapter 4 Test

1. $5^{-4} = \dfrac{1}{5^4} = \dfrac{1}{625}$

2. $(-3)^0 + 4^0 = 1 + 1 = 2$

3. $4^{-1} + 3^{-1} = \dfrac{1}{4^1} + \dfrac{1}{3^1} = \dfrac{3}{12} + \dfrac{4}{12} = \dfrac{7}{12}$

4.
$$\begin{aligned} \frac{(3x^2y)^2(xy^3)^2}{(xy)^3} &= \frac{3^2(x^2)^2y^2x^2(y^3)^2}{x^3y^3} \\ &= \frac{9x^4y^2x^2y^6}{x^3y^3} \\ &= \frac{9x^6y^8}{x^3y^3} \\ &= 9x^3y^5 \end{aligned}$$

5. $\dfrac{8^{-1} \cdot 8^4}{8^{-2}} = \dfrac{8^{(-1)+4}}{8^{-2}} = \dfrac{8^3}{8^{-2}} = 8^{3-(-2)} = 8^5$

6.
$$\begin{aligned} \frac{(x^{-3})^{-2}(x^{-1}y)^2}{(xy^{-2})^2} &= \frac{(x^{-3})^{-2}(x^{-1})^2(y)^2}{(x)^2(y^{-2})^2} \\ &= \frac{x^6x^{-2}y^2}{x^2y^{-4}} \\ &= \frac{x^4y^2}{x^2y^{-4}} \\ &= x^{4-2}y^{2-(-4)} \\ &= x^2y^6 \end{aligned}$$

7. **(a)** $3^{-4} = \dfrac{1}{3^4} = \dfrac{1}{81}$, which is *positive.*

A negative exponent indicates a reciprocal, not a negative number.

(b) $(-3)^4 = 81$, which is *positive.*

(c) $-3^4 = -1 \cdot 3^4 = -81$, which is *negative.*

(d) $3^0 = 1$, which is *positive.*

(e) $(-3)^0 - 3^0 = 1 - 1 = 0$ *(zero)*

(f) $(-3)^{-3} = \dfrac{1}{(-3)^3} = \dfrac{1}{-27}$, which is *negative.*

8. **(a)** $45{,}000{,}000{,}000 = 4.5 \times 10^{10}$

Move the decimal point left 10 places so it is to the right of the first nonzero digit. 45,000,000,000 is *greater* than 4.5, so the power is *positive.*

(b) $3.6 \times 10^{-6} = 0.000\,003\,6$

Move the decimal point 6 places to the left.

(c)
$$\begin{aligned} \frac{9.5 \times 10^{-1}}{5 \times 10^3} &= \frac{9.5}{5} \times \frac{10^{-1}}{10^3} \\ &= 1.9 \times 10^{-1-3} \\ &= 1.9 \times 10^{-4} \\ &= 0.00019 \end{aligned}$$

9. **(a)** $1000 = 1 \times 10^3$

$5{,}890{,}000{,}000{,}000 = 5.89 \times 10^{12}$

(b)
$$\begin{aligned} &(1 \times 10^3 \text{ light-years})\left(5.89 \times 10^{12}\ \frac{\text{miles}}{\text{light-year}}\right) \\ &= 5.89 \times 10^{3+12} \text{ miles} \\ &= 5.89 \times 10^{15} \text{ miles} \end{aligned}$$

10.
$$\begin{aligned} 5x^2 + 8x - 12x^2 &= 5x^2 - 12x^2 + 8x \\ &= -7x^2 + 8x \end{aligned}$$

degree 2; binomial (2 terms)

11.
$$\begin{aligned} 13n^3 - n^2 + n^4 + 3n^4 - 9n^2 &= n^4 + 3n^4 + 13n^3 - n^2 - 9n^2 \\ &= 4n^4 + 13n^3 - 10n^2 \end{aligned}$$

degree 4; trinomial (3 terms)

12. $y = 2x^2 - 4$

$$\begin{aligned} x = -2 &: y = 2(-2)^2 - 4 = 2 \cdot 4 - 4 = 4 \\ x = -1 &: y = 2(-1)^2 - 4 = 2 \cdot 1 - 4 = -2 \\ x = 0 &: y = 2(0)^2 - 4 = 2 \cdot 0 - 4 = -4 \\ x = 1 &: y = 2(1)^2 - 4 = 2 \cdot 1 - 4 = -2 \\ x = 2 &: y = 2(2)^2 - 4 = 2 \cdot 4 - 4 = 4 \end{aligned}$$

x	-2	-1	0	1	2
y	4	-2	-4	-2	4

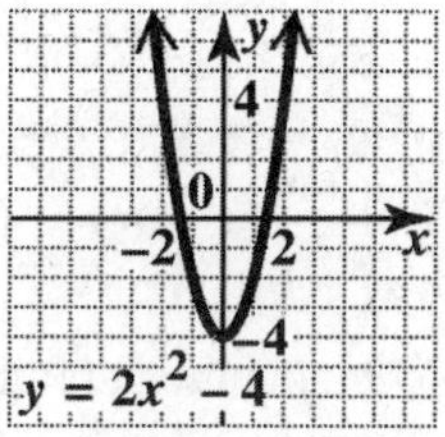

13. $(2y^2 - 8y + 8) + (-3y^2 + 2y + 3) - (y^2 + 3y - 6)$

$$\begin{aligned} &= (2y^2 - 8y + 8) + (-3y^2 + 2y + 3) + (-y^2 - 3y + 6) \\ &= (2y^2 - 3y^2 - y^2) + (-8y + 2y - 3y) + (8 + 3 + 6) \\ &= -2y^2 - 9y + 17 \end{aligned}$$

14. $(-9a^3b^2 + 13ab^5 + 5a^2b^2) - (6ab^5 + 12a^3b^2 + 10a^2b^2)$

$$\begin{aligned} &= (-9a^3b^2 + 13ab^5 + 5a^2b^2) + (-6ab^5 - 12a^3b^2 - 10a^2b^2) \\ &= (-9a^3b^2 - 12a^3b^2) + (13ab^5 - 6ab^5) + (5a^2b^2 - 10a^2b^2) \\ &= -21a^3b^2 + 7ab^5 - 5a^2b^2 \end{aligned}$$

15. Subtract.

$$\begin{array}{rrrr} 9t^3 & -\ 4t^2 & +\ 2t & +\ 2 \\ \underline{9t^3} & \underline{+\ 8t^2} & \underline{-\ 3t} & \underline{-\ 6} \end{array}$$

Change all signs in the second row and then add.

$$\begin{array}{rrrr} 9t^3 & -\ \ 4t^2 & +\ 2t & +\ 2 \\ -9t^3 & -\ \ 8t^2 & +\ 3t & +\ 6 \\ \hline & -\ 12t^2 & +\ 5t & +\ 8 \end{array}$$

16. $3x^2(-9x^3 + 6x^2 - 2x + 1)$

$$\begin{aligned} &= 3x^2(-9x^3) + 3x^2(6x^2) + 3x^2(-2x) + 3x^2(1) \\ &= -27x^5 + 18x^4 - 6x^3 + 3x^2 \end{aligned}$$

17. $(t - 8)(t + 3)$

F O I L

$$\begin{aligned} &= t^2 + 3t - 8t - 24 \\ &= t^2 - 5t - 24 \end{aligned}$$

18. $(4x + 3y)(2x - y)$

F O I L

$$\begin{aligned} &= 8x^2 - 4xy + 6xy - 3y^2 \\ &= 8x^2 + 2xy - 3y^2 \end{aligned}$$

19. $(5x - 2y)^2 = (5x)^2 - 2(5x)(2y) + (2y)^2$

$$= 25x^2 - 20xy + 4y^2$$

20. $(10v + 3w)(10v - 3w) = (10v)^2 - (3w)^2$

$$= 100v^2 - 9w^2$$

21. $(2r - 3)(r^2 + 2r - 5)$

Multiply vertically.

$$\begin{array}{rrrr} & r^2 & +\ \ 2r & -\ \ 5 \\ & & 2r & -\ \ 3 \\ \hline & -\ 3r^2 & -\ \ 6r & +\ 15 \\ 2r^3 & +\ 4r^2 & -\ 10r & \\ \hline 2r^3 & +\ \ r^2 & -\ 16r & +\ 15 \end{array}$$

22. Use the formula for the perimeter of a square, $P = 4s$, with $s = 3x + 9$.

$$\begin{aligned} P &= 4s \\ &= 4(3x + 9) \\ &= 4(3x) + 4(9) \\ &= 12x + 36 \end{aligned}$$

The perimeter of the square is $(12x + 26)$ units.

Use the formula for the area of a square, $\mathcal{A} = s^2$, with $s = 3x + 9$.

$$\begin{aligned} \mathcal{A} &= s^2 \\ &= (3x + 9)^2 \\ &= (3x)^2 + 2(3x)(9) + 9^2 \\ &= 9x^2 + 54x + 81 \end{aligned}$$

The area of the square is $(9x^2 + 54x + 81)$ square units.

23. $\dfrac{8y^3 - 6y^2 + 4y + 10}{2y}$

$$\begin{aligned} &= \frac{8y^3}{2y} - \frac{6y^2}{2y} + \frac{4y}{2y} + \frac{10}{2y} \\ &= 4y^2 - 3y + 2 + \frac{5}{y} \end{aligned}$$

24. $(-9x^2y^3 + 6x^4y^3 + 12xy^3) \div (3xy)$

$$\begin{aligned} &= \frac{-9x^2y^3 + 6x^4y^3 + 12xy^3}{3xy} \\ &= \frac{-9x^2y^3}{3xy} + \frac{6x^4y^3}{3xy} + \frac{12xy^3}{3xy} \\ &= -3xy^2 + 2x^3y^2 + 4y^2 \end{aligned}$$

25. $\dfrac{5x^2 - x - 18}{5x + 9}$

$$\begin{array}{r|l} & x \;-\; 2 \\ \hline 5x+9 & 5x^2 \;-\; x \;-\; 18 \\ & 5x^2 \;+9x \\ \hline & \quad -10x \;-\; 18 \\ & \quad -10x \;-\; 18 \\ \hline & \qquad\qquad 0 \end{array}$$

The remainder is 0. The answer is the quotient, $x - 2$.

26. $(3x^3 - x + 4) \div (x - 2)$

$$\begin{array}{r|l} & 3x^2 \;+\; 6x \;+11 \\ \hline x-2 & 3x^3 \;+0x^2 \;-\; x \;+\; 4 \\ & 3x^3 \;-6x^2 \\ \hline & \quad 6x^2 \;-\; x \\ & \quad 6x^2 \;-12x \\ \hline & \qquad 11x \;+\; 4 \\ & \qquad 11x \;-22 \\ \hline & \qquad\qquad 26 \end{array}$$

The answer is $3x^2 + 6x + 11 + \dfrac{26}{x-2}$.

Cumulative Review Exercises (Chapters 1–4)

1. $\dfrac{28}{16} = \dfrac{7 \cdot 4}{4 \cdot 4} = \dfrac{7}{4}$

2. $\dfrac{55}{11} = \dfrac{5 \cdot 11}{1 \cdot 11} = \dfrac{5}{1} = 5$

3. Each shed requires $1\frac{1}{4}$ cubic yards of concrete, so the total amount of concrete needed for 25 sheds would be

$$\begin{aligned} 25 \times 1\tfrac{1}{4} &= 25 \times \tfrac{5}{4} \\ &= \tfrac{125}{4} \\ &= 31\tfrac{1}{4} \text{ cubic yards.} \end{aligned}$$

4. Use the formula for simple interest, $I = Prt$, with $P = \$34{,}000$, $r = 5.4\%$, and $t = 1$.

$$\begin{aligned} I &= Prt \\ &= (34{,}000)(0.054)(1) \\ &= 1836 \end{aligned}$$

She earned \$1836 in interest.

5. The positive integer factors of 45 are

$$1, 3, 5, 9, 15, \text{ and } 45.$$

6. $\dfrac{4x - 2y}{x + y} = \dfrac{4(-2) - 2(4)}{(-2) + 4}$ *Let $x = -2$, $y = 4$.*

$$= \dfrac{-8 - 8}{2} = \dfrac{-16}{2} = -8$$

7. $\dfrac{(-13 + 15) - (3 + 2)}{6 - 12} = \dfrac{2 - 5}{-6} = \dfrac{-3}{-6} = \dfrac{1}{2}$

8. $$\begin{aligned} -7 - 3[2 + (5 - 8)] &= -7 - 3[2 + (-3)] \\ &= -7 - 3[-1] \\ &= -7 + 3 = -4 \end{aligned}$$

9. $(9 + 2) + 3 = 9 + (2 + 3)$

The numbers are in the same order but grouped differently, so this is an example of the associative property of addition.

10. $6(4 + 2) = 6(4) + 6(2)$

The number 6 outside the parentheses is "distributed" over the 4 and the 2. This is an example of the distributive property.

11. $$\begin{aligned} &-3(2x^2 - 8x + 9) - (4x^2 + 3x + 2) \\ &= -6x^2 + 24x - 27 - 4x^2 - 3x - 2 \\ &= -10x^2 + 21x - 29 \end{aligned}$$

12. $$\begin{aligned} 2 - 3(t - 5) &= 4 + t \\ 2 - 3t + 15 &= 4 + t \\ -3t + 17 &= 4 + t \\ -4t + 17 &= 4 \\ -4t &= -13 \\ t &= \tfrac{-13}{-4} = \tfrac{13}{4} \end{aligned}$$

The solution set is $\{\frac{13}{4}\}$.

13. $$\begin{aligned} 2(5x + 1) &= 10x + 4 \\ 10x + 2 &= 10x + 4 \\ 2 &= 4 \qquad \textit{False} \end{aligned}$$

The false statement indicates that the equation has no solution, symbolized by $\emptyset$.

14. Solve $d = rt$ for r.

$$\frac{d}{t} = \frac{rt}{t} \qquad \textit{Divide by } t.$$

$$\frac{d}{t} = r$$

15. $$\begin{aligned} \frac{x}{5} &= \frac{x - 2}{7} \\ 7x &= 5(x - 2) \qquad \textit{Cross products are equal.} \\ 7x &= 5x - 10 \\ 2x &= -10 \\ x &= -5 \end{aligned}$$

The solution set is $\{-5\}$.

16. $\frac{1}{3}p - \frac{1}{6}p = -2$

To clear fractions, multiply both sides of the equation by the least common denominator, which is 6.

$$\begin{aligned} 6(\tfrac{1}{3}p - \tfrac{1}{6}p) &= (6)(-2) \\ 6(\tfrac{1}{3}p) - 6(\tfrac{1}{6}p) &= -12 \\ 2p - p &= -12 \\ p &= -12 \end{aligned}$$

The solution set is $\{-12\}$.

17. $0.05x + 0.15(50 - x) = 5.50$

To clear decimals, multiply both sides of the equation by 100.

$$\begin{aligned} 100[0.05x + 0.15(50 - x)] &= 100(5.50) \\ 100(0.05x) + 100[0.15(50 - x)] &= 100(5.50) \\ 5x + 15(50 - x) &= 550 \\ 5x + 750 - 15x &= 550 \\ -10x + 750 &= 550 \\ -10x &= -200 \\ x &= 20 \end{aligned}$$

The solution set is $\{20\}$.

18. $4 - (3x + 12) = (2x - 9) - (5x - 1)$

$$\begin{aligned} 4 - 3x - 12 &= 2x - 9 - 5x + 1 \\ -3x - 8 &= -3x - 8 \qquad \textit{True} \end{aligned}$$

The true statement indicates that the solution set is {all real numbers}.

19. Let $x =$ the number of calories burned in thermoregulation.

Then $5\frac{3}{8}x = \frac{43}{8}x =$ the number of calories burned in exertion.

$$\begin{aligned} x + \tfrac{43}{8}x &= 11{,}200 \\ 8(x + \tfrac{43}{8}x) &= 8(11{,}200) \\ 8x + 43x &= 89{,}600 \\ 51x &= 89{,}600 \\ x &= \tfrac{89{,}600}{51} \approx 1757 \end{aligned}$$

A husky burns approximately 1757 calories for thermoregulation and $\frac{43}{8}(\frac{89{,}600}{51}) \approx 9443$ calories for exertion.

20. Let $x =$ one side of the triangle.
Then $2x =$ the other (unknown) side.

The third side is 17 feet. The perimeter of the triangle cannot be more than 50 feet. This is equivalent to stating that the sum of the lengths of the sides must be less than or equal to 50 feet. Write this statement as an inequality and solve.

$$\begin{aligned} x + 2x + 17 &\le 50 \\ 3x + 17 &\le 50 \\ 3x &\le 33 \\ x &\le 11 \end{aligned}$$

One side cannot be more than 11 feet. The other side cannot be more than $2 \cdot 11 = 22$ feet.

21. $-2(x + 4) > 3x + 6$

$$\begin{aligned} -2x - 8 &> 3x + 6 \\ -2x &> 3x + 14 \\ -5x &> 14 \\ \frac{-5x}{-5} &< \frac{14}{-5} \qquad \textit{Divide by } -5\textit{; reverse the symbol.} \\ x &< -\tfrac{14}{5} \end{aligned}$$

The solution set is $(-\infty, -\frac{14}{5})$.

22. $-3 \le 2x + 5 < 9$

$$\begin{aligned} -8 \le \ &2x < 4 \qquad \textit{Subtract 5.} \\ \frac{-8}{2} \le \ &\frac{2x}{2} < \frac{4}{2} \qquad \textit{Divide by 2.} \\ -4 \le \ &x < 2 \end{aligned}$$

The solution set is $[-4, 2)$.

23. We recognize $y = -3x + 6$ as the equation of a line with y-intercept $(0, 6)$ and slope -3. From the point $(0, 6)$, we can move right 1 unit and down 3 units to the point $(1, 3)$ to get another point on the graph.

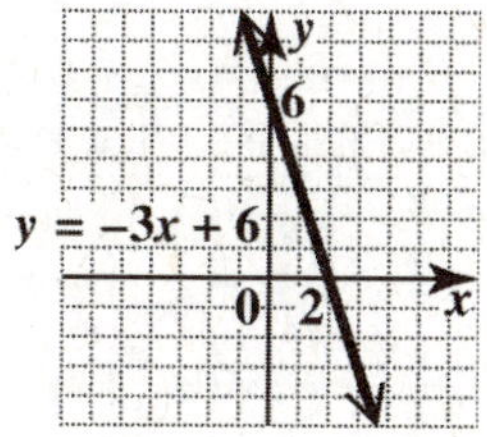

24. **(a)** Use the definition of slope with $(x_1, y_1) = (-1, 5)$ and $(x_2, y_2) = (2, 8)$.

$$m = \frac{\text{change in } y}{\text{change in } x} = \frac{y_2 - y_1}{x_2 - x_1} = \frac{8-5}{2-(-1)} = \frac{3}{3} = 1$$

(b) Use the point-slope form of the equation of a line with $m = 1$ and $(x_1, y_1) = (-1, 5)$.

$$\begin{aligned} y - y_1 &= m(x - x_1) \\ y - 5 &= 1[x - (-1)] \\ y - 5 &= x + 1 \\ y &= x + 6 \end{aligned}$$

25. $4^{-1} + 3^0 = \frac{1}{4^1} + 1 = 1\frac{1}{4}$, or $\frac{5}{4}$

26. $\frac{8^{-5} \cdot 8^7}{8^2} = \frac{8^{-5+7}}{8^2} = \frac{8^2}{8^2} = 1$

27.
$$\begin{aligned} \frac{(a^{-3}b^2)^2}{(2a^{-4}b^{-3})^{-1}} &= \frac{(a^{-3})^2(b^2)^2}{2^{-1}(a^{-4})^{-1}(b^{-3})^{-1}} \\ &= \frac{a^{-6}b^4}{2^{-1}a^4b^3} \\ &= \frac{2b^4}{a^6a^4b^3} = \frac{2b}{a^{10}} \end{aligned}$$

28. $(3.6 \times 10^1 \text{ sec})\left(3.0 \times 10^5 \frac{\text{km}}{\text{sec}}\right)$

$= 3.6 \times 3.0 \times 10^{1+5}$ km

$= 10.8 \times 10^6$ km

Venus is about 10,800,000 km from the sun.

29. $y = (x + 4)^2$

$x = -6 : y = (-6 + 4)^2 = (-2)^2 = 4$
$x = -5 : y = (-5 + 4)^2 = (-1)^2 = 1$
$x = -4 : y = (-4 + 4)^2 = (0)^2 = 0$
$x = -3 : y = (-3 + 4)^2 = (1)^2 = 1$
$x = -2 : y = (-2 + 4)^2 = (2)^2 = 4$

x	-6	-5	-4	-3	-2
y	4	1	0	1	4

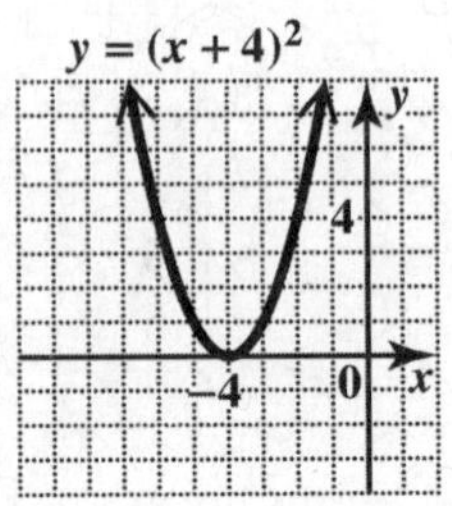

30.
$$\begin{aligned} &(7x^3 - 12x^2 - 3x + 8) + (6x^2 + 4) \\ &\quad - (-4x^3 + 8x^2 - 2x - 2) \\ &= (7x^3 - 12x^2 - 3x + 8) + (6x^2 + 4) \\ &\quad + (4x^3 - 8x^2 + 2x + 2) \\ &= (7 + 4)x^3 + (-12 + 6 - 8)x^2 \\ &\quad + (-3 + 2)x + (8 + 4 + 2) \\ &= 11x^3 - 14x^2 - x + 14 \end{aligned}$$

31.
$$\begin{aligned} &(7x + 4)(9x + 3) \\ &= 63x^2 + 21x + 36x + 12 \quad \textit{FOIL} \\ &= 63x^2 + 57x + 12 \end{aligned}$$

32. $\dfrac{y^3 - 3y^2 + 8y - 6}{y - 1}$

$$\begin{array}{r|l} & y^2 - 2y + 6 \\ \hline y - 1 & y^3 - 3y^2 + 8y - 6 \\ & \underline{y^3 - y^2} \\ & \quad -2y^2 + 8y \\ & \quad \underline{-2y^2 + 2y} \\ & \qquad 6y - 6 \\ & \qquad \underline{6y - 6} \\ & \qquad\qquad 0 \end{array}$$

The remainder is 0. The answer is the quotient,

$$y^2 - 2y + 6.$$

CHAPTER 5 FACTORING AND APPLICATIONS

5.1 The Greatest Common Factor; Factoring by Grouping

5.1 Now Try Exercises

N1. **(a)** 24, 36

Write each number in prime factored form.

$$24 = 2 \cdot 2 \cdot 2 \cdot 3, \quad 36 = 2 \cdot 2 \cdot 3 \cdot 3$$

Use each prime the *least* number of times it appears in *all* the factored forms. The least number of times 2 appears in all the factored forms is 2, and the least number of times 3 appears is 1. The greatest common factor (**GCF**) is $2 \cdot 2 \cdot 3 = 12$.

(b) 54, 90, 108

Write each number in prime factored form.

$$54 = 2 \cdot 3 \cdot 3 \cdot 3, \qquad 90 = 2 \cdot 3 \cdot 3 \cdot 5,$$
$$108 = 2 \cdot 2 \cdot 3 \cdot 3 \cdot 3$$

Use each prime the least number of times it appears. The greatest common factor is $2 \cdot 3 \cdot 3 = 18$.

(c) 15, 19, 25

Write each number in prime factored form.

$$15 = 3 \cdot 5, \quad 19 = 19, \quad 25 = 5 \cdot 5$$

Use each prime the least number of times it appears. Since no prime appears in all the factored forms, the GCF is 1.

N2. **(a)** $25k^3, 15k^2, 35k^5$

Write each number in prime factored form.

$$25k^3 = 5^2 \cdot k^3, \quad 15k^2 = 3 \cdot 5 \cdot k^2,$$
$$35k^5 = 5 \cdot 7 \cdot k^5$$

The greatest common factor of the coefficients 25, 15, and 35 is 5. The greatest common factor of the terms k^3, k^2, and k^5 is k^2 since 2 is the least exponent on k. Thus, the GCF of these terms is the product of 5 and k^2, that is, $5k^2$.

(b) m^3n^5, m^4n^4, m^5n^2

The least exponent on m is 3, and the least exponent on n is 2. Thus, the GCF is m^3n^2.

N3. **(a)** $7t^4 - 14t^3$

7 is the greatest common factor of 7 and 14, and 3 is the least exponent on t. Hence, $7t^3$ is the GCF.

$$\begin{aligned} 7t^4 - 14t^3 &= 7t^3(t) - 7t^3(2) \\ &= 7t^3(t-2) \end{aligned}$$

(b) $8x^6 - 20x^5 + 28x^4$

4 is the greatest common factor of 8, 20, and 28, so 4 can be factored out. Since x occurs in every term and the least exponent on x is 4, x^4 can be factored out. Hence, $4x^4$ is the GCF.

$$\begin{aligned} &8x^6 - 20x^5 + 28x^4 \\ &\quad = 4x^4(2x^2) - 4x^4(5x) + 4x^4(7) \\ &\quad = 4x^4(2x^2 - 5x + 7) \end{aligned}$$

(c) $30m^4n^3 - 42m^2n^2$

6 is the greatest common factor of 30 and 42. The least exponents on m and n are 2 and 2, respectively. Hence, $6m^2n^2$ is the GCF.

$$\begin{aligned} &30m^4n^3 - 42m^2n^2 \\ &\quad = (6m^2n^2)(5m^2n) - (6m^2n^2)(7) \\ &\quad = 6m^2n^2(5m^2n - 7) \end{aligned}$$

N4. **(a)** $x(x+2) + 5(x+2)$

The binomial $x + 2$ is the greatest common factor here.

$$x(x+2) + 5(x+2) = (x+2)(x+5)$$

(b) $a(t+10) - b(t+10)$

The binomial $t + 10$ is the greatest common factor.

$$a(t+10) - b(t+10) = (t+10)(a-b)$$

N5. **(a)** $ab + 3a + 5b + 15$

$= (ab + 3a) + (5b + 15)$ *Group terms.*

$= a(b+3) + 5(b+3)$ *Factor each group.*

$= (b+3)(a+5)$ *Factor out $b + 3$.*

(b) $12xy + 3x + 4y + 1$

$= (12xy + 3x) + (4y + 1)$ *Group terms.*

$= 3x(4y+1) + 1(4y+1)$ *Factor each group; remember the 1.*

$= (4y+1)(3x+1)$ *Factor out $4y + 1$.*

(c) $x^3 + 5x^2 - 8x - 40$

$= (x^3 + 5x^2) + (-8x - 40)$ *Group terms.*

$= x^2(x+5) - 8(x+5)$ *Factor each group.*

$= (x+5)(x^2 - 8)$ *Factor out $x + 5$.*

N6. **(a)** $12p^2 - 28q - 16pq + 21p$

Factoring out the common factor 4 from the first two terms and the common factor p from the last two terms gives

$$\begin{aligned} &12p^2 - 28q - 16pq + 21p \\ &= 4(3p^2 - 7q) + p(-16q + 21). \end{aligned}$$

This does not lead to a common factor, so we try rearranging the terms.

$= (12p^2 - 16pq) + (21p - 28q)$ *Rearrange.*

$= 4p(3p - 4q) + 7(3p - 4q)$ *Factor each group.*

$= (3p - 4q)(4p + 7)$ *Factor out* $3p - 4q$.

Here's another rearrangement.

$$\begin{aligned} &= (12p^2 + 21p) + (-16pq - 28q) \\ &= 3p(4p + 7) - 4q(4p + 7) \\ &= (4p + 7)(3p - 4q) \end{aligned}$$

This is an equivalent answer.

(b) $5xy - 6 - 15x + 2y$

$$\begin{aligned} &= (5xy + 2y) + (-15x - 6) \\ &= y(5x + 2) - 3(5x + 2) \\ &= (5x + 2)(y - 3) \end{aligned}$$

5.1 Section Exercises

1. Find the prime factored form of each number.

$$\begin{aligned} 40 &= 2 \cdot 2 \cdot 2 \cdot 5 \\ 20 &= 2 \cdot 2 \cdot 5 \\ 4 &= 2 \cdot 2 \end{aligned}$$

The least number of times 2 appears in all the factored forms is 2. There is no 5 in the prime factored form of 4, so the

$$\text{GCF} = 2^2 = 4.$$

3. Find the prime factored form of each number.

$$\begin{aligned} 18 &= 2 \cdot 3 \cdot 3 \\ 24 &= 2 \cdot 2 \cdot 2 \cdot 3 \\ 36 &= 2 \cdot 2 \cdot 3 \cdot 3 \\ 48 &= 2 \cdot 2 \cdot 2 \cdot 2 \cdot 3 \end{aligned}$$

The least number of times the primes 2 and 3 appear in all four factored forms is once, so

$$\text{GCF} = 2 \cdot 3 = 6.$$

5. 6, 8, 9

Find the prime factored form of each number.

$$\begin{aligned} 6 &= 2 \cdot 3 \\ 8 &= 2 \cdot 2 \cdot 2 \\ 9 &= 3 \cdot 3 \end{aligned}$$

There are no primes common to all three numbers, so the GCF is 1.

7. Write each term in prime factored form.

$$\begin{aligned} 16y &= 2^4 \cdot y \\ 24 &= 2^3 \cdot 3 \end{aligned}$$

There is no y in the second term, so y will not appear in the GCF. Thus, the GCF of $16y$ and 24 is

$$2^3 = 8.$$

9.
$$\begin{aligned} 30x^3 &= 2 \cdot 3 \cdot 5 \cdot x^3 \\ 40x^6 &= 2^3 \cdot 5 \cdot x^6 \\ 50x^7 &= 2 \cdot 5^2 \cdot x^7 \end{aligned}$$

The GCF of the coefficients, 30, 40, and 50, is $2^1 \cdot 5^1 = 10$. The smallest exponent on the variable x is 3. Thus, the GCF of the given terms is $10x^3$.

11. $x^4y^3 = x^4 \cdot y^3$; $\quad xy^2 = x \cdot y^2$

The GCF is xy^2.

13.
$$\begin{aligned} 12m^3n^2 &= 2^2 \cdot 3 \cdot m^3 \cdot n^2 \\ 18m^5n^4 &= 2 \cdot 3^2 \cdot m^5 \cdot n^4 \\ 36m^8n^3 &= 2^2 \cdot 3^2 \cdot m^8 \cdot n^3 \end{aligned}$$

The GCF is $2 \cdot 3 \cdot m^3 \cdot n^2 = 6m^3n^2$.

15. $2k^2(5k)$ is written as a product of $2k^2$ and $5k$ and hence, it is *factored.*

17. $2k^2 + (5k + 1)$ is written as a sum of $2k^2$ and $(5k + 1)$, and hence, it is *not factored.*

19. $9m^4 = 3m^2(3m^2)$

Factor out $3m^2$ from $9m^4$ to obtain $3m^2$.

21. $-8z^9 = -4z^5(2z^4)$

Factor out $-4z^5$ from $-8z^9$ to obtain $2z^4$.

23. $6m^4n^5 = 3m^3n(2mn^4)$

Factor out $3m^3n$ from $6m^4n^5$ to obtain $2mn^4$.

25.
$$\begin{aligned} 12y + 24 &= 12 \cdot y + 12 \cdot 2 \\ &= 12(y + 2) \end{aligned}$$

27.
$$\begin{aligned} 10a^2 - 20a &= 10a(a) - 10a(2) \\ &= 10a(a - 2) \end{aligned}$$

29.
$$\begin{aligned} 8x^2y + 12x^3y^2 &= 4x^2y(2) + 4x^2y(3xy) \\ &= 4x^2y(2 + 3xy) \end{aligned}$$

31. First, verify that you have factored completely. Then multiply the factors. The product should be the original polynomial.

33. The greatest common factor for $x^2 - 4x$ is x.

$$\begin{aligned} x^2 - 4x &= x(x) + x(-4) \\ &= x(x - 4) \end{aligned}$$

35. The greatest common factor for $6t^2 + 15t$ is $3t$.

$$\begin{aligned} 6t^2 + 15t &= 3t(2t) + 3t(5) \\ &= 3t(2t + 5) \end{aligned}$$

37. $27m^3 - 9m$
The GCF is $9m$.

$$\begin{aligned} 27m^3 - 9m &= 9m(3m^2) + 9m(-1) \\ &= 9m(3m^2 - 1) \end{aligned}$$

39. $16z^4 + 24z^2$
The GCF is $8z^2$.

$$\begin{aligned} 16z^4 + 24z^2 &= 8z^2(2z^2) + 8z^2(3) \\ &= 8z^2(2z^2 + 3) \end{aligned}$$

41. $12x^3 + 6x^2$
The GCF is $6x^2$.

$$\begin{aligned} 12x^3 + 6x^2 &= 6x^2(2x) + 6x^2(1) \\ &= 6x^2(2x + 1) \end{aligned}$$

43. $65y^{10} + 35y^6$
The GCF is $5y^6$.

$$\begin{aligned} 65y^{10} + 35y^6 &= (5y^6)(13y^4) + (5y^6)(7) \\ &= 5y^6(13y^4 + 7) \end{aligned}$$

45. $11w^3 - 100$

The two terms of this expression have no common factor (except 1), so $11w^3 - 100$ is in factored form.

47. $8mn^3 + 24m^2n^3$
The GCF is $8mn^3$.

$$\begin{aligned} &8mn^3 + 24m^2n^3 \\ &= (8mn^3)(1) + (8mn^3)(3m) \\ &= 8mn^3(1 + 3m) \end{aligned}$$

49. $13y^8 + 26y^4 - 39y^2$
The GCF is $13y^2$.

$$\begin{aligned} &13y^8 + 26y^4 - 39y^2 \\ &= 13y^2(y^6) + 13y^2(2y^2) + 13y^2(-3) \\ &= 13y^2(y^6 + 2y^2 - 3) \end{aligned}$$

51. $36p^6q + 45p^5q^4 + 81p^3q^2$
The GCF is $9p^3q$.

$$\begin{aligned} &36p^6q + 45p^5q^4 + 81p^3q^2 \\ &= 9p^3q(4p^3) + 9p^3q(5p^2q^3) + 9p^3q(9q) \\ &= 9p^3q(4p^3 + 5p^2q^3 + 9q) \end{aligned}$$

53. $a^5 + 2a^3b^2 - 3a^5b^2 + 4a^4b^3$
The GCF is a^3.

$$\begin{aligned} &a^5 + 2a^3b^2 - 3a^5b^2 + 4a^4b^3 \\ &= a^3(a^2) + a^3(2b^2) + a^3(-3a^2b^2) \\ &\quad + a^3(4ab^3) \\ &= a^3(a^2 + 2b^2 - 3a^2b^2 + 4ab^3) \end{aligned}$$

55. The GCF of the terms of $c(x + 2) - d(x + 2)$ is the binomial $x + 2$.

$$\begin{aligned} &c(x + 2) - d(x + 2) \\ &= (x + 2)(c) + (x + 2)(-d) \\ &= (x + 2)(c - d) \end{aligned}$$

57. The GCF of the terms of $m(m + 2n) + n(m + 2n)$ is the binomial $m + 2n$.

$$\begin{aligned} &m(m + 2n) + n(m + 2n) \\ &= (m + 2n)(m) + (m + 2n)(n) \\ &= (m + 2n)(m + n) \end{aligned}$$

59. The GCF for $q^2(p - 4) + 1(p - 4)$ is $p - 4$.

$$q^2(p - 4) + 1(p - 4) = (p - 4)(q^2 + 1)$$

61. $8(7t + 4) + x(7t + 4)$

This expression is the *sum* of two terms, $8(7t + 4)$ and $x(7t + 4)$, so it is not in factored form. We can factor out $7t + 4$.

$$\begin{aligned} &8(7t + 4) + x(7t + 4) \\ &= (7t + 4)(8) + (7t + 4)(x) \\ &= (7t + 4)(8 + x) \end{aligned}$$

63. $(8 + x)(7t + 4)$

This expression is the *product* of two factors, $8 + x$ and $7t + 4$, so it is in factored form.

65. $18x^2(y + 4) + 7(y - 4)$

This expression is the *sum* of two terms, $18x^2(y + 4)$ and $7(y - 4)$, so it is not in factored form.

67. It is not possible to factor the expression in Exercise 65 because the two terms, $18x^2(y + 4)$ and $7(y - 4)$, do not have a common factor.

69. $p^2 + 4p + pq + 4q$

The first two terms have a common factor of p, and the last two terms have a common factor of q. Thus,

$$\begin{aligned} &p^2 + 4p + pq + 4q \\ &= (p^2 + 4p) + (pq + 4q) \\ &= p(p + 4) + q(p + 4). \end{aligned}$$

Now we have two terms which have a common binomial factor of $p + 4$. Thus,

$$\begin{aligned} &p^2 + 4p + pq + 4q \\ &= p(p + 4) + q(p + 4) \\ &= (p + 4)(p + q). \end{aligned}$$

71. $a^2 - 2a + ab - 2b$

$= (a^2 - 2a) + (ab - 2b)$ *Group the terms.*

$= a(a - 2) + b(a - 2)$ *Factor each group.*

$= (a - 2)(a + b)$ *Factor out $a - 2$.*

73. $7z^2 + 14z - az - 2a$

$= (7z^2 + 14z) + (-az - 2a)$ *Group the terms.*

$= 7z(z + 2) - a(z + 2)$ *Factor each group.*

$= (z + 2)(7z - a)$ *Factor out $z + 2$.*

75. $18r^2 + 12ry - 3xr - 2xy$

$= (18r^2 + 12ry) + (-3xr - 2xy)$ *Group the terms.*

$= 6r(3r + 2y) - x(3r + 2y)$ *Factor each group.*

$= (3r + 2y)(6r - x)$ *Factor out $3r + 2y$.*

77. $3a^3 + 3ab^2 + 2a^2b + 2b^3$

$= (3a^3 + 3ab^2) + (2a^2b + 2b^3)$ *Group the terms.*

$= 3a(a^2 + b^2) + 2b(a^2 + b^2)$ *Factor each group.*

$= (a^2 + b^2)(3a + 2b)$ *Factor out $a^2 + b^2$.*

79. $12 - 4a - 3b + ab$

$= (12 - 4a) + (-3b + ab)$ *Group the terms.*

$= 4(3 - a) - b(3 - a)$ *Factor each group.*

$= (3 - a)(4 - b)$ *Factor out $3 - a$.*

81. $16m^3 - 4m^2p^2 - 4mp + p^3$

$= (16m^3 - 4m^2p^2) + (-4mp + p^3)$

$= 4m^2(4m - p^2) - p(4m - p^2)$

$= (4m - p^2)(4m^2 - p)$

83. $y^2 + 3x + 3y + xy$

$= y^2 + 3y + xy + 3x$ *Rearrange.*

$= (y^2 + 3y) + (xy + 3x)$

$= y(y + 3) + x(y + 3)$

$= (y + 3)(y + x)$

85. $5m - 6p - 2mp + 15$

We need to rearrange these terms to get two groups that each have a common factor. We could group $5m$ with either $-2mp$ or 15.

$5m + 15 - 2mp - 6p$ *Rearrange.*

$= (5m + 15) + (-2mp - 6p)$ *Group the terms.*

$= 5(m + 3) - 2p(m + 3)$ *Factor each group.*

$= (m + 3)(5 - 2p)$ *Factor out $m + 3$.*

87. $18r^2 - 2ty + 12ry - 3rt$

We'll rearrange the terms so that $18r^2$ is grouped with another term containing r.

$18r^2 + 12ry - 3rt - 2ty$ *Rearrange.*

$= (18r^2 + 12ry) + (-3rt - 2ty)$ *Group the terms.*

$= 6r(3r + 2y) - t(3r + 2y)$ *Factor each group.*

$= (3r + 2y)(6r - t)$ *Factor out $3r + 2y$.*

89. $a^5 - 3 + 2a^5b - 6b$

$= a^5 + 2a^5b - 3 - 6b$ *Rearrange.*

$= (a^5 + 2a^5b) + (-3 - 6b)$ *Group the terms.*

$= a^5(1 + 2b) - 3(1 + 2b)$ *Factor each group.*

$= (1 + 2b)(a^5 - 3)$ *Factor out $1 + 2b$.*

♦♦♦ Relating Concepts 91–94 ♦♦♦

91. In order to rewrite

$$2xy + 12 - 3y - 8x$$

as $2xy - 8x - 3y + 12$,

we must change the order of the terms. The property that allows us to do this is the commutative property of addition.

92. After we group both pairs of terms in the rearranged polynomial, we have

$$(2xy - 8x) + (-3y + 12).$$

The greatest common factor for the first pair of terms is $2x$. The GCF for the second pair is -3. Factoring each group gives us

$$2x(y - 4) - 3(y - 4).$$

93. The expression obtained in Exercise 92 is the *difference* between two terms, $2x(y - 4)$ and $3(y - 4)$, so it is *not* in factored form.

94. $2x(y - 4) - 3(y - 4)$

$= (y - 4)(2x - 3)$

or $= (2x - 3)(y - 4)$

Yes, this is the same result as the one shown in Example 6(b), even though the terms were grouped in a different way.

95. $(x + 6)(x - 9) = x^2 - 9x + 6x - 54$

$= x^2 - 3x - 54$

97. $(x + 2)(x + 7) = x^2 + 7x + 2x + 14$

$= x^2 + 9x + 14$

99. $2x^2(x^2 + 3x + 5) = 2x^4 + 6x^3 + 10x^2$

5.2 Factoring Trinomials

5.2 Now Try Exercises

N1. $p^2 + 7p + 10$

Factors of 10	Sums of Factors
10, 1	$10 + 1 = 11$
5, 2	$5 + 2 = 7$ ←

The pair of integers whose product is 10 and whose sum is 7 is 5 and 2. Thus,

$$p^2 + 7p + 10 = (p + 2)(p + 5).$$

N2. $t^2 - 9t + 18$

Factors of 18	Sums of Factors
−18, −1	$-18 + (-1) = -19$
−9, −2	$-9 + (-2) = -11$
−6, −3	$-6 + (-3) = -9$ ←

The pair of integers whose product is 18 and whose sum is −9 is −6 and −3. Thus,

$$t^2 - 9t + 18 = (t - 3)(t - 6).$$

N3. $x^2 + x - 42$, or $x^2 + 1x - 42$

Find the two integers whose product is −42 and whose sum is 1. Because the last term is negative, the pair must include one positive and one negative integer.

Factors of −42	Sums of Factors
42, −1	$42 + (-1) = 41$
21, −2	$21 + (-2) = 19$
14, −3	$14 + (-3) = 11$
7, −6	$7 + (-6) = 1$ ←
−42, 1	$-42 + 1 = -41$
−21, 2	$-21 + 2 = -19$
−14, 3	$-14 + 3 = -11$
−7, 6	$-7 + 6 = -1$

The required integers are −6 and 7, so

$$x^2 + x - 42 = (x - 6)(x + 7).$$

N4. $x^2 - 4x - 21$

Find the two integers whose product is −21 and whose sum is −4. Because the last term is negative, the pair must include one positive and one negative integer.

Factors of −21	Sums of Factors
21, −1	$21 + (-1) = 20$
7, −3	$7 + (-3) = 4$
−21, 1	$-21 + 1 = -20$
−7, 3	$-7 + 3 = -4$ ←

The required integers are −7 and 3, so

$$x^2 - 4x - 21 = (x - 7)(x + 3).$$

N5. **(a)** $m^2 + 5m + 8$

There is no pair of integers whose product is 8 and whose sum is 5, so $m^2 + 5m + 8$ is a prime polynomial.

(b) $t^2 + 11t - 24$

There is no pair of integers whose product is −24 and whose sum is 11, so $t^2 + 11t - 24$ is a prime polynomial.

N6. $a^2 + 2ab - 15b^2$

Two expressions whose product is $-15b^2$ and whose sum is $2b$ are $5b$ and $-3b$, so

$$a^2 + 2ab - 15b^2 = (a + 5b)(a - 3b).$$

N7. $3y^4 - 27y^3 + 60y^2 = 3y^2(y^2 - 9y + 20)$

Factor $y^2 - 9y + 20$. The integers −5 and −4 have a product of 20 and a sum of −9, so

$$y^2 - 9y + 20 = (y - 5)(y - 4).$$

The completely factored form is

$$3y^4 - 27y^3 + 60y^2 = 3y^2(y - 5)(y - 4).$$

5.2 Section Exercises

1. Product: 48 Sum: −19

Factors of 48	Sums of Factors
1, 48	$1 + 48 = 49$
−1, −48	$-1 + (-48) = -49$
2, 24	$2 + 24 = 26$
−2, −24	$-2 + (-24) = -26$
3, 16	$3 + 16 = 19$
−3, −16	$-3 + (-16) = -19$ ←
4, 12	$4 + 12 = 16$
−4, −12	$-4 + (-12) = -16$
6, 8	$6 + 8 = 14$
−6, −8	$-6 + (-8) = -14$

The pair of integers whose product is 48 and whose sum is −19 is −3 and −16.

3. Product: −24 Sum: −5

Factors of −24	Sums of Factors
1, −24	$1 + (-24) = -23$
−1, 24	$-1 + 24 = 23$
2, −12	$2 + (-12) = -10$
−2, 12	$-2 + 12 = 10$
3, −8	$3 + (-8) = -5$ ←
−3, 8	$-3 + 8 = 5$
4, −6	$4 + (-6) = -2$
−4, 6	$-4 + 6 = 2$

The pair of integers whose product is −24 and whose sum is −5 is 3 and −8.

5. If the coefficient of the last term of the trinomial is negative, then a and b must have different signs, one positive and one negative.

7. A *prime polynomial* is one that cannot be factored using only integers in the factors.

9. $x^2 - 12x + 32$

Multiply each of the given pairs of factors to determine which one gives the required product.

A. $(x-8)(x+4) = x^2 - 4x - 32$

B. $(x+8)(x-4) = x^2 + 4x - 32$

C. $(x-8)(x-4) = x^2 - 12x + 32$

D. $(x+8)(x+4) = x^2 + 12x + 32$

Choice **C** is the correct factored form.

11. Multiply the factors using FOIL to determine the polynomial.

$$\begin{aligned}&(a+9)(a+4)\\ &\quad\;\;\textbf{F}\qquad\;\textbf{O}\qquad\;\textbf{I}\qquad\;\textbf{L}\\ &= a(a) + a(4) + 9(a) + 9(4)\\ &= a^2 + 4a + 9a + 36\\ &= a^2 + 13a + 36\end{aligned}$$

13. $p^2 + 11p + 30 = (p+5)(\_\_\_\_)$

Look for an integer whose product with 5 is 30 and whose sum with 5 is 11. That integer is 6.

$$p^2 + 11p + 30 = (p+5)(p+6)$$

15. $x^2 + 15x + 44 = (x+4)(\_\_\_\_)$

Look for an integer whose product with 4 is 44 and whose sum with 4 is 15. That integer is 11.

$$x^2 + 15x + 44 = (x+4)(x+11)$$

17. $x^2 - 9x + 8 = (x-1)(\_\_\_\_)$

Look for an integer whose product with -1 is 8 and whose sum with -1 is -9. That integer is -8.

$$x^2 - 9x + 8 = (x-1)(x-8)$$

19. $y^2 - 2y - 15 = (y+3)(\_\_\_\_)$

Look for an integer whose product with 3 is -15 and whose sum with 3 is -2. That integer is -5.

$$y^2 - 2y - 15 = (y+3)(y-5)$$

21. $x^2 + 9x - 22 = (x-2)(\_\_\_\_)$

Look for an integer whose product with -2 is -22 and whose sum with -2 is 9. That integer is 11.

$$x^2 + 9x - 22 = (x-2)(x+11)$$

23. $y^2 - 7y - 18 = (y+2)(\_\_\_\_)$

Look for an integer whose product with 2 is -18 and whose sum with 2 is -7. That integer is -9.

$$y^2 - 7y - 18 = (y+2)(y-9)$$

25. $y^2 + 9y + 8$

Look for two integers whose product is 8 and whose sum is 9. Both integers must be positive because b and c are both positive.

Factors of 8	Sums of Factors
1, 8	9 ←
2, 4	6

Thus,

$$y^2 + 9y + 8 = (y+8)(y+1).$$

27. $b^2 + 8b + 15$

Look for two integers whose product is 15 and whose sum is 8. Both integers must be positive because b and c are both positive.

Factors of 15	Sums of Factors
1, 15	16
3, 5	8 ←

Thus,

$$b^2 + 8b + 15 = (b+3)(b+5).$$

29. $m^2 + m - 20$

Look for two integers whose product is -20 and whose sum is 1. Since c is negative, one integer must be positive and one must be negative.

Factors of -20	Sums of Factors
$-1, 20$	19
$1, -20$	-19
$-2, 10$	8
$2, -10$	-8
$-4, 5$	1 ←
$4, -5$	-1

Thus,

$$m^2 + m - 20 = (m-4)(m+5).$$

31. $y^2 - 8y + 15$

Find two integers whose product is 15 and whose sum is -8. Since c is positive and b is negative, both integers must be negative.

Factors of 15	Sums of Factors
$-1, -15$	-16
$-3, -5$	-8 ←

Thus,

$$y^2 - 8y + 15 = (y-5)(y-3).$$

33. $x^2 + 4x + 5$

Look for two integers whose product is 5 and whose sum is 4. Both integers must be positive since b and c are both positive.

Product	*Sum*
$5 \cdot 1 = 5$	$5 + 1 = 6$

There is no other pair of positive integers whose product is 5. Since there is no pair of integers whose product is 5 and whose sum is 4, $x^2 + 4x + 5$ is a *prime* polynomial.

35. $z^2 - 15z + 56$

Find two integers whose product is 56 and whose sum is -15. Since c is positive and b is negative, both integers must be negative.

Factors of 56	Sums of Factors
$-1, -56$	-57
$-2, -28$	-30
$-4, -14$	-18
$-7, -8$	-15 ←

Thus,

$$z^2 - 15z + 56 = (z - 7)(z - 8).$$

37. $r^2 - r - 30$

Look for two integers whose product is -30 and whose sum is -1. Because c is negative, one integer must be positive and the other must be negative.

Factors of -30	Sums of Factors
$-1, 30$	29
$1, -30$	-29
$-2, 15$	13
$2, -15$	-13
$-3, 10$	7
$3, -10$	-7
$-5, 6$	1
$5, -6$	-1 ←

Thus,

$$r^2 - r - 30 = (r + 5)(r - 6).$$

39. $a^2 - 8a - 48$

Find two integers whose product is -48 and whose sum is -8. Since c is negative, one integer must be positive and one must be negative.

Factors of -48	Sums of Factors
$-1, 48$	47
$1, -48$	-47
$-2, 24$	22
$2, -24$	-22
$-3, 16$	13
$3, -16$	-13
$-4, 12$	8
$4, -12$	-8 ←
$-6, 8$	2
$6, -8$	-2

Thus,

$$a^2 - 8a - 48 = (a + 4)(a - 12).$$

41. $x^2 + 3x - 39$

Look for two integers whose product is -39 and whose sum is 3. Because c is negative, one integer must be positive and one must be negative.

Factors of -39	Sums of Factors
$-1, 39$	38
$1, -39$	-38
$-3, 13$	10
$3, -13$	-10

This list does not produce the required integers, and there are no other possibilities to try. Therefore, $x^2 + 3x - 39$ is *prime*.

43. $-32 + 14x + x^2$, or $x^2 + 14x - 32$

Look for two integers whose product is -32 and whose sum is 14. Since c is negative, one integer must be positive and one must be negative.

Factors of -32	Sums of Factors
$-1, 32$	31
$1, -32$	-31
$-2, 16$	14 ←
$2, -16$	-14
$-4, 8$	4
$4, -8$	-4

Thus,

$$-32 + 14x + x^2 = (x - 2)(x + 16).$$

45. $r^2 + 3ra + 2a^2$

Look for two expressions whose product is $2a^2$ and whose sum is $3a$. They are $2a$ and a, so

$$r^2 + 3ra + 2a^2 = (r + 2a)(r + a).$$

47. $t^2 - tz - 6z^2$

Look for two expressions whose product is $-6z^2$ and whose sum is $-z$. They are $2z$ and $-3z$, so

$$t^2 - tz - 6z^2 = (t + 2z)(t - 3z).$$

49. $x^2 + 4xy + 3y^2$

Look for two expressions whose product is $3y^2$ and whose sum is $4y$. The expressions are $3y$ and y, so

$$x^2 + 4xy + 3y^2 = (x + 3y)(x + y).$$

51. $v^2 - 11vw + 30w^2$

Factors of $30w^2$	Sums of Factors
$-30w, -w$	$-31w$
$-15w, -2w$	$-17w$
$-10w, -3w$	$-13w$
$-5w, -6w$	$-11w$ ←

The completely factored form is

$$v^2 - 11vw + 30w^2 = (v - 5w)(v - 6w).$$

53. $4x^2 + 12x - 40$

First, factor out the GCF, 4.

$$4x^2 + 12x - 40 = 4(x^2 + 3x - 10)$$

Now factor $x^2 + 3x - 10$.

Factors of -10	Sums of Factors
$-1, 10$	9
$1, -10$	-9
$2, -5$	-3
$-2, 5$	3 ←

Thus,

$$x^2 + 3x - 10 = (x - 2)(x + 5).$$

The completely factored form is

$$4x^2 + 12x - 40 = 4(x - 2)(x + 5).$$

55. $2t^3 + 8t^2 + 6t$

First, factor out the GCF, $2t$.

$$2t^3 + 8t^2 + 6t = 2t(t^2 + 4t + 3)$$

Then factor $t^2 + 4t + 3$.

$$t^2 + 4t + 3 = (t + 1)(t + 3)$$

The completely factored form is

$$2t^3 + 8t^2 + 6t = 2t(t + 1)(t + 3).$$

57. $2x^6 + 8x^5 - 42x^4$

First, factor out the GCF, $2x^4$.

$$2x^6 + 8x^5 - 42x^4 = 2x^4(x^2 + 4x - 21)$$

Now factor $x^2 + 4x - 21$.

Factors of -21	Sums of Factors
$1, -21$	-20
$-1, 21$	20
$3, -7$	-4
$-3, 7$	4 ←

Thus,

$$x^2 + 4x - 21 = (x - 3)(x + 7).$$

The completely factored form is

$$2x^6 + 8x^5 - 42x^4 = 2x^4(x - 3)(x + 7).$$

59. $5m^5 + 25m^4 - 40m^2$

Factor out the GCF, $5m^2$.

$$5m^5 + 25m^4 - 40m^2 = 5m^2(m^3 + 5m^2 - 8)$$

61. $m^3n - 10m^2n^2 + 24mn^3$

First, factor out the GCF, mn.

$$\begin{aligned} &m^3n - 10m^2n^2 + 24mn^3 \\ &= mn(m^2 - 10mn + 24n^2) \end{aligned}$$

The expressions $-6n$ and $-4n$ have a product of $24n^2$ and a sum of $-10n$. The completely factored form is

$$\begin{aligned} &m^3n - 10m^2n^2 + 24mn^3 \\ &= mn(m - 6n)(m - 4n). \end{aligned}$$

63. $a^5 + 3a^4b - 4a^3b^2$

The GCF is a^3, so

$$\begin{aligned} &a^5 + 3a^4b - 4a^3b^2 \\ &= a^3(a^2 + 3ab - 4b^2). \end{aligned}$$

Now factor $a^2 + 3ab - 4b^2$. The expressions $4b$ and $-b$ have a product of $-4b^2$ and a sum of $3b$. The completely factored form is

$$a^5 + 3a^4b - 4a^3b^2 = a^3(a + 4b)(a - b).$$

65. $y^3z + y^2z^2 - 6yz^3$

The GCF is yz, so

$$y^3z + y^2z^2 - 6yz^3 = yz(y^2 + yz - 6z^2).$$

Now factor $y^2 + yz - 6z^2$. The expressions $3z$ and $-2z$ have a product of $-6z^2$ and a sum of z. The completely factored form is

$$y^3z + y^2z^2 - 6yz^3 = yz(y + 3z)(y - 2z).$$

67. $z^{10} - 4z^9y - 21z^8y^2$

$$\begin{aligned} &= z^8(z^2 - 4zy - 21y^2) \quad \textit{GCF is } z^8. \\ &= z^8(z - 7y)(z + 3y) \end{aligned}$$

69. $(a + b)x^2 + (a + b)x - 12(a + b)$

The GCF is $(a + b)$, so

$$\begin{aligned} &(a + b)x^2 + (a + b)x - 12(a + b) \\ &= (a + b)(x^2 + x - 12). \end{aligned}$$

Now factor $x^2 + x - 12$.

$$x^2 + x - 12 = (x + 4)(x - 3)$$

The completely factored form is

$$\begin{aligned} &(a + b)x^2 + (a + b)x - 12(a + b) \\ &= (a + b)(x + 4)(x - 3). \end{aligned}$$

71. $(2p+q)r^2 - 12(2p+q)r + 27(2p+q)$

The GCF is $(2p+q)$, so

$$\begin{aligned}&(2p+q)r^2 - 12(2p+q)r + 27(2p+q)\\ &= (2p+q)(r^2 - 12r + 27).\end{aligned}$$

Now factor $r^2 - 12r + 27$.

$$r^2 - 12r + 27 = (r-9)(r-3)$$

The completely factored form is

$$\begin{aligned}&(2p+q)r^2 - 12(2p+q)r + 27(2p+q)\\ &= (2p+q)(r-9)(r-3).\end{aligned}$$

73. $$\begin{aligned}(2y-7)(y+4) &= 2y^2 + 8y - 7y - 28\\ &= 2y^2 + y - 28\end{aligned}$$

75. $$\begin{aligned}(5z+2)(3z-2) &= 15z^2 - 10z + 6z - 4\\ &= 15z^2 - 4z - 4\end{aligned}$$

5.3 More on Factoring Trinomials

5.3 Now Try Exercises

N1. **(a)** $2z^2 + 5z + 3$

Find the two integers whose product is $2(3) = 6$ and whose sum is 5. The integers are 2 and 3. Write the middle term, $5z$, as $2z + 3z$.

$$\begin{aligned}2z^2 + 5z + 3 &= 2z^2 + 2z + 3z + 3\\ &= (2z^2 + 2z) + (3z + 3)\\ &= 2z(z+1) + 3(z+1)\\ &= (z+1)(2z+3)\end{aligned}$$

(b) $15m^2 + m - 2$

Find two integers whose product is $15(-2) = -30$ and whose sum is 1. The integers are 6 and -5.

$$\begin{aligned}15m^2 + m - 2 &= 15m^2 + 6m - 5m - 2\\ &= (15m^2 + 6m) + (-5m - 2)\\ &= 3m(5m+2) - 1(5m+2)\\ &= (5m+2)(3m-1)\end{aligned}$$

Note that if we had written $-5m + 6m$ instead of $6m - 5m$, the order of the factors in the final expression would be reversed; that is, $(3m-1)(5m+2)$.

(c) $8x^2 - 2xy - 3y^2$

Find two integers whose product is $8(-3) = -24$ and whose sum is -2. The integers are -6 and 4.

$$\begin{aligned}&8x^2 - 2xy - 3y^2\\ &= 8x^2 - 6xy + 4xy - 3y^2\\ &= (8x^2 - 6xy) + (4xy - 3y^2)\\ &= 2x(4x - 3y) + y(4x - 3y)\\ &= (4x - 3y)(2x + y)\end{aligned}$$

N2. $15z^6 + 18z^5 - 24z^4$

First factor out the greatest common factor, $3z^4$.

$$15z^6 + 18z^5 - 24z^4 = 3z^4(5z^2 + 6z - 8)$$

Find two integers whose product is $5(-8) = -40$ and whose sum is 6. The integers are -4 and 10.

$$\begin{aligned}5z^2 + 6z - 8 &= 5z^2 - 4z + 10z - 8\\ &= (5z^2 - 4z) + (10z - 8)\\ &= z(5z-4) + 2(5z-4)\\ &= (5z-4)(z+2)\end{aligned}$$

The completely factored form is

$$15z^6 + 18z^5 - 24z^4 = 3z^4(5z-4)(z+2).$$

N3. $8y^2 + 22y + 5$

The factors of $8y^2$ are $2y$ and $4y$, or $8y$ and y. The factors of 5 are 5 and 1. Try various combinations, checking to see if the middle term is $22y$ in each case.

$(4y+5)(2y+1) = 8y^2 + 14y + 5$ *Incorrect*
$(4y+1)(2y+5) = 8y^2 + 22y + 5$ *Correct*

Thus, $8y^2 + 22y + 5 = (4y+1)(2y+5)$.

N4. $10x^2 - 9x + 2$

The factors of $10x^2$ are $10x$ and x, or $5x$ and $2x$. Try $5x$ and $2x$. Since the last term is positive and the coefficient of the middle term is negative, only negative factors of 2 should be considered. The factors of 2 are -2 and -1.

$(5x-1)(2x-2)$
$= 10x^2 - 12x + 2$ *Incorrect*

$(5x-2)(2x-1)$
$= 10x^2 - 9x + 2$ *Correct*

Thus, $10x^2 - 9x + 2 = (5x-2)(2x-1)$.

N5. $10a^2 + 31a - 14$

The factors of $10a^2$ are $5a$ and $2a$, or $10a$ and a. Some factors of -14 are 14 and -1, or 7 and -2. Try various possibilities.

$(5a-14)(2a+1) = 10a^2 - 23a - 14$ *Incorrect*
$(5a-2)(2a+7) = 10a^2 + 31a - 14$ *Correct*

Thus, $10a^2 + 31a - 14 = (5a-2)(2a+7)$.

N6. $8z^2 + 2wz - 15w^2$

Try various possibilities.

$(4z+5w)(2z-3w)$
$= 8z^2 - 2wz - 15w^2$ *Incorrect*

The middle terms differ only in sign, so reverse the signs of the two factors.

$$(4z-5w)(2z+3w)$$
$$=8z^2+2wz-15w^2 \quad \textit{Correct}$$

Thus,

$$8z^2+2wz-15w^2=(4z-5w)(2z+3w).$$

N7. $-10x^3-45x^2+90x$

The common factor could be $5x$ or $-5x$. If we factor out $-5x$, the first term of the trinomial factor will be positive, which makes it easier to factor.

$$-10x^3-45x^2+90x$$
$$=-5x(2x^2+9x-18)$$

Factor $2x^2+9x-18$ by trial and error to obtain

$$2x^2+9x-18=(2x-3)(x+6).$$

The completely factored form is

$$-10x^3-45x^2+90x$$
$$=-5x(2x-3)(x+6).$$

5.3 Section Exercises

1. $10t^2+5t+4t+2$

$=(10t^2+5t)+(4t+2)$ *Group terms.*
$=5t(2t+1)+2(2t+1)$ *Factor each group.*
$=(2t+1)(5t+2)$ *Factor out 2t + 1.*

3. $15z^2-10z-9z+6$

$=(15z^2-10z)+(-9z+6)$ *Group terms.*
$=5z(3z-2)-3(3z-2)$ *Factor each group.*
$=(3z-2)(5z-3)$ *Factor out 3z − 2.*

5. $8s^2-4st+6st-3t^2$

$=(8s^2-4st)+(6st-3t^2)$ *Group terms.*
$=4s(2s-t)+3t(2s-t)$ *Factor each group.*
$=(2s-t)(4s+3t)$ *Factor out 2s − t.*

7. $2m^2+11m+12$

(a) Find two integers whose product is $\underline{2}\cdot\underline{12}=\underline{24}$ and whose sum is $\underline{11}$.

(b) The required integers are $\underline{3}$ and $\underline{8}$. (Order is irrelevant.)

(c) Write the middle term, $11m$, as $\underline{3m}+\underline{8m}$.

(d) Rewrite the given trinomial as $\underline{2m^2+3m+8m+12}$.

(e) $(2m^2+3m)+(8m+12)$ *Group terms.*
$=m(2m+3)+4(2m+3)$ *Factor each group.*
$=(2m+3)(m+4)$ *Factor out 2m + 3.*

(f) $(2m+3)(m+4)$

F O I L

$=2m(m)+2m(4)+3(m)+3(4)$
$=2m^2+8m+3m+12$
$=2m^2+11m+12$

9. To factor $12y^2+5y-2$, we must find two integers with a product of $12(-2)=-24$ and a sum of 5. The only pair of integers satisfying those conditions is 8 and -3, choice **B**.

11. $2x^2-x-1$

Multiply the factors in the choices together to see which ones give the correct product. Since

$$(2x-1)(x+1)=2x^2+x-1$$
$$\text{and } (2x+1)(x-1)=2x^2-x-1,$$

the correct factored form is choice **B**, $(2x+1)(x-1)$.

13. $4y^2+17y-15$

Multiply the factors in the choices together to see which ones give the correct product. Since

$$(y+5)(4y-3)=4y^2+17y-15$$
$$\text{and } (2y-5)(2y+3)=4y^2-4y-15,$$

the correct factored form is choice **A**, $(y+5)(4y-3)$.

15. $6a^2+7ab-20b^2=(3a-4b)(\underline{\qquad})$

The first term in the missing expression must be $2a$ since

$$(3a)(2a)=6a^2.$$

The second term in the missing expression must be $5b$ since

$$(-4b)(5b)=-20b^2.$$

Checking our answer by multiplying, we see that $(3a-4b)(\underline{2a+5b})=6a^2+7ab-20b^2$, as desired.

17. $2x^2+6x-8=2(\underline{x^2+3x-4})$

To factor x^2+3x-4, we look for two integers whose product is -4 and whose sum is 3. The integers are 4 and -1. Thus,

$$2x^2+6x-8=2(\underline{x+4})(\underline{x-1}).$$

19. $4z^3-10z^2-6z=2z(\underline{2z^2-5z-3})$
$=2z(\underline{2z+1})(\underline{z-3})$

21. The binomial $2x-6$ cannot be a factor of $12x^2+7x-12$ because its terms have a common factor of 2, which the polynomial terms do not have.

In Exercises 23–78, either the trial and error method (which uses FOIL in reverse) or the grouping method can be used to factor each polynomial.

23. $3a^2+10a+7$

Factor by the grouping method. Look for two integers whose product is $3(7)=21$ and whose sum is 10. The integers are 3 and 7. Use these integers to rewrite the middle term, $10a$, as $3a + 7a$, and then factor the resulting four-term polynomial by grouping.

$$\begin{aligned} &3a^2 + 10a + 7 \\ &= 3a^2 + 3a + 7a + 7 && \textit{10a = 3a + 7a} \\ &= (3a^2 + 3a) + (7a + 7) && \textit{Group the terms.} \\ &= 3a(a + 1) + 7(a + 1) && \textit{Factor each group.} \\ &= (a + 1)(3a + 7) && \textit{Factor out a + 1.} \end{aligned}$$

25. $2y^2 + 7y + 6$

Factor by the grouping method. Look for two integers whose product is $2(6) = 12$ and whose sum is 7. The integers are 3 and 4.

$$\begin{aligned} &2y^2 + 7y + 6 \\ &= 2y^2 + 3y + 4y + 6 && \textit{7y = 3y + 4y} \\ &= (2y^2 + 3y) + (4y + 6) && \textit{Group terms.} \\ &= y(2y + 3) + 2(2y + 3) && \textit{Factor each group.} \\ &= (2y + 3)(y + 2) && \textit{Factor out 2y + 3.} \end{aligned}$$

27. $15m^2 + m - 2$

Factor by the grouping method. Look for two integers whose product is $15(-2) = -30$ and whose sum is 1. The integers are 6 and -5.

$$\begin{aligned} &15m^2 + m - 2 \\ &= 15m^2 + 6m - 5m - 2 && \textit{m = 6m – 5m} \\ &= (15m^2 + 6m) + (-5m - 2) && \textit{Group the terms.} \\ &= 3m(5m + 2) - 1(5m + 2) && \textit{Factor each group.} \\ &= (5m + 2)(3m - 1) && \textit{Factor out 5m + 2.} \end{aligned}$$

29. $12s^2 + 11s - 5$

Factor by trial and error.

Possible factors of $12s^2$ are s and $12s$, $2s$ and $6s$, or $3s$ and $4s$.

Factors of -5 are -1 and 5 or -5 and 1.

$$\begin{aligned} (2s - 1)(6s + 5) &= 12s^2 + 4s - 5 && \textit{Incorrect} \\ (2s + 1)(6s - 5) &= 12s^2 - 4s - 5 && \textit{Incorrect} \\ (3s - 1)(4s + 5) &= 12s^2 + 11s - 5 && \textit{Correct} \end{aligned}$$

31. $10m^2 - 23m + 12$

Factor by the grouping method. Look for two integers whose product is $10(12) = 120$ and whose sum is -23. The integers are -8 and -15.

$$\begin{aligned} &10m^2 - 23m + 12 \\ &= 10m^2 - 8m - 15m + 12 && \textit{–23m = –8m – 15m} \\ &= (10m^2 - 8m) + (-15m + 12) && \textit{Group the terms.} \\ &= 2m(5m - 4) - 3(5m - 4) && \textit{Factor each group.} \\ &= (5m - 4)(2m - 3) && \textit{Factor out 5m – 4.} \end{aligned}$$

33. $8w^2 - 14w + 3$

Factor by trial and error. Possible factors of $8w^2$ are w and $8w$ or $2w$ and $4w$.
Factors of 3 are -1 and -3 (since $b = -14$ is negative).

$$\begin{aligned} (4w - 3)(2w - 1) &= 8w^2 - 10w + 3 && \textit{Incorrect} \\ (4w - 1)(2w - 3) &= 8w^2 - 14w + 3 && \textit{Correct} \end{aligned}$$

35. $20y^2 - 39y - 11$

Factor by the grouping method. Look for two integers whose product is $20(-11) = -220$ and whose sum is -39. The integers are -44 and 5.

$$\begin{aligned} &20y^2 - 39y - 11 \\ &= 20y^2 - 44y + 5y - 11 && \textit{–39y = –44y + 5y} \\ &= (20y^2 - 44y) + (5y - 11) && \textit{Group the terms.} \\ &= 4y(5y - 11) + 1(5y - 11) && \textit{Factor each group.} \\ &= (5y - 11)(4y + 1) && \textit{Factor out 5y – 11.} \end{aligned}$$

37. $3x^2 - 15x + 16$

Factor by the grouping method. Look for two integers whose product is $3(16) = 48$ and whose sum is -15. The negative factors and their sums are:

$$\begin{aligned} -1 + (-48) &= -49 \\ -2 + (-24) &= -26 \\ -3 + (-16) &= -19 \\ -4 + (-12) &= -16 \\ -6 + (-8) &= -14 \end{aligned}$$

So there are no integers satisfying the conditions and the polynomial is *prime.*

39. First, factor out the greatest common factor, 2.

$$20x^2 + 22x + 6 = 2(10x^2 + 11x + 3)$$

Now factor $10x^2 + 11x + 3$ by trial and error to obtain

$$10x^2 + 11x + 3 = (5x + 3)(2x + 1).$$

The complete factorization is

$$20x^2 + 22x + 6 = 2(5x + 3)(2x + 1).$$

41. First, factor out the GCF, 3.

$$24x^2 - 42x + 9 = 3(8x^2 - 14x + 3)$$

Use the grouping method to factor $8x^2 - 14x + 3$. Look for two integers whose product is $8(3) = 24$ and whose sum is -14. The integers are -12 and -2.

$$\begin{aligned} &24x^2 - 42x + 9 \\ &= 3(8x^2 - 12x - 2x + 3) && \textit{-14x = -12x - 2x} \\ &= 3[(8x^2 - 12x) + (-2x + 3)] && \textit{Group the terms.} \\ &= 3[4x(2x - 3) - 1(2x - 3)] && \textit{Factor each group.} \\ &= 3(2x - 3)(4x - 1) && \textit{Factor out } 2x - 3. \end{aligned}$$

43. First, factor out the GCF, q.

$$40m^2q + mq - 6q = q(40m^2 + m - 6)$$

Now factor $40m^2 + m - 6$ by trial and error to obtain

$$40m^2 + m - 6 = (5m + 2)(8m - 3).$$

The complete factorization is

$$40m^2q + mq - 6q = q(5m + 2)(8m - 3).$$

45. First, factor out the GCF, $3n^2$.

$$15n^4 - 39n^3 + 18n^2 = 3n^2(5n^2 - 13n + 6)$$

Factor $5n^2 - 13n + 6$ by the trial and error method. Possible factors of $5n^2$ are $5n$ and n.

Possible factors of 6 are -6 and -1, or -3 and -2.

$(5n - 6)(n - 1) = 5n^2 - 11n + 6$ *Incorrect*
$(5n - 3)(n - 2) = 5n^2 - 13n + 6$ *Correct*

The completely factored form is

$$15n^4 - 39n^3 + 18n^2 = 3n^2(5n - 3)(n - 2).$$

47. First, factor out the GCF, y^2.

$$15x^2y^2 - 7xy^2 - 4y^2 = y^2(15x^2 - 7x - 4)$$

Factor $15x^2 - 7x - 4$ by the grouping method. Look for two integers whose product is $15(-4) = -60$ and whose sum is -7. The integers are -12 and 5.

$$\begin{aligned} &15x^2y^2 - 7xy^2 - 4y^2 \\ &= y^2(15x^2 - 12x + 5x - 4) \\ &= y^2[3x(5x - 4) + 1(5x - 4)] \\ &= y^2(5x - 4)(3x + 1) \end{aligned}$$

49. $5a^2 - 7ab - 6b^2$

Factor by the grouping method. Look for two integers whose product is $5(-6) = -30$ and whose sum is -7. The integers are -10 and 3.

$$\begin{aligned} &5a^2 - 7ab - 6b^2 \\ &= 5a^2 - 10ab + 3ab - 6b^2 \\ &= (5a^2 - 10ab) + (3ab - 6b^2) \\ &= 5a(a - 2b) + 3b(a - 2b) \\ &= (a - 2b)(5a + 3b) \end{aligned}$$

51. $12s^2 + 11st - 5t^2$

Factor by the grouping method. Look for two integers whose product is $12(-5) = -60$ and whose sum is 11. The integers are 15 and -4.

$$\begin{aligned} &12s^2 + 11st - 5t^2 \\ &= 12s^2 + 15st - 4st - 5t^2 \\ &= (12s^2 + 15st) + (-4st - 5t^2) \\ &= 3s(4s + 5t) - t(4s + 5t) \\ &= (4s + 5t)(3s - t) \end{aligned}$$

53. $6m^6n + 7m^5n^2 + 2m^4n^3$
$= m^4n(6m^2 + 7mn + 2n^2)$ *GCF* $= m^4n$

Now factor $6m^2 + 7mn + 2n^2$ by trial and error.

Possible factors of $6m^2$ are $6m$ and m or $3m$ and $2m$. Possible factors of $2n^2$ are $2n$ and n.

$(3m + 2n)(2m + n) = 6m^2 + 7mn + 2n^2$
Correct

The completely factored form is

$$\begin{aligned} &6m^6n + 7m^5n^2 + 2m^4n^3 \\ &= m^4n(3m + 2n)(2m + n). \end{aligned}$$

55. $5 - 6x + x^2$

$$\begin{aligned} &= x^2 - 6x + 5 && \textit{Re-order.} \\ &= x^2 - 5x - x + 5 && \textit{-6x = -5x - x} \\ &= (x^2 - 5x) + (-x + 5) && \textit{Group the terms.} \\ &= x(x - 5) - 1(x - 5) && \textit{Factor each group.} \\ &= (x - 5)(x - 1) && \textit{Factor out } x - 5. \end{aligned}$$

57. $16 + 16x + 3x^2 = 3x^2 + 16x + 16$

Factor by the grouping method. Find two integers whose product is $(3)(16) = 48$ and whose sum is 16. The numbers are 4 and 12.

$$\begin{aligned} &3x^2 + 16x + 16 \\ &= 3x^2 + 4x + 12x + 16 \\ &= (3x^2 + 4x) + (12x + 16) \\ &= x(3x + 4) + 4(3x + 4) \\ &= (3x + 4)(x + 4) \end{aligned}$$

59. $-10x^3 + 5x^2 + 140x$

First, factor out $-5x$; then complete the factoring by trial and error, using FOIL to test various possibilities until the correct one is found.

$$-10x^3 + 5x^2 + 140x$$
$$= -5x(2x^2 - x - 28)$$
$$= -5x(2x + 7)(x - 4)$$

61. $12x^2 - 47x - 4$

Factor by the grouping method. Find two integers whose product is $12(-4) = -48$ and whose sum is -47. The numbers are 1 and -48.

$$12x^2 - 47x - 4$$
$$= 12x^2 + x - 48x - 4$$
$$= (12x^2 + x) + (-48x - 4)$$
$$= x(12x + 1) - 4(12x + 1)$$
$$= (12x + 1)(x - 4)$$

63. $24y^2 - 41xy - 14x^2$

Factor by using trial and error. Only positive factors of 24 should be considered: 1 and 24, 2 and 12, 3 and 8, or 4 and 6. The factors of -14 include 7 and -2. The correct factorization is

$$24y^2 - 41xy - 14x^2 = (24y + 7x)(y - 2x).$$

65. $36x^4 - 64x^2y + 15y^2$

Factor by using trial and error. For $15y^2$, consider only $-y$ and $-15y$, and $-3y$ and $-5y$.

$$36x^4 - 64x^2y + 15y^2 = (18x^2 - 5y)(2x^2 - 3y)$$

67. $48a^2 - 94ab - 4b^2$
$= 2(24a^2 - 47ab - 2b^2)$ $\quad GCF = 2$

Now factor $24a^2 - 47ab - 2b^2$ by the grouping method. Look for two integers whose product is $24(-2) = -48$ and whose sum is -47. The integers are 1 and -48.

$$24a^2 - 47ab - 2b^2$$
$$= 24a^2 + ab - 48ab - 2b^2$$
$$= (24a^2 + ab) + (-48ab - 2b^2)$$
$$= a(24a + b) - 2b(24a + b)$$
$$= (24a + b)(a - 2b)$$

The completely factored form is

$$48a^2 - 94ab - 4b^2 = 2(24a + b)(a - 2b).$$

69. $10x^4y^5 + 39x^3y^5 - 4x^2y^5$
$= x^2y^5(10x^2 + 39x - 4)$ $\quad GCF = x^2y^5$

Now factor $10x^2 + 39x - 4$ by the grouping method. Look for two integers whose product is $10(-4) = -40$ and whose sum is 39. The integers are -1 and 40.

$$10x^2 + 39x - 4$$
$$= 10x^2 - x + 40x - 4$$
$$= (10x^2 - x) + (40x - 4)$$
$$= x(10x - 1) + 4(10x - 1)$$
$$= (10x - 1)(x + 4)$$

The completely factored form is

$$10x^4y^5 + 39x^3y^5 - 4x^2y^5$$
$$= x^2y^5(10x - 1)(x + 4).$$

71. $36a^3b^2 - 104a^2b^2 - 12ab^2$
$= 4ab^2(9a^2 - 26a - 3)$ $\quad GCF = 4ab^2$

Now factor $9a^2 - 26a - 3$ by the grouping method. Look for two integers whose product is $9(-3) = -27$ and whose sum is -26. The integers are 1 and -27.

$$9a^2 - 26a - 3$$
$$= 9a^2 + a - 27a - 3$$
$$= (9a^2 + a) + (-27a - 3)$$
$$= a(9a + 1) - 3(9a + 1)$$
$$= (9a + 1)(a - 3)$$

The completely factored form is

$$36a^3b^2 - 104a^2b^2 - 12ab^2$$
$$= 4ab^2(9a + 1)(a - 3).$$

73. $24x^2 - 46x + 15$

Factor by the grouping method. Look for two integers whose product is $24(15) = 360$ and whose sum is -46. The integers are -10 and -36.

$$24x^2 - 46x + 15$$
$$= 24x^2 - 10x - 36x + 15$$
$$= (24x^2 - 10x) + (-36x + 15)$$
$$= 2x(12x - 5) - 3(12x - 5)$$
$$= (12x - 5)(2x - 3)$$

75. $24x^4 + 55x^2 - 24$

Factor by the grouping method. Look for two integers whose product is $24(-24) = -576$ and whose sum is 55. The integers are -9 and 64. Note the x^2-term.

$$24x^4 + 55x^2 - 24$$
$$= 24x^4 - 9x^2 + 64x^2 - 24$$
$$= (24x^4 - 9x^2) + (64x^2 - 24)$$
$$= 3x^2(8x^2 - 3) + 8(8x^2 - 3)$$
$$= (8x^2 - 3)(3x^2 + 8)$$

77. $24x^2 + 38xy + 15y^2$

Factor by the grouping method. Look for two integers whose product is $24(15) = 360$ and whose sum is 38. The integers are 18 and 20. Note the xy-term.

$$24x^2 + 38xy + 15y^2$$
$$= 24x^2 + 18xy + 20xy + 15y^2$$
$$= (24x^2 + 18xy) + (20xy + 15y^2)$$
$$= 6x(4x + 3y) + 5y(4x + 3y)$$
$$= (4x + 3y)(6x + 5y)$$

79. $-x^2 - 4x + 21 = -1(x^2 + 4x - 21)$
$= -1(x + 7)(x - 3)$

81. $-3x^2 - x + 4 = -1(3x^2 + x - 4)$
$= -1(3x + 4)(x - 1)$

83. $-2a^2 - 5ab - 2b^2$
$= -1(2a^2 + 5ab + 2b^2)$
Factor out –1.
$= -1(2a^2 + 4ab + ab + 2b^2)$
5ab = 4ab + ab
$= -1[(2a^2 + 4ab) + (ab + 2b^2)]$
Group the terms.
$= -1[2a(a + 2b) + b(a + 2b)]$
Factor each group.
$= -1(a + 2b)(2a + b)$
Factor out a + 2b.

85. $25q^2(m + 1)^3 - 5q(m + 1)^3 - 2(m + 1)^3$

First, factor out the GCF, $(m + 1)^3$; then factor the resulting trinomial by trial and error.

$25q^2(m + 1)^3 - 5q(m + 1)^3 - 2(m + 1)^3$
$= (m + 1)^3(25q^2 - 5q - 2)$
$= (m + 1)^3(5q - 2)(5q + 1)$

87. $9x^2(r + 3)^3 + 12xy(r + 3)^3 + 4y^2(r + 3)^3$
$= (r + 3)^3(9x^2 + 12xy + 4y^2)$
$= (r + 3)^3(3x + 2y)(3x + 2y)$
$= (r + 3)^3(3x + 2y)^2$

89. $5x^2 + kx - 1$

Look for two integers whose product is $5(-1) = -5$ and whose sum is k.

Factors of -5	Sums of Factors
$-5, 1$	-4
$5, -1$	4

Thus, there are two possible integer values for k: -4 and 4.

91. $2m^2 + km + 5$

Look for two integers whose product is $2(5) = 10$ and whose sum is k.

Factors of 10	Sums of Factors
$-10, -1$	-11
$10, 1$	11
$-5, -2$	-7
$5, 2$	7

Thus, there are four possible integer values for k: $-11, -7, 7,$ and 11.

93. $(7p + 3)(7p - 3) = (7p)^2 - 3^2$
$= 49p^2 - 9$

95. $(x + 6)^2 = x^2 + 2(x)(6) + 6^2$
$= x^2 + 12x + 36$

5.4 Special Factoring Techniques

5.4 Now Try Exercises

N1. (a) $x^2 - 100 = x^2 - 10^2$
$= (x + 10)(x - 10)$

(b) $x^2 + 49$
This binomial is the sum of two squares and the terms have no common factor. Unlike the difference of two squares, it cannot be factored. It is a *prime* polynomial.

N2. (a) $9t^2 - 100 = (3t)^2 - 10^2$
$= (3t + 10)(3t - 10)$

(b) $36a^2 - 49b^2 = (6a)^2 - (7b)^2$
$= (6a + 7b)(6a - 7b)$

N3. (a) $16k^2 - 64$
$= 16(k^2 - 4)$ *Factor out 16.*
$= 16(k^2 - 2^2)$
$= 16(k + 2)(k - 2)$ *Difference of squares*

(b) $m^4 - 144 = (m^2)^2 - 12^2$
$= (m^2 + 12)(m^2 - 12)$

(c) $v^4 - 625 = (v^2)^2 - 25^2$
$= (v^2 + 25)(v^2 - 25)$
$= (v^2 + 25)(v^2 - 5^2)$
$= (v^2 + 25)(v + 5)(v - 5)$

N4. $y^2 + 14y + 49$
The term y^2 is a perfect square, and so is 49. Try to factor the trinomial as

$$y^2 + 14y + 49 = (y + 7)^2.$$

To check, take twice the product of the two terms in the squared binomial.

$$2 \cdot y \cdot 7 = 14y$$

Since $14y$ is the middle term of the trinomial, the trinomial is a perfect square and can be factored as $(y + 7)^2$. Thus,

$$y^2 + 14y + 49 = (y + 7)^2.$$

N5. (a) $t^2 - 18t + 81 \stackrel{?}{=} (t - 9)^2$

$2 \cdot t \cdot 9 = 18t$, so this is a perfect square trinomial, and

$$t^2 - 18t + 81 = (t - 9)^2.$$

(b) $4p^2 - 28p + 49 \stackrel{?}{=} (2p - 7)^2$

$2 \cdot 2p \cdot 7 = 28p$, so this is a perfect square trinomial, and

$$4p^2 - 28p + 49 = (2p - 7)^2.$$

(c) $9x^2 + 6x + 4 \stackrel{?}{=} (3x+2)^2$

$2 \cdot 3x \cdot 2 = 12x \neq 6x$, so this is not a perfect square trinomial. The trinomial cannot be factored even with the methods of the previous sections. It is *prime*.

(d) $80x^3 + 120x^2 + 45x$

Factor out the greatest common factor, $5x$.

$$80x^3 + 120x^2 + 45x = 5x(16x^2 + 24x + 9) \stackrel{?}{=} 5x(4x+3)^2$$

$2 \cdot 4x \cdot 3 = 24x$, so this is a perfect square trinomial, and

$$5x(16x^2 + 24x + 9) = 5x(4x+3)^2.$$

N6. **(a)** $a^3 - 27 = a^3 - 3^3$

Let $x = a$ and $y = 3$ in the pattern for the difference of cubes.

$$\begin{aligned} x^3 - y^3 &= (x-y)(x^2 + xy + y^2) \\ a^3 - 3^3 &= (a-3)(a^2 + a \cdot 3 + 3^2) \\ &= (a-3)(a^2 + 3a + 9) \end{aligned}$$

(b) $8t^3 - 125 = (2t)^3 - 5^3$

Let $x = 2t$ and $y = 5$ in the pattern for the difference of cubes.

$$\begin{aligned} x^3 - y^3 &= (x-y)(x^2 + xy + y^2) \\ (2t)^3 - 5^3 &= (2t-5)[(2t)^2 + 2t \cdot 5 + 5^2] \\ &= (2t-5)(4t^2 + 10t + 25) \end{aligned}$$

(c) $$\begin{aligned} 3k^3 - 192 &= 3(k^3 - 64) \\ &= 3(k^3 - 4^3) \\ &= 3(k-4)(k^2 + k \cdot 4 + 4^2) \\ &= 3(k-4)(k^2 + 4k + 16) \end{aligned}$$

(d) $125x^3 - 343y^6$

$$\begin{aligned} &= (5x)^3 - (7y^2)^3 \\ &= (5x - 7y^2)[(5x)^2 + 5x(7y^2) + (7y^2)^2] \\ &= (5x - 7y^2)(25x^2 + 35xy^2 + 49y^4) \end{aligned}$$

N7. **(a)** $x^3 + 125 = x^3 + 5^3$

Let $x = x$ and $y = 5$ in the pattern for the sum of cubes.

$$\begin{aligned} x^3 + y^3 &= (x+y)(x^2 - xy + y^2) \\ x^3 + 5^3 &= (x+5)(x^2 - x \cdot 5 + 5^2) \\ &= (x+5)(x^2 - 5x + 25) \end{aligned}$$

(b) $27a^3 + 8b^3$

$$\begin{aligned} &= (3a)^3 + (2b)^3 \\ &= (3a + 2b)[(3a)^2 - 3a(2b) + (2b)^2] \\ &= (3a + 2b)(9a^2 - 6ab + 4b^2) \end{aligned}$$

5.4 Section Exercises

1.

$1^2 = \underline{1}$	$2^2 = \underline{4}$	$3^2 = \underline{9}$
$4^2 = \underline{16}$	$5^2 = \underline{25}$	$6^2 = \underline{36}$
$7^2 = \underline{49}$	$8^2 = \underline{64}$	$9^2 = \underline{81}$
$10^2 = \underline{100}$	$11^2 = \underline{121}$	$12^2 = \underline{144}$
$13^2 = \underline{169}$	$14^2 = \underline{196}$	$15^2 = \underline{225}$
$16^2 = \underline{256}$	$17^2 = \underline{289}$	$18^2 = \underline{324}$
$19^2 = \underline{361}$	$20^2 = \underline{400}$	

3.

$1^3 = \underline{1}$	$2^3 = \underline{8}$	$3^3 = \underline{27}$
$4^3 = \underline{64}$	$5^3 = \underline{125}$	$6^3 = \underline{216}$
$7^3 = \underline{343}$	$8^3 = \underline{512}$	$9^3 = \underline{729}$
$10^3 = \underline{1000}$		

5. **(a)** $64x^6y^{12} = (8x^3y^6)^2$, so $64x^6y^{12}$ is a perfect square.

$64x^6y^{12} = (4x^2y^4)^3$, so $64x^6y^{12}$ is also a perfect cube.

Therefore, the answer is "both of these."

(b) $125t^6 = (5t^2)^3$, so $125t^6$ is a perfect cube. Since 125 is not a perfect square, $125t^6$ is not a perfect square.

(c) $49x^{12} = (7x^6)^2$, so $49x^{12}$ is a perfect square. Since 49 is not a perfect cube, $49x^{12}$ is not a perfect cube.

(d) $81r^{10} = (9r^5)^2$, so $81r^{10}$ is a perfect square. It is not a perfect cube.

7. $y^2 - 25$

To factor this binomial, use the rule for factoring a difference of squares.

$$\begin{aligned} a^2 - b^2 &= (a+b)(a-b) \\ \downarrow\quad \downarrow\quad &\quad\ \downarrow\ \downarrow\ \ \downarrow\ \downarrow \\ y^2 - 25 = y^2 - 5^2 &= (y+5)(y-5) \end{aligned}$$

9. $$\begin{aligned} x^2 - 144 &= x^2 - 12^2 \\ &= (x+12)(x-12) \end{aligned}$$

11. $m^2 + 64$

This binomial is the *sum* of squares and the terms have no common factor. Unlike the *difference* of squares, it cannot be factored. It is a *prime* polynomial.

13. $4m^2 + 16$

Factor out the GCF, 4. The resulting sum of squares cannot be factored further.

$$4m^2 + 16 = 4(m^2 + 4)$$

15. $$\begin{aligned} 9r^2 - 4 &= (3r)^2 - 2^2 \\ &= (3r+2)(3r-2) \end{aligned}$$

17. $36x^2 - 16$

First factor out the GCF, 4; then use the rule for factoring the difference of squares.

$$\begin{aligned}36x^2 - 16 &= 4(9x^2 - 4)\\ &= 4[(3x)^2 - 2^2]\\ &= 4(3x+2)(3x-2)\end{aligned}$$

19. $$\begin{aligned}196p^2 - 225 &= (14p)^2 - 15^2\\ &= (14p+15)(14p-15)\end{aligned}$$

21. $$\begin{aligned}16r^2 - 25a^2 &= (4r)^2 - (5a)^2\\ &= (4r+5a)(4r-5a)\end{aligned}$$

23. $100x^2 + 49$

This binomial is the *sum* of squares and the terms have no common factor. Unlike the *difference* of squares, it cannot be factored. It is a *prime* polynomial.

25. $$\begin{aligned}p^4 - 49 &= (p^2)^2 - 7^2\\ &= (p^2+7)(p^2-7)\end{aligned}$$

27. $x^4 - 1$

To factor this binomial completely, factor the difference of squares twice.

$$\begin{aligned}x^4 - 1 &= (x^2)^2 - 1^2\\ &= (x^2+1)(x^2-1)\\ &= (x^2+1)(x^2-1^2)\\ &= (x^2+1)(x+1)(x-1)\end{aligned}$$

29. $p^4 - 256$

To factor this binomial completely, factor the difference of squares twice.

$$\begin{aligned}p^4 - 256 &= (p^2)^2 - 16^2\\ &= (p^2+16)(p^2-16)\\ &= (p^2+16)(p^2-4^2)\\ &= (p^2+16)(p+4)(p-4)\end{aligned}$$

31. The student's answer is not a complete factorization because $x^2 - 9$ can be factored further. The correct complete factorization is

$$k^4 - 81 = (k^2+9)(k+3)(k-3).$$

In Exercises 33–50, use the rules for factoring perfect square trinomials:

$$\begin{aligned}a^2 + 2ab + b^2 &= (a+b)^2\\ a^2 - 2ab + b^2 &= (a-b)^2.\end{aligned}$$

33. Find b so that

$$x^2 + bx + 25 = (x+5)^2.$$

Since $(x+5)^2 = x^2 + 10x + 25$, $b = 10$.

35. Find a so that

$$ay^2 - 12y + 4 = (3y-2)^2.$$

Since $(3y-2)^2 = 9y^2 - 12y + 4$, $a = 9$.

37. $w^2 + 2w + 1$

The first and last terms are perfect squares, w^2 and 1^2. This trinomial is a perfect square, since the middle term is twice the product of w and 1, or

$$2 \cdot w \cdot 1 = 2w.$$

Therefore,

$$w^2 + 2w + 1 = (w+1)^2.$$

39. $x^2 - 8x + 16$

The first and last terms are perfect squares, x^2 and $(-4)^2$. This trinomial is a perfect square, since the middle term is twice the product of x and -4, or

$$2 \cdot x \cdot (-4) = -8x.$$

Therefore,

$$x^2 - 8x + 16 = (x-4)^2.$$

41. $2x^2 + 24x + 72$

First, factor out the GCF, 2.

$$2x^2 + 24x + 72 = 2(x^2 + 12x + 36)$$

Now factor $x^2 + 12x + 36$ as a perfect square trinomial.

$$x^2 + 12x + 36 = (x+6)^2$$

The final factored form is

$$2x^2 + 24x + 72 = 2(x+6)^2.$$

43. $16x^2 - 40x + 25$

The first and last terms are perfect squares, $(4x)^2$ and $(-5)^2$. The middle term is

$$2(4x)(-5) = -40x.$$

Therefore,

$$16x^2 - 40x + 25 = (4x-5)^2.$$

45. $49x^2 - 28xy + 4y^2$

The first and last terms are perfect squares, $(7x)^2$ and $(-2y)^2$. The middle term is

$$2(7x)(-2y) = -28xy.$$

Therefore,

$$49x^2 - 28xy + 4y^2 = (7x-2y)^2.$$

47. $$\begin{aligned}&64x^2 + 48xy + 9y^2\\ &= (8x)^2 + 2(8x)(3y) + (3y)^2\\ &= (8x+3y)^2\end{aligned}$$

49. $$\begin{aligned}&50h^2 - 40hy + 8y^2\\ &= 2(25h^2 - 20hy + 4y^2)\\ &= 2[(5h)^2 - 2(5h)(2y) + (2y)^2]\\ &= 2(5h-2y)^2\end{aligned}$$

51. $4k^3 - 4k^2 + 9k$

First, factor out the GCF, k.

$$4k^3 - 4k^2 + 9k = k(4k^2 - 4k + 9)$$

Since $4k^2 - 4k + 9$ cannot be factored, $k(4k^2 - 4k + 9)$ is the final factored form.

53. $25z^4 + 5z^3 + z^2$

First, factor out the GCF, z^2.

$$25z^4 + 5z^3 + z^2 = z^2(25z^2 + 5z + 1)$$

Since $25z^2 + 5z + 1$ cannot be factored, $z^2(25z^2 + 5z + 1)$ is the final factored form.

55. $a^3 - 1$

Let $x = a$ and $y = 1$ in the pattern for the difference of cubes.

$$\begin{aligned} x^3 - y^3 &= (x - y)(x^2 + xy + y^2) \\ a^3 - 1 = a^3 - 1^3 &= (a - 1)(a^2 + a \cdot 1 + 1^2) \\ &= (a - 1)(a^2 + a + 1) \end{aligned}$$

57. $m^3 + 8$

Let $x = m$ and $y = 2$ in the pattern for the sum of cubes.

$$\begin{aligned} x^3 + y^3 &= (x + y)(x^2 - xy + y^2) \\ m^3 + 8 = m^3 + 2^3 &= (m + 2)(m^2 - m \cdot 2 + 2^2) \\ &= (m + 2)(m^2 - 2m + 4) \end{aligned}$$

59. Factor $k^3 + 1000$ as the sum of cubes.

$$\begin{aligned} k^3 + 1000 &= k^3 + 10^3 \\ &= (k + 10)(k^2 - 10k + 10^2) \\ &= (k + 10)(k^2 - 10k + 100) \end{aligned}$$

61. Factor $27x^3 - 64$ as the difference of cubes.

$$\begin{aligned} 27x^3 - 64 &= (3x)^3 - 4^3 \\ &= (3x - 4)[(3x)^2 + 3x \cdot 4 + 4^2] \\ &= (3x - 4)(9x^2 + 12x + 16) \end{aligned}$$

63. $$\begin{aligned} 6p^3 + 6 &= 6(p^3 + 1) && \textit{GCF} = 6 \\ &= 6(p^3 + 1^3) && \textit{Sum of cubes} \\ &= 6(p + 1)(p^2 - p \cdot 1 + 1^2) \\ &= 6(p + 1)(p^2 - p + 1) \end{aligned}$$

65. $$\begin{aligned} 5x^3 + 40 &= 5(x^3 + 8) && \textit{GCF} = 5 \\ &= 5(x^3 + 2^3) && \textit{Sum of cubes} \\ &= 5(x + 2)(x^2 - x \cdot 2 + 2^2) \\ &= 5(x + 2)(x^2 - 2x + 4) \end{aligned}$$

67. Factor $y^3 - 8x^3$ as the difference of cubes.

$$\begin{aligned} y^3 - 8x^3 &= y^3 - (2x)^3 \\ &= (y - 2x)[y^2 + y(2x) + (2x)^2] \\ &= (y - 2x)(y^2 + 2yx + 4x^2) \end{aligned}$$

69. $$\begin{aligned} &2x^3 - 16y^3 \\ &= 2(x^3 - 8y^3) && \textit{GCF} = 2 \\ &= 2[x^3 - (2y)^3] && \textit{Difference of cubes} \\ &= 2(x - 2y)[x^2 + x \cdot 2y + (2y)^2] \\ &= 2(x - 2y)(x^2 + 2xy + 4y^2) \end{aligned}$$

71. Factor $8p^3 + 729q^3$ as the sum of cubes.

$$\begin{aligned} &8p^3 + 729q^3 \\ &= (2p)^3 + (9q)^3 \\ &= (2p + 9q)[(2p)^2 - 2p \cdot 9q + (9q)^2] \\ &= (2p + 9q)(4p^2 - 18pq + 81q^2) \end{aligned}$$

73. Factor $27a^3 + 64b^3$ as the sum of cubes.

$$\begin{aligned} &27a^3 + 64b^3 \\ &= (3a)^3 + (4b)^3 \\ &= (3a + 4b)[(3a)^2 - 3a \cdot 4b + (4b)^2] \\ &= (3a + 4b)(9a^2 - 12ab + 16b^2) \end{aligned}$$

75. Factor $125t^3 + 8s^3$ as the sum of cubes.

$$\begin{aligned} &125t^3 + 8s^3 \\ &= (5t)^3 + (2s)^3 \\ &= (5t + 2s)[(5t)^2 - 5t \cdot 2s + (2s)^2] \\ &= (5t + 2s)(25t^2 - 10ts + 4s^2) \end{aligned}$$

77. Factor $8x^3 - 125y^6$ as the difference of cubes.

$$\begin{aligned} &8x^3 - 125y^6 \\ &= (2x)^3 - (5y^2)^3 \\ &= (2x - 5y^2)[(2x)^2 + 2x(5y^2) + (5y^2)^2] \\ &= (2x - 5y^2)(4x^2 + 10xy^2 + 25y^4) \end{aligned}$$

79. Factor $27m^6 + 8n^3$ as the sum of cubes.

$$\begin{aligned} &27m^6 + 8n^3 \\ &= (3m^2)^3 + (2n)^3 \\ &= (3m^2 + 2n)[(3m^2)^2 - 3m^2(2n) + (2n)^2] \\ &= (3m^2 + 2n)(9m^4 - 6m^2n + 4n^2) \end{aligned}$$

81. Factor $x^9 + y^9$ as the sum of cubes.

$$\begin{aligned} &x^9 + y^9 \\ &= (x^3)^3 + (y^3)^3 \\ &= (x^3 + y^3)[(x^3)^2 - x^3(y^3) + (y^3)^2] \\ &= (x + y)(x^2 - xy + y^2)(x^6 - x^3y^3 + y^6) \end{aligned}$$

83. $$\begin{aligned} p^2 - \tfrac{1}{9} &= p^2 - (\tfrac{1}{3})^2 \\ &= (p + \tfrac{1}{3})(p - \tfrac{1}{3}) \end{aligned}$$

85. $$\begin{aligned} 36m^2 - \tfrac{16}{25} &= (6m)^2 - (\tfrac{4}{5})^2 \\ &= (6m + \tfrac{4}{5})(6m - \tfrac{4}{5}) \end{aligned}$$

87. $$\begin{aligned} x^2 - 0.64 &= x^2 - (0.8)^2 \\ &= (x + 0.8)(x - 0.8) \end{aligned}$$

89. $t^2 + t + \frac{1}{4}$

t^2 is a perfect square, and $\frac{1}{4}$ is a perfect square since $\frac{1}{2} \cdot \frac{1}{2} = \frac{1}{4}$. The middle term is twice the product of t and $\frac{1}{2}$, or

$$t = 2(t)(\tfrac{1}{2}).$$

Therefore,

$$t^2 + t + \tfrac{1}{4} = (t + \tfrac{1}{2})^2.$$

91. $x^2 - 1.0x + 0.25$

The first and last terms are perfect squares, x^2 and $(-0.5)^2$. The trinomial is a perfect square, since the middle term is

$$2 \cdot x \cdot (-0.5) = -1.0x.$$

Therefore,

$$x^2 - 1.0x + 0.25 = (x - 0.5)^2.$$

93. Factor $x^3 + \frac{1}{8}$ as the sum of cubes.

$$\begin{aligned} x^3 + \tfrac{1}{8} &= x^3 + (\tfrac{1}{2})^3 \\ &= (x + \tfrac{1}{2})[x^2 - \tfrac{1}{2}x + (\tfrac{1}{2})^2] \\ &= (x + \tfrac{1}{2})(x^2 - \tfrac{1}{2}x + \tfrac{1}{4}) \end{aligned}$$

95. $(m + n)^2 - (m - n)^2$

Factor as the difference of squares. Substitute into the rule using $x = m + n$ and $y = m - n$.

$$\begin{aligned} &(m + n)^2 - (m - n)^2 \\ &\quad = [(m + n) + (m - n)] \cdot [(m + n) - (m - n)] \\ &\quad = (2m)(2n) \\ &\quad = 4mn \end{aligned}$$

97. $m^2 - p^2 + 2m + 2p$

This expression can be factored by grouping.

$= (m + p)(m - p) + 2(m + p)$ *Factor each group.*

$= (m + p)(m - p + 2)$ *Factor out* $m + p$.

99. $m - 4 = 0$

$m = 4$ *Add 4.*

The solution set is $\{4\}$.

101. $2t + 10 = 0$

$2t = -10$ *Subtract 10.*

$t = -5$ *Divide by 2.*

The solution set is $\{-5\}$.

Summary Exercises on Factoring

1. $$\begin{aligned} 12x^2 + 20x + 8 &= 4(3x^2 + 5x + 2) \\ &= 4(3x + 2)(x + 1) \end{aligned}$$

Match with choice **G**. [Factor out the GCF; then factor a trinomial by grouping or trial and error.]

3. $$\begin{aligned} &16m^2n + 24mn - 40mn^2 \\ &\quad = 8mn(2m + 3 - 5n) \end{aligned}$$

Match with choice **A**. [Factor out the GCF; no further factoring is possible.]

5. $36p^2 - 60pq + 25q^2$

The first and last terms are perfect squares, $(6p)^2$ and $(-5q)^2$. The middle term is

$$2(6p)(-5q) = -60pq.$$

Therefore,

$$36p^2 - 60pq + 25q^2 = (6p - 5q)^2.$$

Match with choice **E**. [Factor a perfect square trinomial.]

7. Factor $8r^3 - 125$ as the difference of cubes.

$$\begin{aligned} &8r^3 - 125 \\ &\quad = (2r)^3 - 5^3 \\ &\quad = (2r - 5)[(2r)^2 + 2r(5) + 5^2] \\ &\quad = (2r - 5)(4r^2 + 10r + 25) \end{aligned}$$

Match with choice **C**. [Factor a difference of cubes.]

9. $4w^2 + 49$

Match with choice **I**. [The polynomial is prime.]

11. $a^2 - 4a - 12 = (a - 6)(a + 2)$

13. $$\begin{aligned} 6y^2 - 6y - 12 &= 6(y^2 - y - 2) \\ &= 6(y - 2)(y + 1) \end{aligned}$$

15. $$\begin{aligned} &6a + 12b + 18c \\ &\quad = 6(a + 2b + 3c) \end{aligned}$$

17. $p^2 - 17p + 66 = (p - 11)(p - 6)$

19. $10z^2 - 7z - 6$

Use the grouping method.
Look for two integers whose product is $10(-6) = -60$ and whose sum is -7.
The integers are -12 and 5.

$10z^2 - 7z - 6$

$= 10z^2 - 12z + 5z - 6$

$= 2z(5z - 6) + 1(5z - 6)$ *Factor each group.*

$= (5z - 6)(2z + 1)$ *Factor out* $5z - 6$.

21. $17x^3y^2 + 51xy = 17xy(x^2y + 3)$

23. $$\begin{aligned} &8a^5 - 8a^4 - 48a^3 \\ &\quad = 8a^3(a^2 - a - 6) \\ &\quad = 8a^3(a - 3)(a + 2) \end{aligned}$$

25. $z^2 - 3za - 10a^2 = (z - 5a)(z + 2a)$

27. $x^2 - 4x - 5x + 20$

$= x(x-4) - 5(x-4)$ *Factor each group.*

$= (x-4)(x-5)$ *Factor out $x - 4$.*

29. $6n^2 - 19n + 10 = (3n-2)(2n-5)$

31. $16x + 20 = 4(4x+5)$

33. $6y^2 - 5y - 4$

Factor by grouping. Find two integers whose product is $6(-4) = -24$ and whose sum is -5. The integers are -8 and 3.

$6y^2 - 5y - 4$
$= 6y^2 - 8y + 3y - 4$
$= 2y(3y-4) + 1(3y-4)$
$= (3y-4)(2y+1)$

35. $6z^2 + 31z + 5 = (6z+1)(z+5)$

37. $4k^2 - 12k + 9$
$= (2k)^2 - 2 \cdot 2k \cdot 3 + 3^2$
$= (2k-3)^2$ *Perfect square trinomial*

39. $54m^2 - 24z^2$
$= 6(9m^2 - 4z^2)$
$= 6[(3m)^2 - (2z)^2]$
$= 6(3m+2z)(3m-2z)$

41. $3k^2 + 4k - 4 = (3k-2)(k+2)$

43. $14k^3 + 7k^2 - 70k$
$= 7k(2k^2 + k - 10)$
$= 7k(2k+5)(k-2)$

45. $y^4 - 16$
$= (y^2)^2 - 4^2$
$= (y^2+4)(y^2-4)$ *Difference of squares*
$= (y^2+4)(y+2)(y-2)$ *Difference of squares*

47. $8m - 16m^2 = 8m(1-2m)$

49. Factor $z^3 - 8$ as the difference of cubes.

$z^3 - 8 = z^3 - 2^3$
$= (z-2)(z^2 + z \cdot 2 + 2^2)$
$= (z-2)(z^2 + 2z + 4)$

51. $k^2 + 9$ cannot be factored because it is the sum of squares with no GCF. The expression is prime.

53. $32m^9 + 16m^5 + 24m^3$
$= 8m^3(4m^6 + 2m^2 + 3)$

55. $16r^2 + 24rm + 9m^2$
$= (4r)^2 + 2 \cdot 4r \cdot 3m + (3m)^2$
$= (4r+3m)^2$ *Perfect square trinomial*

57. $15h^2 + 11hg - 14g^2$

Factor by grouping. Look for two integers whose product is $15(-14) = -210$ and whose sum is 11. The integers are 21 and -10.

$15h^2 + 11hg - 14g^2$
$= 15h^2 + 21hg - 10hg - 14g^2$
$= 3h(5h+7g) - 2g(5h+7g)$ *Factor each group.*
$= (5h+7g)(3h-2g)$ *Factor out $5h + 7g$.*

59. $k^2 - 11k + 30 = (k-5)(k-6)$

61. $3k^3 - 12k^2 - 15k$
$= 3k(k^2 - 4k - 5)$
$= 3k(k-5)(k+1)$

63. $1000p^3 + 27$
$= (10p)^3 + 3^3$ *Sum of Cubes*
$= (10p+3)[(10p)^2 - 10p \cdot 3 + 3^2]$
$= (10p+3)(100p^2 - 30p + 9)$

65. $6 + 3m + 2p + mp$
$= (6+3m) + (2p+mp)$
$= 3(2+m) + p(2+m)$
$= (2+m)(3+p)$

67. $16z^2 - 8z + 1$
$= (4z)^2 - 2 \cdot 4z \cdot 1 + 1^2$
$= (4z-1)^2$ *Perfect square trinomial*

69. $108m^2 - 36m + 3$
$= 3(36m^2 - 12m + 1)$
$= 3(6m-1)^2$ *Perfect square trinomial*

71. $x^2 - xy + y^2$ is *prime*. The middle term would have to be $+2xy$ or $-2xy$ in order to make this a perfect square trinomial.

73. $32z^3 + 56z^2 - 16z$
$= 8z(4z^2 + 7z - 2)$
$= 8z(4z-1)(z+2)$

75. $20 + 5m + 12n + 3mn$
$= 5(4+m) + 3n(4+m)$
$= (4+m)(5+3n)$

77. $6a^2 + 10a - 4$
$= 2(3a^2 + 5a - 2)$
$= 2(3a-1)(a+2)$

79. $a^3 - b^3 + 2a - 2b$

Factor by grouping. The first two terms form a difference of cubes.

$a^3 - b^3 + 2a - 2b$
$= (a^3 - b^3) + (2a - 2b)$
$= (a-b)(a^2 + ab + b^2) + 2(a-b)$
$= (a-b)(a^2 + ab + b^2 + 2)$ *Factor out $(a - b)$.*

81. $64m^2 - 80mn + 25n^2$
$= (8m)^2 - 2(8m)(5n) + (5n)^2$
$= (8m - 5n)^2$ *Perfect square trinomial*

83. $8k^2 - 2kh - 3h^2$
$= 8k^2 - 6kh + 4kh - 3h^2$
$= 2k(4k - 3h) + h(4k - 3h)$ *Factor each group.*
$= (4k - 3h)(2k + h)$ *Factor out 4k – 3h.*

85. $2x^3 + 128$
$= 2(x^3 + 64)$ *GCF = 2*
$= 2(x^3 + 4^3)$ *Sum of Cubes*
$= 2(x + 4)(x^2 - x \cdot 4 + 4^2)$
$= 2(x + 4)(x^2 - 4x + 16)$

87. $10y^2 - 7yz - 6z^2$
$= 10y^2 - 12yz + 5yz - 6z^2$
$= 2y(5y - 6z) + z(5y - 6z)$ *Factor each group.*
$= (5y - 6z)(2y + z)$ *Factor out 5y – 6z.*

89. $8a^2 + 23ab - 3b^2$
$= 8a^2 + 24ab - ab - 3b^2$
$= 8a(a + 3b) - b(a + 3b)$ *Factor each group.*
$= (a + 3b)(8a - b)$ *Factor out a + 3b.*

♦♦♦ Relating Concepts 91–98 ♦♦♦

91. $x^6 - 1 = (x^3)^2 - 1^2$
$= (x^3 + 1)(x^3 - 1)$
or $(x^3 - 1)(x^3 + 1)$

92. From Exercise 91, we have

$$x^6 - 1 = (x^3 - 1)(x^3 + 1).$$

Use the rules for the difference and sum of cubes to factor further.

Since

$$x^3 - 1 = (x - 1)(x^2 + x + 1)$$

and

$$x^3 + 1 = (x + 1)(x^2 - x + 1),$$

we obtain the factorization

$$x^6 - 1 = (x - 1)(x^2 + x + 1) \cdot (x + 1)(x^2 - x + 1).$$

93. $x^6 - 1 = (x^2)^3 - 1^3$
$= (x^2 - 1)[(x^2)^2 + x^2 \cdot 1 + 1^2]$
$= (x^2 - 1)(x^4 + x^2 + 1)$

94. From Exercise 93, we have

$$x^6 - 1 = (x^2 - 1)(x^4 + x^2 + 1).$$

Use the rule for the difference of two squares to factor the binomial.

$$x^2 - 1 = (x - 1)(x + 1)$$

Thus, we obtain the factorization

$$x^6 - 1 = (x - 1)(x + 1)(x^4 + x^2 + 1).$$

95. The result in Exercise 92 is the completely factored form.

96. Multiply the trinomials from the factored form in Exercise 92 vertically.

$$\begin{array}{rrrrr} & & x^2 & +x & +1 \\ & & x^2 & -x & +1 \\ \hline & & x^2 & +x & +1 \\ & -x^3 & -x^2 & -x & \\ x^4 & +x^3 & +x^2 & & \\ \hline x^4 & & +x^2 & & +1 \end{array}$$

97. In general, if I must choose between factoring first using the method for difference of squares or the method for difference of cubes, I should choose the *difference of squares* method to eventually obtain the completely factored form.

98. $x^6 - 729 = (x^3)^2 - 27^2$
$= (x^3 - 27)(x^3 + 27)$
$= (x^3 - 3^3)(x^3 + 3^3)$
$= (x - 3)(x^2 + 3x + 9) \cdot (x + 3)(x^2 - 3x + 9)$

5.5 Solving Quadratic Equations by Factoring

5.5 Now Try Exercises

N1. **(a)** $(x - 4)(3x + 1) = 0$

By the zero-factor property, either $x - 4 = 0$ or $3x + 1 = 0$. The equation $x - 4 = 0$ gives the solution $x = 4$; $3x + 1 = 0$ gives $x = -\frac{1}{3}$. Check both solutions by substituting first 4 and then $-\frac{1}{3}$ for x in the original equation. The solution set is $\{-\frac{1}{3}, 4\}$.

(b) $y(4y - 5) = 0$

By the zero-factor property, either $y = 0$ or $4y - 5 = 0$, so the solutions are $y = 0$ and $y = \frac{5}{4}$. Check each solution in the original equation. The solution set is $\{0, \frac{5}{4}\}$.

N2. $t^2 = -3t + 18$

Rewrite in standard form by adding $3t$ to each side and subtracting 18 from each side.

$$t^2 + 3t - 18 = 0$$
$$(t-3)(t+6) = 0$$
$$t - 3 = 0 \quad \text{or} \quad t + 6 = 0$$
$$t = 3 \quad \text{or} \quad t = -6$$

The solution set is $\{-6, 3\}$.

N3.
$$10p^2 + 65p = 35$$
$$10p^2 + 65p - 35 = 0 \quad \textit{Standard form}$$
$$5(2p^2 + 13p - 7) = 0 \quad \textit{Factor out 5.}$$
$$2p^2 + 13p - 7 = 0 \quad \textit{Divide by 5.}$$
$$(2p-1)(p+7) = 0 \quad \textit{Factor.}$$
$$2p - 1 = 0 \quad \text{or} \quad p + 7 = 0$$
$$p = \tfrac{1}{2} \quad \text{or} \quad p = -7$$

Check that the solution set is $\{-7, \frac{1}{2}\}$.

N4. (a)
$$9x^2 - 64 = 0$$
$$(3x+8)(3x-8) = 0$$
$$3x + 8 = 0 \quad \text{or} \quad 3x - 8 = 0$$
$$3x = -8 \quad \text{or} \quad 3x = 8$$
$$x = -\tfrac{8}{3} \quad \text{or} \quad x = \tfrac{8}{3}$$

Check that the solution set is $\{-\frac{8}{3}, \frac{8}{3}\}$.

(b)
$$m^2 = 5m$$
$$m^2 - 5m = 0 \quad \textit{Standard form}$$
$$m(m-5) = 0 \quad \textit{Factor.}$$
$$m = 0 \quad \text{or} \quad m - 5 = 0 \quad \textit{Zero-factor prop.}$$
$$m = 5 \quad \textit{Solve.}$$

Check that the solution set is $\{0, 5\}$.

(c) $p(6p-1) = 2$

Write the equation in standard form.

$$6p^2 - p = 2$$
$$6p^2 - p - 2 = 0$$
$$(3p-2)(2p+1) = 0$$
$$3p - 2 = 0 \quad \text{or} \quad 2p + 1 = 0$$
$$3p = 2 \quad \text{or} \quad 2p = -1$$
$$p = \tfrac{2}{3} \qquad p = -\tfrac{1}{2}$$

Check that the solution set is $\{-\frac{1}{2}, \frac{2}{3}\}$.

N5. $4x^2 - 4x + 1 = 0$

Factor $4x^2 - 4x + 1$ as a perfect square trinomial.

$$(2x-1)^2 = 0$$

Set the factor $2x - 1$ equal to 0 and solve.

$$2x - 1 = 0$$
$$2x = 1$$
$$x = \tfrac{1}{2}$$

Check this solution by substituting it in the original equation. The solution set is $\{\frac{1}{2}\}$.

N6. (a)
$$3x^3 - 27x = 0$$
$$3x(x^2 - 9) = 0$$
$$3x(x+3)(x-3) = 0$$
$$3x = 0 \quad \text{or} \quad x + 3 = 0 \quad \text{or} \quad x - 3 = 0$$
$$x = 0 \quad \text{or} \quad x = -3 \quad \text{or} \quad x = 3$$

Check that the solution set is $\{-3, 0, 3\}$.

(b) $(3a-1)(2a^2 - 5a - 12) = 0$

We know that $a = \frac{1}{3}$ is the solution of $3a - 1 = 0$, so we need to find the solutions of $2a^2 - 5a - 12 = 0$.

$$2a^2 - 5a - 12 = 0$$
$$(2a+3)(a-4) = 0 \quad \textit{Factor.}$$
$$2a + 3 = 0 \quad \text{or} \quad a - 4 = 0 \quad \textit{Zero-factor property}$$
$$a = -\tfrac{3}{2} \quad \text{or} \quad a = 4 \quad \textit{Solve.}$$

The solutions of the original equation are $-\frac{3}{2}$, $\frac{1}{3}$, and 4. Check each solution to verify that the solution set is $\{-\frac{3}{2}, \frac{1}{3}, 4\}$.

N7.
$$x(4x-9) = (x-2)^2 + 24$$
$$4x^2 - 9x = x^2 - 4x + 4 + 24 \quad \textit{Multiply.}$$
$$3x^2 - 5x - 28 = 0 \quad \textit{St. form}$$
$$(3x+7)(x-4) = 0 \quad \textit{Factor.}$$
$$3x + 7 = 0 \quad \text{or} \quad x - 4 = 0 \quad \textit{Zero-factor property}$$
$$x = -\tfrac{7}{3} \quad \text{or} \quad x = 4 \quad \textit{Solve.}$$

Check that the solution set is $\{-\frac{7}{3}, 4\}$.

5.5 Section Exercises

1. A quadratic equation is an equation that can be put in the form $\underline{ax^2 + bx + c} = 0$. $(a \neq 0)$

3. If a quadratic equation is in standard form, to solve the equation we should begin by attempting to *factor* the polynomial.

5. If a quadratic equation $ax^2 + bx + c = 0$ has $c = 0$, then $\underline{0}$ *must* be a solution because $\underline{x}$ is a factor of the polynomial. Note that if $c = 0$, the equation becomes $ax^2 + bx = 0$, which can be factored as $x(ax + b) = 0$, and has solutions $x = 0$ and $x = -b/a$.

7. $2x(3x-4)$ is the product of two factors, $2x$ and $3x - 4$. Applying the zero-factor property yields two equations,

$$2x = 0 \quad \text{or} \quad 3x - 4 = 0.$$

Dividing each side of $2x = 0$ by 2 gives us $x = 0$, so we end up with the two solutions, $x = 0$ and $x = \frac{4}{3}$.

We conclude that multiplying a polynomial by a constant does not affect the solutions of the corresponding equation.

9. The variable x is another factor to set equal to 0, so the solution set is $\{0, \frac{1}{7}\}$.

Not including 0 as a solution is *WHAT WENT WRONG.*

For all equations in this section, answers should be checked by substituting into the original equation. These checks will be shown here for only a few of the exercises.

11. $(x+5)(x-2)=0$

By the zero-factor property, the only way that the product of these two factors can be zero is if at least one of the factors is zero.

$$x+5=0 \quad \text{or} \quad x-2=0$$

Solve each of these linear equations.

$$x=-5 \quad \text{or} \quad x=2$$

Check $x=-5$: $0(-7)=0$ *True*
Check $x=2$: $7(0)=0$ *True*

The solution set is $\{-5, 2\}$.

13. $(2m-7)(m-3)=0$

Set each factor equal to zero and solve the resulting linear equations.

$$\begin{aligned} 2m-7&=0 \quad \text{or} \quad & m-3&=0\\ 2m&=7 \quad \text{or} \quad & m&=3\\ m&=\tfrac{7}{2} \end{aligned}$$

The solution set is $\{3, \frac{7}{2}\}$.

15. $(2x+1)(6x-1)=0$

Set each factor equal to zero and solve the resulting linear equations.

$$\begin{aligned} 2x+1&=0 \quad \text{or} \quad & 6x-1&=0\\ 2x&=-1 \quad \text{or} \quad & 6x&=1\\ x&=-\tfrac{1}{2} & x&=\tfrac{1}{6} \end{aligned}$$

The solution set is $\{-\frac{1}{2}, \frac{1}{6}\}$.

17. $t(6t+5)=0$

Set each factor equal to zero and solve the resulting linear equations.

$$\begin{aligned} t=0 \quad \text{or} \quad 6t+5&=0\\ 6t&=-5\\ t&=-\tfrac{5}{6} \end{aligned}$$

The solution set is $\{-\frac{5}{6}, 0\}$.

19. $2x(3x-4)=0$

Set each factor equal to zero and solve the resulting linear equations.

$$\begin{aligned} 2x&=0 \quad \text{or} \quad & 3x-4&=0\\ x&=0 \quad \text{or} \quad & 3x&=4\\ & & x&=\tfrac{4}{3} \end{aligned}$$

The solution set is $\{0, \frac{4}{3}\}$.

21. $(x-6)(x-6)=0$

Set each factor equal to zero and solve the resulting linear equations.

$$\begin{aligned} x-6&=0 \quad \text{or} \quad & x-6&=0\\ x&=6 \quad \text{or} \quad & x&=6 \end{aligned}$$

6 is called a *double solution* for $(x-6)^2=0$ because it occurs twice when the equation is solved.

The solution set is $\{6\}$.

23. $y^2+3y+2=0$

Factor the polynomial.

$$(y+2)(y+1)=0$$

Set each factor equal to 0.

$$y+2=0 \quad \text{or} \quad y+1=0$$

Solve each equation.

$$y=-2 \quad \text{or} \quad y=-1$$

Check these solutions by substituting -2 for y and then -1 for y in the original equation.

$$\begin{aligned} y^2+3y+2&=0\\ (-2)^2+3(-2)+2&\stackrel{?}{=}0 \quad \textit{Let } y=-2.\\ 4-6+2&\stackrel{?}{=}0\\ -2+2&=0 \quad \textit{True} \end{aligned}$$

$$\begin{aligned} y^2+3y+2&=0\\ (-1)^2+3(-1)+2&\stackrel{?}{=}0 \quad \textit{Let } y=-1.\\ 1-3+2&\stackrel{?}{=}0\\ -2+2&=0 \quad \textit{True} \end{aligned}$$

The solution set is $\{-2, -1\}$.

25. $y^2-3y+2=0$

Factor the polynomial.

$$(y-1)(y-2)=0$$

Set each factor equal to 0.

$$y-1=0 \quad \text{or} \quad y-2=0$$

Solve each equation.

$$y=1 \quad \text{or} \quad y=2$$

The solution set is $\{1, 2\}$.

27. $x^2 = 24 - 5x$

Write the equation in standard form.

$$x^2 + 5x - 24 = 0$$

Factor the polynomial.

$$(x + 8)(x - 3) = 0$$

Set each factor equal to 0.

$$x + 8 = 0 \quad \text{or} \quad x - 3 = 0$$

Solve each equation.

$$x = -8 \quad \text{or} \quad x = 3$$

The solution set is $\{-8, 3\}$.

29. $x^2 = 3 + 2x$

Write the equation in standard form.

$$x^2 - 2x - 3 = 0$$
$$(x + 1)(x - 3) = 0$$

$$x + 1 = 0 \quad \text{or} \quad x - 3 = 0$$
$$x = -1 \quad \text{or} \quad x = 3$$

The solution set is $\{-1, 3\}$.

31. $z^2 + 3z = -2$

Write the equation in standard form.

$$z^2 + 3z + 2 = 0$$

Factor the polynomial.

$$(z + 2)(z + 1) = 0$$

Set each factor equal to 0.

$$z + 2 = 0 \quad \text{or} \quad z + 1 = 0$$
$$z = -2 \quad \text{or} \quad z = -1$$

The solution set is $\{-2, -1\}$.

33. $m^2 + 8m + 16 = 0$

Factor $m^2 + 8m + 16$ as a perfect square trinomial.

$$(m + 4)^2 = 0$$

Set the factor $m + 4$ equal to 0 and solve.

$$m + 4 = 0$$
$$m = -4$$

The solution set is $\{-4\}$.

35. $3x^2 + 5x - 2 = 0$

Factor the polynomial.

$$(3x - 1)(x + 2) = 0$$

Set each factor equal to 0.

$$3x - 1 = 0 \quad \text{or} \quad x + 2 = 0$$
$$3x = 1 \quad \text{or} \quad x = -2$$
$$x = \tfrac{1}{3}$$

The solution set is $\{-2, \frac{1}{3}\}$.

37. $12p^2 = 8 - 10p$

$$12p^2 + 10p - 8 = 0 \qquad \textit{Standard form}$$
$$2(6p^2 + 5p - 4) = 0 \qquad \textit{Factor out 2.}$$
$$2(3p + 4)(2p - 1) = 0$$

$$3p + 4 = 0 \quad \text{or} \quad 2p - 1 = 0$$
$$3p = -4 \quad \text{or} \quad 2p = 1$$
$$p = -\tfrac{4}{3} \quad \text{or} \quad p = \tfrac{1}{2}$$

The solution set is $\{-\frac{4}{3}, \frac{1}{2}\}$.

39. $9s^2 + 12s = -4$

$$9s^2 + 12s + 4 = 0$$
$$(3s + 2)^2 = 0$$

Set the factor $3s + 2$ equal to 0 and solve.

$$3s + 2 = 0$$
$$3s = -2$$
$$s = -\tfrac{2}{3}$$

The solution set is $\{-\frac{2}{3}\}$.

41. $y^2 - 9 = 0$

$$(y + 3)(y - 3) = 0$$

$$y + 3 = 0 \quad \text{or} \quad y - 3 = 0$$
$$y = -3 \quad \text{or} \quad y = 3$$

The solution set is $\{-3, 3\}$.

43. $16x^2 - 49 = 0$

$$(4x + 7)(4x - 7) = 0$$

$$4x + 7 = 0 \quad \text{or} \quad 4x - 7 = 0$$
$$4x = -7 \quad \text{or} \quad 4x = 7$$
$$x = -\tfrac{7}{4} \quad \text{or} \quad x = \tfrac{7}{4}$$

The solution set is $\{-\frac{7}{4}, \frac{7}{4}\}$.

45. $n^2 = 121$

$$n^2 - 121 = 0$$
$$(n + 11)(n - 11) = 0$$

$$n + 11 = 0 \quad \text{or} \quad n - 11 = 0$$
$$n = -11 \quad \text{or} \quad n = 11$$

The solution set is $\{-11, 11\}$.

47. $x^2 = 7x$

$$x^2 - 7x = 0$$
$$x(x - 7) = 0$$

$$x = 0 \quad \text{or} \quad x - 7 = 0$$
$$x = 0 \quad \text{or} \quad x = 7$$

Check $x = 0$: $0 = 0$ *True*
Check $x = 7$: $49 = 49$ *True*

The solution set is $\{0, 7\}$.

49.
$$6r^2 = 3r$$
$$6r^2 - 3r = 0$$
$$3r(2r - 1) = 0$$

$3r = 0$ or $2r - 1 = 0$
$r = 0$ or $2r = 1$
$r = \frac{1}{2}$

The solution set is $\{0, \frac{1}{2}\}$.

51.
$$x(x-7) = -10$$
$$x^2 - 7x = -10$$
$$x^2 - 7x + 10 = 0$$
$$(x-2)(x-5) = 0$$

$x - 2 = 0$ or $x - 5 = 0$
$x = 2$ or $x = 5$

The solution set is $\{2, 5\}$.

53.
$$3z(2z+7) = 12$$
$z(2z+7) = 4$ *Divide by 3.*
$$2z^2 + 7z = 4$$
$$2z^2 + 7z - 4 = 0$$
$$(2z-1)(z+4) = 0$$

$2z - 1 = 0$ or $z + 4 = 0$
$2z = 1$ or $z = -4$
$z = \frac{1}{2}$

The solution set is $\{-4, \frac{1}{2}\}$.

55.
$$2y(y+13) = 136$$
$y(y+13) = 68$ *Divide by 2.*
$y^2 + 13y = 68$ *Multiply.*
$y^2 + 13y - 68 = 0$ *Standard form*
$(y+17)(y-4) = 0$ *Factor.*

$x + 17 = 0$ or $y - 4 = 0$ *Zero-factor property*
$y = -17$ or $y = 4$ *Solve.*

Check $y = -17$:

$$2y(y+13) = 136$$
$2(-17)(-17+13) \stackrel{?}{=} 136$ *Let y = −17.*
$(-34)(-4) = 136$ *True*

Check $y = 4$:

$2(4)(4+13) \stackrel{?}{=} 136$ *Let y = 4.*
$8(17) = 136$ *True*

The solution set is $\{-17, 4\}$.

57. $(2r+5)(3r^2 - 16r + 5) = 0$

Begin by factoring $3r^2 - 16r + 5$.

$(2r+5)(3r-1)(r-5) = 0$

Set each of the three factors equal to 0 and solve the resulting equations.

$2r + 5 = 0$ or $3r - 1 = 0$ or $r - 5 = 0$
$2r = -5$ $\quad 3r = 1$
$r = -\frac{5}{2}$ or $r = \frac{1}{3}$ or $r = 5$

The solution set is $\{-\frac{5}{2}, \frac{1}{3}, 5\}$.

59.
$$(2x+7)(x^2 + 2x - 3) = 0$$
$$(2x+7)(x+3)(x-1) = 0$$

$2x + 7 = 0$ or $x + 3 = 0$ or $x - 1 = 0$
$2x = -7$
$x = -\frac{7}{2}$ or $x = -3$ or $x = 1$

The solution set is $\{-\frac{7}{2}, -3, 1\}$.

61. $9y^3 - 49y = 0$

To factor the polynomial, begin by factoring out the greatest common factor.

$$y(9y^2 - 49) = 0$$

Now factor $9y^2 - 49$ as the difference of squares.

$$y(3y+7)(3y-7) = 0$$

Set each of the three factors equal to 0 and solve.

$y = 0$ or $3y + 7 = 0$ or $3y - 7 = 0$
$3y = -7$ $\quad 3y = 7$
$y = 0$ or $y = -\frac{7}{3}$ or $y = \frac{7}{3}$

The solution set is $\{-\frac{7}{3}, 0, \frac{7}{3}\}$.

63.
$$r^3 - 2r^2 - 8r = 0$$
$r(r^2 - 2r - 8) = 0$ *Factor out r.*
$r(r-4)(r+2) = 0$ *Factor.*

Set each factor equal to zero and solve.

$r = 0$ or $r - 4 = 0$ or $r + 2 = 0$
$r = 0$ or $r = 4$ or $r = -2$

The solution set is $\{-2, 0, 4\}$.

65.
$$x^3 + x^2 - 20x = 0$$
$x(x^2 + x - 20) = 0$ *Factor out x.*
$x(x+5)(x-4) = 0$ *Factor.*

Set each factor equal to zero and solve.

$x = 0$ or $x + 5 = 0$ or $x - 4 = 0$
$x = 0$ or $x = -5$ or $x = 4$

The solution set is $\{-5, 0, 4\}$.

67. $r^4 = 2r^3 + 15r^2$

Rewrite with all terms on the left side.

$r^4 - 2r^3 - 15r^2 = 0$
$r^2(r^2 - 2r - 15) = 0$ *Factor out r^2.*
$r^2(r-5)(r+3) = 0$ *Factor.*

Set each factor equal to zero and solve.

$$r^2 = 0 \quad \text{or} \quad r - 5 = 0 \quad \text{or} \quad r + 3 = 0$$
$$r = 0 \quad \text{or} \quad r = 5 \quad \text{or} \quad r = -3$$

The solution set is $\{-3, 0, 5\}$.

69.
$$3x(x+1) = (2x+3)(x+1)$$
$$3x^2 + 3x = 2x^2 + 5x + 3$$
$$x^2 - 2x - 3 = 0$$
$$(x+1)(x-3) = 0$$

$$x + 1 = 0 \quad \text{or} \quad x - 3 = 0$$
$$x = -1 \quad \text{or} \quad x = 3$$

The solution set is $\{-1, 3\}$.

Alternatively, we could begin by moving all the terms to the left side and then factoring out $x + 1$.

$$3x(x+1) - (2x+3)(x+1) = 0$$
$$(x+1)[3x - (2x+3)] = 0$$
$$(x+1)(x-3) = 0$$

The rest of the solution is the same.

71.
$$x^2 + (x+1)^2 = (x+2)^2$$
$$x^2 + x^2 + 2x + 1 = x^2 + 4x + 4$$
$$x^2 - 2x - 3 = 0$$
$$(x+1)(x-3) = 0$$

$$x + 1 = 0 \quad \text{or} \quad x - 3 = 0$$
$$x = -1 \quad \text{or} \quad x = 3$$

The solution set is $\{-1, 3\}$.

73.
$$(2x)^2 = (2x+4)^2 - (x+5)^2$$
$$4x^2 = 4x^2 + 16x + 16 - (x^2 + 10x + 25)$$
$$4x^2 = 4x^2 + 16x + 16 - x^2 - 10x - 25$$
$$4x^2 = 3x^2 + 6x - 9$$

$$x^2 - 6x + 9 = 0$$
$$(x-3)(x-3) = 0$$

Set the factor $x - 3$ equal to 0 and solve.

$$x - 3 = 0$$
$$x = 3$$

The solution set is $\{3\}$.

75.
$$(x+3)^2 - (2x-1)^2 = 0$$
$$(x^2 + 6x + 9) - (4x^2 - 4x + 1) = 0$$
Square the binomials.
$$x^2 + 6x + 9 - 4x^2 + 4x - 1 = 0$$
$$-3x^2 + 10x + 8 = 0$$
Combine like terms.
$$-1(3x^2 - 10x - 8) = 0$$
Factor out –1.
$$-1(3x+2)(x-4) = 0$$
Factor.

Set each factor equal to zero and solve.

$$3x + 2 = 0 \quad \text{or} \quad x - 4 = 0$$
$$3x = -2$$
$$x = -\tfrac{2}{3} \quad \text{or} \quad x = 4$$

The solution set is $\{-\frac{2}{3}, 4\}$.

Alternatively we could begin by factoring the left side as the difference of two squares.

$$(x+3)^2 - (2x-1)^2 = 0$$
$$[(x+3) + (2x-1)][(x+3) - (2x-1)] = 0$$
$$[3x+2][-x+4] = 0$$

The same solution set is obtained.

77.
$$6p^2(p+1) = 4(p+1) - 5p(p+1)$$
$$6p^2(p+1) + 5p(p+1) - 4(p+1) = 0$$
Rewrite with all terms on the left.
$$(p+1)(6p^2 + 5p - 4) = 0$$
Factor out p + 1.
$$(p+1)(2p-1)(3p+4) = 0$$
Factor.

Set each factor equal to zero and solve.

$$p + 1 = 0 \quad \text{or} \quad 2p - 1 = 0 \quad \text{or} \quad 3p + 4 = 0$$
$$2p = 1 \qquad 3p = -4$$
$$p = -1 \quad \text{or} \quad p = \tfrac{1}{2} \quad \text{or} \quad p = -\tfrac{4}{3}$$

The solution set is $\{-\frac{4}{3}, -1, \frac{1}{2}\}$.

79. **(a)** $d = 16t^2$

$t = 2$: $d = 16(2)^2 = 16(4) = 64$
$t = 3$: $d = 16(3)^2 = 16(9) = 144$
$d = 256$: $256 = 16t^2$; $16 = t^2$; $t = 4$
$d = 576$: $576 = 16t^2$; $36 = t^2$; $t = 6$

t in seconds	0	1	2	3	4	6
d in feet	0	16	64	144	256	576

(b) When $t = 0$, $d = 0$, no time has elapsed, so the object hasn't fallen (been released) yet.

81. From the calculator screens, we see that the solution set of

$$x^2 + 0.4x - 0.05 = 0$$

is $\{-0.5, 0.1\}$. Substituting -0.5 for x gives us

$$(-0.5)^2 + 0.4(-0.5) - 0.05 \stackrel{?}{=} 0$$
$$0.25 - 0.20 - 0.05 \stackrel{?}{=} 0$$
$$0.25 - 0.25 = 0 \quad \textit{True}$$

This shows that -0.5 is a solution of the given quadratic equation.

A similar check shows that 0.1 is also a solution.

The solution set is $\{-0.5, 0.1\}$.

83. Let $x =$ the number.
"If a number is doubled and 6 is subtracted from this result, the answer is 3684" translates to

$$\begin{aligned} 2x - 6 &= 3684. \\ 2x &= 3690 \\ x &= 1845 \end{aligned}$$

The number is 1845, so that is when Texas was admitted to the Union.

85. Let $x =$ the first consecutive integer. Then $x + 1 =$ the second consecutive integer.
"Twice the sum of two consecutive integers is 28 more than the greater integer" is translated as

$$2[x + (x + 1)] = 28 + (x + 1).$$

Solve this equation.

$$\begin{aligned} 2(2x + 1) &= 28 + x + 1 \\ 4x + 2 &= x + 29 \\ 3x + 2 &= 29 \\ 3x &= 27 \\ x &= 9 \end{aligned}$$

Since $x = 9$, $x + 1 = 10$, so the integers are 9 and 10.

5.6 Applications of Quadratic Equations

5.6 Now Try Exercises

N1. *Step 2*
Let $x =$ the length of one leg.
Then $x - 4 =$ the length of the other leg.

Step 3 $\quad \mathcal{A} = \frac{1}{2}bh$

$$6 = \tfrac{1}{2}x(x - 4)$$

Step 4
Solve the equation.

$$\begin{aligned} 12 &= x(x - 4) && \textit{Multiply by 2.} \\ 12 &= x^2 - 4x && \textit{Distributive prop.} \\ x^2 - 4x - 12 &= 0 && \textit{Standard form} \\ (x + 2)(x - 6) &= 0 && \textit{Factor.} \end{aligned}$$

$$\begin{aligned} x + 2 &= 0 \quad \text{or} \quad x - 6 = 0 \quad \textit{Zero-factor prop.} \\ x &= -2 \quad \text{or} \quad x = 6 \end{aligned}$$

Step 5
Because a triangle's side cannot be negative, $x = 6$ and $x - 4 = 2$. The lengths of the legs are 2 feet and 6 feet.

Step 6
The length of one leg is 4 meters less than the length of the other leg, and the area is $\frac{1}{2}(6)(2) = 6 \text{ m}^2$, as required.

N2. Let $x =$ the first integer.
Then $x + 1 =$ the second integer
and $x + 2 =$ the third integer.

From the given information, we write the equation

$$x(x + 1) = 8(x + 2) + 2.$$

Solve this equation.

$$\begin{aligned} x^2 + x &= 8x + 16 + 2 \\ x^2 - 7x - 18 &= 0 \\ (x + 2)(x - 9) &= 0 \end{aligned}$$

$$\begin{aligned} x + 2 &= 0 \quad \text{or} \quad x - 9 = 0 \\ x &= -2 \quad \text{or} \quad x = 9 \end{aligned}$$

If $x = -2$, then $x + 1 = -1$ and $x + 2 = 0$.
If $x = 9$, then $x + 1 = 10$ and $x + 2 = 11$. There are two sets of integers that satisfy the statement of the problem: $-2, -1, 0$, and $9, 10, 11$.

N3. Let $x =$ the length of the shorter leg of the right triangle.
Then $x + 7 =$ the length of the longer leg
and $x + 8 =$ the length of the hypotenuse.

Use the Pythagorean theorem, substituting x for a, $x + 7$ for b, and $x + 8$ for c.

$$\begin{aligned} a^2 + b^2 &= c^2 \\ x^2 + (x + 7)^2 &= (x + 8)^2 \\ x^2 + x^2 + 14x + 49 &= x^2 + 16x + 64 \\ 2x^2 + 14x + 49 &= x^2 + 16x + 64 \\ x^2 - 2x - 15 &= 0 \\ (x + 3)(x - 5) &= 0 \end{aligned}$$

$$\begin{aligned} x + 3 &= 0 \quad \text{or} \quad x - 5 = 0 \\ x &= -3 \quad \text{or} \quad x = 5 \end{aligned}$$

Discard -3 since the length of a side of a triangle cannot be negative. The length of the shorter leg is 5 feet, the length of the longer leg is $x + 7 = 5 + 7 = 12$ feet, and the length of the hypotenuse is $x + 8 = 5 + 8 = 13$ feet.

N4. $h = -16t^2 + 180t + 6$

To find how long it will take for the ball to reach a height of 50 feet, let $h = 50$ and solve the resulting equation. (For convenience, we reverse the sides of the equation.)

$$\begin{aligned} -16t^2 + 180t + 6 &= 50 \\ -16t^2 + 180t - 44 &= 0 && \textit{Standard form} \\ 4t^2 - 45t + 11 &= 0 && \textit{Divide by } -4. \\ (4t - 1)(t - 11) &= 0 && \textit{Factor.} \end{aligned}$$

$$\begin{aligned} 4t - 1 &= 0 \quad \text{or} \quad t - 11 = 0 \\ t &= \tfrac{1}{4} \quad \text{or} \quad t = 11 \end{aligned}$$

The ball will reach a height of 50 feet after $\frac{1}{4}$ second (on the way up) and 11 seconds (on the way down).

N5. For 2000, $x = 2000 - 1930 = 70$.

$$\begin{aligned} y &= 0.01048x^2 - 0.5400x + 15.43 \\ y &= 0.01048(70)^2 - 0.5400(70) + 15.43 \quad \textit{Let } x = 70. \\ &= 28.982 \approx 29.0 \end{aligned}$$

According to the model, the foreign-born population of the United States in 2000, to the nearest tenth of a million, was about 29.0 million. The actual value from the table is 28.4 million, so the answer using the model is slightly high.

5.6 Section Exercises

1. Read; variable; equation; Solve; answer; Check, original

3. $\mathcal{A} = bh$; $\mathcal{A} = 45$, $b = 2x + 1$, $h = x + 1$

Step 3

$$\begin{aligned} \mathcal{A} &= bh \\ 45 &= (2x + 1)(x + 1) \end{aligned}$$

Step 4 Solve the equation.

$$\begin{aligned} 45 &= 2x^2 + 3x + 1 \\ 0 &= 2x^2 + 3x - 44 \\ 0 &= (2x + 11)(x - 4) \end{aligned}$$

$$\begin{aligned} 2x + 11 &= 0 \quad \text{or} \quad x - 4 = 0 \\ 2x &= -11 \\ x &= -\tfrac{11}{2} \quad \text{or} \quad x = 4 \end{aligned}$$

Step 5
The solution cannot be $x = -\frac{11}{2}$ since, when substituted, $2(-\frac{11}{2}) + 1$ and $-\frac{11}{2} + 1$ are negative numbers and base and height cannot be negative. Thus, $x = 4$ and

$$\begin{aligned} b &= 2x + 1 = 8 + 1 = 9 \\ h &= x + 1 = 4 + 1 = 5. \end{aligned}$$

The base is 9 units and the height is 5 units.

Step 6
$bh = 9 \cdot 5 = 45$, the desired value of $\mathcal{A}$.

5. $\mathcal{A} = LW$; $\mathcal{A} = 80$, $L = x + 8$, $W = x - 8$

Step 3

$$\begin{aligned} \mathcal{A} &= LW \\ 80 &= (x + 8)(x - 8) \end{aligned}$$

Step 4
Solve the equation.

$$\begin{aligned} 80 &= x^2 - 64 \\ 0 &= x^2 - 144 \\ 0 &= (x + 12)(x - 12) \end{aligned}$$

$$\begin{aligned} x + 12 &= 0 \quad \text{or} \quad x - 12 = 0 \\ x &= -12 \quad \text{or} \quad x = 12 \end{aligned}$$

Step 5
The solution cannot be $x = -12$ since, when substituted, $-12 + 8$ and $-12 - 8$ are negative numbers and length and width cannot be negative. Thus, $x = 12$ and

$$\begin{aligned} L &= x + 8 = 12 + 8 = 20 \\ W &= x - 8 = 12 - 8 = 4. \end{aligned}$$

The length is 20 units, and the width is 4 units.

Step 6
$LW = 20 \cdot 4 = 80$, the desired value of $\mathcal{A}$.

7. *Step 2*
Let $x =$ the width of the case.
Then $x + 2 =$ the length of the case.

Step 3

$$\begin{aligned} \mathcal{A} &= LW \\ 168 &= (x + 2)x \quad \textit{Substitute.} \end{aligned}$$

Step 4

$$\begin{aligned} 168 &= x^2 + 2x && \textit{Multiply.} \\ x^2 + 2x - 168 &= 0 && \textit{Standard form} \\ (x + 14)(x - 12) &= 0 && \textit{Factor.} \end{aligned}$$

$$\begin{aligned} x + 14 &= 0 \quad \text{or} \quad x - 12 = 0 \quad \textit{Zero-factor prop.} \\ x &= -14 \quad \text{or} \quad x = 12 \quad \textit{Solve.} \end{aligned}$$

Step 5
Because a width cannot be negative, $x = 12$ and $x + 2 = 14$. The width is 12 cm and the length is 14 cm.

Step 6
The length is 2 cm more than the width and the area is $14(12) = 168 \text{ cm}^2$, as required.

9. Let $h =$ the height of the triangle.
Then $2h + 2 =$ the base of the triangle.

The area of the triangle is 30 square inches.

$$\begin{aligned} \mathcal{A} &= \tfrac{1}{2}bh \\ 30 &= \tfrac{1}{2}(2h + 2) \cdot h \\ 60 &= (2h + 2)h \\ 60 &= 2h^2 + 2h \\ 0 &= 2h^2 + 2h - 60 \\ 0 &= 2(h^2 + h - 30) \\ 0 &= 2(h + 6)(h - 5) \end{aligned}$$

$$\begin{aligned} h + 6 &= 0 \quad \text{or} \quad h - 5 = 0 \\ h &= -6 \quad \text{or} \quad h = 5 \end{aligned}$$

The solution $h = -6$ must be discarded since a triangle cannot have a negative height. Thus,

$$h = 5 \text{ and } 2h + 2 = 2(5) + 2 = 12.$$

The height is 5 inches, and the base is 12 inches.

11. Let $x =$ the width of the aquarium.
Then $x + 3 =$ the height of the aquarium.

Use the formula for the volume of a rectangular box.

$$\begin{aligned} V &= LWH \\ 2730 &= 21x(x+3) \\ 130 &= x(x+3) \quad \textit{Divide by 21.} \\ 130 &= x^2 + 3x \\ 0 &= x^2 + 3x - 130 \\ 0 &= (x+13)(x-10) \end{aligned}$$

$$x + 13 = 0 \quad \text{or} \quad x - 10 = 0$$
$$x = -13 \quad \text{or} \quad x = 10$$

We discard -13 because the width cannot be negative. The width is 10 inches. The height is $10 + 3 = 13$ inches.

13. *Step 2*
Let $x =$ the width of the monitor.
Then $x + 3 =$ the length of the monitor.
The area is $x(x + 3)$.

Step 3
If the length were doubled $[2(x + 3)]$ and if the width were decreased by 1 in. $[x - 1]$, the area would be increased by 150 in.^{2} $[x(x + 3) + 150]$. Write an equation.

$$\begin{aligned} LW &= A \\ [2(x+3)](x-1) &= x(x+3) + 150 \end{aligned}$$

Step 4

$$\begin{aligned} (2x+6)(x-1) &= x^2 + 3x + 150 \\ 2x^2 + 4x - 6 &= x^2 + 3x + 150 \\ x^2 + x - 156 &= 0 \\ (x+13)(x-12) &= 0 \end{aligned}$$

$$x + 13 = 0 \quad \text{or} \quad x - 12 = 0$$
$$x = -13 \quad \text{or} \quad x = 12$$

Step 5
Reject -13, so the width is 12 inches and the length is $12 + 3 = 15$ inches.

Step 6
The length is 3 inches more than the width. The area of the monitor is $15(12) = 180$ in.^{2}. Doubling the length and decreasing the width by 1 inch gives us an area of $30(11) = 330$ in.^{2}, which is 150 in.^{2} more than the area of the original monitor, as required.

15. Let $x =$ the length of a side of the square painting.
Then $x - 2 =$ the length of a side of the square mirror.

Since the formula for the area of a square is $A = s^2$, the area of the painting is x^2, and the area of the mirror is $(x - 2)^2$. The difference between their areas is 32, so

$$\begin{aligned} x^2 - (x-2)^2 &= 32 \\ x^2 - (x^2 - 4x + 4) &= 32 \\ x^2 - x^2 + 4x - 4 &= 32 \\ 4x - 4 &= 32 \\ 4x &= 36 \\ x &= 9. \end{aligned}$$

The length of a side of the painting is 9 feet. The length of a side of the mirror is $9 - 2 = 7$ feet.

Check: $9^2 - 7^2 = 81 - 49 = 32$

17. Let $x =$ the first volume number.
Then $x + 1 =$ the second volume number.
The product of the numbers is 420.

$$\begin{aligned} x(x+1) &= 420 \\ x^2 + x - 420 &= 0 \\ (x-20)(x+21) &= 0 \end{aligned}$$

$$x - 20 = 0 \quad \text{or} \quad x + 21 = 0$$
$$x = 20 \quad \text{or} \quad x = -21$$

The volume number cannot be negative, so we reject -21. The volume numbers are 20 and $x + 1 = 20 + 1 = 21$.

19. Let $x =$ the first integer.
Then $x + 1 =$ the second integer,
and $x + 2 =$ the third integer.

The product of the second and third	is	2	more than	10 times the first
↓	↓	↓	↓	↓
$(x+1)(x+2)$	$=$	2	$+$	$10x$

$$\begin{aligned} x^2 + 3x + 2 &= 2 + 10x \\ x^2 - 7x &= 0 \\ x(x-7) &= 0 \end{aligned}$$

$$x = 0 \quad \text{or} \quad x - 7 = 0$$
$$x = 7$$

If $x = 0$, then $x + 1 = 1$, and $x + 2 = 2$.
If $x = 7$, then $x + 1 = 8$, and $x + 2 = 9$.
So there are two sets of consecutive integers that satisfy the condition: 0, 1, 2 and 7, 8, 9.

Check 0, 1, 2: $1(2) \stackrel{?}{=} 2 + 10(0)$
$2 = 2 + 0$ *True*

Check 7, 8, 9: $8(9) \stackrel{?}{=} 2 + 10(7)$
$72 = 2 + 70$ *True*

21. Let $x =$ the first odd integer.
Then $x + 2 =$ the second odd integer and
$x + 4 =$ the third odd integer.

$$\begin{aligned} 3[x + (x + 2) + (x + 4)] &= x(x + 2) + 18 \\ 3(3x + 6) &= x^2 + 2x + 18 \\ 9x + 18 &= x^2 + 2x + 18 \\ 0 &= x^2 - 7x \\ 0 &= x(x - 7) \end{aligned}$$

$$x = 0 \quad \text{or} \quad x - 7 = 0$$
$$x = 7$$

We must discard 0 because it is even and the problem requires the integers to be odd. If $x = 7$, $x + 2 = 9$, and $x + 4 = 11$. The three integers are 7, 9, and 11.

23. Let $x =$ the first even integer. Then $x + 2$ and $x + 4$ are the next two even integers.

$$\begin{aligned} x^2 + (x + 2)^2 &= (x + 4)^2 \\ x^2 + x^2 + 4x + 4 &= x^2 + 8x + 16 \\ x^2 - 4x - 12 &= 0 \\ (x - 6)(x + 2) &= 0 \end{aligned}$$

$$x - 6 = 0 \quad \text{or} \quad x + 2 = 0$$
$$x = 6 \quad \text{or} \quad x = -2$$

If $x = 6$, $x + 2 = 8$, and $x + 4 = 10$.
If $x = -2$, $x + 2 = 0$, and $x + 4 = 2$.

The three integers are 6, 8, and 10 or -2, 0, and 2.

25. Let $x =$ the length of the longer leg of the right triangle.
Then $x + 1 =$ the length of the hypotenuse and
$x - 7 =$ the length of the shorter leg.

Refer to the figure in the text.
Use the Pythagorean theorem with $a = x$, $b = x - 7$, and $c = x + 1$.

$$\begin{aligned} a^2 + b^2 &= c^2 \\ x^2 + (x - 7)^2 &= (x + 1)^2 \\ x^2 + (x^2 - 14x + 49) &= x^2 + 2x + 1 \\ 2x^2 - 14x + 49 &= x^2 + 2x + 1 \\ x^2 - 16x + 48 &= 0 \\ (x - 12)(x - 4) &= 0 \end{aligned}$$

$$x - 12 = 0 \quad \text{or} \quad x - 4 = 0$$
$$x = 12 \quad \text{or} \quad x = 4$$

Discard 4 because if the length of the longer leg is 4 centimeters, by the conditions of the problem, the length of the shorter leg would be $4 - 7 = -3$ centimeters, which is impossible. The length of the longer leg is 12 centimeters.

Check: $12^2 + 5^2 = 13^2$; $169 = 169$ *True*

27. Let $x =$ Alan's distance from home. Then $x + 1 =$ the distance between Tram and Alan.

Refer to the diagram in the textbook.
Use the Pythagorean theorem.

$$\begin{aligned} a^2 + b^2 &= c^2 \\ x^2 + 5^2 &= (x + 1)^2 \\ x^2 + 25 &= x^2 + 2x + 1 \\ 24 &= 2x \\ 12 &= x \end{aligned}$$

Alan is 12 miles from home.

Check: $12^2 + 5^2 = 13^2$; $169 = 169$ *True*

29. Let $x =$ the length of the ladder. Then
$x - 4 =$ the distance from the bottom of the ladder to the building and
$x - 2 =$ the distance on the side of the building to the top of the ladder.

Substitute into the Pythagorean theorem.

$$\begin{aligned} a^2 + b^2 &= c^2 \\ (x - 2)^2 + (x - 4)^2 &= x^2 \\ x^2 - 4x + 4 + x^2 - 8x + 16 &= x^2 \\ x^2 - 12x + 20 &= 0 \\ (x - 10)(x - 2) &= 0 \end{aligned}$$

$$x - 10 = 0 \quad \text{or} \quad x - 2 = 0$$
$$x = 10 \quad \text{or} \quad x = 2$$

The solution cannot be 2 because then a negative distance results. Thus, $x = 10$ and the top of the ladder reaches $x - 2 = 10 - 2 = 8$ feet up the side of the building.

Check: $8^2 + 6^2 = 10^2$; $100 = 100$ *True*

31.
$$\begin{aligned} h &= -16t^2 + 128t \\ h &= -16(1)^2 + 128(1) \quad \textit{Let } t = 1. \\ &= -16 + 128 \\ &= 112 \end{aligned}$$

After 1 second, the height is 112 feet.

33.
$$\begin{aligned} h &= -16t^2 + 128t \\ h &= -16(4)^2 + 128(4) \quad \textit{Let } t = 4. \\ &= -16(16) + 512 \\ &= -256 + 512 \\ &= 256 \end{aligned}$$

After 4 seconds, the height is 256 feet.

35. **(a)** Let $h = 64$ in the given formula and solve for t.

$$h = -16t^2 + 32t + 48$$
$$64 = -16t^2 + 32t + 48$$
$$16t^2 - 32t + 16 = 0$$
$$16(t^2 - 2t + 1) = 0$$
$$16(t-1)^2 = 0$$
$$t - 1 = 0$$
$$t = 1$$

The height of the object will be 64 feet after 1 second.

(b) To find the time when the height is 60 feet, let $h = 60$ in the given equation and solve for t.

$$h = -16t^2 + 32t + 48$$
$$60 = -16t^2 + 32t + 48$$
$$16t^2 - 32t + 12 = 0$$
$$4(4t^2 - 8t + 3) = 0$$
$$4(2t-1)(2t-3) = 0$$

$$2t - 1 = 0 \quad \text{or} \quad 2t - 3 = 0$$
$$2t = 1 \quad \text{or} \quad 2t = 3$$
$$t = \tfrac{1}{2} \quad \text{or} \quad t = \tfrac{3}{2}$$

The height of the object is 60 feet after $\frac{1}{2}$ second (on the way up) and after $\frac{3}{2}$ or $1\frac{1}{2}$ seconds (on the way down).

(c) To find the time when the object hits the ground, let $h = 0$ and solve for t.

$$h = -16t^2 + 32t + 48$$
$$0 = -16t^2 + 32t + 48$$
$$16t^2 - 32t - 48 = 0$$
$$16(t^2 - 2t - 3) = 0$$
$$16(t+1)(t-3) = 0$$

$$t + 1 = 0 \quad \text{or} \quad t - 3 = 0$$
$$t = -1 \quad \text{or} \quad t = 3$$

We discard -1 because time cannot be negative. The object will hit the ground after 3 seconds.

(d) The negative solution, -1, does not make sense, since *t* represents time, which cannot be negative.

37. **(a)** $x = 10$ in 2000.

$$y = 0.590x^2 + 4.523x + 0.136$$
$$y = 0.590(10)^2 + 4.523(10) + 0.136$$
$$y = 104.366 \approx 104.4$$

The model indicates there were about 104.4 million cellular phone subscribers in 2000. The result using the model is less than 109 million, the actual number for 2000.

(b) $x = 2008 - 1990 = 18$
$x = 18$ corresponds to 2008.

(c) $y = 0.590(18)^2 + 4.523(18) + 0.136$
$y = 272.71 \approx 272.7$

The model indicates there were about 272.7 million cellular phone subscribers in 2008. The result using the model is more than 263 million, the actual number for 2008.

(d) $x = 2010 - 1990 = 20$
$y = 0.590(20)^2 + 4.523(20) + 0.136$
$y = 326.596 \approx 326.6$

The model gives an estimate of 326.6 million cellular phone subscribers in 2010.

39. $\dfrac{50}{72} = \dfrac{25 \cdot 2}{36 \cdot 2} = \dfrac{25}{36}$

41. $\dfrac{48}{-27} = \dfrac{16 \cdot 3}{-9 \cdot 3} = \dfrac{16}{-9}$, or $-\dfrac{16}{9}$

Chapter 5 Review Exercises

1. $7t + 14 = 7 \cdot t + 7 \cdot 2 = 7(t+2)$

2. $60z^3 + 30z = 30z \cdot 2z^2 + 30z \cdot 1$
$= 30z(2z^2 + 1)$

3. $2xy - 8y + 3x - 12$
$= (2xy - 8y) + (3x - 12)$ *Group terms.*
$= 2y(x-4) + 3(x-4)$ *Factor each group.*
$= (x-4)(2y+3)$ *Factor out x − 4.*

4. $6y^2 + 9y + 4xy + 6x$
$= (6y^2 + 9y) + (4xy + 6x)$
$= 3y(2y+3) + 2x(2y+3)$
$= (2y+3)(3y+2x)$

5. $x^2 + 5x + 6$

Find two integers whose product is 6 and whose sum is 5. The integers are 3 and 2. Thus,

$$x^2 + 5x + 6 = (x+3)(x+2).$$

6. $y^2 - 13y + 40$

Find two integers whose product is 40 and whose sum is -13.

Factors of 40	Sums of Factors
$-1, -40$	-41
$-2, -20$	-22
$-4, -10$	-14
$-5, -8$	-13 ←

The integers are -5 and -8, so

$$y^2 - 13y + 40 = (y-5)(y-8).$$

7. $q^2 + 6q - 27$

Find two integers whose product is -27 and whose sum is 6. The integers are -3 and 9, so

$$q^2 + 6q - 27 = (q - 3)(q + 9).$$

8. $r^2 - r - 56$

Find two integers whose product is -56 and whose sum is -1. The integers are 7 and -8, so

$$r^2 - r - 56 = (r + 7)(r - 8).$$

9. $r^2 - 4rs - 96s^2$

Find two expressions whose product is $-96s^2$ and whose sum is $-4s$. The expressions are $8s$ and $-12s$, so

$$r^2 - 4rs - 96s^2 = (r + 8s)(r - 12s).$$

10. $p^2 + 2pq - 120q^2$

Find two expressions whose product is $-120q^2$ and whose sum is $2q$. The expressions are $12q$ and $-10q$, so

$$p^2 + 2pq - 120q^2 = (p + 12q)(p - 10q).$$

11. $8p^3 - 24p^2 - 80p$

First, factor out the GCF, $8p$.

$$8p^3 - 24p^2 - 80p = 8p(p^2 - 3p - 10)$$

Now factor $p^2 - 3p - 10$.

$$p^2 - 3p - 10 = (p + 2)(p - 5)$$

The completely factored form is

$$8p^3 - 24p^2 - 80p = 8p(p + 2)(p - 5).$$

12. $3x^4 + 30x^3 + 48x^2$
$= 3x^2(x^2 + 10x + 16)$
$= 3x^2(x + 2)(x + 8)$

13. $p^7 - p^6q - 2p^5q^2$
$= p^5(p^2 - pq - 2q^2)$
$= p^5(p + q)(p - 2q)$

14. $3r^5 - 6r^4s - 45r^3s^2$
$= 3r^3(r^2 - 2rs - 15s^2)$
$= 3r^3(r + 3s)(r - 5s)$

15. $9x^4y - 9x^3y - 54x^2y$
$= 9x^2y(x^2 - x - 6)$
$= 9x^2y(x + 2)(x - 3)$

16. $2x^7 + 2x^6y - 12x^5y^2$
$= 2x^5(x^2 + xy - 6y^2)$
$= 2x^5(x - 2y)(x + 3y)$

17. To begin factoring $6r^2 - 5r - 6$, the possible first terms of the two binomial factors are r and $6r$, or $2r$ and $3r$, if we consider only positive integer coefficients.

18. When factoring $2z^3 + 9z^2 - 5z$, the first step is to factor out the GCF, z.

In Exercises 19–27, either the trial and error method or the grouping method can be used to factor each polynomial.

19. Factor $2k^2 - 5k + 2$ by trial and error.

$$2k^2 - 5k + 2 = (2k - 1)(k - 2)$$

20. Factor $3r^2 + 11r - 4$ by grouping. Look for two integers whose product is $3(-4) = -12$ and whose sum is 11. The integers are 12 and -1.

$$\begin{aligned} 3r^2 + 11r - 4 &= 3r^2 + 12r - r - 4 \\ &= (3r^2 + 12r) + (-r - 4) \\ &= 3r(r + 4) - 1(r + 4) \\ &= (r + 4)(3r - 1) \end{aligned}$$

21. Factor $6r^2 - 5r - 6$ by grouping. Find two integers whose product is $6(-6) = -36$ and whose sum is -5. The integers are -9 and 4.

$$\begin{aligned} 6r^2 - 5r - 6 &= 6r^2 - 9r + 4r - 6 \\ &= (6r^2 - 9r) + (4r - 6) \\ &= 3r(2r - 3) + 2(2r - 3) \\ &= (2r - 3)(3r + 2) \end{aligned}$$

22. Factor $10z^2 - 3z - 1$ by trial and error.

$$10z^2 - 3z - 1 = (5z + 1)(2z - 1)$$

23. Factor $8v^2 + 17v - 21$ by grouping. Look for two integers whose product is $8(-21) = -168$ and whose sum is 17. The integers are 24 and -7.

$$\begin{aligned} 8v^2 + 17v - 21 &= 8v^2 + 24v - 7v - 21 \\ &= (8v^2 + 24v) + (-7v - 21) \\ &= 8v(v + 3) - 7(v + 3) \\ &= (v + 3)(8v - 7) \end{aligned}$$

24. $24x^5 - 20x^4 + 4x^3$

Factor out the GCF, $4x^3$. Then complete the factoring by trial and error.

$$\begin{aligned} &24x^5 - 20x^4 + 4x^3 \\ &= 4x^3(6x^2 - 5x + 1) \\ &= 4x^3(3x - 1)(2x - 1) \end{aligned}$$

25. $-6x^2 + 3x + 30$
$= -3(2x^2 - x - 10)$
$= -3(2x - 5)(x + 2)$

26. $10r^3s + 17r^2s^2 + 6rs^3$
$= rs(10r^2 + 17rs + 6s^2)$
$= rs(5r + 6s)(2r + s)$

27. $48x^4y + 4x^3y^2 - 4x^2y^3$
$= 4x^2y(12x^2 + xy - y^2)$
$= 4x^2y(3x + y)(4x - y)$

28. The student stopped too soon. He needs to factor out the common factor $4x - 1$ to get $(4x - 1)(4x - 5)$ as the correct answer.

29. Only choice **B**, $4x^2y^2 - 25z^2$, is the difference of squares. In **A**, 32 is not a perfect square. In **C**, we have a sum, not a difference. In **D**, y^3 is not a square. The correct choice is **B**.

30. Only choice **D**, $x^2 - 20x + 100$, is a perfect square trinomial because $x^2 = x \cdot x$, $100 = 10 \cdot 10$, and $-20x = -2(x)(10)$.

In Exercises 31–34, use the rule for factoring a difference of squares.

31. $n^2 - 49 = n^2 - 7^2 = (n + 7)(n - 7)$

32. $$\begin{aligned} 25b^2 - 121 &= (5b)^2 - 11^2 \\ &= (5b + 11)(5b - 11) \end{aligned}$$

33. $$\begin{aligned} 49y^2 - 25w^2 &= (7y)^2 - (5w)^2 \\ &= (7y + 5w)(7y - 5w) \end{aligned}$$

34. $$\begin{aligned} 144p^2 - 36q^2 &= 36(4p^2 - q^2) \\ &= 36[(2p)^2 - q^2] \\ &= 36(2p + q)(2p - q) \end{aligned}$$

35. $x^2 + 100$

This polynomial is *prime* because it is the sum of squares and the two terms have no common factor.

In Exercises 36–37, use the rules for factoring a perfect square trinomial.

36. $$\begin{aligned} r^2 - 12r + 36 &= r^2 - 2(6)(r) + 6^2 \\ &= (r - 6)^2 \end{aligned}$$

37. $$\begin{aligned} 9t^2 - 42t + 49 &= (3t)^2 - 2(3t)(7) + 7^2 \\ &= (3t - 7)^2 \end{aligned}$$

In Exercises 38–39, use the rule for factoring a sum of cubes.

38. $$\begin{aligned} &m^3 + 1000 \\ &= m^3 + 10^3 \\ &= (m + 10)(m^2 - 10 \cdot m + 10^2) \\ &= (m + 10)(m^2 - 10m + 100) \end{aligned}$$

39. $$\begin{aligned} &125k^3 + 64x^3 \\ &= (5k)^3 + (4x)^3 \\ &= (5k + 4x)[(5k)^2 - 5k \cdot 4x + (4x)^2] \\ &= (5k + 4x)(25k^2 - 20kx + 16x^2) \end{aligned}$$

In Exercises 40–41, use the rule for factoring a difference of cubes.

40. $$\begin{aligned} &343x^3 - 64 \\ &= (7x)^3 - 4^3 \\ &= (7x - 4)[(7x)^2 + 7x \cdot 4 + 4^2] \\ &= (7x - 4)(49x^2 + 28x + 16) \end{aligned}$$

41. $$\begin{aligned} &1000 - 27x^6 \\ &= 10^3 - (3x^2)^3 \\ &= (10 - 3x^2)[10^2 + 10(3x^2) + (3x^2)^2] \\ &= (10 - 3x^2)(100 + 30x^2 + 9x^4) \end{aligned}$$

42. $$\begin{aligned} &x^6 - y^6 \\ &= (x^3)^2 - (y^3)^2 \quad \textit{Difference of squares} \\ &= (x^3 + y^3)(x^3 - y^3) \end{aligned}$$

Now factor as the sum and difference of cubes.

$$= (x + y)(x^2 - xy + y^2)(x - y)(x^2 + xy + y^2)$$

In Exercises 43–56, all solutions should be checked by substituting in the original equations. The checks will not be shown here.

43. $(4t + 3)(t - 1) = 0$

$$\begin{aligned} 4t + 3 = 0 \quad &\text{or} \quad t - 1 = 0 \\ 4t = -3 \quad &\text{or} \quad t = 1 \\ t = -\tfrac{3}{4} \quad & \end{aligned}$$

The solution set is $\left\{-\frac{3}{4}, 1\right\}$.

44. $(x + 7)(x - 4)(x + 3) = 0$

$$\begin{aligned} x + 7 = 0 \quad \text{or} \quad x - 4 = 0 \quad \text{or} \quad x + 3 = 0 \\ x = -7 \quad \text{or} \quad x = 4 \quad \text{or} \quad x = -3 \end{aligned}$$

The solution set is $\{-7, -3, 4\}$.

45. $x(2x - 5) = 0$

$$\begin{aligned} x = 0 \quad \text{or} \quad 2x - 5 &= 0 \\ 2x &= 5 \\ x &= \tfrac{5}{2} \end{aligned}$$

The solution set is $\left\{0, \frac{5}{2}\right\}$.

46. $$\begin{aligned} z^2 + 4z + 3 &= 0 \\ (z + 3)(z + 1) &= 0 \end{aligned}$$

$$\begin{aligned} z + 3 = 0 \quad &\text{or} \quad z + 1 = 0 \\ z = -3 \quad &\text{or} \quad z = -1 \end{aligned}$$

The solution set is $\{-3, -1\}$.

47. $$\begin{aligned} m^2 - 5m + 4 &= 0 \\ (m - 1)(m - 4) &= 0 \end{aligned}$$

$$\begin{aligned} m - 1 = 0 \quad &\text{or} \quad m - 4 = 0 \\ m = 1 \quad &\text{or} \quad m = 4 \end{aligned}$$

The solution set is $\{1, 4\}$.

48. $$\begin{aligned} x^2 &= -15 + 8x \\ x^2 - 8x + 15 &= 0 \\ (x - 3)(x - 5) &= 0 \end{aligned}$$

$$\begin{aligned} x - 3 = 0 \quad &\text{or} \quad x - 5 = 0 \\ x = 3 \quad &\text{or} \quad x = 5 \end{aligned}$$

The solution set is $\{3, 5\}$.

49.
$$3z^2 - 11z - 20 = 0$$
$$(3z+4)(z-5) = 0$$
$$3z + 4 = 0 \quad \text{or} \quad z - 5 = 0$$
$$3z = -4 \quad \text{or} \quad z = 5$$
$$z = -\tfrac{4}{3}$$

The solution set is $\{-\frac{4}{3}, 5\}$.

50.
$$81t^2 - 64 = 0$$
$$(9t+8)(9t-8) = 0$$
$$9t + 8 = 0 \quad \text{or} \quad 9t - 8 = 0$$
$$9t = -8 \quad \text{or} \quad 9t = 8$$
$$t = -\tfrac{8}{9} \quad \text{or} \quad t = \tfrac{8}{9}$$

The solution set is $\{-\frac{8}{9}, \frac{8}{9}\}$.

51.
$$y^2 = 8y$$
$$y^2 - 8y = 0$$
$$y(y-8) = 0$$
$$y = 0 \quad \text{or} \quad y - 8 = 0$$
$$y = 8$$

The solution set is $\{0, 8\}$.

52.
$$n(n-5) = 6$$
$$n^2 - 5n = 6$$
$$n^2 - 5n - 6 = 0$$
$$(n+1)(n-6) = 0$$
$$n + 1 = 0 \quad \text{or} \quad n - 6 = 0$$
$$n = -1 \quad \text{or} \quad n = 6$$

The solution set is $\{-1, 6\}$.

53.
$$t^2 - 14t + 49 = 0$$
$$(t-7)^2 = 0$$
$$t - 7 = 0$$
$$t = 7$$

The solution set is $\{7\}$.

54.
$$t^2 = 12(t-3)$$
$$t^2 = 12t - 36$$
$$t^2 - 12t + 36 = 0$$
$$(t-6)^2 = 0$$
$$t - 6 = 0$$
$$t = 6$$

The solution set is $\{6\}$.

55.
$$(5z+2)(z^2+3z+2) = 0$$
$$(5z+2)(z+2)(z+1) = 0$$
$$5z + 2 = 0 \quad \text{or} \quad z + 2 = 0 \quad \text{or} \quad z + 1 = 0$$
$$5z = -2$$
$$z = -\tfrac{2}{5} \quad \text{or} \quad z = -2 \quad \text{or} \quad z = -1$$

The solution set is $\{-2, -1, -\frac{2}{5}\}$.

56.
$$x^2 = 9$$
$$x^2 - 9 = 0$$
$$(x+3)(x-3) = 0$$
$$x + 3 = 0 \quad \text{or} \quad x - 3 = 0$$
$$x = -3 \quad \text{or} \quad x = 3$$

The solution set is $\{-3, 3\}$.

57. Let $x =$ the width of the rug.
Then $x + 6 =$ the length of the rug.

$$A = LW$$
$$40 = (x+6)x$$
$$40 = x^2 + 6x$$
$$0 = x^2 + 6x - 40$$
$$0 = (x+10)(x-4)$$
$$x + 10 = 0 \quad \text{or} \quad x - 4 = 0$$
$$x = -10 \quad \text{or} \quad x = 4$$

Reject -10 since the width cannot be negative. The width of the rug is 4 feet and the length is $4 + 6$ or 10 feet.

58. From the figure, we have $L = 20$, $W = x$, and $H = x + 4$.

$$S = 2WH + 2WL + 2LH$$
$$650 = 2x(x+4) + 2x(20) + 2(20)(x+4)$$
$$650 = 2x^2 + 8x + 40x + 40(x+4)$$
$$650 = 2x^2 + 48x + 40x + 160$$
$$0 = 2x^2 + 88x - 490$$
$$0 = 2(x^2 + 44x - 245)$$
$$0 = 2(x+49)(x-5)$$
$$x + 49 = 0 \quad \text{or} \quad x - 5 = 0$$
$$x = -49 \quad \text{or} \quad x = 5$$

Reject -49 because the width cannot be negative. The width of the chest is 5 feet.

59. Let $x =$ the first integer.
Then $x + 1 =$ the next integer.

The product of the integers is 29 more than their sum, so

$$x(x+1) = 29 + [x + (x+1)].$$

Solve this equation.

$$x^2 + x = 29 + 2x + 1$$
$$x^2 - x - 30 = 0$$
$$(x-6)(x+5) = 0$$
$$x - 6 = 0 \quad \text{or} \quad x + 5 = 0$$
$$x = 6 \quad \text{or} \quad x = -5$$

If $x = 6$, $x + 1 = 6 + 1 = 7$.
If $x = -5$, $x + 1 = -5 + 1 = -4$.

The consecutive integers are 6 and 7 or -5 and -4.

60. Let $x =$ the distance traveled west.
Then $x - 14 =$ the distance traveled south,
and $(x - 14) + 16 = x + 2 =$ the distance between the cars.

These three distances form a right triangle with x and $x - 14$ representing the lengths of the legs and $x + 2$ representing the length of the hypotenuse. Use the Pythagorean theorem.

$$\begin{aligned} a^2 + b^2 &= c^2 \\ x^2 + (x-14)^2 &= (x+2)^2 \\ x^2 + x^2 - 28x + 196 &= x^2 + 4x + 4 \\ x^2 - 32x + 192 &= 0 \\ (x-8)(x-24) &= 0 \end{aligned}$$

$$x - 8 = 0 \quad \text{or} \quad x - 24 = 0$$
$$x = 8 \quad \text{or} \quad x = 24$$

If $x = 8$, then $x - 14 = -6$, which is not possible because a distance cannot be negative.

If $x = 24$, then $x - 14 = 10$ and $x + 2 = 26$. The cars were 26 miles apart.

61. **(a)** $d = 16t^2$

$$t = 4\colon\ d = 16(4)^2 = 256$$

In 4 seconds, the object would fall 256 feet.

(b) $t = 8\colon\ d = 16(8)^2 = 1024$

In 8 seconds, the object would fall 1024 feet.

62. **(a)** In 2005, $x = 5$

$$\begin{aligned} y &= -2.84x^2 + 61.1x + 366 \\ y &= -2.84(5)^2 + 61.1(5) + 366 \\ y &= 600.5 \approx 601 \end{aligned}$$

If $x = 5$, then $y = 601$, so the prediction is 601,000 vehicles. The result is slightly higher than the actual number for 2005.

(b) In 2007, $x = 7$

$$\begin{aligned} y &= -2.84x^2 + 61.1x + 366 \\ y &= -2.84(7)^2 + 61.1(7) + 366 \\ y &= 654.54 \approx 655 \end{aligned}$$

If $x = 7$, then $y = 655$, so the prediction is 655,000 vehicles.

(c) The estimate may be unreliable because the conditions that prevailed in the years 2001–2006 may have changed, causing either a greater increase or a decrease predicted by the model for the number of alternative-fueled vehicles.

63. **[5.1]** **D** is not factored completely.
$3(7t + 4) + x(7t + 4) = (7t + 4)(3 + x)$

64. **[5.1]** The factor $2x + 8$ has a factor of 2. The completely factored form is $2(x + 4)(3x - 4)$.

65. **[5.3]** $3k^2 + 11k + 10$

Two integers with product $3(10) = 30$ and sum 11 are 5 and 6.

$$\begin{aligned} &3k^2 + 11k + 10 \\ &= 3k^2 + 5k + 6k + 10 \\ &= (3k^2 + 5k) + (6k + 10) \\ &= k(3k + 5) + 2(3k + 5) \\ &= (3k + 5)(k + 2) \end{aligned}$$

66. **[5.2]** $z^2 - 11zx + 10x^2 = (z - x)(z - 10x)$

67. **[5.4]** $y^4 - 625$

$$\begin{aligned} &= (y^2)^2 - 25^2 \\ &= (y^2 + 25)(y^2 - 25) \quad \textit{Difference of squares} \\ &= (y^2 + 25)(y + 5)(y - 5) \quad \textit{Difference of squares} \end{aligned}$$

68. **[5.1]** $15m^2 + 20m - 12mp - 16p$

$$\begin{aligned} &= 5m(3m + 4) - 4p(3m + 4) \quad \textit{Factor by grouping.} \\ &= (3m + 4)(5m - 4p) \end{aligned}$$

69. **[5.1]** $24ab^3c^2 - 56a^2bc^3 + 72a^2b^2c$

$$= 8abc(3b^2c - 7ac^2 + 9ab)$$

70. **[5.3]** $6m^3 - 21m^2 - 45m$

$$\begin{aligned} &= 3m(2m^2 - 7m - 15) \\ &= 3m[(2m^2 - 10m) + (3m - 15)] \quad \textit{Factor by grouping.} \\ &= 3m[2m(m - 5) + 3(m - 5)] \\ &= 3m(m - 5)(2m + 3) \end{aligned}$$

71. **[5.1]** $12x^2yz^3 + 12xy^2z - 30x^3y^2z^4$

$$= 6xyz(2xz^2 + 2y - 5x^2yz^3)$$

72. **[5.3]** $25a^2 + 15ab + 9b^2$ is a *prime* polynomial.

73. **[5.1]** $12r^2 + 18rq - 10r - 15q$

$$\begin{aligned} &= 6r(2r + 3q) - 5(2r + 3q) \quad \textit{Factor by grouping.} \\ &= (2r + 3q)(6r - 5) \end{aligned}$$

74. **[5.2]** $2a^5 - 8a^4 - 24a^3$

$$\begin{aligned} &= 2a^3(a^2 - 4a - 12) \\ &= 2a^3(a - 6)(a + 2) \end{aligned}$$

75. **[5.4]** $49t^2 + 56t + 16$

$$\begin{aligned} &= (7t)^2 + 2(7t)(4) + 4^2 \\ &= (7t + 4)^2 \end{aligned}$$

76. **[5.4]** $1000a^3 + 27$

$$\begin{aligned} &= (10a)^3 + 3^3 \\ &= (10a + 3)[(10a)^2 - 10a \cdot 3 + 3^2] \\ &= (10a + 3)(100a^2 - 30a + 9) \end{aligned}$$

77. **[5.5]** $t(t-7)=0$

$$t=0 \quad \text{or} \quad t-7=0$$
$$t=7$$

The solution set is $\{0,7\}$.

78. **[5.5]**

$$x^2+3x=10$$
$$x^2+3x-10=0$$
$$(x+5)(x-2)=0$$

$$x+5=0 \quad \text{or} \quad x-2=0$$
$$x=-5 \quad \text{or} \quad x=2$$

The solution set is $\{-5,2\}$.

79. **[5.5]**

$$25x^2+20x+4=0$$
$$(5x)^2+2(5x)(2)+2^2=0$$
$$(5x+2)^2=0$$
$$5x+2=0$$
$$5x=-2$$
$$x=-\tfrac{2}{5}$$

The solution set is $\{-\frac{2}{5}\}$.

80. **[5.6]** Let $x=$ the first integer. Then $x+1$ and $x+2$ are the next two integers.

The product of the first two of three consecutive integers is equal to 23 plus the third.

$$x(x+1)=23+(x+2)$$
$$x^2+x=23+x+2$$
$$x^2-25=0$$
$$(x+5)(x-5)=0$$

$$x+5=0 \quad \text{or} \quad x-5=0$$
$$x=-5 \quad \text{or} \quad x=5$$

If $x=-5$, then $x+1=-4$ and $x+2=-3$.
If $x=5$, then $x+1=6$ and $x+2=7$.
The integers are -5, -4, and -3, or 5, 6, and 7.

81. **[5.6]** Let $x=$ the width of the base.
Then $x+2=$ the length of the base.

The area of the base, B, is given by LW, so

$$B=x(x+2).$$

Use the formula for the volume of a pyramid,

$$V=\tfrac{1}{3}\cdot B\cdot h.$$
$$48=\tfrac{1}{3}x(x+2)(6)$$
$$48=2x(x+2)$$
$$24=x^2+2x$$
$$x^2+2x-24=0$$
$$(x+6)(x-4)=0$$

$$x+6=0 \quad \text{or} \quad x-4=0$$
$$x=-6 \quad \text{or} \quad x=4$$

Reject -6. The width of the base is 4 meters and the length is $4+2$ or 6 meters.

82. **[5.6]** Let $x=$ the length of the shorter leg.
Then $2x+6=$ the length of the longer leg and $(2x+6)+3=2x+9=$ the length of the hypotenuse.

Use the Pythagorean theorem, $a^2+b^2=c^2$.

$$x^2+(2x+6)^2=(2x+9)^2$$
$$x^2+4x^2+24x+36=4x^2+36x+81$$
$$x^2-12x-45=0$$
$$(x-15)(x+3)=0$$

$$x-15=0 \quad \text{or} \quad x+3=0$$
$$x=15 \quad \text{or} \quad x=-3$$

Reject -3 because a length cannot be negative.
The sides of the lot are 15 meters, $2(15)+6=36$ meters, and $36+3=39$ meters.

83. **[5.6]** Let $b=$ the base of the sail.
Then $b+4=$ the height of the sail.

Use the formula for the area of a triangle.

$$\mathcal{A}=\tfrac{1}{2}bh$$
$$30=\tfrac{1}{2}(b)(b+4) \qquad \textit{Let } \mathcal{A}=30.$$
$$60=b^2+4b$$
$$0=b^2+4b-60$$
$$0=(b+10)(b-6)$$

$$b+10=0 \quad \text{or} \quad b-6=0$$
$$b=-10 \quad \text{or} \quad b=6$$

Discard -10 since the base of a triangle cannot be negative. The base of the triangular sail is 6 meters.

84. **[5.6]** Let $x=$ the width of the house.
Then $x+7=$ the length of the house.

Use $\mathcal{A}=LW$ with 170 for $\mathcal{A}$, $x+7$ for L, and x for W.

$$170=(x+7)(x)$$
$$170=x^2+7x$$
$$0=x^2+7x-170$$
$$0=(x+17)(x-10)$$

$$x+17=0 \quad \text{or} \quad x-10=0$$
$$x=-17 \quad \text{or} \quad x=10$$

Discard -17 because the width cannot be negative. If $x=10$, $x+7=10+7=17$.

The width is 10 meters and the length is 17 meters.

Chapter 5 Test

1. $$2x^2 - 2x - 24 = 2(x^2 - x - 12) = 2(x+3)(x-4)$$

The correct completely factored form is choice **D**. Note that the factored forms **A**, $(2x+6)(x-4)$, and **B**, $(x+3)(2x-8)$, also can be multiplied to give a product of $2x^2 - 2x - 24$, but neither of these is completely factored because $2x+6$ and $2x-8$ both contain a common factor of 2.

2. $$12x^2 - 30x = 6x(2x-5)$$

3. $$2m^3n^2 + 3m^3n - 5m^2n^2 = m^2n(2mn + 3m - 5n)$$

4. $$\begin{aligned} &2ax - 2bx + ay - by \\ &= 2x(a-b) + y(a-b) \\ &= (a-b)(2x+y) \end{aligned}$$

5. $x^2 - 5x - 24$

Find two integers whose product is -24 and whose sum is -5. The integers are 3 and -8.

$$x^2 - 5x - 24 = (x+3)(x-8)$$

6. Factor $2x^2 + x - 3$ by trial and error.

$$2x^2 + x - 3 = (2x+3)(x-1)$$

7. Factor $10z^2 - 17z + 3$ by trial and error.

$$10z^2 - 17z + 3 = (2z-3)(5z-1)$$

8. $t^2 + 2t + 3$

We cannot find two integers whose product is 3 and whose sum is 2. This polynomial is prime.

9. $x^2 + 36$

This polynomial is *prime* because the sum of squares cannot be factored and the two terms have no common factor.

10. $$\begin{aligned} &12 - 6a + 2b - ab \\ &= (12 - 6a) + (2b - ab) \\ &= 6(2-a) + b(2-a) \\ &= (2-a)(6+b) \end{aligned}$$

11. $$\begin{aligned} &9y^2 - 64 \\ &= (3y)^2 - 8^2 \\ &= (3y+8)(3y-8) \end{aligned}$$

12. $$\begin{aligned} &4x^2 - 28xy + 49y^2 \\ &= (2x)^2 - 2(2x)(7y) + (7y)^2 \\ &= (2x-7y)^2 \end{aligned}$$

13. $$\begin{aligned} &-2x^2 - 4x - 2 \\ &= -2(x^2 + 2x + 1) \\ &= -2(x^2 + 2\cdot x\cdot 1 + 1^2) \\ &= -2(x+1)^2 \end{aligned}$$

14. $$\begin{aligned} &6t^4 + 3t^3 - 108t^2 \\ &= 3t^2(2t^2 + t - 36) \\ &= 3t^2(2t+9)(t-4) \end{aligned}$$

15. $$\begin{aligned} r^3 - 125 &= r^3 - 5^3 \\ &= (r-5)(r^2 + 5\cdot r + 5^2) \\ &= (r-5)(r^2 + 5r + 25) \end{aligned}$$

16. $$\begin{aligned} 8k^3 + 64 &= 8(k^3 + 8) \\ &= 8(k^3 + 2^3) \\ &= 8(k+2)(k^2 - 2\cdot k + 2^2) \\ &= 8(k+2)(k^2 - 2k + 4) \end{aligned}$$

17. $$\begin{aligned} x^4 - 81 &= (x^2)^2 - 9^2 \\ &= (x^2+9)(x^2-9) \\ &= (x^2+9)(x+3)(x-3) \end{aligned}$$

18. $$\begin{aligned} 81x^4 - 16y^4 &= (9x^2)^2 - (4y^2)^2 \\ &= (9x^2 + 4y^2)(9x^2 - 4y^2) \\ &= (9x^2 + 4y^2)[(3x)^2 - (2y)^2] \\ &= (9x^2 + 4y^2)[(3x+2y)(3x-2y)] \\ &= (3x+2y)(3x-2y)(9x^2+4y^2) \end{aligned}$$

19. $$\begin{aligned} &9x^6y^4 + 12x^3y^2 + 4 \\ &= (3x^3y^2)^2 + 2(3x^3y^2)(2) + (2)^2 \\ &= (3x^3y^2 + 2)^2 \end{aligned}$$

20. $$\begin{aligned} 2r^2 - 13r + 6 &= 0 \\ (2r-1)(r-6) &= 0 \end{aligned}$$

$$2r - 1 = 0 \quad \text{or} \quad r - 6 = 0$$
$$2r = 1 \quad \text{or} \quad r = 6$$
$$r = \tfrac{1}{2}$$

The solution set is $\left\{\frac{1}{2}, 6\right\}$.

21. $$\begin{aligned} 25x^2 - 4 &= 0 \\ (5x+2)(5x-2) &= 0 \end{aligned}$$

$$5x + 2 = 0 \quad \text{or} \quad 5x - 2 = 0$$
$$5x = -2 \quad \text{or} \quad 5x = 2$$
$$x = -\tfrac{2}{5} \quad \text{or} \quad x = \tfrac{2}{5}$$

The solution set is $\left\{-\frac{2}{5}, \frac{2}{5}\right\}$.

22. $$\begin{aligned} t^2 &= 9t \\ t^2 - 9t &= 0 \\ t(t-9) &= 0 \end{aligned}$$

$$t = 0 \quad \text{or} \quad t - 9 = 0$$
$$t = 0 \quad \text{or} \quad t = 9$$

The solution set is $\{0, 9\}$.

23. $$\begin{aligned} x(x-20) &= -100 \\ x^2 - 20x &= -100 \\ x^2 - 20x + 100 &= 0 \\ (x-10)^2 &= 0 \\ x - 10 &= 0 \\ x &= 10 \end{aligned}$$

The solution set is $\{10\}$.

24. $(s+8)(6s^2+13s-5)=0$

If $s+8=0$, then $s=-8$.

$$\begin{aligned} 6s^2+13s-5&=0\\ (2s+5)(3s-1)&=0 \end{aligned}$$

$$\begin{aligned} 2s+5&=0 &\text{or}\quad 3s-1&=0\\ 2s&=-5 &\text{or}\quad 3s&=1\\ s&=-\tfrac{5}{2} & s&=\tfrac{1}{3} \end{aligned}$$

The solution set is $\{-8, -\frac{5}{2}, \frac{1}{3}\}$.

25. Let $x=$ the width of the flower bed.
Then $2x-3=$ the length of the flower bed.

Use the formula $A=LW$.

$$\begin{aligned} x(2x-3)&=54\\ 2x^2-3x&=54\\ 2x^2-3x-54&=0\\ (2x+9)(x-6)&=0 \end{aligned}$$

$$\begin{aligned} 2x+9&=0 &\text{or}\quad x-6&=0\\ 2x&=-9 &\text{or}\quad x&=6\\ x&=-\tfrac{9}{2} \end{aligned}$$

Reject $-\frac{9}{2}$. If $x=6$, $2x-3=2(6)-3=9$.
The dimensions of the flower bed are 6 feet by 9 feet.

26. Let $x=$ the first integer.
Then $x+1=$ the second integer.

The square of the sum of the two integers is 11 more than the first integer.

$$\begin{aligned} [x+(x+1)]^2&=x+11\\ (2x+1)^2&=x+11\\ 4x^2+4x+1&=x+11\\ 4x^2+3x-10&=0\\ (4x-5)(x+2)&=0 \end{aligned}$$

$$\begin{aligned} 4x-5&=0 &\text{or}\quad x+2&=0\\ 4x&=5 &\text{or}\quad x&=-2\\ x&=\tfrac{5}{4} \end{aligned}$$

Reject $\frac{5}{4}$ because it is not an integer. If $x=-2$, $x+1=-1$. The integers are -2 and -1.

27. Let $x=$ the length of the stud.
Then $3x-7=$ the length of the brace.

The figure shows that a right triangle is formed with the brace as the hypotenuse. Use the Pythagorean theorem, $a^2+b^2=c^2$.

$$\begin{aligned} x^2+15^2&=(3x-7)^2\\ x^2+225&=9x^2-42x+49\\ 0&=8x^2-42x-176\\ 0&=2(4x^2-21x-88)\\ 0&=2(4x+11)(x-8) \end{aligned}$$

$$\begin{aligned} 4x+11&=0 &\text{or}\quad x-8&=0\\ 4x&=-11 &\text{or}\quad x&=8\\ x&=-\tfrac{11}{4} \end{aligned}$$

Reject $-\frac{11}{4}$. If $x=8$, $3x-7=24-7=17$, so the brace should be 17 feet long.

28. For 2006, $x=2006-2000=6$.

$$\begin{aligned} y&=29.92x^2+305.8x+5581\\ y&=29.92(6)^2+305.8(6)+5581\\ &=8492.92\approx 8493 \end{aligned}$$

The model estimates that the public debt of the United States was \$8493 billion in 2006.

Cumulative Review Exercises (Chapters 1–5)

1.
$$\begin{aligned} 3x+2(x-4)&=4(x-2)\\ 3x+2x-8&=4x-8\\ 5x-8&=4x-8\\ x-8&=-8\\ x&=0 \end{aligned}$$

The solution set is $\{0\}$.

2. $0.3x+0.9x=0.06$

Multiply both sides by 100 to clear decimals.

$$\begin{aligned} 100(0.3x+0.9x)&=100(0.06)\\ 30x+90x&=6\\ 120x&=6\\ x&=\tfrac{6}{120}=\tfrac{1}{20}=0.05 \end{aligned}$$

The solution set is $\{0.05\}$.

3. $\frac{2}{3}m-\frac{1}{2}(m-4)=3$

To clear fractions, multiply both sides by the least common denominator, which is 6.

$$\begin{aligned} 6[\tfrac{2}{3}m-\tfrac{1}{2}(m-4)]&=6(3)\\ 4m-3(m-4)&=18\\ 4m-3m+12&=18\\ m+12&=18\\ m&=6 \end{aligned}$$

The solution set is $\{6\}$.

4.
$$\begin{aligned} A&=P+Prt\\ A&=P(1+rt) && \textit{Factor out P.}\\ \frac{A}{1+rt}&=\frac{P(1+rt)}{1+rt} && \textit{Divide by 1 + rt.}\\ \frac{A}{1+rt}&=P \quad\text{or}\quad P=\frac{A}{1+rt} \end{aligned}$$

5. The angles are supplementary, so the sum of the angles is 180°.

$$\begin{aligned}(2x+16)+(x+23)&=180\\3x+39&=180\\3x&=141\\x&=47\end{aligned}$$

Since $x = 47$, $2x + 16 = 2(47) + 16 = 110$ and $x + 23 = 47 + 23 = 70$.
The angles are 110° and 70°.

6. Let x = number of bronze medals.
Then $x + 5$ = number of gold medals,
and $(x + 5) + 1 = x + 6$ = the number of silver medals.

The total number of medals was 29.

$$\begin{aligned}x+(x+5)+(x+6)&=29\\3x+11&=29\\3x&=18\\x&=6\end{aligned}$$

Since $x = 6$, $x + 5 = 11$, and $x + 6 = 12$.
Germany won 11 gold medals, 12 silver medals, and 6 bronze medals.

7. 46% of 500 is what number?

$$\begin{aligned}\frac{\text{part}}{\text{whole}}&=\frac{p}{100}\\\frac{a}{500}&=\frac{46}{100}\\100a&=46(500)\\a&=230\end{aligned}$$

41% of 500 is what number?

$$\begin{aligned}\frac{\text{part}}{\text{whole}}&=\frac{p}{100}\\\frac{a}{500}&=\frac{41}{100}\\100a&=500(41)\\a&=205\end{aligned}$$

What percent of 500 is 190?

$$\begin{aligned}\frac{\text{part}}{\text{whole}}&=\frac{p}{100}\\\frac{190}{500}&=\frac{p}{100}\\500p&=190(100)\\p&=38\end{aligned}$$

What percent of 500 is 60?

$$\begin{aligned}\frac{\text{part}}{\text{whole}}&=\frac{p}{100}\\\frac{60}{500}&=\frac{p}{100}\\500p&=60(100)\\p&=12\end{aligned}$$

Item	Percent	Number
Personal computer	46%	230
Cell phone	41%	205
High-speed internet	38%	190
MP3 player	12%	60

8. The point with coordinates (a, b) is in

(a) quadrant II if a is *negative* and b is *positive*.

(b) quadrant III if a is *negative* and b is *negative*.

9. **(a)** The equation $y = 12x + 3$ is in slope-intercept form, so the y-intercept is $(0, 3)$.

Let $y = 0$ to find the x-intercept.

$$\begin{aligned}0&=12x+3\\-3&=12x\\-\tfrac{1}{4}&=x\end{aligned}$$

The x-intercept is $(-\frac{1}{4}, 0)$.

(b) The equation $y = 12x + 3$ is in slope-intercept form, so the slope is the coefficient of x, that is, 12.

(c)

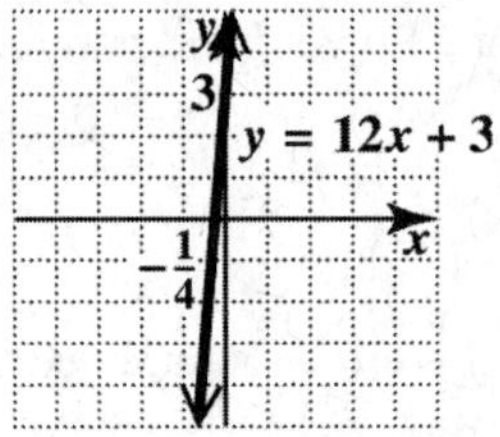

10. **(a)** $(2001, 161), (2007, 259)$

$$\begin{aligned}m&=\frac{y_2-y_1}{x_2-x_1}\\&=\frac{259-161}{2007-2001}\\&=\frac{98}{6}\approx 16\text{ (to the nearest whole number)}\end{aligned}$$

A slope of (approximately) 16 means that the retail sales of prescription drugs increased by about \$16 billion per year.

(b) The graph of the line will fall about 16 units from 2007 to 2006 and also from 2006 to 2005, so the y-value is $259 - 16 - 16 = 227$. The ordered pair is $(2005, 227)$, or about $(2005, 230)$.

11. $(\frac{3}{4})^{-2} = (\frac{4}{3})^2 = \frac{16}{9}$

12. $$\left(\frac{4^{-3}\cdot 4^4}{4^5}\right)^{-1}=\left(\frac{4^5}{4^{-3}\cdot 4^4}\right)^{1}=\frac{4^5}{4^1}=4^4=256$$

13. $\dfrac{(p^2)^3p^{-4}}{(p^{-3})^{-1}p} = \dfrac{p^{2\cdot 3}p^{-4}}{p^{(-3)(-1)}p}$

$= \dfrac{p^6p^{-4}}{p^3p^1}$

$= \dfrac{p^{6-4}}{p^{3+1}}$

$= \dfrac{p^2}{p^4} = \dfrac{1}{p^2}$

14. $\dfrac{(m^{-2})^3m}{m^5m^{-4}} = \dfrac{m^{-2(3)}m^1}{m^{5+(-4)}}$

$= \dfrac{m^{-6+1}}{m^1}$

$= \dfrac{m^{-5}}{m^1} = \dfrac{1}{m^6}$

15. $(2k^2 + 4k) - (5k^2 - 2) - (k^2 + 8k - 6)$
$= (2k^2 + 4k) + (-5k^2 + 2)$
$+ (-k^2 - 8k + 6)$
$= 2k^2 + 4k - 5k^2 + 2 - k^2 - 8k + 6$
$= -4k^2 - 4k + 8$

16. $(9x + 6)(5x - 3)$

F **O** **I** **L**

$= 9x(5x) + 9x(-3) + 6(5x) + 6(-3)$
$= 45x^2 - 27x + 30x - 18$
$= 45x^2 + 3x - 18$

17. $(3p + 2)^2 = (3p)^2 + 2\cdot 3p\cdot 2 + 2^2$
$= 9p^2 + 12p + 4$

18. $\dfrac{8x^4 + 12x^3 - 6x^2 + 20x}{2x}$

$= \dfrac{8x^4}{2x} + \dfrac{12x^3}{2x} - \dfrac{6x^2}{2x} + \dfrac{20x}{2x}$

$= 4x^3 + 6x^2 - 3x + 10$

19. $55{,}000 = 5.5 \times 10^4$

Move the decimal point left 4 places so it is to the right of the first nonzero digit. 55,000 is *greater* than 5.5, so the power is *positive.*

$2{,}000{,}000 = 2.0 \times 10^6$

Move the decimal point left 6 places so it is to the right of the first nonzero digit. 2,000,000 is *greater* than 2, so the power is *positive.*

20. Factor $2a^2 + 7a - 4$ by trial and error.

$2a^2 + 7a - 4 = (a + 4)(2a - 1)$

21. $10m^2 + 19m + 6$

To factor by grouping, find two integers whose product is $10(6) = 60$ and whose sum is 19. The integers are 15 and 4.

$10m^2 + 19m + 6 = 10m^2 + 15m + 4m + 6$
$= 5m(2m + 3) + 2(2m + 3)$
$= (2m + 3)(5m + 2)$

22. Factor $8t^2 + 10tv + 3v^2$ by trial and error.

$8t^2 + 10tv + 3v^2 = (4t + 3v)(2t + v)$

23. $4p^2 - 12p + 9 = (2p - 3)(2p - 3)$
$= (2p - 3)^2$

24. $25r^2 - 81t^2 = (5r)^2 - (9t)^2$
$= (5r + 9t)(5r - 9t)$

25. $2pq + 6p^3q + 8p^2q$
$= 2pq(1 + 3p^2 + 4p)$
$= 2pq(3p^2 + 4p + 1)$
$= 2pq(3p + 1)(p + 1)$

26. $6m^2 + m - 2 = 0$
$(3m + 2)(2m - 1) = 0$

$3m + 2 = 0$ or $2m - 1 = 0$
$3m = -2$ or $2m = 1$
$m = -\frac{2}{3}$ or $m = \frac{1}{2}$

The solution set is $\{-\frac{2}{3}, \frac{1}{2}\}$.

27. $8x^2 = 64x$
$8x^2 - 64x = 0$
$8x(x - 8) = 0$

$8x = 0$ or $x - 8 = 0$
$x = 0$ or $x = 8$

The solution set is $\{0, 8\}$.

28. Let $x =$ the length of the shorter leg.
Then $x + 7 =$ the length of the longer leg, and
$2x + 3 =$ the length of the hypotenuse.

Use the Pythagorean theorem.

$x^2 + (x + 7)^2 = (2x + 3)^2$
$x^2 + (x^2 + 14x + 49) = 4x^2 + 12x + 9$
$2x^2 + 14x + 49 = 4x^2 + 12x + 9$
$0 = 2x^2 - 2x - 40$
$0 = 2(x^2 - x - 20)$
$0 = (x - 5)(x + 4)$

$x - 5 = 0$ or $x + 4 = 0$
$x = 5$ or $x = -4$

Reject -4 because the length of a leg cannot be negative. Since $x = 5$, $x + 7 = 12$, and $2x + 3 = 2(5) + 3 = 13$. The length of the sides are 5 meters, 12 meters, and 13 meters.

CHAPTER 6 RATIONAL EXPRESSIONS AND APPLICATIONS

6.1 The Fundamental Property of Rational Expressions

6.1 Now Try Exercises

N1. $\frac{2x-1}{x+4} = \frac{2(-3)-1}{-3+4}$ *Let x = −3.*

$= \frac{-7}{1} = -7$

N2. (a) $\frac{k-4}{2k-1}$

Solve $2k - 1 = 0$.
Since $k = \frac{1}{2}$ will make the denominator zero, the expression is undefined for $k = \frac{1}{2}$. We write the answer as $k \neq \frac{1}{2}$.

(b) $\frac{2x}{x^2+5x-14}$

Solve $x^2 + 5x - 14 = 0$.
$(x+7)(x-2) = 0$, so the denominator is zero when either $x + 7 = 0$ or $x - 2 = 0$, that is, when $x = -7$ or $x = 2$. The expression is undefined for $x = -7$ and $x = 2$. We write the answer as $x \neq -7, x \neq 2$.

(c) $\frac{y+10}{y^2+10}$

This denominator will not equal 0 for any value of y, because y^2 is always greater than or equal to 0, and adding 10 makes the sum greater than 0. Thus, there are no values for which this rational expression is undefined.

N3. $\frac{21y^5}{7y^2} = \frac{3 \cdot 7 \cdot y \cdot y \cdot y \cdot y \cdot y}{7 \cdot y \cdot y}$ *Factor.*

$= \frac{3 \cdot y \cdot y \cdot y(7 \cdot y \cdot y)}{1(7 \cdot y \cdot y)}$ *Group.*

$= \frac{3 \cdot y \cdot y \cdot y}{1}$ *Fundamental property*

$= 3y^3$

N4. (a) $\frac{3x+15}{5x+25} = \frac{3(x+5)}{5(x+5)}$ *Factor.*

$= \frac{3}{5}$ *Fundamental property*

(b) $\frac{k^2-36}{k^2+8k+12}$

$= \frac{(k+6)(k-6)}{(k+6)(k+2)}$ *Factor.*

$= \frac{k-6}{k+2}$ *Fundamental property*

N5. $\frac{10-a^2}{a^2-10} = \frac{1(10-a^2)}{-1(10-a^2)} = \frac{1}{-1} = -1$

N6. (a) $\frac{p-4}{4-p}$

Since $p - 4$ and $4 - p$ are opposites, this expression equals -1.

(b) $\frac{4m^2-n^2}{2n-4m} = \frac{(2m)^2-n^2}{2(n-2m)}$

$= \frac{(2m+n)(2m-n)}{2(-1)(2m-n)}$

$= \frac{2m+n}{2(-1)}$

$= \frac{2m+n}{-2}$, or $-\frac{2m+n}{2}$

(c) $\frac{x+y}{x-y}$

$x - y = -1(-x+y) \neq -1(x+y)$

The expressions $x + y$ and $x - y$ are not opposites of each other. They do not have any common factors (other than 1), so the rational expression is already in lowest terms.

N7. To write four equivalent expressions for $-\frac{4k-9}{k+3}$, we will follow the outline in Example 7.
Applying the negative sign to the numerator we have

$$\frac{-(4k-9)}{k+3}.$$

Distributing the negative sign gives us

$$\frac{-4k+9}{k+3}.$$

Applying the negative sign to the denominator yields

$$\frac{4k-9}{-(k+3)}.$$

Again, we distribute to get

$$\frac{4k-9}{-k-3}.$$

6.1 Section Exercises

1. (a) $\frac{3x+1}{5x} = \frac{3(2)+1}{5(2)}$ *Let x = 2.*

$= \frac{7}{10}$

(b) $\frac{3x+1}{5x} = \frac{3(-3)+1}{5(-3)}$ *Let x = −3.*

$= \frac{-8}{-15} = \frac{8}{15}$

3. **(a)** $\frac{x^2-4}{2x+1} = \frac{(2)^2-4}{2(2)+1}$ *Let x = 2.*

$= \frac{0}{5} = 0$

(b) $\frac{x^2-4}{2x+1} = \frac{(-3)^2-4}{2(-3)+1}$ *Let x = −3.*

$= \frac{5}{-5} = -1$

5. **(a)** $\frac{(-2x)^3}{3x+9} = \frac{(-2\cdot 2)^3}{3\cdot 2+9}$ *Let x = 2.*

$= \frac{-64}{15} = -\frac{64}{15}$

(b) $\frac{(-2x)^3}{3x+9} = \frac{[-2(-3)]^3}{3(-3)+9}$ *Let x = −3.*

$= \frac{216}{0}$

Since substituting -3 for x makes the denominator zero, the given rational expression is undefined when $x = -3$.

7. **(a)** $\frac{7-3x}{3x^2-7x+2}$

$= \frac{7-3(2)}{3(2)^2-7(2)+2}$ *Let x = 2.*

$= \frac{7-6}{12-14+2} = \frac{1}{0}$

Since substituting 2 for x makes the denominator zero, the given rational expression is undefined when $x = 2$.

(b) $\frac{7-3x}{3x^2-7x+2}$

$= \frac{7-3(-3)}{3(-3)^2-7(-3)+2}$ *Let x = −3.*

$= \frac{7+9}{27+21+2} = \frac{16}{50} = \frac{8}{25}$

9. **(a)** $\frac{(x+3)(x-2)}{500x}$

$= \frac{(2+3)(2-2)}{500(2)}$ *Let x = 2.*

$= \frac{5(0)}{1000} = \frac{0}{1000} = 0$

(b) $\frac{(x+3)(x-2)}{500x}$

$= \frac{(-3+3)(-3-2)}{500(-3)}$ *Let x = −3.*

$= \frac{0(-5)}{-1500} = \frac{0}{-1500} = 0$

11. **(a)** $\frac{x^2-4}{x^2-9} = \frac{2^2-4}{2^2-9}$ *Let x = 2.*

$= \frac{4-4}{4-9} = \frac{0}{-5} = 0$

(b) $\frac{x^2-4}{x^2-9} = \frac{(-3)^2-4}{(-3)^2-9}$ *Let x = −3.*

$= \frac{9-4}{9-9} = \frac{5}{0}$

Since substituting -3 for x makes the denominator zero, the given rational expression is undefined when $x = -3$.

13. A rational expression is a quotient of two polynomials, such as $\frac{x^2+3x-6}{x+4}$. One can think of this as an algebraic fraction.

15. Division by 0 is undefined. If the denominator of a rational expression equals 0, the expression is undefined.

17. $\frac{12}{5y}$

The denominator $5y$ will be zero when $y = 0$, so the given expression is undefined for $y = 0$. We write the answer as $y \neq 0$.

19. $\frac{x+1}{x-6}$

To find the values for which this expression is undefined, set the denominator equal to zero and solve for x.

$$x - 6 = 0$$
$$x = 6$$

Because $x = 6$ will make the denominator zero, the given expression is undefined for 6. We write the answer as $x \neq 6$.

21. $\frac{4x^2}{3x+5}$

To find the values for which this expression is undefined, set the denominator equal to zero and solve for x.

$$3x + 5 = 0$$
$$3x = -5$$
$$x = -\tfrac{5}{3}$$

Because $x = -\frac{5}{3}$ will make the denominator zero, the given expression is undefined for $-\frac{5}{3}$. We write the answer as $x \neq -\frac{5}{3}$.

23. $\dfrac{5m+2}{m^2+m-6}$

To find the numbers that make the denominator 0, we must solve

$$m^2+m-6=0$$
$$(m+3)(m-2)=0$$
$$m+3=0 \quad \text{or} \quad m-2=0$$
$$m=-3 \quad \text{or} \quad m=2$$

The given expression is undefined for $m=-3$ and for $m=2$. We write the answer as $m \neq -3$ and $m \neq 2$.

25. $\dfrac{x^2+3x}{4}$ is never undefined since the denominator is never zero.

27. $\dfrac{3x-1}{x^2+2}$

This denominator cannot equal zero for any value of x because x^2 is always greater than or equal to zero, and adding 2 makes the sum greater than zero. Thus, the given rational expression is never undefined.

29. **(a)** $\dfrac{x^2+4x}{x+4}$

The two terms in the numerator are x^2 and $4x$. The two terms in the denominator are x and 4.

(b) To express the rational expression in lowest terms, factor the numerator and denominator and divide both by the common factor $x+4$ to get x.

$$\frac{x^2+4x}{x+4}=\frac{x(x+4)}{x+4}=\frac{x}{1}=x$$

31. $\dfrac{18r^3}{6r}=\dfrac{3r^2(6r)}{1(6r)}$ *Factor.*

$=3r^2$ *Fundamental property*

33. $\dfrac{4(y-2)}{10(y-2)}=\dfrac{2\cdot 2(y-2)}{5\cdot 2(y-2)}$ *Factor.*

$=\dfrac{2}{5}$ *Fundamental property*

35. $\dfrac{(x+1)(x-1)}{(x+1)^2}=\dfrac{(x+1)(x-1)}{(x+1)(x+1)}$

$=\dfrac{x-1}{x+1}$ *Fundamental property*

37. $\dfrac{7m+14}{5m+10}=\dfrac{7(m+2)}{5(m+2)}$ *Factor.*

$=\dfrac{7}{5}$ *Fundamental property*

39. $\dfrac{6m-18}{7m-21}=\dfrac{6(m-3)}{7(m-3)}$ *Factor.*

$=\dfrac{6}{7}$ *Fundamental property*

41. $\dfrac{m^2-n^2}{m+n}=\dfrac{(m+n)(m-n)}{m+n}$

$=m-n$

43. $\dfrac{2t+6}{t^2-9}=\dfrac{2(t+3)}{(t+3)(t-3)}$

$=\dfrac{2}{t-3}$

45. $\dfrac{12m^2-3}{8m-4}=\dfrac{3(4m^2-1)}{4(2m-1)}$

$=\dfrac{3(2m+1)(2m-1)}{4(2m-1)}$

$=\dfrac{3(2m+1)}{4}$

47. $\dfrac{3m^2-3m}{5m-5}=\dfrac{3m(m-1)}{5(m-1)}$

$=\dfrac{3m}{5}$

49. $\dfrac{9r^2-4s^2}{9r+6s}=\dfrac{(3r+2s)(3r-2s)}{3(3r+2s)}$

$=\dfrac{3r-2s}{3}$

51. $\dfrac{5k^2-13k-6}{5k+2}=\dfrac{(5k+2)(k-3)}{5k+2}$

$=k-3$

53. $\dfrac{x^2+2x-15}{x^2+6x+5}=\dfrac{(x+5)(x-3)}{(x+5)(x+1)}$

$=\dfrac{x-3}{x+1}$

55. $\dfrac{2x^2-3x-5}{2x^2-7x+5}=\dfrac{(2x-5)(x+1)}{(2x-5)(x-1)}$

$=\dfrac{x+1}{x-1}$

57. $\dfrac{3x^3+13x^2+14x}{3x^3-5x^2-28x}$

$=\dfrac{x(3x^2+13x+14)}{x(3x^2-5x-28)}$

$=\dfrac{x(3x+7)(x+2)}{x(3x+7)(x-4)}$

$=\dfrac{x+2}{x-4}$ *Fundamental property*

59. $\dfrac{-3t+6t^2-3t^3}{7t^2-14t^3+7t^4}$

$=\dfrac{-3t(1-2t+t^2)}{7t^2(1-2t+t^2)}$

$=-\dfrac{3}{7t}$ *Fundamental property*

61. $\dfrac{zw+4z-3w-12}{zw+4z+5w+20}$

$= \dfrac{z(w+4)-3(w+4)}{z(w+4)+5(w+4)}$ *Factor by grouping.*

$= \dfrac{(w+4)(z-3)}{(w+4)(z+5)}$

$= \dfrac{z-3}{z+5}$ *Fundamental property*

63. $\dfrac{pr+qr+ps+qs}{pr+qr-ps-qs}$

$= \dfrac{r(p+q)+s(p+q)}{r(p+q)-s(p+q)}$ *Factor by grouping.*

$= \dfrac{(p+q)(r+s)}{(p+q)(r-s)}$

$= \dfrac{r+s}{r-s}$ *Fundamental property*

65. $\dfrac{ac-ad+bc-bd}{ac-ad-bc+bd}$

$= \dfrac{a(c-d)+b(c-d)}{a(c-d)-b(c-d)}$ *Factor by grouping.*

$= \dfrac{(c-d)(a+b)}{(c-d)(a-b)}$

$= \dfrac{a+b}{a-b}$ *Fundamental property*

67. $\dfrac{m^2-n^2-4m-4n}{2m-2n-8}$

$= \dfrac{(m+n)(m-n)-4(m+n)}{2(m-n-4)}$ *Factor by grouping.*

$= \dfrac{(m+n)(m-n-4)}{2(m-n-4)}$

$= \dfrac{m+n}{2}$ *Fundamental property*

69. $\dfrac{x^2y+y+x^2z+z}{xy+xz}$

$= \dfrac{y(x^2+1)+z(x^2+1)}{x(y+z)}$ *Factor by grouping.*

$= \dfrac{(x^2+1)(y+z)}{x(y+z)}$

$= \dfrac{x^2+1}{x}$ *Fundamental property*

71. The numerator is the sum of cubes.

$$\frac{1+p^3}{1+p} = \frac{1^3+p^3}{1+p} = \frac{(1+p)(1-p+p^2)}{1+p} = 1-p+p^2$$

73. The numerator is the difference of cubes.

$$\frac{x^3-27}{x-3} = \frac{x^3-3^3}{x-3} = \frac{(x-3)(x^2+3x+9)}{x-3} = x^2+3x+9$$

75. The numerator is the difference of cubes and the denominator is the difference of squares.

$$\frac{b^3-a^3}{a^2-b^2} = \frac{(b-a)(b^2+ba+a^2)}{(a-b)(a+b)}$$

$$= (-1)\cdot\frac{(b^2+ba+a^2)}{(a+b)} \qquad \frac{(b-a)}{(a-b)} = -1$$

$$= -\frac{b^2+ba+a^2}{a+b}$$

77. The numerator is the sum of cubes and the denominator is the difference of squares.

$$\frac{k^3+8}{k^2-4} = \frac{k^3+2^3}{(k+2)(k-2)} = \frac{(k+2)(k^2-2k+4)}{(k+2)(k-2)} = \frac{k^2-2k+4}{k-2}$$

79. The numerator is the sum of cubes. The denominator has a common factor of z.

$$\frac{z^3+27}{z^3-3z^2+9z} = \frac{z^3+3^3}{z(z^2-3z+9)} = \frac{(z+3)(z^2-3z+9)}{z(z^2-3z+9)} = \frac{z+3}{z}$$

81. The numerator is the difference of cubes. The denominator has a common factor of 2.

$$\frac{1-8r^3}{8r^2+4r+2} = \frac{1^3-(2r)^3}{2(4r^2+2r+1)} = \frac{(1-2r)(1+2r+4r^2)}{2(4r^2+2r+1)} = \frac{1-2r}{2}$$

83. **A.** $\dfrac{2x+3}{2x-3} \neq -1$

B. $\dfrac{2x-3}{3-2x} = \dfrac{-1(3-2x)}{3-2x} = -1$

C. $\dfrac{2x+3}{3+2x} = 1 \neq -1$

D. $\dfrac{2x+3}{-2x-3} = \dfrac{2x+3}{-1(2x+3)} = -1$

B and **D** are equal to -1.

85. $\dfrac{6-t}{t-6}=\dfrac{-1(t-6)}{1(t-6)}=\dfrac{-1}{1}=-1$

Note that $6-t$ and $t-6$ are opposites, so we know that their quotient will be -1.

87. $\dfrac{m^2-1}{1-m}=\dfrac{(m+1)(m-1)}{-1(m-1)}$

$=\dfrac{m+1}{-1}$

$=-(m+1)$ or $-m-1$

89. $\dfrac{q^2-4q}{4q-q^2}=\dfrac{q(q-4)}{q(4-q)}$

$=\dfrac{q-4}{4-q}=-1$

$q-4$ and $4-q$ are opposites.

91. In the expression $\dfrac{p+6}{p-6}$, neither numerator nor denominator can be factored. It is already in lowest terms. *Note:* $(p+6)$ and $(p-6)$ are not opposites.

93. **A.** $\dfrac{3-x}{x-4}=\dfrac{-1(3-x)}{-1(x-4)}=\dfrac{-3+x}{-x+4}=\dfrac{x-3}{4-x}$

C. $-\dfrac{3-x}{4-x}=\dfrac{-1(3-x)}{4-x}=\dfrac{-3+x}{4-x}=\dfrac{x-3}{4-x}$

D. $-\dfrac{x-3}{x-4}=\dfrac{x-3}{-1(x-4)}=\dfrac{x-3}{-x+4}=\dfrac{x-3}{4-x}$

Since **A**, **C**, and **D** are equivalent to $\dfrac{x-3}{4-x}$, **B** is the one that is not.

There are many possible answers for Exercises 95–100.

95. To write four equivalent expressions for $-\dfrac{x+4}{x-3}$, we will follow the outline in Example 7. Applying the negative sign to the numerator we have

$$\frac{-(x+4)}{x-3}.$$

Distributing the negative sign gives us

$$\frac{-x-4}{x-3}.$$

Applying the negative sign to the denominator yields

$$\frac{x+4}{-(x-3)}.$$

Again, we distribute to get

$$\frac{x+4}{-x+3}.$$

97. $-\dfrac{2x-3}{x+3}$ is equivalent to each of the following:

$$\frac{-(2x-3)}{x+3},\quad \frac{-2x+3}{x+3},$$
$$\frac{2x-3}{-(x+3)},\quad \frac{2x-3}{-x-3}$$

99. $-\dfrac{3x-1}{5x-6}$ is equivalent to each of the following:

$$\frac{-(3x-1)}{5x-6},\quad \frac{-3x+1}{5x-6},$$
$$\frac{3x-1}{-(5x-6)},\quad \frac{3x-1}{-5x+6}$$

101. $L \cdot W=\mathcal{A}$

$W=\dfrac{\mathcal{A}}{L}$

$W=\dfrac{x^4+10x^2+21}{x^2+7}$

$=\dfrac{(x^2+7)(x^2+3)}{x^2+7}$

$=x^2+3$

Note: If it is not apparent that we can factor $\mathcal{A}$ as $x^4+10x^2+21=(x^2+7)(x^2+3)$, we may use "long division" to find the quotient $\dfrac{\mathcal{A}}{L}$. Remember to insert zeros for the coefficients of the missing terms in the dividend and divisor.

$$\begin{array}{r|lllll}
 & x^2 & & & & +\ 3 \\ \hline
x^2+0x+7 & x^4 & +0x^3 & +10x^2 & +0x & +21 \\
 & x^4 & +0x^3 & +\ 7x^2 & & \\ \hline
 & & & 3x^2 & +0x & +21 \\
 & & & 3x^2 & +0x & +21 \\ \hline
 & & & & & 0
\end{array}$$

The width of the rectangle is x^2+3.

103. Let $w=\dfrac{x^2}{2(1-x)}$.

(a) $w=\dfrac{(0.1)^2}{2(1-0.1)}$ *Let $x=0.1$.*

$=\dfrac{0.01}{2(0.9)}\approx 0.006$

To the nearest tenth, w is 0.

(b) $w=\dfrac{(0.8)^2}{2(1-0.8)}$ *Let $x=0.8$.*

$=\dfrac{0.64}{2(0.2)}=1.6$

(c) $w = \dfrac{(0.9)^2}{2(1-0.9)}$ *Let x = 0.9.*

$= \dfrac{0.81}{2(0.1)} = 4.05 \approx 4.1$

(d) Based on the answers in (a), (b), and (c), we see that as the traffic intensity increases, the waiting time also increases.

105. $\dfrac{2}{3} \cdot \dfrac{5}{6} = \dfrac{2 \cdot 5}{3 \cdot 2 \cdot 3} = \dfrac{5}{3 \cdot 3} = \dfrac{5}{9}$

107. $\dfrac{10}{3} \div \dfrac{5}{6} = \dfrac{10}{3} \cdot \dfrac{6}{5}$

$= \dfrac{2 \cdot 5 \cdot 2 \cdot 3}{3 \cdot 5}$

$= \dfrac{2 \cdot 2}{1} = 4$

6.2 Multiplying and Dividing Rational Expressions

6.2 Now Try Exercises

N1. $\dfrac{4k^2}{7} \cdot \dfrac{14}{11k} = \dfrac{2 \cdot 2 \cdot k \cdot k \cdot 2 \cdot 7}{7 \cdot 11 \cdot k}$

$= \dfrac{2 \cdot 2 \cdot k \cdot 2}{11} = \dfrac{8k}{11}$

N2. $\dfrac{m-3}{3m} \cdot \dfrac{9m^2}{8(m-3)^2}$

$= \dfrac{(m-3) \cdot 9m^2}{3m \cdot 8(m-3)^2}$

$= \dfrac{(m-3) \cdot 3 \cdot 3 \cdot m \cdot m}{3 \cdot m \cdot 2 \cdot 2 \cdot 2(m-3)^2} = \dfrac{3m}{8(m-3)}$

N3. $\dfrac{y^2-3y-28}{y^2-9y+14} \cdot \dfrac{y^2-7y+10}{y^2+4y}$

$= \dfrac{(y-7)(y+4)}{(y-7)(y-2)} \cdot \dfrac{(y-2)(y-5)}{y(y+4)}$ *Factor.*

$= \dfrac{(y-7)(y+4)(y-2)(y-5)}{(y-7)(y-2)y(y+4)}$ *Multiply.*

$= \dfrac{y-5}{y}$ *Lowest terms*

N4. $\dfrac{2x-5}{3x^2} \div \dfrac{2x-5}{12x}$

$= \dfrac{2x-5}{3x^2} \cdot \dfrac{12x}{2x-5}$ *Multiply by reciprocal.*

$= \dfrac{2x-5}{3 \cdot x \cdot x} \cdot \dfrac{2 \cdot 2 \cdot 3 \cdot x}{2x-5}$ *Factor.*

$= \dfrac{2 \cdot 2 \cdot 3 \cdot x \cdot (2x-5)}{3 \cdot x \cdot x \cdot (2x-5)}$ *Multiply.*

$= \dfrac{4}{x}$ *Lowest terms*

N5. $\dfrac{(3k)^3}{2j^4} \div \dfrac{9k^2}{6j}$

$= \dfrac{(3k)^3}{2j^4} \cdot \dfrac{6j}{9k^2}$ *Multiply by reciprocal.*

$= \dfrac{27k^3 \cdot 6j}{2j^4 \cdot 9k^2}$ *Multiply.*

$= \dfrac{9k}{j^3}$ *Lowest terms*

N6. $\dfrac{(t+2)(t-5)}{-4t} \div \dfrac{t^2-25}{(t+5)(t+2)}$

$= \dfrac{(t+2)(t-5)}{-4t} \cdot \dfrac{(t+5)(t+2)}{t^2-25}$ *Multiply by reciprocal.*

$= \dfrac{(t+2)(t-5)}{-4t} \cdot \dfrac{(t+5)(t+2)}{(t+5)(t-5)}$ *Factor.*

$= \dfrac{(t+2)(t-5)(t+5)(t+2)}{-4t(t+5)(t-5)}$ *Multiply.*

$= \dfrac{(t+2)^2}{-4t}$, or $-\dfrac{(t+2)^2}{4t}$ *Lowest terms*

N7. $\dfrac{7-x}{2x+6} \div \dfrac{x^2-49}{x^2+6x+9}$

$= \dfrac{7-x}{2x+6} \cdot \dfrac{x^2+6x+9}{x^2-49}$

$= \dfrac{(7-x)(x^2+6x+9)}{(2x+6)(x^2-49)}$

$= \dfrac{(7-x)(x+3)(x+3)}{2(x+3)(x+7)(x-7)}$

$= \dfrac{(-1)(x+3)}{2(x+7)}$

$= \dfrac{-x-3}{2(x+7)}$, or $-\dfrac{x+3}{2(x+7)}$

6.2 Section Exercises

1. (a) $\dfrac{5x^3}{10x^4} \cdot \dfrac{10x^7}{4x} = \dfrac{5 \cdot 10 \cdot x^3 \cdot x^7}{10 \cdot 4 \cdot x^4 \cdot x}$

$= \dfrac{5x^{10}}{4x^5}$

$= \dfrac{5x^5}{4}$ **(B)**

(b) $\dfrac{10x^4}{5x^3} \cdot \dfrac{10x^7}{4x} = \dfrac{5 \cdot 2 \cdot 5 \cdot 2 \cdot x^4 \cdot x^7}{5 \cdot 2 \cdot 2 \cdot x^3 \cdot x}$

$= \dfrac{5x^{11}}{1x^4}$

$= 5x^7$ **(D)**

(c) $\dfrac{5x^3}{10x^4} \cdot \dfrac{4x}{10x^7} = \dfrac{5 \cdot 2 \cdot 2 \cdot x^3 \cdot x}{5 \cdot 2 \cdot 5 \cdot 2 \cdot x^4 \cdot x^7}$

$= \dfrac{1x^4}{5x^{11}}$

$= \dfrac{1}{5x^7}$ **(C)**

(d) $\dfrac{10x^4}{5x^3} \cdot \dfrac{4x}{10x^7} = \dfrac{10 \cdot 4 \cdot x^4 \cdot x}{5 \cdot 10 \cdot x^3 \cdot x^7}$

$= \dfrac{4x^5}{5x^{10}}$

$= \dfrac{4}{5x^5}$ **(A)**

3. $\dfrac{15a^2}{14} \cdot \dfrac{7}{5a} = \dfrac{3 \cdot 5 \cdot a \cdot a \cdot 7}{2 \cdot 7 \cdot 5 \cdot a}$ *Multiply and factor.*

$= \dfrac{3 \cdot a(5 \cdot 7 \cdot a)}{2(5 \cdot 7 \cdot a)}$

$= \dfrac{3a}{2}$ *Lowest terms*

5. $\dfrac{12x^4}{18x^3} \cdot \dfrac{-8x^5}{4x^2} = \dfrac{-96x^9}{72x^5}$ *Multiply numerators; multiply denominators.*

$= \dfrac{-4x^4(24x^5)}{3(24x^5)}$ *Group common factors.*

$= -\dfrac{4x^4}{3}$ *Lowest terms*

7. $\dfrac{2(c+d)}{3} \cdot \dfrac{18}{6(c+d)^2}$

$= \dfrac{3 \cdot 3 \cdot 2 \cdot 2(c+d)}{3 \cdot 3 \cdot 2(c+d)(c+d)}$ *Multiply and factor.*

$= \dfrac{2}{c+d}$ *Lowest terms*

9. $\dfrac{(x-y)^2}{2} \cdot \dfrac{24}{3(x-y)}$

$= \dfrac{6 \cdot 4(x-y)(x-y)}{6(x-y)}$

$= 4(x-y)$

11. $\dfrac{t-4}{8} \cdot \dfrac{4t^2}{t-4}$

$= \dfrac{4t^2(t-4)}{2 \cdot 4(t-4)}$

$= \dfrac{t^2}{2}$

13. $\dfrac{3x}{x+3} \cdot \dfrac{(x+3)^2}{6x^2}$

$= \dfrac{3x(x+3)(x+3)}{2 \cdot 3 \cdot x \cdot x(x+3)}$

$= \dfrac{x+3}{2x}$

15. $\dfrac{9z^4}{3z^5} \div \dfrac{3z^2}{5z^3} = \dfrac{9z^4}{3z^5} \cdot \dfrac{5z^3}{3z^2}$

$= \dfrac{9 \cdot 5z^7}{3 \cdot 3z^7}$

$= 5$

17. $\dfrac{4t^4}{2t^5} \div \dfrac{(2t)^3}{-6} = \dfrac{4t^4}{2t^5} \cdot \dfrac{-6}{(2t)^3}$

$= \dfrac{4t^4}{2t^5} \cdot \dfrac{-6}{8t^3}$

$= \dfrac{-24t^4}{16t^8}$

$= \dfrac{-3(8t^4)}{2t^4(8t^4)}$

$= \dfrac{-3}{2t^4} = -\dfrac{3}{2t^4}$

19. $\dfrac{3}{2y-6} \div \dfrac{6}{y-3} = \dfrac{3}{2y-6} \cdot \dfrac{y-3}{6}$

$= \dfrac{3}{2(y-3)} \cdot \dfrac{y-3}{6}$

$= \dfrac{3(y-3)}{2 \cdot 2 \cdot 3(y-3)}$

$= \dfrac{1}{2 \cdot 2} = \dfrac{1}{4}$

21. $\dfrac{7t+7}{-6} \div \dfrac{4t+4}{15}$

$= \dfrac{7t+7}{-6} \cdot \dfrac{15}{4t+4}$

$= \dfrac{7(t+1)}{-2 \cdot 3} \cdot \dfrac{3 \cdot 5}{4(t+1)}$

$= \dfrac{3 \cdot 5 \cdot 7(t+1)}{-2 \cdot 3 \cdot 4(t+1)} = -\dfrac{35}{8}$

23. $\dfrac{2x}{x-1} \div \dfrac{x^2}{x+2}$

$= \dfrac{2x}{x-1} \cdot \dfrac{x+2}{x^2}$

$= \dfrac{2x(x+2)}{x \cdot x(x-1)} = \dfrac{2(x+2)}{x(x-1)}$

25. $\dfrac{(x-3)^2}{6x} \div \dfrac{x-3}{x^2}$

$= \dfrac{(x-3)^2}{6x} \cdot \dfrac{x^2}{x-3}$

$= \dfrac{x \cdot x(x-3)(x-3)}{6x(x-3)} = \dfrac{x(x-3)}{6}$

27. $\dfrac{5x-15}{3x+9} \cdot \dfrac{4x+12}{6x-18}$

$= \dfrac{5(x-3)}{3(x+3)} \cdot \dfrac{4(x+3)}{6(x-3)}$

$= \dfrac{5 \cdot 4 \cdot (x-3)(x+3)}{3 \cdot 6 \cdot (x-3)(x+3)}$

$= \dfrac{10}{9}$

29. $\frac{2-t}{8} \div \frac{t-2}{6}$

$= \frac{2-t}{8} \cdot \frac{6}{t-2}$ *Multiply by reciprocal.*

$= \frac{6(2-t)}{8(t-2)}$ *Multiply numerators; multiply denominators.*

$= \frac{6(-1)}{8}$ $\frac{2-t}{t-2} = -1$

$= -\frac{3}{4}$ *Lowest terms*

31. $\frac{27-3z}{4} \cdot \frac{12}{2z-18}$

$= \frac{3(9-z)}{4} \cdot \frac{3 \cdot 4}{2(z-9)}$

$= \frac{3 \cdot 3 \cdot 4(9-z)}{4 \cdot 2(z-9)}$

$= \frac{3 \cdot 3 \cdot (-1)}{2}$

$= -\frac{9}{2}$

33. $\frac{p^2+4p-5}{p^2+7p+10} \div \frac{p-1}{p+4}$

$= \frac{p^2+4p-5}{p^2+7p+10} \cdot \frac{p+4}{p-1}$

$= \frac{(p+5)(p-1) \cdot (p+4)}{(p+5)(p+2) \cdot (p-1)}$

$= \frac{p+4}{p+2}$

35. $\frac{m^2-4}{16-8m} \div \frac{m+2}{8}$

$= \frac{(m+2)(m-2)}{8(2-m)} \cdot \frac{8}{m+2}$

$= \frac{8(m+2)(m-2)}{8(m+2)(2-m)}$

$= -1$

37. $\frac{2x^2-7x+3}{x-3} \cdot \frac{x+2}{x-1}$

$= \frac{(2x-1)(x-3)}{x-3} \cdot \frac{x+2}{x-1}$

$= \frac{(2x-1)(x+2)}{x-1}$

39. $\frac{2k^2-k-1}{2k^2+5k+3} \div \frac{4k^2-1}{2k^2+k-3}$

$= \frac{2k^2-k-1}{2k^2+5k+3} \cdot \frac{2k^2+k-3}{4k^2-1}$

$= \frac{(2k+1)(k-1)(2k+3)(k-1)}{(2k+3)(k+1)(2k+1)(2k-1)}$

$= \frac{(k-1)(k-1)}{(k+1)(2k-1)}$

$= \frac{(k-1)^2}{(k+1)(2k-1)}$

41. $\frac{2k^2+3k-2}{6k^2-7k+2} \cdot \frac{4k^2-5k+1}{k^2+k-2}$

$= \frac{(2k-1)(k+2)}{(3k-2)(2k-1)} \cdot \frac{(4k-1)(k-1)}{(k+2)(k-1)}$

$= \frac{(2k-1)(k+2)(4k-1)(k-1)}{(3k-2)(2k-1)(k+2)(k-1)}$

$= \frac{4k-1}{3k-2}$

43. $\frac{m^2+2mp-3p^2}{m^2-3mp+2p^2} \div \frac{m^2+4mp+3p^2}{m^2+2mp-8p^2}$

$= \frac{m^2+2mp-3p^2}{m^2-3mp+2p^2} \cdot \frac{m^2+2mp-8p^2}{m^2+4mp+3p^2}$

$= \frac{(m+3p)(m-p)(m+4p)(m-2p)}{(m-2p)(m-p)(m+3p)(m+p)}$

$= \frac{m+4p}{m+p}$

45. $\frac{m^2+3m+2}{m^2+5m+4} \cdot \frac{m^2+10m+24}{m^2+5m+6}$

$= \frac{(m+2)(m+1)}{(m+4)(m+1)} \cdot \frac{(m+6)(m+4)}{(m+3)(m+2)}$

$= \frac{m+6}{m+3}$ *Multiply and use fundamental property.*

47. $\frac{y^2+y-2}{y^2+3y-4} \div \frac{y+2}{y+3}$

$= \frac{y^2+y-2}{y^2+3y-4} \cdot \frac{y+3}{y+2}$

$= \frac{(y+2)(y-1)}{(y+4)(y-1)} \cdot \frac{y+3}{y+2}$

$= \frac{y+3}{y+4}$ *Multiply and use fundamental property.*

49. $\frac{2m^2+7m+3}{m^2-9} \cdot \frac{m^2-3m}{2m^2+11m+5}$

$= \frac{(2m+1)(m+3)}{(m-3)(m+3)} \cdot \frac{m(m-3)}{(2m+1)(m+5)}$

$= \frac{(2m+1)(m+3)m(m-3)}{(m-3)(m+3)(2m+1)(m+5)}$

$= \frac{m}{m+5}$

51. $\dfrac{r^2+rs-12s^2}{r^2-rs-20s^2} \div \dfrac{r^2-2rs-3s^2}{r^2+rs-30s^2}$

$$= \frac{r^2+rs-12s^2}{r^2-rs-20s^2} \cdot \frac{r^2+rs-30s^2}{r^2-2rs-3s^2}$$

$$= \frac{(r-3s)(r+4s)(r+6s)(r-5s)}{(r-5s)(r+4s)(r-3s)(r+s)}$$

$$= \frac{r+6s}{r+s}$$

53. $\dfrac{(q-3)^4(q+2)}{q^2+3q+2} \div \dfrac{q^2-6q+9}{q^2+4q+4}$

$$= \frac{(q-3)^4(q+2)}{q^2+3q+2} \cdot \frac{q^2+4q+4}{q^2-6q+9}$$

$$= \frac{(q-3)^4(q+2)(q+2)^2}{(q+2)(q+1)(q-3)^2}$$

$$= \frac{(q-3)^2(q+2)^2}{q+1}$$

55. $\dfrac{x+5}{x+10} \div \left(\dfrac{x^2+10x+25}{x^2+10x} \cdot \dfrac{10x}{x^2+15x+50}\right)$

$$= \frac{x+5}{x+10} \div \left[\frac{(x+5)^2 \cdot 10x}{x(x+10)(x+5)(x+10)}\right]$$

$$= \frac{x+5}{x+10} \div \left[\frac{10(x+5)}{(x+10)^2}\right]$$

$$= \frac{x+5}{x+10} \cdot \frac{(x+10)^2}{10(x+5)}$$

$$= \frac{x+10}{10}$$

57. $\dfrac{3a-3b-a^2+b^2}{4a^2-4ab+b^2} \cdot \dfrac{4a^2-b^2}{2a^2-ab-b^2}$

Factor $3a-3b-a^2+b^2$ by grouping.

$$\begin{aligned} &3a-3b-a^2+b^2 \\ &= 3(a-b)-(a^2-b^2) \\ &= 3(a-b)-(a-b)(a+b) \\ &= (a-b)[3-(a+b)] \\ &= (a-b)(3-a-b) \end{aligned}$$

Thus,

$$\frac{3a-3b-a^2+b^2}{4a^2-4ab+b^2} \cdot \frac{4a^2-b^2}{2a^2-ab-b^2}$$

$$= \frac{(a-b)(3-a-b)}{(2a-b)(2a-b)} \cdot \frac{(2a-b)(2a+b)}{(2a+b)(a-b)}$$

$$= \frac{(a-b)(3-a-b)(2a-b)(2a+b)}{(2a-b)(2a-b)(2a+b)(a-b)}$$

$$= \frac{3-a-b}{2a-b}.$$

59. $\dfrac{-x^3-y^3}{x^2-2xy+y^2} \div \dfrac{3y^2-3xy}{x^2-y^2}$

$$= \frac{-1(x^3+y^3)}{x^2-2xy+y^2} \cdot \frac{x^2-y^2}{3y^2-3xy}$$

$$= \frac{-1(x+y)(x^2-xy+y^2)}{(x-y)(x-y)} \cdot \frac{(x-y)(x+y)}{3y(y-x)}$$

$$= \frac{-1(x+y)(x^2-xy+y^2)(x-y)(x+y)}{-1(x-y)(x-y)(3y)(x-y)}$$

$$= \frac{(x+y)^2(x^2-xy+y^2)}{3y(x-y)^2}$$

If we had not changed $y-x$ to $-1(x-y)$ in the denominator, we would have obtained an alternate form of the answer,

$$-\frac{(x+y)^2(x^2-xy+y^2)}{3y(y-x)(x-y)}.$$

61. Use the formula for the area of a rectangle with $\mathcal{A} = \dfrac{5x^2y^3}{2pq}$ and $L = \dfrac{2xy}{p}$ to solve for W.

$$\mathcal{A} = L \cdot W$$

$$\frac{5x^2y^3}{2pq} = \frac{2xy}{p} \cdot W$$

$$W = \frac{5x^2y^3}{2pq} \div \frac{2xy}{p}$$

$$= \frac{5x^2y^3}{2pq} \cdot \frac{p}{2xy}$$

$$= \frac{5x^2y^3p}{4pqxy}$$

$$= \frac{5xy^2}{4q}$$

Thus, the rational expression $\dfrac{5xy^2}{4q}$ represents the width of the rectangle.

63. $18 = 2 \cdot 9$
$\quad = 2 \cdot 3 \cdot 3$

The prime factored form of 18 is $2 \cdot 3^2$.

65. $108 = 2 \cdot 54$
$\quad = 2 \cdot 2 \cdot 27$
$\quad = 2 \cdot 2 \cdot 3 \cdot 9$
$\quad = 2 \cdot 2 \cdot 3 \cdot 3 \cdot 3$

The prime factored form of 108 is $2^2 \cdot 3^3$.

67.
$$24m = 2^3 \cdot 3 \cdot m$$
$$18m^2 = 2 \cdot 3^2 \cdot m^2$$
$$6 = 2 \cdot 3$$

The GCF is $2 \cdot 3 = 6$.

69.
$$84q^3 = 2^2 \cdot 3 \cdot 7 \cdot q^3$$
$$90q^6 = 2 \cdot 3^2 \cdot 5 \cdot q^6$$

The GCF is $2 \cdot 3 \cdot q^3 = 6q^3$.

6.3 Least Common Denominators

6.3 Now Try Exercises

N1. (a) $\frac{5}{48}, \frac{1}{30}$

Factor each denominator.

$$48 = 2 \cdot 2 \cdot 2 \cdot 2 \cdot 3, \qquad 30 = 2 \cdot 3 \cdot 5$$

Take each different factor the *greatest* number of times it appears as a factor in any of the denominators, and use it to form the **least common denominator (LCD)**.

$$\text{LCD} = 2 \cdot 2 \cdot 2 \cdot 2 \cdot 3 \cdot 5 = 240$$

(b) $\frac{3}{10y}, \frac{1}{6y}$

Factor each denominator.

$$10y = 2 \cdot 5 \cdot y$$
$$6y = 2 \cdot 3 \cdot y$$

Take each factor the greatest number of times it appears in any denominator; then multiply.

$$\text{LCD} = 2 \cdot 3 \cdot 5 \cdot y = 30y$$

N2. $\frac{5}{6x^4}, \frac{7}{8x^3}$

Factor each denominator.

$$6x^4 = 2 \cdot 3 \cdot x^4$$
$$8x^3 = 2 \cdot 2 \cdot 2 \cdot x^3$$

Take each factor the greatest number of times it appears in any denominator; then multiply.

$$\text{LCD} = 2 \cdot 2 \cdot 2 \cdot 3 \cdot x^4 = 24x^4$$

N3. (a) $\frac{3t}{2t^2 - 10t}, \frac{t+4}{t^2 - 25}$

Factor each denominator.

$$2t^2 - 10t = 2t(t-5)$$
$$t^2 - 25 = (t+5)(t-5)$$

Take each factor the greatest number of times it appears in any denominator; then multiply.

$$\text{LCD} = 2t(t-5)(t+5)$$

(b) $\frac{1}{x^2 + 7x + 12}, \frac{2}{x^2 + 6x + 9}, \frac{5}{x^2 + 2x - 8}$

Factor each denominator.

$$x^2 + 7x + 12 = (x+3)(x+4)$$
$$x^2 + 6x + 9 = (x+3)(x+3)$$
$$x^2 + 2x - 8 = (x+4)(x-2)$$

Take each factor the greatest number of times it appears in any denominator; then multiply.

$$\text{LCD} = (x+3)^2(x+4)(x-2)$$

(c) $\frac{2}{a-4}, \frac{1}{4-a}$

The expressions $a - 4$ and $4 - a$ are opposites of each other because

$$-(a-4) = -a + 4 = 4 - a.$$

Therefore, either $a - 4$ or $4 - a$ can be used as the LCD.

N4. (a) $\frac{2}{9} = \frac{?}{27}$

First factor the denominator on the right. Then compare the denominator on the left with the one on the right to decide what factors are missing.

$$\frac{2}{9} = \frac{?}{3 \cdot 9}$$

A factor of 3 is missing, so multiply $\frac{2}{9}$ by $\frac{3}{3}$, which is equal to 1.

$$\frac{2}{9} = \frac{2}{9} \cdot \frac{3}{3} = \frac{6}{27}$$

(b) $\frac{4t}{11} = \frac{?}{33t}$

Factor the denominator on the right; then compare it to the denominator on the left.

$$\frac{4t}{11} = \frac{?}{3 \cdot 11 \cdot t}$$

The factors 3 and t are missing on the left, so multiply by $\frac{3t}{3t}$.

$$\frac{4t}{11} = \frac{4t}{11} \cdot \frac{3t}{3t} = \frac{12t^2}{33t}$$

N5. (a) $\frac{8k}{5k-2} = \frac{?}{25k-10}$

Factor the denominator on the right.

$$\frac{8k}{5k-2} = \frac{?}{5(5k-2)}$$

The factor 5 is missing on the left, so multiply by $\frac{5}{5}$.

$$\frac{8k}{5k-2} = \frac{8k}{5k-2} \cdot \frac{5}{5} = \frac{40k}{25k-10}$$

(b) $\dfrac{2t-1}{t^2+4t}=\dfrac{?}{t^3+12t^2+32t}$

Factor and compare the denominators.

$$\frac{2t-1}{t(t+4)}=\frac{?}{t(t+4)(t+8)}$$

The factor $t+8$ is missing on the left, so multiply by $\frac{t+8}{t+8}$.

$$\begin{aligned}\frac{2t-1}{t^2+4t}&=\frac{2t-1}{t(t+4)}\cdot\frac{t+8}{t+8}\\&=\frac{(2t-1)(t+8)}{t(t+4)(t+8)}\end{aligned}$$

6.3 Section Exercises

1. The factor x appears at most one time in any denominator as does the factor y. Thus, the LCD is the product of the two factors, xy. The correct response is **C**.

3. Since $20=2^2\cdot 5$, the LCD of $\frac{9}{20}$ and $\frac{1}{2}$ must have 5 as a factor and 2^2 as a factor. Because 2 appears twice in $2^2\cdot 5$, we don't have to include another 2 in the LCD for the number $\frac{1}{2}$. Thus, the LCD is just $2^2\cdot 5=20$. Note that this is a specific case of Exercise 2 since 2 is a factor of 20. The correct response is **C**.

5. $\dfrac{7}{15},\dfrac{21}{20}$

Factor each denominator.

$$\begin{aligned}15&=3\cdot 5\\20&=2\cdot 2\cdot 5=2^2\cdot 5\end{aligned}$$

Take each factor the greatest number of times it appears as a factor in any one of the denominators.

$$\text{LCD}=2^2\cdot 3\cdot 5=60$$

7. $\dfrac{17}{100},\dfrac{23}{120},\dfrac{43}{180}$

Factor each denominator.

$$\begin{aligned}100&=2^2\cdot 5^2\\120&=2^3\cdot 3\cdot 5\\180&=2^2\cdot 3^2\cdot 5\end{aligned}$$

Take each factor the greatest number of times it appears as a factor in any one of the denominators.

$$\text{LCD}=2^3\cdot 3^2\cdot 5^2=1800$$

9. $\dfrac{9}{x^2},\dfrac{8}{x^5}$

The greatest number of times x appears as a factor in any denominator is the greatest exponent on x, which is 5.

$$\text{LCD}=x^5$$

11. $\dfrac{-2}{5p},\dfrac{13}{6p}$

Factor each denominator.

$$\begin{aligned}5p&=5\cdot p\\6p&=2\cdot 3\cdot p\end{aligned}$$

Take each factor the greatest number of times it appears; then multiply.

$$\text{LCD}=2\cdot 3\cdot 5\cdot p=30p$$

13. $\dfrac{17}{15y^2},\dfrac{55}{36y^4}$

Factor each denominator.

$$\begin{aligned}15y^2&=3\cdot 5\cdot y^2\\36y^4&=2^2\cdot 3^2\cdot y^4\end{aligned}$$

Take each factor the greatest number of times it appears; then multiply.

$$\text{LCD}=2^2\cdot 3^2\cdot 5\cdot y^4=180y^4$$

15. $\dfrac{5}{21r^3},\dfrac{7}{12r^5}$

Factor each denominator.

$$\begin{aligned}21r^3&=3\cdot 7\cdot r^3\\12r^5&=2^2\cdot 3\cdot r^5\end{aligned}$$

Take each factor the greatest number of times it appears; then multiply.

$$\text{LCD}=2^2\cdot 3\cdot 7\cdot r^5=84r^5$$

17. $\dfrac{13}{5a^2b^3},\dfrac{29}{15a^5b}$

Factor each denominator.

$$\begin{aligned}5a^2b^3&=5\cdot a^2\cdot b^3\\15a^5b&=3\cdot 5\cdot a^5\cdot b\end{aligned}$$

Take each factor the greatest number of times it appears; then multiply.

$$\text{LCD}=3\cdot 5\cdot a^5\cdot b^3=15a^5b^3$$

19. $\dfrac{7}{6p},\dfrac{15}{4p-8}$

Factor each denominator.

$$\begin{aligned}6p&=2\cdot 3\cdot p\\4p-8&=4(p-2)=2^2(p-2)\end{aligned}$$

Take each factor the greatest number of times it appears; then multiply.

$$\text{LCD}=2^2\cdot 3\cdot p(p-2)=12p(p-2)$$

21. $\frac{9}{28m^2}, \frac{3}{12m-20}$

Factor each denominator.

$$28m^2 = 2^2 \cdot 7 \cdot m^2$$
$$12m - 20 = 4(3m-5) = 2^2(3m-5)$$

Take each factor the greatest number of times it appears; then multiply.

$$\text{LCD} = 2^2 \cdot 7m^2(3m-5) = 28m^2(3m-5)$$

23. $\frac{7}{5b-10}, \frac{11}{6b-12}$

Factor each denominator.

$$5b - 10 = 5(b-2)$$
$$6b - 12 = 6(b-2) = 2 \cdot 3(b-2)$$

Take each factor the greatest number of times it appears; then multiply.

$$\text{LCD} = 2 \cdot 3 \cdot 5(b-2) = 30(b-2)$$

25. $\frac{37}{6r-12}, \frac{25}{9r-18}$

Factor each denominator.

$$6r - 12 = 6(r-2) = 2 \cdot 3(r-2)$$
$$9r - 18 = 9(r-2) = 3^2(r-2)$$

Take each factor the greatest number of times it appears; then multiply.

$$\text{LCD} = 2 \cdot 3^2(r-2) = 18(r-2)$$

27. $\frac{5}{12p+60}, \frac{-17}{p^2+5p}, \frac{-16}{p^2+10p+25}$

Factor each denominator.

$$12p + 60 = 12(p+5) = 2^2 \cdot 3(p+5)$$
$$p^2 + 5p = p(p+5)$$
$$p^2 + 10p + 25 = (p+5)(p+5)$$
$$\text{LCD} = 2^2 \cdot 3 \cdot p(p+5)^2 = 12p(p+5)^2$$

29. $\frac{-3}{8y+16}, \frac{-22}{y^2+3y+2}$

Factor each denominator.

$$8y + 16 = 8(y+2) = 2^3(y+2)$$
$$y^2 + 3y + 2 = (y+2)(y+1)$$
$$\text{LCD} = 8(y+2)(y+1)$$

31. $\frac{5}{c-d}, \frac{8}{d-c}$

The denominators, $c - d$ and $d - c$, are opposites of each other since

$$-(c-d) = -c + d = d - c.$$

Therefore, either $c - d$ or $d - c$ can be used as the LCD.

33. $\frac{12}{m-3}, \frac{-4}{3-m}$

The expression $3 - m$ can be written as $-1(m-3)$, since

$$-1(m-3) = -m + 3 = 3 - m.$$

Because of this, either $m - 3$ or $3 - m$ can be used as the LCD.

35. $\frac{29}{p-q}, \frac{18}{q-p}$

The expression $q - p$ can be written as $-1(p-q)$, since

$$-1(p-q) = -p + q = q - p.$$

Because of this, either $p - q$ or $q - p$ can be used as the LCD.

37. $\frac{3}{k^2+5k}, \frac{2}{k^2+3k-10}$

Factor each denominator.

$$k^2 + 5k = k(k+5)$$
$$k^2 + 3k - 10 = (k+5)(k-2)$$

$$\text{LCD} = k(k+5)(k-2)$$

39. $\frac{6}{a^2+6a}, \frac{-5}{a^2+3a-18}$

Factor each denominator.

$$a^2 + 6a = a(a+6)$$
$$a^2 + 3a - 18 = (a+6)(a-3)$$

$$\text{LCD} = a(a+6)(a-3)$$

41. $\frac{5}{p^2+8p+15}, \frac{3}{p^2-3p-18}, \frac{12}{p^2-p-30}$

Factor each denominator.

$$p^2 + 8p + 15 = (p+5)(p+3)$$
$$p^2 - 3p - 18 = (p-6)(p+3)$$
$$p^2 - p - 30 = (p-6)(p+5)$$

$$\text{LCD} = (p+3)(p+5)(p-6)$$

43. $\frac{-5}{k^2+2k-35}, \frac{-8}{k^2+3k-40}, \frac{19}{k^2-2k-15}$

Factor each denominator.

$$k^2 + 2k - 35 = (k+7)(k-5)$$
$$k^2 + 3k - 40 = (k+8)(k-5)$$
$$k^2 - 2k - 15 = (k-5)(k+3)$$

$$\text{LCD} = (k+7)(k-5)(k+8)(k+3)$$

◆◆◆ Relating Concepts 45–50 ◆◆◆

45. $\frac{3}{4} = \frac{?}{28}$

To change 4 into 28, multiply by 7. If you multiply the denominator by 7, you must multiply the numerator by 7.

46. $\frac{3}{4} = \frac{3}{4} \cdot \frac{7}{7} = \frac{21}{28}$

Note that numerator and denominator are being multiplied by 7, so $\frac{3}{4}$ is being multiplied by the fraction $\frac{7}{7}$, which is equal to 1.

47. Since $\frac{7}{7}$ has a value of 1, the multiplier is 1. The *identity property of multiplication* is being used when we write a common fraction as an equivalent one with a larger denominator.

48. $\frac{2x+5}{x-4} = \frac{?}{7x-28} = \frac{?}{7(x-4)}$

The expression $7x - 28$ is factored as $7(x-4)$, so the multiplier is 7.

49. $\frac{2x+5}{x-4} = \frac{?}{7x-28} = \frac{?}{7(x-4)}$

To form the new denominator, 7 must be used as the multiplier for the denominator. To form an equivalent fraction, the same multiplier must be used for numerator and denominator. Thus, the multiplier is $\frac{7}{7}$, which is equal to 1.

50. The *identity property of multiplication* is being used when we write an algebraic fraction as an equivalent one with a larger denominator.

51. $\frac{4}{11} = \frac{?}{55}$

First factor the denominator on the right. Then compare the denominator on the left with the one on the right to decide what factors are missing.

$$\frac{4}{11} = \frac{?}{11 \cdot 5}$$

A factor of 5 is missing, so multiply $\frac{4}{11}$ by $\frac{5}{5}$, which is equal to 1.

$$\frac{4}{11} \cdot \frac{5}{5} = \frac{20}{55}$$

53. $\frac{-5}{k} = \frac{?}{9k}$

A factor of 9 is missing in the first fraction, so multiply numerator and denominator by 9.

$$\frac{-5}{k} \cdot \frac{9}{9} = \frac{-45}{9k}$$

55. $\frac{15m^2}{8k} = \frac{?}{32k^4}$

$32k^4 = (8k)(4k^3)$, so we must multiply the numerator and the denominator by $4k^3$.

$$\frac{15m^2}{8k} = \frac{15m^2}{8k} \cdot \frac{4k^3}{4k^3} \quad \textit{Multiplicative identity property}$$

$$= \frac{60m^2k^3}{32k^4}$$

57. $\frac{19z}{2z-6} = \frac{?}{6z-18}$

Begin by factoring each denominator.

$$2z - 6 = 2(z-3)$$
$$6z - 18 = 6(z-3)$$

The fractions may now be written as follows.

$$\frac{19z}{2(z-3)} = \frac{?}{6(z-3)}$$

Comparing the two factored forms, we see that the denominator of the fraction on the left side must be multiplied by 3; the numerator must also be multiplied by 3.

$$\frac{19z}{2z-6}$$

$$= \frac{19z}{2(z-3)} \cdot \frac{3}{3} \quad \textit{Multiplicative identity property}$$

$$= \frac{19z(3)}{2(z-3)(3)} \quad \textit{Multiplication of rational expressions}$$

$$= \frac{57z}{6z-18} \quad \textit{Multiply the factors}$$

59. $\frac{-2a}{9a-18} = \frac{?}{18a-36}$

$$\frac{-2a}{9(a-2)} = \frac{?}{18(a-2)} \quad \textit{Factor each denominator}$$

$$\frac{-2a}{9a-18} = \frac{-2a}{9(a-2)} \cdot \frac{2}{2} \quad \textit{Multiplicative identity property}$$

$$= \frac{-4a}{18a-36} \quad \textit{Multiply}$$

61. $\frac{6}{k^2-4k} = \frac{?}{k(k-4)(k+1)}$

$$\frac{6}{k(k-4)} = \frac{?}{k(k-4)(k+1)} \quad \textit{Factor first denominator}$$

$$\frac{6}{k^2-4k} = \frac{6}{k(k-4)} \cdot \frac{(k+1)}{(k+1)} \quad \textit{Multiplicative identity property}$$

$$= \frac{6(k+1)}{k(k-4)(k+1)} \quad \textit{Multiply}$$

63. $\dfrac{36r}{r^2-r-6} = \dfrac{?}{(r-3)(r+2)(r+1)}$

$\dfrac{36r}{(r-3)(r+2)} = \dfrac{?}{(r-3)(r+2)(r+1)}$

Factor first denominator

$\dfrac{36r}{r^2-r-6} = \dfrac{36r}{(r-3)(r+2)} \cdot \dfrac{(r+1)}{(r+1)}$

Multiplicative identity property

$= \dfrac{36r(r+1)}{(r-3)(r+2)(r+1)}$

65. $\dfrac{a+2b}{2a^2+ab-b^2} = \dfrac{?}{2a^3b+a^2b^2-ab^3}$

$\dfrac{a+2b}{2a^2+ab-b^2} = \dfrac{?}{ab(2a^2+ab-b^2)}$

Factor second denominator

$\dfrac{a+2b}{2a^2+ab-b^2} = \dfrac{(a+2b)}{(2a^2+ab-b^2)} \cdot \dfrac{ab}{ab}$

Multiplicative identity property

$= \dfrac{ab(a+2b)}{2a^3b+a^2b^2-ab^3}$

67. $\dfrac{4r-t}{r^2+rt+t^2} = \dfrac{?}{t^3-r^3}$

Factor the second denominator as the difference of cubes.

$t^3-r^3 = (t-r)(t^2+rt+r^2)$

$\dfrac{4r-t}{r^2+rt+t^2} = \dfrac{(4r-t)}{(r^2+rt+t^2)} \cdot \dfrac{(t-r)}{(t-r)}$

Multiplicative identity property

$= \dfrac{(4r-t)(t-r)}{t^3-r^3}$

Multiply the factors

69. $\dfrac{2(z-y)}{y^2+yz+z^2} = \dfrac{?}{y^4-z^3y}$

Factor the second denominator.

$y^4-z^3y = y(y^3-z^3) \quad GCF = y$

$= y(y-z)(y^2+yz+z^2)$

Difference of cubes

$\dfrac{2(z-y)}{y^2+yz+z^2} = \dfrac{?}{y(y-z)(y^2+yz+z^2)}$

$\dfrac{2(z-y)}{y^2+yz+z^2} = \dfrac{2(z-y)}{(y^2+yz+z^2)} \cdot \dfrac{y(y-z)}{y(y-z)}$

$= \dfrac{2y(z-y)(y-z)}{y(y-z)(y^2+yz+z^2)}$

Multiplicative identity property

$= \dfrac{2y(z-y)(y-z)}{y(y^3-z^3)}$

$= \dfrac{2y(z-y)(y-z)}{y^4-z^3y}$,

or $\dfrac{-2y(y-z)^2}{y^4-z^3y}$, since $z-y=-1(y-z)$.

71. $\dfrac{1}{2}+\dfrac{7}{8} = \dfrac{4}{8}+\dfrac{7}{8} = \dfrac{4+7}{8} = \dfrac{11}{8}$

73. $\dfrac{7}{5}-\dfrac{3}{4} = \dfrac{28}{20}-\dfrac{15}{20} = \dfrac{28-15}{20} = \dfrac{13}{20}$

6.4 Adding and Subtracting Rational Expressions

6.4 Now Try Exercises

N1. **(a)** $\dfrac{2}{7k}+\dfrac{4}{7k} = \dfrac{2+4}{7k}$

$= \dfrac{6}{7k}$

(b) $\dfrac{4y}{y+3}+\dfrac{12}{y+3} = \dfrac{4y+12}{y+3}$

$= \dfrac{4(y+3)}{y+3}$

$= 4$

N2. **(a)** $\dfrac{5}{12}+\dfrac{3}{20}$

Step 1

$\text{LCD} = 2\cdot 2\cdot 3\cdot 5 = 60$

Step 2

$\dfrac{5}{12} = \dfrac{5(5)}{12(5)} = \dfrac{25}{60}$

$\dfrac{3}{20} = \dfrac{3(3)}{20(3)} = \dfrac{9}{60}$

Step 3

$\dfrac{25}{60}+\dfrac{9}{60} = \dfrac{25+9}{60}$

$= \dfrac{34}{60}$

Step 4

$\dfrac{34}{60} = \dfrac{17(2)}{30(2)} = \dfrac{17}{30}$

(b) $\dfrac{3}{5x}+\dfrac{2}{7x}$

Step 1

$\text{LCD} = 5\cdot 7\cdot x = 35x$

Step 2

$\dfrac{3}{5x} = \dfrac{3(7)}{5x(7)} = \dfrac{21}{35x}$

$\dfrac{2}{7x} = \dfrac{2(5)}{7x(5)} = \dfrac{10}{35x}$

Step 3

$\dfrac{21}{35x}+\dfrac{10}{35x} = \dfrac{31}{35x}$

N3. $\dfrac{6t}{t^2-9}+\dfrac{-3}{t+3}$

$=\dfrac{6t}{(t+3)(t-3)}+\dfrac{-3}{t+3}$ *Factor denominator.*

$=\dfrac{6t}{(t+3)(t-3)}+\dfrac{-3(t-3)}{(t+3)(t-3)}$

LCD = (t + 3)(t − 3)

$=\dfrac{6t+[-3(t-3)]}{(t+3)(t-3)}$ *Add numerators.*

$=\dfrac{6t-3t+9}{(t+3)(t-3)}$ *Dist. property*

$=\dfrac{3t+9}{(t+3)(t-3)}$ *Combine terms.*

$=\dfrac{3(t+3)}{(t+3)(t-3)}$ *Factor numerator.*

$=\dfrac{3}{t-3}$ *Lowest terms*

N4. $\dfrac{x-1}{x^2+6x+8}+\dfrac{4x}{x^2+x-12}$

$=\dfrac{x-1}{(x+2)(x+4)}+\dfrac{4x}{(x+4)(x-3)}$ *Factor.*

$=\dfrac{(x-1)(x-3)}{(x+2)(x+4)(x-3)}$

$+\dfrac{4x(x+2)}{(x+4)(x-3)(x+2)}$

LCD = (x + 4)(x + 2)(x − 3)

$=\dfrac{(x-1)(x-3)+4x(x+2)}{(x+4)(x+2)(x-3)}$ *Add numerators.*

$=\dfrac{x^2-4x+3+4x^2+8x}{(x+4)(x+2)(x-3)}$ *Multiply.*

$=\dfrac{5x^2+4x+3}{(x+4)(x+2)(x-3)}$ *Combine terms.*

N5. $\dfrac{2k}{k-7}+\dfrac{5}{7-k}$

$=\dfrac{2k}{k-7}+\dfrac{5(-1)}{(7-k)(-1)}$

$=\dfrac{2k}{k-7}+\dfrac{-5}{k-7}$

$=\dfrac{2k-5}{k-7}$, or $\dfrac{5-2k}{7-k}$

N6. $\dfrac{2x}{x+5}-\dfrac{x+1}{x+5}$

$=\dfrac{2x-(x+1)}{x+5}$ *Use parentheses.*

$=\dfrac{2x-x-1}{x+5}$

$=\dfrac{x-1}{x+5}$

N7. $\dfrac{6}{y-6}-\dfrac{2}{y}$

$=\dfrac{6(y)}{(y-6)(y)}-\dfrac{2(y-6)}{y(y-6)}$ *LCD = y(y − 6)*

$=\dfrac{6y-2(y-6)}{y(y-6)}$

$=\dfrac{6y-2y+12}{y(y-6)}$

$=\dfrac{4y+12}{y(y-6)}$

$=\dfrac{4(y+3)}{y(y-6)}$

N8. $\dfrac{2m}{m-4}-\dfrac{m-12}{4-m}$

The denominators are opposites, so either may be used as the common denominator. We will choose $m-4$.

$\dfrac{2m}{m-4}-\dfrac{m-12}{4-m}$

$=\dfrac{2m}{m-4}-\dfrac{(m-12)(-1)}{(4-m)(-1)}$

$=\dfrac{2m}{m-4}-\dfrac{-m+12}{m-4}$

$=\dfrac{2m-(-m+12)}{m-4}$

$=\dfrac{2m+m-12}{m-4}$

$=\dfrac{3m-12}{m-4}$

$=\dfrac{3(m-4)}{m-4}=3$

N9. $\dfrac{5}{t^2-6t+9}-\dfrac{2t}{t^2-9}$

$=\dfrac{5}{(t-3)^2}-\dfrac{2t}{(t+3)(t-3)}$ *Factor.*

$=\dfrac{5(t+3)}{(t-3)^2(t+3)}-\dfrac{2t(t-3)}{(t-3)^2(t+3)}$

LCD = $(t-3)^2(t+3)$

$=\dfrac{5(t+3)-2t(t-3)}{(t-3)^2(t+3)}$

$=\dfrac{5t+15-2t^2+6t}{(t-3)^2(t+3)}$

$=\dfrac{-2t^2+11t+15}{(t-3)^2(t+3)}$

6.4 Section Exercises

1. $\dfrac{x}{x+8}+\dfrac{8}{x+8}$

The denominators are the same, so the sum is found by adding the two numerators and keeping the same (common) denominator.

$$\frac{x}{x+8}+\frac{8}{x+8}=\frac{x+8}{x+8}=1$$

Choice **E** is correct.

3. $\dfrac{8}{x-8}-\dfrac{x}{x-8}$

The denominators are the same, so the difference is found by subtracting the two numerators and keeping the same (common) denominator.

$$\begin{aligned}\frac{8}{x-8}-\frac{x}{x-8}&=\frac{8-x}{x-8}\\&=\frac{-1(x-8)}{x-8}=-1\end{aligned}$$

Choice **C** is correct.

5. $\dfrac{x}{x+8}-\dfrac{8}{x+8}$

The denominators are the same, so the difference is found by subtracting the two numerators and keeping the same (common) denominator.

$$\frac{x}{x+8}-\frac{8}{x+8}=\frac{x-8}{x+8}$$

Choice **B** is correct.

7. $\dfrac{1}{8}-\dfrac{1}{x}$

The LCD is $8x$. Now rewrite each rational expression as a fraction with the LCD as its denominator.

$$\frac{1}{8}\cdot\frac{x}{x}=\frac{x}{8x}$$
$$\frac{1}{x}\cdot\frac{8}{8}=\frac{8}{8x}$$

Since the fractions now have a common denominator, subtract the numerators and use the LCD as the denominator of the sum.

$$\frac{1}{8}-\frac{1}{x}=\frac{x}{8x}-\frac{8}{8x}=\frac{x-8}{8x}$$

Choice **G** is correct.

9. $\dfrac{4}{m}+\dfrac{7}{m}$

The denominators are the same, so the sum is found by adding the two numerators and keeping the same (common) denominator.

$$\frac{4}{m}+\frac{7}{m}=\frac{4+7}{m}=\frac{11}{m}$$

11. $\dfrac{5}{y+4}-\dfrac{1}{y+4}$

The denominators are the same, so the difference is found by subtracting the two numerators and keeping the same (common) denominator.

$$\frac{5}{y+4}-\frac{1}{y+4}=\frac{5-1}{y+4}=\frac{4}{y+4}$$

13. $\dfrac{x}{x+y}+\dfrac{y}{x+y}$

The denominators are the same, so the sum is found by adding the two numerators and keeping the same (common) denominator.

$$\frac{x}{x+y}+\frac{y}{x+y}=\frac{x+y}{x+y}=1$$

15. $\dfrac{5m}{m+1}-\dfrac{1+4m}{m+1}$

The denominators are the same, so the difference is found by subtracting the two numerators and keeping the same (common) denominator. Don't forget the parentheses on the second numerator.

$$\begin{aligned}\frac{5m}{m+1}-\frac{1+4m}{m+1}&=\frac{5m-(1+4m)}{m+1}\\&=\frac{5m-1-4m}{m+1}\\&=\frac{m-1}{m+1}\end{aligned}$$

17. $\dfrac{a+b}{2}-\dfrac{a-b}{2}$

The denominators are the same, so the difference is found by subtracting the two numerators and keeping the same (common) denominator. Don't forget the parentheses on the second numerator.

$$\begin{aligned}\frac{a+b}{2}-\frac{a-b}{2}&=\frac{(a+b)-(a-b)}{2}\\&=\frac{a+b-a+b}{2}\\&=\frac{2b}{2}=b\end{aligned}$$

19.
$$\begin{aligned}\frac{x^2}{x+5}+\frac{5x}{x+5}&=\frac{x^2+5x}{x+5} && \textit{Add numerators.}\\&=\frac{x(x+5)}{x+5} && \textit{Factor numerator.}\\&=x && \textit{Lowest terms}\end{aligned}$$

21.
$$\begin{aligned}\frac{y^2-3y}{y+3}+\frac{-18}{y+3}&=\frac{y^2-3y-18}{y+3}\\&=\frac{(y-6)(y+3)}{y+3}\\&=y-6\end{aligned}$$

23. $\dfrac{x}{x^2-9}-\dfrac{-3}{x^2-9}$

The denominators are the same, so the sum is found by adding the two numerators and keeping the same (common) denominator.

$$\frac{x}{x^2-9}-\frac{-3}{x^2-9}=\frac{x+3}{x^2-9}$$
$$=\frac{x+3}{(x+3)(x-3)}$$
$$=\frac{1}{x-3}$$

25. $\dfrac{z}{5}+\dfrac{1}{3}$

The LCD is 15. Now rewrite each rational expression as a fraction with the LCD as its denominator.

$$\frac{z}{5}\cdot\frac{3}{3}=\frac{3z}{15}$$
$$\frac{1}{3}\cdot\frac{5}{5}=\frac{5}{15}$$

Since the fractions now have a common denominator, add the numerators and use the LCD as the denominator of the sum.

$$\frac{z}{5}+\frac{1}{3}=\frac{3z}{15}+\frac{5}{15}=\frac{3z+5}{15}$$

27. $\dfrac{5}{7}-\dfrac{r}{2}=\dfrac{5}{7}\cdot\dfrac{2}{2}-\dfrac{r}{2}\cdot\dfrac{7}{7}$ *LCD = 14*

$$=\frac{10}{14}-\frac{7r}{14}$$
$$=\frac{10-7r}{14}$$

29. $-\dfrac{3}{4}-\dfrac{1}{2x}=-\dfrac{3\cdot x}{4\cdot x}-\dfrac{1\cdot 2}{2x\cdot 2}$ *LCD = 4x*

$$=\frac{-3x-2}{4x}$$

31. $\dfrac{6}{5x}+\dfrac{9}{2x}=\dfrac{6}{5x}\cdot\dfrac{2}{2}+\dfrac{9}{2x}\cdot\dfrac{5}{5}$ *LCD = 10x*

$$=\frac{12}{10x}+\frac{45}{10x}$$
$$=\frac{12+45}{10x}=\frac{57}{10x}$$

33. $\dfrac{x+1}{6}+\dfrac{3x+3}{9}$

First reduce the second fraction.

$$\frac{3x+3}{9}=\frac{3(x+1)}{9}=\frac{x+1}{3}$$

Now the LCD of $\dfrac{x+1}{6}$ and $\dfrac{x+1}{3}$ is 6. Thus,

$$\frac{x+1}{6}+\frac{x+1}{3}=\frac{x+1}{6}+\frac{x+1}{3}\cdot\frac{2}{2}$$
$$=\frac{x+1+2x+2}{6}$$
$$=\frac{3x+3}{6}$$
$$=\frac{3(x+1)}{6}=\frac{x+1}{2}.$$

35. $\dfrac{x+3}{3x}+\dfrac{2x+2}{4x}=\dfrac{x+3}{3x}+\dfrac{2(x+1)}{4x}$

$$=\frac{x+3}{3x}+\frac{x+1}{2x}\quad \textit{Reduce.}$$
$$=\frac{x+3}{3x}\cdot\frac{2}{2}+\frac{x+1}{2x}\cdot\frac{3}{3}$$

LCD = 6x

$$=\frac{2x+6+3x+3}{6x}$$
$$=\frac{5x+9}{6x}$$

37. $\dfrac{7}{3p^2}-\dfrac{2}{p}=\dfrac{7}{3p^2}-\dfrac{2}{p}\cdot\dfrac{3p}{3p}$ *LCD = $3p^2$*

$$=\frac{7-6p}{3p^2}$$

39. $\dfrac{1}{k+4}-\dfrac{2}{k}=\dfrac{1}{k+4}\cdot\dfrac{k}{k}-\dfrac{2}{k}\cdot\dfrac{k+4}{k+4}$ *LCD = k(k + 4)*

$$=\frac{k}{k(k+4)}-\frac{2(k+4)}{k(k+4)}$$
$$=\frac{k-2k-8}{k(k+4)}$$
$$=\frac{-k-8}{k(k+4)}$$

41. $\dfrac{x}{x-2}+\dfrac{-8}{x^2-4}$

$$=\frac{x}{x-2}+\frac{-8}{(x+2)(x-2)}$$
$$=\frac{x}{x-2}\cdot\frac{x+2}{x+2}+\frac{-8}{(x+2)(x-2)}$$

LCD = (x + 2)(x − 2)

$$=\frac{x(x+2)-8}{(x+2)(x-2)}$$
$$=\frac{x^2+2x-8}{(x+2)(x-2)}$$
$$=\frac{(x+4)(x-2)}{(x+2)(x-2)}=\frac{x+4}{x+2}$$

43. $\frac{4m}{m^2+3m+2}+\frac{2m-1}{m^2+6m+5}$

$=\frac{4m}{(m+2)(m+1)}+\frac{2m-1}{(m+1)(m+5)}$

$=\frac{4m(m+5)}{(m+2)(m+1)(m+5)}$

$+\frac{(2m-1)(m+2)}{(m+1)(m+5)(m+2)}$

LCD $=(m+2)(m+1)(m+5)$

$=\frac{(4m^2+20m)+(2m^2+3m-2)}{(m+2)(m+1)(m+5)}$

$=\frac{6m^2+23m-2}{(m+2)(m+1)(m+5)}$

45. $\frac{4y}{y^2-1}-\frac{5}{y^2+2y+1}$

$=\frac{4y}{(y+1)(y-1)}-\frac{5}{(y+1)(y+1)}$

$=\frac{4y(y+1)}{(y+1)^2(y-1)}-\frac{5(y-1)}{(y+1)^2(y-1)}$

LCD $=(y+1)^2(y-1)$

$=\frac{(4y^2+4y)-(5y-5)}{(y+1)^2(y-1)}$

$=\frac{4y^2-y+5}{(y+1)^2(y-1)}$

47. $\frac{t}{t+2}+\frac{5-t}{t}-\frac{4}{t^2+2t}$

$=\frac{t}{t+2}+\frac{5-t}{t}-\frac{4}{t(t+2)}$

$=\frac{t}{t+2}\cdot\frac{t}{t}+\frac{5-t}{t}\cdot\frac{t+2}{t+2}$

$-\frac{4}{t(t+2)}$ *LCD* $=t(t+2)$

$=\frac{t\cdot t+(5-t)(t+2)-4}{t(t+2)}$

$=\frac{t^2+5t+10-t^2-2t-4}{t(t+2)}$

$=\frac{3t+6}{t(t+2)}$

$=\frac{3(t+2)}{t(t+2)}=\frac{3}{t}$

49. $\frac{10}{m-2}+\frac{5}{2-m}$

Since

$$2-m=-1(m-2),$$

either $m-2$ or $2-m$ could be used as the LCD.

51. $\frac{4}{x-5}+\frac{6}{5-x}$

The two denominators, $x-5$ and $5-x$, are opposites of each other, so either one may be used as the common denominator. We will work the exercise both ways and compare the answers.

$\frac{4}{x-5}+\frac{6}{5-x}=\frac{4}{x-5}+\frac{6(-1)}{(5-x)(-1)}$

LCD $=x-5$

$=\frac{4}{x-5}+\frac{-6}{x-5}$

$=\frac{-2}{x-5}$

$\frac{4}{x-5}+\frac{6}{5-x}=\frac{4(-1)}{(x-5)(-1)}+\frac{6}{5-x}$

LCD $=5-x$

$=\frac{-4}{5-x}+\frac{6}{5-x}$

$=\frac{2}{5-x}$

The two answers are equivalent, since

$$\frac{-2}{x-5}\cdot\frac{-1}{-1}=\frac{2}{5-x}.$$

53. $\frac{-1}{1-y}-\frac{4y-3}{y-1}$

The LCD is either $1-y$ or $y-1$. We'll use $y-1$.

$\frac{-1}{1-y}-\frac{4y-3}{y-1}=\frac{-1\cdot-1}{-1\cdot(1-y)}-\frac{4y-3}{y-1}$

$=\frac{1-(4y-3)}{y-1}$

$=\frac{1-4y+3}{y-1}$

$=\frac{4-4y}{y-1}$

$=\frac{4(1-y)}{y-1}=-4$

55. $\frac{2}{x-y^2}+\frac{7}{y^2-x}$

LCD $=x-y^2$ or y^2-x

We will use $x-y^2$.

$\frac{2}{x-y^2}+\frac{7}{y^2-x}$

$=\frac{2}{x-y^2}+\frac{-1(7)}{-1(y^2-x)}$

$=\frac{2}{x-y^2}+\frac{-7}{-y^2+x}$

$$= \frac{2}{x-y^2} + \frac{-7}{x-y^2}$$

$$= \frac{2+(-7)}{x-y^2} = \frac{-5}{x-y^2}$$

If $y^2 - x$ is used as the LCD, we will obtain the equivalent answer

$$\frac{5}{y^2-x}.$$

57. $\dfrac{x}{5x-3y} - \dfrac{y}{3y-5x}$

LCD $= 5x - 3y$ or $3y - 5x$
We will use $5x - 3y$.

$$\frac{x}{5x-3y} - \frac{y}{3y-5x}$$

$$= \frac{x}{5x-3y} - \frac{-1(y)}{-1(3y-5x)}$$

$$= \frac{x}{5x-3y} - \frac{-y}{-3y+5x}$$

$$= \frac{x}{5x-3y} - \frac{-y}{5x-3y}$$

$$= \frac{x-(-y)}{5x-3y} = \frac{x+y}{5x-3y}$$

If $3y - 5x$ is used as the LCD, we will obtain the equivalent answer

$$\frac{-x-y}{3y-5x}.$$

59. $\dfrac{3}{4p-5} + \dfrac{9}{5-4p}$

LCD $= 4p - 5$ or $5 - 4p$

We will use $4p - 5$.

$$\frac{3}{4p-5} + \frac{9}{5-4p}$$

$$= \frac{3}{4p-5} + \frac{-1(9)}{-1(5-4p)}$$

$$= \frac{3}{4p-5} + \frac{-9}{-5+4p}$$

$$= \frac{3}{4p-5} + \frac{-9}{4p-5}$$

$$= \frac{3+(-9)}{4p-5} = \frac{-6}{4p-5}$$

If $5 - 4p$ is used as the LCD, we will obtain the equivalent answer

$$\frac{6}{5-4p}.$$

61. $\dfrac{2m}{m-n} - \dfrac{5m+n}{2m-2n}$

$$= \frac{2m}{m-n} - \frac{5m+n}{2(m-n)} \quad \textit{Factor second denominator.}$$

$$= \frac{2m}{m-n} \cdot \frac{2}{2} - \frac{5m+n}{2(m-n)} \quad LCD = 2(m-n)$$

$$= \frac{4m-(5m+n)}{2(m-n)}$$

$$= \frac{4m-5m-n}{2(m-n)}$$

$$= \frac{-m-n}{2(m-n)}$$

63. $\dfrac{5}{x^2-9} - \dfrac{x+2}{x^2+4x+3}$

To find the LCD, factor the denominators.

$$x^2 - 9 = (x+3)(x-3)$$
$$x^2 + 4x + 3 = (x+3)(x+1)$$

The LCD is $(x+3)(x-3)(x+1)$.

$$\frac{5}{x^2-9} - \frac{x+2}{x^2+4x+3}$$

$$= \frac{5 \cdot (x+1)}{(x+3)(x-3) \cdot (x+1)} - \frac{(x+2) \cdot (x-3)}{(x+3)(x+1) \cdot (x-3)}$$

$$= \frac{5x+5}{(x+3)(x-3)(x+1)} - \frac{x^2-x-6}{(x+3)(x+1)(x-3)}$$

$$= \frac{(5x+5)-(x^2-x-6)}{(x+3)(x-3)(x+1)}$$

$$= \frac{5x+5-x^2+x+6}{(x+3)(x-3)(x+1)}$$

$$= \frac{-x^2+6x+11}{(x+3)(x-3)(x+1)}$$

65. $\dfrac{2q+1}{3q^2+10q-8} - \dfrac{3q+5}{2q^2+5q-12}$

$$= \frac{2q+1}{(3q-2)(q+4)} - \frac{3q+5}{(2q-3)(q+4)}$$

$$= \frac{(2q+1) \cdot (2q-3)}{(3q-2)(q+4) \cdot (2q-3)} - \frac{(3q+5) \cdot (3q-2)}{(2q-3)(q+4) \cdot (3q-2)}$$

$$LCD = (3q-2)(q+4)(2q-3)$$

$$= \frac{(4q^2-4q-3)-(9q^2+9q-10)}{(3q-2)(q+4)(2q-3)}$$

$$= \frac{4q^2-4q-3-9q^2-9q+10}{(3q-2)(q+4)(2q-3)}$$

$$= \frac{-5q^2-13q+7}{(3q-2)(q+4)(2q-3)}$$

67. $\frac{4}{r^2-r}+\frac{6}{r^2+2r}-\frac{1}{r^2+r-2}$

$$=\frac{4}{r(r-1)}+\frac{6}{r(r+2)}-\frac{1}{(r+2)(r-1)}$$

$$=\frac{4\cdot(r+2)}{r(r-1)\cdot(r+2)}+\frac{6\cdot(r-1)}{r(r+2)\cdot(r-1)}$$

$$-\frac{1\cdot r}{r\cdot(r+2)(r-1)}$$

LCD $= r(r+2)(r-1)$

$$=\frac{4r+8+6r-6-r}{r(r+2)(r-1)}$$

$$=\frac{9r+2}{r(r+2)(r-1)}$$

69. $\frac{x+3y}{x^2+2xy+y^2}+\frac{x-y}{x^2+4xy+3y^2}$

$$=\frac{x+3y}{(x+y)(x+y)}+\frac{x-y}{(x+3y)(x+y)}$$

$$=\frac{(x+3y)\cdot(x+3y)}{(x+y)(x+y)\cdot(x+3y)}$$

$$+\frac{(x-y)\cdot(x+y)}{(x+3y)(x+y)\cdot(x+y)}$$

LCD $= (x+y)(x+y)(x+3y)$

$$=\frac{(x^2+6xy+9y^2)+(x^2-y^2)}{(x+y)(x+y)(x+3y)}$$

$$=\frac{2x^2+6xy+8y^2}{(x+y)(x+y)(x+3y)}$$

$$=\frac{2(x^2+3xy+4y^2)}{(x+y)(x+y)(x+3y)},$$

or $\frac{2(x^2+3xy+4y^2)}{(x+y)^2(x+3y)}$

71. $\frac{r+y}{18r^2+9ry-2y^2}+\frac{3r-y}{36r^2-y^2}$

Factor the first denominator by grouping. Find integers whose product is $18(-2)=-36$ and whose sum is 9. The integers are 12 and -3.

$$\begin{aligned}&18r^2+9ry-2y^2\\&=18r^2+12ry-3ry-2y^2\\&=6r(3r+2y)-y(3r+2y)\\&=(3r+2y)(6r-y)\end{aligned}$$

Factor the second denominator.

$$\frac{r+y}{(3r+2y)(6r-y)}+\frac{3r-y}{(6r-y)(6r+y)}$$

Rewrite fractions with the LCD, $(3r+2y)(6r-y)(6r+y)$.

$$=\frac{(r+y)\cdot(6r+y)}{(3r+2y)(6r-y)\cdot(6r+y)}$$

$$+\frac{(3r-y)\cdot(3r+2y)}{(6r-y)(6r+y)\cdot(3r+2y)}$$

$$=\frac{6r^2+7ry+y^2}{(3r+2y)(6r-y)(6r+y)}$$

$$+\frac{9r^2+3ry-2y^2}{(3r+2y)(6r-y)(6r+y)}$$

$$=\frac{6r^2+7ry+y^2+9r^2+3ry-2y^2}{(3r+2y)(6r-y)(6r+y)}$$

$$=\frac{15r^2+10ry-y^2}{(3r+2y)(6r-y)(6r+y)}$$

73. **(a)** $P=2L+2W$

$$=2\left(\frac{3k+1}{10}\right)+2\left(\frac{5}{6k+2}\right)$$

$$=2\left(\frac{3k+1}{2\cdot 5}\right)+2\left(\frac{5}{2(3k+1)}\right)$$

$$=\frac{3k+1}{5}+\frac{5}{3k+1}$$

To add the two fractions on the right, use $5(3k+1)$ as the LCD.

$$P=\frac{(3k+1)(3k+1)}{5(3k+1)}+\frac{(5)(5)}{5(3k+1)}$$

$$=\frac{(3k+1)(3k+1)+(5)(5)}{5(3k+1)}$$

$$=\frac{9k^2+6k+1+25}{5(3k+1)}$$

$$=\frac{9k^2+6k+26}{5(3k+1)}$$

(b) $\mathcal{A}=L\cdot W$

$$\mathcal{A}=\frac{3k+1}{10}\cdot\frac{5}{6k+2}$$

$$=\frac{3k+1}{5\cdot 2}\cdot\frac{5}{2(3k+1)}$$

$$=\frac{1}{2\cdot 2}=\frac{1}{4}$$

75. $\frac{1010}{49(101-x)}-\frac{10}{49}$

$$=\frac{1010}{49(101-x)}-\frac{10(101-x)}{49(101-x)}$$

$$=\frac{1010-1010+10x}{49(101-x)}$$

$$=\frac{10x}{49(101-x)}$$

77. $\frac{\frac{5}{6}}{\frac{2}{3}} = \frac{5}{6} \div \frac{2}{3} = \frac{5}{6} \cdot \frac{3}{2} = \frac{5}{\cancel{6}_2} \cdot \frac{\cancel{3}^1}{2} = \frac{5}{4}$

79. $\frac{\frac{3}{2}}{\frac{7}{4}} = \frac{3}{2} \div \frac{7}{4} = \frac{3}{2} \cdot \frac{4}{7} = \frac{3}{\cancel{2}_1} \cdot \frac{\cancel{4}^2}{7} = \frac{6}{7}$

6.5 Complex Fractions

6.5 Now Try Exercises

N1. (a) $$\frac{\frac{2}{5}+\frac{1}{4}}{\frac{1}{6}+\frac{3}{8}} = \frac{\frac{2(4)}{5(4)}+\frac{1(5)}{4(5)}}{\frac{1(4)}{6(4)}+\frac{3(3)}{8(3)}} = \frac{\frac{8+5}{20}}{\frac{4+9}{24}} = \frac{13}{20} \div \frac{13}{24}$$

$$= \frac{13}{20} \cdot \frac{24}{13} \quad \textit{Multiply by reciprocal.}$$

$$= \frac{13 \cdot 4 \cdot 6}{4 \cdot 5 \cdot 13} = \frac{6}{5}$$

(b) $$\frac{2+\frac{4}{x}}{\frac{5}{6}+\frac{5x}{12}} = \frac{\frac{2}{1}+\frac{4}{x}}{\frac{5}{6}+\frac{5x}{12}} = \frac{\frac{2(x)}{1(x)}+\frac{4}{x}}{\frac{5(2)}{6(2)}+\frac{5x}{12}}$$

$$= \frac{\frac{2x+4}{x}}{\frac{10+5x}{12}} = \frac{2x+4}{x} \div \frac{10+5x}{12}$$

$$= \frac{2x+4}{x} \cdot \frac{12}{10+5x} = \frac{2(x+2) \cdot 12}{x \cdot 5(2+x)} = \frac{24}{5x}$$

N2. $$\frac{\frac{a^2b}{c}}{\frac{ab^2}{c^3}} = \frac{a^2b}{c} \div \frac{ab^2}{c^3} = \frac{a^2b}{c} \cdot \frac{c^3}{ab^2} = \frac{ac^2}{b}$$

N3. $$\frac{5+\frac{2}{a-3}}{\frac{1}{a-3}-2} = \frac{\frac{5(a-3)}{1(a-3)}+\frac{2}{a-3}}{\frac{1}{a-3}-\frac{2(a-3)}{1(a-3)}}$$

$$= \frac{\frac{5(a-3)+2}{a-3}}{\frac{1-2(a-3)}{a-3}} = \frac{\frac{5a-15+2}{a-3}}{\frac{1-2a+6}{a-3}} = \frac{\frac{5a-13}{a-3}}{\frac{7-2a}{a-3}}$$

$$= \frac{5a-13}{a-3} \div \frac{7-2a}{a-3} = \frac{5a-13}{a-3} \cdot \frac{a-3}{7-2a} = \frac{5a-13}{7-2a}$$

N4. (a) $$\frac{\frac{3}{5}-\frac{1}{4}}{\frac{1}{8}+\frac{3}{20}}$$

$$= \frac{40\left(\frac{3}{5}-\frac{1}{4}\right)}{40\left(\frac{1}{8}+\frac{3}{20}\right)} \quad \textit{LCD} = 2^3 \cdot 5 = 40$$

$$= \frac{40\left(\frac{3}{5}\right)-40\left(\frac{1}{4}\right)}{40\left(\frac{1}{8}\right)+40\left(\frac{3}{20}\right)} = \frac{24-10}{5+6} = \frac{14}{11}$$

(b) $$\frac{\frac{2}{x}-3}{7+\frac{x}{5}} \quad \textit{LCD} = 5x$$

$$= \frac{5x\left(\frac{2}{x}-3\right)}{5x\left(7+\frac{x}{5}\right)} \quad \textit{Multiply numerator and denominator by } 5x.$$

$$= \frac{5x\left(\frac{2}{x}\right)-5x(3)}{5x(7)+5x\left(\frac{x}{5}\right)} = \frac{10-15x}{35x+x^2}, \text{ or } \frac{10-15x}{x^2+35x}$$

N5. $\dfrac{\frac{1}{y}+\frac{2}{3y^2}}{\frac{5}{4y^2}-\frac{3}{2y^3}}$ *LCD = $12y^3$*

$$=\frac{12y^3\left(\frac{1}{y}+\frac{2}{3y^2}\right)}{12y^3\left(\frac{5}{4y^2}-\frac{3}{2y^3}\right)}$$

$$=\frac{12y^3\left(\frac{1}{y}\right)+12y^3\left(\frac{2}{3y^2}\right)}{12y^3\left(\frac{5}{4y^2}\right)-12y^3\left(\frac{3}{2y^3}\right)}$$

$$=\frac{12y^2+8y}{15y-18}$$

N6. (a) $\dfrac{1-\frac{2}{x}-\frac{15}{x^2}}{1+\frac{5}{x}+\frac{6}{x^2}}$ *LCD = x^2*

$$=\frac{x^2\left(1-\frac{2}{x}-\frac{15}{x^2}\right)}{x^2\left(1+\frac{5}{x}+\frac{6}{x^2}\right)}\quad\text{Use Method 2.}$$

$$=\frac{x^2(1)-x^2\left(\frac{2}{x}\right)-x^2\left(\frac{15}{x^2}\right)}{x^2(1)+x^2\left(\frac{5}{x}\right)+x^2\left(\frac{6}{x^2}\right)}$$

$$=\frac{x^2-2x-15}{x^2+5x+6}$$

$$=\frac{(x-5)(x+3)}{(x+2)(x+3)}$$

$$=\frac{x-5}{x+2}$$

(b) $$\frac{\frac{9y^2-16}{y^2-100}}{\frac{3y-4}{y+10}}=\frac{9y^2-16}{y^2-100}\div\frac{3y-4}{y+10}$$

$$=\frac{9y^2-16}{y^2-100}\cdot\frac{y+10}{3y-4}$$

$$=\frac{(3y+4)(3y-4)}{(y+10)(y-10)}\cdot\frac{y+10}{3y-4}$$

$$=\frac{3y+4}{y-10}$$

N7. $\dfrac{2y^{-1}-3y^{-2}}{y^{-2}+3x^{-1}}=\dfrac{\frac{2}{y}-\frac{3}{y^2}}{\frac{1}{y^2}+\frac{3}{x}}$ *LCD = xy^2*

$$=\frac{xy^2\left(\frac{2}{y}-\frac{3}{y^2}\right)}{xy^2\cdot\left(\frac{1}{y^2}+\frac{3}{x}\right)}$$

$$=\frac{xy^2\cdot\frac{2}{y}-xy^2\cdot\frac{3}{y^2}}{xy^2\cdot\frac{1}{y^2}+xy^2\cdot\frac{3}{x}}$$

$$=\frac{2xy-3x}{x+3y^2}$$

6.5 Section Exercises

1. **(a)** The LCD of $\frac{3}{2}$ and $\frac{4}{3}$ is $2\cdot3=6$. The simplified form of the numerator is

$$\frac{3}{2}-\frac{4}{3}=\frac{9}{6}-\frac{8}{6}=\frac{1}{6}.$$

(b) The LCD of $\frac{1}{6}$ and $\frac{5}{12}$ is 12 since 12 is a multiple of 6. The simplified form of the denominator is

$$\frac{1}{6}-\frac{5}{12}=\frac{2}{12}-\frac{5}{12}=-\frac{3}{12}=-\frac{1}{4}.$$

(c) $$\frac{\frac{1}{6}}{-\frac{1}{4}}=\frac{1}{6}\div\left(-\frac{1}{4}\right)$$

(d) $$\frac{1}{6}\div\left(-\frac{1}{4}\right)=\frac{1}{6}\cdot\left(-\frac{4}{1}\right)$$

$$=-\frac{2\cdot2}{2\cdot3}=-\frac{2}{3}$$

3. $$\frac{2-\frac{1}{4}}{3-\frac{1}{2}}=\frac{-2+\frac{1}{4}}{-3+\frac{1}{2}}$$

Choice **D** is equivalent to the given fraction. Each term of the numerator and denominator has been multiplied by -1. Since $\frac{-1}{-1}=1$, the fraction has been multiplied by the identity element, so its value is unchanged.

In Exercises 5–40, either Method 1 or Method 2 can be used to simplify each complex fraction. Only one method will be shown for each exercise.

5. To use Method 1, divide the numerator of the complex fraction by the denominator.

$$\frac{-\frac{4}{3}}{\frac{2}{9}}=-\frac{4}{3}\div\frac{2}{9}$$

$$=-\frac{4}{3}\cdot\frac{9}{2}=-\frac{36}{6}=-6$$

7. To use Method 2, multiply the numerator and denominator of the complex fraction by the LCD, y^2.

$$\frac{\frac{x}{y^2}}{\frac{x^2}{y}}=\frac{y^2\left(\frac{x}{y^2}\right)}{y^2\left(\frac{x^2}{y}\right)}$$

$$=\frac{x}{yx^2}=\frac{1}{xy}$$

9. $\dfrac{\frac{4a^4b^3}{3a}}{\frac{2ab^4}{b^2}} = \dfrac{4a^4b^3}{3a} \div \dfrac{2ab^4}{b^2}$ *Method 1*

$$= \frac{4a^4b^3}{3a} \cdot \frac{b^2}{2ab^4}$$
$$= \frac{4a^4b^3 \cdot b^2}{3a \cdot 2ab^4}$$
$$= \frac{4a^4b^5}{6a^2b^4}$$
$$= \frac{2a^2b}{3}$$

11. To use Method 2, multiply the numerator and denominator of the complex fraction by the LCD, $3m$.

$$\frac{\frac{m+2}{3}}{\frac{m-4}{m}} = \frac{3m\left(\frac{m+2}{3}\right)}{3m\left(\frac{m-4}{m}\right)}$$
$$= \frac{m(m+2)}{3(m-4)}$$

13. $\dfrac{\frac{2}{x}-3}{\frac{2-3x}{2}} = \dfrac{2x\left(\frac{2}{x}-3\right)}{2x\left(\frac{2-3x}{2}\right)}$ *Method 2; LCD = 2x*

$$= \frac{2x\left(\frac{2}{x}\right) - 2x(3)}{x(2-3x)}$$
$$= \frac{4-6x}{x(2-3x)}$$
$$= \frac{2(2-3x)}{x(2-3x)} \quad \textit{Factor.}$$
$$= \frac{2}{x} \quad \textit{Lowest terms}$$

15. $\dfrac{\frac{1}{x}+x}{\frac{x^2+1}{8}} = \dfrac{8x\left(\frac{1}{x}+x\right)}{8x\left(\frac{x^2+1}{8}\right)}$ *Method 2; LCD = 8x*

$$= \frac{8+8x^2}{x(x^2+1)} \quad \textit{Distributive property}$$
$$= \frac{8(1+x^2)}{x(x^2+1)} \quad \textit{Factor.}$$
$$= \frac{8}{x} \quad \textit{Lowest terms}$$

17. $\dfrac{a-\frac{5}{a}}{a+\frac{1}{a}} = \dfrac{a\left(a-\frac{5}{a}\right)}{a\left(a+\frac{1}{a}\right)}$ *Method 2; LCD = a*

$$= \frac{a^2-5}{a^2+1}$$

19. $\dfrac{\frac{5}{8}+\frac{2}{3}}{\frac{7}{3}-\frac{1}{4}} = \dfrac{24\left(\frac{5}{8}+\frac{2}{3}\right)}{24\left(\frac{7}{3}-\frac{1}{4}\right)}$ *Method 2; LCD = 24*

$$= \frac{24\left(\frac{5}{8}\right)+24\left(\frac{2}{3}\right)}{24\left(\frac{7}{3}\right)-24\left(\frac{1}{4}\right)}$$
$$= \frac{15+16}{56-6} = \frac{31}{50}$$

21. $\dfrac{\frac{1}{x^2}+\frac{1}{y^2}}{\frac{1}{x}-\frac{1}{y}}$

$$= \frac{x^2y^2\left(\frac{1}{x^2}+\frac{1}{y^2}\right)}{x^2y^2\left(\frac{1}{x}-\frac{1}{y}\right)} \quad \textit{Method 2; LCD} = x^2y^2$$
$$= \frac{x^2y^2\left(\frac{1}{x^2}\right)+x^2y^2\left(\frac{1}{y^2}\right)}{x^2y^2\left(\frac{1}{x}\right)-x^2y^2\left(\frac{1}{y}\right)}$$
$$= \frac{y^2+x^2}{xy^2-x^2y} = \frac{y^2+x^2}{xy(y-x)}$$

23. $\dfrac{\frac{2}{p^2}-\frac{3}{5p}}{\frac{4}{p}+\frac{1}{4p}} = \dfrac{20p^2\left(\frac{2}{p^2}-\frac{3}{5p}\right)}{20p^2\left(\frac{4}{p}+\frac{1}{4p}\right)}$ *Method 2; LCD = $20p^2$*

$$= \frac{20p^2\left(\frac{2}{p^2}\right)-20p^2\left(\frac{3}{5p}\right)}{20p^2\left(\frac{4}{p}\right)+20p^2\left(\frac{1}{4p}\right)}$$
$$= \frac{40-12p}{80p+5p} = \frac{40-12p}{85p}$$

25. $\dfrac{\frac{5}{x^2y}-\frac{2}{xy^2}}{\frac{3}{x^2y^2}+\frac{4}{xy}}$

$$= \frac{x^2y^2\left(\frac{5}{x^2y}-\frac{2}{xy^2}\right)}{x^2y^2\left(\frac{3}{x^2y^2}+\frac{4}{xy}\right)} \quad \textit{Method 2; LCD} = x^2y^2$$
$$= \frac{x^2y^2\left(\frac{5}{x^2y}\right)-x^2y^2\left(\frac{2}{xy^2}\right)}{x^2y^2\left(\frac{3}{x^2y^2}\right)+x^2y^2\left(\frac{4}{xy}\right)}$$
$$= \frac{5y-2x}{3+4xy}$$

27. $\dfrac{\frac{1}{4}-\frac{1}{a^2}}{\frac{1}{2}+\frac{1}{a}}$

$= \dfrac{4a^2\left(\frac{1}{4}-\frac{1}{a^2}\right)}{4a^2\left(\frac{1}{2}+\frac{1}{a}\right)}$ *Method 2; LCD = $4a^2$*

$= \dfrac{a^2-4}{2a^2+4a}$ *Distributive property*

$= \dfrac{(a-2)(a+2)}{2a(a+2)}$ *Factor numerator and denominator.*

$= \dfrac{a-2}{2a}$ *Fundamental property*

29. $\dfrac{\frac{1}{z+5}}{\frac{4}{z^2-25}}$

$= \dfrac{1}{z+5} \div \dfrac{4}{z^2-25}$ *Method 1*

$= \dfrac{1}{z+5} \cdot \dfrac{z^2-25}{4}$ *Multiply by reciprocal.*

$= \dfrac{1 \cdot (z^2-25)}{(z+5) \cdot 4}$ *Multiply.*

$= \dfrac{(z+5)(z-5)}{(z+5) \cdot 4}$ *Factor numerator.*

$= \dfrac{z-5}{4}$ *Fundamental property*

31. $\dfrac{\frac{1}{m+1}-1}{\frac{1}{m+1}+1}$

$= \dfrac{(m+1)\left(\frac{1}{m+1}-1\right)}{(m+1)\left(\frac{1}{m+1}+1\right)}$ *Method 2; LCD = m + 1*

$= \dfrac{1-1(m+1)}{1+1(m+1)}$ *Distributive property*

$= \dfrac{1-m-1}{1+m+1}$ *Distributive property*

$= \dfrac{-m}{m+2}$

33. $\dfrac{\frac{1}{m-1}+\frac{2}{m+2}}{\frac{2}{m+2}-\frac{1}{m-3}}$

$= \dfrac{(m-1)(m+2)(m-3)\left(\frac{1}{m-1}+\frac{2}{m+2}\right)}{(m-1)(m+2)(m-3)\left(\frac{2}{m+2}-\frac{1}{m-3}\right)}$

Method 2; LCD = (m – 1)(m + 2)(m – 3)

$= \dfrac{(m+2)(m-3)+2(m-1)(m-3)}{2(m-1)(m-3)-(m-1)(m+2)}$

Distributive property

$= \dfrac{(m-3)[(m+2)+2(m-1)]}{(m-1)[2(m-3)-(m+2)]}$

Factor out m – 3 in numerator and m – 1 in denominator.

$= \dfrac{(m-3)[m+2+2m-2]}{(m-1)[2m-6-m-2]}$

Distributive property

$= \dfrac{3m(m-3)}{(m-1)(m-8)}$ *Combine like terms.*

35. $\dfrac{2+\frac{1}{x}-\frac{28}{x^2}}{3+\frac{13}{x}+\frac{4}{x^2}}$ *LCD = x^2*

$= \dfrac{x^2\left(2+\frac{1}{x}-\frac{28}{x^2}\right)}{x^2\left(3+\frac{13}{x}+\frac{4}{x^2}\right)}$ *Method 2*

$= \dfrac{2x^2+x-28}{3x^2+13x+4}$

$= \dfrac{(2x-7)(x+4)}{(3x+1)(x+4)}$

$= \dfrac{2x-7}{3x+1}$

37. $\dfrac{\frac{y+8}{y-4}}{\frac{y^2-64}{y^2-16}}$

$= \dfrac{y+8}{y-4} \div \dfrac{y^2-64}{y^2-16}$ *Method 1*

$= \dfrac{y+8}{y-4} \cdot \dfrac{y^2-16}{y^2-64}$ *Multiply by reciprocal.*

$= \dfrac{(y+8) \cdot (y^2-16)}{(y-4) \cdot (y^2-64)}$ *Multiply.*

$= \dfrac{(y+8)(y+4)(y-4)}{(y-4)(y+8)(y-8)}$ *Factor.*

$= \dfrac{y+4}{y-8}$ *Fundamental property*

39. $\dfrac{1}{x^{-2}+y^{-2}} = \dfrac{1}{\frac{1}{x^2}+\frac{1}{y^2}}$

$= \dfrac{x^2y^2(1)}{x^2y^2\left(\frac{1}{x^2}+\frac{1}{y^2}\right)}$ *LCD = x^2y^2*

$= \dfrac{x^2y^2}{y^2+x^2}$

41. $\dfrac{x^{-2}+y^{-2}}{x^{-1}+y^{-1}}$

$$=\frac{\frac{1}{x^2}+\frac{1}{y^2}}{\frac{1}{x}+\frac{1}{y}}$$

Multiply the numerator and denominator by x^2y^2, the LCD of all the fractions.

$$=\frac{x^2y^2\left(\frac{1}{x^2}+\frac{1}{y^2}\right)}{x^2y^2\left(\frac{1}{x}+\frac{1}{y}\right)}$$

$$=\frac{x^2y^2\cdot\frac{1}{x^2}+x^2y^2\cdot\frac{1}{y^2}}{x^2y^2\cdot\frac{1}{x}+x^2y^2\cdot\frac{1}{y}}$$

$$=\frac{y^2+x^2}{xy^2+x^2y},\text{ or }\frac{y^2+x^2}{xy(y+x)}$$

43. $\dfrac{x^{-1}+2y^{-1}}{2y+4x}=\dfrac{\frac{1}{x}+\frac{2}{y}}{2y+4x}$

Multiply the numerator and denominator by xy, the LCD of all the fractions.

$$=\frac{xy\left(\frac{1}{x}+\frac{2}{y}\right)}{xy(2y+4x)}$$

$$=\frac{y+2x}{2xy(y+2x)}$$

$$=\frac{1}{2xy}$$

45. $\dfrac{1+x^{-1}-12x^{-2}}{1-x^{-1}-20x^{-2}}$

$$=\frac{1+\frac{1}{x}-\frac{12}{x^2}}{1-\frac{1}{x}-\frac{20}{x^2}}\quad\textit{LCD} = x^2$$

$$=\frac{x^2\left(1+\frac{1}{x}-\frac{12}{x^2}\right)}{x^2\left(1-\frac{1}{x}-\frac{20}{x^2}\right)}\quad\textit{Method 2}$$

$$=\frac{x^2+x-12}{x^2-x-20}\quad\textit{Multiply.}$$

$$=\frac{(x+4)(x-3)}{(x-5)(x+4)}\quad\textit{Factor.}$$

$$=\frac{x-3}{x-5}\quad\textit{Fundamental property}$$

47. In a fraction, the fraction bar represents division. For example, $\frac{3}{5}$ can be read "3 divided by 5."

◆◆◆ Relating Concepts 49–52 ◆◆◆

49. "The sum of $\frac{3}{8}$ and $\frac{5}{6}$, divided by 2" is written

$$\frac{\frac{3}{8}+\frac{5}{6}}{2}.$$

50. $\dfrac{\frac{3}{8}+\frac{5}{6}}{2}=\dfrac{\frac{9}{24}+\frac{20}{24}}{2}$ *Method 1*

$$=\frac{\frac{29}{24}}{\frac{2}{1}}=\frac{29}{24}\cdot\frac{1}{2}=\frac{29}{48}$$

51. $\dfrac{\frac{3}{8}+\frac{5}{6}}{2}=\dfrac{24\left(\frac{3}{8}+\frac{5}{6}\right)}{24(2)}$ *Method 2; LCD = 24*

$$=\frac{24\left(\frac{3}{8}\right)+24\left(\frac{5}{6}\right)}{24(2)}$$

$$=\frac{9+20}{48}=\frac{29}{48}$$

52. Method 2 is usually shorter for more complex problems because the problem can be worked without adding and subtracting rational expressions, which can be complicated and time-consuming.

53. $1+\dfrac{1}{1+\frac{1}{1+1}}=1+\dfrac{1}{1+\frac{1}{2}}$

$$=1+\frac{1}{\frac{2}{2}+\frac{1}{2}}$$

$$=1+\frac{1}{\frac{3}{2}}$$

$$=1+1\cdot\tfrac{2}{3}$$

$$=1+\tfrac{2}{3}$$

$$=\tfrac{3}{3}+\tfrac{2}{3}=\tfrac{5}{3}$$

55. $7-\dfrac{3}{5+\frac{2}{4-2}}=7-\dfrac{3}{5+\frac{2}{2}}$

$$=7-\frac{3}{5+1}$$

$$=7-\tfrac{3}{6}$$

$$=7-\tfrac{1}{2}$$

$$=\tfrac{14}{2}-\tfrac{1}{2}=\tfrac{13}{2}$$

57. $r+\dfrac{r}{4-\frac{2}{6+2}}=r+\dfrac{r}{4-\frac{2}{8}}$

$$=r+\frac{r}{4-\frac{1}{4}}$$

$$=r+\frac{r}{\frac{16}{4}-\frac{1}{4}}$$

$$=r+\frac{r}{\frac{15}{4}}$$

$$=r+r\cdot\tfrac{4}{15}$$

$$=r+\frac{4r}{15}$$

$$=\frac{15r}{15}+\frac{4r}{15}$$

$$=\frac{19r}{15}$$

59. $9\left(\frac{4x}{3}+\frac{2}{9}\right)=9\left(\frac{4x}{3}\right)+9\left(\frac{2}{9}\right)$
$=3(4x)+2$
$=12x+2$

61. $-12\left(\frac{11p^2}{3}-\frac{9p}{4}\right)$
$=-12\left(\frac{11p^2}{3}\right)-12\left(-\frac{9p}{4}\right)$
$=-4(11p^2)+3(9p)$
$=-44p^2+27p$

63. $3x+5=7x+3$
$5=4x+3$ *Subtract 3x.*
$2=4x$ *Subtract 3.*
$x=\frac{2}{4}=\frac{1}{2}$ *Divide by 4.*

The solution set is $\{\frac{1}{2}\}$.

65. $6(z-3)+5=8z-3$
$6z-18+5=8z-3$
$6z-13=8z-3$
$-13=2z-3$
$-10=2z$
$z=\frac{-10}{2}=-5$

The solution set is $\{-5\}$.

6.6 Solving Equations with Rational Expressions

6.6 Now Try Exercises

N1. **(a)** $\frac{3}{2}t-\frac{5}{7}t=\frac{11}{7}$ has an equals sign, so this is an *equation* to be solved. Use the multiplication property of equality to clear fractions. The LCD is 14.

$\frac{3}{2}t-\frac{5}{7}t=\frac{11}{7}$
$14(\frac{3}{2}t-\frac{5}{7}t)=14(\frac{11}{7})$ *Multiply by 14.*
$14(\frac{3}{2}t)-14(\frac{5}{7}t)=14(\frac{11}{7})$ *Distributive property*
$21t-10t=22$ *Multiply.*
$11t=22$ *Combine terms.*
$t=2$ *Divide by 11.*

Check $t=2$: $\frac{21}{7}-\frac{10}{7}=\frac{11}{7}$ *True*

The solution set is $\{2\}$.

(b) $\frac{3}{2}t-\frac{5}{7}t$ is the difference of two terms, so it is an *expression* to be simplified. Simplify by finding the LCD, writing each coefficient with this LCD, and combining like terms.

$\frac{3}{2}t-\frac{5}{7}t=\frac{3\cdot 7}{2\cdot 7}t-\frac{5\cdot 2}{7\cdot 2}t$ *LCD = 14*
$=\frac{21}{14}t-\frac{10}{14}t$ *Multiply.*
$=\frac{11}{14}t$ *Combine terms.*

N2. $\frac{x+5}{5}-\frac{x}{7}=\frac{3}{7}$
$35\left(\frac{x+5}{5}-\frac{x}{7}\right)=35\left(\frac{3}{7}\right)$
Multiply by the LCD, 35.
$7(x+5)-5x=5(3)$
$7x+35-5x=15$
$2x+35=15$ *Combine like terms.*
$2x=-20$ *Subtract 35.*
$x=-10$ *Divide by 2.*

Check $x=-10$: $-1+\frac{10}{7}=\frac{3}{7}$ *True*

The solution set is $\{-10\}$.

N3. $4+\frac{6}{x-3}=\frac{2x}{x-3}$
$(x-3)\left(4+\frac{6}{x-3}\right)=(x-3)\frac{2x}{x-3}$
Multiply by the LCD, x – 3.
$4(x-3)+6=2x$
$4x-12+6=2x$
$4x-6=2x$
$2x=6$
$x=3$

Check $x=3$: $4+\frac{6}{0}=\frac{6}{0}$

The fractions are undefined, so the equation has no solution. The solution set is $\emptyset$.

N4. $\frac{3}{2x^2-8x}=\frac{1}{x^2-16}$
$\frac{3}{2x(x-4)}=\frac{1}{(x+4)(x-4)}$ *Factor.*
Multiply by the LCD, 2x(x + 4)(x – 4).
$2x(x+4)(x-4)\left(\frac{3}{2x(x-4)}\right)$
$=2x(x+4)(x-4)\left(\frac{1}{(x+4)(x-4)}\right)$
$3(x+4)=2x(1)$
$3x+12=2x$
$x=-12$

Check $x=-12$: $\frac{3}{384}=\frac{1}{128}$ *True*

The solution set is $\{-12\}$.

N5. $\frac{2y}{y^2-25}=\frac{8}{y+5}-\frac{1}{y-5}$
$\frac{2y}{(y+5)(y-5)}=\frac{8}{y+5}-\frac{1}{y-5}$ *Factor.*
Multiply by the LCD, (y + 5)(y – 5).
$(y+5)(y-5)\left(\frac{2y}{(y+5)(y-5)}\right)$
$=(y+5)(y-5)\left(\frac{8}{y+5}-\frac{1}{y-5}\right)$

$$2y = 8(y-5) - 1(y+5)$$
$$2y = 8y - 40 - y - 5$$
$$2y = 7y - 45$$
$$45 = 5y$$
$$9 = y$$

Check $y = 9$: $\frac{18}{56} = \frac{8}{14} - \frac{1}{4}$ *True*

The solution set is $\{9\}$.

N6.
$$\frac{3}{m^2-9} = \frac{1}{2(m-3)} - \frac{1}{4}$$
$$\frac{3}{(m+3)(m-3)} = \frac{1}{2(m-3)} - \frac{1}{4} \quad \textit{Factor.}$$

Multiply by the LCD, 4(m − 3)(m + 3).

$$4(m-3)(m+3)\left(\frac{3}{(m+3)(m-3)}\right)$$
$$= 4(m-3)(m+3)\left(\frac{1}{2(m-3)} - \frac{1}{4}\right)$$
$$4(3) = 2(m+3) - (m-3)(m+3)$$
$$12 = 2m + 6 - (m^2 - 9)$$
$$12 = 2m + 6 - m^2 + 9$$
$$0 = -m^2 + 2m + 3$$
$$0 = m^2 - 2m - 3$$
$$0 = (m-3)(m+1)$$

$m - 3 = 0$ or $m + 1 = 0$
$m = 3$ or $m = -1$

Check $m = 3$: $\frac{3}{0} = \frac{1}{0} - \frac{1}{4}$ (undefined)

Check $m = -1$: $-\frac{3}{8} = -\frac{1}{8} - \frac{1}{4}$ *True*

The solution set is $\{-1\}$.

N7.
$$\frac{5}{k^2+k-2} = \frac{1}{3k-3} - \frac{1}{k+2}$$
$$\frac{5}{(k+2)(k-1)} = \frac{1}{3(k-1)} - \frac{1}{k+2} \quad \textit{Factor.}$$

Multiply by the LCD, 3(k + 2)(k − 1).

$$3(k+2)(k-1)\left(\frac{5}{(k+2)(k-1)}\right)$$
$$= 3(k+2)(k-1)\left(\frac{1}{3(k-1)} - \frac{1}{k+2}\right)$$
$$3(5) = k + 2 - 3(k-1)$$
$$15 = k + 2 - 3k + 3$$
$$15 = -2k + 5$$
$$10 = -2k$$
$$-5 = k$$

Check $k = -5$: $\frac{5}{18} = -\frac{1}{18} + \frac{1}{3}$ *True*

The solution set is $\{-5\}$.

N8. **(a)** Solve $p = \frac{x-y}{z}$ for x.

$$pz = \left(\frac{x-y}{z}\right)(z) \quad \textit{Multiply by the LCD, z.}$$
$$pz = x - y$$
$$pz + y = x \quad \textit{Isolate x.}$$

(b) Solve $a = \frac{b}{c+d}$ for d.

Multiply by the LCD, $c + d$.

$$a(c+d) = \left(\frac{b}{c+d}\right)(c+d)$$
$$ac + ad = b$$
$$ad = b - ac \quad \textit{Isolate ad.}$$
$$d = \frac{b-ac}{a} \quad \textit{Divide by a.}$$

Another solution method is to multiply by $c + d$, divide by a, and then subtract c to get $y = \frac{b}{a} - c$.

N9. Solve the formula $\frac{2}{w} = \frac{1}{x} - \frac{3}{y}$ for x.

Multiply by the LCD, wxy.

$$wxy\left(\frac{2}{w}\right) = wxy\left(\frac{1}{x} - \frac{3}{y}\right)$$
$$wxy\left(\frac{2}{w}\right) = wxy\left(\frac{1}{x}\right) - wxy\left(\frac{3}{y}\right)$$
$$2xy = wy - 3wx$$

Get the x-terms on one side.

$$2xy + 3wx = wy$$
$$x(2y + 3w) = wy \quad \textit{Factor.}$$
$$x = \frac{wy}{2y+3w}$$

6.6 Section Exercises

1. $\frac{7}{8}x + \frac{1}{5}x$ is the sum of two terms, so it is an *expression* to be simplified. Simplify by finding the LCD, writing each coefficient with this LCD, and combining like terms.

$$\frac{7}{8}x + \frac{1}{5}x = \frac{35}{40}x + \frac{8}{40}x \quad \textit{LCD = 40}$$
$$= \frac{43}{40}x \quad \textit{Combine like terms.}$$

3. $\frac{7}{8}x + \frac{1}{5}x = 1$ has an equals sign, so this is an *equation* to be solved. Use the multiplication property of equality to clear fractions. The LCD is 40.

$$\frac{7}{8}x + \frac{1}{5}x = 1$$
$$40\left(\frac{7}{8}x + \frac{1}{5}x\right) = 40 \cdot 1 \quad \textit{Multiply by 40.}$$
$$40\left(\frac{7}{8}x\right) + 40\left(\frac{1}{5}x\right) = 40 \cdot 1 \quad \textit{Distributive property}$$
$$35x + 8x = 40 \quad \textit{Multiply.}$$
$$43x = 40 \quad \textit{Combine like terms.}$$
$$x = \frac{40}{43} \quad \textit{Divide by 43.}$$

The solution set is $\left\{\frac{40}{43}\right\}$.

5. $\frac{3}{5}x - \frac{7}{10}x$ is the difference of two terms, so it is an *expression* to be simplified.

$$\frac{3}{5}x - \frac{7}{10}x = \frac{6}{10}x - \frac{7}{10}x \quad \textit{LCD = 10}$$
$$= -\frac{1}{10}x \quad \textit{Combine like terms.}$$

7. $\frac{3}{5}x - \frac{7}{10}x = 1$ has an equals sign, so it is an *equation* to be solved.

$$\frac{3}{5}x - \frac{7}{10}x = 1$$
$$10(\tfrac{3}{5}x - \tfrac{7}{10}x) = 10 \cdot 1 \quad \textit{LCD = 10}$$
$$10(\tfrac{3}{5}x) - 10(\tfrac{7}{10}x) = 10 \cdot 1 \quad \textit{Distributive property}$$
$$6x - 7x = 10 \quad \textit{Multiply.}$$
$$-x = 10 \quad \textit{Combine like terms.}$$
$$x = -10 \quad \textit{Divide by –1.}$$

The solution set is $\{-10\}$.

9. $\frac{3}{4}x - \frac{1}{2}x = 0$ has an equals sign, so it is an *equation* to be solved.

$$\frac{3}{4}x - \frac{1}{2}x = 0$$
$$4(\tfrac{3}{4}x - \tfrac{1}{2}x) = 4(0) \quad \textit{LCD = 4}$$
$$4(\tfrac{3}{4}x) - 4(\tfrac{1}{2}x) = 4(0) \quad \textit{Distributive property}$$
$$3x - 2x = 0 \quad \textit{Multiply.}$$
$$x = 0 \quad \textit{Combine like terms.}$$

The solution set is $\{0\}$.

11. $$\frac{3}{x+2} - \frac{5}{x} = 1$$

The denominators, $x+2$ and x, are equal to 0 for the values -2 and 0, respectively. Thus, $x \neq -2, 0$.

13. $$\frac{-1}{(x+3)(x-4)} = \frac{1}{2x+1}$$

The denominators, $(x+3)(x-4)$ and $2x+1$, are equal to 0 for the values -3, 4, and $-\frac{1}{2}$, respectively. Thus, $x \neq -3, 4, -\frac{1}{2}$.

15. $$\frac{4}{x^2+8x-9} + \frac{1}{x^2-4} = 0$$

The denominators, $x^2+8x-9 = (x+9)(x-1)$ and $x^2-4 = (x+2)(x-2)$, are equal to 0 for the values -9, 1, -2, and 2, respectively. Thus, $x \neq -9, 1, -2, 2$.

17. We cannot solve

$$\frac{2}{3x} + \frac{1}{5x}$$

because it is an expression, not an equation. An expression cannot be solved. Only equations and inequalities are "solved."

Note: In Exercises 19–76, all proposed solutions should be checked by substituting in the original equation. It is essential to determine whether a proposed solution will make any denominator in the original equation equal to zero. Full checks will be shown here for only a few of the exercises.

19. $$\frac{5}{m} - \frac{3}{m} = 8$$

Multiply each side by the LCD, m.

$$m\left(\frac{5}{m} - \frac{3}{m}\right) = m \cdot 8$$

Use the distributive property to remove parentheses; then solve.

$$m\left(\frac{5}{m}\right) - m\left(\frac{3}{m}\right) = 8m$$
$$5 - 3 = 8m$$
$$2 = 8m$$
$$m = \tfrac{2}{8} = \tfrac{1}{4}$$

Check this proposed solution by replacing m with $\frac{1}{4}$ in the original equation.

$$\frac{5}{\frac{1}{4}} - \frac{3}{\frac{1}{4}} \stackrel{?}{=} 8 \quad \textit{Let } m = \frac{1}{4}.$$
$$5 \cdot 4 - 3 \cdot 4 \stackrel{?}{=} 8 \quad \textit{Multiply by reciprocals.}$$
$$20 - 12 \stackrel{?}{=} 8$$
$$8 = 8 \quad \textit{True}$$

Thus, the solution set is $\{\frac{1}{4}\}$.

21. $$\frac{5}{y} + 4 = \frac{2}{y}$$

$$y\left(\frac{5}{y} + 4\right) = y\left(\frac{2}{y}\right) \quad \textit{Multiply by LCD, y.}$$
$$y\left(\frac{5}{y}\right) + y(4) = y\left(\frac{2}{y}\right) \quad \textit{Distributive property}$$
$$5 + 4y = 2$$
$$4y = -3$$
$$y = -\tfrac{3}{4}$$

Check $y = -\frac{3}{4}$: $-\frac{8}{3} = -\frac{8}{3}$ *True*

Thus, the solution set is $\{-\frac{3}{4}\}$.

23.
$$\frac{3x}{5} - 6 = x$$
$$5\left(\frac{3x}{5} - 6\right) = 5(x) \quad \textit{Multiply by LCD, 5.}$$
$$5\left(\frac{3x}{5}\right) - 5(6) = 5x \quad \textit{Distributive property}$$
$$3x - 30 = 5x$$
$$-30 = 2x$$
$$-15 = x$$

Check $x = -15$: $-15 = -15$ *True*

Thus, the solution set is $\{-15\}$.

25.
$$\frac{4m}{7} + m = 11$$
$$7\left(\frac{4m}{7} + m\right) = 7(11) \quad \textit{Multiply by LCD, 7.}$$
$$7\left(\frac{4m}{7}\right) + 7(m) = 77 \quad \textit{Distributive property}$$
$$4m + 7m = 77$$
$$11m = 77$$
$$m = 7$$

Check $m = 7$: $11 = 11$ *True*

Thus, the solution set is $\{7\}$.

27.
$$\frac{z-1}{4} = \frac{z+3}{3}$$
$$12\left(\frac{z-1}{4}\right) = 12\left(\frac{z+3}{3}\right) \quad \textit{Multiply by LCD, 12.}$$
$$3(z-1) = 4(z+3)$$
$$3z - 3 = 4z + 12 \quad \textit{Dist. prop.}$$
$$-15 = z$$

Check $z = -15$: $-4 = -4$ *True*

Thus, the solution set is $\{-15\}$.

29.
$$\frac{3p+6}{8} = \frac{3p-3}{16}$$
$$16\left(\frac{3p+6}{8}\right) = 16\left(\frac{3p-3}{16}\right) \quad \textit{Multiply by LCD, 16.}$$
$$2(3p+6) = 3p - 3$$
$$6p + 12 = 3p - 3 \quad \textit{Dist. prop.}$$
$$3p = -15$$
$$p = -5$$

Check $p = -5$: $-\frac{9}{8} = -\frac{9}{8}$ *True*

Thus, the solution set is $\{-5\}$.

31.
$$\frac{2x+3}{x} = \frac{3}{2}$$
$$2x\left(\frac{2x+3}{x}\right) = 2x\left(\frac{3}{2}\right) \quad \textit{Multiply by LCD, 2x.}$$
$$2(2x+3) = 3x$$
$$4x + 6 = 3x \quad \textit{Dist. prop.}$$
$$x = -6$$

Check $x = -6$: $\frac{3}{2} = \frac{3}{2}$ *True*

Thus, the solution set is $\{-6\}$.

33.
$$\frac{k}{k-4} - 5 = \frac{4}{k-4}$$
$$(k-4)\left(\frac{k}{k-4} - 5\right) = (k-4)\left(\frac{4}{k-4}\right)$$
Multiply by LCD, k – 4.
$$(k-4)\left(\frac{k}{k-4}\right) - 5(k-4) = 4 \quad \textit{Distributive property}$$
$$k - 5k + 20 = 4$$
$$-4k = -16$$
$$k = 4$$

The proposed solution is 4. However, 4 cannot be a solution because it makes the denominator $k - 4$ equal 0. Therefore, the solution set is $\emptyset$.

35.
$$\frac{q+2}{3} + \frac{q-5}{5} = \frac{7}{3}$$
$$15\left(\frac{q+2}{3} + \frac{q-5}{5}\right) = 15\left(\frac{7}{3}\right) \quad \textit{Mult. by LCD, 15.}$$
$$15\left(\frac{q+2}{3}\right) + 15\left(\frac{q-5}{5}\right) = 5 \cdot 7$$
$$5(q+2) + 3(q-5) = 35$$
$$5q + 10 + 3q - 15 = 35$$
$$8q - 5 = 35$$
$$8q = 40$$
$$q = 5$$

Check $q = 5$: $\frac{7}{3} = \frac{7}{3}$ *True*

Thus, the solution set is $\{5\}$.

37.
$$\frac{x}{2} = \frac{5}{4} + \frac{x-1}{4}$$
$$4\left(\frac{x}{2}\right) = 4\left(\frac{5}{4} + \frac{x-1}{4}\right) \quad \textit{Multiply by LCD, 4.}$$
$$2(x) = 4\left(\frac{5}{4}\right) + 4\left(\frac{x-1}{4}\right)$$
$$2x = 5 + x - 1$$
$$x = 4$$

Check $x = 4$: $2 = 2$ *True*

Thus, the solution set is $\{4\}$.

39.
$$x + \frac{17}{2} = \frac{x}{2} + x + 6$$
$$2\left(x + \frac{17}{2}\right) = 2\left(\frac{x}{2} + x + 6\right) \quad \textit{Multiply by LCD, 2.}$$
$$2(x) + 2\left(\frac{17}{2}\right) = 2\left(\frac{x}{2}\right) + 2(x) + 2(6)$$
$$2x + 17 = x + 2x + 12$$
$$5 = x$$

Check $x = 5$: $\frac{27}{2} = \frac{27}{2}$ *True*

Thus, the solution set is $\{5\}$.

41. $\dfrac{9}{3x+4}=\dfrac{36-27x}{16-9x^2}$

$\dfrac{9}{3x+4}=\dfrac{9(4-3x)}{(4+3x)(4-3x)}$ *Factor.*

$\dfrac{9}{3x+4}=\dfrac{9}{4+3x}$ *Reduce.* $x \neq 4/3$

This is an identity, which is true for all real numbers for which the original equation is defined. We must exclude $\pm\frac{4}{3}$. Thus, the solution set is $\{x|x\neq\pm\frac{4}{3}\}$.

43. $\dfrac{a+7}{8}-\dfrac{a-2}{3}=\dfrac{4}{3}$

$24\left(\dfrac{a+7}{8}-\dfrac{a-2}{3}\right)=24\left(\dfrac{4}{3}\right)$

Multiply by LCD, 24.

$$24\left(\frac{a+7}{8}\right)-24\left(\frac{a-2}{3}\right)=8(4)$$
$$3(a+7)-8(a-2)=32$$
$$3a+21-8a+16=32$$
$$-5a+37=32$$
$$-5a=-5$$
$$a=1$$

Check $a=1$: $\frac{4}{3}=\frac{4}{3}$ *True*

Thus, the solution set is $\{1\}$.

45. $\dfrac{p}{2}-\dfrac{p-1}{4}=\dfrac{5}{4}$

$4\left(\dfrac{p}{2}-\dfrac{p-1}{4}\right)=4\left(\dfrac{5}{4}\right)$ *Multiply by LCD, 4.*

$$4\left(\frac{p}{2}\right)-4\left(\frac{p-1}{4}\right)=5$$
$$2p-1(p-1)=5$$
$$2p-p+1=5$$
$$p=4$$

Check $p=4$: $\frac{5}{4}=\frac{5}{4}$ *True*

Thus, the solution set is $\{4\}$.

47. $\dfrac{3x}{5}-\dfrac{x-5}{7}=3$

$35\left(\dfrac{3x}{5}-\dfrac{x-5}{7}\right)=35(3)$ *Multiply by LCD, 35.*

$$35\left(\frac{3x}{5}\right)-35\left(\frac{x-5}{7}\right)=105$$
$$7(3x)-5(x-5)=105$$
$$21x-5x+25=105$$
$$16x=80$$
$$x=5$$

Check $x=5$: $3=3$ *True*

Thus, the solution set is $\{5\}$.

49. $\dfrac{4}{x^2-3x}=\dfrac{1}{x^2-9}$

$\dfrac{4}{x(x-3)}=\dfrac{1}{(x+3)(x-3)}$ *Factor denominators.*

$$x(x+3)(x-3)\cdot\frac{4}{x(x-3)}=x(x+3)(x-3)\cdot\frac{1}{(x+3)(x-3)}$$

Multiply by LCD, x(x + 3)(x − 3).

$$4(x+3)=x\cdot 1$$
$$4x+12=x$$
$$3x=-12$$
$$x=-4$$

Check $x=-4$: $\frac{1}{7}=\frac{1}{7}$ *True*

Thus, the solution set is $\{-4\}$.

51. $\dfrac{2}{m}=\dfrac{m}{5m+12}$

$$m(5m+12)\left(\frac{2}{m}\right)=m(5m+12)\left(\frac{m}{5m+12}\right)$$

Multiply by LCD, m(5m + 12).

$$(5m+12)(2)=m(m)$$
$$10m+24=m^2$$
$$-m^2+10m+24=0$$
$$m^2-10m-24=0 \quad \textit{Multiply by } -1.$$
$$(m-12)(m+2)=0$$
$$m-12=0 \quad \text{or} \quad m+2=0$$
$$m=12 \quad \text{or} \quad m=-2$$

Check $m=12$: $\frac{1}{6}=\frac{1}{6}$ *True*

Check $m=-2$: $-1=-1$ *True*

Thus, the solution set is $\{-2,12\}$.

53. $\dfrac{-2}{z+5}+\dfrac{3}{z-5}=\dfrac{20}{z^2-25}$

$\dfrac{-2}{z+5}+\dfrac{3}{z-5}=\dfrac{20}{(z+5)(z-5)}$

$$(z+5)(z-5)\left(\frac{-2}{z+5}+\frac{3}{z-5}\right)=(z+5)(z-5)\left(\frac{20}{(z+5)(z-5)}\right)$$

Multiply by LCD, (z + 5)(z − 5).

$$(z+5)(z-5)\left(\frac{-2}{z+5}\right)+(z+5)(z-5)\left(\frac{3}{z-5}\right)=20$$
$$-2(z-5)+3(z+5)=20$$
$$-2z+10+3z+15=20$$
$$z+25=20$$
$$z=-5$$

The proposed solution, -5, cannot be a solution because it would make the denominators $z+5$ and z^2-25 equal 0 and the corresponding fractions undefined. Since -5 cannot be a solution, the solution set is $\emptyset$.

55.
$$\frac{3}{x-1}+\frac{2}{4x-4}=\frac{7}{4}$$
$$\frac{3}{x-1}+\frac{2}{4(x-1)}=\frac{7}{4}$$
$$4(x-1)\left(\frac{3}{x-1}+\frac{2}{4(x-1)}\right)=4(x-1)\left(\frac{7}{4}\right)$$
Multiply by LCD, 4(x – 1).
$$4(3)+2=(x-1)(7)$$
$$14=7x-7$$
$$21=7x$$
$$3=x$$

Check $x=3$: $\frac{7}{4}=\frac{7}{4}$ *True*

Thus, the solution set is $\{3\}$.

57.
$$\frac{x}{3x+3}=\frac{2x-3}{x+1}-\frac{2x}{3x+3}$$
$$\frac{x}{3(x+1)}=\frac{2x-3}{x+1}-\frac{2x}{3(x+1)}$$
$$3(x+1)\left(\frac{x}{3(x+1)}\right)=$$
$$3(x+1)\left[\frac{2x-3}{x+1}-\frac{2x}{3(x+1)}\right]$$
Multiply by LCD, 3(x + 1).
$$x=3(x+1)\left(\frac{2x-3}{x+1}\right)$$
$$-3(x+1)\left(\frac{2x}{3(x+1)}\right)$$
$$x=3(2x-3)-2x$$
$$x=6x-9-2x$$
$$x=4x-9$$
$$-3x=-9$$
$$x=3$$

Check $x=3$: $\frac{1}{4}=\frac{1}{4}$ *True*

Thus, the solution set is $\{3\}$.

59.
$$\frac{2p}{p^2-1}=\frac{2}{p+1}-\frac{1}{p-1}$$
$$\frac{2p}{(p+1)(p-1)}=\frac{2}{p+1}-\frac{1}{p-1}$$
$$(p+1)(p-1)\left[\frac{2p}{(p+1)(p-1)}\right]$$
$$=(p+1)(p-1)\left(\frac{2}{p+1}\right)$$
$$-(p+1)(p-1)\left(\frac{1}{p-1}\right)$$
Multiply by LCD, (p + 1)(p – 1).
$$2p=2(p-1)-1(p+1)$$
$$2p=2p-2-p-1$$
$$p=-3$$

Check $p=-3$: $-\frac{3}{4}=-1+\frac{1}{4}$ *True*

Thus, the solution set is $\{-3\}$.

61.
$$\frac{5x}{14x+3}=\frac{1}{x}$$
$$x(14x+3)\left(\frac{5x}{14x+3}\right)=x(14x+3)\left(\frac{1}{x}\right)$$
Multiply by LCD, x(14x + 3).
$$x(5x)=(14x+3)(1)$$
$$5x^2=14x+3$$
$$5x^2-14x-3=0$$
$$(5x+1)(x-3)=0$$

Note to reader: We may skip writing out the zero-factor property since this step can be easily performed mentally.

$$x=-\tfrac{1}{5}\quad\text{or}\quad x=3$$

Check $x=-\frac{1}{5}$: $-5=-5$ *True*
Check $x=3$: $\frac{1}{3}=\frac{1}{3}$ *True*

Thus, the solution set is $\left\{-\frac{1}{5},3\right\}$.

63.
$$\frac{2}{x-1}-\frac{2}{3}=\frac{-1}{x+1}$$
$$3(x-1)(x+1)\left(\frac{2}{x-1}-\frac{2}{3}\right)$$
$$=3(x-1)(x+1)\left(\frac{-1}{x+1}\right)$$
Multiply by LCD, 3(x – 1)(x + 1).
$$3(x-1)(x+1)\left(\frac{2}{x-1}\right)$$
$$-3(x-1)(x+1)\left(\frac{2}{3}\right)$$
$$=3(x-1)(x+1)\left(\frac{-1}{x+1}\right)$$
$$3(x+1)(2)-(x-1)(x+1)(2)$$
$$=3(x-1)(-1)$$
$$6(x+1)-2(x^2-1)=-3(x-1)$$
$$6x+6-2x^2+2=-3x+3$$
$$-2x^2+9x+5=0$$
$$2x^2-9x-5=0$$
$$(2x+1)(x-5)=0$$
$$x=-\tfrac{1}{2}\quad\text{or}\quad x=5$$

Check $x=-\frac{1}{2}$: $-2=-2$ *True*
Check $x=5$: $-\frac{1}{6}=-\frac{1}{6}$ *True*

Thus, the solution set is $\left\{-\frac{1}{2},5\right\}$.

65.
$$\frac{x}{2x+2}=\frac{-2x}{4x+4}+\frac{2x-3}{x+1}$$
$$\frac{x}{2(x+1)}=\frac{-2x}{4(x+1)}+\frac{2x-3}{x+1}$$
$$4(x+1)\left(\frac{x}{2(x+1)}\right)=4(x+1)\left(\frac{-2x}{4(x+1)}\right)$$
$$+4(x+1)\left(\frac{2x-3}{x+1}\right)$$
Multiply by LCD, 4(x + 1).

$$2(x) = -2x + 4(2x - 3)$$
$$2x = -2x + 8x - 12$$
$$-4x = -12$$
$$x = 3$$

Check $x = 3$: $\frac{3}{8} = \frac{3}{8}$ *True*

Thus, the solution set is $\{3\}$.

67. $$\frac{8x+3}{x} = 3x$$
$$x\left(\frac{8x+3}{x}\right) = x(3x) \quad \textit{Multiply by LCD, x.}$$
$$8x + 3 = 3x^2$$
$$0 = 3x^2 - 8x - 3$$
$$0 = (3x+1)(x-3)$$
$$x = -\tfrac{1}{3} \quad \text{or} \quad x = 3$$

Check $x = -\frac{1}{3}$: $-1 = -1$ *True*

Check $x = 3$: $9 = 9$ *True*

Thus, the solution set is $\left\{-\frac{1}{3}, 3\right\}$.

69. $$\frac{1}{x+4} + \frac{x}{x-4} = \frac{-8}{x^2-16}$$
$$(x+4)(x-4)\left(\frac{1}{x+4}\right) + (x+4)(x-4)\left(\frac{x}{x-4}\right) = (x+4)(x-4)\left(\frac{-8}{x^2-16}\right)$$
Multiply by LCD, (x + 4)(x – 4).
$$1(x-4) + x(x+4) = -8$$
$$x - 4 + x^2 + 4x = -8$$
$$x^2 + 5x + 4 = 0$$
$$(x+4)(x+1) = 0$$
$$x = -4 \quad \text{or} \quad x = -1$$

$x = -4$ cannot be a solution because it would make the denominators $x + 4$ and $x^2 - 16$ equal 0 and the corresponding fractions undefined.

Check $x = -1$: $\frac{1}{3} + \frac{1}{5} = \frac{8}{15}$ *True*

Thus, the solution set is $\{-1\}$.

71. $$\frac{4}{3x+6} - \frac{3}{x+3} = \frac{8}{x^2+5x+6}$$
$$\frac{4}{3(x+2)} - \frac{3}{x+3} = \frac{8}{(x+2)(x+3)}$$
$$3(x+2)(x+3)\cdot\frac{4}{3(x+2)} - 3(x+2)(x+3)\cdot\frac{3}{x+3} = 3(x+2)(x+3)\cdot\frac{8}{(x+2)(x+3)}$$
Multiply by LCD, 3(x + 2)(x + 3).
$$4(x+3) - 3(x+2)(3) = 3(8)$$
$$4x + 12 - 9x - 18 = 24$$
$$-5x = 30$$
$$x = -6$$

Check $x = -6$: $-\frac{1}{3} - (-1) = \frac{2}{3}$ *True*

Thus, the solution set is $\{-6\}$.

73. $$\frac{3x}{x^2+5x+6} = \frac{5x}{x^2+2x-3} - \frac{2}{x^2+x-2}$$
$$\frac{3x}{(x+2)(x+3)} = \frac{5x}{(x+3)(x-1)} - \frac{2}{(x-1)(x+2)}$$
$$(x+2)(x+3)(x-1)\cdot\left[\frac{3x}{(x+2)(x+3)}\right] = (x+2)(x+3)(x-1)\cdot\left[\frac{5x}{(x+3)(x-1)}\right] - (x+2)(x+3)(x-1)\cdot\left[\frac{2}{(x-1)(x+2)}\right]$$
Multiply by LCD, (x + 2)(x + 3)(x – 1).
$$3x(x-1) = 5x(x+2) - 2(x+3)$$
$$3x^2 - 3x = 5x^2 + 10x - 2x - 6$$
$$0 = 2x^2 + 11x - 6$$
$$0 = (2x-1)(x+6)$$
$$x = \tfrac{1}{2} \quad \text{or} \quad x = -6$$

Check $x = \frac{1}{2}$: $\frac{6}{35} = -\frac{10}{7} - \left(-\frac{8}{5}\right)$ *True*

Check $x = -6$: $-\frac{3}{2} = -\frac{10}{7} - \frac{1}{14}$ *True*

Thus, the solution set is $\left\{-6, \frac{1}{2}\right\}$.

75. $$\frac{x+4}{x^2-3x+2} - \frac{5}{x^2-4x+3} = \frac{x-4}{x^2-5x+6}$$
$$\frac{x+4}{(x-2)(x-1)} - \frac{5}{(x-3)(x-1)} = \frac{x-4}{(x-3)(x-2)}$$
$$(x-2)(x-1)(x-3)\cdot\left[\frac{x+4}{(x-2)(x-1)} - \frac{5}{(x-3)(x-1)}\right] = (x-2)(x-1)(x-3)\left[\frac{x-4}{(x-3)(x-2)}\right]$$
Multiply by LCD, (x – 2)(x – 1)(x – 3).
$$(x+4)(x-3) - 5(x-2) = (x-1)(x-4)$$
$$x^2 + x - 12 - 5x + 10 = x^2 - 5x + 4$$
$$-4x - 2 = -5x + 4$$
$$x = 6$$

Check $x = 6$: $\frac{1}{2} - \frac{1}{3} = \frac{1}{6}$ *True*

Thus, the solution set is $\{6\}$.

77. $kr - mr = km$

If you are solving for k, put both terms with k on one side and the remaining term on the other side.

$$kr - km = mr$$

79. $m = \frac{kF}{a}$ for F

We need to isolate F on one side of the equation.

$$m \cdot a = \left(\frac{kF}{a}\right)(a) \quad \textit{Multiply by a.}$$
$$ma = kF$$
$$\frac{ma}{k} = \frac{kF}{k} \quad \textit{Divide by k.}$$
$$\frac{ma}{k} = F$$

81. $m = \frac{kF}{a}$ for a

$$m \cdot a = \left(\frac{kF}{a}\right)(a) \quad \textit{Multiply by a.}$$
$$ma = kF$$
$$\frac{ma}{m} = \frac{kF}{m} \quad \textit{Divide by m.}$$
$$a = \frac{kF}{m}$$

83. $I = \frac{E}{R+r}$ for R

We need to isolate R on one side of the equation.

$$I(R+r) = \left(\frac{E}{R+r}\right)(R+r) \quad \textit{Multiply by } R + r.$$
$$IR + Ir = E \quad \textit{Distributive property}$$
$$IR = E - Ir \quad \textit{Subtract Ir.}$$
$$R = \frac{E - Ir}{I}, \quad \textit{Divide by I.}$$
$$\text{or } R = \frac{E}{I} - r$$

85. $h = \frac{2\mathcal{A}}{B+b}$ for $\mathcal{A}$

$$(B+b)h = (B+b) \cdot \frac{2\mathcal{A}}{B+b}$$
Multiply by B + b.
$$h(B+b) = 2\mathcal{A}$$
$$\frac{h(B+b)}{2} = \mathcal{A} \quad \textit{Divide by 2.}$$

87. $d = \frac{2S}{n(a+L)}$ for a

We need to isolate a on one side of the equation.

$$d \cdot n(a+L) = \frac{2S}{n(a+L)} \cdot n(a+L)$$
Multiply by n(a + L).
$$nd(a+L) = 2S$$
$$and + ndL = 2S$$
$$and = 2S - ndL \quad \textit{Subtract ndL.}$$
$$a = \frac{2S - ndL}{nd}, \quad \textit{Divide by nd.}$$
$$\text{or } a = \frac{2S}{nd} - L$$

89. $\frac{1}{x} = \frac{1}{y} - \frac{1}{z}$ for y

The LCD of all the fractions in the equation is xyz, so multiply both sides by xyz.

$$xyz\left(\frac{1}{x}\right) = xyz\left(\frac{1}{y} - \frac{1}{z}\right)$$
$$xyz\left(\frac{1}{x}\right) = xyz\left(\frac{1}{y}\right) - xyz\left(\frac{1}{z}\right)$$
Distributive property
$$yz = xz - xy$$

Since we are solving for y, get all terms with y on one side of the equation.

$$xy + yz = xz \quad \textit{Add xy.}$$

Factor out the common factor y on the left.

$$y(x+z) = xz$$

Finally, divide both sides by the coefficient of y, which is $x + z$.

$$y = \frac{xz}{x+z}$$

91. $\frac{2}{r} + \frac{3}{s} + \frac{1}{t} = 1$ for t

The LCD of all the fractions in the equation is rst, so multiply both sides by rst.

$$rst\left(\frac{2}{r} + \frac{3}{s} + \frac{1}{t}\right) = rst(1)$$
$$rst\left(\frac{2}{r}\right) + rst\left(\frac{3}{s}\right) + rst\left(\frac{1}{t}\right) = rst$$
$$2st + 3rt + rs = rst$$

Since we are solving for t, get all terms with t on one side of the equation.

$$2st + 3rt - rst = -rs$$

Factor out the common factor t on the left.

$$t(2s + 3r - rs) = -rs$$

Finally, divide both sides by the coefficient of t, which is $2s + 3r - rs$.

$$t = \frac{-rs}{2s + 3r - rs}, \quad \text{or} \quad t = \frac{rs}{-2s - 3r + rs}$$

93. $9x + \frac{3}{z} = \frac{5}{y}$ for z

$$yz\left(9x + \frac{3}{z}\right) = yz\left(\frac{5}{y}\right) \quad \textit{Multiply by LCD, yz.}$$
$$yz(9x) + yz\left(\frac{3}{z}\right) = yz\left(\frac{5}{y}\right) \quad \textit{Distributive property}$$
$$9xyz + 3y = 5z$$
$$9xyz - 5z = -3y \quad \textit{Get the z terms on one side.}$$

$z(9xy-5)=-3y$ *Factor out z.*

$z=\dfrac{-3y}{9xy-5}$, *Divide by 9xy – 5.*

or $z=\dfrac{3y}{5-9xy}$

95. $\dfrac{t}{x-1}-\dfrac{2}{x+1}=\dfrac{1}{x^2-1}$ for t

$$(x+1)(x-1)\left(\frac{t}{x-1}-\frac{2}{x+1}\right)=(x+1)(x-1)\left(\frac{1}{x^2-1}\right)$$

Multiply by LCD, (x + 1)(x – 1).

$$t(x+1)-2(x-1)=1$$
$$t(x+1)=1+2(x-1)$$
$$t(x+1)=1+2x-2$$
$$t=\frac{2x-1}{x+1},\text{ or } t=\frac{-2x+1}{-x-1}$$

97. Using $d=rt$, we get

$$r=\frac{d}{t}=\frac{288}{t}.$$

His rate is $\frac{288}{t}$ mph.

99. Using $d=rt$, we get

$$t=\frac{d}{r}=\frac{289}{z}.$$

His time is $\frac{289}{z}$ hr.

Summary Exercises on Rational Expressions and Equations

1. No equals sign appears so this is an *expression.*

$$\frac{4}{p}+\frac{6}{p}=\frac{4+6}{p}=\frac{10}{p}$$

3. No equals sign appears so this is an *expression.*

$$\frac{1}{x^2+x-2}\div\frac{4x^2}{2x-2}$$
$$=\frac{1}{x^2+x-2}\cdot\frac{2x-2}{4x^2}$$
$$=\frac{1}{(x+2)(x-1)}\cdot\frac{2(x-1)}{2\cdot 2x^2}$$
$$=\frac{1}{2x^2(x+2)}$$

5. No equals sign appears so this is an *expression.*

$$\frac{2y^2+y-6}{2y^2-9y+9}\cdot\frac{y^2-2y-3}{y^2-1}$$
$$=\frac{(2y-3)(y+2)(y-3)(y+1)}{(2y-3)(y-3)(y+1)(y-1)}$$
$$=\frac{y+2}{y-1}$$

7. $\dfrac{x-4}{5}=\dfrac{x+3}{6}$

There is an equals sign, so this is an *equation.*

$$30\left(\frac{x-4}{5}\right)=30\left(\frac{x+3}{6}\right)\quad \textit{Multiply by LCD, 30.}$$
$$6(x-4)=5(x+3)$$
$$6x-24=5x+15$$
$$x=39$$

Check $x=39$: $7=7$ *True*

The solution is set is $\{39\}$.

9. No equals sign appears so this is an *expression.*

$$\frac{4}{p+2}+\frac{1}{3p+6}$$
$$=\frac{4}{p+2}+\frac{1}{3(p+2)}$$
$$=\frac{3\cdot 4}{3(p+2)}+\frac{1}{3(p+2)}\quad \textit{LCD = 3(p+2)}$$
$$=\frac{12+1}{3(p+2)}$$
$$=\frac{13}{3(p+2)}$$

11. $\dfrac{3}{t-1}+\dfrac{1}{t}=\dfrac{7}{2}$

There is an equals sign, so this is an *equation.*

$$2t(t-1)\left(\frac{3}{t-1}+\frac{1}{t}\right)=2t(t-1)\left(\frac{7}{2}\right)$$

Multiply by LCD, 2t(t – 1).

$$2t(t-1)\left(\frac{3}{t-1}\right)+2t(t-1)\left(\frac{1}{t}\right)=7t(t-1)$$
$$2t(3)+2(t-1)=7t(t-1)$$
$$6t+2t-2=7t^2-7t$$
$$0=7t^2-15t+2$$
$$0=(7t-1)(t-2)$$
$$t=\tfrac{1}{7}\quad\text{or}\quad t=2$$

Check $t=\frac{1}{7}$: $-\frac{7}{2}+7=\frac{7}{2}$ *True*
Check $t=2$: $3+\frac{1}{2}=\frac{7}{2}$ *True*
The solution set is $\left\{\frac{1}{7},2\right\}$.

13. No equals sign appears so this is an *expression.*

$$\frac{5}{4z}-\frac{2}{3z}=\frac{3\cdot 5}{3\cdot 4z}-\frac{4\cdot 2}{4\cdot 3z}\quad \textit{LCD = 12z}$$
$$=\frac{15}{12z}-\frac{8}{12z}$$
$$=\frac{15-8}{12z}=\frac{7}{12z}$$

15. No equals sign appears so this is an *expression.*

$$\frac{1}{m^2+5m+6}+\frac{2}{m^2+4m+3}$$
$$=\frac{1}{(m+2)(m+3)}+\frac{2}{(m+1)(m+3)}$$
$$=\frac{1(m+1)}{(m+2)(m+3)(m+1)}$$
$$+\frac{2(m+2)}{(m+1)(m+3)(m+2)}$$

LCD = (m + 1)(m + 2)(m + 3)

$$=\frac{(m+1)+(2m+4)}{(m+1)(m+2)(m+3)}$$
$$=\frac{3m+5}{(m+1)(m+2)(m+3)}$$

17. $\frac{2}{x+1}+\frac{5}{x-1}=\frac{10}{x^2-1}$

There is an equals sign, so this is an *equation.*

$$\frac{2}{x+1}+\frac{5}{x-1}=\frac{10}{(x+1)(x-1)}$$
$$(x+1)(x-1)\left(\frac{2}{x+1}+\frac{5}{x-1}\right)$$
$$=(x+1)(x-1)\left[\frac{10}{(x+1)(x-1)}\right]$$

Multiply by LCD, (x + 1)(x – 1).

$$(x+1)(x-1)\left(\frac{2}{x+1}\right)$$
$$+(x+1)(x-1)\left(\frac{5}{x-1}\right)=10$$

Distributive property

$$\begin{aligned}2(x-1)+5(x+1)&=10\\2x-2+5x+5&=10\\3+7x&=10\\7x&=7\\x&=1\end{aligned}$$

Replacing x by 1 in the original equation makes the denominators $x-1$ and x^2-1 equal to 0, so the solution set is $\emptyset$.

19. No equals sign appears so this is an *expression.*

$$\frac{4t^2-t}{6t^2+10t}\div\frac{8t^2+2t-1}{3t^2+11t+10}$$
$$=\frac{4t^2-t}{6t^2+10t}\cdot\frac{3t^2+11t+10}{8t^2+2t-1}$$

Multiply by reciprocal.

$$=\frac{t(4t-1)}{2t(3t+5)}\cdot\frac{(3t+5)(t+2)}{(4t-1)(2t+1)}$$

Factor numerators and denominators.

$$=\frac{t+2}{2(2t+1)}$$

6.7 Applications of Rational Expressions

6.7 Now Try Exercises

N1. *Step 2*
Let x = the denominator of the original fraction, so that the numerator is $x-4$.

Step 3
If 7 is added to both the numerator and denominator, the resulting fraction is equivalent to $\frac{7}{8}$ translates to

$$\frac{(x-4)+7}{x+7}=\frac{7}{8}.$$

Step 4
Multiply by the LCD, $8(x+7)$.

$$\begin{aligned}8(x+7)\frac{(x-4)+7}{x+7}&=8(x+7)\frac{7}{8}\\8(x+3)&=7(x+7)\\8x+24&=7x+49\\x&=25\end{aligned}$$

Step 5
The denominator is 25 and the numerator is $25-4=21$, so the original fraction is $\frac{21}{25}$.

Step 6
If 7 is added to the numerator and the denominator of $\frac{21}{25}$, the result is $\frac{28}{32}$, which is equal to $\frac{7}{8}$.

N2. Let x = the rate of the boat with no current. Complete a table.

	d	r	t
With the Current	12	$x+2$	$\frac{12}{x+2}$
Against the Current	4	$x-2$	$\frac{4}{x-2}$

Since the times are equal, we get the following equation.

$$\frac{12}{x+2}=\frac{4}{x-2}$$

Multiply by the LCD, $(x+2)(x-2)$.

$$\begin{aligned}12(x-2)&=4(x+2)\\12x-24&=4x+8\\8x-24&=8\\8x&=32\\x&=4\end{aligned}$$

The rate of the boat with no current is 4 miles per hour.

N3.

	Rate	Time Working Together	Fractional Part of the Job Done when Working Together
Sarah	$\frac{1}{10}$	x	$\frac{1}{10}x$
Joyce	$\frac{1}{12}$	x	$\frac{1}{12}x$

Since together Sarah and Joyce complete 1 whole job, we must add their individual parts and set the sum equal to 1.

$$\frac{1}{10}x + \frac{1}{12}x = 1$$

$$60(\tfrac{1}{10}x + \tfrac{1}{12}x) = 60(1) \quad \textit{Multiply by the LCD, 60.}$$

$$60(\tfrac{1}{10}x) + 60(\tfrac{1}{12}x) = 60$$

$$6x + 5x = 60$$

$$11x = 60$$

$$x = \tfrac{60}{11}, \text{ or } 5\tfrac{5}{11}$$

Working together, Sarah and Joyce can proofread the manuscript in $5\frac{5}{11}$ hours.

6.7 Section Exercises

1. **(a)** Let $x =$ the amount.

(b) An expression for "the numerator of the fraction $\frac{5}{6}$ is increased by an amount" is $5 + x$. We could also use $\dfrac{5+x}{6}$.

(c) An equation that can be used to solve the problem is

$$\frac{5+x}{6} = \frac{13}{3}.$$

3. *Step 2*
Let $x =$ the numerator of the original fraction. Then $x + 6 =$ the denominator of the original fraction.

Step 3
If 3 is added to both the numerator and denominator, the resulting fraction is equivalent to $\frac{5}{7}$ translates to

$$\frac{x+3}{(x+6)+3} = \frac{5}{7}.$$

Step 4
Since we have a fraction equal to another fraction, we can use cross multiplication.

$$7(x+3) = 5[(x+6)+3]$$

$$7x + 21 = 5x + 45$$

$$2x = 24$$

$$x = 12$$

Step 5
The original fraction is

$$\frac{x}{x+6} = \frac{12}{12+6} = \frac{12}{18}.$$

Step 6
Adding 3 to both the numerator and the denominator gives us

$$\frac{12+3}{18+3} = \frac{15}{21},$$

which is equivalent to $\frac{5}{7}$.

5. *Step 2*
Let $x =$ the denominator of the original fraction. Then $4x =$ the numerator of the original fraction.

Step 3
If 6 is added to both the numerator and the denominator, the resulting fraction is equivalent to 2 translates to

$$\frac{4x+6}{x+6} = 2.$$

Step 4

$$4x + 6 = 2(x+6)$$

$$4x + 6 = 2x + 12$$

$$2x = 6$$

$$x = 3$$

Step 5
The original fraction is

$$\frac{4x}{x} = \frac{4(3)}{3} = \frac{12}{3}.$$

Step 6
Adding 6 to both the numerator and the denominator gives us

$$\frac{12+6}{3+6} = \frac{18}{9} = 2.$$

7. *Step 2*
Let $x =$ the number.

Step 3
One-third of a number is 2 greater than one-sixth of the same number translates to

$$\tfrac{1}{3}x = 2 + \tfrac{1}{6}x.$$

Step 4
Multiply both sides by the LCD, 6.

$$6(\tfrac{1}{3}x) = 6(2 + \tfrac{1}{6}x)$$

$$2x = 12 + x$$

$$x = 12$$

Step 5 The number is 12.

Step 6
One-third of 12 is 4 and one-sixth of 12 is 2. So $x = 12$ checks since 4 is 2 more than 2.

9. *Step 2*
Let $x =$ the quantity.
Then $\frac{2}{3}$ of it, $\frac{1}{2}$ of it, and $\frac{1}{7}$ of it are

$$\tfrac{2}{3}x,\ \tfrac{1}{2}x,\ \text{and } \tfrac{1}{7}x.$$

Step 3
Added together equals 33 translates to

$$x + \tfrac{2}{3}x + \tfrac{1}{2}x + \tfrac{1}{7}x = 33.$$

Step 4
Multiply both sides by the LCD of 3, 2, and 7, which is 42.

$$\begin{aligned} 42(x + \tfrac{2}{3}x + \tfrac{1}{2}x + \tfrac{1}{7}x) &= 42(33) \\ 42x + 42(\tfrac{2}{3}x) + 42(\tfrac{1}{2}x) + 42(\tfrac{1}{7}x) &= 42(33) \\ 42x + 28x + 21x + 6x &= 1386 \\ 97x &= 1386 \\ x &= \tfrac{1386}{97} \end{aligned}$$

Step 5
The quantity is $\frac{1386}{97}$. (Note that this fraction is already in lowest terms since 97 is a prime number and is not a factor of 1386.)

Step 6
Check $\frac{1386}{97}$ in the original problem.

$$x = \tfrac{1386}{97},\ \tfrac{2}{3}x = \tfrac{924}{97},\ \tfrac{1}{2}x = \tfrac{693}{97},\ \tfrac{1}{7}x = \tfrac{198}{97}$$

Adding gives us

$$\frac{1386 + 924 + 693 + 198}{97} = \frac{3201}{97} = 33,$$

as desired.

11. We are asked to find the *time*, so we'll use the distance, rate, and time relationship $t = d/r$.

$$\begin{aligned} t = \frac{d}{r} &= \frac{0.6 \text{ miles}}{0.0319 \text{ miles per minute}} \\ &\approx 18.809 \text{ minutes} \end{aligned}$$

13. We are asked to find the average *rate*, so we'll use the distance, rate, and time relationship $r = d/t$.

$$\begin{aligned} r = \frac{d}{t} &= \frac{5000 \text{ meters}}{15.911 \text{ minutes}} \\ &\approx 314.248 \text{ meters per minute} \end{aligned}$$

15. We are asked to find the *time*, so we'll use the distance, rate, and time relationship $t = d/r$.

$$\begin{aligned} t = \frac{d}{r} &= \frac{500 \text{ miles}}{152.672 \text{ miles per hour}} \\ &\approx 3.275 \text{ hours} \end{aligned}$$

17. Use time $= \frac{\text{distance}}{\text{rate}}$. Since we know that the times for Stephanie and Wally are the same,

$$\text{time}_{\text{Stephanie}} = \text{time}_{\text{Wally}}$$

or

$$\frac{D}{R} = \frac{d}{r}$$

19. Let $x =$ rate of the plane in still air. Then the rate against the wind is $x - 10$ and the rate with the wind is $x + 10$. The time flying against the wind is

$$t = \frac{d}{r} = \frac{500}{x - 10},$$

and the time flying with the wind is

$$t = \frac{d}{r} = \frac{600}{x + 10}.$$

Now complete the chart.

	d	**r**	**t**
Against the Wind	500	$x - 10$	$\frac{500}{x-10}$
With the Wind	600	$x + 10$	$\frac{600}{x+10}$

Since the problem states that the two times are equal, we have

$$\frac{500}{x - 10} = \frac{600}{x + 10}.$$

We would use this equation to solve the problem.

21. Let x represent the rate of the boat in still water. Then $x - 4$ is the rate against the current and $x + 4$ is the rate with the current. We fill in the chart as follows, realizing that the time column is filled in by using the formula $t = d/r$.

	d	**r**	**t**
Against the Current	20	$x - 4$	$\frac{20}{x-4}$
With the Current	60	$x + 4$	$\frac{60}{x+4}$

Since the times are equal, we get the following equation.

$$\begin{aligned} \frac{20}{x - 4} &= \frac{60}{x + 4} \\ (x + 4)(x - 4)\frac{20}{x - 4} &= (x + 4)(x - 4)\frac{60}{x + 4} \\ &\textit{Multiply by LCD, (x + 4)(x − 4)} \\ 20(x + 4) &= 60(x - 4) \\ 20x + 80 &= 60x - 240 \\ 320 &= 40x \\ 8 &= x \end{aligned}$$

The rate of the boat in still water is 8 miles per hour.

23. Let x = rate of the bird in still air. Then the rate against the wind is $x - 8$ and the rate with the wind is $x + 8$. Use $t = d/r$ to complete the chart.

	d	**r**	**t**
Against the Wind	18	$x-8$	$\frac{18}{x-8}$
With the Wind	30	$x+8$	$\frac{30}{x+8}$

Since the problem states that the two times are equal, we get the following equation.

$$\frac{18}{x-8} = \frac{30}{x+8}$$
$$(x+8)(x-8)\frac{18}{x-8} = (x+8)(x-8)\frac{30}{x+8}$$
Multiply by LCD, (x + 8)(x − 8)
$$18(x+8) = 30(x-8)$$
$$18x + 144 = 30x - 240$$
$$384 = 12x$$
$$32 = x$$

The rate of the bird in still air is 32 miles per hour.

25. Let x = rate of the plane in still air. Then the rate against the wind is $x - 15$ and the rate with the wind is $x + 15$. Use $t = d/r$ to complete the chart.

	d	**r**	**t**
Against the Wind	375	$x-15$	$\frac{375}{x-15}$
With the Wind	450	$x+15$	$\frac{450}{x+15}$

Since the problem states that the two times are equal, we get the following equation.

$$\frac{375}{x-15} = \frac{450}{x+15}$$
$$(x+15)(x-15)\frac{375}{x-15} = (x+15)(x-15)\frac{450}{x+15}$$
Multiply by LCD, (x + 15)(x − 15)
$$375(x+15) = 450(x-15)$$
$$375x + 5625 = 450x - 6750$$
$$12{,}375 = 75x$$
$$165 = x$$

The rate of the plane in still air is 165 miles per hour.

27. Let x represent the rate of the current of the river. Then $12 - x$ is the rate upstream (against the current) and $12 + x$ is the rate downstream (with the current). Use $t = d/r$ to complete the table.

	d	**r**	**t**
Upstream	6	$12-x$	$\frac{6}{12-x}$
Downstream	10	$12+x$	$\frac{10}{12+x}$

Since the times are equal, we get the following equation.

$$\frac{6}{12-x} = \frac{10}{12+x}$$
$$(12+x)(12-x)\frac{6}{12-x} = (12+x)(12-x)\frac{10}{12+x}$$
Multiply by LCD, (12 + x)(12 − x)
$$6(12+x) = 10(12-x)$$
$$72 + 6x = 120 - 10x$$
$$16x = 48$$
$$x = 3$$

The rate of the current of the river is 3 miles per hour.

29. Let x = the average rate of the ferry.

Use the formula $t = d/r$ to make a chart.

	d	**r**	**t**
Seattle-Victoria	148	x	$\frac{148}{x}$
Victoria-Vancouver	74	x	$\frac{74}{x}$

Since the time for the Victoria-Vancouver trip is 4 hours less than the time for the Seattle-Victoria trip, solve the equation

$$\frac{74}{x} = \frac{148}{x} - 4.$$
$$x\left(\frac{74}{x}\right) = x\left(\frac{148}{x} - 4\right)$$ *Multiply by LCD, x.*
$$74 = 148 - 4x$$
$$4x = 74$$
$$x = \tfrac{74}{4} = \tfrac{37}{2}, \text{ or } 18\tfrac{1}{2}$$

The average rate of the ferry is $18\frac{1}{2}$ miles per hour.

31. If it takes Elayn 10 hours to do a job, her rate is

$\frac{1}{10}$ job per hour.

33. Let $t =$ the number of hours it will take Edward and Abdalla to paint the room working together.

	Rate	Time Working Together	Fractional Part of the Job Done when Working Together
Edward	$\frac{1}{8}$	t	$\frac{1}{8}t$
Abdalla	$\frac{1}{6}$	t	$\frac{1}{6}t$

Working together, they complete 1 whole job, so add their individual fractional parts and set the sum equal to 1.

part done by Edward	+	part done by Abdalla	=	1 whole job
↓	↓	↓	↓	↓
$\frac{1}{8}t$	+	$\frac{1}{6}t$	=	1

An equation that can be used to solve this problem is

$$\tfrac{1}{8}t + \tfrac{1}{6}t = 1.$$

Alternatively, we can compare the hourly rates of completion. In one hour, Edward will complete $\frac{1}{8}$ of the job, Abdalla will complete $\frac{1}{6}$ of the job, and together they will complete $\frac{1}{t}$ of the job. So another equation that can be used to solve this problem is

$$\tfrac{1}{8} + \tfrac{1}{6} = \tfrac{1}{t}.$$

35. Let x represent the number of hours it will take for Heather and Courtney to grade the tests working together. Since Heather can grade the test in 4 hours, her rate alone is $\frac{1}{4}$ job per hour. Also, since Courtney can do the job alone in 6 hours, her rate is $\frac{1}{6}$ job per hour.

	Rate	Time Working Together	Fractional Part of the Job Done when Working Together
Heather	$\frac{1}{4}$	x	$\frac{1}{4}x$
Courtney	$\frac{1}{6}$	x	$\frac{1}{6}x$

Since together Heather and Courtney complete 1 whole job, we must add their individual fractional parts and set the sum equal to 1.

$$\begin{aligned} \tfrac{1}{4}x + \tfrac{1}{6}x &= 1 \\ 12(\tfrac{1}{4}x + \tfrac{1}{6}x) &= 12(1) \quad \textit{LCD = 12} \\ 12(\tfrac{1}{4}x) + 12(\tfrac{1}{6}x) &= 12 \\ 3x + 2x &= 12 \\ 5x &= 12 \\ x &= \tfrac{12}{5}, \text{ or } 2\tfrac{2}{5} \end{aligned}$$

It will take Heather and Courtney $2\frac{2}{5}$ hours to grade the tests if they work together.

37. Let $x =$ the number of hours to pump the water using both pumps.

	Rate	Time Working Together	Fractional Part of the Job Done when Working Together
Pump 1	$\frac{1}{10}$	x	$\frac{1}{10}x$
Pump 2	$\frac{1}{12}$	x	$\frac{1}{12}x$

Since together the two pumps complete 1 whole job, we must add their individual fractional parts and set the sum equal to 1.

$$\begin{aligned} \tfrac{1}{10}x + \tfrac{1}{12}x &= 1 \\ 60(\tfrac{1}{10}x + \tfrac{1}{12}x) &= 60(1) \quad \textit{LCD = 60} \\ 60(\tfrac{1}{10}x) + 60(\tfrac{1}{12}x) &= 60 \\ 6x + 5x &= 60 \\ 11x &= 60 \\ x &= \tfrac{60}{11}, \text{ or } 5\tfrac{5}{11} \end{aligned}$$

It would take $5\frac{5}{11}$ hours to pump out the basement if both pumps were used.

39. Let x represent the number of hours it will take the experienced employee to enter the data. Then $2x$ represents the number of hours it will take the new employee (the experienced employee takes less time). The experienced employee's rate is $\dfrac{1}{x}$ job per hour and the new employee's rate is $\dfrac{1}{2x}$ job per hour.

	Rate	Time Working Together	Fractional Part of the Job Done when Working Together
Experienced employee	$\frac{1}{x}$	2	$\frac{1}{x} \cdot 2 = \frac{2}{x}$
New employee	$\frac{1}{2x}$	2	$\frac{1}{2x} \cdot 2 = \frac{1}{x}$

Since together the two employees complete the whole job, we must add their individual fractional parts and set the sum equal to 1.

$$\begin{aligned} \frac{2}{x} + \frac{1}{x} &= 1 \\ x\left(\frac{2}{x} + \frac{1}{x}\right) &= x(1) \quad \textit{Multiply by the LCD, x.} \\ x\left(\frac{2}{x}\right) + x\left(\frac{1}{x}\right) &= x \\ 2 + 1 &= x \\ 3 &= x \end{aligned}$$

Working alone, it will take the experienced employee 3 hours to enter the data.

41. Let $x =$ the number of hours to fill the pool $\frac{3}{4}$ full with both pipes working together.

	Rate	Time Working Together	Fractional Part of the Job Done when Working Together
First pipe	$\frac{1}{6}$	x	$\frac{1}{6}x$
Second pipe	$\frac{1}{9}$	x	$\frac{1}{9}x$

Part done by first pipe + Part done by second pipe = $\frac{3}{4}$ full

$$\frac{1}{6}x + \frac{1}{9}x = \frac{3}{4}$$

$$36\left(\tfrac{1}{6}x + \tfrac{1}{9}x\right) = 36\left(\tfrac{3}{4}\right) \quad \textit{LCD = 36}$$
$$36\left(\tfrac{1}{6}x\right) + 36\left(\tfrac{1}{9}x\right) = 36\left(\tfrac{3}{4}\right)$$
$$6x + 4x = 27$$
$$10x = 27$$
$$x = \tfrac{27}{10}, \text{ or } 2\tfrac{7}{10}$$

It takes $2\frac{7}{10}$ hours to fill the pool $\frac{3}{4}$ full using both pipes.

Alternatively, we could solve $\frac{1}{6}x + \frac{1}{9}x = 1$ (filling the whole pool) and then multiply that answer by $\frac{3}{4}$. In this case, x would represent the number of hours to fill the pool with both pipes working together.

43. Let $x =$ the number of minutes it takes to fill the sink.

In 1 minute, the cold water faucet (alone) can fill $\frac{1}{12}$ of the sink. In the same time, the hot water faucet (alone) can fill $\frac{1}{15}$ of the sink. In 1 minute, the drain (alone) empties $\frac{1}{25}$ of the sink. Together, they fill $\frac{1}{x}$ of the sink in one minute, so solve the equation

$$\tfrac{1}{12} + \tfrac{1}{15} - \tfrac{1}{25} = \tfrac{1}{x}.$$
$$300x\left(\tfrac{1}{12} + \tfrac{1}{15} - \tfrac{1}{25}\right) = 300x\left(\tfrac{1}{x}\right)$$

Multiply by the LCD, 300x.

$$25x + 20x - 12x = 300$$
$$33x = 300$$
$$x = \tfrac{300}{33} = \tfrac{100}{11}, \text{ or } 9\tfrac{1}{11}$$

It will take $9\frac{1}{11}$ minutes to fill the sink.

45. $\dfrac{6-2}{5-3} = \dfrac{4}{2} = 2$

47. $\dfrac{4-(-1)}{-3-(-5)} = \dfrac{4+1}{-3+5} = \dfrac{5}{2}$

49. $\dfrac{-5-(-5)}{3-2} = \dfrac{-5+5}{1} = \dfrac{0}{1} = 0$

51.
$$3x + 2y = 8$$
$$2y = -3x + 8$$
$$y = -\tfrac{3}{2}x + 4$$

Chapter 6 Review Exercises

1. **(a)** $\dfrac{4x-3}{5x+2} = \dfrac{4(-2)-3}{5(-2)+2}$ *Let x = –2.*

$$= \frac{-8-3}{-10+2} = \frac{-11}{-8} = \frac{11}{8}$$

(b) $\dfrac{4x-3}{5x+2} = \dfrac{4(4)-3}{5(4)+2}$ *Let x = 4.*

$$= \frac{16-3}{20+2} = \frac{13}{22}$$

2. **(a)** $\dfrac{3x}{x^2-4} = \dfrac{3(-2)}{(-2)^2-4}$ *Let x = –2.*

$$= \frac{-6}{4-4} = \frac{-6}{0}$$

Substituting -2 for x makes the denominator zero, so the given expression is undefined when $x = -2$.

(b) $\dfrac{3x}{x^2-4} = \dfrac{3(4)}{(4)^2-4}$ *Let x = 4.*

$$= \frac{12}{16-4} = \frac{12}{12} = 1$$

3. $\dfrac{4}{x-3}$

To find the values for which this expression is undefined, set the denominator equal to zero and solve for x.

$$x - 3 = 0$$
$$x = 3$$

Because $x = 3$ will make the denominator zero, the given expression is undefined for 3. Thus, $x \neq 3$.

4. $\dfrac{y+3}{2y}$

Set the denominator equal to zero and solve for y.

$$2y = 0$$
$$y = 0$$

The given expression is undefined for 0. Thus, $y \neq 0$.

5. $\dfrac{2k+1}{3k^2+17k+10}$

Set the denominator equal to zero and solve for k.

$$3k^2+17k+10=0$$
$$(3k+2)(k+5)=0$$
$$k=-\tfrac{2}{3} \text{ or } k=-5$$

The given expression is undefined for -5 and $-\frac{2}{3}$. Thus, $k \neq -5, -\frac{2}{3}$.

6. Set the denominator equal to 0 and solve the equation. Any solutions are values for which the rational expression is undefined.

7. $\dfrac{5a^3b^3}{15a^4b^2}=\dfrac{b\cdot 5a^3b^2}{3a\cdot 5a^3b^2}=\dfrac{b}{3a}$

8. $\dfrac{m-4}{4-m}=\dfrac{-1(4-m)}{4-m}=-1$

9. $\dfrac{4x^2-9}{6-4x}=\dfrac{(2x+3)(2x-3)}{-2(2x-3)}$

$$=\frac{2x+3}{-2}=\frac{-1(2x+3)}{2}$$
$$=\frac{-(2x+3)}{2}$$

10. $\dfrac{4p^2+8pq-5q^2}{10p^2-3pq-q^2}=\dfrac{(2p-q)(2p+5q)}{(5p+q)(2p-q)}$

$$=\frac{2p+5q}{5p+q}$$

11. $-\dfrac{4x-9}{2x+3}$

Apply the negative sign to the numerator:

$$\frac{-(4x-9)}{2x+3}$$

Now distribute the negative sign:

$$\frac{-4x+9}{2x+3}$$

Apply the negative sign to the denominator:

$$\frac{4x-9}{-(2x+3)}$$

Again, distribute the negative sign:

$$\frac{4x-9}{-2x-3}$$

12. $-\dfrac{8-3x}{3-6x}$

Four equivalent forms are:

$$\frac{-(8-3x)}{3-6x},\quad \frac{-8+3x}{3-6x},$$
$$\frac{8-3x}{-(3-6x)},\quad \frac{8-3x}{-3+6x}$$

13. $\dfrac{18p^3}{6}\cdot\dfrac{24}{p^4}=\dfrac{6\cdot 3\cdot 24p^3}{6p^4}=\dfrac{72}{p}$

14. $\dfrac{8x^2}{12x^5}\cdot\dfrac{6x^4}{2x}=\dfrac{2\cdot 4}{3\cdot 4x^3}\cdot\dfrac{3x^3}{1}=2$

15. $\dfrac{x-3}{4}\cdot\dfrac{5}{2x-6}=\dfrac{x-3}{4}\cdot\dfrac{5}{2(x-3)}=\dfrac{5}{8}$

16. $\dfrac{2r+3}{r-4}\cdot\dfrac{r^2-16}{6r+9}$

$$=\frac{2r+3}{r-4}\cdot\frac{(r+4)(r-4)}{3(2r+3)}$$
$$=\frac{r+4}{3}$$

17. $\dfrac{6a^2+7a-3}{2a^2-a-6}\div\dfrac{a+5}{a-2}$

$$=\frac{6a^2+7a-3}{2a^2-a-6}\cdot\frac{a-2}{a+5}$$
$$=\frac{(3a-1)(2a+3)}{(2a+3)(a-2)}\cdot\frac{a-2}{a+5}$$
$$=\frac{3a-1}{a+5}$$

18. $\dfrac{y^2-6y+8}{y^2+3y-18}\div\dfrac{y-4}{y+6}$

$$=\frac{y^2-6y+8}{y^2+3y-18}\cdot\frac{y+6}{y-4}$$
$$=\frac{(y-4)(y-2)}{(y+6)(y-3)}\cdot\frac{y+6}{y-4}$$
$$=\frac{y-2}{y-3}$$

19. $\dfrac{2p^2+13p+20}{p^2+p-12}\cdot\dfrac{p^2+2p-15}{2p^2+7p+5}$

$$=\frac{(2p+5)(p+4)}{(p+4)(p-3)}\cdot\frac{(p+5)(p-3)}{(2p+5)(p+1)}$$
$$=\frac{p+5}{p+1}$$

20. $\dfrac{3z^2+5z-2}{9z^2-1}\cdot\dfrac{9z^2+6z+1}{z^2+5z+6}$

$$=\frac{(3z-1)(z+2)}{(3z-1)(3z+1)}\cdot\frac{(3z+1)^2}{(z+3)(z+2)}$$
$$=\frac{3z+1}{z+3}$$

21. $\dfrac{4}{9y},\ \dfrac{7}{12y^2},\ \dfrac{5}{27y^4}$

Factor each denominator.

$$9y=3^2y$$
$$12y^2=2^2\cdot 3\cdot y^2$$
$$27y^4=3^3\cdot y^4$$

$\text{LCD}=2^2\cdot 3^3\cdot y^4=108y^4$

22. $\dfrac{3}{x^2+4x+3}, \dfrac{5}{x^2+5x+4}$

Factor each denominator.

$$x^2+4x+3=(x+3)(x+1)$$
$$x^2+5x+4=(x+1)(x+4)$$

LCD $=(x+3)(x+1)(x+4)$

23. $\dfrac{3}{2a^3}=\dfrac{?}{10a^4}$

$\dfrac{3}{2a^3}=\dfrac{3}{2a^3}\cdot\dfrac{5a}{5a}=\dfrac{15a}{10a^4}$

24. $\dfrac{9}{x-3}=\dfrac{?}{18-6x}=\dfrac{?}{-6(x-3)}$

$$\dfrac{9}{x-3}=\dfrac{9}{x-3}\cdot\dfrac{-6}{-6}=\dfrac{-54}{-6x+18}=\dfrac{-54}{18-6x}$$

25. $\dfrac{-3y}{2y-10}=\dfrac{?}{50-10y}=\dfrac{?}{-5(2y-10)}$

$$\dfrac{-3y}{2y-10}=\dfrac{-3y}{2y-10}\cdot\dfrac{-5}{-5}=\dfrac{15y}{-10y+50}=\dfrac{15y}{50-10y}$$

26. $\dfrac{4b}{b^2+2b-3}=\dfrac{?}{(b+3)(b-1)(b+2)}$

$$\dfrac{4b}{b^2+2b-3}=\dfrac{4b}{(b+3)(b-1)}=\dfrac{4b}{(b+3)(b-1)}\cdot\dfrac{b+2}{b+2}=\dfrac{4b(b+2)}{(b+3)(b-1)(b+2)}$$

27. $\dfrac{10}{x}+\dfrac{5}{x}=\dfrac{10+5}{x}=\dfrac{15}{x}$

28. $\dfrac{6}{3p}-\dfrac{12}{3p}=\dfrac{6-12}{3p}=\dfrac{-6}{3p}=-\dfrac{2}{p}$

29. $\dfrac{9}{k}-\dfrac{5}{k-5}$

$$=\dfrac{9(k-5)}{k(k-5)}-\dfrac{5\cdot k}{(k-5)k} \quad \textit{LCD} = k(k-5)$$
$$=\dfrac{9(k-5)-5k}{k(k-5)}=\dfrac{9k-45-5k}{k(k-5)}=\dfrac{4k-45}{k(k-5)}$$

30. $\dfrac{4}{y}+\dfrac{7}{7+y}$

$$=\dfrac{4(7+y)}{y(7+y)}+\dfrac{7\cdot y}{(7+y)y} \quad \textit{LCD} = y(7+y)$$
$$=\dfrac{28+4y+7y}{y(7+y)}=\dfrac{28+11y}{y(7+y)}$$

31. $\dfrac{m}{3}-\dfrac{2+5m}{6}$

$$=\dfrac{m\cdot 2}{3\cdot 2}-\dfrac{2+5m}{6} \quad \textit{LCD} = 6$$
$$=\dfrac{2m-(2+5m)}{6}=\dfrac{2m-2-5m}{6}=\dfrac{-2-3m}{6}$$

32. $\dfrac{12}{x^2}-\dfrac{3}{4x}$

$$=\dfrac{12\cdot 4}{x^2\cdot 4}-\dfrac{3\cdot x}{4x\cdot x} \quad \textit{LCD} = 4x^2$$
$$=\dfrac{48-3x}{4x^2}=\dfrac{3(16-x)}{4x^2}$$

33. $\dfrac{5}{a-2b}+\dfrac{2}{a+2b}$

$$=\dfrac{5(a+2b)}{(a-2b)(a+2b)}+\dfrac{2(a-2b)}{(a+2b)(a-2b)}$$

LCD $=(a-2b)(a+2b)$

$$=\dfrac{5(a+2b)+2(a-2b)}{(a-2b)(a+2b)}=\dfrac{5a+10b+2a-4b}{(a-2b)(a+2b)}=\dfrac{7a+6b}{(a-2b)(a+2b)}$$

34. $\dfrac{4}{k^2-9}-\dfrac{k+3}{3k-9}$

$$=\dfrac{4}{(k+3)(k-3)}-\dfrac{k+3}{3(k-3)}$$

LCD $=3(k+3)(k-3)$

$$=\dfrac{4\cdot 3}{(k+3)(k-3)\cdot 3}-\dfrac{(k+3)(k+3)}{3(k-3)(k+3)}$$
$$=\dfrac{12-(k+3)(k+3)}{3(k+3)(k-3)}$$

$$= \frac{12-(k^2+6k+9)}{3(k+3)(k-3)}$$

$$= \frac{12-k^2-6k-9}{3(k+3)(k-3)}$$

$$= \frac{-k^2-6k+3}{3(k+3)(k-3)}$$

35. $\frac{8}{z^2+6z} - \frac{3}{z^2+4z-12}$

$$= \frac{8}{z(z+6)} - \frac{3}{(z+6)(z-2)}$$

LCD = z(z + 6)(z − 2)

$$= \frac{8(z-2)}{z(z+6)(z-2)} - \frac{3 \cdot z}{(z+6)(z-2) \cdot z}$$

$$= \frac{8(z-2)-3z}{z(z+6)(z-2)}$$

$$= \frac{8z-16-3z}{z(z+6)(z-2)}$$

$$= \frac{5z-16}{z(z+6)(z-2)}$$

36. $\frac{11}{2p-p^2} - \frac{2}{p^2-5p+6}$

$$= \frac{11}{p(2-p)} - \frac{2}{(p-3)(p-2)}$$

LCD = p(p − 3)(p − 2)

$$= \frac{11(-1)(p-3)}{p(2-p)(-1)(p-3)}$$

$$- \frac{2 \cdot p}{(p-3)(p-2)p}$$

$$= \frac{-11(p-3)-2p}{p(p-2)(p-3)}$$

$$= \frac{-11p+33-2p}{p(p-2)(p-3)}$$

$$= \frac{-13p+33}{p(p-2)(p-3)}$$

37. $\frac{\frac{y-3}{y}}{\frac{y+3}{4y}} = \frac{y-3}{y} \cdot \frac{4y}{y+3} = \frac{4(y-3)}{y+3}$

38. $\frac{\frac{2}{3}-\frac{1}{6}}{\frac{1}{4}+\frac{2}{5}} = \frac{60(\frac{2}{3}-\frac{1}{6})}{60(\frac{1}{4}+\frac{2}{5})}$ *Multiply by LCD, 60*

$$= \frac{60 \cdot \frac{2}{3} - 60 \cdot \frac{1}{6}}{60 \cdot \frac{1}{4} + 60 \cdot \frac{2}{5}}$$

$$= \frac{40-10}{15+24} = \frac{30}{39} = \frac{10}{13}$$

39. $\frac{x+\frac{1}{w}}{x-\frac{1}{w}}$

$$= \frac{(x+\frac{1}{w}) \cdot w}{(x-\frac{1}{w}) \cdot w}$$ *Multiply by LCD, w.*

$$= \frac{xw+(\frac{1}{w})w}{xw-(\frac{1}{w})w}$$

$$= \frac{xw+1}{xw-1}$$

40. $\frac{\frac{1}{p}-\frac{1}{q}}{\frac{1}{q-p}}$

$$= \frac{\left(\frac{1}{p}-\frac{1}{q}\right)pq(q-p)}{\left(\frac{1}{q-p}\right)pq(q-p)}$$ *Multiply by LCD, pq(q − p).*

$$= \frac{\frac{1}{p}[pq(q-p)] - \frac{1}{q}[pq(q-p)]}{pq}$$

$$= \frac{q(q-p)-p(q-p)}{pq}$$

$$= \frac{q^2-pq-pq+p^2}{pq}$$

$$= \frac{q^2-2pq+p^2}{pq}$$

$$= \frac{(q-p)^2}{pq}$$

41. $\frac{\frac{x^2-25}{x+3}}{\frac{x+5}{x^2-9}}$

$$= \frac{x^2-25}{x+3} \cdot \frac{x^2-9}{x+5}$$

$$= \frac{(x+5)(x-5)}{x+3} \cdot \frac{(x+3)(x-3)}{x+5}$$

$$= \frac{x-5}{1} \cdot \frac{x-3}{1}$$

$= (x-5)(x-3)$, or $x^2-8x+15$

42. $\frac{x^{-2}-y^{-2}}{x^{-1}-y^{-1}} = \frac{\frac{1}{x^2}-\frac{1}{y^2}}{\frac{1}{x}-\frac{1}{y}}$

Multiply the numerator and denominator by the LCD of all the fractions, x^2y^2.

$$= \frac{x^2y^2\left(\frac{1}{x^2}-\frac{1}{y^2}\right)}{x^2y^2\left(\frac{1}{x}-\frac{1}{y}\right)}$$

$$= \frac{x^2y^2\left(\frac{1}{x^2} - \frac{1}{y^2}\right)}{x^2y^2\left(\frac{1}{x} - \frac{1}{y}\right)} \text{ [repeated]}$$

$$= \frac{y^2 - x^2}{xy^2 - x^2y}$$

$$= \frac{(y+x)(y-x)}{xy(y-x)}$$

$$= \frac{y+x}{xy}$$

43. $\frac{3x-1}{x-2} = \frac{5}{x-2} + 1$

$(x-2)\left(\frac{3x-1}{x-2}\right) = (x-2)\left(\frac{5}{x-2} + 1\right)$

Multiply by LCD, x – 2.

$(x-2)\left(\frac{3x-1}{x-2}\right) = (x-2)\left(\frac{5}{x-2}\right) + (x-2)(1)$

Distributive property

$$3x - 1 = 5 + x - 2$$
$$3x - 1 = 3 + x$$
$$2x = 4$$
$$x = 2$$

The solution set is $\emptyset$ because $x = 2$ makes the original denominators equal to zero.

44. $\frac{4-z}{z} + \frac{3}{2} = \frac{-4}{z}$

Multiply each side by the LCD, $2z$.

$$2z\left(\frac{4-z}{z} + \frac{3}{2}\right) = 2z\left(-\frac{4}{z}\right)$$
$$2z\left(\frac{4-z}{z}\right) + 2z\left(\frac{3}{2}\right) = -8$$
$$2(4-z) + 3z = -8$$
$$8 - 2z + 3z = -8$$
$$8 + z = -8$$
$$z = -16$$

Check $z = -16$: $-\frac{5}{4} + \frac{3}{2} = \frac{1}{4}$ *True*

Thus, the solution set is $\{-16\}$.

45. $\frac{3}{x+4} - \frac{2x}{5} = \frac{3}{x+4}$

$-\frac{2x}{5} = 0$ *Subtract* $\frac{3}{x+4}$.

$x = 0$ *Multiply by* $-\frac{5}{2}$.

Check $x = 0$: $\frac{3}{4} - 0 = \frac{3}{4}$ *True*

Thus, the solution set is $\{0\}$.

46. $\frac{3}{m-2} + \frac{1}{m-1} = \frac{7}{m^2 - 3m + 2}$

$\frac{3}{m-2} + \frac{1}{m-1} = \frac{7}{(m-2)(m-1)}$

$(m-2)(m-1)\left(\frac{3}{m-2} + \frac{1}{m-1}\right) = (m-2)(m-1) \cdot \frac{7}{(m-2)(m-1)}$

Multiply by LCD, (m – 2)(m – 1).

$$3(m-1) + 1(m-2) = 7$$
$$3m - 3 + m - 2 = 7$$
$$4m - 5 = 7$$
$$4m = 12$$
$$m = 3$$

Check $m = 3$: $3 + \frac{1}{2} = \frac{7}{2}$ *True*

Thus, the solution set is $\{3\}$.

47. $m = \frac{Ry}{t}$ for t

$t \cdot m = t\left(\frac{Ry}{t}\right)$ *Multiply by t.*

$tm = Ry$

$t = \frac{Ry}{m}$ *Divide by m.*

48. $x = \frac{3y-5}{4}$ for y

$$4x = 4\left(\frac{3y-5}{4}\right)$$
$$4x = 3y - 5$$
$$4x + 5 = 3y$$
$$\frac{4x+5}{3} = y$$

49. $p^2 = \frac{4}{3m-q}$ for m

$$(3m-q)p^2 = (3m-q)\left(\frac{4}{3m-q}\right)$$
$$3mp^2 - p^2q = 4$$
$$3mp^2 = 4 + p^2q$$
$$m = \frac{4 + p^2q}{3p^2}$$

50. Let $x =$ the numerator. Then $x - 5 =$ the denominator. Adding 5 to both the numerator and the denominator gives us a fraction that is equivalent to $\frac{5}{4}$.

$$\frac{x+5}{x-5+5} = \frac{5}{4}$$
$$\frac{x+5}{x} = \frac{5}{4}$$
$$4x\left(\frac{x+5}{x}\right) = 4x\left(\frac{5}{4}\right)$$
$$4(x+5) = x(5)$$
$$4x + 20 = 5x$$
$$20 = x$$

The numerator is 20 and the denominator is $20 - 5 = 15$, so the original fraction is $\frac{20}{15}$.

51. Let $x =$ the numerator. Then $6x =$ the denominator. Adding 3 to the numerator and subtracting 3 from the denominator gives us a fraction equivalent to $\frac{2}{5}$.

$$\frac{x+3}{6x-3} = \frac{2}{5}$$
$$5(6x-3)\left(\frac{x+3}{6x-3}\right) = 5(6x-3)\left(\frac{2}{5}\right)$$
$$5(x+3) = 2(6x-3)$$
$$5x + 15 = 12x - 6$$
$$21 = 7x$$
$$3 = x$$

The numerator is 3 and the denominator is $6 \cdot 3 = 18$, so the original fraction is $\frac{3}{18}$.

52. Let $x =$ the rate of the wind. Then the rate against the wind is $165 - x$ and the rate with the wind is $165 + x$. Complete the chart using $t = d/r$.

	d	r	t
Against the Wind	310	$165 - x$	$\frac{310}{165-x}$
With the Wind	350	$165 + x$	$\frac{350}{165+x}$

Since the times are equal, we get the following equation.

$$\frac{310}{165-x} = \frac{350}{165+x}$$
$$(165+x)(165-x)\frac{310}{165-x} = (165+x)(165-x)\frac{350}{165+x}$$
$$310(165+x) = 350(165-x)$$
$$51{,}150 + 310x = 57{,}750 - 350x$$
$$660x = 6600$$
$$x = 10$$

The rate of the wind is 10 miles per hour.

53. *Step 2*
Let $x =$ the number of hours it takes them to do the job working together.

	Rate	Time Working Together	Fractional Part of the Job Done when Working Together
Susan	$\frac{1}{5}$	x	$\frac{1}{5}x$
Friend	$\frac{1}{8}$	x	$\frac{1}{8}x$

Step 3
Working together, they do 1 whole job, so

$$\tfrac{1}{5}x + \tfrac{1}{8}x = 1.$$

Step 4
Solve this equation by multiplying both sides by the LCD, 40.

$$40(\tfrac{1}{5}x + \tfrac{1}{8}x) = 40(1)$$
$$8x + 5x = 40$$
$$13x = 40$$
$$x = \tfrac{40}{13}, \text{ or } 3\tfrac{1}{13}$$

Step 5
Working together, it takes them $3\frac{1}{13}$ hours.

Step 6
Susan does $\frac{1}{5}$ of the job per hour for $\frac{40}{13}$ hours:

$$\tfrac{1}{5} \cdot \tfrac{40}{13} = \tfrac{8}{13} \text{ of the job}$$

Her friend does $\frac{1}{8}$ of the job per hour for $\frac{40}{13}$ hours:

$$\tfrac{1}{8} \cdot \tfrac{40}{13} = \tfrac{5}{13} \text{ of the job}$$

Together, they have done

$$\tfrac{8}{13} + \tfrac{5}{13} = \tfrac{13}{13} = 1 \text{ total job.}$$

54. *Step 2*
Let $x =$ the time needed by the head gardener to mow the lawns.
Then $2x =$ the time needed by the assistant to mow the lawns.

	Rate	Time Working Together	Fractional Part of the Job Done when Working Together
Head gardener	$\frac{1}{x}$	$1\frac{1}{3} = \frac{4}{3}$	$\frac{1}{x} \cdot \frac{4}{3} = \frac{4}{3x}$
Assistant	$\frac{1}{2x}$	$1\frac{1}{3} = \frac{4}{3}$	$\frac{1}{2x} \cdot \frac{4}{3} = \frac{2}{3x}$

Step 3 $\quad \frac{4}{3x} + \frac{2}{3x} = 1$

Step 4
$$\frac{4+2}{3x}=1$$
$$\frac{6}{3x}=1$$
$$3x\left(\frac{6}{3x}\right)=3x(1)$$
$$6=3x$$
$$2=x$$

Step 5
It takes the head gardener 2 hours to mow the lawns.

Step 6
The head gardener does $\frac{1}{2}$ of the job per hour for $\frac{4}{3}$ hours:

$$\tfrac{1}{2}\cdot\tfrac{4}{3}=\tfrac{2}{3}\text{ of the job}$$

The assistant does $\frac{1}{4}$ of the job per hour for $\frac{4}{3}$ hours:

$$\tfrac{1}{4}\cdot\tfrac{4}{3}=\tfrac{1}{3}\text{ of the job}$$

Together, they have done

$$\tfrac{2}{3}+\tfrac{1}{3}=\tfrac{3}{3}=1\text{ total job.}$$

55. [6.4] $\dfrac{4}{m-1}-\dfrac{3}{m+1}$

To perform the indicated subtraction, use $(m-1)(m+1)$ as the LCD.

$$\begin{aligned}\frac{4}{m-1}-\frac{3}{m+1}&=\frac{4(m+1)}{(m-1)(m+1)}-\frac{3(m-1)}{(m+1)(m-1)}\\&=\frac{4(m+1)-3(m-1)}{(m-1)(m+1)}\\&=\frac{4m+4-3m+3}{(m-1)(m+1)}\\&=\frac{m+7}{(m-1)(m+1)}\end{aligned}$$

56. [6.2] $\dfrac{8p^5}{5}\div\dfrac{2p^3}{10}$

To perform the indicated division, multiply the first rational expression by the reciprocal of the second.

$$\begin{aligned}\frac{8p^5}{5}\div\frac{2p^3}{10}&=\frac{8p^5}{5}\cdot\frac{10}{2p^3}\\&=\frac{80p^5}{10p^3}\\&=8p^2\end{aligned}$$

57. [6.2]
$$\begin{aligned}\frac{r-3}{8}\div\frac{3r-9}{4}&=\frac{r-3}{8}\cdot\frac{4}{3r-9}\\&=\frac{r-3}{8}\cdot\frac{4}{3(r-3)}\\&=\frac{4}{24}=\frac{1}{6}\end{aligned}$$

58. [6.5]
$$\frac{t^{-2}+s^{-2}}{t^{-1}-s^{-1}}=\frac{\frac{1}{t^2}+\frac{1}{s^2}}{\frac{1}{t}-\frac{1}{s}}$$

Multiply the numerator and denominator by the LCD of all the fractions, t^2s^2.

$$\begin{aligned}&=\frac{t^2s^2\left(\frac{1}{t^2}+\frac{1}{s^2}\right)}{t^2s^2\left(\frac{1}{t}-\frac{1}{s}\right)}\\&=\frac{s^2+t^2}{ts^2-t^2s}\\&=\frac{s^2+t^2}{st(s-t)}\end{aligned}$$

59. [6.5]
$$\frac{\frac{5}{x}-1}{\frac{5-x}{3x}}=\frac{\left(\frac{5}{x}-1\right)3x}{\left(\frac{5-x}{3x}\right)3x}$$ *Multiply by LCD, 3x.*

$$\begin{aligned}&=\frac{\frac{5}{x}(3x)-1(3x)}{5-x}\\&=\frac{15-3x}{5-x}\\&=\frac{3(5-x)}{5-x}=3\end{aligned}$$

60. [6.4]
$$\begin{aligned}&\frac{4}{z^2-2z+1}-\frac{3}{z^2-1}\\&=\frac{4}{(z-1)^2}-\frac{3}{(z+1)(z-1)}\end{aligned}$$

LCD $= (z+1)(z-1)^2$

$$\begin{aligned}&=\frac{4(z+1)}{(z-1)^2(z+1)}\\&\quad-\frac{3(z-1)}{(z+1)(z-1)(z-1)}\\&=\frac{4(z+1)-3(z-1)}{(z+1)(z-1)^2}\\&=\frac{4z+4-3z+3}{(z+1)(z-1)^2}\\&=\frac{z+7}{(z+1)(z-1)^2}\end{aligned}$$

61. **[6.4]** $\frac{1}{t^2-4}+\frac{1}{2-t}$

$$=\frac{1}{(t+2)(t-2)}+\frac{1}{2-t}$$

LCD = (t + 2)(t − 2)

$$=\frac{1}{(t+2)(t-2)}-\frac{1(t+2)}{(t-2)(t+2)}$$

$$=\frac{1-t-2}{(t+2)(t-2)}$$

$$=\frac{-t-1}{(t+2)(t-2)}, \text{ or } \frac{t+1}{(t+2)(2-t)}$$

62. **[6.6]** $\frac{2}{z}-\frac{z}{z+3}=\frac{1}{z+3}$

Multiply each side of the equation by the LCD, $z(z+3)$.

$$z(z+3)\left(\frac{2}{z}-\frac{z}{z+3}\right)=z(z+3)\left(\frac{1}{z+3}\right)$$

$$z(z+3)\left(\frac{2}{z}\right)-z(z+3)\left(\frac{z}{z+3}\right)=z(1)$$

$$2(z+3)-z^2=z$$

$$2z+6-z^2=z$$

$$0=z^2-z-6$$

$$0=(z-3)(z+2)$$

$$z-3=0 \quad \text{or} \quad z+2=0$$

$$z=3 \quad \text{or} \quad z=-2$$

Check $z=-2$: $-1-(-2)=1$ *True*
Check $z=3$: $\frac{2}{3}-\frac{1}{2}=\frac{1}{6}$ *True*
Thus, the solution set is $\{-2,3\}$.

63. **[6.6]** Solve $a=\frac{v-w}{t}$ for v.

$t\cdot a=v-w$ *Multiply by t.*
$at+w=v$ *Add w.*

64. **[6.7]** Let $x=$ the rate of the plane in still air. Then the rate of the plane with the wind is $x+50$, and the rate of the plane against the wind is $x-50$. Use $t=d/r$ to complete the chart.

	d	***r***	***t***
With the Wind	400	$x+50$	$\frac{400}{x+50}$
Against the Wind	200	$x-50$	$\frac{200}{x-50}$

The times are the same, so

$$\frac{400}{x+50}=\frac{200}{x-50}.$$

To solve this equation, multiply both sides by the LCD, $(x+50)(x-50)$.

$$(x+50)(x-50)\cdot\frac{400}{x+50}$$
$$=(x+50)(x-50)\cdot\frac{200}{x-50}$$

$$400(x-50)=200(x+50)$$
$$400x-20{,}000=200x+10{,}000$$
$$200x=30{,}000$$
$$x=150$$

The rate of the plane is 150 kilometers per hour.

65. **[6.7]** Let $x=$ the number of hours it takes them to do the job working together.

	Rate	**Time Working Together**	**Fractional Part of the Job Done when Working Together**
Lizette	$\frac{1}{8}$	x	$\frac{1}{8}x$
Seyed	$\frac{1}{14}$	x	$\frac{1}{14}x$

Working together, they do 1 whole job, so

$$\tfrac{1}{8}x+\tfrac{1}{14}x=1.$$

To clear fractions, multiply both sides by the LCD, 56.

$$56(\tfrac{1}{8}x+\tfrac{1}{14}x)=56(1)$$
$$7x+4x=56$$
$$11x=56$$
$$x=\tfrac{56}{11}, \text{ or } 5\tfrac{1}{11}$$

Working together, they can paint the woodwork in $5\frac{1}{11}$ hours.

♦♦♦ Relating Concepts 66–75 ♦♦♦

66. **(a)** $x+3=0$
$x=-3$

This value of x makes the value of the denominator zero, so P will be undefined when $x=-3$.

(b) $x+1=0$
$x=-1$

This value of x makes the value of the denominator zero, so Q will be undefined when $x=-1$.

(c) $x^2+4x+3=0$
$(x+3)(x+1)=0$
$x=-3$ or $x=-1$

These values of x make the value of the denominator zero, so the values for which R is undefined are -3 and -1.

67. $(P \cdot Q) \div R$

$$= \left(\frac{6}{x+3} \cdot \frac{5}{x+1}\right) \div \frac{4x}{x^2+4x+3}$$

$$= \frac{30}{(x+3)(x+1)} \cdot \frac{x^2+4x+3}{4x}$$

$$= \frac{30}{(x+3)(x+1)} \cdot \frac{(x+3)(x+1)}{4x}$$

$$= \frac{30}{4x} = \frac{15}{2x}$$

68. If $x = 0$, the divisor R is equal to 0, and division by 0 is undefined.

69. List the three denominators and factor if possible.

$x+3,\ x+1,\ x^2+4x+3 = (x+3)(x+1)$

The LCD for P, Q, and R is $(x+3)(x+1)$.

70. $P + Q - R$

$$= \frac{6}{x+3} + \frac{5}{x+1} - \frac{4x}{x^2+4x+3}$$

$$= \frac{6}{x+3} + \frac{5}{x+1} - \frac{4x}{(x+3)(x+1)}$$

LCD = (x + 3)(x + 1)

$$= \frac{6(x+1)}{(x+3)(x+1)} + \frac{5(x+3)}{(x+3)(x+1)} - \frac{4x}{(x+3)(x+1)}$$

$$= \frac{6(x+1)+5(x+3)-4x}{(x+3)(x+1)}$$

$$= \frac{6x+6+5x+15-4x}{(x+3)(x+1)}$$

$$= \frac{7x+21}{(x+3)(x+1)}$$

$$= \frac{7(x+3)}{(x+3)(x+1)} = \frac{7}{x+1}$$

71. $$\frac{P+Q}{R} = \frac{\frac{6}{x+3}+\frac{5}{x+1}}{\frac{4x}{x^2+4x+3}} = \frac{\frac{6}{x+3}+\frac{5}{x+1}}{\frac{4x}{(x+3)(x+1)}}$$

To simplify this complex fraction, use Method 2. Multiply numerator and denominator by the LCD for all the fractions, $(x+3)(x+1)$.

$$\frac{(x+3)(x+1)\left(\frac{6}{x+3}+\frac{5}{x+1}\right)}{(x+3)(x+1)\left(\frac{4x}{(x+3)(x+1)}\right)}$$

$$= \frac{(x+3)(x+1)\left(\frac{6}{x+3}\right) + (x+3)(x+1)\left(\frac{5}{x+1}\right)}{4x}$$

$$= \frac{6(x+1)+5(x+3)}{4x}$$

$$= \frac{6x+6+5x+15}{4x}$$

$$= \frac{11x+21}{4x}$$

72.

$$P + Q = R$$

$$\frac{6}{x+3} + \frac{5}{x+1} = \frac{4x}{x^2+4x+3}$$

$$\frac{6}{x+3} + \frac{5}{x+1} = \frac{4x}{(x+3)(x+1)}$$

Multiply by LCD, (x + 3)(x + 1).

$$(x+3)(x+1)\left(\frac{6}{x+3}\right) + (x+3)(x+1)\left(\frac{5}{x+1}\right) = (x+3)(x+1)\left(\frac{4x}{(x+3)(x+1)}\right)$$

$$6(x+1) + 5(x+3) = 4x$$

$$6x+6+5x+15 = 4x$$

$$7x = -21$$

$$x = -3$$

To check, substitute -3 for x. The denominators of the first and third fractions become zero. Reject -3 as a solution. There is no solution to the equation, so the solution set is $\emptyset$.

73. We know that -3 is not allowed because P and R are undefined for $x = -3$.

74. Solving $d = rt$ for r gives us $r = d/t$. If $d = 6$ miles and $t = (x+3)$ minutes, then $r = \frac{d}{t} = \frac{6}{x+3}$ miles per minute. Thus,

$$P = \frac{6}{x+3}$$

represents the rate of the car (in miles per minute).

75.

$$R = \frac{4x}{x^2+4x+3}$$

$$\frac{40}{77} = \frac{4x}{x^2+4x+3}$$

$$40(x^2+4x+3) = (4x)(77) \quad \textit{Cross products are equal.}$$

$$40x^2+160x+120 = 308x$$

$$40x^2-148x+120 = 0$$

$$10x^2-37x+30 = 0$$

$$(5x-6)(2x-5) = 0$$

$$x = \tfrac{6}{5} \quad \text{or} \quad x = \tfrac{5}{2}$$

Chapter 6 Test

1. **(a)** $\dfrac{6r+1}{2r^2-3r-20}$

$= \dfrac{6(-2)+1}{2(-2)^2-3(-2)-20}$ *Let r = −2.*

$= \dfrac{-12+1}{2\cdot 4+6-20}$

$= \dfrac{-11}{8+6-20}$

$= \dfrac{-11}{-6} = \dfrac{11}{6}$

(b) $\dfrac{6r+1}{2r^2-3r-20}$

$= \dfrac{6(4)+1}{2(4)^2-3(4)-20}$ *Let r = 4.*

$= \dfrac{24+1}{2\cdot 16-12-20}$

$= \dfrac{25}{32-12-20}$

$= \dfrac{25}{20-20} = \dfrac{25}{0}$

The expression is undefined when $r = 4$ because the denominator is 0.

2. $\dfrac{3x-1}{x^2-2x-8}$

Set the denominator equal to zero and solve for x.

$$x^2-2x-8=0$$
$$(x+2)(x-4)=0$$

$x+2=0$ or $x-4=0$

$x=-2$ or $x=4$

The expression is undefined for -2 and 4, so $x \neq -2, 4$.

3. $-\dfrac{6x-5}{2x+3}$

Apply the negative sign to the numerator:

$$\frac{-(6x-5)}{2x+3}$$

Now distribute the negative sign:

$$\frac{-6x+5}{2x+3}$$

Apply the negative sign to the denominator:

$$\frac{6x-5}{-(2x+3)}$$

Again, distribute the negative sign:

$$\frac{6x-5}{-2x-3}$$

4. $\dfrac{-15x^6y^4}{5x^4y} = \dfrac{(5x^4y)(-3x^2y^3)}{(5x^4y)(1)}$

$= \dfrac{5x^4y}{5x^4y} \cdot \dfrac{-3x^2y^3}{1} = -3x^2y^3$

5. $\dfrac{6a^2+a-2}{2a^2-3a+1} = \dfrac{(3a+2)(2a-1)}{(2a-1)(a-1)}$

$= \dfrac{3a+2}{a-1}$

6. $\dfrac{5(d-2)}{9} \div \dfrac{3(d-2)}{5}$

$= \dfrac{5(d-2)}{9} \cdot \dfrac{5}{3(d-2)}$

$= \dfrac{5\cdot 5}{9\cdot 3} = \dfrac{25}{27}$

7. $\dfrac{6k^2-k-2}{8k^2+10k+3} \cdot \dfrac{4k^2+7k+3}{3k^2+5k+2}$

$= \dfrac{(3k-2)(2k+1)}{(4k+3)(2k+1)} \cdot \dfrac{(4k+3)(k+1)}{(3k+2)(k+1)}$

$= \dfrac{3k-2}{3k+2}$

8. $\dfrac{4a^2+9a+2}{3a^2+11a+10} \div \dfrac{4a^2+17a+4}{3a^2+2a-5}$

$= \dfrac{4a^2+9a+2}{3a^2+11a+10} \cdot \dfrac{3a^2+2a-5}{4a^2+17a+4}$

$= \dfrac{(4a+1)(a+2)}{(3a+5)(a+2)} \cdot \dfrac{(3a+5)(a-1)}{(4a+1)(a+4)}$

$= \dfrac{a-1}{a+4}$

9. $\dfrac{x^2-10x+25}{9-6x+x^2} \cdot \dfrac{x-3}{5-x}$

$= \dfrac{(x-5)(x-5)}{(3-x)(3-x)} \cdot \dfrac{x-3}{5-x}$ *Factor.*

$= \dfrac{(x-5)(x-5)(x-3)}{(3-x)(3-x)(5-x)}$ *Multiply.*

$= \dfrac{-1(x-5)}{-1(3-x)}$ *Lowest terms*

$= \dfrac{x-5}{3-x}$

10. $\dfrac{-3}{10p^2}, \dfrac{21}{25p^3}, \dfrac{-7}{30p^5}$

Factor each denominator.

$$10p^2 = 2\cdot 5\cdot p^2$$
$$25p^3 = 5^2\cdot p^3$$
$$30p^5 = 2\cdot 3\cdot 5\cdot p^5$$

$\text{LCD} = 2\cdot 3\cdot 5^2\cdot p^5 = 150p^5$

11. $\frac{r+1}{2r^2+7r+6}, \frac{-2r+1}{2r^2-7r-15}$

Factor each denominator.

$$2r^2+7r+6=(2r+3)(r+2)$$
$$2r^2-7r-15=(2r+3)(r-5)$$

LCD $=(2r+3)(r+2)(r-5)$

12. $\frac{15}{4p}=\frac{?}{64p^3}=\frac{?}{4p\cdot 16p^2}$

$$\frac{15}{4p}=\frac{15\cdot 16p^2}{4p\cdot 16p^2}=\frac{240p^2}{64p^3}$$

13. $\frac{3}{6m-12}=\frac{?}{42m-84}=\frac{?}{7(6m-12)}$

$$\frac{3}{6m-12}=\frac{3\cdot 7}{(6m-12)7}=\frac{21}{42m-84}$$

14. $\frac{4x+2}{x+5}+\frac{-2x+8}{x+5}$

$$=\frac{(4x+2)+(-2x+8)}{x+5}$$
$$=\frac{2x+10}{x+5}$$
$$=\frac{2(x+5)}{x+5}=2$$

15. $\frac{-4}{y+2}+\frac{6}{5y+10}$

$$=\frac{-4}{y+2}+\frac{6}{5(y+2)} \quad LCD = 5(y+2)$$
$$=\frac{-4\cdot 5}{(y+2)\cdot 5}+\frac{6}{5(y+2)}$$
$$=\frac{-20+6}{5(y+2)}=\frac{-14}{5(y+2)}$$

16. Using LCD $=3-x$,

$$\frac{x+1}{3-x}+\frac{x^2}{x-3}=\frac{x+1}{3-x}+\frac{-1(x^2)}{-1(x-3)}$$
$$=\frac{x+1}{3-x}+\frac{-x^2}{-x+3}$$
$$=\frac{x+1}{3-x}+\frac{-x^2}{3-x}$$
$$=\frac{(x+1)+(-x^2)}{3-x}$$
$$=\frac{-x^2+x+1}{3-x}.$$

If we use $x-3$ for the LCD, we obtain the equivalent answer

$$\frac{x^2-x-1}{x-3}.$$

17. $\frac{3}{2m^2-9m-5}-\frac{m+1}{2m^2-m-1}$

$$=\frac{3}{(2m+1)(m-5)}-\frac{m+1}{(2m+1)(m-1)}$$

LCD $=(2m+1)(m-5)(m-1)$

$$=\frac{3(m-1)}{(2m+1)(m-5)(m-1)}$$
$$-\frac{(m+1)(m-5)}{(2m+1)(m-1)(m-5)}$$
$$=\frac{3(m-1)-(m+1)(m-5)}{(2m+1)(m-5)(m-1)}$$
$$=\frac{(3m-3)-(m^2-4m-5)}{(2m+1)(m-5)(m-1)}$$
$$=\frac{3m-3-m^2+4m+5}{(2m+1)(m-5)(m-1)}$$
$$=\frac{-m^2+7m+2}{(2m+1)(m-5)(m-1)}$$

18. $\frac{\frac{2p}{k^2}}{\frac{3p^2}{k^3}}=\frac{2p}{k^2}\div\frac{3p^2}{k^3}$

$$=\frac{2p}{k^2}\cdot\frac{k^3}{3p^2}$$
$$=\frac{2k^3p}{3k^2p^2}=\frac{2k}{3p}$$

19. $\frac{\frac{1}{x+3}-1}{1+\frac{1}{x+3}}$

Start by multiplying the numerator and the denominator by the LCD, $x+3$.

$$=\frac{(x+3)\left(\frac{1}{x+3}-1\right)}{(x+3)\left(1+\frac{1}{x+3}\right)}$$
$$=\frac{(x+3)\left(\frac{1}{x+3}\right)-(x+3)(1)}{(x+3)(1)+(x+3)\left(\frac{1}{x+3}\right)}$$
$$=\frac{1-(x+3)}{(x+3)+1}$$
$$=\frac{1-x-3}{x+4}$$
$$=\frac{-2-x}{x+4}$$

20. $\dfrac{2x^{-2}+y^{-2}}{x^{-1}-y^{-1}} = \dfrac{\frac{2}{x^2}+\frac{1}{y^2}}{\frac{1}{x}-\frac{1}{y}}$

Multiply the numerator and denominator by the LCD of all the fractions, x^2y^2.

$$= \frac{x^2y^2\left(\frac{2}{x^2}+\frac{1}{y^2}\right)}{x^2y^2\left(\frac{1}{x}-\frac{1}{y}\right)}$$

$$= \frac{2y^2+x^2}{xy^2-x^2y}$$

$$= \frac{2y^2+x^2}{xy(y-x)}$$

21. $\dfrac{3x}{x+1} = \dfrac{3}{2x}$

$$2x(x+1)\frac{3x}{x+1} = 2x(x+1)\frac{3}{2x}$$

Multiply by LCD, 2x(x + 1).

$$2x(3x) = 3(x+1)$$
$$6x^2 = 3x+3$$
$$6x^2-3x-3=0$$
$$3(2x^2-x-1)=0$$
$$3(2x+1)(x-1)=0$$
$$x=-\tfrac{1}{2} \quad \text{or} \quad x=1$$

Check $x=-\frac{1}{2}$: $-3=-3$ *True*
Check $x=1$: $\frac{3}{2}=\frac{3}{2}$ *True*

Thus, the solution set is $\{-\frac{1}{2}, 1\}$.

22. $\dfrac{2x}{x-3}+\dfrac{1}{x+3}=\dfrac{-6}{x^2-9}$

$$\frac{2x}{x-3}+\frac{1}{x+3}=\frac{-6}{(x+3)(x-3)}$$

$$(x+3)(x-3)\left(\frac{2x}{x-3}+\frac{1}{x+3}\right)$$
$$= (x+3)(x-3)\left(\frac{-6}{(x+3)(x-3)}\right)$$

Multiply by LCD, (x + 3)(x − 3).

$$2x(x+3)+1(x-3)=-6$$
$$2x^2+6x+x-3=-6$$
$$2x^2+7x+3=0$$
$$(2x+1)(x+3)=0$$
$$x=-\tfrac{1}{2} \quad \text{or} \quad x=-3$$

x cannot equal -3 because the denominator $x+3$ would equal 0.

Check $x=-\frac{1}{2}$: $\frac{2}{7}+\frac{2}{5}=\frac{24}{35}$ *True*

Thus, the solution set is $\{-\frac{1}{2}\}$.

23. Solve $F=\dfrac{k}{d-D}$ for D.

$$(d-D)(F)=(d-D)\left(\frac{k}{d-D}\right) \quad \textit{Multiply by LCD, } d-D.$$
$$(d-D)(F)=k$$
$$dF-DF=k$$
$$-DF=k-dF$$
$$DF=dF-k$$
$$D=\frac{dF-k}{F}, \text{ or } D=d-\frac{k}{F}$$

24. Let x = the rate of the current.

	d	r	t
Upstream	20	$7-x$	$\frac{20}{7-x}$
Downstream	50	$7+x$	$\frac{50}{7+x}$

The times are equal, so

$$\frac{20}{7-x}=\frac{50}{7+x}.$$

$$(7-x)(7+x)\left(\frac{20}{7-x}\right)=(7-x)(7+x)\left(\frac{50}{7+x}\right)$$

Multiply by LCD, (7 − x)(7 + x).

$$20(7+x)=50(7-x)$$
$$140+20x=350-50x$$
$$70x=210$$
$$x=3$$

The rate of the current is 3 miles per hour.

25. Let x = the time required for Sanford and his neighbor to paint the room working together.

	Rate	Time Working Together	Fractional Part of the Job Done when Working Together
Sanford	$\frac{1}{5}$	x	$\frac{1}{5}x$
Neighbor	$\frac{1}{4}$	x	$\frac{1}{4}x$

Working together, they do 1 whole job, so

$$\tfrac{1}{5}x+\tfrac{1}{4}x=1.$$
$$20(\tfrac{1}{5}x+\tfrac{1}{4}x)=20(1) \quad \textit{Multiply by LCD, 20.}$$
$$20(\tfrac{1}{5}x)+20(\tfrac{1}{4}x)=20$$
$$4x+5x=20$$
$$9x=20$$
$$x=\tfrac{20}{9}, \text{ or } 2\tfrac{2}{9}$$

They can paint the room in $2\frac{2}{9}$ hours.

Cumulative Review Exercises (Chapters 1–6)

1. $3+4(\frac{1}{2}-\frac{3}{4})$
$=3+4(\frac{2}{4}-\frac{3}{4})$
$=3+4(-\frac{1}{4})$ *Parentheses*
$=3+(-1)$ *Multiplication*
$=2$ *Addition*

2. $3(2y-5)=2+5y$
$6y-15=2+5y$
$y=17$

The solution set is $\{17\}$.

3. Solve $\mathcal{A}=\frac{1}{2}bh$ for b.

$2\cdot\mathcal{A}=2\cdot\frac{1}{2}bh$ *Multiply by 2.*
$2\mathcal{A}=bh$
$\frac{2\mathcal{A}}{h}=b$ *Divide by h.*

4. $\frac{2+m}{2-m}=\frac{3}{4}$
$4(2+m)=3(2-m)$ *Cross multiply.*
$8+4m=6-3m$
$7m=-2$
$m=-\frac{2}{7}$

The solution set is $\{-\frac{2}{7}\}$.

5. $5y\le 6y+8$
$-y\le 8$
$y\ge -8$ *Reverse the inequality symbol.*

The solution set is $[-8,\infty)$.

6. $4x+3y=-12$

(a) Let $y=0$ to find the x-intercept.

$4x+3(0)=-12$
$4x=-12$
$x=-3$

The x-intercept is $(-3,0)$.

(b) Let $x=0$ to find the y-intercept.

$4(0)+3y=-12$
$3y=-12$
$y=-4$

The y-intercept is $(0,-4)$.

7. $y=-3x+2$

This is an equation of a line.

If $x=0,\ y=2$; so the y-intercept is $(0,2)$.
If $x=1,\ y=-1$; and if $x=2,\ y=-4$.

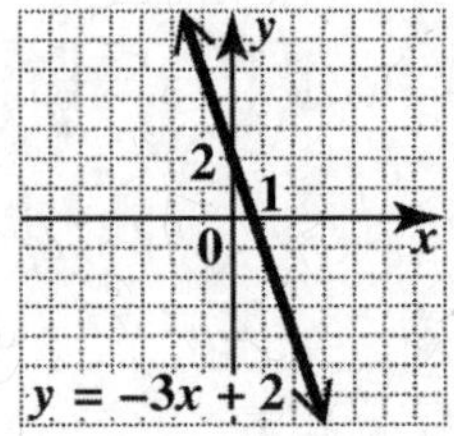

8. $y=-x^2+1$

This is the equation of a parabola opening downward with a y-intercept $(0,1)$.
If $x=+2$ or -2, $y=-3$.

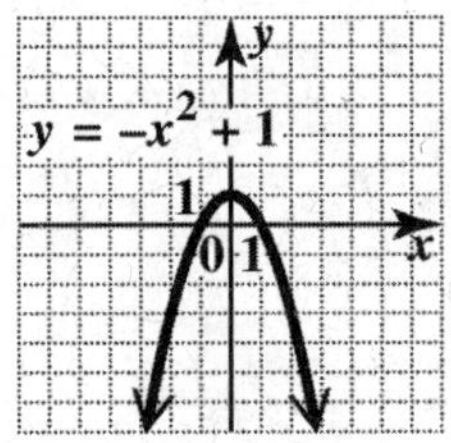

9. $\frac{(2x^3)^{-1}\cdot x}{2^3x^5}=\frac{2^{-1}(x^3)^{-1}\cdot x}{2^3x^5}=\frac{2^{-1}x^{-3}x}{2^3x^5}$
$=\frac{2^{-1}x^{-2}}{2^3x^5}=\frac{1}{2^1\cdot 2^3\cdot x^2\cdot x^5}$
$=\frac{1}{2^4x^7}$

10. $\frac{(m^{-2})^3m}{m^5m^{-4}}=\frac{m^{-6}m}{m^5m^{-4}}=\frac{m\cdot m^4}{m^5\cdot m^6}$
$=\frac{m^5}{m^{11}}=\frac{1}{m^6}$

11. $(2k^2+3k)-(k^2+k-1)$
$=2k^2+3k-k^2-k+1$
$=k^2+2k+1$

12. $(2a-b)^2=(2a)^2-2(2a)(b)+(b)^2$
$=4a^2-4ab+b^2$

13. $(y^2+3y+5)(3y-1)$

Multiply vertically.

$$\begin{array}{rrrr} & y^2 & +\ 3y & +\ 5 \\ & & 3y & -\ 1 \\ \hline & -\ y^2 & -\ 3y & -\ 5 \\ 3y^3 & +\ 9y^2 & +\ 15y & \\ \hline 3y^3 & +\ 8y^2 & +\ 12y & -\ 5 \end{array}$$

14. $\dfrac{12p^3+2p^2-12p+4}{2p-2}$

$$\begin{array}{r|l} & 6p^2 + 7p + 1 \\ 2p-2 & 12p^3 + 2p^2 - 12p + 4 \\ & \underline{12p^3 - 12p^2} \\ & \quad 14p^2 - 12p \\ & \quad \underline{14p^2 - 14p} \\ & \qquad 2p + 4 \\ & \qquad \underline{2p - 2} \\ & \qquad\quad 6 \end{array}$$

The result is

$$6p^2+7p+1+\frac{6}{2p-2}$$
$$=6p^2+7p+1+\frac{2\cdot 3}{2(p-1)}$$
$$=6p^2+7p+1+\frac{3}{p-1}.$$

15. $8t^2+10tv+3v^2$

$$=8t^2+6tv+4tv+3v^2 \qquad \textit{6 • 4 = 24; 6 + 4 = 10}$$
$$=(8t^2+6tv)+(4tv+3v^2)$$
$$=2t(4t+3v)+v(4t+3v)$$
$$=(4t+3v)(2t+v)$$

16. $8r^2-9rs+12s^2$

To factor this polynomial by the grouping method, we must find two integers whose product is $(8)(12)=96$ and whose sum is -9. There is no pair of integers that satisfies both of these conditions, so the polynomial is *prime*.

17. $16x^4-1$

$$=(4x^2)^2-(1)^2$$
$$=(4x^2+1)(4x^2-1)$$
$$=(4x^2+1)[(2x)^2-(1)^2]$$
$$=(4x^2+1)(2x+1)(2x-1)$$

18.
$$r^2=2r+15$$
$$r^2-2r-15=0$$
$$(r+3)(r-5)=0$$

$$r+3=0 \quad \text{or} \quad r-5=0$$
$$r=-3 \quad \text{or} \quad r=5$$

The solution set is $\{-3,5\}$.

19. $(r-5)(2r+1)(3r-2)=0$

$$r-5=0 \quad \text{or} \quad 2r+1=0 \quad \text{or} \quad 3r-2=0$$
$$2r=-1 \quad \text{or} \quad 3r=2$$
$$r=5 \quad \text{or} \quad r=-\tfrac{1}{2} \quad \text{or} \quad r=\tfrac{2}{3}$$

The solution set is $\{5,-\frac{1}{2},\frac{2}{3}\}$.

20. Let x = the lesser number
Then $x+4$ = the greater number.

The product of the numbers is 2 less than the lesser number translates to

$$x(x+4)=x-2.$$
$$x^2+4x=x-2$$
$$x^2+3x+2=0$$
$$(x+2)(x+1)=0$$
$$x=-2 \quad \text{or} \quad x=-1$$

The lesser number can be either -2 or -1.

21. Let w = the width of the rectangle.
Then $2w-2$ = the length of the rectangle.

Use the formula $\mathcal{A}=LW$ with the area $=60$.

$$60=(2w-2)w$$
$$60=2w^2-2w$$
$$0=2w^2-2w-60$$
$$0=2(w^2-w-30)$$
$$0=2(w-6)(w+5)$$
$$w=6 \quad \text{or} \quad w=-5$$

Discard -5 because the width cannot be negative. The width of the rectangle is 6 meters.

22. All of the given expressions are equal to 1 for all real numbers for which they are defined. However, expressions **B**, **C**, and **D** all have one or more values for which the expression is undefined and therefore cannot be equal to 1 at these values. Since k^2+2 is *always* positive, the denominator in expression **A** is never equal to zero. This expression is defined and equal to 1 for all real numbers, so the correct choice is **A**.

23. The appropriate choice is **D** since

$$\frac{-(3x+4)}{7}=\frac{-3x-4}{7}$$
$$\neq\frac{4-3x}{7}.$$

24. $\dfrac{5}{q}-\dfrac{1}{q}=\dfrac{5-1}{q}=\dfrac{4}{q}$

25. $\dfrac{3}{7}+\dfrac{4}{r}=\dfrac{3\cdot r}{7\cdot r}+\dfrac{4\cdot 7}{r\cdot 7} \qquad \textit{LCD = 7r}$

$$=\frac{3r+28}{7r}$$

26. $\dfrac{4}{5q-20}-\dfrac{1}{3q-12}$

$=\dfrac{4}{5(q-4)}-\dfrac{1}{3(q-4)}$

$=\dfrac{4\cdot 3}{5(q-4)\cdot 3}-\dfrac{1\cdot 5}{3(q-4)\cdot 5}$

LCD $= 5\cdot 3\cdot (q-4) = 15(q-4)$

$=\dfrac{12-5}{15(q-4)}=\dfrac{7}{15(q-4)}$

27. $\dfrac{2}{k^2+k}-\dfrac{3}{k^2-k}$

$=\dfrac{2}{k(k+1)}-\dfrac{3}{k(k-1)}$

$=\dfrac{2(k-1)}{k(k+1)(k-1)}-\dfrac{3(k+1)}{k(k-1)(k+1)}$

LCD $= k(k+1)(k-1)$

$=\dfrac{2(k-1)-3(k+1)}{k(k+1)(k-1)}$

$=\dfrac{2k-2-3k-3}{k(k+1)(k-1)}$

$=\dfrac{-k-5}{k(k+1)(k-1)}$

28. $\dfrac{7z^2+49z+70}{16z^2+72z-40}\div\dfrac{3z+6}{4z^2-1}$

$=\dfrac{7z^2+49z+70}{16z^2+72z-40}\cdot\dfrac{4z^2-1}{3z+6}$

$=\dfrac{7(z^2+7z+10)}{8(2z^2+9z-5)}\cdot\dfrac{(2z+1)(2z-1)}{3(z+2)}$

$=\dfrac{7(z+5)(z+2)}{8(2z-1)(z+5)}\cdot\dfrac{(2z+1)(2z-1)}{3(z+2)}$

$=\dfrac{7(2z+1)}{8\cdot 3}=\dfrac{7(2z+1)}{24}$

29. $\dfrac{\frac{4}{a}+\frac{5}{2a}}{\frac{7}{6a}-\frac{1}{5a}}$

$=\dfrac{\left(\frac{4}{a}+\frac{5}{2a}\right)\cdot 30a}{\left(\frac{7}{6a}-\frac{1}{5a}\right)\cdot 30a}$ *Multiply by LCD, 30a.*

$=\dfrac{\frac{4}{a}(30a)+\frac{5}{2a}(30a)}{\frac{7}{6a}(30a)-\frac{1}{5a}(30a)}$

$=\dfrac{4\cdot 30+5\cdot 15}{7\cdot 5-1\cdot 6}$

$=\dfrac{120+75}{35-6}=\dfrac{195}{29}$

30. $\dfrac{r+2}{5}=\dfrac{r-3}{3}$

$15\left(\dfrac{r+2}{5}\right)=15\left(\dfrac{r-3}{3}\right)$ *Multiply by LCD, 15.*

$3(r+2)=5(r-3)$

$3r+6=5r-15$

$21=2r$

$\frac{21}{2}=r$

Check $r=\frac{21}{2}$: $\frac{5}{2}=\frac{5}{2}$ *True*

The solution set is $\left\{\frac{21}{2}\right\}$.

31. $\dfrac{1}{x}=\dfrac{1}{x+1}+\dfrac{1}{2}$

$2x(x+1)\left(\dfrac{1}{x}\right)=2x(x+1)\left(\dfrac{1}{x+1}+\dfrac{1}{2}\right)$

Multiply by LCD, 2x(x + 1).

$2(x+1)=2x(x+1)\left(\dfrac{1}{x+1}\right)$

$+\,2x(x+1)\left(\frac{1}{2}\right)$

$2(x+1)=2x+x(x+1)$

$2x+2=2x+x^2+x$

$0=x^2+x-2$

$0=(x+2)(x-1)$

$x=-2$ or $x=1$

Check $x=-2$: $-\frac{1}{2}=-1+\frac{1}{2}$ *True*
Check $x=1$: $1=\frac{1}{2}+\frac{1}{2}$ *True*

Thus, the solution set is $\{-2, 1\}$.

32. Let $x=$ the number of hours it will take Jody and Pat to weed the yard working together.

	Rate	**Time Working Together**	**Fractional Part of the Job Done when Working Together**
Jody	$\frac{1}{3}$	x	$\frac{1}{3}x$
Pat	$\frac{1}{2}$	x	$\frac{1}{2}x$

Working together, they can do 1 whole job, so

$\frac{1}{3}x+\frac{1}{2}x=1.$

$6\left(\frac{1}{3}x+\frac{1}{2}x\right)=6\cdot 1$ *LCD = 6*

$6\left(\frac{1}{3}x\right)+6\left(\frac{1}{2}x\right)=6$

$2x+3x=6$

$5x=6$

$x=\frac{6}{5}$, or $1\frac{1}{5}$

If Jody and Pat worked together, it would take them $1\frac{1}{5}$ hours to weed the yard.

CHAPTER 7 GRAPHS, LINEAR EQUATIONS, AND FUNCTIONS

7.1 Review of Graphs and Slopes of Lines

7.1 Now Try Exercises

N1. $x - 2y = 4$
To find the x-intercept, let $y = 0$.

$$\begin{aligned} x - 2y &= 4 \\ x - 2(0) &= 4 \\ x &= 4 \end{aligned}$$

The x-intercept is $(4, 0)$.

To find the y-intercept, let $x = 0$.

$$\begin{aligned} x - 2y &= 4 \\ 0 - 2y &= 4 \\ -2y &= 4 \\ y &= -2 \end{aligned}$$

The y-intercept is $(0, -2)$.

Plot the intercepts, and draw the line through them.

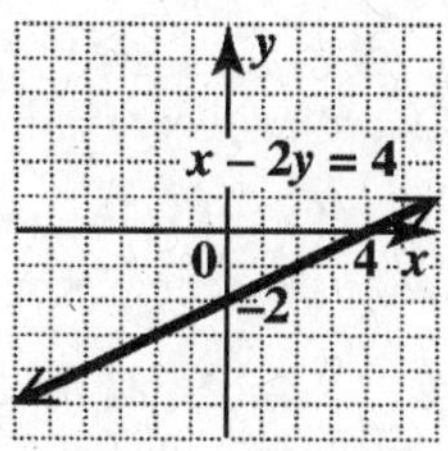

N2. $y = -2$

In standard form, the equation is $0x + y = -2$. Every value of x leads to $y = -2$, so the y-intercept is $(0, -2)$. There is no x-intercept. The graph is the horizontal line through $(0, -2)$.

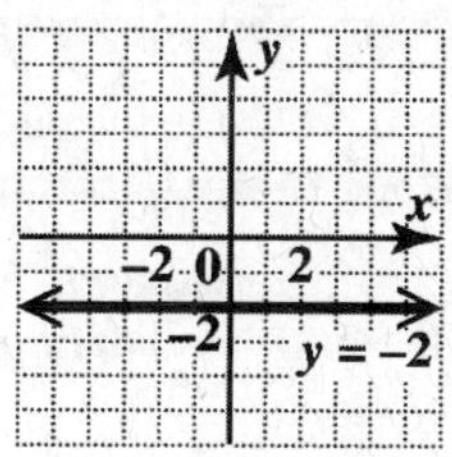

N3. $x + 3 = 0$

In standard form, the equation is $x + 0y = -3$. Every value of y leads to $x = -3$, the x-intercept is $(-3, 0)$. There is no y-intercept. The graph is the vertical line through $(-3, 0)$.

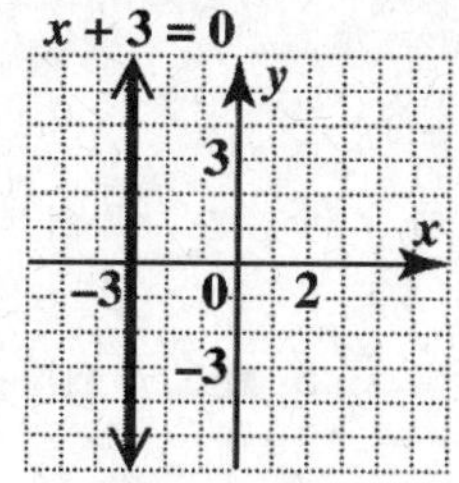

N4. $2x + 3y = 0$
To find the x-intercept, let $y = 0$.

$$\begin{aligned} 2x + 3(0) &= 0 \\ 2x &= 0 \\ x &= 0 \end{aligned}$$

Since the x-intercept is $(0, 0)$, the y-intercept is also $(0, 0)$.

Find another point. Let $x = 3$.

$$\begin{aligned} 2(3) + 3y &= 0 \\ 6 + 3y &= 0 \\ 3y &= -6 \\ y &= -2 \end{aligned}$$

This gives the ordered pair $(3, -2)$. Plot $(3, -2)$ and $(0, 0)$ and draw the line through them.

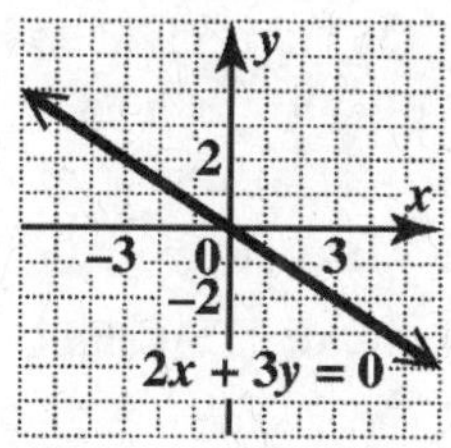

N5. By the midpoint formula, the midpoint of the segment with endpoints $(2, -5)$ and $(-4, 7)$ is

$$\left(\frac{2 + (-4)}{2}, \frac{-5 + 7}{2}\right) = \left(\frac{-2}{2}, \frac{2}{2}\right) = (-1, 1).$$

N6. If $(x_1, y_1) = (2, -6)$ and $(x_2, y_2) = (-3, 5)$, then

$$m = \frac{y_2 - y_1}{x_2 - x_1} = \frac{5 - (-6)}{-3 - 2} = \frac{11}{-5} = -\frac{11}{5}.$$

The slope is $-\frac{11}{5}$.

N7. To find the slope of the line with equation

$$3x - 7y = 21,$$

first find the intercepts. The x-intercept is $(7, 0)$, and the y-intercept is $(0, -3)$. The slope is then

$$m = \frac{-3 - 0}{0 - 7} = \frac{-3}{-7} = \frac{3}{7}.$$

The slope is $\frac{3}{7}$.

N8. Solve the equation for y.

$$\begin{aligned} 5x - 4y &= 7 \\ -4y &= -5x + 7 \quad \textit{Subtract 5x.} \\ y &= \tfrac{5}{4}x - \tfrac{7}{4} \quad \textit{Divide by } -4. \end{aligned}$$

The slope is given by the coefficient of x, so the slope is $\frac{5}{4}$.

N9. Through $(-4, 1)$; $m = -\frac{2}{3}$

Locate the point $(-4, 1)$ on the graph. Use the slope formula to find a second point on the line.

$$m = \frac{\text{change in } y}{\text{change in } x} = \frac{-2}{3}$$

From $(-4, 1)$, move *down* 2 units and then 3 units to the *right* to $(-1, -1)$. Draw the line through the two points.

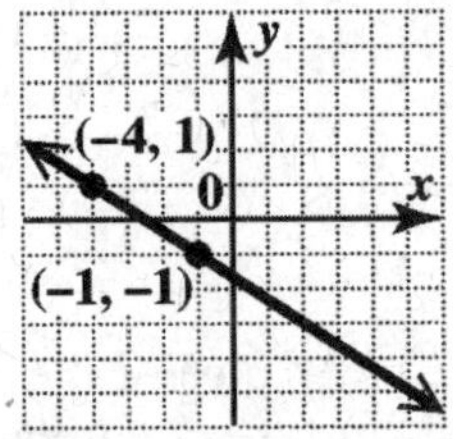

N10. (a) Find the slope of each line.
The line L_1 through $(2, 5)$ and $(4, 8)$ has slope

$$m_1 = \frac{8 - 5}{4 - 2} = \frac{3}{2}.$$

The line L_2 through $(2, 0)$ and $(-1, -2)$ has slope

$$m_2 = \frac{-2 - 0}{-1 - 2} = \frac{-2}{-3} = \frac{2}{3}.$$

The slopes of the lines are neither equal nor negative reciprocals. Therefore, the lines are *neither* parallel nor perpendicular.

(b) Solve each equation for y.

$$\begin{aligned} x + 2y &= 7 \\ 2y &= -x + 7 \\ y &= -\tfrac{1}{2}x + \tfrac{7}{5} \\ (\text{so } m_1 &= -\tfrac{1}{2}) \end{aligned} \qquad \begin{aligned} 2x - y &= -4 \\ -y &= -2x - 4 \\ y &= 2x + 4 \\ (\text{so } m_2 &= 2) \end{aligned}$$

Since $m_1 m_2 = (-\frac{1}{2})(2) = -1$, the lines are *perpendicular.*

(c) Solve each equation for y.

$$\begin{aligned} 2x - y &= 4 \\ -y &= -2x + 4 \\ y &= 2x - 4 \\ (\text{so } m_1 &= 2) \end{aligned} \qquad \begin{aligned} -2x + y &= 6 \\ y &= 2x + 6 \\ (\text{so } m_2 &= 2) \end{aligned}$$

Since $m_1 = m_2$, the lines are *parallel.*

N11. (a) $(x_1, y_1) = (2000, 690)$ and $(x_2, y_2) = (2002, 828)$.

$$\begin{aligned} \text{average rate of change} &= \frac{y_2 - y_1}{x_2 - x_1} \\ &= \frac{828 - 690}{2002 - 2000} \\ &= \frac{138}{2} = 69 \end{aligned}$$

The average rate of change is 69 hr per yr.

(b) It is greater than 58, which is the average rate of change from 2000 to 2005 found in Example 11.

N12. $(x_1, y_1) = (2000, 2838)$ and $(x_2, y_2) = (2008, 1885)$.

$$\begin{aligned} \text{average rate of change} &= \frac{y_2 - y_1}{x_2 - x_1} \\ &= \frac{1885 - 2838}{2008 - 2000} \\ &= \frac{-953}{8} = -119.125 \end{aligned}$$

Thus, the average rate of change in sales of digital camcorders in the United States from 2000 to 2008 was about $-\$119$ million per year.

7.1 Section Exercises

1. **(a)** The point $(1, 6)$ is located in quadrant I, since the x- and y-coordinates are both positive.

(b) The point $(-4, -2)$ is located in quadrant III, since the x- and y-coordinates are both negative.

(c) The point $(-3, 6)$ is located in quadrant II, since the x-coordinate is negative and the y-coordinate is positive.

(d) The point $(7, -5)$ is located in quadrant IV, since the x-coordinate is positive and the y-coordinate is negative.

(e) The point $(-3, 0)$ is located on the x-axis, so it does not belong to any quadrant.

(f) The point $(0, -0.5)$ is located on the y-axis, so it does not belong to any quadrant.

3. **(a)** To plot $(2, 3)$, go two units from zero to the right along the x-axis, and then go three units up parallel to the y-axis.

(b) To plot $(-3, -2)$, go three units from zero to the left along the x-axis, and then go two units down parallel to the y-axis.

(c) To plot $(0, 5)$, do not move along the x-axis at all since the x-coordinate is 0. Move five units up along the y-axis.

(d) To plot $(-2, 4)$, go two units from zero to the left along the x-axis, and then go four units up parallel to the y-axis.

(e) To plot $(-2, 0)$, go two units to the left along the x-axis. Do not move up or down since the y-coordinate is 0.

(f) To plot $(3, -3)$, go 3 units right on the x-axis, then 3 units down.

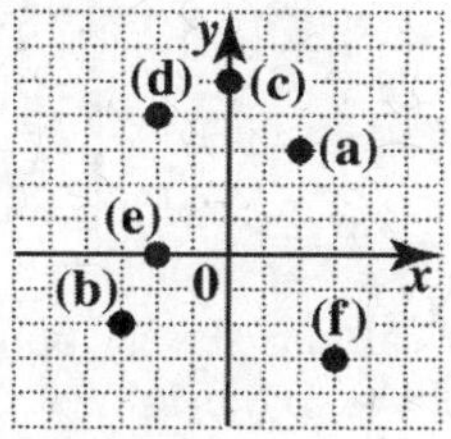

5. $x - y = 3$

(a) To complete the table, substitute the given values for x and y in the equation.

For $x = 0$:
$$x - y = 3$$
$$0 - y = 3$$
$$y = -3 \quad \mathbf{(0, -3)}$$

For $y = 0$:
$$x - y = 3$$
$$x - 0 = 3$$
$$x = 3 \quad \mathbf{(3, 0)}$$

For $x = 5$:
$$x - y = 3$$
$$5 - y = 3$$
$$-y = -2$$
$$y = 2 \quad \mathbf{(5, 2)}$$

For $x = 2$:
$$x - y = 3$$
$$2 - y = 3$$
$$-y = 1$$
$$y = -1 \quad \mathbf{(2, -1)}$$

(b) Plot the ordered pairs and draw the line through them.

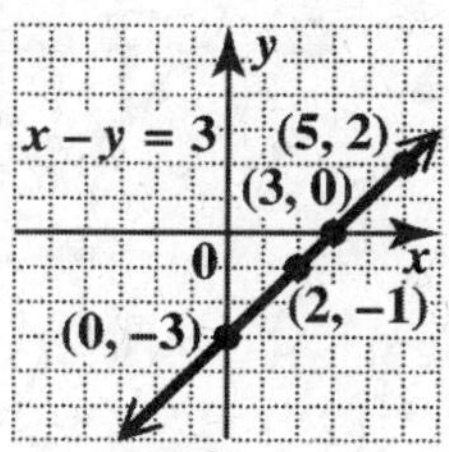

7. $y = -2x + 3$

(a) To complete the table, substitute the given values for x or y in the equation.

For $x = 0$:
$$y = -2x + 3$$
$$y = -2(0) + 3$$
$$y = 0 + 3$$
$$y = 3 \quad \mathbf{(0, 3)}$$

For $y = 1$:
$$y = -2x + 3$$
$$1 = -2x + 3$$
$$-2 = -2x$$
$$1 = x \quad \mathbf{(1, 1)}$$

For $x = 2$:
$$y = -2x + 3$$
$$y = -2(2) + 3$$
$$y = -4 + 3$$
$$y = -1 \quad \mathbf{(2, -1)}$$

For $y = -3$:
$$y = -2x + 3$$
$$-3 = -2x + 3$$
$$-6 = -2x$$
$$3 = x \quad \mathbf{(3, -3)}$$

(b) Plot the ordered pairs and draw the line through them.

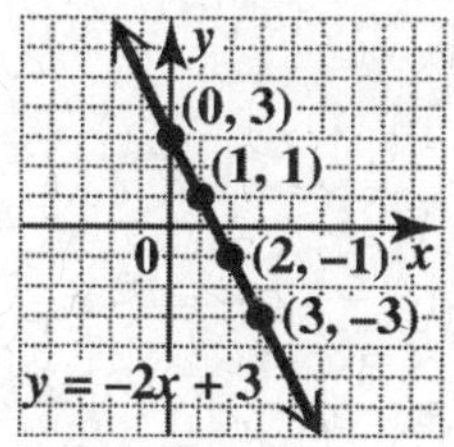

9. $2x + 3y = 12$

To find the x-intercept, let $y = 0$.

$$2x + 3y = 12$$
$$2x + 3(0) = 12$$
$$2x = 12$$
$$x = 6$$

The x-intercept is $(6, 0)$.
To find the y-intercept, let $x = 0$.

$$2x + 3y = 12$$
$$2(0) + 3y = 12$$
$$3y = 12$$
$$y = 4$$

The y-intercept is $(0, 4)$.
Plot the intercepts and draw the line through them.

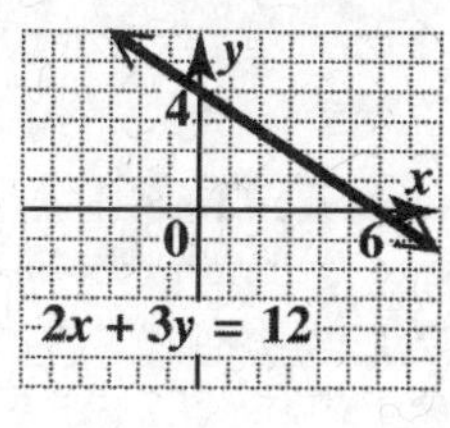

11. $x - 3y = 6$

To find the x-intercept, let $y = 0$.

$$x - 3y = 6$$
$$x - 3(0) = 6$$
$$x - 0 = 6$$
$$x = 6$$

The x-intercept is $(6, 0)$.
To find the y-intercept, let $x = 0$.

$$\begin{aligned} x - 3y &= 6 \\ 0 - 3y &= 6 \\ -3y &= 6 \\ y &= -2 \end{aligned}$$

The y-intercept is $(0, -2)$.
Plot the intercepts and draw the line through them.

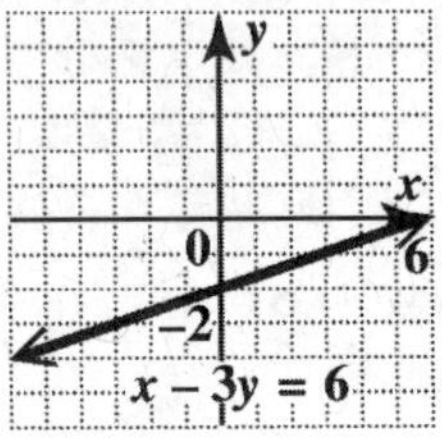

13. $5x + 6y = -10$
To find the x-intercept, let $y = 0$.

$$\begin{aligned} 5x + 6(0) &= -10 \\ 5x &= -10 \\ x &= -2 \end{aligned}$$

The x-intercept is $(-2, 0)$.
To find the y-intercept, let $x = 0$.

$$\begin{aligned} 5(0) + 6y &= -10 \\ 6y &= -10 \\ y &= -\tfrac{10}{6} = -\tfrac{5}{3} \end{aligned}$$

The y-intercept is $(0, -\frac{5}{3})$.

Plot the intercepts and draw the line through them.

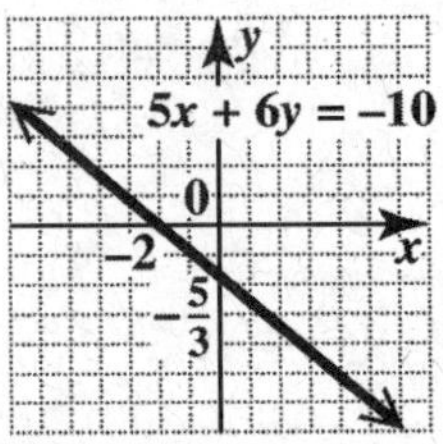

15. $y = 5$
This is a horizontal line. Every point has y-coordinate 5, so no point has y-coordinate 0. There is no x-intercept.
Since every point of the line has y-coordinate 5, the y-intercept is $(0, 5)$. Draw the horizontal line through $(0, 5)$.

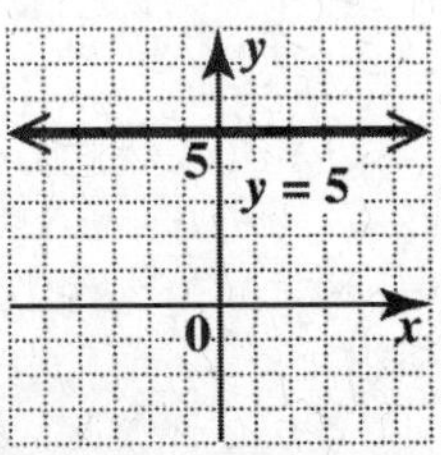

17. $x = 2$
This is a vertical line. Every point has x-coordinate 2, so the x-intercept is $(2, 0)$.
Since every point of the line has x-coordinate 2, no point has x-coordinate 0. There is no y-intercept. Draw the vertical line through $(2, 0)$.

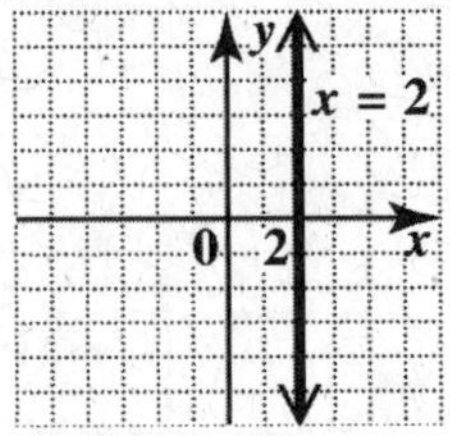

19. $x + 4 = 0 \quad (x = -4)$
This is a vertical line. Every point has x-coordinate -4, so the x-intercept is $(-4, 0)$.
Since every point of the line has x-coordinate -4, no point has x-coordinate 0. There is no y-intercept. Draw the vertical line through $(-4, 0)$.

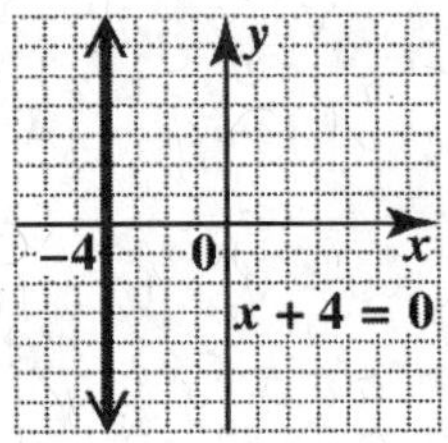

21. $y + 2 = 0 \quad (y = -2)$
This is a horizontal line. Every point has y-coordinate -2, so no point has y-coordinate 0. There is no x-intercept.
Since every point of the line has y-coordinate -2, the y-intercept is $(0, -2)$. Draw the horizontal line through $(0, -2)$.

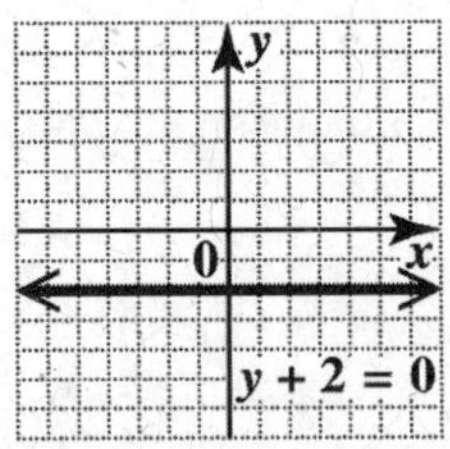

23. $x + 5y = 0$
To find the x-intercept, let $y = 0$.

$$\begin{aligned} x + 5y &= 0 \\ x + 5(0) &= 0 \\ x &= 0 \end{aligned}$$

The x-intercept is $(0, 0)$, and since $x = 0$, this is also the y-intercept. Since the intercepts are the same, another point is needed to graph the line. Choose any number for y, say $y = -1$, and solve the equation for x.

$$\begin{aligned} x + 5y &= 0 \\ x + 5(-1) &= 0 \\ x &= 5 \end{aligned}$$

This gives the ordered pair $(5,-1)$. Plot $(5,-1)$ and $(0,0)$, and draw the line through them.

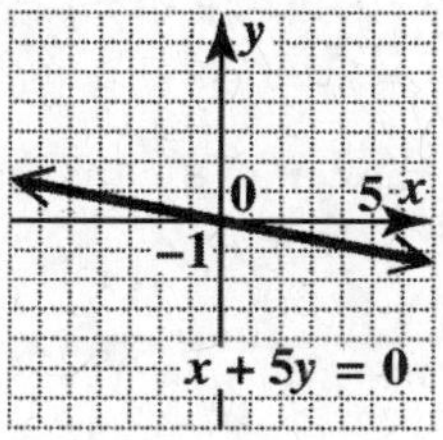

25. $2x = 3y$
If $x = 0$, then $y = 0$, so the x- and y-intercepts are $(0,0)$. To get another point, let $x = 3$.

$$2(3) = 3y$$
$$2 = y$$

Plot $(3,2)$ and $(0,0)$, and draw the line through them.

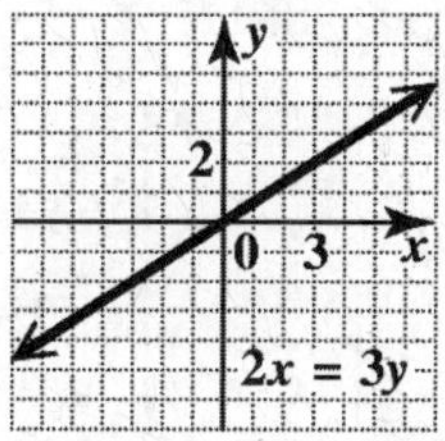

27. By the Midpoint Formula, the midpoint of the segment with endpoints $(-8,4)$ and $(-2,-6)$ is

$$\left(\frac{-8+(-2)}{2}, \frac{4+(-6)}{2}\right) = \left(\frac{-10}{2}, \frac{-2}{2}\right) = (-5,-1).$$

29. By the Midpoint Formula, the midpoint of the segment with endpoints $(3,-6)$ and $(6,3)$ is

$$\left(\frac{3+6}{2}, \frac{-6+3}{2}\right) = \left(\frac{9}{2}, \frac{-3}{2}\right) = \left(\frac{9}{2}, -\frac{3}{2}\right).$$

31. By the Midpoint Formula, the midpoint of the segment with endpoints $(-9,3)$ and $(9,8)$ is

$$\left(\frac{-9+9}{2}, \frac{3+8}{2}\right) = \left(\frac{0}{2}, \frac{11}{2}\right) = \left(0, \frac{11}{2}\right).$$

33. By the Midpoint Formula, the midpoint of the segment with endpoints $(2.5, 3.1)$ and $(1.7, -1.3)$ is

$$\left(\frac{2.5+1.7}{2}, \frac{3.1+(-1.3)}{2}\right) = \left(\frac{4.2}{2}, \frac{1.8}{2}\right) = (2.1, 0.9).$$

35. By the Midpoint Formula, the midpoint of the segment with endpoints $\left(\frac{1}{2}, \frac{1}{3}\right)$ and $\left(\frac{3}{2}, \frac{5}{3}\right)$ is

$$\left(\frac{\frac{1}{2}+\frac{3}{2}}{2}, \frac{\frac{1}{3}+\frac{5}{3}}{2}\right) = \left(\frac{\frac{4}{2}}{2}, \frac{\frac{6}{3}}{2}\right) = \left(\frac{2}{2}, \frac{2}{2}\right) = (1,1).$$

37. By the Midpoint Formula, the midpoint of the segment with endpoints $\left(-\frac{1}{3}, \frac{2}{7}\right)$ and $\left(-\frac{1}{2}, \frac{1}{14}\right)$ is

$$\left(\frac{-\frac{1}{3}+\left(-\frac{1}{2}\right)}{2}, \frac{\frac{2}{7}+\frac{1}{14}}{2}\right) = \left(\frac{-\frac{5}{6}}{2}, \frac{\frac{5}{14}}{2}\right) = \left(-\frac{5}{12}, \frac{5}{28}\right).$$

39. The graph goes through the point $(-2,0)$ which satisfies only equations **B** and **C**. The graph also goes through the point $(0,3)$ which satisfies only equations **A** and **B**. Therefore, the correct equation is **B**.

41.
$$\text{slope} = \frac{\text{change in vertical position}}{\text{change in horizontal position}} = \frac{30 \text{ feet}}{100 \text{ feet}}$$

Choices **A**, 0.3, **B**, $\frac{3}{10}$, **D**, $\frac{30}{100}$, and **G**, 30%, are all correct.

43. **(a)** Graph **C** indicates that sales leveled off during the second quarter.

(b) Graph **A** indicates that sales leveled off during the fourth quarter.

(c) Graph **D** indicates that sales rose sharply during the first quarter, and then fell to the original level during the second quarter.

(d) Graph **B** is the only graph that indicates that sales fell during the first two quarters.

45. **(a)** To get to B from A, we must go up 2 units and move right 1 unit. Thus,

$$\text{slope of } AB = \frac{\text{rise}}{\text{run}} = \frac{2}{1} = 2.$$

(b) slope of $BC = \frac{\text{rise}}{\text{run}} = \frac{0}{-4} = 0$

(c) slope of $CD = \frac{\text{rise}}{\text{run}} = \frac{-7}{0}$, which is undefined.

(d) slope of $DE = \frac{\text{rise}}{\text{run}} = \frac{-1}{3} = -\frac{1}{3}$

(e) slope of $EF = \frac{\text{rise}}{\text{run}} = \frac{3}{3} = 1$

(f) slope of $FG = \frac{\text{rise}}{\text{run}} = \frac{-4}{1} = -4$

47. **(a)** Let $(x_1, y_1) = (-2,-3)$ and $(x_2, y_2) = (-1,5)$. Then

$$m = \frac{y_2 - y_1}{x_2 - x_1} = \frac{5-(-3)}{-1-(-2)} = \frac{8}{1} = 8.$$

The slope is 8.

(b) The slope is positive, so the line rises.

49. **(a)** Let $(x_1, y_1) = (-4, 1)$ and $(x_2, y_2) = (2, 6)$. Then

$$m = \frac{y_2 - y_1}{x_2 - x_1} = \frac{6 - 1}{2 - (-4)} = \frac{5}{6}.$$

The slope is $\frac{5}{6}$.

(b) The slope is positive, so the line rises.

51. **(a)** Let $(x_1, y_1) = (2, 4)$ and $(x_2, y_2) = (-4, 4)$. Then

$$m = \frac{y_2 - y_1}{x_2 - x_1} = \frac{4 - 4}{-4 - 2} = \frac{0}{-6} = 0.$$

The slope is 0.

(b) The slope is zero, so the line is horizontal.

53. **(a)** Let $(x_1, y_1) = (-2, 2)$ and $(x_2, y_2) = (4, -1)$. Then

$$m = \frac{y_2 - y_1}{x_2 - x_1} = \frac{-1 - 2}{4 - (-2)} = \frac{-3}{6} = -\frac{1}{2}.$$

The slope is $-\frac{1}{2}$.

(b) The slope is negative, so the line falls.

55. **(a)** Let $(x_1, y_1) = (5, -3)$ and $(x_2, y_2) = (5, 2)$. Then

$$m = \frac{y_2 - y_1}{x_2 - x_1} = \frac{2 - (-3)}{5 - 5} = \frac{5}{0}.$$

The slope is undefined.

(b) The slope is undefined, so the line is vertical.

57. **(a)** Let $(x_1, y_1) = (1.5, 2.6)$ and $(x_2, y_2) = (0.5, 3.6)$. Then

$$m = \frac{y_2 - y_1}{x_2 - x_1} = \frac{3.6 - 2.6}{0.5 - 1.5} = \frac{1}{-1} = -1.$$

The slope is -1.

(b) The slope is negative, so the line falls.

59. Let $(x_1, y_1) = (\frac{1}{6}, \frac{1}{2})$ and $(x_2, y_2) = (\frac{5}{6}, \frac{9}{2})$. Then

$$m = \frac{y_2 - y_1}{x_2 - x_1} = \frac{\frac{9}{2} - \frac{1}{2}}{\frac{5}{6} - \frac{1}{6}} = \frac{\frac{8}{2}}{\frac{4}{6}} = 4 \cdot \tfrac{3}{2} = 6.$$

The slope is 6.

61. Let $(x_1, y_1) = (-\frac{2}{9}, \frac{5}{18})$ and $(x_2, y_2) = (\frac{1}{18}, -\frac{5}{9})$. Then

$$m = \frac{y_2 - y_1}{x_2 - x_1} = \frac{-\frac{5}{9} - \frac{5}{18}}{\frac{1}{18} - (-\frac{2}{9})} = \frac{-\frac{15}{18}}{\frac{5}{18}}$$
$$= -\tfrac{15}{18} \cdot \tfrac{18}{5} = -3.$$

The slope is -3.

63. The points shown on the line are $(-3, 3)$ and $(-1, -2)$. The slope is

$$m = \frac{-2 - 3}{-1 - (-3)} = \frac{-5}{2} = -\frac{5}{2}.$$

65. The points shown on the line are $(3, 3)$ and $(3, -3)$. The slope is

$$m = \frac{-3 - 3}{3 - 3} = \frac{-6}{0}, \text{ which is } \textit{undefined}.$$

67. To find the slope of

$$x + 2y = 4,$$

first find the intercepts. Replace y with 0 to find that the x-intercept is $(4, 0)$; replace x with 0 to find that the y-intercept is $(0, 2)$. The slope is then

$$m = \frac{2 - 0}{0 - 4} = -\frac{2}{4} = -\frac{1}{2}.$$

To sketch the graph, plot the intercepts and draw the line through them.

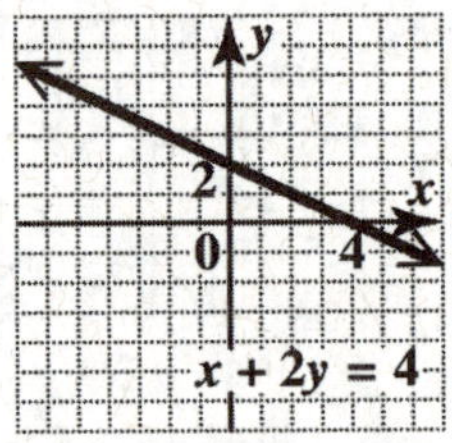

69. To find the slope of

$$5x - 2y = 10,$$

first find the intercepts. Replace y with 0 to find that the x-intercept is $(2, 0)$; replace x with 0 to find that the y-intercept is $(0, -5)$. The slope is then

$$m = \frac{-5 - 0}{0 - 2} = \frac{-5}{-2} = \frac{5}{2}.$$

To sketch the graph, plot the intercepts and draw the line through them.

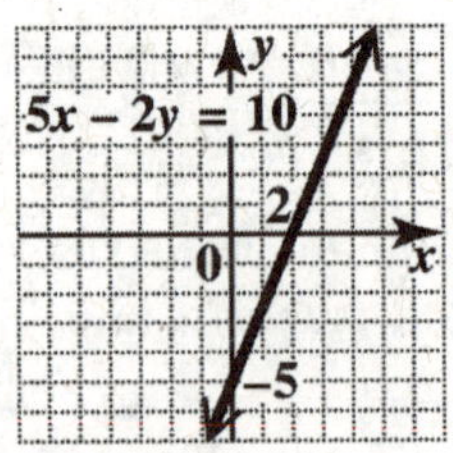

71. In the equation

$$y = 4x,$$

replace x with 0 and then x with 1 to get the ordered pairs $(0, 0)$ and $(1, 4)$, respectively. (There are other possibilities for ordered pairs.) The slope is then

$$m = \frac{4-0}{1-0} = \frac{4}{1} = 4.$$

To sketch the graph, plot the two points and draw the line through them.

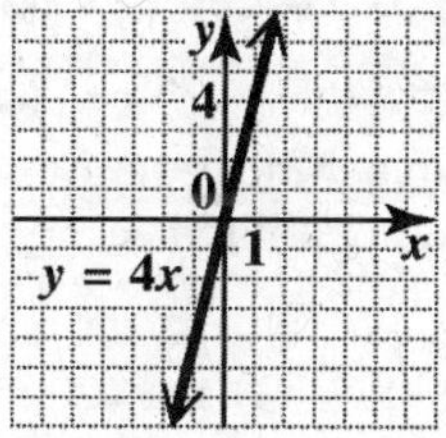

73. $x - 3 = 0 \quad (x = 3)$

The graph of $x = 3$ is the vertical line with x-intercept $(3, 0)$. The slope of a vertical line is undefined.

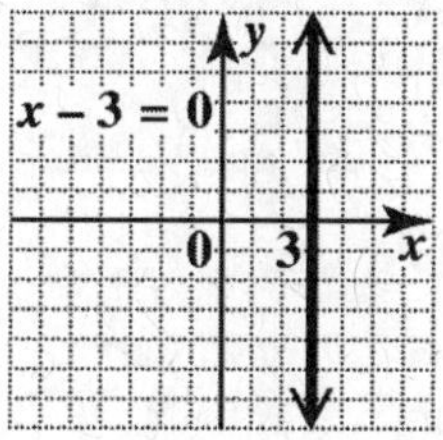

75. $y = -5$

The graph of $y = -5$ is the horizontal line with y-intercept $(0, -5)$. The slope of a horizontal line is 0.

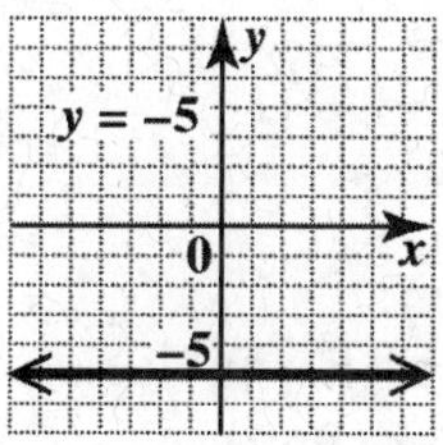

77. To graph the line through $(-4, 2)$ with slope $m = \frac{1}{2}$, locate $(-4, 2)$ on the graph. To find a second point, use the definition of slope.

$$m = \frac{\text{change in } y}{\text{change in } x} = \frac{1}{2}$$

From $(-4, 2)$, go up 1 unit. Then go 2 units to the right to get to $(-2, 3)$. Draw the line through $(-4, 2)$ and $(-2, 3)$.

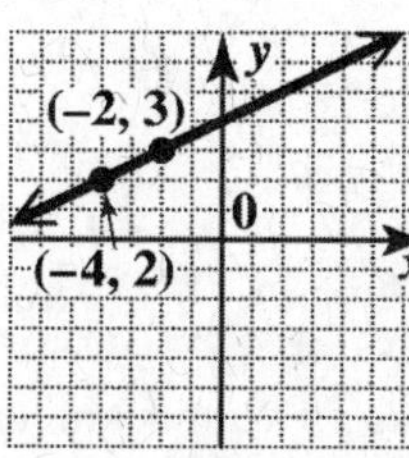

79. To graph the line through $(0, -2)$ with slope $m = -\frac{2}{3}$, locate the point $(0, -2)$ on the graph. To find a second point on the line, use the definition of slope, writing $-\frac{2}{3}$ as $\frac{-2}{3}$.

$$m = \frac{\text{change in } y}{\text{change in } x} = \frac{-2}{3}$$

From $(0, -2)$, move 2 units down and then 3 units to the right. Draw a line through this second point and $(0, -2)$. (Note that the slope could also be written as $\frac{2}{-3}$. In this case, move 2 units up and 3 units to the left to get another point on the same line.)

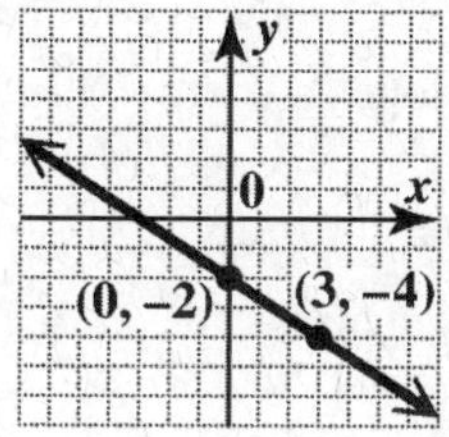

81. Locate $(-1, -2)$. Then use $m = 3 = \frac{3}{1}$ to go 3 units up and 1 unit right to $(0, 1)$. Draw the line through $(-1, -2)$ and $(0, 1)$.

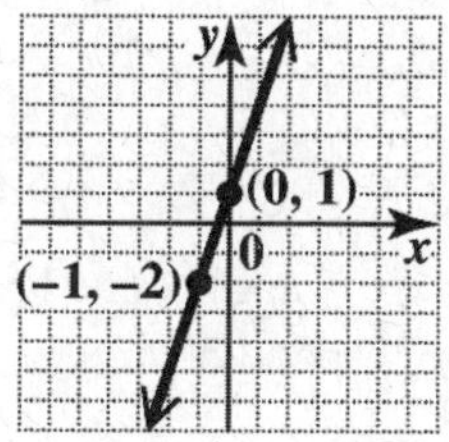

83. Locate $(2, -5)$. A slope of 0 means that the line is horizontal, so $y = -5$ at every point. Draw the horizontal line through $(2, -5)$.

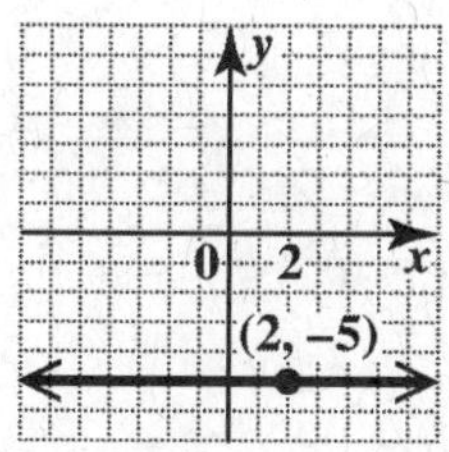

85. Locate $(-3, 1)$. Since the slope is undefined, the line is vertical. The x-value of every point is -3. Draw the vertical line through $(-3, 1)$.

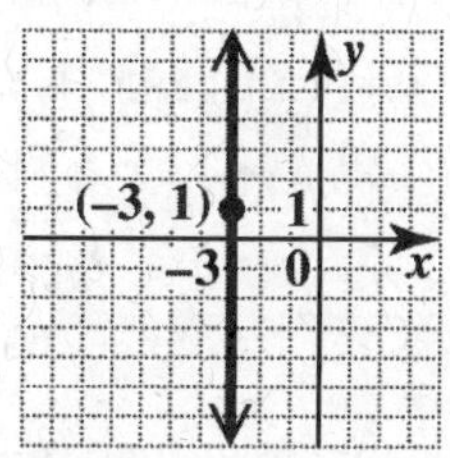

87. If a line has slope $-\frac{4}{9}$, then any line parallel to it has slope $\underline{-\frac{4}{9}}$ (the slope must be the same), and any line perpendicular to it has slope $\underline{\frac{9}{4}}$ (the slope must be the negative reciprocal).

89. The slope of the line through $(15, 9)$ and $(12, -7)$ is

$$m = \frac{-7-9}{12-15} = \frac{-16}{-3} = \frac{16}{3}.$$

The slope of the line through $(8, -4)$ and $(5, -20)$ is

$$m = \frac{-20-(-4)}{5-8} = \frac{-16}{-3} = \frac{16}{3}.$$

Since the slopes are equal, the two lines are *parallel.*

91. $x + 4y = 7$ and $4x - y = 3$
Solve the equations for y.

$$4y = -x + 7 \qquad -y = -4x + 3$$
$$y = -\tfrac{1}{4}x + \tfrac{7}{4} \qquad y = 4x - 3$$

The slopes, $-\frac{1}{4}$ and 4, are negative reciprocals of one another, so the lines are *perpendicular.*

93. $4x - 3y = 6$ and $3x - 4y = 2$
Solve the equations for y.

$$-3y = -4x + 6 \qquad -4y = -3x + 2$$
$$y = \tfrac{4}{3}x - 2 \qquad y = \tfrac{3}{4}x - \tfrac{1}{2}$$

The slopes are $\frac{4}{3}$ and $\frac{3}{4}$. The lines are *neither* parallel nor perpendicular.

95. $x = 6$ and $6 - x = 8$

The second equation can be simplified as $x = -2$. Both lines are vertical lines, so they are *parallel.*

97. $4x + y = 0$ and $5x - 8 = 2y$
Solve the equations for y.

$$y = -4x \qquad \tfrac{5}{2}x - 4 = y$$

The slopes are -4 and $\frac{5}{2}$. The lines are *neither* parallel nor perpendicular.

99. $2x = y + 3$ and $2y + x = 3$
Solve the equations for y.

$$2x - 3 = y \qquad 2y = -x + 3$$
$$y = -\tfrac{1}{2}x + \tfrac{3}{2}$$

The slopes, 2 and $-\frac{1}{2}$, are negative reciprocals of one another, so the lines are *perpendicular.*

101. Use the points $(0, 20)$ and $(4, 4)$.

average rate of change

$$= \frac{\text{change in } y}{\text{change in } x} = \frac{4-20}{4-0} = \frac{-16}{4} = -4$$

The average rate of change is $-\$4000$ per year, that is, the value of the machine is decreasing \$4000 each year during these years.

103. We can see that there is no change in the percent of pay raise. Thus, the average rate of change is 0% per year, that is, the percent of pay raise is not changing—it is 3% each year during these years.

105. (a) $m = \dfrac{270.9 - 207.3}{2008 - 2005} = \dfrac{63.6}{3} = 21.2$

(b) The number of subscribers increased by an average of 21.2 million each year from 2005 to 2008.

107. (a) Use $(2000, 443)$ and $(2007, 405)$.

$$m = \frac{405 - 443}{2007 - 2000} = \frac{-38}{7} \approx -5$$

The average rate of change is -5 theaters per year.

(b) The negative slope means that the number of drive-in theaters *decreased* by an average of 5 each year from 2000 to 2007.

109. Use $(2003, 1590)$ and $(2006, 5705)$.

average rate of change

$$= \frac{5705 - 1590}{2006 - 2003} = \frac{4115}{3} \approx 1371.67$$

The average rate of change in sales is about \$1371.67 million per year, that is, the sales of plasma TVs in the United States increased by an average of \$1371.67 million each year from 2003 to 2006.

111.
$$y - (-2) = \tfrac{3}{2}(x - 5)$$
$$2(y + 2) = 3(x - 5)$$
$$2y + 4 = 3x - 15$$
$$19 = 3x - 2y$$

7.2 Review of Equations of Lines; Linear Models

7.2 Now Try Exercises

N1. Slope $\frac{2}{3}$; y-intercept $(0, 1)$
Here $m = \frac{2}{3}$ and $b = 1$. Substitute these values into the slope-intercept form.

$$y = mx + b$$
$$y = \tfrac{2}{3}x + 1$$

N2. $4x + 3y = 6$
Solve the equation for y.

$$3y = -4x + 6$$
$$y = -\tfrac{4}{3}x + 2$$

Plot the y-intercept $(0, 2)$. The slope can be interpreted as either $\frac{-4}{3}$ or $\frac{4}{-3}$. Using $\frac{-4}{3}$, move from $(0, 2)$ *down* 4 units and to the *right* 3 units to locate the point $(3, -2)$. Draw a line through the two points.

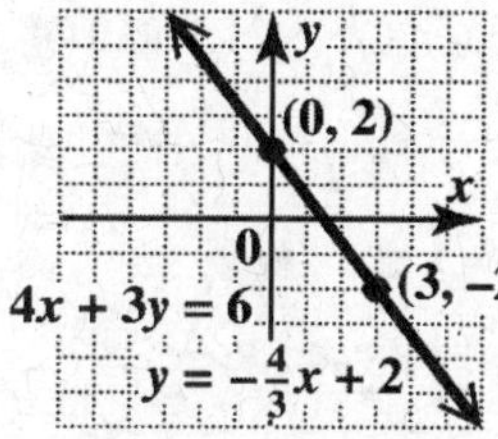

N3. Through $(5, -3)$; slope $= m = -\frac{1}{5}$
Use the point-slope form with $(x_1, y_1) = (5, -3)$ and $m = -\frac{1}{5}$.

$$\begin{aligned} y - y_1 &= m(x - x_1) \\ y - (-3) &= -\tfrac{1}{5}(x - 5) \\ y + 3 &= -\tfrac{1}{5}(x - 5) \\ y + 3 &= -\tfrac{1}{5}x + 1 \\ y &= -\tfrac{1}{5}x - 2 \quad \textit{Subtract 3.} \end{aligned}$$

N4. **(a)** Through $(4, -4)$; m undefined

This is a vertical line since the slope is undefined. A vertical line through the point (a, b) has equation $x = a$. Here the x-coordinate is 4, so the equation is $x = 4$.

(b) Through $(4, -4)$; $m = 0$

Since the slope is 0, this is a horizontal line. A horizontal line through the point (a, b) has equation $y = b$. Here the y-coordinate is -4, so the equation is $y = -4$.

N5. Through $(3, -4)$ and $(-2, -1)$

$$m = \frac{-1 - (-4)}{-2 - 3} = \frac{3}{-5} = -\frac{3}{5}$$

Let $(x_1, y_1) = (3, -4)$.

$$\begin{aligned} y - y_1 &= m(x - x_1) \\ y - (-4) &= -\tfrac{3}{5}(x - 3) \\ y + 4 &= -\tfrac{3}{5}(x - 3) \\ 5y + 20 &= -3x + 9 \quad \textit{Multiply by 5.} \\ 3x + 5y &= -11 \quad \textit{Standard form} \end{aligned}$$

N6. **(a)** Through $(6, -1)$; parallel to the line $3x - 5y = 7$

Find the slope of the given line.

$$\begin{aligned} 3x - 5y &= 7 \\ -5y &= -3x + 7 \\ y &= \tfrac{3}{5}x - \tfrac{7}{5} \end{aligned}$$

The slope is $\frac{3}{5}$, so a line parallel to it also has slope $\frac{3}{5}$. Use $m = \frac{3}{5}$ and $(x_1, y_1) = (6, -1)$ in the point-slope form.

$$\begin{aligned} y - y_1 &= m(x - x_1) \\ y - (-1) &= \tfrac{3}{5}(x - 6) \\ y + 1 &= \tfrac{3}{5}x - \tfrac{18}{5} \\ y &= \tfrac{3}{5}x - \tfrac{18}{5} - \tfrac{5}{5} \\ y &= \tfrac{3}{5}x - \tfrac{23}{5} \end{aligned}$$

(b) Through $(6, -1)$; perpendicular to $3x - 5y = 7$

The slope of $3x - 5y = 7$ is $\frac{3}{5}$. The negative reciprocal of $\frac{3}{5}$ is $-\frac{5}{3}$, so the slope of the line through $(6, -1)$ is $-\frac{5}{3}$.

$$\begin{aligned} y - y_1 &= m(x - x_1) \\ y - (-1) &= -\tfrac{5}{3}(x - 6) \\ y + 1 &= -\tfrac{5}{3}x + 10 \\ y &= -\tfrac{5}{3}x + 9 \end{aligned}$$

N7. Since the price you pay is \$85 per month plus a flat fee of \$100, an equation for x months is

$$y = 85x + 100.$$

N8. **(a)** Use $(2, 183)$ and $(6, 251)$.

$$m = \frac{251 - 183}{6 - 2} = \frac{68}{4} = 17$$

Use the point-slope form with $(x_1, y_1) = (2, 183)$.

$$\begin{aligned} y - y_1 &= m(x - x_1) \\ y - 183 &= 17(x - 2) \\ y - 183 &= 17x - 34 \\ y &= 17x + 149 \end{aligned}$$

(b) For 2011, $x = 2011 - 2000 = 11$.

$$\begin{aligned} y &= 17x + 149 \\ y &= 17(11) + 149 \quad \textit{Let } x = 11. \\ &= 187 + 149 \\ &= 336 \end{aligned}$$

Using the equation from part (a), we estimate the retail spending on prescription drugs in 2011 to be \$336 billion.

7.2 Section Exercises

1. Choice **A**, $3x - 2y = 5$, is in the form $Ax + By = C$ with $A \geq 0$ and integers A, B, and C having no common factor (except 1).

3. Choice **A**, $y = 6x + 2$, is in the form $y = mx + b$.

5.
$$\begin{aligned} y + 2 &= -3(x - 4) \\ y + 2 &= -3x + 12 \\ 3x + y &= 10 \quad \textit{Standard form} \end{aligned}$$

7. $y = 2x + 3$
This line is in slope-intercept form with slope $m = 2$ and y-intercept $(0, b) = (0, 3)$. The only graph with positive slope and with a positive y-coordinate of its y-intercept is **A**.

9. $y = -2x - 3$
This line is in slope-intercept form with slope $m = -2$ and y-intercept $(0, b) = (0, -3)$. The only graph with negative slope and with a negative y-coordinate of its y-intercept is **C**.

11. $y = 2x$
This line has slope $m = 2$ and y-intercept $(0, b) = (0, 0)$. The only graph with positive slope and with y-intercept $(0, 0)$ is **H**.

13. $y = 3$
This line is a horizontal line with y-intercept $(0, 3)$. Its y-coordinate is positive. The only graph that has these characteristics is **B**.

15. $m = 5; b = 15$
Substitute these values in the slope-intercept form.

$$y = mx + b$$
$$y = 5x + 15$$

17. $m = -\frac{2}{3}; b = \frac{4}{5}$
Substitute these values in the slope-intercept form.

$$y = mx + b$$
$$y = -\tfrac{2}{3}x + \tfrac{4}{5}$$

19. Slope 1; y-intercept $(0, -1)$
Here, $m = 1$ and $b = -1$. Substitute these values in the slope-intercept form.

$$y = mx + b$$
$$y = 1x - 1, \text{ or } y = x - 1$$

21. Slope $\frac{2}{5}$; y-intercept $(0, 5)$
Here, $m = \frac{2}{5}$ and $b = 5$. Substitute these values in the slope-intercept form.

$$y = mx + b$$
$$y = \tfrac{2}{5}x + 5$$

23. To get to the point $(3, 3)$ from the y-intercept $(0, 1)$, we must go up 2 units and to the right 3 units, so the slope is $\frac{2}{3}$. The slope-intercept form is

$$y = \tfrac{2}{3}x + 1.$$

25. To get to the point $(-3, 1)$ from the y-intercept $(0, -2)$, we must go up 3 units and to the left 3 units, so the slope is $\frac{3}{-3} = -1$. The slope-intercept form is

$$y = -1x - 2, \text{ or } y = -x - 2.$$

27. $-x + y = 4$
(a) Solve for y to get the equation in slope-intercept form.

$$-x + y = 4$$
$$y = x + 4$$

(b) The slope is the coefficient of x, 1.

(c) The y-intercept is the point $(0, b)$, or $(0, 4)$.

(d)

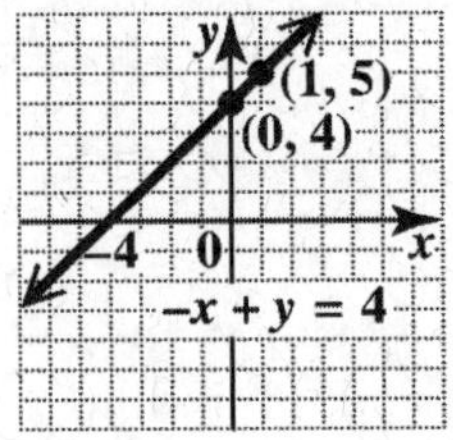

29. $6x + 5y = 30$
(a) Solve for y to get the equation in slope-intercept form.

$$6x + 5y = 30$$
$$5y = -6x + 30$$
$$y = -\tfrac{6}{5}x + 6$$

(b) The slope is the coefficient of x, $-\frac{6}{5}$.

(c) The y-intercept is the point $(0, b)$, or $(0, 6)$.

(d)

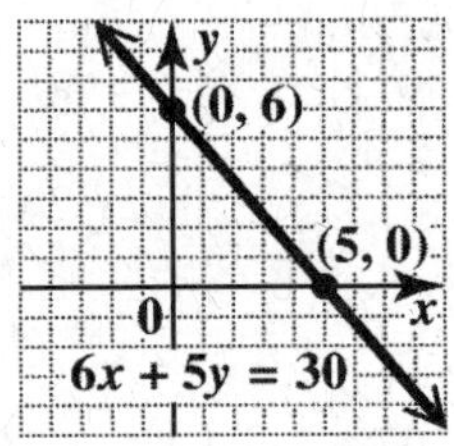

31. $4x - 5y = 20$
(a) Solve for y to get the equation in slope-intercept form.

$$4x - 5y = 20$$
$$-5y = -4x + 20$$
$$y = \tfrac{4}{5}x - 4$$

(b) The slope is the coefficient of x, $\frac{4}{5}$.

(c) The y-intercept is the point $(0, b)$, or $(0, -4)$.

(d)

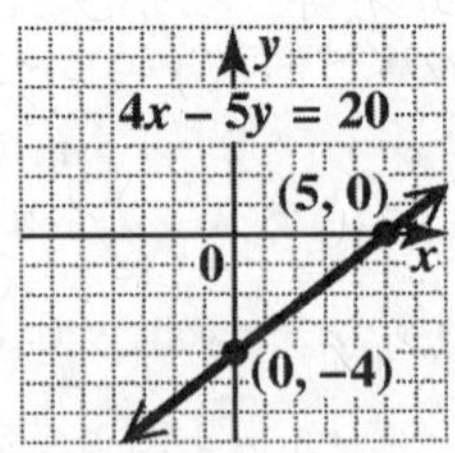

33. $x + 2y = -4$
(a) Solve for y to get the equation in slope-intercept form.

$$x + 2y = -4$$
$$2y = -x - 4$$
$$y = -\tfrac{1}{2}x - 2$$

(b) The slope is the coefficient of x, $-\frac{1}{2}$.

(c) The y-intercept is the point $(0, b)$, or $(0, -2)$.

(d)

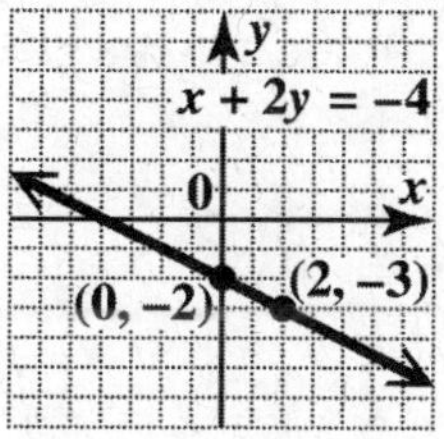

35. **(a)** Through $(5, 8)$; slope -2
Use the point-slope form with $(x_1, y_1) = (5, 8)$ and $m = -2$.

$$\begin{aligned} y - y_1 &= m(x - x_1) \\ y - 8 &= -2(x - 5) \\ y - 8 &= -2x + 10 \\ 2x + y &= 18 \end{aligned}$$

(b) Solve the last equation from part (a) for y.

$$\begin{aligned} 2x + y &= 18 \\ y &= -2x + 18 \end{aligned}$$

37. **(a)** Through $(-2, 4)$; slope $-\frac{3}{4}$
Use the point-slope form with $(x_1, y_1) = (-2, 4)$ and $m = -\frac{3}{4}$.

$$\begin{aligned} y - y_1 &= m(x - x_1) \\ y - 4 &= -\tfrac{3}{4}[x - (-2)] \\ 4(y - 4) &= -3(x + 2) \\ 4y - 16 &= -3x - 6 \\ 3x + 4y &= 10 \end{aligned}$$

(b) Solve the last equation from part (a) for y.

$$\begin{aligned} 3x + 4y &= 10 \\ 4y &= -3x + 10 \\ y &= -\tfrac{3}{4}x + \tfrac{10}{4} \\ y &= -\tfrac{3}{4}x + \tfrac{5}{2} \end{aligned}$$

39. **(a)** Through $(-5, 4)$; slope $\frac{1}{2}$
Use the point-slope form with $(x_1, y_1) = (-5, 4)$ and $m = \frac{1}{2}$.

$$\begin{aligned} y - y_1 &= m(x - x_1) \\ y - 4 &= \tfrac{1}{2}[x - (-5)] \\ 2(y - 4) &= 1(x + 5) \\ 2y - 8 &= x + 5 \\ -x + 2y &= 13 \\ x - 2y &= -13 \end{aligned}$$

(b) Solve the last equation from part (a) for y.

$$\begin{aligned} x - 2y &= -13 \\ -2y &= -x - 13 \\ y &= \tfrac{1}{2}x + \tfrac{13}{2} \end{aligned}$$

41. **(a)** Through $(3, 0)$; slope 4
Use the point-slope form with $(x_1, y_1) = (3, 0)$ and $m = 4$.

$$\begin{aligned} y - y_1 &= m(x - x_1) \\ y - 0 &= 4(x - 3) \\ y &= 4x - 12 \\ -4x + y &= -12 \\ 4x - y &= 12 \end{aligned}$$

(b) Solve the last equation from part (a) for y.

$$\begin{aligned} 4x - y &= 12 \\ -y &= -4x + 12 \\ y &= 4x - 12 \end{aligned}$$

43. **(a)** Through $(2, 6.8)$; slope 1.4
Use the point-slope form with $(x_1, y_1) = (2, 6.8)$ and $m = 1.4$.

$$\begin{aligned} y - y_1 &= m(x - x_1) \\ y - 6.8 &= 1.4(x - 2) \\ y - 6.8 &= \tfrac{7}{5}(x - 2) \\ 5(y - 6.8) &= 7(x - 2) \\ 5y - 34 &= 7x - 14 \\ -7x + 5y &= 20 \\ 7x - 5y &= -20 \end{aligned}$$

(b) Solve the last equation from part (a) for y.

$$\begin{aligned} 7x - 5y &= -20 \\ -5y &= -7x - 20 \\ y &= \tfrac{7}{5}x + 4, \quad \text{or} \quad y = 1.4x + 4 \end{aligned}$$

45. Through $(9, 5)$; slope 0
A line with slope 0 is a horizontal line. A horizontal line through the point (x, k) has equation $y = k$. Here $k = 5$, so an equation is $y = 5$.

47. Through $(9, 10)$; undefined slope
A vertical line has undefined slope and equation $x = c$. Since the x-value in $(9, 10)$ is 9, the equation is $x = 9$.

49. Through $(-\frac{3}{4}, -\frac{3}{2})$; slope 0
The equation of this line is $y = -\frac{3}{2}$.

51. Through $(-7, 8)$; horizontal
A horizontal line through the point (x, k) has equation $y = k$, so the equation is $y = 8$.

53. Through $(0.5, 0.2)$; vertical
A vertical line through the point (k, y) has equation $x = k$. Here $k = 0.5$, so the equation is $x = 0.5$.

55. **(a)** $(3, 4)$ and $(5, 8)$
Find the slope.

$$m = \frac{8-4}{5-3} = \frac{4}{2} = 2$$

Use the point-slope form with $(x_1, y_1) = (3, 4)$ and $m = 2$.

$$\begin{aligned} y - y_1 &= m(x - x_1) \\ y - 4 &= 2(x - 3) \\ y - 4 &= 2x - 6 \\ -2x + y &= -2 \\ 2x - y &= 2 \end{aligned}$$

(b) Solve the last equation from part (a) for y.

$$\begin{aligned} 2x - y &= 2 \\ -y &= -2x + 2 \\ y &= 2x - 2 \end{aligned}$$

Note: You could use *any* of the equations in part (a) to solve for y. In this exercise, choosing

$$y - 4 = 2x - 6 \quad \text{or} \quad -2x + y = -2,$$

easily leads to the equation $y = 2x - 2$.

57. **(a)** $(6, 1)$ and $(-2, 5)$
Find the slope.

$$m = \frac{5-1}{-2-6} = \frac{4}{-8} = -\frac{1}{2}$$

Use the point-slope form with $(x_1, y_1) = (6, 1)$ and $m = -\frac{1}{2}$.

$$\begin{aligned} y - y_1 &= m(x - x_1) \\ y - 1 &= -\tfrac{1}{2}(x - 6) \\ 2(y - 1) &= -1(x - 6) \\ 2y - 2 &= -x + 6 \\ x + 2y &= 8 \end{aligned}$$

(b) Solve the last equation from part (a) for y.

$$\begin{aligned} x + 2y &= 8 \\ 2y &= -x + 8 \\ y &= -\tfrac{1}{2}x + 4 \end{aligned}$$

59. **(a)** $(2, 5)$ and $(1, 5)$
Find the slope.

$$m = \frac{5-5}{1-2} = \frac{0}{-1} = 0$$

A line with slope 0 is horizontal. A horizontal line through the point (x, k) has equation $y = k$, so the equation is $y = 5$.

(b) $y = 5$ is already in the slope-intercept form.

61. **(a)** $(7, 6)$ and $(7, -8)$
Find the slope.

$$m = \frac{-8-6}{7-7} = \frac{-14}{0} \quad \textit{Undefined}$$

A line with undefined slope is a vertical line. The equation of a vertical line is $x = k$, where k is the common x-value. So the equation is $x = 7$.

(b) It is not possible to write $x = 7$ in slope-intercept form.

63. **(a)** $(\frac{1}{2}, -3)$ and $(-\frac{2}{3}, -3)$
Find the slope.

$$m = \frac{-3-(-3)}{-\frac{2}{3}-\frac{1}{2}} = \frac{0}{-\frac{7}{6}} = 0$$

A line with slope 0 is horizontal. A horizontal line through the point (x, k) has equation $y = k$, so the equation is $y = -3$.

(b) $y = -3$ is already in the slope-intercept form.

65. **(a)** $(-\frac{2}{5}, \frac{2}{5})$ and $(\frac{4}{3}, \frac{2}{3})$
Find the slope.

$$\begin{aligned} m &= \frac{\frac{2}{3}-\frac{2}{5}}{\frac{4}{3}-(-\frac{2}{5})} = \frac{\frac{10-6}{15}}{\frac{20+6}{15}} \\ &= \frac{\frac{4}{15}}{\frac{26}{15}} = \frac{4}{26} = \frac{2}{13} \end{aligned}$$

Use the point-slope form with $(x_1, y_1) = (-\frac{2}{5}, \frac{2}{5})$ and $m = \frac{2}{13}$.

$$\begin{aligned} y - \tfrac{2}{5} &= \tfrac{2}{13}\left[x - (-\tfrac{2}{5})\right] \\ 13(y - \tfrac{2}{5}) &= 2(x + \tfrac{2}{5}) \\ 13y - \tfrac{26}{5} &= 2x + \tfrac{4}{5} \\ -2x + 13y &= \tfrac{30}{5} \\ 2x - 13y &= -6 \end{aligned}$$

(b) Solve the last equation from part (a) for y.

$$\begin{aligned} 2x - 13y &= -6 \\ -13y &= -2x - 6 \\ y &= \tfrac{2}{13}x + \tfrac{6}{13} \end{aligned}$$

67. **(a)** Through $(7, 2)$; parallel to the graph of the line having equation $3x - y = 8$
Find the slope of $3x - y = 8$.

$$\begin{aligned} -y &= -3x + 8 \\ y &= 3x - 8 \end{aligned}$$

The slope is 3, so a line parallel to it also has slope 3. Use $m = 3$ and $(x_1, y_1) = (7, 2)$ in the point-slope form.

$$\begin{aligned} y - y_1 &= m(x - x_1) \\ y - 2 &= 3(x - 7) \\ y - 2 &= 3x - 21 \\ y &= 3x - 19 \end{aligned}$$

(b)
$$\begin{aligned} y &= 3x - 19 \\ -3x + y &= -19 \\ 3x - y &= 19 \end{aligned}$$

69. **(a)** Through $(-2,-2)$; parallel to $-x+2y=10$
Find the slope of $-x+2y=10$.

$$2y = x + 10$$
$$y = \tfrac{1}{2}x + 5$$

The slope is $\frac{1}{2}$, so a line parallel to it also has slope $\frac{1}{2}$. Use $m=\frac{1}{2}$ and $(x_1, y_1) = (-2,-2)$ in the point-slope form.

$$\begin{aligned} y - y_1 &= m(x - x_1) \\ y - (-2) &= \tfrac{1}{2}[x - (-2)] \\ y + 2 &= \tfrac{1}{2}(x + 2) \\ y + 2 &= \tfrac{1}{2}x + 1 \\ y &= \tfrac{1}{2}x - 1 \end{aligned}$$

(b)
$$\begin{aligned} y &= \tfrac{1}{2}x - 1 \\ 2y &= x - 2 \quad \textit{Multiply by 2.} \\ -x + 2y &= -2 \\ x - 2y &= 2 \end{aligned}$$

71. **(a)** Through $(8,5)$; perpendicular to $2x - y = 7$
Find the slope of $2x - y = 7$.

$$-y = -2x + 7$$
$$y = 2x - 7$$

The slope of the line is 2. Therefore, the slope of the line perpendicular to it is $-\frac{1}{2}$ since $2(-\frac{1}{2}) = -1$. Use $m = -\frac{1}{2}$ and $(x_1, y_1) = (8,5)$ in the point-slope form.

$$\begin{aligned} y - y_1 &= m(x - x_1) \\ y - 5 &= -\tfrac{1}{2}(x - 8) \\ y - 5 &= -\tfrac{1}{2}x + 4 \\ y &= -\tfrac{1}{2}x + 9 \end{aligned}$$

(b)
$$\begin{aligned} y &= -\tfrac{1}{2}x + 9 \\ 2y &= -x + 18 \quad \textit{Multiply by 2.} \\ x + 2y &= 18 \end{aligned}$$

73. **(a)** Through $(-2,7)$; perpendicular to $x = 9$
$x = 9$ is a vertical line so a line perpendicular to it will be a horizontal line. It goes through $(-2,7)$, so its equation is

$$y = 7.$$

(b) $y = 7$ is already in standard form.

75. Distance = (rate)(time), so

$$y = 45x.$$

x	$y = 45x$	Ordered Pair
0	$45(0) = 0$	$(0,0)$
5	$45(5) = 225$	$(5,225)$
10	$45(10) = 450$	$(10,450)$

77. Total cost = (cost/gal)(number of gallons), so

$$y = 3.10x.$$

x	$y = 3.10x$	Ordered Pair
0	$3.10(0) = 0$	$(0,0)$
5	$3.10(5) = 15.50$	$(5,15.50)$
10	$3.10(10) = 31.00$	$(10,31.00)$

79. Total cost = (cost/credit)(number of credits), so

$$y = 111x.$$

x	$y = 111x$	Ordered Pair
0	$111(0) = 0$	$(0,0)$
5	$111(5) = 555$	$(5,555)$
10	$111(10) = 1110$	$(10,1110)$

81. **(a)** The fixed cost is \$12, so that is the value of b. The variable cost is \$112.50, so

$$y = mx + b = 112.50x + 12.$$

(b) If $x = 5$, $y = 112.50(5) + 12 = 574.50$. The ordered pair is $(5, 574.50)$. The cost of 5 tickets and a parking pass is \$574.50.

(c) If $x = 2$, $y = 112.50(2) + 12 = 237$. The cost of 2 tickets and a parking pass is \$237.

83. **(a)** The fixed cost is \$99, so that is the value of b. The variable cost is \$41, so

$$y = mx + b = 41x + 99.$$

(b) If $x = 5$, $y = 41(5) + 99 = 304$. The ordered pair is $(5, 304)$. The cost for a 5-month membership is \$304.

(c) If $x = 12$, $y = 41(12) + 99 = 591$. The cost for the first year's membership is \$591.

85. **(a)** The fixed cost is \$36, so that is the value of b. The variable cost is \$60, so

$$y = mx + b = 60x + 36.$$

(b) If $x = 5$, $y = 60(5) + 36 = 336$. The ordered pair is $(5, 336)$. The cost of the plan for 5 months is \$336.

(c) For a 1-year contract, $x = 12$, so $y = 60(12) + 36 = 756$. The cost of the plan for 1 year is \$756.

87. **(a)** The fixed cost is \$30, so that is the value of b. The variable cost is \$6, so

$$y = mx + b = 6x + 30.$$

(b) If $x = 5$, $y = 6(5) + 30 = 60$. The ordered pair is $(5, 60)$. It costs \$60 to rent the saw for 5 days.

(c)
$$\begin{aligned} 138 &= 6x + 30 \quad \textit{Let } y = 138. \\ 108 &= 6x \\ x &= \frac{108}{6} = 18 \end{aligned}$$

The saw is rented for 18 days.

89. **(a)** Use $(0, 3921)$ and $(3, 7805)$.

$$m = \frac{7805 - 3921}{3 - 0} = \frac{3884}{3} \approx 1294.7$$

The equation is $y = 1294.7x + 3921$. The slope tells us that the sales of digital cameras in the United States increased by \$1294.7 million per year from 2003 to 2006.

(b) The year 2007 corresponds to $x = 4$, so $y = 1294.7(4) + 3921 = \$9099.8$ million.

91. **(a)** Use $(3, 38.0)$ and $(7, 59.0)$.

$$m = \frac{59.0 - 38.0}{7 - 3} = \frac{21}{4} = 5.25$$

Now use the point-slope form.

$$\begin{aligned} y - 38 &= 5.25(x - 3) \\ y - 38 &= 5.25x - 15.75 \\ y &= 5.25x + 22.25 \end{aligned}$$

(b) The year 2005 corresponds to $x = 5$, so $y = 5.25(5) + 22.25 = 48.5$. This value is greater than the actual value.

♦♦♦ Relating Concepts 93–100 ♦♦♦

93. When $C = 0°$, $F = \underline{32°}$, and when $C = 100°$, $F = \underline{212°}$. These are the freezing and boiling temperatures for water.

94. The two points of the form (C, F) would be $(0, 32)$ and $(100, 212)$.

95. $m = \dfrac{212 - 32}{100 - 0} = \dfrac{180}{100} = \dfrac{9}{5}$

96. Let $m = \frac{9}{5}$ and $(x_1, y_1) = (0, 32)$.

$$\begin{aligned} y - y_1 &= m(x - x_1) \\ F - 32 &= \tfrac{9}{5}(C - 0) \\ F - 32 &= \tfrac{9}{5}C \\ F &= \tfrac{9}{5}C + 32 \end{aligned}$$

97.
$$\begin{aligned} F &= \tfrac{9}{5}C + 32 \\ F - 32 &= \tfrac{9}{5}C \\ \tfrac{5}{9}(F - 32) &= C \end{aligned}$$

98. $F = \frac{9}{5}C + 32$

If $C = 30$, $\quad F = \frac{9}{5}(30) + 32 = 54 + 32 = 86.$

Thus, when $C = 30°$, $F = 86°$.

99. $C = \frac{5}{9}(F - 32)$

When $F = 50$, $\quad C = \frac{5}{9}(50 - 32) = \frac{5}{9}(18) = 10.$

Thus, when $F = 50°$, $C = 10°$.

100. Let $F = C$ in the equation obtained in Exercise 96.

$$\begin{aligned} F &= \tfrac{9}{5}C + 32 \\ C &= \tfrac{9}{5}C + 32 && \textit{Let } F = C. \\ 5C &= 5(\tfrac{9}{5}C + 32) && \textit{Multiply by 5.} \\ 5C &= 9C + 160 \\ -4C &= 160 && \textit{Subtract 9C.} \\ C &= -40 && \textit{Divide by } -4. \end{aligned}$$

(The same result may be found by using either form of the equation obtained in Exercise 97.) The Celsius and Fahrenheit temperatures are equal $(F = C)$ at -40 degrees.

101. $x \geq 0$ written in interval notation is $[0, \infty)$.

103. $-4 \leq x \leq 4$ written in interval notation is $[-4, 4]$.

Summary Exercises on Slopes and Equations of Lines

1. Solve the equation for y.

$$\begin{aligned} 3x + 5y &= 9 \\ 5y &= -3x + 9 \\ y &= -\tfrac{3}{5}x + \tfrac{9}{5} \end{aligned}$$

The slope is $-\frac{3}{5}$.

3. $y = 2x - 5$ is already in the slope-intercept form, so the slope is 2.

5. The graph of $x - 4 = 0$, or $x = 4$, is a vertical line with x-intercept $(4, 0)$. The slope of a vertical line is *undefined* because the denominator equals zero in the slope formula.

7. Through $(-2, 6)$ and $(4, 1)$

(a) The slope is

$$m = \frac{1 - 6}{4 - (-2)} = \frac{-5}{6} = -\frac{5}{6}.$$

Use the point-slope form.

$$\begin{aligned} y - y_1 &= m(x - x_1) \\ y - 6 &= -\tfrac{5}{6}[x - (-2)] \\ y - 6 &= -\tfrac{5}{6}x - \tfrac{5}{3} \\ y &= -\tfrac{5}{6}x - \tfrac{5}{3} + \tfrac{18}{3} \\ y &= -\tfrac{5}{6}x + \tfrac{13}{3} \end{aligned}$$

(b)
$$\begin{aligned} y &= -\tfrac{5}{6}x + \tfrac{13}{3} \\ 6y &= -5x + 26 && \textit{Multiply by 6.} \\ 5x + 6y &= 26 \end{aligned}$$

9. Through $(0, 0)$; perpendicular to $2x - 5y = 6$

(a) Find the slope of $2x - 5y = 6$.

$$-5y = -2x + 6$$
$$y = \frac{2}{5}x - \frac{6}{5}$$

The slope of the line is $\frac{2}{5}$. Therefore, the slope of the line perpendicular to it is $-\frac{5}{2}$ since $\frac{2}{5}(-\frac{5}{2}) = -1$. Use $m = -\frac{5}{2}$ and $(x_1, y_1) = (0, 0)$ in the point-slope form.

$$y - y_1 = m(x - x_1)$$
$$y - 0 = -\frac{5}{2}(x - 0)$$
$$y = -\frac{5}{2}x$$

(b)
$$y = -\frac{5}{2}x$$
$$2y = -5x$$
$$5x + 2y = 0$$

11. Through $(\frac{3}{4}, -\frac{7}{9})$; perpendicular to $x = \frac{2}{3}$

(a) $x = \frac{2}{3}$ is a vertical line so a line perpendicular to it will be a horizontal line. It goes through $(\frac{3}{4}, -\frac{7}{9})$, so its equation is

$$y = -\frac{7}{9}.$$

(b) $y = -\frac{7}{9}$

$9y = -7$ *Multiply by 9.*

13. Through $(-4, 2)$; parallel to the line through $(3, 9)$ and $(6, 11)$

(a) The slope of the line through $(3, 9)$ and $(6, 11)$ is

$$m = \frac{11 - 9}{6 - 3} = \frac{2}{3}.$$

Use the point-slope form with $(x_1, y_1) = (-4, 2)$ and $m = \frac{2}{3}$ (since the slope of the desired line must equal the slope of the given line).

$$y - y_1 = m(x - x_1)$$
$$y - 2 = \frac{2}{3}[x - (-4)]$$
$$y - 2 = \frac{2}{3}(x + 4)$$
$$y - 2 = \frac{2}{3}x + \frac{8}{3}$$
$$y = \frac{2}{3}x + \frac{8}{3} + \frac{6}{3}$$
$$y = \frac{2}{3}x + \frac{14}{3}$$

(b)
$$y = \frac{2}{3}x + \frac{14}{3}$$
$$3y = 2x + 14$$
$$-2x + 3y = 14$$
$$2x - 3y = -14$$

15. Through $(4, -8)$ and $(-4, 12)$

(a) The slope is

$$m = \frac{12 - (-8)}{-4 - 4} = \frac{20}{-8} = -\frac{5}{2}.$$

Use the point-slope form.

$$y - y_1 = m(x - x_1)$$
$$y - (-8) = -\frac{5}{2}(x - 4)$$
$$y + 8 = -\frac{5}{2}x + 10$$
$$y = -\frac{5}{2}x + 2$$

(b)
$$y = -\frac{5}{2}x + 2$$

$2y = -5x + 4$ *Multiply by 2.*

$$5x + 2y = 4$$

17. Through $(0, 3)$ and the midpoint of the segment with endpoints $(2, 8)$ and $(-4, 12)$

(a) The midpoint of the segment with endpoints $(2, 8)$ and $(-4, 12)$ is

$$\left(\frac{2 + (-4)}{2}, \frac{8 + 12}{2}\right) = \left(\frac{-2}{2}, \frac{20}{2}\right) = (-1, 10).$$

The slope of the line through $(0, 3)$ and $(-1, 10)$ is

$$m = \frac{10 - 3}{-1 - 0} = \frac{7}{-1} = -7.$$

Since we are given the y-intercept, $(0, 3)$, the slope-intercept form is

$$y = -7x + 3.$$

(b)
$$y = -7x + 3$$
$$7x + y = 3$$

7.3 Introduction to Relations and Functions

7.3 Now Try Exercises

N1. **(a)** $\{(1, 5), (3, 5), (5, 5)\}$
The relation *is a function* because for each different x-value there is exactly one y-value. It is acceptable to have different x-values paired with the same y-value.

(b) $\{(-1, -3), (0, 2), (-1, 6)\}$
The first and last ordered pairs have the *same* x-value paired with *two different* y-values (-1 is paired with both -3 and 6), so this relation *is not a function.*

N2. **(a)** $\{(2, 2), (2, 5), (4, 8), (6, 5)\}$
The first two ordered pairs have the *same* x-value paired with *two different* y-values (2 is paired with both 2 and 5), so this relation *is not a function.*

(b) The domain of this relation is the set of all first components, that is, $\{1, 10, 15\}$. The range of this relation is the set of all second components, that is, $\{\$1.39, \$10.39, \$20.85\}$. This relation *is a function* because for each different first component, there is exactly one second component.

N3. The arrowheads indicate that the graph extends indefinitely left and right, as well as upward. The domain includes all real numbers, written $(-\infty, \infty)$. Because there is a least y-value, -2, the range includes all numbers greater than or equal to -2, $[-2, \infty)$.

N4. A vertical line intersects the graph more than once, so the relation is *not* a function.

N5. **(a)** $y = 4x - 3$ is a function because each value of x corresponds to exactly one value of y. Its domain is the set of all real numbers, $(-\infty, \infty)$.

(b) $y = \dfrac{1}{x-8}$
Given any value of x in the domain, we find y by subtracting 8, and then dividing the result into 1. This process produces exactly one value of y for each value in the domain, so the given equation defines a function. The domain includes all real numbers except those that make the denominator 0. We find those numbers by setting the denominator equal to 0 and solving for x.

$$x - 8 = 0$$
$$x = 8$$

The domain includes all real numbers *except* 8, written as $(-\infty, 8) \cup (8, \infty)$.

7.3 Section Exercises

1. Answers will vary. A function is a set of ordered pairs in which each first component corresponds to exactly one second component. For example, $\{(0,1), (1,2), (2,3), (3,4), \ldots\}$ is a function.

3. In an ordered pair of a relation, the first element is the independent variable.

5. $\{(0,2), (2,4), (4,6)\}$
We can represent this set of ordered pairs by plotting them on a graph.

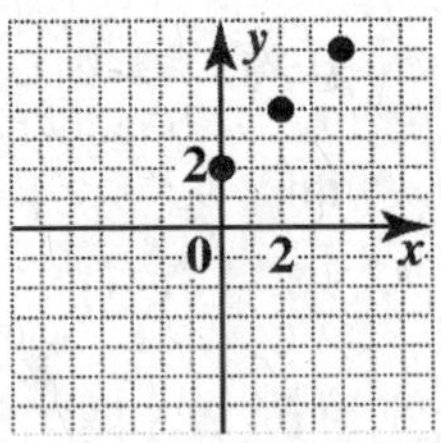

7. We can represent this diagram in table form.

x	y
-3	-4
-3	1
2	0

9. $\{(5,1), (3,2), (4,9), (7,6)\}$
The relation is a function since for each x-value, there is only one y-value.

The domain is the set of x-values: $\{5, 3, 4, 7\}$.
The range is the set of y-values: $\{1, 2, 9, 6\}$.

11. $\{(2,4), (0,2), (2,5)\}$
The relation is not a function since the x-value 2 has two different y-values associated with it, 4 and 5.

The domain is the set of x-values: $\{2, 0\}$.
The range is the set of y-values: $\{4, 2, 5\}$.

13. $\{(-3,1), (4,1), (-2,7)\}$
The relation is a function since for each x-value, there is only one y-value.

The domain is the set of x-values: $\{-3, 4, -2\}$.
The range is the set of y-values: $\{1, 7\}$.

15. $\{(1,1), (1,-1), (0,0), (2,4), (2,-4)\}$
The relation is not a function since the x-value 1 has two different y-values associated with it, 1 and -1. (A similar statement can be made for $x = 2$.)

The domain is the set of x-values: $\{1, 0, 2\}$.
The range is the set of y-values: $\{1, -1, 0, 4, -4\}$.

17. The relation can be described by the set of ordered pairs

$$\{(2,1), (5,1), (11,7), (17,20), (3,20)\}.$$

The relation is a function since for each x-value, there is only one y-value.

The domain is the set of x-values: $\{2, 5, 11, 17, 3\}$.
The range is the set of y-values: $\{1, 7, 20\}$.

19. The relation can be described by the set of ordered pairs

$$\{(1,5), (1,2), (1,-1), (1,-4)\}.$$

The relation is not a function since the x-value 1 has four different y-values associated with it, 5, 2, -1, and -4.

The domain is the set of x-values: $\{1\}$.
The range is the set of y-values: $\{5, 2, -1, -4\}$.

21. The relation can be described by the set of ordered pairs

$$\{(4,-3), (2,-3), (0,-3), (-2,-3)\}.$$

The relation is a function since for each x-value, there is only one y-value.

The domain is the set of x-values: $\{4, 2, 0, -2\}$.
The range is the set of y-values: $\{-3\}$.

23. The relation can be described by the set of ordered pairs

$$\{(-2,2), (0,3), (3,2)\}.$$

The relation is a function since for each x-value, there is only one y-value.

The domain is the set of x-values: $\{-2, 0, 3\}$.
The range is the set of y-values: $\{2, 3\}$.

25. Using the vertical line test, we find any vertical line will intersect the graph at most once. This indicates that the graph represents a function. This graph extends indefinitely to the left $(-\infty)$ and indefinitely to the right (∞). Therefore, the domain is $(-\infty, \infty)$. This graph extends indefinitely downward $(-\infty)$, and indefinitely upward (∞). Thus, the range is $(-\infty, \infty)$.

27. Since a vertical line, such as $x = -4$, intersects the graph in two points, the relation is not a function. The domain is $(-\infty, 0]$, and the range is $(-\infty, \infty)$.

29. Using the vertical line test, we find any vertical line will intersect the graph at most once. This indicates that the graph represents a function. This graph extends indefinitely to the left $(-\infty)$ and indefinitely to the right (∞). Therefore, the domain is $(-\infty, \infty)$. This graph extends indefinitely downward $(-\infty)$, and reaches a high point at $y = 4$. Therefore, the range is $(-\infty, 4]$.

31. Since a vertical line can intersect the graph of the relation in more than one point, the relation is not a function. The domain, the x-values of the points on the graph, is $[-4, 4]$. The range, the y-values of the points on the graph, is $[-3, 3]$.

33. $y = -6x$
Each value of x corresponds to one y-value. For example, if $x = 3$, then $y = -6(3) = -18$. Therefore, $y = -6x$ defines y as a function of x. Since any x-value, positive, negative, or zero, can be multiplied by -6, the domain is $(-\infty, \infty)$.

35. $y = 2x - 6$
For any value of x, there is exactly one value of y, so this equation defines a function. The domain is the set of all real numbers, $(-\infty, \infty)$.

37. $y = x^2$
Each value of x corresponds to one y-value. For example, if $x = 3$, then $y = 3^2 = 9$. Therefore, $y = x^2$ defines y as a function of x.
Since any x-value, positive, negative, or zero, can be squared, the domain is $(-\infty, \infty)$.

39. $x = y^6$
The ordered pairs $(64, 2)$ and $(64, -2)$ both satisfy the equation. Since one value of x, 64, corresponds to two values of y, 2 and -2, the relation does not define a function. Because x is equal to the sixth power of y, the values of x must always be nonnegative. The domain is $[0, \infty)$.

41. $y = \dfrac{x+4}{5}$
Given any value of x, y is found by adding 4, then dividing the result by 5. This process produces exactly one value of y for each x-value in the domain, so the relation represents a function. The denominator is never 0, so the domain is $(-\infty, \infty)$.

43. $y = -\dfrac{2}{x}$
Given any value of x, y is found by dividing that value into 2, and negating that result. This process produces exactly one value of y for each x-value in the domain, so the relation represents a function. The domain includes all real numbers except those that make the denominator 0, namely 0. The domain is $(-\infty, 0) \cup (0, \infty)$.

45. $y = \dfrac{2}{x-4}$
Given any value of x, y is found by subtracting 4, then dividing the result into 2. This process produces exactly one value of y for each x-value in the domain, so the relation represents a function. The domain includes all real numbers except those that make the denominator 0, namely 4. The domain is $(-\infty, 4) \cup (4, \infty)$.

47. $y = \dfrac{1}{4x+3}$ defines y as a function of x. The domain includes all real numbers except those that make the denominator 0, namely $-\frac{3}{4}$. The domain is $\left(-\infty, -\frac{3}{4}\right) \cup \left(-\frac{3}{4}, \infty\right)$.

49. $x = y^2 + 1$
The ordered pairs $(2, 1)$ and $(2, -1)$ both satisfy the equation. Since one value of x, 2, corresponds to two values of y, 1 and -1, the relation does not define a function. Because x is equal to the second power of y, plus 1, the values of x must always be at least 1. The domain is $[1, \infty)$.

51. $xy = 1$
Rewrite $xy = 1$ as $y = \frac{1}{x}$. Note that x can never equal 0, otherwise the denominator would equal 0. The domain is $(-\infty, 0) \cup (0, \infty)$.
Each nonzero x-value gives exactly one y-value. Therefore, $xy = 1$ defines y as a function of x.

53. **(a)** Each year corresponds to exactly one percentage, so the table defines a function.

(b) The domain is $\{2004, 2005, 2006, 2007, 2008\}$.
The range is $\{42.3, 42.8, 43.7, 43.8\}$.

(c) Answers will vary. Two possible answers are $(2005, 42.3)$ and $(2008, 43.8)$.

55.
$$\begin{aligned} y &= -7x + 12 \\ y &= -7(3) + 12 \quad \textit{Let } x = 3. \\ y &= -9 \end{aligned}$$

57.
$$\begin{aligned} y &= 3x - 8 \\ y &= 3(3) - 8 \quad \textit{Let } x = 3. \\ y &= 1 \end{aligned}$$

59.
$$\begin{aligned} 2x - 4y &= 7 \\ -4y &= -2x + 7 \\ y &= \tfrac{1}{2}x - \tfrac{7}{4} \end{aligned}$$

7.4 Function Notation and Linear Functions

7.4 Now Try Exercises

N1. $f(x) = 4x + 3$

$f(-2) = 4(-2) + 3$ *Replace x with −2.*

$= -8 + 3$ *Multiply.*

$= -5$ *Add.*

N2. $f(x) = 2x^2 - 4x + 1$

(a) $f(-2) = 2(-2)^2 - 4(-2) + 1 = 8 + 8 + 1 = 17$

(b) $f(a) = 2a^2 - 4a + 1$

N3.
$$\begin{aligned} g(x) &= 8x - 5 \\ g(a-2) &= 8(a-2) - 5 \\ &= 8a - 16 - 5 \\ &= 8a - 21 \end{aligned}$$

N4. (a) When $x = -1$, $y = 4$, so $f(-1) = 4$.

(b) $f(x) = x^2 - 12$

$f(-1) = (-1)^2 - 12 = 1 - 12 = -11$

N5. (a) When $x = -1$, $y = 0$, so $f(-1) = 0$.

(b) $f(x) = 2$ is equivalent to $y = 2$, and $y = 2$ when $x = 1$.

N6. $-4x^2 + y = 5$

Solving for y gives us $y = 4x^2 + 5$.

$f(x) = 4x^2 + 5$ *Replace y with f(x).*

$f(-3) = 4(-3)^2 + 5 = 36 + 5 = 41$

$f(h) = 4h^2 + 5$

N7. The graph of $g(x) = \frac{1}{3}x - 2$ is a line with slope $\frac{1}{3}$ and y-intercept -2.

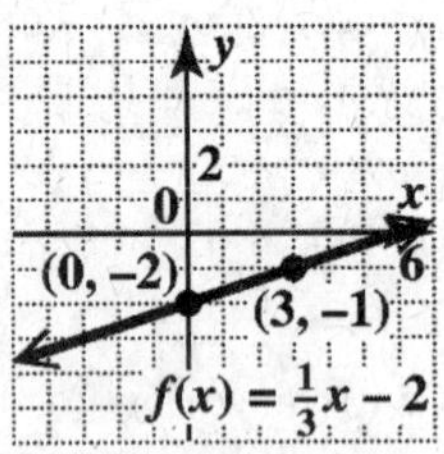

The domain and range are $(-\infty, \infty)$.

7.4 Section Exercises

1. $f(3)$ is the value of the dependent variable when the independent variable is 3—choice **B**.

3.
$$\begin{aligned} f(x) &= -3x + 4 \\ f(0) &= -3(0) + 4 \\ &= 0 + 4 \\ &= 4 \end{aligned}$$

5.
$$\begin{aligned} g(x) &= -x^2 + 4x + 1 \\ g(-2) &= -(-2)^2 + 4(-2) + 1 \\ &= -(4) - 8 + 1 \\ &= -11 \end{aligned}$$

7.
$$\begin{aligned} f(x) &= -3x + 4 \\ f(\tfrac{1}{3}) &= -3(\tfrac{1}{3}) + 4 \\ &= -1 + 4 \\ &= 3 \end{aligned}$$

9.
$$\begin{aligned} g(x) &= -x^2 + 4x + 1 \\ g(0.5) &= -(0.5)^2 + 4(0.5) + 1 \\ &= -0.25 + 2 + 1 \\ &= 2.75 \end{aligned}$$

11.
$$\begin{aligned} f(x) &= -3x + 4 \\ f(p) &= -3(p) + 4 \\ &= -3p + 4 \end{aligned}$$

13.
$$\begin{aligned} f(x) &= -3x + 4 \\ f(-x) &= -3(-x) + 4 \\ &= 3x + 4 \end{aligned}$$

15.
$$\begin{aligned} f(x) &= -3x + 4 \\ f(x+2) &= -3(x+2) + 4 \\ &= -3x - 6 + 4 \\ &= -3x - 2 \end{aligned}$$

17.
$$\begin{aligned} g(x) &= -x^2 + 4x + 1 \\ g(\pi) &= -\pi^2 + 4\pi + 1 \end{aligned}$$

19.
$$\begin{aligned} f(x) &= -3x + 4 \\ f(x+h) &= -3(x+h) + 4 \\ &= -3x - 3h + 4 \end{aligned}$$

21. $f(x) = -3x + 4$, $g(x) = -x^2 + 4x + 1$

$$\begin{aligned} &f(4) - g(4) \\ &= [-3(4) + 4] - [-(4)^2 + 4(4) + 1] \\ &= [-8] - [1] \\ &= -9 \end{aligned}$$

23. $f = \{(-2, 2), (-1, -1), (2, -1)\}$

(a) When $x = 2$, $y = -1$, so $f(2) = -1$.

(b) When $x = -1$, $y = -1$, so $f(-1) = -1$.

25. (a) When $x = 2$, $y = 2$, so $f(2) = 2$.

(b) When $x = -1$, $y = 3$, so $f(-1) = 3$.

27. (a) When $x = 2$, $y = 15$, so $f(2) = 15$.

(b) When $x = -1$, $y = 10$, so $f(-1) = 10$.

29. (a) When $x = 2$, $y = 4$, so $f(2) = 4$.

(b) When $x = -1$, $y = 1$, so $f(-1) = 1$.

31. **(a)** The point $(2, 3)$ is on the graph of f, so $f(2) = 3$.

(b) The point $(-1, -3)$ is on the graph of f, so $f(-1) = -3$.

33. **(a)** The point $(2, -3)$ is on the graph of f, so $f(2) = -3$.

(b) The point $(-1, 2)$ is on the graph of f, so $f(-1) = 2$.

35. **(a)** $f(x) = 3$: when $y = 3$, $x = 2$

(b) $f(x) = -1$: when $y = -1$, $x = 0$

(c) $f(x) = -3$: when $y = -3$, $x = -1$

37. **(a)** Solve the equation for y.

$$\begin{aligned} x + 3y &= 12 \\ 3y &= 12 - x \\ y &= \frac{12 - x}{3} \end{aligned}$$

Since $y = f(x)$,

$$f(x) = \frac{12 - x}{3} = -\frac{1}{3}x + 4.$$

(b) $f(3) = \dfrac{12 - 3}{3} = \dfrac{9}{3} = 3$

39. **(a)** Solve the equation for y.

$$\begin{aligned} y + 2x^2 &= 3 \\ y &= 3 - 2x^2 \end{aligned}$$

Since $y = f(x)$,

$$f(x) = 3 - 2x^2.$$

(b) $$\begin{aligned} f(3) &= 3 - 2(3)^2 \\ &= 3 - 2(9) \\ &= -15 \end{aligned}$$

41. **(a)** Solve the equation for y.

$$\begin{aligned} 4x - 3y &= 8 \\ -3y &= 8 - 4x \\ y &= \frac{8 - 4x}{-3} \end{aligned}$$

Since $y = f(x)$,

$$f(x) = \frac{8 - 4x}{-3} = \frac{4}{3}x - \frac{8}{3}.$$

(b) $$\begin{aligned} f(3) &= \frac{8 - 4(3)}{-3} = \frac{8 - 12}{-3} \\ &= \frac{-4}{-3} = \frac{4}{3} \end{aligned}$$

43. The equation $2x + y = 4$ has a straight __line__ as its graph. To find y in $(3, y)$, let $x = 3$ in the equation.

$$\begin{aligned} 2x + y &= 4 \\ 2(3) + y &= 4 \\ 6 + y &= 4 \\ y &= -2 \end{aligned}$$

So one point that lies on the graph is $(3, \underline{-2})$. To use functional notation for $2x + y = 4$, solve for y to get

$$y = -2x + 4.$$

Replace y with $f(x)$ to get

$$\begin{aligned} f(x) &= \underline{-2x + 4}. \\ f(3) &= -2(3) + 4 = \underline{-2} \end{aligned}$$

Because $y = -2$ when $x = 3$, the point $(\underline{3}, \underline{-2})$ lies on the graph of the function.

45. $f(x) = -2x + 5$
The graph will be a line. The intercepts are $(0, 5)$ and $(\frac{5}{2}, 0)$.
The domain is $(-\infty, \infty)$. The range is $(-\infty, \infty)$.

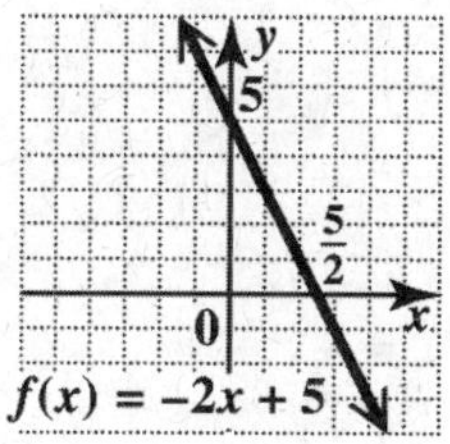

47. $h(x) = \frac{1}{2}x + 2$
The graph will be a line. The intercepts are $(0, 2)$ and $(-4, 0)$.
The domain is $(-\infty, \infty)$. The range is $(-\infty, \infty)$.

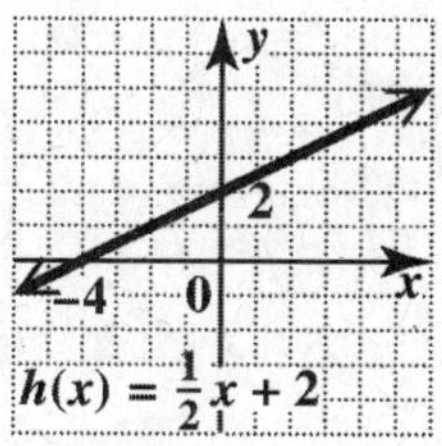

49. $G(x) = 2x$
This line includes the points $(0, 0)$, $(1, 2)$, and $(2, 4)$. The domain is $(-\infty, \infty)$. The range is $(-\infty, \infty)$.

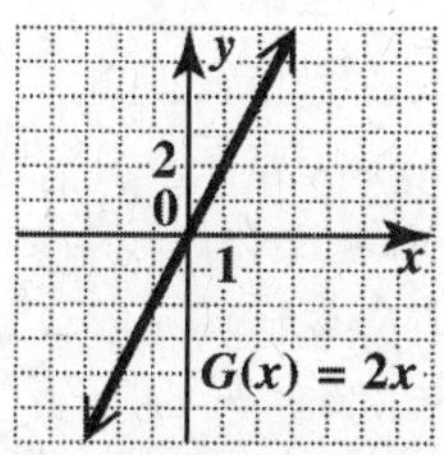

51. $g(x) = -4$
Using a y-intercept of $(0, -4)$ and a slope of $m = 0$, we graph the horizontal line. From the graph we see that the domain is $(-\infty, \infty)$. The range is $\{-4\}$.

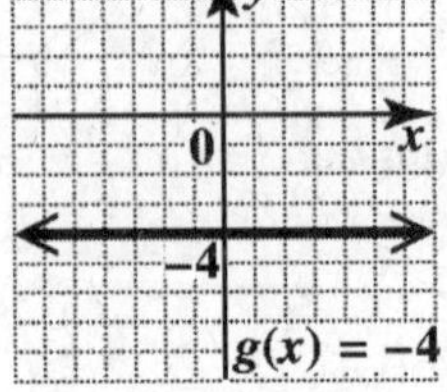

53. $f(x) = 0$
Draw the horizontal line through the point $(0, 0)$. On the horizontal line the value of x can be any real number, so the domain is $(-\infty, \infty)$. The range is $\{0\}$.

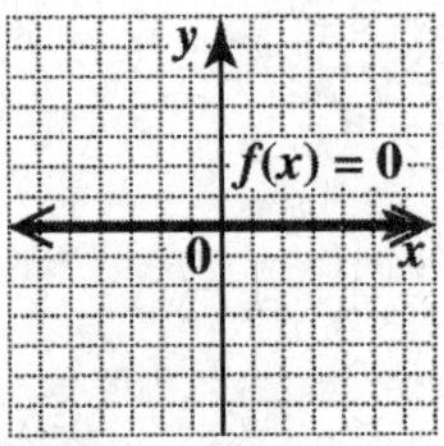

55. $f(x) = 0$, or $y = 0$, is the x-axis.

57. **(a)** $f(x) = 3.75x$
$$f(3) = 3.75(3) = 11.25 \text{ (dollars)}$$

(b) 3 is the value of the independent variable, which represents a package weight of 3 pounds; $f(3)$ is the value of the dependent variable representing the cost to mail a 3-pound package.

(c) $3.75(5) = \$18.75$, the cost to mail a 5-lb package. Using function notation, we have $f(5) = 18.75$.

59. **(a)** Since the length of a man's femur is given, use the formula $h(r) = 69.09 + 2.24r$.

$$h(56) = 69.09 + 2.24(56) \quad \textit{Let } r = 56.$$
$$= 194.53$$

The man is 194.53 cm tall.

(b) Use the formula $h(t) = 81.69 + 2.39t$.

$$h(40) = 81.69 + 2.39(40) \quad \textit{Let } t = 40.$$
$$= 177.29$$

The man is 177.29 cm tall.

(c) Since the length of a woman's femur is given, use the formula $h(r) = 61.41 + 2.32r$.

$$h(50) = 61.41 + 2.32(50) \quad \textit{Let } r = 50.$$
$$= 177.41$$

The woman is 177.41 cm tall.

(d) Use the formula $h(t) = 72.57 + 2.53t$.

$$h(36) = 72.57 + 2.53(36) \quad \textit{Let } t = 36.$$
$$= 163.65$$

The woman is 163.65 cm tall.

61. **(a)** $f(x) = 12x + 100$

(b) $f(125) = 12(125) + 100 = 1500 + 100 = 1600$

The cost to print 125 t-shirts is \$1600.

(c)
$$f(x) = 1000$$
$$12x + 100 = 1000$$
$$12x = 900 \quad \textit{Subtract 100.}$$
$$x = 75 \quad \textit{Divide by 12.}$$

In function notation, $f(75) = 1000$.

The cost to print 75 t-shirts is \$1000.

63. **(a)** $f(2) = 1.1$

(b) y is -2.5 when x is 5.
So, if $f(x) = -2.5$, then $x = 5$.

(c) Let $(x_1, y_1) = (0, 3.5)$ and $(x_2, y_2) = (1, 2.3)$. Then

$$m = \frac{2.3 - 3.5}{1 - 0} = \frac{-1.2}{1} = -1.2.$$

The slope is -1.2.

(d) When $x = 0$, y is 3.5, so the y-intercept is $(0, 3.5)$.

(e) Use the slope-intercept form of the equation of a line and the information found in parts (c) and (d).

$$f(x) = mx + b$$
$$f(x) = -1.2x + 3.5$$

65. **(a)** The independent variable is t, the number of hours, and the possible values are in the set $[0, 100]$. The dependent variable is g, the number of gallons, and the possible values are in the set $[0, 3000]$.

(b) The graph rises for the first 25 hours, so the water level increases for 25 hours. The graph falls for $t = 50$ to $t = 75$, so the water level decreases for 25 hours.

(c) There are 2000 gallons in the pool when $t = 90$.

(d) $f(0)$ is the number of gallons in the pool at time $t = 0$. Here, $f(0) = 0$, which means the pool is empty at time 0.

(e) $f(25) = 3000$; After 25 hours, there are 3000 gallons of water in the pool.

67. $(15x^2 - 2x) + (x - 4) = 15x^2 - 2x + x - 4$
$$= 15x^2 - x - 4$$

69. $(4x-5)(3x+1)$
$= 4x(3x)+4x(1)-5(3x)-5(1)$
$= 12x^2+4x-15x-5$
$= 12x^2-11x-5$

71. $\dfrac{27x^3-18x^2}{9x} = \dfrac{27x^3}{9x} - \dfrac{18x^2}{9x}$
$= 3x^2-2x$

7.5 Operations on Functions and Composition

7.5 Now Try Exercises

N1. $f(x) = x^3-2x^2+7$
$f(-3) = (-3)^3 - 2(-3)^2 + 7$
$= -27 - 2(9) + 7$
$= -38$

N2. $f(x) = x^3-3x^2+4,\ g(x) = -2x^3+x^2-12$

(a) $(f+g)(x)$
$= f(x)+g(x)$
$= (x^3-3x^2+4)+(-2x^3+x^2-12)$
$= x^3-3x^2+4-2x^3+x^2-12$
$= -x^3-2x^2-8$

(b) $(f-g)(x)$
$= f(x)-g(x)$
$= (x^3-3x^2+4)-(-2x^3+x^2-12)$
$= x^3-3x^2+4+2x^3-x^2+12$
$= 3x^3-4x^2+16$

N3. $f(x) = x^2-4,\ g(x) = -6x^2$

(a) $(f+g)(x)$
$= f(x)+g(x)$
$= (x^2-4)+(-6x^2)$
$= -5x^2-4$

(b) $(f-g)(x)$
$= f(x)-g(x)$
$= (x^2-4)-(-6x^2)$
$= 7x^2-4$
$(f-g)(-4)$
$= 7(-4)^2-4$
$= 7(16)-4$
$= 108$

N4. $f(x) = 3x^2-1,\quad g(x) = 8x+7$
$(fg)(x) = f(x)\cdot g(x)$
$= (3x^2-1)(8x+7)$
$= 24x^3+21x^2-8x-7$

Using the last result,
$(fg)(-2)$
$= 24(-2)^3+21(-2)^2-8(-2)-7$
$= -192+84+16-7$
$= -99.$

N5. $f(x) = 8x^2+2x-3,\ g(x) = 2x-1$

$$\left(\frac{f}{g}\right)(x) = \frac{f(x)}{g(x)} = \frac{8x^2+2x-3}{2x-1}$$

$$\begin{array}{r|l} & 4x + 3 \\ 2x-1 & 8x^2+2x-3 \\ & \underline{8x^2-4x} \\ & 6x-3 \\ & \underline{6x-3} \\ & 0 \end{array}$$

First step: $\dfrac{8x^2}{2x} = 4x$

We conclude that $\left(\frac{f}{g}\right)(x) = 4x+3$, provided the denominator, $2x-1$, is *not* equal to zero; that is, $x \neq \frac{1}{2}$.

Using the above result, $\left(\frac{f}{g}\right)(x) = 4x+3$, we have

$$\left(\frac{f}{g}\right)(8) = 32+3 = 35.$$

N6. $f(x) = 3x+7$ and $g(x) = x-2$

$(f \circ g)(7) = f(g(7))$ *Definition*
$= f(7-2)$ *g(x) = x − 2*
$= f(5)$ *Subtract.*
$= 3(5)+7$ *f(x) = 3x + 7*
$= 22$

N7. $f(x) = x-5$ and $g(x) = -x^2+2$

(a) $(g \circ f)(-1) = g(f(-1))$ *Definition*
$= g(-1-5)$ *f(x) = x − 5*
$= f(-6)$ *Subtract.*
$= -(-6)^2+2$ *g(x) = −x² + 2*
$= -34$

(b) $(f \circ g)(x) = f(g(x))$ *Definition*
$= f(-x^2+2)$ *g(x) = −x² + 2*
$= (-x^2+2)-5$ *f(x) = x − 5*
$= -x^2-3$

7.5 Section Exercises

1. $f(x) = 6x-4$

(a) $f(-1) = 6(-1)-4$
$= -6-4 = -10$

(b) $f(2) = 6(2)-4$
$= 12-4 = 8$

3. $f(x) = x^2-3x+4$

(a) $f(-1) = (-1)^2-3(-1)+4$
$= 1+3+4 = 8$

(b) $f(2) = (2)^2-3(2)+4$
$= 4-6+4 = 2$

5. $f(x) = 5x^4 - 3x^2 + 6$

(a) $f(-1) = 5(-1)^4 - 3(-1)^2 + 6$
$= 5 \cdot 1 - 3 \cdot 1 + 6$
$= 5 - 3 + 6 = 8$

(b) $f(2) = 5(2)^4 - 3(2)^2 + 6$
$= 5 \cdot 16 - 3 \cdot 4 + 6$
$= 80 - 12 + 6 = 74$

7. $f(x) = -x^2 + 2x^3 - 8$

(a) $f(-1) = -(-1)^2 + 2(-1)^3 - 8$
$= -(1) + 2(-1) - 8$
$= -1 - 2 - 8 = -11$

(b) $f(2) = -(2)^2 + 2(2)^3 - 8$
$= -(4) + 2 \cdot 8 - 8$
$= -4 + 16 - 8 = 4$

9. (a) $(f+g)(x) = f(x) + g(x)$
$= (5x - 10) + (3x + 7)$
$= 8x - 3$

(b) $(f-g)(x) = f(x) - g(x)$
$= (5x - 10) - (3x + 7)$
$= (5x - 10) + (-3x - 7)$
$= 2x - 17$

11. (a) $(f+g)(x)$
$= f(x) + g(x)$
$= (4x^2 + 8x - 3) + (-5x^2 + 4x - 9)$
$= -x^2 + 12x - 12$

(b) $(f-g)(x)$
$= f(x) - g(x)$
$= (4x^2 + 8x - 3) - (-5x^2 + 4x - 9)$
$= (4x^2 + 8x - 3) + (5x^2 - 4x + 9)$
$= 9x^2 + 4x + 6$

For Exercises 13–28, let $f(x) = x^2 - 9$, $g(x) = 2x$, and $h(x) = x - 3$.

13. $(f+g)(x) = f(x) + g(x)$
$= (x^2 - 9) + (2x)$
$= x^2 + 2x - 9$

15. $(f+g)(3) = f(3) + g(3)$
$= (3^2 - 9) + 2(3)$
$= 0 + 6 = 6$

Alternatively, we could evaluate the polynomial in Exercise 13, $x^2 + 2x - 9$, using $x = 3$.

17. $(f-h)(x) = f(x) - h(x)$
$= (x^2 - 9) - (x - 3)$
$= x^2 - 9 - x + 3$
$= x^2 - x - 6$

19. $(f-h)(-3) = f(-3) - h(-3)$
$= [(-3)^2 - 9] - [(-3) - 3]$
$= (9 - 9) - (-6)$
$= 0 + 6 = 6$

21. $(g+h)(-10) = g(-10) + h(-10)$
$= 2(-10) + [(-10) - 3]$
$= -20 + (-13)$
$= -33$

23. $(g-h)(-3) = g(-3) - h(-3)$
$= 2(-3) - [(-3) - 3]$
$= -6 - (-6)$
$= -6 + 6 = 0$

25. $(g+h)(\frac{1}{4}) = g(\frac{1}{4}) + h(\frac{1}{4})$
$= 2(\frac{1}{4}) + (\frac{1}{4} - 3)$
$= \frac{1}{2} + (-\frac{11}{4})$
$= -\frac{9}{4}$

27. $(g+h)(-\frac{1}{2}) = g(-\frac{1}{2}) + h(-\frac{1}{2})$
$= 2(-\frac{1}{2}) + (-\frac{1}{2} - 3)$
$= -1 + (-\frac{7}{2})$
$= -\frac{9}{2}$

29. Answers will vary. Let $f(x) = x^3$ and $g(x) = x^4$.

$(f-g)(x) = f(x) - g(x)$
$= x^3 - x^4$

$(g-f)(x) = g(x) - f(x)$
$= x^4 - x^3$

Because the two differences are not equal, subtraction of polynomial functions is not commutative.

31. $f(x) = 2x$, $g(x) = 5x - 1$

$(fg)(x) = f(x) \cdot g(x)$
$= 2x(5x - 1)$
$= 10x^2 - 2x$

33. $f(x) = x + 1$, $g(x) = 2x - 3$

$(fg)(x) = f(x) \cdot g(x)$
$= (x + 1)(2x - 3)$
$= 2x^2 - 3x + 2x - 3$
$= 2x^2 - x - 3$

35. $f(x) = 2x - 3$, $g(x) = 4x^2 + 6x + 9$

$(fg)(x) = f(x) \cdot g(x)$
$= (2x - 3)(4x^2 + 6x + 9)$

Multiply vertically.

$$\begin{array}{r} 4x^2 + 6x + 9 \\ 2x - 3 \\ \hline -12x^2 - 18x - 27 \\ 8x^3 + 12x^2 + 18x \\ \hline 8x^3 \qquad\qquad - 27 \end{array}$$

For Exercises 37–48, let $f(x) = x^2 - 9$, $g(x) = 2x$, and $h(x) = x - 3$.

37. $$\begin{aligned}(fg)(x) &= f(x)\cdot g(x)\\ &= (x^2-9)(2x)\\ &= 2x^3 - 18x\end{aligned}$$

39. $$\begin{aligned}(fg)(2) &= f(2)\cdot g(2)\\ &= (2^2-9)[2(2)]\\ &= -5\cdot 4 = -20\end{aligned}$$

41. $$\begin{aligned}(gh)(x) &= g(x)\cdot h(x)\\ &= (2x)(x-3)\\ &= 2x^2 - 6x\end{aligned}$$

43. $$\begin{aligned}(gh)(-3) &= g(-3)\cdot h(-3)\\ &= [2(-3)]\cdot[(-3)-3]\\ &= -6(-6) = 36\end{aligned}$$

45. $$\begin{aligned}(fg)(-\tfrac{1}{2}) &= f(-\tfrac{1}{2})\cdot g(-\tfrac{1}{2})\\ &= \left[(-\tfrac{1}{2})^2 - 9\right]\cdot\left[2(-\tfrac{1}{2})\right]\\ &= (\tfrac{1}{4} - \tfrac{36}{4})(-1)\\ &= -\tfrac{35}{4}(-1) = \tfrac{35}{4}\end{aligned}$$

47. $$\begin{aligned}(fh)(-\tfrac{1}{4}) &= f(-\tfrac{1}{4})\cdot h(-\tfrac{1}{4})\\ &= \left[(-\tfrac{1}{4})^2 - 9\right]\cdot\left[-\tfrac{1}{4} - 3\right]\\ &= (\tfrac{1}{16} - \tfrac{144}{16})(-\tfrac{1}{4} - \tfrac{12}{4})\\ &= -\tfrac{143}{16}(-\tfrac{13}{4}) = \tfrac{1859}{64}\end{aligned}$$

49. $$\begin{aligned}\left(\frac{f}{g}\right)(x) &= \frac{f(x)}{g(x)} = \frac{10x^2 - 2x}{2x}\\ &= \frac{10x^2}{2x} - \frac{2x}{2x}\\ &= 5x - 1\end{aligned}$$

The x-values that are not in the domain of the quotient function are found by solving $g(x) = 0$.

$$\begin{aligned}2x &= 0\\ x &= 0\end{aligned}$$

Exclude $x = 0$ from the domain.

51.
$$\begin{array}{r} 2x - 3 \\ x + 1 \overline{)\,2x^2 - x - 3} \\ \underline{2x^2 + 2x} \\ -3x - 3 \\ \underline{-3x - 3} \\ 0 \end{array}$$

Quotient: $2x - 3$

$$\begin{aligned}g(x) &= 0\\ x + 1 &= 0\\ x &= -1\end{aligned}$$

Exclude $x = -1$ from the domain.

53.
$$\begin{array}{r} 4x^2 + 6x + 9 \\ 2x - 3 \overline{)\,8x^3 + 0x^2 + 0x - 27} \\ \underline{8x^3 - 12x^2} \\ 12x^2 + 0x \\ \underline{12x^2 - 18x} \\ 18x - 27 \\ \underline{18x - 27} \\ 0 \end{array}$$

Quotient: $4x^2 + 6x + 9$

$$\begin{aligned}g(x) &= 0\\ 2x - 3 &= 0\\ 2x &= 3\\ x &= \tfrac{3}{2}\end{aligned}$$

Exclude $x = \frac{3}{2}$ from the domain.

For Exercises 55–66, let $f(x) = x^2 - 9$, $g(x) = 2x$, and $h(x) = x - 3$.

55. $$\left(\frac{f}{g}\right)(x) = \frac{f(x)}{g(x)} = \frac{x^2 - 9}{2x}$$

We must exclude any values of x that make the denominator equal to zero, that is, $x \neq 0$.

57. $$\left(\frac{f}{g}\right)(2) = \frac{f(2)}{g(2)} = \frac{2^2 - 9}{2(2)} = \frac{-5}{4} = -\frac{5}{4}$$

59. $$\left(\frac{h}{g}\right)(x) = \frac{h(x)}{g(x)} = \frac{x - 3}{2x},\ x \neq 0$$

61. $$\left(\frac{h}{g}\right)(3) = \frac{h(3)}{g(3)} = \frac{(3) - 3}{2(3)} = \frac{0}{6} = 0$$

63. $$\begin{aligned}\left(\frac{f}{g}\right)\left(\frac{1}{2}\right) &= \frac{f(\frac{1}{2})}{g(\frac{1}{2})} = \frac{(\frac{1}{2})^2 - 9}{2(\frac{1}{2})} = \frac{\frac{1}{4} - \frac{36}{4}}{1}\\ &= -\tfrac{35}{4}\end{aligned}$$

65. $$\left(\frac{h}{g}\right)\left(-\frac{1}{2}\right) = \frac{h(-\frac{1}{2})}{g(-\frac{1}{2})} = \frac{-\frac{1}{2} - 3}{2(-\frac{1}{2})} = \frac{-\frac{7}{2}}{-1} = \frac{7}{2}$$

For Exercises 67–82,
$f(x) = x^2 + 4$, $g(x) = 2x + 3$, and $h(x) = x - 5$.

67. $$\begin{aligned}(h \circ g)(4) &= h(g(4))\\ &= h(2\cdot 4 + 3)\\ &= h(11)\\ &= 11 - 5\\ &= 6\end{aligned}$$

69. $(g \circ f)(6) = g(f(6))$
$= g(6^2 + 4)$
$= g(40)$
$= 2 \cdot 40 + 3$
$= 83$

71. $(f \circ h)(-2) = f(h(-2))$
$= f(-2 - 5)$
$= f(-7)$
$= (-7)^2 + 4$
$= 53$

73. $(f \circ g)(0) = f(g(0))$
$= f(2(0) + 3)$
$= f(3)$
$= 3^2 + 4$
$= 13$

75. $(g \circ f)(x) = g(f(x))$
$= g(x^2 + 4)$
$= 2(x^2 + 4) + 3$
$= 2x^2 + 8 + 3$
$= 2x^2 + 11$

77. $(h \circ g)(x) = h(g(x))$
$= h(2x + 3)$
$= 2x + 3 - 5$
$= 2x - 2$

79. $(f \circ h)(\frac{1}{2}) = f(h(\frac{1}{2}))$
$= f(\frac{1}{2} - 5)$
$= f(-\frac{9}{2})$
$= (-\frac{9}{2})^2 + 4$
$= \frac{81}{4} + \frac{16}{4} = \frac{97}{4}$

81. $(f \circ g)(-\frac{1}{2}) = f(g(-\frac{1}{2}))$
$= f[2(-\frac{1}{2}) + 3]$
$= f(2)$
$= 2^2 + 4$
$= 4 + 4 = 8$

83. $f(x) = 12x,\ g(x) = 5280x$
$(f \circ g)(x) = f(g(x))$
$= f(5280x)$
$= 12(5280x)$
$= 63{,}360x$

$(f \circ g)(x)$ computes the number of inches in x mi.

85. $r(t) = 2t,\ \mathcal{A}(r) = \pi r^2$
$(\mathcal{A} \circ r)(t) = \mathcal{A}(r(t))$
$= \mathcal{A}(2t)$
$= \pi(2t)^2$
$= 4\pi t^2$

This is the area of the circular layer as a function of time.

87. $y = kx$
$1 = k(3)$ *Let y = 1, x = 3.*
$\frac{1}{3} = k$

89. $y = \frac{k}{x}$
$1 = \frac{k}{3}$ *Let y = 1, x = 3.*
$1(3) = k$
$3 = k$

7.6 Variation

7.6 Now Try Exercises

N1. Let C represent the cost of d dozen eggs. C varies directly as d, so

$$C = kd,$$

where k represents the cost of one dozen eggs. Since $C = 20$ when $d = 8$,

$$20 = k \cdot 8$$
$$k = \tfrac{20}{8} = 2.50$$

One dozen eggs costs \$2.50, and C and d are related by

$$C = 2.50d.$$

N2. Let $\mathcal{A}$ represent the area of a parallelogram with base b and constant height. Use $\mathcal{A} = kb$ with $\mathcal{A} = 20$ and $b = 4$ to find k.

$$\mathcal{A} = kb$$
$$20 = k(4)$$
$$\tfrac{20}{4} = k$$
$$5 = k$$

So $\mathcal{A} = 5b$.

Let $b = 7$. Find $\mathcal{A}$.

$$\mathcal{A} = 5(7) = 35$$

The area is 35 cm^2.

N3. y varies directly as the square of x, so

$$y = kx^2.$$

$y = 200$ when $x = 5$, so

$$200 = k(5)^2.$$
$$200 = k(25)$$
$$8 = k$$

Thus, $y = 8x^2$. When $x = 7$,

$$y = 8(7)^2 = 8(49) = 392.$$

N4. Let h represent the height of a triangle with base b and constant area. h varies inversely as b, so

$$h = \frac{k}{b},$$

for some constant k. Since $h = 7$ when $b = 8$,

$$7 = \frac{k}{8}.$$
$$56 = k$$

Thus, $h = \frac{56}{b}$. When $b = 14$,

$$h = \tfrac{56}{14} = 4.$$

The height is 4 centimeters.

N5. Let w represent the weight of an object above Earth at a distance d from the center of Earth. w varies inversely as the square of d, so

$$w = \frac{k}{d^2}.$$

$w = 150$ when $d = 3960$, so

$$150 = \frac{k}{3960^2}.$$
$$150(3960^2) = k$$

Thus, $w = \dfrac{150(3960^2)}{d^2}$.

When $d = 3960 + 1000 = 4960$,

$$w = \frac{150(3960^2)}{4960^2} \approx 95.6.$$

The object weighs about 96 pounds.

N6. Let V represent the volume of a right pyramid with height h and base area $\mathcal{A}$. V varies jointly as h and $\mathcal{A}$, so

$$V = kh\mathcal{A}.$$

$V = 100$ when $h = 10$ and $\mathcal{A} = 30$, so

$$100 = k(10)(30).$$
$$100 = k(300)$$
$$k = \tfrac{100}{300} = \tfrac{1}{3}$$

Thus, $V = \frac{1}{3}h\mathcal{A}$. When $h = 20$ and $\mathcal{A} = 90$,

$$V = \tfrac{1}{3}(20)(90) = 20(30) = 600.$$

The volume is 600 cubic feet.

N7. Let S represent the sample size, v the variance, and e the maximum error of the estimate. S varies directly as v and inversely as e^2, so

$$S = \frac{kv}{e^2}.$$

$S = 200$ when $v = 25$ and $e = 0.5$, so

$$200 = \frac{k(25)}{(0.5)^2}$$
$$200 = 100k$$
$$k = \tfrac{200}{100} = 2$$

Thus, $S = \dfrac{2v}{e^2}$. When $v = 25$ and $e = 0.1$,

$$S = \frac{2(25)}{(0.1)^2} = 5000.$$

7.6 Section Exercises

1. This suggests *direct* variation.
As the number of tickets you buy increases, so does the probability that you will win.

3. This suggests *direct* variation.
As the pressure put on the accelerator increases, so does the speed of the car.

5. This suggests *inverse* variation.
As your age gets larger, the probability that you believe in Santa Claus gets smaller.

7. This suggests *inverse* variation.
As the number of days gets smaller, the number of home runs increases.

9. The equation $y = \frac{3}{x}$ represents *inverse* variation.
y varies inversely as x because x is in the denominator.

11. The equation $y = 10x^2$ represents *direct* variation. The number 10 is the constant of variation, and y varies directly as the square of x.

13. The equation $y = 3xz^4$ represents *joint* variation. y varies directly as x and z^4.

15. The equation $y = \dfrac{4x}{wz}$ represents *combined* variation. In the numerator, 4 is the constant of variation, and y varies directly as x. In the denominator, y varies inversely as w and z.

17. For $k > 0$, if y varies directly as x (then $y = kx$), when x increases, y *increases*, and when x decreases, y *decreases*.

19. $P = 4s$ The perimeter of a square varies directly as the length of its side.

21. $S = 4\pi r^2$ The surface area of a sphere varies directly as the square of its radius.

23. $\mathcal{A} = \frac{1}{2}bh$ The area of a triangle varies jointly as the length of its base and its height.

25. For $P = 4s$, $d = 2r$, $S = 4\pi r^2$, $V = \frac{4}{3}\pi r^3$, $\mathcal{A} = \frac{1}{2}bh$, and $V = \frac{1}{3}\pi r^2 h$, the constants of variation are 4, 2, 4π, $\frac{4}{3}\pi$, $\frac{1}{2}$, and $\frac{1}{3}\pi$, respectively.

27. "x varies directly as y" means $x = ky$ for some constant k. Let $x = 9$ and $y = 3$ to find k.

$$x = ky$$
$$9 = k(3)$$
$$k = \tfrac{9}{3} = 3$$

So $x = 3y$. To find x when $y = 12$, substitute 12 for y in the equation.

$$x = 3y$$
$$x = 3(12)$$
$$x = 36$$

29. "a varies directly as the square of b" means

$$a = kb^2$$

for some constant k. Substitute $a = 4$ and $b = 3$ in the equation and solve for k.

$$a = kb^2$$
$$4 = k(3)^2$$
$$k = \tfrac{4}{9}$$

So $a = \frac{4}{9}b^2$. To find a when $b = 2$, substitute 2 for b in the equation.

$$a = \tfrac{4}{9}b^2$$
$$a = \tfrac{4}{9}(2)^2$$
$$a = \frac{4 \cdot 4}{9} = \frac{16}{9}$$

31. "z varies inversely as w" means

$$z = \frac{k}{w}$$

for some constant k. Since $z = 10$ when $w = 0.5$, substitute these values in the equation and solve for k.

$$z = \frac{k}{w}$$
$$10 = \frac{k}{0.5}$$
$$k = 10(0.5) = 5$$

So $z = \frac{5}{w}$. To find z when $w = 8$, substitute 8 for w in the equation.

$$z = \tfrac{5}{w}$$
$$z = \tfrac{5}{8} \text{ or } 0.625$$

33. "m varies inversely as p^2" means

$$m = \frac{k}{p^2}$$

for some constant k. Since $m = 20$ when $p = 2$, substitute these values in the equation and solve for k.

$$m = \frac{k}{p^2}$$
$$20 = \frac{k}{2^2}$$
$$k = 20(4) = 80$$

So $m = \dfrac{80}{p^2}$. Now let $p = 5$.

$$m = \frac{80}{p^2}$$
$$m = \frac{80}{5^2} = \frac{16}{5}$$

35. "p varies jointly as q and r^2" means

$$p = kqr^2$$

for some constant k. Given that $p = 200$ when $q = 2$ and $r = 3$, solve for k.

$$p = kqr^2$$
$$200 = k(2)(3)^2$$
$$200 = 18k$$
$$k = \tfrac{200}{18} = \tfrac{100}{9}$$

So $p = \frac{100}{9}qr^2$. Now let $q = 5$ and $r = 2$.

$$p = \tfrac{100}{9}qr^2$$
$$p = \tfrac{100}{9}(5)(2)^2$$
$$= \tfrac{100}{9}(20)$$
$$= \tfrac{2000}{9} \text{ or } 222\tfrac{2}{9}$$

37. Let $x =$ the number of gallons he bought and let $C =$ the cost.
C varies directly as x, so

$$C = kx.$$

Since $C = 43.79$ when $x = 15$,

$$43.79 = k(15)$$
$$k = \tfrac{43.79}{15} \approx 2.919.$$

The price per gallon is \$2.919, or \$2.91$\frac{9}{10}$.

39. Let $y =$ the weight of an object on earth and $x =$ the weight of the object on the moon.
y varies directly as x, so

$$y = kx$$

for some constant k. Since $y = 200$ when $x = 32$, substitute these values in the equation and solve for k.

$$y = kx$$
$$200 = k(32)$$
$$k = \tfrac{200}{32} = 6.25$$

So $y = 6.25x$. To find x when $y = 50$, substitute 50 for y in the equation

$$y = 6.25x.$$
$$50 = 6.25x$$
$$x = \tfrac{50}{6.25} = 8$$

The dog would weigh 8 lb on the moon.

41. Let $V =$ the volume of the can and let $h =$ the height of the can. V varies directly as h, so

$$V = kh.$$

Since $V = 300$ when $h = 10.62$,

$$300 = k(10.62)$$
$$k = \tfrac{300}{10.62} \approx 28.25.$$

So $V = 28.25h$. Now let $h = 15.92$.

$$V = 28.25h$$
$$V = 28.25(15.92) = 449.74$$

The volume is about 450 cubic centimeters.

43. Let d = the distance and t = the time. d varies directly as the square of t, so $d = kt^2$. Let $d = -576$ and $t = 6$. (You could also use $d = 576$, but the negative sign indicates the direction of the body.)

$$-576 = k(6)^2$$
$$-576 = 36k$$
$$-16 = k$$

So $d = -16t^2$. Now let $t = 4$.

$$d = -16(4)^2 = -256$$

The object fell 256 feet in the first 4 seconds.

45. Let s = the rate and t = the time. The rate varies inversely with time, so there is a constant k such that $s = k/t$. Find the value of k by replacing s with 160 and t with $\frac{1}{2}$.

$$s = \frac{k}{t}$$
$$160 = \frac{k}{\frac{1}{2}}$$
$$k = 160(\tfrac{1}{2}) = 80$$

So $s = \frac{80}{t}$. Now let $t = \frac{3}{4}$.

$$s = \frac{80}{\frac{3}{4}}$$
$$s = \frac{80}{1} \cdot \frac{4}{3} = \frac{320}{3} \quad \text{or} \quad 106\frac{2}{3}$$

A rate of $106\frac{2}{3}$ miles per hour is needed to go the same distance in $\frac{3}{4}$ minute.

47. Let f = the frequency of a string in cycles per second and s = the length in feet. f varies inversely as s, so

$$f = \frac{k}{s}$$

for some constant k. Since $f = 250$ when $s = 2$, substitute these values in the equation and solve for k.

$$f = \frac{k}{s}$$
$$250 = \frac{k}{2}$$
$$k = 250(2) = 500$$

So $f = \dfrac{500}{s}$. Now let $s = 5$.

$$f = \tfrac{500}{5} = 100.$$

The string would have a frequency of 100 cycles per second.

49. Let I = the illumination produced by a light source and d = the distance from the source. I varies inversely as d^2, so

$$I = \frac{k}{d^2}$$

for some constant k. Since $I = 768$ when $d = 1$, substitute these values in the equation and solve for k.

$$I = \frac{k}{d^2}$$
$$768 = \frac{k}{1^2}$$
$$768 = k$$

So $I = \dfrac{768}{d^2}$. Now let $d = 6$.

$$I = \frac{768}{d^2}$$
$$I = \frac{768}{6^2} = \frac{768}{36} = \frac{64}{3} \quad \text{or} \quad 21\frac{1}{3}$$

The illumination produced by the light source is $21\frac{1}{3}$ foot-candles.

51. Let I = the simple interest, P the principal, and t the time.
Since I varies jointly as the principal and time, there is a constant k such that $I = kPt$. Find k by replacing I with 280, P with 2000, and t with 4.

$$I = kPt$$
$$280 = k(2000)(4)$$
$$k = \tfrac{280}{8000} = 0.035$$

So $I = 0.035Pt$. Now let $t = 6$.

$$I = 0.035(2000)(6)$$
$$= 420$$

The interest would be \$420.

53. The weight W of a bass varies jointly as its girth G and the square of its length L, so

$$W = kGL^2$$

for some constant k. Substitute 22.7 for W, 21 for G, and 36 for L.

$$22.7 = k(21)(36)^2$$
$$k = \frac{22.7}{27{,}216} \approx 0.000834$$

So $W = 0.000834GL^2$. Now let $G = 18$ and $L = 28$.

$$W = 0.000834GL^2$$
$$= 0.000834(18)(28)^2$$
$$\approx 11.8$$

The bass would weigh about 11.8 pounds.

55. Let F = the force, w = the weight of the car, s = the speed, and r = the radius.
The force varies inversely as the radius and jointly as the weight and the square of the speed, so

$$F = \frac{kws^2}{r}.$$

Let $F = 242$, $w = 2000$, $r = 500$, and $s = 30$.

$$242 = \frac{k(2000)(30)^2}{500}$$
$$k = \frac{242(500)}{2000(900)} = \frac{121}{1800}$$

So $F = \dfrac{121ws^2}{1800r}$.
Let $r = 750$, $s = 50$, and $w = 2000$.

$$F = \frac{121(2000)(50)^2}{1800(750)} \approx 448.1$$

Approximately 448.1 pounds of force would be needed.

57. Let N = the number of long distance calls,
p_1 = the population of City 1,
p_2 = the population of City 2,
and d = the distance between them.

$$N = \frac{kp_1p_2}{d}$$

Let $N = 80{,}000$, $p_1 = 70{,}000$, $p_2 = 100{,}000$, and $d = 400$.

$$80{,}000 = \frac{k(70{,}000)(100{,}000)}{400}$$
$$80{,}000 = 17{,}500{,}000k$$
$$k = \frac{80{,}000}{17{,}500{,}000} = \frac{4}{875}$$

So $N = \dfrac{4}{875}\left(\dfrac{p_1p_2}{d}\right)$.

Let $p_1 = 50{,}000$, $p_2 = 75{,}000$, and $d = 250$.

$$N = \frac{4}{875}\left(\frac{50{,}000 \cdot 75{,}000}{250}\right)$$
$$= \frac{480{,}000}{7} = 68{,}571\tfrac{3}{7}$$

Rounded to the nearest hundred, there are about 68,600 calls.

59. Use the BMI from Example 7, with $k = 694$.

$$B = \frac{694w}{h^2}$$

Substitute your weight in pounds for w and your height in inches for h to determine your BMI. Answers will vary. A BMI from 19 to 25 is considered desirable.

61. $2x + 3y = 12$

If $x = 0$, $y = 4$. If $y = 0$, $x = 6$. Graph a line through the intercepts, $(0, 4)$ and $(6, 0)$.

$4x - 2y = 8$

If $x = 0$, $y = -4$. If $y = 0$, $x = 2$. Graph a line through the intercepts, $(0, -4)$ and $(2, 0)$.

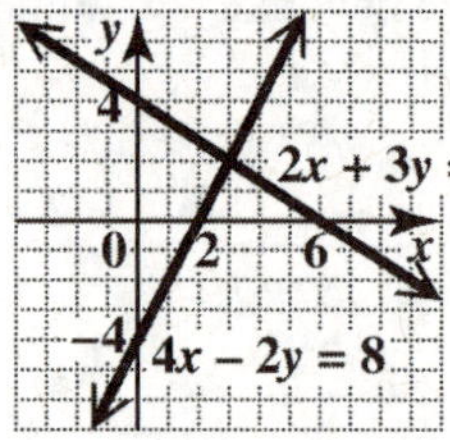

63. $-5x + 2y = 10$

If $x = 0$, $y = 5$. If $y = 0$, $x = -2$. Graph a line through the intercepts, $(0, 5)$ and $(-2, 0)$.

$2y = -4 + 5x$

If $x = 0$, $y = -2$. If $y = 0$, $x = \frac{4}{5}$. Graph a line through the intercepts, $(0, -2)$ and $(\frac{4}{5}, 0)$.

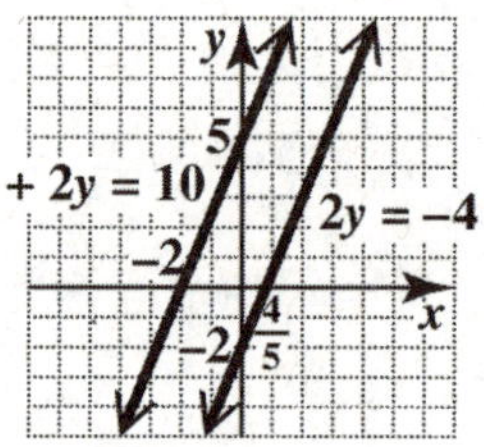

Chapter 7 Review Exercises

1. $3x + 2y = 10$

For $x = 0$:

$$3(0) + 2y = 10$$
$$2y = 10$$
$$y = 5 \quad \mathbf{(0, 5)}$$

For $y = 0$:

$$3x + 2(0) = 10$$
$$3x = 10$$
$$x = \tfrac{10}{3} \quad \mathbf{(\tfrac{10}{3}, 0)}$$

For $x = 2$:

$$3(2) + 2y = 10$$
$$6 + 2y = 10$$
$$2y = 4$$
$$y = 2 \quad \mathbf{(2, 2)}$$

For $y = -2$:

$$3x + 2(-2) = 10$$
$$3x - 4 = 10$$
$$3x = 14$$
$$x = \tfrac{14}{3} \quad \mathbf{(\tfrac{14}{3}, -2)}$$

Plot the ordered pairs, and draw the line through them.

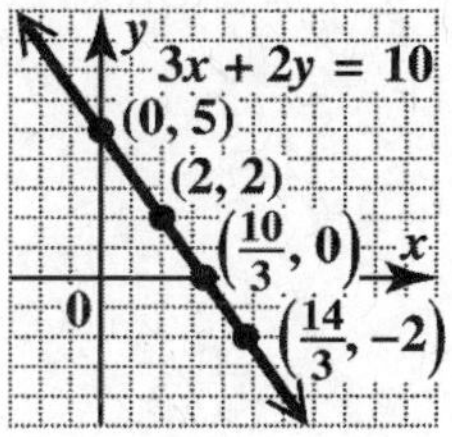

2. $x - y = 8$
For $x = 2$:

$$\begin{aligned} 2 - y &= 8 \\ -y &= 6 \\ y &= -6 \quad \mathbf{(2, -6)} \end{aligned}$$

For $y = -3$:

$$\begin{aligned} x - (-3) &= 8 \\ x + 3 &= 8 \\ x &= 5 \quad \mathbf{(5, -3)} \end{aligned}$$

For $x = 3$:

$$\begin{aligned} 3 - y &= 8 \\ -y &= 5 \\ y &= -5 \quad \mathbf{(3, -5)} \end{aligned}$$

For $y = -2$:

$$\begin{aligned} x - (-2) &= 8 \\ x + 2 &= 8 \\ x &= 6 \quad \mathbf{(6, -2)} \end{aligned}$$

Plot the ordered pairs, and draw the line through them.

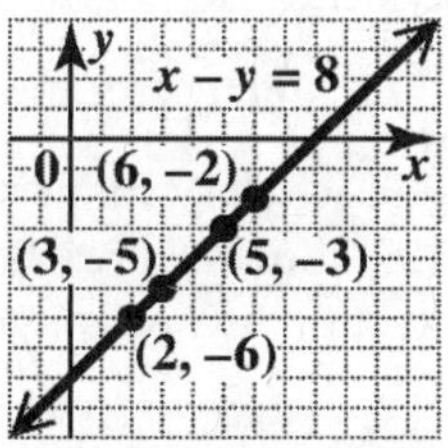

3. $4x - 3y = 12$
To find the x-intercept, let $y = 0$.

$$\begin{aligned} 4x - 3y &= 12 \\ 4x - 3(0) &= 12 \\ 4x &= 12 \\ x &= 3 \end{aligned}$$

The x-intercept is $(3, 0)$.
To find the y-intercept, let $x = 0$.

$$\begin{aligned} 4x - 3y &= 12 \\ 4(0) - 3y &= 12 \\ -3y &= 12 \\ y &= -4 \end{aligned}$$

The y-intercept is $(0, -4)$.
Plot the intercepts and draw the line through them.

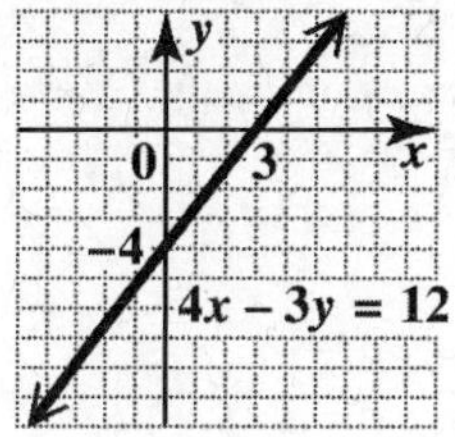

4. $5x + 7y = 28$
To find the x-intercept, let $y = 0$.

$$\begin{aligned} 5x + 7y &= 28 \\ 5x + 7(0) &= 28 \\ 5x &= 28 \\ x &= \tfrac{28}{5} \end{aligned}$$

The x-intercept is $(\frac{28}{5}, 0)$.
To find the y-intercept, let $x = 0$.

$$\begin{aligned} 5x + 7y &= 28 \\ 5(0) + 7y &= 28 \\ 7y &= 28 \\ y &= 4 \end{aligned}$$

The y-intercept is $(0, 4)$.
Plot the intercepts and draw the line through them.

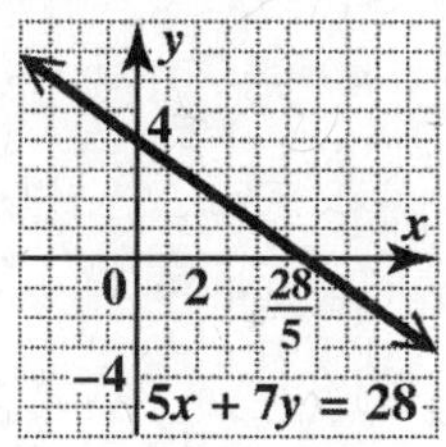

5. $2x + 5y = 20$
To find the x-intercept, let $y = 0$.

$$\begin{aligned} 2x + 5y &= 20 \\ 2x + 5(0) &= 20 \\ 2x &= 20 \\ x &= 10 \end{aligned}$$

The x-intercept is $(10, 0)$.
To find the y-intercept, let $x = 0$.

$$\begin{aligned} 2x + 5y &= 20 \\ 2(0) + 5y &= 20 \\ 5y &= 20 \\ y &= 4 \end{aligned}$$

The y-intercept is $(0, 4)$.
Plot the intercepts and draw the line through them.

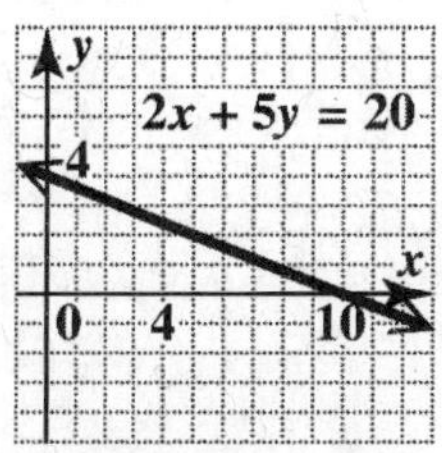

6. $x - 4y = 8$
To find the x-intercept, let $y = 0$.

$$\begin{aligned} x - 4y &= 8 \\ x - 4(0) &= 8 \\ x &= 8 \end{aligned}$$

The x-intercept is $(8, 0)$.
To find the y-intercept, let $x = 0$.

$$\begin{aligned} 0 - 4y &= 8 \\ -4y &= 8 \\ y &= -2 \end{aligned}$$

The y-intercept is $(0, -2)$.
Plot the intercepts and draw the line through them.

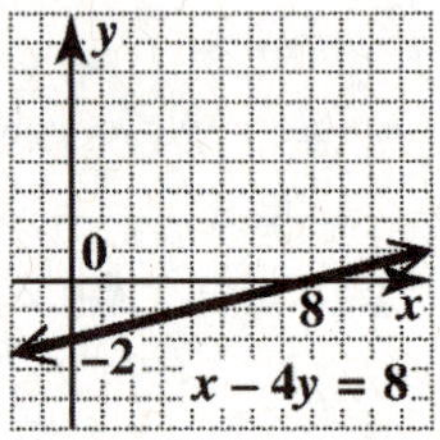

7. By the midpoint formula, the midpoint of the segment with endpoints $(-8, -12)$ and $(8, 16)$ is

$$\left(\frac{-8+8}{2}, \frac{-12+16}{2}\right) = \left(\frac{0}{2}, \frac{4}{2}\right) = (0, 2).$$

8. By the midpoint formula, the midpoint of the segment with endpoints $(0, -5)$ and $(-9, 8)$ is

$$\left(\frac{0+(-9)}{2}, \frac{-5+8}{2}\right) = \left(\frac{-9}{2}, \frac{3}{2}\right) = \left(-\frac{9}{2}, \frac{3}{2}\right).$$

9. Through $(-1, 2)$ and $(4, -5)$

$$m = \frac{\text{change in } y}{\text{change in } x} = \frac{-5-2}{4-(-1)} = \frac{-7}{5} = -\frac{7}{5}$$

10. Through $(0, 3)$ and $(-2, 4)$

Let $(x_1, y_1) = (0, 3)$ and $(x_2, y_2) = (-2, 4)$.

$$m = \frac{y_2 - y_1}{x_2 - x_1} = \frac{4-3}{-2-0} = \frac{1}{-2} = -\frac{1}{2}$$

11. The slope of $y = 2x + 3$ is 2, the coefficient of x.

12. $3x - 4y = 5$
Write the equation in slope-intercept form.

$$\begin{aligned} -4y &= -3x + 5 \\ y &= \tfrac{3}{4}x - \tfrac{5}{4} \end{aligned}$$

The slope is $\frac{3}{4}$.

13. $x = 5$ is a vertical line and has *undefined* slope.

14. Parallel to $3y = 2x + 5$
Write the equation in slope-intercept form.

$$\begin{aligned} 3y &= 2x + 5 \\ y &= \tfrac{2}{3}x + \tfrac{5}{3} \end{aligned}$$

The slope of $3y = 2x + 5$ is $\frac{2}{3}$; all lines parallel to it will also have a slope of $\frac{2}{3}$.

15. Perpendicular to $3x - y = 4$
Solve for y.

$$y = 3x - 4$$

The slope is 3; the slope of a line perpendicular to it is $-\frac{1}{3}$ since

$$3(-\tfrac{1}{3}) = -1.$$

16. Through $(-1, 5)$ and $(-1, -4)$

$$m = \frac{\Delta y}{\Delta x} = \frac{-4-5}{-1-(-1)} = \frac{-9}{0} \quad \textit{Undefined}$$

This is a vertical line; it has undefined slope.

17. Through $(3, -1)$ and $(-3, 1)$

$$m = \frac{\Delta y}{\Delta x} = \frac{1-(-1)}{-3-3} = \frac{2}{-6} = -\frac{1}{3}.$$

18. The x-intercept is $(2, 0)$ and the y-intercept is $(0, 2)$. The slope is

$$m = \frac{\text{change in } y}{\text{change in } x} = \frac{2-0}{0-2} = \frac{2}{-2} = -1.$$

19. The line goes up from left to right, so it has positive slope.

20. The line goes down from left to right, so it has negative slope.

21. The line is vertical, so it has *undefined* slope.

22. The line is horizontal, so it has 0 slope.

23. To rise 1 foot, we must move 4 feet in the horizontal direction. To rise 3 feet, we must move $3(4) = 12$ feet in the horizontal direction.

24. Let $(x_1, y_1) = (1980, 21{,}000)$ and $(x_2, y_2) = (2007, 61{,}400)$. Then

$$m = \frac{\Delta y}{\Delta x} = \frac{61{,}400 - 21{,}000}{2007 - 1980} = \frac{40{,}400}{27} \approx 1496.$$

The average rate of change is \$1496 per year (to the nearest dollar).

25. (a) Slope $-\frac{1}{3}$, y-intercept $(0, -1)$
Use the slope-intercept form with $m = -\frac{1}{3}$ and $b = -1$.

$$\begin{aligned} y &= mx + b \\ y &= -\tfrac{1}{3}x - 1 \end{aligned}$$

(b)

$$\begin{aligned} y &= -\tfrac{1}{3}x - 1 \\ 3y &= -x - 3 \\ x + 3y &= -3 \end{aligned}$$

26. **(a)** Slope 0, y-intercept $(0, -2)$
Use the slope-intercept form with $m = 0$ and $b = -2$.

$$\begin{aligned} y &= mx + b \\ y &= (0)x - 2 \\ y &= -2 \end{aligned}$$

(b) $y = -2$ is already in standard form.

27. **(a)** Slope $-\frac{4}{3}$, through $(2, 7)$
Use the point-slope form with $m = -\frac{4}{3}$ and $(x_1, y_1) = (2, 7)$.

$$\begin{aligned} y - y_1 &= m(x - x_1) \\ y - 7 &= -\tfrac{4}{3}(x - 2) \\ y - 7 &= -\tfrac{4}{3}x + \tfrac{8}{3} \\ y &= -\tfrac{4}{3}x + \tfrac{29}{3} \end{aligned}$$

(b)
$$\begin{aligned} y &= -\tfrac{4}{3}x + \tfrac{29}{3} \\ 3y &= -4x + 29 \\ 4x + 3y &= 29 \end{aligned}$$

28. **(a)** Slope 3, through $(-1, 4)$
Use the point-slope form with $m = 3$ and $(x_1, y_1) = (-1, 4)$.

$$\begin{aligned} y - y_1 &= m(x - x_1) \\ y - 4 &= 3[x - (-1)] \\ y - 4 &= 3(x + 1) \\ y - 4 &= 3x + 3 \\ y &= 3x + 7 \end{aligned}$$

(b)
$$\begin{aligned} y &= 3x + 7 \\ -3x + y &= 7 \\ 3x - y &= -7 \end{aligned}$$

29. **(a)** Vertical, through $(2, 5)$
The equation of any vertical line is in the form $x = k$. Since the line goes through $(2, 5)$, the equation is $x = 2$. (Slope-intercept form is not possible.)

(b) $x = 2$ is already in standard form.

30. **(a)** Through $(2, -5)$ and $(1, 4)$
Find the slope.

$$m = \frac{\Delta y}{\Delta x} = \frac{4 - (-5)}{1 - 2} = \frac{9}{-1} = -9$$

Use the point-slope form with $m = -9$ and $(x_1, y_1) = (2, -5)$.

$$\begin{aligned} y - y_1 &= m(x - x_1) \\ y - (-5) &= -9(x - 2) \\ y + 5 &= -9x + 18 \\ y &= -9x + 13 \end{aligned}$$

(b)
$$\begin{aligned} y &= -9x + 13 \\ 9x + y &= 13 \end{aligned}$$

31. **(a)** Through $(-3, -1)$ and $(2, 6)$
Find the slope.

$$m = \frac{\Delta y}{\Delta x} = \frac{6 - (-1)}{2 - (-3)} = \frac{7}{5}$$

Use the point-slope form with $m = \frac{7}{5}$ and $(x_1, y_1) = (2, 6)$.

$$\begin{aligned} y - y_1 &= m(x - x_1) \\ y - 6 &= \tfrac{7}{5}(x - 2) \\ y - 6 &= \tfrac{7}{5}x - \tfrac{14}{5} \\ y &= \tfrac{7}{5}x + \tfrac{16}{5} \end{aligned}$$

(b)
$$\begin{aligned} y &= \tfrac{7}{5}x + \tfrac{16}{5} \\ 5y &= 7x + 16 \\ -7x + 5y &= 16 \\ 7x - 5y &= -16 \end{aligned}$$

32. **(a)** From Exercise 18, we have $m = -1$ and a y-intercept of $(0, 2)$. The slope-intercept form is

$$y = -1x + 2 \quad \text{or} \quad y = -x + 2.$$

(b)
$$\begin{aligned} y &= -x + 2 \\ x + y &= 2 \end{aligned}$$

33. **(a)** Parallel to $4x - y = 3$ and through $(7, -1)$
Writing $4x - y = 3$ in slope-intercept form gives us $y = 4x - 3$, which has slope 4. Lines parallel to it will also have slope 4. The line with slope 4 through $(7, -1)$ is:

$$\begin{aligned} y - y_1 &= m(x - x_1) \\ y - (-1) &= 4(x - 7) \\ y + 1 &= 4x - 28 \\ y &= 4x - 29 \end{aligned}$$

(b)
$$\begin{aligned} y &= 4x - 29 \\ -4x + y &= -29 \\ 4x - y &= 29 \end{aligned}$$

34. **(a)** Perpendicular to $2x - 5y = 7$ and through $(4, 3)$
Write the equation in slope-intercept form.

$$\begin{aligned} 2x - 5y &= 7 \\ -5y &= -2x + 7 \\ y &= \tfrac{2}{5}x - \tfrac{7}{5} \end{aligned}$$

$y = \frac{2}{5}x - \frac{7}{5}$ has slope $\frac{2}{5}$ and is perpendicular to lines with slope $-\frac{5}{2}$.
The line with slope $-\frac{5}{2}$ through $(4, 3)$ is

$$\begin{aligned} y - y_1 &= m(x - x_1) \\ y - 3 &= -\tfrac{5}{2}(x - 4) \\ y - 3 &= -\tfrac{5}{2}x + 10 \\ y &= -\tfrac{5}{2}x + 13 \end{aligned}$$

(b)
$$\begin{aligned} y &= -\tfrac{5}{2}x + 13 \\ 2y &= -5x + 26 \\ 5x + 2y &= 26 \end{aligned}$$

35. **(a)** The fixed cost is \$159, so that is the value of b. The variable cost is \$57, so

$$y = mx + b = 57x + 159.$$

The cost of a 1-year membership can be found by substituting 12 for x.

$$\begin{aligned} y &= 57(12) + 159 \\ &= 684 + 159 = 843 \end{aligned}$$

The cost is \$843.

(b) As in part (a),

$$\begin{aligned} y &= 47x + 159. \\ y &= 47(12) + 159 \\ &= 564 + 159 = 723 \end{aligned}$$

The cost is \$723.

36. **(a)** Use $(3, 1839)$ and $(7, 2414)$.

$$m = \frac{\Delta y}{\Delta x} = \frac{2414 - 1839}{7 - 3} = \frac{575}{4} \approx 143.75$$

Use the point-slope form of a line.

$$\begin{aligned} y - y_1 &= m(x - x_1) \\ y - 1839 &= 143.75(x - 3) \\ y - 1839 &= 143.75x - 431.25 \\ y &= 143.75x + 1407.75 \end{aligned}$$

The slope, 143.75, indicates that the revenue from skiing facilities increased by an average of \$143.75 million each year from 2003 to 2007.

(b) The year 2008 corresponds to $x = 8$.

$$\begin{aligned} y &= 143.75(8) + 1407.75 \\ &= 1150 + 1407.75 = 2557.75 \end{aligned}$$

According to the equation from part (a), we estimate the revenue from skiing facilities to be \$2558 million (to the nearest million).

37. $\{(-4, 2), (-4, -2), (1, 5), (1, -5)\}$
The domain, the set of x-values, is $\{-4, 1\}$.
The range, the set of y-values, is $\{2, -2, 5, -5\}$.
Since each x-value has more than one y-value, the relation is not a function.

38. The relation can be described by the set of ordered pairs

$$\{(9, 32), (11, 47), (4, 47), (17, 69), (25, 14)\}.$$

The relation is a function since for each x-value, there is only one y-value.

The domain is the set of x-values: $\{9, 11, 4, 17, 25\}$.
The range is the set of y-values: $\{32, 47, 69, 14\}$.

39. The domain, the x-values of the points on the graph, is $[-4, 4]$. The range, the y-values of the points on the graph, is $[0, 2]$. Since a vertical line intersects the graph of the relation in at most one point, the relation is a function.

40. The x-values are negative or zero, so the domain is $(-\infty, 0]$. The y-values can be any real number, so the range is $(-\infty, \infty)$. A vertical line, such as $x = -3$, will intersect the graph twice, so by the vertical line test, the relation is not a function.

41. $y = 3x - 3$
For any value of x, there is exactly one value of y, so the equation defines a function, actually a linear function. The domain is the set of all real numbers, $(-\infty, \infty)$.

42. $x = y^2$
The ordered pairs $(4, 2)$ and $(4, -2)$ both satisfy the equation. Since one value of x, 4, corresponds to two values of y, 2 and -2, the equation does not define a function. Because x is equal to the square of y, the values of x must always be nonnegative. The domain is $[0, \infty)$.

43. $y = \dfrac{7}{x - 6}$
Given any value of x, y is found by subtracting 6, then dividing the result into 7. This process produces exactly one value of y for each x-value in the domain, so the equation defines a function. The domain includes all real numbers except those that make the denominator 0, namely 6. The domain is $(-\infty, 6) \cup (6, \infty)$.

In Exercises 44–47, use

$$f(x) = -2x^2 + 3x - 6.$$

44. $f(0) = -2(0)^2 + 3(0) - 6 = -6$

45.
$$\begin{aligned} f(2.1) &= -2(2.1)^2 + 3(2.1) - 6 \\ &= -8.82 + 6.3 - 6 = -8.52 \end{aligned}$$

46.
$$\begin{aligned} f(-\tfrac{1}{2}) &= -2(-\tfrac{1}{2})^2 + 3(-\tfrac{1}{2}) - 6 \\ &= -\tfrac{1}{2} - \tfrac{3}{2} - 6 = -8 \end{aligned}$$

47. $f(k) = -2k^2 + 3k - 6$

48.
$$\begin{aligned} 2x^2 - y &= 0 \\ -y &= -2x^2 \\ y &= 2x^2 \end{aligned}$$

Since $y = f(x)$,
$f(x) = 2x^2$,
and $f(3) = 2(3)^2 = 2(9) = 18$.

49. Solve for y in terms of x.

$$2x - 5y = 7$$
$$2x - 7 = 5y$$
$$\tfrac{2}{5}x - \tfrac{7}{5} = y$$

Thus, choice **C** is correct.

50. **(a)** For each year, there is exactly one life expectancy associated with the year, so the table defines a function.

(b) The domain is the set of years, that is, $\{1960, 1970, 1980, 1990, 2000, 2009\}$.
The range is the set of life expectancies, that is, $\{69.7, 70.8, 73.7, 75.4, 76.8, 78.1\}$.

(c) Answers will vary. Two possible answers are $(1960, 69.7)$ and $(2009, 78.1)$.

(d) $f(1980) = 73.7$. In 1980, life expectancy at birth was 73.7 yr.

(e) Since $f(2000) = 76.8$, $x = 2000$.

51. $f(x) = -2x^2 + 5x + 7$

(a) $f(-2) = -2(-2)^2 + 5(-2) + 7$
$= -2(4) - 10 + 7$
$= -8 - 10 + 7$
$= -18 + 7 = -11$

(b) $f(3) = -2(3)^2 + 5(3) + 7$
$= -2(9) + 15 + 7$
$= -18 + 15 + 7$
$= -3 + 7 = 4$

52. $f(x) = 2x + 3$, $g(x) = 5x^2 - 3x + 2$

(a) $(f + g)(x) = f(x) + g(x)$
$= (2x + 3) + (5x^2 - 3x + 2)$
$= 5x^2 + 2x - 3x + 3 + 2$
$= 5x^2 - x + 5$

(b) $(f - g)(x) = f(x) - g(x)$
$= (2x + 3) - (5x^2 - 3x + 2)$
$= 2x + 3 - 5x^2 + 3x - 2$
$= -5x^2 + 5x + 1$

(c) $(f + g)(-1)$
$= f(-1) + g(-1)$
$= [2(-1) + 3] + [5(-1)^2 - 3(-1) + 2]$
$= [1] + [10]$
$= 11$

(d) $(f - g)(-1) = f(-1) - g(-1)$
$= 1 - 10$ *from part (c)*
$= -9$

53. **(a)** $(fg)(x) = f(x) \cdot g(x)$
$= (12x^2 - 3x)(3x)$
$= 12x^2(3x) - 3x(3x)$
$= 36x^3 - 9x^2$

(b) $\left(\dfrac{f}{g}\right)(x) = \dfrac{f(x)}{g(x)} = \dfrac{12x^2 - 3x}{3x}$
$= \dfrac{12x^2}{3x} - \dfrac{3x}{3x}$
$= 4x - 1, x \neq 0$

(c) $(fg)(-1)$
$= 36(-1)^3 - 9(-1)^2$ *from part (a)*
$= -36 - 9$
$= -45$

(d) $\left(\dfrac{f}{g}\right)(2) = 4(2) - 1$ *from part (b)*
$= 8 - 1 = 7$

54. $f(x) = 3x^2 + 2x - 1$, $g(x) = 5x + 7$

(a) $(g \circ f)(3) = g(f(3))$
$= g(3 \cdot 3^2 + 2 \cdot 3 - 1)$
$= g(32)$
$= 5 \cdot 32 + 7$
$= 167$

(b) $(f \circ g)(3) = f(g(3))$
$= f(5 \cdot 3 + 7)$
$= f(22)$
$= 3 \cdot 22^2 + 2 \cdot 22 - 1$
$= 1495$

(c) $(f \circ g)(-2) = f(g(-2))$
$= f[5(-2) + 7]$
$= f(-3)$
$= 3(-3)^2 + 2(-3) - 1$
$= 20$

(d) $(g \circ f)(-2) = g(f(-2))$
$= g[3(-2)^2 + 2(-2) - 1]$
$= g(7)$
$= 5 \cdot 7 + 7$
$= 42$

(e) $(f \circ g)(x)$
$= f(g(x))$
$= f(5x + 7)$
$= 3(5x + 7)^2 + 2(5x + 7) - 1$
$= 3(25x^2 + 70x + 49) + 10x + 14 - 1$
$= 75x^2 + 210x + 147 + 10x + 13$
$= 75x^2 + 220x + 160$

(f) $(g \circ f)(x) = g(f(x))$
$= g(3x^2 + 2x - 1)$
$= 5(3x^2 + 2x - 1) + 7$
$= 15x^2 + 10x - 5 + 7$
$= 15x^2 + 10x + 2$

55. If y varies inversely as x, then $y = \frac{k}{x}$, for some constant k. This form fits choice **C**.

56. Let $v =$ the viewing distance and $e =$ the amount of enlargement. v varies directly as e, so

$$v = ke$$

for some constant k. Since $v = 250$ when $e = 5$, substitute these values in the equation and solve for k.

$$\begin{aligned} v &= ke \\ 250 &= k(5) \\ 50 &= k \end{aligned}$$

So $v = 50e$. Now let $e = 8.6$.

$$v = 50(8.6) = 430$$

It should be viewed from 430 mm.

57. The frequency f of a vibrating guitar string varies inversely as its length L, so

$$f = \frac{k}{L}$$

for some constant k. Substitute 0.65 for L and 4.3 for f.

$$\begin{aligned} 4.3 &= \frac{k}{0.65} \\ k &= 4.3(0.65) = 2.795 \end{aligned}$$

So $f = \frac{2.795}{L}$. Now let $L = 0.5$.

$$f = \frac{2.795}{0.5} = 5.59$$

The frequency would be 5.59 vibrations per second.

58. The volume V of a rectangular box of a given height is proportional to its width W and length L, so

$$V = kWL$$

for some constant k. Substitute 2 for W, 4 for L, and 12 for V.

$$\begin{aligned} 12 &= k(2)(4) \\ k &= \tfrac{12}{8} = \tfrac{3}{2} \end{aligned}$$

So $V = \frac{3}{2}WL$. Now let $W = 3$ and $L = 5$.

$$V = \tfrac{3}{2}(3)(5) = \tfrac{45}{2}$$

The volume is 22.5 cubic feet.

Chapter 7 Test

1. $2x - 3y = 12$

For $x = 1$:

$$\begin{aligned} 2(1) - 3y &= 12 \\ 2 - 3y &= 12 \\ -3y &= 10 \\ y &= -\tfrac{10}{3} \quad \mathbf{(1, -\tfrac{10}{3})} \end{aligned}$$

For $x = 3$:

$$\begin{aligned} 2(3) - 3y &= 12 \\ 6 - 3y &= 12 \\ -3y &= 6 \\ y &= -2 \quad \mathbf{(3, -2)} \end{aligned}$$

For $y = -4$:

$$\begin{aligned} 2x - 3(-4) &= 12 \\ 2x + 12 &= 12 \\ 2x &= 0 \\ x &= 0 \quad \mathbf{(0, -4)} \end{aligned}$$

2. $3x - 2y = 20$
To find the x-intercept, let $y = 0$.

$$\begin{aligned} 3x - 2(0) &= 20 \\ 3x &= 20 \\ x &= \tfrac{20}{3} \end{aligned}$$

The x-intercept is $\left(\frac{20}{3}, 0\right)$.
To find the y-intercept, let $x = 0$.

$$\begin{aligned} 3(0) - 2y &= 20 \\ -2y &= 20 \\ y &= -10 \end{aligned}$$

The y-intercept is $(0, -10)$.
Draw the line through these two points.

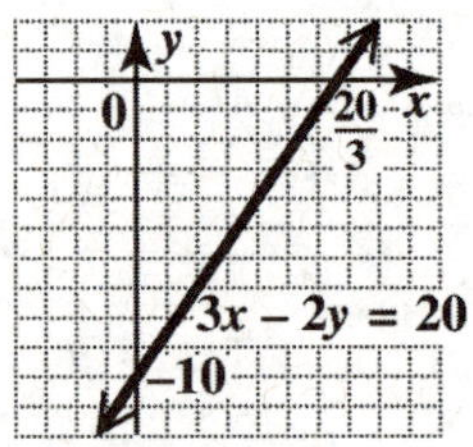

3. The graph of $y = 5$ is the horizontal line with slope 0 and y-intercept $(0, 5)$. There is no x-intercept.

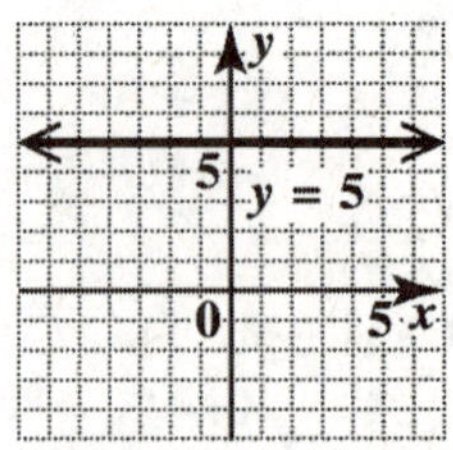

4. The graph of $x = 2$ is the vertical line with x-intercept at $(2, 0)$. There is no y-intercept.

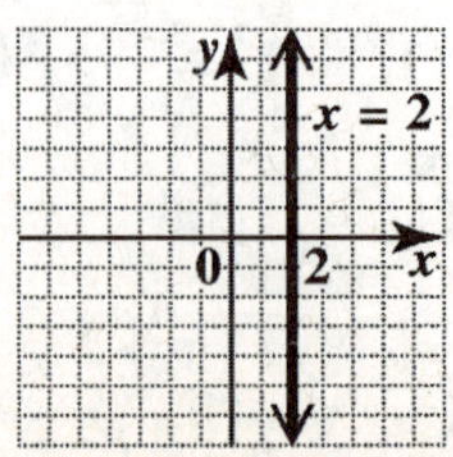

5. Through $(6,4)$ and $(-4,-1)$

$$m=\frac{\Delta y}{\Delta x}=\frac{-1-4}{-4-6}=\frac{-5}{-10}=\frac{1}{2}$$

The slope of the line is $\frac{1}{2}$.

6. The graph of a line with undefined slope is the graph of a vertical line.

7. Find the slope of each line.

$$\begin{aligned}5x-y&=8\\-y&=-5x+8\\y&=5x-8\end{aligned}$$

The slope is 5.

$$\begin{aligned}5y&=-x+3\\y&=-\tfrac{1}{5}x+\tfrac{3}{5}\end{aligned}$$

The slope is $-\frac{1}{5}$.
Since $5(-\frac{1}{5})=-1$, the two slopes are negative reciprocals and the lines are perpendicular.

8. Find the slope of each line.

$$\begin{aligned}2y&=3x+12\\y&=\tfrac{3}{2}x+6\end{aligned}$$

The slope is $\frac{3}{2}$.

$$\begin{aligned}3y&=2x-5\\y&=\tfrac{2}{3}x-\tfrac{5}{3}\end{aligned}$$

The slope is $\frac{2}{3}$.
The lines are neither parallel nor perpendicular.

9. Use the points $(1980, 119{,}000)$ and $(2008, 93{,}000)$.

average rate of change

$$\begin{aligned}&=\frac{\text{change in } y}{\text{change in } x}=\frac{93{,}000-119{,}000}{2008-1980}\\&=\frac{-26{,}000}{28}\approx -929\end{aligned}$$

The average rate of change is about -929 farms per year, that is, the number of farms decreased by about 929 each year from 1980 to 2008.

10. Through $(4,-1)$; $m=-5$

(a) Let $m=-5$ and $(x_1,y_1)=(4,-1)$ in the point-slope form.

$$\begin{aligned}y-y_1&=m(x-x_1)\\y-(-1)&=-5(x-4)\\y+1&=-5x+20\\y&=-5x+19\end{aligned}$$

(b)

$y=-5x+19$	*From part (a)*
$5x+y=19$	*Standard form*

11. Through $(-3,14)$; horizontal

(a) A horizontal line has equation $y=k$. Here $k=14$, so the line has equation $y=14$.

(b) $y=14$ is already in standard form.

12. Through $(-2,3)$ and $(6,-1)$

(a) First find the slope.

$$m=\frac{\Delta y}{\Delta x}=\frac{-1-3}{6-(-2)}=\frac{-4}{8}=-\frac{1}{2}$$

Use $m=-\frac{1}{2}$ and $(x_1,y_1)=(-2,3)$ in the point-slope form.

$$\begin{aligned}y-y_1&=m(x-x_1)\\y-3&=-\tfrac{1}{2}[x-(-2)]\\y-3&=-\tfrac{1}{2}(x+2)\\y-3&=-\tfrac{1}{2}x-1\\y&=-\tfrac{1}{2}x+2\end{aligned}$$

(b)

$y=-\frac{1}{2}x+2$	*From part (a)*
$\frac{1}{2}x+y=2$	*Variable terms on one side*
$2(\frac{1}{2}x+y)=2(2)$	*Multiply by 2.*
$x+2y=4$	*Standard form*

13. Through $(5,-6)$; vertical

(a) The equation of any vertical line is in the form $x=k$. Since the line goes through $(5,-6)$, the equation is $x=5$. Writing $x=5$ in slope-intercept form is *not possible* since there is no y-term.

(b) From part (a), the standard form is $x=5$.

14. Through $(-7,2)$ and parallel to $3x+5y=6$

(a) To find the slope of $3x+5y=6$, write the equation in slope-intercept form by solving for y.

$$\begin{aligned}3x+5y&=6\\5y&=-3x+6\\y&=-\tfrac{3}{5}x+\tfrac{6}{5}\end{aligned}$$

The slope is $-\frac{3}{5}$, so a line parallel to it also has slope $-\frac{3}{5}$. Let $m=-\frac{3}{5}$ and $(x_1,y_1)=(-7,2)$ in the point-slope form.

$$\begin{aligned}y-y_1&=m(x-x_1)\\y-2&=-\tfrac{3}{5}[x-(-7)]\\y-2&=-\tfrac{3}{5}(x+7)\\y-2&=-\tfrac{3}{5}x-\tfrac{21}{5}\\y&=-\tfrac{3}{5}x-\tfrac{11}{5}\end{aligned}$$

(b)

$y=-\frac{3}{5}x-\frac{11}{5}$	*From part (a)*
$\frac{3}{5}x+y=-\frac{11}{5}$	*Variable terms on one side*
$5(\frac{3}{5}x+y)=5(-\frac{11}{5})$	*Multiply by 5.*
$3x+5y=-11$	*Standard form*

15. Through $(-7, 2)$ and perpendicular to $y = 2x$

(a) Since $y = 2x$ is in slope-intercept form $(b = 0)$, the slope, m, of $y = 2x$ is 2. A line perpendicular to it has a slope that is the negative reciprocal of 2, that is, $-\frac{1}{2}$. Let $m = -\frac{1}{2}$ and $(x_1, y_1) = (-7, 2)$ in the point-slope form.

$$\begin{aligned} y - y_1 &= m(x - x_1) \\ y - 2 &= -\tfrac{1}{2}(x + 7) \\ y - 2 &= -\tfrac{1}{2}x - \tfrac{7}{2} \\ y &= -\tfrac{1}{2}x - \tfrac{3}{2} \end{aligned}$$

(b)

$$\begin{aligned} y &= -\tfrac{1}{2}x - \tfrac{3}{2} && \textit{From part (a)} \\ \tfrac{1}{2}x + y &= -\tfrac{3}{2} && \textit{Variable terms on one side} \\ 2(\tfrac{1}{2}x + y) &= 2(-\tfrac{3}{2}) && \textit{Multiply by 2.} \\ x + 2y &= -3 && \textit{Standard form} \end{aligned}$$

16. Positive slope means that the line goes up from left to right. The only line that has positive slope and a negative y-coordinate for its y-intercept is choice **B**.

17. **(a)** Use the points $(1, 53{,}635)$ and $(7, 66{,}103)$.

$$m = \frac{\Delta y}{\Delta x} = \frac{66{,}103 - 53{,}635}{7 - 1} = \frac{12{,}468}{6} = 2078$$

Use the point-slope form with $m = 2078$ and $(x_1, y_1) = (1, 53{,}635)$.

$$\begin{aligned} y - 53{,}635 &= 2078(x - 1) \\ y - 53{,}635 &= 2078x - 2078 \\ y &= 2078x + 51{,}557 \end{aligned}$$

(b) $y = 2078(5) + 51{,}557$ *Let $x = 5$.*
$= \$61{,}947$,
which is more than the actual value.

18. Choice **D** is the only graph that passes the vertical line test.

19. Choice **D** does not define a function, since its domain (input) element 0 is paired with two different range (output) elements, 1 and 2.

20. **(a)** The x-values are greater than or equal to zero, so the domain is $[0, \infty)$. Since y can be any value, the range is $(-\infty, \infty)$.

(b) The domain is the set of x-values: $\{0, -2, 4\}$. The range is the set of y-values: $\{1, 3, 8\}$.

21. $f(x) = -x^2 + 2x - 1$

(a)
$$\begin{aligned} f(1) &= -(1)^2 + 2(1) - 1 \\ &= -1 + 2 - 1 \\ &= 0 \end{aligned}$$

(b) $f(a) = -a^2 + 2a - 1$

22. $f(x) = \frac{2}{3}x - 1$
This function represents a line with y-intercept $(0, -1)$ and x-intercept $(\frac{3}{2}, 0)$.
Draw the line through these two points.
The domain is $(-\infty, \infty)$, and the range is $(-\infty, \infty)$.

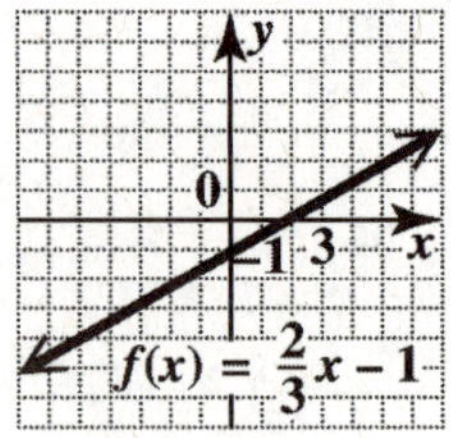

23. $f(x) = -2x^2 + 5x - 6$, $g(x) = 7x - 3$

(a)
$$\begin{aligned} f(x) &= -2x^2 + 5x - 6 \\ f(4) &= -2(4)^2 + 5(4) - 6 \\ &= -2 \cdot 16 + 20 - 6 \\ &= -32 + 20 - 6 \\ &= -12 - 6 = -18 \end{aligned}$$

(b)
$$\begin{aligned} (f + g)(x) &= f(x) + g(x) \\ &= (-2x^2 + 5x - 6) + (7x - 3) \\ &= -2x^2 + 12x - 9 \end{aligned}$$

(c)
$$\begin{aligned} (f - g)(x) &= f(x) - g(x) \\ &= (-2x^2 + 5x - 6) - (7x - 3) \\ &= -2x^2 + 5x - 6 - 7x + 3 \\ &= -2x^2 - 2x - 3 \end{aligned}$$

(d) Using the answer in part (c), we have

$$\begin{aligned} (f - g)(-2) &= -2(-2)^2 - 2(-2) - 3 \\ &= -8 + 4 - 3 = -7. \end{aligned}$$

24. $f(x) = x^2 + 3x + 2$, $g(x) = x + 1$

(a)
$$\begin{aligned} (fg)(x) &= f(x) \cdot g(x) \\ &= (x^2 + 3x + 2)(x + 1) \\ &= (x^2 + 3x + 2)(x) \\ &\quad + (x^2 + 3x + 2)(1) \\ &= x^3 + 3x^2 + 2x + x^2 + 3x + 2 \\ &= x^3 + 4x^2 + 5x + 2 \end{aligned}$$

(b)
$$\begin{aligned} (fg)(-2) &= f(-2) \cdot g(-2) \\ &= [(-2)^2 + 3(-2) + 2] \cdot [(-2) + 1] \\ &= [4 - 6 + 2] \cdot [-1] \\ &= 0(-1) = 0 \end{aligned}$$

Alternatively, we could have substituted -2 for x into our answer from part (a).

25. **(a)** $\left(\dfrac{f}{g}\right)(x) = \dfrac{f(x)}{g(x)} = \dfrac{x^2 + 3x + 2}{x + 1}$

$$\begin{array}{r} x+2 \\ x+1\,\overline{)\,x^2+3x+2} \\ \underline{x^2+x} \\ 2x+2 \\ \underline{2x+2} \\ 0 \end{array}$$

Thus, $\left(\frac{f}{g}\right)(x) = x+2$ if $x+1 \neq 0$, that is, $x \neq -1$.

(b) Using our answer from part (a),

$$\left(\tfrac{f}{g}\right)(-2) = (-2)+2 = 0.$$

26. $f(x) = 3x+5$, $g(x) = x^2+2$

(a) $$\begin{aligned}(f \circ g)(-2) &= f(g(-2)) \\ &= f\left((-2)^2+2\right) \\ &= f(6) \\ &= 3 \cdot 6 + 5 \\ &= 23\end{aligned}$$

(b) $$\begin{aligned}(f \circ g)(x) &= f(g(x)) \\ &= f(x^2+2) \\ &= 3(x^2+2)+5 \\ &= 3x^2+6+5 \\ &= 3x^2+11\end{aligned}$$

(c) $$\begin{aligned}(g \circ f)(x) &= g(f(x)) \\ &= g(3x+5) \\ &= (3x+5)^2+2 \\ &= 9x^2+30x+25+2 \\ &= 9x^2+30x+27\end{aligned}$$

27. The current I is inversely proportional to the resistance R, so

$$I = \frac{k}{R}$$

for some constant k. Let $I = 80$ and $R = 30$. Find k.

$$\begin{aligned}80 &= \frac{k}{30} \\ k &= 80(30) = 2400\end{aligned}$$

So $I = \frac{2400}{R}$. Now let $R = 12$.

$$I = \tfrac{2400}{12} = 200$$

The current is 200 amperes.

28. The force F of the wind blowing on a vertical surface varies jointly as the area $\mathcal{A}$ of the surface and the square of the velocity V, so

$$F = k\mathcal{A}V^2$$

for some constant k. Let $F = 50$, $\mathcal{A} = 500$, and $V = 40$. Find k.

$$\begin{aligned}50 &= k(500)(40)^2 \\ k &= \frac{50}{500(1600)} = \frac{1}{16{,}000}\end{aligned}$$

So $F = \frac{1}{16{,}000}\mathcal{A}V^2$. Now let $\mathcal{A} = 2$ and $V = 80$.

$$F = \frac{1}{16{,}000}(2)(80)^2 = 0.8$$

The force of the wind is 0.8 pounds.

Cumulative Review Exercises (Chapters 1–7)

1. The absolute value of a negative number is a positive number and the additive inverse of the same negative number is the same positive number. For example, suppose the negative number is -5:

$$\begin{aligned}|-5| &= -(-5) = 5 \\ \text{and} \quad -(-5) &= 5\end{aligned}$$

The statement is *always true*.

2. The sum of two negative numbers is another negative number, so the statement is *never true*.

3. The statement is *sometimes true*. For example,

$$\begin{aligned}&3+(-3) = 0, \\ \text{but} \quad &3+(-1) = 2 \neq 0.\end{aligned}$$

4. $$\begin{aligned}-|-2|-4+|-3|+7 &= -2-4+3+7 \\ &= -6+3+7 \\ &= -3+7 \\ &= 4\end{aligned}$$

5. $$\begin{aligned}-(-4m+3) &= -(-4m)-3 \\ &= 4m-3\end{aligned}$$

6. $$\begin{aligned}\frac{(4^2-4)-(-1)7}{4+(-6)} &= \frac{(16-4)-(-7)}{-2} \\ &= \frac{12+7}{-2} = -\frac{19}{2}\end{aligned}$$

7. Let $p = -4$ and $q = \frac{1}{2}$.

$$\begin{aligned}-3(2q-3p) &= -3\left[2\left(\tfrac{1}{2}\right)-3(-4)\right] \\ &= -3(1+12) \\ &= -3(13) \\ &= -39\end{aligned}$$

8. Let $p = -4$ and $r = 16$.

$$\begin{aligned}\frac{r}{8p+2r} &= \frac{16}{8(-4)+2(16)} \\ &= \frac{16}{-32+32} \\ &= \frac{16}{0}, \text{ which is } \textit{undefined}.\end{aligned}$$

9.

$$\begin{aligned} 2z - 5 + 3z &= 2 - z \\ 5z - 5 &= 2 - z \\ 6z &= 7 \\ z &= \tfrac{7}{6} \end{aligned}$$

The solution set is $\{\frac{7}{6}\}$.

10.

$$\frac{3x-1}{5} + \frac{x+2}{2} = -\frac{3}{10}$$

Multiply both sides by the LCD, 10.

$$\begin{aligned} 10\left(\frac{3x-1}{5} + \frac{x+2}{2}\right) &= 10\left(-\frac{3}{10}\right) \\ 2(3x-1) + 5(x+2) &= -3 \\ 6x - 2 + 5x + 10 &= -3 \\ 11x + 8 &= -3 \\ 11x &= -11 \\ x &= -1 \end{aligned}$$

The solution set is $\{-1\}$.

11. Let x denote the side of the original square and $4x$ the perimeter. Now $x + 4$ is the side of the new square and $4(x+4)$ is its perimeter.
"The perimeter would be 8 inches less than twice the perimeter of the original square " translates as

$$\begin{aligned} 4(x+4) &= 2(4x) - 8. \\ 4x + 16 &= 8x - 8 \\ 24 &= 4x \\ 6 &= x \end{aligned}$$

The length of a side of the original square is 6 inches.

12. Let $x =$ the time it takes for the planes to be 2100 miles apart.
Make a table. Use the formula $d = rt$.

	r	t	d
Eastbound Plane	550	x	$550x$
Westbound Plane	500	x	$500x$

The total distance is 2100 miles.

$$\begin{aligned} 550x + 500x &= 2100 \\ 1050x &= 2100 \\ x &= 2 \end{aligned}$$

It will take 2 hr for the planes to be 2100 mi apart.

13.

$$\begin{aligned} -4 < 3 - 2k &< 9 \\ -7 < -2k &< 6 \\ \frac{7}{2} > k &> -3 \quad \textit{Divide by } -2, \textit{ reverse inequalities} \\ -3 < k &< \tfrac{7}{2} \quad \textit{Equivalent inequality} \end{aligned}$$

The solution set is $(-3, \frac{7}{2})$.

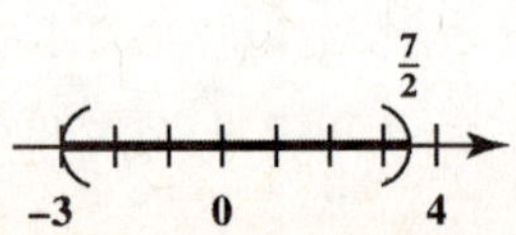

14.

$$\begin{aligned} -0.3x + 2.1(x-4) &\le -6.6 \\ -3x + 21(x-4) &\le -66 \quad \textit{Multiply by 10.} \\ -3x + 21x - 84 &\le -66 \\ 18x - 84 &\le -66 \\ 18x &\le 18 \\ x &\le 1 \end{aligned}$$

The solution set is $(-\infty, 1]$.

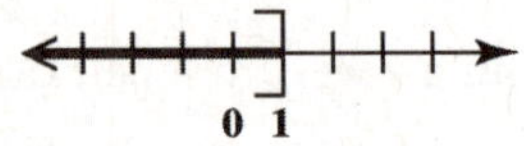

15. $3x + 5y = 12$
To find the x-intercept, let $y = 0$.

$$\begin{aligned} 3x + 5(0) &= 12 \\ 3x &= 12 \\ x &= 4 \end{aligned}$$

The x-intercept is $(4, 0)$.
To find the y-intercept, let $x = 0$.

$$\begin{aligned} 3(0) + 5y &= 12 \\ 5y &= 12 \\ y &= \tfrac{12}{5} \end{aligned}$$

The y-intercept is $(0, \frac{12}{5})$.
Plot the intercepts and draw the line through them.

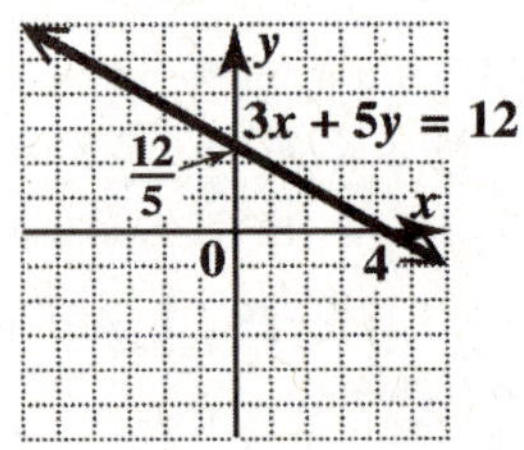

16. $A(-2, 1)$ and $B(3, -5)$
(a) The slope of line AB is

$$m = \frac{\Delta y}{\Delta x} = \frac{-5-1}{3-(-2)} = \frac{-6}{5} = -\frac{6}{5}.$$

(b) The slope of a line perpendicular to line AB is the negative reciprocal of $-\frac{6}{5}$, which is $\frac{5}{6}$.

17. **(a)** Slope $-\frac{3}{4}$; y-intercept $(0, -1)$
To write an equation of this line, let $m = -\frac{3}{4}$ and $b = -1$ in the slope-intercept form.

$$\begin{aligned} y &= mx + b \\ y &= -\tfrac{3}{4}x - 1 \end{aligned}$$

(b)

$$\begin{aligned} y &= -\tfrac{3}{4}x - 1 \\ 4y &= -3x - 4 \\ 3x + 4y &= -4 \end{aligned}$$

18. **(a)** Through $(4, -3)$ and $(1, 1)$
First find the slope of the line.

$$m = \frac{\Delta y}{\Delta x} = \frac{1-(-3)}{1-4} = \frac{4}{-3} = -\frac{4}{3}$$

Now substitute $(x_1, y_1) = (4, -3)$ and $m = -\frac{4}{3}$ in the point-slope form. Then solve for y.

$$\begin{aligned} y - y_1 &= m(x - x_1) \\ y - (-3) &= -\tfrac{4}{3}(x - 4) \\ y + 3 &= -\tfrac{4}{3}x + \tfrac{16}{3} \\ y &= -\tfrac{4}{3}x + \tfrac{7}{3} \end{aligned}$$

(b)
$$\begin{aligned} y &= -\tfrac{4}{3}x + \tfrac{7}{3} \\ 3y &= -4x + 7 \\ 4x + 3y &= 7 \end{aligned}$$

19. $(3x^2y^{-1})^{-2}(2x^{-3}y)^{-1}$
$= 3^{-2}x^{-4}y^2 \cdot 2^{-1}x^3y^{-1}$
$= 3^{-2}2^{-1}x^{-1}y$
$= \dfrac{y}{3^2 \cdot 2x} = \dfrac{y}{18x}$

20. $(7x + 3y)^2 = (7x)^2 + 2(7x)(3y) + (3y)^2$
$= 49x^2 + 42xy + 9y^2$

21. $(3x^3 + 4x^2 - 7) - (2x^3 - 8x^2 + 3x)$
$= 3x^3 + 4x^2 - 7 - 2x^3 + 8x^2 - 3x$
$= x^3 + 12x^2 - 3x - 7$

22. $\dfrac{m^3 - 3m^2 + 5m - 3}{m - 1}$

$$\begin{array}{r} m^2 - 2m + 3 \\ m - 1 \overline{\smash{)}\, m^3 - 3m^2 + 5m - 3} \\ \underline{m^3 - m^2 \qquad\qquad} \\ -2m^2 + 5m \qquad \\ \underline{-2m^2 + 2m \qquad} \\ 3m - 3 \\ \underline{3m - 3} \\ 0 \end{array}$$

The quotient is $m^2 - 2m + 3$.

23. $16w^2 + 50wz - 21z^2$

Two integer factors whose product is $(16)(-21) = -336$ and whose sum is 50 are 56 and -6.

$= 16w^2 + 56wz - 6wz - 21z^2$
$= 8w(2w + 7z) - 3z(2w + 7z)$
$= (2w + 7z)(8w - 3z)$

24. $4x^2 - 4x + 1 - y^2$
Group the first three terms.
$= (4x^2 - 4x + 1) - y^2$
$= (2x - 1)^2 - y^2$
$= [(2x - 1) + y][(2x - 1) - y]$
$= (2x - 1 + y)(2x - 1 - y)$

25. $8p^3 + 27$
$= (2p)^3 + 3^3$
$= (2p + 3)[(2p)^2 - (2p)(3) + 3^2]$
$= (2p + 3)(4p^2 - 6p + 9)$

26.
$$\begin{aligned} 9x^2 &= 6x - 1 \\ 9x^2 - 6x + 1 &= 0 \\ (3x - 1)^2 &= 0 \\ 3x - 1 &= 0 \\ 3x &= 1 \\ x &= \tfrac{1}{3} \end{aligned}$$

The solution set is $\left\{\frac{1}{3}\right\}$.

27. Let x be the length of the base. Then $x + 3$ will be the height. The area is 14 square feet. Use the formula $\mathcal{A} = \frac{1}{2}bh$, and substitute 14 for $\mathcal{A}$, x for b, and $x + 3$ for h.

$$\begin{aligned} \tfrac{1}{2}bh &= \mathcal{A} \\ \tfrac{1}{2}(x)(x + 3) &= 14 \\ x(x + 3) &= 28 \quad \textit{Multiply by 2.} \\ x^2 + 3x &= 28 \\ x^2 + 3x - 28 &= 0 \\ (x + 7)(x - 4) &= 0 \end{aligned}$$

$x + 7 = 0$ or $x - 4 = 0$
$x = -7$ or $x = 4$

The length cannot be negative, so reject -7 as a solution. The only possible solution is 4. The base is 4 feet long.

28. Let x be the distance between the longer sides. (This is actually the width.) Then $x+2$ will be the length of the longer side. The area of the rectangle is 288 in^2. Use the formula $LW = \mathcal{A}$. Substitute 288 for $\mathcal{A}$, $x+2$ for L, and x for W.

$$\begin{aligned}(x+2)x &= 288\\ x^2+2x &= 288\\ x^2+2x-288 &= 0\\ (x+18)(x-16) &= 0\end{aligned}$$

$$x+18=0 \quad \text{or} \quad x-16=0$$
$$x=-18 \quad \text{or} \quad x=16$$

The distance cannot be negative, so reject -18 as a solution. The only possible solution is 16. The distance between the longer sides is 16 inches, and the length of the longer sides is $16+2=18$ inches.

29. $$\frac{8}{x+1}-\frac{2}{x+3} = \frac{8(x+3)}{(x+1)(x+3)}-\frac{2(x+1)}{(x+3)(x+1)} = \frac{8x+24-2x-2}{(x+1)(x+3)} = \frac{6x+22}{(x+1)(x+3)}$$

30. $$\frac{x^2+5x+6}{3x} \div \frac{x^2-4}{x^2+x-6} = \frac{(x+2)(x+3)}{3x}\cdot\frac{(x+3)(x-2)}{(x+2)(x-2)} = \frac{(x+3)^2}{3x}$$

31. $$\frac{\frac{12}{x+6}}{\frac{4}{2x+12}} = \frac{12}{x+6}\div\frac{4}{2x+12} = \frac{12}{x+6}\cdot\frac{2(x+6)}{4} = \frac{12\cdot 2}{4} = 6$$

32. $$\frac{2}{x-1} = \frac{5}{x-1}-\frac{3}{4}$$

Multiply each side by the LCD, $4(x-1)$.

$$\begin{aligned}4(x-1)\left(\frac{2}{x-1}\right) &= 4(x-1)\left(\frac{5}{x-1}-\frac{3}{4}\right)\\ 4\cdot 2 &= 4\cdot 5-3(x-1)\\ 8 &= 20-3x+3\\ 3x &= 15\\ x &= 5\end{aligned}$$

Check $x=5$: $\frac{2}{4}=\frac{5}{4}-\frac{3}{4}$ *True*

Solution set: $\{5\}$

33. The domain of the relation consists of the elements in the leftmost figure; that is, $\{14, 91, 75, 23\}$.

The range of the relation consists of the elements in the rightmost figure; that is, $\{9, 70, 56, 5\}$.

Since the element 75 in the domain is paired with two different values, 70 and 56, in the range, the relation is not a function.

34. $f(x) = -4x+10$

(a) The variable x can be any real number, so the domain is $(-\infty, \infty)$. The function is a non-constant linear function, so its range is $(-\infty, \infty)$.

(b) $f(-3) = -4(-3)+10 = 12+10 = 22$

(c)
$$\begin{aligned}f(x) &= 6\\ -4x+10 &= 6\\ -4x &= -4 && \textit{Subtract 10.}\\ x &= 1 && \textit{Divide by } -4.\end{aligned}$$

35. Use $(2003, 46.8)$ and $(2008, 36.7)$.

$$m = \frac{\Delta y}{\Delta x} = \frac{36.7-46.8}{2008-2003} = \frac{-10.1}{5} = -2.02$$

So the average rate of change is -2.02 per year; that is, the per capita consumption of potatoes in the United States decreased by an average of 2.02 pounds per year from 2003 to 2008.

CHAPTER 8 SYSTEMS OF LINEAR EQUATIONS

8.1 Solving Systems of Linear Equations by Graphing

8.1 Now Try Exercises

N1. **(a)** $(5, 2)$

$$2x + 5y = 20$$
$$x - y = 7$$

To decide whether $(5, 2)$ is a solution, substitute 5 for x and 2 for y in each equation.

$$2x + 5y = 20$$
$$2(5)+5(2) \stackrel{?}{=} 20$$
$$10 + 10 \stackrel{?}{=} 20$$
$$20 = 20 \quad \textit{True}$$

$$x - y = 7$$
$$5 - 2 \stackrel{?}{=} 7$$
$$3 = 7 \quad \textit{False}$$

Since $(5, 2)$ does not satisfy the second equation, it is not a solution of the system. **No**

(b) $(5, 2)$

$$3x - y = 13$$
$$2x + y = 12$$

To decide whether $(5, 2)$ is a solution, substitute 5 for x and 2 for y in each equation.

$$3x - y = 13$$
$$3(5) - 2 \stackrel{?}{=} 13$$
$$15 - 2 \stackrel{?}{=} 13 \quad \textit{Multiply.}$$
$$13 = 13 \quad \textit{True}$$

$$2x + y = 12$$
$$2(5) + 2 \stackrel{?}{=} 12$$
$$10 + 2 \stackrel{?}{=} 12 \quad \textit{Multiply.}$$
$$12 = 12 \quad \textit{True}$$

Since $(5, 2)$ satisfies both equations, it is a solution of the system. **Yes**

N2.

$$x - 2y = 4$$
$$2x + y = 3$$

Graph each equation by plotting several points for each line. The intercepts are good choices.

Graph $x - 2y = 4$.

x	y
0	-2
2	-1
4	0

Graph $2x + y = 3$.

x	y
0	3
1	1
$\frac{3}{2}$	0

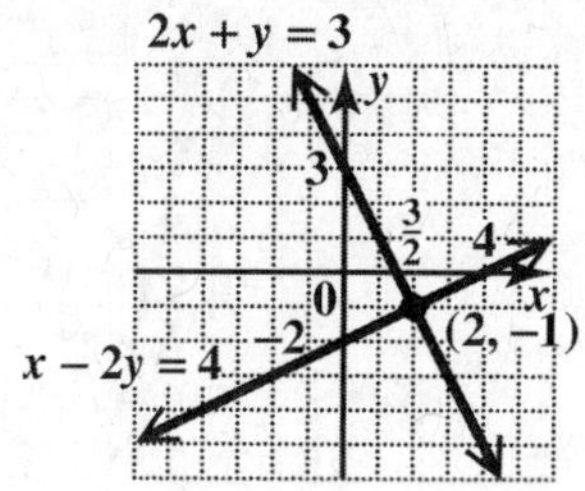

As suggested by the figure, the point at which the graphs of the two lines intersect is $(2, -1)$. Check by substituting 2 for x and -1 for y in both equations of the system. The solution set is $\{(2, -1)\}$.

N3. **(a)**

$$5x - 3y = 2$$
$$10x - 6y = 4$$

Notice that if we multiply each side of the first equation by 2, we obtain the second equation. Thus, the graphs of these two equations are the same line.

Graph $5x - 3y = 2$.

x	y
0	$-\frac{2}{3}$
$\frac{2}{5}$	0
1	1

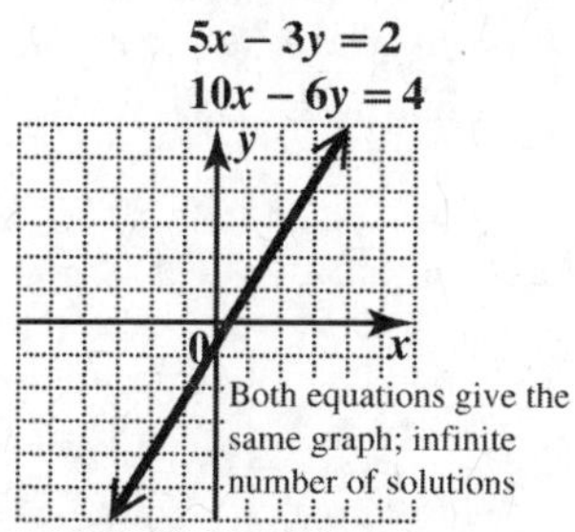

In this case, every point on the line is a solution of the system, and the solution set contains an infinite number of ordered pairs, each of which satisfies both equations of the system. We write the solution set as $\{(x, y) \mid 5x - 3y = 2\}$.

(b)

$$4x + y = 7$$
$$12x + 3y = 10$$

Graph $4x + y = 7$.

x	y
$\frac{7}{4}$	0
0	7
1	3

Graph $12x + 3y = 10$.

x	y
$\frac{5}{6}$	0
0	$\frac{10}{3}$
1	$-\frac{2}{3}$

graph on next page

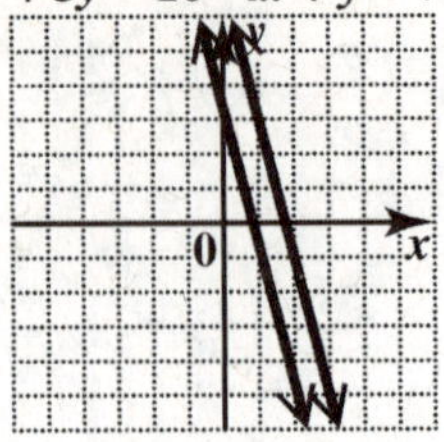

The lines each have slope -4 and hence, are parallel. Since they do not intersect, there is *no solution*. We write the solution set as $\emptyset$.

N4. (a) Solve each equation for y.

$$5x - 8y = 4 \qquad x - \tfrac{8}{5}y = \tfrac{4}{5}$$
$$-8y = -5x + 4 \qquad -\tfrac{8}{5}y = -x + \tfrac{4}{5}$$
$$y = \tfrac{5}{8}x - \tfrac{1}{2} \qquad y = \tfrac{5}{8}x - \tfrac{1}{2}$$

The slope-intercept forms are the same, so the equations represent the same line. The system has an infinite number of solutions.

(b) Solve each equation for y.

$$2x + y = 7 \qquad 3y = -6x - 12$$
$$y = -2x + 7 \qquad y = -2x - 4$$

The equations have the same slope, -2, but different y-intercepts, 7 and -4. The equations represent parallel lines. The system has no solution.

(c) Solve each equation for y.

$$y - 3x = 7 \qquad 3y - x = 0$$
$$y = 3x + 7 \qquad 3y = x$$
$$y = \tfrac{1}{3}x$$

The slopes, 3 and $\frac{1}{3}$, are different, so the graphs of these equations are neither parallel nor the same line. The system has exactly one solution.

8.1 Section Exercises

1. From the graph, the ordered pair that is a solution of the system is in the second quadrant. Choice **A**, $(-4, -4)$, is the only ordered pair given that is not in quadrant II, so it is the only valid choice.

3. $(2, -3)$ $\quad x + y = -1$
$\quad 2x + 5y = 19$

To decide whether $(2, -3)$ is a solution of the system, substitute 2 for x and -3 for y in each equation.

$$x + y = -1$$
$$2 + (-3) \stackrel{?}{=} -1$$
$$-1 = -1 \quad \textit{True}$$

$$2x + 5y = 19$$
$$2(2) + 5(-3) \stackrel{?}{=} 19$$
$$4 + (-15) \stackrel{?}{=} 19$$
$$-11 = 19 \quad \textit{False}$$

The ordered pair $(2, -3)$ satisfies the first equation but not the second. Because it does not satisfy *both* equations, it is not a solution of the system.

5. $(-1, -3)$ $\quad 3x + 5y = -18$
$\quad 4x + 2y = -10$

Substitute -1 for x and -3 for y in each equation.

$$3x + 5y = -18$$
$$3(-1) + 5(-3) \stackrel{?}{=} -18$$
$$-3 - 15 \stackrel{?}{=} -18$$
$$-18 = -18 \quad \textit{True}$$

$$4x + 2y = -10$$
$$4(-1) + 2(-3) \stackrel{?}{=} -10$$
$$-4 - 6 \stackrel{?}{=} -10$$
$$-10 = -10 \quad \textit{True}$$

Since $(-1, -3)$ satisfies both equations, it is a solution of the system.

7. $(7, -2)$ $\quad 4x = 26 - y$
$\quad 3x = 29 + 4y$

Substitute 7 for x and -2 for y in each equation.

$$4x = 26 - y$$
$$4(7) \stackrel{?}{=} 26 - (-2)$$
$$28 \stackrel{?}{=} 26 + 2$$
$$28 = 28 \quad \textit{True}$$

$$3x = 29 + 4y$$
$$3(7) \stackrel{?}{=} 29 + 4(-2)$$
$$21 \stackrel{?}{=} 29 - 8$$
$$21 = 21 \quad \textit{True}$$

Since $(7, -2)$ satisfies both equations, it is a solution of the system.

9. $(6, -8)$ $\quad -2y = x + 10$
$\quad 3y = 2x + 30$

Substitute 6 for x and -8 for y in each equation.

$$-2y = x + 10$$
$$-2(-8) \stackrel{?}{=} 6 + 10$$
$$16 = 16 \quad \textit{True}$$

$$3y = 2x + 30$$
$$3(-8) \stackrel{?}{=} 2(6) + 30$$
$$-24 \stackrel{?}{=} 12 + 30$$
$$-24 = 42 \quad \textit{False}$$

The ordered pair $(6, -8)$ satisfies the first equation but not the second. Because it does not satisfy *both* equations, it is not a solution of the system.

11. $(0,0)$ $\quad 4x + 2y = 0$
$\quad x + y = 0$

Substitute 0 for x and 0 for y in each equation.

$$\begin{aligned} 4x + 2y &= 0 \\ 4(0) + 2(0) &\stackrel{?}{=} 0 \\ 0 &= 0 \quad \textit{True} \end{aligned}$$

$$\begin{aligned} x + y &= 0 \\ 0 + 0 &\stackrel{?}{=} 0 \\ 0 &= 0 \quad \textit{True} \end{aligned}$$

Since $(0,0)$ satisfies both equations, it is a solution of the system.

13. **(a)** $(3,4)$ is in quadrant I—choice **B** is correct.

(b) $(-2,3)$ is in quadrant II—choice **C** is correct since choice **D** is the only other choice with intersection point in quadrant II, but its x-coordinate of the intersection point is -3, not -2.

(c) $(-3,2)$ is in quadrant II—choice **D** is correct.

(d) $(5,-2)$ is in quadrant IV—choice **A** is correct.

15. $x - y = 2$
$x + y = 6$

To graph the equations, find the intercepts.

$\boldsymbol{x - y = 2}$: Let $y = 0$; then $x = 2$.
Let $x = 0$; then $y = -2$.

Plot the intercepts, $(2,0)$ and $(0,-2)$, and draw the line through them.

$\boldsymbol{x + y = 6}$: Let $y = 0$; then $x = 6$.
Let $x = 0$; then $y = 6$.

Plot the intercepts, $(6,0)$ and $(0,6)$, and draw the line through them.

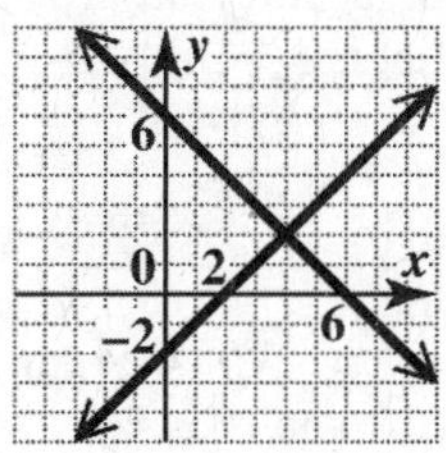

It appears that the lines intersect at the point $(4,2)$. Check this by substituting 4 for x and 2 for y in both equations. Since $(4,2)$ satisfies both equations, the solution set of this system is $\{(4,2)\}$.

17. $x + y = 4$
$y - x = 4$

To graph the equations, find the intercepts.

$\boldsymbol{x + y = 4}$: Let $y = 0$; then $x = 4$.
Let $x = 0$; then $y = 4$.

Plot the intercepts, $(4,0)$ and $(0,4)$, and draw the line through them.

$\boldsymbol{y - x = 4}$: Let $y = 0$; then $x = -4$.
Let $x = 0$; then $y = 4$.

Plot the intercepts, $(-4,0)$ and $(0,4)$, and draw the line through them.

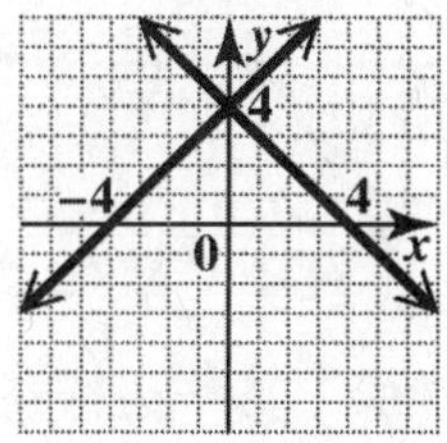

The lines intersect at their common y-intercept, $(0,4)$, so $\{(0,4)\}$ is the solution set of the system.

19. $x - 2y = 6$
$x + 2y = 2$

To graph the equations, find the intercepts.

$\boldsymbol{x - 2y = 6}$: Let $y = 0$; then $x = 6$.
Let $x = 0$; then $y = -3$.

Plot the intercepts, $(6,0)$ and $(0,-3)$, and draw the line through them.

$\boldsymbol{x + 2y = 2}$: Let $y = 0$; then $x = 2$.
Let $x = 0$; then $y = 1$.

Plot the intercepts, $(2,0)$ and $(0,1)$, and draw the line through them.

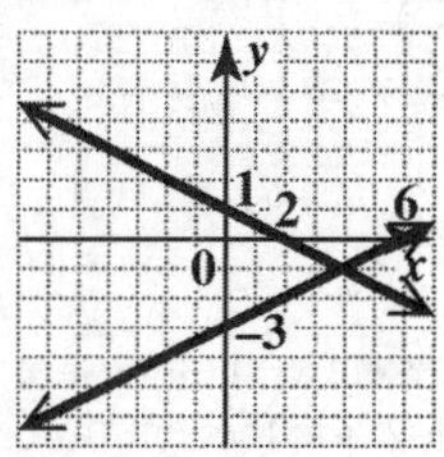

It appears that the lines intersect at the point $(4,-1)$. Since $(4,-1)$ satisfies both equations, the solution set of this system is $\{(4,-1)\}$.

21. $3x - 2y = -3$
$-3x - y = -6$

To graph the equations, find the intercepts.

$\mathbf{3x - 2y = -3}$: Let $y = 0$; then $x = -1$.
Let $x = 0$; then $y = \frac{3}{2}$.

Plot the intercepts, $(-1, 0)$ and $(0, \frac{3}{2})$, and draw the line through them.

$\mathbf{-3x - y = -6}$: Let $y = 0$; then $x = 2$.
Let $x = 0$; then $y = 6$.

Plot the intercepts, $(2, 0)$ and $(0, 6)$, and draw the line through them.

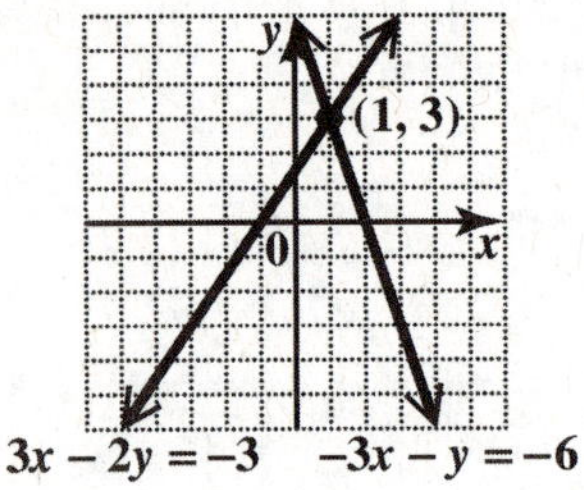

It appears that the lines intersect at the point $(1, 3)$. Since $(1, 3)$ satisfies both equations, the solution set of this system is $\{(1, 3)\}$.

23. $2x - 3y = -6$
$y = -3x + 2$

To graph the first line, find the intercepts.

$\mathbf{2x - 3y = -6}$: Let $y = 0$; then $x = -3$.
Let $x = 0$; then $y = 2$.

Plot the intercepts, $(-3, 0)$ and $(0, 2)$, and draw the line through them.

To graph the second line, start by plotting the y-intercept, $(0, 2)$. From this point, go 3 units down and 1 unit to the right (because the slope is -3) to reach the point $(1, -1)$. Draw the line through $(0, 2)$ and $(1, -1)$.

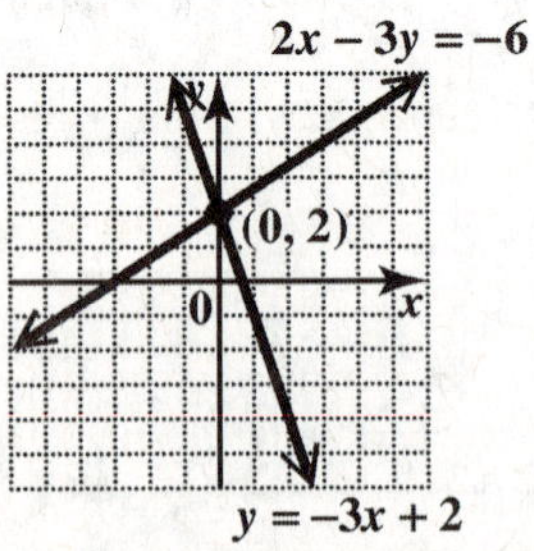

The lines intersect at their common y-intercept, $(0, 2)$, so $\{(0, 2)\}$ is the solution set of the system.

25. $2x - y = 6$
$4x - 2y = 8$

Graph the line $2x - y = 6$ through its intercepts, $(3, 0)$ and $(0, -6)$.

Graph the line $4x - 2y = 8$ through its intercepts, $(2, 0)$ and $(0, -4)$.

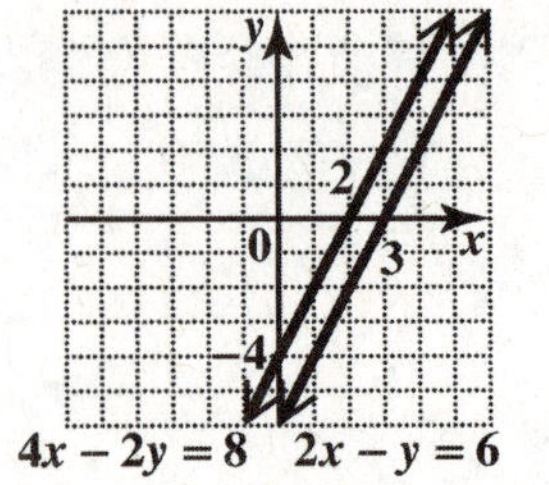

Solving each equation for y gives us $y = 2x - 6$ and $y = 2x - 4$, so the lines each have slope 2, and hence, are parallel. Since they do not intersect, there is no solution. This is an inconsistent system and the solution set is $\emptyset$.

27. $3x + y = 5$
$6x + 2y = 10$

$\mathbf{3x + y = 5}$: Let $y = 0$; then $x = \frac{5}{3}$.
Let $x = 0$; then $y = 5$.

Plot these intercepts, $(\frac{5}{3}, 0)$ and $(0, 5)$, and draw the line through them.

$\mathbf{6x + 2y = 10}$: Let $y = 0$; then $x = \frac{5}{3}$.
Let $x = 0$; then $y = 5$.

Plot these intercepts, $(\frac{5}{3}, 0)$ and $(0, 5)$, and draw the line through them.

Since both equations have the same intercepts, they are equations of the same line.

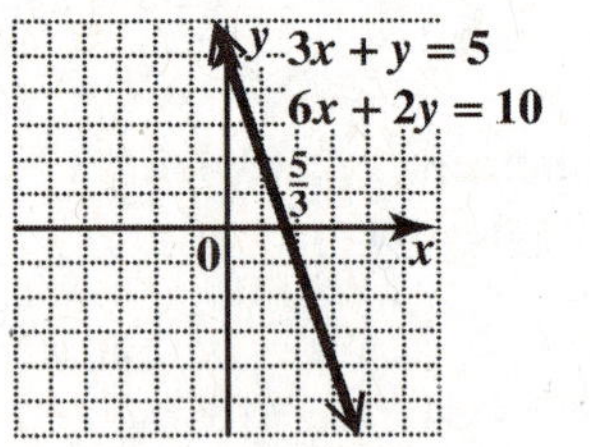

There is an infinite number of solutions. The equations are dependent equations and the solution set contains an infinite number of ordered pairs. The solution set is $\{(x, y) \mid 3x + y = 5\}$.

29. $3x - 4y = 24$
$y = -\frac{3}{2}x + 3$

Graph the line $3x - 4y = 24$ through its intercepts, $(8, 0)$ and $(0, -6)$.

To graph the line $y = -\frac{3}{2}x + 3$, plot the y-intercept $(0, 3)$ and then go 3 units down and 2 units to the right (because the slope is $-\frac{3}{2}$) to reach the point $(2, 0)$. Draw the line through $(0, 3)$ and $(2, 0)$.

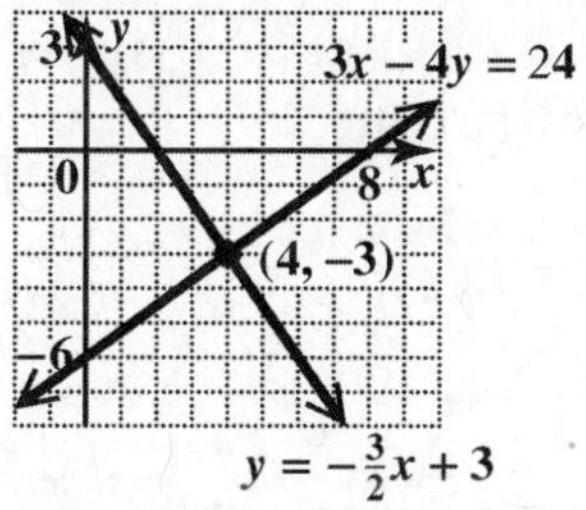

It appears that the lines intersect at the point $(4, -3)$. Since $(4, -3)$ satisfies both equations, the solution set of this system is $\{(4, -3)\}$.

31. $2x = y - 4$
$4x + 4 = 2y$

Graph the line $2x = y - 4$ through its intercepts, $(-2, 0)$ and $(0, 4)$.

Graph the line $4x + 4 = 2y$ through its intercepts, $(-1, 0)$ and $(0, 2)$.

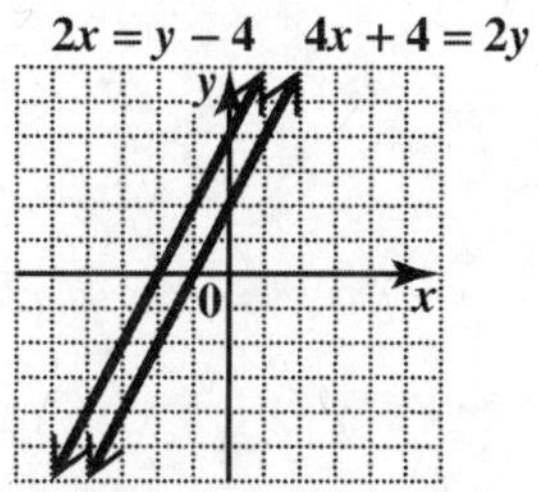

Solving each equation for y gives us $y = 2x + 4$ and $y = 2x + 2$, so the lines each have slope 2, and hence, are parallel. Since they do not intersect, there is no solution. This is an inconsistent system and the solution set is $\emptyset$.

33. Graph both lines on the same axes. For $2x + 3y = 6$, use the intercepts $(3, 0)$ and $(0, 2)$. For $x - 3y = 5$, use the intercepts $(5, 0)$ and $(0, -\frac{5}{3})$. The two lines do not intersect at integer coordinates. It is difficult to read the exact coordinates of the solution. Thus, the solution cannot be checked.

35. $y - x = -5$
$x + y = 1$

Write the equations in slope-intercept form.

$y - x = -5$	$x + y = 1$
$y = x - 5$	$y = -x + 1$
$m = 1$	$m = -1$

The lines have different slopes.

(a) The system is consistent because it has a solution. The equations are independent because they have different graphs. Therefore, the answer is "neither."

(b) The graph is a pair of intersecting lines.

(c) The system has one solution.

37. $x + 2y = 0$
$4y = -2x$

Write the equations in slope-intercept form.

$x + 2y = 0$	$4y = -2x$
$2y = -x$	$y = -\frac{1}{2}x$
$y = -\frac{1}{2}x$	

For both lines, $m = -\frac{1}{2}$ and $b = 0$.

(a) Since the equations have the same slope and y-intercept, they are dependent.

(b) The graph is one line.

(c) The system has an infinite number of solutions.

39. $x - 3y = 5$
$2x + y = 8$

Write the equations in slope-intercept form.

$x - 3y = 5$	$2x + y = 8$
$-3y = -x + 5$	$y = -2x + 8$
$y = \frac{1}{3}x - \frac{5}{3}$	$m = -2$
$m = \frac{1}{3}$	

The lines have different slopes.

(a) The system is consistent because it has a solution. The equations are independent because they have different graphs. Therefore, the answer is "neither."

(b) The graph is a pair of intersecting lines.

(c) The system has one solution.

41. $5x + 4y = 7$

$10x + 8y = 4$

Write the equations in slope-intercept form.

$5x + 4y = 7$	$10x + 8y = 4$
$4y = -5x + 7$	$8y = -10x + 4$
$y = -\frac{5}{4}x + \frac{7}{4}$	$y = -\frac{10}{8}x + \frac{4}{8}$
$m = -\frac{5}{4},\ b = \frac{7}{4}$	$y = -\frac{5}{4}x + \frac{1}{2}$
	$m = -\frac{5}{4},\ b = \frac{1}{2}$

The lines have the same slope but different y-intercepts.

(a) The system is inconsistent because it has no solution.

(b) The graph is a pair of parallel lines.

(c) The system has no solution.

43. **(a)** The graph representing evening dailies is above the graph representing morning dailies for the years 1980 to 2000, so there were more evening dailies than morning dailies during the years 1980–2000.

(b) The graphs intersect when the year is about 2001. The intersection point is about $\frac{3}{4}$ of the way between 600 and 800, so there were about 750 newspapers of each type in 2001.

45. **(a)** The graph representing ATM use is above the graphs representing credit card *and* debit card use for the years 1997–2002.

(b) The graph representing debit card use rises above the graph for credit card use in about 2001.

(c) The graph representing debit card use rises above the graph for ATM use in about 2002.

(d) The y-value is approximately 30, so the ordered pair is $(1998, 30)$.

47. Supply equals demand at the point where the two lines intersect, or when $p = 30$.

49. For $x + y = 4$, the x-intercept is $(4, 0)$ and the y-intercept is $(0, 4)$. The only screen with a graph having those intercepts is choice **B**. The point of intersection, $(3, 1)$ is listed at the bottom of the screen. Check $(3, 1)$ in the system of equations. Since it satisfies the equations, $\{(3, 1)\}$ is the solution set.

51. The graph of $x - y = 0$ goes through the origin. The only screen with a graph that goes through the origin is choice **A**.

53.
$$\begin{aligned} 3x + y &= 4 \\ y &= -3x + 4 \end{aligned}$$

55.
$$\begin{aligned} 9x - 2y &= 4 \\ -2y &= -9x + 4 \\ y &= \tfrac{9}{2}x - 2 \end{aligned}$$

57.
$$\begin{aligned} -2(x - 2) + 5x &= 10 \\ -2x + 4 + 5x &= 10 \\ 3x + 4 &= 10 \\ 3x &= 6 \\ x &= 2 \end{aligned}$$

Check $x = 2$: $0 + 10 = 10$

The solution set is $\{2\}$.

59.
$$\begin{aligned} 4x - 2(1 - 3x) &= 6 \\ 4x - 2 + 6x &= 6 \\ 10x - 2 &= 6 \\ 10x &= 8 \\ x &= \tfrac{8}{10} = \tfrac{4}{5} \end{aligned}$$

Check $x = \frac{4}{5}$: $\frac{16}{5} - 2(-\frac{7}{5}) = 6$

$\frac{16}{5} + \frac{14}{5} = 6$

The solution set is $\{\frac{4}{5}\}$.

8.2 Solving Systems of Linear Equations by Substitution

8.2 Now Try Exercises

N1.
$$\begin{aligned} 2x - 4y &= 28 \\ y &= -3x \end{aligned}$$

The second equation is already solved for y. Substitute $-3x$ for x in the first equation.

$$\begin{aligned} 2x - 4y &= 28 \\ 2x - 4(-3x) &= 28 && \textit{Let } y = -3x. \\ 2x + 12x &= 28 && \textit{Multiply.} \\ 14x &= 28 && \textit{Combine like terms.} \\ x &= 2 && \textit{Divide by 14.} \end{aligned}$$

Now substitute 2 for x in the second equation.

$$y = -3x = -3(2) = -6$$

Check $x = 2, y = -6$:

$$\begin{aligned} 2x - 4y &= 28 \\ 2(2) - 4(-6) &\stackrel{?}{=} 28 \\ 4 + 24 &\stackrel{?}{=} 28 \\ 28 &= 28 \quad \textit{True} \end{aligned}$$

Check $x = 2, y = -6$:

$$\begin{aligned} y &= -3x \\ -6 &\stackrel{?}{=} -3(2) \\ -6 &= -6 \quad \textit{True} \end{aligned}$$

Since $(2, -6)$ satisfies both equations, the solution set is $\{(2, -6)\}$.

N2.
$$\begin{aligned} 4x + 9y &= 1 \\ x &= y - 3 \end{aligned}$$

Substitute $y - 3$ for x in the first equation, and solve for y.

$$\begin{aligned} 4x + 9y &= 1 \\ 4(y - 3) + 9y &= 1 && \textit{Let } x = y - 3. \\ 4y - 12 + 9y &= 1 \\ 13y - 12 &= 1 \\ 13y &= 13 \\ y &= 1 \end{aligned}$$

To find x, use $x = y - 3$ and $y = 1$.

$$x = y - 3 = 1 - 3 = -2$$

The solution set is $\{(-2, 1)\}$.

N3.
$$\begin{aligned} 2y &= x - 2 && (1) \\ 4x - 5y &= -4 && (2) \end{aligned}$$

Solve (1) for x.

$$\begin{aligned} 2y &= x - 2 \\ 2y + 2 &= x && (3) \end{aligned}$$

Substitute $2y + 2$ for x in equation (2) and solve for y.

$$\begin{aligned} 4x - 5y &= -4 \quad \textit{Let } x = 2y + 2. \\ 4(2y + 2) - 5y &= -4 \\ 8y + 8 - 5y &= -4 \\ 3y + 8 &= -4 \\ 3y &= -12 \\ y &= -4 \end{aligned}$$

To find x, let $y = -4$ in equation (3).

$$\begin{aligned} x = 2y + 2 &= 2(-4) + 2 \\ &= -8 + 2 = -6 \end{aligned}$$

Check that $(-6, -4)$ is the solution.

$$\begin{aligned} 2y &= x - 2 \quad (1) \\ 2(-4) &\stackrel{?}{=} -6 - 2 \\ -8 &= -8 \quad \textit{True} \end{aligned}$$

$$\begin{aligned} 4x - 5y &= -4 \quad (2) \\ 4(-6) - 5(-4) &\stackrel{?}{=} -4 \\ -4 &= -4 \quad \textit{True} \end{aligned}$$

The solution set of the system is $\{(-6, -4)\}$.

N4. $8x - 2y = 1 \quad (1)$
$y = 4x - 8 \quad (2)$

Substitute $4x - 8$ for y in equation (1).

$$\begin{aligned} 8x - 2y &= 1 \\ 8x - 2(4x - 8) &= 1 \quad \textit{Let } y = 4x - 8. \\ 8x - 8x + 16 &= 8 \\ 16 &= 8 \quad \textit{False} \end{aligned}$$

This false result indicates that the system has no solution, so the solution set is $\emptyset$.

N5. $5x - y = 6 \quad (1)$
$-10x + 2y = -12 \quad (2)$

Solve equation (1) for y.

$$\begin{aligned} 5x - y &= 6 \\ 5x - 6 &= y \end{aligned}$$

Substitute $5x - 6$ for y in equation (2) and solve the resulting equation.

$$\begin{aligned} -10x + 2y &= -12 \quad (2) \\ -10x + 2(5x - 6) &= -12 \quad \textit{Let } y = 5x - 6. \\ -10x + 10x - 12 &= -12 \quad \textit{Distributive property} \\ 0 &= 0 \quad \textit{Add 12; combine like terms.} \end{aligned}$$

This true result means that every solution of one equation is also a solution of the other, so the system has an infinite number of solutions. The solution set is $\{(x, y) \mid 5x - y = 6\}$.

N6. $x + \frac{1}{2}y = \frac{1}{2} \quad (1)$
$\frac{1}{6}x - \frac{1}{3}y = \frac{4}{3} \quad (2)$

First, clear all fractions.

Equation (1):

$$\begin{aligned} 2(x + \tfrac{1}{2}y) &= 2(\tfrac{1}{2}) \quad \textit{Multiply by the LCD, 2.} \\ 2(x) + 2(\tfrac{1}{2}y) &= 1 \quad \textit{Distributive property} \\ 2x + y &= 1 \quad (3) \end{aligned}$$

Equation (2):

$$\begin{aligned} 6(\tfrac{1}{6}x - \tfrac{1}{3}y) &= 6(\tfrac{4}{3}) \quad \textit{Multiply by the LCD, 6.} \\ 6(\tfrac{1}{6}x) - 6(\tfrac{1}{3}y) &= 8 \quad \textit{Distributive property} \\ x - 2y &= 8 \quad (4) \end{aligned}$$

The system has been simplified to

$$\begin{aligned} 2x + y &= 1 \quad (3) \\ x - 2y &= 8 \quad (4) \end{aligned}$$

Solve this system by the substitution method.

$$x = 2y + 8 \quad (5) \quad \textit{Solve (4) for x.}$$

$$\begin{aligned} 2(2y + 8) + y &= 1 \quad \textit{Substitute for x in (3).} \\ 4y + 16 + y &= 1 \\ 5y + 16 &= 1 \\ 5y &= -15 \\ y &= -3 \end{aligned}$$

To find x, let $y = -3$ in equation (5).

$$x = 2(-3) + 8 = -6 + 8 = 2$$

The solution set is $\{(2, -3)\}$.

N7. $0.2x + 0.3y = 0.5 \quad (1)$
$0.3x - 0.1y = 1.3 \quad (2)$

First, clear all decimals.

Equation (1):

$$\begin{aligned} 10(0.2x + 0.3y) &= 10(0.5) \quad \textit{Multiply by 10.} \\ 10(0.2x) + 10(0.3y) &= 10(0.5) \quad \textit{Dist. property} \\ 2x + 3y &= 5 \quad (3) \end{aligned}$$

Equation (2):

$$\begin{aligned} 10(0.3x - 0.1y) &= 10(1.3) \quad \textit{Multiply by 10.} \\ 10(0.3x) - 10(0.1y) &= 10(1.3) \quad \textit{Dist. property} \\ 3x - y &= 13 \quad (4) \end{aligned}$$

The system has been simplified to

$$\begin{aligned} 2x + 3y &= 5 \quad (3) \\ 3x - y &= 13 \quad (4) \end{aligned}$$

Solve this system by the substitution method.

$$y = 3x - 13 \quad (5) \quad \textit{Solve (4) for y.}$$

$$\begin{aligned} 2x + 3(3x - 13) &= 5 \quad \textit{Substitute for y in (3).} \\ 2x + 9x - 39 &= 5 \\ 11x - 39 &= 5 \\ 11x &= 44 \\ x &= 4 \end{aligned}$$

To find y, let $x = 4$ in equation (5).

$$y = 3(4) - 13 = 12 - 13 = -1$$

The solution set is $\{(4, -1)\}$.

8.2 Section Exercises

1. The student must find the value of y and write the solution as an ordered pair. Substituting 3 for x in the first equation, $5x - y = 15$, gives us $15 - y = 15$, so $y = 0$. The correct solution set is $\{(3, 0)\}$.

In this section, all solutions should be checked by substituting in *both* equations of the original system. Checks will not be shown here.

3. $x + y = 12 \quad (1)$
$y = 3x \quad (2)$

Equation (2) is already solved for y. Substitute $3x$ for y in equation (1) and solve the resulting equation for x.

$$\begin{aligned} x + y &= 12 \\ x + 3x &= 12 \quad \textit{Let } y = 3x. \\ 4x &= 12 \\ x &= 3 \end{aligned}$$

To find the y-value of the solution, substitute 3 for x in equation (2).

$$\begin{aligned} y &= 3x \\ y &= 3(3) \quad \textit{Let } x = 3. \\ &= 9 \end{aligned}$$

The solution set is $\{(3, 9)\}$.

To check this solution, substitute 3 for x and 9 for y in both equations of the given system.

5. $3x + 2y = 27 \quad (1)$
$x = y + 4 \quad (2)$

Equation (2) is already solved for x. Substitute $y + 4$ for x in equation (1).

$$\begin{aligned} 3x + 2y &= 27 \\ 3(y + 4) + 2y &= 27 \\ 3y + 12 + 2y &= 27 \\ 5y &= 15 \\ y &= 3 \end{aligned}$$

To find x, substitute 3 for y in equation (2).

$$\begin{aligned} x &= y + 4 \\ x &= 3 + 4 = 7 \end{aligned}$$

The solution set is $\{(7, 3)\}$.

7. $3x + 4 = -y \quad (1)$
$2x + y = 0 \quad (2)$

Solve equation (1) for y.

$$\begin{aligned} 3x + 4 &= -y \\ y &= -3x - 4 \quad (3) \end{aligned}$$

Substitute $-3x - 4$ for y in equation (2) and solve for x.

$$\begin{aligned} 2x + y &= 0 \\ 2x + (-3x - 4) &= 0 \\ -x - 4 &= 0 \\ -x &= 4 \\ x &= -4 \end{aligned}$$

To find y, substitute -4 for x in equation (3).

$$\begin{aligned} y &= -3x - 4 \\ &= -3(-4) - 4 = 8 \end{aligned}$$

The solution set is $\{(-4, 8)\}$.

9. $7x + 4y = 13 \quad (1)$
$x + y = 1 \quad (2)$

Solve equation (2) for y.

$$\begin{aligned} x + y &= 1 \\ y &= 1 - x \quad (3) \end{aligned}$$

Substitute $1 - x$ for y in equation (1).

$$\begin{aligned} 7x + 4y &= 13 \\ 7x + 4(1 - x) &= 13 \\ 7x + 4 - 4x &= 13 \\ 3x + 4 &= 13 \\ 3x &= 9 \\ x &= 3 \end{aligned}$$

To find y, substitute 3 for x in equation (3).

$$\begin{aligned} y &= 1 - x \\ y &= 1 - 3 = -2 \end{aligned}$$

The solution set is $\{(3, -2)\}$.

11. $3x + 5y = 25 \quad (1)$
$x - 2y = -10 \quad (2)$

Solve equation (2) for x since its coefficient is 1.

$$\begin{aligned} x - 2y &= -10 \\ x &= 2y - 10 \quad (3) \end{aligned}$$

Substitute $2y - 10$ for x in equation (1) and solve for y.

$$\begin{aligned} 3x + 5y &= 25 \\ 3(2y - 10) + 5y &= 25 \\ 6y - 30 + 5y &= 25 \\ 11y &= 55 \\ y &= 5 \end{aligned}$$

To find x, substitute 5 for y in equation (3).

$$\begin{aligned} x &= 2y - 10 \\ x &= 2(5) - 10 = 0 \end{aligned}$$

The solution set is $\{(0, 5)\}$.

13. $3x - y = 5$ (1)
$y = 3x - 5$ (2)

Equation (2) is already solved for y, so we substitute $3x - 5$ for y in equation (1).

$$\begin{aligned} 3x - (3x - 5) &= 5 \\ 3x - 3x + 5 &= 5 \\ 5 &= 5 \quad \textit{True} \end{aligned}$$

This true result means that every solution of one equation is also a solution of the other, so the system has an infinite number of solutions. The solution set is $\{(x, y) \mid 3x - y = 5\}$.

15. $2x + y = 0$ (1)
$4x - 2y = 2$ (2)

Solve equation (1) for y.

$$\begin{aligned} 2x + y &= 0 \\ y &= -2x \quad (3) \end{aligned}$$

Substitute $-2x$ for y in equation (2) and solve for x.

$$\begin{aligned} 4x - 2(-2x) &= 2 \\ 4x + 4x &= 2 \\ 8x &= 2 \\ x &= \tfrac{1}{4} \end{aligned}$$

To find y, let $x = \frac{1}{4}$ in equation (3).

$$\begin{aligned} y &= -2x \\ y &= -2\left(\tfrac{1}{4}\right) \\ &= -\tfrac{1}{2} \end{aligned}$$

The solution set is $\left\{\left(\frac{1}{4}, -\frac{1}{2}\right)\right\}$.

17. $2x + 8y = 3$ (1)
$x = 8 - 4y$ (2)

Equation (2) is already solved for x, so substitute $8 - 4y$ for x in equation (1).

$$\begin{aligned} 2(8 - 4y) + 8y &= 3 \\ 16 - 8y + 8y &= 3 \\ 16 &= 3 \quad \textit{False} \end{aligned}$$

This false result means that the system is inconsistent and its solution set is $\emptyset$.

19. $2y = 4x + 24$ (1)
$2x - y = -12$ (2)

Solve equation (2) for y.

$$\begin{aligned} 2x - y &= -12 \\ 2x &= y - 12 \\ 2x + 12 &= y \end{aligned}$$

Substitute $2x + 12$ for y in (1) and solve for x.

$$\begin{aligned} 2(2x + 12) &= 4x + 24 \\ 4x + 24 &= 4x + 24 \quad \textit{True} \end{aligned}$$

This true result means that every solution of one equation is also a solution of the other, so the system has an infinite number of solutions. The solution set is $\{(x, y) \mid 2x - y = -12\}$.

21. $\frac{1}{2}x + \frac{1}{3}y = 3$ (1)
$y = 3x$ (2)

First, clear all fractions in equation (1).

$$\begin{aligned} 6\left(\tfrac{1}{2}x + \tfrac{1}{3}y\right) &= 6(3) && \textit{Multiply by the LCD, 6.} \\ 6\left(\tfrac{1}{2}x\right) + 6\left(\tfrac{1}{3}y\right) &= 18 && \textit{Distributive property} \\ 3x + 2y &= 18 && (3) \end{aligned}$$

From equation (2), substitute $3x$ for y in equation (3).

$$\begin{aligned} 3x + 2(3x) &= 18 \\ 3x + 6x &= 18 \\ 9x &= 18 \\ x &= 2 \end{aligned}$$

To find y, let $x = 2$ in equation (2).

$$y = 3(2) = 6$$

The solution set is $\{(2, 6)\}$.

23. $\frac{1}{2}x + \frac{1}{3}y = -\frac{1}{3}$ (1)
$\frac{1}{2}x + 2y = -7$ (2)

First, clear all fractions.

Equation (1):

$$\begin{aligned} 6\left(\tfrac{1}{2}x + \tfrac{1}{3}y\right) &= 6\left(-\tfrac{1}{3}\right) && \textit{Multiply by the LCD, 6.} \\ 6\left(\tfrac{1}{2}x\right) + 6\left(\tfrac{1}{3}y\right) &= -2 && \textit{Distributive property} \\ 3x + 2y &= -2 && (3) \end{aligned}$$

Equation (2):

$$\begin{aligned} 2\left(\tfrac{1}{2}x + 2y\right) &= 2(-7) && \textit{Multiply by 2.} \\ x + 4y &= -14 && (4) \end{aligned}$$

The system has been simplified to

$$3x + 2y = -2 \quad (3)$$
$$x + 4y = -14 \quad (4)$$

Solve this system by the substitution method.

$$x = -4y - 14 \quad (5) \quad \textit{Solve (4) for x.}$$

$$\begin{aligned} 3(-4y - 14) + 2y &= -2 \quad \textit{Substitute for x in (3).} \\ -12y - 42 + 2y &= -2 \\ -10y - 42 &= -2 \\ -10y &= 40 \\ y &= -4 \end{aligned}$$

To find x, let $y = -4$ in equation (5).

$$x = -4(-4) - 14 = 16 - 14 = 2$$

The solution set is $\{(2, -4)\}$.

25. $\dfrac{x}{5} + 2y = \dfrac{8}{5} \quad (1)$

$\dfrac{3x}{5} + \dfrac{y}{2} = -\dfrac{7}{10} \quad (2)$

First, clear all fractions.

Equation (1):

$$5\left(\frac{x}{5} + 2y\right) = 5\left(\frac{8}{5}\right) \quad \textit{Multiply by 5.}$$
$$x + 10y = 8 \quad (3)$$

Equation (2):

$$10\left(\frac{3x}{5} + \frac{y}{2}\right) = 10\left(-\frac{7}{10}\right) \quad \textit{Multiply by 10.}$$
$$6x + 5y = -7 \quad (4)$$

The system has been simplified to

$$x + 10y = 8 \quad (3)$$
$$6x + 5y = -7. \quad (4)$$

Solve this system by the substitution method. Solve equation (3) for x.

$$x = 8 - 10y \quad (5)$$

Substitute $8 - 10y$ for x in equation (4).

$$\begin{aligned} 6(8 - 10y) + 5y &= -7 \\ 48 - 60y + 5y &= -7 \\ -55y &= -55 \\ y &= 1 \end{aligned}$$

To find x, let $y = 1$ in equation (5).

$$x = 8 - 10(1) = -2$$

The solution set is $\{(-2, 1)\}$.

27. $\frac{1}{6}x + \frac{1}{3}y = 8 \quad (1)$

$\frac{1}{4}x + \frac{1}{2}y = 12 \quad (2)$

Multiply equation (1) by 6.

$$6\left(\tfrac{1}{6}x + \tfrac{1}{3}y\right) = 6(8)$$
$$x + 2y = 48 \quad (3)$$

Multiply equation (2) by 4.

$$4\left(\tfrac{1}{4}x + \tfrac{1}{2}y\right) = 4(12)$$
$$x + 2y = 48 \quad (4)$$

Equations (3) and (4) are identical.

This means that every solution of one equation is also a solution of the other, so the system has an infinite number of solutions. The solution set is $\{(x, y) \mid x + 2y = 48\}$.

29. $0.2x - 1.3y = -3.2 \quad (1)$

$-0.1x + 2.7y = 9.8 \quad (2)$

Multiply equation (1) and equation (2) by 10 to eliminate the decimals. Then solve the resulting equations by the substitution method.

$$2x - 13y = -32 \quad (3)$$
$$-x + 27y = 98 \quad (4)$$

Solve equation (4) for x.

$$27y - 98 = x \quad (5)$$

Substitute $27y - 98$ for x in equation (3).

$$\begin{aligned} 2x - 13y &= -32 \\ 2(27y - 98) - 13y &= -32 \quad \textit{Let } x = 27y - 98. \\ 54y - 196 - 13y &= -32 \\ 41y &= 164 \\ y &= 4 \end{aligned}$$

To find x, let $y = 4$ in equation (5).

$$x = 27(4) - 98 = 10$$

The solution set is $\{(10, 4)\}$. Check in both of the original equations.

31. $0.3x - 0.1y = 2.1 \quad (1)$

$0.6x + 0.3y = -0.3 \quad (2)$

Multiply equation (1) and equation (2) by 10 to eliminate the decimals. Then solve the resulting equations by the substitution method.

$$3x - y = 21 \quad (3)$$
$$6x + 3y = -3 \quad (4)$$

Solve equation (3) for y.

$$y = 3x - 21 \quad (5)$$

Substitute $3x - 21$ for y in equation (4).

$$\begin{aligned} 6x + 3(3x - 21) &= -3 \\ 6x + 9x - 63 &= -3 \quad \textit{Let } y = 3x - 21. \\ 15x - 63 &= -3 \\ 15x &= 60 \\ x &= 4 \end{aligned}$$

To find y, let $x = 4$ in equation (5).

$$y = 3(4) - 21 = -9$$

The solution set is $\{(4, -9)\}$. Check in both of the original equations.

◆◆◆ Relating Concepts 33–36 ◆◆◆

33. To find the total cost, multiply the number of bicycles (x) by the cost per bicycle ($400), and add the fixed cost ($5000). Thus, $y_1 = 400x + 5000$ gives the total cost (in dollars).

34. Since each bicycle sells for $600, the total revenue for selling x bicycles is $600x$ (in dollars). Thus, $y_2 = 600x$ gives the total revenue.

35. $y_1 = 400x + 5000$ (1)
$y_2 = 600x$ (2)

To solve this system by the substitution method, substitute $600x$ for y_1 in equation (1).

$$\begin{aligned} 600x &= 400x + 5000 \\ 200x &= 5000 \\ x &= 25 \end{aligned}$$

If $x = 25$, $y_2 = 600(25) = 15{,}000$.

The solution set is $\{(25, 15{,}000)\}$.

36. The value of x from Exercise 35 is the number of bikes it takes to break even. When <u>25</u> bikes are sold, the break-even point is reached. At that point, you have spent <u>15,000</u> dollars and taken in <u>15,000</u> dollars.

37. $y = 6 - x$ (1)
$y = 2x$ (2)

Substitute $2x$ for y in equation (1).

$$\begin{aligned} 2x &= 6 - x \\ 3x &= 6 \\ x &= 2 \end{aligned}$$

Substituting 2 for x in either of the original equations gives $y = 4$.

The solution set is $\{(2, 4)\}$.

Input the equations $Y_1 = 6 - X$ and $Y_2 = 2X$. Next, graph the equations using a standard window. On the TI-82/3/4, just press [ZOOM] [6].

Now we'll let the calculator find the coordinates of the point of intersection of the graphs. Press [2nd] [CALC] [5] [ENTER] [ENTER] to indicate the graphs for which we're trying to find the point of intersection. Now press the right cursor key, [▷], nine times to get close to the point of intersection. Lastly, press [ENTER] to produce the following graph.

10

−10 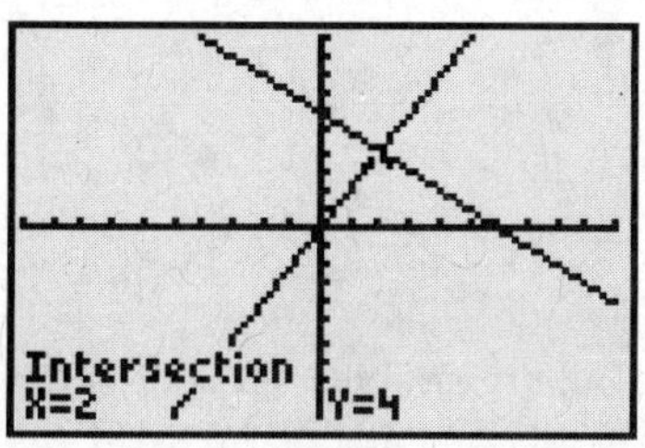 10

−10

The display at the bottom of the last figure indicates that the solution set is $\{(2, 4)\}$.

39. $y = -\frac{4}{3}x + \frac{19}{3}$ (1)
$y = \frac{15}{2}x - \frac{5}{2}$ (2)

Substitute the expression from equation (2) into equation (1).

$$\tfrac{15}{2}x - \tfrac{5}{2} = -\tfrac{4}{3}x + \tfrac{19}{3}$$

Multiply by 6 to clear fractions.

$$\begin{aligned} 6(\tfrac{15}{2}x - \tfrac{5}{2}) &= 6(-\tfrac{4}{3}x + \tfrac{19}{3}) \\ 45x - 15 &= -8x + 38 \\ 53x &= 53 \\ x &= 1 \end{aligned}$$

To find y, let $x = 1$ in equation (1).

$$\begin{aligned} y &= -\tfrac{4}{3}(1) + \tfrac{19}{3} \\ &= -\tfrac{4}{3} + \tfrac{19}{3} = \tfrac{15}{3} = 5 \end{aligned}$$

The solution set is $\{(1, 5)\}$.

10

−10 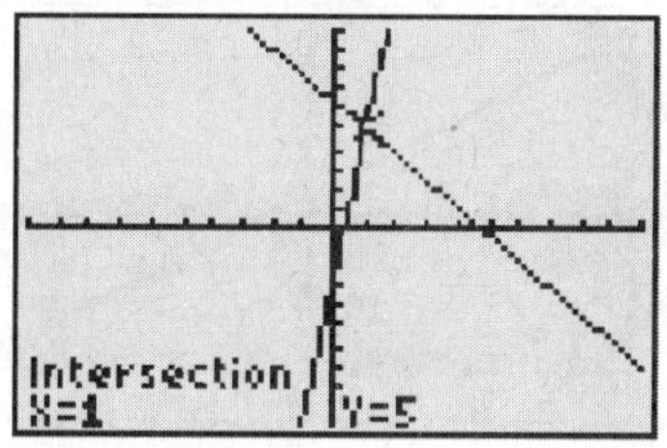 10

−10

41. $4x + 5y = 5$ (1)
$2x + 3y = 1$ (2)

Solve equation (2) for y.

$$\begin{aligned} 2x + 3y &= 1 \\ 3y &= 1 - 2x \\ y &= \frac{1 - 2x}{3} \quad (3) \end{aligned}$$

Substitute for y in equation (1).

$$4x + 5\left(\frac{1 - 2x}{3}\right) = 5$$

$$4x + \frac{5(1 - 2x)}{3} = 5$$

Multiply by 3 to clear fractions.

$$\begin{aligned} 3\left[4x + \frac{5(1 - 2x)}{3}\right] &= 3(5) \\ 12x + 5(1 - 2x) &= 15 \\ 12x + 5 - 10x &= 15 \\ 2x &= 10 \\ x &= 5 \end{aligned}$$

To find y, let $x = 5$ in equation (3).

$$y = \frac{1-2(5)}{3} = \frac{-9}{3} = -3$$

The solution set is $\{(5, -3)\}$.

To graph the original system on a graphing calculator, each equation must be solved for y.

Equation (1):

$$\begin{aligned} 4x + 5y &= 5 \\ 5y &= 5 - 4x \\ y &= \frac{5-4x}{5} \end{aligned}$$

Equation (2) was solved for y — see (3).

Thus, the equations to input are

$$Y_1 = \frac{5-4X}{5} = (5-4X)/5$$

and

$$Y_2 = \frac{1-2X}{3} = (1-2X)/3.$$

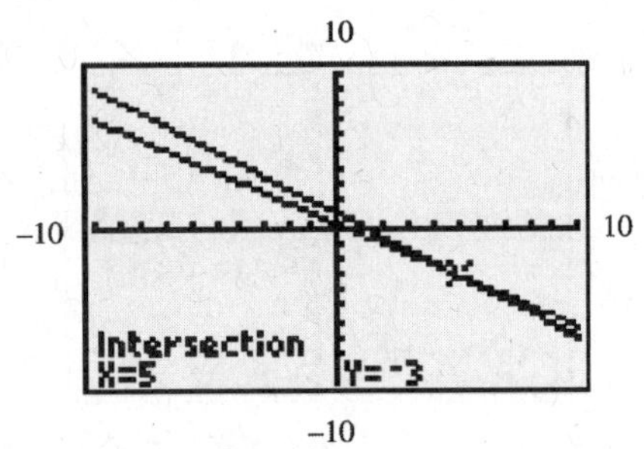

43. $(14x - 3y) + (2x + 3y)$
$= 14x - 3y + 2x + 3y$
$= (14 + 2)x + (-3 + 3)y$
$= 16x + 0y = 16x$

45. $(-x + 7y) + (3y + x)$
$= -x + 7y + 3y + x$
$= (-1 + 1)x + (7 + 3)y$
$= 0x + 10y = 10y$

47. To get a sum of 0, we must add the additive inverse of $-4x$, that is, $4x$.

49. The desired term times $4y$ must equal $-8y$, which is the additive inverse of $8y$. By trial and error, -2 times $4y$ equals $-8y$, so $4y$ must be multiplied by -2.

8.3 Solving Systems of Linear Equations by Elimination

8.3 Now Try Exercises

N1. $x - y = 4$
$3x + y = 8$

To eliminate y, add equations (1) and (2).

$$\begin{array}{rcrcrl} x & - & y & = & 4 & (1) \\ 3x & + & y & = & 8 & (2) \\ \hline 4x & & & = & 12 & \textit{Add (1) and (2).} \\ & & x & = & 3 & \end{array}$$

This result gives the x-value of the solution. To find the y-value of the solution, substitute 2 for x in either equation. We will use equation (2).

$$\begin{aligned} 3x + y &= 8 && (2) \\ 3(3) + y &= 8 && \textit{Let } x = 3. \\ 9 + y &= 8 \\ y &= -1 \end{aligned}$$

Check by substituting 3 for x and -1 for y in both equations of the original system.

Check Equation (1)

$$\begin{aligned} x - y &= 4 \\ 3 - (-1) &\stackrel{?}{=} 4 && \textit{Let } x = 3,\ y = -1. \\ 3 + 1 &\stackrel{?}{=} 4 \\ 4 &= 4 && \textit{True} \end{aligned}$$

Check Equation (2)

$$\begin{aligned} 3x + y &= 8 \\ 3(3) + (-1) &\stackrel{?}{=} 8 && \textit{Let } x = 3,\ y = -1. \\ 9 - 1 &\stackrel{?}{=} 8 \\ 8 &= 8 && \textit{True} \end{aligned}$$

Since both results are true, the solution set of the system is $\{(3, -1)\}$.

N2. $2x - 6 = -3y$
$5x - 3y = -27$

First, rewrite the first equation in the standard form, $Ax + By = C$.

$$\begin{array}{rcrcrl} 2x & + & 3y & = & 6 & (1) \\ 5x & - & 3y & = & -27 & (2) \\ \hline 7x & & & = & -21 & \textit{Add (1) and (2).} \\ & & x & = & -3 & \end{array}$$

Substitute -3 for x in equation (1).

$$\begin{aligned} 2x + 3y &= 6 \\ 2(-3) + 3y &= 6 \\ -6 + 3y &= 6 \\ 3y &= 12 \\ y &= 4 \end{aligned}$$

The solution set is $\{(-3, 4)\}$.

N3. $3x - 5y = 25$ (1)
$2x + 8y = -6$ (2)

If we simply add the equations, we will not eliminate either variable. To eliminate y, multiply equation (1) by 8, equation (2) by 5, and add the resulting equations.

$$\begin{array}{rcrcrl} 24x & - & 40y & = & 200 & (3)\;\; 8 \times \text{Eq. } (1) \\ 10x & + & 40y & = & -30 & (4)\;\; 5 \times \text{Eq. } (2) \\ \hline 34x & & & = & 170 & \textit{Add } (3) \textit{ and } (4). \\ & & x & = & 5 & \end{array}$$

Substitute 5 for x in equation (1).

$$\begin{aligned} 3x - 5y &= 25 \\ 3(5) - 5y &= 25 \\ 15 - 5y &= 25 \\ -5y &= 10 \\ y &= -2 \end{aligned}$$

Check $x = 5$, $y = -2$:

$$\begin{array}{lll} \text{Eq. } (1)\text{:} & 15 - (-10) = 25 & \textit{True} \\ \text{Eq. } (2)\text{:} & 10 + (-16) = -6 & \textit{True} \end{array}$$

The solution set is $\{(5, -2)\}$.

N4. $4x + 9y = 3$ (A)
$5y = 6 - 3x$ (B)

Rewrite in standard form.

$$\begin{array}{ll} 4x + 9y = 3 & (1) \\ 3x + 5y = 6 & (2) \end{array}$$

Solve for y by using elimination. The least common multiple of 4 and 3 is 12.

$$\begin{array}{rcrcrl} -12x & - & 27y & = & -9 & (3)\;\; -3 \times \text{Eq. } (1) \\ 12x & + & 20y & = & 24 & (4)\;\; 4 \times \text{Eq. } (2) \\ \hline & & -7y & = & 15 & \textit{Add } (3) \textit{ and } (4). \\ & & y & = & -\frac{15}{7} & \end{array}$$

Solve for x by using elimination. The least common multiple of 9 and 5 is 45.

$$\begin{array}{rcrcrl} 20x & + & 45y & = & 15 & (5)\;\; 5 \times \text{Eq. } (1) \\ -27x & - & 45y & = & -54 & (6)\;\; -9 \times \text{Eq. } (2) \\ \hline -7x & & & = & -39 & \textit{Add } (5) \textit{ and } (6). \\ & & x & = & \frac{39}{7} & \end{array}$$

Check $x = \frac{39}{7}$, $y = -\frac{15}{7}$:

$$\begin{array}{lll} \text{Eq. (A):} & \frac{156}{7} - \frac{135}{7} = \frac{21}{7} & \textit{True} \\ \text{Eq. (B):} & -\frac{75}{7} = \frac{42}{7} - \frac{117}{7} & \textit{True} \end{array}$$

The solution set is $\left\{\left(\frac{39}{7}, -\frac{15}{7}\right)\right\}$.

N5. (a) $x - y = 2$ (1)
$5x - 5y = 10$ (2)

$$\begin{array}{rcrcrl} -5x & + & 5y & = & -10 & (3)\;\; -5 \times \text{Eq. } (1) \\ 5x & - & 5y & = & 10 & (2) \\ \hline & & 0 & = & 0 & \textit{Add } (3) \textit{ and } (2). \end{array}$$

The true statement, $0 = 0$, indicates that there are an infinite number of solutions and the solution set is $\{(x, y) \mid x - y = 2\}$.

(b)

$$\begin{array}{rcrcrl} 4x & + & 3y & = & 0 & (1) \\ -4x & - & 3y & = & -1 & (2) \\ \hline & & 0 & = & -1 & \textit{Add } (1) \textit{ and } (2). \end{array}$$

This false statement, $0 = -1$, indicates that the given system has solution set $\emptyset$.

8.3 Section Exercises

1. The statement is *false*; if the elimination method leads to the equation $0 = -1$, the solution set of the system is $\emptyset$.

In Exercises 3–42, check your answers by substituting into *both* of the original equations. The check will be shown only for Exercise 3.

3. To eliminate y, add equations (1) and (2).

$$\begin{array}{rcrcrl} x & - & y & = & -2 & (1) \\ x & + & y & = & 10 & (2) \\ \hline 2x & & & = & 8 & \textit{Add } (1) \textit{ and } (2). \\ & & x & = & 4 & \end{array}$$

This result gives the x-value of the solution. To find the y-value of the solution, substitute 4 for x in either equation. We will use equation (2).

$$\begin{array}{rll} x + y &= 10 & \\ 4 + y &= 10 & \textit{Let } x = 4. \\ y &= 6 & \end{array}$$

Check by substituting 4 for x and 6 for y in both equations of the original system.

Check Equation (1)

$$\begin{array}{rll} x - y &= -2 & \\ 4 - 6 &\stackrel{?}{=} -2 & \textit{Let } x = 4,\ y = 6. \\ -2 &= -2 & \textit{True} \end{array}$$

Check Equation (2)

$$\begin{array}{rll} x + y &= 10 & \\ 4 + 6 &\stackrel{?}{=} 10 & \textit{Let } x = 4,\ y = 6. \\ 10 &= 10 & \textit{True} \end{array}$$

The solution set is $\{(4, 6)\}$.

5.

$$\begin{array}{rcrcrl} 2x & + & y & = & -5 & (1) \\ x & - & y & = & 2 & (2) \\ \hline 3x & & & = & -3 & \textit{Add } (1) \textit{ and } (2). \\ & & x & = & -1 & \end{array}$$

Substitute -1 for x in equation (1) to find the y-value of the solution.

$$\begin{array}{rll} 2x + y &= -5 & \\ 2(-1) + y &= -5 & \textit{Let } x = -1. \\ -2 + y &= -5 & \\ y &= -3 & \end{array}$$

The solution set is $\{(-1, -3)\}$.

7. First, rewrite both equations in the standard form, $Ax + By = C$. We can write the first equation, $2y = -3x$, as $3x + 2y = 0$ by adding $3x$ to each side.

$$\begin{array}{rcrcrl} 3x & + & 2y & = & 0 & (1) \\ -3x & - & y & = & 3 & (2) \\ \hline & & y & = & 3 & \textit{Add } (1) \textit{ and } (2). \end{array}$$

Substitute 3 for y in equation (1).

$$\begin{aligned} 3x + 2y &= 0 \\ 3x + 2(3) &= 0 \\ 3x + 6 &= 0 \\ 3x &= -6 \\ x &= -2 \end{aligned}$$

The solution set is $\{(-2, 3)\}$.

9. $6x - y = -1$
$5y = 17 + 6x$

Rewrite in standard form.

$$\begin{array}{rcrcrl} 6x & - & y & = & -1 & (1) \\ -6x & + & 5y & = & 17 & (2) \\ \hline & & 4y & = & 16 & \textit{Add } (1) \textit{ and } (2). \\ & & y & = & 4 & \textit{Solve for } y. \end{array}$$

Substitute 4 for y in equation (1).

$$\begin{aligned} 6x - y &= -1 \\ 6x - 4 &= -1 \\ 6x &= 3 \\ x &= \tfrac{3}{6} = \tfrac{1}{2} \end{aligned}$$

The solution set is $\left\{\left(\frac{1}{2}, 4\right)\right\}$.

11. $2x - y = 12$ (1)
$3x + 2y = -3$ (2)

If we simply add the equations, we will not eliminate either variable. To eliminate y, multiply equation (1) by 2 and add the result to equation (2).

$$\begin{array}{rcrcrl} 4x & - & 2y & = & 24 & (3) \\ 3x & + & 2y & = & -3 & (2) \\ \hline 7x & & & = & 21 & \textit{Add } (3) \textit{ and } (2). \\ & & x & = & 3 & \end{array}$$

Substitute 3 for x in equation (1).

$$\begin{aligned} 2x - y &= 12 \\ 2(3) - y &= 12 \\ -y &= 6 \\ y &= -6 \end{aligned}$$

The solution set is $\{(3, -6)\}$.

13. $x + 4y = 16$ (1)
$3x + 5y = 20$ (2)

To eliminate x, multiply equation (1) by -3 and add the result to equation (2).

$$\begin{array}{rcrcrl} -3x & - & 12y & = & -48 & (3) \\ 3x & + & 5y & = & 20 & (2) \\ \hline & & -7y & = & -28 & \textit{Add } (3) \textit{ and } (2). \\ & & y & = & 4 & \end{array}$$

Substitute 4 for y in equation (1).

$$\begin{aligned} x + 4y &= 16 \\ x + 4(4) &= 16 \\ x + 16 &= 16 \\ x &= 0 \end{aligned}$$

The solution set is $\{(0, 4)\}$.

15. $2x - 8y = 0$ (1)
$4x + 5y = 0$ (2)

To eliminate x, multiply equation (1) by -2 and add the result to equation (2).

$$\begin{array}{rcrcrl} -4x & + & 16y & = & 0 & (3) \\ 4x & + & 5y & = & 0 & (2) \\ \hline & & 21y & = & 0 & \textit{Add } (3) \textit{ and } (2). \\ & & y & = & 0 & \end{array}$$

Substitute 0 for y in equation (1).

$$\begin{aligned} 2x - 8y &= 0 \\ 2x - 8(0) &= 0 \\ 2x &= 0 \\ x &= 0 \end{aligned}$$

The solution set is $\{(0, 0)\}$.

17. $3x + 3y = 33$ (1)
$5x - 2y = 27$ (2)

To eliminate y, multiply equation (1) by 2 and equation (2) by 3.

$$\begin{array}{rcrcrl} 6x & + & 6y & = & 66 & (3) \\ 15x & - & 6y & = & 81 & (4) \\ \hline 21x & & & = & 147 & \textit{Add } (3) \textit{ and } (4). \\ & & x & = & 7 & \end{array}$$

Substitute 7 for x in equation (2).

$$\begin{aligned} 5x - 2y &= 27 \\ 5(7) - 2y &= 27 \\ 35 - 2y &= 27 \\ -2y &= -8 \\ y &= 4 \end{aligned}$$

The solution set is $\{(7, 4)\}$.

Note that we could have reduced equation (1) by dividing by 3 in the beginning.

19. $5x + 4y = 12$ (1)
$3x + 5y = 15$ (2)

To eliminate x, we could multiply equation (1) by $-\frac{3}{5}$, but that would introduce fractions and make the solution more complicated. Instead, we'll work with the least common multiple of the coefficients of x, which is 15, and choose suitable multipliers of these coefficients so that the new coefficients are opposites.

In this case, we could pick -3 times equation (1) and 5 times equation (2) *or* 3 times equation (1) and -5 times equation (2). If we wanted to eliminate y, we could multiply equation (1) by -5 and equation (2) by 4 *or* equation (1) by 5 and equation (2) by -4.

$$\begin{array}{rcrcrl} -15x & - & 12y & = & -36 & (3)\ -3 \times \text{Eq.}(1) \\ 15x & + & 25y & = & 75 & (4)\ \ 5 \times \text{Eq.}(2) \\ \hline & & 13y & = & 39 & \textit{Add } (3) \textit{ and } (4). \\ & & y & = & 3 & \end{array}$$

Substitute 3 for y in equation (1).

$$\begin{aligned} 5x + 4y &= 12 \\ 5x + 4(3) &= 12 \\ 5x + 12 &= 12 \\ 5x &= 0 \\ x &= 0 \end{aligned}$$

The solution set is $\{(0, 3)\}$.

21. $5x - 4y = 15$ (1)
$-3x + 6y = -9$ (2)

$$\begin{array}{rcrcrl} 15x & - & 12y & = & 45 & (3)\ \ 3 \times \text{Eq.}(1) \\ -15x & + & 30y & = & -45 & (4)\ \ 5 \times \text{Eq.}(2) \\ \hline & & 18y & = & 0 & \textit{Add } (3) \textit{ and } (4). \\ & & y & = & 0 & \end{array}$$

Substitute 0 for y in equation (1).

$$\begin{aligned} 5x - 4y &= 15 \\ 5x - 4(0) &= 15 \\ 5x &= 15 \\ x &= 3 \end{aligned}$$

The solution set is $\{(3, 0)\}$.

23. $-x + 3y = 4$ (1)
$-2x + 6y = 8$ (2)

$$\begin{array}{rcrcrl} 2x & - & 6y & = & -8 & (3)\ \ -2 \times \text{Eq.}(1) \\ -2x & + & 6y & = & 8 & (2) \\ \hline & & 0 & = & 0 & \textit{Add } (3) \textit{ and } (2). \end{array}$$

Since $0 = 0$ is a *true* statement, the equations are equivalent. This result indicates that every solution of one equation is also a solution of the other; there are an *infinite number of solutions.* The solution set is $\{(x, y) \mid x - 3y = -4\}$.

25. $5x - 2y = 3$ (1)
$10x - 4y = 5$ (2)

$$\begin{array}{rcrcrl} -10x & + & 4y & = & -6 & (3)\ \ -2 \times \text{Eq.}(1) \\ 10x & - & 4y & = & 5 & (2) \\ \hline & & 0 & = & -1 & \textit{Add } (3) \textit{ and } (2). \end{array}$$

Since $0 = -1$ is a *false* statement, there are no solutions of the system and the solution set is $\emptyset$.

27. $6x - 2y = -22$ (1)
$-3x + 4y = 17$ (2)

$$\begin{array}{rcrcrl} 6x & - & 2y & = & -22 & (1) \\ -6x & + & 8y & = & 34 & (3)\ \ 2 \times \text{Eq.}(2) \\ \hline & & 6y & = & 12 & \textit{Add } (1) \textit{ and } (3). \\ & & y & = & 2 & \end{array}$$

Substitute 2 for y in equation (1).

$$\begin{aligned} 6x - 2y &= -22 \\ 6x - 2(2) &= -22 \\ 6x - 4 &= -22 \\ 6x &= -18 \\ x &= -3 \end{aligned}$$

The solution set is $\{(-3, 2)\}$.

29. $3x = 3 + 2y$ (1)
$-\frac{4}{3}x + y = \frac{1}{3}$ (2)

Rewrite equation (1) in standard form and multiply equation (2) by 3 to clear fractions.

$3x - 2y = 3$ (3)
$-4x + 3y = 1$ (4)

$$\begin{array}{rcrcrl} 12x & - & 8y & = & 12 & (5)\ \ 4 \times \text{Eq.}(3) \\ -12x & + & 9y & = & 3 & (6)\ \ 3 \times \text{Eq.}(4) \\ \hline & & y & = & 15 & \textit{Add } (5) \textit{ and } (6). \end{array}$$

$$\begin{array}{rcrcrl} 9x & - & 6y & = & 9 & (7)\ \ 3 \times \text{Eq.}(3) \\ -8x & + & 6y & = & 2 & (8)\ \ 2 \times \text{Eq.}(4) \\ \hline x & & & = & 11 & \textit{Add } (7) \textit{ and } (8). \end{array}$$

The solution set is $\{(11, 15)\}$.

31. $\frac{1}{5}x + y = \frac{6}{5}$ (1)
$\frac{1}{10}x + \frac{1}{3}y = \frac{5}{6}$ (2)

First, clear all fractions.

Equation (1):

$5(\frac{1}{5}x + y) = 5(\frac{6}{5})$ *Multiply by LCD, 5.*
$x + 5y = 6$ (3)

Equation (2):

$30(\frac{1}{10}x + \frac{1}{3}y) = 30(\frac{5}{6})$ *Multiply by LCD, 30.*
$3x + 10y = 25$ (4)

The system has been simplified to

$$\begin{aligned} x + 5y &= 6 \quad &(3)\\ 3x + 10y &= 25 \quad &(4)\end{aligned}$$

To eliminate y, add -2 times equation (3) to (4).

$$\begin{array}{rcrcrl} -2x & - & 10y & = & -12 & (5) \quad -2 \times \text{Eq.}(3)\\ 3x & + & 10y & = & 25 & (4)\\ \hline x & & & = & 13 & \textit{Add } (5) \textit{ and } (4).\end{array}$$

Substitute 13 for x in equation (3).

$$\begin{aligned} x + 5y &= 6 \quad (3)\\ 13 + 5y &= 6\\ 5y &= -7\\ y &= -\tfrac{7}{5}\end{aligned}$$

The solution set is $\left\{\left(13, -\frac{7}{5}\right)\right\}$.

33. $2.4x + 1.7y = 7.6 \quad (1)$
$1.2x - 0.5y = 9.2 \quad (2)$

Multiply equation (1) and equation (2) by 10 to eliminate the decimals. Then solve the resulting equations by the elimination method.

$$\begin{aligned} 24x + 17y &= 76 \quad (3)\\ 12x - \ \ 5y &= 92 \quad (4)\end{aligned}$$

Multiply equation (2) by -2.

$$\begin{array}{rcrcrl} 24x & + & 17y & = & 76 & (3)\\ -24x & + & 10y & = & -184 & (5)\\ \hline & & 27y & = & -108 & \textit{Add.}\\ & & y & = & -4 & \end{array}$$

Substitute -4 for y in equation (4).

$$\begin{aligned} 12x - 5y &= 92\\ 12x - 5(-4) &= 92 \quad \textit{Let } y = -4.\\ 12x + 20 &= 92\\ 12x &= 72\\ x &= 6 \quad \textit{Divide by 12.}\end{aligned}$$

The solution set is $\{(6, -4)\}$.

35. $x + 3y = 6$
$-2x + 12 = 6y$

Rewrite in standard form.

$$\begin{aligned} x + 3y &= 6 \quad &(1)\\ -2x - 6y &= -12 \quad &(2)\end{aligned}$$

To eliminate x, multiply equation (1) by 2 and add the result to equation (2).

$$\begin{array}{rcrcrl} 2x & + & 6y & = & 12 & (3)\\ -2x & - & 6y & = & -12 & (2)\\ \hline & & 0 & = & 0 & \textit{Add } (3) \textit{ and } (2).\end{array}$$

The true statement, $0 = 0$, indicates that there are an infinite number of solutions and the solution set is $\{(x, y) \mid x + 3y = 6\}$.

37. $4x - 3y = 1$
$8x = 3 + 6y$

Rewrite in standard form.

$$\begin{aligned} 4x - 3y &= 1 \quad &(1)\\ 8x - 6y &= 3 \quad &(2)\end{aligned}$$

To eliminate x, multiply equation (1) by -2 and add the result to equation (2).

$$\begin{array}{rcrcrl} -8x & + & 6y & = & -2 & (3)\\ 8x & - & 6y & = & 3 & (2)\\ \hline & & 0 & = & 1 & \textit{Add } (3) \textit{ and } (2).\end{array}$$

The false statement, $0 = 1$, indicates that the given system has solution set $\emptyset$.

39. $4x = 3y - 2$
$5x + 3 = 2y$

Rewrite in standard form.

$$\begin{aligned} -4x + 3y &= 2 \quad &(1)\\ 5x - 2y &= -3 \quad &(2)\end{aligned}$$

$$\begin{array}{rcrcrl} -8x & + & 6y & = & 4 & (3) \quad 2 \times \text{Eq.}(1)\\ 15x & - & 6y & = & -9 & (4) \quad 3 \times \text{Eq.}(2)\\ \hline 7x & & & = & -5 & \textit{Add } (3) \textit{ and } (4).\\ & & x & = & -\frac{5}{7} & \end{array}$$

Rather than substitute $-\frac{5}{7}$ for x in (1) or (2), we will eliminate x by multiplying equation (1) by 5, equation (2) by 4, and adding the results.

$$\begin{array}{rcrcrl} -20x & + & 15y & = & 10 & (5) \quad 5 \times \text{Eq.}(1)\\ 20x & - & 8y & = & -12 & (6) \quad 4 \times \text{Eq.}(2)\\ \hline & & 7y & = & -2 & \textit{Add } (5) \textit{ and } (6).\\ & & y & = & -\frac{2}{7} & \end{array}$$

Solving the system in this fashion reduces the chance of making an arithmetic error.

The solution set is $\left\{\left(-\frac{5}{7}, -\frac{2}{7}\right)\right\}$.

When you get a solution that has non-integer components, it is sometimes more difficult to check the problem than it was to solve it. A graphing calculator can be very helpful in this case. Just store the values for x and y in their respective memory locations, and then type the expressions as shown in the following screen. The results 2 and -3 (the right sides of the *equations* (1) and (2)) indicate that we have found the correct solution.

41. $24x + 12y = -7$ (1)
$16x - 18y = 17$ (2)

$$\begin{array}{rcrl} 48x + 24y &=& -14 & (3) \quad 2 \times \text{Eq.}(1) \\ -48x + 54y &=& -51 & (4) \quad -3 \times \text{Eq.}(2) \\ \hline 78y &=& -65 & \textit{Add (3) and (4).} \\ y &=& \frac{-65}{78} = -\frac{5}{6} & \end{array}$$

$$\begin{array}{rcrl} 72x + 36y &=& -21 & (5) \quad 3 \times \text{Eq.}(1) \\ 32x - 36y &=& 34 & (6) \quad 2 \times \text{Eq.}(2) \\ \hline 104x &=& 13 & \textit{Add (5) and (6).} \\ x &=& \frac{13}{104} = \frac{1}{8} & \end{array}$$

The solution set is $\left\{\left(\frac{1}{8}, -\frac{5}{6}\right)\right\}$.

◆◆◆ Relating Concepts 43–48 ◆◆◆

43. $y = ax + b$
$5.39 = a(2000) + b$ *Let x = 2000, y = 5.39.*
$5.39 = 2000a + b$

44. As in Exercise 43,

$$7.18 = 2008a + b.$$

45. $2000a + b = 5.39$ (1)
$2008a + b = 7.18$ (2)

Multiply equation (1) by -1 and add the result to equation (2),

$$\begin{array}{rcr} -2000a - b &=& -5.39 \\ 2008a + b &=& 7.18 \\ \hline 8a &=& 1.79 \\ a &=& 0.22375 \end{array}$$

Substitute 0.22375 for a in equation (1).

$$2000(0.22375) + b = 5.39$$
$$447.5 + b = 5.39$$
$$b = -442.11$$

The solution set is $\{(0.22375, -442.11)\}$.

46. **(a)** An equation of the segment *PQ* is

$$y = 0.22375x - 442.11$$

for $2000 \le x \le 2008$.

(b) $y = 0.22375x - 442.11$
$y = 0.22375(2007) - 442.11$ *Let x = 2007.*
$= 449.06625 - 442.11$
$= 6.95625 \approx 6.96$

This is a bit more ($0.08) than the actual figure.

47. $4(2x - 3y + z) = 4(5)$
$8x - 12y + 4z = 20$

48. $-3(-3x + 8y - z) = -3(0)$
$9x - 24y + 3z = 0$

49. $x + 2y + 3z = 9$
$1 + 2(-2) + 3z = 9$ *Let x = 1, y = −2.*
$1 - 4 + 3z = 9$
$3z = 12$
$z = 4$

51. Multiplying the first equation by -3 will give us $-3x$, which when added to $3x$ in the second equation, gives us 0.

Summary Exercises on Solving Systems of Linear Equations

1. **(a)** $3x + 5y = 69$
$y = 4x$

Use substitution since the second equation is solved for y.

(b) $3x + y = -7$
$x - y = -5$

Use elimination since the coefficients of the y-terms are opposites.

(c) $3x - 2y = 0$
$9x + 8y = 7$

Use elimination since the equations are in standard form with no coefficients of 1 or -1. Solving by substitution would involve fractions.

3. $4x - 3y = -8$ (1)
$x + 3y = 13$ (2)

(a) Solve the system by the elimination method.

$$\begin{array}{rcrl} 4x - 3y &=& -8 & (1) \\ x + 3y &=& 13 & (2) \\ \hline 5x &=& 5 & \textit{Add (1) and (2).} \\ x &=& 1 & \end{array}$$

To find y, let $x = 1$ in equation (2).

$$x + 3y = 13$$
$$1 + 3y = 13$$
$$3y = 12$$
$$y = 4$$

The solution set is $\{(1, 4)\}$.

(b) To solve this system by the substitution method, begin by solving equation (2) for x.

$$x + 3y = 13$$
$$x = -3y + 13$$

Substitute $-3y + 13$ for x in equation (1).

$$4(-3y + 13) - 3y = -8$$
$$-12y + 52 - 3y = -8$$
$$-15y = -60$$
$$y = 4$$

To find x, let $y = 4$ in equation (2).

$$\begin{aligned} x + 3y &= 13 \\ x + 3(4) &= 13 \\ x + 12 &= 13 \\ x &= 1 \end{aligned}$$

The solution set is $\{(1, 4)\}$.

(c) For this particular system, the elimination method is preferable because both equations are already written in the form $Ax + By = C$, and the equations can be added without multiplying either by a constant. Comparing solutions by the two methods, we see that the elimination method requires fewer steps than the substitution method for this system.

5.

$$\begin{aligned} 3x + 5y &= 69 \quad (1) \\ y &= 4x \quad (2) \end{aligned}$$

Equation (2) is already solved for y, so we'll use the substitution method. Substitute $4x$ for y in equation (1) and solve the resulting equation for x.

$$\begin{aligned} 3x + 5y &= 69 \\ 3x + 5(4x) &= 69 \quad \textit{Let } y = 4x. \\ 3x + 20x &= 69 \\ 23x &= 69 \\ x &= \tfrac{69}{23} = 3 \end{aligned}$$

To find the y-value of the solution, substitute 3 for x in equation (2).

$$\begin{aligned} y &= 4x \\ y &= 4(3) \quad \textit{Let } x = 3. \\ &= 12 \end{aligned}$$

The solution set is $\{(3, 12)\}$.

7.

$$\begin{aligned} 3x - 2y &= 0 \quad (1) \\ 9x + 8y &= 7 \quad (2) \end{aligned}$$

$$\begin{array}{rcrcrl} 12x & - & 8y & = & 0 & (3) \quad 4 \times \text{Eq. } (1) \\ 9x & + & 8y & = & 7 & (2) \\ \hline 21x & & & = & 7 & \textit{Add } (3) \textit{ and } (2). \\ & & x & = & \tfrac{7}{21} = \tfrac{1}{3} & \end{array}$$

$$\begin{array}{rcrcrl} -9x & + & 6y & = & 0 & (4) \quad -3 \times \text{Eq.}(1) \\ 9x & + & 8y & = & 7 & (2) \\ \hline & & 14y & = & 7 & \textit{Add } (4) \textit{ and } (2). \\ & & y & = & \tfrac{7}{14} = \tfrac{1}{2} & \end{array}$$

The solution set is $\left\{\left(\frac{1}{3}, \frac{1}{2}\right)\right\}$.

9.

$$\begin{aligned} 6x + 7y &= 4 \quad (1) \\ 5x + 8y &= -1 \quad (2) \end{aligned}$$

$$\begin{array}{rcrcrl} -30x & - & 35y & = & -20 & (3) \quad -5 \times \text{Eq.}(1) \\ 30x & + & 48y & = & -6 & (4) \quad 6 \times \text{Eq.}(2) \\ \hline & & 13y & = & -26 & \textit{Add } (3) \textit{ and } (4). \\ & & y & = & -2 & \end{array}$$

Substitute -2 for y in equation (1).

$$\begin{aligned} 6x + 7y &= 4 \\ 6x + 7(-2) &= 4 \quad \textit{Let } y = -2. \\ 6x - 14 &= 4 \\ 6x &= 18 \\ x &= 3 \end{aligned}$$

The solution set is $\{(3, -2)\}$.

11.

$$\begin{aligned} 4x - 6y &= 10 \quad (1) \\ -10x + 15y &= -25 \quad (2) \end{aligned}$$

Divide equation (1) by 2 and equation (2) by 5.

$$\begin{array}{rcrcrl} 2x & - & 3y & = & 5 & (3) \\ -2x & + & 3y & = & -5 & (4) \\ \hline & & 0 & = & 0 & \textit{Add } (3) \textit{ and } (4). \end{array}$$

Since $0 = 0$ is a *true* statement, there are an *infinite number of solutions*. The solution set is $\{(x, y) \mid 2x - 3y = 5\}$.

13.

$$\begin{aligned} 5x &= 7 + 2y \\ 5y &= 5 - 3x \end{aligned}$$

Rewrite in standard form.

$$\begin{aligned} 5x - 2y &= 7 \quad (1) \\ 3x + 5y &= 5 \quad (2) \end{aligned}$$

$$\begin{array}{rcrcrl} 25x & - & 10y & = & 35 & (3) \quad 5 \times \text{Eq.}(1) \\ 6x & + & 10y & = & 10 & (4) \quad 2 \times \text{Eq.}(2) \\ \hline 31x & & & = & 45 & \textit{Add } (3) \textit{ and } (4). \\ & & x & = & \tfrac{45}{31} & \end{array}$$

$$\begin{array}{rcrcrl} -15x & + & 6y & = & -21 & (5) \quad -3 \times \text{Eq.}(1) \\ 15x & + & 25y & = & 25 & (6) \quad 5 \times \text{Eq.}(2) \\ \hline & & 31y & = & 4 & \textit{Add } (5) \textit{ and } (6). \\ & & y & = & \tfrac{4}{31} & \end{array}$$

The solution set is $\left\{\left(\frac{45}{31}, \frac{4}{31}\right)\right\}$.

15.

$$\begin{aligned} 2x - 3y &= 7 \quad (1) \\ -4x + 6y &= 14 \quad (2) \end{aligned}$$

$$\begin{array}{rcrcrl} 4x & - & 6y & = & 14 & (3) \quad 2 \times \text{Eq.}(1) \\ -4x & + & 6y & = & 14 & (2) \\ \hline & & 0 & = & 28 & \textit{Add } (3) \textit{ and } (2). \end{array}$$

Since $0 = 28$ is *false*, the solution set is $\emptyset$.

17. $2x + 5y = 4$ (1)
$x + y = -1$ (2)

Solve equation (2) for x.

$$x = -y - 1 \quad (3)$$

Substitute $-y - 1$ for x in equation (1).

$$\begin{aligned} 2(-y-1) + 5y &= 4 \quad \textit{Let } x = -y - 1. \\ -2y - 2 + 5y &= 4 \\ 3y - 2 &= 4 \\ 3y &= 6 \\ y &= 2 \end{aligned}$$

Substitute 2 for y in equation (3).

$$\begin{aligned} x &= -(2) - 1 \quad \textit{Let } y = 2. \\ &= -3 \end{aligned}$$

The solution set is $\{(-3, 2)\}$.

19. $7x - 4y = 0$
$3x = 2y$

Rewrite in standard form.

$$\begin{aligned} 7x - 4y &= 0 \quad (1) \\ 3x - 2y &= 0 \quad (2) \end{aligned}$$

$$\begin{array}{rcrcrl} 7x & - & 4y & = & 0 & (1) \\ -6x & + & 4y & = & 0 & (3) \quad -2 \times \text{Eq.}(2) \\ \hline x & & & = & 0 & \textit{Add } (1) \textit{ and } (3). \end{array}$$

Substitute 0 for x in equation (1).

$$\begin{aligned} 7x - 4y &= 0 \\ 7(0) - 4y &= 0 \\ -4y &= 0 \\ y &= 0 \end{aligned}$$

The solution set is $\{(0, 0)\}$.

21. $\frac{1}{6}x + \frac{1}{6}y = 2$ (1)
$-\frac{1}{2}x - \frac{1}{3}y = -8$ (2)

Multiply each side of equation (1) by 6 to clear fractions.

$$\begin{aligned} 6(\tfrac{1}{6}x + \tfrac{1}{6}y) &= 6(2) \\ 6(\tfrac{1}{6}x) + 6(\tfrac{1}{6}y) &= 6(2) \\ x + y &= 12 \quad (3) \end{aligned}$$

Multiply each side of equation (2) by the LCD, 6, to clear fractions.

$$\begin{aligned} 6(-\tfrac{1}{2}x - \tfrac{1}{3}y) &= 6(-8) \\ 6(-\tfrac{1}{2}x) + 6(-\tfrac{1}{3}y) &= 6(-8) \\ -3x - 2y &= -48 \quad (4) \end{aligned}$$

The given system of equations has been simplified as follows.

$$\begin{aligned} x + y &= 12 \quad (3) \\ -3x - 2y &= -48 \quad (4) \end{aligned}$$

Multiply equation (3) by 3 and add the result to equation (4).

$$\begin{array}{rcrcr} 3x & + & 3y & = & 36 \\ -3x & - & 2y & = & -48 \\ \hline & & y & = & -12 \end{array}$$

To find x, let $y = -12$ in equation (3).

$$\begin{aligned} x + y &= 12 \\ x + (-12) &= 12 \\ x - 12 &= 12 \\ x &= 24 \end{aligned}$$

The solution set is $\{(24, -12)\}$.

23. $\frac{x}{2} - \frac{y}{3} = 9$ (1)
$\frac{x}{5} - \frac{y}{4} = 5$ (2)

To clear fractions, multiply each side of equation (1) by the LCD, 6.

$$\begin{aligned} 6\left(\frac{x}{2} - \frac{y}{3}\right) &= 6(9) \\ 3x - 2y &= 54 \quad (3) \end{aligned}$$

To clear fractions, multiply each side of equation (2) by the LCD, 20.

$$\begin{aligned} 20\left(\frac{x}{5} - \frac{y}{4}\right) &= 20(5) \\ 4x - 5y &= 100 \quad (4) \end{aligned}$$

We now have the simplified system

$$\begin{aligned} 3x - 2y &= 54 \quad (3) \\ 4x - 5y &= 100. \quad (4) \end{aligned}$$

To solve this system by the elimination method, multiply equation (3) by 5 and equation (4) by -2; then add the results.

$$\begin{array}{rcrcr} 15x & - & 10y & = & 270 \\ -8x & + & 10y & = & -200 \\ \hline 7x & & & = & 70 \\ & & x & = & 10 \end{array}$$

To find y, let $x = 10$ in equation (3).

$$\begin{aligned} 3x - 2y &= 54 \\ 3(10) - 2y &= 54 \\ 30 - 2y &= 54 \\ -2y &= 24 \\ y &= -12 \end{aligned}$$

The solution set is $\{(10, -12)\}$.

25. $0.2x - 0.3y = 0.1$ (1)
$0.3x - 0.2y = 0.9$ (2)

Multiply equation (1) and equation (2) by 10 to eliminate the decimals. Then solve the resulting equations by the elimination method.

$$2x - 3y = 1 \quad (3)$$
$$3x - 2y = 9 \quad (4)$$

Multiply equation (3) by −3 and equation (4) by 2.

$$\begin{array}{rcrcrl} -6x & + & 9y & = & -3 & (5) \\ 6x & - & 4y & = & 18 & (6) \\ \hline & & 5y & = & 15 & \textit{Add.} \\ & & y & = & 3 & \end{array}$$

Substitute 3 for y in equation (4).

$$\begin{aligned} 3x - 2y &= 9 \\ 3x - 2(3) &= 9 && \textit{Let } y = 3. \\ 3x - 6 &= 9 \\ 3x &= 15 \\ x &= 5 && \textit{Divide by 3.} \end{aligned}$$

The solution set is $\{(5, 3)\}$.

8.4 Solving Systems of Linear Equations in Three Variables

8.4 Now Try Exercises

N1.
$$\begin{aligned} x - y + 2z &= 1 && (1) \\ 3x + 2y + 7z &= 8 && (2) \\ -3x - 4y + 9z &= -10 && (3) \end{aligned}$$

Step 1
Eliminate x by adding equations (2) and (3).

$$\begin{array}{rcrcrcrl} 3x & + & 2y & + & 7z & = & 8 & (2) \\ -3x & - & 4y & + & 9z & = & -10 & (3) \\ \hline & - & 2y & + & 16z & = & -2 & (4) \end{array}$$

Step 2
To eliminate x again, multiply equation (1) by 3 and add the result to equation (3).

$$\begin{array}{rcrcrcrl} 3x & - & 3y & + & 6z & = & 3 & 3 \times (1) \\ -3x & - & 4y & + & 9z & = & -10 & (3) \\ \hline & - & 7y & + & 15z & = & -7 & (5) \end{array}$$

Step 3
Use equations (4) and (5) to eliminate y. Multiply equation (4) by −7 and add the result to 2 times equation (5).

$$\begin{array}{rcrcrl} 14y & - & 112z & = & 14 & -7 \times (4) \\ -14y & + & 30z & = & -14 & 2 \times (5) \\ \hline & - & 82z & = & 0 & \\ & & z & = & 0 & \end{array}$$

Step 4
Substitute 0 for z in equation (5) to find y.

$$\begin{aligned} -7y + 15z &= -7 && (5) \\ -7y + 15(0) &= -7 && \textit{Let } z = 0. \\ -7y &= -7 \\ y &= 1 \end{aligned}$$

Step 5
Substitute 0 for z and 1 for y in equation (1) to find x.

$$\begin{aligned} x - y + 2z &= 1 && (1) \\ x - 1 + 0 &= 1 && \textit{Let } y = 1, z = 0. \\ x &= 2 \end{aligned}$$

Step 6
Check $(2, 1, 0)$

Equation (1):	$2 - 1 + 0 = 1$	*True*
Equation (2):	$6 + 2 + 0 = 8$	*True*
Equation (3):	$-6 - 4 + 0 = -10$	*True*

The solution set is $\{(2, 1, 0)\}$.

N2.
$$\begin{array}{rcrcrcrl} 3x & & & - & z & = & -10 & (1) \\ & & 4y & + & 5z & = & 24 & (2) \\ x & - & 6y & & & = & -8 & (3) \end{array}$$

Since equation (1) is missing y, eliminate y again from equations (2) and (3). Multiply equation (2) by 3 and add the result to 2 times equation (3).

$$\begin{array}{rcrcrcrl} & & 12y & + & 15z & = & 72 & 3 \times (2) \\ 2x & - & 12y & & & = & -16 & 2 \times (3) \\ \hline 2x & & & + & 15z & = & 56 & (4) \end{array}$$

Use equation (4) together with equation (1) to eliminate z. Multiply equation (1) by 15 and add it to equation (4).

$$\begin{array}{rcrcrl} 45x & - & 15z & = & -150 & 15 \times (1) \\ 2x & + & 15z & = & 56 & (4) \\ \hline 47x & & & = & -94 & \\ & & x & = & -2 & \end{array}$$

Substitute −2 for x in equation (1) to find z.

$$\begin{aligned} 3x - z &= -10 && (1) \\ 3(-2) - z &= -10 && \textit{Let } x = -2. \\ -6 - z &= -10 \\ -z &= -4 \\ z &= 4 \end{aligned}$$

Substitute 4 for z in equation (2) to find y.

$$\begin{aligned} 4y + 5z &= 24 && (2) \\ 4y + 5(4) &= 24 && \textit{Let } z = 4. \\ 4y + 20 &= 24 \\ 4y &= 4 \\ y &= 1 \end{aligned}$$

Check $(-2, 1, 4)$

Equation (1): $-6 - 4 = -10$ *True*
Equation (2): $4 + 20 = 24$ *True*
Equation (3): $-2 - 6 = -8$ *True*

The solution set is $\{(-2, 1, 4)\}$.

N3.
$$\begin{aligned} x - 5y + 2z &= 4 \quad (1)\\ 3x + y - z &= 6 \quad (2)\\ -2x + 10y - 4z &= 7 \quad (3) \end{aligned}$$

Multiply equation (1) by 2 and add the result to equation (3).

$$\begin{array}{rcll} 2x - 10y + 4z &=& 8 & 2 \times (1)\\ -2x + 10y - 4z &=& 7 & (3)\\ \hline 0 &=& 15 & \textit{False} \end{array}$$

Since a false statement results, the system is inconsistent.

The solution set is $\emptyset$.

N4.
$$\begin{aligned} x - 3y + 2z &= 10 \quad (1)\\ -2x + 6y - 4z &= -20 \quad (2)\\ \tfrac{1}{2}x - \tfrac{3}{2}y + z &= 5 \quad (3) \end{aligned}$$

Since equation (2) is -2 times equation (1) and equation (3) is $\frac{1}{2}$ times equation (1), the three equations are dependent. All three have the same graph.

The solution set is $\{(x, y, z) \mid x - 3y + 2z = 10\}$.

N5.
$$\begin{aligned} x - 3y + 2z &= 4 \quad (1)\\ \tfrac{1}{3}x - y + \tfrac{2}{3}z &= 7 \quad (2)\\ \tfrac{1}{2}x - \tfrac{3}{2}y + z &= 2 \quad (3) \end{aligned}$$

Eliminate the fractions in equations (2) and (3).

$$\begin{array}{rll} x - 3y + 2z = 4 & (1) & \\ x - 3y + 2z = 24 & (4) & 3 \times (2)\\ x - 3y + 2z = 4 & (5) & 2 \times (3) \end{array}$$

Equations (1) and (5) are dependent (they have the same graph). Equations (1) and (4) are not equivalent. Since they have the same coefficients, but a different constant term, their graphs have no points in common (the planes are parallel). Thus, the system is inconsistent and the solution set is $\emptyset$.

8.4 Section Exercises

1. Substitute 1 for x, 2 for y, and 3 for z in $3x + 2y - z$, which is the left side of each equation.

$$\begin{aligned} 3(1) + 2(2) - (3) &= 3 + 4 - 3\\ &= 7 - 3\\ &= 4 \end{aligned}$$

Choice **B** is correct since its right side is 4.

3.
$$\begin{aligned} 2x - 5y + 3z &= -1 \quad (1)\\ x + 4y - 2z &= 9 \quad (2)\\ x - 2y - 4z &= -5 \quad (3) \end{aligned}$$

Eliminate x by adding equation (1) to -2 times equation (2).

$$\begin{array}{rcll} 2x - 5y + 3z &=& -1 & (1)\\ -2x - 8y + 4z &=& -18 & -2 \times (2)\\ \hline -13y + 7z &=& -19 & (4) \end{array}$$

Now eliminate x by adding equation (2) to -1 times equation (3).

$$\begin{array}{rcll} x + 4y - 2z &=& 9 & (2)\\ -x + 2y + 4z &=& 5 & -1 \times (3)\\ \hline 6y + 2z &=& 14 & (5) \end{array}$$

Use equations (4) and (5) to eliminate z. Multiply equation (4) by -2 and add the result to 7 times equation (5).

$$\begin{array}{rcll} 26y - 14z &=& 38 & -2 \times (4)\\ 42y + 14z &=& 98 & 7 \times (5)\\ \hline 68y &=& 136 & \\ y &=& 2 & \end{array}$$

Substitute 2 for y in equation (5) to find z.

$$\begin{aligned} 6y + 2z &= 14 \quad (5)\\ 6(2) + 2z &= 14\\ 12 + 2z &= 14\\ 2z &= 2\\ z &= 1 \end{aligned}$$

Substitute 2 for y and 1 for z in equation (3) to find x.

$$\begin{aligned} x - 2y - 4z &= -5 \quad (3)\\ x - 2(2) - 4(1) &= -5\\ x - 4 - 4 &= -5\\ x - 8 &= -5\\ x &= 3 \end{aligned}$$

The solution $(3, 2, 1)$ checks in all three of the original equations.
The solution set is $\{(3, 2, 1)\}$.

5.
$$\begin{aligned} 3x + 2y + z &= 8 \quad (1)\\ 2x - 3y + 2z &= -16 \quad (2)\\ x + 4y - z &= 20 \quad (3) \end{aligned}$$

Eliminate z by adding equations (1) and (3).

$$\begin{array}{rcll} 3x + 2y + z &=& 8 & (1)\\ x + 4y - z &=& 20 & (3)\\ \hline 4x + 6y &=& 28 & (4) \end{array}$$

To get another equation without z, multiply equation (3) by 2 and add the result to equation (2).

$$\begin{array}{rcll} 2x + 8y - 2z &=& 40 & 2 \times (3) \\ 2x - 3y + 2z &=& -16 & (2) \\ \hline 4x + 5y &=& 24 & (5) \end{array}$$

Use equations (4) and (5) to eliminate x. Multiply equation (4) by -1 and add the result to equation (5).

$$\begin{array}{rcll} -4x - 6y &=& -28 & -1 \times (4) \\ 4x + 5y &=& 24 & (5) \\ \hline -y &=& -4 & \\ y &=& 4 & \end{array}$$

Substitute 4 for y in equation (5) to find x.

$$\begin{aligned} 4x + 5y &= 24 \quad (5) \\ 4x + 5(4) &= 24 \\ 4x + 20 &= 24 \\ 4x &= 4 \\ x &= 1 \end{aligned}$$

Substitute 1 for x and 4 for y in equation (3) to find z.

$$\begin{aligned} x + 4y - z &= 20 \quad (3) \\ 1 + 4(4) - z &= 20 \\ 1 + 16 - z &= 20 \\ 17 - z &= 20 \\ -z &= 3 \\ z &= -3 \end{aligned}$$

The solution $(1, 4, -3)$ checks in all three of the original equations.
The solution set is $\{(1, 4, -3)\}$.

7.

$$\begin{array}{rcll} 2x + 5y + 2z &=& 0 & (1) \\ 4x - 7y - 3z &=& 1 & (2) \\ 3x - 8y - 2z &=& -6 & (3) \end{array}$$

Add equations (1) and (3) to eliminate z.

$$\begin{array}{rcll} 2x + 5y + 2z &=& 0 & (1) \\ 3x - 8y - 2z &=& -6 & (3) \\ \hline 5x - 3y &=& -6 & (4) \end{array}$$

Multiply equation (1) by 3 and equation (2) by 2. Then add the results to eliminate z again.

$$\begin{array}{rcll} 6x + 15y + 6z &=& 0 & 3 \times (1) \\ 8x - 14y - 6z &=& 2 & 2 \times (2) \\ \hline 14x + y &=& 2 & (5) \end{array}$$

Solve the system

$$\begin{array}{rcll} 5x - 3y &=& -6 & (4) \\ 14x + y &=& 2. & (5) \end{array}$$

Multiply equation (5) by 3 then add this result to (4).

$$\begin{array}{rcll} 5x - 3y &=& -6 & (4) \\ 42x + 3y &=& 6 & 3 \times (5) \\ \hline 47x &=& 0 & \\ x &=& 0 & \end{array}$$

To find y, substitute $x = 0$ into equation (4).

$$\begin{aligned} 5x - 3y &= -6 \quad (4) \\ 5(0) - 3y &= -6 \\ y &= 2 \end{aligned}$$

To find z, substitute $x = 0$ and $y = 2$ in equation (1).

$$\begin{aligned} 2x + 5y + 2z &= 0 \quad (1) \\ 2(0) + 5(2) + 2z &= 0 \\ 10 + 2z &= 0 \\ 2z &= -10 \\ z &= -5 \end{aligned}$$

The solution set is $\{(0, 2, -5)\}$.

9.

$$\begin{array}{rcll} x + 2y + z &=& 4 & (1) \\ 2x + y - z &=& -1 & (2) \\ x - y - z &=& -2 & (3) \end{array}$$

Add (1) and (2) to eliminate z.

$$\begin{array}{rcll} x + 2y + z &=& 4 & (1) \\ 2x + y - z &=& -1 & (2) \\ \hline 3x + 3y &=& 3 & (4) \end{array}$$

Add (1) and (3) to eliminate z.

$$\begin{array}{rcll} x + 2y + z &=& 4 & (1) \\ x - y - z &=& -2 & (3) \\ \hline 2x + y &=& 2 & (5) \end{array}$$

Multiply equation (5) by -3.

$$\begin{array}{rcll} -6x - 3y &=& -6 & (6)\ -3 \times (5) \\ 3x + 3y &=& 3 & (4) \\ \hline -3x &=& -3 & \\ x &=& 1 & \end{array}$$

Substitute 1 for x in equation (5) and solve for y.

$$\begin{aligned} 2(1) + y &= 2 \\ y &= 0 \end{aligned}$$

Substitute 1 for x and 0 for y in (1).

$$\begin{aligned} (1) + 2(0) + z &= 4 \\ z &= 3 \end{aligned}$$

The solution set is $\{(1, 0, 3)\}$.

11.

$$\begin{array}{rcll} -x + 2y + 6z &=& 2 & (1) \\ 3x + 2y + 6z &=& 6 & (2) \\ x + 4y - 3z &=& 1 & (3) \end{array}$$

Eliminate y and z by adding equation (1) to -1 times equation (2).

$$\begin{array}{rcll} -x + 2y + 6z &=& 2 & (1) \\ -3x - 2y - 6z &=& -6 & -1 \times (2) \\ \hline -4x &=& -4 & \\ x &=& 1 & \end{array}$$

Eliminate z by adding equation (2) to 2 times equation (3).

$$\begin{array}{rcrcrcrll} 3x & + & 2y & + & 6z & = & 6 & (2) & \\ 2x & + & 8y & - & 6z & = & 2 & & 2\times(3) \\ \hline 5x & + & 10y & & & = & 8 & (4) & \end{array}$$

Substitute 1 for x in equation (4).

$$\begin{aligned} 5x + 10y &= 8 \qquad (4) \\ 5(1) + 10y &= 8 \\ 10y &= 3 \\ y &= \tfrac{3}{10} \end{aligned}$$

Substitute 1 for x and $\frac{3}{10}$ for y in equation (1).

$$\begin{aligned} -x + 2y + 6z &= 2 \qquad (1) \\ -1 + 2(\tfrac{3}{10}) + 6z &= 2 \\ \tfrac{3}{5} + 6z &= 3 \\ 6z &= \tfrac{12}{5} \\ z &= \tfrac{2}{5} \end{aligned}$$

The solution set is $\left\{\left(1, \frac{3}{10}, \frac{2}{5}\right)\right\}$.

13.
$$\begin{array}{rcrcrcrl} x & + & y & - & z & = & -2 & (1) \\ 2x & - & y & + & z & = & -5 & (2) \\ -x & + & 2y & - & 3z & = & -4 & (3) \end{array}$$

Eliminate y and z by adding equations (1) and (2).

$$\begin{array}{rcrcrcrl} x & + & y & - & z & = & -2 & (1) \\ 2x & - & y & + & z & = & -5 & (2) \\ \hline 3x & & & & & = & -7 & \\ & & & & x & = & -\frac{7}{3} & \end{array}$$

To get another equation without y, multiply equation (2) by 2 and add the result to equation (3).

$$\begin{array}{rcrcrcrll} 4x & - & 2y & + & 2z & = & -10 & & 2\times(2) \\ -x & + & 2y & - & 3z & = & -4 & (3) & \\ \hline 3x & & & - & z & = & -14 & (4) & \end{array}$$

Substitute $-\frac{7}{3}$ for x in equation (4) to find z.

$$\begin{aligned} 3x - z &= -14 \qquad (4) \\ 3(-\tfrac{7}{3}) - z &= -14 \\ -7 - z &= -14 \\ -z &= -7 \\ z &= 7 \end{aligned}$$

Substitute $-\frac{7}{3}$ for x and 7 for z in equation (1) to find y.

$$\begin{aligned} x + y - z &= -2 \qquad (1) \\ -\tfrac{7}{3} + y - 7 &= -2 \\ -7 + 3y - 21 &= -6 \qquad \textit{Multiply by 3.} \\ 3y - 28 &= -6 \\ 3y &= 22 \\ y &= \tfrac{22}{3} \end{aligned}$$

The solution set is $\left\{\left(-\frac{7}{3}, \frac{22}{3}, 7\right)\right\}$.

A calculator check reduces the probability of making any arithmetic errors and is highly recommended. The following screen shows the substitution of the solution for x, y, and z along with the left sides of the three original equations. The evaluation of the three expressions, -2, -5, and -4 (the right sides of the three equations), indicates that we have found the correct solution.

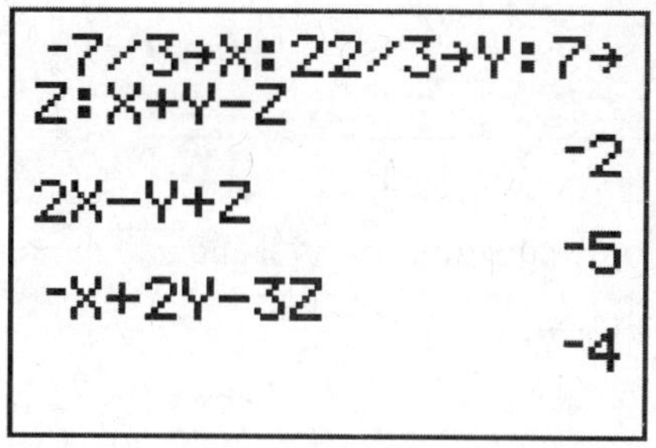

15.
$$\begin{array}{rcrcrcrl} \frac{1}{3}x & + & \frac{1}{6}y & - & \frac{2}{3}z & = & -1 & (1) \\ -\frac{3}{4}x & - & \frac{1}{3}y & - & \frac{1}{4}z & = & 3 & (2) \\ \frac{1}{2}x & + & \frac{3}{2}y & + & \frac{3}{4}z & = & 21 & (3) \end{array}$$

Eliminate all fractions.

$$\begin{array}{rcrcrcrll} 2x & + & y & - & 4z & = & -6 & (4) & 6\times(1) \\ -9x & - & 4y & - & 3z & = & 36 & (5) & 12\times(2) \\ 2x & + & 6y & + & 3z & = & 84 & (6) & 4\times(3) \end{array}$$

Multiply (4) by 4.

$$\begin{array}{rcrcrcrl} 8x & + & 4y & - & 16z & = & -24 & \\ -9x & - & 4y & - & 3z & = & 36 & (5) \\ \hline -x & & & - & 19z & = & 12 & (7) \end{array}$$

Multiply (4) by -6.

$$\begin{array}{rcrcrcrl} -12x & - & 6y & + & 24z & = & 36 & \\ 2x & + & 6y & + & 3z & = & 84 & (6) \\ \hline -10x & & & + & 27z & = & 120 & (8) \end{array}$$

Multiply (7) by -10.

$$\begin{array}{rcrcrl} 10x & + & 190z & = & -120 & \\ -10x & + & 27z & = & 120 & (8) \\ \hline & & 217z & = & 0 & \\ & & z & = & 0 & \end{array}$$

Substituting 0 for z in (7) gives us $x = -12$. Substitute -12 for x and 0 for z in (4).

$$\begin{aligned} 2(-12) + y - 4(0) &= -6 \\ -24 + y &= -6 \\ y &= 18 \end{aligned}$$

The solution set is $\{(-12, 18, 0)\}$.

17.
$$\begin{aligned} 5.5x - 2.5y + 1.6z &= 11.83 \quad (1)\\ 2.2x + 5.0y - 0.1z &= -5.97 \quad (2)\\ 3.3x - 7.5y + 3.2z &= 21.25 \quad (3) \end{aligned}$$

Multiply the equations by 10 to eliminate the decimals for the coefficients of the variables.

$$\begin{aligned} 55x - 25y + 16z &= 118.3 \quad (4)\\ 22x + 50y - z &= -59.7 \quad (5)\\ 33x - 75y + 32z &= 212.5 \quad (6) \end{aligned}$$

Multiply equation (4) by -3 and add the result to equation (6) to eliminate y.

$$\begin{aligned} -165x + 75y - 48z &= -354.9 \quad -3 \times (4)\\ 33x - 75y + 32z &= 212.5 \quad (6)\\ \hline -132x \qquad - 16z &= -142.4 \quad (7) \end{aligned}$$

Multiply equation (4) by 2 and add the result to equation (5) to eliminate y.

$$\begin{aligned} 110x - 50y + 32z &= 236.6 \quad 2 \times (4)\\ 22x + 50y - z &= -59.7 \quad (5)\\ \hline 132x \qquad + 31z &= 176.9 \quad (8) \end{aligned}$$

Now add equations (7) and (8) to eliminate x.

$$\begin{aligned} -132x - 16z &= -142.4 \quad (7)\\ 132x + 31z &= 176.9 \quad (8)\\ \hline 15z &= 34.5\\ z &= 2.3 \end{aligned}$$

Substitute 2.3 for z in equation (8) and solve for x.

$$\begin{aligned} 132x + 31(2.3) &= 176.9\\ 132x + 71.3 &= 176.9\\ 132x &= 105.6\\ x &= 0.8 \end{aligned}$$

Use equation (5) to solve for y.

$$\begin{aligned} 22x + 50y - z &= -59.7\\ 22(0.8) + 50y - 2.3 &= -59.7\\ 50y &= -75\\ y &= -1.5 \end{aligned}$$

The solution set is $\{(0.8, -1.5, 2.3)\}$.

19.
$$\begin{aligned} 2x - 3y + 2z &= -1 \quad (1)\\ x + 2y + z &= 17 \quad (2)\\ 2y - z &= 7 \quad (3) \end{aligned}$$

Multiply equation (2) by -2, and add the result to equation (1).

$$\begin{aligned} 2x - 3y + 2z &= -1 \quad (1)\\ -2x - 4y - 2z &= -34 \quad -2 \times (2)\\ \hline -7y &= -35\\ y &= 5 \end{aligned}$$

To find z, substitute 5 for y in equation (3).

$$\begin{aligned} 2y - z &= 7 \quad (3)\\ 2(5) - z &= 7\\ 10 - z &= 7\\ -z &= -3\\ z &= 3 \end{aligned}$$

To find x, substitute $y = 5$ and $z = 3$ into equation (1).

$$\begin{aligned} 2x - 3y + 2z &= -1 \quad (1)\\ 2x - 3(5) + 2(3) &= -1\\ 2x - 9 &= -1\\ 2x &= 8\\ x &= 4 \end{aligned}$$

The solution set is $\{(4, 5, 3)\}$.

21.
$$\begin{aligned} 4x + 2y - 3z &= 6 \quad (1)\\ x - 4y + z &= -4 \quad (2)\\ -x \qquad + 2z &= 2 \quad (3) \end{aligned}$$

Equation (3) is missing y. Eliminate y again by multiplying equation (1) by 2 and adding the result to equation (2).

$$\begin{aligned} 8x + 4y - 6z &= 12 \quad 2 \times (1)\\ x - 4y + z &= -4 \quad (2)\\ \hline 9x \qquad - 5z &= 8 \quad (4) \end{aligned}$$

Use equations (3) and (4) to eliminate x. Multiply equation (3) by 9 and add the result to equation (4).

$$\begin{aligned} -9x + 18z &= 18 \quad 9 \times (3)\\ 9x - 5z &= 8 \quad (4)\\ \hline 13z &= 26\\ z &= 2 \end{aligned}$$

Substitute 2 for z in equation (3) to find x.

$$\begin{aligned} -x + 2z &= 2 \quad (3)\\ -x + 2(2) &= 2\\ -x + 4 &= 2\\ -x &= -2\\ x &= 2 \end{aligned}$$

Substitute 2 for x and 2 for z in equation (2) to find y.

$$\begin{aligned} x - 4y + z &= -4 \quad (2)\\ 2 - 4y + 2 &= -4\\ -4y + 4 &= -4\\ -4y &= -8\\ y &= 2 \end{aligned}$$

The solution set is $\{(2, 2, 2)\}$.

23.
$$\begin{aligned} 2x + y &= 6 \quad (1) \\ 3y - 2z &= -4 \quad (2) \\ 3x - 5z &= -7 \quad (3) \end{aligned}$$

To eliminate y, multiply equation (1) by -3 and add the result to equation (2).

$$\begin{aligned} -6x - 3y &= -18 \quad -3 \times (1) \\ 3y - 2z &= -4 \quad (2) \\ \hline -6x - 2z &= -22 \quad (4) \end{aligned}$$

Since equation (3) does not have a y-term, we can multiply equation (3) by 2 and add the result to equation (4) to eliminate x and solve for z.

$$\begin{aligned} 6x - 10z &= -14 \quad 2 \times (3) \\ -6x - 2z &= -22 \quad (4) \\ \hline -12z &= -36 \\ z &= 3 \end{aligned}$$

To find x, substitute 3 for z into equation (3).

$$\begin{aligned} 3x - 5z &= -7 \quad (3) \\ 3x - 5(3) &= -7 \\ 3x &= 8 \\ x &= \tfrac{8}{3} \end{aligned}$$

To find y, substitute 3 for z into equation (2).

$$\begin{aligned} 3y - 2z &= -4 \quad (2) \\ 3y - 2(3) &= -4 \\ 3y &= 2 \\ y &= \tfrac{2}{3} \end{aligned}$$

The solution set is $\left\{\left(\frac{8}{3}, \frac{2}{3}, 3\right)\right\}$.

25.
$$\begin{aligned} -5x + 2y + z &= 5 \quad (1) \\ -3x - 2y - z &= 3 \quad (2) \\ -x + 6y &= 1 \quad (3) \end{aligned}$$

Add (1) and (2) to eliminate y and z.

$$\begin{aligned} -8x &= 8 \\ x &= -1 \end{aligned}$$

Substitute -1 for x in equation (3).

$$\begin{aligned} -x + 6y &= 1 \quad (3) \\ -(-1) + 6y &= 1 \\ 6y &= 0 \\ y &= 0 \end{aligned}$$

Substitute -1 for x and 0 for y in (1).

$$\begin{aligned} -5x + 2y + z &= 5 \quad (1) \\ -5(-1) + 2(0) + z &= 5 \\ 5 + z &= 5 \\ z &= 0 \end{aligned}$$

The solution set is $\{(-1, 0, 0)\}$.

27.
$$\begin{aligned} 7x - 3z &= -34 \quad (1) \\ 2y + 4z &= 20 \quad (2) \\ \tfrac{3}{4}x + \tfrac{1}{6}y &= -2 \quad (3) \end{aligned}$$

Eliminate fractions from equation (3) by multiplying by -12.

$$-9x - 2y = 24 \quad (4) \quad -12 \times (3)$$

Adding equations (2) and (4) gives us

$$-9x + 4z = 44. \quad (5)$$

Eliminate z by adding 4 times (1) to 3 times (5).

$$\begin{aligned} 28x - 12z &= -136 \quad 4 \times (1) \\ -27x + 12z &= 132 \quad 3 \times (5) \\ \hline x &= -4 \end{aligned}$$

Substitute -4 for x in (5).

$$\begin{aligned} -9(-4) + 4z &= 44 \\ 36 + 4z &= 44 \\ 4z &= 8 \\ z &= 2 \end{aligned}$$

Substitute 2 for z in (2).

$$\begin{aligned} 2y + 4(2) &= 20 \\ 2y + 8 &= 20 \\ 2y &= 12 \\ y &= 6 \end{aligned}$$

The solution set is $\{(-4, 6, 2)\}$.

29.
$$\begin{aligned} 4x \phantom{+ \tfrac{3}{5}y} - z &= -6 \quad (1) \\ \tfrac{3}{5}y + \tfrac{1}{2}z &= 0 \quad (2) \\ \tfrac{1}{3}x \phantom{+ \tfrac{3}{5}y} + \tfrac{2}{3}z &= -5 \quad (3) \end{aligned}$$

Eliminate fractions first.

$$\begin{aligned} 6y + 5z &= 0 \quad (4) \quad 10 \times (2) \\ x + 2z &= -15 \quad (5) \quad 3 \times (3) \end{aligned}$$

Eliminate z by adding (5) to 2 times (1).

$$\begin{aligned} 8x - 2z &= -12 \quad 2 \times (1) \\ x + 2z &= -15 \quad (5) \\ \hline 9x &= -27 \\ x &= -3 \end{aligned}$$

Substitute -3 for x in (5).

$$\begin{aligned} x + 2z &= -15 \quad (5) \\ -3 + 2z &= -15 \\ 2z &= -12 \\ z &= -6 \end{aligned}$$

Substitute -6 for z in (4).

$$\begin{aligned} 6y + 5z &= 0 \quad (4) \\ 6y + 5(-6) &= 0 \\ 6y &= 30 \\ y &= 5 \end{aligned}$$

The solution set is $\{(-3, 5, -6)\}$.

31.
$$\begin{aligned} 2x + 2y - 6z &= 5 \quad (1) \\ -3x + y - z &= -2 \quad (2) \\ -x - y + 3z &= 4 \quad (3) \end{aligned}$$

Multiply equation (3) by 2 and add the result to equation (1).

$$\begin{array}{rl} 2x + 2y - 6z = 5 & (1) \\ -2x - 2y + 6z = 8 & 2 \times (3) \\ \hline 0 = 13 & \textit{False} \end{array}$$

The solution set is $\emptyset$; inconsistent system.

33.
$$\begin{aligned} -5x + 5y - 20z &= -40 \quad (1) \\ x - y + 4z &= 8 \quad (2) \\ 3x - 3y + 12z &= 24 \quad (3) \end{aligned}$$

Dividing equation (1) by -5 gives equation (2). Dividing equation (3) by 3 also gives equation (2). The resulting equations are the same, so the three equations are dependent.

The solution set is $\{(x, y, z) \mid x - y + 4z = 8\}$.

35.
$$\begin{aligned} x + 5y - 2z &= -1 \quad (1) \\ -2x + 8y + z &= -4 \quad (2) \\ 3x - y + 5z &= 19 \quad (3) \end{aligned}$$

Eliminate x by adding (2) to 2 times (1).

$$\begin{array}{rl} 2x + 10y - 4z = -2 & 2 \times (1) \\ -2x + 8y + z = -4 & (2) \\ \hline 18y - 3z = -6 & (4) \end{array}$$

Eliminate x by adding (3) to -3 times (1).

$$\begin{array}{rl} -3x - 15y + 6z = 3 & -3 \times (1) \\ 3x - y + 5z = 19 & (3) \\ \hline -16y + 11z = 22 & (5) \end{array}$$

To eliminate z from equations (4) and (5), we could first divide equation (4) by 3 and then multiply the resulting equation by 11, or we could simply multiply equation (4) by $\frac{11}{3}$ and add it to equation (5).

$$\begin{array}{rl} 66y - 11z = -22 & \frac{11}{3} \times (4) \\ -16y + 11z = 22 & (5) \\ \hline 50y = 0 & \\ y = 0 & \end{array}$$

Substitute 0 for y in equation (5).

$$\begin{aligned} -16y + 11z &= 22 \quad (5) \\ -16(0) + 11z &= 22 \\ 11z &= 22 \\ z &= 2 \end{aligned}$$

Substitute 0 for y and 2 for z in equation (1).

$$\begin{aligned} x + 5y - 2z &= -1 \quad (1) \\ x + 5(0) - 2(2) &= -1 \\ x - 4 &= -1 \\ x &= 3 \end{aligned}$$

The solution set is $\{(3, 0, 2)\}$.

37.
$$\begin{aligned} 2x + y - z &= 6 \quad (1) \\ 4x + 2y - 2z &= 12 \quad (2) \\ -x - \tfrac{1}{2}y + \tfrac{1}{2}z &= -3 \quad (3) \end{aligned}$$

Multiplying equation (1) by 2 gives equation (2). Multiplying equation (3) by -4 also gives equation (2). The resulting equations are the same, so the three equations are dependent.

The solution set is $\{(x, y, z) \mid 2x + y - z = 6\}$.

39.
$$\begin{aligned} x + y - 2z &= 0 \quad (1) \\ 3x - y + z &= 0 \quad (2) \\ 4x + 2y - z &= 0 \quad (3) \end{aligned}$$

Eliminate z by adding equations (2) and (3).

$$\begin{array}{rl} 3x - y + z = 0 & (2) \\ 4x + 2y - z = 0 & (3) \\ \hline 7x + y = 0 & (4) \end{array}$$

To get another equation without z, multiply equation (2) by 2 and add the result to equation (1).

$$\begin{array}{rl} 6x - 2y + 2z = 0 & 2 \times (2) \\ x + y - 2z = 0 & (1) \\ \hline 7x - y = 0 & (5) \end{array}$$

Add equations (4) and (5) to find x.

$$\begin{array}{rl} 7x + y = 0 & (4) \\ 7x - y = 0 & (5) \\ \hline 14x = 0 & \\ x = 0 & \end{array}$$

Substitute 0 for x in equation (4) to find y.

$$\begin{aligned} 7x + y &= 0 \quad (4) \\ 7(0) + y &= 0 \\ 0 + y &= 0 \\ y &= 0 \end{aligned}$$

Substitute 0 for x and 0 for y in equation (1) to find z.

$$\begin{aligned} x + y - 2z &= 0 \quad (1) \\ 0 + 0 - 2z &= 0 \\ -2z &= 0 \\ z &= 0 \end{aligned}$$

The solution set is $\{(0, 0, 0)\}$.

41.
$$\begin{aligned} x - 2y + \tfrac{1}{3}z &= 4 \quad (1) \\ 3x - 6y + z &= 12 \quad (2) \\ -6x + 12y - 2z &= -3 \quad (3) \end{aligned}$$

The coefficients of z are $\frac{1}{3}$, 1, and -2. We'll multiply the equations by values that will eliminate fractions and make the coefficients of z easy to compare.

$$\begin{aligned} -6x + 12y - 2z &= -24 \quad (4) \quad -6 \times (1) \\ -6x + 12y - 2z &= -24 \quad (5) \quad -2 \times (2) \\ -6x + 12y - 2z &= -3 \quad (3) \end{aligned}$$

We can now easily see that (4) and (5) are dependent equations (they have the same graph—in fact, their graph is the same plane). Equation (3) has the same coefficients, but a different constant term, so its graph is a plane parallel to the other plane—that is, there are no points in common. Thus, the system is inconsistent and the solution set is $\emptyset$.

43.
$$\begin{aligned} x + y + z - w &= 5 \quad (1) \\ 2x + y - z + w &= 3 \quad (2) \\ x - 2y + 3z + w &= 18 \quad (3) \\ -x - y + z + 2w &= 8 \quad (4) \end{aligned}$$

Eliminate w. Add equations (1) and (2).

$$\begin{aligned} x + y + z - w &= 5 \quad (1) \\ 2x + y - z + w &= 3 \quad (2) \\ \hline 3x + 2y &= 8 \quad (5) \end{aligned}$$

Eliminate w again. Add equations (1) and (3).

$$\begin{aligned} x + y + z - w &= 5 \quad (1) \\ x - 2y + 3z + w &= 18 \quad (3) \\ \hline 2x - y + 4z &= 23 \quad (6) \end{aligned}$$

Eliminate w again. Multiply equation (2) by -2. Add the result to equation (4).

$$\begin{aligned} -4x - 2y + 2z - 2w &= -6 \quad -2 \times (2) \\ -x - y + z + 2w &= 8 \quad (4) \\ \hline -5x - 3y + 3z &= 2 \quad (7) \end{aligned}$$

Equations (5), (6), and (7) do not contain a w-term. Since (5) does not have a z-term, we will find another equation without a z-term.

Eliminate z. Multiply equation (6) by 3 and equation (7) by -4. Then add the results.

$$\begin{aligned} 6x - 3y + 12z &= 69 \quad 3 \times (6) \\ 20x + 12y - 12z &= -8 \quad -4 \times (7) \\ \hline 26x + 9y &= 61 \quad (8) \end{aligned}$$

Eliminate y. Multiply equation (5) by 9 and equation (8) by -2. Then add the results.

$$\begin{aligned} 27x + 18y &= 72 \quad 9 \times (5) \\ -52x - 18y &= -122 \quad -2 \times (8) \\ \hline -25x &= -50 \\ x &= 2 \end{aligned}$$

To find y, substitute $x = 2$ into equation (5).

$$\begin{aligned} 3x + 2y &= 8 \quad (5) \\ 3(2) + 2y &= 8 \\ 2y &= 2 \\ y &= 1 \end{aligned}$$

To find z, substitute $x = 2$ and $y = 1$ into equation (6).

$$\begin{aligned} 2x - y + 4z &= 23 \quad (6) \\ 2(2) - 1 + 4z &= 23 \\ 4z &= 20 \\ z &= 5 \end{aligned}$$

To find w, substitute $x = 2$, $y = 1$, and $z = 5$ into equation (1).

$$\begin{aligned} x + y + z - w &= 5 \quad (1) \\ 2 + 1 + 5 - w &= 5 \\ -w &= -3 \\ w &= 3 \end{aligned}$$

The solution set is $\{(2, 1, 5, 3)\}$.

45.
$$\begin{aligned} 3x + y - z + w &= -3 \quad (1) \\ 2x + 4y + z - w &= -7 \quad (2) \\ -2x + 3y - 5z + w &= 3 \quad (3) \\ 5x + 4y - 5z + 2w &= -7 \quad (4) \end{aligned}$$

Eliminate w. Add equations (1) and (2).

$$\begin{aligned} 3x + y - z + w &= -3 \quad (1) \\ 2x + 4y + z - w &= -7 \quad (2) \\ \hline 5x + 5y &= -10 \quad (5) \end{aligned}$$

Eliminate w again. Add equations (2) and (3).

$$\begin{aligned} 2x + 4y + z - w &= -7 \quad (2) \\ -2x + 3y - 5z + w &= 3 \quad (3) \\ \hline 7y - 4z &= -4 \quad (6) \end{aligned}$$

Eliminate w again. Multiply equation (2) by 2. Add the result to equation (4).

$$\begin{aligned} 4x + 8y + 2z - 2w &= -14 \quad 2 \times (2) \\ 5x + 4y - 5z + 2w &= -7 \quad (4) \\ \hline 9x + 12y - 3z &= -21 \quad (7) \end{aligned}$$

Equations (5), (6), and (7) do not contain a w-term. Since (5) does not have a z-term, we will find another equation without a z-term.

Eliminate z. Multiply equation (6) by 3 and equation (7) by -4. Then add the results.

$$\begin{aligned} 21y - 12z &= -12 \quad 3 \times (6) \\ -36x - 48y + 12z &= 84 \quad -4 \times (7) \\ \hline -36x - 27y &= 72 \quad (8) \end{aligned}$$

Eliminate y using (5) and (8).

$$\begin{aligned} 3x + 3y &= -6 \quad \tfrac{3}{5} \times (5) \\ -4x - 3y &= 8 \quad (8) \div 9 \\ \hline -x &= 2 \\ x &= -2 \end{aligned}$$

Substitute -2 for x in (5).

$$\begin{aligned} 5(-2) + 5y &= -10 \\ 5y &= 0 \\ y &= 0 \end{aligned}$$

Substitute 0 for y in (6).

$$\begin{aligned} 7(0) - 4z &= -4 \\ -4z &= -4 \\ z &= 1 \end{aligned}$$

Substitute -2 for x, 0 for y, and 1 for z in (1).

$$\begin{aligned} 3(-2) + (0) - (1) + w &= -3 \\ -6 - 1 + w &= -3 \\ w &= 4 \end{aligned}$$

The solution set is $\{(-2, 0, 1, 4)\}$.

47. Let $x =$ the length of the longest side.
Then $\frac{5}{6}x =$ the length of the shortest side,
and $x - 17 =$ the length of the medium side.

The perimeter is 323, so

$$\begin{aligned} x + \tfrac{5}{6}x + (x - 17) &= 323. \\ 6x + 5x + 6(x - 17) &= 6(323) \quad \textit{Multiply by 6.} \\ 6x + 5x + 6x - 102 &= 1938 \\ 17x &= 2040 \\ x &= 120 \end{aligned}$$

Since $x = 120$, $\frac{5}{6}x = 100$, and $x - 17 = 103$. The lengths of the sides are 100 inches, 103 inches, and 120 inches.

49. Let $x =$ the least number.
Then $-3x =$ the greatest number,
and $-3x - 4 =$ the middle number.

The sum is 16, so

$$\begin{aligned} x + (-3x) + (-3x - 4) &= 16. \\ -5x - 4 &= 16 \\ -5x &= 20 \\ x &= -4 \end{aligned}$$

Since $x = -4$, $-3x = 12$, and $-3x - 4 = 8$. The three numbers are -4, 8, and 12.

8.5 Applications of Systems of Linear Equations

8.5 Now Try Exercises

N1. *Step 2*
Let $W =$ the width of the field
and $L =$ the length of the field.

Step 3
Since the length is 10 feet more than twice its width,

$$L = 2W + 10. \quad (1)$$

The perimeter of a rectangle is given by

$$2W + 2L = P.$$

With perimeter $P = 620$ feet,

$$2W + 2L = 620. \quad (2)$$

Step 4
Substitute $2W + 10$ for L in equation (2).

$$\begin{aligned} 2W + 2(2W + 10) &= 620 \\ 2W + 4W + 20 &= 620 \\ 6W &= 600 \\ W &= 100 \end{aligned}$$

Substitute $W = 100$ into equation (1).

$$L = 2W + 10 = 200 + 10 = 210$$

Step 5
The length of the field is 210 feet and the width is 100 feet.

Step 6
Check 210 is 10 more than 2 times 100 and $2(210) + 2(100) = 620$, as required.

N2. *Step 2*
Let $x =$ the cost of a general admission ticket, and $y =$ the cost of a ticket for children under 48 inches tall.

Step 3
From the given information,

$$\begin{aligned} 2x + 3y &= 172.98 \quad (1) \\ x + 4y &= 163.99. \quad (2) \end{aligned}$$

Step 4
Multiply equation (2) by -2 and add to equation (1).

$$\begin{array}{rcrl} 2x + 3y & = & 172.98 & (1) \\ -2x - 8y & = & -327.98 & -2 \times (2) \\ \hline -5y & = & -155.00 & \\ y & = & 31.00 & \end{array}$$

Let $y = 31.00$ in equation (2).

$$\begin{aligned} x + 4(31.00) &= 163.99 \\ x + 124.00 &= 163.99 \\ x &= 39.99 \end{aligned}$$

Step 5
The cost of a general admission ticket is \$39.99 and the cost of a ticket for children under 48 inches tall is \$31.00.

Step 6
Check $2(39.99) + 3(31.00) = 172.98$ and $39.99 + 4(31.00) = 163.99$.

N3. *Step 2*
Let $x =$ the amount of 15% acid solution and $y =$ the amount of 25% acid solution.

Liters of Solution	Percent (as a decimal)	Liters of Pure Acid
x	$15\% = 0.15$	$0.15x$
y	$25\% = 0.25$	$0.25y$
30	$18\% = 0.18$	$0.18(30) = 5.4$

Step 3
Write the system from the columns.

$$\begin{aligned} x + \quad y &= 30 \quad (1) \\ 0.15x + 0.25y &= 5.4 \quad (2) \end{aligned}$$

Step 4
Multiply (2) by 100 to clear the equation of decimals, and multiply (1) by -15.

$$\begin{array}{rcrl} -15x - 15y &=& -450 & -15 \times (1) \\ 15x + 25y &=& 540 & 100 \times (2) \\ \hline 10y &=& 90 & \\ y &=& 9 & \end{array}$$

Substitute 9 for y in (1).

$$\begin{aligned} x + 9 &= 30 \\ x &= 30 - 9 = 21 \end{aligned}$$

Step 5
21 L of 15% and 9 L of 25% should be mixed.

Step 6
Check $21 + 9 = 30$ and
$0.15(21) + 0.25(9) = 5.4$.

N4. *Step 2*
Let x = Vann's rate and y = Ivy's rate.

Make a table. Use $t = \frac{d}{r}$.

	Distance	Rate	Time
Vann	50	x	$\frac{50}{x}$
Ivy	40	y	$\frac{40}{y}$

Step 3
Since the times are the same,

$$\begin{aligned} \frac{50}{x} &= \frac{40}{y} \\ 50y &= 40x \quad \textit{Multiply by xy.} \\ -40x + 50y &= 0. \quad (1) \end{aligned}$$

Since Vann's rate, x, is 2 mph faster than Ivy's rate, y,

$$x = y + 2. \quad (2)$$

Step 4
Substitute $y + 2$ for x in equation (1) to find y.

$$\begin{aligned} -40x + 50y &= 0 \quad (1) \\ -40(y + 2) + 50y &= 0 \\ -40y - 80 + 50y &= 0 \\ 10y &= 80 \\ y &= 8 \end{aligned}$$

Since $y = 8$ and $x = y + 2$, $x = 10$.

Step 5
Vann's rate is 10 mph, and Ivy's rate is 8 mph.

Step 6
Check 10 mph is 2 mph faster than 8 mph. It would take Vann 5 hours to travel 50 miles at 10 mph, which is the same amount of time it would take Ivy to travel 40 miles at 8 mph.

N5. Let x = the number of bottles of \$2.00 Gatorade sold, y = the number of \$1.50 pretzels sold, and z = the number of \$1.00 candy items sold.

The number of pretzels was four times the number of candy items, so

$$y = 4z. \quad (1)$$

The number of Gatorades was 310 more than the number of pretzels, so

$$x = 310 + y. \quad (2)$$

Sales for these items totaled \$1520, so

$$2x + 1.5y + 1z = 1520. \quad (3)$$

Substitute $310 + y$ for x in (3).

$$2(310 + y) + 1.5y + 1z = 1520 \quad (4)$$

Now substitute $4z$ for y in (4).

$$\begin{aligned} 2(310 + 4z) + 1.5(4z) + 1z &= 1520 \\ 620 + 8z + 6z + 1z &= 1520 \\ 15z &= 900 \\ z &= 60 \end{aligned}$$

From (1), $y = 4(60) = 240$.
From (2), $x = 310 + 240 = 550$.
Workers sold 60 candy items, 240 pretzels, and 550 bottles of Gatorade.

N6. Let x = the number of lone star quilts, y = the number of bandana quilts, and z = the number of log cabin quilts.

The time allocated for piecework is 74 hours, so

$$8x + 2y + 10z = 74. \quad (1)$$

The time allocated for machine quilting is 42 hours, so

$$4x + 2y + 5z = 42. \quad (2)$$

The time allocated for finishing is 24 hours, so

$$2x + 2y + 2z = 24, \text{ or equivalently,}$$

$$x + y + z = 12. \quad (3)$$

Solve the system of equations (1), (2), and (3). To eliminate z, multiply equation (3) by -10 and add the result to equation (1).

$$\begin{array}{rcrl} -10x - 10y - 10z &=& -120 & -10 \times (3) \\ 8x + 2y + 10z &=& 74 & (1) \\ \hline -2x - 8y &=& -46 & \end{array}$$

Divide the last equation by -2 to get

$$x + 4y = 23. \quad (4)$$

To eliminate z again, multiply equation (3) by -5 and add the result to equation (2).

$$\begin{array}{rcll} -5x - 5y - 5z &=& -60 & -5 \times (3) \\ 4x + 2y + 5z &=& 42 & (2) \\ \hline -x - 3y &=& -18 & (5) \end{array}$$

Add equations (4) and (5) to eliminate y.

$$\begin{array}{rcll} x + 4y &=& 23 & (4) \\ -x - 3y &=& -18 & (5) \\ \hline y &=& 5 & \end{array}$$

Substitute 5 for y in equation (4).

$$\begin{aligned} x + 4y &= 23 \qquad (4) \\ x + 4(5) &= 23 \\ x + 20 &= 23 \\ x &= 3 \end{aligned}$$

Substitute 3 for x and 5 for y in equation (3).

$$\begin{aligned} x + y + z &= 12 \qquad (3) \\ 3 + 5 + z &= 12 \\ 8 + z &= 12 \\ z &= 4 \end{aligned}$$

Each month, the quilting shop should make 3 lone star quilts, 5 bandana quilts, and 4 log cabin quilts.

8.5 Section Exercises

1. *Step 2*
Let $x =$ the number of games that the Dodgers won and let $y =$ the number of games that they lost.

Step 3
They played 162 games, so

$$x + y = 162. \quad (1)$$

They won 28 more games than they lost, so

$$x = 28 + y. \quad (2)$$

Step 4
Substitute $28 + y$ for x in (1).

$$\begin{aligned} (28 + y) + y &= 162 \\ 28 + 2y &= 162 \\ 2y &= 134 \\ y &= 67 \end{aligned}$$

Substitute 67 for y in (2).

$$x = 28 + 67 = 95$$

Step 5
The Dodgers win-loss record was 95 wins and 67 losses.

Step 6
95 is 28 more than 67 and the sum of 95 and 67 is 162.

3. Let $W =$ the width of the tennis court and $L =$ the length of the court.

Since the length is 42 ft more than the width,

$$L = W + 42. \quad (1)$$

The perimeter of a rectangle is given by

$$2W + 2L = P.$$

With perimeter $P = 228$ ft,

$$2W + 2L = 228. \quad (2)$$

Substitute $W + 42$ for L in equation (2).

$$\begin{aligned} 2W + 2(W + 42) &= 228 \\ 2W + 2W + 84 &= 228 \\ 4W &= 144 \\ W &= 36 \end{aligned}$$

Substitute $W = 36$ into equation (1).

$$\begin{aligned} L &= W + 42 \qquad (1) \\ L &= 36 + 42 \\ &= 78 \end{aligned}$$

The length is 78 ft and the width is 36 ft.

5. Let $x =$ the revenue for Verizon and $y =$ the revenue for AT&T (both in billions of dollars).

The total revenue was \$221.4 billion.

$$x + y = 221.4 \quad (1)$$

AT&T's revenue was \$26.6 billion more than that of Verizon.

$$y = x + 26.6 \quad (2)$$

Substitute $x + 26.6$ for y in equation (1).

$$\begin{aligned} x + (x + 26.6) &= 221.4 \\ 2x + 26.6 &= 221.4 \\ 2x &= 194.8 \\ x &= 97.4 \end{aligned}$$

Substitute 97.4 for x in equation (2).

$$y = x + 26.6 = 97.4 + 26.6 = 124.0$$

Verizon's revenue was \$97.4 billion and AT&T's revenue was \$124.0 billion.

7. From the figure in the text, the angles marked y and $3x + 10$ are supplementary, so

$$(3x + 10) + y = 180. \quad (1)$$

Also, the angles x and y are complementary, so

$$x + y = 90. \quad (2)$$

Solve equation (2) for y to get

$$y = 90 - x. \quad (3)$$

Substitute $90 - x$ for y in equation (1).

$$\begin{aligned}(3x + 10) + (90 - x) &= 180\\ 2x + 100 &= 180\\ 2x &= 80\\ x &= 40\end{aligned}$$

Substitute $x = 40$ into equation (3) to get

$$y = 90 - x = 90 - 40 = 50.$$

The angles measure $40°$ and $50°$.

9. Let $x =$ the hockey FCI
and $y =$ the basketball FCI.

The sum is $\$580.16$, so

$$x + y = 580.16. \quad (1)$$

The hockey FCI was $\$3.70$ less than the basketball FCI, so

$$x = y - 3.70. \quad (2)$$

From (2), substitute $y - 3.70$ for x in (1).

$$\begin{aligned}(y - 3.70) + y &= 580.16\\ 2y - 3.70 &= 580.16\\ 2y &= 583.86\\ y &= 291.93\end{aligned}$$

From (2),

$$x = y - 3.70 = 291.93 - 3.70 = 288.23.$$

The hockey FCI was $\$288.23$ and the basketball FCI was $\$291.93$.

11. Let $x =$ the cost of a single Junior Roast Beef sandwich, and $y =$ the cost of a single Big Montana sandwich.

15 Junior Roast Beef sandwiches and 10 Big Montana sandwiches cost $\$75.25$.

$$15x + 10y = 75.25 \quad (1)$$

30 Junior Roast Beef sandwiches and 5 Big Montana sandwiches cost $\$84.65$.

$$30x + 5y = 84.65 \quad (2)$$

Multiply equation (1) by -2 and add to equation (2).

$$\begin{array}{rcrcrl} -30x & - & 20y & = & -150.50 & \quad -2 \times (1)\\ 30x & + & 5y & = & 84.65 & \quad (2)\\ \hline & - & 15y & = & -65.85 & \\ & & y & = & 4.39 & \end{array}$$

Substitute 4.39 for y in equation (1).

$$\begin{aligned}15x + 10y &= 75.25\\ 15x + 10(4.39) &= 75.25\\ 15x + 43.90 &= 75.25\\ 15x &= 31.35\\ x &= 2.09\end{aligned}$$

A single Junior Roast Beef sandwich costs $\$2.09$ and a single Big Montana sandwich costs $\$4.39$.

13. Use the formula (rate of percent) • (base amount) $=$ amount (percentage) of pure acid to compute parts (a) – (d).

(a) $0.10(60) = 6$ oz

(b) $0.25(60) = 15$ oz

(c) $0.40(60) = 24$ oz

(d) $0.50(60) = 30$ oz

15. The cost is the price per pound, $\$2.29$, times the number of pounds, x, or $\$2.29x$.

17. Let $x =$ the amount of 25% alcohol solution, and $y =$ the amount of 35% alcohol solution.

Make a table. The percent times the amount of solution gives the amount of pure alcohol in the third column.

Gallons of Solution	Percent (as a decimal)	Gallons of Pure Alcohol
x	$25\% = 0.25$	$0.25x$
y	$35\% = 0.35$	$0.35y$
20	$32\% = 0.32$	$0.32(20) = 6.4$

The third row gives the total amounts of solution and pure alcohol. From the columns in the table, write a system of equations.

$$\begin{aligned} x + \quad y &= 20 \quad (1)\\ 0.25x + 0.35y &= 6.4 \quad (2)\end{aligned}$$

Solve the system. Multiply equation (1) by -25 and equation (2) by 100. Then add the results.

$$\begin{array}{rcrcrl} -25x & - & 25y & = & -500 & \quad -25 \times (1)\\ 25x & + & 35y & = & 640 & \quad 100 \times (2)\\ \hline & & 10y & = & 140 & \\ & & y & = & 14 & \end{array}$$

Substitute $y = 14$ into equation (1).

$$\begin{aligned}x + y &= 20 \quad (1)\\ x + 14 &= 20\\ x &= 6\end{aligned}$$

Mix 6 gal of 25% solution and 14 gal of 35% solution.

19. Let $x =$ the amount of pure acid and $y =$ the amount of 10% acid.

Make a table.

Liters of Solution	Percent (as a decimal)	Liters of Pure Acid
x	$100\% = 1$	$1.00x = x$
y	$10\% = 0.10$	$0.10y$
54	$20\% = 0.20$	$0.20(54) = 10.8$

Solve the following system.

$$\begin{aligned} x + \quad y &= 54 \quad (1) \\ x + 0.10y &= 10.8 \quad (2) \end{aligned}$$

Multiply equation (2) by 10 to clear the decimals.

$$10x + y = 108 \quad (3)$$

To eliminate y, multiply equation (1) by -1 and add the result to equation (3).

$$\begin{array}{rcrcrl} -x & - & y & = & -54 & \quad -1 \times (1) \\ 10x & + & y & = & 108 & \quad (3) \\ \hline 9x & & & = & 54 & \\ & & x & = & 6 & \end{array}$$

Since $x = 6$,

$$\begin{aligned} x + y &= 54 \quad (1) \\ 6 + y &= 54 \\ y &= 48. \end{aligned}$$

Use 6 L of pure acid and 48 L of 10% acid.

21. Complete the table.

	Number of Kilograms	Price per Kilogram	Value
Nuts	x	2.50	$2.50x$
Cereal	y	1.00	$1.00y$
Mixture	30	1.70	$1.70(30) = 51$

From the "Number of Kilograms" column,

$$x + y = 30. \quad (1)$$

From the "Value" column,

$$2.50x + 1.00y = 51. \quad (2)$$

Solve the system.

$$\begin{array}{rcrcrl} -10x & - & 10y & = & -300 & \quad -10 \times (1) \\ 25x & + & 10y & = & 510 & \quad 10 \times (2) \\ \hline 15x & & & = & 210 & \\ & & x & = & 14 & \end{array}$$

From (1), $14 + y = 30$, so $y = 16$.
The party mix should be made from 14 kg of nuts and 16 kg of cereal.

23. From the "Principal" column in the text,

$$x + y = 3000. \quad (1)$$

From the "Interest" column in the text,

$$0.02x + 0.04y = 100. \quad (2)$$

Multiply equation (2) by 100 to clear the decimals.

$$2x + 4y = 10{,}000 \quad (3)$$

To eliminate x, multiply equation (1) by -2 and add the result to equation (3).

$$\begin{array}{rcrcrl} -2x & - & 2y & = & -6000 & \quad -2 \times (1) \\ 2x & + & 4y & = & 10{,}000 & \quad (3) \\ \hline & & 2y & = & 4000 & \\ & & y & = & 2000 & \end{array}$$

From (1), $x + 2000 = 3000$, so $x = 1000$.
\$1000 is invested at 2%, and \$2000 is invested at 4%.

25. **(a)** The rate of the boat going upstream is *decreased* by the rate of the current, so it is $(10 - x)$ mph.

(b) The rate of the boat going downstream is *increased* by the rate of the current, so it is $(10 + x)$ mph.

27. Let $x =$ the rate of the train and $y =$ the rate of the plane.

	r	t	d
Train	x	$\frac{150}{x}$	150
Plane	y	$\frac{400}{y}$	400

The times are equal, so

$$\frac{150}{x} = \frac{400}{y}. \quad (1)$$

The rate of the plane is 20 km per hour less than 3 times the rate of the train, so

$$3x - 20 = y. \quad (2)$$

Multiply (1) by xy (or use cross products).

$$150y = 400x \quad (3)$$

From (2), substitute $3x - 20$ for y in (3).

$$\begin{aligned} 150(3x - 20) &= 400x \\ 450x - 3000 &= 400x \\ 50x &= 3000 \\ x &= 60 \end{aligned}$$

From (2), $y = 3(60) - 20 = 160$.
The rate of the train is 60 km/hr, and the rate of the plane is 160 km/hr.

29. Let $x =$ the rate of the boat in still water and $y =$ the rate of the current.

Furthermore,

$$\text{rate upstream} = x - y$$
$$\text{and} \quad \text{rate downstream} = x + y.$$

Use these rates and the information in the problem to make a table.

	r	t	d
Upstream	$x - y$	2	36
Downstream	$x + y$	1.5	36

From the table, use the formula $d = rt$ to write a system of equations.

$$36 = 2(x - y)$$
$$36 = 1.5(x + y)$$

Remove the parentheses and move the variables to the left side.

$$2x - 2y = 36 \quad (1)$$
$$1.5x + 1.5y = 36 \quad (2)$$

Solve the system. Multiply equation (1) by -3 and equation (2) by 4. Then add the results.

$$\begin{array}{rcrl} -6x + 6y &=& -108 & -3 \times (1) \\ 6x + 6y &=& 144 & 4 \times (2) \\ \hline 12y &=& 36 & \\ y &=& 3 & \end{array}$$

Substitute $y = 3$ into equation (1).

$$2x - 2y = 36 \quad (1)$$
$$2x - 2(3) = 36$$
$$2x = 42$$
$$x = 21$$

The rate of the boat is 21 mph, and the rate of the current is 3 mph.

31. Let $x =$ the number of pounds of the \$0.75-per-lb candy and $y =$ the number of pounds of the \$1.25-per-lb candy.

Make a table.

	Price per Pound	Number of Pounds	Value
Less Expensive Candy	\$0.75	x	$\$0.75x$
More Expensive Candy	\$1.25	y	$\$1.25y$
Mixture	\$0.96	9	$\$0.96(9) = \$8.64$

From the "Number of Pounds" column,

$$x + y = 9. \quad (1)$$

From the "Value" column,

$$0.75x + 1.25y = 8.64. \quad (2)$$

Solve the system.

$$\begin{array}{rcrl} -75x - 75y &=& -675 & -75 \times (1) \\ 75x + 125y &=& 864 & 100 \times (2) \\ \hline 50y &=& 189 & \\ y &=& 3.78 & \end{array}$$

From (1), $x + 3.78 = 9$, so $x = 5.22$.
Mix 5.22 pounds of the \$0.75-per-lb candy with 3.78 pounds of the \$1.25-per-lb candy to obtain 9 pounds of a mixture that sells for \$0.96 per pound.

33. Let $x =$ the number of general admission tickets and $y =$ the number of student tickets.

Make a table.

Ticket	Number	Value of Tickets
General	x	$5 \cdot x = 5x$
Student	y	$4 \cdot y = 4y$
Totals	184	812

Solve the system.

$$x + y = 184 \quad (1)$$
$$5x + 4y = 812 \quad (2)$$

To eliminate y, multiply equation (1) by -4 and add the result to equation (2).

$$\begin{array}{rcrl} -4x - 4y &=& -736 & -4 \times (1) \\ 5x + 4y &=& 812 & (2) \\ \hline x &=& 76 & \end{array}$$

From (1), $76 + y = 184$, so $y = 108$.

76 general admission tickets and 108 student tickets were sold.

35. Let $x =$ the price for a citron and let $y =$ the price for a wood apple.

"9 citrons and 7 fragrant wood apples is 107" gives us

$$9x + 7y = 107. \quad (1)$$

"7 citrons and 9 fragrant wood apples is 101" gives us

$$7x + 9y = 101. \quad (2)$$

Multiply equation (1) by -7 and equation (2) by 9. Then add.

$$\begin{array}{rcrl} -63x - 49y &=& -749 & -7 \times (1) \\ 63x + 81y &=& 909 & 9 \times (2) \\ \hline 32y &=& 160 & \\ y &=& 5 & \end{array}$$

Substitute 5 for y in equation (1).

$$9x + 7(5) = 107$$
$$9x + 35 = 107$$
$$9x = 72$$
$$x = 8$$

The prices are 8 for a citron and 5 for a wood apple.

37. Let $x =$ the measure of one angle,
$y =$ the measure of another angle,
and $z =$ the measure of the last angle.

Two equations are given, so

$$z = x + 10$$
$$\text{or} \quad -x + z = 10 \qquad (1)$$
$$\text{and} \quad x + y = 100. \qquad (2)$$

Since the sum of the measures of the angles of a triangle is 180°, the third equation of the system is

$$x + y + z = 180. \quad (3)$$

Equation (1) is missing y. To eliminate y again, multiply equation (2) by -1 and add the result to equation (3).

$$\begin{array}{rcl} -x - y & = & -100 \quad -1 \times (2) \\ x + y + z & = & 180 \quad (3) \\ \hline z & = & 80 \end{array}$$

Since $z = 80$,

$$\begin{aligned} -x + z &= 10 \qquad (1) \\ -x + 80 &= 10 \\ -x &= -70 \\ x &= 70. \end{aligned}$$

From (2), $70 + y = 100$, so $y = 30$.
The measures of the angles are 70°, 30°, and 80°.

39. Let $x =$ the measure of the first angle,
$y =$ the measure of the second angle, and
$z =$ the measure of the third angle.

The sum of the angles in a triangle equals 180°, so

$$x + y + z = 180. \quad (1)$$

The measure of the second angle is 10° greater than 3 times that of the first angle, so

$$y = 3x + 10. \quad (2)$$

The third angle is equal to the sum of the other two, so

$$z = x + y. \quad (3)$$

Solve the system. Substitute z for $x + y$ in equation (1).

$$\begin{aligned} (x + y) + z &= 180 \qquad (1) \\ z + z &= 180 \\ 2z &= 180 \\ z &= 90 \end{aligned}$$

Substitute 90 for z and $3x + 10$ for y in equation (3).

$$\begin{aligned} z &= x + y \qquad (3) \\ 90 &= x + (3x + 10) \\ 80 &= 4x \\ 20 &= x \end{aligned}$$

Substitute $x = 20$ and $z = 90$ into equation (3).

$$\begin{aligned} z &= x + y \qquad (3) \\ 90 &= 20 + y \\ 70 &= y \end{aligned}$$

The three angles have measures of 20°, 70°, and 90°.

41. Let $x =$ the length of the longest side,
$y =$ the length of the middle side,
and $z =$ the length of the shortest side.

Perimeter is the sum of the measures of the sides, so

$$x + y + z = 70. \quad (1)$$

The longest side is 4 cm less than the sum of the other sides, so

$$x = y + z - 4$$
$$\text{or} \quad x - y - z = -4. \qquad (2)$$

Twice the shortest side is 9 cm less than the longest side, so

$$2z = x - 9$$
$$\text{or} \quad -x + 2z = -9 \qquad (3)$$

Add equations (1) and (2) to eliminate y and z.

$$\begin{array}{rcl} x + y + z & = & 70 \quad (1) \\ x - y - z & = & -4 \quad (2) \\ \hline 2x & = & 66 \\ x & = & 33 \end{array}$$

Substitute 33 for x in (3).

$$\begin{aligned} -x + 2z &= -9 \qquad (3) \\ -33 + 2z &= -9 \\ 2z &= 24 \\ z &= 12 \end{aligned}$$

Substitute 33 for x and 12 for z in (1).

$$\begin{aligned} x + y + z &= 70 \qquad (1) \\ 33 + y + 12 &= 70 \\ y + 45 &= 70 \\ y &= 25 \end{aligned}$$

The shortest side is 12 cm long, the middle side is 25 cm long, and the longest side is 33 cm long.

43. Let $x =$ the number of gold medals, $y =$ the number of silver medals, and $z =$ the number of bronze medals.

Russia earned 5 fewer gold medals than bronze, so

$$x = z - 5. \quad (1)$$

The number of silver medals earned was 35 less than twice the number of bronze medals, so

$$y = 2z - 35. \quad (2)$$

The total number of medals earned was 72, so

$$x + y + z = 72. \quad (3)$$

Substitute $z - 5$ for x and $2z - 35$ for y in (3).

$$\begin{aligned}(z-5) + (2z-35) + z &= 72\\ 4z - 40 &= 72\\ 4z &= 112\\ z &= 28\end{aligned}$$

From (1), $x = 28 - 5 = 23$.
From (2), $y = 2(28) - 35 = 56 - 35 = 21$.

Russia earned 23 gold, 21 silver, and 28 bronze medals.

45. Let $x =$ the number of \$16 tickets, $y =$ the number of \$23 tickets, and $z =$ the number of VIP \$40 tickets.

Nine times as many \$16 tickets have been sold as VIP tickets, so

$$x = 9z. \quad (1)$$

The number of \$16 tickets was 55 more than the sum of the number of \$23 tickets and the number of VIP tickets, so

$$x = 55 + y + z. \quad (2)$$

Sales of these tickets totaled \$46,575.

$$16x + 23y + 40z = 46{,}575 \quad (3)$$

Since x is in terms of z in (1), we'll substitute $9z$ for x in (2) and then get y in terms of z.

$$\begin{aligned}9z &= 55 + y + z\\ 8z - 55 &= y \qquad (4)\end{aligned}$$

Substitute $9z$ for x and $8z - 55$ for y in (3).

$$\begin{aligned}16(9z) + 23(8z - 55) + 40z &= 46{,}575\\ 144z + 184z - 1265 + 40z &= 46{,}575\\ 368z &= 47{,}840\\ z &= 130\end{aligned}$$

From (1), $x = 9(130) = 1170$.
From (4), $y = 8(130) - 55 = 985$.

There were 1170 \$16 tickets, 985 \$23 tickets, and 130 \$40 tickets sold.

47. Let $x =$ the number of T-shirts shipped to bookstore A,
$y =$ the number of T-shirts shipped to bookstore B,
and $z =$ the number of T-shirts shipped to bookstore C.

Twice as many T-shirts were shipped to bookstore B as to bookstore A, so

$$y = 2x. \quad (1)$$

The number shipped to bookstore C was 40 less than the sum of the numbers shipped to the other two bookstores, so

$$z = x + y - 40. \quad (2)$$

The total number of T-shirts shipped was 800, so

$$x + y + z = 800. \quad (3)$$

Substitute $2x$ for y [from (1)] into equation (2) to get z in terms of x.

$$\begin{aligned}z &= x + y - 40 \qquad (2)\\ z &= x + (2x) - 40\\ z &= 3x - 40 \qquad (4)\end{aligned}$$

Substitute $2x$ for y and $3x - 40$ for z in (3).

$$\begin{aligned}x + (2x) + (3x - 40) &= 800\\ 6x - 40 &= 800\\ 6x &= 840\\ x &= 140\end{aligned}$$

From (1), $y = 2(140) = 280$.
From (4), $z = 3(140) - 40 = 380$.
The number of T-shirts shipped to bookstores A, B, and C was 140, 280, and 380, respectively.

49. Let x, y, and z denote the number of kilograms of the first, second, and third chemicals, respectively. The mix must include 60% of the first and second chemicals, so

$$x + y = 0.60(750) = 450. \quad (1)$$

The second and third chemicals must be in a ratio of 4 to 3 by weight, so

$$y = \tfrac{4}{3}z. \quad (2)$$

The total is 750, so

$$x + y + z = 750. \quad (3)$$

From (1), $x = 450 - y$. (4)

From (2), $z = \frac{3}{4}y$. (5)

Substitute $450 - y$ for x and $\frac{3}{4}y$ for z in (3).

$$\begin{aligned}(450 - y) + y + \tfrac{3}{4}y &= 750\\ \tfrac{3}{4}y &= 300\\ y &= 400\end{aligned}$$

From (4), $x = 450 - 400 = 50$.
From (5), $z = \frac{3}{4}(400) = 300$.

Use 50 kg of the first chemical, 400 kg of the second chemical, and 300 kg of the third chemical to make the plant food.

51. *Step 2*
Let $x =$ the number of wins,
$y =$ the number of losses in regulation play,
and $z =$ the number of overtime losses.

Step 3
They played 82 games, so

$$x + y + z = 82. \quad (1)$$

The points from their wins and overtime losses totaled 116, so

$$2x + z = 116. \quad (2)$$

They had 9 more losses in regulation play than overtime losses, so

$$y = z + 9. \quad (3)$$

Step 4
Solving equation (2) for x gives us $x = 58 - \frac{1}{2}z$. Now substitute for x and y in (1).

$$\begin{aligned} x + y + z &= 82 \qquad (1) \\ (58 - \tfrac{1}{2}z) + (z + 9) + z &= 82 \\ 1.5z + 67 &= 82 \\ 1.5z &= 15 \\ z &= 10 \end{aligned}$$

Since $y = z + 9$, $y = 10 + 9 = 19$.
Since $x = 58 - \frac{1}{2}z$, $x = 58 - \frac{1}{2}(10) = 53$.

Step 5
The Bruins won 53 games, lost 19 games, and had 10 overtime losses.

Step 6
Adding 53, 19, and 10 gives 82 total games. The points from their wins and overtime losses totaled $2(53) + 10 = 106 + 10 = 116$, and there were 9 more losses in regulation play than overtime losses. The solution is correct.

53. **(a)** The additive inverse of -6 is $-(-6) = 6$.

(b) The multiplicative inverse (reciprocal) of -6 is $\frac{1}{-6} = -\frac{1}{6}$.

55. **(a)** The additive inverse of $\frac{7}{8}$ is $-\frac{7}{8}$.

(b) The multiplicative inverse (reciprocal) of $\frac{7}{8}$ is $\frac{1}{7/8} = \frac{8}{7}$.

8.6 Solving Systems of Linear Equations by Matrix Methods

8.6 Now Try Exercises

N1. $x + 3y = 3$
$2x - 3y = -12$

Write the augmented matrix for this system.

$$\left[\begin{array}{cc|c} 1 & 3 & 3 \\ 2 & -3 & -12 \end{array}\right]$$

$$\left[\begin{array}{cc|c} 1 & 3 & 3 \\ 0 & -9 & -18 \end{array}\right] \quad -2R_1 + R_2$$

$$\left[\begin{array}{cc|c} 1 & 3 & 3 \\ 0 & 1 & 2 \end{array}\right] \quad -\tfrac{1}{9}R_2$$

This matrix gives the system

$$\begin{aligned} x + 3y &= 3 \\ y &= 2. \end{aligned}$$

Substitute 2 for y in the first equation.

$$\begin{aligned} x + 3(2) &= 3 \\ x + 6 &= 3 \\ x &= -3 \end{aligned}$$

The solution set is $\{(-3, 2)\}$.

N2. $x + y - 2z = -5$
$-x + 2y + z = -1$
$2x - y + 3z = 14$

Write the augmented matrix for this system.

$$\left[\begin{array}{ccc|c} 1 & 1 & -2 & -5 \\ -1 & 2 & 1 & -1 \\ 2 & -1 & 3 & 14 \end{array}\right]$$

$$\left[\begin{array}{ccc|c} 1 & 1 & -2 & -5 \\ 0 & 3 & -1 & -6 \\ 2 & -1 & 3 & 14 \end{array}\right] \quad R_1 + R_2$$

$$\left[\begin{array}{ccc|c} 1 & 1 & -2 & -5 \\ 0 & 3 & -1 & -6 \\ 0 & -3 & 7 & 24 \end{array}\right] \quad -2R_1 + R_3$$

$$\left[\begin{array}{ccc|c} 1 & 1 & -2 & -5 \\ 0 & 1 & -\frac{1}{3} & -2 \\ 0 & -3 & 7 & 24 \end{array}\right] \quad \tfrac{1}{3}R_2$$

$$\left[\begin{array}{ccc|c} 1 & 1 & -2 & -5 \\ 0 & 1 & -\frac{1}{3} & -2 \\ 0 & 0 & 6 & 18 \end{array}\right] \quad 3R_2 + R_3$$

$$\left[\begin{array}{ccc|c} 1 & 1 & -2 & -5 \\ 0 & 1 & -\frac{1}{3} & -2 \\ 0 & 0 & 1 & 3 \end{array}\right] \quad \tfrac{1}{6}R_3$$

This matrix gives the system

$$\begin{aligned} x + y - 2z &= -5 \\ y - \tfrac{1}{3}z &= -2 \\ z &= 3. \end{aligned}$$

Substitute 3 for z in the second equation.

$$\begin{aligned} y - \tfrac{1}{3}(3) &= -2 \\ y &= -1 \end{aligned}$$

Substitute -1 for y and 3 for z in the first equation.

$$x+(-1)-2(3)=-5$$
$$x-7=-5$$
$$x=2$$

The solution set is $\{(2,-1,3)\}$.

N3. **(a)** $3x-y=8$
$-6x+2y=4$

Write the augmented matrix for this system.

$$\left[\begin{array}{rr|r} 3 & -1 & 8 \\ -6 & 2 & 4 \end{array}\right]$$

$$\left[\begin{array}{rr|r} 3 & -1 & 8 \\ 0 & 0 & 20 \end{array}\right] \quad 2R_1+R_2$$

This matrix gives the system

$$3x-y=8$$
$$0=20. \quad \textit{False}$$

The false statement indicates that the system is inconsistent and has no solution.

The solution set is $\emptyset$.

(b) $x+2y=7$
$-x-2y=-7$

Write the augmented matrix for this system.

$$\left[\begin{array}{rr|r} 1 & 2 & 7 \\ -1 & -2 & -7 \end{array}\right]$$

$$\left[\begin{array}{rr|r} 1 & 2 & 7 \\ 0 & 0 & 0 \end{array}\right] \quad R_1+R_2$$

This matrix gives the system

$$x+2y=7$$
$$0=0. \quad \textit{True}$$

The true statement indicates that the system has dependent equations.

The solution set is $\{(x,y)\,|\,x+2y=7\}$.

8.6 Section Exercises

1. $$\begin{bmatrix} -2 & 3 & 1 \\ 0 & 5 & -3 \\ 1 & 4 & 8 \end{bmatrix}$$

(a) The elements of the second row are 0, 5, and -3.

(b) The elements of the third column are 1, -3, and 8.

(c) The matrix is square since the number of rows (three) is the same as the number of columns.

(d) The matrix obtained by interchanging the first and third rows is

$$\begin{bmatrix} 1 & 4 & 8 \\ 0 & 5 & -3 \\ -2 & 3 & 1 \end{bmatrix}.$$

(e) The matrix obtained by multiplying the first row by $-\frac{1}{2}$ is

$$\begin{bmatrix} -2(-\frac{1}{2}) & 3(-\frac{1}{2}) & 1(-\frac{1}{2}) \\ 0 & 5 & -3 \\ 1 & 4 & 8 \end{bmatrix} = \begin{bmatrix} 1 & -\frac{3}{2} & -\frac{1}{2} \\ 0 & 5 & -3 \\ 1 & 4 & 8 \end{bmatrix}.$$

(f) The matrix obtained by multiplying the third row by 3 and adding to the first row is

$$\begin{bmatrix} -2+3(1) & 3+3(4) & 1+3(8) \\ 0 & 5 & -3 \\ 1 & 4 & 8 \end{bmatrix} = \begin{bmatrix} 1 & 15 & 25 \\ 0 & 5 & -3 \\ 1 & 4 & 8 \end{bmatrix}.$$

3. There are 3 rows and 2 columns, so the dimensions are written as 3×2.

5. There are 4 rows and 2 columns, so the dimensions are written as 4×2.

7. $x+y=5$
$x-y=3$

Write the augmented matrix for this system.

$$\left[\begin{array}{rr|r} 1 & 1 & 5 \\ 1 & -1 & 3 \end{array}\right]$$

$$\left[\begin{array}{rr|r} 1 & 1 & 5 \\ 0 & -2 & -2 \end{array}\right] \quad -1R_1+R_2$$

$$\left[\begin{array}{rr|r} 1 & 1 & 5 \\ 0 & 1 & 1 \end{array}\right] \quad -\tfrac{1}{2}R_2$$

This matrix gives the system

$$x+y=5$$
$$y=1.$$

Substitute $y=1$ in the first equation.

$$x+y=5$$
$$x+1=5$$
$$x=4$$

The solution set is $\{(4,1)\}$.

9. $2x+4y=6$
$3x-\ y=2$

Write the augmented matrix.

$$\left[\begin{array}{rr|r} 2 & 4 & 6 \\ 3 & -1 & 2 \end{array}\right]$$

The easiest way to get a 1 in the first row, first column position is to multiply the elements in the first row by $\frac{1}{2}$.

$$\left[\begin{array}{rr|r} 1 & 2 & 3 \\ 3 & -1 & 2 \end{array}\right] \quad \tfrac{1}{2}R_1$$

To get a 0 in row two, column 1, we need to subtract 3 from the 3 that is in that position. To do this we will multiply row 1 by -3 and add the result to row 2.

$$\left[\begin{array}{cc|c} 1 & 2 & 3 \\ 0 & -7 & -7 \end{array}\right] \quad -3R_1 + R_2$$

$$\left[\begin{array}{cc|c} 1 & 2 & 3 \\ 0 & 1 & 1 \end{array}\right] \quad -\tfrac{1}{7}R_2$$

This matrix gives the system

$$\begin{aligned} x + 2y &= 3 \\ y &= 1. \end{aligned}$$

Substitute $y = 1$ in the first equation.

$$\begin{aligned} x + 2y &= 3 \\ x + 2(1) &= 3 \\ x &= 1 \end{aligned}$$

The solution set is $\{(1, 1)\}$.

11. $3x + 4y = 13$
$2x - 3y = -14$

Write the augmented matrix.

$$\left[\begin{array}{cc|c} 3 & 4 & 13 \\ 2 & -3 & -14 \end{array}\right]$$

$$\left[\begin{array}{cc|c} 1 & 7 & 27 \\ 2 & -3 & -14 \end{array}\right] \quad -1R_2 + R_1$$

$$\left[\begin{array}{cc|c} 1 & 7 & 27 \\ 0 & -17 & -68 \end{array}\right] \quad -2R_1 + R_2$$

$$\left[\begin{array}{cc|c} 1 & 7 & 27 \\ 0 & 1 & 4 \end{array}\right] \quad -\tfrac{1}{17}R_2$$

This matrix gives the system

$$\begin{aligned} x + 7y &= 27 \\ y &= 4. \end{aligned}$$

Substitute $y = 4$ in the first equation.

$$\begin{aligned} x + 7y &= 27 \\ x + 7(4) &= 27 \\ x + 28 &= 27 \\ x &= -1 \end{aligned}$$

The solution set is $\{(-1, 4)\}$.

13. $-4x + 12y = 36$
$x - 3y = 9$

Write the augmented matrix.

$$\left[\begin{array}{cc|c} -4 & 12 & 36 \\ 1 & -3 & 9 \end{array}\right]$$

$$\left[\begin{array}{cc|c} -1 & 3 & 9 \\ 1 & -3 & 9 \end{array}\right] \quad \tfrac{1}{4}R_1$$

$$\left[\begin{array}{cc|c} -1 & 3 & 9 \\ 0 & 0 & 18 \end{array}\right] \quad R_1 + R_2$$

The corresponding system is

$$\begin{aligned} -x + 3y &= 9 \\ 0 &= 18 \qquad \textit{False} \end{aligned}$$

which is inconsistent and has no solution.

The solution set is $\emptyset$.

15. $2x + y = 4$
$4x + 2y = 8$

Write the augmented matrix.

$$\left[\begin{array}{cc|c} 2 & 1 & 4 \\ 4 & 2 & 8 \end{array}\right]$$

$$\left[\begin{array}{cc|c} 1 & \frac{1}{2} & 2 \\ 4 & 2 & 8 \end{array}\right] \quad \tfrac{1}{2}R_1$$

$$\left[\begin{array}{cc|c} 1 & \frac{1}{2} & 2 \\ 0 & 0 & 0 \end{array}\right] \quad -4R_1 + R_2$$

Row 2, $0 = 0$, indicates that the system has dependent equations.

The solution set is $\{(x, y) \mid 2x + y = 4\}$.

17. $-3x + 2y = 0$
$x - y = 0$

Write the augmented matrix.

$$\left[\begin{array}{cc|c} -3 & 2 & 0 \\ 1 & -1 & 0 \end{array}\right]$$

$$\left[\begin{array}{cc|c} 1 & -1 & 0 \\ -3 & 2 & 0 \end{array}\right] \quad R_1 \leftrightarrow R_2$$

We use $\leftrightarrow$ to represent the interchanging of two rows.

$$\left[\begin{array}{cc|c} 1 & -1 & 0 \\ 0 & -1 & 0 \end{array}\right] \quad 3R_1 + R_2$$

$$\left[\begin{array}{cc|c} 1 & -1 & 0 \\ 0 & 1 & 0 \end{array}\right] \quad -R_2$$

This matrix gives the system

$$\begin{aligned} x - y &= 0 \\ y &= 0. \end{aligned}$$

Substitute $y = 0$ in the first equation.

$$\begin{aligned} x - 0 &= 0 \\ x &= 0 \end{aligned}$$

The solution set is $\{(0, 0)\}$.

19. $x + y - 3z = 1$
$2x - y + z = 9$
$3x + y - 4z = 8$

Write the augmented matrix.

$$\left[\begin{array}{ccc|c} 1 & 1 & -3 & 1 \\ 2 & -1 & 1 & 9 \\ 3 & 1 & -4 & 8 \end{array}\right]$$

$$\left[\begin{array}{ccc|c} 1 & 1 & -3 & 1 \\ 0 & -3 & 7 & 7 \\ 0 & -2 & 5 & 5 \end{array}\right] \quad \begin{array}{l} \\ -2R_1 + R_2 \\ -3R_1 + R_3 \end{array}$$

$$\left[\begin{array}{ccc|c} 1 & 1 & -3 & 1 \\ 0 & 1 & -\frac{7}{3} & -\frac{7}{3} \\ 0 & -2 & 5 & 5 \end{array}\right] \quad -\tfrac{1}{3}R_2$$

$$\left[\begin{array}{ccc|c} 1 & 1 & -3 & 1 \\ 0 & 1 & -\frac{7}{3} & -\frac{7}{3} \\ 0 & 0 & \frac{1}{3} & \frac{1}{3} \end{array}\right] \quad 2R_2 + R_3$$

$$\left[\begin{array}{ccc|c} 1 & 1 & -3 & 1 \\ 0 & 1 & -\frac{7}{3} & -\frac{7}{3} \\ 0 & 0 & 1 & 1 \end{array}\right] \quad 3R_3$$

This matrix gives the system

$$\begin{aligned} x + y - 3z &= 1 \\ y - \tfrac{7}{3}z &= -\tfrac{7}{3} \\ z &= 1. \end{aligned}$$

Substitute $z = 1$ in the second equation.

$$\begin{aligned} y - \tfrac{7}{3}z &= -\tfrac{7}{3} \\ y - \tfrac{7}{3}(1) &= -\tfrac{7}{3} \\ y &= 0 \end{aligned}$$

Substitute $y = 0$ and $z = 1$ in the first equation.

$$\begin{aligned} x + y - 3z &= 1 \\ x + 0 - 3(1) &= 1 \\ x - 3 &= 1 \\ x &= 4 \end{aligned}$$

The solution set is $\{(4, 0, 1)\}$.

21.
$$\begin{aligned} x + y - z &= 6 \\ 2x - y + z &= -9 \\ x - 2y + 3z &= 1 \end{aligned}$$

Write the augmented matrix.

$$\left[\begin{array}{ccc|c} 1 & 1 & -1 & 6 \\ 2 & -1 & 1 & -9 \\ 1 & -2 & 3 & 1 \end{array}\right]$$

$$\left[\begin{array}{ccc|c} 1 & 1 & -1 & 6 \\ 0 & -3 & 3 & -21 \\ 0 & -3 & 4 & -5 \end{array}\right] \quad \begin{array}{l} \\ -2R_1 + R_2 \\ -1R_1 + R_3 \end{array}$$

$$\left[\begin{array}{ccc|c} 1 & 1 & -1 & 6 \\ 0 & 1 & -1 & 7 \\ 0 & -3 & 4 & -5 \end{array}\right] \quad -\tfrac{1}{3}R_2$$

$$\left[\begin{array}{ccc|c} 1 & 1 & -1 & 6 \\ 0 & 1 & -1 & 7 \\ 0 & 0 & 1 & 16 \end{array}\right] \quad 3R_2 + R_3$$

This matrix gives the system

$$\begin{aligned} x + y - z &= 6 \\ y - z &= 7 \\ z &= 16. \end{aligned}$$

Substitute $z = 16$ in the second equation.

$$\begin{aligned} y - 16 &= 7 \\ y &= 23 \end{aligned}$$

Substitute $y = 23$ and $z = 16$ in the first equation.

$$\begin{aligned} x + 23 - 16 &= 6 \\ x + 7 &= 6 \\ x &= -1 \end{aligned}$$

The solution set is $\{(-1, 23, 16)\}$.

23.
$$\begin{aligned} x - y &= 1 \\ y - z &= 6 \\ x + z &= -1 \end{aligned}$$

Write the augmented matrix.

$$\left[\begin{array}{ccc|c} 1 & -1 & 0 & 1 \\ 0 & 1 & -1 & 6 \\ 1 & 0 & 1 & -1 \end{array}\right]$$

$$\left[\begin{array}{ccc|c} 1 & -1 & 0 & 1 \\ 0 & 1 & -1 & 6 \\ 0 & 1 & 1 & -2 \end{array}\right] \quad -1R_1 + R_3$$

$$\left[\begin{array}{ccc|c} 1 & -1 & 0 & 1 \\ 0 & 1 & -1 & 6 \\ 0 & 0 & 2 & -8 \end{array}\right] \quad -1R_2 + R_3$$

$$\left[\begin{array}{ccc|c} 1 & -1 & 0 & 1 \\ 0 & 1 & -1 & 6 \\ 0 & 0 & 1 & -4 \end{array}\right] \quad \tfrac{1}{2}R_3$$

This matrix gives the system

$$\begin{aligned} x - y &= 1 \\ y - z &= 6 \\ z &= -4. \end{aligned}$$

Substitute $z = -4$ in the second equation.

$$\begin{aligned} y - z &= 6 \\ y - (-4) &= 6 \\ y &= 2 \end{aligned}$$

Substitute $y = 2$ in the first equation.

$$\begin{aligned} x - y &= 1 \\ x - 2 &= 1 \\ x &= 3 \end{aligned}$$

The solution set is $\{(3, 2, -4)\}$.

25.
$$\begin{aligned} x - 2y + z &= 4 \\ 3x - 6y + 3z &= 12 \\ -2x + 4y - 2z &= -8 \end{aligned}$$

Write the augmented matrix.

$$\left[\begin{array}{ccc|c} 1 & -2 & 1 & 4 \\ 3 & -6 & 3 & 12 \\ -2 & 4 & -2 & -8 \end{array}\right]$$

$$\left[\begin{array}{ccc|c} 1 & -2 & 1 & 4 \\ 1 & -2 & 1 & 4 \\ -1 & 2 & -1 & -4 \end{array}\right] \quad \begin{array}{l} \\ \tfrac{1}{3}R_2 \\ \tfrac{1}{2}R_3 \end{array}$$

$$\left[\begin{array}{ccc|c} 1 & -2 & 1 & 4 \\ 0 & 0 & 0 & 0 \\ 0 & 0 & 0 & 0 \end{array}\right] \quad \begin{array}{l} \\ -1R_1 + R_2 \\ R_1 + R_3 \end{array}$$

This augmented matrix represents a system of dependent equations.
The solution set is $\{(x, y, z) \mid x - 2y + z = 4\}$.

27.
$$\begin{aligned} x + 2y + 3z &= -2 \\ 2x + 4y + 6z &= -5 \\ x - y + 2z &= 6 \end{aligned}$$

Write the augmented matrix.

$$\left[\begin{array}{ccc|c} 1 & 2 & 3 & -2 \\ 2 & 4 & 6 & -5 \\ 1 & -1 & 2 & 6 \end{array}\right]$$

$$\left[\begin{array}{ccc|c} 1 & 2 & 3 & -2 \\ 0 & 0 & 0 & -1 \\ 0 & -3 & -1 & 8 \end{array}\right] \quad \begin{array}{l} -2R_1 + R_2 \\ -R_1 + R_3 \end{array}$$

From the second row, $0 = -1$, we see that the system is inconsistent.
The solution set is $\emptyset$.

29.
$$\begin{aligned} 4x + y &= 5 \\ 2x + y &= 3 \end{aligned}$$

Enter the augmented matrix as [A].

$$\begin{bmatrix} 4 & 1 & 5 \\ 2 & 1 & 3 \end{bmatrix}$$

The TI-83/4 screen for A follows. (Use MATRX EDIT.)

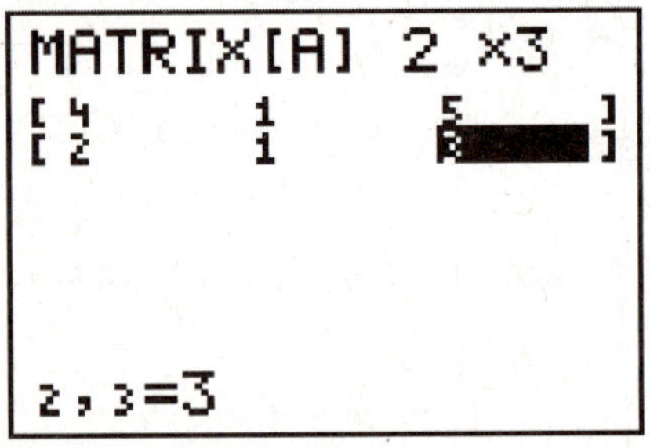

Now use the reduced row echelon form (rref) command to simplify the system. (Use MATRX MATH ALPHA B for rref and MATRX 1 for [A].)

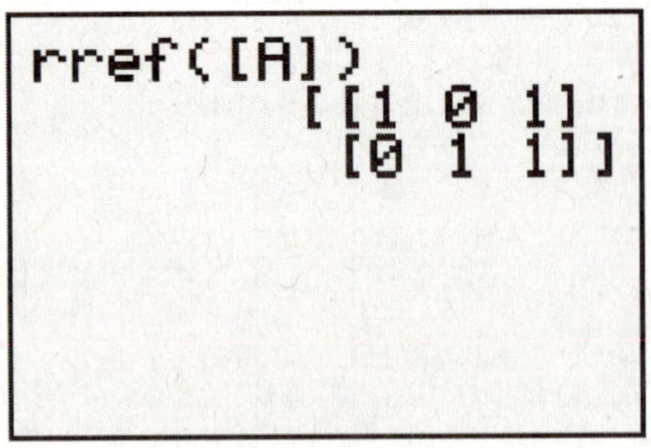

This matrix gives the system

$$\begin{aligned} 1x + 0y &= 1 \\ 0x + 1y &= 1, \end{aligned}$$

or simply, $x = 1$ and $y = 1$.
The solution set is $\{(1, 1)\}$.

31.
$$\begin{aligned} 5x + y - 3z &= -6 \\ 2x + 3y + z &= 5 \\ -3x - 2y + 4z &= 3 \end{aligned}$$

Enter the augmented matrix as [C].

$$\begin{bmatrix} 5 & 1 & -3 & -6 \\ 2 & 3 & 1 & 5 \\ -3 & -2 & 4 & 3 \end{bmatrix}$$

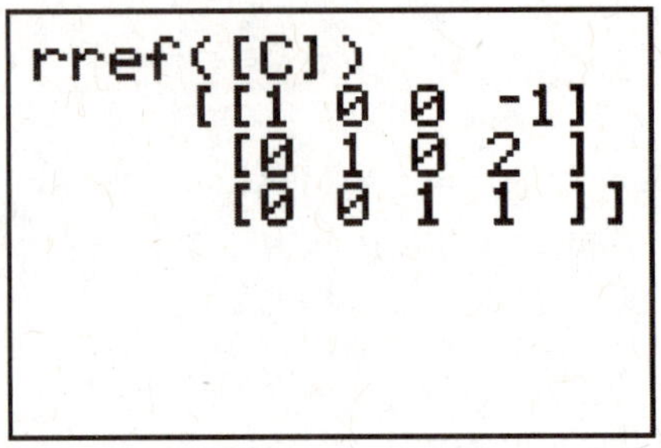

The solution set is $\{(-1, 2, 1)\}$.

33.
$$\begin{aligned} x + z &= -3 \\ y + z &= 3 \\ x + y &= 8 \end{aligned}$$

Enter the augmented matrix as [E].

$$\begin{bmatrix} 1 & 0 & 1 & -3 \\ 0 & 1 & 1 & 3 \\ 1 & 1 & 0 & 8 \end{bmatrix}$$

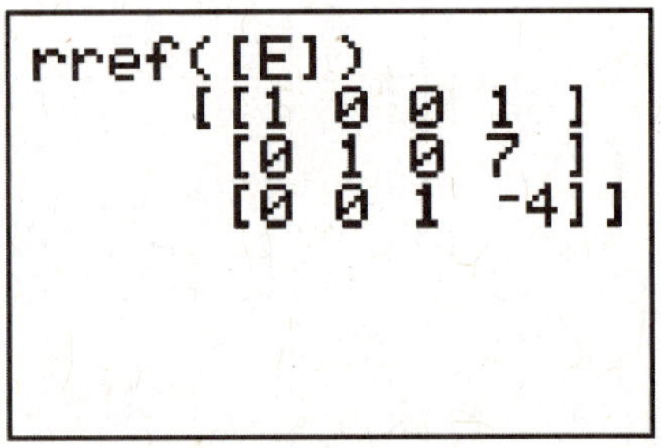

The solution set is $\{(1, 7, -4)\}$.

35. $x - 4 \geq 12$
$x \geq 16$ *Add 4.*

Check that the solution set is the interval $[16, \infty)$.

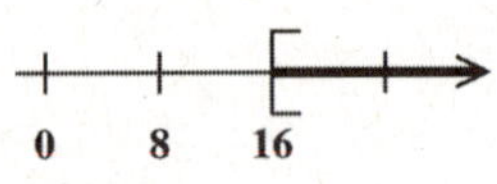

37. $-5z + 18 > -2$
$-5z > -20$
Divide by -5; reverse the inequality symbol.
$z < 4$

Check that the solution set is the interval $(-\infty, 4)$.

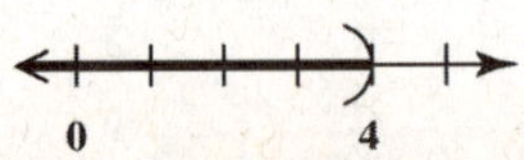

Chapter 8 Review Exercises

1. $(3,4)$ $\quad 4x - 2y = 4$
$5x + y = 19$

To decide whether $(3,4)$ is a solution of the system, substitute 3 for x and 4 for y in each equation.

$$\begin{aligned} 4x - 2y &= 4 \\ 4(3) - 2(4) &\overset{?}{=} 4 \\ 12 - 8 &\overset{?}{=} 4 \\ 4 &= 4 \quad \textit{True} \end{aligned}$$

$$\begin{aligned} 5x + y &= 19 \\ 5(3) + 4 &\overset{?}{=} 19 \\ 15 + 4 &\overset{?}{=} 19 \\ 19 &= 19 \quad \textit{True} \end{aligned}$$

Since $(3,4)$ satisfies both equations, it is a solution of the system.

2. $(-5,2)$ $\quad x - 4y = -13 \quad (1)$
$2x + 3y = 4 \quad (2)$

Substitute -5 for x and 2 for y in equation (2).

$$\begin{aligned} 2x + 3y &= 4 \\ 2(-5) + 3(2) &\overset{?}{=} 4 \\ -10 + 6 &\overset{?}{=} 4 \\ -4 &= 4 \quad \textit{False} \end{aligned}$$

Since $(-5,2)$ is not a solution of the second equation, it cannot be a solution of the system.

3. $x + y = 4$
$2x - y = 5$

To graph the equations, find the intercepts.

$\boldsymbol{x + y = 4}$**:** Let $y = 0$; then $x = 4$.
Let $x = 0$; then $y = 4$.

Plot the intercepts, $(4,0)$ and $(0,4)$, and draw the line through them.

$\boldsymbol{2x - y = 5}$**:** Let $y = 0$; then $x = \frac{5}{2}$.
Let $x = 0$; then $y = -5$.

Plot the intercepts, $(\frac{5}{2}, 0)$ and $(0,-5)$, and draw the line through them.

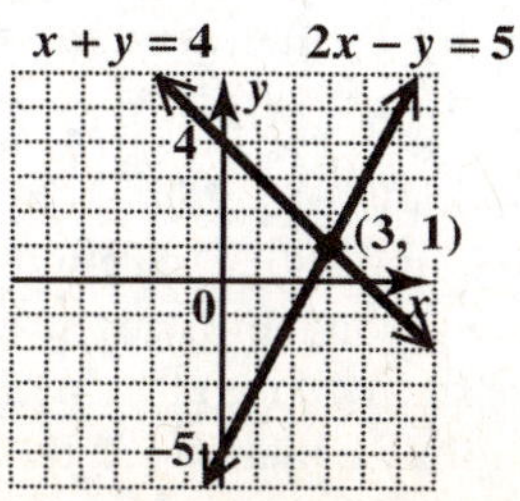

It appears that the lines intersect at the point $(3,1)$. Check this by substituting 3 for x and 1 for y in both equations. Since $(3,1)$ satisfies both equations, the solution set of this system is $\{(3,1)\}$.

4. $x - 2y = 4$
$2x + y = -2$

To graph the equations, find the intercepts.

$\boldsymbol{x - 2y = 4}$**:** Let $y = 0$; then $x = 4$.
Let $x = 0$; then $y = -2$.

Plot the intercepts, $(4,0)$ and $(0,-2)$, and draw the line through them.

$\boldsymbol{2x + y = -2}$**:** Let $y = 0$; then $x = -1$.
Let $x = 0$; then $y = -2$.

Plot the intercepts, $(-1,0)$ and $(0,-2)$, and draw the line through them.

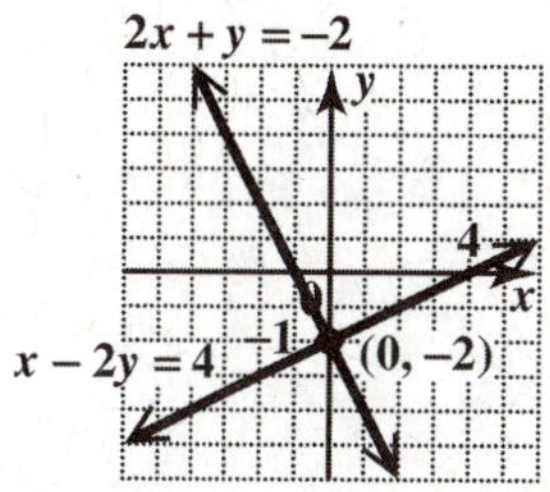

The lines intersect at their common y-intercept, $(0,-2)$, so $\{(0,-2)\}$ is the solution set of the system.

5. $2x + 4 = 2y$
$y - x = -3$

Graph the line $2x + 4 = 2y$ through its intercepts, $(-2,0)$ and $(0,2)$. Graph the line $y - x = -3$ through its intercepts, $(3,0)$ and $(0,-3)$.

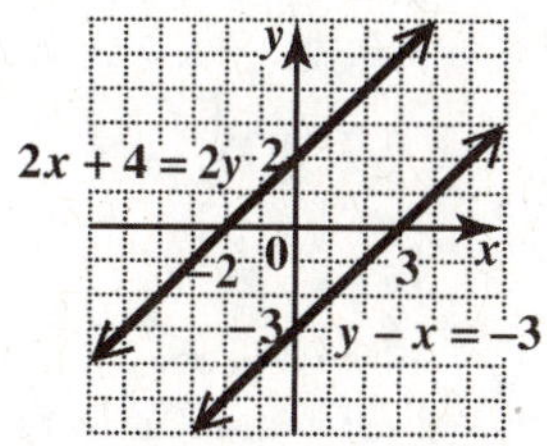

Solving each equation for y gives us $y = x + 2$ and $y = x - 3$, so the lines are parallel (both lines have slope 1). Since they do not intersect, there is no solution. This is an inconsistent system and the solution set is $\emptyset$.

6. $x - 2 = 2y$
$2x - 4y = 4$

Graph the line $x - 2 = 2y$ through its intercepts, $(2,0)$ and $(0,-1)$.

Graph the line $2x - 4y = 4$ through its intercepts, $(2,0)$ and $(0,-1)$.

Since both equations have the same intercepts, they are equations of the same line.

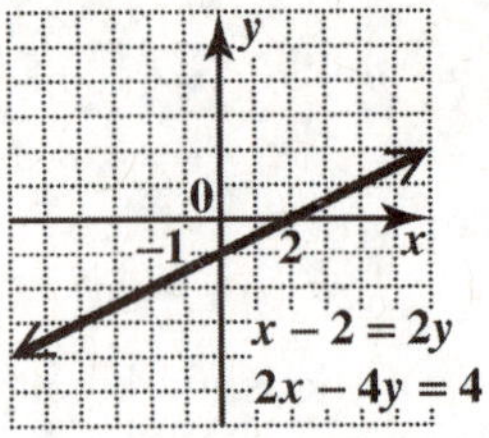

There is an infinite number of solutions. The equations are dependent equations and the solution set contains an infinite number of ordered pairs. The solution set is $\{(x, y) \mid x - 2 = 2y\}$.

7. It would be easiest to solve for x in the second equation because its coefficient is -1. No fractions would be involved.

8. The true statement $0 = 0$ is an indication that the system has an infinite number of solutions. Write the solution set using set-builder notation and the equation of the system that is in standard form with integer coefficients having greatest common factor 1.

9. $3x + y = 7 \quad (1)$
$x = 2y \quad (2)$

Substitute $2y$ for x in equation (1) and solve the resulting equation for y.

$$\begin{aligned} 3x + y &= 7 \\ 3(2y) + y &= 7 \\ 6y + y &= 7 \\ 7y &= 7 \\ y &= 1 \end{aligned}$$

To find x, let $y = 1$ in equation (2).

$$\begin{aligned} x &= 2y \\ x &= 2(1) = 2 \end{aligned}$$

The solution set is $\{(2, 1)\}$.

10. $2x - 5y = -19 \quad (1)$
$y = x + 2 \quad (2)$

Substitute $x + 2$ for y in equation (1).

$$\begin{aligned} 2x - 5y &= -19 \\ 2x - 5(x + 2) &= -19 \\ 2x - 5x - 10 &= -19 \\ -3x - 10 &= -19 \\ -3x &= -9 \\ x &= 3 \end{aligned}$$

To find y, let $x = 3$ in equation (2).

$$\begin{aligned} y &= x + 2 \\ y &= 3 + 2 = 5 \end{aligned}$$

The solution set is $\{(3, 5)\}$.

11. $4x + 5y = 44 \quad (1)$
$x + 2 = 2y \quad (2)$

Solve equation (2) for x.

$$\begin{aligned} x + 2 &= 2y \\ x &= 2y - 2 \quad (3) \end{aligned}$$

Substitute $2y - 2$ for x in equation (1) and solve the resulting equation for y.

$$\begin{aligned} 4x + 5y &= 44 \\ 4(2y - 2) + 5y &= 44 \\ 8y - 8 + 5y &= 44 \\ 13y - 8 &= 44 \\ 13y &= 52 \\ y &= 4 \end{aligned}$$

To find x, let $y = 4$ in equation (3).

$$\begin{aligned} x &= 2y - 2 \\ x &= 2(4) - 2 = 6 \end{aligned}$$

The solution set is $\{(6, 4)\}$.

12. $5x + 15y = 30 \quad (1)$
$x + 3y = 6 \quad (2)$

Solve equation (2) for x.

$$\begin{aligned} x + 3y &= 6 \\ x &= 6 - 3y \quad (3) \end{aligned}$$

Substitute $6 - 3y$ for x in equation (1).

$$\begin{aligned} 5x + 15y &= 30 \\ 5(6 - 3y) + 15y &= 30 \\ 30 - 15y + 15y &= 30 \\ 30 &= 30 \quad \textit{True} \end{aligned}$$

This true result means that every solution of one equation is also a solution of the other, so the system has an infinite number of solutions. The solution set is $\{(x, y) \mid x + 3y = 6\}$.

13. If we simply add the given equations without first multiplying one or both equations by a constant, choice **C** is the only system in which a variable will be eliminated. If we add the equations in **C** we get $3x = 17$. (The variable y was eliminated.)

14. $2x + 12y = 7 \quad (1)$
$3x + 4y = 1 \quad (2)$

(a) If we multiply equation (1) by -3, the first term will become $-6x$. To eliminate x, we need to change the first term on the left side of equation (2) from $3x$ to $6x$. In order to do this, we must multiply equation (2) by 2.

(b) If we multiply equation (1) by -3, the second term will become $-36y$. To eliminate y, we need to change the second term on the left side of equation (2) from $4y$ to $36y$. In order to do this, we must multiply equation (2) by 9.

15.
$$\begin{array}{rcrcrl} 2x & - & y & = & 13 & (1) \\ x & + & y & = & 8 & (2) \\ \hline 3x & & & = & 21 & \textit{Add (1) and (2).} \\ & & x & = & 7 & \end{array}$$

From (2), $y = 1$.

The solution set is $\{(7, 1)\}$.

16.
$$\begin{aligned} -4x + 3y &= 25 \quad (1) \\ 6x - 5y &= -39 \quad (2) \end{aligned}$$

Multiply equation (1) by 3 and equation (2) by 2; then add the results.

$$\begin{array}{rcrcr} -12x & + & 9y & = & 75 \\ 12x & - & 10y & = & -78 \\ \hline & & -y & = & -3 \\ & & y & = & 3 \end{array}$$

To find x, let $y = 3$ in equation (1).

$$\begin{aligned} -4x + 3y &= 25 \\ -4x + 3(3) &= 25 \\ -4x + 9 &= 25 \\ -4x &= 16 \\ x &= -4 \end{aligned}$$

The solution set is $\{(-4, 3)\}$.

17.
$$\begin{aligned} 3x - 4y &= 9 \quad (1) \\ 6x - 8y &= 18 \quad (2) \end{aligned}$$

Multiply equation (1) by -2 and add the result to equation (2).

$$\begin{array}{rcrcrl} -6x & + & 8y & = & -18 & \\ 6x & - & 8y & = & 18 & \\ \hline & & 0 & = & 0 & \textit{True} \end{array}$$

This result indicates that all solutions of equation (1) are also solutions of equation (2). The given system has an infinite number of solutions. The solution set is $\{(x, y) \mid 3x - 4y = 9\}$.

18.
$$\begin{aligned} 2x + y &= 3 \quad (1) \\ -4x - 2y &= 6 \quad (2) \end{aligned}$$

Multiply equation (1) by 2 and add the result to equation (2).

$$\begin{array}{rcrcrl} 4x & + & 2y & = & 6 & \\ -4x & - & 2y & = & 6 & \\ \hline & & 0 & = & 12 & \textit{False} \end{array}$$

This result indicates that the given system has solution set $\emptyset$.

19.
$$\begin{aligned} 2x + 3y &= -5 \quad (1) \\ 3x + 4y &= -8 \quad (2) \end{aligned}$$

Multiply equation (1) by -3 and equation (2) by 2; then add the results.

$$\begin{array}{rcrcr} -6x & - & 9y & = & 15 \\ 6x & + & 8y & = & -16 \\ \hline & & -y & = & -1 \\ & & y & = & 1 \end{array}$$

To find x, let $y = 1$ in equation (1).

$$\begin{aligned} 2x + 3y &= -5 \\ 2x + 3(1) &= -5 \\ 2x + 3 &= -5 \\ 2x &= -8 \\ x &= -4 \end{aligned}$$

The solution set is $\{(-4, 1)\}$.

20.
$$\begin{aligned} 6x - 9y &= 0 \quad (1) \\ 2x - 3y &= 0 \quad (2) \end{aligned}$$

Multiply equation (2) by -3 and add the result to equation (1).

$$\begin{array}{rcrcrl} 6x & - & 9y & = & 0 & \\ -6x & + & 9y & = & 0 & \\ \hline & & 0 & = & 0 & \textit{True} \end{array}$$

This result indicates that the system has an infinite number of solutions. The solution set is $\{(x, y) \mid 2x - 3y = 0\}$.

21.
$$\begin{aligned} x - 2y &= 5 \quad (1) \\ y &= x - 7 \quad (2) \end{aligned}$$

From (2), substitute $x - 7$ for y in equation (1).

$$\begin{aligned} x - 2y &= 5 \\ x - 2(x - 7) &= 5 \\ x - 2x + 14 &= 5 \\ -x &= -9 \\ x &= 9 \end{aligned}$$

Let $x = 9$ in equation (2) to find y.

$$y = 9 - 7 = 2$$

The solution set is $\{(9, 2)\}$.

22.
$$\frac{x}{2} + \frac{y}{3} = 7 \quad (1)$$

$$\frac{x}{4} + \frac{2y}{3} = 8 \quad (2)$$

Multiply equation (1) by 6 to clear fractions.

$$\begin{aligned} 6\left(\frac{x}{2} + \frac{y}{3}\right) &= 6(7) \\ 3x + 2y &= 42 \quad (3) \end{aligned}$$

Multiply equation (2) by 12 to clear fractions.

$$\begin{aligned} 12\left(\frac{x}{4} + \frac{2y}{3}\right) &= 12(8) \\ 3x + 8y &= 96 \quad (4) \end{aligned}$$

To solve this system by the elimination method, multiply equation (3) by -1 and add the result to equation (4).

$$\begin{array}{rcrcr} -3x & - & 2y & = & -42 \\ 3x & + & 8y & = & 96 \\ \hline & & 6y & = & 54 \\ & & y & = & 9 \end{array}$$

To find x, let $y = 9$ in equation (3).

$$\begin{aligned} 3x + 2y &= 42 \\ 3x + 2(9) &= 42 \\ 3x + 18 &= 42 \\ 3x &= 24 \\ x &= 8 \end{aligned}$$

The solution set is $\{(8, 9)\}$.

23. $\frac{3}{4}x - \frac{1}{3}y = \frac{7}{6}$ (1)
$\frac{1}{2}x + \frac{2}{3}y = \frac{5}{3}$ (2)

Multiply equation (1) by 12 to clear fractions.

$$\begin{aligned} 12(\tfrac{3}{4}x) - 12(\tfrac{1}{3}y) &= 12(\tfrac{7}{6}) \\ 9x - 4y &= 14 \qquad (3) \end{aligned}$$

Multiply equation (2) by 6 to clear fractions.

$$\begin{aligned} 6(\tfrac{1}{2}x) + 6(\tfrac{2}{3}y) &= 6(\tfrac{5}{3}) \\ 3x + 4y &= 10 \qquad (4) \end{aligned}$$

Add equations (3) and (4) to eliminate y.

$$\begin{array}{rcrcr} 9x & - & 4y & = & 14 \\ 3x & + & 4y & = & 10 \\ \hline 12x & & & = & 24 \\ & & x & = & 2 \end{array}$$

To find y, let $x = 2$ in equation (4).

$$\begin{aligned} 3(2) + 4y &= 10 \\ 6 + 4y &= 10 \\ 4y &= 4 \\ y &= 1 \end{aligned}$$

The solution set is $\{(2, 1)\}$.

24. $0.4x - 0.5y = -2.2$ (1)
$0.3x + 0.2y = -0.5$ (2)

Multiply equation (1) and equation (2) by 10 to eliminate the decimals. Then solve the resulting equations by the substitution method.

$$\begin{aligned} 4x - 5y &= -22 \quad (3) \\ 3x + 2y &= \;\; -5 \quad (4) \end{aligned}$$

To solve this system by the elimination method, multiply equation (3) by 2, multiply equation (4) by 5, and add the resulting equations.

$$\begin{array}{rcrcr} 8x & - & 10y & = & -44 \\ 15x & + & 10y & = & -25 \\ \hline 23x & & & = & -69 \\ & & x & = & -3 \end{array}$$

To find y, let $x = -3$ in equation (4).

$$\begin{aligned} 3x + 2y &= -5 \\ 3(-3) + 2y &= -5 \\ -9 + 2y &= -5 \\ 2y &= 4 \\ y &= 2 \end{aligned}$$

The solution set is $\{(-3, 2)\}$.

25.
$$\begin{array}{rcrcrcrl} 2x & + & 3y & - & z & = & -16 & (1) \\ x & + & 2y & + & 2z & = & -3 & (2) \\ -3x & + & y & + & z & = & -5 & (3) \end{array}$$

To eliminate z, add equations (1) and (3).

$$\begin{array}{rcrcrcrl} 2x & + & 3y & - & z & = & -16 & (1) \\ -3x & + & y & + & z & = & -5 & (3) \\ \hline -x & + & 4y & & & = & -21 & (4) \end{array}$$

To eliminate z again, multiply equation (1) by 2 and add the result to equation (2).

$$\begin{array}{rcrcrcrl} 4x & + & 6y & - & 2z & = & -32 & 2 \times (1) \\ x & + & 2y & + & 2z & = & -3 & (2) \\ \hline 5x & + & 8y & & & = & -35 & (5) \end{array}$$

Use equations (4) and (5) to eliminate x. Multiply equation (4) by 5 and add the result to equation (5).

$$\begin{array}{rcrcrl} -5x & + & 20y & = & -105 & 5 \times (4) \\ 5x & + & 8y & = & -35 & (5) \\ \hline & & 28y & = & -140 & \\ & & y & = & -5 & \end{array}$$

Substitute -5 for y in equation (4) to find x.

$$\begin{aligned} -x + 4y &= -21 \qquad (4) \\ -x + 4(-5) &= -21 \\ -x - 20 &= -21 \\ -x &= -1 \\ x &= 1 \end{aligned}$$

Substitute 1 for x and -5 for y in equation (2) to find z.

$$\begin{aligned} x + 2y + 2z &= -3 \qquad (2) \\ 1 + 2(-5) + 2z &= -3 \\ 1 - 10 + 2z &= -3 \\ 2z &= 6 \\ z &= 3 \end{aligned}$$

The solution set is $\{(1, -5, 3)\}$.

26.
$$\begin{aligned} 4x - y \quad &= 2 \quad (1)\\ 3y + z &= 9 \quad (2)\\ x \quad + 2z &= 7 \quad (3)\end{aligned}$$

To eliminate y, multiply equation (1) by 3 and add the result to equation (2).

$$\begin{array}{rcll} 12x - 3y \quad &=& 6 & 3\times(1)\\ 3y + z &=& 9 & (2)\\ \hline 12x \quad + z &=& 15 & (4)\end{array}$$

To eliminate z, multiply equation (4) by -2 and add the result to equation (3).

$$\begin{array}{rcll} -24x - 2z &=& -30 & -2\times(4)\\ x + 2z &=& 7 & (3)\\ \hline -23x &=& -23 & \\ x &=& 1 & \end{array}$$

Substitute 1 for x in equation (3) to find z.

$$\begin{aligned} x + 2z &= 7 \quad (3)\\ 1 + 2z &= 7\\ 2z &= 6\\ z &= 3\end{aligned}$$

Substitute 1 for x in equation (1) to find y.

$$\begin{aligned} 4x - y &= 2 \quad (1)\\ 4(1) - y &= 2\\ 4 - y &= 2\\ -y &= -2\\ y &= 2\end{aligned}$$

The solution set is $\{(1, 2, 3)\}$.

27.
$$\begin{aligned} 3x - y - z &= -8 \quad (1)\\ 4x + 2y + 3z &= 15 \quad (2)\\ -6x + 2y + 2z &= 10 \quad (3)\end{aligned}$$

To eliminate y, multiply equation (1) by 2 and add the result to equation (3).

$$\begin{array}{rcll} 6x - 2y - 2z &=& -16 & 2\times(1)\\ -6x + 2y + 2z &=& 10 & (3)\\ \hline 0 &=& -6 & \textit{False}\end{array}$$

Since a false statement results, equations (1) and (3) have no common solution. The system is *inconsistent*. The solution set is $\emptyset$.

28. Let x = the width of the rink, and y = the length of the rink.

The length is 30 ft longer than twice the width.

$$y = 2x + 30 \quad (1)$$

The perimeter is 570 ft.

$$2x + 2y = 570 \quad (2)$$

Substitute $2x + 30$ for y in equation (2).

$$\begin{aligned} 2x + 2y &= 570 \quad (2)\\ 2x + 2(2x + 30) &= 570\\ 2x + 4x + 60 &= 570\\ 6x &= 510\\ x &= 85\end{aligned}$$

From (1), $y = 2(85) + 30 = 200$.

The width is 85 ft and the length is 200 ft.

29. Let x = the average price for a Yankees ticket and y = the average price for a Red Sox ticket.

From the given information, we get the following system of equations:

$$\begin{aligned} 2x + 3y &= 296.66 \quad (1)\\ 3x + 2y &= 319.39 \quad (2)\end{aligned}$$

To eliminate y, multiply equation (1) by 2 and add the result to -3 times equation (2).

$$\begin{array}{rcll} 4x + 6y &=& 593.32 & 2\times(1)\\ -9x - 6y &=& -958.17 & -3\times(2)\\ \hline -5x &=& -364.85 & \\ x &=& 72.97 & \end{array}$$

Substitute 72.97 for x in equation (1).

$$\begin{aligned} 2x + 3y &= 296.66 \quad (1)\\ 2(72.97) + 3y &= 296.66\\ 145.94 + 3y &= 296.66\\ 3y &= 150.72\\ y &= 50.24\end{aligned}$$

The average price for a Yankees ticket was \$72.97 and the average price for a Red Sox ticket was \$50.24.

30. Let x = the speed of the plane and y = the speed of the wind.

Complete the chart.

	r	t	d
With Wind	$x + y$	1.75	$1.75(x + y)$
Against Wind	$x - y$	2	$2(x - y)$

The distance each way is 560 miles. From the chart,

$$1.75(x + y) = 560.$$

Divide by 1.75.

$$x + y = 320 \quad (1)$$

From the chart,

$$\begin{aligned} 2(x - y) &= 560\\ x - y &= 280. \quad (2)\end{aligned}$$

Solve the system by adding equations (1) and (2) to eliminate y.

$$\begin{aligned} x + y &= 320 \quad (1)\\ x - y &= 280 \quad (2)\\ \hline 2x \quad &= 600\\ x &= 300 \end{aligned}$$

From (1), $300 + y = 320$, so $y = 20$.

The speed of the plane was 300 mph, and the speed of the wind was 20 mph.

31. Let $x =$ the amount of \$2-per-pound nuts and $y =$ the amount of \$1-per-pound chocolate candy.

Complete the chart.

	Number of Pounds	Price per Pound	Value
Nuts	x	2	$2x$
Chocolate	y	1	$1y = y$
Mixture	100	1.30	$1.30(100) = 130$

Solve the system formed from the second and fourth columns.

$$\begin{aligned} x + y &= 100 \quad (1)\\ 2x + y &= 130 \quad (2) \end{aligned}$$

Solve equation (1) for y.

$$y = 100 - x \quad (3)$$

Substitute $100 - x$ for y in equation (2).

$$\begin{aligned} 2x + (100 - x) &= 130\\ x &= 30 \end{aligned}$$

From (3), $y = 100 - 30 = 70$.

She should use 30 lb of \$2-per-pound nuts and 70 lb of \$1-per-pound chocolate candy.

32. Let $x =$ the measure of the largest angle, $y =$ the measure of the middle-sized angle, and $z =$ the measure of the smallest angle.

Since the sum of the measures of the angles of a triangle is 180°,

$$x + y + z = 180. \quad (1)$$

Since the largest angle measures 10° less than the sum of the other two,

$$\begin{aligned} x &= y + z - 10\\ \text{or} \quad x - y - z &= -10. \quad (2) \end{aligned}$$

Since the measure of the middle-sized angle is the average of the other two,

$$\begin{aligned} y &= \frac{x+z}{2}\\ 2y &= x + z\\ -x + 2y - z &= 0. \quad (3) \end{aligned}$$

Solve the system.

$$\begin{aligned} x + y + z &= 180 \quad (1)\\ x - y - z &= -10 \quad (2)\\ -x + 2y - z &= 0 \quad (3) \end{aligned}$$

Add equations (1) and (2) to find x.

$$\begin{aligned} x + y + z &= 180 \quad (1)\\ x - y - z &= -10 \quad (2)\\ \hline 2x \quad &= 170\\ x &= 85 \end{aligned}$$

Add equations (1) and (3), to find y.

$$\begin{aligned} x + y + z &= 180 \quad (1)\\ -x + 2y - z &= 0 \quad (3)\\ \hline 3y &= 180\\ y &= 60 \end{aligned}$$

Substitute 85 for x and 60 for y in equation (1) to find z.

$$\begin{aligned} x + y + z &= 180 \quad (1)\\ 85 + 60 + z &= 180\\ 145 + z &= 180\\ z &= 35 \end{aligned}$$

The three angle measures are 85°, 60°, and 35°.

33. Let $x =$ the value of sales at 10%, $y =$ the value of sales at 6%, and $z =$ the value of sales at 5%.

Since her total sales were \$280,000,

$$x + y + z = 280{,}000 \quad (1)$$

Since her commissions on the sales totaled \$17,000,

$$0.10x + 0.06y + 0.05z = 17{,}000.$$

Multiply by 100 to clear the decimals, so

$$10x + 6y + 5z = 1{,}700{,}000. \quad (2)$$

Since the 5% sale amounted to the sum of the other two sales,

$$z = x + y. \quad (3)$$

Solve the system.

$$\begin{aligned} x + y + z &= 280{,}000 \quad (1)\\ 10x + 6y + 5z &= 1{,}700{,}000 \quad (2)\\ z &= x + y \quad (3) \end{aligned}$$

Since equation (3) is given in terms of z, substitute $x + y$ for z in equations (1) and (2).

$$\begin{aligned} x + y + z &= 280{,}000 \quad (1)\\ x + y + (x + y) &= 280{,}000\\ 2x + 2y &= 280{,}000\\ x + y &= 140{,}000 \quad (4) \end{aligned}$$

$$\begin{aligned} 10x + 6y + 5z &= 1{,}700{,}000 \quad (2) \\ 10x + 6y + 5(x + y) &= 1{,}700{,}000 \\ 10x + 6y + 5x + 5y &= 1{,}700{,}000 \\ 15x + 11y &= 1{,}700{,}000 \quad (5) \end{aligned}$$

To eliminate x, multiply equation (4) by -11 and add the result to equation (5).

$$\begin{array}{rcrl} -11x - 11y & = & -1{,}540{,}000 & \quad -11 \times (4) \\ 15x + 11y & = & 1{,}700{,}000 & \quad (5) \\ \hline 4x & = & 160{,}000 & \\ x & = & 40{,}000 & \end{array}$$

From (4), $y = 100{,}000$.
From (3), $z = 40{,}000 + 100{,}000 = 140{,}000$.
He sold \$40,000 at 10%, \$100,000 at 6%, and \$140,000 at 5%.

34. Let $x =$ the number of liters of 8% solution,
$y =$ the number of liters of 10% solution,
and $z =$ the number of liters of 20% solution.

Since the amount of the mixture will be 8 L,

$$x + y + z = 8. \quad (1)$$

Since the final solution will be 12.5% hydrogen peroxide,

$$0.08x + 0.10y + 0.20z = 0.125(8).$$

Multiply by 100 to clear the decimals.

$$8x + 10y + 20z = 100 \quad (2)$$

Since the amount of 8% solution used must be 2 L more than the amount of 20% solution,

$$x = z + 2. \quad (3)$$

Solve the system.

$$\begin{array}{rcll} x + y + z & = & 8 & (1) \\ 8x + 10y + 20z & = & 100 & (2) \\ x & = & z + 2 & (3) \end{array}$$

Since equation (3) is given in terms of x, substitute $z + 2$ for x in equations (1) and (2).

$$\begin{aligned} x + y + z &= 8 \quad (1) \\ (z + 2) + y + z &= 8 \\ y + 2z &= 6 \quad (4) \end{aligned}$$

$$\begin{aligned} 8x + 10y + 20z &= 100 \quad (2) \\ 8(z + 2) + 10y + 20z &= 100 \\ 8z + 16 + 10y + 20z &= 100 \\ 10y + 28z &= 84 \quad (5) \end{aligned}$$

To eliminate y, multiply equation (4) by -10 and add the result to equation (5).

$$\begin{array}{rcrl} -10y - 20z & = & -60 & \quad -10 \times (4) \\ 10y + 28z & = & 84 & \quad (5) \\ \hline 8z & = & 24 & \\ z & = & 3 & \end{array}$$

From (3), $x = z + 2 = 3 + 2 = 5$.
From (4), $y = 6 - 2z = 6 - 2(3) = 0$.
Mix 5 L of 8% solution, none of 10% solution, and 3 L of 20% solution.

35. Let $x =$ the number of home runs hit by Mantle,
$y =$ the number of home runs hit by Maris,
and $z =$ the number of home runs hit by Berra.

They combined for 137 home runs, so

$$x + y + z = 137. \quad (1)$$

Mantle hit 7 fewer than Maris, so

$$x = y - 7. \quad (2)$$

Maris hit 39 more than Berra, so

$$y = z + 39 \quad \text{or} \quad z = y - 39. \quad (3)$$

Substitute $y - 7$ for x and $y - 39$ for z in (1).

$$\begin{aligned} x + y + z &= 137 \quad (1) \\ (y - 7) + y + (y - 39) &= 137 \\ 3y - 46 &= 137 \\ 3y &= 183 \\ y &= 61 \end{aligned}$$

From (2), $x = y - 7 = 61 - 7 = 54$.
From (3), $z = y - 39 = 61 - 39 = 22$.
Mantle hit 54 home runs, Maris hit 61 home runs, and Berra hit 22 home runs.

36.
$$\begin{aligned} 2x + 5y &= -4 \\ 4x - y &= 14 \end{aligned}$$

Write the augmented matrix.

$$\left[\begin{array}{rr|r} 2 & 5 & -4 \\ 4 & -1 & 14 \end{array}\right]$$

$$\left[\begin{array}{rr|r} 2 & 5 & -4 \\ 0 & -11 & 22 \end{array}\right] \quad -2R_1 + R_2$$

$$\left[\begin{array}{rr|r} 2 & 5 & -4 \\ 0 & 1 & -2 \end{array}\right] \quad -\tfrac{1}{11}R_2$$

This matrix gives the system

$$\begin{aligned} 2x + 5y &= -4 \\ y &= -2. \end{aligned}$$

Substitute $y = -2$ in the first equation.

$$\begin{aligned} 2x + 5y &= -4 \\ 2x + 5(-2) &= -4 \\ 2x - 10 &= -4 \\ 2x &= 6 \\ x &= 3 \end{aligned}$$

The solution set is $\{(3, -2)\}$.

37. $$\begin{aligned} 6x + 3y &= 9 \\ -7x + 2y &= 17 \end{aligned}$$

Write the augmented matrix.

$$\left[\begin{array}{rr|r} 6 & 3 & 9 \\ -7 & 2 & 17 \end{array}\right]$$

$$\left[\begin{array}{rr|r} 1 & -5 & -26 \\ -7 & 2 & 17 \end{array}\right] \quad -R_1 - R_2$$

$$\left[\begin{array}{rr|r} 1 & -5 & -26 \\ 0 & -33 & -165 \end{array}\right] \quad 7R_1 + R_2$$

$$\left[\begin{array}{rr|r} 1 & -5 & -26 \\ 0 & 1 & 5 \end{array}\right] \quad -\tfrac{1}{33}R_2$$

This matrix gives the system

$$\begin{aligned} x - 5y &= -26 \\ y &= 5. \end{aligned}$$

Substitute $y = 5$ in the first equation.

$$\begin{aligned} x - 5y &= -26 \\ x - 5(5) &= -26 \\ x - 25 &= -26 \\ x &= -1 \end{aligned}$$

The solution set is $\{(-1, 5)\}$.

38. $$\begin{aligned} x + 2y - z &= 1 \\ 3x + 4y + 2z &= -2 \\ -2x - y + z &= -1 \end{aligned}$$

$$\left[\begin{array}{rrr|r} 1 & 2 & -1 & 1 \\ 3 & 4 & 2 & -2 \\ -2 & -1 & 1 & -1 \end{array}\right]$$

$$\left[\begin{array}{rrr|r} 1 & 2 & -1 & 1 \\ 0 & -2 & 5 & -5 \\ 0 & 3 & -1 & 1 \end{array}\right] \quad \begin{array}{l} -3R_1 + R_2 \\ 2R_1 + R_3 \end{array}$$

$$\left[\begin{array}{rrr|r} 1 & 2 & -1 & 1 \\ 0 & 1 & 4 & -4 \\ 0 & 3 & -1 & 1 \end{array}\right] \quad R_3 + R_2$$

$$\left[\begin{array}{rrr|r} 1 & 2 & -1 & 1 \\ 0 & 1 & 4 & -4 \\ 0 & 0 & -13 & 13 \end{array}\right] \quad -3R_2 + R_3$$

$$\left[\begin{array}{rrr|r} 1 & 2 & -1 & 1 \\ 0 & 1 & 4 & -4 \\ 0 & 0 & 1 & -1 \end{array}\right] \quad -\tfrac{1}{13}R_3$$

This matrix gives the system

$$\begin{aligned} x + 2y - z &= 1 \\ y + 4z &= -4 \\ z &= -1. \end{aligned}$$

Substitute $z = -1$ in the second equation.

$$\begin{aligned} y + 4z &= -4 \\ y + 4(-1) &= -4 \\ y &= 0 \end{aligned}$$

Substitute $y = 0$ and $z = -1$ in the first equation.

$$\begin{aligned} x + 2y - z &= 1 \\ x + 2(0) - (-1) &= 1 \\ x + 1 &= 1 \\ x &= 0 \end{aligned}$$

The solution set is $\{(0, 0, -1)\}$.

39. $$\begin{aligned} x + 3y \quad &= 7 \\ 3x \quad + z &= 2 \\ y - 2z &= 4 \end{aligned}$$

$$\left[\begin{array}{rrr|r} 1 & 3 & 0 & 7 \\ 3 & 0 & 1 & 2 \\ 0 & 1 & -2 & 4 \end{array}\right]$$

$$\left[\begin{array}{rrr|r} 1 & 3 & 0 & 7 \\ 0 & -9 & 1 & -19 \\ 0 & 1 & -2 & 4 \end{array}\right] \quad -3R_1 + R_2$$

$$\left[\begin{array}{rrr|r} 1 & 3 & 0 & 7 \\ 0 & 1 & -2 & 4 \\ 0 & -9 & 1 & -19 \end{array}\right] \quad R_2 \leftrightarrow R_3$$

We use $\leftrightarrow$ to represent the interchanging of two rows.

$$\left[\begin{array}{rrr|r} 1 & 3 & 0 & 7 \\ 0 & 1 & -2 & 4 \\ 0 & 0 & -17 & 17 \end{array}\right] \quad 9R_2 + R_3$$

$$\left[\begin{array}{rrr|r} 1 & 3 & 0 & 7 \\ 0 & 1 & -2 & 4 \\ 0 & 0 & 1 & -1 \end{array}\right] \quad -\tfrac{1}{17}R_3$$

This matrix gives the system

$$\begin{aligned} x + 3y &= 7 \\ y - 2z &= 4 \\ z &= -1. \end{aligned}$$

Substitute $z = -1$ in the second equation.

$$\begin{aligned} y - 2z &= 4 \\ y - 2(-1) &= 4 \\ y + 2 &= 4 \\ y &= 2 \end{aligned}$$

Substitute $y = 2$ in the first equation.

$$\begin{aligned} x + 3y &= 7 \\ x + 3(2) &= 7 \\ x + 6 &= 7 \\ x &= 1 \end{aligned}$$

The solution set is $\{(1, 2, -1)\}$.

40. **[8.2]** System **B** would be easier to solve using the substitution method because the second equation is already solved for y.

41. **[8.3]** $\frac{2}{3}x + \frac{1}{6}y = \frac{19}{2}$ (1)

$\frac{1}{3}x - \frac{2}{9}y = 2$ (2)

Multiply equation (1) by 6 and equation (2) by 9 to clear the fractions.

$$\begin{aligned} 4x + y &= 57 \quad (3) \quad 6 \times (1) \\ 3x - 2y &= 18 \quad (4) \quad 9 \times (2) \end{aligned}$$

To eliminate y, multiply equation (3) by 2 and add the result to equation (4).

$$\begin{aligned} 8x + 2y &= 114 \qquad 2 \times (3) \\ 3x - 2y &= 18 \quad (4) \\ \hline 11x &= 132 \\ x &= 12 \end{aligned}$$

Substitute 12 for x in equation (3) to find y.

$$\begin{aligned} 4x + y &= 57 \quad (3) \\ 4(12) + y &= 57 \\ 48 + y &= 57 \\ y &= 9 \end{aligned}$$

The solution set is $\{(12, 9)\}$.

42. **[8.4]**

$$\begin{aligned} 2x + 5y - z &= 12 \quad (1) \\ -x + y - 4z &= -10 \quad (2) \\ -8x - 20y + 4z &= 31 \quad (3) \end{aligned}$$

Multiply equation (1) by 4 and add the result to equation (3).

$$\begin{aligned} 8x + 20y - 4z &= 48 \qquad 4 \times (1) \\ -8x - 20y + 4z &= 31 \quad (3) \\ \hline 0 &= 79 \quad \textit{False} \end{aligned}$$

Since a false statement results, the system is *inconsistent*. The solution set is $\emptyset$.

43. **[8.2]** $x = 7y + 10$ (1)

$2x + 3y = 3$ (2)

Since equation (1) is given in terms of x, substitute $7y + 10$ for x in equation (2) and solve for y.

$$\begin{aligned} 2(7y + 10) + 3y &= 3 \\ 14y + 20 + 3y &= 3 \\ 17y &= -17 \\ y &= -1 \end{aligned}$$

From (1), $x = 7(-1) + 10 = 3$.

The solution set is $\{(3, -1)\}$.

44. **[8.3]**

$$\begin{aligned} x + 4y &= 17 \quad (1) \\ -3x + 2y &= -9 \quad (2) \end{aligned}$$

To eliminate x, multiply equation (1) by 3 and add the result to equation (2).

$$\begin{aligned} 3x + 12y &= 51 \qquad 3 \times (1) \\ -3x + 2y &= -9 \quad (2) \\ \hline 14y &= 42 \\ y &= 3 \end{aligned}$$

Substitute 3 for y in equation (1) to find x.

$$\begin{aligned} x + 4y &= 17 \quad (1) \\ x + 4(3) &= 17 \\ x + 12 &= 17 \\ x &= 5 \end{aligned}$$

The solution set is $\{(5, 3)\}$.

45. **[8.3]**

$$\begin{aligned} -7x + 3y &= 12 \quad (1) \\ 5x + 2y &= 8 \quad (2) \end{aligned}$$

To eliminate y, multiply equation (1) by 2 and equation (2) by -3. Then add the results.

$$\begin{aligned} -14x + 6y &= 24 \quad 2 \times (1) \\ -15x - 6y &= -24 \quad -3 \times (2) \\ \hline -29x &= 0 \\ x &= 0 \end{aligned}$$

Substitute 0 for x in equation (1) to find y.

$$\begin{aligned} -7x + 3y &= 12 \quad (1) \\ -7(0) + 3y &= 12 \\ 3y &= 12 \\ y &= 4 \end{aligned}$$

The solution set is $\{(0, 4)\}$.

46. **[8.3]**

$$\begin{aligned} 2x - 5y &= 8 \quad (1) \\ 3x + 4y &= 10 \quad (2) \end{aligned}$$

To eliminate y, multiply equation (1) by 4 and equation (2) by 5 and add the results.

$$\begin{aligned} 8x - 20y &= 32 \quad 4 \times (1) \\ 15x + 20y &= 50 \quad 5 \times (2) \\ \hline 23x &= 82 \\ x &= \tfrac{82}{23} \end{aligned}$$

Instead of substituting to find y, we'll choose different multipliers and eliminate x from the original system.

$$\begin{aligned} 6x - 15y &= 24 \quad 3 \times (1) \\ -6x - 8y &= -20 \quad -2 \times (2) \\ \hline -23y &= 4 \\ y &= -\tfrac{4}{23} \end{aligned}$$

The solution set is $\{(\frac{82}{23}, -\frac{4}{23})\}$.

47. **[8.5]** Let $x =$ the number of liters of 5% solution and $y =$ the number of liters of 10% solution.

Liters of Solution	Percent (as a decimal)	Liters of Pure Acid
x	0.05	$0.05x$
10	0.20	$0.20(10) = 2$
y	0.10	$0.10y$

Solve the system formed from the first and third columns.

$$\begin{aligned} x + 10 &= y & (1) \\ 0.05x + 2 &= 0.10y & (2) \end{aligned}$$

Multiply equation (2) by 100 to clear the decimals.

$$5x + 200 = 10y \quad (3)$$

Substitute $x + 10$ for y in equation (3) and solve for x.

$$\begin{aligned} 5x + 200 &= 10y & (3) \\ 5x + 200 &= 10(x + 10) \\ 5x + 200 &= 10x + 100 \\ 100 &= 5x \\ 20 &= x \end{aligned}$$

He should use 20 L of 5% solution.

48. **[8.5]** Let x, y, and z denote the number of medals won by Germany, the United States, and Canada, respectively.

The total number of medals won was 93, so

$$x + y + z = 93. \quad (1)$$

Germany won seven fewer medals than the United States, so

$$x = y - 7. \quad (2)$$

Canada won eleven fewer medals than the United States, so

$$z = y - 11. \quad (3)$$

Substitute $y - 7$ for x and $y - 11$ for z in (1).

$$\begin{aligned} (y - 7) + y + (y - 11) &= 93 \\ 3y - 18 &= 93 \\ 3y &= 111 \\ y &= 37 \end{aligned}$$

From (2), $x = 37 - 7 = 30$.
From (3), $z = 37 - 11 = 26$.

Germany won 30 medals, the United States won 37 medals, and Canada won 26 medals.

Chapter 8 Test

1. The lines intersect at $(8, 3000)$, so the cost equals the revenue at $x = 8$ (which is 800 items). The revenue is $3000.

2. On the sale of 1100 parts ($x = 11$), the revenue is about $4100 and the cost is about $3700.

$$\begin{aligned} \text{Profit} &= \text{Revenue} - \text{Cost} \\ &\approx 4100 - 3700 \\ &= 400 \end{aligned}$$

The profit is about $400.

3.
$$\begin{aligned} 2x + y &= -3 & (1) \\ x - y &= -9 & (2) \end{aligned}$$

(a) $(1, -5)$

Substitute 1 for x and -5 for y in (1) and (2).

$$\begin{aligned} (1) \quad 2(1) + (-5) &\stackrel{?}{=} -3 \\ -3 &= -3 \quad \textit{True} \end{aligned}$$

$$\begin{aligned} (2) \quad (1) - (-5) &\stackrel{?}{=} -9 \\ 6 &= -9 \quad \textit{False} \end{aligned}$$

Since $(1, -5)$ does not satisfy both equations, it *is not* a solution of the system.

(b) $(1, 10)$

$$\begin{aligned} (1) \quad 2(1) + (10) &\stackrel{?}{=} -3 \\ 12 &= -3 \quad \textit{False} \end{aligned}$$

The false result indicates that $(1, 10)$ *is not* a solution of the system.

(c) $(-4, 5)$

$$\begin{aligned} (1) \quad 2(-4) + (5) &\stackrel{?}{=} -3 \\ -3 &= -3 \quad \textit{True} \end{aligned}$$

$$\begin{aligned} (2) \quad (-4) - (5) &\stackrel{?}{=} -9 \\ -9 &= -9 \quad \textit{True} \end{aligned}$$

Since $(-4, 5)$ satisfies both equations, it *is* a solution of the system.

4. When each equation of the system

$$\begin{aligned} x + y &= 7 \\ x - y &= 5 \end{aligned}$$

is graphed, the point of intersection appears to be $(6, 1)$. To check, substitute 6 for x and 1 for y in each of the equations. Since $(6, 1)$ makes both equations true, the solution set of the system is $\{(6, 1)\}$.

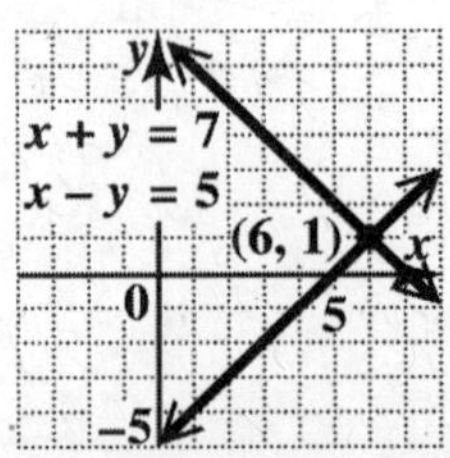

5. $2x - 3y = 24$ $y = 2/3x$ $2/3 \times 3/1$ 6

$3[2x - 3(2/3x)] = 24(3)$

$6x - 9(2x) = 72$

$6x - 18x = 72$

$\frac{-12x}{-12} = \frac{72}{-12}$ $x = -6$

$-12 = 2/3(-6)$

$y = -6(2/3)$ $-12/3$

$2(-6) - 3y = 24$

$-12 - 3y = 24$

$+12 \quad +12$

$\frac{-3y}{-3} = \frac{36}{-3}$ $y = -4$

$2(-6) - 3(-4) = 24$

-12

$\frac{2x}{2} = \frac{3y + 24}{2}$ $x = 3/2y + 12$

$y = 2/3(3/2)y + 12$

$y =$

$2x - (3 \cdot 2/3x) = 24$

$2x - 6/3x = 24$

inconsistent

6.) $3x - y = -8 \Rightarrow -y = -3x - 8 \Rightarrow y = 3x + 8$

$2x + 6y = 3$

$2x + 6(3x + 8) = 3$

$2x + 18x + 48 = 3$

$-48 \quad -48$

$\frac{20x}{20} = \frac{-45}{20}$ $x = -45/20$

$3(-45/20) + \frac{45}{20} = -8$

$-135/20 + 45/20$ $-90/20$

135
-45
90

560 mi in 1.75 hr with wind $d = rt$

560 mi in 2 hr against wind

	d =	r ×	t
trip 1	560 m	x	1.75 hr
trip 2	560 m	x + y	2 hr

$r = d/t$

speed = x
wind = y

$x = 560/1.75$
$x = 320$

$560 = (320 + y) \times 2$
$560 = 640 + 2y$
$-640 \quad -640$
$\frac{-80}{2} = \frac{2y}{2}$
$y = -40$

$560 = (320 - 40)2$

$2/lb + $1/lb = 1.30 100 lb

Ch 8 test

1.) (8, 3000)

2.) $ 300   4000 profit   3700 cost

3.) a.) $2x + y = -3$ $2 \cdot 1 + (-5)$ $2 + 10$ $2 \cdot (-4) + 5 = -3$
-8

b.) $x - y = -9$ -4 -5

(1, -5) no
(1, 10) no
(-4, 5) yes

4.) $x + y = 7 \Rightarrow y = 7 - x$
$x - y = -5 \Rightarrow -y = -5 - x$
$y = 5 + x$

$x + (5 + x) = 7 \Rightarrow \frac{2x}{2} = \frac{2}{2}$
-5   -5

$x = 1$

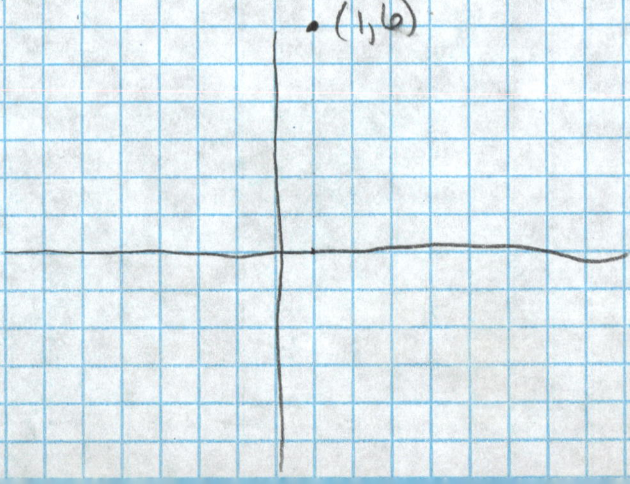

5. $$\begin{aligned} 2x - 3y &= 24 \quad (1) \\ y &= -\tfrac{2}{3}x \quad (2) \end{aligned}$$

Since equation (2) is solved for y, substitute $-\frac{2}{3}x$ for y in equation (1) and solve for x.

$$\begin{aligned} 2x - 3y &= 24 \quad (1) \\ 2x - 3(-\tfrac{2}{3}x) &= 24 \\ 2x + 2x &= 24 \\ 4x &= 24 \\ x &= 6 \end{aligned}$$

From (2), $y = -\frac{2}{3}(6) = -4$.

The solution set is $\{(6, -4)\}$.

6. $$\begin{aligned} 3x - y &= -8 \quad (1) \\ 2x + 6y &= 3 \quad (2) \end{aligned}$$

To eliminate y, multiply equation (1) by 6. Then add that equation and equation (2).

$$\begin{aligned} 18x - 6y &= -48 \quad 6 \times (1) \\ 2x + 6y &= 3 \quad (2) \\ \hline 20x &= -45 \\ x &= -\tfrac{45}{20} = -\tfrac{9}{4} \end{aligned}$$

To find y, substitute $-\frac{9}{4}$ for x in equation (2). (We could also use elimination by adding $2 \times (1)$ and $-3 \times (2)$.)

$$\begin{aligned} 2x + 6y &= 3 \quad (2) \\ 2(-\tfrac{9}{4}) + 6y &= 3 \\ -\tfrac{9}{2} + 6y &= 3 \\ 6y &= \tfrac{15}{2} \\ y &= \tfrac{15}{12} = \tfrac{5}{4} \end{aligned}$$

The solution set is $\{-\frac{9}{4}, \frac{5}{4}\}$.

7. $$\begin{aligned} 12x - 5y &= 8 \quad (1) \\ 3x &= \tfrac{5}{4}y + 2 \\ \text{or} \quad x &= \tfrac{5}{12}y + \tfrac{2}{3} \quad (2) \end{aligned}$$

Substitute $\frac{5}{12}y + \frac{2}{3}$ for x in equation (1) and solve for y.

$$\begin{aligned} 12x - 5y &= 8 \quad (1) \\ 12(\tfrac{5}{12}y + \tfrac{2}{3}) - 5y &= 8 \\ 5y + 8 - 5y &= 8 \\ 8 &= 8 \quad \textit{True} \end{aligned}$$

Equations (1) and (2) are dependent.
The solution set is $\{(x, y) \mid 12x - 5y = 8\}$.

8. $$\begin{aligned} 3x + y &= 12 \quad (1) \\ 2x - y &= 3 \quad (2) \end{aligned}$$

To eliminate y, add equations (1) and (2).

$$\begin{aligned} 3x + y &= 12 \quad (1) \\ 2x - y &= 3 \quad (2) \\ \hline 5x &= 15 \\ x &= 3 \end{aligned}$$

Substitute 3 for x in equation (1) to find y.

$$\begin{aligned} 3x + y &= 12 \quad (1) \\ 3(3) + y &= 12 \\ 9 + y &= 12 \\ y &= 3 \end{aligned}$$

The solution set is $\{(3, 3)\}$.

9. $$\begin{aligned} -5x + 2y &= -4 \quad (1) \\ 6x + 3y &= -6 \quad (2) \end{aligned}$$

To eliminate x, multiply equation (1) by 6 and equation (2) by 5. Then add the results.

$$\begin{aligned} -30x + 12y &= -24 \quad 6 \times (1) \\ 30x + 15y &= -30 \quad 5 \times (2) \\ \hline 27y &= -54 \\ y &= -2 \end{aligned}$$

Substitute -2 for y in equation (1) to find x.

$$\begin{aligned} -5x + 2y &= -4 \quad (1) \\ -5x + 2(-2) &= -4 \\ -5x - 4 &= -4 \\ -5x &= 0 \\ x &= 0 \end{aligned}$$

The solution set is $\{(0, -2)\}$.

10. $$\begin{aligned} 3x + 4y &= 8 \quad (1) \\ 8y &= 7 - 6x \\ \text{or} \quad 6x + 8y &= 7 \quad (2) \end{aligned}$$

Multiply equation (1) by -2 and add the result to equation (2).

$$\begin{aligned} -6x - 8y &= -16 \quad -2 \times (1) \\ 6x + 8y &= 7 \quad (2) \\ \hline 0 &= -9 \quad \textit{False} \end{aligned}$$

Since a false statement results, the system is *inconsistent*. The solution set is $\emptyset$.

11. $$\begin{aligned} \tfrac{6}{5}x - \tfrac{1}{3}y &= -20 \quad (1) \\ -\tfrac{2}{3}x + \tfrac{1}{6}y &= 11 \quad (2) \end{aligned}$$

To clear fractions, multiply by the LCD for each equation. Multiply equation (1) by 15.

$$\begin{aligned} 15(\tfrac{6}{5}x - \tfrac{1}{3}y) &= 15(-20) \\ 18x - 5y &= -300 \quad (3) \end{aligned}$$

Multiply equation (2) by 6.

$$\begin{aligned} 6(-\tfrac{2}{3}x + \tfrac{1}{6}y) &= 6(11) \\ -4x + y &= 66 \quad (4) \end{aligned}$$

To solve this system by the elimination method, multiply equation (4) by 5 and add the result to equation (3).

$$\begin{array}{rcrcr} 18x & - & 5y & = & -300 \\ -20x & + & 5y & = & 330 \\ \hline -2x & & & = & 30 \\ & & x & = & -15 \end{array}$$

To find y, let $x = -15$ in equation (4).

$$\begin{aligned} -4x + y &= 66 \\ -4(-15) + y &= 66 \\ 60 + y &= 66 \\ y &= 6 \end{aligned}$$

The solution set is $\{(-15, 6)\}$.

12. $$\begin{array}{rcrcrcrl} 3x & + & 5y & + & 3z & = & 2 & (1) \\ 6x & + & 5y & + & z & = & 0 & (2) \\ 3x & + & 10y & - & 2z & = & 6 & (3) \end{array}$$

To eliminate x, multiply equation (1) by -1 and add the result to equation (3).

$$\begin{array}{rcrcrcrll} -3x & - & 5y & - & 3z & = & -2 & & -1 \times (1) \\ 3x & + & 10y & - & 2z & = & 6 & (3) & \\ \hline & & 5y & - & 5z & = & 4 & (4) & \end{array}$$

To eliminate x again, multiply equation (1) by -2 and add the result to equation (2).

$$\begin{array}{rcrcrcrll} -6x & - & 10y & - & 6z & = & -4 & & -2 \times (1) \\ 6x & + & 5y & + & z & = & 0 & (2) & \\ \hline & - & 5y & - & 5z & = & -4 & (5) & \end{array}$$

To eliminate y, add equations (4) and (5).

$$\begin{array}{rcrcrl} 5y & - & 5z & = & 4 & (4) \\ -5y & - & 5z & = & -4 & (5) \\ \hline & - & 10z & = & 0 & \\ & & z & = & 0 & \end{array}$$

Substitute 0 for z in equation (4) to find y.

$$\begin{aligned} 5y - 5z &= 4 \qquad (4) \\ 5y - 5(0) &= 4 \\ 5y - 0 &= 4 \\ 5y &= 4 \\ y &= \tfrac{4}{5} \end{aligned}$$

Substitute $\frac{4}{5}$ for y and 0 for z in equation (1) to find x.

$$\begin{aligned} 3x + 5y + 3z &= 2 \qquad (1) \\ 3x + 5(\tfrac{4}{5}) + 3(0) &= 2 \\ 3x + 4 + 0 &= 2 \\ 3x &= -2 \\ x &= -\tfrac{2}{3} \end{aligned}$$

The solution set is $\{(-\frac{2}{3}, \frac{4}{5}, 0)\}$.

13. $$\begin{array}{rcrcrcrl} 4x & + & y & + & z & = & 11 & (1) \\ x & - & y & - & z & = & 4 & (2) \\ & & y & + & 2z & = & 0 & (3) \end{array}$$

To eliminate x, multiply equation (2) by -4 and add the result to equation (1).

$$\begin{array}{rcrcrcrll} 4x & + & y & + & z & = & 11 & (1) & \\ -4x & + & 4y & + & 4z & = & -16 & & -4 \times (2) \\ \hline & & 5y & + & 5z & = & -5 & (4) & \end{array}$$

To eliminate y, divide equation (4) by -5 and add the result to equation (3).

$$\begin{array}{rcrcrll} -y & - & z & = & 1 & & (4) \div (-5) \\ y & + & 2z & = & 0 & (3) & \\ \hline & & z & = & 1 & & \end{array}$$

From (3), $y + 2(1) = 0$, so $y = -2$.
From (2), $x - (-2) - 1 = 4$, so $x = 3$.

The solution set is $\{(3, -2, 1)\}$.

14. Let $x =$ the gross (in millions of dollars) for *Star Wars Episode IV: A New Hope*, and $y =$ the gross (in millions of dollars) for *Indiana Jones and the Kingdom of the Crystal Skull.*

Together the movies grossed \$778.0 million, so

$$x + y = 778.0. \quad (1)$$

Indiana Jones and the Kingdom of the Crystal Skull grossed \$144.0 million less than *Star Wars Episode IV: A New Hope*, so

$$y = x - 144.0. \quad (2)$$

Substitute $x - 144.0$ for y in equation (1).

$$\begin{aligned} x + y &= 778.0 \qquad (1) \\ x + (x - 144.0) &= 778.0 \\ 2x &= 922.0 \\ x &= 461.0 \end{aligned}$$

From (2), $y = 461.0 - 144.0 = 317.0$.

Star Wars Episode IV: A New Hope grossed \$461.0 million and *Indiana Jones and the Kingdom of the Crystal Skull* grossed \$317.0 million.

15. Let $x =$ the rate of the faster car and $y =$ the rate of the slower car.

Make a table.

	r	t	d
Faster Car	x	3.5	$3.5x$
Slower Car	y	3.5	$3.5y$

Since the slow car travels 30 mph slower than the fast car,

$$x - y = 30. \quad (1)$$

Since the cars travel a total of 420 miles,

$$3.5x + 3.5y = 420.$$

Multiply by 10 to clear the decimals.

$$35x + 35y = 4200 \quad (2)$$

To eliminate y, multiply equation (1) by 35 and add the result to equation (2).

$$\begin{array}{rcrcrl} 35x & - & 35y & = & 1050 & \quad 35 \times (1) \\ 35x & + & 35y & = & 4200 & (2) \\ \hline 70x & & & = & 5250 & \\ & & x & = & 75 & \end{array}$$

Substitute 75 for x in equation (1) to find y.

$$\begin{aligned} x - y &= 30 \quad (1) \\ 75 - y &= 30 \\ -y &= -45 \\ y &= 45 \end{aligned}$$

The faster car is traveling at 75 mph, and the slower car is traveling at 45 mph.

16. Let x = the number of liters of 20% solution and y = the number of liters of 50% solution.

Make a table.

Liters of Solution	Percent (as a decimal)	Liters of Pure Alcohol
x	0.20	$0.20x$
y	0.50	$0.50y$
12	0.40	$0.40(12) = 4.8$

Since 12 L of the mixture are needed,

$$x + y = 12. \quad (1)$$

Since the amount of pure alcohol in the 20% solution plus the amount of pure alcohol in the 50% solution must equal the amount of alcohol in the mixture,

$$0.2x + 0.5y = 4.8.$$

Multiply by 10 to clear the decimals.

$$2x + 5y = 48 \quad (2)$$

Multiply equation (1) by -2 and add the result to equation (2).

$$\begin{array}{rcrcrl} -2x & - & 2y & = & -24 & \quad -2 \times (1) \\ 2x & + & 5y & = & 48 & (2) \\ \hline & & 3y & = & 24 & \\ & & y & = & 8 & \end{array}$$

From (1), $x + 8 = 12$, so $x = 4$.
4 L of 20% solution and 8 L of 50% solution are needed.

17. Let x = the price of an AC adaptor and y = the price of a rechargeable flashlight.

Since 7 AC adaptors and 2 rechargeable flashlights cost \$86,

$$7x + 2y = 86. \quad (1)$$

Since 3 AC adaptors and 4 rechargeable flashlights cost \$84,

$$3x + 4y = 84. \quad (2)$$

Solve the system.

$$\begin{array}{rcrcrl} 7x & + & 2y & = & 86 & (1) \\ 3x & + & 4y & = & 84 & (2) \end{array}$$

To eliminate y, multiply equation (1) by -2 and add the result to equation (2).

$$\begin{array}{rcrcrl} -14x & - & 4y & = & -172 & \quad -2 \times (1) \\ 3x & + & 4y & = & 84 & (2) \\ \hline -11x & & & = & -88 & \\ & & x & = & 8 & \end{array}$$

Substitute 8 for x in equation (1) to find y.

$$\begin{aligned} 7x + 2y &= 86 \quad (1) \\ 7(8) + 2y &= 86 \\ 56 + 2y &= 86 \\ 2y &= 30 \\ y &= 15 \end{aligned}$$

An AC adaptor costs \$8, and a rechargeable flashlight costs \$15.

18. Let x = the amount of Orange Pekoe, y = the amount of Irish Breakfast, and z = the amount of Earl Grey.

The owner wants 100 oz of tea, so

$$x + y + z = 100. \quad (1)$$

An equation which relates the prices of the tea is

$$0.80x + 0.85y + 0.95z = 0.83(100).$$

Multiply by 100 to clear the decimals.

$$80x + 85y + 95z = 8300 \quad (2)$$

The mixture must contain twice as much Orange Pekoe as Irish Breakfast, so

$$x = 2y. \quad (3)$$

To eliminate z, multiply equation (1) by -95 and add the result to equation (2).

$$\begin{array}{rcrcrcrl} -95x & - & 95y & - & 95z & = & -9500 & \quad -95 \times (1) \\ 80x & + & 85y & + & 95z & = & 8300 & (2) \\ \hline -15x & - & 10y & & & = & -1200 & (4) \end{array}$$

Divide equation (4) by -5.

$$3x + 2y = 240 \quad (5)$$

Substitute $2y$ for x in equation (5) to find y.

$$\begin{aligned} 3x + 2y &= 240 \qquad (5) \\ 3(2y) + 2y &= 240 \\ 8y &= 240 \\ y &= 30 \end{aligned}$$

From (3), $x = 2(30) = 60$.
Substitute 60 for x and 30 for y in equation (1) to find z.

$$\begin{aligned} x + y + z &= 100 \qquad (1) \\ 60 + 30 + z &= 100 \\ z &= 10 \end{aligned}$$

He should use 60 oz of Orange Pekoe, 30 oz of Irish Breakfast, and 10 oz of Earl Grey.

19. $3x + 2y = 4$
$5x + 5y = 9$

Write the augmented matrix.

$$\left[\begin{array}{cc|c} 3 & 2 & 4 \\ 5 & 5 & 9 \end{array}\right]$$

We could divide row 1 by 3, but to avoid working with fractions, we'll multiply row 1 by 2 and then multiply row 2 by -1 and add to row 1 to obtain a "1" in the first row.

$$\left[\begin{array}{cc|c} 6 & 4 & 8 \\ 5 & 5 & 9 \end{array}\right] \quad 2R_1$$

$$\left[\begin{array}{cc|c} 1 & -1 & -1 \\ 5 & 5 & 9 \end{array}\right] \quad -R_2 + R_1$$

$$\left[\begin{array}{cc|c} 1 & -1 & -1 \\ 0 & 10 & 14 \end{array}\right] \quad -5R_1 + R_2$$

$$\left[\begin{array}{cc|c} 1 & -1 & -1 \\ 0 & 1 & \frac{7}{5} \end{array}\right] \quad \tfrac{1}{10}R_2$$

This matrix gives the system

$$\begin{aligned} x - y &= -1 \\ y &= \tfrac{7}{5}. \end{aligned}$$

Substitute $y = \frac{7}{5}$ in the first equation.

$$\begin{aligned} x - \tfrac{7}{5} &= -1 \\ x &= -1 + \tfrac{7}{5} \\ &= -\tfrac{5}{5} + \tfrac{7}{5} = \tfrac{2}{5} \end{aligned}$$

The solution set is $\left\{\left(\frac{2}{5}, \frac{7}{5}\right)\right\}$.

20. $x + 3y + 2z = 11$
$3x + 7y + 4z = 23$
$5x + 3y - 5z = -14$

Write the augmented matrix.

$$\left[\begin{array}{ccc|c} 1 & 3 & 2 & 11 \\ 3 & 7 & 4 & 23 \\ 5 & 3 & -5 & -14 \end{array}\right]$$

$$\left[\begin{array}{ccc|c} 1 & 3 & 2 & 11 \\ 0 & -2 & -2 & -10 \\ 0 & -12 & -15 & -69 \end{array}\right] \quad \begin{array}{l} \\ -3R_1 + R_2 \\ -5R_1 + R_3 \end{array}$$

$$\left[\begin{array}{ccc|c} 1 & 3 & 2 & 11 \\ 0 & 1 & 1 & 5 \\ 0 & -12 & -15 & -69 \end{array}\right] \quad -\tfrac{1}{2}R_2$$

$$\left[\begin{array}{ccc|c} 1 & 3 & 2 & 11 \\ 0 & 1 & 1 & 5 \\ 0 & 0 & -3 & -9 \end{array}\right] \quad 12R_2 + R_3$$

$$\left[\begin{array}{ccc|c} 1 & 3 & 2 & 11 \\ 0 & 1 & 1 & 5 \\ 0 & 0 & 1 & 3 \end{array}\right] \quad -\tfrac{1}{3}R_3$$

This matrix gives the system

$$\begin{aligned} x + 3y + 2z &= 11 \\ y + z &= 5 \\ z &= 3. \end{aligned}$$

Substitute $z = 3$ in the second equation.

$$\begin{aligned} y + z &= 5 \\ y + 3 &= 5 \\ y &= 2 \end{aligned}$$

Substitute $y = 2$ and $z = 3$ in the first equation.

$$\begin{aligned} x + 3y + 2z &= 11 \\ x + 3(2) + 2(3) &= 11 \\ x + 6 + 6 &= 11 \\ x &= -1 \end{aligned}$$

The solution set is $\{(-1, 2, 3)\}$.

Cumulative Review Exercises (Chapters 1–8)

1. The integer factors of 40 are $-1, 1, -2, 2, -4, 4, -5, 5, -8, 8, -10, 10, -20, 20, -40$, and 40.

2.
$$\begin{aligned} -2 + 6[3 - (4 - 9)] &= -2 + 6[3 - (-5)] \\ &= -2 + 6(8) \\ &= -2 + 48 \\ &= 46 \end{aligned}$$

3. $\dfrac{3x^2+2y^2}{10y+3} = \dfrac{3\cdot 1^2+2\cdot 5^2}{10(5)+3}$ *Let x = 1, y = 5.*

$= \dfrac{3\cdot 1+2\cdot 25}{50+3}$

$= \dfrac{3+50}{50+3} = \dfrac{53}{53} = 1$

4. $r(s-k) = rs - rk$

This is an example of the distributive property.

5.
$$\begin{aligned} 2-3(6x+2) &= 4(x+1)+18 \\ 2-18x-6 &= 4x+4+18 \\ -18x-4 &= 4x+22 \\ -22x &= 26 \\ x &= \tfrac{26}{-22} = -\tfrac{13}{11} \end{aligned}$$

The solution set is $\{-\frac{13}{11}\}$.

6. $\frac{3}{2}\left(\frac{1}{3}x+4\right) = 6\left(\frac{1}{4}+x\right)$

Multiply each side by 2.

$$\begin{aligned} 2\left[\tfrac{3}{2}\left(\tfrac{1}{3}x+4\right)\right] &= 2\left[6\left(\tfrac{1}{4}+x\right)\right] \\ 3\left(\tfrac{1}{3}x+4\right) &= 12\left(\tfrac{1}{4}+x\right) \end{aligned}$$

Use the distributive property to remove parentheses.

$$\begin{aligned} x+12 &= 3+12x \\ -11x &= -9 \\ x &= \tfrac{9}{11} \end{aligned}$$

The solution set is $\{\frac{9}{11}\}$.

7. $P = \dfrac{kT}{V}$ for T

$PV = kT$ *Multiply by V.*

$\dfrac{PV}{k} = \dfrac{kT}{k}$ *Divide by k.*

$\dfrac{PV}{k} = T$ or $T = \dfrac{PV}{k}$

8. $-\frac{5}{6}x < 15$

Multiply each side by the reciprocal of $-\frac{5}{6}$, which is $-\frac{6}{5}$, and reverse the direction of the inequality symbol.

$$\begin{aligned} -\tfrac{6}{5}\left(-\tfrac{5}{6}x\right) &> -\tfrac{6}{5}(15) \\ x &> -18 \end{aligned}$$

The solution set is $(-18, \infty)$.

9.
$$\begin{aligned} -8 &< 2x+3 \\ -11 &< 2x \\ -\tfrac{11}{2} &< x \quad \text{or} \quad x > -\tfrac{11}{2} \end{aligned}$$

The solution set is $(-\frac{11}{2}, \infty)$.

10. 80.4% of 2500 is $0.804(2500) = 2010$
72.5% of 2500 is $0.725(2500) = 1812.5 \approx 1813$

$\dfrac{1570}{2500} = 0.628$ or 62.8%

$\dfrac{1430}{2500} = 0.572$ or 57.2%

Product or Company	Percent	Actual Number
Charmin	80.4%	2010
Wheaties	72.5%	1813
Budweiser	62.8%	1570
State Farm	57.2%	1430

11. Let x = number of "against" votes;
y = number of "in favor" votes.

There were 99 total votes and there were 37 more "in favor" votes than "against" votes, so we have:

$$\begin{aligned} x+y &= 99 \qquad (1) \\ y &= x+37 \qquad (2) \end{aligned}$$

Substitute $x+37$ for y in equation (1).

$$\begin{aligned} x+(x+37) &= 99 \\ 2x+37 &= 99 \\ 2x &= 62 \\ x &= 31 \end{aligned}$$

From (2), $y = 31+37 = 68$.
There were 68 "in favor" votes and 31 "against" votes.

12. Let x = the measure of the equal angles.
Then $2x-4$ = the measure of the third angle.

The sum of the measures of the angles in a triangle is 180°.

$$\begin{aligned} x+x+(2x-4) &= 180 \\ 4x-4 &= 180 \\ 4x &= 184 \\ x &= 46 \end{aligned}$$

The third angle has measure

$2x-4 = 2(46)-4 = 92-4 = 88.$

The measures of the angles are 46°, 46°, and 88°.

13. $x - y = 4$

To graph this line, find the intercepts.

If $y = 0$, $x = 4$, so the x-intercept is $(4, 0)$.
If $x = 0$, $y = -4$, so the y-intercept is $(0, -4)$.

Graph the line through these intercepts. A third point, such as $(5, 1)$, may be used as a check.

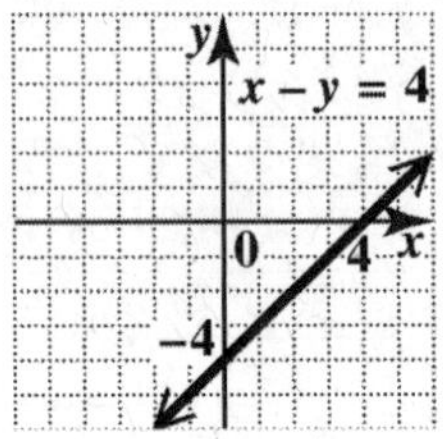

14. $3x + y = 6$

If $y = 0$, $x = 2$, so the x-intercept is $(2, 0)$.
If $x = 0$, $y = 6$, so the y-intercept is $(0, 6)$.

Graph the line through these intercepts. A third point, such as $(1, 3)$, may be used as a check.

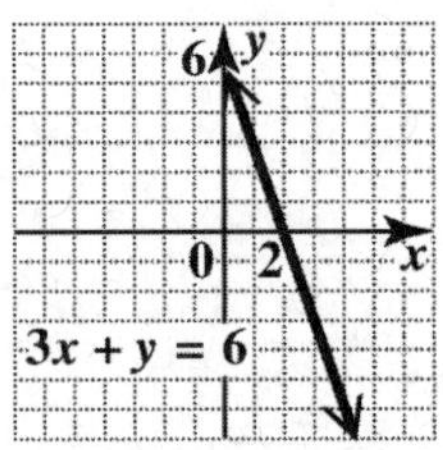

15. The slope m of the line passing through the points $(-5, 6)$ and $(1, -2)$ is

$$m = \frac{y_2 - y_1}{x_2 - x_1} = \frac{-2 - 6}{1 - (-5)} = \frac{-8}{6} = -\frac{4}{3}.$$

16. The slope of the line $y = 4x - 3$ is 4. The slope of the line whose graph is perpendicular to that of $y = 4x - 3$ is the negative reciprocal of 4, namely $-\frac{1}{4}$.

17. Through $(-4, 1)$ with slope $\frac{1}{2}$
Use the point-slope form of a line.

$$\begin{aligned} y - y_1 &= m(x - x_1) \\ y - 1 &= \tfrac{1}{2}[x - (-4)] \\ y - 1 &= \tfrac{1}{2}(x + 4) \\ y - 1 &= \tfrac{1}{2}x + 2 \\ y &= \tfrac{1}{2}x + 3 \quad \textit{Slope-intercept form} \end{aligned}$$

18. Through the points $(1, 3)$ and $(-2, -3)$
Find the slope.

$$m = \frac{y_2 - y_1}{x_2 - x_1} = \frac{-3 - 3}{-2 - 1} = \frac{-6}{-3} = 2$$

Use the point-slope form of a line.

$$\begin{aligned} y - y_1 &= m(x - x_1) \\ y - 3 &= 2(x - 1) \\ y - 3 &= 2x - 2 \\ y &= 2x + 1 \quad \textit{Slope-intercept form} \end{aligned}$$

19. **(a)** On the vertical line through $(9, -2)$, the x-coordinate of every point is 9. Therefore, an equation of this line is $x = 9$.

(b) On the horizontal line through $(4, -1)$, the y-coordinate of every point is -1. Therefore, an equation of this line is $y = -1$.

20.
$$\begin{aligned} &\left(\frac{m^{-4}n^2}{m^2n^{-3}}\right) \cdot \left(\frac{m^5n^{-1}}{m^{-2}n^5}\right) \\ &= \left(\frac{n^2n^3}{m^2m^4}\right) \cdot \left(\frac{m^5m^2}{n^1n^5}\right) \\ &= \left(\frac{n^5}{m^6}\right) \cdot \left(\frac{m^7}{n^6}\right) \\ &= \frac{n^5m^7}{n^6m^6} = \frac{m}{n} \end{aligned}$$

21.
$$\begin{aligned} &(3y^2 - 2y + 6) - (-y^2 + 5y + 12) \\ &= 3y^2 - 2y + 6 + y^2 - 5y - 12 \\ &= 4y^2 - 7y - 6 \end{aligned}$$

22.
$$\begin{aligned} (4f + 3)(3f - 1) &= 12f^2 - 4f + 9f - 3 \\ &= 12f^2 + 5f - 3 \end{aligned}$$

23. $\left(\frac{1}{4}x + 5\right)^2$
Use the formula for the square of a binomial, $(a + b)^2 = a^2 + 2ab + b^2$.

$$\begin{aligned} \left(\tfrac{1}{4}x + 5\right)^2 &= \left(\tfrac{1}{4}x\right)^2 + 2\left(\tfrac{1}{4}x\right)(5) + 5^2 \\ &= \tfrac{1}{16}x^2 + \tfrac{5}{2}x + 25 \end{aligned}$$

24. $(3x^3 + 13x^2 - 17x - 7) \div (3x + 1)$

$$\begin{array}{r|l} & x^2 + 4x - 7 \\ \hline 3x + 1 & 3x^3 + 13x^2 - 17x - 7 \\ & \underline{3x^3 + x^2} \\ & 12x^2 - 17x \\ & \underline{12x^2 + 4x} \\ & -21x - 7 \\ & \underline{-21x - 7} \\ & 0 \end{array}$$

Answer: $x^2 + 4x - 7$

25. $2x^2 - 13x - 45 = (2x + 5)(x - 9)$

26. $100t^4 - 25 = 25(4t^4 - 1)$
$$= 25[(2t^2)^2 - 1^2]$$
$$= 25(2t^2 + 1)(2t^2 - 1)$$

27. Use the sum of cubes formula,
$$x^3 + y^3 = (x + y)(x^2 - xy + y^2).$$
$$8p^3 + 125 = (2p)^3 + 5^3$$
$$= (2p + 5)[(2p)^2 - (2p)(5) + 5^2]$$
$$= (2p + 5)(4p^2 - 10p + 25)$$

28.
$$3x^2 + 4x = 7$$
$$3x^2 + 4x - 7 = 0$$
$$(3x + 7)(x - 1) = 0$$
$$3x + 7 = 0 \quad \text{or} \quad x - 1 = 0$$
$$3x = -7$$
$$x = -\tfrac{7}{3} \quad \text{or} \quad x = 1$$

The solution set is $\{-\frac{7}{3}, 1\}$.

29.
$$\frac{y^2 - 16}{y^2 - 8y + 16} = \frac{(y + 4)(y - 4)}{(y - 4)(y - 4)}$$
$$= \frac{y + 4}{y - 4}$$

30.
$$\frac{2a^2}{a + b} \cdot \frac{a - b}{4a} = \frac{2a^2(a - b)}{4a(a + b)}$$
$$= \frac{a(a - b)}{2(a + b)}$$

31.
$$\frac{x + 4}{x - 2} + \frac{2x - 10}{x - 2} = \frac{x + 4 + 2x - 10}{x - 2}$$
$$= \frac{3x - 6}{x - 2}$$
$$= \frac{3(x - 2)}{x - 2} = 3$$

32.
$$\frac{-3x}{x + 1} + \frac{4x + 1}{x} = \frac{-3}{x^2 + x}$$
$$\frac{-3x}{x + 1} + \frac{4x + 1}{x} = \frac{-3}{x(x + 1)}$$

Multiply by the LCD, $x(x + 1)$. $(x \neq -1, 0)$
$$x(x + 1)\left(\frac{-3x}{x + 1} + \frac{4x + 1}{x}\right)$$
$$= x(x + 1)\left(\frac{-3}{x(x + 1)}\right)$$
$$x(-3x) + (x + 1)(4x + 1) = -3$$
$$-3x^2 + 4x^2 + x + 4x + 1 = -3$$
$$x^2 + 5x + 4 = 0$$
$$(x + 4)(x + 1) = 0$$
$$x + 4 = 0 \quad \text{or} \quad x + 1 = 0$$
$$x = -4 \quad \text{or} \quad x = -1$$

The number -1 is not allowed as a solution because substituting it in the original equation results in division by 0.

Check $x = -4$: $\quad -4 + \frac{15}{4} = -\frac{1}{4}$ *True*

The solution set is $\{-4\}$.

33. (a) Solve the equation for y.
$$5x - 3y = 8$$
$$5x - 8 = 3y$$
$$\frac{5x - 8}{3} = y$$

So $f(x) = \dfrac{5x - 8}{3}$ or $f(x) = \dfrac{5}{3}x - \dfrac{8}{3}$.

(b) $f(1) = \dfrac{5(1) - 8}{3} = \dfrac{-3}{3} = -1$

34. (a) $(f + g)(x) = f(x) + g(x)$
$$= (x^2 + 2x - 3)$$
$$+ (2x^3 - 3x^2 + 4x - 1)$$
$$= 2x^3 - 2x^2 + 6x - 4$$

(b) $(g - f)(x) = g(x) - f(x)$
$$= (2x^3 - 3x^2 + 4x - 1)$$
$$- (x^2 + 2x - 3)$$
$$= 2x^3 - 3x^2 + 4x - 1$$
$$- x^2 - 2x + 3$$
$$= 2x^3 - 4x^2 + 2x + 2$$

(c) Using part (a),
$$(f + g)(-1)$$
$$= 2(-1)^3 - 2(-1)^2 + 6(-1) - 4$$
$$= -2 - 2 - 6 - 4$$
$$= -14$$

(d) $(f \circ h)(x) = f(h(x))$
$$= f(x^2)$$
$$= (x^2)^2 + 2(x^2) - 3$$
$$= x^4 + 2x^2 - 3$$

35.
$$\begin{aligned} -2x + 3y &= -15 \quad (1) \\ 4x - y &= 15 \quad (2) \end{aligned}$$

To eliminate x, multiply equation (1) by 2 and add the result to equation (2).
$$\begin{aligned} -4x + 6y &= -30 \quad 2 \times (1) \\ 4x - y &= 15 \quad (2) \\ \hline 5y &= -15 \\ y &= -3 \end{aligned}$$

Substitute -3 for y in equation (2) to find x.
$$4x - y = 15 \quad (2)$$
$$4x - (-3) = 15$$
$$4x + 3 = 15$$
$$4x = 12$$
$$x = 3$$

The solution set is $\{(3, -3)\}$.

36.

$$\begin{aligned} x - 3y &= 7 \quad &(1) \\ 2x - 6y &= 14 \quad &(2) \end{aligned}$$

Multiplying equation (1) by 2 gives equation (2). Since those equations are the same, the two equations are dependent.

The solution set is $\{(x, y) \mid x - 3y = 7\}$.

37.

$$\begin{aligned} x + y + z &= 10 \quad &(1) \\ x - y - z &= 0 \quad &(2) \\ -x + y - z &= -4 \quad &(3) \end{aligned}$$

Add equations (1) and (2) to eliminate y and z. The result is

$$\begin{aligned} 2x &= 10 \\ x &= 5. \end{aligned}$$

Add equations (2) and (3) to eliminate x and y. The result is

$$\begin{aligned} -2z &= -4 \\ z &= 2. \end{aligned}$$

Substitute 5 for x and 2 for z in equation (1) to find y.

$$\begin{aligned} x + y + z &= 10 \quad (1) \\ 5 + y + 2 &= 10 \\ y + 7 &= 10 \\ y &= 3 \end{aligned}$$

The solution set is $\{(5, 3, 2)\}$.

38. Let $x =$ the average cost of the original Tickle Me Elmo
and $y =$ the recommended cost of T.M.X.

The original's cost was $12.37 less than T.M.X.'s, so

$$x = y - 12.37. \quad (1)$$

Together they cost $67.63, so

$$x + y = 67.63. \quad (2)$$

From (1), substitute $y - 12.37$ for x in equation (2).

$$\begin{aligned} (y - 12.37) + y &= 67.63 \\ 2y - 12.37 &= 67.63 \\ 2y &= 80 \\ y &= 40 \end{aligned}$$

From (1), $x = y - 12.37 = 40 - 12.37 = 27.63$. The average cost of the original Elmo was $27.63 and the recommended cost of a T.M.X. is $40.

CHAPTER 9 INEQUALITIES AND ABSOLUTE VALUE

9.1 Set Operations and Compound Inequalities

9.1 Now Try Exercises

N1. Let $A = \{2, 4, 6, 8\}$ and $B = \{0, 2, 6, 8\}$.

The set $A \cap B$, the intersection of A and B, contains those elements that belong to both A *and* B; that is, the numbers 2, 6, and 8. Therefore,

$$A \cap B = \{2, 6, 8\}.$$

N2. $x - 2 \leq 5$ and $x + 5 \geq 9$

Solve each inequality.

$$\begin{array}{rcr} x - 2 \leq 5 & \text{and} & x + 5 \geq 9 \\ x - 2 + 2 \leq 5 + 2 & \text{and} & x + 5 - 5 \geq 9 - 5 \\ x \leq 7 & \text{and} & x \geq 4 \end{array}$$

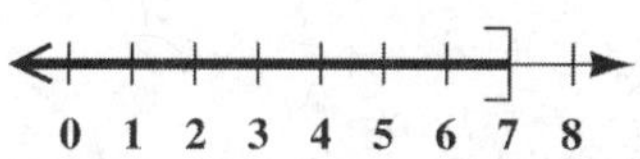

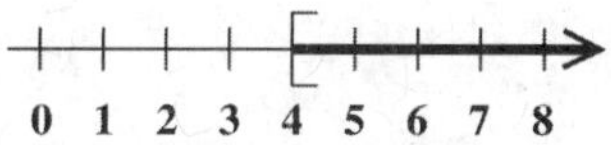

The values that satisfy both inequalities are the numbers between 4 and 7, including 4 and 7.

The solution set is $[4, 7]$.

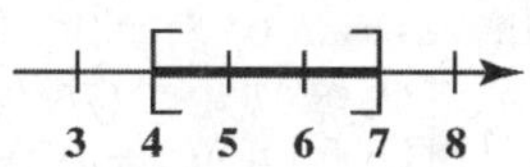

N3.
$$\begin{array}{rcr} -4x - 1 < 7 & \text{and} & 3x + 4 \geq -5 \\ -4x < 8 & \text{and} & 3x \geq -9 \\ x > -2 & \text{and} & x \geq -3 \end{array}$$

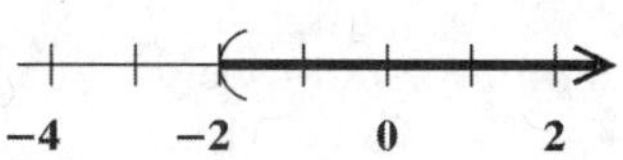

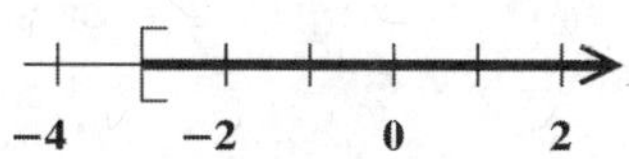

The overlap of the two graphs consists of the numbers that are greater than -2 and are also greater than or equal to -3; that is, the numbers greater than -2.

The solution set is $(-2, \infty)$.

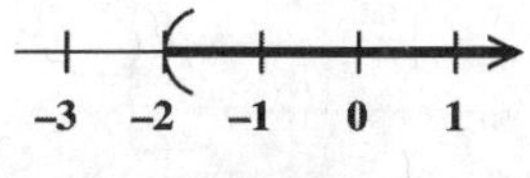

N4.
$$\begin{array}{rcr} x - 7 < -12 & \text{and} & 2x + 1 > 5 \\ & & 2x > 4 \\ x < -5 & \text{and} & x > 2 \end{array}$$

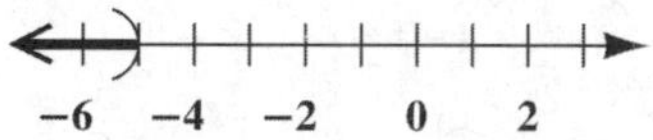

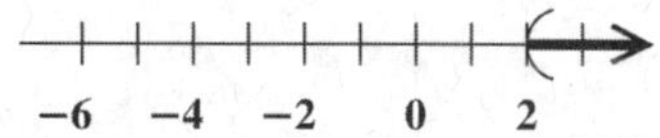

The two graphs do not overlap. Therefore, there is no number that is both greater than 2 *and* less than -5, so the given compound inequality has no solution. The solution set is $\emptyset$.

N5. Let $A = \{5, 10, 15, 20\}$ and $B = \{5, 15, 25\}$.

The set $A \cup B$, the union of A and B, consists of all elements in either A *or* B (or both). Start by listing the elements of set A: 5, 10, 15, 20. Then list any additional elements from set B. In this case, the elements 5 and 15 are already listed, so the only additional element is 25. Therefore,

$$A \cup B = \{5, 10, 15, 20, 25\}.$$

N6.
$$\begin{array}{rcr} -12x \leq -24 & \text{or} & x + 9 < 8 \\ x \geq 2 & \text{or} & x < -1 \end{array}$$

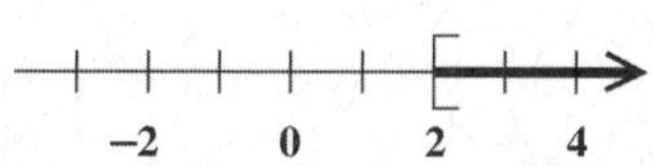

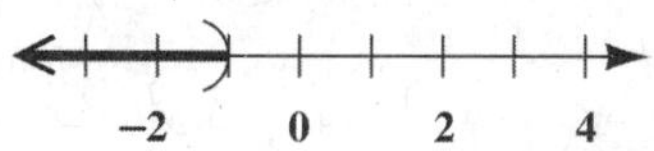

The graph of the solution set consists of all numbers greater than or equal to 2 *or* less than -1.

The solution set is $(-\infty, -1) \cup [2, \infty)$.

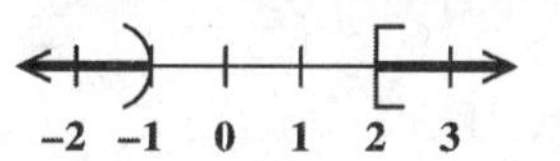

N7.
$$\begin{array}{rcr} -x + 2 < 6 & \text{or} & 6x - 8 \geq 10 \\ -x < 4 & \text{or} & 6x \geq 18 \\ x > -4 & \text{or} & x \geq 3 \end{array}$$

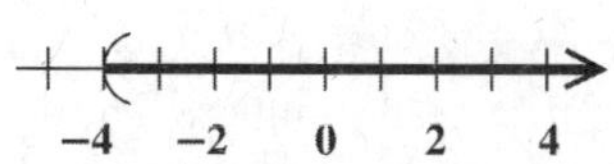

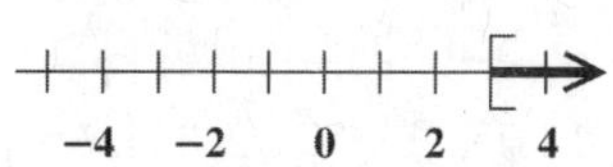

The solution set is all numbers that are either greater than -4 *or* greater than or equal to 3. All real numbers greater than -4 are included.

The solution set is $(-4, \infty)$.

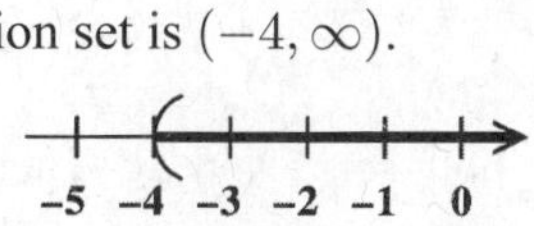

N8.
$$\begin{aligned} 8x-4 &\geq 20 \quad \text{or} \quad -2x+1 > -9 \\ 8x &\geq 24 \quad \text{or} \quad -2x > -10 \\ x &\geq 3 \quad \text{or} \quad x < 5 \end{aligned}$$

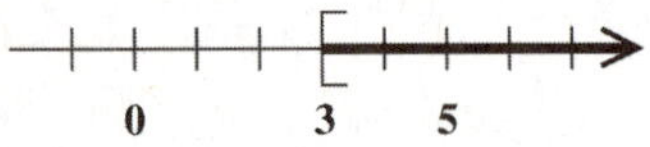

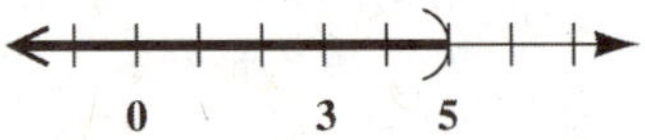

The solution set is all numbers that are either greater than or equal to 3 *or* less than 5. All real numbers are included.

The solution set is $(-\infty, \infty)$.

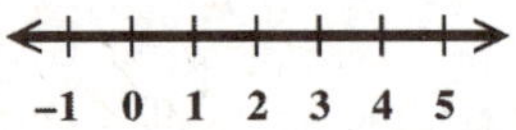

N9. **(a)** All films except *The Ten Commandments* had admissions greater than 140,000,000. *The Sound of Music, E.T.,* and *The Ten Commandments* had a gross income of less than \$1,200,000,000. Thus, the set of elements (films) that satisfy both sets is {*The Sound of Music, E.T.*}.

(b) *Star Wars, The Sound of Music, E.T.,* and *The Ten Commandments* had admissions less than 200,000,000. *The Sound of Music, E.T.,* and *The Ten Commandments* had a gross income of less than \$1,200,000,000. Thus, the set of elements (films) that satisfy either set is {*Star Wars, The Sound of Music, E.T., The Ten Commandments*}.

9.1 Section Exercises

1. This statement is *true*. The solution set of $x+1=6$ is $\{5\}$. The solution set of $x+1>6$ is $(5,\infty)$. The solution set of $x+1<6$ is $(-\infty,5)$. Taken together we have the set of real numbers. (See Section 2.8, Exercises 59–62, for a discussion of this concept.)

3. This statement is *false*. The union is $(-\infty,7)\cup(7,\infty)$. The only real number that is *not* in the union is 7.

5. This statement is *false* since 0 is a rational number but not an irrational number. The sets of rational numbers and irrational numbers have no common elements so their intersection is $\emptyset$.

In Exercises 7–14, let $A=\{1,2,3,4,5,6\}$, $B=\{1,3,5\}$, $C=\{1,6\}$, and $D=\{4\}$.

7. The intersection of sets B and A contains only those elements in both sets B and A.

$$B\cap A=\{1,3,5\} \text{ or set } B$$

9. The intersection of sets A and D is the set of all elements in both set A and D. Therefore,

$$A\cap D=\{4\} \text{ or set } D.$$

11. The intersection of set B and the set of no elements (empty set), $B\cap\emptyset$, is the set of no elements or $\emptyset$.

13. The union of sets A and B is the set of all elements that are in either set A or set B or both sets A and B. Since all numbers in set B are also in set A, the set $A\cup B$ will be the same as set A.

$$A\cup B=\{1,2,3,4,5,6\} \text{ or set } A$$

15. The first graph represents the set $(-\infty,2)$. The second graph represents the set $(-3,\infty)$. The intersection includes the elements common to both sets, that is, $(-3,2)$.

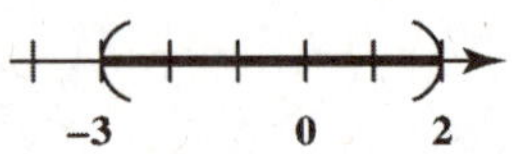

17. The first graph represents the set $(-\infty,5]$. The second graph represents the set $(-\infty,2]$. The intersection includes the elements common to both sets, that is, $(-\infty,2]$.

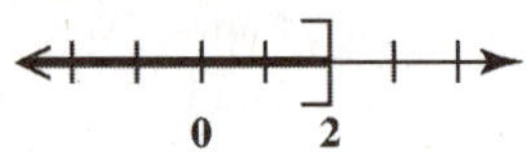

19. $x<2$ and $x>-3$

The graph of the solution set will be all numbers that are both less than 2 and greater than -3. The solution set is $(-3,2)$.

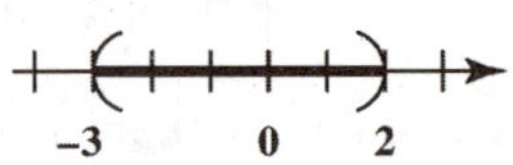

21. $x\leq 2$ and $x\leq 5$

The graph of the solution set will be all numbers that are both less than or equal to 2 and less than or equal to 5. The overlap is the numbers less than or equal to 2. The solution set is $(-\infty,2]$.

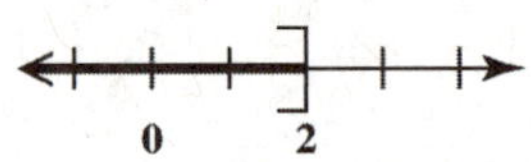

23. $x\leq 3$ and $x\geq 6$

The graph of the solution set will be all numbers that are both less than or equal to 3 and greater than or equal to 6. There are no such numbers. The solution set is $\emptyset$.

25.
$$\begin{aligned} x-3\leq 6 &\quad \text{and} \quad x+2\geq 7 \\ x\leq 9 &\quad \text{and} \quad x\geq 5 \end{aligned}$$

The graph of the solution set is all numbers that are both less than or equal to 9 and greater than or equal to 5. This is the intersection. The elements common to both sets are the numbers between 5 and 9, including the endpoints. The solution set is $[5,9]$.

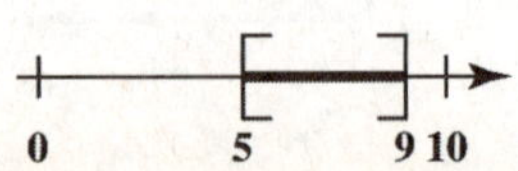

27. $-3x > 3$ and $x + 3 > 0$
$x < -1$ and $x > -3$

The graph of the solution set is all numbers that are both less than -1 and greater than -3. This is the intersection. The elements common to both sets are the numbers between -3 and -1, not including the endpoints. The solution set is $(-3, -1)$.

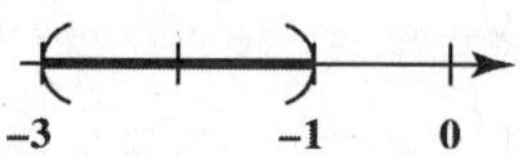

29. $3x - 4 \leq 8$ and $-4x + 1 \geq -15$
$3x \leq 12$ and $-4x \geq -16$
$x \leq 4$ and $x \leq 4$

Since both inequalities are identical, the graph of the solution set is the same as the graph of one of the inequalities. The solution set is $(-\infty, 4]$.

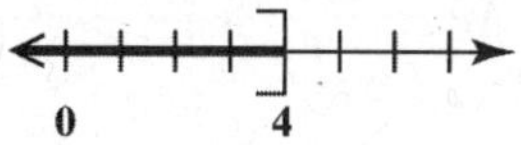

31. The first graph represents the set $(-\infty, 2]$. The second graph represents the set $[4, \infty)$. The union includes all elements in either set, or in both, that is, $(-\infty, 2] \cup [4, \infty)$.

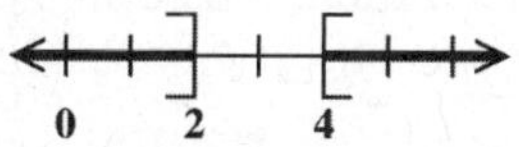

33. The first graph represents the set $[1, \infty)$. The second graph represents the set $(-\infty, 8]$. The union includes all elements in either set, or in both, that is, $(-\infty, \infty)$.

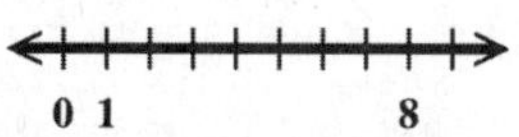

35. $x \leq 1$ or $x \leq 8$

The word "or" means to take the union of both sets. The graph of the solution set is all numbers that are either less than or equal to 1 *or* less than or equal to 8, or both. This is all numbers less than or equal to 8. The solution set is $(-\infty, 8]$.

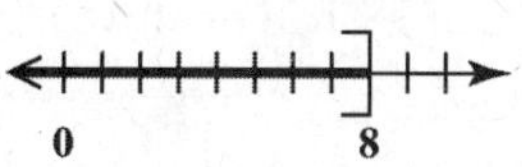

37. $x \geq -2$ or $x \geq 5$

The graph of the solution set will be all numbers that are either greater than or equal to -2 or greater than or equal to 5. The solution set is $[-2, \infty)$.

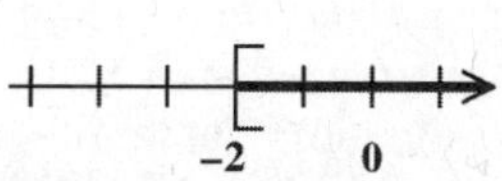

39. $x \geq -2$ or $x \leq 4$

The graph of the solution set will be all numbers that are either greater than or equal to -2 or less than or equal to 4. This is the set of all real numbers. The solution set is $(-\infty, \infty)$.

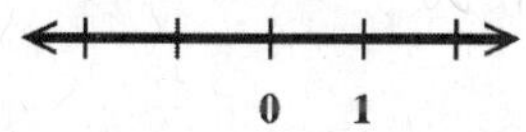

41. $x + 2 > 7$ or $1 - x > 6$
$-x > 5$
$x > 5$ or $x < -5$

The graph of the solution set is all numbers either greater than 5 or less than -5. This is the union. The solution set is $(-\infty, -5) \cup (5, \infty)$.

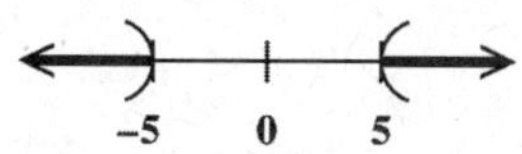

43. $x + 1 > 3$ or $-4x + 1 > 5$
$-4x > 4$
$x > 2$ or $x < -1$

The graph of the solution set is all numbers either less than -1 or greater than 2. This is the union. The solution set is $(-\infty, -1) \cup (2, \infty)$.

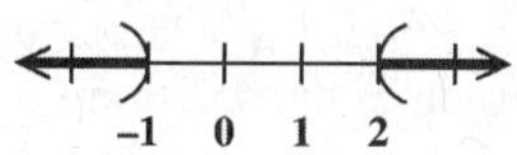

45. $4x + 1 \geq -7$ or $-2x + 3 \geq 5$
$4x \geq -8$ or $-2x \geq 2$
$x \geq -2$ or $x \leq -1$

The graph of the solution set is all numbers either greater than or equal to -2 or less than or equal to -1. This is the set of all real numbers. The solution set is $(-\infty, \infty)$.

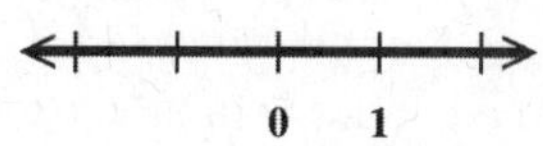

47. $(-\infty, -1] \cap [-4, \infty)$

The intersection is the set of numbers less than or equal to -1 and greater than or equal to -4. The numbers common to both original sets are between, and including, -4 and -1. The simplest interval form is $[-4, -1]$.

49. $(-\infty, -6] \cap [-9, \infty)$

The intersection is the set of numbers less than or equal to -6 and greater than or equal to -9. The numbers common to both original sets are between, and including, -9 and -6. The simplest interval form is $[-9, -6]$.

51. $(-\infty, 3) \cup (-\infty, -2)$

The union is the set of numbers that are either less than 3 or less than -2, or both. This is all numbers less than 3. The simplest interval form is $(-\infty, 3)$.

53. $[3, 6] \cup (4, 9)$

The union is the set of numbers between, and including, 3 and 6, or between, but not including, 4 and 9. This is the set of numbers greater than or equal to 3 and less than 9. The simplest interval form is $[3, 9)$.

55. $x < -1$ and $x > -5$

The word "and" means to take the intersection of both sets. $x < -1$ and $x > -5$ is true only when

$$-5 < x < -1.$$

The graph of the solution set is all numbers greater than -5 *and* less than -1. This is all numbers between -5 and -1, not including -5 or -1. The solution set is $(-5, -1)$.

57. $x < 4$ or $x < -2$

The word "or" means to take the union of both sets. The graph of the solution set is all numbers that are either less than 4 *or* less than -2, or both. This is all numbers less than 4. The solution set is $(-\infty, 4)$.

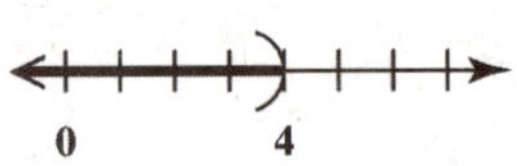

59. $-3x \leq -6$ or $-3x \geq 0$

$x \geq 2$ or $x \leq 0$

The word "or" means to take the union of both sets. The graph of the solution set is all numbers that are either greater than or equal to 2 *or* less than or equal to 0. The solution set is $(-\infty, 0] \cup [2, \infty)$.

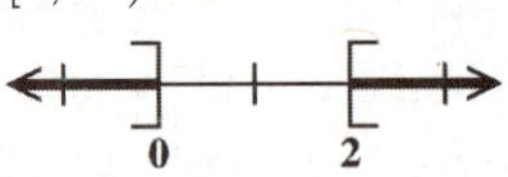

61. $x + 1 \geq 5$ and $x - 2 \leq 10$

$x \geq 4$ and $x \leq 12$

The word "and" means to take the intersection of both sets. The graph of the solution set is all numbers that are both greater than or equal to 4 *and* less than or equal to 12. This is all numbers between, and including, 4 and 12. The solution set is $[4, 12]$.

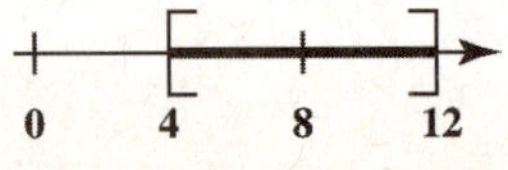

63. The set of expenses that are less than \$6500 for public schools *and* are greater than \$10,000 for private schools is {Tuition and fees}.

65. The set of expenses that are less than \$6500 for public schools *or* are greater than \$10,000 for private schools is {Tuition and fees, Board rates, Dormitory charges}.

♦♦♦ Relating Concepts 67–72 ♦♦♦

67. Find "the yard can be fenced *and* the yard can be sodded."

A yard that can be fenced has $P \leq 150$. Maria and Joe qualify.

A yard that can be sodded has $A \leq 1400$. Again, Maria and Joe qualify.

Find the intersection. Maria's and Joe's yards are common to both sets, so Maria and Joe can have their yards both fenced and sodded.

68. Find "the yard can be fenced *and* the yard cannot be sodded."

A yard that can be fenced has $P \leq 150$. Maria and Joe qualify.

A yard that cannot be sodded has $A > 1400$. Luigi and Than qualify.

Find the intersection. There are no yards common to both sets, so none of them qualify.

69. Find "the yard cannot be fenced *and* the yard can be sodded."

A yard that cannot be fenced has $P > 150$. Luigi and Than qualify.

A yard that can be sodded has $A \leq 1400$. Maria and Joe qualify.

Find the intersection. There are no yards common to both sets, so none of them qualify.

70. Find "the yard cannot be fenced *and* the yard cannot be sodded."

A yard that cannot be fenced has $P > 150$. Luigi and Than qualify.

A yard that cannot be sodded has $A > 1400$. Again, Luigi and Than qualify.

Find the intersection. Luigi's and Than's yards are common to both sets, so Luigi and Than qualify.

71. Find "the yard can be fenced *or* the yard can be sodded." From Exercise 67, Maria's and Joe's yards qualify for both conditions, so the union is Maria and Joe.

72. Find "the yard cannot be fenced *or* the yard can be sodded." From Exercise 69, Luigi's and Than's yards cannot be fenced, and Maria's and Joe's yards can be sodded. The union includes all of them.

73.
$$\begin{aligned} -5 < 2x+1 &< 5 \\ -6 < 2x &< 4 \\ -3 < x &< 2 \end{aligned}$$

The solution set is $(-3, 2)$.

75.
$$\begin{aligned} -|6| - |-11| + (-4) &= -6 - (11) + (-4) \\ &= -17 + (-4) \\ &= -21 \end{aligned}$$

77. The absolute value of 0 is 0, which is not a positive number, so the statement is *false*.

9.2 Absolute Value Equations and Inequalities

9.2 Now Try Exercises

N1. $|4x - 1| = 11$

$$\begin{aligned} 4x - 1 &= 11 \quad \text{or} \quad 4x - 1 = -11 \\ 4x &= 12 \quad \text{or} \quad 4x = -10 \\ x &= 3 \quad \text{or} \quad x = -\tfrac{5}{2} \end{aligned}$$

Check $x = 3$: $|11| = 11$ *True*

Check $x = -\frac{5}{2}$: $|-11| = 11$ *True*

The solution set is $\{-\frac{5}{2}, 3\}$.

N2. $|4x - 1| > 11$

$$\begin{aligned} 4x - 1 &> 11 \quad \text{or} \quad 4x - 1 < -11 \\ 4x &> 12 \quad \text{or} \quad 4x < -10 \\ x &> 3 \quad \text{or} \quad x < -\tfrac{5}{2} \end{aligned}$$

Check $x = -3$, 0, and 6 in $|4x - 1| > 11$.

Check $x = -3$: $|-13| > 11$ *True*
Check $x = 0$: $|-1| > 11$ *False*
Check $x = 6$: $|23| > 11$ *True*

The solution set is $(-\infty, -\frac{5}{2}) \cup (3, \infty)$.

N3. $|4x - 1| < 11$

$$\begin{aligned} -11 < 4x - 1 &< 11 \\ -10 < 4x &< 12 \\ -\tfrac{5}{2} < x &< 3 \end{aligned}$$

Check $x = -5$, 0, and 5 in $|4x - 1| < 11$.

Check $x = -5$: $|-21| < 11$ *False*
Check $x = 0$: $|-1| < 11$ *True*
Check $x = 5$: $|19| < 11$ *False*

The solution set is $(-\frac{5}{2}, 3)$.

N4. $|10x - 2| - 2 = 12$

We first *isolate* the absolute value expression, that is, rewrite the equation so that the absolute value expression is alone on one side of the equals symbol.

$$|10x - 2| = 14$$

$$\begin{aligned} 10x - 2 &= 14 \quad \text{or} \quad 10x - 2 = -14 \\ 10x &= 16 \quad \text{or} \quad 10x = -12 \\ x &= \tfrac{8}{5} \quad \text{or} \quad x = -\tfrac{6}{5} \end{aligned}$$

Check $x = \frac{8}{5}$: $|14| - 2 = 12$ *True*
Check $x = -\frac{6}{5}$: $|-14| - 2 = 12$ *True*

The solution set is $\{-\frac{6}{5}, \frac{8}{5}\}$.

N5. (a) $|x - 1| - 4 \le 2$

$|x - 1| \le 6$ *Isolate.*

$$\begin{aligned} -6 \le x - 1 &\le 6 \\ -5 \le x &\le 7 \end{aligned}$$

The solution set is $[-5, 7]$.

(b) $|x - 1| - 4 \ge 2$

$|x - 1| \ge 6$ *Isolate.*

$$\begin{aligned} x - 1 &\ge 6 \quad \text{or} \quad x - 1 \le -6 \\ x &\ge 7 \quad \text{or} \quad x \le -5 \end{aligned}$$

The solution set is $(-\infty, -5] \cup [7, \infty)$.

N6. $|3x - 4| = |5x + 12|$

$$\begin{aligned} 3x - 4 &= 5x + 12 \quad \text{or} \quad 3x - 4 = -(5x + 12) \\ -2x &= 16 \quad \text{or} \quad 3x - 4 = -5x - 12 \\ x &= -8 \qquad\qquad 8x = -8 \\ &\qquad\qquad\qquad x = -1 \end{aligned}$$

Check $x = -8$: $|-28| = |-28|$ *True*
Check $x = -1$: $|-7| = |7|$ *True*

The solution set is $\{-8, -1\}$.

N7. (a) $|3x - 8| = -2$

Since the absolute value of an expression can never be negative, there are no solutions for this equation.

The solution set is $\emptyset$.

(b) $|7x + 12| = 0$

The expression $7x + 12$ will equal 0 *only* if $7x + 12 = 0$.

$7x = -12$ *Subtract 12.*

$x = -\frac{12}{7}$ *Divide by 7.*

The solution set is $\{-\frac{12}{7}\}$.

N8. (a) $|x| > -10$

The absolute value of a number is always greater than or equal to 0. Therefore, the inequality is true for all real numbers.

The solution set is $(-\infty, \infty)$.

(b) $|4x + 1| + 5 < 4$

$|4x + 1| < -1$

There is no number whose absolute value is less than a negative number, so this inequality has no solution.

The solution set is $\emptyset$.

(c) $|x-2|-3 \le -3$

$|x-2| \le 0$ *Isolate abs. value.*

The value of $|x-2|$ will never be less than 0. However, $|x-2|$ will equal 0 when $x=2$.

The solution set is $\{2\}$.

9.2 Section Exercises

1. $|x|=5$ has two solutions, $x=5$ or $x=-5$. The graph is Choice **E**.

$|x|<5$ is written $-5<x<5$. Notice that -5 and 5 are not included. The graph is Choice **C**, which uses parentheses.

$|x|>5$ is written $x<-5$ or $x>5$. The graph is Choice **D**, which uses parentheses.

$|x|\le 5$ is written $-5\le x\le 5$. This time -5 and 5 are included. The graph is Choice **B**, which uses brackets.

$|x|\ge 5$ is written $x\le -5$ or $x\ge 5$. The graph is Choice **A**, which uses brackets.

3. **(a)** $|ax+b|=k,\ k=0$
This means the distance from $ax+b$ to 0 is 0, so $ax+b=0$, which has one solution.

(b) $|ax+b|=k,\ k>0$
This means the distance from $ax+b$ to 0 is a positive number, so $ax+b=k$ or $ax+b=-k$. There are two solutions.

(c) $|ax+b|=k,\ k<0$
This means the distance from $ax+b$ to 0 is a negative number, which is impossible because distance is always positive. There are no solutions.

5. $|x|=12$

$x=12$ or $x=-12$

The solution set is $\{-12,12\}$.

7. $|4x|=20$

$4x=20$ or $4x=-20$

$x=5$ or $x=-5$

The solution set is $\{-5,5\}$.

9. $|x-3|=9$

$x-3=9$ or $x-3=-9$

$x=12$ or $x=-6$

The solution set is $\{-6,12\}$.

11. $|2x-1|=11$

$2x-1=11$ or $2x-1=-11$

$2x=12$ $2x=-10$

$x=6$ or $x=-5$

The solution set is $\{-5,6\}$.

13. $|4x-5|=17$

$4x-5=17$ or $4x-5=-17$

$4x=22$ $4x=-12$

$x=\frac{22}{4}=\frac{11}{2}$ or $x=-3$

The solution set is $\left\{-3,\frac{11}{2}\right\}$.

15. $|2x+5|=14$

$2x+5=14$ or $2x+5=-14$

$2x=9$ $2x=-19$

$x=\frac{9}{2}$ or $x=-\frac{19}{2}$

The solution set is $\left\{-\frac{19}{2},\frac{9}{2}\right\}$.

17. $\left|\frac{1}{2}x+3\right|=2$

$\frac{1}{2}x+3=2$ or $\frac{1}{2}x+3=-2$

$\frac{1}{2}x=-1$ $\frac{1}{2}x=-5$

$x=-2$ or $x=-10$

The solution set is $\{-10,-2\}$.

19. $\left|1+\frac{3}{4}x\right|=7$

$1+\frac{3}{4}x=7$ or $1+\frac{3}{4}x=-7$

Multiply each side by 4.

$4+3x=28$ or $4+3x=-28$

$3x=24$ $3x=-32$

$x=8$ or $x=\frac{-32}{3}$

The solution set is $\left\{-\frac{32}{3},8\right\}$.

21. $|0.02x-1|=2.50$

$0.02x-1=2.50$ or $0.02x-1=-2.50$

$0.02x=3.50$ $0.02x=-1.50$

$x=50(3.5)$ $x=50(-1.5)$

$x=175$ or $x=-75$

The solution set is $\{-75,175\}$.

23. $|x|>3 \Leftrightarrow x>3$ or $x<-3$

The solution set is $(-\infty,-3)\cup(3,\infty)$.

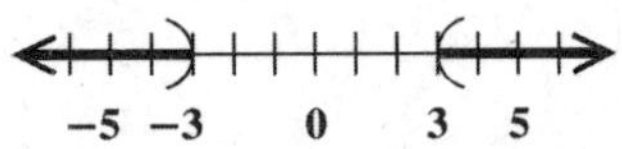

25. $|x|\ge 4 \Leftrightarrow x\ge 4$ or $x\le -4$

The solution set is $(-\infty,-4]\cup[4,\infty)$.

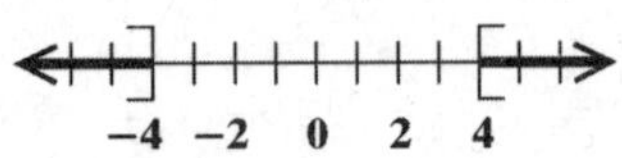

27. $|r+5|\ge 20$

$r+5\le -20$ or $r+5\ge 20$

$r\le -25$ or $r\ge 15$

The solution set is $(-\infty,-25]\cup[15,\infty)$.

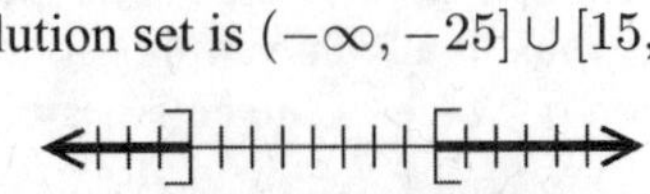

29. $|x+2| > 10$

$$x+2>10 \quad \text{or} \quad x+2<-10$$
$$x>8 \quad \text{or} \quad x<-12$$

The solution set is $(-\infty,-12)\cup(8,\infty)$.

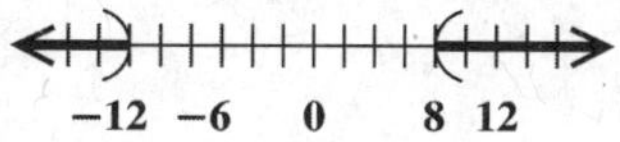

31. $|3-x|>5$

$$3-x>5 \quad \text{or} \quad 3-x<-5$$
$$-x>2 \quad \text{or} \quad -x<-8$$

Multiply by -1, and reverse the inequality symbols.

$$x<-2 \quad \text{or} \quad x>8$$

The solution set is $(-\infty,-2)\cup(8,\infty)$.

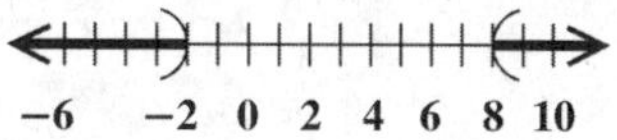

33. $|-5x+3|\geq 12$

$$-5x+3\geq 12 \quad \text{or} \quad -5x+3\leq -12$$
$$-5x\geq 9 \qquad\qquad -5x\leq -15$$
$$x\leq -\tfrac{9}{5} \quad \text{or} \quad x\geq 3$$

The solution set is $(-\infty,-\frac{9}{5}]\cup[3,\infty)$.

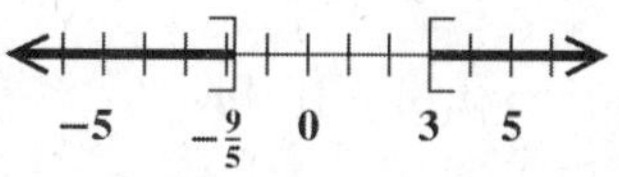

35. **(a)** $|2x+1|<9$

The graph of the solution set will be all numbers between -5 and 4, since the absolute value is less than 9.

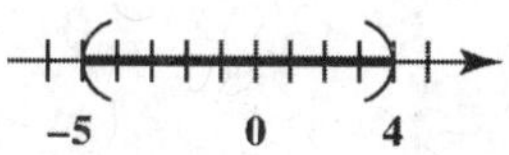

(b) $|2x+1|>9$

The graph of the solution set will be all numbers less than -5 or greater than 4, since the absolute value is greater than 9.

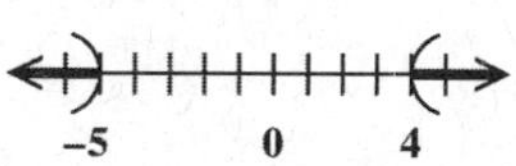

37. $|x|\leq 3 \Leftrightarrow -3\leq x\leq 3$

The solution set is $[-3,3]$.

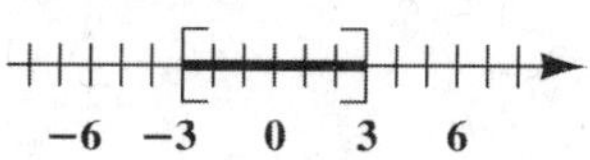

39. $|x|<4 \Leftrightarrow -4<x<4$

The solution set is $(-4,4)$.

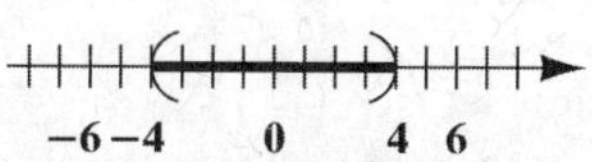

41. $|r+5|<20$

$$-20<r+5<20$$
$$-25<r\quad<15 \qquad \textit{Subtract 5.}$$

The solution set is $(-25,15)$.

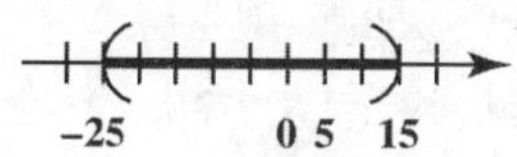

43. $|x+2|\leq 10$

$$-10\leq x+2\leq 10$$
$$-12\leq x\quad\leq 8$$

The solution set is $[-12,8]$.

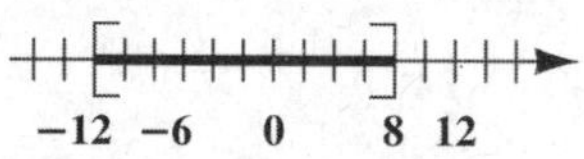

45. $|3-x|\leq 5$

$$-5\leq 3-x\leq 5$$
$$-8\leq -x\leq 2$$
$$8\geq x\geq -2 \qquad \textit{Multiply by } -1\textit{, reverse inequalities}$$
$$-2\leq x\leq 8 \qquad \textit{Equivalent inequality}$$

The solution set is $[-2,8]$.

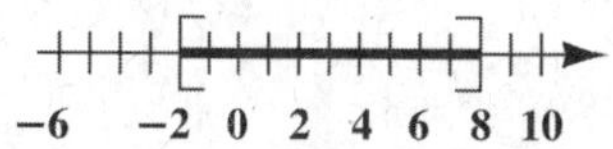

47. $|-5x+3|<12$

$$-12<-5x+3<12$$
$$-15<-5x<9$$
$$3>x>-\frac{9}{5} \qquad \textit{Divide by } -5\textit{, reverse inequalities}$$
$$-\tfrac{9}{5}<x<3 \qquad \textit{Equivalent inequality}$$

The solution set is $(-\frac{9}{5},3)$.

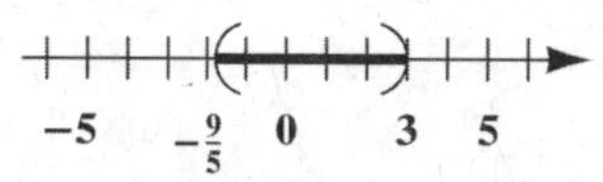

49. $|-4+x|>9$

$$-4+x>9 \quad \text{or} \quad -4+x<-9$$
$$x>13 \quad \text{or} \quad x<-5$$

The solution set is $(-\infty,-5)\cup(13,\infty)$.

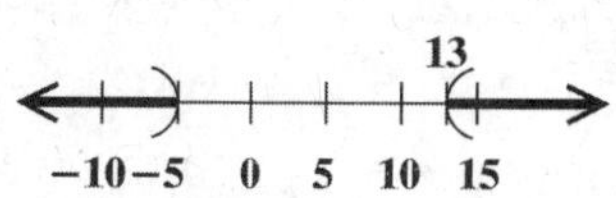

51. $|x+5|>20$

$$x+5>20 \quad \text{or} \quad x+5<-20$$
$$x>15 \quad \text{or} \quad x<-25$$

The solution set is $(-\infty,-25)\cup(15,\infty)$.

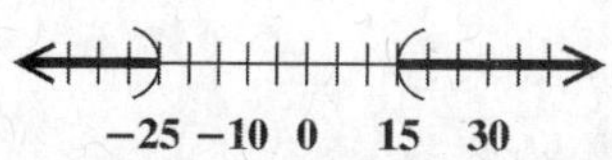

53. $|7+2x|=5$

$$\begin{aligned} 7+2x&=5 &\text{or}\quad 7+2x&=-5\\ 2x&=-2 & 2x&=-12\\ x&=-1 &\text{or}\quad x&=-6 \end{aligned}$$

The solution set is $\{-6,-1\}$.

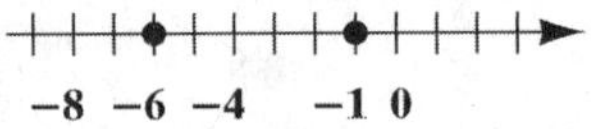

55. $|3x-1|\le 11$

$$\begin{aligned} -11 &\le 3x-1 \le 11\\ -10 &\le 3x \le 12\\ -\tfrac{10}{3} &\le x \le 4 \end{aligned}$$

The solution set is $[-\frac{10}{3},4]$.

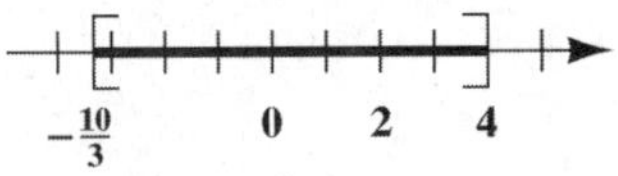

57. $|-6x-6|\le 1$

$$\begin{aligned} -1 &\le -6x-6 \le 1\\ 5 &\le -6x \le 7\\ -\frac{5}{6} &\ge x \ge -\frac{7}{6} && \textit{Divide by } -6. \textit{ Reverse inequalities.}\\ -\tfrac{7}{6} &\le x \le -\tfrac{5}{6} && \textit{Equivalent inequality} \end{aligned}$$

The solution set is $[-\frac{7}{6},-\frac{5}{6}]$.

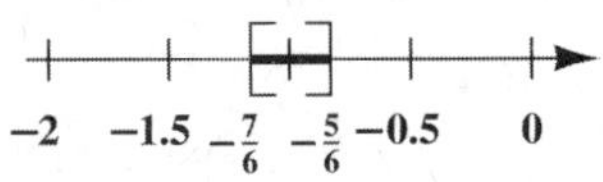

59. $|2x-1|\ge 7$

$$\begin{aligned} 2x-1&\ge 7 &\text{or}\quad 2x-1&\le -7\\ 2x&\ge 8 &\text{or}\quad 2x&\le -6\\ x&\ge 4 &\text{or}\quad x&\le -3 \end{aligned}$$

The solution set is $(-\infty,-3]\cup[4,\infty)$.

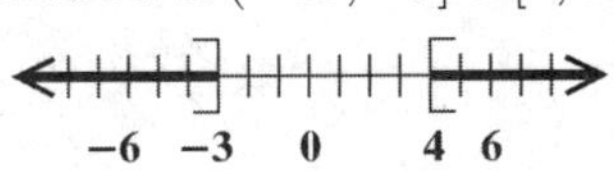

61. $|x+2|=3$

$$\begin{aligned} x+2&=3 &\text{or}\quad x+2&=-3\\ x&=1 &\text{or}\quad x&=-5 \end{aligned}$$

The solution set is $\{-5,1\}$.

63. $|x-6|=3$

$$\begin{aligned} x-6&=3 &\text{or}\quad x-6&=-3\\ x&=9 &\text{or}\quad x&=3 \end{aligned}$$

The solution set is $\{3,9\}$.

65. $|2-0.2x|=2$

$$\begin{aligned} 2-0.2x&=2 &\text{or}\quad 2-0.2x&=-2\\ -0.2x&=0 & -0.2x&=-4\\ x&=0 &\text{or}\quad x&=20 \end{aligned}$$

The solution set is $\{0,20\}$.

67. $|x|-1=4$

$$|x|=5$$

$$x=5 \quad\text{or}\quad x=-5$$

The solution set is $\{-5,5\}$.

69. $|x+4|+1=2$

$$|x+4|=1$$

$$\begin{aligned} x+4&=1 &\text{or}\quad x+4&=-1\\ x&=-3 &\text{or}\quad x&=-5 \end{aligned}$$

The solution set is $\{-5,-3\}$.

71. $|2x+1|+3>8$

$$|2x+1|>5$$

$$\begin{aligned} 2x+1&>5 &\text{or}\quad 2x+1&<-5\\ 2x&>4 & 2x&<-6\\ x&>2 &\text{or}\quad x&<-3 \end{aligned}$$

The solution set is $(-\infty,-3)\cup(2,\infty)$.

73. $|x+5|-6\le -1$

$$\begin{aligned} |x+5| &\le 5\\ -5 \le x+5 &\le 5\\ -10 \le x &\le 0 \end{aligned}$$

The solution set is $[-10,0]$.

75. $\left|\frac{1}{2}x+\frac{1}{3}\right|+\frac{1}{4}=\frac{3}{4}$

$$\left|\tfrac{1}{2}x+\tfrac{1}{3}\right|=\tfrac{1}{2}$$

$$\begin{aligned} \tfrac{1}{2}x+\tfrac{1}{3}&=\tfrac{1}{2} &\text{or}\quad \tfrac{1}{2}x+\tfrac{1}{3}&=-\tfrac{1}{2}\\ 6(\tfrac{1}{2}x+\tfrac{1}{3})&=6(\tfrac{1}{2}) & 6(\tfrac{1}{2}x+\tfrac{1}{3})&=6(-\tfrac{1}{2})\\ 3x+2&=3 & 3x+2&=-3\\ 3x&=1 & 3x&=-5\\ x&=\tfrac{1}{3} &\text{or}\quad x&=-\tfrac{5}{3} \end{aligned}$$

The solution set is $\left\{-\frac{5}{3},\frac{1}{3}\right\}$.

77. $|0.1x-2.5|+0.3\ge 0.8$

$$|0.1x-2.5|\ge 0.5$$

$$\begin{aligned} 0.1x-2.5&\ge 0.5 &\text{or}\quad 0.1x-2.5&\le -0.5\\ 0.1x&\ge 3 &\text{or}\quad 0.1x&\le 2\\ x&\ge 30 &\text{or}\quad x&\le 20 \end{aligned}$$

The solution set is $(-\infty,20]\cup[30,\infty)$.

79. $|3x+1|=|2x+4|$

$$\begin{aligned} 3x+1&=2x+4 &\text{or}\quad 3x+1&=-(2x+4)\\ & & 3x+1&=-2x-4\\ & & 5x&=-5\\ x&=3 &\text{or}\quad x&=-1 \end{aligned}$$

The solution set is $\{-1,3\}$.

81. $\left|x-\frac{1}{2}\right|=\left|\frac{1}{2}x-2\right|$

$$x-\tfrac{1}{2}=\tfrac{1}{2}x-2 \quad \text{or} \quad x-\tfrac{1}{2}=-\left(\tfrac{1}{2}x-2\right)$$

Multiply by 2. $\quad x-\frac{1}{2}=-\frac{1}{2}x+2$

Multiply by 2.

$$2x-1=x-4 \qquad 2x-1=-x+4$$
$$3x=5$$
$$x=-3 \quad \text{or} \quad x=\tfrac{5}{3}$$

The solution set is $\left\{-3,\frac{5}{3}\right\}$.

83. $|6x|=|9x+1|$

$$6x=9x+1 \quad \text{or} \quad 6x=-(9x+1)$$
$$-3x=1 \qquad 6x=-9x-1$$
$$15x=-1$$
$$x=-\tfrac{1}{3} \quad \text{or} \quad x=-\tfrac{1}{15}$$

The solution set is $\left\{-\frac{1}{3},-\frac{1}{15}\right\}$.

85. $|2x-6|=|2x+11|$

$$2x-6=2x+11 \quad \text{or} \quad 2x-6=-(2x+11)$$
$$-6=11 \quad \textit{False} \qquad 2x-6=-2x-11$$
$$4x=-5$$
$$\textit{No solution} \quad \text{or} \quad x=-\tfrac{5}{4}$$

The solution set is $\left\{-\frac{5}{4}\right\}$.

87. $|x|\geq-10$

The absolute value of a number is always greater than or equal to 0. Therefore, the inequality is true for all real numbers.

The solution set is $(-\infty,\infty)$.

89. $|12t-3|=-8$

Since the absolute value of an expression can never be negative, there are no solutions for this equation.

The solution set is $\emptyset$.

91. $|4x+1|=0$

The expression $4x+1$ will equal 0 *only* for the solution of the equation

$$4x+1=0.$$
$$4x=-1$$
$$x=\tfrac{-1}{4} \text{ or } -\tfrac{1}{4}$$

The solution set is $\left\{-\frac{1}{4}\right\}$.

93. $|2x-1|=-6$

Since the absolute value of an expression can never be negative, there are no solutions for this equation.

The solution set is $\emptyset$.

95. $|x+5|>-9$

Since the absolute value of an expression is always nonnegative (positive or zero), the inequality is true for any real number x.

The solution set is $(-\infty,\infty)$.

97. $|7x+3|\leq 0$

The absolute value of an expression is always nonnegative (positive or zero), so this inequality is true only when

$$7x+3=0$$
$$7x=-3$$
$$x=-\tfrac{3}{7}.$$

The solution set is $\left\{-\frac{3}{7}\right\}$.

99. $|5x-2|=0$

The expression $5x-2$ will equal 0 *only* for the solution of the equation

$$5x-2=0.$$
$$5x=2$$
$$x=\tfrac{2}{5}$$

The solution set is $\left\{\frac{2}{5}\right\}$.

101. $|x-2|+3\geq 2$
$|x-2|\geq -1$

Since the absolute value of an expression is always nonnegative (positive or zero), the inequality is true for any real number x.

The solution set is $(-\infty,\infty)$.

103. $|10x+7|+3<1$
$|10x+7|<-2$

There is no number whose absolute value is less than -2, so this inequality has no solution.

The solution set is $\emptyset$.

105. Let x represent the calcium intake for a specific female. For x to be within 100 mg of 1000 mg, we must have

$$|x-1000|\leq 100.$$
$$-100\leq x-1000\leq 100$$
$$900\leq x\leq 1100$$

♦ ♦ ♦ Relating Concepts 107–110 ♦ ♦ ♦

107. Add the given heights with a calculator to get 8105. There are 10 numbers, so divide the sum by 10.

$$\frac{8105}{10}=810.5$$

The average height is 810.5 ft.

108. $|x - k| < 50$

Substitute 810.5 for k and solve the inequality.

$$|x - 810.5| < 50$$
$$-50 < x - 810.5 < 50$$
$$760.5 < x < 860.5$$

The buildings with heights between 760.5 ft and 860.5 ft are Bank of America Center and Texaco Heritage Plaza.

109. $|x - k| < 95$

Substitute 810.5 for k and solve the inequality.

$$|x - 810.5| < 95$$
$$-95 < x - 810.5 < 95$$
$$715.5 < x < 905.5$$

The buildings with heights between 715.5 ft and 905.5 ft are Williams Tower, Bank of America Center, Texaco Heritage Plaza, Enterprise Plaza, Centerpoint Energy Plaza, Continental Center I, and Fulbright Tower.

110. **(a)** This would be the opposite of the inequality in Exercise 109, that is,

$$|x - 810.5| \geq 95.$$
$$x - 810.5 \geq 95 \quad \text{or} \quad x - 810.5 \leq -95$$
$$x \geq 905.5 \quad \text{or} \quad x \leq 715.5$$

(b) The buildings that are not within 95 ft of the average have height less than or equal to 715.5 or greater than or equal to 905.5. They are JPMorgan Chase Tower, Wells Fargo Plaza, and One Shell Plaza.

The answer makes sense because it includes all the buildings *not* listed in the answer to Exercise 109, which had heights within 95 ft of the average.

111. $x - y = 5$

To find the x-intercept, let $y = 0$.

$$x - y = 5$$
$$x - 0 = 5$$
$$x = 5$$

The x-intercept is $(5, 0)$.
To find the y-intercept, let $x = 0$.

$$x - y = 5$$
$$0 - y = 5$$
$$-y = 5$$
$$y = -5$$

The y-intercept is $(0, -5)$.
Plot the intercepts and draw the line through them.

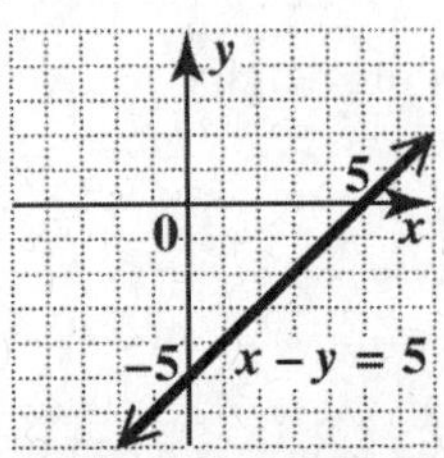

113. $(-4, 3)$

$$3x - 4y = 12$$
$$3(-4) - 4(3) \stackrel{?}{=} 12 \quad \textit{Let } x = -4,\ y = 3.$$
$$-12 - 12 \stackrel{?}{=} 12$$
$$-24 = 12 \quad \textit{False}$$

No, $(-4, 3)$ is not a solution.

Summary Exercises on Solving Linear and Absolute Value Equations and Inequalities

1. $4x + 1 = 49$
$$4x = 48$$
$$x = 12$$

The solution set is $\{12\}$.

3. $6x - 9 = 12 + 3x$
$$3x = 21$$
$$x = 7$$

The solution set is $\{7\}$.

5. $|x + 3| = -4$

Since the absolute value of an expression is always nonnegative, there is no number that makes this statement true. Therefore, the solution set is $\emptyset$.

7. $8x + 2 \geq 5x$
$$3x \geq -2$$
$$x \geq -\tfrac{2}{3}$$

The solution set is $[-\frac{2}{3}, \infty)$.

9. $2x - 1 = -7$
$$2x = -6$$
$$x = -3$$

The solution set is $\{-3\}$.

11. $6x - 5 \leq 3x + 10$
$$3x \leq 15$$
$$x \leq 5$$

The solution set is $(-\infty, 5]$.

13. $9x - 3(x + 1) = 8x - 7$
$$9x - 3x - 3 = 8x - 7$$
$$6x - 3 = 8x - 7$$
$$4 = 2x$$
$$2 = x$$

The solution set is $\{2\}$.

15. $9x - 5 \geq 9x + 3$
$$-5 \geq 3 \quad \textit{False}$$

This is a false statement, so the inequality is a contradiction.

The solution set is $\emptyset$.

17. $|x| < 5.5$
$-5.5 < x < 5.5$

The solution set is $(-5.5, 5.5)$.

19. $\frac{2}{3}x + 8 = \frac{1}{4}x$
$8x + 96 = 3x$ *Multiply by 12.*
$5x = -96$
$x = -\frac{96}{5}$

The solution set is $\left\{-\frac{96}{5}\right\}$.

21. $\frac{1}{4}x < -6$
$4(\frac{1}{4}x) < 4(-6)$
$x < -24$

The solution set is $(-\infty, -24)$.

23. $\frac{3}{5}x - \frac{1}{10} = 2$
$6x - 1 = 20$ *Multiply by 10.*
$6x = 21$
$x = \frac{21}{6} = \frac{7}{2}$

The solution set is $\left\{\frac{7}{2}\right\}$.

25. $x + 9 + 7x = 4(3 + 2x) - 3$
$8x + 9 = 12 + 8x - 3$
$8x + 9 = 8x + 9$
$0 = 0$ *True*

The last statement is true for any real number x.

The solution set is {all real numbers}.

27. $|2x - 3| > 11$

$2x - 3 > 11$ or $2x - 3 < -11$
$2x > 14$ $\quad$ $2x < -8$
$x > 7$ or $x < -4$

The solution set is $(-\infty, -4) \cup (7, \infty)$.

29. $|5x + 1| \leq 0$

The expression $|5x + 1|$ is never less than 0 since an absolute value expression must be nonnegative. However, $|5x + 1| = 0$ if

$$5x + 1 = 0$$
$$5x = -1$$
$$x = \frac{-1}{5} = -\frac{1}{5}$$

The solution set is $\left\{-\frac{1}{5}\right\}$.

31. $-2 \leq 3x - 1 \leq 8$
$-1 \leq 3x \leq 9$
$-\frac{1}{3} \leq x \leq 3$

The solution set is $[-\frac{1}{3}, 3]$.

33. $|7x - 1| = |5x + 3|$
$7x - 1 = 5x + 3$ or $7x - 1 = -(5x + 3)$
$2x = 4$ $\quad$ $7x - 1 = -5x - 3$
$12x = -2$
$x = 2$ or $x = \frac{-2}{12} = -\frac{1}{6}$
The solution set is $\left\{-\frac{1}{6}, 2\right\}$.

35. $|1 - 3x| \geq 4$

$1 - 3x \geq 4$ or $1 - 3x \leq -4$
$-3x \geq 3$ $\quad$ $-3x \leq -5$
$x \leq -1$ or $x \geq \frac{5}{3}$

The solution set is $(-\infty, -1] \cup [\frac{5}{3}, \infty)$.

37. $-(x + 4) + 2 = 3x + 8$
$-x - 4 + 2 = 3x + 8$
$-x - 2 = 3x + 8$
$-10 = 4x$
$x = \frac{-10}{4} = -\frac{5}{2}$

The solution set is $\left\{-\frac{5}{2}\right\}$.

39. $-6 \leq \frac{3}{2} - x \leq 6$
$-\frac{15}{2} \leq -x \leq \frac{9}{2}$ *Subtract $\frac{3}{2}$.*
$\frac{15}{2} \geq x \geq -\frac{9}{2}$ *Multiply by −1. Reverse inequalities.*
$-\frac{9}{2} \leq x \leq \frac{15}{2}$ *Equivalent inequality*

The solution set is $[-\frac{9}{2}, \frac{15}{2}]$.

41. $|x - 1| \geq -6$

The absolute value of an expression is always nonnegative, so the inequality is true for any real number x.

The solution set is $(-\infty, \infty)$.

43. $8x - (1 - x) = 3(1 + 3x) - 4$
$8x - 1 + x = 3 + 9x - 4$
$9x - 1 = 9x - 1$ *True*

This is an identity.

The solution set is {all real numbers}.

45. $|x - 5| = |x + 9|$

$x - 5 = x + 9$ or $x - 5 = -(x + 9)$
$-5 = 9$ *False* $\quad$ $x - 5 = -x - 9$
$2x = -4$
No solution or $x = -2$

The solution set is $\{-2\}$.

47. $2x + 1 > 5$ or $3x + 4 < 1$
$2x > 4$ $\quad$ $3x < -3$
$x > 2$ or $x < -1$

The solution set is $(-\infty, -1) \cup (2, \infty)$.

9.3 Linear Inequalities in Two Variables

9.3 Now Try Exercises

N1. $-x + 2y \geq 4$

Step 1
Graph the line, $-x + 2y = 4$, which has intercepts $(-4, 0)$ and $(0, 2)$, as a solid line since the inequality involves " $\leq$ ".

Step 2
Test $(0, 0)$.

$$x + y \geq 4$$
$$0 + 0 \overset{?}{\geq} 4$$
$$0 \geq 4 \quad \textit{False}$$

Step 3
Since the result is false, shade the region that does not contain $(0, 0)$.

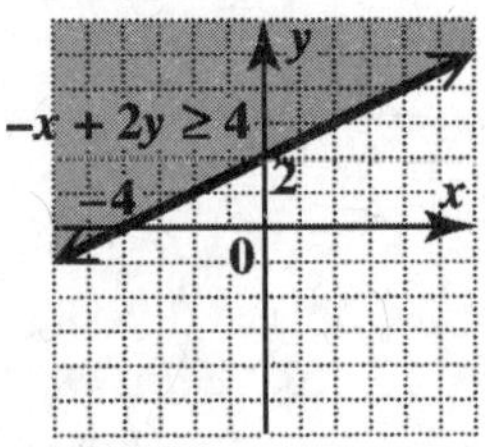

N2. $3x - y < 6$

Solve the inequality for y.

$$-y < -3x + 6 \quad \textit{Subtract 3x.}$$
$$y > 3x - 6 \quad \textit{Multiply by } -1.$$

Graph the boundary line, $y = 3x - 6$ [which has slope 3 and y-intercept $(0, -6)$], as a dashed line because the inequality symbol is $>$. Since the inequality is solved for y and the inequality symbol is $>$, we shade the half-plane above the boundary line.

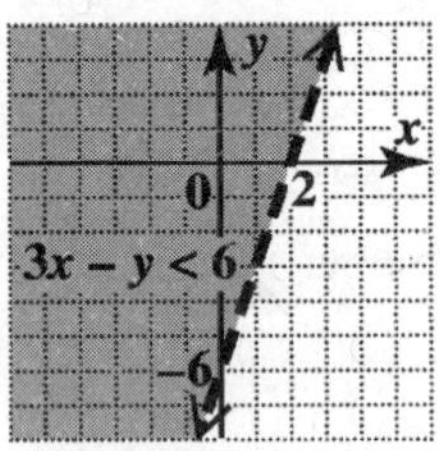

N3. $x > 2$

First graph $x = 2$, a vertical line through the point $(2, 0)$. Use a dashed line because of the $>$ symbol. Choose $(0, 0)$ as a test point.

$$x > 2$$
$$0 > 2 \quad \textit{Let } x = 0. \quad \textit{False}$$

Since the last statement is false, shade the region that does *not* include the test point $(0, 0)$, that is, the region to the right of line. The dashed line shows that the boundary is not part of the graph.

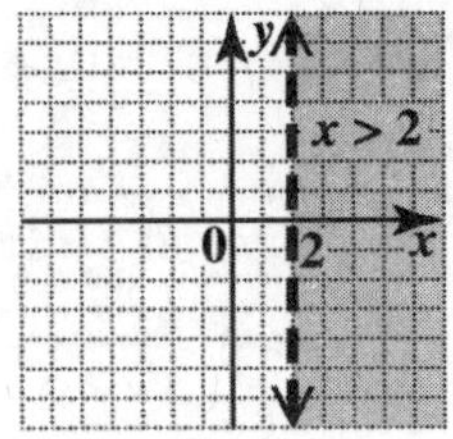

N4. $y \leq -2x$

Graph $y = -2x$ as a solid line through $(0, 0)$ and $(1, -2)$. We cannot use $(0, 0)$ as a test point because $(0, 0)$ is on the line $y = -2x$. Instead, we choose a test point off the line, $(1, 0)$.

$$y \leq -2x$$
$$0 \overset{?}{\leq} -2(1) \quad \textit{Let } x = 1, y = 0.$$
$$0 \leq -2 \quad \textit{False}$$

Since the last statement is false, shade the region that does *not* include the test point $(1, 0)$, that is, the region below the line. The shaded region, along with the boundary, is the desired graph.

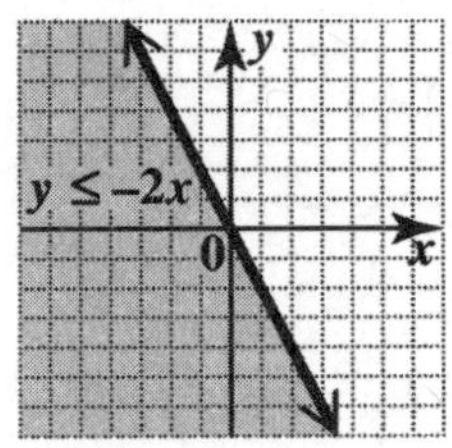

N5. $x + y < 3$ and $y \leq 2$

Graph $x + y = 3$, which has intercepts $(3, 0)$ and $(0, 3)$, as a dashed line since the inequality involves " $<$ ". Test $(0, 0)$, which yields $0 < 3$, a true statement. Shade the region that includes $(0, 0)$.

$x + y < 3$

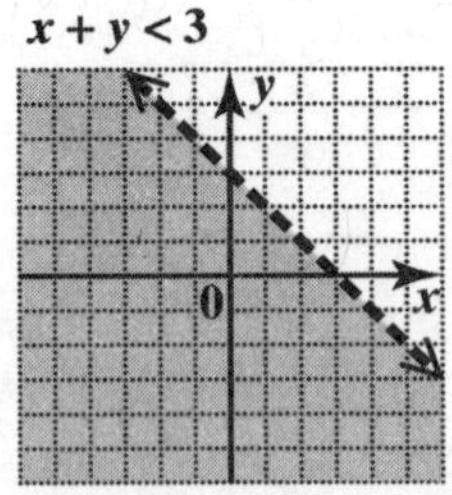

Graph $y = 2$ as a solid horizontal line through $(0, 2)$. Shade the region below $y = 2$.

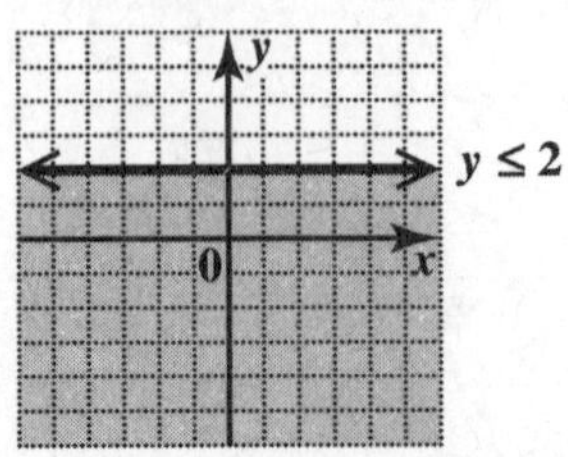

The graph of the intersection is the region common to both graphs.

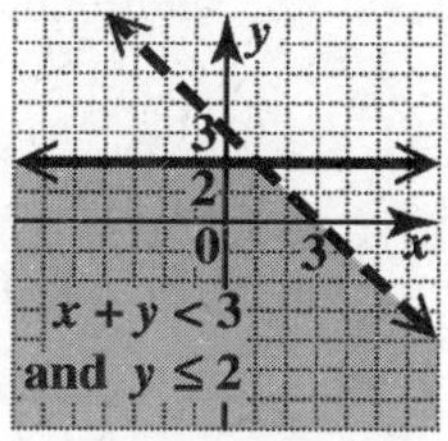

N6. $3x - 5y < 15$ or $x > 4$

Graph $3x - 5y = 15$ as a dashed line through its intercepts $(5, 0)$ and $(0, -3)$. Test $(0, 0)$, which yields $0 < 15$, a true statement. Shade the region that includes $(0, 0)$.

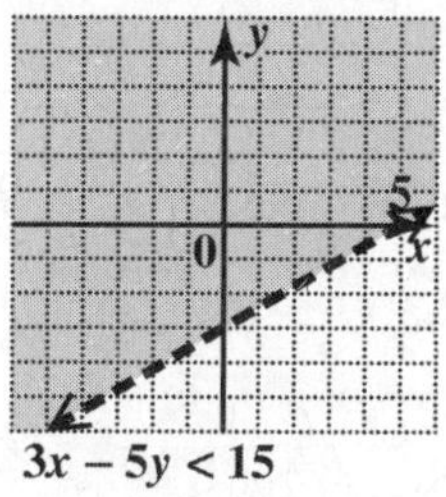

Graph $x = 4$ as a dashed vertical line through $(4, 0)$. Shade the region to the right of $x = 4$.

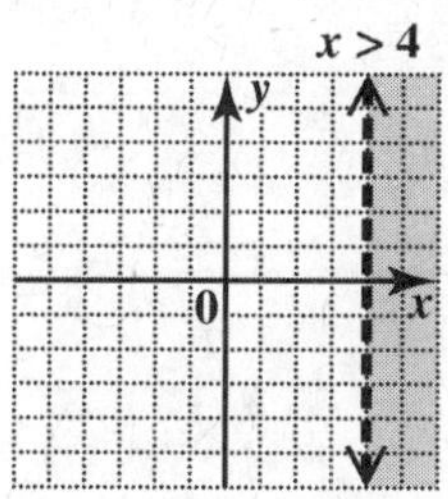

The graph of the union is the region that includes all the points in both graphs.

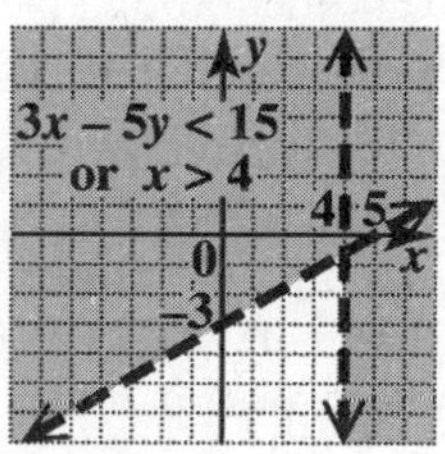

9.3 Section Exercises

1. The boundary of the graph of $y \le -x + 2$ will be a <u>solid</u> line (since the inequality involves $\le$), and the shading will be <u>below</u> the line (since the inequality sign is $\le$ or $<$).

3. The boundary of the graph of $y > -x + 2$ will be a <u>dashed</u> line (since the inequality involves $>$), and the shading will be <u>above</u> the line (since the inequality sign is $\ge$ or $>$).

5. $x + 2y \ge 7$

Use $(0, 0)$ as a test point.

$$x + 2y \ge 7$$
$$0 + 2(0) \overset{?}{\ge} 7 \quad \textit{Let } x = 0,\ y = 0.$$
$$0 \ge 7 \quad \textit{False}$$

Because the last statement is *false*, we shade the region that does *not* include the test point $(0, 0)$. This is the region above the line.

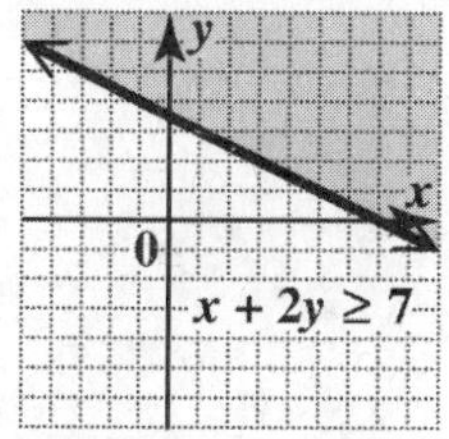

7. $-3x + 4y > 12$

Use $(0, 0)$ as a test point.

$$-3x + 4y > 12$$
$$-3(0) + 4(0) \overset{?}{>} 12 \quad \textit{Let } x = 0,\ y = 0.$$
$$0 > 12 \quad \textit{False}$$

Because the last statement is *false*, we shade the region that does *not* include the test point $(0, 0)$. This is the region above the line.

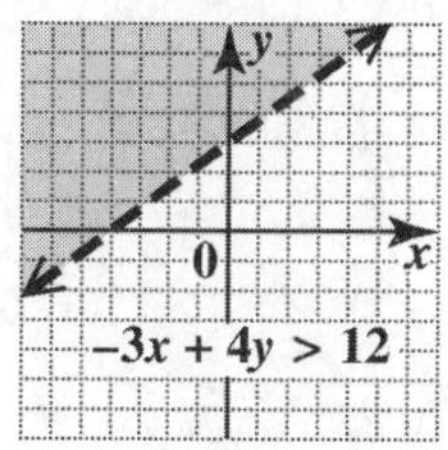

9. $y < -1$

Use $(0, 0)$ as a test point.

$$y < -1$$
$$0 < -1 \quad \textit{Let } y = 0;\ \textit{False}$$

Because $0 < -1$ is *false*, shade the region *not* containing $(0, 0)$. This is the region below the line.

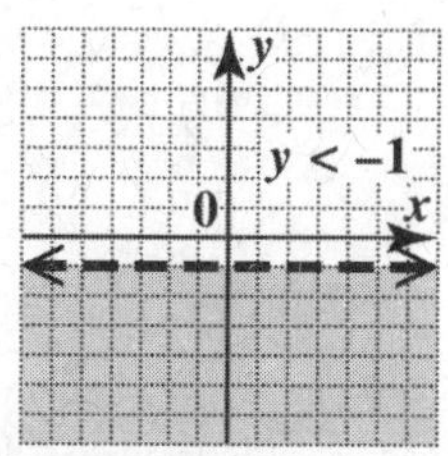

11. Use a *dashed* line if the symbol is $<$ or $>$. Use a *solid* line if the symbol is $\le$ or $\ge$.

13. $x + y \leq 2$
Graph the line $x + y = 2$ by drawing a solid line (since the inequality involves $\leq$) through the intercepts $(2, 0)$ and $(0, 2)$.
Test a point not on this line, such as $(0, 0)$.

$$x + y \leq 2$$
$$0 + 0 \overset{?}{\leq} 2$$
$$0 \leq 2 \quad \textit{True}$$

Shade the side of the line containing the test point $(0, 0)$.

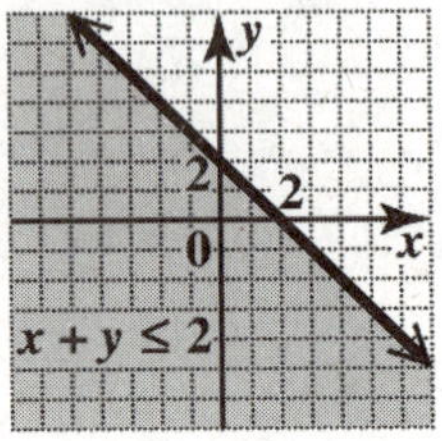

15. $4x - y < 4$
Graph the line $4x - y = 4$ by drawing a dashed line (since the inequality involves $<$) through the intercepts $(1, 0)$ and $(0, -4)$. Instead of using a test point, we will solve the inequality for y.

$$-y < -4x + 4$$
$$y > 4x - 4$$

Since we have "$y >$ " in the last inequality, shade the region *above* the boundary line.

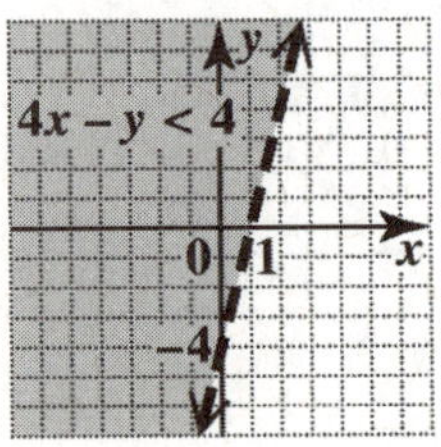

17. $x + 3y \geq -2$
Graph the solid line $x + 3y = -2$ (since the inequality involves $\geq$) through the intercepts $(-2, 0)$ and $(0, -\frac{2}{3})$.
Test a point not on this line such as $(0, 0)$.

$$0 + 3(0) \overset{?}{\geq} -2$$
$$0 \geq -2 \quad \textit{True}$$

Shade the side of the line containing the test point $(0, 0)$.

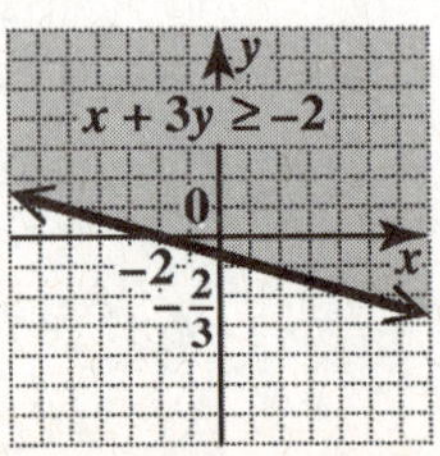

19. $2x + 3y \geq 6$
Graph the solid line $2x + 3y = 6$ (since the inequality involves $\geq$) through the intercepts $(3, 0)$ and $(0, 2)$.
Test a point not on this line such as $(0, 0)$.

$$2(0) + 3(0) \overset{?}{\geq} 6$$
$$0 \geq 6 \quad \textit{False}$$

Shade the side of the line *not* containing the test point $(0, 0)$.

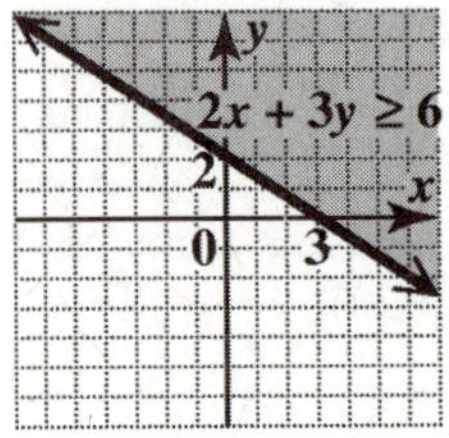

21. $5x - 3y > 15$
Graph the dashed line $5x - 3y = 15$ (since the inequality involves $>$) through the intercepts $(3, 0)$ and $(0, -5)$.
Test a point not on this line such as $(0, 0)$.

$$5(0) - 3(0) \overset{?}{>} 15$$
$$0 > 15 \quad \textit{False}$$

Shade the side of the line *not* containing the test point $(0, 0)$.

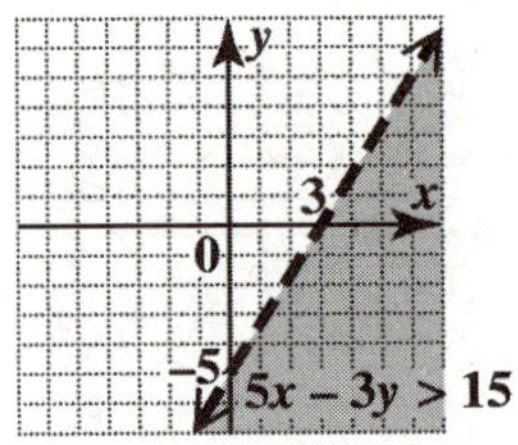

23. $x < -2$
The boundary is the line with equation $x = -2$. This is a vertical line through $(-2, 0)$. Make this line dashed because of the $<$ sign.
Using $(0, 0)$ as a test point will result in the inequality $0 < -2$, which is false. Shade the region *not* containing the origin. This is the region to the left of the boundary. The dashed line shows that the boundary is not part of the graph.

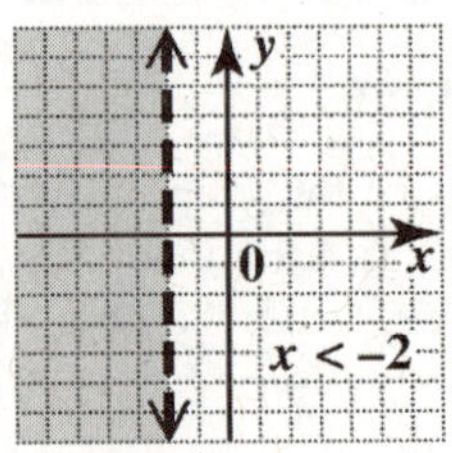

25. $y \leq 5$
The boundary is the line with equation $y = 5$. This is the horizontal line through $(0, 5)$. Make this line solid because of the $\leq$ sign.

Using $(0, 0)$ as a test point will result in the inequality $0 \leq 5$, which is true. Shade the region containing the origin, that is, the region below the line. The solid line shows that the boundary is part of the graph.

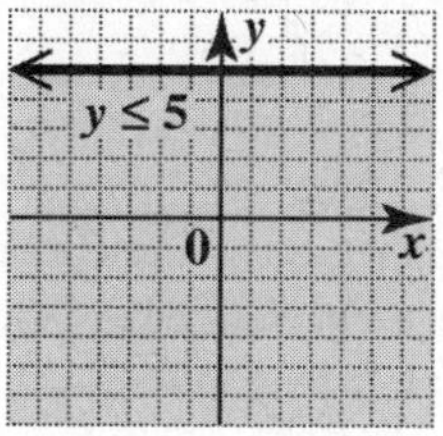

27. $x + y > 0$
Graph the line $x + y = 0$, which includes the points $(0, 0)$ and $(2, -2)$, as a dashed line (since the inequality involves $>$). Solving the inequality for y gives us

$$y > -x,$$

So shade the region above the boundary line.

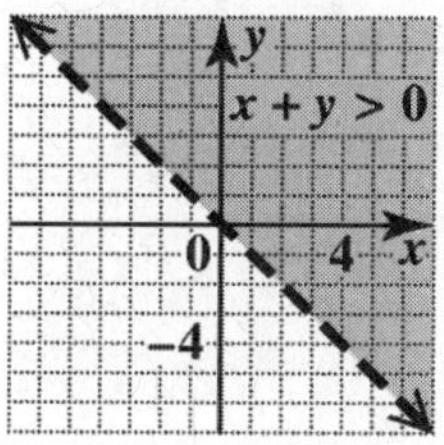

29. $x - 3y \leq 0$
Graph the solid line $x - 3y = 0$ through the points $(0, 0)$ and $(3, 1)$.
Solve the inequality for y.

$$-3y \leq -x$$
$$y \geq \tfrac{1}{3}x$$

Shade the region above the boundary line.

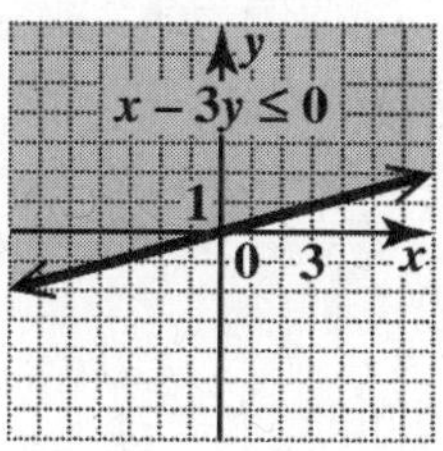

31. $y < x$
Graph the dashed line $y = x$ through $(0, 0)$ and $(2, 2)$. Since we have "$y <$ " in the inequality, shade the region *below* the boundary line.

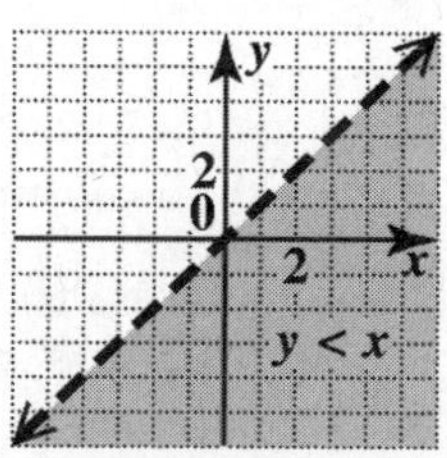

33. $x + y \leq 1$ and $x \geq 1$
Graph the solid line $x + y = 1$ through $(0, 1)$ and $(1, 0)$. The inequality $x + y \leq 1$ can be written as $y \leq -x + 1$, so shade the region below the boundary line.
Graph the solid vertical line $x = 1$ through $(1, 0)$ and shade the region to the right. The required graph is the common shaded area as well as the portions of the lines that bound it.

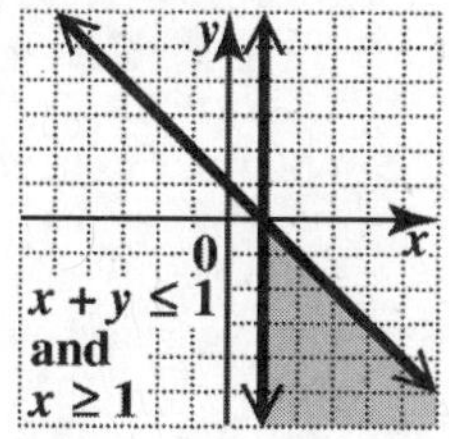

35. $2x - y \geq 2$ and $y < 4$
Graph the solid line $2x - y = 2$ through the intercepts $(1, 0)$ and $(0, -2)$. Test $(0, 0)$ to get $0 \geq 2$, a false statement. Shade the side of the line not containing $(0, 0)$. To graph $y < 4$ on the same axes, graph the dashed horizontal line through $(0, 4)$. Test $(0, 0)$ to get $0 < 4$, a true statement. Shade the side of the dashed line containing $(0, 0)$. The word "and" indicates the intersection of the two graphs. The final solution set consists of the region where the two shaded regions overlap.

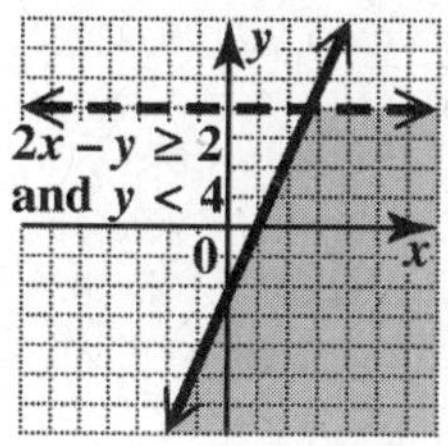

37. $x + y > -5$ and $y < -2$
Graph $x + y = -5$, which has intercepts $(-5, 0)$ and $(0, -5)$, as a dashed line. Test $(0, 0)$, which yields $0 > -5$, a true statement. Shade the region that includes $(0, 0)$.
Graph $y = -2$ as a dashed horizontal line. Shade the region below $y = -2$. The required graph of the intersection is the region common to both graphs.

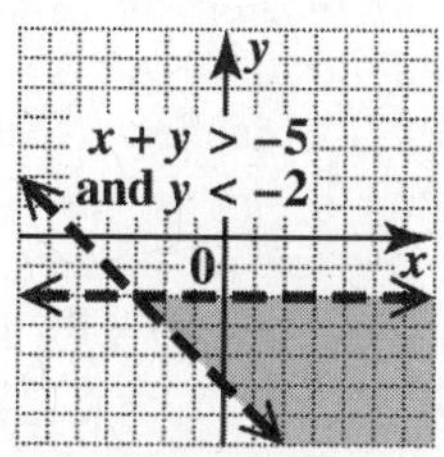

39. $|x| < 3$ can be rewritten as $-3 < x < 3$. The boundaries are the dashed vertical lines $x = -3$ and $x = 3$. Since x is between -3 and 3, the graph includes all points between the lines.

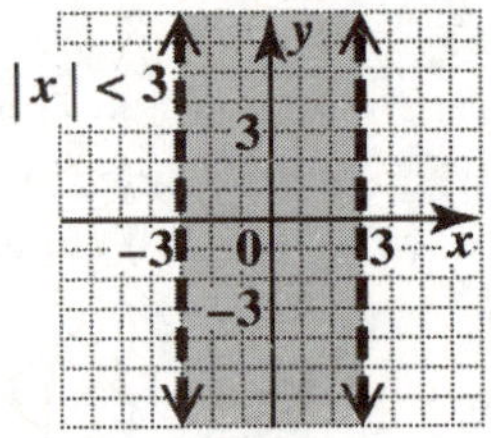

41. $|x + 1| < 2$ can be rewritten as

$$\begin{aligned} -2 &< x + 1 < 2 \\ -3 &< x \quad\;\; < 1. \end{aligned}$$

The boundaries are the dashed vertical lines $x = -3$ and $x = 1$. Since x is between -3 and 1, the graph includes all points between the lines.

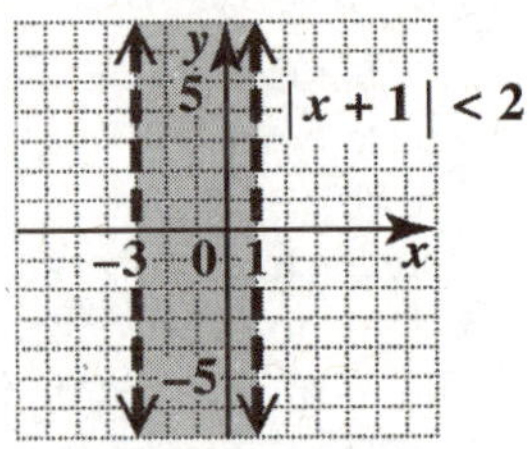

43. $x - y \geq 1$ or $y \geq 2$
Graph the solid line $x - y = 1$, which crosses the y-axis at -1 and the x-axis at 1. Use $(0, 0)$ as a test point, which yields $0 \geq 1$, a false statement. Shade the region that does not include $(0, 0)$.
Now graph the solid line $y = 2$. Since the inequality is $y \geq 2$, shade above this line.
The required graph of the union includes all the shaded regions, that is, all the points that satisfy either inequality.

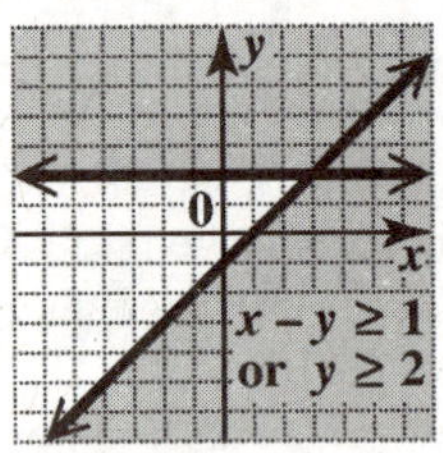

45. $x - 2 > y$ or $x < 1$
Graph $x - 2 = y$, which has intercepts $(2, 0)$ and $(0, -2)$, as a dashed line. Test $(0, 0)$, which yields $-2 > 0$, a false statement. Shade the region that does not include $(0, 0)$.
Graph $x = 1$ as a dashed vertical line. Shade the region to the left of $x = 1$.
The required graph of the union includes all the shaded regions, that is, all the points that satisfy either inequality.

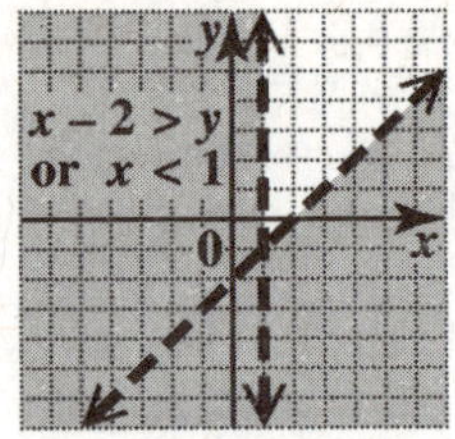

47. $3x + 2y < 6$ or $x - 2y > 2$
Graph $3x + 2y = 6$, which has intercepts $(2, 0)$ and $(0, 3)$, as a dashed line. Test $(0, 0)$, which yields $0 < 6$, a true statement. Shade the region that includes $(0, 0)$.
Graph $x - 2y = 2$, which has intercepts $(2, 0)$ and $(0, -1)$, as a dashed line. Test $(0, 0)$, which yields $0 > 2$, a false statement. Shade the region that does not include $(0, 0)$.
The required graph of the union includes all the shaded regions, that is, all the points that satisfy either inequality.

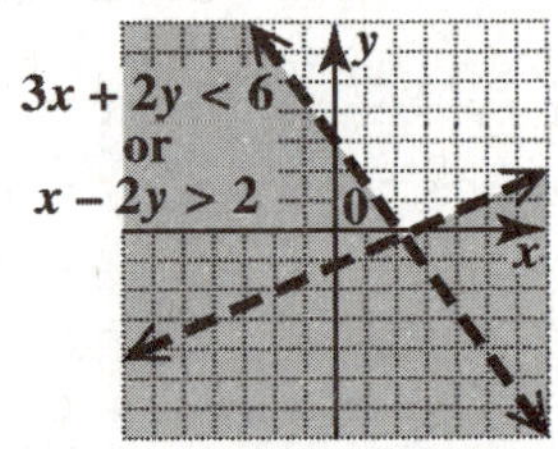

49. $y \leq 3x - 6$
The boundary line, $y = 3x - 6$, has slope 3 and y-intercept -6. This would be graph **B** or graph **C**. Since we want the region less than or equal to $3x - 6$, we want the region on or below the boundary line. The answer is graph **C**.

51. $y \leq -3x - 6$
The slope of the boundary line $y = -3x - 6$ is -3, and the y-intercept is -6. This would be graph **A** or graph **D**. The inequality sign is $\leq$, so we want the region on or below the boundary line. The answer is graph **A**.

53. **(a)** The x-intercept is $(-4, 0)$, so the solution set for $y = 0$ is $\{-4\}$.

(b) The solution set for $y < 0$ is $(-\infty, -4)$, since the graph is below the x-axis for these values of x.

(c) The solution set for $y > 0$ is $(-4, \infty)$, since the graph is above the x-axis for these values of x.

55. **(a)** The x-intercept is $(3.5, 0)$, so the solution set for $y = 0$ is $\{3.5\}$.

(b) The solution set for $y < 0$ is $(3.5, \infty)$, since the graph is below the x-axis for these values of x.

(c) The solution set for $y > 0$ is $(-\infty, 3.5)$, since the graph is above the x-axis for these values of x.

♦♦♦ Relating Concepts 57–62 ♦♦♦

57. "A factory can have *no more than* 200 workers on a shift, but must have *at least* 100" can be translated as $x \leq 200$ and $x \geq 100$. "Must manufacture *at least* 3000 units" can be translated as $y \geq 3000$.

58.

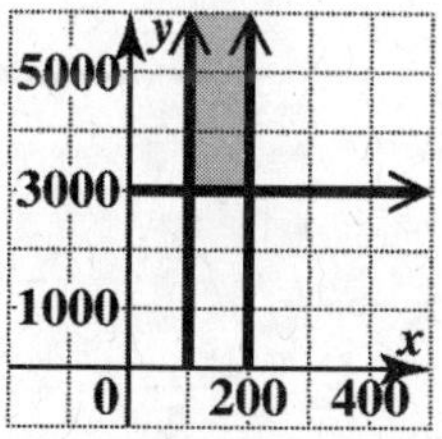

59. The total daily cost C consists of \$50 per worker and \$100 to manufacture 1 unit, so $C = 50x + 100y$.

60. Some examples of points in the shaded region are $(150, 4000)$, $(120, 3500)$, and $(180, 6000)$. Some examples of points on the boundary are $(100, 5000)$, $(150, 3000)$, and $(200, 4000)$. The corner points are $(100, 3000)$ and $(200, 3000)$.

61.

(x, y)	$50x + 100y = C$
$(150, 4000)$	$50(150) + 100(4000) = 407{,}500$
$(120, 3500)$	$50(120) + 100(3500) = 356{,}000$
$(180, 6000)$	$50(180) + 100(6000) = 609{,}000$
$(100, 5000)$	$50(100) + 100(5000) = 505{,}000$
$(150, 3000)$	$50(150) + 100(3000) = 307{,}500$
$(200, 4000)$	$50(200) + 100(4000) = 410{,}000$
$(100, 3000)$	$50(100) + 100(3000) = 305{,}000$ (least value)
$(200, 3000)$	$50(200) + 100(3000) = 310{,}000$

62. The company should use 100 workers and manufacture 3000 units to achieve the least possible cost.

63. $8^2 = 8 \cdot 8 = 64$

65. $-12^2 = -1 \cdot 12 \cdot 12 = -144$

Chapter 9 Review Exercises

For Exercises 1–4, let $A = \{a, b, c, d\}$, $B = \{a, c, e, f\}$, and $C = \{a, e, f, g\}$.

1. $A \cap B = \{a, b, c, d\} \cap \{a, c, e, f\}$
$= \{a, c\}$

2. $A \cap C = \{a, b, c, d\} \cap \{a, e, f, g\}$
$= \{a\}$

3. $B \cup C = \{a, c, e, f\} \cup \{a, e, f, g\}$
$= \{a, c, e, f, g\}$

4. $A \cup C = \{a, b, c, d\} \cup \{a, e, f, g\}$
$= \{a, b, c, d, e, f, g\}$

5. $x > 6$ and $x < 9$

The graph of the solution set will be all numbers which are both greater than 6 and less than 9. The overlap is the numbers between 6 and 9, not including the endpoints.

The solution set is $(6, 9)$.

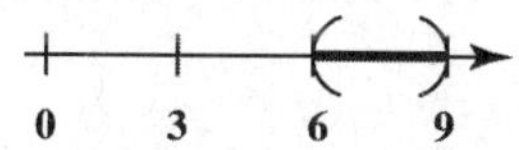

6.
$$\begin{aligned} x + 4 > 12 &\quad \text{and} \quad x - 2 < 12 \\ x > 8 &\quad \text{and} \quad x < 14 \end{aligned}$$

The graph of the solution set will be all numbers between 8 and 14, not including the endpoints.

The solution set is $(8, 14)$.

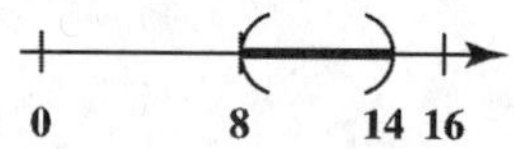

7. $x > 5$ or $x \leq -3$

The graph of the solution set will be all numbers that are either greater than 5 or less than or equal to -3.

The solution set is $(-\infty, -3] \cup (5, \infty)$.

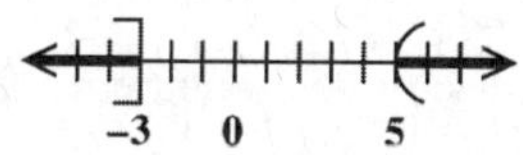

8. $x \geq -2$ or $x < 2$

The graph of the solution set will be all numbers that are either greater than or equal to -2 or less than 2. All real numbers satisfy these criteria.

The solution set is $(-\infty, \infty)$.

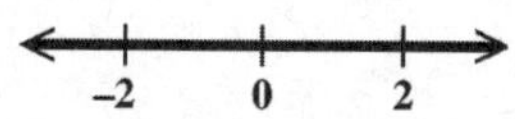

9.
$$\begin{aligned} x - 4 > 6 &\quad \text{and} \quad x + 3 \leq 10 \\ x > 10 &\quad \text{and} \quad x \leq 7 \end{aligned}$$

The graph of the solution set will be all numbers that are both greater than 10 and less than or equal to 7. There are no real numbers satisfying these criteria.

The solution set is $\emptyset$.

10.
$$\begin{aligned} -5x + 1 \geq 11 &\quad \text{or} \quad 3x + 5 \geq 26 \\ -5x \geq 10 &\quad \phantom{\text{or}} \quad 3x \geq 21 \\ x \leq -2 &\quad \text{or} \quad x \geq 7 \end{aligned}$$

The graph of the solution set will be all numbers that are either less than or equal to -2 or greater than or equal to 7.

The solution set is $(-\infty, -2] \cup [7, \infty)$.

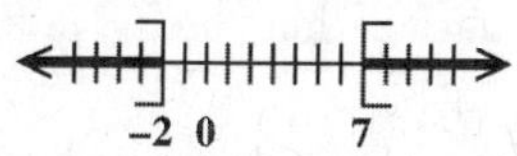

11. $(-3, \infty) \cap (-\infty, 4)$
$(-3, \infty)$ includes all real numbers greater than -3.
$(-\infty, 4)$ includes all real numbers less than 4.
Find the intersection. The numbers common to both sets are greater than -3 and less than 4.

$$-3 < x < 4$$

The solution set is $(-3, 4)$.

12. $(-\infty, 6) \cap (-\infty, 2)$
$(-\infty, 6)$ includes all real numbers less than 6.
$(-\infty, 2)$ includes all real numbers less than 2.
Find the intersection. The numbers common to both sets are less than 2.

The solution set is $(-\infty, 2)$.

13. $(4, \infty) \cup (9, \infty)$
$(4, \infty)$ includes all real numbers greater than 4.
$(9, \infty)$ includes all real numbers greater than 9.
Find the union. The numbers in the first set, the second set, or in both sets are all the real numbers that are greater than 4.

The solution set is $(4, \infty)$.

14. $(1, 2) \cup (1, \infty)$
$(1, 2)$ includes the real numbers between 1 and 2, not including 1 and 2.
$(1, \infty)$ includes all real numbers greater than 1.
Find the union. The numbers in the first set, the second set, or in both sets are all real numbers greater than 1.

The solution set is $(1, \infty)$.

15. $|x| = 7 \Leftrightarrow x = 7 \text{ or } x = -7$

The solution set is $\{-7, 7\}$.

16. $|x + 2| = 9$

$$\begin{aligned} x + 2 &= 9 \quad \text{or} \quad x + 2 = -9 \\ x &= 7 \quad \text{or} \quad x = -11 \end{aligned}$$

The solution set is $\{-11, 7\}$.

17. $|3x - 7| = 8$

$$\begin{aligned} 3x - 7 &= 8 \quad \text{or} \quad 3x - 7 = -8 \\ 3x &= 15 \qquad\quad 3x = -1 \\ x &= 5 \quad \text{or} \quad x = -\tfrac{1}{3} \end{aligned}$$

The solution set is $\left\{-\frac{1}{3}, 5\right\}$.

18. $|x - 4| = -12$

Since the absolute value of an expression can never be negative, there are no solutions for this equation.

The solution set is $\emptyset$.

19. $|2x - 7| + 4 = 11$
$|2x - 7| = 7$

$$\begin{aligned} 2x - 7 &= 7 \quad \text{or} \quad 2x - 7 = -7 \\ 2x &= 14 \qquad\quad 2x = 0 \\ x &= 7 \quad \text{or} \quad x = 0 \end{aligned}$$

The solution set is $\{0, 7\}$.

20. $|4x + 2| - 7 = -3$
$|4x + 2| = 4$

$$\begin{aligned} 4x + 2 &= 4 \quad \text{or} \quad 4x + 2 = -4 \\ 4x &= 2 \qquad\quad 4x = -6 \\ x &= \tfrac{2}{4} \qquad\quad x = -\tfrac{6}{4} \\ x &= \tfrac{1}{2} \quad \text{or} \quad x = -\tfrac{3}{2} \end{aligned}$$

The solution set is $\left\{-\frac{3}{2}, \frac{1}{2}\right\}$.

21. $|3x + 1| = |x + 2|$

$$\begin{aligned} 3x + 1 &= x + 2 \quad \text{or} \quad 3x + 1 = -(x + 2) \\ 2x &= 1 \qquad\qquad 3x + 1 = -x - 2 \\ & \qquad\qquad\qquad 4x = -3 \\ x &= \tfrac{1}{2} \quad \text{or} \quad x = -\tfrac{3}{4} \end{aligned}$$

The solution set is $\left\{-\frac{3}{4}, \frac{1}{2}\right\}$.

22. $|2x - 1| = |2x + 3|$

$$\begin{aligned} 2x - 1 &= 2x + 3 \quad \text{or} \quad 2x - 1 = -(2x + 3) \\ -1 &= 3 \quad \textit{False} \qquad 2x - 1 = -2x - 3 \\ & \qquad\qquad\qquad 4x = -2 \\ & \textit{No solution} \quad \text{or} \quad x = -\tfrac{2}{4} = -\tfrac{1}{2} \end{aligned}$$

The solution set is $\left\{-\frac{1}{2}\right\}$.

23. $|x| < 14 \Leftrightarrow -14 < x < 14$

The solution set is $(-14, 14)$.

24. $|-x + 6| \le 7$

$$\begin{aligned} -7 \le -x + 6 &\le 7 \\ -13 \le -x \quad &\le 1 \quad \textit{Subtract 6.} \\ 13 \ge x \quad &\ge -1 \quad \textit{Multiply by } -1\textit{. Reverse inequalities.} \\ -1 \le x \quad &\le 13 \quad \textit{Equivalent inequality} \end{aligned}$$

The solution set is $[-1, 13]$.

25. $|2x + 5| \le 1$

$$\begin{aligned} -1 \le 2x + 5 &\le 1 \\ -6 \le 2x \quad &\le -4 \\ -3 \le x \quad &\le -2 \end{aligned}$$

The solution set is $[-3, -2]$.

26. $|x + 1| \ge -3$

Since the absolute value of an expression is always nonnegative (positive or zero), the inequality is *true* for any real number x.

The solution set is $(-\infty, \infty)$.

27. $|5x - 1| > 9$

$$\begin{aligned} 5x - 1 &> 9 \quad \text{or} \quad & 5x - 1 &< -9 \\ 5x &> 10 & 5x &< -8 \\ x &> 2 \quad \text{or} & x &< -\tfrac{8}{5} \end{aligned}$$

The solution set is $\left(-\infty, -\frac{8}{5}\right) \cup (2, \infty)$.

28. $|11x - 3| \le -2$

There is no number whose absolute value is less than or equal to -2, so this inequality has no solution.

The solution set is $\emptyset$.

29. $|11x - 3| \ge -2$

Since the absolute value of an expression is always nonnegative (positive or zero), the inequality is true for any real number x.

The solution set is $(-\infty, \infty)$.

30. $|11x - 3| \le 0$

The absolute value of an expression is always nonnegative (positive or zero), so this inequality is true only when

$$\begin{aligned} 11x - 3 &= 0 \\ 11x &= 3 \\ x &= \tfrac{3}{11}. \end{aligned}$$

The solution set is $\left\{\frac{3}{11}\right\}$.

31. $3x - 2y \le 12$
Graph $3x - 2y = 12$ as a solid line through $(0, -6)$ and $(4, 0)$. Use $(0, 0)$ as a test point. Since $(0, 0)$ satisfies the inequality, shade the region on the side of the line containing $(0, 0)$.

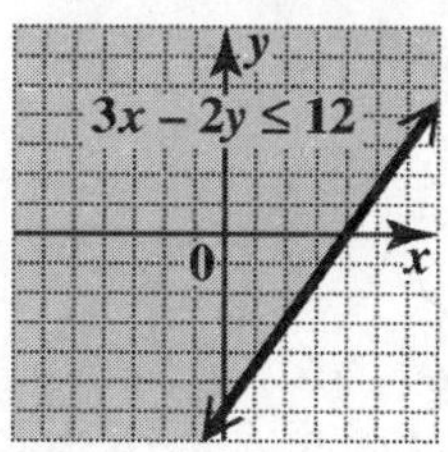

32. $5x - y > 6$
Graph $5x - y = 6$ as a dashed line through $(0, -6)$ and $(\frac{6}{5}, 0)$. Use $(0, 0)$ as a test point. Since $(0, 0)$ does not satisfy the inequality, shade the region on the side of the line that does not contain $(0, 0)$.

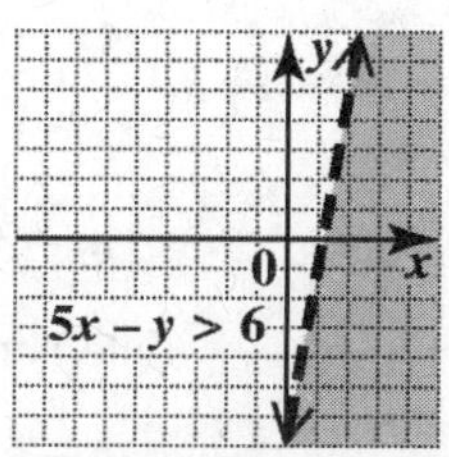

33. $3x + 2y < 0$

Graph the boundary, $3x + 2y = 0$, as a dashed line through $(0, 0)$, $(-2, 3)$, and $(2, -3)$. Choose a test point not on the line. Using $(1, 1)$ results in the false statement $5 < 0$, so shade the region *not* containing $(1, 1)$. This is the region below the line. The dashed line shows that the boundary is not part of the graph.

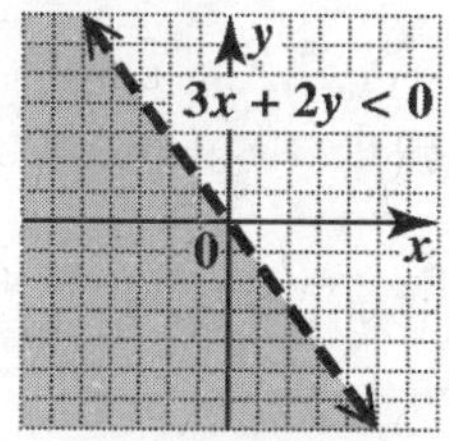

34. $2x + y \le 1$ and $x \ge 2y$
Graph $2x + y = 1$ as a solid line through $(\frac{1}{2}, 0)$ and $(0, 1)$, and shade the region on the side containing $(0, 0)$ since it satisfies the inequality. Next, graph $x = 2y$ as a solid line through $(0, 0)$ and $(2, 1)$, and shade the region on the side containing $(2, 0)$ since $2 > 2(0)$ or $2 > 0$ is true. The intersection is the region where the graphs overlap.

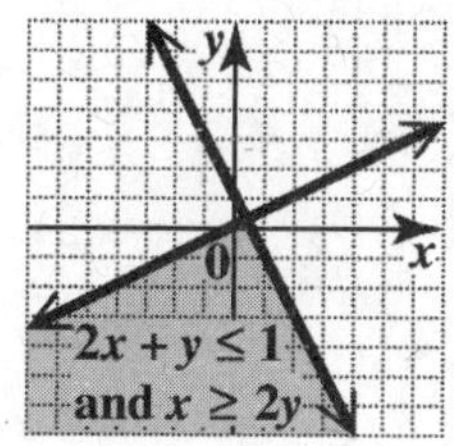

35. $x \ge 2$ or $y \ge 2$

Graph $x = 2$ as a solid vertical line through $(2, 0)$. Shade the region to the right of $x = 2$.

Graph $y = 2$ as a solid horizontal line through $(0, 2)$. Shade the region above $y = 2$. The graph of

$$x \ge 2 \quad \text{or} \quad y \ge 2$$

includes all the shaded regions.

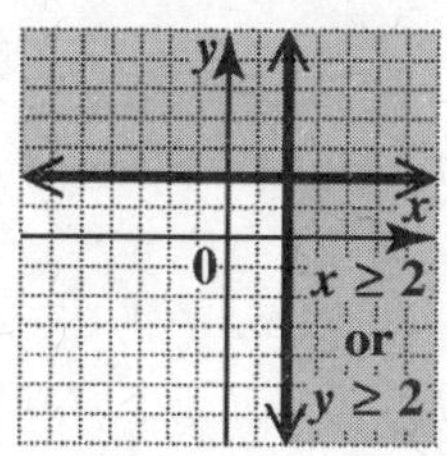

36. In $y < 4x + 3$, the " $<$ " symbol indicates that the graph has a dashed boundary line and that the shading is below the line, so the correct choice is **D**.

37. **[9.1]** $x < 3$ and $x \geq -2$

The real numbers that are common to both sets are the numbers greater than or equal to -2 and less than 3.

$$-2 \leq x < 3$$

The solution set is $[-2, 3)$.

38. **[9.2]** $|3x + 6| \geq 0$

The absolute value of an expression is always nonnegative, so the inequality is true for any real number k.

The solution set is $(-\infty, \infty)$.

39. **[9.2]** $|3x + 2| + 4 = 9$

$$|3x + 2| = 5$$

$$\begin{aligned} 3x + 2 &= 5 \quad \text{or} \quad & 3x + 2 &= -5 \\ 3x &= 3 & 3x &= -7 \\ x &= 1 \quad \text{or} & x &= -\tfrac{7}{3} \end{aligned}$$

The solution set is $\left\{-\frac{7}{3}, 1\right\}$.

40. **[9.2]** $|x + 3| \leq 13$

$$\begin{aligned} -13 \leq x + 3 &\leq 13 \\ -16 \leq \quad x \quad &\leq 10 \end{aligned}$$

The solution set is $[-16, 10]$.

41. **[9.2]** $|5x - 1| > 14$

$$\begin{aligned} 5x - 1 &> 14 \quad \text{or} \quad & 5x - 1 &< -14 \\ 5x &> 15 & 5x &< -13 \\ x &> 3 \quad \text{or} & x &< -\tfrac{13}{5} \end{aligned}$$

The solution set is $(-\infty, -\frac{13}{5}) \cup (3, \infty)$.

42. **[9.1]** $x \geq -2$ or $x < 4$

The solution set includes all numbers either greater than or equal to -2 or all numbers less than 4. This is the union and is the set of all real numbers.

The solution set is $(-\infty, \infty)$.

43. **[9.2]** $|x - 1| = |2x + 3|$

$$\begin{aligned} x - 1 &= 2x + 3 \quad \text{or} \quad & x - 1 &= -(2x + 3) \\ & & x - 1 &= -2x - 3 \\ & & 3x &= -2 \\ -4 &= x \quad \text{or} & x &= -\tfrac{2}{3} \end{aligned}$$

The solution set is $\left\{-4, -\frac{2}{3}\right\}$.

44. **[9.2]** $|x + 3| \leq 1$

$$\begin{aligned} -1 \leq x + 3 &\leq 1 \\ -4 \leq x \quad &\leq -2 \end{aligned}$$

The solution set is $[-4, -2]$.

45. **[9.2]** $|3x - 7| = 4$

$$\begin{aligned} 3x - 7 &= 4 \quad \text{or} \quad & 3x - 7 &= -4 \\ 3x &= 11 & 3x &= 3 \\ x &= \tfrac{11}{3} \quad \text{or} & x &= 1 \end{aligned}$$

The solution set is $\left\{1, \frac{11}{3}\right\}$.

46. **[9.2]** **(a)** $|2x - 5| = k$

If $k < 0$, then $|2x - 5|$ would be equal to a negative number. Since the absolute value of an expression is always nonnegative (positive or zero), the solution set is $\emptyset$.

(b) $|2x - 5| < k$

If $k < 0$, then $|2x - 5|$ would be less than a negative number. Since the absolute value of an expression is always nonnegative (positive or zero), the solution set is $\emptyset$.

(c) $|2x - 5| > k$

If $k < 0$, then $|2x - 5|$ would be greater than a negative number. Since the absolute value of an expression is always nonnegative (positive or zero), the solution set is the set of all real numbers, $(-\infty, \infty)$.

47. **[9.1]** $x > 6$ and $x < 8$

The graph of the solution set is all numbers both greater than 6 *and* less than 8. This is the intersection. The elements common to both sets are the numbers between 6 and 8, not including the endpoints. The solution set is $(6, 8)$.

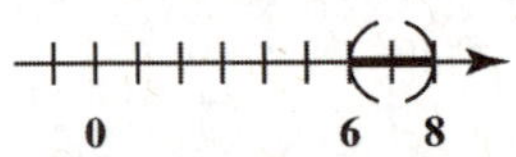

48. **[9.1]** $-5x + 1 \geq 11$ or $3x + 5 \geq 26$

$$\begin{aligned} -5x + 1 &\geq 11 \quad \text{or} \quad & 3x + 5 &\geq 26 \\ -5x &\geq 10 & 3x &\geq 21 \\ x &\leq -2 \quad \text{or} & x &\geq 7 \end{aligned}$$

The graph of the solution set is all numbers either less than or equal to -2 *or* greater than or equal to 7. This is the union. The solution set is $(-\infty, -2] \cup [7, \infty)$.

49. **[9.3]** $2x - 3y > -6$

Use intercepts to graph the boundary, $2x - 3y = -6$.

If $y = 0$, $x = -3$, so the x-intercept is $(-3, 0)$.

If $x = 0$, $y = 2$, so the y-intercept is $(0, 2)$.

Draw a dashed line through $(-3, 0)$ and $(0, 2)$.

Using $(0, 0)$ as a test point will result in the inequality $0 > -6$, which is true. Shade the region containing the origin. This is the region below the line. The dashed line shows that the boundary is not part of the graph.

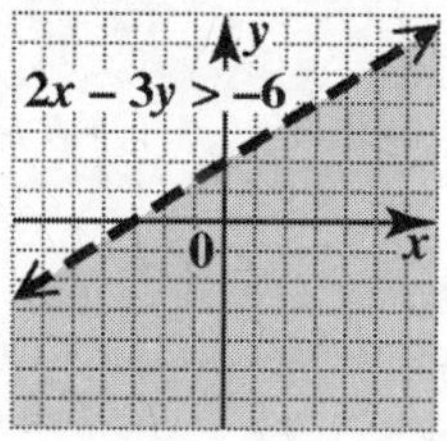

50. **[9.3]** $3x + 5y > 9$

To graph the boundary, which is the line $3x + 5y = 9$, find its intercepts.

$$\begin{aligned} 3x + 5y &= 9 \\ 3x + 5(0) &= 9 \quad \textit{Let } y = 0. \\ 3x &= 9 \\ x &= 3 \end{aligned} \qquad \begin{aligned} 3x + 5y &= 9 \\ 3(0) + 5y &= 9 \quad \textit{Let } x = 0. \\ 5y &= 9 \\ y &= \tfrac{9}{5} \end{aligned}$$

The x-intercept is $(3, 0)$ and the y-intercept is $\left(0, \frac{9}{5}\right)$. (A third point may be used as a check.) Draw a dashed line through these points. In order to determine which side of the line should be shaded, use $(0, 0)$ as a test point. Substituting 0 for x and 0 for y will result in the inequality $0 > 9$, which is false. Shade the region *not* containing the origin. This is the region above the line. The dashed line shows that the boundary is not part of the graph.

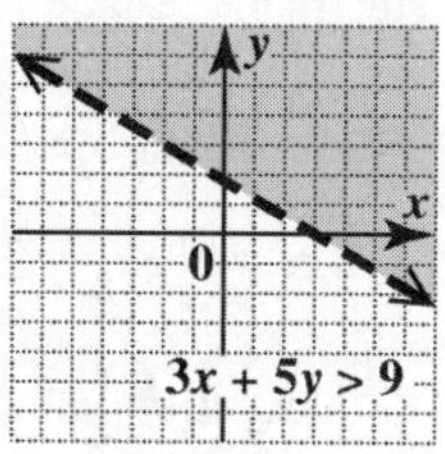

51. **[9.1]** **(a)** All states have less than 3 million female workers. The set of states with more than 3 million male workers is {Illinois}. Illinois is the only state in both sets, so the set of states with less than 3 million female workers *and* more than 3 million male workers is {Illinois}.

(b) The set of states with less than 1 million female workers *or* more than 2 million male workers is {Illinois, Maine, North Carolina, Oregon, Utah}.

(c) It is easy to see that the sum of the female and male workers for each state doesn't exceed 7 million, so the set of states with a total of more than 7 million civilian workers is {}, or $\emptyset$.

52. **[9.2]** **(a)** The endpoints on the number line are $-\frac{11}{3}$ and 1. The solution set of $|3x + 4| \geq 7$ would be

$$\left(-\infty, -\tfrac{11}{3}\right] \cup [1, \infty).$$

(b) The solution set of $|3x + 4| \leq 7$ would be

$$\left[-\tfrac{11}{3}, 1\right].$$

Chapter 9 Test

1. There are *no* occupations with median earnings for men less than \$900 and for women greater than \$500. This set is denoted by $\emptyset$.

2. *All* of the listed occupations have median earnings for men greater than \$600 or for women less than \$400. They are: Managerial/Professional, Technical/Sales/Administrative Support, Service, Operators/Fabricators/Laborers.

3. $$\begin{aligned} A \cap B &= \{1, 2, 5, 7\} \cap \{1, 5, 9, 12\} \\ &= \{1, 5\} \end{aligned}$$

4. $$\begin{aligned} A \cup B &= \{1, 2, 5, 7\} \cup \{1, 5, 9, 12\} \\ &= \{1, 2, 5, 7, 9, 12\} \end{aligned}$$

5. $$\begin{aligned} 3k \geq 6 \quad &\text{and} \quad k - 4 < 5 \\ k \geq 2 \quad &\text{and} \quad k < 9 \end{aligned}$$

The solution set is all numbers both greater than or equal to 2 *and* less than 9. This is the intersection. The numbers common to both sets are between 2 and 9, including 2 but not 9. The solution set is $[2, 9)$.

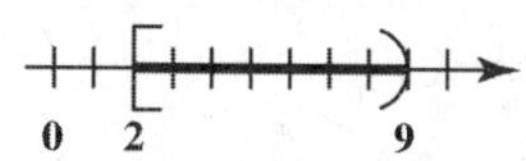

6. $$\begin{aligned} -4x \leq -24 \quad &\text{or} \quad 4x - 2 < 10 \\ &\qquad\quad 4x < 12 \\ x \geq 6 \quad &\text{or} \quad x < 3 \end{aligned}$$

The solution set is all numbers less than 3 or greater than or equal to 6. This is the union. The solution set is $(-\infty, 3) \cup [6, \infty)$.

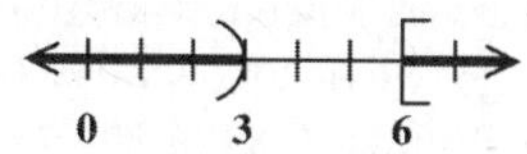

7. $|4x - 3| = 7$

$$\begin{aligned} 4x - 3 &= 7 & \text{or} \quad 4x - 3 &= -7 \\ 4x &= 10 & 4x &= -4 \\ x &= \tfrac{10}{4} = \tfrac{5}{2} & \text{or} \quad x &= -1 \end{aligned}$$

The solution set is $\left\{-1, \frac{5}{2}\right\}$.

8. $|5 - 6x| > 12$

$$\begin{aligned} 5 - 6x &> 12 & \text{or} \quad 5 - 6x &< -12 \\ -6x &> 7 & -6x &< -17 \\ x &< -\tfrac{7}{6} & \text{or} \quad x &> \tfrac{17}{6} \end{aligned}$$

The solution set is $\left(-\infty, -\frac{7}{6}\right) \cup \left(\frac{17}{6}, \infty\right)$.

9. $|7 - x| \le -1$

Since the absolute value of an expression is always nonnegative (positive or zero), the inequality is *false* for any real number x.

The solution set is $\emptyset$.

10. $|3 - 5x| = |2x + 8|$

$$\begin{aligned} 3 - 5x &= 2x + 8 & \text{or} \quad 3 - 5x &= -(2x + 8) \\ -7x &= 5 & 3 - 5x &= -2x - 8 \\ & & -3x &= -11 \\ x &= -\tfrac{5}{7} & \text{or} \quad x &= \tfrac{11}{3} \end{aligned}$$

The solution set is $\left\{-\frac{5}{7}, \frac{11}{3}\right\}$.

11. $|-3x + 4| - 4 < -1$

$$\begin{aligned} |-3x + 4| &< 3 && \textit{Isolate.} \\ -3 < -3x + 4 &< 3 && \textit{Equivalent inequality} \\ -7 < -3x &< -1 && \textit{Subtract 4.} \\ \frac{7}{3} > x &> \frac{1}{3} && \textit{Divide by } -3, \textit{ reverse inequalities} \\ \tfrac{1}{3} < x &< \tfrac{7}{3} && \textit{Equivalent inequality} \end{aligned}$$

The solution set is $\left(\frac{1}{3}, \frac{7}{3}\right)$.

12. $|12t + 7| \ge 0$

Since the absolute value of an expression is always nonnegative (positive or zero), the inequality is true for any real number t.

The solution set is $(-\infty, \infty)$.

13. **(a)** $|5x + 3| < k$

If $k < 0$, then $|5x + 3|$ would be less than a negative number. Since the absolute value of an expression is always nonnegative (positive or zero), the solution set is $\emptyset$.

(b) $|5x + 3| > k$

If $k < 0$, then $|5x + 3|$ would be greater than a negative number. Since the absolute value of an expression is always nonnegative (positive or zero), the solution set is the set of all real numbers, $(-\infty, \infty)$.

(c) $|5x + 3| = k$

If $k < 0$, then $|5x + 3|$ would be equal to a negative number. Since the absolute value of an expression is always nonnegative (positive or zero), the solution set is $\emptyset$.

14. $3x - 2y > 6$
Graph the line $3x - 2y = 6$, which has intercepts $(2, 0)$ and $(0, -3)$, as a dashed line since the inequality involves $>$. Test $(0, 0)$, which yields $0 > 6$, a false statement. Shade the region that does not include $(0, 0)$.

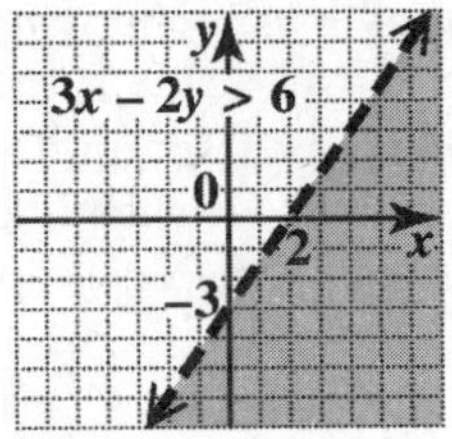

15. $3x - y > 0$

The boundary, $3x - y = 0$, goes through the origin, so both intercepts are $(0, 0)$. Two other points on this line are $(1, 3)$ and $(-1, -3)$. Draw the boundary as a dashed line.

Choose a test point which is not on the boundary. Using $(3, 0)$ results in the true statement $9 > 0$, so shade the region containing $(3, 0)$. This is the region below the line. The dashed line shows that the boundary is not part of the graph.

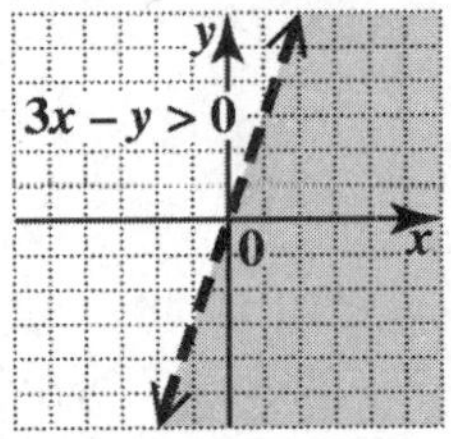

16. $y < 2x - 1$ and $x - y < 3$
First graph $y = 2x - 1$ as a dashed line through $(2, 3)$ and $(0, -1)$. Test $(0, 0)$, which yields $0 < -1$, a false statement. Shade the side of the line not containing $(0, 0)$.
Next, graph $x - y = 3$ as a dashed line through $(3, 0)$ and $(0, -3)$. Test $(0, 0)$, which yields $0 < 3$, a true statement. Shade the side of the line containing $(0, 0)$. The intersection is the region where the graphs overlap.

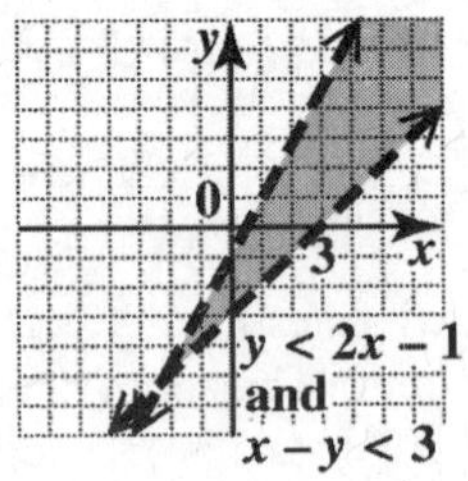

17. $x - 2 \ge y$ or $y \ge 3$

Graph $x - 2 = y$, which has intercepts $(2, 0)$ and $(0, -2)$, as a solid line. Test $(0, 0)$, which yields $-2 \ge 0$, a false statement. Shade the region that does not include $(0, 0)$.

Graph $y = 3$ as a solid horizontal line through $(0, 3)$. Shade the region above the line.

The required graph of the union includes all the shaded regions, that is, all the points that satisfy either inequality.

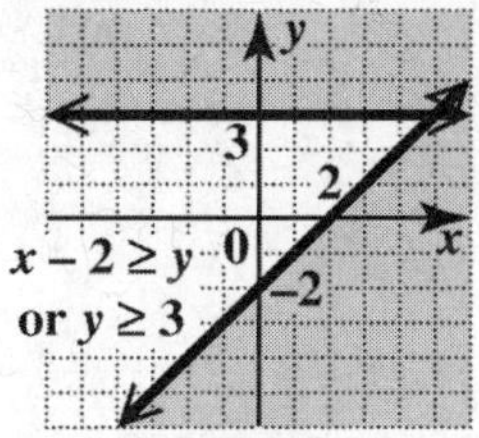

Cumulative Review Exercises (Chapters 1–9)

1. **(a)** 34 is a natural number, so it is also a whole number, an integer, a rational number, and a real number. **A, B, C, D, F**

   (b) 0 is a whole number, so it is also an integer, a rational number, and a real number. **B, C, D, F**

   (c) 2.16 is a rational number, so it is also a real number. **D, F**

   (d) $-\sqrt{36} = -6$ is an integer, so it is also a rational number and a real number. **C, D, F**

   (e) $\sqrt{13}$ is an irrational number, so it is also a real number. **E, F**

   (f) $-\frac{4}{5}$ is a rational number, so it is also a real number. **D, F**

2. $9 \cdot 4 - 16 \div 4 = (9 \cdot 4) - (16 \div 4) = 36 - 4 = 32$

3. $$\begin{aligned} -|8-13| - |-4| + |-9| &= -|-5| - 4 + 9 \\ &= -5 - 4 + 9 \\ &= -9 + 9 = 0 \end{aligned}$$

4. $$\begin{aligned} -5(8-2z) + 4(7-z) &= 7(8+z) - 3 \\ -40 + 10z + 28 - 4z &= 56 + 7z - 3 \quad \textit{Distributive property} \\ 6z - 12 &= 7z + 53 \quad \textit{Combine like terms.} \\ -65 &= z \quad \textit{Subtract 6z; 53.} \end{aligned}$$

   Thus, the solution set is $\{-65\}$.

5. $$\begin{aligned} 3(x+2) - 5(x+2) &= -2x - 4 \\ 3x + 6 - 5x - 10 &= -2x - 4 \\ -2x - 4 &= -2x - 4 \end{aligned}$$

   The last statement is true for all real numbers, so the solution set is {all real numbers}.

6. Solve $A = p + prt$ for t.

   $$\begin{aligned} A &= p + prt \\ A - p &= prt \quad \textit{Subtract p.} \\ \frac{A-p}{pr} &= t \quad \textit{Divide by pr.} \end{aligned}$$

7. $$\begin{aligned} 2(m+5) - 3m + 1 &> 5 \\ 2m + 10 - 3m + 1 &> 5 \\ -m + 11 &> 5 \\ -m &> -6 \\ m &< 6 \end{aligned}$$

   The solution set is $(-\infty, 6)$.

8. Personal computer: $\frac{480}{1500} = 32\%$

   Pacemaker:

   $$26\% \text{ of } 1500 = 0.26(1500) = 390$$

   Wireless communication:

   $$18\% \text{ of } 1500 = 0.18(1500) = 270$$

   Television: $\frac{150}{1500} = 10\%$

9. The sum of the measures of the angles of any triangle is 180°, so

   $$(x + 15) + (6x + 10) + (x - 5) = 180.$$

   Solve this equation.

   $$\begin{aligned} 8x + 20 &= 180 \\ 8x &= 160 \\ x &= 20 \end{aligned}$$

   Substitute 20 for x to find the measures of the angles.

   $$\begin{aligned} x - 5 &= 20 - 5 = 15 \\ x + 15 &= 20 + 15 = 35 \\ 6x + 10 &= 6(20) + 10 = 130 \end{aligned}$$

   The measures of the angles of the triangle are 15°, 35°, and 130°.

10. Through $(-4, 5)$ and $(2, -3)$

    Use the definition of slope with $x_1 = -4$, $y_1 = 5$, $x_2 = 2$, and $y_2 = -3$.

    $$m = \frac{y_2 - y_1}{x_2 - x_1} = \frac{-3-5}{2-(-4)} = \frac{-8}{6} = -\frac{4}{3}$$

11. Horizontal, through $(4, 5)$

    The slope of every horizontal line is 0.

12. Through $(4, -1)$, $m = -4$

    (a) Use the point-slope form with $x_1 = 4$, $y_1 = -1$, and $m = -4$.

    $$\begin{aligned} y - y_1 &= m(x - x_1) \\ y - (-1) &= -4(x - 4) \\ y + 1 &= -4x + 16 \\ y &= -4x + 15 \end{aligned}$$

    (b) The standard form is $Ax + By = C$.

    $$\begin{aligned} y &= -4x + 15 \\ 4x + y &= 15 \end{aligned}$$

13. Through $(0, 0)$ and $(1, 4)$

    (a) Find the slope.

    $$m = \frac{4-0}{1-0} = 4$$

    Because the slope is 4 and the y-intercept is 0, the equation of the line in slope-intercept form is

    $$y = 4x.$$

(b) The standard form is $Ax + By = C$.

$$y = 4x$$
$$0 = 4x - y, \text{ or } 4x - y = 0$$

14. $-3x + 4y = 12$

If $y = 0$, $x = -4$, so the x-intercept is $(-4, 0)$.

If $x = 0$, $y = 3$, so the y-intercept is $(0, 3)$.

Draw a line through these intercepts. A third point may be used as a check.

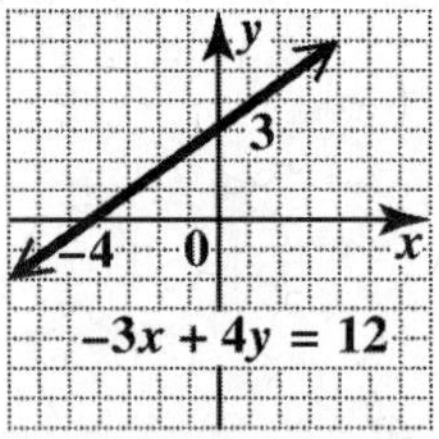

15. $\left(\dfrac{2m^3n}{p^2}\right)^3 = \dfrac{2^3(m^3)^3n^3}{(p^2)^3} = \dfrac{8m^9n^3}{p^6}$

16. $\dfrac{x^{-6}y^3z^{-1}}{x^7y^{-4}z} = \dfrac{y^4y^3}{x^6x^7z^1z} = \dfrac{y^7}{x^{13}z^2}$

17. $(3x^2 - 8x + 1) - (x^2 - 3x - 9)$

$= 3x^2 - 8x + 1 - x^2 + 3x + 9$ *Distributive property*

$= (3x^2 - x^2) + (-8x + 3x) + (1 + 9)$ *Combine like terms.*

$= 2x^2 - 5x + 10$

18. $(x + 2y)(x^2 - 2xy + 4y^2)$

Multiply vertically.

$$\begin{array}{rrrr} & x^2 & -\ 2xy & +\ 4y^2 \\ & & x & +\ 2y \\ \hline & 2x^2y & -\ 4xy^2 & +\ 8y^3 \\ x^3 & -\ 2x^2y & +\ 4xy^2 & \\ \hline x^3 & & & +\ 8y^3 \end{array}$$

Thus,

$$(x + 2y)(x^2 - 2xy + 4y^2) = x^3 + 8y^3.$$

19. $(3x + 2y)(5x - y)$

F O I L

$= 3x(5x) + 3x(-y) + 2y(5x) + 2y(-y)$

$= 15x^2 - 3xy + 10xy - 2y^2$

$= 15x^2 + 7xy - 2y^2$

20. $\dfrac{16x^3y^5 - 8x^2y^2 + 4}{4x^2y}$

$= \dfrac{16x^3y^5}{4x^2y} - \dfrac{8x^2y^2}{4x^2y} + \dfrac{4}{4x^2y}$

$= 4xy^4 - 2y + \dfrac{1}{x^2y}$

21. $m^2 + 12m + 32 = (m + 8)(m + 4)$

22. $25t^4 - 36 = (5t^2)^2 - (6)^2$

$= (5t^2 + 6)(5t^2 - 6)$

23. $81z^2 + 72z + 16$

$= (9z)^2 + 2(9z)(4) + 4^2$

$= (9z + 4)^2$

24. $(x + 4)(x - 1) = -6$

$x^2 + 3x - 4 = -6$

$x^2 + 3x + 2 = 0$

$(x + 2)(x + 1) = 0$

$x = -2$ or $x = -1$

The solution set is $\{-2, -1\}$.

25. $\dfrac{3}{x^2 + 5x - 14} = \dfrac{3}{(x + 7)(x - 2)}$

The expression is undefined when x is -7 or 2, because those values make the denominator equal zero.

26. $\dfrac{x^2 - 3x - 4}{x^2 + 3x} \cdot \dfrac{x^2 + 2x - 3}{x^2 - 5x + 4}$

$= \dfrac{(x - 4)(x + 1)}{x(x + 3)} \cdot \dfrac{(x - 1)(x + 3)}{(x - 4)(x - 1)}$ *Factor*

$= \dfrac{x + 1}{x}$ *Lowest terms*

27. $\dfrac{t^2 + 4t - 5}{t + 5} \div \dfrac{t - 1}{t^2 + 8t + 15}$

$= \dfrac{t^2 + 4t - 5}{t + 5} \cdot \dfrac{t^2 + 8t + 15}{t - 1}$

Multiply by the reciprocal

$= \dfrac{(t + 5)(t - 1)}{t + 5} \cdot \dfrac{(t + 5)(t + 3)}{t - 1}$ *Factor*

$= (t + 5)(t + 3),$ *Lowest terms*

or $t^2 + 8t + 15$

28. $\dfrac{2}{x + 3} - \dfrac{4}{x - 1}$

$= \dfrac{2(x - 1)}{(x + 3)(x - 1)} - \dfrac{4(x + 3)}{(x - 1)(x + 3)}$

$= \dfrac{2(x - 1) - 4(x + 3)}{(x + 3)(x - 1)}$

$= \dfrac{2x - 2 - 4x - 12}{(x + 3)(x - 1)}$

$= \dfrac{-2x - 14}{(x + 3)(x - 1)}$

29. $$\frac{\frac{2}{3}+\frac{1}{2}}{\frac{1}{9}-\frac{1}{6}}=\frac{\frac{4}{6}+\frac{3}{6}}{\frac{2}{18}-\frac{3}{18}}$$

$$=\frac{\frac{7}{6}}{\frac{-1}{18}}=\frac{7}{6}\div\frac{-1}{18}$$

$$=\frac{7}{6}\cdot\frac{18}{-1}=7\cdot(-3)=-21$$

30. $$\frac{x}{x+8}-\frac{3}{x-8}=\frac{128}{x^2-64}$$

Multiply by the LCD, (x + 8)(x − 8)

$$x(x-8)-3(x+8)=128$$
$$x^2-8x-3x-24=128$$
$$x^2-11x-152=0$$
$$(x+8)(x-19)=0$$
$$x=-8 \quad\text{or}\quad x=19$$

Check $x=19$: $\frac{19}{27}-\frac{3}{11}=\frac{128}{297}$ *True*

We cannot have -8 for a solution because that would result in division by zero. Thus, the solution set is $\{19\}$.

31. (a) $$m=\frac{261{,}122-322{,}486}{5-0}=\frac{-61{,}364}{5}$$
$$=-12{,}272.8$$

The average rate of change is $-12{,}272.8$ thousand pounds per year, that is, the number of pounds of shrimp caught in the United States decreased an average of $-12{,}272.8$ thousand pounds per year during the selected years.

(b) $y=mx+b$
$y=-12{,}272.8x+322{,}486$

(c) For the year 2003, $x=2003-2000=3$.

$$y=-12{,}272.8(3)+322{,}486$$
$$=285{,}667.6\approx 285{,}668$$

The model approximates about 285,668 thousand pounds of shrimp caught in 2003.

32. $\{(-4,-2),\ (-1,0),\ (2,0),\ (5,2)\}$

The domain is the set of first components, that is, $\{-4,-1,2,5\}$.
The range is the set of second components, that is, $\{-2,0,2\}$.
The relation is a function since each first component is paired with a unique second component.

33. $$g(x)=-x^2-2x+6$$
$$g(3)=-3^2-2(3)+6$$
$$=-9-6+6$$
$$=-9$$

34. $3x-4y=1 \quad (1)$
$2x+3y=12 \quad (2)$

To eliminate y, multiply equation (1) by 3 and equation (2) by 4. Then add the results.

$$\begin{array}{rcll} 9x-12y &=& 3 & 3\times(1)\\ 8x+12y &=& 48 & 4\times(2)\\ \hline 17x &=& 51 & \\ x &=& 3 & \end{array}$$

Since $x=3$,

$$3x-4y=1 \quad (1)$$
$$3(3)-4y=1$$
$$9-4y=1$$
$$-4y=-8$$
$$y=2$$

The solution set is $\{(3,2)\}$.

35. $3x-2y=4 \quad (1)$
$-6x+4y=7 \quad (2)$

Multiply equation (1) by 2 and add the result to equation (2).

$$\begin{array}{rcll} 6x-4y &=& 8 & 2\times(1)\\ -6x+4y &=& 7 & (2)\\ \hline 0 &=& 15 & \textit{False} \end{array}$$

Since a false statement results, the system is *inconsistent*. The solution set is $\emptyset$.

36. $x+3y-6z=7 \quad (1)$
$2x-y+z=1 \quad (2)$
$x+2y+2z=-1 \quad (3)$

To eliminate x, multiply equation (1) by -2 and add the result to equation (2).

$$\begin{array}{rcll} -2x-6y+12z &=& -14 & -2\times(1)\\ 2x-y+z &=& 1 & (2)\\ \hline -7y+13z &=& -13 & (4) \end{array}$$

To eliminate x again, multiply equation (3) by -2 and add the result to equation (2).

$$\begin{array}{rcll} 2x-y+z &=& 1 & (2)\\ -2x-4y-4z &=& 2 & -2\times(3)\\ \hline -5y-3z &=& 3 & (5) \end{array}$$

Use equations (4) and (5) to eliminate z. Multiply equation (4) by 3 and add the result to 13 times equation (5).

$$\begin{array}{rcll} -21y+39z &=& -39 & 3\times(4)\\ -65y-39z &=& 39 & 13\times(5)\\ \hline -86y &=& 0 & \\ y &=& 0 & \end{array}$$

From (5), $-3z=3$, so $z=-1$.
From (3), $x-2=-1$, so $x=1$.

The solution set is $\{(1,0,-1)\}$.

37. The length L of the rectangular flag measured 12 feet more than its width W, so

$$L = W + 12. \quad (1)$$

The perimeter is 144 feet.

$$P = 2L + 2W \quad (2)$$

Substitute $W + 12$ for L into equation (2).

$$144 = 2(W + 12) + 2W$$
$$144 = 2W + 24 + 2W$$
$$120 = 4W$$
$$30 = W$$

From (1), $L = 30 + 12 = 42$.

The length is 42 feet and the width is 30 feet.

38. Make a chart.

Number of Liters of Solution	Percent (as a decimal)	Pure Liters of Alcohol
x	0.15	$0.15x$
y	0.30	$0.30y$
9	0.20	$0.20(9) = 1.8$

From the first and third columns, we have the following system:

$$x + y = 9 \quad (1)$$
$$0.15x + 0.30y = 1.8 \quad (2)$$

To eliminate x, multiply equation (1) by -15 and add the result to 100 times equation (2).

$$-15x - 15y = -135$$
$$15x + 30y = 180$$
$$15y = 45$$
$$y = 3$$

From (1), $x + 3 = 9$, so $x = 6$.

Agbe should use 6 L of 15% solution and 3 L of 30% solution.

39. $x > -4$ and $x < 4$

The numbers that are greater than -4 *and* less than 4 are in the interval $(-4, 4)$.

The solution set is $(-4, 4)$.

40. $2x + 1 > 5$ or $2 - x \geq 2$

$2x > 4$ $\quad -x \geq 0$

$x > 2$ or $x \leq 0$

The solution set is $(-\infty, 0] \cup (2, \infty)$.

41. $|3x - 1| = 2$

$3x - 1 = 2$ or $3x - 1 = -2$

$3x = 3$ $\quad 3x = -1$

$x = 1$ or $x = -\frac{1}{3}$

The solution set is $\left\{-\frac{1}{3}, 1\right\}$.

42. $|3z + 1| \geq 7$

$3z + 1 \geq 7$ or $3z + 1 \leq -7$

$3z \geq 6$ $\quad 3z \leq -8$

$z \geq 2$ or $z \leq -\frac{8}{3}$

The solution set is $(-\infty, -\frac{8}{3}] \cup [2, \infty)$.

43. $y \leq 2x - 6$

Graph the boundary, $y = 2x - 6$, as a solid line through the intercepts $(3, 0)$ and $(0, -6)$. A third point such as $(1, -4)$ can be used as a check. Using $(0, 0)$ as a test point results in the false inequality $0 \leq -6$, so shade the region *not* containing the origin. This is the region below the line. The solid line shows that the boundary is part of the graph.

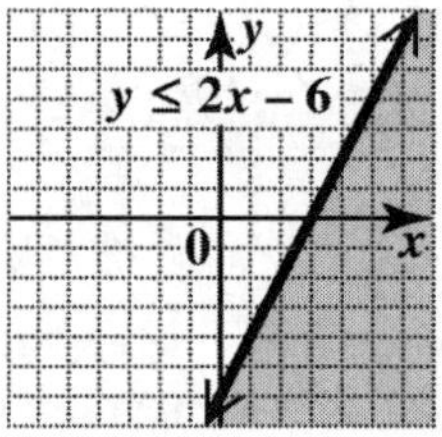

44. $x - y \geq 3$ and $3x + 4y \leq 12$

Graph the solid line $x - y = 3$ through $(3, 0)$ and $(0, -3)$. The inequality $x - y \geq 3$ can be written as $y \leq x - 3$, so shade the region below the boundary line.
Graph the solid line $3x + 4y = 12$ through $(4, 0)$ and $(0, 3)$. The inequality $3x + 4y \leq 12$ can be written as $y \leq -\frac{3}{4}x + 3$, so shade the region below the boundary line.
The required graph is the common shaded area as well as the portions of the lines that bound it.

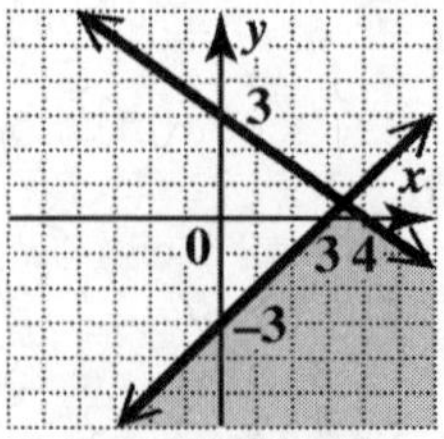

$x - y \geq 3$ and $3x + 4y \leq 12$

CHAPTER 10 ROOTS, RADICALS, AND ROOT FUNCTIONS

10.1 Radical Expressions and Graphs

10.1 Now Try Exercises

N1. The square roots of 81 are 9 and -9 because

$$9 \cdot 9 = 81 \quad \text{and} \quad (-9)(-9) = 81.$$

N2. **(a)** $\sqrt{400}$ is the positive square root of 400.

$$\sqrt{400} = 20$$

(b) $-\sqrt{169}$ is the negative square root of 169.

$$-\sqrt{169} = -13$$

(c) $\sqrt{\frac{100}{121}}$ is the positive square root of $\frac{100}{121}$.

$$\sqrt{\frac{100}{121}} = \frac{10}{11}$$

N3. **(a)** The square of $\sqrt{15}$ is

$$\left(\sqrt{15}\right)^2 = 15.$$

(b) The square of $-\sqrt{23}$ is

$$\left(-\sqrt{23}\right)^2 = \left(\sqrt{23}\right)^2 = 23.$$

(c) The square of $\sqrt{2k^2+5}$ is

$$\left(\sqrt{2k^2+5}\right)^2 = 2k^2+5.$$

N4. **(a)** 31 is not a perfect square, so $\sqrt{31}$ is irrational.

(b) 900 is a perfect square, so $\sqrt{900} = 30$ is rational.

(c) There is no real number whose square is -16, so $\sqrt{-16}$ is not a real number.

N5. **(a)** $\sqrt[3]{343} = 7$ because $7^3 = 343$.

(b) $\sqrt[3]{-1000} = -10$ because $(-10)^3 = -1000$.

(c) $\sqrt[3]{27} = 3$ because $3^3 = 27$.

N6. **(a)** $\sqrt[4]{625} = 5$ because $5^4 = 625$.

(b) $\sqrt[4]{-625}$ is not a real number because a fourth power of a real number cannot be negative.

(c) $\sqrt[5]{3125} = 5$ because $5^5 = 3125$.

(d) $\sqrt[5]{-3125} = -5$ because $(-5)^5 = -3125$.

N7. **(a)** $f(x) = \sqrt{x+1}$

For the radicand to be nonnegative, we must have $x + 1 \geq 0$, or, equivalently $x \geq -1$. Therefore, the domain is $[-1, \infty)$. The function values are at least 0 since $\sqrt{x+1} \geq 0$, so the range is $[0, \infty)$.

x	$f(x) = \sqrt{x+1}$
-1	$\sqrt{-1+1} = 0$
0	$\sqrt{0+1} = 1$
3	$\sqrt{3+1} = 2$
8	$\sqrt{8+1} = 3$

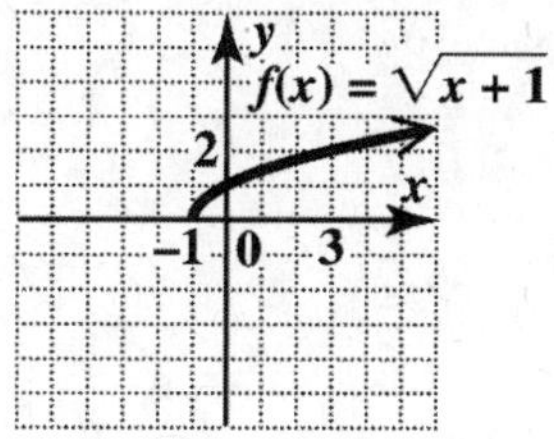

(b) $f(x) = \sqrt[3]{x} - 1$

Since we can take the cube root of any real number, the domain is $(-\infty, \infty)$.
The result of a cube root can be any real number, so the range is $(-\infty, \infty)$ (subtracting 1 doesn't change the range).

x	$f(x) = \sqrt[3]{x} - 1$
-8	$\sqrt[3]{-8} - 1 = -3$
-1	$\sqrt[3]{-1} - 1 = -2$
0	$\sqrt[3]{0} - 1 = -1$
1	$\sqrt[3]{1} - 1 = 0$
8	$\sqrt[3]{8} - 1 = 1$

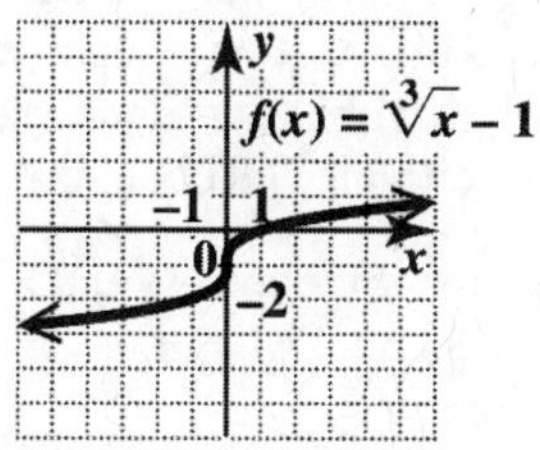

N8. **(a)** $\sqrt{11^2} = |11| = 11$

(b) $\sqrt{(-11)^2} = |-11| = 11$

(c) $\sqrt{z^2} = |z|$

(d) $\sqrt{(-z)^2} = |-z| = |z|$

N9. **(a)** $\sqrt[8]{(-2)^8} = |-2| = 2$ (n is even)

(b) $\sqrt[3]{(-9)^3} = -9$ (n is odd)

(c) $-\sqrt[4]{(-10)^4} = -|-10| = -10$ (n is even)

(d) $-\sqrt{m^8} = -\sqrt{(m^4)^2} = -m^4$ (n is even)
No absolute value bars are needed here because m^4 is nonnegative for any real number value of m.

(e) $\sqrt[3]{x^{18}} = \sqrt[3]{(x^6)^3} = x^6$

(f) $\sqrt[4]{t^{20}} = \sqrt[4]{(t^5)^4} = |t^5|$
Since t^5 could be negative, we need the absolute value bars.

N10. Use a calculator, and round to three decimal places.

(a) $-\sqrt{92} \approx -9.592$

(b) $\sqrt[4]{39} \approx 2.499$

(c) $\sqrt[5]{33} \approx 2.012$

N11. $$f = \frac{1}{2\pi\sqrt{LC}} = \frac{1}{2\pi\sqrt{(7 \times 10^{-5})(3 \times 10^{-9})}} \approx 347{,}305$$
or about 347,000 cycles per second.

10.1 Section Exercises

1. Every positive number has two real square roots. This statement is *true*. One of the real square roots is a positive number and the other is its opposite.

3. Every nonnegative number has two real square roots. This statement is *false* since zero is a nonnegative number that has only one square root, namely 0.

5. The cube root of every nonzero real number has the same sign as the number itself. This statement is *true*. The cube root of a positive real number is positive and the cube root of a negative real number is negative. The cube root of 0 is 0.

7. The square roots of 9 are -3 and 3 because $(-3)(-3) = 9$ and $3 \cdot 3 = 9$.

9. The square roots of 64 are -8 and 8 because $(-8)(-8) = 64$ and $8 \cdot 8 = 64$.

11. The square roots of 169 are -13 and 13 because $(-13)(-13) = 169$ and $13 \cdot 13 = 169$.

13. The square roots of $\frac{25}{196}$ are $-\frac{5}{14}$ and $\frac{5}{14}$ because $(-\frac{5}{14})(-\frac{5}{14}) = \frac{25}{196}$ and $\frac{5}{14} \cdot \frac{5}{14} = \frac{25}{196}$.

15. The square roots of 900 are -30 and 30 because $(-30)(-30) = 900$ and $30 \cdot 30 = 900$.

17. $\sqrt{1}$ represents the positive square root of 1. Since $1 \cdot 1 = 1$, $\sqrt{1} = 1$.

19. $\sqrt{49}$ represents the positive square root of 49. Since $7 \cdot 7 = 49$, $\sqrt{49} = 7$.

21. $-\sqrt{256}$ represents the negative square root of 256. Since $16 \cdot 16 = 256$, $-\sqrt{256} = -16$.

23. $-\sqrt{\frac{144}{121}}$ represents the negative square root of $\frac{144}{121}$. Since $\frac{12}{11} \cdot \frac{12}{11} = \frac{144}{121}$, $-\sqrt{\frac{144}{121}} = -\frac{12}{11}$.

25. $\sqrt{0.64}$ represents the positive square root of 0.64. Since $0.8 \cdot 0.8 = 0.64$, $\sqrt{0.64} = 0.8$.

27. $\sqrt{-121}$ is not a real number because there is no real number whose square is -121.

29. $-\sqrt{-49}$ is not a real number because there is no real number whose square is -49. The leading negative sign is irrelevant.

31. The square of $\sqrt{19}$ is
$$\left(\sqrt{19}\right)^2 = 19,$$
by the definition of square root.

33. The square of $-\sqrt{19}$ is
$$\left(-\sqrt{19}\right)^2 = 19,$$
by the definition of square root and since the square of a negative number is positive.

35. The square of $\sqrt{\frac{2}{3}}$ is
$$\left(\sqrt{\tfrac{2}{3}}\right)^2 = \tfrac{2}{3},$$
by the definition of square root.

37. The square of $\sqrt{3x^2 + 4}$ is
$$\left(\sqrt{3x^2 + 4}\right)^2 = 3x^2 + 4.$$

39. For the statement "$\sqrt{a}$ represents a positive number" to be true, a must be positive because the square root of a negative number is not a real number and $\sqrt{0} = 0$.

41. For the statement "$\sqrt{a}$ is not a real number" to be true, a must be negative.

43. $\sqrt{25}$
The number 25 is a perfect square, 5^2, so $\sqrt{25}$ is a *rational* number.
$$\sqrt{25} = 5$$

45. $\sqrt{29}$
Because 29 is not a perfect square, $\sqrt{29}$ is *irrational*. Using a calculator, we obtain
$$\sqrt{29} \approx 5.385.$$

47. $-\sqrt{64}$
The number 64 is a perfect square, 8^2, so $-\sqrt{64}$ is *rational.*

$$-\sqrt{64} = -8$$

49. $-\sqrt{300}$
The number 300 is not a perfect square, so $-\sqrt{300}$ is *irrational.* Using a calculator, we obtain

$$-\sqrt{300} \approx -17.321.$$

51. $\sqrt{-29}$
There is no real number whose square is -29. Therefore, $\sqrt{-29}$ is *not a real number.*

53. $\sqrt{1200}$
Because 1200 is not a perfect square, $\sqrt{1200}$ is *irrational.* Using a calculator, we obtain

$$\sqrt{1200} \approx 34.641.$$

55. Since $81 < 94 < 100$, $\sqrt{81} = 9$, and $\sqrt{100} = 10$, we conclude that $\sqrt{94}$ is between 9 and 10.

57. Since $49 < 51 < 64$, $\sqrt{49} = 7$, and $\sqrt{64} = 8$, we conclude that $\sqrt{51}$ is between 7 and 8.

59. Since $36 < 40 < 49$, $\sqrt{36} = 6$, and $\sqrt{49} = 7$, we conclude that $-\sqrt{40}$ is between -7 and -6.

61. Since $16 < 23.2 < 25$, $\sqrt{16} = 4$, and $\sqrt{25} = 5$, we conclude that $\sqrt{23.2}$ is between 4 and 5.

63.

$1^3 = \underline{1}$	$6^3 = \underline{216}$
$2^3 = \underline{8}$	$7^3 = \underline{343}$
$3^3 = \underline{27}$	$8^3 = \underline{512}$
$4^3 = \underline{64}$	$9^3 = \underline{729}$
$5^3 = \underline{125}$	$10^3 = \underline{1000}$

65. **(a)** $-\sqrt{25} = -(5) = -5$ **(E)**

(b) $\sqrt{-25}$ is not a real number (the index, 2, is even and the radicand, -25, is negative). **(F)**

(c) $\sqrt[3]{-27} = -3$, because $(-3)^3 = -27$. **(D)**

(d) $\sqrt[5]{-32} = -2$, because $(-2)^5 = -32$. **(B)**

(e) $\sqrt[4]{81} = 3$, because $3^4 = 81$. **(A)**

(f) $\sqrt[3]{8} = 2$, because $2^3 = 8$. **(C)**

67. $\sqrt{81} = 9$, because $9^2 = 81$. Notice that -9 is not an answer because the symbol $\sqrt{81}$ means the *nonnegative* square root of 81. However, the negative in front of the radical does lead to a negative answer since

$$-\sqrt{81} = -(9) = -9.$$

69. $\sqrt[3]{216} = 6$, because $6^3 = 216$.

71. $\sqrt[3]{-64} = -4$, because $(-4)^3 = -64$.

73. $\sqrt[3]{512} = 8$, because $8^3 = 512$, so $-\sqrt[3]{512} = -8$.

75. $\sqrt[4]{1296} = 6$, because $6^4 = 1296$.

77. $\sqrt[4]{16} = 2$, because $2^4 = 16$, so $-\sqrt[4]{16} = -2$.

79. $\sqrt[4]{-625}$ is not a real number since no real number to the fourth power equals -625. Any real number raised to the fourth power is 0 or positive.

81. $\sqrt[6]{64} = 2$, because $2^6 = 64$.

83. $\sqrt[6]{-32}$ is not a real number, because the index, 6, is even and the radicand, -32, is negative.

85. $\sqrt{\frac{64}{81}} = \frac{8}{9}$, because $\left(\frac{8}{9}\right)^2 = \frac{64}{81}$.

87. $\sqrt[3]{\frac{64}{27}} = \frac{4}{3}$, because $\left(\frac{4}{3}\right)^3 = \frac{64}{27}$.

89. $\sqrt[6]{\frac{1}{64}} = \frac{1}{2}$, because $\left(\frac{1}{2}\right)^6 = \frac{1}{64}$, so $-\sqrt[6]{\frac{1}{64}} = -\frac{1}{2}$.

91. $\sqrt[3]{-27} = -3$, because $(-3)^3 = -27$, so $-\sqrt[3]{-27} = -(-3) = 3$.

93. $\sqrt{0.25} = 0.5$, because $(0.5)^2 = 0.25$.

95. $\sqrt{0.49} = 0.7$, because $(0.7)^2 = 0.49$, so $-\sqrt{0.49} = -0.7$.

97. $\sqrt[3]{0.001} = 0.1$, because $(0.1)^3 = 0.001$.

99. $f(x) = \sqrt{x+3}$
For the radicand to be nonnegative, we must have

$$x + 3 \geq 0 \quad \text{or} \quad x \geq -3.$$

Thus, the domain is $[-3, \infty)$.
The function values are positive or zero (the result of the radical), so the range is $[0, \infty)$.

x	$f(x) = \sqrt{x+3}$
-3	$\sqrt{-3+3} = 0$
-2	$\sqrt{-2+3} = 1$
1	$\sqrt{1+3} = 2$

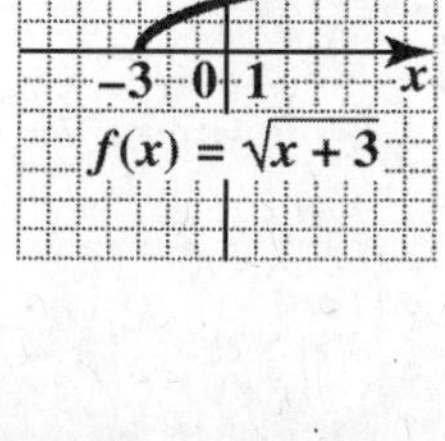

101. $f(x) = \sqrt{x} - 2$
For the radicand to be nonnegative, we must have $x \geq 0$. Note that the "-2" does not affect the domain, which is $[0, \infty)$.

The result of the radical is positive or zero, but the function values are 2 less than those values, so the range is $[-2, \infty)$.

x	$f(x) = \sqrt{x} - 2$
0	$\sqrt{0} - 2 = -2$
1	$\sqrt{1} - 2 = -1$
4	$\sqrt{4} - 2 = 0$

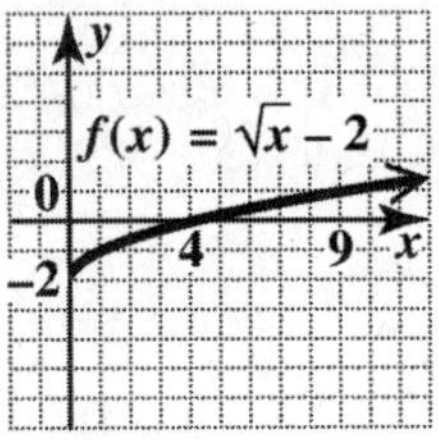

103. $f(x) = \sqrt[3]{x} - 3$
Since we can take the cube root of any real number, the domain is $(-\infty, \infty)$.
The result of a cube root can be any real number, so the range is $(-\infty, \infty)$. (The "-3" does not affect the range.)

x	$f(x) = \sqrt[3]{x} - 3$
-8	$\sqrt[3]{-8} - 3 = -5$
-1	$\sqrt[3]{-1} - 3 = -4$
0	$\sqrt[3]{0} - 3 = -3$
1	$\sqrt[3]{1} - 3 = -2$
8	$\sqrt[3]{8} - 3 = -1$

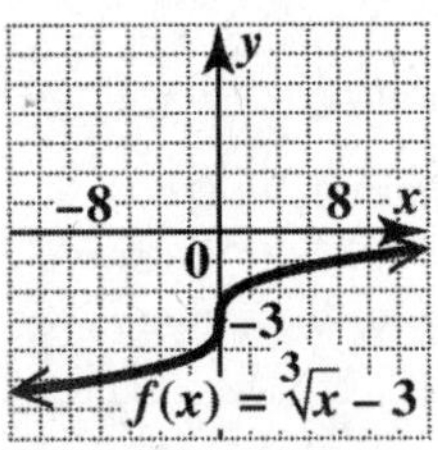

105. $f(x) = \sqrt[3]{x - 3}$
Both the domain and range are $(-\infty, \infty)$.

x	$f(x) = \sqrt[3]{x - 3}$
-5	$\sqrt[3]{-5 - 3} = -2$
2	$\sqrt[3]{2 - 3} = -1$
3	$\sqrt[3]{3 - 3} = 0$
4	$\sqrt[3]{4 - 3} = 1$
11	$\sqrt[3]{11 - 3} = 2$

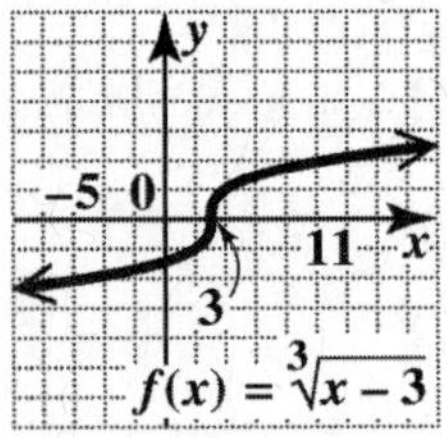

107. $\sqrt{12^2} = |12| = 12$

109. $\sqrt{(-10)^2} = |-10| = -(-10) = 10$

111. Since 6 is an even positive integer, $\sqrt[6]{a^6} = |a|$, so $\sqrt[6]{(-2)^6} = |-2| = 2$.

113. Since 5 is odd, $\sqrt[5]{a^5} = a$, so $\sqrt[5]{(-9)^5} = -9$.

115. $\sqrt[6]{(-5)^6} = |-5| = 5$, so $-\sqrt[6]{(-5)^6} = -5$.

117. Since the index is even, $\sqrt{x^2} = |x|$.

119. $\sqrt{(-z)^2} = |-z| = |z|$

121. Since the index is odd, $\sqrt[3]{x^3} = x$.

123. $\sqrt[3]{x^{15}} = \sqrt[3]{(x^5)^3} = x^5$ (3 is odd)

125. $\sqrt[6]{x^{30}} = \sqrt[6]{(x^5)^6} = |x^5|$, or $|x|^5$ (6 is even)

127. $\sqrt{123.5} \approx \sqrt{121} = 11$, because $11^2 = 121$. **(C)**

In Exercises 129–140, use a calculator and round to three decimal places

129. $\sqrt{9483} \approx 97.381$

131. $\sqrt{284.361} \approx 16.863$

133. $-\sqrt{82} \approx -9.055$

135. $\sqrt[3]{423} \approx 7.507$

137. $\sqrt[4]{100} \approx 3.162$

139. $\sqrt[5]{23.8} \approx 1.885$

141. The length $\sqrt{98}$ is closer to $\sqrt{100} = 10$ than to $\sqrt{81} = 9$. The width $\sqrt{26}$ is closer to $\sqrt{25} = 5$ than to $\sqrt{36} = 6$. Use the estimates $L = 10$ and $W = 5$ in $\mathcal{A} = LW$ to find an estimate of the area.

$$\mathcal{A} \approx 10 \cdot 5 = 50$$

Choice **A** is the best estimate.

143. $f = \dfrac{1}{2\pi\sqrt{LC}}$

$$= \frac{1}{2\pi\sqrt{(7.237 \times 10^{-5})(2.5 \times 10^{-10})}}$$

$\approx 1{,}183{,}235$
or about 1,183,000 cycles per second.

145. Since $H = 44 + 6 = 50$ ft, substitute 50 for H in the formula.

$$D = \sqrt{2H} = \sqrt{2 \cdot 50} = \sqrt{100} = 10$$

She will be able to see about 10 miles.

147. Let $a = 850$ mi, $b = 925$ mi, and $c = 1300$ mi. First find s.

$$\begin{aligned} s &= \tfrac{1}{2}(a+b+c) \\ &= \tfrac{1}{2}(850 + 925 + 1300) \\ &= \tfrac{3075}{2} = 1537.5 \end{aligned}$$

Now find the area using Heron's formula.

$$\begin{aligned} \mathcal{A} &= \sqrt{s(s-a)(s-b)(s-c)} \\ &= \sqrt{1537.5(687.5)(612.5)(237.5)} \\ &\approx 392{,}128.8 \end{aligned}$$

The area of the Bermuda Triangle is about 392,000 square miles.

149. **(a)** $I = \sqrt{\dfrac{2P}{L}} = \sqrt{\dfrac{2(120)}{80}}$ *Let P = 120, L = 80.*

$= \sqrt{3} \approx 1.732$ amps

(b) $I = \sqrt{\dfrac{2P}{L}} = \sqrt{\dfrac{2(100)}{40}}$ *Let P = 100, L = 40.*

$= \sqrt{5} \approx 2.236$ amps

151. $x^5 \cdot x^{-1} \cdot x^{-3} = x^{5+(-1)+(-3)} = x^1$, or x

153. $\left(\dfrac{2}{3}\right)^{-3} = \left(\dfrac{3}{2}\right)^{3} = \dfrac{3^3}{2^3}$, or $\dfrac{27}{8}$

10.2 Rational Exponents

10.2 Now Try Exercises

N1. **(a)** $81^{1/2} = \sqrt[2]{81} = \sqrt{81} = 9 \quad (9^2 = 81)$

(b) $125^{1/3} = \sqrt[3]{125} = 5 \quad (5^3 = 125)$

(c) $-625^{1/4} = -\sqrt[4]{625} = -5 \quad (5^4 = 625)$

(d) $(-625)^{1/4} = \sqrt[4]{-625}$ is *not a real number* because the radicand, -625, is negative and the index, 4, is even.

(e) $(-125)^{1/3} = \sqrt[3]{-125} = -5 \quad [(-5)^3 = -125]$

(f) $\left(\frac{1}{16}\right)^{1/4} = \sqrt[4]{\frac{1}{16}} = \frac{1}{2} \quad \left[\left(\frac{1}{2}\right)^4 = \frac{1}{16}\right]$

N2. **(a)** $32^{2/5} = (32^{1/5})^2 = \left(\sqrt[5]{32}\right)^2 = 2^2 = 4$

(b) $8^{5/3} = (8^{1/3})^5 = \left(\sqrt[3]{8}\right)^5 = 2^5 = 32$

(c) $-100^{3/2} = -(100^{1/2})^3 = -\left(\sqrt{100}\right)^3 = -10^3 = -1000$

(d) $(-121)^{3/2}$ is *not a real number*, since $(-121)^{1/2}$, or $\sqrt{-121}$, is not a real number.

(e) $(-125)^{4/3} = \left[(-125)^{1/3}\right]^4 = \left(\sqrt[3]{-125}\right)^4 = (-5)^4 = 625$

N3. **(a)** $243^{-3/5} = \dfrac{1}{243^{3/5}} = \dfrac{1}{(243^{1/5})^3} = \dfrac{1}{\left(\sqrt[5]{243}\right)^3} = \dfrac{1}{3^3} = \dfrac{1}{27}$

(b) $4^{-5/2} = \dfrac{1}{4^{5/2}} = \dfrac{1}{(4^{1/2})^5} = \dfrac{1}{\left(\sqrt{4}\right)^5} = \dfrac{1}{2^5} = \dfrac{1}{32}$

(c) $\left(\dfrac{216}{125}\right)^{-2/3} = \left(\dfrac{125}{216}\right)^{2/3} = \left(\sqrt[3]{\dfrac{125}{216}}\right)^2 = \left(\dfrac{5}{6}\right)^2 = \dfrac{25}{36}$

N4. **(a)** $21^{1/2} = \left(\sqrt[2]{21}\right)^1 = \sqrt{21}$

(b) $17^{5/4} = \left(\sqrt[4]{17}\right)^5$

(c) $4t^{3/5} + (4t)^{2/3} = 4\left(\sqrt[5]{t}\right)^3 + \left(\sqrt[3]{4t}\right)^2$

(d) $w^{-2/5} = \dfrac{1}{w^{2/5}} = \dfrac{1}{\left(\sqrt[5]{w}\right)^2}$

(e) $(a^2 - b^2)^{1/4} = \sqrt[4]{a^2 - b^2}$

(f) $\sqrt[3]{15} = 15^{1/3}$

(g) $\sqrt[4]{4^2} = 4^{2/4} = 4^{1/2} = \sqrt{4} = 2$

(h) $\sqrt[4]{x^4} = x$, since x is assumed to be positive.

N5. **(a)** $5^{1/4} \cdot 5^{2/3} = 5^{1/4+2/3} = 5^{3/12+8/12} = 5^{11/12}$

(b) $\dfrac{9^{3/5}}{9^{7/5}} = 9^{3/5-7/5} = 9^{-4/5} = \dfrac{1}{9^{4/5}}$

(c) $\dfrac{(r^{2/3}t^{1/4})^8}{t} = \dfrac{(r^{2/3})^8(t^{1/4})^8}{t} = \dfrac{r^{(2/3)8}t^{(1/4)8}}{t} = \dfrac{r^{16/3}t^2}{t} = r^{16/3}t$

(d) $$\begin{aligned} &\left(\frac{2x^{1/2}y^{-2/3}}{x^{-3/5}y^{-1/5}}\right)^{-3} \\ &= \left(2x^{1/2-(-3/5)}y^{-2/3-(-1/5)}\right)^{-3} \\ &= \left(2x^{5/10+6/10}y^{-10/15+3/15}\right)^{-3} \\ &= \left(2x^{11/10}y^{-7/15}\right)^{-3} \\ &= (2)^{-3}(x^{11/10})^{-3}(y^{-7/15})^{-3} \\ &= \frac{1}{2^3}x^{-33/10}y^{21/15} \\ &= \frac{y^{7/5}}{8x^{33/10}} \end{aligned}$$

(e) $y^{2/3}(y^{1/3}+y^{5/3}) = y^{2/3}\cdot y^{1/3} + y^{2/3}\cdot y^{5/3}$
$= y^{2/3+1/3} + y^{2/3+5/3}$
$= y^{3/3} + y^{7/3}$
$= y + y^{7/3}$

N6. (a) $\sqrt[5]{y^3}\cdot\sqrt[3]{y} = y^{3/5}\cdot y^{1/3}$
$= y^{3/5+1/3}$
$= y^{9/15+5/15}$
$= y^{14/15}$

(b) $\dfrac{\sqrt[4]{y^3}}{\sqrt{y^5}} = \dfrac{y^{3/4}}{y^{5/2}} = y^{3/4-5/2}$
$= y^{3/4-10/4} = y^{-7/4} = \dfrac{1}{y^{7/4}}$

(c) $\sqrt{\sqrt[3]{y}} = \sqrt{y^{1/3}} = (y^{1/3})^{1/2}$
$= y^{(1/3)(1/2)} = y^{1/6}$

10.2 Section Exercises

1. $3^{1/2} = \sqrt[2]{3^1} = \sqrt{3}$ **(C)**

3. $-16^{1/2} = -\sqrt{16} = -(4) = -4$ **(A)**

5. $(-32)^{1/5} = \sqrt[5]{-32} = -2$ **(H)**

7. $4^{3/2} = (4^{1/2})^3 = 2^3 = 8$ **(B)**

9. $-6^{2/4} = -(6^{1/2}) = -\sqrt{6}$ **(D)**

11. $169^{1/2} = \sqrt{169} = 13$

13. $729^{1/3} = \sqrt[3]{729} = 9$

15. $16^{1/4} = \sqrt[4]{16} = 2$

17. $\left(\frac{64}{81}\right)^{1/2} = \sqrt{\frac{64}{81}} = \frac{8}{9}$

19. $(-27)^{1/3} = \sqrt[3]{-27} = -3$

21. $(-144)^{1/2}$ is not a real number, because no real number squared equals -144.

23. $100^{3/2} = (100^{1/2})^3 = 10^3 = 1000$

25. $81^{3/4} = (81^{1/4})^3 = (\sqrt[4]{81})^3 = 3^3 = 27$

27. $-16^{5/2} = -(16^{1/2})^5 = -(4)^5 = -1024$

29. $(-8)^{4/3} = [(-8)^{1/3}]^4 = \left(\sqrt[3]{-8}\right)^4$
$= (-2)^4 = 16$

31. $32^{-3/5} = \dfrac{1}{32^{3/5}} = \dfrac{1}{(32^{1/5})^3} = \dfrac{1}{(\sqrt[5]{32})^3}$
$= \dfrac{1}{2^3} = \dfrac{1}{8}$

33. $64^{-3/2} = \dfrac{1}{64^{3/2}} = \dfrac{1}{(64^{1/2})^3} = \dfrac{1}{8^3} = \dfrac{1}{512}$

35. $\left(\frac{125}{27}\right)^{-2/3} = \left(\frac{27}{125}\right)^{2/3}$
$= \left[\left(\dfrac{27}{125}\right)^{1/3}\right]^2$
$= \left(\frac{3}{5}\right)^2 = \frac{9}{25}$

37. $\left(\frac{16}{81}\right)^{-3/4} = \left(\frac{81}{16}\right)^{3/4}$
$= \left[\left(\frac{81}{16}\right)^{1/4}\right]^3$
$= \left(\frac{3}{2}\right)^3 = \frac{27}{8}$

39. $10^{1/2} = (\sqrt[2]{10})^1 = \sqrt{10}$

41. $8^{3/4} = (8^{1/4})^3 = \left(\sqrt[4]{8}\right)^3$

43. $(9q)^{5/8} - (2x)^{2/3}$
$= [(9q)^{1/8}]^5 - [(2x)^{1/3}]^2$
$= (\sqrt[8]{9q})^5 - (\sqrt[3]{2x})^2$

45. $(2m)^{-3/2} = [(2m)^{1/2}]^{-3}$
$= (\sqrt{2m})^{-3}$
$= \dfrac{1}{(\sqrt{2m})^3}$

47. $(2y+x)^{2/3} = [(2y+x)^{1/3}]^2$
$= (\sqrt[3]{2y+x})^2$

49. $(3m^4+2k^2)^{-2/3}$
$= \dfrac{1}{(3m^4+2k^2)^{2/3}}$
$= \dfrac{1}{[(3m^4+2k^2)^{1/3}]^2}$
$= \dfrac{1}{(\sqrt[3]{3m^4+2k^2})^2}$

51. $\sqrt{2^{12}} = (2^{12})^{1/2} = 2^{12/2} = 2^6 = 64$

53. $\sqrt[3]{4^9} = 4^{9/3} = 4^3 = 64$

55. $\sqrt{x^{20}} = x^{20/2} = x^{10}$

57. $\sqrt[3]{x}\cdot\sqrt{x} = x^{1/3}\cdot x^{1/2}$
$= x^{1/3+1/2} = x^{2/6+3/6}$
$= x^{5/6} = \sqrt[6]{x^5}$

59. $\dfrac{\sqrt[3]{t^4}}{\sqrt[5]{t^4}} = \dfrac{t^{4/3}}{t^{4/5}} = \dfrac{t^{20/15}}{t^{12/15}} = t^{20/15-12/15}$
$= t^{8/15} = \sqrt[15]{t^8}$

61. $3^{1/2}\cdot 3^{3/2} = 3^{1/2+3/2} = 3^{4/2} = 3^2 = 9$

63. $\dfrac{64^{5/3}}{64^{4/3}} = 64^{5/3-4/3} = 64^{1/3} = \sqrt[3]{64} = 4$

65. $y^{7/3}\cdot y^{-4/3} = y^{7/3+(-4/3)}$
$= y^{3/3} = y^1 = y$

67. $x^{2/3} \cdot x^{-1/4} = x^{2/3+(-1/4)}$
$= x^{8/12-3/12} = x^{5/12}$

69. $\dfrac{k^{1/3}}{k^{2/3} \cdot k^{-1}} = \dfrac{k^{1/3}}{k^{-1/3}} = k^{1/3-(-1/3)} = k^{2/3}$

71. $\dfrac{(x^{1/4}y^{2/5})^{20}}{x^2} = \dfrac{x^{(1/4) \cdot 20}y^{(2/5) \cdot 20}}{x^2}$
$= \dfrac{x^5y^8}{x^2} = x^3y^8$

73. $\dfrac{(x^{2/3})^2}{(x^2)^{7/3}} = \dfrac{x^{4/3}}{x^{14/3}}$
$= x^{4/3-14/3} = x^{-10/3} = \dfrac{1}{x^{10/3}}$

75. $\dfrac{m^{3/4}n^{-1/4}}{(m^2n)^{1/2}} = \dfrac{m^{3/4}n^{-1/4}}{m^1n^{1/2}}$
$= m^{3/4-1}n^{-1/4-1/2}$
$= m^{3/4-4/4}n^{-1/4-2/4}$
$= m^{-1/4}n^{-3/4}$
$= \dfrac{1}{m^{1/4}n^{3/4}}$

77. $\dfrac{p^{1/5}p^{7/10}p^{1/2}}{(p^3)^{-1/5}} = \dfrac{p^{2/10+7/10+5/10}}{p^{-3/5}}$
$= \dfrac{p^{14/10}}{p^{-6/10}}$
$= p^{14/10-(-6/10)}$
$= p^{20/10} = p^2$

79. $\left(\dfrac{b^{-3/2}}{c^{-5/3}}\right)^2 (b^{-1/4}c^{-1/3})^{-1}$
$= \left(\dfrac{c^{5/3}}{b^{3/2}}\right)^2 (b^{1/4}c^{1/3})$
$= \dfrac{c^{10/3}}{b^3}(b^{1/4}c^{1/3})$
$= \dfrac{c^{10/3}b^{1/4}c^{1/3}}{b^3}$
$= c^{10/3+1/3}b^{1/4-3}$
$= c^{11/3}b^{-11/4}$
$= \dfrac{c^{11/3}}{b^{11/4}}$

81. $\left(\dfrac{p^{-1/4}q^{-3/2}}{3^{-1}p^{-2}q^{-2/3}}\right)^{-2}$
$= \dfrac{p^{1/2}q^3}{3^2p^4q^{4/3}}$
$= \dfrac{q^{3-4/3}}{9p^{4-1/2}}$
$= \dfrac{q^{5/3}}{9p^{7/2}}$

83. $p^{2/3}(p^{1/3} + 2p^{4/3})$
$= p^{2/3}p^{1/3} + p^{2/3}(2p^{4/3})$
$= p^{2/3+1/3} + 2p^{2/3+4/3}$
$= p^{3/3} + 2p^{6/3}$
$= p^1 + 2p^2$
$= p + 2p^2$

85. $k^{1/4}(k^{3/2} - k^{1/2})$
$= k^{1/4+3/2} - k^{1/4+1/2}$
$= k^{1/4+6/4} - k^{1/4+2/4}$
$= k^{7/4} - k^{3/4}$

87. $6a^{7/4}(a^{-7/4} + 3a^{-3/4})$
$= 6a^{7/4+(-7/4)} + 18a^{7/4+(-3/4)}$
$= 6a^0 + 18a^{4/4}$
$= 6(1) + 18a^1 = 6 + 18a$

89. $-5x^{7/6}(x^{5/6} - x^{-1/6})$
$= -5x^{7/6+5/6} + 5x^{7/6+(-1/6)}$
$= -5x^{12/6} + 5x^{6/6}$
$= -5x^2 + 5x$

91. $\sqrt[5]{x^3} \cdot \sqrt[4]{x} = x^{3/5} \cdot x^{1/4}$
$= x^{3/5+1/4}$
$= x^{12/20+5/20}$
$= x^{17/20}$

93. $\dfrac{\sqrt{x^5}}{\sqrt{x^8}} = \dfrac{(x^5)^{1/2}}{(x^8)^{1/2}}$
$= x^{5/2-(8/2)}$
$= x^{-3/2} = \dfrac{1}{x^{3/2}}$

95. $\sqrt{y} \cdot \sqrt[3]{yz} = y^{1/2} \cdot (yz)^{1/3}$
$= y^{1/2}y^{1/3}z^{1/3}$
$= y^{1/2+1/3}z^{1/3}$
$= y^{3/6+2/6}z^{1/3}$
$= y^{5/6}z^{1/3}$

97. $\sqrt[4]{\sqrt[3]{m}} = \sqrt[4]{m^{1/3}} = (m^{1/3})^{1/4} = m^{1/12}$

99. $\sqrt{\sqrt{\sqrt{x}}} = \sqrt{\sqrt{x^{1/2}}} = \sqrt{(x^{1/2})^{1/2}}$
$= \sqrt{x^{1/4}} = (x^{1/4})^{1/2} = x^{1/8}$

101. $\sqrt{\sqrt[3]{\sqrt[4]{x}}} = \sqrt{\sqrt[3]{x^{1/4}}}$
$= \sqrt{(x^{1/4})^{1/3}}$
$= (x^{1/12})^{1/2}$
$= x^{1/24}$

103. We are to show that, in general,

$$\sqrt{a^2+b^2} \neq a+b.$$

When $a=3$ and $b=4$,

$$\begin{aligned}\sqrt{a^2+b^2} &= \sqrt{3^2+4^2}\\ &= \sqrt{9+16}\\ &= \sqrt{25}=5,\end{aligned}$$

but $a+b=3+4=7$.

Since $5\neq 7$, $\sqrt{a^2+b^2}\neq a+b$.

105. Use $T(D)=0.07D^{3/2}$ with $D=16$.

$$\begin{aligned}T(16) &= 0.07(16)^{3/2} = 0.07(16^{1/2})^3\\ &= 0.07(4)^3 = 0.07(64) = 4.48\end{aligned}$$

To the nearest tenth of an hour, time T is 4.5 hours.

107. Use a calculator and the formula

$$W = 35.74 + 0.6215T - 35.75V^{4/25} + 0.4275TV^{4/25}.$$

For $T=30°$F and $V=15$ mph, $W\approx 19.0°$F. The table gives 19.0°F.

109. As in Exercise 107, for $T=20°$F and $V=20$ mph, $W\approx 4.2°$F. The table gives 4°F.

111. $\sqrt{25}\cdot\sqrt{36}=5\cdot 6=30$

$\sqrt{25\cdot 36}=\sqrt{900}=30$

The results are the same.

10.3 Simplifying Radical Expressions

10.3 Now Try Exercises

N1. **(a)** $\sqrt{7}\cdot\sqrt{11}=\sqrt{7\cdot 11}=\sqrt{77}$

(b) $\sqrt{2mn}\cdot\sqrt{15}=\sqrt{2mn\cdot 15}=\sqrt{30mn}$

N2. **(a)** $\sqrt[3]{4}\cdot\sqrt[3]{5}=\sqrt[3]{4\cdot 5}=\sqrt[3]{20}$

(b) $\sqrt[4]{5t}\cdot\sqrt[4]{6r^3}=\sqrt[4]{5t\cdot 6r^3}=\sqrt[4]{30tr^3}$

(c) $\sqrt[7]{20x}\cdot\sqrt[7]{3xy^3}=\sqrt[7]{20x\cdot 3xy^3}=\sqrt[7]{60x^2y^3}$

(d) $\sqrt[3]{5}\cdot\sqrt[4]{9}$ cannot be simplified using the product rule for radicals, because the indexes (3 and 4) are different.

N3. **(a)** $\sqrt{\dfrac{49}{36}}=\dfrac{\sqrt{49}}{\sqrt{36}}=\dfrac{7}{6}$

(b) $\sqrt{\dfrac{5}{144}}=\dfrac{\sqrt{5}}{\sqrt{144}}=\dfrac{\sqrt{5}}{12}$

(c) $\sqrt[3]{-\dfrac{27}{1000}}=\sqrt[3]{\dfrac{-27}{1000}}=\dfrac{\sqrt[3]{-27}}{\sqrt[3]{1000}}$

$$=\frac{-3}{10}=-\frac{3}{10}$$

(d) $\sqrt[4]{\dfrac{t}{16}}=\dfrac{\sqrt[4]{t}}{\sqrt[4]{16}}=\dfrac{\sqrt[4]{t}}{2}$

(e) $-\sqrt[5]{\dfrac{m^{15}}{243}}=-\dfrac{\sqrt[5]{m^{15}}}{\sqrt[5]{243}}=-\dfrac{m^3}{3}$

N4. **(a)** $\sqrt{50}=\sqrt{25\cdot 2}=\sqrt{25}\cdot\sqrt{2}=5\sqrt{2}$

(b) $\sqrt{192}=\sqrt{64\cdot 3}=\sqrt{64}\cdot\sqrt{3}=8\sqrt{3}$

(c) $\sqrt{42}$

No perfect square (other than 1) divides into 42, so $\sqrt{42}$ *cannot be simplified further.*

(d) $\sqrt[3]{108}=\sqrt[3]{27\cdot 4}=\sqrt[3]{27}\cdot\sqrt[3]{4}=3\sqrt[3]{4}$

(e) $-\sqrt[4]{80}=-\sqrt[4]{16\cdot 5}=-\sqrt[4]{16}\cdot\sqrt[4]{5}=-2\sqrt[4]{5}$

N5. **(a)**
$$\begin{aligned}\sqrt{36x^5} &= \sqrt{36x^4\cdot x}\\ &= \sqrt{36x^4}\cdot\sqrt{x}=6x^2\sqrt{x}\end{aligned}$$

(b)
$$\begin{aligned}\sqrt{32m^5n^4} &= \sqrt{16m^4n^4\cdot 2m}\\ &= \sqrt{16m^4n^4}\cdot\sqrt{2m}\\ &= 4m^2n^2\sqrt{2m}\end{aligned}$$

(c)
$$\begin{aligned}\sqrt[3]{-125k^3p^7} &= \sqrt[3]{-125k^3p^6\cdot p}\\ &= \sqrt[3]{-125k^3p^6}\cdot\sqrt[3]{p}\\ &= -5kp^2\sqrt[3]{p}\end{aligned}$$

(d)
$$\begin{aligned}-\sqrt[4]{162x^7y^8} &= -\sqrt[4]{81x^4y^8\cdot 2x^3}\\ &= -\sqrt[4]{81x^4y^8}\cdot\sqrt[4]{2x^3}\\ &= -3xy^2\sqrt[4]{2x^3}\end{aligned}$$

N6. **(a)** $\sqrt[6]{7^2}=7^{2/6}=7^{1/3}=\sqrt[3]{7}$

(b) $\sqrt[6]{y^4}=y^{4/6}=y^{2/3}=\sqrt[3]{y^2}$

N7. $\sqrt[3]{3}\cdot\sqrt{6}$

The least common index of 3 and 2 is 6. Write each radical as a sixth root.

$$\sqrt[3]{3}=3^{1/3}=3^{2/6}=\sqrt[6]{3^2}=\sqrt[6]{9}$$
$$\sqrt{6}=6^{1/2}=6^{3/6}=\sqrt[6]{6^3}=\sqrt[6]{216}$$

Therefore,

$$\sqrt[3]{3}\cdot\sqrt{6}=\sqrt[6]{9}\cdot\sqrt[6]{216}=\sqrt[6]{1944}.$$

N8. **(a)** Substitute 5 for a and 8 for b in the Pythagorean theorem to find c.

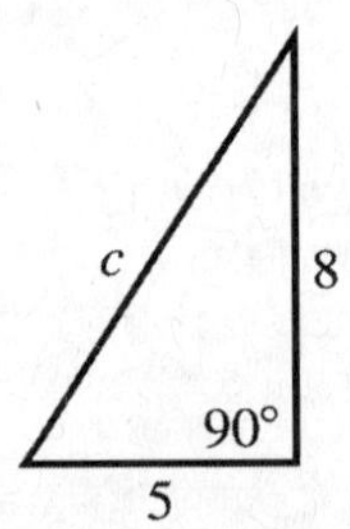

$$c^2 = a^2 + b^2$$
$$c^2 = 5^2 + 8^2$$
$$c^2 = 89$$
$$c = \sqrt{89}$$

The length of the hypotenuse is $\sqrt{89}$.

(b) Substitute 6 for a and 12 for c to find b.

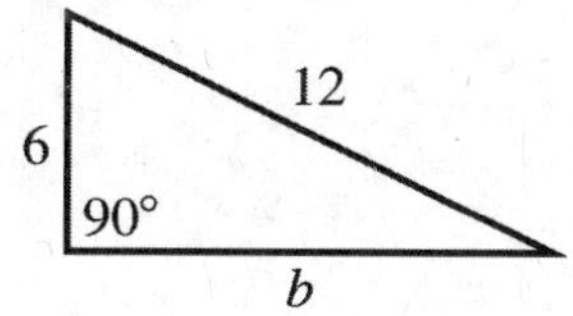

$$a^2 + b^2 = c^2$$
$$b^2 = c^2 - a^2$$
$$b^2 = 12^2 - 6^2$$
$$b^2 = 108$$
$$b = \sqrt{108}$$
$$b = \sqrt{36 \cdot 3}$$
$$b = \sqrt{36} \cdot \sqrt{3}$$
$$b = 6\sqrt{3}$$

The length of the unknown side is $6\sqrt{3}$.

N9. Use the distance formula,

$$d = \sqrt{(x_2 - x_1)^2 + (y_2 - y_1)^2}.$$

$(x_1, y_1) = (-4, -3)$ and $(x_2, y_2) = (-8, 6)$

$$d = \sqrt{[-8 - (-4)]^2 + [6 - (-3)]^2}$$
$$= \sqrt{(-4)^2 + 9^2}$$
$$= \sqrt{16 + 81}$$
$$= \sqrt{97}$$

10.3 Section Exercises

1. $\sqrt{3} \cdot \sqrt{3} = \sqrt{3 \cdot 3} = \sqrt{9}$, or 3

3. $\sqrt{18} \cdot \sqrt{2} = \sqrt{18 \cdot 2} = \sqrt{36}$, or 6

5. $\sqrt{5} \cdot \sqrt{6} = \sqrt{5 \cdot 6} = \sqrt{30}$

7. $\sqrt{14} \cdot \sqrt{x} = \sqrt{14 \cdot x} = \sqrt{14x}$

9. $\sqrt{14} \cdot \sqrt{3pqr} = \sqrt{14 \cdot 3pqr} = \sqrt{42pqr}$

11. $\sqrt[3]{2} \cdot \sqrt[3]{5} = \sqrt[3]{2 \cdot 5} = \sqrt[3]{10}$

13. $\sqrt[3]{7x} \cdot \sqrt[3]{2y} = \sqrt[3]{7x \cdot 2y} = \sqrt[3]{14xy}$

15. $\sqrt[4]{11} \cdot \sqrt[4]{3} = \sqrt[4]{11 \cdot 3} = \sqrt[4]{33}$

17. $\sqrt[4]{2x} \cdot \sqrt[4]{3x^2} = \sqrt[4]{2x \cdot 3x^2} = \sqrt[4]{6x^3}$

19. $\sqrt[3]{7} \cdot \sqrt[4]{3}$ cannot be multiplied using the product rule, because the indexes (3 and 4) are different.

21. $\sqrt{\frac{64}{121}} = \frac{\sqrt{64}}{\sqrt{121}} = \frac{8}{11}$

23. $\sqrt{\frac{3}{25}} = \frac{\sqrt{3}}{\sqrt{25}} = \frac{\sqrt{3}}{5}$

25. $\sqrt{\frac{x}{25}} = \frac{\sqrt{x}}{\sqrt{25}} = \frac{\sqrt{x}}{5}$

27. $\sqrt{\frac{p^6}{81}} = \frac{\sqrt{p^6}}{\sqrt{81}} = \frac{\sqrt{(p^3)^2}}{9} = \frac{p^3}{9}$

29. $\sqrt[3]{-\frac{27}{64}} = \sqrt[3]{\frac{-27}{64}} = \frac{\sqrt[3]{-27}}{\sqrt[3]{64}} = \frac{-3}{4} = -\frac{3}{4}$

31. $\sqrt[3]{\frac{r^2}{8}} = \frac{\sqrt[3]{r^2}}{\sqrt[3]{8}} = \frac{\sqrt[3]{r^2}}{2}$

33. $-\sqrt[4]{\frac{81}{x^4}} = -\frac{\sqrt[4]{3^4}}{\sqrt[4]{x^4}} = -\frac{3}{x}$

35. $\sqrt[5]{\frac{1}{x^{15}}} = \frac{\sqrt[5]{1}}{\sqrt[5]{(x^3)^5}} = \frac{1}{x^3}$

37. $\sqrt{12} = \sqrt{4 \cdot 3} = \sqrt{4} \cdot \sqrt{3} = 2\sqrt{3}$

39. $\sqrt{288} = \sqrt{144 \cdot 2} = \sqrt{144} \cdot \sqrt{2} = 12\sqrt{2}$

41. $-\sqrt{32} = -\sqrt{16 \cdot 2} = -\sqrt{16} \cdot \sqrt{2} = -4\sqrt{2}$

43. $-\sqrt{28} = -\sqrt{4 \cdot 7} = -\sqrt{4} \cdot \sqrt{7} = -2\sqrt{7}$

45. $\sqrt{30}$ cannot be simplified further.

47. $\sqrt[3]{128} = \sqrt[3]{64 \cdot 2} = \sqrt[3]{64} \cdot \sqrt[3]{2} = 4\sqrt[3]{2}$

49. $\sqrt[3]{-16} = \sqrt[3]{-8 \cdot 2} = \sqrt[3]{-8} \cdot \sqrt[3]{2} = -2\sqrt[3]{2}$

51. $\sqrt[3]{40} = \sqrt[3]{8 \cdot 5} = \sqrt[3]{8} \cdot \sqrt[3]{5} = 2\sqrt[3]{5}$

53. $-\sqrt[4]{512} = -\sqrt[4]{256 \cdot 2} = -\sqrt[4]{4^4} \cdot \sqrt[4]{2} = -4\sqrt[4]{2}$

55. $\sqrt[5]{64} = \sqrt[5]{32 \cdot 2} = \sqrt[5]{2^5} \cdot \sqrt[5]{2} = 2\sqrt[5]{2}$

57. $-\sqrt[5]{486} = -\sqrt[5]{243 \cdot 2} = -\sqrt[5]{3^5} \cdot \sqrt[5]{2} = -3\sqrt[5]{2}$

59. $\sqrt[6]{128} = \sqrt[6]{64 \cdot 2} = \sqrt[6]{2^6} \cdot \sqrt[6]{2} = 2\sqrt[6]{2}$

61. His reasoning was incorrect. The radicand 14 must be written as a product of two factors (not a sum of two terms) where one of the two factors is a perfect cube.

63. $\sqrt{72k^2} = \sqrt{36k^2 \cdot 2} = \sqrt{36k^2} \cdot \sqrt{2} = 6k\sqrt{2}$

65.
$$\sqrt{144x^3y^9} = \sqrt{144x^2y^8 \cdot xy}$$
$$= \sqrt{(12xy^4)^2} \cdot \sqrt{xy}$$
$$= 12xy^4\sqrt{xy}$$

67. $\sqrt{121x^6} = \sqrt{(11x^3)^2} = 11x^3$

69. $-\sqrt[3]{27t^{12}} = -\sqrt[3]{(3t^4)^3} = -3t^4$

71. $-\sqrt{100m^8z^4} = -\sqrt{(10m^4z^2)^2} = -10m^4z^2$

73. $-\sqrt[3]{-125a^6b^9c^{12}} = -\sqrt[3]{(-5a^2b^3c^4)^3}$
$= -(-5a^2b^3c^4)$
$= 5a^2b^3c^4$

75. $\sqrt[4]{\frac{1}{16}r^8t^{20}} = \sqrt[4]{(\frac{1}{2}r^2t^5)^4} = \frac{1}{2}r^2t^5$

77. $\sqrt{50x^3} = \sqrt{25x^2 \cdot 2x} = \sqrt{(5x)^2} \cdot \sqrt{2x}$
$= 5x\sqrt{2x}$

79. $-\sqrt{500r^{11}} = -\sqrt{100r^{10} \cdot 5r}$
$= -\sqrt{(10r^5)^2} \cdot \sqrt{5r} = -10r^5\sqrt{5r}$

81. $\sqrt{13x^7y^8} = \sqrt{(x^6y^8)(13x)}$
$= \sqrt{x^6y^8} \cdot \sqrt{13x}$
$= x^3y^4\sqrt{13x}$

83. $\sqrt[3]{8z^6w^9} = \sqrt[3]{(2z^2w^3)^3} = 2z^2w^3$

85. $\sqrt[3]{-16z^5t^7} = \sqrt[3]{(-2^3z^3t^6)(2z^2t)}$
$= -2zt^2\sqrt[3]{2z^2t}$

87. $\sqrt[4]{81x^{12}y^{16}} = \sqrt[4]{(3x^3y^4)^4} = 3x^3y^4$

89. $-\sqrt[4]{162r^{15}s^{10}} = -\sqrt[4]{81r^{12}s^8(2r^3s^2)}$
$= -\sqrt[4]{81r^{12}s^8} \cdot \sqrt[4]{2r^3s^2}$
$= -3r^3s^2\sqrt[4]{2r^3s^2}$

91. $\sqrt{\frac{y^{11}}{36}} = \frac{\sqrt{y^{11}}}{\sqrt{36}} = \frac{\sqrt{y^{10} \cdot y}}{6} = \frac{y^5\sqrt{y}}{6}$

93. $\sqrt[3]{\frac{x^{16}}{27}} = \frac{\sqrt[3]{x^{15} \cdot x}}{\sqrt[3]{27}} = \frac{x^5\sqrt[3]{x}}{3}$

95. $\sqrt[4]{48^2} = 48^{2/4} = 48^{1/2} = \sqrt{48} = \sqrt{16 \cdot 3}$
$= \sqrt{16} \cdot \sqrt{3} = 4\sqrt{3}$

97. $\sqrt[4]{25} = 25^{1/4} = (5^2)^{1/4} = 5^{2/4} = 5^{1/2} = \sqrt{5}$

99. $\sqrt[10]{x^{25}} = x^{25/10} = x^{5/2} = \sqrt{x^5}$
$= \sqrt{x^4 \cdot x} = x^2\sqrt{x}$

101. $\sqrt[3]{4} \cdot \sqrt{3}$
The least common index of 3 and 2 is 6. Write each radical as a sixth root.

$$\sqrt[3]{4} = 4^{1/3} = 4^{2/6} = \sqrt[6]{4^2} = \sqrt[6]{16}$$
$$\sqrt{3} = 3^{1/2} = 3^{3/6} = \sqrt[6]{3^3} = \sqrt[6]{27}$$

Therefore,

$$\sqrt[3]{4} \cdot \sqrt{3} = \sqrt[6]{16} \cdot \sqrt[6]{27}$$
$$= \sqrt[6]{16 \cdot 27} = \sqrt[6]{432}.$$

103. $\sqrt[4]{3} \cdot \sqrt[3]{4}$
The least common index of 4 and 3 is 12. Write each radical as a twelfth root.

$$\sqrt[4]{3} = 3^{1/4} = 3^{3/12} = \sqrt[12]{3^3} = \sqrt[12]{27}$$
$$\sqrt[3]{4} = 4^{1/3} = 4^{4/12} = \sqrt[12]{4^4} = \sqrt[12]{256}$$

Therefore,

$$\sqrt[4]{3} \cdot \sqrt[3]{4} = \sqrt[12]{27} \cdot \sqrt[12]{256}$$
$$= \sqrt[12]{27 \cdot 256} = \sqrt[12]{6912}.$$

105. $\sqrt{x} \cdot \sqrt[3]{x}$

$$\sqrt{x} = x^{1/2} = x^{3/6} = \sqrt[6]{x^3}$$
$$\sqrt[3]{x} = x^{1/3} = x^{2/6} = \sqrt[6]{x^2}$$
So $\sqrt{x} \cdot \sqrt[3]{x} = \sqrt[6]{x^3} \cdot \sqrt[6]{x^2}$
$= \sqrt[6]{x^3 \cdot x^2} = \sqrt[6]{x^5}.$

107. Substitute 3 for a and 4 for b in the Pythagorean theorem to find c.

$$c^2 = a^2 + b^2$$
$$c = \sqrt{a^2 + b^2} = \sqrt{3^2 + 4^2}$$
$$= \sqrt{9 + 16} = \sqrt{25} = 5$$

The length of the hypotenuse is 5.

109. Substitute 4 for a and 12 for c in the Pythagorean theorem to find b.

$$a^2 + b^2 = c^2$$
$$b = \sqrt{c^2 - a^2} = \sqrt{12^2 - 4^2}$$
$$= \sqrt{144 - 16} = \sqrt{128}$$
$$= \sqrt{64}\sqrt{2} = 8\sqrt{2}$$

The length of the unknown leg is $8\sqrt{2}$.

111. Substitute 5 for b and 9 for c in the Pythagorean theorem to find a.

$$a^2 + b^2 = c^2$$
$$a = \sqrt{c^2 - b^2} = \sqrt{9^2 - 5^2}$$
$$= \sqrt{81 - 25} = \sqrt{56}$$
$$= \sqrt{4}\sqrt{14} = 2\sqrt{14}$$

The length of the unknown leg is $2\sqrt{14}$.

In Exercises 113–124, use the distance formula

$$d = \sqrt{(x_2 - x_1)^2 + (y_2 - y_1)^2}.$$

113. $(6, 13)$ and $(1, 1)$

$$d = \sqrt{(6-1)^2 + (13-1)^2}$$
$$= \sqrt{5^2 + 12^2} = \sqrt{25 + 144}$$
$$= \sqrt{169} = 13$$

115. $(-6, 5)$ and $(3, -4)$

$$d = \sqrt{(-6-3)^2 + [5-(-4)]^2}$$
$$= \sqrt{(-9)^2 + (9)^2} = \sqrt{81 + 81}$$
$$= \sqrt{162} = \sqrt{81} \cdot \sqrt{2} = 9\sqrt{2}$$

117. $(-8, 2)$ and $(-4, 1)$

$$\begin{aligned} d &= \sqrt{[-8-(-4)]^2 + (2-1)^2} \\ &= \sqrt{(-4)^2 + 1^2} \\ &= \sqrt{16+1} = \sqrt{17} \end{aligned}$$

119. $(4.7, 2.3)$ and $(1.7, -1.7)$

$$\begin{aligned} d &= \sqrt{(4.7-1.7)^2 + [2.3-(-1.7)]^2} \\ &= \sqrt{3^2+4^2} = \sqrt{9+16} \\ &= \sqrt{25} = 5 \end{aligned}$$

121. $(\sqrt{2}, \sqrt{6})$ and $(-2\sqrt{2}, 4\sqrt{6})$

$$\begin{aligned} d &= \sqrt{\left[\sqrt{2}-\left(-2\sqrt{2}\right)\right]^2 + \left(\sqrt{6}-4\sqrt{6}\right)^2} \\ &= \sqrt{(3\sqrt{2})^2 + (-3\sqrt{6})^2} \\ &= \sqrt{9 \cdot 2 + 9 \cdot 6} = \sqrt{18+54} \\ &= \sqrt{72} = \sqrt{36} \cdot \sqrt{2} = 6\sqrt{2} \end{aligned}$$

123. $(x+y, y)$ and $(x-y, x)$

$$\begin{aligned} d &= \sqrt{[(x+y)-(x-y)]^2 + (y-x)^2} \\ &= \sqrt{(2y)^2 + (y-x)^2} \\ &= \sqrt{4y^2 + y^2 - 2xy + x^2} \\ &= \sqrt{5y^2 - 2xy + x^2} \end{aligned}$$

125. To find the lengths of the three sides of the triangle, use the distance formula to find the distance between each pair of points. Then add the distances to find the perimeter.

$$\begin{aligned} P &= \sqrt{(-3-2)^2 + (-3-6)^2} \\ &\quad + \sqrt{(2-6)^2 + (6-2)^2} \\ &\quad + \sqrt{[6-(-3)]^2 + [2-(-3)]^2} \\ &= \sqrt{(-5)^2 + (-9)^2} + \sqrt{(-4)^2 + 4^2} \\ &\quad + \sqrt{9^2 + 5^2} \\ &= \sqrt{25+81} + \sqrt{16+16} + \sqrt{81+25} \\ &= \sqrt{106} + \sqrt{32} + \sqrt{106} \\ &= 2\sqrt{106} + \sqrt{16} \cdot \sqrt{2} \\ &= 2\sqrt{106} + 4\sqrt{2} \end{aligned}$$

127.
$$\begin{aligned} d &= 1.224\sqrt{h} \\ &= 1.224\sqrt{156} \approx 15.3 \text{ miles} \end{aligned}$$

129. Substitute 21.7 for a and 16 for b in the Pythagorean theorem to find c.

$$\begin{aligned} c^2 &= a^2 + b^2 = (21.7)^2 + 16^2 \\ c &= \sqrt{726.89} \approx 26.96 \end{aligned}$$

To the nearest tenth of an inch, the length of the diagonal of the screen is 27.0 inches.

131. Use $f_1 = f_2\sqrt{\frac{F_1}{F_2}}$ with $F_1 = 300$, $F_2 = 60$, and $f_2 = 260$ to find f_1.

$$f_1 = 260\sqrt{\tfrac{300}{60}} = 260\sqrt{5} \approx 581$$

133. $\sqrt{9} + \sqrt{9} = 3 + 3 = 6$, and $\sqrt{4} = 2$; $6 \neq 2$, so the statement is false.

135. For the model MSX-77, we have a right triangle with legs x and x, and hypotenuse 26. Substitute x for a, x for b, and 26 for c in the Pythagorean theorem, then solve for x.

$$\begin{aligned} a^2 + b^2 &= c^2 \\ x^2 + x^2 &= 26^2 \\ 2x^2 &= 676 \\ x^2 &= 338 \\ x &= \sqrt{338} \approx 18.4 \end{aligned}$$

The length of the leg is about 18.4 inches. Similarly, for the model MSX-83 with $c = 24$, we get $\sqrt{288} \approx 17.0$ inches, and for the model MSX-60 with $c = 20$, we get $\sqrt{200} \approx 14.1$ inches.

137.
$$\begin{aligned} &13x^4 - 12x^3 + 9x^4 + 2x^3 \\ &= 13x^4 + 9x^4 - 12x^3 + 2x^3 \\ &= 22x^4 - 10x^3 \end{aligned}$$

139.
$$\begin{aligned} &9q^2 + 2q - 5q - q^2 \\ &= 9q^2 - q^2 + 2q - 5q \\ &= 8q^2 - 3q \end{aligned}$$

10.4 Adding and Subtracting Radical Expressions

10.4 Now Try Exercises

N1. **(a)**
$$\begin{aligned} &\sqrt{12} + \sqrt{75} \\ &= \sqrt{4} \cdot \sqrt{3} + \sqrt{25} \cdot \sqrt{3} \\ &= 2\sqrt{3} + 5\sqrt{3} \\ &= (2+5)\sqrt{3} \\ &= 7\sqrt{3} \end{aligned}$$

(b)
$$\begin{aligned} &-\sqrt{63t} + 3\sqrt{28t},\ t \geq 0 \\ &= -\sqrt{9} \cdot \sqrt{7t} + 3\sqrt{4} \cdot \sqrt{7t} \\ &= -3\sqrt{7t} + 3 \cdot 2 \cdot \sqrt{7t} \\ &= -3\sqrt{7t} + 6\sqrt{7t} \\ &= (-3+6)\sqrt{7t} \\ &= 3\sqrt{7t} \end{aligned}$$

(c) $6\sqrt{7} - 2\sqrt{3}$

Here the radicals differ and are already simplified, so this expression *cannot be simplified further.*

N2. **(a)** $3\sqrt[3]{2000} - 4\sqrt[3]{128}$
$= 3\sqrt[3]{1000 \cdot 2} - 4\sqrt[3]{64 \cdot 2}$
$= 3\sqrt[3]{1000}\sqrt[3]{2} - 4\sqrt[3]{64}\sqrt[3]{2}$
$= 3 \cdot 10\sqrt[3]{2} - 4 \cdot 4\sqrt[3]{2}$
$= 30\sqrt[3]{2} - 16\sqrt[3]{2}$
$= (30 - 16)\sqrt[3]{2}$
$= 14\sqrt[3]{2}$

(b) $5\sqrt[4]{a^5b^3} + \sqrt[4]{81ab^7}$
$= 5\sqrt[4]{a^4 \cdot ab^3} + \sqrt[4]{81b^4 \cdot ab^3}$
$= 5\sqrt[4]{a^4} \cdot \sqrt[4]{ab^3} + \sqrt[4]{81b^4} \cdot \sqrt[4]{ab^3}$
$= 5a\sqrt[4]{ab^3} + 3b\sqrt[4]{ab^3}$
$= (5a + 3b)\sqrt[4]{ab^3}$

(c) $\sqrt[3]{128t^4} - 2\sqrt{72t^3}$
$= \sqrt[3]{64t^3 \cdot 2t} - 2\sqrt{36t^2 \cdot 2t}$
$= \sqrt[3]{64t^3} \cdot \sqrt[3]{2t} - 2\sqrt{36t^2} \cdot \sqrt{2t}$
$= 4t\sqrt[3]{2t} - 2 \cdot 6t\sqrt{2t}$
$= 4t\sqrt[3]{2t} - 12t\sqrt{2t}$

N3. **(a)** $5\frac{\sqrt{5}}{\sqrt{45}} - 4\sqrt{\frac{28}{9}}$
$= 5\frac{\sqrt{5}}{\sqrt{9 \cdot 5}} - 4\frac{\sqrt{4 \cdot 7}}{\sqrt{9}}$
$= 5\left(\frac{\sqrt{5}}{3\sqrt{5}}\right) - 4\left(\frac{2\sqrt{7}}{3}\right)$
$= \frac{5}{3} - \frac{8\sqrt{7}}{3}$
$= \frac{5 - 8\sqrt{7}}{3}$

(b) $6\sqrt[3]{\frac{16}{x^{12}}} + 7\sqrt[3]{\frac{9}{x^9}}$
$= 6\frac{\sqrt[3]{8 \cdot 2}}{\sqrt[3]{x^{12}}} + 7\frac{\sqrt[3]{9}}{\sqrt[3]{x^9}}$
$= \frac{6\sqrt[3]{8}\sqrt[3]{2}}{x^4} + \frac{7\sqrt[3]{9}}{x^3}$
$= \frac{6 \cdot 2\sqrt[3]{2}}{x^4} + \frac{7x\sqrt[3]{9}}{x \cdot x^3}$
$= \frac{12\sqrt[3]{2} + 7x\sqrt[3]{9}}{x^4}$

10.4 Section Exercises

1. Simplify each radical and subtract.

$$\sqrt{36} - \sqrt{100} = 6 - 10 = -4$$

3. $-2\sqrt{48} + 3\sqrt{75}$
$= -2\sqrt{16 \cdot 3} + 3\sqrt{25 \cdot 3}$
$= -2 \cdot 4\sqrt{3} + 3 \cdot 5\sqrt{3}$
$= -8\sqrt{3} + 15\sqrt{3} = 7\sqrt{3}$

5. $\sqrt[3]{16} + 4\sqrt[3]{54}$
$= \sqrt[3]{8 \cdot 2} + 4\sqrt[3]{27 \cdot 2}$
$= \sqrt[3]{8}\sqrt[3]{2} + 4\sqrt[3]{27}\sqrt[3]{2}$
$= 2\sqrt[3]{2} + 4 \cdot 3\sqrt[3]{2}$
$= 2\sqrt[3]{2} + 12\sqrt[3]{2} = 14\sqrt[3]{2}$

7. $\sqrt[4]{32} + 3\sqrt[4]{2}$
$= \sqrt[4]{16 \cdot 2} + 3\sqrt[4]{2}$
$= \sqrt[4]{16}\sqrt[4]{2} + 3\sqrt[4]{2}$
$= 2\sqrt[4]{2} + 3\sqrt[4]{2} = 5\sqrt[4]{2}$

9. $6\sqrt{18} - \sqrt{32} + 2\sqrt{50}$
$= 6\sqrt{9 \cdot 2} - \sqrt{16 \cdot 2} + 2\sqrt{25 \cdot 2}$
$= 6 \cdot 3\sqrt{2} - 4\sqrt{2} + 2 \cdot 5\sqrt{2}$
$= 18\sqrt{2} - 4\sqrt{2} + 10\sqrt{2}$
$= 24\sqrt{2}$

11. $5\sqrt{6} + 2\sqrt{10}$

The radicals differ and are already simplified, so the expression cannot be simplified further.

13. $2\sqrt{5} + 3\sqrt{20} + 4\sqrt{45}$
$= 2\sqrt{5} + 3\sqrt{4 \cdot 5} + 4\sqrt{9 \cdot 5}$
$= 2\sqrt{5} + 3 \cdot 2\sqrt{5} + 4 \cdot 3\sqrt{5}$
$= 2\sqrt{5} + 6\sqrt{5} + 12\sqrt{5}$
$= 20\sqrt{5}$

15. $\sqrt{72x} - \sqrt{8x}$
$= \sqrt{36 \cdot 2x} - \sqrt{4 \cdot 2x}$
$= 6\sqrt{2x} - 2\sqrt{2x}$
$= 4\sqrt{2x}$

17. $3\sqrt{72m^2} - 5\sqrt{32m^2} - 3\sqrt{18m^2}$
$= 3\sqrt{36m^2 \cdot 2} - 5\sqrt{16m^2 \cdot 2} - 3\sqrt{9m^2 \cdot 2}$
$= 3 \cdot 6m\sqrt{2} - 5 \cdot 4m\sqrt{2} - 3 \cdot 3m\sqrt{2}$
$= 18m\sqrt{2} - 20m\sqrt{2} - 9m\sqrt{2}$
$= (18m - 20m - 9m)\sqrt{2} = -11m\sqrt{2}$

19. $2\sqrt[3]{16} + \sqrt[3]{54} = 2\sqrt[3]{8 \cdot 2} + \sqrt[3]{27 \cdot 2}$
$= 2 \cdot 2\sqrt[3]{2} + 3\sqrt[3]{2}$
$= 4\sqrt[3]{2} + 3\sqrt[3]{2}$
$= 7\sqrt[3]{2}$

21. $2\sqrt[3]{27x} - 2\sqrt[3]{8x}$
$= 2\sqrt[3]{27 \cdot x} - 2\sqrt[3]{8 \cdot x}$
$= 2 \cdot 3\sqrt[3]{x} - 2 \cdot 2\sqrt[3]{x}$
$= 6\sqrt[3]{x} - 4\sqrt[3]{x}$
$= 2\sqrt[3]{x}$

23. $3\sqrt[3]{x^2y} - 5\sqrt[3]{8x^2y}$
$= 3\sqrt[3]{x^2y} - 5\sqrt[3]{8}\sqrt[3]{x^2y}$
$= 3\sqrt[3]{x^2y} - 5\cdot 2\sqrt[3]{x^2y}$
$= (3-10)\sqrt[3]{x^2y}$
$= -7\sqrt[3]{x^2y}$

25. $3x\sqrt[3]{xy^2} - 2\sqrt[3]{8x^4y^2}$
$= 3x\sqrt[3]{xy^2} - 2\sqrt[3]{8x^3}\cdot\sqrt[3]{xy^2}$
$= 3x\sqrt[3]{xy^2} - 2\cdot 2x\cdot\sqrt[3]{xy^2}$
$= (3x-4x)\sqrt[3]{xy^2} = -x\sqrt[3]{xy^2}$

27. $5\sqrt[4]{32} + 3\sqrt[4]{162}$
$= 5\sqrt[4]{16\cdot 2} + 3\sqrt[4]{81\cdot 2}$
$= 5\cdot 2\sqrt[4]{2} + 3\cdot 3\sqrt[4]{2}$
$= 10\sqrt[4]{2} + 9\sqrt[4]{2}$
$= 19\sqrt[4]{2}$

29. $3\sqrt[4]{x^5y} - 2x\sqrt[4]{xy}$
$= 3\sqrt[4]{x^4\cdot xy} - 2x\sqrt[4]{xy}$
$= 3x\sqrt[4]{xy} - 2x\sqrt[4]{xy}$
$= (3x-2x)\sqrt[4]{xy} = x\sqrt[4]{xy}$

31. $2\sqrt[4]{32a^3} + 5\sqrt[4]{2a^3}$
$= 2\sqrt[4]{16}\cdot\sqrt[4]{2a^3} + 5\sqrt[4]{2a^3}$
$= 2\cdot 2\cdot\sqrt[4]{2a^3} + 5\sqrt[4]{2a^3}$
$= (4+5)\sqrt[4]{2a^3} = 9\sqrt[4]{2a^3}$

33. $\sqrt[3]{64xy^2} + \sqrt[3]{27x^4y^5}$
$= \sqrt[3]{64\cdot xy^2} + \sqrt[3]{27x^3y^3\cdot xy^2}$
$= 4\sqrt[3]{xy^2} + 3xy\sqrt[3]{xy^2}$
$= (4+3xy)\sqrt[3]{xy^2}$

35. $\sqrt[3]{192st^4} - \sqrt{27s^3t}$
$= \sqrt[3]{64t^3\cdot 3st} - \sqrt{9s^2\cdot 3st}$
$= 4t\sqrt[3]{3st} - 3s\sqrt{3st}$

37. $2\sqrt[3]{8x^4} + 3\sqrt[4]{16x^5}$
$= 2\sqrt[3]{8x^3\cdot x} + 3\sqrt[4]{16x^4\cdot x}$
$= 2\cdot 2x\sqrt[3]{x} + 3\cdot 2x\sqrt[4]{x}$
$= 4x\sqrt[3]{x} + 6x\sqrt[4]{x}$

39. $\sqrt{8} - \dfrac{\sqrt{64}}{\sqrt{16}} = \sqrt{4\cdot 2} - \dfrac{8}{4}$
$= \sqrt{4}\cdot\sqrt{2} - 2$
$= 2\sqrt{2} - 2$

41. $\dfrac{2\sqrt{5}}{3} + \dfrac{\sqrt{5}}{6} = \dfrac{4\sqrt{5}}{6} + \dfrac{1\sqrt{5}}{6}$
$= \dfrac{4\sqrt{5} + 1\sqrt{5}}{6}$
$= \dfrac{5\sqrt{5}}{6}$

43. $\sqrt{\dfrac{8}{9}} + \sqrt{\dfrac{18}{36}} = \dfrac{\sqrt{8}}{\sqrt{9}} + \dfrac{\sqrt{18}}{\sqrt{36}}$
$= \dfrac{\sqrt{4}\sqrt{2}}{3} + \dfrac{\sqrt{9}\sqrt{2}}{6}$
$= \dfrac{2\sqrt{2}}{3} + \dfrac{3\sqrt{2}}{6}$
$= \dfrac{4\sqrt{2}}{6} + \dfrac{3\sqrt{2}}{6}$
$= \dfrac{4\sqrt{2} + 3\sqrt{2}}{6} = \dfrac{7\sqrt{2}}{6}$

45. $\dfrac{\sqrt{32}}{3} + \dfrac{2\sqrt{2}}{3} - \dfrac{\sqrt{2}}{\sqrt{9}}$
$= \dfrac{\sqrt{16}\sqrt{2}}{3} + \dfrac{2\sqrt{2}}{3} - \dfrac{\sqrt{2}}{3}$
$= \dfrac{4\sqrt{2} + 2\sqrt{2} - \sqrt{2}}{3} = \dfrac{5\sqrt{2}}{3}$

47. $3\sqrt{\dfrac{50}{9}} + 8\dfrac{\sqrt{2}}{\sqrt{8}} = 3\dfrac{\sqrt{50}}{\sqrt{9}} + 8\dfrac{\sqrt{2}}{2\sqrt{2}}$
$= 3\cdot\dfrac{5\sqrt{2}}{3} + 8\cdot\dfrac{1}{2}$
$= 5\sqrt{2} + 4$

49. $\sqrt{\dfrac{25}{x^8}} + \sqrt{\dfrac{9}{x^6}} = \dfrac{\sqrt{25}}{\sqrt{x^8}} + \dfrac{\sqrt{9}}{\sqrt{x^6}}$
$= \dfrac{5}{x^4} + \dfrac{3}{x^3}$
$= \dfrac{5}{x^4} + \dfrac{3\cdot x}{x^3\cdot x}$ $\quad LCD = x^4$
$= \dfrac{5+3x}{x^4}$

51. $3\sqrt[3]{\dfrac{m^5}{27}} - 2m\sqrt[3]{\dfrac{m^2}{64}}$
$= \dfrac{3\sqrt[3]{m^5}}{\sqrt[3]{27}} - \dfrac{2m\sqrt[3]{m^2}}{\sqrt[3]{64}}$
$= \dfrac{3\sqrt[3]{m^3}\sqrt[3]{m^2}}{3} - \dfrac{2m\sqrt[3]{m^2}}{4}$
$= \dfrac{m\sqrt[3]{m^2}}{1} - \dfrac{m\sqrt[3]{m^2}}{2}$
$= \dfrac{2m\sqrt[3]{m^2} - m\sqrt[3]{m^2}}{2} = \dfrac{m\sqrt[3]{m^2}}{2}$

53. $3\sqrt[3]{\dfrac{2}{x^6}} - 4\sqrt[3]{\dfrac{5}{x^9}} = 3\dfrac{\sqrt[3]{2}}{\sqrt[3]{x^6}} - 4\dfrac{\sqrt[3]{5}}{\sqrt[3]{x^9}}$
$= 3\dfrac{\sqrt[3]{2}}{x^2} - 4\dfrac{\sqrt[3]{5}}{x^3}$
$= \dfrac{3\cdot x\cdot\sqrt[3]{2}}{x^2\cdot x} - \dfrac{4\sqrt[3]{5}}{x^3}$
$LCD = x^3$
$= \dfrac{3x\sqrt[3]{2} - 4\sqrt[3]{5}}{x^3}$

55. Only choice **B** has like radical terms, so it can be simplified without first simplifying the individual radical expressions.

$$3\sqrt{6}+9\sqrt{6}=12\sqrt{6}$$

57. $$\begin{aligned}\sqrt{64}+\sqrt[3]{125}+\sqrt[4]{16}&=\sqrt{8^2}+\sqrt[3]{5^3}+\sqrt[4]{2^4}\\&=8+5+2=15\end{aligned}$$

This sum can be found easily since each radicand has a whole number power corresponding to the index of the radical; that is, each radical expression simplifies to a whole number.

59. Let $L=\sqrt{192}\approx\sqrt{196}=14$ and $W=\sqrt{48}\approx\sqrt{49}=7$. An estimate of the perimeter is $2L+2W=2(14)+2(7)=42$ meters. **(A)**

61. The perimeter, P, of a triangle is the sum of the measures of the sides.

$$\begin{aligned}P&=3\sqrt{20}+2\sqrt{45}+\sqrt{75}\\&=3\sqrt{4\cdot5}+2\sqrt{9\cdot5}+\sqrt{25\cdot3}\\&=3\cdot2\sqrt{5}+2\cdot3\sqrt{5}+5\sqrt{3}\\&=6\sqrt{5}+6\sqrt{5}+5\sqrt{3}\\&=12\sqrt{5}+5\sqrt{3}\end{aligned}$$

The perimeter is $(12\sqrt{5}+5\sqrt{3})$ inches.

63. To find the perimeter, add the lengths of the sides.

$$\begin{aligned}&4\sqrt{18}+\sqrt{108}+2\sqrt{72}+3\sqrt{12}\\&=4\sqrt{9}\sqrt{2}+\sqrt{36}\sqrt{3}+2\sqrt{36}\sqrt{2}+3\sqrt{4}\sqrt{3}\\&=4\cdot3\sqrt{2}+6\sqrt{3}+2\cdot6\sqrt{2}+3\cdot2\sqrt{3}\\&=12\sqrt{2}+6\sqrt{3}+12\sqrt{2}+6\sqrt{3}\\&=24\sqrt{2}+12\sqrt{3}\end{aligned}$$

The perimeter is $(24\sqrt{2}+12\sqrt{3})$ inches.

65. $$\begin{aligned}5xy(2x^2y^3-4x)&=5xy(2x^2y^3)-5xy(4x)\\&=10x^3y^4-20x^2y\end{aligned}$$

67. $$\begin{aligned}(a^2+b)(a^2-b)&=(a^2)^2-b^2\\&=a^4-b^2\end{aligned}$$

69. $$\begin{aligned}(4x^3+3)^2&=(4x^3)^2+2\cdot4x^3\cdot3+3^2\\&=16x^6+24x^3+9\end{aligned}$$

Now multiply by $4x^3+3$.

$$\begin{array}{rrrr} & 16x^6 + & 24x^3 + & 9\\ & & 4x^3 + & 3\\ \hline & 48x^6 + & 72x^3 + & 27\\ 64x^9 + & 96x^6 + & 36x^3 & \\ \hline 64x^9 + & 144x^6 + & 108x^3 + & 27\end{array}$$

Thus, $(4x^3+3)^3=64x^9+144x^6+108x^3+27$.

71. $$\begin{aligned}\frac{8x^2-10x}{6x^2}&=\frac{2x(4x-5)}{2x\cdot3x}\\&=\frac{4x-5}{3x}\end{aligned}$$

10.5 Multiplying and Dividing Radical Expressions

10.5 Now Try Exercises

N1. Use the FOIL method and multiply as two binomials.

(a) $(8-\sqrt{5})(9-\sqrt{2})$

F O I L

$$\begin{aligned}&=8\cdot9-8\sqrt{2}-9\sqrt{5}+\sqrt{5}\cdot\sqrt{2}\\&=72-8\sqrt{2}-9\sqrt{5}+\sqrt{10}\end{aligned}$$

(b) $(\sqrt{7}+\sqrt{5})(\sqrt{7}-\sqrt{5})$

$$\begin{aligned}&=\sqrt{7}\cdot\sqrt{7}-\sqrt{7}\cdot\sqrt{5}\\&\quad+\sqrt{5}\cdot\sqrt{7}-\sqrt{5}\cdot\sqrt{5}\\&=7-5\\&=2\end{aligned}$$

(c) $(\sqrt{15}-4)^2$

$$\begin{aligned}&=(\sqrt{15}-4)(\sqrt{15}-4)\\&=\sqrt{15}\cdot\sqrt{15}-4\sqrt{15}-4\sqrt{15}+4\cdot4\\&=15-8\sqrt{15}+16\\&=31-8\sqrt{15}\end{aligned}$$

(d) $(8+\sqrt[3]{5})(8-\sqrt[3]{5})$

$$\begin{aligned}&=8\cdot8-8\sqrt[3]{5}+8\sqrt[3]{5}-\sqrt[3]{5}\cdot\sqrt[3]{5}\\&=64-\sqrt[3]{25}\end{aligned}$$

(e) For this one, we use the difference of squares.

$$\begin{aligned}&(\sqrt{m}-\sqrt{n})(\sqrt{m}+\sqrt{n})\\&=(\sqrt{m})^2-(\sqrt{n})^2\\&=m-n,\quad m\ge0\text{ and }n\ge0\end{aligned}$$

N2. **(a)** $$\frac{8}{\sqrt{13}}=\frac{8\cdot\sqrt{13}}{\sqrt{13}\cdot\sqrt{13}}=\frac{8\sqrt{13}}{13}$$

(b) $$\frac{9\sqrt{7}}{\sqrt{3}}=\frac{9\sqrt{7}\cdot\sqrt{3}}{\sqrt{3}\cdot\sqrt{3}}=\frac{9\sqrt{21}}{3}=3\sqrt{21}$$

(c) $$\begin{aligned}\frac{-10}{\sqrt{20}}&=\frac{-10}{\sqrt{4\cdot5}}=\frac{-10}{2\sqrt{5}}=\frac{-10\cdot\sqrt{5}}{2\sqrt{5}\cdot\sqrt{5}}\\&=\frac{-10\sqrt{5}}{2\cdot5}=\frac{-10\sqrt{5}}{10}=-\sqrt{5}\end{aligned}$$

N3. **(a)** $$\begin{aligned}-\sqrt{\frac{27}{80}}&=-\frac{\sqrt{27}}{\sqrt{80}}=-\frac{\sqrt{9\cdot3}}{\sqrt{16\cdot5}}\\&=-\frac{3\sqrt{3}}{4\sqrt{5}}=-\frac{3\sqrt{3}\cdot\sqrt{5}}{4\sqrt{5}\cdot\sqrt{5}}\\&=-\frac{3\sqrt{15}}{4\cdot5}=-\frac{3\sqrt{15}}{20}\end{aligned}$$

(b) $\sqrt{\frac{48x^8}{y^3}} = \frac{\sqrt{16x^8 \cdot 3}}{\sqrt{y^2 \cdot y}}$

$= \frac{4x^4\sqrt{3}}{y\sqrt{y}}$

$= \frac{4x^4\sqrt{3} \cdot \sqrt{y}}{y\sqrt{y} \cdot \sqrt{y}}$

$= \frac{4x^4\sqrt{3y}}{y \cdot y}$

$= \frac{4x^4\sqrt{3y}}{y^2} \quad (y > 0)$

N4. (a) $\sqrt[3]{\frac{8}{81}} = \frac{\sqrt[3]{8}}{\sqrt[3]{81}} = \frac{2}{\sqrt[3]{27} \cdot \sqrt[3]{3}} = \frac{2}{3\sqrt[3]{3}}$

$= \frac{2 \cdot \sqrt[3]{9}}{3\sqrt[3]{3} \cdot \sqrt[3]{9}} = \frac{2\sqrt[3]{9}}{3\sqrt[3]{27}}$

$= \frac{2\sqrt[3]{9}}{3 \cdot 3} = \frac{2\sqrt[3]{9}}{9}$

(b) $\sqrt[4]{\frac{7x}{y}} = \frac{\sqrt[4]{7x}}{\sqrt[4]{y}} \cdot \frac{\sqrt[4]{y^3}}{\sqrt[4]{y^3}}$

$= \frac{\sqrt[4]{7xy^3}}{\sqrt[4]{y^4}}$

$= \frac{\sqrt[4]{7xy^3}}{y} \quad (x \geq 0, y > 0)$

N5. (a) $\frac{4}{1+\sqrt{3}}$

Multiply both the numerator and denominator by the conjugate of the denominator, $1 - \sqrt{3}$.

$= \frac{4(1-\sqrt{3})}{(1+\sqrt{3})(1-\sqrt{3})}$

$= \frac{4(1-\sqrt{3})}{1-3} = \frac{4(1-\sqrt{3})}{-2}$

$= -2(1-\sqrt{3})$, or $-2+2\sqrt{3}$

(b) $\frac{4}{5+\sqrt{7}}$

Multiply both the numerator and denominator by the conjugate of the denominator, $5 - \sqrt{7}$.

$= \frac{4(5-\sqrt{7})}{(5+\sqrt{7})(5-\sqrt{7})}$

$= \frac{4(5-\sqrt{7})}{25-7}$

$= \frac{4(5-\sqrt{7})}{18}$

$= \frac{2(5-\sqrt{7})}{9}$

(c) $\frac{\sqrt{3}+\sqrt{7}}{\sqrt{5}-\sqrt{2}}$

$= \frac{(\sqrt{3}+\sqrt{7})(\sqrt{5}+\sqrt{2})}{(\sqrt{5}-\sqrt{2})(\sqrt{5}+\sqrt{2})}$

$= \frac{\sqrt{15}+\sqrt{6}+\sqrt{35}+\sqrt{14}}{5-2}$

$= \frac{\sqrt{15}+\sqrt{6}+\sqrt{35}+\sqrt{14}}{3}$

(d) $\frac{8}{\sqrt{3x}-\sqrt{y}}$

$= \frac{8\left(\sqrt{3x}+\sqrt{y}\right)}{\left(\sqrt{3x}-\sqrt{y}\right)\left(\sqrt{3x}+\sqrt{y}\right)}$

$= \frac{8\left(\sqrt{3x}+\sqrt{y}\right)}{3x-y}$

$(3x \neq y,\ x > 0, y > 0)$

N6. (a) $\frac{15-6\sqrt{2}}{18} = \frac{3(5-2\sqrt{2})}{18} = \frac{5-2\sqrt{2}}{6}$

(b) $\frac{15k+\sqrt{50k^2}}{20k} = \frac{15k+\sqrt{25k^2 \cdot 2}}{20k}$

$= \frac{15k+5k\sqrt{2}}{20k} \quad (k > 0)$

$= \frac{5k(3+\sqrt{2})}{5k \cdot 4}$

$= \frac{3+\sqrt{2}}{4}$

10.5 Section Exercises

1. $\left(A+\sqrt{B}\right)\left(A-\sqrt{B}\right)$

$= A^2 - \left(\sqrt{B}\right)^2$

$= A^2 - B$ **(E)**

3. $\left(\sqrt{A}+\sqrt{B}\right)\left(\sqrt{A}-\sqrt{B}\right)$

$= \left(\sqrt{A}\right)^2 - \left(\sqrt{B}\right)^2$

$= A - B$ **(A)**

5. $\left(\sqrt{A}-\sqrt{B}\right)^2$

$= \left(\sqrt{A}\right)^2 - 2\sqrt{A}\sqrt{B} + \left(\sqrt{B}\right)^2$

$= A - 2\sqrt{AB} + B$ **(D)**

7. $\sqrt{6}(3+\sqrt{2}) = 3\sqrt{6}+\sqrt{6} \cdot \sqrt{2}$

$= 3\sqrt{6}+\sqrt{12}$

$= 3\sqrt{6}+\sqrt{4} \cdot \sqrt{3}$

$= 3\sqrt{6}+2\sqrt{3}$

9. $5(\sqrt{72}-\sqrt{8}) = 5\sqrt{72}-5\sqrt{8}$
$= 5\sqrt{36}\cdot\sqrt{2}-5\sqrt{4}\cdot\sqrt{2}$
$= 5\cdot 6\cdot\sqrt{2}-5\cdot 2\cdot\sqrt{2}$
$= 30\sqrt{2}-10\sqrt{2}$
$= 20\sqrt{2}$

11. $(\sqrt{7}+3)(\sqrt{7}-3) = (\sqrt{7})^2-3^2$
$= 7-9 = -2$

13. $(\sqrt{2}-\sqrt{3})(\sqrt{2}+\sqrt{3})$
$= (\sqrt{2})^2-(\sqrt{3})^2$
$= 2-3 = -1$

15. $(\sqrt{8}-\sqrt{2})(\sqrt{8}+\sqrt{2})$
$= (\sqrt{8})^2-(\sqrt{2})^2$
$= 8-2 = 6$

17. $(\sqrt{2}+1)(\sqrt{3}-1)$
F O I L
$= \sqrt{2}\cdot\sqrt{3}-1\sqrt{2}+1\sqrt{3}-1\cdot 1$
$= \sqrt{6}-\sqrt{2}+\sqrt{3}-1$

19. $(\sqrt{11}-\sqrt{7})(\sqrt{2}+\sqrt{5})$
F O I L
$= \sqrt{11}\cdot\sqrt{2}+\sqrt{11}\cdot\sqrt{5}-\sqrt{7}\cdot\sqrt{2}-\sqrt{7}\cdot\sqrt{5}$
$= \sqrt{22}+\sqrt{55}-\sqrt{14}-\sqrt{35}$

21. $(2\sqrt{3}+\sqrt{5})(3\sqrt{3}-2\sqrt{5})$
$= (2\sqrt{3})(3\sqrt{3})+(2\sqrt{3})(-2\sqrt{5})$
$+ (\sqrt{5})(3\sqrt{3})+(\sqrt{5})(-2\sqrt{5})$
$= 2\cdot 3\cdot 3-2\cdot 2\sqrt{3\cdot 5}+3\sqrt{5\cdot 3}-2\cdot 5$
$= 18-4\sqrt{15}+3\sqrt{15}-10$
$= 8-\sqrt{15}$

23. $(\sqrt{5}+2)^2 = (\sqrt{5})^2+2\cdot\sqrt{5}\cdot 2+2^2$
$= 5+4\sqrt{5}+4$
$= 9+4\sqrt{5}$

25. $(\sqrt{21}-\sqrt{5})^2$
$= (\sqrt{21})^2-2\cdot\sqrt{21}\cdot\sqrt{5}+(\sqrt{5})^2$
$= 21-2\sqrt{105}+5$
$= 26-2\sqrt{105}$

27. $(2+\sqrt[3]{6})(2-\sqrt[3]{6}) = 2^2-(\sqrt[3]{6})^2$
$= 4-\sqrt[3]{6}\cdot\sqrt[3]{6}$
$= 4-\sqrt[3]{36}$

29. $(2+\sqrt[3]{2})(4-2\sqrt[3]{2}+\sqrt[3]{4})$
$= 2\cdot 4-2\cdot 2\sqrt[3]{2}+2\sqrt[3]{4}$
$+ 4\sqrt[3]{2}-2\sqrt[3]{2}\cdot\sqrt[3]{2}+\sqrt[3]{2}\cdot\sqrt[3]{4}$
$= 8-4\sqrt[3]{2}+2\sqrt[3]{4}+4\sqrt[3]{2}-2\sqrt[3]{4}+\sqrt[3]{8}$
$= 8+2$
$= 10$

31. $(3\sqrt{x}-\sqrt{5})(2\sqrt{x}+1)$
$= (3\sqrt{x})(2\sqrt{x})+3\sqrt{x}-2\sqrt{5}\sqrt{x}-\sqrt{5}$
$= 6x+3\sqrt{x}-2\sqrt{5x}-\sqrt{5}$

33. $(3\sqrt{r}-\sqrt{s})(3\sqrt{r}+\sqrt{s})$
$= (3\sqrt{r})^2-(\sqrt{s})^2$
$= 9r-s$

35. $(\sqrt[3]{2y}-5)(4\sqrt[3]{2y}+1)$
$= \sqrt[3]{2y}\cdot 4\sqrt[3]{2y}+\sqrt[3]{2y}-5\cdot 4\sqrt[3]{2y}-5$
$= 4\sqrt[3]{4y^2}+\sqrt[3]{2y}-20\sqrt[3]{2y}-5$
$= 4\sqrt[3]{4y^2}-19\sqrt[3]{2y}-5$

37. $(\sqrt{3x}+2)(\sqrt{3x}-2) = (\sqrt{3x})^2-2^2$
$= 3x-4$

39. $(2\sqrt{x}+\sqrt{y})(2\sqrt{x}-\sqrt{y})$
$= (2\sqrt{x})^2-(\sqrt{y})^2$
$= 2^2(\sqrt{x})^2-y$
$= 4x-y$

41. $[(\sqrt{2}+\sqrt{3})-\sqrt{6}][(\sqrt{2}+\sqrt{3})+\sqrt{6}]$
$= (\sqrt{2}+\sqrt{3})^2-(\sqrt{6})^2$
$= [(\sqrt{2})^2+2\sqrt{2}\sqrt{3}+(\sqrt{3})^2]-6$
$= (2+2\sqrt{6}+3)-6$
$= 2\sqrt{6}-1$

43. $\dfrac{7}{\sqrt{7}} = \dfrac{7\cdot\sqrt{7}}{\sqrt{7}\cdot\sqrt{7}} = \dfrac{7\sqrt{7}}{7} = \sqrt{7}$

45. $\dfrac{15}{\sqrt{3}} = \dfrac{15\cdot\sqrt{3}}{\sqrt{3}\cdot\sqrt{3}} = \dfrac{15\sqrt{3}}{3} = 5\sqrt{3}$

47. $\dfrac{\sqrt{3}}{\sqrt{2}} = \dfrac{\sqrt{3}\cdot\sqrt{2}}{\sqrt{2}\cdot\sqrt{2}} = \dfrac{\sqrt{6}}{2}$

49. $\dfrac{9\sqrt{3}}{\sqrt{5}} = \dfrac{9\sqrt{3}\cdot\sqrt{5}}{\sqrt{5}\cdot\sqrt{5}} = \dfrac{9\sqrt{15}}{5}$

51. $\dfrac{-7}{\sqrt{48}} = \dfrac{-7}{4\sqrt{3}} = \dfrac{-7\cdot\sqrt{3}}{4\sqrt{3}\cdot\sqrt{3}}$
$= \dfrac{-7\sqrt{3}}{4\cdot 3} = -\dfrac{7\sqrt{3}}{12}$

53. $\sqrt{\dfrac{7}{2}} = \dfrac{\sqrt{7}}{\sqrt{2}} = \dfrac{\sqrt{7}\cdot\sqrt{2}}{\sqrt{2}\cdot\sqrt{2}} = \dfrac{\sqrt{14}}{2}$

55. $-\sqrt{\dfrac{7}{50}} = -\dfrac{\sqrt{7}}{\sqrt{50}} = -\dfrac{\sqrt{7}}{\sqrt{25\cdot 2}} = -\dfrac{\sqrt{7}}{5\sqrt{2}}$
$= -\dfrac{\sqrt{7}\cdot\sqrt{2}}{5\sqrt{2}\cdot\sqrt{2}} = -\dfrac{\sqrt{14}}{5\cdot 2} = -\dfrac{\sqrt{14}}{10}$

57. $\sqrt{\dfrac{24}{x}} = \dfrac{\sqrt{24}}{\sqrt{x}} = \dfrac{\sqrt{4\cdot 6}}{\sqrt{x}} = \dfrac{2\sqrt{6}}{\sqrt{x}}$
$= \dfrac{2\sqrt{6}\cdot\sqrt{x}}{\sqrt{x}\cdot\sqrt{x}} = \dfrac{2\sqrt{6x}}{x}$

59. $\dfrac{-8\sqrt{3}}{\sqrt{k}}=\dfrac{-8\sqrt{3}\cdot\sqrt{k}}{\sqrt{k}\cdot\sqrt{k}}=\dfrac{-8\sqrt{3k}}{k}$

61. $-\sqrt{\dfrac{150m^5}{n^3}}$

$=\dfrac{-\sqrt{150m^5}}{\sqrt{n^3}}=\dfrac{-\sqrt{25m^4\cdot 6m}}{\sqrt{n^2\cdot n}}$

$=\dfrac{-5m^2\sqrt{6m}}{n\sqrt{n}}=\dfrac{-5m^2\sqrt{6m}\cdot\sqrt{n}}{n\sqrt{n}\cdot\sqrt{n}}$

$=\dfrac{-5m^2\sqrt{6mn}}{n\cdot n}=\dfrac{-5m^2\sqrt{6mn}}{n^2}$

63. $\sqrt{\dfrac{288x^7}{y^9}}=\dfrac{\sqrt{288x^7}}{\sqrt{y^9}}=\dfrac{\sqrt{144x^6\cdot 2x}}{\sqrt{y^8\cdot y}}$

$=\dfrac{12x^3\sqrt{2x}}{y^4\sqrt{y}}=\dfrac{12x^3\sqrt{2x}\cdot\sqrt{y}}{y^4\sqrt{y}\cdot\sqrt{y}}$

$=\dfrac{12x^3\sqrt{2xy}}{y^4\cdot y}=\dfrac{12x^3\sqrt{2xy}}{y^5}$

65. $\dfrac{5\sqrt{2m}}{\sqrt{y^3}}=\dfrac{5\sqrt{2m}}{\sqrt{y^3}}\cdot\dfrac{\sqrt{y}}{\sqrt{y}}$

$=\dfrac{5\sqrt{2my}}{\sqrt{y^4}}$

$=\dfrac{5\sqrt{2my}}{y^2}$

67. $-\sqrt{\dfrac{48k^2}{z}}=-\dfrac{\sqrt{16k^2}\cdot\sqrt{3}}{\sqrt{z}}\cdot\dfrac{\sqrt{z}}{\sqrt{z}}$

$=-\dfrac{4k\sqrt{3z}}{z}$

69. $\sqrt[3]{\dfrac{2}{3}}=\dfrac{\sqrt[3]{2}}{\sqrt[3]{3}}=\dfrac{\sqrt[3]{2}\cdot\sqrt[3]{9}}{\sqrt[3]{3}\cdot\sqrt[3]{9}}=\dfrac{\sqrt[3]{18}}{\sqrt[3]{27}}=\dfrac{\sqrt[3]{18}}{3}$

71. $\sqrt[3]{\dfrac{4}{9}}=\dfrac{\sqrt[3]{4}}{\sqrt[3]{9}}=\dfrac{\sqrt[3]{4}}{\sqrt[3]{3^2}}=\dfrac{\sqrt[3]{4}\cdot\sqrt[3]{3}}{\sqrt[3]{3^2}\cdot\sqrt[3]{3}}$

$=\dfrac{\sqrt[3]{12}}{\sqrt[3]{3^3}}=\dfrac{\sqrt[3]{12}}{3}$

73. $\sqrt[3]{\dfrac{9}{32}}=\dfrac{\sqrt[3]{9}}{\sqrt[3]{32}}=\dfrac{\sqrt[3]{9}}{\sqrt[3]{8}\cdot\sqrt[3]{4}}\cdot\dfrac{\sqrt[3]{2}}{\sqrt[3]{2}}$

$=\dfrac{\sqrt[3]{9\cdot 2}}{2\cdot\sqrt[3]{8}}$

$=\dfrac{\sqrt[3]{18}}{4}$

75. $-\sqrt[3]{\dfrac{2p}{r^2}}=-\dfrac{\sqrt[3]{2p}}{\sqrt[3]{r^2}}=-\dfrac{\sqrt[3]{2p}\cdot\sqrt[3]{r}}{\sqrt[3]{r^2}\cdot\sqrt[3]{r}}$

$=-\dfrac{\sqrt[3]{2pr}}{\sqrt[3]{r^3}}=-\dfrac{\sqrt[3]{2pr}}{r}$

77. $\sqrt[3]{\dfrac{x^6}{y}}=\dfrac{\sqrt[3]{x^6}}{\sqrt[3]{y}}=\dfrac{x^2}{\sqrt[3]{y}}\cdot\dfrac{\sqrt[3]{y^2}}{\sqrt[3]{y^2}}$

$=\dfrac{x^2\sqrt[3]{y^2}}{\sqrt[3]{y^3}}$

$=\dfrac{x^2\sqrt[3]{y^2}}{y}$

79. $\sqrt[4]{\dfrac{16}{x}}=\dfrac{\sqrt[4]{16}}{\sqrt[4]{x}}=\dfrac{2}{\sqrt[4]{x}}=\dfrac{2\cdot\sqrt[4]{x^3}}{\sqrt[4]{x}\cdot\sqrt[4]{x^3}}$

$=\dfrac{2\sqrt[4]{x^3}}{\sqrt[4]{x^4}}=\dfrac{2\sqrt[4]{x^3}}{x}$

81. $\sqrt[4]{\dfrac{2y}{z}}=\dfrac{\sqrt[4]{2y}}{\sqrt[4]{z}}\cdot\dfrac{\sqrt[4]{z^3}}{\sqrt[4]{z^3}}$

$=\dfrac{\sqrt[4]{2yz^3}}{z}$

83. $\dfrac{3}{4+\sqrt{5}}$

Multiply both the numerator and denominator by the conjugate of the denominator, $4-\sqrt{5}$.

$=\dfrac{3(4-\sqrt{5})}{(4+\sqrt{5})(4-\sqrt{5})}$

$=\dfrac{3(4-\sqrt{5})}{16-5}=\dfrac{3(4-\sqrt{5})}{11}$

85. $\dfrac{\sqrt{8}}{3-\sqrt{2}}$

Multiply both the numerator and denominator by the conjugate of the denominator, $3+\sqrt{2}$.

$=\dfrac{\sqrt{4}\sqrt{2}(3+\sqrt{2})}{(3-\sqrt{2})(3+\sqrt{2})}$

$=\dfrac{2\sqrt{2}(3+\sqrt{2})}{3^2-2}$

$=\dfrac{2\cdot 3\sqrt{2}+2\cdot 2}{7}$

$=\dfrac{6\sqrt{2}+4}{7}$

87. $\dfrac{2}{3\sqrt{5}+2\sqrt{3}}$

Multiply both the numerator and denominator by the conjugate of the denominator, $3\sqrt{5}-2\sqrt{3}$.

$=\dfrac{2(3\sqrt{5}-2\sqrt{3})}{(3\sqrt{5}+2\sqrt{3})(3\sqrt{5}-2\sqrt{3})}$

$=\dfrac{2(3\sqrt{5}-2\sqrt{3})}{3^2\cdot 5-2^2\cdot 3}$

$=\dfrac{2(3\sqrt{5}-2\sqrt{3})}{45-12}=\dfrac{2(3\sqrt{5}-2\sqrt{3})}{33}$

89. $\dfrac{\sqrt{2}-\sqrt{3}}{\sqrt{6}-\sqrt{5}}$

Multiply both the numerator and denominator by the conjugate of the denominator, $\sqrt{6}+\sqrt{5}$.

$$=\frac{(\sqrt{2}-\sqrt{3})(\sqrt{6}+\sqrt{5})}{(\sqrt{6}-\sqrt{5})(\sqrt{6}+\sqrt{5})}$$
$$=\frac{\sqrt{12}+\sqrt{10}-\sqrt{18}-\sqrt{15}}{(\sqrt{6})^2-(\sqrt{5})^2}$$
$$=\frac{\sqrt{4}\cdot\sqrt{3}+\sqrt{10}-\sqrt{9}\cdot\sqrt{2}-\sqrt{15}}{6-5}$$
$$=2\sqrt{3}+\sqrt{10}-3\sqrt{2}-\sqrt{15}$$

91. $\dfrac{m-4}{\sqrt{m}+2}$

Multiply both the numerator and denominator by the conjugate of the denominator, $\sqrt{m}-2$.

$$=\frac{(m-4)(\sqrt{m}-2)}{(\sqrt{m}+2)(\sqrt{m}-2)}$$
$$=\frac{(m-4)(\sqrt{m}-2)}{m-4}$$
$$=\sqrt{m}-2$$

93. $\dfrac{4}{\sqrt{x}-2\sqrt{y}}$

Multiply both the numerator and denominator by the conjugate of the denominator, $\sqrt{x}+2\sqrt{y}$.

$$=\frac{4(\sqrt{x}+2\sqrt{y})}{(\sqrt{x}-2\sqrt{y})(\sqrt{x}+2\sqrt{y})}$$
$$=\frac{4(\sqrt{x}+2\sqrt{y})}{x-4y}$$

95. $\dfrac{\sqrt{x}-\sqrt{y}}{\sqrt{x}+\sqrt{y}}$

Multiply both the numerator and denominator by the conjugate of the denominator, $\sqrt{x}-\sqrt{y}$.

$$=\frac{(\sqrt{x}-\sqrt{y})(\sqrt{x}-\sqrt{y})}{(\sqrt{x}+\sqrt{y})(\sqrt{x}-\sqrt{y})}$$
$$=\frac{(\sqrt{x})^2-2\sqrt{x}\sqrt{y}+(\sqrt{y})^2}{(\sqrt{x})^2-(\sqrt{y})^2}$$
$$=\frac{x-2\sqrt{xy}+y}{x-y}$$

97. $\dfrac{5\sqrt{k}}{2\sqrt{k}+\sqrt{q}}$

Multiply both the numerator and denominator by the conjugate of the denominator, $2\sqrt{k}-\sqrt{q}$.

$$=\frac{5\sqrt{k}(2\sqrt{k}-\sqrt{q})}{(2\sqrt{k}+\sqrt{q})(2\sqrt{k}-\sqrt{q})}$$
$$=\frac{5\sqrt{k}(2\sqrt{k}-\sqrt{q})}{4k-q}$$

99. $\dfrac{30-20\sqrt{6}}{10}=\dfrac{10(3-2\sqrt{6})}{10}=3-2\sqrt{6}$

101. $\dfrac{3-3\sqrt{5}}{3}=\dfrac{3(1-\sqrt{5})}{3}=1-\sqrt{5}$

103. $\dfrac{16-4\sqrt{8}}{12}=\dfrac{16-4(2\sqrt{2})}{12}=\dfrac{16-8\sqrt{2}}{12}$

$$=\frac{4(4-2\sqrt{2})}{4\cdot 3}=\frac{4-2\sqrt{2}}{3}$$

105. $\dfrac{6p+\sqrt{24p^3}}{3p}$

$$=\frac{6p+\sqrt{4p^2\cdot 6p}}{3p}=\frac{6p+2p\sqrt{6p}}{3p}$$
$$=\frac{p(6+2\sqrt{6p})}{3p}=\frac{6+2\sqrt{6p}}{3}$$

107. $\dfrac{3}{\sqrt{x+y}}=\dfrac{3}{\sqrt{x+y}}\cdot\dfrac{\sqrt{x+y}}{\sqrt{x+y}}$

$$=\frac{3\cdot\sqrt{x+y}}{(\sqrt{x+y})^2}$$
$$=\frac{3\sqrt{x+y}}{x+y}$$

109. $\dfrac{p}{\sqrt{p+2}}=\dfrac{p}{\sqrt{p+2}}\cdot\dfrac{\sqrt{p+2}}{\sqrt{p+2}}$

$$=\frac{p\sqrt{p+2}}{p+2}$$

111. $\dfrac{1}{\sqrt{2}}\cdot\dfrac{\sqrt{3}}{2}-\dfrac{1}{\sqrt{2}}\cdot\dfrac{1}{2}=\dfrac{\sqrt{3}}{2\sqrt{2}}-\dfrac{1}{2\sqrt{2}}$

$$=\frac{\sqrt{3}-1}{2\sqrt{2}}=\frac{(\sqrt{3}-1)\sqrt{2}}{(2\sqrt{2})\sqrt{2}}$$
$$=\frac{\sqrt{6}-\sqrt{2}}{2\cdot 2}=\frac{\sqrt{6}-\sqrt{2}}{4}$$

Using a calculator,

$$\frac{1}{\sqrt{2}}\cdot\frac{\sqrt{3}}{2}-\frac{1}{\sqrt{2}}\cdot\frac{1}{2}\approx 0.2588190451 \text{ and}$$
$$\frac{\sqrt{6}-\sqrt{2}}{4}\approx 0.2588190451.$$

113. $\dfrac{6-\sqrt{3}}{8}=\dfrac{6-\sqrt{3}}{8}\cdot\dfrac{6+\sqrt{3}}{6+\sqrt{3}}$

$$=\frac{6^2-(\sqrt{3})^2}{8(6+\sqrt{3})}$$
$$=\frac{36-3}{8(6+\sqrt{3})}$$
$$=\frac{33}{8(6+\sqrt{3})}$$

115. $\dfrac{2\sqrt{x}-\sqrt{y}}{3x} = \dfrac{2\sqrt{x}-\sqrt{y}}{3x}\cdot\dfrac{2\sqrt{x}+\sqrt{y}}{2\sqrt{x}+\sqrt{y}}$

$= \dfrac{(2\sqrt{x})^2-(\sqrt{y})^2}{3x(2\sqrt{x}+\sqrt{y})}$

$= \dfrac{4x-y}{3x(2\sqrt{x}+\sqrt{y})}$

117. $-8x+7=4$

$-8x=-3$

$x=\frac{3}{8}$

The solution set is $\left\{\frac{3}{8}\right\}$.

119. $6x^2-7x=3$

$6x^2-7x-3=0$

$(3x+1)(2x-3)=0$

$3x+1=0$ or $2x-3=0$

$3x=-1$ or $2x=3$

$x=-\frac{1}{3}$ or $x=\frac{3}{2}$

The solution set is $\left\{-\frac{1}{3},\frac{3}{2}\right\}$.

Summary Exercises on Operations with Radicals and Rational Exponents

1. $6\sqrt{10}-12\sqrt{10}=(6-12)\sqrt{10}$

$=-6\sqrt{10}$

3. $(1-\sqrt{3})(2+\sqrt{6})$

F O I L

$=2+\sqrt{6}-2\sqrt{3}-\sqrt{18}$

$=2+\sqrt{6}-2\sqrt{3}-\sqrt{9}\cdot\sqrt{2}$

$=2+\sqrt{6}-2\sqrt{3}-3\sqrt{2}$

5. $(3\sqrt{5}+2\sqrt{7})^2$

$=(3\sqrt{5})^2+2(3\sqrt{5})(2\sqrt{7})+(2\sqrt{7})^2$

$=9\cdot 5+12\sqrt{35}+4\cdot 7$

$=45+12\sqrt{35}+28$

$=73+12\sqrt{35}$

7. $\dfrac{8}{\sqrt{7}+\sqrt{5}}=\dfrac{8}{\sqrt{7}+\sqrt{5}}\cdot\dfrac{\sqrt{7}-\sqrt{5}}{\sqrt{7}-\sqrt{5}}$

$=\dfrac{8(\sqrt{7}-\sqrt{5})}{7-5}$

$=\dfrac{8(\sqrt{7}-\sqrt{5})}{2}$

$=4(\sqrt{7}-\sqrt{5})$

9. $(\sqrt{5}+7)(\sqrt{5}-7)=(\sqrt{5})^2-7^2$

$=5-49$

$=-44$

11. $\sqrt[3]{8a^3b^5c^9}=\sqrt[3]{8a^3b^3c^9}\cdot\sqrt[3]{b^2}$

$=2abc^3\sqrt[3]{b^2}$

13. $\dfrac{3}{\sqrt{5}+2}=\dfrac{3}{\sqrt{5}+2}\cdot\dfrac{\sqrt{5}-2}{\sqrt{5}-2}$

$=\dfrac{3(\sqrt{5}-2)}{5-4}$

$=3(\sqrt{5}-2)$

15. $\dfrac{16\sqrt{3}}{5\sqrt{12}}=\dfrac{16\sqrt{3}}{5\cdot\sqrt{4}\cdot\sqrt{3}}$

$=\dfrac{16}{5\cdot 2}$

$=\frac{8}{5}$

17. $\dfrac{-10}{\sqrt[3]{10}}=\dfrac{-10}{\sqrt[3]{10}}\cdot\dfrac{\sqrt[3]{100}}{\sqrt[3]{100}}$

$=\dfrac{-10\sqrt[3]{100}}{\sqrt[3]{1000}}$

$=\dfrac{-10\sqrt[3]{100}}{10}=-\sqrt[3]{100}$

19. $\sqrt{12x}-\sqrt{75x}=\sqrt{4}\cdot\sqrt{3x}-\sqrt{25}\cdot\sqrt{3x}$

$=2\sqrt{3x}-5\sqrt{3x}$

$=-3\sqrt{3x}$

21. $\sqrt[3]{\dfrac{13}{81}}=\dfrac{\sqrt[3]{13}}{\sqrt[3]{81}}$

$=\dfrac{\sqrt[3]{13}}{\sqrt[3]{27}\cdot\sqrt[3]{3}}\cdot\dfrac{\sqrt[3]{9}}{\sqrt[3]{9}}$

$=\dfrac{\sqrt[3]{13\cdot 9}}{3\cdot\sqrt[3]{27}}$

$=\dfrac{\sqrt[3]{117}}{3\cdot 3}=\dfrac{\sqrt[3]{117}}{9}$

23. $\dfrac{6}{\sqrt[4]{3}}=\dfrac{6}{\sqrt[4]{3}}\cdot\dfrac{\sqrt[4]{3^3}}{\sqrt[4]{3^3}}$

$=\dfrac{6\sqrt[4]{27}}{3}=2\sqrt[4]{27}$

25. $\sqrt[3]{\dfrac{x^2y}{x^{-3}y^4}}=\sqrt[3]{x^{2-(-3)}y^{1-4}}$

$=\sqrt[3]{x^5y^{-3}}$

$=\dfrac{\sqrt[3]{x^5}}{\sqrt[3]{y^3}}$

$=\dfrac{\sqrt[3]{x^3\cdot x^2}}{y}=\dfrac{x\sqrt[3]{x^2}}{y}$

27. $\dfrac{x^{-2/3}y^{4/5}}{x^{-5/3}y^{-2/5}}=x^{-2/3-(-5/3)}y^{4/5-(-2/5)}$

$=x^{3/3}y^{6/5}$

$=xy^{6/5}$

29. $$(125x^3)^{-2/3} = \frac{1}{(125x^3)^{2/3}} = \frac{1}{\left(\sqrt[3]{125x^3}\right)^2} = \frac{1}{(5x)^2} = \frac{1}{25x^2}$$

31. $$\sqrt[3]{16x^2} - \sqrt[3]{54x^2} + \sqrt[3]{128x^2}$$
$$= \sqrt[3]{8}\cdot\sqrt[3]{2x^2} - \sqrt[3]{27}\cdot\sqrt[3]{2x^2} + \sqrt[3]{64}\cdot\sqrt[3]{2x^2}$$
$$= 2\sqrt[3]{2x^2} - 3\sqrt[3]{2x^2} + 4\sqrt[3]{2x^2}$$
$$= (2-3+4)\sqrt[3]{2x^2}$$
$$= 3\sqrt[3]{2x^2}$$

33. **(a)** $\sqrt{64} = 8$

(b) $x^2 = 64$

$x = -\sqrt{64}$ or $x = \sqrt{64}$
$x = -8$ or $x = 8$

The solution set is $\{-8, 8\}$.

35. **(a)** $x^2 = 16$

$x = -\sqrt{16}$ or $x = \sqrt{16}$
$x = -4$ or $x = 4$

The solution set is $\{-4, 4\}$.

(b) $-\sqrt{16} = -\left(\sqrt{16}\right) = -4$

37. **(a)** Since $\left(\frac{9}{11}\right)^2 = \frac{81}{121}$, $-\sqrt{\frac{81}{121}} = -\frac{9}{11}$.

(b) $x^2 = \frac{81}{121}$

$x = -\sqrt{\frac{81}{121}}$ or $x = \sqrt{\frac{81}{121}}$
$x = -\frac{9}{11}$ or $x = \frac{9}{11}$

The solution set is $\left\{-\frac{9}{11}, \frac{9}{11}\right\}$.

39. **(a)** $x^2 = 0.04$

$x = -\sqrt{0.04}$ or $x = \sqrt{0.04}$
$x = -0.2$ or $x = 0.2$

The solution set is $\{-0.2, 0.2\}$.

(b) Since $(0.2)^2 = 0.04$, $\sqrt{0.04} = 0.2$.

10.6 Solving Equations with Radicals

10.6 Now Try Exercises

N1. $$\sqrt{9x+7} = 5$$
$$\left(\sqrt{9x+7}\right)^2 = 5^2 \quad \textit{Square each side.}$$
$$9x + 7 = 25$$
$$9x = 18$$
$$x = 2$$

Check 2 in the original equation.
$$\sqrt{9x+7} = 5$$
$$\sqrt{9\cdot 2+7} \stackrel{?}{=} 5 \quad \textit{Let } x = 2.$$
$$\sqrt{25} \stackrel{?}{=} 5$$
$$5 = 5 \quad \textit{True}$$

The solution set is $\{2\}$.

N2. $\sqrt{3x+4} + 5 = 0$

Step 1

Isolate the radical on one side of the equation.
$$\sqrt{3x+4} = -5$$

The equation has no solution, because the square root of a real number must be nonnegative.

The solution set is $\emptyset$.

N3. $$\sqrt{16-x} = x+4$$
$$\left(\sqrt{16-x}\right)^2 = (x+4)^2 \quad \textit{Square.}$$
$$16 - x = x^2 + 8x + 16$$
$$0 = x^2 + 9x$$
$$0 = x(x+9)$$

$x = 0$ or $x + 9 = 0$
or $x = -9$

Check $x = 0$: $\sqrt{16} = 4$ *True*

Check $x = -9$: $\sqrt{25} = -5$ *False*

The solution set is $\{0\}$.

N4. $$\sqrt{x^2 - 3x + 18} = x + 3$$
$$\left(\sqrt{x^2-3x+18}\right)^2 = (x+3)^2 \quad \textit{Square.}$$
$$x^2 - 3x + 18 = x^2 + 6x + 9$$
$$-9x = -9$$
$$x = 1$$

Check $x = 1$: $\sqrt{16} = 4$ *True*

The solution set is $\{1\}$.

N5. $\sqrt{3x+1} - \sqrt{x+4} = 1$

Get one radical on each side of the equals sign.
$$\sqrt{3x+1} = 1 + \sqrt{x+4}$$

Square each side.
$$\left(\sqrt{3x+1}\right)^2 = \left(1+\sqrt{x+4}\right)^2$$
$$3x + 1 = 1 + 2\sqrt{x+4} + (x+4)$$

Isolate the remaining radical.
$$3x = 2\sqrt{x+4} + x + 4$$
$$2x - 4 = 2\sqrt{x+4}$$
$$x - 2 = \sqrt{x+4}$$

Square each side again.
$$(x-2)^2 = \left(\sqrt{x+4}\right)^2$$
$$x^2 - 4x + 4 = x + 4$$
$$x^2 - 5x = 0$$
$$x(x-5) = 0$$

$x = 0$ or $x - 5 = 0$
or $x = 5$

Check $x = 0$: $\sqrt{1} - \sqrt{4} = 1$ *False*

Check $x = 5$: $\sqrt{16} - \sqrt{9} = 1$ *True*

The solution set is $\{5\}$.

N6.
$$\sqrt[3]{4x-5} = \sqrt[3]{3x+2}$$
$$\left(\sqrt[3]{4x-5}\right)^3 = \left(\sqrt[3]{3x+2}\right)^3 \quad \textit{Cube each side.}$$
$$4x - 5 = 3x + 2$$
$$x = 7$$

Check $x = 7$: $\sqrt[3]{23} = \sqrt[3]{23}$ *True*
The solution set is $\{7\}$.

N7. Solve $x = \sqrt{\dfrac{y+2}{a}}$ for a.
$$x^2 = \frac{y+2}{a} \quad \textit{Square each side.}$$
$$ax^2 = y + 2 \quad \textit{Multiply by a.}$$
$$a = \frac{y+2}{x^2} \quad \textit{Divide by } x^2.$$

10.6 Section Exercises

1. $\sqrt{3x+18} - x = 0$

(a) Check $x = 6$.
$$\sqrt{3(6)+18} - 6 \stackrel{?}{=} 0$$
$$\sqrt{18+18} - 6 \stackrel{?}{=} 0$$
$$\sqrt{36} - 6 \stackrel{?}{=} 0$$
$$6 - 6 = 0 \quad \textit{True}$$

The number 6 is a solution.

(b) Check $x = -3$.

$\sqrt{3(-3)+18} - (-3) = 0$ is a false statement since the principal square root of a number is nonnegative, and adding 3 to the radical makes the left side at least 3. The number -3 is not a solution.

3. $\sqrt{x+2} - \sqrt{9x-2} = -2\sqrt{x-1}$

(a) Check $x = 2$.
$$\sqrt{2+2} - \sqrt{9(2)-2} \stackrel{?}{=} -2\sqrt{2-1}$$
$$\sqrt{4} - \sqrt{16} \stackrel{?}{=} -2\sqrt{1}$$
$$2 - 4 \stackrel{?}{=} -2$$
$$-2 = -2 \quad \textit{True}$$

The number 2 is a solution.

(b) Check $x = 7$.
$$\sqrt{7+2} - \sqrt{9(7)-2} \stackrel{?}{=} -2\sqrt{7-1}$$
$$\sqrt{9} - \sqrt{61} \stackrel{?}{=} -2\sqrt{6}$$
$$3 - \sqrt{61} = -2\sqrt{6} \quad \textit{False}$$

The number 7 is not a solution.

5. $\sqrt{9} = 3$, not -3. There is no solution of $\sqrt{x} = -3$ since the value of a principal square root cannot equal a negative number.

In Exercises 7–34, check each solution in the original equation.

7.
$$\sqrt{x-2} = 3$$
$$\left(\sqrt{x-2}\right)^2 = 3^2 \quad \textit{Square each side.}$$
$$x - 2 = 9$$
$$x = 11$$

Check the proposed solution, 11.
Check $x = 11$: $\sqrt{9} = 3$ *True*
The solution set is $\{11\}$.

9.
$$\sqrt{6k-1} = 1$$
$$\left(\sqrt{6k-1}\right)^2 = 1^2 \quad \textit{Square each side.}$$
$$6k - 1 = 1$$
$$6k = 2$$
$$k = \tfrac{2}{6} = \tfrac{1}{3}$$

Check the proposed solution, $\frac{1}{3}$.
Check $k = \frac{1}{3}$: $\sqrt{1} = 1$ *True*
The solution set is $\{\frac{1}{3}\}$.

11.
$$\sqrt{4r+3} + 1 = 0$$
$$\sqrt{4r+3} = -1 \quad \textit{Isolate the radical.}$$

This equation has no solution, because $\sqrt{4r+3}$ cannot be negative.

The solution set is $\emptyset$.

13.
$$\sqrt{3x+1} - 4 = 0$$
$$\sqrt{3x+1} = 4 \quad \textit{Isolate the radical.}$$
$$\left(\sqrt{3x+1}\right)^2 = 4^2 \quad \textit{Square each side.}$$
$$3x + 1 = 16$$
$$3x = 15$$
$$x = 5$$

Check $x = 5$: $\sqrt{16} - 4 = 0$ *True*
The solution set is $\{5\}$.

15.
$$4 - \sqrt{x-2} = 0$$
$$4 = \sqrt{x-2} \quad \textit{Isolate.}$$
$$4^2 = \left(\sqrt{x-2}\right)^2 \quad \textit{Square.}$$
$$16 = x - 2$$
$$18 = x$$

Check $x = 18$: $4 - \sqrt{16} = 0$ *True*
The solution set is $\{18\}$.

17.
$$\sqrt{9x-4} = \sqrt{8x+1}$$
$$\left(\sqrt{9x-4}\right)^2 = \left(\sqrt{8x+1}\right)^2 \quad \textit{Square.}$$
$$9x - 4 = 8x + 1$$
$$x = 5$$

Check $x = 5$: $\sqrt{41} = \sqrt{41}$ *True*
The solution set is $\{5\}$.

19. $2\sqrt{x} = \sqrt{3x+4}$

$(2\sqrt{x})^2 = (\sqrt{3x+4})^2$ *Square.*

$4x = 3x + 4$

$x = 4$

Check $x = 4$: $4 = \sqrt{16}$ *True*

The solution set is $\{4\}$.

21. $3\sqrt{x-1} = 2\sqrt{2x+2}$

$(3\sqrt{x-1})^2 = (2\sqrt{2x+2})^2$ *Square.*

$9(x-1) = 4(2x+2)$

$9x - 9 = 8x + 8$

$x = 17$

Check $x = 17$: $3(4) = 2(6)$ *True*

The solution set is $\{17\}$.

23. $x = \sqrt{x^2+4x-20}$

$x^2 = (\sqrt{x^2+4x-20})^2$ *Square.*

$x^2 = x^2 + 4x - 20$

$20 = 4x$

$5 = x$

Check $x = 5$: $5 = \sqrt{25}$ *True*

The solution set is $\{5\}$.

25. $x = \sqrt{x^2+3x+9}$

$x^2 = (\sqrt{x^2+3x+9})^2$ *Square.*

$x^2 = x^2 + 3x + 9$

$-3x = 9$

$x = -3$

Substituting -3 for x makes the left side of the original equation negative, but the right side is nonnegative, so the solution set is $\emptyset$.

27. $\sqrt{9-x} = x + 3$

$(\sqrt{9-x})^2 = (x+3)^2$ *Square.*

$9 - x = x^2 + 6x + 9$

$0 = x^2 + 7x$

$0 = x(x+7)$

$x = 0$ or $x + 7 = 0$

$x = -7$

Check $x = -7$: $\sqrt{16} = -4$ *False*

Check $x = 0$: $\sqrt{9} = 3$ *True*

The solution set is $\{0\}$.

29. $\sqrt{k^2+2k+9} = k + 3$

$(\sqrt{k^2+2k+9})^2 = (k+3)^2$ *Square.*

$k^2 + 2k + 9 = k^2 + 6k + 9$

$0 = 4k$

$0 = k$

Check $k = 0$: $\sqrt{9} = 3$ *True*

The solution set is $\{0\}$.

31. $\sqrt{x^2+12x-4} = x - 4$

$(\sqrt{x^2+12x-4})^2 = (x-4)^2$ *Square.*

$x^2 + 12x - 4 = x^2 - 8x + 16$

$20x = 20$

$x = 1$

Substituting 1 for x makes the left side of the original equation positive, but the right side is zero, so the solution set is $\emptyset$.

33. $\sqrt{r^2+9r+15} - r - 4 = 0$

$\sqrt{r^2+9r+15} = r + 4$ *Isolate.*

$(\sqrt{r^2+9r+15})^2 = (r+4)^2$ *Square.*

$r^2 + 9r + 15 = r^2 + 8r + 16$

$r = 1$

Check $r = 1$: $\sqrt{25} - 1 - 4 = 0$ *True*

The solution set is $\{1\}$.

35. $\sqrt{3x+4} = 8 - x$

$(8-x)^2$ equals $64 - 16x + x^2$, not $64 + x^2$. The first step should be

$3x + 4 = 64 - 16x + x^2.$

Then we have

$0 = x^2 - 19x + 60.$

$0 = (x-4)(x-15)$

$x - 4 = 0$ or $x - 15 = 0$

$x = 4$ or $x = 15$

Check $x = 4$: $\sqrt{16} = 8 - 4$ *True*

Check $x = 15$: $\sqrt{49} = 8 - 15$ *False*

The solution set is $\{4\}$.

37. $\sqrt[3]{2x+5} = \sqrt[3]{6x+1}$

$(\sqrt[3]{2x+5})^3 = \left(\sqrt[3]{6x+1}\right)^3$ *Cube each side.*

$2x + 5 = 6x + 1$

$4 = 4x$

$1 = x$

Check $x = 1$: $\sqrt[3]{7} = \sqrt[3]{7}$ *True*

The solution set is $\{1\}$.

39. $\sqrt[3]{x^2+5x+1} = \sqrt[3]{x^2+4x}$

$(\sqrt[3]{x^2+5x+1})^3 = (\sqrt[3]{x^2+4x})^3$ *Cube.*

$x^2 + 5x + 1 = x^2 + 4x$

$x = -1$

Check $x = -1$: $\sqrt[3]{-3} = \sqrt[3]{-3}$ *True*

The solution set is $\{-1\}$.

41. $\sqrt[3]{2m-1} = \sqrt[3]{m+13}$

$(\sqrt[3]{2m-1})^3 = (\sqrt[3]{m+13})^3$ *Cube.*

$2m - 1 = m + 13$

$m = 14$

Check $m = 14$: $\sqrt[3]{27} = \sqrt[3]{27}$ *True*

The solution set is $\{14\}$.

43. $\sqrt[4]{x+12} = \sqrt[4]{3x-4}$

Raise each side to the fourth power.

$$\left(\sqrt[4]{x+12}\right)^4 = \left(\sqrt[4]{3x-4}\right)^4$$
$$x+12 = 3x-4$$
$$16 = 2x$$
$$8 = x$$

Check $x=8$: $\sqrt[4]{20} = \sqrt[4]{20}$ *True*

The solution set is $\{8\}$.

45. $\sqrt[3]{x-8}+2=0$

$$\sqrt[3]{x-8} = -2 \quad \textit{Isolate.}$$
$$\left(\sqrt[3]{x-8}\right)^3 = (-2)^3 \quad \textit{Cube.}$$
$$x-8=-8$$
$$x=0$$

Check $x=0$: $\sqrt[3]{-8}+2=0$ *True*

The solution set is $\{0\}$.

47. $\sqrt[4]{2k-5}+4=0$

$$\sqrt[4]{2k-5} = -4 \quad \textit{Isolate.}$$

This equation has no solution, because $\sqrt[4]{2k-5}$ cannot be negative.

The solution set is $\emptyset$.

49. $\sqrt{k+2}-\sqrt{k-3}=1$

Get one radical on each side of the equals sign.

$$\sqrt{k+2} = 1+\sqrt{k-3}$$
$$\left(\sqrt{k+2}\right)^2 = \left(1+\sqrt{k-3}\right)^2 \quad \textit{Square.}$$
$$k+2 = 1+2\sqrt{k-3}+k-3$$
$$4 = 2\sqrt{k-3} \quad \textit{Isolate.}$$
$$2 = \sqrt{k-3} \quad \textit{Divide by 2.}$$
$$2^2 = \left(\sqrt{k-3}\right)^2 \quad \textit{Square again.}$$
$$4 = k-3$$
$$7 = k$$

Check $k=7$: $\sqrt{9}-\sqrt{4}=1$ *True*

The solution set is $\{7\}$.

51. $\sqrt{2r+11}-\sqrt{5r+1}=-1$

Get one radical on each side of the equals sign.

$$\sqrt{2r+11} = -1+\sqrt{5r+1}$$

Square each side.

$$\left(\sqrt{2r+11}\right)^2 = \left(-1+\sqrt{5r+1}\right)^2$$
$$2r+11 = 1-2\sqrt{5r+1}+5r+1$$

Isolate the remaining radical.

$$2\sqrt{5r+1} = 3r-9$$

Square each side again.

$$\left(2\sqrt{5r+1}\right)^2 = (3r-9)^2$$
$$4(5r+1) = 9r^2-54r+81$$
$$20r+4 = 9r^2-54r+81$$
$$0 = 9r^2-74r+77$$
$$0 = (9r-11)(r-7)$$

$$9r-11=0 \quad \text{or} \quad r-7=0$$
$$r=\tfrac{11}{9} \qquad\qquad r=7$$

Check $r=\frac{11}{9}$: $\frac{11}{3}-\frac{8}{3}=-1$ *False*

Check $r=7$: $5-6=-1$ *True*

The solution set is $\{7\}$.

53. $\sqrt{3p+4}-\sqrt{2p-4}=2$

Get one radical on each side of the equals sign.

$$\sqrt{3p+4} = 2+\sqrt{2p-4}$$

Square each side.

$$\left(\sqrt{3p+4}\right)^2 = \left(2+\sqrt{2p-4}\right)^2$$
$$3p+4 = 4+4\sqrt{2p-4}+2p-4$$

Isolate the remaining radical.

$$p+4 = 4\sqrt{2p-4}$$

Square each side again.

$$(p+4)^2 = \left(4\sqrt{2p-4}\right)^2$$
$$p^2+8p+16 = 16(2p-4)$$
$$p^2+8p+16 = 32p-64$$
$$p^2-24p+80 = 0$$
$$(p-4)(p-20) = 0$$

$$p-4=0 \quad \text{or} \quad p-20=0$$
$$p=4 \qquad\qquad p=20$$

Check $p=4$: $\sqrt{16}-\sqrt{4}=2$ *True*

Check $p=20$: $\sqrt{64}-\sqrt{36}=2$ *True*

The solution set is $\{4, 20\}$.

55. $\sqrt{3-3p}-3=\sqrt{3p+2}$

Square each side.

$$\left(\sqrt{3-3p}-3\right)^2 = \left(\sqrt{3p+2}\right)^2$$
$$3-3p-6\sqrt{3-3p}+9 = 3p+2$$

Isolate the remaining radical.

$$-6\sqrt{3-3p} = 6p-10$$
$$-3\sqrt{3-3p} = 3p-5$$

Square each side again.

$$\left(-3\sqrt{3-3p}\right)^2 = (3p-5)^2$$
$$9(3-3p) = 9p^2-30p+25$$
$$27-27p = 9p^2-30p+25$$
$$0 = 9p^2-3p-2$$
$$0 = (3p+1)(3p-2)$$

$$3p+1=0 \quad \text{or} \quad 3p-2=0$$
$$p=-\tfrac{1}{3} \qquad\qquad p=\tfrac{2}{3}$$

Check $p=-\frac{1}{3}$: $\sqrt{4}-3=\sqrt{1}$ *False*

Check $p=\frac{2}{3}$: $\sqrt{1}-3=\sqrt{4}$ *False*

The solution set is $\emptyset$.

57. $\sqrt{2\sqrt{x+11}} = \sqrt{4x+2}$

$$2\sqrt{x+11} = 4x+2 \quad \textit{Square.}$$
$$\left(2\sqrt{x+11}\right)^2 = (4x+2)^2 \quad \textit{Square again.}$$
$$4(x+11) = 16x^2+16x+4$$
$$4x+44 = 16x^2+16x+4$$
$$0 = 16x^2+12x-40$$
$$0 = 4x^2+3x-10$$
$$0 = (x+2)(4x-5)$$

$x+2=0$ or $4x-5=0$

$x=-2$ $\qquad x=\frac{5}{4}$

Check $x=-2$: $\sqrt{6}=\sqrt{-6}$ *False*

Check $x=\frac{5}{4}$: $\sqrt{7}=\sqrt{7}$ *True*

The solution set is $\left\{\frac{5}{4}\right\}$.

59. $(2x-9)^{1/2} = 2+(x-8)^{1/2}$

$$\sqrt{2x-9} = 2+\sqrt{x-8}$$
$$\left(\sqrt{2x-9}\right)^2 = \left(2+\sqrt{x-8}\right)^2$$
$$2x-9 = 4+4\sqrt{x-8}+x-8$$
$$x-5 = 4\sqrt{x-8}$$
$$(x-5)^2 = \left(4\sqrt{x-8}\right)^2$$
$$x^2-10x+25 = 16(x-8)$$
$$x^2-10x+25 = 16x-128$$
$$x^2-26x+153 = 0$$
$$(x-9)(x-17) = 0$$

$x-9=0$ or $x-17=0$

$x=9$ $\qquad x=17$

Check $x=9$: $9^{1/2} \stackrel{?}{=} 2+1^{1/2}$

$3=2+1$ *True*

Check $x=17$: $25^{1/2} \stackrel{?}{=} 2+9^{1/2}$

$5=2+3$ *True*

The solution set is $\{9, 17\}$.

61. $(2w-1)^{2/3} - w^{1/3} = 0$

$$\sqrt[3]{(2w-1)^2} = \sqrt[3]{w}$$
$$\left[\sqrt[3]{(2w-1)^2}\right]^3 = \left(\sqrt[3]{w}\right)^3$$
$$(2w-1)^2 = w$$
$$4w^2-4w+1 = w$$
$$4w^2-5w+1 = 0$$
$$(4w-1)(w-1) = 0$$

$4w-1=0$ or $w-1=0$

$w=\frac{1}{4}$ or $w=1$

Check $w=\frac{1}{4}$: $\left(-\frac{1}{2}\right)^{2/3} - \left(\frac{1}{4}\right)^{1/3} \stackrel{?}{=} 0$

$\left(\frac{1}{4}\right)^{1/3} - \left(\frac{1}{4}\right)^{1/3} = 0$ *True*

Check $w=1$: $1^{2/3} - 1^{1/3} = 0$ *True*

The solution set is $\left\{\frac{1}{4}, 1\right\}$.

63. Solve $Z = \sqrt{\frac{L}{C}}$ for L.

$$(Z)^2 = \left(\sqrt{\frac{L}{C}}\right)^2 \quad \textit{Square.}$$
$$Z^2 = \frac{L}{C}$$
$$CZ^2 = L \quad \textit{Multiply by C.}$$

65. Solve $V = \sqrt{\frac{2K}{m}}$ for K.

$$(V)^2 = \left(\sqrt{\frac{2K}{m}}\right)^2 \quad \textit{Square.}$$
$$V^2 = \frac{2K}{m} \quad \textit{Multiply by } \frac{m}{2}.$$
$$\frac{V^2 m}{2} = K$$

67. Solve $r = \sqrt{\frac{Mm}{F}}$ for M.

$$(r)^2 = \left(\sqrt{\frac{Mm}{F}}\right)^2 \quad \textit{Square.}$$
$$r^2 = \frac{Mm}{F}$$
$$Fr^2 = Mm$$
$$\frac{Fr^2}{m} = M$$

69. Solve $N = \frac{1}{2\pi}\sqrt{\frac{a}{r}}$ for r.

$$2\pi N = \sqrt{\frac{a}{r}}$$
$$(2\pi N)^2 = \left(\sqrt{\frac{a}{r}}\right)^2$$
$$4\pi^2 N^2 = \frac{a}{r}$$
$$4\pi^2 N^2 r = a$$
$$r = \frac{a}{4\pi^2 N^2}$$

71. $(5+9x)+(-4-8x)$

$= 5-4+9x-8x$

$= 1+x$

73. $(x+3)(2x-5) = 2x^2-5x+6x-15$

$= 2x^2+x-15$

75. $\frac{-7}{5-\sqrt{2}} = \frac{-7}{5-\sqrt{2}} \cdot \frac{5+\sqrt{2}}{5+\sqrt{2}}$

$$= \frac{-7\left(5+\sqrt{2}\right)}{5^2-\left(\sqrt{2}\right)^2}$$
$$= \frac{-7\left(5+\sqrt{2}\right)}{25-2} = \frac{-7\left(5+\sqrt{2}\right)}{23}$$

10.7 Complex Numbers

10.7 Now Try Exercises

N1. **(a)** $\sqrt{-49} = i\sqrt{49} = 7i$

(b) $-\sqrt{-121} = -i\sqrt{121} = -11i$

(c) $\sqrt{-3} = i\sqrt{3}$

(d) $\sqrt{-32} = i\sqrt{32} = i\sqrt{16 \cdot 2} = 4i\sqrt{2}$

N2. **(a)** $\sqrt{-4} \cdot \sqrt{-16} = i\sqrt{4} \cdot i\sqrt{16}$
$= i \cdot 2 \cdot i \cdot 4$
$= 8i^2$
$= 8(-1)$
$= -8$

(b) $\sqrt{-5} \cdot \sqrt{-11} = i\sqrt{5} \cdot i\sqrt{11}$
$= i^2\sqrt{5 \cdot 11}$
$= (-1)\sqrt{55}$
$= -\sqrt{55}$

(c) $\sqrt{-3} \cdot \sqrt{-12} = i\sqrt{3} \cdot i\sqrt{12}$
$= i^2\sqrt{3 \cdot 12}$
$= (-1)\sqrt{36}$
$= -6$

(d) $\sqrt{13} \cdot \sqrt{-2} = \sqrt{13} \cdot i\sqrt{2}$
$= i\sqrt{13 \cdot 2}$
$= i\sqrt{26}$

N3. **(a)** $\dfrac{\sqrt{-72}}{\sqrt{-8}} = \dfrac{i\sqrt{72}}{i\sqrt{8}} = \sqrt{\dfrac{72}{8}} = \sqrt{9} = 3$

(b) $\dfrac{\sqrt{-48}}{\sqrt{3}} = \dfrac{i\sqrt{48}}{\sqrt{3}} = i\sqrt{\dfrac{48}{3}} = i\sqrt{16} = 4i$

N4. To add complex numbers, add the real parts and add the imaginary parts.

(a) $(-3 + 2i) + (4 + 7i)$
$= (-3 + 4) + (2 + 7)i$
$= 1 + 9i$

(b) $(5 - i) + (-3 + 3i) + (6 - 4i)$
$= [5 + (-3) + 6] + [(-1) + 3 + (-4)]i$
$= 8 - 2i$

N5. To subtract complex numbers, subtract the real parts and subtract the imaginary parts.

(a) $(7 + 10i) - (3 + 5i)$
$= (7 - 3) + (10 - 5)i$
$= 4 + 5i$

(b) $(5 - 2i) - (9 - 7i)$
$= (5 - 9) + [-2 - (-7)]i$
$= -4 + 5i$

(c) $(-1 + 12i) - (-1 - i)$
$= [-1 - (-1)] + [12 - (-1)]i$
$= 0 + 13i = 13i$

N6. **(a)** $8i(3 - 5i) = 8i(3) - 8i(5i)$
$= 24i - 40i^2$
$= 24i - 40(-1)$
$= 40 + 24i$

(b) $(7 - 2i)(4 + 3i)$
F O I L
$= 7(4) + 7(3i) + 4(-2i) + 3i(-2i)$
$= 28 + 21i - 8i - 6i^2$
$= 28 + 13i - 6(-1)$
$= 28 + 13i + 6$
$= 34 + 13i$

N7. In each case, multiply the numerator and denominator by the conjugate of the denominator.

(a) $\dfrac{4 + 2i}{1 + 3i} = \dfrac{(4 + 2i)(1 - 3i)}{(1 + 3i)(1 - 3i)}$
$= \dfrac{4 - 12i + 2i - 6i^2}{1^2 - (3i)^2}$
$= \dfrac{4 - 10i - 6(-1)}{1 - (-9)}$
$= \dfrac{4 - 10i + 6}{10}$
$= \dfrac{10}{10} - \dfrac{10}{10}i = 1 - i$

(b) $\dfrac{5 - 4i}{i}$ The conjugate of i is $-i$.
$= \dfrac{(5 - 4i)(-i)}{i(-i)}$
$= \dfrac{-5i + 4i^2}{-i^2}$
$= \dfrac{-5i + 4(-1)}{-(-1)}$
$= \dfrac{-5i - 4}{1}$
$= -4 - 5i$

N8. **(a)** $i^{16} = (i^4)^4 = 1^4 = 1$

(b) $i^{21} = i^{20} \cdot i^1 = (i^4)^5 \cdot i = 1^5 \cdot i = i$

(c) $i^{-6} = \dfrac{1}{i^6} = \dfrac{1}{i^4 \cdot i^2} = \dfrac{1}{1 \cdot (-1)} = \dfrac{1}{-1} = -1$

(d) $i^{-13} = \dfrac{1}{i^{13}} = \dfrac{1}{i^{12} \cdot i} = \dfrac{1}{(i^4)^3 \cdot i} = \dfrac{1}{1^3 \cdot i} = \dfrac{1}{i}$

Now multiply by $-i$, the conjugate of i.

$\dfrac{1}{i} = \dfrac{1(-i)}{i(-i)} = \dfrac{-i}{-i^2} = \dfrac{-i}{-(-1)} = \dfrac{-i}{1} = -i$

10.7 Section Exercises

1. $\sqrt{-1} = i$

3. $i^2 = -1$

5. $\frac{1}{i} = \frac{1}{i} \cdot \frac{-i}{-i}$ *−i is the conjugate of i*
$= \frac{-i}{-i^2} = \frac{i}{i^2} = \frac{i}{-1} = -i$

7. $\sqrt{-169} = i\sqrt{169} = 13i$

9. $-\sqrt{-144} = -i\sqrt{144} = -12i$

11. $\sqrt{-5} = i\sqrt{5}$

13. $\sqrt{-48} = i\sqrt{48} = i\sqrt{16 \cdot 3} = 4i\sqrt{3}$

15. $\sqrt{-7} \cdot \sqrt{-15} = i\sqrt{7} \cdot i\sqrt{15} = i^2\sqrt{7 \cdot 15}$
$= -1\sqrt{105} = -\sqrt{105}$

17. $\sqrt{-4} \cdot \sqrt{-25} = i\sqrt{4} \cdot i\sqrt{25} = 2i \cdot 5i = 10i^2$
$= 10(-1) = -10$

19. $\sqrt{-3} \cdot \sqrt{11} = i\sqrt{3} \cdot \sqrt{11}$
$= i\sqrt{33}$

21. $\frac{\sqrt{-300}}{\sqrt{-100}} = \frac{i\sqrt{300}}{i\sqrt{100}} = \sqrt{\frac{300}{100}} = \sqrt{3}$

23. $\frac{\sqrt{-75}}{\sqrt{3}} = \frac{i\sqrt{75}}{\sqrt{3}} = i\sqrt{\frac{75}{3}} = i\sqrt{25} = 5i$

25. $\frac{-\sqrt{-64}}{\sqrt{-16}} = \frac{-i\sqrt{64}}{i\sqrt{16}} = -\frac{8}{4} = -2$

27. Any real number a can be written as $a + 0i$, a complex number with imaginary part 0.

29. $(3 + 2i) + (-4 + 5i)$
$= [(3 + (-4)] + (2 + 5)i$
$= -1 + 7i$

31. $(5 - i) + (-5 + i)$
$= (5 - 5) + (-1 + 1)i$
$= 0$

33. $(4 + i) - (-3 - 2i)$
$= [(4 - (-3)] + [(1 - (-2)]i$
$= 7 + 3i$

35. $(-3 - 4i) - (-1 - 4i)$
$= [-3 - (-1)] + [-4 - (-4)]i$
$= -2$

37. $(-4 + 11i) + (-2 - 4i) + (7 + 6i)$
$= (-4 - 2 + 7) + (11 - 4 + 6)i$
$= 1 + 13i$

39. $[(7 + 3i) - (4 - 2i)] + (3 + i)$
Work inside the brackets first.
$= [(7 - 4) + (3 + 2)i] + (3 + i)$
$= (3 + 5i) + (3 + i)$
$= (3 + 3) + (5 + 1)i$
$= 6 + 6i$

41. If $a - c = b$, then $b + c = a$.
So, $(4 + 2i) - (3 + i) = 1 + i$ implies that
$$(1 + i) + (3 + i) = \underline{4 + 2i}.$$

43. $(3i)(27i) = 81i^2 = 81(-1) = -81$

45. $(-8i)(-2i) = 16i^2 = 16(-1) = -16$

47. $5i(-6 + 2i) = (5i)(-6) + (5i)(2i)$
$= -30i + 10i^2$
$= -30i + 10(-1)$
$= -10 - 30i$

49. $(4 + 3i)(1 - 2i)$
F O I L
$= (4)(1) + 4(-2i) + (3i)(1) + (3i)(-2i)$
$= 4 - 8i + 3i - 6i^2$
$= 4 - 5i - 6(-1)$
$= 4 - 5i + 6 = 10 - 5i$

51. $(4 + 5i)^2 = 4^2 + 2(4)(5i) + (5i)^2$
$= 16 + 40i + 25i^2$
$= 16 + 40i + 25(-1)$
$= 16 + 40i - 25$
$= -9 + 40i$

53. $2i(-4 - i)^2$
$= 2i[(-4)^2 - 2(-4)(i) + i^2]$
$= 2i(16 + 8i - 1)$
$= 2i(15 + 8i) = 30i + 16i^2$
$= 30i + 16(-1) = -16 + 30i$

55. $(12 + 3i)(12 - 3i)$
$= 12^2 - (3i)^2 = 144 - 9i^2$
$= 144 - 9(-1) = 144 + 9 = 153$

57. $(4 + 9i)(4 - 9i)$
$= 4^2 - (9i)^2 = 16 - 81i^2$
$= 16 - 81(-1) = 16 + 81 = 97$

59. $(1 + i)^2(1 - i)^2$
$= [(1 + i)(1 - i)]^2$
$= (1^2 - i^2)^2$
$= [1 - (-1)]^2$
$= 2^2 = 4$

61. The conjugate of $a + bi$ is $a - bi$.

63. $\frac{2}{1 - i}$
Multiply the numerator and the denominator by the conjugate of the denominator, $1 + i$.
$= \frac{2(1 + i)}{(1 - i)(1 + i)} = \frac{2(1 + i)}{1^2 - i^2}$
$= \frac{2(1 + i)}{1 - (-1)} = \frac{2(1 + i)}{2} = 1 + i$

65. $\dfrac{8i}{2+2i}$

Write in lowest terms.

$= \dfrac{2 \cdot 4i}{2(1+i)} = \dfrac{4i}{1+i}$

Multiply the numerator and the denominator by the conjugate of the denominator, $1 - i$.

$= \dfrac{4i(1-i)}{(1+i)(1-i)}$

In the denominator, we make use of the fact that $(a+bi)(a-bi) = a^2 + b^2$.

$= \dfrac{4(i-i^2)}{1^2+1^2} = \dfrac{4(i+1)}{2}$

$= 2(i+1) = 2+2i$

67. $\dfrac{-7+4i}{3+2i}$

Multiply the numerator and the denominator by the conjugate of the denominator, $3 - 2i$.

$= \dfrac{(-7+4i)(3-2i)}{(3+2i)(3-2i)}$

In the denominator, we make use of the fact that $(a+bi)(a-bi) = a^2 + b^2$.

$= \dfrac{-21+14i+12i+8}{3^2+2^2}$

$= \dfrac{-13+26i}{13} = \dfrac{13(-1+2i)}{13} = -1+2i$

69. $\dfrac{2-3i}{2+3i}$

Multiply the numerator and the denominator by the conjugate of the denominator, $2 - 3i$.

$= \dfrac{(2-3i)(2-3i)}{(2+3i)(2-3i)} = \dfrac{2^2 - 2(2)(3i) + (3i)^2}{2^2+3^2}$

$= \dfrac{4-12i+9i^2}{4+9} = \dfrac{4-12i-9}{13}$

$= \dfrac{-5-12i}{13} = -\dfrac{5}{13} - \dfrac{12}{13}i$

71. $\dfrac{3+i}{i}$

Multiply the numerator and the denominator by the conjugate of the denominator, $-i$.

$= \dfrac{(3+i)(-i)}{i(-i)} = \dfrac{-3i-i^2}{-i^2}$

$= \dfrac{-3i-(-1)}{-(-1)} = \dfrac{-3i+1}{1}$

$= 1-3i$

73. $\dfrac{3-i}{-i}$

Multiply the numerator and the denominator by the conjugate of the denominator, i.

$= \dfrac{(3-i)(i)}{-i(i)} = \dfrac{3i-i^2}{-i^2} = \dfrac{3i-(-1)}{-(-1)}$

$= \dfrac{3i+1}{1} = 1+3i$

75. $i^{18} = i^{16} \cdot i^2 = (i^4)^4 \cdot i^2$

$= 1^4 \cdot (-1) = 1 \cdot (-1) = -1$

77. $i^{89} = i^{88} \cdot i = (i^4)^{22} \cdot i = 1^{22} \cdot i$

$= 1 \cdot i = i$

79. $i^{38} = i^{36} \cdot i^2$

$= (i^4)^9 \cdot i^2 = 1^9(-1) = -1$

81. $i^{43} = i^{40} \cdot i^2 \cdot i$

$= (i^4)^{10} \cdot (-1) \cdot i$

$= 1^{10} \cdot (-i) = -i$

83. $i^{-5} = \dfrac{1}{i^5} = \dfrac{1}{i^4 \cdot i} = \dfrac{1}{1 \cdot i} = \dfrac{1}{i}$

From Exercise 5, $\dfrac{1}{i} = -i$.

85. Since $i^{20} = (i^4)^5 = 1^5 = 1$, the student multiplied by 1, which is justified by the identity property for multiplication.

87. $I = \dfrac{E}{R + (X_L - X_c)i}$

Substitute $2+3i$ for E, 5 for R, 4 for X_L, and 3 for X_c.

$I = \dfrac{2+3i}{5+(4-3)i} = \dfrac{2+3i}{5+i}$

$= \dfrac{(2+3i)(5-i)}{(5+i)(5-i)} = \dfrac{10-2i+15i-3i^2}{5^2+1^2}$

$= \dfrac{10+3+13i}{25+1} = \dfrac{13+13i}{26}$

$= \dfrac{13(1+i)}{13 \cdot 2} = \dfrac{1+i}{2} = \dfrac{1}{2} + \dfrac{1}{2}i$

89. To show that $1+5i$ is a solution of the equation $x^2 - 2x + 26 = 0$, substitute $1+5i$ for x.

$$x^2 - 2x + 26 = 0$$
$$(1+5i)^2 - 2(1+5i) + 26 \stackrel{?}{=} 0$$
$$(1+10i+25i^2) - 2 - 10i + 26 \stackrel{?}{=} 0$$
$$1 + 10i - 25 - 2 - 10i + 26 \stackrel{?}{=} 0$$
$$(1-25-2+26) + (10-10)i \stackrel{?}{=} 0$$
$$0 = 0 \quad \textit{True}$$

Thus, $1+5i$ is a solution of the given equation.

Now substitute $1-5i$ for x.

$$(1-5i)^2 - 2(1-5i) + 26 \stackrel{?}{=} 0$$
$$(1-10i+25i^2) - 2 + 10i + 26 \stackrel{?}{=} 0$$
$$1 - 10i - 25 - 2 + 10i + 26 \stackrel{?}{=} 0$$
$$(1-25-2+26) + (-10+10)i \stackrel{?}{=} 0$$
$$0 = 0 \quad \textit{True}$$

Thus, $1-5i$ is also a solution of the given equation.

91. $\dfrac{3}{2-i}+\dfrac{5}{1+i}$

$=\dfrac{3(2+i)}{(2-i)(2+i)}+\dfrac{5(1-i)}{(1+i)(1-i)}$

$=\dfrac{6+3i}{4+1}+\dfrac{5-5i}{1+1}$

$=\dfrac{2(6+3i)}{2\cdot 5}+\dfrac{5(5-5i)}{5\cdot 2}$

$=\dfrac{12+6i}{10}+\dfrac{25-25i}{10}$

$=\dfrac{37-19i}{10}=\dfrac{37}{10}-\dfrac{19}{10}i$

93. $\dfrac{2+i}{2-i}+\dfrac{i}{1+i}$

$=\dfrac{(2+i)(2+i)}{(2-i)(2+i)}+\dfrac{i(1-i)}{(1+i)(1-i)}$

$=\dfrac{4+4i+i^2}{4+1}+\dfrac{i-i^2}{1+1}$

$=\dfrac{3+4i}{5}+\dfrac{1+i}{2}$

$=\dfrac{2(3+4i)}{2\cdot 5}+\dfrac{5(1+i)}{5\cdot 2}$

$=\dfrac{6+8i}{10}+\dfrac{5+5i}{10}$

$=\dfrac{11+13i}{10}=\dfrac{11}{10}+\dfrac{13}{10}i$

Multiplying by i gives us

$\left(\dfrac{2+i}{2-i}+\dfrac{i}{1+i}\right)i=\left(\dfrac{11}{10}+\dfrac{13}{10}i\right)i$

$=\dfrac{11}{10}i+\dfrac{13}{10}i^2$

$=-\dfrac{13}{10}+\dfrac{11}{10}i.$

95. $6x+13=0$

$6x=-13$

$x=-\frac{13}{6}$

The solution set is $\{-\frac{13}{6}\}$.

97. $x(x+3)=40$

$x^2+3x=40$

$x^2+3x-40=0$

$(x+8)(x-5)=0$

$x+8=0$ or $x-5=0$

$x=-8$ or $x=5$

Check $x=-8$: $-8(-5)=40$ *True*

Check $x=5$: $5(8)=40$ *True*

The solution set is $\{-8,5\}$.

99. $5x^2-3x=2$

$5x^2-3x-2=0$

$(5x+2)(x-1)=0$

$5x+2=0$ or $x-1=0$

$x=-\frac{2}{5}$ or $x=1$

Check $x=-\frac{2}{5}$: $\frac{4}{5}+\frac{6}{5}=2$ *True*

Check $x=1$: $5-3=2$ *True*

The solution set is $\{-\frac{2}{5},1\}$.

Chapter 10 Review Exercises

1. $\sqrt{1764}=42$, because $42^2=1764$.

2. $-\sqrt{289}=-(17)=-17$, since $17^2=289$.

3. $\sqrt[3]{216}=6$, because $6^3=216$.

4. $\sqrt[3]{-125}=-5$, because $(-5)^3=-125$.

5. $-\sqrt[3]{27}=-(3)=-3$, since $3^3=27$.

6. $\sqrt[5]{-32}=-2$, because $(-2)^5=-32$.

7. $\sqrt[n]{a}$ is not a real number if n is even and a is negative.

8. **(a)** $\sqrt{x^2}=|x|$

(b) $-\sqrt{x^2}=-|x|$

(c) $\sqrt[3]{x^3}=x$

9. $-\sqrt{47}\approx -6.856$

10. $\sqrt[3]{-129}\approx -5.053$

11. $\sqrt[4]{605}\approx 4.960$

12. $\sqrt[4]{500^{-3}}\approx 0.009$

13. $-\sqrt[3]{500^4}\approx -3968.503$

14. $-\sqrt{28^{-1}}\approx -0.189$

15. $f(x)=\sqrt{x-1}$

For the radicand to be nonnegative, we must have

$$x-1\ge 0 \quad\text{or}\quad x\ge 1.$$

Thus, the domain is $[1,\infty)$.

The function values are positive or zero (the result of the radical), so the range is $[0,\infty)$.

x	$f(x)=\sqrt{x-1}$
1	$\sqrt{1-1}=0$
2	$\sqrt{2-1}=1$
5	$\sqrt{5-1}=2$

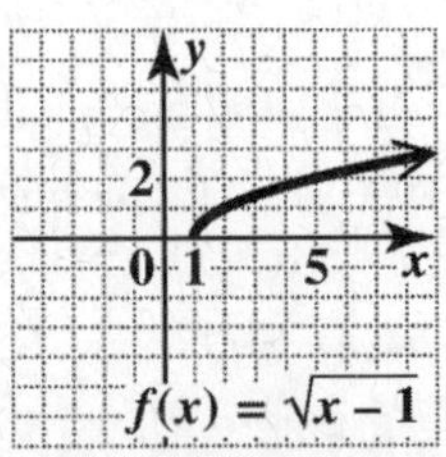

16. $f(x) = \sqrt[3]{x} + 4$

Since we can take the cube root of any real number, the domain is $(-\infty, \infty)$.

The result of a cube root can be any real number, so the range is $(-\infty, \infty)$. (The "+4" does not affect that range.)

x	$f(x) = \sqrt[3]{x} + 4$
-8	$\sqrt[3]{-8} + 4 = 2$
-1	$\sqrt[3]{-1} + 4 = 3$
0	$\sqrt[3]{0} + 4 = 4$
1	$\sqrt[3]{1} + 4 = 5$
8	$\sqrt[3]{8} + 4 = 6$

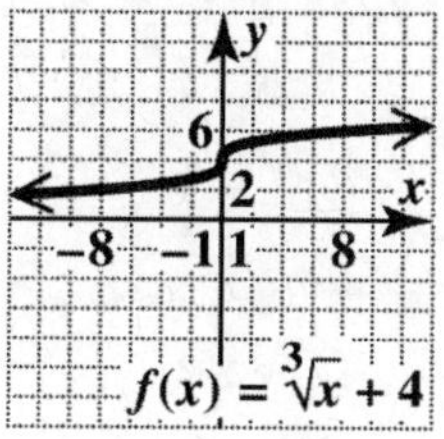

17. The base $\sqrt{38}$ is closest to $\sqrt{36} = 6$. The height $\sqrt{99}$ is closest to $\sqrt{100} = 10$. Use the estimates $b = 6$ and $h = 10$ in $\mathcal{A} = \frac{1}{2}bh$ to find an estimate of the area.

$$\mathcal{A} \approx \tfrac{1}{2}(6)(10) = 30$$

Choice **B** is the best estimate.

18. One way to evaluate $8^{2/3}$ is to first find the _cube (or third)_ root of _8_, which is _2_. Then raise that result to the _second_ power, to get an answer of _4_. Therefore, $8^{2/3} =$ _4_.

19. **A.** $(-27)^{2/3} = \left[(-27)^{1/3}\right]^2$

This number is a square, so it is a positive number.

B. $(-64)^{5/3} = \left[(-64)^{1/3}\right]^5$

This number is the odd power of an odd root of a negative number, so it is a negative number.

C. $(-100)^{1/2}$

This number is the square root of a negative number, so it is not a real number.

D. $(-32)^{1/5}$

This number is an odd root of a negative number, so it is a negative number.

The only positive number is choice **A**.

20. $a^{m/n} = \sqrt[n]{a^m}$

Since n is odd, $\sqrt[n]{a^m}$ is positive if a^m is positive and negative if a^m is negative. Since a is negative, a^m is positive if m is even and negative if m is odd.

(a) If a is negative and n is odd, then $a^{m/n}$ is positive if m is even.

(b) If a is negative and n is odd, then $a^{m/n}$ is negative if m is odd.

21. If a is negative and n is even, then $a^{1/n}$ is *not* a real number. An example is $(-4)^{1/2}$, which is *not* a real number.

22. $49^{1/2} = \sqrt{49} = 7$

23. $-121^{1/2} = -\sqrt{121} = -11$

24. $16^{5/4} = \left(16^{1/4}\right)^5 = 2^5 = 32$

25. $-8^{2/3} = -\left(8^{1/3}\right)^2 = -2^2 = -4$

26. $-\left(\frac{36}{25}\right)^{3/2} = -\left(\sqrt{\frac{36}{25}}\right)^3 = -\left(\frac{6}{5}\right)^3 = -\frac{216}{125}$

27. $\left(-\frac{1}{8}\right)^{-5/3} = (-8)^{5/3} = \left[(-8)^{1/3}\right]^5 = (-2)^5 = -32$

28. $\left(\frac{81}{10{,}000}\right)^{-3/4} = \left(\frac{10{,}000}{81}\right)^{3/4} = \left[\left(\frac{10{,}000}{81}\right)^{1/4}\right]^3 = \left(\frac{10}{3}\right)^3 = \frac{1000}{27}$

29. The base, -16, is negative, and $(-16)^3$ is still negative. The index, 4, is even, so $(-16)^{3/4}$ is not a real number.

30. First way: $8^{2/3} = \left(8^{1/3}\right)^2 = \left(\sqrt[3]{8}\right)^2$

Second way: $8^{2/3} = \left(8^2\right)^{1/3} = \sqrt[3]{8^2}$

31. The expression with fractional exponents, $a^{m/n}$, is equivalent to the radical expression, $\sqrt[n]{a^m}$. The denominator of the exponent is the index of the radical. For example, $\sqrt[3]{8^2} = \sqrt[3]{64} = 4$, and $8^{2/3} = (8^{1/3})^2 = 2^2 = 4$.

32. $(m + 3n)^{1/2} = \sqrt{m + 3n}$

33. $(3a + b)^{-5/3} = \dfrac{1}{(3a+b)^{5/3}} = \dfrac{1}{((3a+b)^{1/3})^5} = \dfrac{1}{\left(\sqrt[3]{3a+b}\right)^5}$ or $\dfrac{1}{\sqrt[3]{(3a+b)^5}}$

34. $\sqrt{7^9} = \left(7^9\right)^{1/2} = 7^{9/2}$

35. $\sqrt[5]{p^4} = \left(p^4\right)^{1/5} = p^{4/5}$

36. $5^{1/4} \cdot 5^{7/4} = 5^{1/4+7/4} = 5^{8/4} = 5^2$, or 25

37. $\dfrac{96^{2/3}}{96^{-1/3}} = 96^{2/3-(-1/3)} = 96^1 = 96$

38. $\dfrac{(a^{1/3})^4}{a^{2/3}} = \dfrac{a^{4/3}}{a^{2/3}} = a^{4/3-2/3} = a^{2/3}$

39. $\dfrac{y^{-1/3} \cdot y^{5/6}}{y} = \dfrac{y^{-2/6}y^{5/6}}{y^{6/6}}$
$= y^{-2/6+5/6-6/6}$
$= y^{-3/6} = y^{-1/2} = \dfrac{1}{y^{1/2}}$

40. $\left(\dfrac{z^{-1}x^{-3/5}}{2^{-2}z^{-1/2}x}\right)^{-1} = \dfrac{z^1x^{3/5}}{2^2z^{1/2}x^{-1}}$
$= \frac{1}{4}z^{1-1/2}x^{3/5-(-1)}$
$= \dfrac{z^{1/2}x^{8/5}}{4}$

41. $r^{-1/2}(r + r^{3/2})$
$= r^{-1/2}(r) + r^{-1/2}(r^{3/2})$
$= r^{-1/2+1} + r^{-1/2+3/2}$
$= r^{1/2} + r^{2/2}$
$= r^{1/2} + r$

42. $\sqrt[8]{s^4} = (s^4)^{1/8} = s^{4/8} = s^{1/2}$

43. $\sqrt[6]{r^9} = (r^9)^{1/6} = r^{9/6} = r^{3/2}$

44. $\dfrac{\sqrt{p^5}}{p^2} = \dfrac{p^{5/2}}{p^2} = p^{5/2-2}$
$= p^{5/2-4/2} = p^{1/2}$

45. $\sqrt[4]{k^3} \cdot \sqrt{k^3} = (k^3)^{1/4}(k^3)^{1/2}$
$= k^{3/4}k^{3/2} = k^{3/4+3/2}$
$= k^{3/4+6/4} = k^{9/4}$

46. $\sqrt[3]{m^5} \cdot \sqrt[3]{m^8} = (m^5)^{1/3}(m^8)^{1/3}$
$= m^{5/3}m^{8/3} = m^{5/3+8/3} = m^{13/3}$

47. $\sqrt[4]{\sqrt[3]{z}} = \sqrt[4]{z^{1/3}} = (z^{1/3})^{1/4} = z^{1/12}$

48. $\sqrt{\sqrt{\sqrt{x}}} = \sqrt{\sqrt{x^{1/2}}} = \sqrt{(x^{1/2})^{1/2}}$
$= \sqrt{x^{1/4}} = (x^{1/4})^{1/2} = x^{1/8}$

49. $\sqrt[3]{\sqrt[5]{x}} = \sqrt[3]{x^{1/5}} = (x^{1/5})^{1/3} = x^{1/15}$

50. $\sqrt{\sqrt[6]{\sqrt[3]{x}}} = \sqrt{\sqrt[6]{x^{1/3}}} = \sqrt{(x^{1/3})^{1/6}}$
$= \sqrt{x^{1/18}} = (x^{1/18})^{1/2} = x^{1/36}$

51. The product rule for exponents applies only if the bases are the same. The bases in $3^{1/4} \cdot 2^{1/5}$ are 3 and 2, respectively.

52. $\sqrt{6} \cdot \sqrt{11} = \sqrt{6 \cdot 11} = \sqrt{66}$

53. $\sqrt{5} \cdot \sqrt{r} = \sqrt{5 \cdot r} = \sqrt{5r}$

54. $\sqrt[3]{6} \cdot \sqrt[3]{5} = \sqrt[3]{6 \cdot 5} = \sqrt[3]{30}$

55. $\sqrt[4]{7} \cdot \sqrt[4]{3} = \sqrt[4]{7 \cdot 3} = \sqrt[4]{21}$

56. $\sqrt{20} = \sqrt{4 \cdot 5} = \sqrt{4}\sqrt{5} = 2\sqrt{5}$

57. $\sqrt{75} = \sqrt{25 \cdot 3} = \sqrt{25}\sqrt{3} = 5\sqrt{3}$

58. $-\sqrt{125} = -\sqrt{25 \cdot 5} = -5\sqrt{5}$

59. $\sqrt[3]{-108} = \sqrt[3]{-27 \cdot 4} = -3\sqrt[3]{4}$

60. $\sqrt{100y^7} = \sqrt{100y^6 \cdot y} = 10y^3\sqrt{y}$

61. $\sqrt[3]{64p^4q^6} = \sqrt[3]{64p^3q^6 \cdot p} = 4pq^2\sqrt[3]{p}$

62. $\sqrt[3]{108a^8b^5} = \sqrt[3]{27a^6b^3 \cdot 4a^2b^2}$
$= 3a^2b\sqrt[3]{4a^2b^2}$

63. $\sqrt[3]{632r^8t^4} = \sqrt[3]{8r^6t^3 \cdot 79r^2t}$
$= 2r^2t\sqrt[3]{79r^2t}$

64. $\sqrt{\dfrac{y^3}{144}} = \dfrac{\sqrt{y^3}}{\sqrt{144}} = \dfrac{\sqrt{y^2 \cdot y}}{12} = \dfrac{y\sqrt{y}}{12}$

65. $\sqrt[3]{\dfrac{m^{15}}{27}} = \dfrac{\sqrt[3]{m^{15}}}{\sqrt[3]{27}} = \dfrac{\sqrt[3]{(m^5)^3}}{\sqrt[3]{3^3}} = \dfrac{m^5}{3}$

66. $\sqrt[3]{\dfrac{r^2}{8}} = \dfrac{\sqrt[3]{r^2}}{\sqrt[3]{8}} = \dfrac{\sqrt[3]{r^2}}{2}$

67. $\sqrt[4]{\dfrac{a^9}{81}} = \dfrac{\sqrt[4]{a^9}}{\sqrt[4]{81}} = \dfrac{\sqrt[4]{a^8 \cdot a}}{3} = \dfrac{a^2\sqrt[4]{a}}{3}$

68. $\sqrt[6]{15^3} = 15^{3/6} = 15^{1/2} = \sqrt{15}$

69. $\sqrt[4]{p^6} = (p^6)^{1/4} = p^{6/4} = p^{3/2}$
$= p^{2/2}p^{1/2} = p\sqrt{p}$

70. $\sqrt[3]{2} \cdot \sqrt[4]{5} = 2^{1/3} \cdot 5^{1/4}$
$= 2^{4/12} \cdot 5^{3/12}$
$= (2^4 \cdot 5^3)^{1/12}$
$= \sqrt[12]{16 \cdot 125} = \sqrt[12]{2000}$

71. $\sqrt{x} \cdot \sqrt[5]{x} = x^{1/2} \cdot x^{1/5}$
$= x^{5/10} \cdot x^{2/10}$
$= x^{7/10} = \sqrt[10]{x^7}$

72. Substitute 8 for a and 6 for b in the Pythagorean theorem to find the hypotenuse, x.

$$x^2 = a^2 + b^2$$
$$x = \sqrt{a^2 + b^2} = \sqrt{8^2 + 6^2}$$
$$= \sqrt{64 + 36} = \sqrt{100} = 10$$

The length of the hypotenuse is 10.

73. The given points are $(-4, 7)$ and $(10, 6)$.

$$\begin{aligned} d &= \sqrt{(x_2 - x_1)^2 + (y_2 - y_1)^2} \\ &= \sqrt{[10 - (-4)]^2 + (6 - 7)^2} \\ &= \sqrt{14^2 + (-1)^2} \\ &= \sqrt{196 + 1} = \sqrt{197} \end{aligned}$$

74. $$\begin{aligned} 2\sqrt{8} - 3\sqrt{50} &= 2\sqrt{4\cdot 2} - 3\sqrt{25\cdot 2} \\ &= 2\cdot 2\sqrt{2} - 3\cdot 5\sqrt{2} \\ &= 4\sqrt{2} - 15\sqrt{2} = -11\sqrt{2} \end{aligned}$$

75. $$\begin{aligned} 8\sqrt{80} - 3\sqrt{45} &= 8\sqrt{16\cdot 5} - 3\sqrt{9\cdot 5} \\ &= 8\cdot 4\sqrt{5} - 3\cdot 3\sqrt{5} \\ &= 32\sqrt{5} - 9\sqrt{5} = 23\sqrt{5} \end{aligned}$$

76. $$\begin{aligned} -\sqrt{27y} + 2\sqrt{75y} &= -\sqrt{9\cdot 3y} + 2\sqrt{25\cdot 3y} \\ &= -3\sqrt{3y} + 2\cdot 5\sqrt{3y} \\ &= -3\sqrt{3y} + 10\sqrt{3y} \\ &= 7\sqrt{3y} \end{aligned}$$

77. $$\begin{aligned} &2\sqrt{54m^3} + 5\sqrt{96m^3} \\ &= 2\sqrt{9m^2\cdot 6m} + 5\sqrt{16m^2\cdot 6m} \\ &= 2\cdot 3m\sqrt{6m} + 5\cdot 4m\sqrt{6m} \\ &= 6m\sqrt{6m} + 20m\sqrt{6m} = 26m\sqrt{6m} \end{aligned}$$

78. $$\begin{aligned} 3\sqrt[3]{54} + 5\sqrt[3]{16} &= 3\sqrt[3]{27\cdot 2} + 5\sqrt[3]{8\cdot 2} \\ &= 3\cdot 3\sqrt[3]{2} + 5\cdot 2\sqrt[3]{2} \\ &= 9\sqrt[3]{2} + 10\sqrt[3]{2} = 19\sqrt[3]{2} \end{aligned}$$

79. $$\begin{aligned} -6\sqrt[4]{32} + \sqrt[4]{512} &= -6\sqrt[4]{16\cdot 2} + \sqrt[4]{256\cdot 2} \\ &= -6\cdot 2\sqrt[4]{2} + 4\sqrt[4]{2} \\ &= -12\sqrt[4]{2} + 4\sqrt[4]{2} = -8\sqrt[4]{2} \end{aligned}$$

80. Add the measures of the sides.

$$\begin{aligned} P &= a + b + c + d \\ P &= 4\sqrt{8} + 6\sqrt{12} + 8\sqrt{2} + 3\sqrt{48} \\ &= 4\sqrt{4\cdot 2} + 6\sqrt{4\cdot 3} + 8\sqrt{2} + 3\sqrt{16\cdot 3} \\ &= 4\cdot 2\sqrt{2} + 6\cdot 2\sqrt{3} + 8\sqrt{2} + 3\cdot 4\sqrt{3} \\ &= 8\sqrt{2} + 12\sqrt{3} + 8\sqrt{2} + 12\sqrt{3} \\ &= 16\sqrt{2} + 24\sqrt{3} \end{aligned}$$

The perimeter is $(16\sqrt{2} + 24\sqrt{3})$ feet.

81. Add the measures of the sides.

$$\begin{aligned} P &= a + b + c \\ P &= 2\sqrt{27} + \sqrt{108} + \sqrt{50} \\ &= 2\sqrt{9\cdot 3} + \sqrt{36\cdot 3} + \sqrt{25\cdot 2} \\ &= 2\cdot 3\sqrt{3} + 6\sqrt{3} + 5\sqrt{2} \\ &= 6\sqrt{3} + 6\sqrt{3} + 5\sqrt{2} \\ &= 12\sqrt{3} + 5\sqrt{2} \end{aligned}$$

The perimeter is $(12\sqrt{3} + 5\sqrt{2})$ feet.

82. $$\begin{aligned} (\sqrt{3}+1)(\sqrt{3}-2) &= 3 - 2\sqrt{3} + \sqrt{3} - 2 \\ &= 1 - \sqrt{3} \end{aligned}$$

83. $$\begin{aligned} (\sqrt{7}+\sqrt{5})(\sqrt{7}-\sqrt{5}) &= (\sqrt{7})^2 - (\sqrt{5})^2 \\ &= 7 - 5 = 2 \end{aligned}$$

84. $$\begin{aligned} &(3\sqrt{2}+1)(2\sqrt{2}-3) \\ &= 6\cdot 2 - 9\sqrt{2} + 2\sqrt{2} - 3 \\ &= 12 - 7\sqrt{2} - 3 = 9 - 7\sqrt{2} \end{aligned}$$

85. $$\begin{aligned} &(\sqrt{13}-\sqrt{2})^2 \\ &= (\sqrt{13})^2 - 2\cdot\sqrt{13}\cdot\sqrt{2} + (\sqrt{2})^2 \\ &= 13 - 2\sqrt{26} + 2 = 15 - 2\sqrt{26} \end{aligned}$$

86. $$\begin{aligned} &(\sqrt[3]{2}+3)(\sqrt[3]{4} - 3\sqrt[3]{2} + 9) \\ &= \sqrt[3]{2}\cdot\sqrt[3]{4} - \sqrt[3]{2}\cdot 3\sqrt[3]{2} + 9\sqrt[3]{2} \\ &\quad + 3\sqrt[3]{4} - 3\cdot 3\sqrt[3]{2} + 27 \\ &= \sqrt[3]{8} - 3\sqrt[3]{4} + 9\sqrt[3]{2} + 3\sqrt[3]{4} - 9\sqrt[3]{2} + 27 \\ &= 2 + 27 = 29 \end{aligned}$$

87. $$\begin{aligned} &(\sqrt[3]{4y}-1)(\sqrt[3]{4y}+3) \\ &= \sqrt[3]{16y^2} + 3\sqrt[3]{4y} - \sqrt[3]{4y} - 3 \\ &= \sqrt[3]{8\cdot 2y^2} + 2\sqrt[3]{4y} - 3 \\ &= 2\sqrt[3]{2y^2} + 2\sqrt[3]{4y} - 3 \end{aligned}$$

88. Show that $15 - 2\sqrt{26} \neq 13\sqrt{26}$.

Find a calculator approximation of each term.

$$4.801960973 \neq 66.28725368$$

Therefore, $15 - 2\sqrt{26} \neq 13\sqrt{26}$.

89. Multiplying by $\sqrt[3]{6}$ still results in an expression with a radical in the denominator.

$$\frac{5\sqrt[3]{6}}{\sqrt[3]{6}\cdot\sqrt[3]{6}} = \frac{5\sqrt[3]{6}}{\sqrt[3]{36}}$$

To rationalize the denominator, multiply by $\sqrt[3]{6^2}$ or $\sqrt[3]{36}$.

$$\frac{5\sqrt[3]{6^2}}{\sqrt[3]{6}\cdot\sqrt[3]{6^2}} = \frac{5\sqrt[3]{36}}{\sqrt[3]{6^3}} = \frac{5\sqrt[3]{36}}{6}$$

90. $$\frac{\sqrt{6}}{\sqrt{5}} = \frac{\sqrt{6}\cdot\sqrt{5}}{\sqrt{5}\cdot\sqrt{5}} = \frac{\sqrt{30}}{5}$$

91. $$\frac{-6\sqrt{3}}{\sqrt{2}} = \frac{-6\sqrt{3}\cdot\sqrt{2}}{\sqrt{2}\cdot\sqrt{2}} = \frac{-6\sqrt{6}}{2} = -3\sqrt{6}$$

92. $$\frac{3\sqrt{7p}}{\sqrt{y}} = \frac{3\sqrt{7p}\cdot\sqrt{y}}{\sqrt{y}\cdot\sqrt{y}} = \frac{3\sqrt{7py}}{y}$$

93. $$\begin{aligned} \sqrt{\frac{11}{8}} &= \frac{\sqrt{11}}{\sqrt{8}} = \frac{\sqrt{11}}{\sqrt{4\cdot 2}} = \frac{\sqrt{11}}{2\sqrt{2}} = \frac{\sqrt{11}\cdot\sqrt{2}}{2\sqrt{2}\cdot\sqrt{2}} \\ &= \frac{\sqrt{22}}{2\cdot 2} = \frac{\sqrt{22}}{4} \end{aligned}$$

94. $-\sqrt[3]{\dfrac{9}{25}} = -\dfrac{\sqrt[3]{9}}{\sqrt[3]{5^2}} = -\dfrac{\sqrt[3]{9}\cdot\sqrt[3]{5}}{\sqrt[3]{5^2}\cdot\sqrt[3]{5}}$

$= -\dfrac{\sqrt[3]{45}}{\sqrt[3]{5^3}} = -\dfrac{\sqrt[3]{45}}{5}$

95. $\sqrt[3]{\dfrac{108m^3}{n^5}} = \dfrac{\sqrt[3]{108m^3}}{\sqrt[3]{n^5}} = \dfrac{\sqrt[3]{27m^3\cdot 4}}{\sqrt[3]{n^3\cdot n^2}}$

$= \dfrac{3m\sqrt[3]{4}}{n\sqrt[3]{n^2}} = \dfrac{3m\sqrt[3]{4}\cdot\sqrt[3]{n}}{n\sqrt[3]{n^2}\cdot\sqrt[3]{n}}$

$= \dfrac{3m\sqrt[3]{4n}}{n\cdot n} = \dfrac{3m\sqrt[3]{4n}}{n^2}$

96. $\dfrac{1}{\sqrt{2}+\sqrt{7}}$

Multiply the numerator and denominator by the conjugate of the denominator, $\sqrt{2}-\sqrt{7}$.

$= \dfrac{1(\sqrt{2}-\sqrt{7})}{(\sqrt{2}+\sqrt{7})(\sqrt{2}-\sqrt{7})}$

$= \dfrac{\sqrt{2}-\sqrt{7}}{2-7} = \dfrac{\sqrt{2}-\sqrt{7}}{-5}$

97. $\dfrac{-5}{\sqrt{6}-3}$

Multiply the numerator and denominator by the conjugate of the denominator, $\sqrt{6}+3$.

$= \dfrac{-5(\sqrt{6}+3)}{(\sqrt{6}-3)(\sqrt{6}+3)} = \dfrac{-5(\sqrt{6}+3)}{6-9}$

$= \dfrac{-5(\sqrt{6}+3)}{-3} = \dfrac{5(\sqrt{6}+3)}{3}$

98. $\dfrac{2-2\sqrt{5}}{8} = \dfrac{2(1-\sqrt{5})}{2\cdot 4} = \dfrac{1-\sqrt{5}}{4}$

99. $\dfrac{4-8\sqrt{8}}{12} = \dfrac{4(1-2\sqrt{8})}{3\cdot 4} = \dfrac{1-2\sqrt{8}}{3}$

$= \dfrac{1-2\sqrt{4\cdot 2}}{3} = \dfrac{1-4\sqrt{2}}{3}$

100. $\dfrac{-18+\sqrt{27}}{6} = \dfrac{-18+\sqrt{9}\cdot\sqrt{3}}{6} = \dfrac{-18+3\sqrt{3}}{6}$

$= \dfrac{3(-6+\sqrt{3})}{3\cdot 2} = \dfrac{-6+\sqrt{3}}{2}$

101. $\sqrt{8x+9} = 5$

$(\sqrt{8x+9})^2 = 5^2$ *Square.*

$8x+9 = 25$

$8x = 16$

$x = 2$

Check $x=2$: $\sqrt{25}=5$ *True*

The solution set is $\{2\}$.

102. $\sqrt{2x-3}-3=0$

$\sqrt{2x-3}=3$ *Isolate.*

$(\sqrt{2x-3})^2 = 3^2$ *Square.*

$2x-3=9$

$2x=12$

$x=6$

Check $x=6$: $\sqrt{9}-3=0$ *True*

The solution set is $\{6\}$.

103. $\sqrt{3x+1}-2=-3$

$\sqrt{3x+1}=-1$ *Isolate.*

This equation has no solution, because $\sqrt{3x+1}$ cannot be negative.

The solution set is $\emptyset$.

104. $\sqrt{7x+1}=x+1$

$(\sqrt{7x+1})^2=(x+1)^2$ *Square.*

$7x+1=x^2+2x+1$

$0=x^2-5x$

$0=x(x-5)$

$x=0$ or $x=5$

Check $x=0$: $\sqrt{1}=1$ *True*

Check $x=5$: $\sqrt{36}=6$ *True*

The solution set is $\{0,5\}$.

105. $3\sqrt{x}=\sqrt{10x-9}$

$(3\sqrt{x})^2=(\sqrt{10x-9})^2$ *Square.*

$9x=10x-9$

$9=x$

Check $x=9$: $3\sqrt{9}=\sqrt{81}$ *True*

The solution set is $\{9\}$.

106. $\sqrt{x^2+3x+7}=x+2$

$(\sqrt{x^2+3x+7})^2=(x+2)^2$ *Square.*

$x^2+3x+7=x^2+4x+4$

$3=x$

Check $x=3$: $\sqrt{25}=5$ *True*

The solution set is $\{3\}$.

107. $\sqrt{x+2}-\sqrt{x-3}=1$

Get one radical on each side of the equals sign.

$\sqrt{x+2}=1+\sqrt{x-3}$

Square each side.

$(\sqrt{x+2})^2=(1+\sqrt{x-3})^2$

$x+2=1+2\sqrt{x-3}+x-3$

$4=2\sqrt{x-3}$

$2=\sqrt{x-3}$

Square each side again.

$2^2=(\sqrt{x-3})^2$

$4=x-3$

$7=x$

Check $x=7$: $\sqrt{9}-\sqrt{4}=1$ *True*

The solution set is $\{7\}$.

108. $\sqrt[3]{5x-1} = \sqrt[3]{3x-2}$

Cube each side.

$$\left(\sqrt[3]{5x-1}\right)^3 = \left(\sqrt[3]{3x-2}\right)^3$$
$$5x-1 = 3x-2$$
$$2x = -1$$
$$x = -\tfrac{1}{2}$$

Check $x = -\frac{1}{2}$: $\sqrt[3]{-\frac{7}{2}} = \sqrt[3]{-\frac{7}{2}}$ *True*

The solution set is $\{-\frac{1}{2}\}$.

109. $\sqrt[3]{2x^2+3x-7} = \sqrt[3]{2x^2+4x+6}$

Cube each side.

$$\left(\sqrt[3]{2x^2+3x-7}\right)^3 = \left(\sqrt[3]{2x^2+4x+6}\right)^3$$
$$2x^2+3x-7 = 2x^2+4x+6$$
$$-13 = x$$

Check $x = -13$: $\sqrt[3]{292} = \sqrt[3]{292}$ *True*

The solution set is $\{-13\}$.

110. $\sqrt[3]{3x^2-4x+6} = \sqrt[3]{3x^2-2x+8}$

$$\left(\sqrt[3]{3x^2-4x+6}\right)^3 = \left(\sqrt[3]{3x^2-2x+8}\right)^3$$
$$3x^2-4x+6 = 3x^2-2x+8$$
$$-2 = 2x$$
$$-1 = x$$

Check $x = -1$: $\sqrt[3]{13} = \sqrt[3]{13}$ *True*

The solution set is $\{-1\}$.

111. $\sqrt[3]{1-2x} - \sqrt[3]{-x-13} = 0$

$$\sqrt[3]{1-2x} = \sqrt[3]{-x-13}$$

Cube each side.

$$\left(\sqrt[3]{1-2x}\right)^3 = \left(\sqrt[3]{-x-13}\right)^3$$
$$1-2x = -x-13$$
$$14 = x$$

Check $x = 14$: $\sqrt[3]{-27} - \sqrt[3]{-27} = 0$ *True*

The solution set is $\{14\}$.

112. $\sqrt[3]{11-2x} - \sqrt[3]{-1-5x} = 0$

$$\sqrt[3]{11-2x} = \sqrt[3]{-1-5x}$$
$$\left(\sqrt[3]{11-2x}\right)^3 = \left(\sqrt[3]{-1-5x}\right)^3$$
$$11-2x = -1-5x$$
$$3x = -12$$
$$x = -4$$

Check $x = -4$: $\sqrt[3]{19} - \sqrt[3]{19} = 0$ *True*

The solution set is $\{-4\}$.

113. $\sqrt[4]{x-1} + 2 = 0$

$$\sqrt[4]{x-1} = -2$$

This equation has no solution, because $\sqrt[4]{x-1}$ cannot be negative.

The solution set is $\emptyset$.

114. $\sqrt[4]{2x+3} + 1 = 0$

$$\sqrt[4]{2x+3} = -1$$

This equation has no solution, because $\sqrt[4]{2x+3}$ cannot be negative.

The solution set is $\emptyset$.

115. $\sqrt[4]{x+7} = \sqrt[4]{2x}$

Raise each side to the fourth power.

$$\left(\sqrt[4]{x+7}\right)^4 = \left(\sqrt[4]{2x}\right)^4$$
$$x+7 = 2x$$
$$7 = x$$

Check $x = 7$: $\sqrt[4]{14} = \sqrt[4]{14}$ *True*

The solution set is $\{7\}$.

116. $\sqrt[4]{x+8} = \sqrt[4]{3x}$

Raise each side to the fourth power.

$$\left(\sqrt[4]{x+8}\right)^4 = \left(\sqrt[4]{3x}\right)^4$$
$$x+8 = 3x$$
$$8 = 2x$$
$$4 = x$$

Check $x = 4$: $\sqrt[4]{12} = \sqrt[4]{12}$ *True*

The solution set is $\{4\}$.

117. (a) Solve $L = \sqrt{H^2+W^2}$ for H.

$$L^2 = H^2 + W^2$$
$$L^2 - W^2 = H^2$$
$$\sqrt{L^2-W^2} = H$$

(b) Substitute 12 for L and 9 for W in $H = \sqrt{L^2-W^2}$.

$$H = \sqrt{12^2-9^2}$$
$$= \sqrt{144-81} = \sqrt{63}$$

To the nearest tenth of a foot, the height is approximately 7.9 feet.

118. $\sqrt{-25} = i\sqrt{25} = 5i$

119. $\sqrt{-200} = i\sqrt{100\cdot 2} = 10i\sqrt{2}$

120. If a is a positive real number, then $-a$ is negative. So, $\sqrt{-a}$ is not a real number. Therefore, $-\sqrt{-a}$ is not a real number either.

121. $(-2+5i)+(-8-7i)$

$$= [-2+(-8)] + [5+(-7)]i$$
$$= -10-2i$$

122. $(5+4i)-(-9-3i)$

$$= [(5-(-9)] + [(4-(-3)]i$$
$$= 14+7i$$

123. $\sqrt{-5}\cdot\sqrt{-7} = i\sqrt{5}\cdot i\sqrt{7} = i^2\sqrt{35}$

$$= -1\left(\sqrt{35}\right) = -\sqrt{35}$$

124. $\sqrt{-25}\cdot\sqrt{-81} = 5i\cdot 9i = 45i^2$
$= 45(-1) = -45$

125. $\dfrac{\sqrt{-72}}{\sqrt{-8}} = \dfrac{i\sqrt{72}}{i\sqrt{8}} = \sqrt{\dfrac{72}{8}} = \sqrt{9} = 3$

126. $(2+3i)(1-i) = 2 - 2i + 3i - 3i^2$
$= 2 + i - 3(-1)$
$= 2 + i + 3$
$= 5 + i$

127. $(6-2i)^2 = 6^2 - 2\cdot 6\cdot 2i + (2i)^2$
$= 36 - 24i + 4i^2$
$= 36 - 24i + 4(-1)$
$= 36 - 24i - 4$
$= 32 - 24i$

128. $\dfrac{3-i}{2+i}$
Multiply by the conjugate of the denominator, $2 - i$.
$= \dfrac{(3-i)(2-i)}{(2+i)(2-i)} = \dfrac{6 - 3i - 2i + i^2}{4 - i^2}$
$= \dfrac{6 - 5i - 1}{4-(-1)} = \dfrac{5-5i}{5} = \dfrac{5(1-i)}{5}$
$= 1 - i$

129. $\dfrac{5+14i}{2+3i}$
Multiply by the conjugate of the denominator, $2 - 3i$.
$= \dfrac{(5+14i)(2-3i)}{(2+3i)(2-3i)} = \dfrac{10 - 15i + 28i - 42i^2}{4 - 9i^2}$
$= \dfrac{10 + 13i - 42(-1)}{4 - 9(-1)} = \dfrac{52 + 13i}{13}$
$= \dfrac{13(4+i)}{13} = 4 + i$

130. $i^{11} = i^8\cdot i^3 = (i^4)^2\cdot i^2\cdot i$
$= 1^2\cdot(-1)\cdot i = -i$

131. $i^{36} = (i^4)^9 = 1^9 = 1$

132. $i^{-10} = \dfrac{1}{i^{10}} = \dfrac{1}{i^8\cdot i^2} = \dfrac{1}{(i^4)^2\cdot(-1)}$
$= \dfrac{1}{1^2\cdot(-1)} = \dfrac{1}{-1} = -1$

Another method:
$i^{-10} = i^{-10}\cdot i^{12} = i^2 = -1$
Note that $i^{12} = (i^4)^3 = 1^3 = 1$.

133. $i^{-8} = \dfrac{1}{i^8} = \dfrac{1}{(i^4)^2} = \dfrac{1}{1^2} = \dfrac{1}{1} = 1$

134. [10.1] $-\sqrt[4]{256} = -\left(\sqrt[4]{4^4}\right) = -(4) = -4$

135. [10.2] $1000^{-2/3} = \dfrac{1}{1000^{2/3}} = \dfrac{1}{\left[(10^3)^{1/3}\right]^2}$
$= \dfrac{1}{10^2} = \dfrac{1}{100}$

136. [10.2] $\dfrac{z^{-1/5}\cdot z^{3/10}}{z^{7/10}} = z^{-2/10+3/10-7/10}$
$= z^{-6/10} = \dfrac{1}{z^{6/10}} = \dfrac{1}{z^{3/5}}$

137. [10.2] $\sqrt[4]{k^{24}} = k^{24/4} = k^6$

138. [10.3] $\sqrt[3]{54z^9t^8} = \sqrt[3]{27z^9t^6\cdot 2t^2} = 3z^3t^2\sqrt[3]{2t^2}$

139. [10.4] $-5\sqrt{18} + 12\sqrt{72}$
$= -5\sqrt{9\cdot 2} + 12\sqrt{36\cdot 2}$
$= -5\cdot 3\sqrt{2} + 12\cdot 6\sqrt{2}$
$= -15\sqrt{2} + 72\sqrt{2} = 57\sqrt{2}$

140. [10.5] $\dfrac{-1}{\sqrt{12}} = \dfrac{-1}{\sqrt{4\cdot 3}} = \dfrac{-1}{2\sqrt{3}} = \dfrac{-1\cdot\sqrt{3}}{2\sqrt{3}\cdot\sqrt{3}}$
$= \dfrac{-\sqrt{3}}{2\cdot 3} = \dfrac{-\sqrt{3}}{6}$

141. [10.5] $\sqrt[3]{\dfrac{12}{25}} = \dfrac{\sqrt[3]{12}}{\sqrt[3]{25}} = \dfrac{\sqrt[3]{12}}{\sqrt[3]{5^2}}$
$= \dfrac{\sqrt[3]{12}\cdot\sqrt[3]{5}}{\sqrt[3]{5^2}\cdot\sqrt[3]{5}} = \dfrac{\sqrt[3]{60}}{\sqrt[3]{5^3}} = \dfrac{\sqrt[3]{60}}{5}$

142. [10.7] $i^{-1000} = \dfrac{1}{i^{1000}} = \dfrac{1}{(i^4)^{250}}$
$= \dfrac{1}{1^{250}} = \dfrac{1}{1} = 1$

143. [10.7] $\sqrt{-49} = i\sqrt{49} = 7i$

144. [10.7] $(4-9i) + (-1+2i)$
$= (4-1) + (-9+2)i$
$= 3 - 7i$

145. [10.7] $\dfrac{\sqrt{50}}{\sqrt{-2}} = \dfrac{\sqrt{25\cdot 2}}{i\sqrt{2}} = \dfrac{5\sqrt{2}}{i\sqrt{2}} = \dfrac{5}{i}$
The conjugate of i is $-i$.
$\dfrac{5(-i)}{i(-i)} = \dfrac{-5i}{-i^2} = \dfrac{-5i}{-(-1)} = -5i$

146. [10.5] $\dfrac{3+\sqrt{54}}{6} = \dfrac{3+\sqrt{9\cdot 6}}{6} = \dfrac{3+3\sqrt{6}}{6}$
$= \dfrac{3(1+\sqrt{6})}{2\cdot 3} = \dfrac{1+\sqrt{6}}{2}$

147. [10.7] $(3+2i)^2 = 3^2 + 2\cdot 3\cdot 2i + (2i)^2$
$= 9 + 12i + 4i^2$
$= 9 + 12i + 4(-1)$
$= 5 + 12i$

148. [10.4] $8\sqrt[3]{x^3y^2} - 2x\sqrt[3]{y^2} = 8x\sqrt[3]{y^2} - 2x\sqrt[3]{y^2}$
$= 6x\sqrt[3]{y^2}$

149. [10.4] $9\sqrt{5} - 4\sqrt{15}$ cannot be simplified further.

150. [10.5] $(\sqrt{5}-\sqrt{3})(\sqrt{7}+\sqrt{3})$
$= \sqrt{35} + \sqrt{15} - \sqrt{21} - 3$

151. [10.6]

$$\sqrt{x+4} = x-2$$
$$\left(\sqrt{x+4}\right)^2 = (x-2)^2 \quad \textit{Square.}$$
$$x+4 = x^2-4x+4$$
$$0 = x^2-5x$$
$$0 = x(x-5)$$

$$x = 0 \quad \text{or} \quad x-5 = 0$$
$$x = 5$$

Check $x=0$: $\sqrt{4} = -2$ *False*

Check $x=5$: $\sqrt{9} = 3$ *True*

The solution set is $\{5\}$.

152. [10.6]

$$\sqrt[3]{2x-9} = \sqrt[3]{5x+3}$$

Cube each side.

$$\left(\sqrt[3]{2x-9}\right)^3 = \left(\sqrt[3]{5x+3}\right)^3$$
$$2x-9 = 5x+3$$
$$-3x = 12$$
$$x = -4$$

Check $x=-4$: $\sqrt[3]{-17} = \sqrt[3]{-17}$ *True*

The solution set is $\{-4\}$.

153. [10.6]

$$\sqrt{6+2x}-1 = \sqrt{7-2x}$$

Square each side.

$$\left(\sqrt{6+2x}-1\right)^2 = \left(\sqrt{7-2x}\right)^2$$
$$6+2x-2\sqrt{6+2x}+1 = 7-2x$$
$$4x = 2\sqrt{6+2x}$$
$$2x = \sqrt{6+2x}$$

Square each side again.

$$(2x)^2 = \left(\sqrt{6+2x}\right)^2$$
$$4x^2 = 6+2x$$
$$4x^2-2x-6 = 0$$
$$2x^2-x-3 = 0$$
$$(2x-3)(x+1) = 0$$

$$2x-3 = 0 \quad \text{or} \quad x+1 = 0$$
$$x = \tfrac{3}{2} \quad \text{or} \quad x = -1$$

Check $x=\frac{3}{2}$: $\sqrt{9}-1 = \sqrt{4}$ *True*

Check $x=-1$: $\sqrt{4}-1 = \sqrt{9}$ *False*

The solution set is $\left\{\frac{3}{2}\right\}$.

154. [10.6]

$$\sqrt{7x+11}-5 = 0$$
$$\sqrt{7x+11} = 5$$
$$\left(\sqrt{7x+11}\right)^2 = 5^2$$
$$7x+11 = 25$$
$$7x = 14$$
$$x = 2$$

Check $x=2$: $\sqrt{25}-5 = 0$ *True*

The solution set is $\{2\}$.

155. [10.6] $\sqrt{6x+2}-\sqrt{5x+3} = 0$

Get one radical on each side of the equals sign.

$$\sqrt{6x+2} = \sqrt{5x+3}$$

Square each side.

$$\left(\sqrt{6x+2}\right)^2 = \left(\sqrt{5x+3}\right)^2$$
$$6x+2 = 5x+3$$
$$x = 1$$

Check $x=1$: $\sqrt{8}-\sqrt{8} = 0$ *True*

The solution set is $\{1\}$.

156. [10.6]

$$\sqrt{3+5x}-\sqrt{x+11} = 0$$
$$\sqrt{3+5x} = \sqrt{x+11}$$
$$\left(\sqrt{3+5x}\right)^2 = \left(\sqrt{x+11}\right)^2$$
$$3+5x = x+11$$
$$4x = 8$$
$$x = 2$$

Check $x=2$: $\sqrt{13}-\sqrt{13} = 0$ *True*

The solution set is $\{2\}$.

157. [10.6]

$$3\sqrt{x} = \sqrt{8x+9}$$
$$\left(3\sqrt{x}\right)^2 = \left(\sqrt{8x+9}\right)^2 \quad \textit{Square.}$$
$$9x = 8x+9$$
$$x = 9$$

Check $x=9$: $3\sqrt{9} = \sqrt{81}$ *True*

The solution set is $\{9\}$.

158. [10.6]

$$6\sqrt{x} = \sqrt{30x+24}$$
$$\left(6\sqrt{x}\right)^2 = \left(\sqrt{30x+24}\right)^2$$
$$36x = 30x+24$$
$$6x = 24$$
$$x = 4$$

Check $x=4$: $6\sqrt{4} = \sqrt{144}$ *True*

The solution set is $\{4\}$.

159. [10.6]

$$\sqrt{11+2x}+1 = \sqrt{5x+1}$$

Square each side.

$$\left(\sqrt{11+2x}+1\right)^2 = \left(\sqrt{5x+1}\right)^2$$
$$\left(\sqrt{11+2x}\right)^2 + 2\cdot\sqrt{11+2x}\cdot 1+1 = 5x+1$$
$$11+2x+2\sqrt{11+2x}+1 = 5x+1$$

Isolate the remaining radical.

$$2\sqrt{11+2x} = 3x-11$$

Square each side again.

$$\left(2\sqrt{11+2x}\right)^2 = (3x-11)^2$$
$$4(11+2x) = 9x^2-66x+121$$
$$44+8x = 9x^2-66x+121$$
$$0 = 9x^2-74x+77$$
$$0 = (9x-11)(x-7)$$

$$9x-11 = 0 \quad \text{or} \quad x-7 = 0$$
$$x = \tfrac{11}{9} \quad \text{or} \quad x = 7$$

Check $x = \frac{11}{9}$: $\sqrt{\frac{121}{9}} + 1 = \sqrt{\frac{64}{9}}$ *False*

Check $x = 7$: $\sqrt{25} + 1 = \sqrt{36}$ *True*

The solution set is $\{7\}$.

160. [10.6]

$\sqrt{5x+6} - \sqrt{x+3} = 3$

Get one radical on each side.

$$\begin{aligned} \sqrt{5x+6} &= 3 + \sqrt{x+3} \\ (\sqrt{5x+6})^2 &= (3+\sqrt{x+3})^2 \\ 5x+6 &= 9 + 6\sqrt{x+3} + x + 3 \end{aligned}$$

Isolate the remaining radical.

$$\begin{aligned} 4x - 6 &= 6\sqrt{x+3} \\ 2x - 3 &= 3\sqrt{x+3} \\ (2x-3)^2 &= (3\sqrt{x+3})^2 \\ 4x^2 - 12x + 9 &= 9(x+3) \\ 4x^2 - 12x + 9 &= 9x + 27 \\ 4x^2 - 21x - 18 &= 0 \\ (4x+3)(x-6) &= 0 \end{aligned}$$

$4x + 3 = 0$ or $x - 6 = 0$

$x = -\frac{3}{4}$ $\qquad x = 6$

Check $x = -\frac{3}{4}$: $\sqrt{\frac{9}{4}} - \sqrt{\frac{9}{4}} = 3$ *False*

Check $x = 6$: $\sqrt{36} - \sqrt{9} = 3$ *True*

The solution set is $\{6\}$.

Chapter 10 Test

1. $\sqrt{841} = 29$, because $29^2 = 841$, so $-\sqrt{841} = -29$.

2. $\sqrt[3]{-512} = -8$, because $(-8)^3 = -512$.

3. $125^{1/3} = \sqrt[3]{125} = 5$, because $5^3 = 125$.

4. $\sqrt{146.25} \approx \sqrt{144} = 12$, because $12^2 = 144$, so choice **C** is the best estimate.

5. $\sqrt{478} \approx 21.863$

6. $\sqrt[3]{-832} \approx -9.405$

7. $f(x) = \sqrt{x+6}$

For the radicand to be nonnegative, we must have

$x + 6 \geq 0$ or $x \geq -6$.

Thus, the domain is $[-6, \infty)$.

The function values are positive or zero (the result of the radical), so the range is $[0, \infty)$.

x	$f(x) = \sqrt{x+6}$
-6	$\sqrt{-6+6} = 0$
-5	$\sqrt{-5+6} = 1$
-2	$\sqrt{-2+6} = 2$

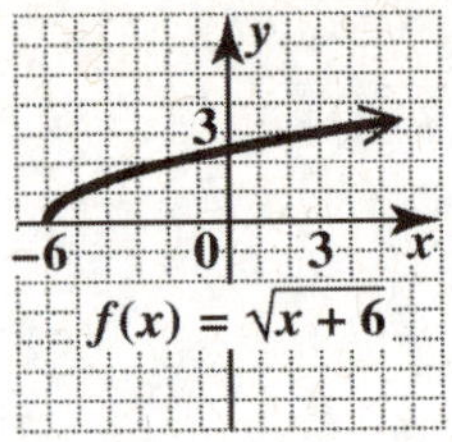

8. $$\left(\frac{16}{25}\right)^{-3/2} = \left(\frac{25}{16}\right)^{3/2} = \left[\left(\frac{25}{16}\right)^{1/2}\right]^3 = \left(\frac{5}{4}\right)^3 = \frac{125}{64}$$

9. $$(-64)^{-4/3} = \frac{1}{(-64)^{4/3}} = \frac{1}{[(-64)^{1/3}]^4} = \frac{1}{(-4)^4} = \frac{1}{256}$$

10. $$\begin{aligned} &\frac{3^{2/5}x^{-1/4}y^{2/5}}{3^{-8/5}x^{7/4}y^{1/10}} \\ &= 3^{2/5-(-8/5)}x^{-1/4-7/4}y^{2/5-1/10} \\ &= 3^{10/5}x^{-8/4}y^{4/10-1/10} \\ &= 3^2x^{-2}y^{3/10} = \frac{9y^{3/10}}{x^2} \end{aligned}$$

11. $$\begin{aligned} \left(\frac{x^{-4}y^{-6}}{x^{-2}y^3}\right)^{-2/3} &= \left(\frac{x^2}{x^4y^6y^3}\right)^{-2/3} \\ &= \left(\frac{1}{x^2y^9}\right)^{-2/3} = (x^2y^9)^{2/3} \\ &= (x^2)^{2/3}(y^9)^{2/3} = x^{4/3}y^6 \end{aligned}$$

12. $$\begin{aligned} 7^{3/4} \cdot 7^{-1/4} &= 7^{3/4+(-1/4)} \\ &= 7^{1/2}, \text{ or } \sqrt{7} \end{aligned}$$

13. $$\begin{aligned} \sqrt[3]{a^4} \cdot \sqrt[3]{a^7} &= \sqrt[3]{a^{4+7}} \\ &= \sqrt[3]{a^{11}} \quad (\text{or } a^{11/3}) \\ &= \sqrt[3]{a^9 \cdot a^2} = a^3\sqrt[3]{a^2} \end{aligned}$$

14. $$\begin{aligned} a^2 + b^2 &= c^2 \\ 12^2 + b^2 &= 17^2 \qquad \textit{Let } a = 12, c = 17. \\ 144 + b^2 &= 289 \\ b^2 &= 145 \\ b &= \sqrt{145} \end{aligned}$$

15. The given points are $(-4, 2)$ and $(2, 10)$.

$$\begin{aligned} d &= \sqrt{(x_2 - x_1)^2 + (y_2 - y_1)^2} \\ &= \sqrt{[2-(-4)]^2 + (10-2)^2} \\ &= \sqrt{6^2 + 8^2} = \sqrt{36+64} = \sqrt{100} = 10 \end{aligned}$$

16. $\sqrt{54x^5y^6} = \sqrt{9x^4y^6 \cdot 6x} = 3x^2y^3\sqrt{6x}$

17. $$\begin{aligned} \sqrt[4]{32a^7b^{13}} &= \sqrt[4]{16a^4b^{12} \cdot 2a^3b} \\ &= 2ab^3\sqrt[4]{2a^3b} \end{aligned}$$

18. $\sqrt{2}\cdot\sqrt[3]{5} = 2^{1/2}\cdot 5^{1/3} = 2^{3/6}\cdot 5^{2/6}$
$= (2^3\cdot 5^2)^{1/6} = \sqrt[6]{2^3\cdot 5^2}$
$= \sqrt[6]{8\cdot 25} = \sqrt[6]{200}$

19. $3\sqrt{20} - 5\sqrt{80} + 4\sqrt{500}$
$= 3\sqrt{4\cdot 5} - 5\sqrt{16\cdot 5} + 4\sqrt{100\cdot 5}$
$= 3\cdot 2\sqrt{5} - 5\cdot 4\sqrt{5} + 4\cdot 10\sqrt{5}$
$= 6\sqrt{5} - 20\sqrt{5} + 40\sqrt{5} = 26\sqrt{5}$

20. $\sqrt[3]{16t^3s^5} - \sqrt[3]{54t^6s^2}$
$= \sqrt[3]{8t^3s^3}\cdot\sqrt[3]{2s^2} - \sqrt[3]{27t^6}\cdot\sqrt[3]{2s^2}$
$= 2ts\sqrt[3]{2s^2} - 3t^2\sqrt[3]{2s^2}$
$= (2ts - 3t^2)\sqrt[3]{2s^2}$, or $t(2s-3t)\sqrt[3]{2s^2}$

21. $(7\sqrt{5}+4)(2\sqrt{5}-1)$
$= 14\cdot 5 - 7\sqrt{5} + 8\sqrt{5} - 4$
$= 70 + \sqrt{5} - 4 = 66 + \sqrt{5}$

22. $(\sqrt{3} - 2\sqrt{5})^2$
$= (\sqrt{3})^2 - 2\cdot\sqrt{3}\cdot 2\sqrt{5} + (2\sqrt{5})^2$
$= 3 - 4\sqrt{15} + 4\cdot 5$
$= 3 - 4\sqrt{15} + 20 = 23 - 4\sqrt{15}$

23. $\dfrac{-5}{\sqrt{40}} = \dfrac{-5}{\sqrt{4\cdot 10}} = \dfrac{-5}{2\sqrt{10}} = \dfrac{-5\cdot\sqrt{10}}{2\sqrt{10}\cdot\sqrt{10}}$
$= \dfrac{-5\sqrt{10}}{2\cdot 10} = \dfrac{-5\sqrt{10}}{20} = -\dfrac{\sqrt{10}}{4}$

24. $\dfrac{2}{\sqrt[3]{5}} = \dfrac{2\cdot\sqrt[3]{5^2}}{\sqrt[3]{5}\sqrt[3]{5^2}} = \dfrac{2\sqrt[3]{25}}{5}$

25. $\dfrac{-4}{\sqrt{7}+\sqrt{5}}$
Multiply the numerator and denominator by the conjugate of the denominator, $\sqrt{7}-\sqrt{5}$.
$= \dfrac{-4(\sqrt{7}-\sqrt{5})}{(\sqrt{7}+\sqrt{5})(\sqrt{7}-\sqrt{5})}$
$= \dfrac{-4(\sqrt{7}-\sqrt{5})}{7-5}$
$= \dfrac{-4(\sqrt{7}-\sqrt{5})}{2} = -2\left(\sqrt{7}-\sqrt{5}\right)$

26. $\dfrac{6+\sqrt{24}}{2} = \dfrac{6+\sqrt{4\cdot 6}}{2} = \dfrac{6+2\sqrt{6}}{2}$
$= \dfrac{2(3+\sqrt{6})}{2} = 3+\sqrt{6}$

27. **(a)** Substitute 50 for V_0, 0.01 for k, and 30 for T in the formula.

$$V = \frac{V_0}{\sqrt{1-kT}}$$
$$V = \frac{50}{\sqrt{1-(0.01)(30)}}$$
$$= \frac{50}{\sqrt{1-0.3}} = \frac{50}{\sqrt{0.7}} \approx 59.8$$

The velocity is about 59.8.

(b)
$$V = \frac{V_0}{\sqrt{1-kT}}$$
$$V^2 = \frac{V_0^2}{1-kT} \quad \textit{Square.}$$
$$V^2(1-kT) = V_0^2$$
$$V^2 - V^2kT = V_0^2$$
$$V^2 - V_0^2 = V^2kT$$
$$T = \frac{V^2 - V_0^2}{V^2k}$$
or
$$T = \frac{V_0^2 - V^2}{-V^2k}$$

28.
$$\sqrt[3]{5x} = \sqrt[3]{2x-3}$$
$$(\sqrt[3]{5x})^3 = (\sqrt[3]{2x-3})^3$$
$$5x = 2x - 3$$
$$3x = -3$$
$$x = -1$$

Check $x=-1$: $\sqrt[3]{-5} = \sqrt[3]{-5}$ *True*
The solution set is $\{-1\}$.

29. $x + \sqrt{x+6} = 9 - x$
Isolate the radical.
$$\sqrt{x+6} = 9 - 2x$$
$$(\sqrt{x+6})^2 = (9-2x)^2 \quad \textit{Square.}$$
$$x + 6 = 81 - 36x + 4x^2$$
$$0 = 4x^2 - 37x + 75$$
$$0 = (x-3)(4x-25)$$

$x - 3 = 0$ or $4x - 25 = 0$
$x = 3$ $\quad\quad x = \frac{25}{4}$

Check $x=3$: $3 + \sqrt{9} = 6$ *True*
Check $x=\frac{25}{4}$: $\frac{25}{4} + \sqrt{\frac{49}{4}} = \frac{11}{4}$ *False*
The solution set is $\{3\}$.

30. $\sqrt{x+4}-\sqrt{1-x}=-1$

$$\begin{aligned}\sqrt{x+4}&=-1+\sqrt{1-x}\\(\sqrt{x+4})^2&=(-1+\sqrt{1-x})^2\\x+4&=1-2\sqrt{1-x}+1-x\\2x+2&=-2\sqrt{1-x}\\x+1&=-\sqrt{1-x}\\(x+1)^2&=(-\sqrt{1-x})^2\\x^2+2x+1&=1-x\\x^2+3x&=0\\x(x+3)&=0\end{aligned}$$

$x=0$ or $x+3=0$
$x=-3$

Check $x=-3$: $1-2=-1$ *True*
Check $x=0$: $2-1=-1$ *False*
The solution set is $\{-3\}$.

31. $(-2+5i)-(3+6i)-7i$
$=(-2-3)+(5-6-7)i$
$=-5-8i$

32. $(1+5i)(3+i)$
$=3+i+15i+5i^2$
$=3+16i+5(-1)$
$=-2+16i$

33. $\dfrac{7+i}{1-i}$
Multiply the numerator and denominator by the conjugate of the denominator, $1+i$.
$=\dfrac{(7+i)(1+i)}{(1-i)(1+i)}$
$=\dfrac{7+7i+i+i^2}{1-i^2}$
$=\dfrac{7+8i-1}{1-(-1)}$
$=\dfrac{6+8i}{2}$
$=\dfrac{2(3+4i)}{2}$
$=3+4i$

34. $i^{37}=i^{36}\cdot i=(i^4)^9\cdot i=1^9\cdot i=1\cdot i=i$

35. **(a)** $i^2=-1$ is *true.*

(b) $i=\sqrt{-1}$ is *true.*

(c) $i=-1$ is *false;* $i=\sqrt{-1}$.

(d) $\sqrt{-3}=i\sqrt{3}$ is *true.*

Cumulative Review Exercises (Chapters 1–10)

In Exercises 1 and 2, $a=-3$, $b=5$, and $c=-4$.

1. $|2a^2-3b+c|$
$=|2(-3)^2-3(5)+(-4)|$
$=|2(9)-15-4|$
$=|18-15-4|=|-1|=1$

2. $\dfrac{(a+b)(a+c)}{3b-6}$
$=\dfrac{(-3+5)[(-3+(-4)]}{3(5)-6}$
$=\dfrac{(2)(-7)}{15-6}=-\dfrac{14}{9}$

3. $3(x+2)-4(2x+3)=-3x+2$

$$\begin{aligned}3x+6-8x-12&=-3x+2\\-5x-6&=-3x+2\\-2x&=8\\x&=-4\end{aligned}$$

Check $x=-4$: $-6+20=14$ *True*
The solution set is $\{-4\}$.

4. $\frac{1}{3}x+\frac{1}{4}(x+8)=x+7$
Multiply by the LCD, 12.

$$\begin{aligned}12\left[\tfrac{1}{3}x+\tfrac{1}{4}(x+8)\right]&=12(x+7)\\4x+3(x+8)&=12x+84\\4x+3x+24&=12x+84\\7x+24&=12x+84\\-5x&=60\\x&=-12\end{aligned}$$

Check $x=-12$: $-4-1=-5$ *True*
The solution set is $\{-12\}$.

5. $0.04x+0.06(100-x)=5.88$
Multiply each side by 100 to clear decimals.

$$\begin{aligned}4x+6(100-x)&=588\\4x+600-6x&=588\\-2x&=-12\\x&=6\end{aligned}$$

Check $x=6$: $0.24+5.64=5.88$ *True*
The solution set is $\{6\}$.

6. $-5-3(x-2)<11-2(x+2)$

$$\begin{aligned}-5-3x+6&<11-2x-4\\1-3x&<7-2x\\-x&<6\end{aligned}$$

Multiply by -1; reverse the inequality.
$x>-6$

The solution set is $(-6,\infty)$.

7. Let $x =$ the number of nickels and then $100 - x =$ the number of quarters.

Value of nickels	+	value of quarters	=	total value.
$0.05x$	+	$0.25(100 - x)$	=	17.80

Multiply by 100 to clear the decimals.

$$\begin{aligned} 5x + 25(100 - x) &= 1780 \\ 5x + 2500 - 25x &= 1780 \\ -20x &= -720 \\ x &= 36 \end{aligned}$$

Since $x = 36$, $100 - x = 100 - 36 = 64$.

There are 36 nickels and 64 quarters.

8. Let $x =$ the amount of pure alcohol. Make a table.

Number of Liters	Percent (as a decimal)	Liters of Pure Alcohol
x	$100\% = 1$	$1 \cdot x = x$
40	$18\% = 0.18$	$0.18(40) = 7.2$
$x + 40$	$22\% = 0.22$	$0.22(x + 40)$

From the last column:

$$\begin{aligned} x + 7.2 &= 0.22x + 8.8 \\ 0.78x &= 1.6 \\ x &= \tfrac{1.6}{0.78} = \tfrac{160}{78} = \tfrac{80}{39} \text{ or } 2\tfrac{2}{39} \end{aligned}$$

The required amount is $\frac{80}{39}$ or $2\frac{2}{39}$ liters of pure alcohol.

9. $4x - 3y = 12$
Let $x = 0$ to find the y-intercept, $(0, -4)$.
Let $y = 0$ to find the x-intercept, $(3, 0)$.
Draw a line through the intercepts.

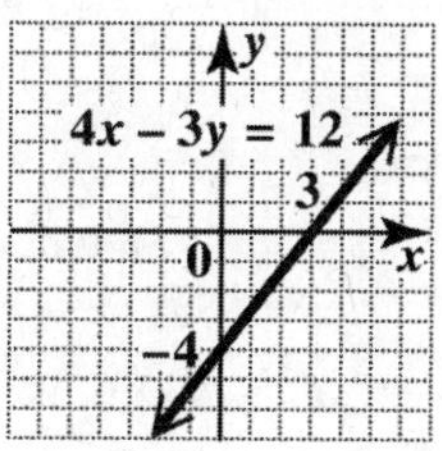

10. The given points are $(-4, 6)$ and $(2, -3)$.

$$m = \frac{-3 - 6}{2 - (-4)} = \frac{-9}{6} = -\frac{3}{2}$$

Use the point-slope form:

$$\begin{aligned} y - 6 &= -\tfrac{3}{2}[x - (-4)] \\ y - 6 &= -\tfrac{3}{2}(x + 4) \\ y - 6 &= -\tfrac{3}{2}x - 6 \\ y &= -\tfrac{3}{2}x \end{aligned}$$

11.
$$\begin{aligned} &(3k^3 - 5k^2 + 8k - 2) - (4k^3 + 11k + 7) \\ &\quad + (2k^2 - 5k) \\ &= 3k^3 - 4k^3 - 5k^2 + 2k^2 \\ &\quad + 8k - 11k - 5k - 2 - 7 \\ &= -k^3 - 3k^2 - 8k - 9 \end{aligned}$$

12.
$$\begin{aligned} &(8x - 7)(x + 3) \\ &= 8x^2 + 24x - 7x - 21 \\ &= 8x^2 + 17x - 21 \end{aligned}$$

13. $\dfrac{6y^4 - 3y^3 + 5y^2 + 6y - 9}{2y + 1}$

$$\begin{array}{r|l} & 3y^3 - 3y^2 + 4y + 1 \\ \hline 2y + 1 & 6y^4 - 3y^3 + 5y^2 + 6y - 9 \\ & 6y^4 + 3y^3 \\ \hline & -6y^3 + 5y^2 \\ & -6y^3 - 3y^2 \\ \hline & 8y^2 + 6y \\ & 8y^2 + 4y \\ \hline & 2y - 9 \\ & 2y + 1 \\ \hline & -10 \end{array}$$

The answer is

$$3y^3 - 3y^2 + 4y + 1 + \frac{-10}{2y + 1}.$$

14. $2p^2 - 5pq + 3q^2 = (2p - 3q)(p - q)$

15.
$$\begin{aligned} 3k^4 + k^2 - 4 &= (3k^2 + 4)(k^2 - 1) \\ &= (3k^2 + 4)(k + 1)(k - 1) \end{aligned}$$

16.
$$\begin{aligned} x^3 + 512 &= x^3 + 8^3 \\ &= (x + 8)(x^2 - 8x + 64) \end{aligned}$$

17. $2x^2 + 11x + 15 = 0$
$(x + 3)(2x + 5) = 0$

$$\begin{aligned} x + 3 = 0 \quad &\text{or} \quad 2x + 5 = 0 \\ x = -3 \quad &\text{or} \quad x = -\tfrac{5}{2} \end{aligned}$$

Check $x = -3$: $18 - 33 + 15 = 0$ *True*
Check $x = -\frac{5}{2}$: $\frac{25}{2} - \frac{55}{2} + \frac{30}{2} = 0$ *True*
The solution set is $\{-3, -\frac{5}{2}\}$.

18.
$$\begin{aligned} 5x(x - 1) &= 2(1 - x) \\ 5x^2 - 5x &= 2 - 2x \\ 5x^2 - 3x - 2 &= 0 \\ (5x + 2)(x - 1) &= 0 \end{aligned}$$

$$\begin{aligned} 5x + 2 = 0 \quad &\text{or} \quad x - 1 = 0 \\ x = -\tfrac{2}{5} \quad &\text{or} \quad x = 1 \end{aligned}$$

Check $x = -\frac{2}{5}$: $-2(-\frac{7}{5}) = 2(\frac{7}{5})$ *True*
Check $x = 1$: $5(0) = 2(0)$ *True*
The solution set is $\{-\frac{2}{5}, 1\}$.

19. $\dfrac{4}{x^2-9} = \dfrac{4}{(x+3)(x-3)}$

The numbers -3 and 3 make the denominator 0 so they are the values of the variable that make the rational expression undefined.

20. $\dfrac{y^2+y-12}{y^3+9y^2+20y} \div \dfrac{y^2-9}{y^3+3y^2}$

$$= \frac{y^2+y-12}{y(y^2+9y+20)} \cdot \frac{y^3+3y^2}{y^2-9}$$

$$= \frac{(y+4)(y-3)}{y(y+4)(y+5)} \cdot \frac{y^2(y+3)}{(y+3)(y-3)}$$

$$= \frac{y}{y+5}$$

21. $\dfrac{1}{x+y} + \dfrac{3}{x-y}$ *The LCD is* $(x+y)(x-y)$.

$$= \frac{1(x-y)}{(x+y)(x-y)} + \frac{3(x+y)}{(x-y)(x+y)}$$

$$= \frac{(x-y)+3(x+y)}{(x+y)(x-y)}$$

$$= \frac{x-y+3x+3y}{(x+y)(x-y)}$$

$$= \frac{4x+2y}{(x+y)(x-y)}$$

22. $\dfrac{\frac{-6}{x-2}}{\frac{8}{3x-6}} = \dfrac{-6}{x-2} \div \dfrac{8}{3x-6}$

$$= \frac{-6}{x-2} \cdot \frac{3x-6}{8}$$

$$= \frac{-6}{x-2} \cdot \frac{3(x-2)}{8}$$

$$= \frac{-2 \cdot 3 \cdot 3}{2 \cdot 4} = -\frac{9}{4}$$

23. $\dfrac{x^{-1}}{y-x^{-1}} = \dfrac{\frac{1}{x}}{y-\frac{1}{x}}$

Multiply the numerator and denominator by x.

$$= \frac{x\left(\frac{1}{x}\right)}{x\left(y-\frac{1}{x}\right)} = \frac{1}{xy-1}$$

24. $\dfrac{x+1}{x-3} = \dfrac{4}{x-3} + 6$

Multiply by the LCD, $x-3$.

$$(x-3)\left(\frac{x+1}{x-3}\right) = (x-3)\left(\frac{4}{x-3}+6\right)$$

$$\begin{aligned} x+1 &= 4+6(x-3) \\ x+1 &= 4+6x-18 \\ x+1 &= -14+6x \\ -5x &= -15 \\ x &= 3 \end{aligned}$$

Substituting 3 in the original equation results in division by zero, so 3 cannot be a solution. The solution set is $\emptyset$.

25. Let $x =$ Richard's speed.
Then $x+4 =$ Danielle's speed.

Use $d = rt$, or $t = \frac{d}{r}$, to make a table.

	Distance	Rate	Time
Richard	24	x	$\frac{24}{x}$
Danielle	48	$x+4$	$\frac{48}{x+4}$

Since the times are the same,

$$\frac{24}{x} = \frac{48}{x+4}.$$

Multiply by the LCD, $x(x+4)$.

$$x(x+4)\left(\frac{24}{x}\right) = x(x+4)\left(\frac{48}{x+4}\right)$$

$$\begin{aligned} 24(x+4) &= 48x \\ 24x+96 &= 48x \\ -24x &= -96 \\ x &= 4 \end{aligned}$$

Since $x = 4$, $x+4 = 4+4 = 8$. Richard's speed is 4 mph; Danielle's speed is 8 mph.

26. $f(x) = 3x-7$
$f(-10) = 3(-10)-7 = -30-7 = -37$

27. Let $C =$ the cost of a pizza and $r =$ the radius of the pizza. C varies directly as r^2, so

$$C = kr^2$$

for some constant k. Since $C = 6$ when $r = 7$, substitute these values in the equation and solve for k.

$$\begin{aligned} C &= kr^2 \\ 6 &= k(7)^2 \\ 6 &= 49k \\ \tfrac{6}{49} &= k \end{aligned}$$

So $C = \frac{6}{49}r^2$. Now let $r = 9$.

$$C = \tfrac{6}{49}(9)^2 = \tfrac{6}{49}(81) = \tfrac{486}{49} \approx 9.92.$$

A pizza with a 9-inch radius should cost \$9.92.

28. $3x - y = 23 \quad (1)$
$2x + 3y = 8 \quad (2)$
To eliminate y, multiply equation (1) by 3 and add the result to equation (2).

$$\begin{array}{rcrl} 9x - 3y &=& 69 & 3 \times (1) \\ 2x + 3y &=& 8 & (2) \\ \hline 11x &=& 77 & \\ x &=& 7 & \end{array}$$

Substitute 7 for x in (2),

$$\begin{aligned} 2(7) + 3y &= 8 \\ 14 + 3y &= 8 \\ 3y &= -6 \\ y &= -2 \end{aligned}$$

The solution set is $\{(7, -2)\}$.

29. $x + y + z = 1$
$x - y - z = -3$
$x + y - z = -1$

Write the augmented matrix.

$$\left[\begin{array}{rrr|r} 1 & 1 & 1 & 1 \\ 1 & -1 & -1 & -3 \\ 1 & 1 & -1 & -1 \end{array}\right]$$

$$\left[\begin{array}{rrr|r} 1 & 1 & 1 & 1 \\ 0 & -2 & -2 & -4 \\ 0 & 0 & -2 & -2 \end{array}\right] \quad \begin{array}{l} \\ -R_1 + R_2 \\ -R_1 + R_3 \end{array}$$

$$\left[\begin{array}{rrr|r} 1 & 1 & 1 & 1 \\ 0 & 1 & 1 & 2 \\ 0 & 0 & 1 & 1 \end{array}\right] \quad \begin{array}{l} \\ -\frac{1}{2}R_2 \\ -\frac{1}{2}R_3 \end{array}$$

This matrix gives the system

$$\begin{aligned} x + y + z &= 1 \quad (1) \\ y + z &= 2 \quad (2) \\ z &= 1. \end{aligned}$$

Substitute 1 for z in (2).

$$\begin{aligned} y + 1 &= 2 \\ y &= 1 \end{aligned}$$

Substitute 1 for y and 1 for z in (1).

$$\begin{aligned} x + 1 + 1 &= 1 \\ x &= -1 \end{aligned}$$

The solution set is $\{(-1, 1, 1)\}$.

30. Let x = the number of 2-ounce letters and
y = the number of 3-ounce letters.

$$\begin{aligned} 5x + 3y &= 5.39 \quad (1) \\ 3x + 5y &= 5.73 \quad (2) \end{aligned}$$

To eliminate x, multiply (1) by -3 and (2) by 5 and add the results.

$$\begin{array}{rcrl} -15x - 9y &=& -16.17 & -3 \times (1) \\ 15x + 25y &=& 28.65 & 5 \times (2) \\ \hline 16y &=& 12.48 & \\ y &=& 0.78 & \end{array}$$

Substitute $y = 0.78$ in (1).

$$\begin{aligned} 5x + 3y &= 5.39 \quad (1) \\ 5x + 3(0.78) &= 5.39 \\ 5x + 2.34 &= 5.39 \\ 5x &= 3.05 \\ x &= 0.61 \end{aligned}$$

The 2010 postage rate for a 2-ounce letter was \$0.61 and for a 3-ounce letter, \$0.78.

31. $2x + 4 < 10$ and $3x - 1 > 5$
$2x < 6 \qquad 3x > 6$
$x < 3$ and $x > 2$

The solution set is $(2, 3)$.

32. $2x + 4 > 10$ or $3x - 1 < 5$
$2x > 6 \qquad 3x < 6$
$x > 3$ or $x < 2$

The solution set is $(-\infty, 2) \cup (3, \infty)$.

33. $|6x + 7| = 13$

$6x + 7 = 13$ or $6x + 7 = -13$
$6x = 6 \qquad 6x = -20$
$x = 1$ or $x = -\frac{20}{6} = -\frac{10}{3}$

Check $x = 1$: $|13| = 13$ *True*
Check $x = -\frac{10}{3}$: $|-13| = 13$ *True*
The solution set is $\{-\frac{10}{3}, 1\}$.

34. $|2p - 5| \geq 9$

$2p - 5 \geq 9$ or $2p - 5 \leq -9$
$2p \geq 14 \qquad 2p \leq -4$
$p \geq 7$ or $p \leq -2$

The solution set is $(-\infty, -2] \cup [7, \infty)$.

35. $\sqrt[3]{16x^2y} \cdot \sqrt[3]{3x^3y}$

$$\begin{aligned} &= \sqrt[3]{8} \cdot \sqrt[3]{2x^2y} \cdot \sqrt[3]{x^3} \cdot \sqrt[3]{3y} \\ &= 2\sqrt[3]{2x^2y} \cdot x\sqrt[3]{3y} \\ &= 2x\sqrt[3]{6x^2y^2} \end{aligned}$$

36. $\sqrt{50} + \sqrt{8} = \sqrt{25 \cdot 2} + \sqrt{4 \cdot 2}$
$= 5\sqrt{2} + 2\sqrt{2} = 7\sqrt{2}$

37. $\dfrac{1}{\sqrt{10}-\sqrt{8}}$

Multiply the numerator and denominator by the conjugate of the denominator, $\sqrt{10}+\sqrt{8}$.

$$= \frac{1(\sqrt{10}+\sqrt{8})}{(\sqrt{10}-\sqrt{8})(\sqrt{10}+\sqrt{8})}$$
$$= \frac{\sqrt{10}+\sqrt{8}}{10-8} = \frac{\sqrt{10}+\sqrt{4\cdot 2}}{2}$$
$$= \frac{\sqrt{10}+2\sqrt{2}}{2}$$

38. The given points are $(-4,4)$ and $(-2,9)$.

$$d = \sqrt{[(-2)-(-4)]^2+(9-4)^2}$$
$$= \sqrt{2^2+5^2}$$
$$= \sqrt{4+25} = \sqrt{29}$$

39. $\sqrt{3x-8} = x-2$

Square each side.

$$(\sqrt{3x-8})^2 = (x-2)^2$$
$$3x-8 = x^2-4x+4$$
$$0 = x^2-7x+12$$
$$0 = (x-3)(x-4)$$

$$x-3=0 \quad \text{or} \quad x-4=0$$
$$x=3 \quad \text{or} \quad x=4$$

Check $x=3$: $\sqrt{1}=1$ *True*
Check $x=4$: $\sqrt{4}=2$ *True*
The solution set is $\{3,4\}$.

40. $\dfrac{6-2i}{1-i}$

Multiply the numerator and denominator by the conjugate of the denominator, $1+i$.

$$= \frac{(6-2i)(1+i)}{(1-i)(1+i)}$$
$$= \frac{6+6i-2i-2i^2}{1^2+1^2}$$
$$= \frac{6+4i+2}{1+1}$$
$$= \frac{8+4i}{2}$$
$$= \frac{2(4+2i)}{2}$$
$$= 4+2i$$

CHAPTER 11 QUADRATIC EQUATIONS, INEQUALITIES, AND FUNCTIONS

11.1 Solving Quadratic Equations by the Square Root Property

11.1 Now Try Exercises

N1. **(a)**
$$x^2 - x - 20 = 0$$
$$(x-5)(x+4) = 0 \quad \textit{Factor.}$$
$$x - 5 = 0 \text{ or } x + 4 = 0 \quad \textit{Zero-factor property}$$
$$x = 5 \text{ or } x = -4 \quad \textit{Solve each equation.}$$

Solution set: $\{-4, 5\}$

(b)
$$x^2 = 36$$
$$x^2 - 36 = 0 \quad \textit{Subtract 36.}$$
$$(x+6)(x-6) = 0 \quad \textit{Factor.}$$
$$x + 6 = 0 \text{ or } x - 6 = 0 \quad \textit{Zero-factor property}$$
$$x = -6 \text{ or } x = 6 \quad \textit{Solve each equation.}$$

Solution set: $\{-6, 6\}$, or $\{\pm 6\}$

N2. **(a)** $t^2 = 25$

Solve by using the square root property.

$$t = \sqrt{25} \quad \text{or} \quad t = -\sqrt{25}$$
$$t = 5 \quad \text{or} \quad t = -5$$

Solution set: $\{-5, 5\}$, or $\{\pm 5\}$

(b) $x^2 = 13$

Use the square root property.

$$x = \sqrt{13} \quad \text{or} \quad x = -\sqrt{13}.$$

Solution set: $\{-\sqrt{13}, \sqrt{13}\}$, or $\{\pm\sqrt{13}\}$

(c) $3x^2 - 54 = 0$
$$3x^2 = 54$$
$$x^2 = 18$$

Use the square root property.

$$x = \sqrt{18} \quad \text{or} \quad x = -\sqrt{18}$$
$$x = 3\sqrt{2} \quad \text{or} \quad x = -3\sqrt{2}$$

Check $x = \pm 3\sqrt{2}$: $3(18) - 54 = 0$ *True*

Solution set: $\left\{3\sqrt{2}, -3\sqrt{2}\right\}$, or $\{\pm 3\sqrt{2}\}$

(d) $2x^2 - 5 = 35$
$$2x^2 = 40$$
$$x^2 = 20$$

Use the square root property.

$$x = \sqrt{20} \quad \text{or} \quad x = -\sqrt{20}$$
$$x = \sqrt{4}\cdot\sqrt{5} \quad \text{or} \quad x = -\sqrt{4}\cdot\sqrt{5}$$
$$x = 2\sqrt{5} \quad \text{or} \quad x = -2\sqrt{5}$$

Solution set: $\{-2\sqrt{5}, 2\sqrt{5}\}$, or $\{\pm 2\sqrt{5}\}$

N3.
$$d = 16t^2$$
$$25 = 16t^2 \quad \textit{Let } d = 25.$$
$$\tfrac{25}{16} = t^2$$

By the square root property,

$$t = \tfrac{5}{4} \quad \text{or} \quad t = -\tfrac{5}{4}.$$

Time cannot be negative, so we discard the negative solution. Therefore, it takes 1.25 sec for the nail to hit the bottom of the bucket.

N4. $(x-2)^2 = 32$

Use the square root property.

$$x - 2 = \sqrt{32} \quad \text{or} \quad x - 2 = -\sqrt{32}$$
$$x = 2 + \sqrt{16}\cdot\sqrt{2} \quad \text{or} \quad x = 2 - \sqrt{16}\cdot\sqrt{2}$$
$$x = 2 + 4\sqrt{2} \quad \text{or} \quad x = 2 - 4\sqrt{2}$$

Check $x = 2 \pm 4\sqrt{2}$:

$$\text{LS} = \left[(2 \pm 4\sqrt{2}) - 2\right]^2$$
$$= \left(\pm 4\sqrt{2}\right)^2 = 16\cdot 2 = 32 = \text{RS}$$

Solution set: $\left\{2 \pm 4\sqrt{2}\right\}$

N5. $(2t-4)^2 = 50$

$$2t - 4 = \sqrt{50} \quad \text{or} \quad 2t - 4 = -\sqrt{50}$$
$$2t - 4 = \sqrt{25}\cdot\sqrt{2} \quad \text{or} \quad 2t - 4 = -\sqrt{25}\cdot\sqrt{2}$$
$$2t = 4 + 5\sqrt{2} \quad \text{or} \quad 2t = 4 - 5\sqrt{2}$$
$$t = \frac{4 + 5\sqrt{2}}{2} \quad \text{or} \quad t = \frac{4 - 5\sqrt{2}}{2}$$

Solution set: $\left\{\frac{4 \pm 5\sqrt{2}}{2}\right\}$

N6. **(a)** $t^2 = -24$

$$t = \sqrt{-24} \quad \text{or} \quad t = -\sqrt{-24}$$
$$t = 2i\sqrt{6} \quad \text{or} \quad t = -2i\sqrt{6}$$

Solution set: $\left\{\pm 2i\sqrt{6}\right\}$

(b) $(x+4)^2 = -36$

$$x + 4 = \sqrt{-36} \quad \text{or} \quad x + 4 = -\sqrt{-36}$$
$$x + 4 = 6i \quad \text{or} \quad x + 4 = -6i$$
$$x = -4 + 6i \quad \text{or} \quad x = -4 - 6i$$

Solution set: $\{-4 \pm 6i\}$

11.1 Section Exercises

1. **B** and **C** are quadratic equations because they are equations with a squared term and no terms of higher degree.

3. The zero-factor property requires a product equal to 0. The first step should have been to rewrite the equation with 0 on one side.

5.
$$x^2 - x - 56 = 0$$
$$(x-8)(x+7) = 0 \quad \textit{Factor.}$$
$$x - 8 = 0 \text{ or } x + 7 = 0 \quad \textit{Zero-factor property}$$
$$x = 8 \text{ or } x = -7 \quad \textit{Solve each equation.}$$

Solution set: $\{-7, 8\}$

7.
$$x^2 = 121$$
$$x^2 - 121 = 0 \quad \textit{Subtract 121.}$$
$$(x+11)(x-11) = 0 \quad \textit{Factor.}$$
$$x + 11 = 0 \text{ or } x - 11 = 0 \quad \textit{Zero-factor prop.}$$
$$x = -11 \text{ or } x = 11 \quad \textit{Solve each eq.}$$

Solution set: $\{-11, 11\}$, or $\{\pm 11\}$

9.
$$3x^2 - 13x = 30$$
$$3x^2 - 13x - 30 = 0$$
$$(x-6)(3x+5) = 0 \quad \textit{Factor.}$$
$$x - 6 = 0 \text{ or } 3x + 5 = 0 \quad \textit{Zero-factor prop.}$$
$$x = 6 \text{ or } x = -\tfrac{5}{3} \quad \textit{Solve each eq.}$$

Solution set: $\{-\frac{5}{3}, 6\}$

11. $x^2 = 81$

Use the square root property to get

$$x = \sqrt{81} = 9 \quad \text{or} \quad x = -\sqrt{81} = -9.$$

Solution set: $\{\pm 9\}$

13. $x^2 = 14$

Use the square root property to get

$$x = \sqrt{14} \quad \text{or} \quad x = -\sqrt{14}.$$

Solution set: $\{\pm\sqrt{14}\}$

15. $t^2 = 48$

$$t = \sqrt{48} \quad \text{or} \quad t = -\sqrt{48}$$

Write $\sqrt{48}$ in simplest form.

$$\sqrt{48} = \sqrt{16} \cdot \sqrt{3} = 4\sqrt{3}$$

Solution set: $\{\pm 4\sqrt{3}\}$

17. $x^2 = \frac{25}{4}$

$$x = \sqrt{\tfrac{25}{4}} \quad \text{or} \quad x = -\sqrt{\tfrac{25}{4}}$$
$$x = \tfrac{5}{2} \quad \text{or} \quad x = -\tfrac{5}{2}$$

Solution set: $\{\pm\frac{5}{2}\}$

19. $x^2 = 2.25$

$$x = \sqrt{2.25} \quad \text{or} \quad x = -\sqrt{2.25}$$
$$x = 1.5 \quad \text{or} \quad x = -1.5$$

Solution set: $\{\pm 1.5\}$

21.
$$r^2 - 3 = 0$$
$$r^2 = 3$$
$$r = \sqrt{3} \text{ or } r = -\sqrt{3}$$

Solution set: $\{\pm\sqrt{3}\}$

23.
$$x^2 - 20 = 0$$
$$x^2 = 20$$
$$x = \sqrt{20} \quad \text{or} \quad x = -\sqrt{20}$$
$$x = 2\sqrt{5} \quad \text{or} \quad x = -2\sqrt{5}$$

Solution set: $\{\pm 2\sqrt{5}\}$

25.
$$7x^2 = 4$$
$$x^2 = \tfrac{4}{7} \quad \textit{Divide by 7.}$$
$$x = \sqrt{\tfrac{4}{7}} \quad \text{or} \quad x = -\sqrt{\tfrac{4}{7}}$$
$$= \frac{\sqrt{4}}{\sqrt{7}} \cdot \frac{\sqrt{7}}{\sqrt{7}} \qquad = -\frac{\sqrt{4}}{\sqrt{7}} \cdot \frac{\sqrt{7}}{\sqrt{7}}$$
$$= \frac{2\sqrt{7}}{7} \qquad = -\frac{2\sqrt{7}}{7}$$

Solution set: $\{\pm\frac{2\sqrt{7}}{7}\}$

27.
$$3n^2 - 72 = 0$$
$$3n^2 = 72$$
$$n^2 = 24$$
$$n = \pm\sqrt{24} = \pm\sqrt{4 \cdot 6} = \pm 2\sqrt{6}$$

Solution set: $\{\pm 2\sqrt{6}\}$

29.
$$5x^2 + 4 = 8$$
$$5x^2 = 4 \quad \textit{Subtract 4.}$$
$$x^2 = \tfrac{4}{5} \quad \textit{Divide by 5.}$$
$$x = \sqrt{\tfrac{4}{5}} \quad \text{or} \quad x = -\sqrt{\tfrac{4}{5}}$$
$$= \frac{\sqrt{4}}{\sqrt{5}} \cdot \frac{\sqrt{5}}{\sqrt{5}} \qquad = -\frac{\sqrt{4}}{\sqrt{5}} \cdot \frac{\sqrt{5}}{\sqrt{5}}$$
$$= \frac{2\sqrt{5}}{5} \qquad = -\frac{2\sqrt{5}}{5}$$

Solution set: $\{\pm\frac{2\sqrt{5}}{5}\}$

31.
$$2t^2 + 7 = 61$$
$$2t^2 = 54$$
$$t^2 = 27$$
$$t = \pm\sqrt{27} = \pm\sqrt{9 \cdot 3} = \pm 3\sqrt{3}$$

Solution set: $\{\pm 3\sqrt{3}\}$

33. $-8x^2 = -64$

$x^2 = 8$ *Divide by -8.*

$x = \pm\sqrt{8} = \pm\sqrt{4 \cdot 2} = \pm 2\sqrt{2}$

Solution set: $\{\pm 2\sqrt{2}\}$

35. $(x-3)^2 = 25$

Use the square root property.

$x - 3 = \sqrt{25}$ or $x - 3 = -\sqrt{25}$

$x - 3 = 5$ or $x - 3 = -5$

$x = 8$ or $x = -2$

Solution set: $\{-2, 8\}$

37. $(x-4)^2 = 3$

$x - 4 = \sqrt{3}$ or $x - 4 = -\sqrt{3}$

$x = 4 + \sqrt{3}$ or $x = 4 - \sqrt{3}$

Solution set: $\{4 \pm \sqrt{3}\}$

39. $(x-8)^2 = 27$

Begin by using the square root property.

$x - 8 = \sqrt{27}$ or $x - 8 = -\sqrt{27}$

Now simplify the radical.

$\sqrt{27} = \sqrt{9} \cdot \sqrt{3} = 3\sqrt{3}$

$x - 8 = 3\sqrt{3}$ or $x - 8 = -3\sqrt{3}$

$x = 8 + 3\sqrt{3}$ or $x = 8 - 3\sqrt{3}$

Solution set: $\{8 \pm 3\sqrt{3}\}$

41. $(3x+2)^2 = 49$

$3x + 2 = \sqrt{49}$ or $3x + 2 = -\sqrt{49}$

$3x + 2 = 7$ or $3x + 2 = -7$

$3x = 5$ or $3x = -9$

$x = \frac{5}{3}$ or $x = -3$

Solution set: $\{-3, \frac{5}{3}\}$

43. $(4x-3)^2 = 9$

$4x - 3 = \sqrt{9}$ or $4x - 3 = -\sqrt{9}$

$4x - 3 = 3$ or $4x - 3 = -3$

$4x = 6$ or $4x = 0$

$x = \frac{6}{4} = \frac{3}{2}$ or $x = 0$

Solution set: $\{0, \frac{3}{2}\}$

45. $(3x-1)^2 = 7$

$3x - 1 = \sqrt{7}$ or $3x - 1 = -\sqrt{7}$

$3x = 1 + \sqrt{7}$ $\quad$ $3x = 1 - \sqrt{7}$

$x = \dfrac{1 + \sqrt{7}}{3}$ or $x = \dfrac{1 - \sqrt{7}}{3}$

Solution set: $\left\{\frac{1 \pm \sqrt{7}}{3}\right\}$

47. $(3k+1)^2 = 18$

$3k + 1 = \sqrt{18}$ or $3k + 1 = -\sqrt{18}$

$3k = -1 + 3\sqrt{2}$ or $3k = -1 - 3\sqrt{2}$

Note that $\sqrt{18} = \sqrt{9 \cdot 2} = 3\sqrt{2}$.

$k = \dfrac{-1 + 3\sqrt{2}}{3}$ or $k = \dfrac{-1 - 3\sqrt{2}}{3}$

Solution set: $\left\{\frac{-1 \pm 3\sqrt{2}}{3}\right\}$

49. $(5-2x)^2 = 30$

$5 - 2x = \sqrt{30}$ or $5 - 2x = -\sqrt{30}$

$-2x = -5 + \sqrt{30}$ or $-2x = -5 - \sqrt{30}$

$x = \dfrac{-5 + \sqrt{30}}{-2}$ or $x = \dfrac{-5 - \sqrt{30}}{-2}$

$x = \dfrac{-5 + \sqrt{30}}{-2} \cdot \dfrac{-1}{-1}$ or $x = \dfrac{-5 - \sqrt{30}}{-2} \cdot \dfrac{-1}{-1}$

$x = \dfrac{5 - \sqrt{30}}{2}$ or $x = \dfrac{5 + \sqrt{30}}{2}$

Solution set: $\left\{\frac{5 \pm \sqrt{30}}{2}\right\}$

51. $(\frac{1}{2}x + 5)^2 = 12$

$\frac{1}{2}x + 5 = \sqrt{12}$ or $\frac{1}{2}x + 5 = -\sqrt{12}$

$\frac{1}{2}x = -5 + 2\sqrt{3}$ or $\frac{1}{2}x = -5 - 2\sqrt{3}$

Note that $\sqrt{12} = \sqrt{4 \cdot 3} = 2\sqrt{3}$.

$x = 2\left(-5 + 2\sqrt{3}\right)$ or $x = 2\left(-5 - 2\sqrt{3}\right)$

$x = -10 + 4\sqrt{3}$ or $x = -10 - 4\sqrt{3}$

Solution set: $\{-10 \pm 4\sqrt{3}\}$

53. $(4x-1)^2 - 48 = 0$

$(4x-1)^2 = 48$

$4x - 1 = \sqrt{48}$ or $4x - 1 = -\sqrt{48}$

$4x - 1 = 4\sqrt{3}$ or $4x - 1 = -4\sqrt{3}$

$4x = 1 + 4\sqrt{3}$ or $4x = 1 - 4\sqrt{3}$

$x = \dfrac{1 + 4\sqrt{3}}{4}$ or $x = \dfrac{1 - 4\sqrt{3}}{4}$

Solution set: $\left\{\frac{1 \pm 4\sqrt{3}}{4}\right\}$

55. $(k + 2.14)^2 = 5.46$

$k + 2.14 = \sqrt{5.46}$ or $k + 2.14 = -\sqrt{5.46}$

$k = -2.14 + \sqrt{5.46}$ or $k = -2.14 - \sqrt{5.46}$

$k \approx 0.20$ or $k \approx -4.48$

To the nearest hundredth, the solution set is $\{-4.48, 0.20\}$.

57. $(2.11p+3.42)^2 = 9.58$

$2.11p+3.42 = \pm\sqrt{9.58}$

Remember that this represents two equations.

$$2.11p = -3.42 \pm \sqrt{9.58}$$
$$p = \frac{-3.42 \pm \sqrt{9.58}}{2.11}$$
$$\approx -3.09 \text{ or } -0.15$$

To the nearest hundredth, the solution set is $\{-3.09, -0.15\}$.

59. $x^2 = -12$

$$x = \sqrt{-12} \quad \text{or} \quad x = -\sqrt{-12}$$
$$x = i\sqrt{12} \qquad x = -i\sqrt{12}$$
$$x = 2i\sqrt{3} \quad \text{or} \quad x = -2i\sqrt{3}$$

Solution set: $\left\{\pm 2i\sqrt{3}\right\}$

61. $(r-5)^2 = -4$

$$r-5 = \sqrt{-4} \quad \text{or} \quad r-5 = -\sqrt{-4}$$
$$r = 5+2i \quad \text{or} \quad r = 5-2i$$

Solution set: $\{5 \pm 2i\}$

63. $(6x-1)^2 = -8$

$$6x-1 = \sqrt{-8} \quad \text{or} \quad 6x-1 = -\sqrt{-8}$$
$$6x-1 = i\sqrt{8} \qquad 6x-1 = -i\sqrt{8}$$
$$6x-1 = 2i\sqrt{2} \qquad 6x-1 = -2i\sqrt{2}$$
$$6x = 1+2i\sqrt{2} \qquad 6x = 1-2i\sqrt{2}$$
$$x = \frac{1+2i\sqrt{2}}{6} \quad \text{or} \quad x = \frac{1-2i\sqrt{2}}{6}$$

Solution set: $\left\{\frac{1}{6} \pm \frac{\sqrt{2}}{3}i\right\}$

65. $d = 16t^2$

$$500 = 16t^2$$
$$t^2 = \tfrac{500}{16} = 31.25$$

$$t = \sqrt{31.25} \quad \text{or} \quad t = -\sqrt{31.25}$$
$$t \approx 5.6 \quad \text{or} \quad t \approx -5.6$$

It would take the rock about 5.6 seconds (time must be positive) to fall to the ground.

67. $A = \pi r^2$

$81\pi = \pi r^2$ *Let $A = 81\pi$.*

$81 = r^2$ *Divide by π.*

$$r = 9 \quad \text{or} \quad r = -9$$

Discard -9 since the radius cannot be negative. The radius is 9 inches.

69. Let $A = 104.04$ and $P = 100$.

$$A = P(1+r)^2$$
$$104.04 = 100(1+r)^2$$
$$(1+r)^2 = \frac{104.04}{100} = 1.0404$$
$$1+r = \pm\sqrt{1.0404}$$
$$1+r = \pm 1.02$$
$$r = -1 \pm 1.02$$

So $r = -1 + 1.02 = 0.02$ or $r = -1 - 1.02 = -2.02$. Reject the solution -2.02. The rate is $r = 0.02$ or 2%.

71. $$\frac{4}{5} + \sqrt{\frac{48}{25}} = \frac{4}{5} + \frac{\sqrt{48}}{\sqrt{25}} = \frac{4}{5} + \frac{\sqrt{16}\cdot\sqrt{3}}{5} = \frac{4+4\sqrt{3}}{5}$$

73. $$\frac{6+\sqrt{24}}{8} = \frac{6+\sqrt{4}\cdot\sqrt{6}}{8} = \frac{6+2\sqrt{6}}{8} = \frac{2\left(3+\sqrt{6}\right)}{2\cdot 4} = \frac{3+\sqrt{6}}{4}$$

75. $$x^2 - 10x + 25 = x^2 - 2\cdot 5\cdot x + 5^2 = (x-5)^2$$

11.2 Solving Quadratic Equations by Completing the Square

11.2 Now Try Exercises

N1. $x^2 + 10x + 8 = 0$

$$x^2 + 10x = -8$$
$$x^2 + 10x + 25 = -8 + 25$$

$2kx = -10x$, so $k = -5$ and $k^2 = 25$.

$$(x-5)^2 = 17$$
$$x - 5 = \sqrt{17} \quad \text{or} \quad x - 5 = -\sqrt{17}$$
$$x = 5 + \sqrt{17} \quad \text{or} \quad x = 5 - \sqrt{17}$$

Solution set: $\left\{5 \pm \sqrt{17}\right\}$

N2. $x^2 - 6x = 9$

Take half of the coefficient of x and square the result.

$$\tfrac{1}{2}(-6) = -3, \quad \text{and} \quad (-3)^2 = 9.$$

Add 9 to each side of the equation, and write the left side as a perfect square.

$$x^2 - 6x + 9 = 9 + 9$$
$$(x-3)^2 = 18 \qquad \textit{Factor.}$$

Use the square root property to solve for x.

$$x - 3 = \sqrt{18} \quad \text{or} \quad x - 3 = -\sqrt{18}$$
$$x - 3 = \sqrt{9}\cdot\sqrt{2} \quad \text{or} \quad x - 3 = -\sqrt{9}\cdot\sqrt{2}$$
$$x = 3 + 3\sqrt{2} \quad \text{or} \quad x = 3 - 3\sqrt{2}$$

Solution set: $\left\{3 \pm 3\sqrt{2}\right\}$

N3. $x^2 + x - 3 = 0$

$x^2 + x = 3$

Complete the square.

$$\left[\tfrac{1}{2}(1)\right]^2 = \left(\tfrac{1}{2}\right)^2 = \tfrac{1}{4}$$

Add $\frac{1}{4}$ to each side.

$$x^2 + x + \tfrac{1}{4} = 3 + \tfrac{1}{4}$$
$$\left(x + \tfrac{1}{2}\right)^2 = \tfrac{13}{4}$$

Use the square root property.

$x + \frac{1}{2} = \sqrt{\frac{13}{4}}$ or $x + \frac{1}{2} = -\sqrt{\frac{13}{4}}$

$x + \frac{1}{2} = \frac{\sqrt{13}}{2}$ or $x + \frac{1}{2} = -\frac{\sqrt{13}}{2}$

$x = \dfrac{-1 + \sqrt{13}}{2}$ or $x = \dfrac{-1 - \sqrt{13}}{2}$

Check that the solution set is $\left\{\frac{-1 \pm \sqrt{13}}{2}\right\}$.

N4. $4t^2 - 4t - 3 = 0$

$4t^2 - 4t = 3$ *Add 3.*

Divide each side by 4 to get a coefficient of 1 for the t^2-term.

$t^2 - t = \frac{3}{4}$ *Divide by 4.*

Take half the coefficient of t, or $\left(\frac{1}{2}\right)(-1) = -\frac{1}{2}$, and square the result: $\left(-\frac{1}{2}\right)^2 = \frac{1}{4}$. Then add $\frac{1}{4}$ to each side.

$t^2 - t + \frac{1}{4} = \frac{3}{4} + \frac{1}{4}$ *Add $\frac{1}{4}$.*

$\left(t - \frac{1}{2}\right)^2 = 1$ *Factor.*

Apply the square root property, and solve for t.

$t - \frac{1}{2} = \sqrt{1}$ or $t - \frac{1}{2} = -\sqrt{1}$

$t - \frac{1}{2} = 1$ or $t - \frac{1}{2} = -1$

$t = \frac{1}{2} + 1$ or $t = \frac{1}{2} - 1$

$t = \frac{3}{2}$ or $t = -\frac{1}{2}$

A check verifies that $-\frac{1}{2}$ and $\frac{3}{2}$ are solutions of the original equation. Using a calculator for your check is highly recommended.

Solution set: $\left\{-\frac{1}{2}, \frac{3}{2}\right\}$

N5. $3x^2 + 12x - 5 = 0$

$3x^2 + 12x = 5$

$x^2 + 4x = \frac{5}{3}$

Complete the square.

$$\left[\tfrac{1}{2}(4)\right]^2 = (2)^2 = 4$$

Add 4 to each side.

$$x^2 + 4x + 4 = \tfrac{5}{3} + 4$$
$$(x + 2)^2 = \tfrac{17}{3}$$

Use the square root property.

$x + 2 = \sqrt{\frac{17}{3}}$ or $x + 2 = -\sqrt{\frac{17}{3}}$

$x = -2 + \sqrt{\frac{17}{3}}$ or $x = -2 - \sqrt{\frac{17}{3}}$

$x = -2 + \dfrac{\sqrt{51}}{3}$ or $x = -2 - \dfrac{\sqrt{51}}{3}$

$x = \dfrac{-6 + \sqrt{51}}{3}$ or $x = \dfrac{-6 - \sqrt{51}}{3}$

Check that the solution set is $\left\{\frac{-6 \pm \sqrt{51}}{3}\right\}$.

N6. $x^2 + 8x + 21 = 0$

$x^2 + 8x = -21$

Complete the square.

$$\left[\tfrac{1}{2}(8)\right]^2 = (4)^2 = 16$$

Add 16 to each side.

$$x^2 + 8x + 16 = -21 + 16$$
$$(x + 4)^2 = -5$$

$x + 4 = \sqrt{-5}$ or $x + 4 = -\sqrt{-5}$

$x + 4 = i\sqrt{5}$ or $x + 4 = -i\sqrt{5}$

$x = -4 + i\sqrt{5}$ or $x = -4 - i\sqrt{5}$

Solution set: $\left\{-4 \pm i\sqrt{5}\right\}$

N7. $(x - 5)(x + 1) = 2$

$x^2 - 4x - 5 = 2$

$x^2 - 4x = 7$

$x^2 - 4x + 4 = 7 + 4$ *Add $\left[\frac{1}{2}(-4)\right]^2 = 4$.*

$(x - 2)^2 = 11$

$x - 2 = \sqrt{11}$ or $x - 2 = -\sqrt{11}$

$x = 2 + \sqrt{11}$ or $x = 2 - \sqrt{11}$

Solution set: $\left\{2 \pm \sqrt{11}\right\}$

11.2 Section Exercises

1. Solve $(2x + 1)^2 = 5$ by the square root property. Solve $x^2 + 4x = 12$ by completing the square.

3. $x^2 + 6x +$ ___

We need to add the square of half the coefficient of x to get a perfect square trinomial.

$$\tfrac{1}{2}(6) = 3 \quad \text{and} \quad 3^2 = 9$$

Add 9 to $x^2 + 6x$ to get a perfect square trinomial.

$$x^2 + 6x + 9 = (x + 3)^2$$

5. $p^2 - 12p +$ ___

$$\tfrac{1}{2}(-12) = -6 \quad \text{and} \quad (-6)^2 = 36$$
$$p^2 - 12p + 36 = (p - 6)^2$$

7. $q^2 + 9q + \_\_\_$

$\frac{1}{2}(9) = \frac{9}{2}$ and $\left(\frac{9}{2}\right)^2 = \frac{81}{4}$

$q^2 + 9q + \frac{81}{4} = \left(q + \frac{9}{2}\right)^2$

9. $x^2 + \frac{1}{4}x + \_\_\_$

$\frac{1}{2}\left(\frac{1}{4}\right) = \frac{1}{8}$ and $\left(\frac{1}{8}\right)^2 = \frac{1}{64}$

$x^2 + \frac{1}{4}x + \frac{1}{64} = \left(x + \frac{1}{8}\right)^2$

11. $x^2 - 0.8x + \_\_\_$

$\frac{1}{2}(-0.8) = -0.4$ and $(-0.4)^2 = 0.16$

$x^2 - 0.8x + 0.16 = (x - 0.4)^2$

13. $x^2 + 4x - 2 = 0$

$x^2 + 4x = 2$

$\left[\frac{1}{2}(4)\right]^2 = 2^2 = 4$

15. $x^2 + 10x + 18 = 0$

$x^2 + 10x = -18$

$\left[\frac{1}{2}(10)\right]^2 = 5^2 = 25$

17. $3w^2 - w - 24 = 0$

$w^2 - \frac{1}{3}w - 8 = 0$ *Divide by 3.*

$w^2 - \frac{1}{3}w = 8$

$\left[\frac{1}{2}\left(-\frac{1}{3}\right)\right]^2 = \left(-\frac{1}{6}\right)^2 = \frac{1}{36}$

19. $x^2 - 4x = -3$

Take half of the coefficient of x and square it. Half of -4 is -2, and $(-2)^2 = 4$. Add 4 to each side of the equation, and write the left side as a perfect square.

$x^2 - 4x + 4 = -3 + 4$

$(x - 2)^2 = 1$

Use the square root property.

$x - 2 = \sqrt{1}$ or $x - 2 = -\sqrt{1}$

$x - 2 = 1$ or $x - 2 = -1$

$x = 3$ or $x = 1$

A check verifies that the solution set is $\{1, 3\}$.

21. $x^2 + 2x - 5 = 0$

Add 5 to each side.

$x^2 + 2x = 5$

Take half the coefficient of x and square it.

$\frac{1}{2}(2) = 1$, and $1^2 = 1$.

Add 1 to each side of the equation, and write the left side as a perfect square.

$x^2 + 2x + 1 = 5 + 1$

$(x + 1)^2 = 6$

Use the square root property.

$x + 1 = \sqrt{6}$ or $x + 1 = -\sqrt{6}$

$x = -1 + \sqrt{6}$ or $x = -1 - \sqrt{6}$

A check verifies that the solution set is $\left\{-1 \pm \sqrt{6}\right\}$. Using a calculator for your check is highly recommended.

23. $x^2 + 4x - 2 = 0$

$x^2 + 4x = 2$

$x^2 + 4x + 4 = 2 + 4$ $\quad\left[\frac{1}{2}(4)\right]^2 = 4$

$(x + 2)^2 = 6$

$x + 2 = \sqrt{6}$ or $x + 2 = -\sqrt{6}$

$x = -2 + \sqrt{6}$ or $x = -2 - \sqrt{6}$

Solution set: $\left\{-2 \pm \sqrt{6}\right\}$

25. $x^2 + 10x + 18 = 0$

$x^2 + 10x = -18$

$x^2 + 10x + 25 = -18 + 25$ $\quad\left[\frac{1}{2}(10)\right]^2 = 25$

$(x + 5)^2 = 7$

$x + 5 = \sqrt{7}$ or $x + 5 = -\sqrt{7}$

$x = -5 + \sqrt{7}$ or $x = -5 - \sqrt{7}$

Solution set: $\left\{-5 \pm \sqrt{7}\right\}$

27. $x^2 - 8x = -4$

$x^2 - 8x + 16 = -4 + 16$ $\quad\left[\frac{1}{2}(-8)\right]^2 = 16$

$(x - 4)^2 = 12$ $\quad$ *Factor.*

$x - 4 = \sqrt{12}$ or $x - 4 = -\sqrt{12}$

$x - 4 = \sqrt{4} \cdot \sqrt{3}$ or $x - 4 = -\sqrt{4} \cdot \sqrt{3}$

$x = 4 + 2\sqrt{3}$ or $x = 4 - 2\sqrt{3}$

Solution set: $\left\{4 \pm 2\sqrt{3}\right\}$

29. $x^2 + 7x - 1 = 0$

$x^2 + 7x = 1$

$x^2 + 7x + \frac{49}{4} = 1 + \frac{49}{4}$ $\quad\left[\frac{1}{2}(7)^2\right] = \frac{49}{4}$

$\left(x + \frac{7}{2}\right)^2 = \frac{53}{4}$

$x + \frac{7}{2} = \sqrt{\frac{53}{4}}$ or $x + \frac{7}{2} = -\sqrt{\frac{53}{4}}$

$x = -\frac{7}{2} + \frac{\sqrt{53}}{2}$ $\quad$ $x = -\frac{7}{2} - \frac{\sqrt{53}}{2}$

$x = \frac{-7+\sqrt{53}}{2}$ or $x = \frac{-7-\sqrt{53}}{2}$

Solution set: $\left\{\frac{-7 \pm \sqrt{53}}{2}\right\}$

31. $4x^2 + 4x = 3$

Divide each side by 4 so that the coefficient of x^2 is 1.

$x^2 + x = \frac{3}{4}$

The coefficient of x is 1. Take half of 1, square the result, and add this square to each side.

$$\tfrac{1}{2}(1) = \tfrac{1}{2} \text{ and } (\tfrac{1}{2})^2 = \tfrac{1}{4}$$

$$x^2 + x + \tfrac{1}{4} = \tfrac{3}{4} + \tfrac{1}{4}$$

The left-hand side can then be written as a perfect square.

$$(x + \tfrac{1}{2})^2 = 1$$

Use the square root property.

$$x + \tfrac{1}{2} = 1 \quad \text{or} \quad x + \tfrac{1}{2} = -1$$
$$x = -\tfrac{1}{2} + 1 \quad \text{or} \quad x = -\tfrac{1}{2} - 1$$
$$x = \tfrac{1}{2} \quad \text{or} \quad x = -\tfrac{3}{2}$$

A check verifies that the solution set is $\{-\frac{3}{2}, \frac{1}{2}\}$.

33. $3w^2 - w = 24$

$w^2 - \frac{1}{3}w = 8$ *Divide by 3.*

Complete the square by taking half of $\frac{1}{3}$, the coefficient of w, and squaring the result.

$$\left[\tfrac{1}{2}(-\tfrac{1}{3})\right]^2 = (-\tfrac{1}{6})^2 = \tfrac{1}{36}$$

Add $\frac{1}{36}$ to each side.

$$w^2 - \tfrac{1}{3}w + \tfrac{1}{36} = 8 + \tfrac{1}{36}$$
$$(w - \tfrac{1}{6})^2 = \tfrac{288}{36} + \tfrac{1}{36}$$
$$(w - \tfrac{1}{6})^2 = \tfrac{289}{36}$$

$$w - \tfrac{1}{6} = \sqrt{\tfrac{289}{36}} \quad \text{or} \quad w - \tfrac{1}{6} = -\sqrt{\tfrac{289}{36}}$$
$$w = \tfrac{1}{6} + \tfrac{\sqrt{289}}{\sqrt{36}} \qquad w = \tfrac{1}{6} - \tfrac{\sqrt{289}}{\sqrt{36}}$$
$$w = \tfrac{1}{6} + \tfrac{17}{6} \qquad w = \tfrac{1}{6} - \tfrac{17}{6}$$
$$w = \tfrac{18}{6} \qquad w = -\tfrac{16}{6}$$
$$w = 3 \quad \text{or} \quad w = -\tfrac{8}{3}$$

Solution set: $\{-\frac{8}{3}, 3\}$

35. $2k^2 + 5k - 2 = 0$

$2k^2 + 5k = 2$

$k^2 + \frac{5}{2}k = 1$ *Divide by 2.*

Complete the square.

$$(\tfrac{1}{2} \cdot \tfrac{5}{2})^2 = (\tfrac{5}{4})^2 = \tfrac{25}{16}$$

Add $\frac{25}{16}$ to each side.

$$k^2 + \tfrac{5}{2}k + \tfrac{25}{16} = 1 + \tfrac{25}{16}$$
$$(k + \tfrac{5}{4})^2 = \tfrac{41}{16}$$

$$k + \tfrac{5}{4} = \sqrt{\tfrac{41}{16}} \quad \text{or} \quad k + \tfrac{5}{4} = -\sqrt{\tfrac{41}{16}}$$
$$k = -\tfrac{5}{4} + \tfrac{\sqrt{41}}{4} \qquad k = -\tfrac{5}{4} - \tfrac{\sqrt{41}}{4}$$
$$k = \tfrac{-5+\sqrt{41}}{4} \quad \text{or} \quad k = \tfrac{-5-\sqrt{41}}{4}$$

Solution set: $\left\{\frac{-5\pm\sqrt{41}}{4}\right\}$

37. $5x^2 - 10x + 2 = 0$

$5x^2 - 10x = -2$

$x^2 - 2x = -\frac{2}{5}$ *Divide by 5.*

Complete the square.

$$\left[\tfrac{1}{2}(-2)\right]^2 = (-1)^2 = 1$$

Add 1 to each side.

$$x^2 - 2x + 1 = -\tfrac{2}{5} + 1$$
$$(x - 1)^2 = \tfrac{3}{5}$$

$$x - 1 = \sqrt{\tfrac{3}{5}} \quad \text{or} \quad x - 1 = -\sqrt{\tfrac{3}{5}}$$
$$x - 1 = \tfrac{\sqrt{3}}{\sqrt{5}} \cdot \tfrac{\sqrt{5}}{\sqrt{5}} \qquad x - 1 = -\tfrac{\sqrt{3}}{\sqrt{5}} \cdot \tfrac{\sqrt{5}}{\sqrt{5}}$$
$$x = \tfrac{5}{5} + \tfrac{\sqrt{15}}{5} \qquad x = \tfrac{5}{5} - \tfrac{\sqrt{15}}{5}$$
$$x = \tfrac{5+\sqrt{15}}{5} \quad \text{or} \quad x = \tfrac{5-\sqrt{15}}{5}$$

Solution set: $\left\{\frac{5\pm\sqrt{15}}{5}\right\}$

39. $9x^2 - 24x = -13$

$x^2 - \frac{24}{9}x = \frac{-13}{9}$ *Divide by 9.*

$x^2 - \frac{8}{3}x = \frac{-13}{9}$

Complete the square.

$$\left[\tfrac{1}{2}(-\tfrac{8}{3})\right]^2 = (-\tfrac{4}{3})^2 = \tfrac{16}{9}$$

Add $\frac{16}{9}$ to each side.

$$x^2 - \tfrac{8}{3}x + \tfrac{16}{9} = \tfrac{-13}{9} + \tfrac{16}{9}$$
$$(x - \tfrac{4}{3})^2 = \tfrac{3}{9}$$

$$x - \tfrac{4}{3} = \sqrt{\tfrac{3}{9}} \quad \text{or} \quad x - \tfrac{4}{3} = -\sqrt{\tfrac{3}{9}}$$
$$x = \tfrac{4}{3} + \tfrac{\sqrt{3}}{3} \qquad x = \tfrac{4}{3} - \tfrac{\sqrt{3}}{3}$$
$$x = \tfrac{4+\sqrt{3}}{3} \quad \text{or} \quad x = \tfrac{4-\sqrt{3}}{3}$$

Solution set: $\left\{\frac{4\pm\sqrt{3}}{3}\right\}$

41. $(x + 3)(x - 1) = 5$

$$x^2 + 2x - 3 = 5$$
$$x^2 + 2x = 8$$
$$x^2 + 2x + 1 = 8 + 1$$
$$(x + 1)^2 = 9$$
$$x + 1 = 3 \quad \text{or} \quad x + 1 = -3$$
$$x = 2 \quad \text{or} \quad x = -4$$

A check verifies that the solution set is $\{-4, 2\}$.

43. $(r - 3)(r - 5) = 2$

$$r^2 - 8r + 15 = 2$$
$$r^2 - 8r = -13$$
$$r^2 - 8r + 16 = -13 + 16$$
$$(r - 4)^2 = 3$$
$$r - 4 = \sqrt{3} \quad \text{or} \quad r - 4 = -\sqrt{3}$$
$$r = 4 + \sqrt{3} \quad \text{or} \quad r = 4 - \sqrt{3}$$

A check verifies that the solution set is $\{4 \pm \sqrt{3}\}$.

45. $-x^2 + 2x = -5$

Divide each side by -1.

$$x^2 - 2x = 5$$

Take half of the coefficient of x and square it. Half of -2 is -1, and $(-1)^2 = 1$. Add 1 to each side of the equation, and write the left side as a perfect square.

$$x^2 - 2x + 1 = 5 + 1$$
$$(x-1)^2 = 6$$

Use the square root property.

$x - 1 = \sqrt{6}$ or $x - 1 = -\sqrt{6}$
$x = 1 + \sqrt{6}$ or $x = 1 - \sqrt{6}$

A check verifies that the solution set is $\{1 \pm \sqrt{6}\}$.

47. $z^2 - \frac{4}{3}z = -\frac{1}{9}$

Complete the square.

$$\left[\tfrac{1}{2}\left(-\tfrac{4}{3}\right)\right]^2 = \left(-\tfrac{2}{3}\right)^2 = \tfrac{4}{9}$$

Add $\frac{4}{9}$ to each side.

$$z^2 - \tfrac{4}{3}z + \tfrac{4}{9} = -\tfrac{1}{9} + \tfrac{4}{9}$$
$$\left(z - \tfrac{2}{3}\right)^2 = \tfrac{3}{9}$$

$z - \frac{2}{3} = \sqrt{\frac{3}{9}}$ or $z - \frac{2}{3} = -\sqrt{\frac{3}{9}}$
$z = \frac{2}{3} + \frac{\sqrt{3}}{3}$ $\quad$ $z = \frac{2}{3} - \frac{\sqrt{3}}{3}$
$z = \frac{2+\sqrt{3}}{3}$ or $z = \frac{2-\sqrt{3}}{3}$

Solution set: $\left\{\frac{2\pm\sqrt{3}}{3}\right\}$

49. $0.1x^2 - 0.2x - 0.1 = 0$

Multiply each side by 10 to clear the decimals.

$$x^2 - 2x - 1 = 0$$
$$x^2 - 2x = 1$$

Complete the square.

$$\left[\tfrac{1}{2}(-2)\right]^2 = (-1)^2 = 1$$

Add 1 to each side.

$$x^2 - 2x + 1 = 1 + 1$$
$$(x-1)^2 = 2$$

$x - 1 = \sqrt{2}$ or $x - 1 = -\sqrt{2}$
$x = 1 + \sqrt{2}$ or $x = 1 - \sqrt{2}$

Solution set: $\{1 \pm \sqrt{2}\}$

51.

$$3r^2 - 2 = 6r + 3$$
$$3r^2 - 6r = 5$$
$$r^2 - 2r = \tfrac{5}{3}$$
$$r^2 - 2r + 1 = \tfrac{5}{3} + 1$$
$$(r-1)^2 = \tfrac{8}{3}$$
$$r - 1 = \pm\sqrt{\tfrac{8}{3}}$$

Simplify the radical.

$$\sqrt{\tfrac{8}{3}} = \tfrac{\sqrt{8}}{\sqrt{3}} = \tfrac{2\sqrt{2}}{\sqrt{3}} \cdot \tfrac{\sqrt{3}}{\sqrt{3}} = \tfrac{2\sqrt{6}}{3}$$
$$r = 1 \pm \tfrac{2\sqrt{6}}{3}$$
$$r = \tfrac{3}{3} \pm \tfrac{2\sqrt{6}}{3} = \tfrac{3\pm2\sqrt{6}}{3}$$

(a) The solution set with *exact* values is $\left\{\frac{3\pm2\sqrt{6}}{3}\right\}$.

(b) $\frac{3+2\sqrt{6}}{3} \approx 2.633$, $\frac{3-2\sqrt{6}}{3} \approx -0.633$

The solution set with *approximate* values is $\{-0.633, 2.633\}$.

53.

$$(x+1)(x+3) = 2$$
$$x^2 + 3x + x + 3 = 2$$
$$x^2 + 4x = -1$$
$$x^2 + 4x + 4 = -1 + 4$$
$$(x+2)^2 = 3$$
$$x + 2 = \pm\sqrt{3}$$
$$x = -2 \pm \sqrt{3}$$

(a) The solution set with *exact* values is $\{-2 \pm \sqrt{3}\}$.

(b) $-2 + \sqrt{3} \approx -0.268$
$-2 - \sqrt{3} \approx -3.732$

The solution set with *approximate* values is $\{-3.732, -0.268\}$.

55. $m^2 + 4m + 13 = 0$

$$m^2 + 4m = -13$$

Complete the square.

$$\left(\tfrac{1}{2} \cdot 4\right)^2 = 2^2 = 4$$

Add 4 to each side.

$$m^2 + 4m + 4 = -13 + 4$$
$$(m+2)^2 = -9$$

$m + 2 = \sqrt{-9}$ or $m + 2 = -\sqrt{-9}$
$m = -2 + 3i$ or $m = -2 - 3i$

Solution set: $\{-2 \pm 3i\}$

57. $3r^2 + 4r + 4 = 0$

$$3r^2 + 4r = -4$$
$$r^2 + \tfrac{4}{3}r = \tfrac{-4}{3} \quad \textit{Divide by 3.}$$

Complete the square.

$$\left(\tfrac{1}{2} \cdot \tfrac{4}{3}\right)^2 = \left(\tfrac{2}{3}\right)^2 = \tfrac{4}{9}$$

Add $\frac{4}{9}$ to each side.

$$r^2 + \tfrac{4}{3}r + \tfrac{4}{9} = \tfrac{-4}{3} + \tfrac{4}{9}$$
$$\left(r + \tfrac{2}{3}\right)^2 = \tfrac{-8}{9}$$

$$r+\frac{2}{3}=\frac{\sqrt{-8}}{\sqrt{9}} \quad \text{or} \quad r+\frac{2}{3}=-\frac{\sqrt{-8}}{\sqrt{9}}$$
$$r=-\frac{2}{3}+\frac{2i\sqrt{2}}{3} \qquad r=-\frac{2}{3}-\frac{2i\sqrt{2}}{3}$$

Solution set: $\left\{-\frac{2}{3}\pm\frac{2\sqrt{2}}{3}i\right\}$

59. $-m^2-6m-12=0$

Multiply each side by -1.

$$m^2+6m+12=0$$
$$m^2+6m=-12$$

Complete the square.

$$\left(\tfrac{1}{2}\cdot 6\right)^2=3^2=9$$

Add 9 to each side.

$$m^2+6m+9=-12+9$$
$$(m+3)^2=-3$$

$$m+3=\sqrt{-3} \quad \text{or} \quad m+3=-\sqrt{-3}$$
$$m=-3+i\sqrt{3} \quad \text{or} \quad m=-3-i\sqrt{3}$$

Solution set: $\left\{-3\pm i\sqrt{3}\right\}$

♦♦♦ Relating Concepts 61–66 ♦♦♦

61. The area of the original square is $x\cdot x$, or x^2.

62. Each rectangular strip has length x and width 1, so each strip has an area of $x\cdot 1$, or x.

63. From Exercise 62, the area of a rectangular strip is x. The area of 6 rectangular strips is $6x$.

64. These are 1 by 1 squares, so each has an area of $1\cdot 1$, or 1.

65. There are 9 small squares, each with area 1 (from Exercise 64), so the total area is $9\cdot 1$, or 9.

66. The area of the larger square is $(x+3)^2$. Using the results from Exercises 61–65,

$$(x+3)^2, \quad \text{or} \quad x^2+6x+9.$$

67. $x^2-b=0$

$$x^2=b$$
$$x=\sqrt{b} \quad \text{or} \quad x=-\sqrt{b}$$

Solution set: $\left\{\pm\sqrt{b}\right\}$

69. $4x^2=b^2+16$

$$x^2=\frac{b^2+16}{4}$$
$$x=\sqrt{\frac{b^2+16}{4}} \quad \text{or} \quad x=-\sqrt{\frac{b^2+16}{4}}$$
$$x=\frac{\sqrt{b^2+16}}{2} \quad \text{or} \quad x=-\frac{\sqrt{b^2+16}}{2}$$

Solution set: $\left\{\pm\frac{\sqrt{b^2+16}}{2}\right\}$

71. $(5x-2b)^2=3a$

$$5x-2b=\sqrt{3a} \quad \text{or} \quad 5x-2b=-\sqrt{3a}$$
$$5x=2b+\sqrt{3a} \qquad 5x=2b-\sqrt{3a}$$
$$x=\frac{2b+\sqrt{3a}}{5} \quad \text{or} \quad x=\frac{2b-\sqrt{3a}}{5}$$

Solution set: $\left\{\frac{2b\pm\sqrt{3a}}{5}\right\}$

73. $a=3, b=1, c=-1$

$$\sqrt{b^2-4ac}=\sqrt{1^2-4(3)(-1)}$$
$$=\sqrt{1+12}$$
$$=\sqrt{13}$$

75. $a=6, b=7, c=2$

$$\sqrt{b^2-4ac}=\sqrt{7^2-4(6)(2)}$$
$$=\sqrt{49-48}$$
$$=\sqrt{1}=1$$

11.3 Solving Quadratic Equations by the Quadratic Formula

11.3 Now Try Exercises

N1. $2x^2+3x-20=0$

Here $a=2$, $b=3$, and $c=-20$.

$$x=\frac{-b\pm\sqrt{b^2-4ac}}{2a}$$
$$x=\frac{-3\pm\sqrt{3^2-4(2)(-20)}}{2(2)}$$
$$=\frac{-3\pm\sqrt{9+160}}{4}$$
$$=\frac{-3\pm\sqrt{169}}{4}$$
$$=\frac{-3\pm 13}{4}$$
$$x=\frac{-3+13}{4}=\frac{10}{4}=\frac{5}{2}$$
$$\text{or} \quad x=\frac{-3-13}{4}=\frac{-16}{4}=-4$$

Solution set: $\left\{-4, \frac{5}{2}\right\}$

N2. $3x^2+1=-5x$

$$3x^2+5x+1=0$$

Here $a=3$, $b=5$, and $c=1$.

$$x=\frac{-b\pm\sqrt{b^2-4ac}}{2a}$$
$$x=\frac{-5\pm\sqrt{5^2-4(3)(1)}}{2(3)}$$
$$=\frac{-5\pm\sqrt{25-12}}{6}=\frac{-5\pm\sqrt{13}}{6}$$

Solution set: $\left\{\frac{-5\pm\sqrt{13}}{6}\right\}$

N3. $(x+5)(x-1)=-18$

$x^2+4x-5=-18$

$x^2+4x+13=0$

Here $a=1$, $b=4$, and $c=13$.

$$x=\frac{-b\pm\sqrt{b^2-4ac}}{2a}$$

$$x=\frac{-4\pm\sqrt{4^2-4(1)(13)}}{2(1)}$$

$$=\frac{-4\pm\sqrt{16-52}}{2}$$

$$=\frac{-4\pm\sqrt{-36}}{2}=\frac{-4\pm 6i}{2}$$

$$=\frac{2(-2\pm 3i)}{2}=-2\pm 3i$$

Solution set: $\{-2\pm 3i\}$

N4. (a) $8x^2-6x-5=0$

Here $a=8$, $b=-6$, and $c=-5$, so the discriminant is

$$\begin{aligned}b^2-4ac&=(-6)^2-4(8)(-5)\\&=36+160\\&=196.\end{aligned}$$

The discriminant is a perfect square and a, b, and c are integers, so there are two different rational solutions and the equation can be solved by factoring.

(b) $9x^2=24x-16$

$9x^2-24x+16=0$

Here $a=9$, $b=-24$, and $c=16$, so the discriminant is

$$\begin{aligned}b^2-4ac&=(-24)^2-4(9)(16)\\&=576-576\\&=0.\end{aligned}$$

The discriminant is zero, so there is one rational solution and the equation can be solved by factoring.

(c) $3x^2+2x=-1$

$3x^2+2x+1=0$

Here $a=3$, $b=2$, and $c=1$, so the discriminant is

$$\begin{aligned}b^2-4ac&=2^2-4(3)(1)\\&=4-12\\&=-8.\end{aligned}$$

The discriminant is negative but not a perfect square and a, b, and c are integers, so there are two nonreal complex solutions and the equation is best solved using the quadratic formula.

N5. $4x^2+kx+25=0$

The equation will have only one rational solution if the discriminant is 0. Since $a=4$, $b=k$, and $c=25$, the discriminant is

$$\begin{aligned}b^2-4ac&=k^2-4(4)(25)\\&=k^2-400.\end{aligned}$$

Set the discriminant equal to 0 and solve for k.

$k^2-400=0$

$k^2=400$ *Add 400.*

$k=20$ or $k=-20$ *Square root property*

The equation will have only one rational solution if $k=20$ or $k=-20$.

11.3 Section Exercises

1. The documentation was incorrect, since the fraction bar should extend under the term $-b$. The correct formula is

$$x=\frac{-b\pm\sqrt{b^2-4ac}}{2a}.$$

3. The last step is wrong. Because 5 is not a common factor in the numerator, the fraction cannot be reduced. The solution set is $\left\{\frac{5\pm\sqrt{5}}{10}\right\}$.

5. $x^2-8x+15=0$

Here $a=1$, $b=-8$, and $c=15$.

$$x=\frac{-b\pm\sqrt{b^2-4ac}}{2a}$$

$$x=\frac{-(-8)\pm\sqrt{(-8)^2-4(1)(15)}}{2(1)}$$

$$=\frac{8\pm\sqrt{64-60}}{2}$$

$$=\frac{8\pm\sqrt{4}}{2}=\frac{8\pm 2}{2}$$

$$x=\frac{8+2}{2}=\frac{10}{2}=5 \text{ or}$$

$$x=\frac{8-2}{2}=\frac{6}{2}=3$$

Solution set: $\{3,5\}$

7. $2x^2+4x+1=0$

Here $a=2$, $b=4$, and $c=1$.

$$x=\frac{-b\pm\sqrt{b^2-4ac}}{2a}$$

$$x=\frac{-4\pm\sqrt{4^2-4(2)(1)}}{2(2)}$$

$$=\frac{-4\pm\sqrt{16-8}}{4}$$

$$=\frac{-4\pm\sqrt{8}}{4}=\frac{-4\pm 2\sqrt{2}}{4}$$

$$=\frac{2\left(-2\pm\sqrt{2}\right)}{2\cdot 2}=\frac{-2\pm\sqrt{2}}{2}$$

Solution set: $\left\{\frac{-2\pm\sqrt{2}}{2}\right\}$

9. $2x^2 - 2x = 1$

$2x^2 - 2x - 1 = 0$

Here $a = 2$, $b = -2$, and $c = -1$.

$$x = \frac{-b \pm \sqrt{b^2 - 4ac}}{2a}$$

$$x = \frac{-(-2) \pm \sqrt{(-2)^2 - 4(2)(-1)}}{2(2)}$$

$$= \frac{2 \pm \sqrt{4 + 8}}{4} = \frac{2 \pm \sqrt{12}}{4}$$

$$= \frac{2 \pm 2\sqrt{3}}{4} = \frac{2\left(1 \pm \sqrt{3}\right)}{2 \cdot 2}$$

$$= \frac{1 \pm \sqrt{3}}{2}$$

Solution set: $\left\{\frac{1 \pm \sqrt{3}}{2}\right\}$

11. $x^2 + 18 = 10x$

$x^2 - 10x + 18 = 0$

Here $a = 1$, $b = -10$, and $c = 18$.

$$x = \frac{-b \pm \sqrt{b^2 - 4ac}}{2a}$$

$$x = \frac{-(-10) \pm \sqrt{(-10)^2 - 4(1)(18)}}{2(1)}$$

$$= \frac{10 \pm \sqrt{100 - 72}}{2} = \frac{10 \pm \sqrt{28}}{2}$$

$$= \frac{10 \pm 2\sqrt{7}}{2} = \frac{2\left(5 \pm \sqrt{7}\right)}{2}$$

$$= 5 \pm \sqrt{7}$$

Solution set: $\left\{5 \pm \sqrt{7}\right\}$

13. $4x^2 + 4x - 1 = 0$

Here $a = 4$, $b = 4$, and $c = -1$.

$$x = \frac{-b \pm \sqrt{b^2 - 4ac}}{2a}$$

$$x = \frac{-4 \pm \sqrt{4^2 - 4(4)(-1)}}{2(4)}$$

$$= \frac{-4 \pm \sqrt{16 + 16}}{8} = \frac{-4 \pm \sqrt{32}}{8}$$

$$= \frac{-4 \pm 4\sqrt{2}}{8} = \frac{4\left(-1 \pm \sqrt{2}\right)}{2 \cdot 4}$$

$$= \frac{-1 \pm \sqrt{2}}{2}$$

Solution set: $\left\{\frac{-1 \pm \sqrt{2}}{2}\right\}$

15. $2 - 2x = 3x^2$

$0 = 3x^2 + 2x - 2$

Here $a = 3$, $b = 2$, and $c = -2$.

$$x = \frac{-b \pm \sqrt{b^2 - 4ac}}{2a}$$

$$x = \frac{-2 \pm \sqrt{2^2 - 4(3)(-2)}}{2(3)}$$

$$= \frac{-2 \pm \sqrt{4 + 24}}{6} = \frac{-2 \pm \sqrt{28}}{6}$$

$$= \frac{-2 \pm 2\sqrt{7}}{6} = \frac{2\left(-1 \pm \sqrt{7}\right)}{2 \cdot 3}$$

$$= \frac{-1 \pm \sqrt{7}}{3}$$

Solution set: $\left\{\frac{-1 \pm \sqrt{7}}{3}\right\}$

17. $\frac{x^2}{4} - \frac{x}{2} = 1$

$\frac{x^2}{4} - \frac{x}{2} - 1 = 0$

$x^2 - 2x - 4 = 0$ *Multiply by 4.*

Here $a = 1$, $b = -2$, and $c = -4$.

$$x = \frac{-b \pm \sqrt{b^2 - 4ac}}{2a}$$

$$x = \frac{-(-2) \pm \sqrt{(-2)^2 - 4(1)(-4)}}{2(1)}$$

$$= \frac{2 \pm \sqrt{4 + 16}}{2} = \frac{2 \pm \sqrt{20}}{2}$$

$$= \frac{2 \pm 2\sqrt{5}}{2} = 1 \pm \sqrt{5}$$

Solution set: $\left\{1 \pm \sqrt{5}\right\}$

19. $-2t(t + 2) = -3$

$-2t^2 - 4t = -3$

$-2t^2 - 4t + 3 = 0$

Here $a = -2$, $b = -4$, and $c = 3$.

$$t = \frac{-b \pm \sqrt{b^2 - 4ac}}{2a}$$

$$t = \frac{-(-4) \pm \sqrt{(-4)^2 - 4(-2)(3)}}{2(-2)}$$

$$= \frac{4 \pm \sqrt{16 + 24}}{-4} = \frac{4 \pm \sqrt{40}}{-4}$$

$$= \frac{4 \pm 2\sqrt{10}}{-4} = \frac{2\left(2 \pm \sqrt{10}\right)}{-2 \cdot 2}$$

$$= \frac{2 \pm \sqrt{10}}{-2} \cdot \frac{-1}{-1} = \frac{-2 \mp \sqrt{10}}{2}$$

$$= \frac{-2 \pm \sqrt{10}}{2}$$

Solution set: $\left\{\frac{-2 \pm \sqrt{10}}{2}\right\}$

21. $(r-3)(r+5)=2$

$r^2+2r-15=2$

$r^2+2r-17=0$

Here $a=1$, $b=2$, and $c=-17$.

$$r=\frac{-b\pm\sqrt{b^2-4ac}}{2a}$$
$$r=\frac{-2\pm\sqrt{2^2-4(1)(-17)}}{2(1)}$$
$$=\frac{-2\pm\sqrt{4+68}}{2}=\frac{-2\pm\sqrt{72}}{2}$$
$$=\frac{-2\pm6\sqrt{2}}{2}=\frac{2\left(-1\pm3\sqrt{2}\right)}{2}$$
$$=-1\pm3\sqrt{2}$$

Solution set: $\left\{-1\pm3\sqrt{2}\right\}$

23. $(x+2)(x-3)=1$

$x^2-x-6=1$

$x^2-x-7=0$

Here $a=1$, $b=-1$, and $c=-7$.

$$x=\frac{-b\pm\sqrt{b^2-4ac}}{2a}$$
$$x=\frac{-(-1)\pm\sqrt{(-1)^2-4(1)(-7)}}{2(1)}$$
$$=\frac{1\pm\sqrt{1+28}}{2}=\frac{1\pm\sqrt{29}}{2}$$

Solution set: $\left\{\frac{1\pm\sqrt{29}}{2}\right\}$

25. $p=\frac{5(5-p)}{3(p+1)}$

$3p(p+1)=5(5-p)$

$3p^2+3p=25-5p$

$3p^2+8p-25=0$

Here $a=3$, $b=8$, and $c=-25$.

$$p=\frac{-b\pm\sqrt{b^2-4ac}}{2a}$$
$$p=\frac{-8\pm\sqrt{8^2-4(3)(-25)}}{2(3)}$$
$$=\frac{-8\pm\sqrt{64+300}}{6}=\frac{-8\pm\sqrt{364}}{6}$$
$$=\frac{-8\pm2\sqrt{91}}{6}=\frac{2\left(-4\pm\sqrt{91}\right)}{2\bullet3}=\frac{-4\pm\sqrt{91}}{3}$$

Solution set: $\left\{\frac{-4\pm\sqrt{91}}{3}\right\}$

27. $(2x+1)^2=x+4$

$4x^2+4x+1=x+4$

$4x^2+3x-3=0$

Here $a=4$, $b=3$, and $c=-3$.

$$x=\frac{-b\pm\sqrt{b^2-4ac}}{2a}$$
$$x=\frac{-3\pm\sqrt{3^2-4(4)(-3)}}{2(4)}$$
$$=\frac{-3\pm\sqrt{9+48}}{8}=\frac{-3\pm\sqrt{57}}{8}$$

Solution set: $\left\{\frac{-3\pm\sqrt{57}}{8}\right\}$

29. $x^2-3x+6=0$

Here $a=1$, $b=-3$, and $c=6$.

$$x=\frac{-b\pm\sqrt{b^2-4ac}}{2a}$$
$$x=\frac{-(-3)\pm\sqrt{(-3)^2-4(1)(6)}}{2(1)}$$
$$=\frac{3\pm\sqrt{9-24}}{2}=\frac{3\pm\sqrt{-15}}{2}$$
$$=\frac{3\pm i\sqrt{15}}{2}=\frac{3}{2}\pm\frac{\sqrt{15}}{2}i$$

Solution set: $\left\{\frac{3}{2}\pm\frac{\sqrt{15}}{2}i\right\}$

31. $r^2-6r+14=0$

Here $a=1$, $b=-6$, and $c=14$.

$$r=\frac{-b\pm\sqrt{b^2-4ac}}{2a}$$
$$r=\frac{-(-6)\pm\sqrt{(-6)^2-4(1)(14)}}{2(1)}$$
$$=\frac{6\pm\sqrt{36-56}}{2}$$
$$=\frac{6\pm\sqrt{-20}}{2}=\frac{6\pm2i\sqrt{5}}{2}$$
$$=\frac{2\left(3\pm i\sqrt{5}\right)}{2}=3\pm i\sqrt{5}$$

Solution set: $\left\{3\pm i\sqrt{5}\right\}$

33. $4x^2-4x=-7$

$4x^2-4x+7=0$

Here $a=4$, $b=-4$, and $c=7$.

$$x=\frac{-b\pm\sqrt{b^2-4ac}}{2a}$$
$$x=\frac{-(-4)\pm\sqrt{(-4)^2-4(4)(7)}}{2(4)}$$
$$=\frac{4\pm\sqrt{16-112}}{8}=\frac{4\pm\sqrt{-96}}{8}$$
$$=\frac{4\pm4i\sqrt{6}}{8}=\frac{4\left(1\pm i\sqrt{6}\right)}{2\bullet4}$$
$$=\frac{1\pm i\sqrt{6}}{2}=\frac{1}{2}\pm\frac{\sqrt{6}}{2}i$$

Solution set: $\left\{\frac{1}{2}\pm\frac{\sqrt{6}}{2}i\right\}$

35. $x(3x+4)=-2$
$3x^2+4x=-2$
$3x^2+4x+2=0$
Here $a=3$, $b=4$, and $c=2$.

$$x=\frac{-b\pm\sqrt{b^2-4ac}}{2a}$$
$$x=\frac{-4\pm\sqrt{4^2-4(3)(2)}}{2(3)}$$
$$=\frac{-4\pm\sqrt{16-24}}{6}=\frac{-4\pm\sqrt{-8}}{6}$$
$$=\frac{-4\pm 2i\sqrt{2}}{6}=\frac{2\left(-2\pm i\sqrt{2}\right)}{2\cdot 3}$$
$$=\frac{-2\pm i\sqrt{2}}{3}=-\frac{2}{3}\pm\frac{\sqrt{2}}{3}i$$

Solution set: $\left\{-\frac{2}{3}\pm\frac{\sqrt{2}}{3}i\right\}$

37. $(2x-1)(8x-4)=-1$
$16x^2-16x+4=-1$
$16x^2-16x+5=0$
Here $a=16$, $b=-16$, and $c=5$.

$$x=\frac{-b\pm\sqrt{b^2-4ac}}{2a}$$
$$x=\frac{-(-16)\pm\sqrt{(-16)^2-4(16)(5)}}{2(16)}$$
$$=\frac{16\pm\sqrt{256-320}}{32}$$
$$=\frac{16\pm\sqrt{-64}}{32}=\frac{16\pm 8i}{32}$$
$$=\tfrac{16}{32}\pm\tfrac{8}{32}i=\tfrac{1}{2}\pm\tfrac{1}{4}i$$

Solution set: $\{\frac{1}{2}\pm\frac{1}{4}i\}$

Note: We could also solve this equation without the quadratic formula as follows:

$$(2x-1)(8x-4)=-1$$
$$(2x-1)4(2x-1)=-1$$
$$(2x-1)^2=-\tfrac{1}{4}$$
$$2x-1=\pm\sqrt{-\tfrac{1}{4}}$$
$$2x=1\pm\tfrac{1}{2}i$$
$$x=\tfrac{1}{2}\pm\tfrac{1}{4}i$$

39. $25x^2+70x+49=0$

Here $a=25$, $b=70$, and $c=49$, so the discriminant is

$$b^2-4ac=70^2-4(25)(49)$$
$$=4900-4900$$
$$=0.$$

Since the discriminant is 0, the quantity under the radical in the quadratic formula is 0, and there is only one rational solution. The answer is **B** and the equation can be solved by factoring.

41. $x^2+4x+2=0$

Here $a=1$, $b=4$, and $c=2$, so the discriminant is

$$b^2-4ac=4^2-4(1)(2)$$
$$=16-8$$
$$=8.$$

Since the discriminant is positive, but not a perfect square, there are two distinct irrational number solutions. The answer is **C** and the equation is best solved using the quadratic formula.

43. $3x^2=5x+2$
$3x^2-5x-2=0$

Here $a=3$, $b=-5$, and $c=-2$, so the discriminant is

$$b^2-4ac=(-5)^2-4(3)(-2)$$
$$=25+24$$
$$=49.$$

Since the discriminant is a perfect square, there are two distinct rational solutions. The answer is **A** and the equation can be solved by factoring.

45. $3m^2-10m+15=0$

Here $a=3$, $b=-10$, and $c=15$, so the discriminant is

$$b^2-4ac=(-10)^2-4(3)(15)$$
$$=100-180$$
$$=-80.$$

Since the discriminant is negative, there are two distinct nonreal complex number solutions. The answer is **D** and the equation is best solved using the quadratic formula.

47. $25x^2+70x+49=0$
$(5x+7)^2=0$
$5x+7=0$
$x=-\frac{7}{5}$

Solution set: $\{-\frac{7}{5}\}$

49. $3x^2=5x+2$
$3x^2-5x-2=0$
$(3x+1)(x-2)=0$

$3x+1=0$ or $x-2=0$
$x=-\frac{1}{3}$ or $x=2$

Solution set: $\{-\frac{1}{3},2\}$

51. **(a)**

$$3x^2 + 13x = -12$$
$$3x^2 + 13x + 12 = 0$$

Here $a = 3$, $b = 13$, and $c = 12$, so the discriminant is

$$\begin{aligned} b^2 - 4ac &= 13^2 - 4(3)(12) \\ &= 169 - 144 \\ &= 25. \end{aligned}$$

The discriminant is a perfect square, so the equation can be solved by factoring.

$$3x^2 + 13x + 12 = 0$$
$$(3x + 4)(x + 3) = 0$$

$$3x + 4 = 0 \quad \text{or} \quad x + 3 = 0$$
$$x = -\tfrac{4}{3} \qquad\qquad x = -3$$

Solution set: $\{-\frac{4}{3}, -3\}$

(b)

$$2x^2 + 19 = 14x$$
$$2x^2 - 14x + 19 = 0$$

Here $a = 2$, $b = -14$, and $c = 19$, so the discriminant is

$$\begin{aligned} b^2 - 4ac &= (-14)^2 - 4(2)(19) \\ &= 196 - 152 \\ &= 44. \end{aligned}$$

The discriminant is not a perfect square, so use the quadratic formula.

$$x = \frac{-b \pm \sqrt{b^2 - 4ac}}{2a}$$
$$x = \frac{-(-14) \pm \sqrt{44}}{2 \cdot 2}$$
$$= \frac{14 \pm 2\sqrt{11}}{2 \cdot 2} = \frac{7 \pm \sqrt{11}}{2}$$

Solution set: $\left\{\frac{7 \pm \sqrt{11}}{2}\right\}$

53. $p^2 + bp + 25 = 0$
For there to be only one rational solution, $b^2 - 4ac$ must equal zero. Since $a = 1$ and $c = 25$,

$$\begin{aligned} b^2 - 4(1)(25) &= 0 \\ b^2 - 100 &= 0 \\ b^2 &= 100 \\ b &= \pm\sqrt{100} \end{aligned}$$
$$b = 10 \quad \text{or} \quad b = -10.$$

55. $am^2 + 8m + 1 = 0$
For there to be only one rational solution, $b^2 - 4ac$ must equal zero. Since $b = 8$ and $c = 1$,

$$\begin{aligned} 8^2 - 4(a)(1) &= 0 \\ 64 - 4a &= 0 \\ -4a &= -64 \\ a &= 16. \end{aligned}$$

57. $9x^2 - 30x + c = 0$
For there to be only one rational solution, $b^2 - 4ac$ must equal zero. Since $a = 9$ and $b = -30$,

$$\begin{aligned} (-30)^2 - 4(9)(c) &= 0 \\ 900 - 36c &= 0 \\ -36c &= -900 \\ c &= 25. \end{aligned}$$

59. Substitute $-\frac{5}{2}$ for x and solve for b.

$$\begin{aligned} 4x^2 + bx - 3 &= 0 \\ 4(-\tfrac{5}{2})^2 + b(-\tfrac{5}{2}) - 3 &= 0 \\ 25 - \tfrac{5}{2}b - 3 &= 0 \\ 22 &= \tfrac{5}{2}b \\ \tfrac{44}{5} &= b \end{aligned}$$

So the equation is

$$4x^2 + \tfrac{44}{5}x - 3 = 0.$$

Now multiply by 5.

$$20x^2 + 44x - 15 = 0$$

Since $-\frac{5}{2}$ is a solution, $2x + 5$ must be one factor.

$$(2x + 5)(10x - 3) = 0$$

So the other solution is $\frac{3}{10}$.

61.

$$\begin{aligned} \tfrac{3}{4}x + \tfrac{1}{2}x &= -10 \\ 3x + 2x &= -40 \quad \textit{Multiply by 4.} \\ 5x &= -40 \\ x &= -8 \end{aligned}$$

Solution set: $\{-8\}$.

63.

$$\begin{aligned} \sqrt{2x + 6} &= x - 1 \\ \left(\sqrt{2x + 6}\right)^2 &= (x - 1)^2 \\ 2x + 6 &= x^2 - 2x + 1 \\ 0 &= x^2 - 4x - 5 \\ 0 &= (x - 5)(x + 1) \end{aligned}$$

$$x - 5 = 0 \quad \text{or} \quad x + 1 = 0$$
$$x = 5 \quad \text{or} \quad x = -1$$

Check $x = 5$: $\sqrt{16} = 4$ *True*
Check $x = -1$: $\sqrt{4} = -2$ *False*

Solution set: $\{5\}$

11.4 Equations Quadratic in Form

11.4 Now Try Exercises

N1. $\dfrac{2}{x} + \dfrac{3}{x + 2} = 1$

Multiply each term by the LCD, $x(x + 2)$.

$$x(x+2)\left(\frac{2}{x}+\frac{3}{x+2}\right)=x(x+2)(1)$$
$$2(x+2)+3x=x^2+2x$$
$$2x+4+3x=x^2+2x$$
$$x^2-3x-4=0$$
$$(x+1)(x-4)=0$$

$$x+1=0 \quad \text{or} \quad x-4=0$$
$$x=-1 \quad \text{or} \quad x=4$$

Check $x=-1$: $-2+3=1$ *True*
Check $x=4$: $\frac{1}{2}+\frac{1}{2}=1$ *True*

Solution set: $\{-1,4\}$

N2. Let $x=$ the rate of the current. Make a table. Use $t=d/r$.

	d	r	t
Up	8	$18-x$	$\frac{8}{18-x}$
Down	8	$18+x$	$\frac{8}{18+x}$

The time going upriver added to the time going downriver is $\frac{9}{10}$ hr.

$$\frac{8}{18-x}+\frac{8}{18+x}=\frac{9}{10}$$

Multiply each term by the LCD, $10(18-x)(18+x)$.

$$10(18-x)(18+x)\frac{8}{18-x}$$
$$+10(18-x)(18+x)\frac{8}{18+x}$$
$$=10(18-x)(18+x)(\tfrac{9}{10})$$
$$80(18+x)+80(18-x)=9(18-x)(18+x)$$
$$1440+80x+1440-80x=9(324-x^2)$$
$$2880=2916-9x^2$$
$$9x^2=36$$
$$x^2=4$$

$$x=-2 \quad \text{or} \quad x=2$$

The rate can't be negative, so the rate of the current is 2 mph.

N3. Let $x=$ the slower electrician's time alone. Then $x-2=$ the faster electrician's time alone. Make a table.

	Rate	**Time Working Together**	**Fractional Part of the Job Done**
Slower Elec.	$\frac{1}{x}$	6	$\frac{6}{x}$
Faster Elec.	$\frac{1}{x-2}$	6	$\frac{6}{x-2}$

Since together they complete 1 whole job,

$$\frac{6}{x}+\frac{6}{x-2}=1.$$

Multiply each term by the LCD, $x(x-2)$.

$$x(x-2)\left(\frac{6}{x}\right)+x(x-2)\left(\frac{6}{x-2}\right)$$
$$=x(x-2)(1)$$
$$6(x-2)+6x=x(x-2)$$
$$6x-12+6x=x^2-2x$$
$$0=x^2-14x+12$$

Here $a=1$, $b=-14$, and $c=12$.

$$x=\frac{-b\pm\sqrt{b^2-4ac}}{2a}$$
$$=\frac{-(-14)\pm\sqrt{(-14)^2-4(1)(12)}}{2(1)}$$
$$=\frac{14\pm\sqrt{196-48}}{2}$$
$$=\frac{14\pm\sqrt{148}}{2}=\frac{14\pm2\sqrt{37}}{2}$$
$$=\frac{2\left(7\pm\sqrt{37}\right)}{2}=7\pm\sqrt{37}$$

$$x=7+\sqrt{37}\approx 13.1 \quad \text{or} \quad x=7-\sqrt{37}\approx 0.9$$

The slower electrician's time cannot be 0.9 since the faster electrician's time would then be $0.9-2$ or -1.1.
So the slower electrician's time working alone is 13.1 hr and the faster electrician's time working alone is $13.1-2=11.1$ hr.

N4. **(a)**

$$x=\sqrt{9x-20}$$
$$x^2=\left(\sqrt{9x-20}\right)^2 \quad \textit{Square.}$$
$$x^2=9x-20$$
$$x^2-9x+20=0$$
$$(x-4)(x-5)=0$$

$$x-4=0 \quad \text{or} \quad x-5=0$$
$$x=4 \quad \text{or} \quad x=5$$

Check $x=4$: $4=\sqrt{16}$ *True*
Check $x=5$: $5=\sqrt{25}$ *True*

Solution set: $\{4,5\}$

(b) $x+\sqrt{x}=20$

$$\sqrt{x}=20-x \quad \textit{Isolate.}$$
$$\left(\sqrt{x}\right)^2=(20-x)^2 \quad \textit{Square.}$$
$$x=400-40x+x^2$$
$$0=x^2-41x+400$$
$$0=(x-16)(x-25)$$

$$x-16=0 \quad \text{or} \quad x-25=0$$
$$x=16 \quad \text{or} \quad x=25$$

Check $x = 16$: $16 + 4 = 20$ *True*
Check $x = 25$: $25 + 5 = 20$ *False*

Solution set: $\{16\}$

N5. **(a)** $x^4 - 10x^2 + 9 = 0$

Since $x^4 = (x^2)^2$, define $u = x^2$, and the original equation becomes $u^2 - 10u + 9 = 0$.

(b) $6(x+2)^2 - 11(x+2) + 4 = 0$

Because this equation involves both $(x+2)^2$ and $(x+2)$, define $u = x + 2$, and the original equation becomes $6u^2 - 11u + 4 = 0$.

N6. **(a)** $x^4 - 17x^2 + 16 = 0$

Let $y = x^2$, so $y^2 = (x^2)^2 = x^4$.

$$y^2 - 17y + 16 = 0$$
$$(y-1)(y-16) = 0$$
$$y - 1 = 0 \quad \text{or} \quad y - 16 = 0$$
$$y = 1 \quad \text{or} \quad y = 16$$

To find x, substitute x^2 for y.

$$x^2 = 1 \quad \text{or} \quad x^2 = 16$$
$$x = \pm 1 \quad \text{or} \quad x = \pm 4$$

Check $x = \pm 1$: $1 - 17 + 16 = 0$ *True*
Check $x = \pm 4$: $256 - 272 + 16 = 0$ *True*

Solution set: $\{\pm 1, \pm 4\}$

(b) $x^4 + 4 = 8x^2$

$$x^4 - 8x^2 + 4 = 0$$

Let $y = x^2$, so $(x^2)^2 = x^4$.

$$y^2 - 8y + 4 = 0$$

Use the quadratic formula with $a = 1, b = -8$, and $c = 4$.

$$y = \frac{-b \pm \sqrt{b^2 - 4ac}}{2a}$$
$$y = \frac{-(-8) \pm \sqrt{(-8)^2 - 4(1)(4)}}{2(1)}$$
$$= \frac{8 \pm \sqrt{64 - 16}}{2} = \frac{8 \pm \sqrt{48}}{2}$$
$$= \frac{8 \pm 4\sqrt{3}}{2} = 4 \pm 2\sqrt{3}$$

To find x, substitute x^2 for y.

$$x^2 = 4 \pm 2\sqrt{3}$$
$$x = \pm\sqrt{4 \pm 2\sqrt{3}}$$

Check $x = \pm\sqrt{4 - 2\sqrt{3}}$:

$$\left(\pm\sqrt{4 - 2\sqrt{3}}\right)^4 + 4 \overset{?}{=} 8\left(\pm\sqrt{4 - 2\sqrt{3}}\right)^2$$
$$\left(4 - 2\sqrt{3}\right)^2 + 4 \overset{?}{=} 8\left(4 - 2\sqrt{3}\right)$$
$$\left(16 - 16\sqrt{3} + 12\right) + 4 \overset{?}{=} 32 - 16\sqrt{3}$$
$$32 - 16\sqrt{3} = 32 - 16\sqrt{3} \quad \textit{True}$$

The check for $x = \pm\sqrt{4 + 2\sqrt{3}}$ is similar.

The solution set contains four numbers:

$$\left\{\pm\sqrt{4 + 2\sqrt{3}}, \pm\sqrt{4 - 2\sqrt{3}}\right\}$$

N7. **(a)** $6(x-4)^2 + 11(x-4) - 10 = 0$

Let $y = x - 4$. The equation becomes:

$$6y^2 + 11y - 10 = 0$$
$$(2y + 5)(3y - 2) = 0$$
$$2y + 5 = 0 \quad \text{or} \quad 3y - 2 = 0$$
$$y = -\tfrac{5}{2} \quad \text{or} \quad y = \tfrac{2}{3}$$

To find x, substitute $x - 4$ for y.

$$x - 4 = -\tfrac{5}{2} \quad \text{or} \quad x - 4 = \tfrac{2}{3}$$
$$x = \tfrac{3}{2} \quad \text{or} \quad x = \tfrac{14}{3}$$

Check $x = \frac{3}{2}$: $\frac{75}{2} - \frac{55}{2} - 10 = 0$ *True*
Check $x = \frac{14}{3}$: $\frac{8}{3} + \frac{22}{3} - 10 = 0$ *True*

Solution set: $\left\{\frac{3}{2}, \frac{14}{3}\right\}$

(b) $2x^{2/3} - 7x^{1/3} + 3 = 0$

Let $y = x^{1/3}$, so $y^2 = (x^{1/3})^2 = x^{2/3}$.

$$2y^2 - 7y + 3 = 0$$
$$(2y - 1)(y - 3) = 0$$
$$2y - 1 = 0 \quad \text{or} \quad y - 3 = 0$$
$$y = \tfrac{1}{2} \quad \text{or} \quad y = 3$$

To find x, substitute $x^{1/3}$ for y.

$$x^{1/3} = \tfrac{1}{2} \quad \text{or} \quad x^{1/3} = 3$$

Cube each side of each equation.

$$(x^{1/3})^3 = (\tfrac{1}{2})^3 \quad \text{or} \quad (x^{1/3})^3 = 3^3$$
$$x = \tfrac{1}{8} \quad \text{or} \quad x = 27$$

Check $x = \frac{1}{8}$: $\frac{1}{2} - \frac{7}{2} + 3 = 0$ *True*
Check $x = 27$: $18 - 21 + 3 = 0$ *True*

Solution set: $\left\{\frac{1}{8}, 27\right\}$

11.4 Section Exercises

1. $\frac{14}{x} = x - 5$

This is a rational equation, so multiply each side by the LCD, x.

3. $(x^2 + x)^2 - 8(x^2 + x) + 12 = 0$

This is quadratic in form, so substitute a variable for $x^2 + x$.

5. The proposed solution -1 does not check. Solution set: $\{4\}$

7. $\frac{14}{x} = x - 5$

Multiply each term by the LCD, x.

$$x\left(\frac{14}{x}\right) = x(x) - 5x$$
$$14 = x^2 - 5x$$

$$-x^2+5x+14=0$$
$$x^2-5x-14=0$$
$$(x+2)(x-7)=0$$

$$x+2=0 \quad \text{or} \quad x-7=0$$
$$x=-2 \quad \text{or} \quad x=7$$

Check $x=-2$: $-7=-7$ *True*
Check $x=7$: $2=2$ *True*

Solution set: $\{-2,7\}$

9. $1-\frac{3}{x}-\frac{28}{x^2}=0$
Multiply each term by the LCD, x^2.

$$x^2(1)-x^2\left(\frac{3}{x}\right)-x^2\left(\frac{28}{x^2}\right)=x^2\bullet 0$$
$$x^2-3x-28=0$$
$$(x+4)(x-7)=0$$

$$x+4=0 \quad \text{or} \quad x-7=0$$
$$x=-4 \quad \text{or} \quad x=7$$

Check $x=-4$: $1+\frac{3}{4}-\frac{7}{4}=0$ *True*
Check $x=7$: $1-\frac{3}{7}-\frac{4}{7}=0$ *True*

Solution set: $\{-4,7\}$

11. $3-\frac{1}{t}=\frac{2}{t^2}$
Multiply each term by the LCD, t^2.

$$t^2(3)-t^2\left(\frac{1}{t}\right)=t^2\left(\frac{2}{t^2}\right)$$
$$3t^2-t=2$$
$$3t^2-t-2=0$$
$$(3t+2)(t-1)=0$$

$$3t+2=0 \quad \text{or} \quad t-1=0$$
$$t=-\tfrac{2}{3} \quad \text{or} \quad t=1$$

Check $t=-\frac{2}{3}$: $3+\frac{3}{2}=\frac{9}{2}$ *True*
Check $t=1$: $3-1=2$ *True*

Solution set: $\{-\frac{2}{3},1\}$

13. $\frac{1}{x}+\frac{2}{x+2}=\frac{17}{35}$
Multiply each term by the LCD, $35x(x+2)$.

$$35x(x+2)\left(\frac{1}{x}\right)+35x(x+2)\left(\frac{2}{x+2}\right)$$
$$=35x(x+2)\left(\tfrac{17}{35}\right)$$
$$35(x+2)+35x(2)=17x(x+2)$$
$$35x+70+70x=17x^2+34x$$
$$70+105x=17x^2+34x$$
$$0=17x^2-71x-70$$
$$0=(17x+14)(x-5)$$

$$17x+14=0 \quad \text{or} \quad x-5=0$$
$$x=-\tfrac{14}{17} \quad \text{or} \quad x=5$$

Check $x=-\frac{14}{17}$: $-\frac{17}{14}+\frac{17}{10}=\frac{17}{35}$ *True*
Check $x=5$: $\frac{1}{5}+\frac{2}{7}=\frac{17}{35}$ *True*

Solution set: $\{-\frac{14}{17},5\}$

15. $\frac{2}{x+1}+\frac{3}{x+2}=\frac{7}{2}$
Multiply each term by the LCD, $2(x+1)(x+2)$.

$$2(x+1)(x+2)\left(\frac{2}{x+1}+\frac{3}{x+2}\right)$$
$$=2(x+1)(x+2)\left(\tfrac{7}{2}\right)$$
$$2(x+2)(2)+2(x+1)(3)$$
$$=(x+1)(x+2)(7)$$
$$4x+8+6x+6=\left(x^2+3x+2\right)(7)$$
$$10x+14=7x^2+21x+14$$
$$0=7x^2+11x$$
$$0=x(7x+11)$$

$$x=0 \quad \text{or} \quad 7x+11=0$$
$$x=-\tfrac{11}{7}$$

Check $x=-\frac{11}{7}$: $-\frac{7}{2}+7=\frac{7}{2}$ *True*
Check $x=0$: $2+\frac{3}{2}=\frac{7}{2}$ *True*

Solution set: $\{-\frac{11}{7},0\}$

17. $\frac{3}{2x}-\frac{1}{2(x+2)}=1$
Multiply each term by the LCD, $2x(x+2)$.

$$2x(x+2)\left(\frac{3}{2x}-\frac{1}{2(x+2)}\right)$$
$$=2x(x+2)\bullet 1$$
$$3(x+2)-x(1)=2x(x+2)$$
$$3x+6-x=2x^2+4x$$
$$0=2x^2+2x-6$$
$$0=x^2+x-3$$

Use $a=1$, $b=1$, $c=-3$ in the quadratic formula.

$$x=\frac{-b\pm\sqrt{b^2-4ac}}{2a}$$
$$x=\frac{-1\pm\sqrt{1^2-4(1)(-3)}}{2(1)}$$
$$=\frac{-1\pm\sqrt{1+12}}{2}=\frac{-1\pm\sqrt{13}}{2}$$

Use a calculator to check both proposed solutions. Both solutions check.
Solution set: $\left\{\frac{-1\pm\sqrt{13}}{2}\right\}$

19. $3 = \frac{1}{t+2} + \frac{2}{(t+2)^2}$

Multiply each term by the LCD, $(t+2)^2$.

$$3(t+2)^2 = 1(t+2) + 2$$
$$3(t^2 + 4t + 4) = t + 2 + 2$$
$$3t^2 + 12t + 12 = t + 4$$
$$3t^2 + 11t + 8 = 0$$
$$(3t+8)(t+1) = 0$$

$3t + 8 = 0$ or $t + 1 = 0$

$t = -\frac{8}{3}$ $\quad$ $t = -1$

Check $t = -\frac{8}{3}$: $3 = -\frac{3}{2} + \frac{9}{2}$ *True*

Check $t = -1$: $3 = 1 + 2$ *True*

Solution set: $\{-\frac{8}{3}, -1\}$

21. $\frac{6}{p} = 2 + \frac{p}{p+1}$

Multiply each term by the LCD, $p(p+1)$.

$$6(p+1) = 2p(p+1) + p \cdot p$$
$$6p + 6 = 2p^2 + 2p + p^2$$
$$0 = 3p^2 - 4p - 6$$

Use $a = 3$, $b = -4$, and $c = -6$ in the quadratic formula.

$$p = \frac{-b \pm \sqrt{b^2 - 4ac}}{2a}$$
$$p = \frac{-(-4) \pm \sqrt{(-4)^2 - 4(3)(-6)}}{2(3)}$$
$$= \frac{4 \pm \sqrt{16 + 72}}{2(3)} = \frac{4 \pm \sqrt{88}}{2(3)}$$
$$= \frac{4 \pm 2\sqrt{22}}{2(3)} = \frac{2 \pm \sqrt{22}}{3}$$

Use a calculator to check both proposed solutions. Both solutions check.

Solution set: $\left\{\frac{2 \pm \sqrt{22}}{3}\right\}$

23. $1 - \frac{1}{2x+1} - \frac{1}{(2x+1)^2} = 0$

Multiply each term by the LCD, $(2x+1)^2$.

$$(2x+1)^2 - 1(2x+1) - 1 = 0$$
$$4x^2 + 4x + 1 - 2x - 1 - 1 = 0$$
$$4x^2 + 2x - 1 = 0$$

Use $a = 4$, $b = 2$, $c = -1$ in the quadratic formula.

$$x = \frac{-b \pm \sqrt{b^2 - 4ac}}{2a}$$
$$x = \frac{-2 \pm \sqrt{2^2 - 4(4)(-1)}}{2(4)}$$
$$= \frac{-2 \pm \sqrt{4 + 16}}{2(4)}$$
$$= \frac{-2 \pm \sqrt{20}}{2(4)}$$
$$= \frac{-2 \pm 2\sqrt{5}}{2(4)}$$
$$= \frac{-1 \pm \sqrt{5}}{4}$$

Use a calculator to check both proposed solutions. Both solutions check.

Solution set: $\left\{\frac{-1 \pm \sqrt{5}}{4}\right\}$

25. Rate in still water: 20 mph
Rate of current: t mph

(a) When the boat travels upstream, the current works against the rate of the boat in still water, so the rate is $(20 - t)$ mph.

(b) When the boat travels downstream, the current works with the rate of the boat in still water, so the rate is $(20 + t)$ mph.

27. Let x = rate of the boat in still water.
With the rate of the current at 15 mph, then
$x - 15$ = rate going upstream and
$x + 15$ = rate going downstream.
Complete a table using the information in the problem, the rates given above, and the formula $d = rt$ or $t = \frac{d}{r}$.

	d	r	t
Upstream	4	$x - 15$	$\frac{4}{x-15}$
Downstream	16	$x + 15$	$\frac{16}{x+15}$

The time, 48 min, is written as $\frac{48}{60} = \frac{4}{5}$ hr. The time upstream plus the time downstream equals $\frac{4}{5}$. So, from the table, the equation is written as

$$\frac{4}{x-15} + \frac{16}{x+15} = \frac{4}{5}.$$

Multiply by the LCD, $5(x-15)(x+15)$.

$$5(x-15)(x+15)\left(\frac{4}{x-15} + \frac{16}{x+15}\right)$$
$$= 5(x-15)(x+15) \cdot \frac{4}{5}$$
$$20(x+15) + 80(x-15)$$
$$= 4(x-15)(x+15)$$
$$20x + 300 + 80x - 1200$$
$$= 4(x^2 - 225)$$
$$100x - 900 = 4x^2 - 900$$
$$0 = 4x^2 - 100x$$
$$0 = 4x(x - 25)$$

$4x = 0$ or $x - 25 = 0$

$x = 0$ or $\quad x = 25$

Reject $x = 0$ mph as a possible boat speed. William's boat had a top speed of 25 mph.

29. Let $x =$ Adrian's rate from Jackson to Lodi. Then $x+10 =$ his rate from Lodi to Manteca.

Make a table. Use $t = \frac{d}{r}$.

	d	r	t
Jackson to Lodi	40	x	$\frac{40}{x}$
Lodi to Manteca	40	$x+10$	$\frac{40}{x+10}$

Driving time for the entire trip was 88 minutes, or $\frac{88}{60} = \frac{22}{15}$ hours.

$$\frac{40}{x} + \frac{40}{x+10} = \frac{22}{15}$$

Multiply each term by the LCD, $15x(x+10)$.

$$15x(x+10)\left(\frac{40}{x} + \frac{40}{x+10}\right) = 15x(x+10)\left(\tfrac{22}{15}\right)$$

$$\begin{aligned} 600(x+10) + 600x &= 22x(x+10) \\ 600x + 6000 + 600x &= 22x^2 + 220x \\ 0 &= 22x^2 - 980x - 6000 \\ 0 &= 11x^2 - 490x - 3000 \\ 0 &= (11x+60)(x-50) \end{aligned}$$

$$11x + 60 = 0 \quad \text{or} \quad x - 50 = 0$$
$$x = -\tfrac{60}{11} \quad \text{or} \quad x = 50$$

Reject $-\frac{60}{11}$ since speed cannot be negative. Adrian's rate from Jackson to Lodi is 50 mph.

31. Let x be the time in hours required for the faster worker to cut the lawn. Then the slower worker requires $x+1$ hours. Complete the chart.

	Rate	Time Working Together	Fractional Part of the Job Done
Faster Worker	$\frac{1}{x}$	2	$\frac{2}{x}$
Slower Worker	$\frac{1}{x+1}$	2	$\frac{2}{x+1}$

Part done by faster worker plus Part done by slower worker equals one whole job.

$$\frac{2}{x} + \frac{2}{x+1} = 1$$

Multiply each side by the LCD, $x(x+1)$.

$$x(x+1)\left(\frac{2}{x} + \frac{2}{x+1}\right) = x(x+1) \cdot 1$$

$$\begin{aligned} 2(x+1) + 2x &= x^2 + x \\ 2x + 2 + 2x &= x^2 + x \\ 0 &= x^2 - 3x - 2 \end{aligned}$$

Solve for x using the quadratic formula with $a = 1$, $b = -3$, and $c = -2$.

$$\begin{aligned} x &= \frac{-(-3) \pm \sqrt{(-3)^2 - 4(1)(-2)}}{2(1)} \\ &= \frac{3 \pm \sqrt{9+8}}{2} = \frac{3 \pm \sqrt{17}}{2} \end{aligned}$$

$$x = \frac{3+\sqrt{17}}{2} \quad \text{or} \quad x = \frac{3-\sqrt{17}}{2}$$
$$x \approx 3.6 \quad \text{or} \quad x \approx -0.6$$

Discard -0.6 as a solution since time cannot be negative.
It would take the faster worker approximately 3.6 hours.

33. Let x represent the time in hours it takes Nancy to plant the flowers. Then $x+2$ is the time it takes Rusty. Organize the information in a chart.

Worker	Rate	Time Working Together	Fractional Part of the Job Done
Nancy	$\frac{1}{x}$	12	$\frac{12}{x}$
Rusty	$\frac{1}{x+2}$	12	$\frac{12}{x+2}$

Part done by Nancy plus part done by Rusty equals one whole job.

$$\frac{12}{x} + \frac{12}{x+2} = 1$$

Multiply each side by the LCD, $x(x+2)$.

$$x(x+2)\left(\frac{12}{x} + \frac{12}{x+2}\right) = x(x+2) \cdot 1$$

$$\begin{aligned} 12(x+2) + 12x &= x^2 + 2x \\ 12x + 24 + 12x &= x^2 + 2x \\ 0 &= x^2 - 22x - 24 \end{aligned}$$

Solve for x using the quadratic formula with $a = 1$, $b = -22$, and $c = -24$.

$$\begin{aligned} x &= \frac{-(-22) \pm \sqrt{(-22)^2 - 4(1)(-24)}}{2(1)} \\ &= \frac{22 \pm \sqrt{580}}{2} \end{aligned}$$

$$x = \frac{22 + \sqrt{580}}{2} \approx 23.0 \quad \text{or}$$
$$x = \frac{22 - \sqrt{580}}{2} \approx -1.0$$

Since x represents time, discard the negative solution.
Nancy takes about 23.0 hours planting flowers alone while Rusty takes about 25.0 hours planting alone.

35. Let $x =$ the number of hours it takes for the faster pipe alone to fill the tank.
$x + 3 =$ the number of hours it takes for the slower pipe alone to fill the tank.

Working together, both pipes can fill the tank in 2 hours. Make a chart.

Pipe	Rate	Time	Fractional Part of Tank Filled
Faster	$\frac{1}{x}$	2	$\frac{2}{x}$
Slower	$\frac{1}{x+3}$	2	$\frac{2}{x+3}$

Since together the faster and slower pipes fill one tank, the sum of their fractional parts is 1; that is,

$$\frac{2}{x} + \frac{2}{x+3} = 1.$$

Multiply each term by the LCD, $x(x+3)$.

$$2(x+3) + 2(x) = x(x+3)$$
$$2x + 6 + 2x = x^2 + 3x$$
$$0 = x^2 - x - 6$$
$$0 = (x-3)(x+2)$$

$$x - 3 = 0 \quad \text{or} \quad x + 2 = 0$$
$$x = 3 \quad \text{or} \quad x = -2$$

Reject -2. The faster pipe takes 3 hours to fill the tank alone and the slower pipe takes 6 hours to fill the tank alone.

37.

$$x = \sqrt{7x - 10}$$
$$(x)^2 = \left(\sqrt{7x-10}\right)^2$$
$$x^2 = 7x - 10$$
$$x^2 - 7x + 10 = 0$$
$$(x-2)(x-5) = 0$$

$$x - 2 = 0 \quad \text{or} \quad x - 5 = 0$$
$$x = 2 \quad \text{or} \quad x = 5$$

Check $x = 2$: $2 = \sqrt{4}$ *True*

Check $x = 5$: $5 = \sqrt{25}$ *True*

Solution set: $\{2, 5\}$

39.

$$2x = \sqrt{11x + 3}$$
$$(2x)^2 = \left(\sqrt{11x+3}\right)^2$$
$$4x^2 = 11x + 3$$
$$4x^2 - 11x - 3 = 0$$
$$(4x+1)(x-3) = 0$$

$$4x + 1 = 0 \quad \text{or} \quad x - 3 = 0$$
$$x = -\tfrac{1}{4} \quad \text{or} \quad x = 3$$

Check $x = -\frac{1}{4}$: $-\frac{1}{2} = \sqrt{\frac{1}{4}}$ *False*

Check $x = 3$: $6 = \sqrt{36}$ *True*

Solution set: $\{3\}$

41.

$$3x = \sqrt{16 - 10x}$$
$$(3x)^2 = \left(\sqrt{16-10x}\right)^2$$
$$9x^2 = 16 - 10x$$
$$9x^2 + 10x - 16 = 0$$
$$(9x-8)(x+2) = 0$$

$$9x - 8 = 0 \quad \text{or} \quad x + 2 = 0$$
$$x = \tfrac{8}{9} \quad \text{or} \quad x = -2$$

Check $x = \frac{8}{9}$: $\frac{8}{3} = \sqrt{\frac{64}{9}}$ *True*

Check $x = -2$: $-6 = \sqrt{36}$ *False*

Solution set: $\{\frac{8}{9}\}$

43.

$$t + \sqrt{t} = 12$$
$$\sqrt{t} = 12 - t$$
$$\left(\sqrt{t}\right)^2 = (12 - t)^2$$
$$t = 144 - 24t + t^2$$
$$0 = t^2 - 25t + 144$$
$$0 = (t-9)(t-16)$$

$$t - 9 = 0 \quad \text{or} \quad t - 16 = 0$$
$$t = 9 \qquad\qquad t = 16$$

Check $t = 9$: $9 + 3 = 12$ *True*

Check $t = 16$: $16 + 4 = 12$ *False*

Solution set: $\{9\}$

45.

$$x = \sqrt{\frac{6-13x}{5}}$$
$$x^2 = \frac{6-13x}{5}$$
$$5x^2 = 6 - 13x$$
$$5x^2 + 13x - 6 = 0$$
$$(5x-2)(x+3) = 0$$

$$5x - 2 = 0 \quad \text{or} \quad x + 3 = 0$$
$$x = \tfrac{2}{5} \quad \text{or} \quad x = -3$$

Check $x = \frac{2}{5}$: $\frac{2}{5} = \sqrt{\frac{4}{25}}$ *True*

Check $x = -3$: $-3 = \sqrt{9}$ *False*

Solution set: $\{\frac{2}{5}\}$

47.

$$-x = \sqrt{\frac{8-2x}{3}}$$
$$(-x)^2 = \left(\sqrt{\frac{8-2x}{3}}\right)^2$$
$$x^2 = \frac{8-2x}{3}$$
$$3x^2 = 8 - 2x$$
$$3x^2 + 2x - 8 = 0$$
$$(3x-4)(x+2) = 0$$

$$3x - 4 = 0 \quad \text{or} \quad x + 2 = 0$$
$$x = \tfrac{4}{3} \quad \text{or} \quad x = -2$$

Check $x = \frac{4}{3}$: $-\frac{4}{3} = \sqrt{\frac{16}{9}}$ *False*

Check $x = -2$: $2 = \sqrt{4}$ *True*

Solution set: $\{-2\}$

49. $x^4 - 29x^2 + 100 = 0$

Let $u = x^2$, so $u^2 = x^4$.

$$u^2 - 29u + 100 = 0$$
$$(u-4)(u-25) = 0$$

$$u - 4 = 0 \quad \text{or} \quad u - 25 = 0$$
$$u = 4 \quad \text{or} \quad u = 25$$

To find x, substitute x^2 for u.

$$x^2 = 4 \quad \text{or} \quad x^2 = 25$$
$$x = \pm 2 \quad \text{or} \quad x = \pm 5$$

Check $x = \pm 2$: $16 - 116 + 100 = 0$ *True*

Check $x = \pm 5$: $625 - 725 + 100 = 0$ *True*

Solution set: $\{\pm 2, \pm 5\}$

51. $4q^4 - 13q^2 + 9 = 0$

Let $u = q^2$, so $u^2 = q^4$.

$$4u^2 - 13u + 9 = 0$$
$$(4u-9)(u-1) = 0$$

$$4u - 9 = 0 \quad \text{or} \quad u - 1 = 0$$
$$u = \tfrac{9}{4} \quad \text{or} \quad u = 1.$$

To find q, substitute q^2 for u.

$$q^2 = \tfrac{9}{4} \quad \text{or} \quad q^2 = 1$$
$$q = \pm\tfrac{3}{2} \quad \text{or} \quad q = \pm 1$$

Check $q = \pm\frac{3}{2}$: $\frac{81}{4} - \frac{117}{4} + 9 = 0$ *True*

Check $q = \pm 1$: $4 - 13 + 9 = 0$ *True*

Solution set: $\{\pm 1, \pm\frac{3}{2}\}$

53. $x^4 + 48 = 16x^2$

$$x^4 - 16x^2 + 48 = 0$$

Let $u = x^2$, so $u^2 = x^4$.

$$u^2 - 16u + 48 = 0$$
$$(u-4)(u-12) = 0$$

$$u - 4 = 0 \quad \text{or} \quad u - 12 = 0$$
$$u = 4 \quad \text{or} \quad u = 12$$

To find x substitute x^2 for u.

$$x^2 = 4 \qquad x^2 = 12$$
$$x = \pm\sqrt{4} \qquad x = \pm\sqrt{12}$$
$$x = \pm 2 \quad \text{or} \quad x = \pm 2\sqrt{3}$$

Check $x = \pm 2$: $16 + 48 = 64$ *True*

Check $x = \pm 2\sqrt{3}$: $144 + 48 = 192$ *True*

Solution set: $\left\{\pm 2, \pm 2\sqrt{3}\right\}$

55. $(x+3)^2 + 5(x+3) + 6 = 0$

Let $u = x + 3$, so $u^2 = (x+3)^2$.

$$u^2 + 5u + 6 = 0$$
$$(u+3)(u+2) = 0$$

$$u + 3 = 0 \quad \text{or} \quad u + 2 = 0$$
$$u = -3 \quad \text{or} \quad u = -2$$

To find x, substitute $x + 3$ for u.

$$x + 3 = -3 \quad \text{or} \quad x + 3 = -2$$
$$x = -6 \quad \text{or} \quad x = -5$$

Check $x = -6$: $9 - 15 + 6 = 0$ *True*

Check $x = -5$: $4 - 10 + 6 = 0$ *True*

Solution set: $\{-6, -5\}$

57. $3(m+4)^2 - 8 = 2(m+4)$

Let $u = m + 4$, so $u^2 = (m+4)^2$.

$$3u^2 - 8 = 2u$$
$$3u^2 - 2u - 8 = 0$$
$$(3u+4)(u-2) = 0$$

$$3u + 4 = 0 \quad \text{or} \quad u - 2 = 0$$
$$u = -\tfrac{4}{3} \qquad u = 2$$

$$m + 4 = -\tfrac{4}{3} \quad \text{or} \quad m + 4 = 2$$
$$m = -\tfrac{16}{3} \qquad m = -2$$

Check $m = -\frac{16}{3}$: $\frac{16}{3} - 8 = -\frac{8}{3}$ *True*

Check $m = -2$: $12 - 8 = 4$ *True*

Solution set: $\{-\frac{16}{3}, -2\}$

59. $x^{2/3} + x^{1/3} - 2 = 0$

Let $u = x^{1/3}$, so $u^2 = x^{2/3}$.

$$u^2 + u - 2 = 0$$
$$(u+2)(u-1) = 0$$

$$u + 2 = 0 \quad \text{or} \quad u - 1 = 0$$
$$u = -2 \quad \text{or} \quad u = 1$$

To find x, substitute $x^{1/3}$ for u.

$$x^{1/3} = -2 \quad \text{or} \quad x^{1/3} = 1$$

Cube each side of each equation.

$$\left(x^{1/3}\right)^3 = (-2)^3 \qquad \left(x^{1/3}\right)^3 = 1^3$$
$$x = -8 \quad \text{or} \quad x = 1$$

Check $x = -8$: $4 - 2 - 2 = 0$ *True*

Check $x = 1$: $1 + 1 - 2 = 0$ *True*

Solution set: $\{-8, 1\}$

61. $r^{2/3} + r^{1/3} - 12 = 0$

Let $u = r^{1/3}$, so $u^2 = r^{2/3}$.

$u^2 + u - 12 = 0$
$(u+4)(u-3) = 0$

$u + 4 = 0$ or $u - 3 = 0$
$u = -4$ or $u = 3$

To find r, substitute $r^{1/3}$ for u.

$r^{1/3} = -4$ or $r^{1/3} = 3$
$(r^{1/3})^3 = (-4)^3$ or $(r^{1/3})^3 = 3^3$
$r = -64$ or $r = 27$

Check $r = -64$: $16 - 4 - 12 = 0$ *True*
Check $r = 27$: $9 + 3 - 12 = 0$ *True*

Solution set: $\{-64, 27\}$

63. $4x^{4/3} - 13x^{2/3} + 9 = 0$

Let $u = x^{2/3}$, so $u^2 = x^{4/3}$.

$4u^2 - 13u + 9 = 0$
$(4u - 9)(u - 1) = 0$

$4u - 9 = 0$ or $u - 1 = 0$
$u = \frac{9}{4}$ or $u = 1$
$x^{2/3} = \frac{9}{4}$ or $x^{2/3} = 1$
$(x^{2/3})^{1/2} = \left(\frac{9}{4}\right)^{1/2}$ or $(x^{2/3})^{1/2} = 1^{1/2}$
$x^{1/3} = \pm\frac{3}{2}$ or $x^{1/3} = \pm 1$
$(x^{1/3})^3 = \left(\pm\frac{3}{2}\right)^3$ or $(x^{1/3})^3 = (\pm 1)^3$
$x = \pm\frac{27}{8}$ or $x = \pm 1$

Check $x = \pm\frac{27}{8}$:
$4\left(\frac{81}{16}\right) - 13\left(\frac{9}{4}\right) + 9 \stackrel{?}{=} 0$
$\frac{81}{4} - \frac{117}{4} + \frac{36}{4} = 0$ *True*
Check $x = \pm 1$:
$4 - 13 + 9 = 0$ *True*

Solution set: $\left\{\pm 1, \pm\frac{27}{8}\right\}$

65. $2 + \dfrac{5}{3x-1} = \dfrac{-2}{(3x-1)^2}$

Let $u = 3x - 1$, so $u^2 = (3x-1)^2$.

$2 + \dfrac{5}{u} = -\dfrac{2}{u^2}$

Multiply each term by the LCD, u^2.

$u^2\left(2 + \dfrac{5}{u}\right) = u^2\left(-\dfrac{2}{u^2}\right)$
$2u^2 + 5u = -2$

$2u^2 + 5u + 2 = 0$
$(2u+1)(u+2) = 0$

$2u + 1 = 0$ or $u + 2 = 0$
$u = -\frac{1}{2}$ $u = -2$

To find x, substitute $3x - 1$ for u.

$3x - 1 = -\frac{1}{2}$ or $3x - 1 = -2$
$3x = \frac{1}{2}$ $3x = -1$
$x = \frac{1}{6}$ or $x = -\frac{1}{3}$

Check $x = \frac{1}{6}$: $2 - 10 = -8$ *True*
Check $x = -\frac{1}{3}$: $2 - \frac{5}{2} = -\frac{1}{2}$ *True*

Solution set: $\left\{-\frac{1}{3}, \frac{1}{6}\right\}$

67. $2 - 6(z-1)^{-2} = (z-1)^{-1}$

Let $u = (z-1)^{-1}$, so $u^2 = (z-1)^{-2}$.

$2 - 6u^{-2} = u^{-1}$
$2 - \dfrac{6}{u^2} = \dfrac{1}{u}.$

Multiply each term by the LCD, u^2.

$2u^2 - 6 = u$
$2u^2 - u - 6 = 0$
$(2u+3)(u-2) = 0$

$2u + 3 = 0$ or $u - 2 = 0$
$u = -\frac{3}{2}$ or $u = 2$

To find z, substitute $z - 1$ for u.

$z - 1 = -\frac{3}{2}$ or $z - 1 = 2$
$z = -\frac{1}{2}$ or $z = 3$

Check $z = -\frac{1}{2}$: $2 - \frac{8}{3} = -\frac{2}{3}$ *True*
Check $z = 3$: $2 - \frac{3}{2} = \frac{1}{2}$ *True*

Solution set: $\left\{-\frac{1}{2}, 3\right\}$

69. $12x^4 - 11x^2 + 2 = 0$

Let $u = x^2$, so $u^2 = x^4$.

$12u^2 - 11u + 2 = 0$
$(4u-1)(3u-2) = 0$

$4u - 1 = 0$ or $3u - 2 = 0$
$u = \frac{1}{4}$ or $u = \frac{2}{3}$
$x^2 = \frac{1}{4}$ $x^2 = \frac{2}{3}$
$x = \pm\frac{1}{2}$ $x = \pm\sqrt{\frac{2}{3}}$

Note: $x = \pm\sqrt{\frac{2}{3}} = \pm\frac{\sqrt{2}}{\sqrt{3}} \cdot \frac{\sqrt{3}}{\sqrt{3}} = \pm\frac{\sqrt{6}}{3}$

Check $x = \pm\frac{1}{2}$: $\frac{3}{4} - \frac{11}{4} + 2 = 0$ *True*
Check $x = \pm\frac{\sqrt{6}}{3}$: $\frac{16}{3} - \frac{22}{3} + 2 = 0$ *True*

Solution set: $\left\{\pm\frac{\sqrt{6}}{3}, \pm\frac{1}{2}\right\}$

71. $\sqrt{2x+3} = 2 + \sqrt{x-2}$

Square each side.

$\left(\sqrt{2x+3}\right)^2 = \left(2 + \sqrt{x-2}\right)^2$
$2x + 3 = 4 + 4\sqrt{x-2} + (x-2)$
$2x + 3 = x + 2 + 4\sqrt{x-2}$

Isolate the radical term on one side.

$$x+1=4\sqrt{x-2}$$

Square each side again.

$$(x+1)^2=\left(4\sqrt{x-2}\right)^2$$
$$x^2+2x+1=16(x-2)$$
$$x^2+2x+1=16x-32$$
$$x^2-14x+33=0$$
$$(x-11)(x-3)=0$$

$$x-11=0 \quad \text{or} \quad x-3=0$$
$$x=11 \quad \text{or} \quad x=3$$

Check $x=11$: $\sqrt{25}=2+\sqrt{9}$ *True*

Check $x=3$: $\sqrt{9}=2+\sqrt{1}$ *True*

Solution set: $\{3, 11\}$

73. $2\left(1+\sqrt{r}\right)^2=13\left(1+\sqrt{r}\right)-6$

Let $u=1+\sqrt{r}$, so $u^2=\left(1+\sqrt{r}\right)^2$.

$$2u^2=13u-6$$
$$2u^2-13u+6=0$$
$$(2u-1)(u-6)=0$$

$$2u-1=0 \quad \text{or} \quad u-6=0$$
$$u=\tfrac{1}{2} \quad \text{or} \quad u=6$$

Replace u with $1+\sqrt{r}$.

$$1+\sqrt{r}=\tfrac{1}{2} \quad \text{or} \quad 1+\sqrt{r}=6$$
$$\sqrt{r}=-\tfrac{1}{2} \qquad \sqrt{r}=5$$

Not possible, $\quad r=25$

since $\sqrt{r}\geq 0$.

Check $r=25$: $72=78-6$ *True*

Solution set: $\{25\}$

75. $2m^6+11m^3+5=0$

Let $y=m^3$, so $y^2=m^6$.

$$2y^2+11y+5=0$$
$$(2y+1)(y+5)=0$$

$$2y+1=0 \quad \text{or} \quad y+5=0$$
$$y=-\tfrac{1}{2} \quad \text{or} \quad y=-5$$

To find m, substitute m^3 for y.

$$m^3=-\tfrac{1}{2} \quad \text{or} \quad m^3=-5$$

Take the cube root of each side of each equation.

$$m=\sqrt[3]{-\tfrac{1}{2}} \quad \text{or} \quad m=\sqrt[3]{-5}$$
$$m=-\sqrt[3]{\tfrac{1}{2}} \qquad m=-\sqrt[3]{5}$$
$$=-\frac{\sqrt[3]{1}}{\sqrt[3]{2}}\cdot\frac{\sqrt[3]{2^2}}{\sqrt[3]{2^2}}$$
$$=-\frac{\sqrt[3]{4}}{2}$$

Check $m=-\frac{\sqrt[3]{4}}{2}$: $\frac{1}{2}-\frac{11}{2}+5=0$ *True*

Check $m=-\sqrt[3]{5}$: $50-55+5=0$ *True*

Solution set: $\left\{-\sqrt[3]{5}, -\frac{\sqrt[3]{4}}{2}\right\}$

77. $6=7(2w-3)^{-1}+3(2w-3)^{-2}$

Let $u=(2w-3)^{-1}$, so $u^2=(2w-3)^{-2}$.

$$6=7u+3u^2$$
$$0=3u^2+7u-6$$
$$0=(3u-2)(u+3)$$

$$3u-2=0 \quad \text{or} \quad u+3=0$$
$$u=\tfrac{2}{3} \quad \text{or} \quad u=-3$$
$$(2w-3)^{-1}=\tfrac{2}{3} \qquad (2w-3)^{-1}=-3$$
$$\frac{1}{2w-3}=\frac{2}{3} \qquad \frac{1}{2w-3}=-3$$
$$3=2(2w-3) \qquad 1=-3(2w-3)$$
$$3=4w-6 \qquad 1=-6w+9$$
$$9=4w \qquad -8=-6w$$
$$\tfrac{9}{4}=w \qquad \tfrac{4}{3}=w$$

Check $w=\frac{4}{3}$: $6=-21+27$ *True*

Check $w=\frac{9}{4}$: $6=\frac{14}{3}+\frac{4}{3}$ *True*

Solution set: $\left\{\frac{4}{3}, \frac{9}{4}\right\}$

79. $2x^4-9x^2=-2$

$$2x^4-9x^2+2=0$$

Let $u=x^2$, so $u^2=x^4$.

$$2u^2-9u+2=0$$

Use $a=2$, $b=-9$, and $c=2$ in the quadratic formula.

$$u=\frac{-b\pm\sqrt{b^2-4ac}}{2a}$$
$$u=\frac{-(-9)\pm\sqrt{(-9)^2-4(2)(2)}}{2(2)}$$
$$=\frac{9\pm\sqrt{81-16}}{4}=\frac{9\pm\sqrt{65}}{4}$$

To find x, substitute x^2 for u.

$$x^2=\frac{9\pm\sqrt{65}}{4}$$
$$x=\pm\sqrt{\frac{9\pm\sqrt{65}}{4}}=\pm\frac{\sqrt{9\pm\sqrt{65}}}{2}$$

Note: The last expression represents four numbers. All four proposed solutions check.

Solution set: $\left\{\pm\frac{\sqrt{9+\sqrt{65}}}{2}, \pm\frac{\sqrt{9-\sqrt{65}}}{2}\right\}$

81. $2x^4 + x^2 - 3 = 0$

Let $m = x^2$, so $m^2 = x^4$.

$2m^2 + m - 3 = 0$

$(2m + 3)(m - 1) = 0$

$2m + 3 = 0$ or $m - 1 = 0$

$m = -\frac{3}{2}$ or $m = 1$

To find x, substitute x^2 for m.

$x^2 = -\frac{3}{2}$ or $x^2 = 1$

$x = \pm\sqrt{-\frac{3}{2}}$ $\quad x = \pm\sqrt{1}$

$x = \pm\frac{\sqrt{3}}{\sqrt{2}} \cdot \frac{\sqrt{2}}{\sqrt{2}} i$ $\quad x = \pm 1$

$x = \pm\frac{\sqrt{6}}{2} i$

Check $x = \pm\frac{\sqrt{6}}{2} i$: $\frac{9}{2} - \frac{3}{2} - 3 = 0$ *True*

Check $x = \pm 1$: $2 + 1 - 3 = 0$ *True*

Solution set: $\left\{\pm 1, \pm\frac{\sqrt{6}}{2} i\right\}$

83. Solve $P = 2L + 2W$ for W.

$P - 2L = 2W$

$\frac{P - 2L}{2} = W$, or $W = \frac{P}{2} - L$

85. Solve $F = \frac{9}{5}C + 32$ for C.

$F - 32 = \frac{9}{5}C$

$\frac{5}{9}(F - 32) = C$

Summary Exercises on Solving Quadratic Equations

1. $(2x + 3)^2 = 4$

Since the equation has the form $(ax + b)^2 = c$, use the *square root property*.

3. $x^2 + 5x - 8 = 0$

The discriminant is

$$b^2 - 4ac = 5^2 - 4(1)(-8) = 25 + 32 = 57.$$

Since the discriminant is not a perfect square, use the *quadratic formula*.

5. $3x^2 = 2 - 5x$

$3x^2 + 5x - 2 = 0$

The discriminant is

$$b^2 - 4ac = 5^2 - 4(3)(-2) = 25 + 24 = 49.$$

Since the discriminant is a perfect square, use *factoring*.

7. $p^2 = 7$

$p = \sqrt{7}$ or $p = -\sqrt{7}$

Solution set: $\left\{\pm\sqrt{7}\right\}$

9. $n^2 + 6n + 4 = 0$

$n^2 + 6n = -4$

$n^2 + 6n + 9 = -4 + 9$ $\quad \left[\frac{1}{2}(6)\right]^2 = 9$

$(n + 3)^2 = 5$

$n + 3 = \sqrt{5}$ or $n + 3 = -\sqrt{5}$

$n = -3 + \sqrt{5}$ or $n = -3 - \sqrt{5}$

Solution set: $\left\{-3 \pm \sqrt{5}\right\}$

11. $\frac{5}{x} + \frac{12}{x^2} = 2$

Multiply each term by the LCD, x^2.

$5x + 12 = 2x^2$

$0 = 2x^2 - 5x - 12$

$0 = (2x + 3)(x - 4)$

$2x + 3 = 0$ or $x - 4 = 0$

$x = -\frac{3}{2}$ or $x = 4$

Solution set: $\left\{-\frac{3}{2}, 4\right\}$

13. $2r^2 - 4r + 1 = 0$

Use $a = 2$, $b = -4$, and $c = 1$ in the quadratic formula.

$$r = \frac{-b \pm \sqrt{b^2 - 4ac}}{2a}$$

$$r = \frac{-(-4) \pm \sqrt{(-4)^2 - 4(2)(1)}}{2(2)}$$

$$= \frac{4 \pm \sqrt{16 - 8}}{2(2)} = \frac{4 \pm \sqrt{8}}{2(2)}$$

$$= \frac{4 \pm 2\sqrt{2}}{2(2)} = \frac{2 \pm \sqrt{2}}{2}$$

Solution set: $\left\{\frac{2 \pm \sqrt{2}}{2}\right\}$

15. $x\sqrt{2} = \sqrt{5x - 2}$

$\left(x\sqrt{2}\right)^2 = \left(\sqrt{5x - 2}\right)^2$

$x^2 \cdot 2 = 5x - 2$

$2x^2 - 5x + 2 = 0$

$(2x - 1)(x - 2) = 0$

$2x - 1 = 0$ or $x - 2 = 0$

$x = \frac{1}{2}$ $\quad x = 2$

Check $x = \frac{1}{2}$: $\frac{1}{2}\sqrt{2} = \sqrt{\frac{1}{2}}$ *True*

Check $x = 2$: $2\sqrt{2} = \sqrt{8}$ *True*

Solution set: $\left\{\frac{1}{2}, 2\right\}$

17. $(2x + 3)^2 = 8$

$2x + 3 = \sqrt{8}$ or $2x + 3 = -\sqrt{8}$

$2x = -3 + 2\sqrt{2}$ $\quad 2x = -3 - 2\sqrt{2}$

$x = \frac{-3 + 2\sqrt{2}}{2}$ or $x = \frac{-3 - 2\sqrt{2}}{2}$

Solution set: $\left\{\frac{-3 \pm 2\sqrt{2}}{2}\right\}$

19.
$$t^4 + 14 = 9t^2$$
$$t^4 - 9t^2 + 14 = 0$$
$$(t^2 - 2)(t^2 - 7) = 0$$
$$t^2 - 2 = 0 \quad \text{or} \quad t^2 - 7 = 0$$
$$t^2 = 2 \qquad\qquad t^2 = 7$$
$$t = \pm\sqrt{2} \quad \text{or} \quad t = \pm\sqrt{7}$$
Solution set: $\left\{\pm\sqrt{2}, \pm\sqrt{7}\right\}$

21. $z^2 + z + 1 = 0$

Use $a = 1$, $b = 1$, and $c = 1$ in the quadratic formula.
$$z = \frac{-b \pm \sqrt{b^2 - 4ac}}{2a}$$
$$z = \frac{-1 \pm \sqrt{1^2 - 4(1)(1)}}{2(1)}$$
$$= \frac{-1 \pm \sqrt{1-4}}{2} = \frac{-1 \pm \sqrt{-3}}{2}$$
$$= \frac{-1 \pm i\sqrt{3}}{2} = -\frac{1}{2} \pm \frac{\sqrt{3}}{2}i$$
Solution set: $\left\{-\frac{1}{2} \pm \frac{\sqrt{3}}{2}i\right\}$

23.
$$4t^2 - 12t + 9 = 0$$
$$(2t - 3)(2t - 3) = 0$$
$$(2t - 3)^2 = 0$$
$$2t - 3 = 0$$
$$t = \tfrac{3}{2}$$
Solution set: $\left\{\frac{3}{2}\right\}$

25. $r^2 - 72 = 0$
$$r^2 = 72$$
$$r = \pm\sqrt{72} = \pm 6\sqrt{2}$$
Solution set: $\left\{\pm 6\sqrt{2}\right\}$

27.
$$x^2 - 5x - 36 = 0$$
$$(x + 4)(x - 9) = 0$$
$$x + 4 = 0 \quad \text{or} \quad x - 9 = 0$$
$$x = -4 \quad \text{or} \quad x = 9$$
Solution set: $\{-4, 9\}$

29.
$$3p^2 = 6p - 4$$
$$3p^2 - 6p + 4 = 0$$
Use $a = 3$, $b = -6$, and $c = 4$ in the quadratic formula.
$$p = \frac{-b \pm \sqrt{b^2 - 4ac}}{2a}$$
$$p = \frac{-(-6) \pm \sqrt{(-6)^2 - 4(3)(4)}}{2(3)}$$
$$= \frac{6 \pm \sqrt{36 - 48}}{2(3)}$$
$$= \frac{6 \pm \sqrt{-12}}{2(3)}$$
$$= \frac{6 \pm 2i\sqrt{3}}{2(3)}$$
$$= \frac{3 \pm i\sqrt{3}}{3}$$
$$= 1 \pm \frac{\sqrt{3}}{3}i$$
Solution set: $\left\{1 \pm \frac{\sqrt{3}}{3}i\right\}$

31.
$$\frac{4}{r^2} + 3 = \frac{1}{r}$$
Multiply each term by the LCD, r^2.
$$4 + 3r^2 = r$$
$$3r^2 - r + 4 = 0$$
Use $a = 3$, $b = -1$, and $c = 4$ in the quadratic formula.
$$r = \frac{-b \pm \sqrt{b^2 - 4ac}}{2a}$$
$$r = \frac{-(-1) \pm \sqrt{(-1)^2 - 4(3)(4)}}{2(3)}$$
$$= \frac{1 \pm \sqrt{1 - 48}}{6} = \frac{1 \pm \sqrt{-47}}{6}$$
$$= \frac{1 \pm i\sqrt{47}}{6} = \frac{1}{6} \pm \frac{\sqrt{47}}{6}i$$
Solution set: $\left\{\frac{1}{6} \pm \frac{\sqrt{47}}{6}i\right\}$

11.5 Formulas and Further Applications

11.5 Now Try Exercises

N1. (a) Solve $n = \dfrac{ab}{E^2}$ for E.

$$E^2 n = ab \qquad \textit{Multiply by } E^2.$$
$$E^2 = \frac{ab}{n} \qquad \textit{Divide by } n.$$
$$E = \pm\sqrt{\frac{ab}{n}} \qquad \textit{Square root property}$$
$$E = \frac{\pm\sqrt{ab}}{\sqrt{n}} \cdot \frac{\sqrt{n}}{\sqrt{n}} \qquad \textit{Rationalize the denominator.}$$
$$E = \frac{\pm\sqrt{abn}}{n}$$

(b) Solve $S = \sqrt{\dfrac{pq}{n}}$ for p.

$$S^2 = \frac{pq}{n} \qquad \textit{Square both sides.}$$
$$nS^2 = pq \qquad \textit{Multiply by } n.$$
$$\frac{nS^2}{q} = p \qquad \textit{Divide by } q.$$

N2. Solve $r^2 + 9r = -c$ for r.

$r^2 + 9r + c = 0$

Use $a = 1$, $b = 9$, and $c = c$ in the quadratic formula.

$$r = \frac{-b \pm \sqrt{b^2 - 4ac}}{2a}$$
$$r = \frac{-9 \pm \sqrt{9^2 - 4(1)c}}{2(1)}$$
$$r = \frac{-9 \pm \sqrt{81 - 4c}}{2}$$

The solutions are

$$r = \frac{-9 + \sqrt{81 - 4c}}{2} \text{ and } r = \frac{-9 - \sqrt{81 - 4c}}{2}.$$

N3. *Step 2*
Let $x =$ the width of the barn. Then $x + 10 =$ the length of the barn.

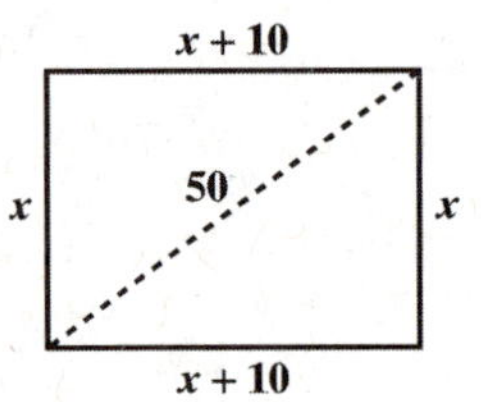

Step 3
The diagonal of the footprint is 50 feet, so a right triangle is formed. Use the Pythagorean theorem.

$$a^2 + b^2 = c^2$$
$$x^2 + (x + 10)^2 = 50^2$$

Step 4

$$x^2 + x^2 + 20x + 100 = 2500$$
$$2x^2 + 20x - 2400 = 0$$
$$x^2 + 10x - 1200 = 0$$
$$(x + 40)(x - 30) = 0$$

$$x + 40 = 0 \quad \text{or} \quad x - 30 = 0$$
$$x = -40 \quad \text{or} \quad x = 30$$

Step 5
The width cannot be -40, so the width of the barn is 30 feet and the length of the barn is $30 + 10 = 40$ feet.

Step 6
$30^2 + 40^2 = 50^2$ and 40 is 10 more than 30, as required.

N4. *Step 2*
Let $x =$ the width of the grass strip.

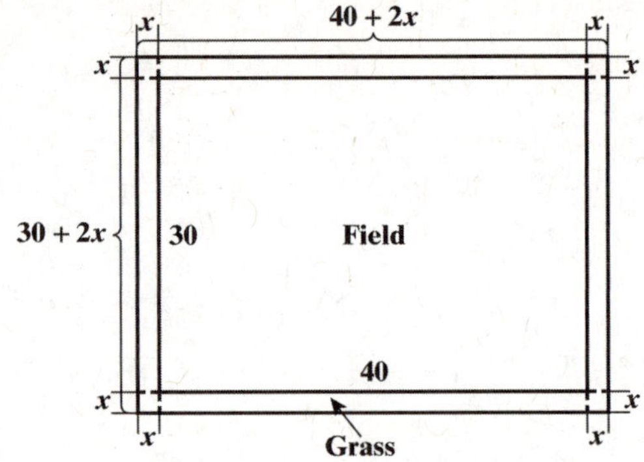

Step 3
The width of the large rectangle is $30 + 2x$, and the length is $40 + 2x$.

$$\text{Area of rectangle} - \text{area of field} = \text{area of grass.}$$
$$(40 + 2x)(30 + 2x) - 40(30) = 296$$

Step 4

$$1200 + 140x + 4x^2 - 1200 = 296$$
$$4x^2 + 140x - 296 = 0$$
$$x^2 + 35x - 74 = 0$$
$$(x + 37)(x - 2) = 0$$

$$x + 37 = 0 \quad \text{or} \quad x - 2 = 0$$
$$x = -37 \quad \text{or} \quad x = 2$$

Step 5
The width cannot be -37, so the grass strip should be 2 yards wide.

Step 6
If $x = 2$, then the area of the large rectangle is

$$(40 + 2 \cdot 2)(30 + 2 \cdot 2) = 44 \cdot 34 = 1496 \text{ yd}^2.$$

The area of the field is $40 \cdot 30 = 1200 \text{ yd}^2$. So, the area of the grass strip is $1496 - 1200 = 296 \text{ yd}^2$, as required. The answer is correct.

N5. $-16t^2 + 60t + 120 = s(t)$

$-16t^2 + 60t + 120 = 0$ *Let s(t) = 0.*

$4t^2 - 15t - 30 = 0$ *Divide by −4.*

Use $a = 4$, $b = -15$, and $c = -30$ in the quadratic formula.

$$t = \frac{-b \pm \sqrt{b^2 - 4ac}}{2a}$$
$$t = \frac{-(-15) \pm \sqrt{(-15)^2 - 4(4)(-30)}}{2(4)}$$
$$= \frac{15 \pm \sqrt{225 + 480}}{8} = \frac{15 \pm \sqrt{705}}{8}$$

$$t = \frac{15 + \sqrt{705}}{8} \approx 5.2 \quad \text{or} \quad t = \frac{15 - \sqrt{705}}{8} \approx -1.4$$

The object will hit the ground about 5.2 seconds after it is projected.

N6. **(a)** For 2005, $x = 2005 - 1980 = 25$.

$$\begin{aligned} f(x) &= -0.065x^2 + 14.8x + 249 \\ f(25) &= -0.065(25)^2 + 14.8(25) + 249 \\ &= 578.375 \end{aligned}$$

The CPI for 2005 was about 578.

(b) Find the value of x that makes $f(x) = 500$.

$$\begin{aligned} f(x) &= -0.065x^2 + 14.8x + 249 \\ 500 &= -0.065x^2 + 14.8x + 249 \\ 0 &= -0.065x^2 + 14.8x - 251 \end{aligned}$$

Now use $a = -0.065$, $b = 14.8$, and $c = -251$ in the quadratic formula.

$$\begin{aligned} x &= \frac{-14.8 \pm \sqrt{14.8^2 - 4(-0.065)(-251)}}{2(-0.065)} \\ &= \frac{-14.8 \pm \sqrt{153.78}}{-0.13} \end{aligned}$$

$x \approx 18.5$ or $x \approx 209.2$

The solution $x \approx 209.2$ is rejected since it corresponds to a year far beyond the period covered by the model. The solution $x \approx 18.5$ is valid for this problem. The corresponding year is 1998.

11.5 Section Exercises

1. The first step in solving a formula that has the specified variable in the denominator is to multiply each side by the LCD to clear the equation of fractions.

3. We must recognize that a formula like

$$gw^2 = kw + 24$$

is quadratic in w. So the first step is to write the formula in standard form (with 0 on one side, in decreasing powers of w). This allows us to apply the quadratic formula to solve for w.

5. Since the triangle is a right triangle, use the Pythagorean theorem with legs m and n and hypotenuse p.

$$\begin{aligned} m^2 + n^2 &= p^2 \\ m^2 &= p^2 - n^2 \\ m &= \sqrt{p^2 - n^2} \end{aligned}$$

Only the positive square root is given since m represents the side of a triangle.

7. Solve $d = kt^2$ for t.

$kt^2 = d$

$t^2 = \frac{d}{k}$ *Divide by k.*

$t = \pm\sqrt{\frac{d}{k}}$ *Use square root property.*

$= \frac{\pm\sqrt{d}}{\sqrt{k}} \cdot \frac{\sqrt{k}}{\sqrt{k}}$ *Rationalize denominator.*

$t = \frac{\pm\sqrt{dk}}{k}$ *Simplify.*

9. Solve $I = \frac{ks}{d^2}$ for d.

$Id^2 = ks$ *Multiply by d^2.*

$d^2 = \frac{ks}{I}$ *Divide by I.*

$d = \pm\sqrt{\frac{ks}{I}}$ *Use square root property.*

$= \pm\frac{\sqrt{ks}}{\sqrt{I}} \cdot \frac{\sqrt{I}}{\sqrt{I}}$ *Rationalize denominator.*

$d = \frac{\pm\sqrt{ksI}}{I}$ *Simplify.*

11. Solve $F = \frac{kA}{v^2}$ for v.

$v^2F = kA$ *Multiply by v^2.*

$v^2 = \frac{kA}{F}$ *Divide by F.*

$v = \pm\sqrt{\frac{kA}{F}}$ *Use square root property.*

$= \frac{\pm\sqrt{kA}}{\sqrt{F}} \cdot \frac{\sqrt{F}}{\sqrt{F}}$ *Rationalize denominator.*

$v = \frac{\pm\sqrt{kAF}}{F}$ *Simplify.*

13. Solve $V = \frac{1}{3}\pi r^2 h$ for r.

$3V = \pi r^2 h$ *Multiply by 3.*

$\frac{3V}{\pi h} = r^2$ *Divide by πh.*

$r = \pm\sqrt{\frac{3V}{\pi h}}$ *Use square root property.*

$= \frac{\pm\sqrt{3V} \cdot \sqrt{\pi h}}{\sqrt{\pi h} \cdot \sqrt{\pi h}}$ *Rationalize denominator.*

$r = \frac{\pm\sqrt{3\pi Vh}}{\pi h}$ *Simplify.*

15. Solve $At^2 + Bt = -C$ for t.

$At^2 + Bt + C = 0$

Use the quadratic formula.

$$t = \frac{-B \pm \sqrt{B^2 - 4AC}}{2A}$$

17. Solve $D = \sqrt{kh}$ for h.

$D^2 = kh$ *Square each side.*

$\frac{D^2}{k} = h$ *Divide by k.*

19. Solve $p = \sqrt{\frac{k\ell}{g}}$ for ℓ.

$$p^2 = \frac{k\ell}{g} \quad \textit{Square each side.}$$
$$p^2 g = k\ell \quad \textit{Multiply by g.}$$
$$\frac{p^2 g}{k} = \ell \quad \textit{Divide by k.}$$

21. Solve $S = 4\pi r^2$ for r.

$$\frac{S}{4\pi} = r^2 \quad \textit{Divide by } 4\pi.$$
$$r = \pm\sqrt{\frac{S}{4\pi}} \quad \textit{Use square root property.}$$
$$= \frac{\pm\sqrt{S}\cdot\sqrt{\pi}}{\sqrt{4\pi}\cdot\sqrt{\pi}} \quad \textit{Rationalize denominator.}$$
$$r = \frac{\pm\sqrt{S\pi}}{2\pi} \quad \textit{Simplify.}$$

23. Solve $p = \frac{E^2 R}{(r+R)^2}$ for R $(E > 0)$.

$$p(r+R)^2 = E^2 R$$
$$p(r^2 + 2rR + R^2) = E^2 R$$
$$pr^2 + 2prR + pR^2 = E^2 R$$
$$pR^2 + 2prR - E^2 R + pr^2 = 0$$
$$pR^2 + (2pr - E^2)R + pr^2 = 0$$

Here $a = p$, $b = 2pr - E^2$, and $c = pr^2$.

$$R = \frac{-(2pr - E^2) \pm \sqrt{(2pr - E^2)^2 - 4p\cdot pr^2}}{2p}$$
$$= \frac{E^2 - 2pr \pm \sqrt{4p^2r^2 - 4prE^2 + E^4 - 4p^2r^2}}{2p}$$
$$= \frac{E^2 - 2pr \pm \sqrt{E^4 - 4prE^2}}{2p}$$
$$= \frac{E^2 - 2pr \pm \sqrt{E^2(E^2 - 4pr)}}{2p}$$
$$R = \frac{E^2 - 2pr \pm E\sqrt{E^2 - 4pr}}{2p}$$

25. Solve $10p^2c^2 + 7pcr = 12r^2$ for r.

$0 = 12r^2 - 7pcr - 10p^2c^2$

Here $a = 12$, $b = -7pc$, and $c = -10p^2c^2$.

$$r = \frac{-(-7pc) \pm \sqrt{(-7pc)^2 - 4(12)(-10p^2c^2)}}{2(12)}$$
$$= \frac{7pc \pm \sqrt{49p^2c^2 + 480p^2c^2}}{24}$$
$$= \frac{7pc \pm \sqrt{529p^2c^2}}{24} = \frac{7pc \pm 23pc}{24}$$
$$r = \frac{7pc + 23pc}{24} = \frac{30pc}{24} = \frac{5pc}{4} \text{ or}$$
$$r = \frac{7pc - 23pc}{24} = \frac{-16pc}{24} = -\frac{2pc}{3}$$

27. Solve $LI^2 + RI + \frac{1}{c} = 0$ for I.

$$cLI^2 + cRI + 1 = 0 \quad \textit{Multiply by c.}$$

Here $a = cL$, $b = cR$, and $c = 1$.

$$I = \frac{-cR \pm \sqrt{(cR)^2 - 4(cL)(1)}}{2(cL)}$$
$$= \frac{-cR \pm \sqrt{c^2R^2 - 4cL}}{2cL}$$

29. Apply the Pythagorean theorem.

$$(x+4)^2 = x^2 + (x+1)^2$$
$$x^2 + 8x + 16 = x^2 + x^2 + 2x + 1$$
$$0 = x^2 - 6x - 15$$

Here $a = 1$, $b = -6$, and $c = -15$.

$$x = \frac{-(-6) \pm \sqrt{(-6)^2 - 4(1)(-15)}}{2(1)}$$
$$= \frac{6 \pm \sqrt{36 + 60}}{2} = \frac{6 \pm \sqrt{96}}{2}$$

$$x = \frac{6 + \sqrt{96}}{2} \approx 7.9 \text{ or}$$
$$x = \frac{6 - \sqrt{96}}{2} \approx -1.9$$

Reject the negative solution.
If $x = 7.9$, then

$$x + 4 = 11.9 \quad \text{and} \quad x + 1 = 8.9.$$

The lengths of the sides of the triangle are approximately 7.9, 8.9, and 11.9.

31. Let $x =$ the distance traveled by the eastbound ship. Then $x + 70 =$ the distance traveled by the southbound ship.

Since the ships are traveling at right angles to one another, the distance d between them can be found using the Pythagorean theorem.

$$c^2 = a^2 + b^2$$
$$d^2 = x^2 + (x+70)^2$$

Let $d = 170$, and solve for x.

$$170^2 = x^2 + (x+70)^2$$
$$28{,}900 = x^2 + x^2 + 140x + 4900$$
$$0 = 2x^2 + 140x - 24{,}000$$
$$0 = x^2 + 70x - 12{,}000$$
$$0 = (x+150)(x-80)$$

$$x + 150 = 0 \quad \text{or} \quad x - 80 = 0$$
$$x = -150 \quad \text{or} \quad x = 80$$

Distance cannot be negative, so reject -150. If $x = 80$, then $x + 70 = 150$. The eastbound ship traveled 80 miles, and the southbound ship traveled 150 miles.

33. Let x = length of the shorter leg;
$2x - 1$ = length of the longer leg;
$2x - 1 + 2$ = length of the hypotenuse.

Use the Pythagorean theorem.

$$\begin{aligned} x^2 + (2x-1)^2 &= (2x+1)^2 \\ x^2 + 4x^2 - 4x + 1 &= 4x^2 + 4x + 1 \\ 5x^2 - 4x + 1 &= 4x^2 + 4x + 1 \\ x^2 - 8x &= 0 \\ x(x-8) &= 0 \end{aligned}$$

$$x = 0 \quad \text{or} \quad x - 8 = 0$$
$$x = 8$$

Since x represents length, discard 0 as a solution. If $x = 8$, then

$$2x - 1 = 2(8) - 1 = 15 \quad \text{and}$$
$$2x - 1 + 2 = 2(8) + 1 = 17.$$

The lengths are 8 inches, 15 inches, and 17 inches.

35. Let x = the width of the rug.
Then $2x + 4$ = the length of the rug.

Use the Pythagorean theorem.

$$\begin{aligned} x^2 + (2x+4)^2 &= 26^2 \\ x^2 + 4x^2 + 16x + 16 &= 676 \\ 5x^2 + 16x - 660 &= 0 \\ (5x+66)(x-10) &= 0 \end{aligned}$$

$$5x + 66 = 0 \quad \text{or} \quad x - 10 = 0$$
$$x = -\tfrac{66}{5} \quad \text{or} \quad x = 10$$

Discard the negative solution. If $x = 10$, then $2x + 4 = 24$. The width of the rug is 10 feet, and the length is 24 feet.

37. Let x = the width of the border. Then the width of the pool and the border is $30 + 2x$ ft and the length of the pool and border is $40 + 2x$ ft.

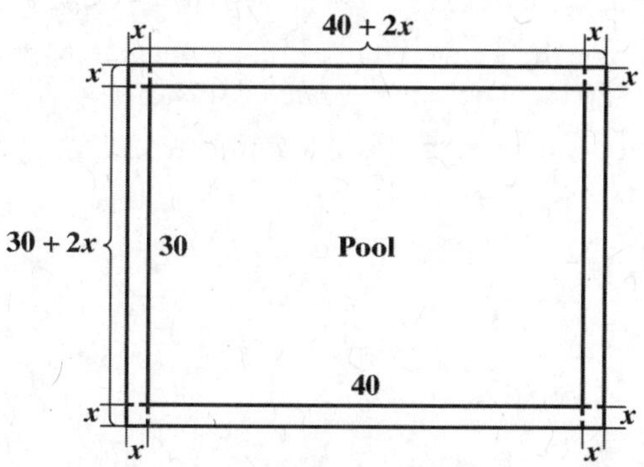

Since the area of the pool is $30 \cdot 40 = 1200$ ft^2, we can write an equation using the total area of the pool and the border.

The area of the pool and border	is	the area of the pool	plus	the area of the border.
$(30+2x)(40+2x)$	$=$	1200	$+$	296

$$\begin{aligned} 1200 + 140x + 4x^2 &= 1496 \\ 4x^2 + 140x - 296 &= 0 \\ x^2 + 35x - 74 &= 0 \\ (x-2)(x+37) &= 0 \end{aligned}$$

$$x - 2 = 0 \quad \text{or} \quad x + 37 = 0$$
$$x = 2 \quad \text{or} \quad x = -37$$

Discard the negative solution. The strip can be 2 feet wide.

39. Let x = original width of the rectangle. Then $2x - 2$ represents the original length of the rectangle.

Now $x + 5$ is the new width that makes the rectangle a square. Thus, the new width must equal the original length since the sides of a square are equal.

$$\begin{aligned} x + 5 &= 2x - 2 \\ 7 &= x \end{aligned}$$

The dimensions of the original rectangle are 7 meters and $2(7) - 2 = 12$ meters. Note that the area of the square is $(12)(12) = 144$ m^2.

41. Let x be the width of the sheet metal. Then the length is $2x - 4$.

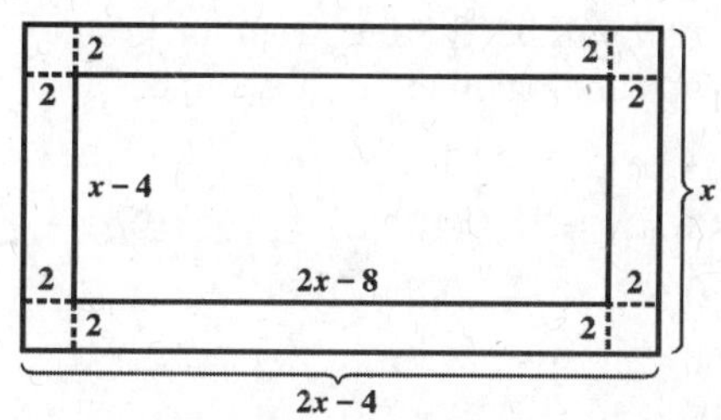

By cutting out 2-inch squares from each corner we get a rectangle with width $x - 4$ and length $(2x - 4) - 4 = 2x - 8$. The uncovered box then has height 2 inches, length $2x - 8$ inches, and width $x - 4$ inches.
Use the formula $V = LWH$ or $V = HLW$.

$$\begin{aligned} 256 &= 2(2x-8)(x-4) \\ 256 &= 4(x-4)(x-4) && \textit{Factor out 2.} \\ 64 &= (x-4)^2 && \textit{Divide by 4.} \end{aligned}$$

Use the square root property.
$\pm 8 = x - 4$

$$x - 4 = 8 \quad \text{or} \quad x - 4 = -8$$
$$x = 12 \quad \text{or} \quad x = -4$$

Since x represents width, discard the negative solution. The width is 12 inches, and the length is $2(12) - 4 = 20$ inches.

43.
$$s = 144t - 16t^2$$
$$128 = 144t - 16t^2 \quad \textit{Let } s = 128.$$
$$0 = -16t^2 + 144t - 128$$
$$0 = t^2 - 9t + 8 \quad \textit{Divide by } -16.$$
$$0 = (t-8)(t-1)$$
$$t - 8 = 0 \quad \text{or} \quad t - 1 = 0$$
$$t = 8 \quad \text{or} \quad t = 1$$

The object will be 128 feet above the ground at two times, going up and coming down, or at 1 second and at 8 seconds.

45.
$$s = -16t^2 + 128t$$
$$213 = -16t^2 + 128t \quad \textit{Let } s = 213.$$
$$0 = -16t^2 + 128t - 213$$

Here $a = -16$, $b = 128$, and $c = -213$.

$$t = \frac{-b \pm \sqrt{b^2 - 4ac}}{2a}$$
$$t = \frac{-128 \pm \sqrt{128^2 - 4(-16)(-213)}}{2(-16)}$$
$$= \frac{-128 \pm \sqrt{16{,}384 - 13{,}632}}{-32}$$
$$= \frac{-128 \pm \sqrt{2752}}{-32}$$
$$t = \frac{-128 + \sqrt{2752}}{-32} \approx 2.4 \quad \text{or}$$
$$t = \frac{-128 - \sqrt{2752}}{-32} \approx 5.6$$

The ball will be 213 feet from the ground after 2.4 seconds and again after 5.6 seconds.

47.
$$D(t) = 13t^2 - 100t$$
$$180 = 13t^2 - 100t \quad \textit{Let } D(t) = 180.$$
$$0 = 13t^2 - 100t - 180$$

Here $a = 13$, $b = -100$, and $c = -180$.

$$t = \frac{-b \pm \sqrt{b^2 - 4ac}}{2a}$$
$$t = \frac{-(-100) \pm \sqrt{(-100)^2 - 4(13)(-180)}}{2(13)}$$
$$= \frac{100 \pm \sqrt{10{,}000 + 9360}}{2(13)}$$
$$= \frac{100 \pm \sqrt{19{,}360}}{2(13)}$$
$$= \frac{100 \pm 44\sqrt{10}}{2(13)} = \frac{50 \pm 22\sqrt{10}}{13}$$
$$t = \frac{50 + 22\sqrt{10}}{13} \approx 9.2 \quad \text{or}$$
$$t = \frac{50 - 22\sqrt{10}}{13} \approx -1.5$$

Discard the negative solution. The car will skid 180 feet in approximately 9.2 seconds.

49.
$$s(t) = -16t^2 + 160t$$
$$400 = -16t^2 + 160t \quad \textit{Let } s(t) = 400.$$
$$0 = -16t^2 + 160t - 400$$
$$0 = t^2 - 10t + 25 \quad \textit{Divide by } -16.$$
$$0 = (t-5)(t-5)$$
$$0 = (t-5)^2$$
$$0 = t - 5$$
$$5 = t$$

The ball reaches a height of 400 feet after 5 seconds. This is its maximum height since this is the only time it reaches 400 feet.

51. Supply and demand are equal when
$$3p - 200 = \frac{3200}{p}.$$

Solve for p.
$$3p^2 - 200p = 3200$$
$$3p^2 - 200p - 3200 = 0$$

Use the quadratic formula with $a = 3$, $b = -200$, and $c = -3200$.

$$p = \frac{-(-200) \pm \sqrt{(-200)^2 - 4(3)(-3200)}}{2(3)}$$
$$= \frac{200 \pm \sqrt{40{,}000 + 38{,}400}}{6}$$
$$= \frac{200 \pm \sqrt{78{,}400}}{6} = \frac{200 \pm 280}{6}$$
$$p = \tfrac{480}{6} = 80 \quad \text{or} \quad p = \tfrac{-80}{6} = -\tfrac{40}{3}$$

Discard the negative solution. The supply and demand are equal when the price is 80 cents or \$0.80.

53. Let $r =$ the interest rate. Let $A = 2142.45$, $P = 2000$, and solve for r.
$$A = P(1+r)^2$$
$$2142.45 = 2000(1+r)^2$$
$$1.071225 = (1+r)^2$$

Use the square root property.
$$1 + r = 1.035 \quad \text{or} \quad 1 + r = -1.035$$
$$r = 0.035 \quad \text{or} \quad r = -2.035$$

Since the interest rate cannot be negative, reject -2.035. The interest rate is 3.5%.

55. Let F denote the Froude number.
$$\text{Solve } F = \frac{v^2}{g\ell} \text{ for } v.$$
$$v^2 = Fg\ell$$
$$v = \pm\sqrt{Fg\ell}$$

v is positive, so
$$v = \sqrt{Fg\ell}.$$

For the rhinoceros, $\ell = 1.2$ and $F = 2.57$.
$$v = \sqrt{(2.57)(9.8)(1.2)} \approx 5.5$$
or 5.5 meters per second.

57. Write a proportion.

$$\frac{x-4}{3x-19} = \frac{4}{x-3}$$

Multiply by the LCD, $(3x-19)(x-3)$.

$$(3x-19)(x-3)\left(\frac{x-4}{3x-19}\right) = (3x-19)(x-3)\left(\frac{4}{x-3}\right)$$

$$(x-3)(x-4) = (3x-19)4$$
$$x^2 - 7x + 12 = 12x - 76$$
$$x^2 - 19x + 88 = 0$$
$$(x-8)(x-11) = 0$$

$$x - 8 = 0 \quad \text{or} \quad x - 11 = 0$$
$$x = 8 \quad \text{or} \quad x = 11$$

If $x = 8$, then $3x - 19 = 3(8) - 19 = 5$.

If $x = 11$, then $3x - 19 = 3(11) - 19 = 14$.

Thus, $AC = 5$ or $AC = 14$.

59. **(a)** From the graph, the spending on physician and clinical services in 2005 appears to be $420 billion (to the nearest $10 billion).

(b) For 2005, $x = 2005 - 2000 = 5$.

$$f(x) = 0.3214x^2 + 25.06x + 288.2$$
$$f(5) = 0.3214(5)^2 + 25.06(5) + 288.2$$
$$= 421.535$$

To the nearest $10 billion, the model gives $420 billion, the same as the estimate in part (a).

61. Use $f(x) = 0.3214x^2 + 25.06x + 288.2$ with $f(x) = 400$.

$$400 = 0.3214x^2 + 25.06x + 288.2$$
$$0 = 0.3214x^2 + 25.06x - 111.8$$

Here $a = 0.3214$, $b = 25.06$, and $c = -111.8$.

$$x = \frac{-b \pm \sqrt{b^2 - 4ac}}{2a}$$

$$x = \frac{-25.06 \pm \sqrt{(25.06)^2 - 4(0.3214)(-111.8)}}{2(0.3214)}$$

$$= \frac{-25.06 \pm \sqrt{771.73368}}{0.6428}$$

$$\approx 4.23 \text{ or } -82.20$$

Reject the negative value. The model indicates that the spending on physician and clinical services reached, and then exceeded, $400 billion in the year $2000 + 4 = 2004$. The graph indicates that spending first exceeded $400 billion in 2005.

63.
$$f(x) = x^2 + 4x - 3$$
$$f(2) = 2^2 + 4(2) - 3$$
$$= 4 + 8 - 3$$
$$= 9$$

65.

x	$f(x) = 2x^2$
-2	$2(-2)^2 = 8$
-1	$2(-1)^2 = 2$
0	$2(0)^2 = 0$
1	$2(1)^2 = 2$
2	$2(2)^2 = 8$

Since the greatest exponent is 2, the graph of f is a parabola.

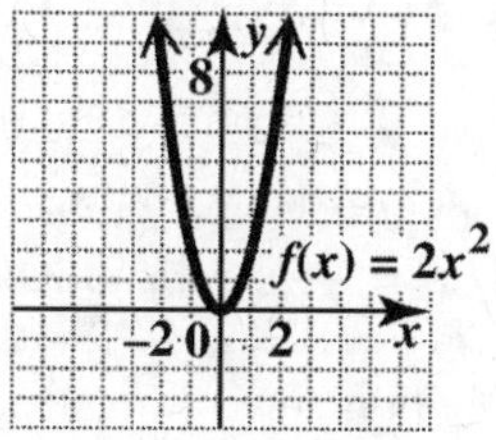

Any x-value can be used, so the domain is $(-\infty, \infty)$. From the graph, we see that the y-values are at least 0, so the range is $[0, \infty)$.

11.6 Graphs of Quadratic Functions

11.6 Now Try Exercises

N1. $f(x) = x^2 - 3$

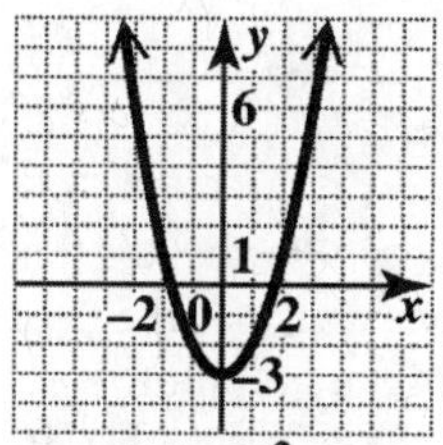

$f(x) = x^2 - 3$

The graph of $f(x) = x^2 - 3$ has the same shape as the graph of $f(x) = x^2$, but the graph is shifted down 3 units and has vertex $(0, -3)$. The graph of this parabola is symmetric about its axis $x = 0$.

Since x can be any real number, the domain is $(-\infty, \infty)$. The value of y is always greater than or equal to -3, so the range is $[-3, \infty)$.

N2. $f(x) = (x+1)^2$

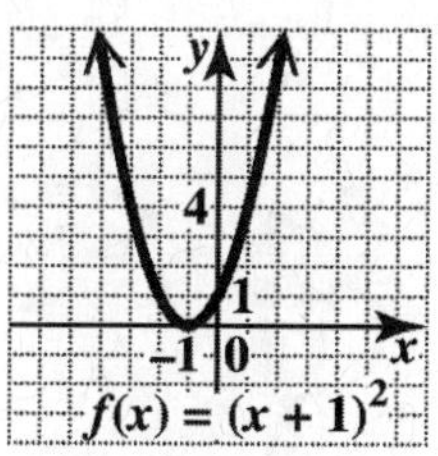

The parabola $f(x) = (x+1)^2$ has the same shape as $f(x) = x^2$ but is shifted 1 unit to the left, since -1 would cause $x + 1$ to equal 0. The vertex is $(-1, 0)$ and the axis is $x = -1$. The domain is $(-\infty, \infty)$ and the range is $[0, \infty)$.

N3. $f(x) = (x+1)^2 - 2$

$f(x) = (x + 1)^2 - 2$

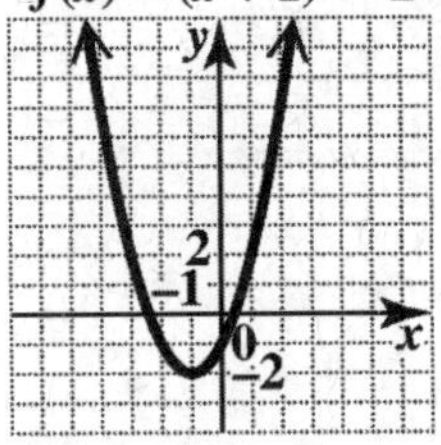

$f(x) = (x+1)^2 - 2$ has the same shape as $f(x) = x^2$ but is shifted 1 unit to the left (since $x + 1 = 0$ if $x = -1$) and 2 units down (because of the -2). The vertex is $(-1, -2)$ and the axis is $x = -1$. The domain is $(-\infty, \infty)$ and the range is $[-2, \infty)$.

N4. $f(x) = -3x^2$

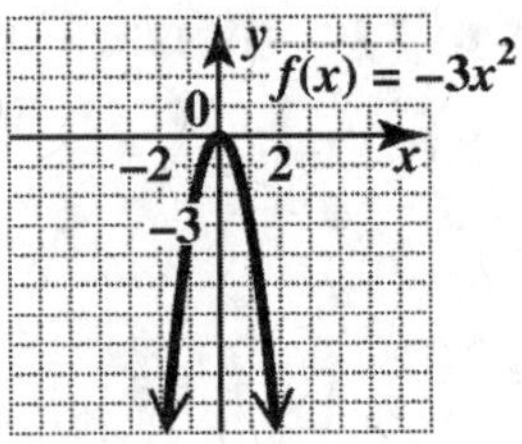

The coefficient -3 indicates that the parabola opens down (because of the negative sign) and is narrower than the graph of $y = x^2$. The vertex is $(0, 0)$ and the axis is $x = 0$. The domain is $(-\infty, \infty)$ and the range is $(-\infty, 0]$.

N5. $f(x) = 2(x-1)^2 + 2$

This equation has a graph like $f(x) = x^2$ but is shifted 1 unit to the right and 2 units up. It has vertex $(1, 2)$. Since $a = 2 > 0$, the parabola opens up. Since $|a| = |2| = 2 > 1$, the parabola is narrower than the graph of $f(x) = x^2$. If $x = 0$, then $f(x) = 4$, so $(0, 4)$ is on the graph. By symmetry about the axis $x = 1$, the point $(2, 4)$ is also on the graph.

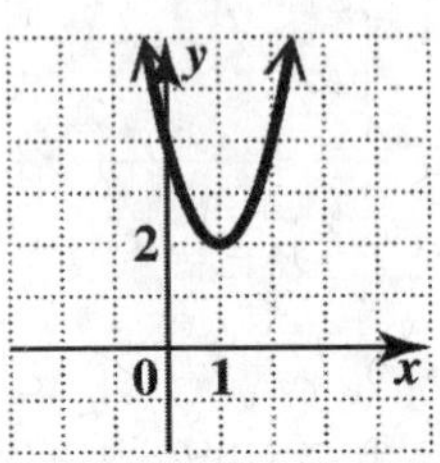

$f(x) = 2(x - 1)^2 + 2$

N6. Use the points with

$$y = ax^2 + bx + c.$$

$$\begin{aligned}
&\text{Use } (0, 4973)\text{: } a(0)^2 + b(0) + c = 4973 \\
&\text{Use } (4, 7321)\text{: } a(4)^2 + b(4) + c = 7321 \\
&\text{Use } (8, 7663)\text{: } a(8)^2 + b(8) + c = 7663
\end{aligned}$$

Simplifying the system gives us:

$$\begin{aligned}
c &= 4973 \quad (1) \\
16a + 4b + c &= 7321 \quad (2) \\
64a + 8b + c &= 7663 \quad (3)
\end{aligned}$$

Substituting 4973 for c and simplifying gives us:

$$\begin{aligned}
16a + 4b &= 2348 \quad (4) \\
64a + 8b &= 2690 \quad (5)
\end{aligned}$$

Dividing (4) by 4 and (5) by 2 gives us:

$$\begin{aligned}
4a + b &= 587 \quad (6) \\
32a + 4b &= 1345 \quad (7)
\end{aligned}$$

From (6) we see that

$$b = 587 - 4a \quad (8),$$

so we substitute $587 - 4a$ for b in (7), and then solve for a.

$$\begin{aligned}
32a + 4(587 - 4a) &= 1345 \\
32a + 2348 - 16a &= 1345 \\
16a &= -1003 \\
a &= -62.6875
\end{aligned}$$

To find b, use (8).

$$\begin{aligned}
b &= 587 - 4a \\
&= 587 - 4(-62.6875) = 837.75
\end{aligned}$$

So using these three points gives us the quadratic model

$$y = -62.6875x^2 + 837.75x + 4973.$$

11.6 Section Exercises

1. A parabola with equation $f(x) = a(x-h)^2 + k$ has vertex $V(h, k)$. We'll identify the vertex for each quadratic function.

(a) $f(x) = (x+2)^2 - 1$

$V(-2, -1)$, choice **B**

(b) $f(x) = (x+2)^2 + 1$

$V(-2, 1)$, choice **C**

(c) $f(x) = (x-2)^2 - 1$

$V(2, -1)$, choice **A**

(d) $f(x) = (x-2)^2 + 1$

$V(2, 1)$, choice **D**

For Exercises 3–12, we write $f(x)$ in the form $f(x) = a(x-h)^2 + k$ and then list the vertex (h, k).

3. $f(x) = -3x^2 = -3(x-0)^2 + 0$
The vertex (h, k) is $(0, 0)$.

5. $f(x) = x^2 + 4 = 1(x-0)^2 + 4$
The vertex (h, k) is $(0, 4)$.

7. $f(x) = (x-1)^2 = 1(x-1)^2 + 0$
The vertex (h, k) is $(1, 0)$.

9. $f(x) = (x+3)^2 - 4 = 1[x-(-3)]^2 - 4$
The vertex (h, k) is $(-3, -4)$.

11. $f(x) = -(x-5)^2 + 6 = -1(x-5)^2 + 6$
The vertex (h, k) is $(5, 6)$.

13. $f(x) = -\frac{2}{5}x^2$

Since $a = -\frac{2}{5} < 0$, the graph opens down. Since $|a| = \left|-\frac{2}{5}\right| = \frac{2}{5} < 1$, the graph is wider than the graph of $f(x) = x^2$.

15. $f(x) = 3x^2 + 1$

Since $a = 3 > 0$, the graph opens up.
Since $|a| = |3| = 3 > 1$, the graph is narrower than the graph of $f(x) = x^2$.

17. $f(x) = -4(x+2)^2 + 5$

Since $a = -4 < 0$, the graph opens down.
Since $|a| = |-4| = 4 > 1$, the graph is narrower than the graph of $f(x) = x^2$.

19. **(a)** $f(x) = (x-4)^2 - 2 = 1(x-4)^2 - 2$ has vertex $(4, -2)$. Because $a = 1 > 0$, the graph opens up. The correct answer is **D**.

(b) $f(x) = (x-2)^2 - 4 = 1(x-2)^2 - 4$ has vertex $(2, -4)$. Because $a = 1 > 0$, the graph opens up. The correct answer is **B**.

(c) $f(x) = -(x-4)^2 - 2 = -1(x-4)^2 - 2$ has vertex $(4, -2)$. Because $a = -1 < 0$, the graph opens down. The correct answer is **C**.

(d) $f(x) = -(x-2)^2 - 4 = -1(x-2)^2 - 4$ has vertex $(2, -4)$. Because $a = -1 < 0$, the graph opens down. The correct answer is **A**.

21. $f(x) = -2x^2$ written in the form
$f(x) = a(x-h)^2 + k$ is
$f(x) = -2(x-0)^2 + 0$.

Here, $h = 0$ and $k = 0$, so the vertex (h, k) is $(0, 0)$. Since $a = -2 < 0$, the graph opens down. Since $|a| = |-2| = 2 > 1$, the graph is narrower than the graph of $f(x) = x^2$. By evaluating the function with $x = 2$ and $x = -2$, we see that the points $(2, -8)$ and $(-2, -8)$ are on the graph.

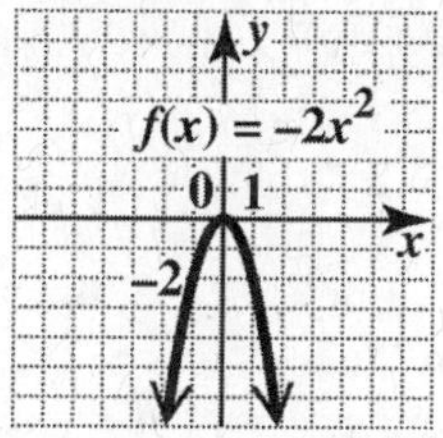

23. $f(x) = x^2 - 1$ written in the form
$f(x) = a(x-h)^2 + k$ is
$f(x) = 1(x-0)^2 + (-1)$.

Here, $h = 0$ and $k = -1$, so the vertex is $(0, -1)$. The graph opens up and has the same shape as $f(x) = x^2$ because $a = 1$. Two other points on the graph are $(-2, 3)$ and $(2, 3)$.

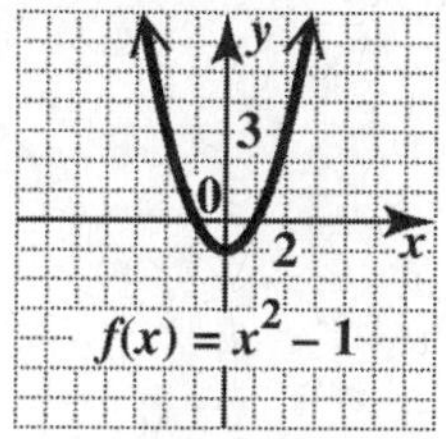

25. $f(x) = -x^2 + 2$ written in the form
$f(x) = a(x-h)^2 + k$ is
$f(x) = -1(x-0)^2 + 2$.

Here, $h = 0$ and $k = 2$, so the vertex (h, k) is $(0, 2)$. Since $a = -1 < 0$, the graph opens down. Since $|a| = |-1| = 1$, the graph has the same shape as $f(x) = x^2$. The points $(2, -2)$ and $(-2, -2)$ are on the graph.

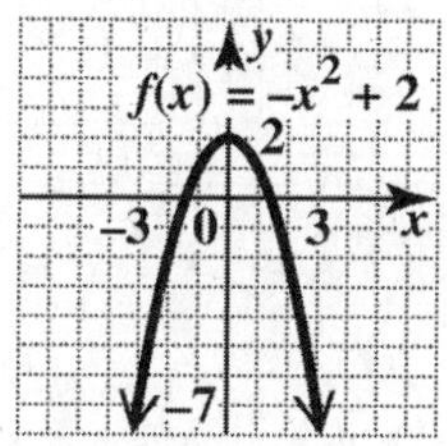

27. $f(x) = (x-4)^2$ written in the form
$f(x) = a(x-h)^2 + k$ is
$f(x) = 1(x-4)^2 + 0$.

Here, $h = 4$ and $k = 0$, so the vertex (h, k) is $(4, 0)$ and the axis is $x = 4$. The graph opens up since a is positive and has the same shape as $f(x) = x^2$ because $|a| = 1$. Two other points on the graph are $(2, 4)$ and $(6, 4)$. We can substitute any value for x, so the domain is $(-\infty, \infty)$. The range is $[0, \infty)$ since the smallest y-value is 0.

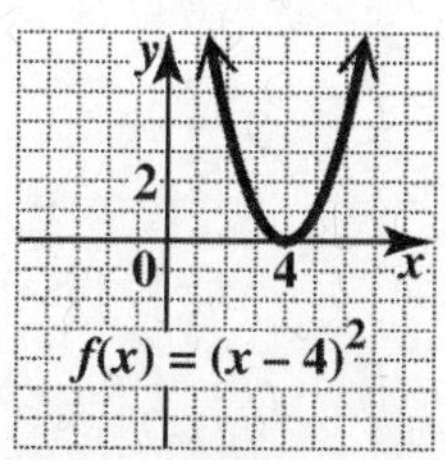

29. $f(x) = (x+2)^2 - 1$ written in the form
$f(x) = a(x-h)^2 + k$ is
$f(x) = 1[x-(-2)]^2 + (-1)$.

Since $h = -2$ and $k = -1$, the vertex (h, k) is $(-2, -1)$ and the axis is $x = -2$. Here, $a = 1$, so the graph opens up and has the same shape as $f(x) = x^2$. The points $(-1, 0)$ and $(-3, 0)$ are on the graph. We can substitute any value for x, so the domain is $(-\infty, \infty)$. The range is $[-1, \infty)$ since the smallest y-value is -1.

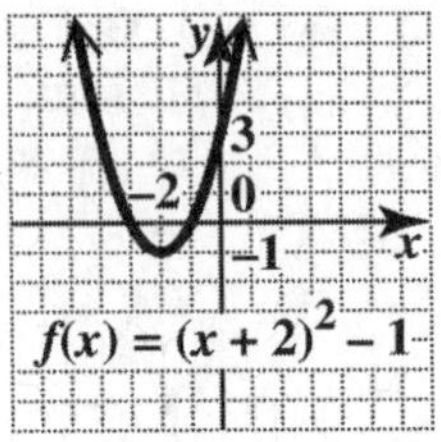

31. $f(x) = 2(x-2)^2 - 4$ written in the form
$f(x) = a(x-h)^2 + k$ is
$f(x) = 2(x-2)^2 + (-4)$.

Here, $h = 2$ and $k = -4$, so the vertex (h, k) is $(2, -4)$ and the axis is $x = 2$. The graph opens up and is narrower than $f(x) = x^2$ because $|a| = |2| > 1$. Two other points on the graph are $(0, 4)$ and $(4, 4)$. We can substitute any value for x, so the domain is $(-\infty, \infty)$. The value of y is greater than or equal to -4, so the range is $[-4, \infty)$.

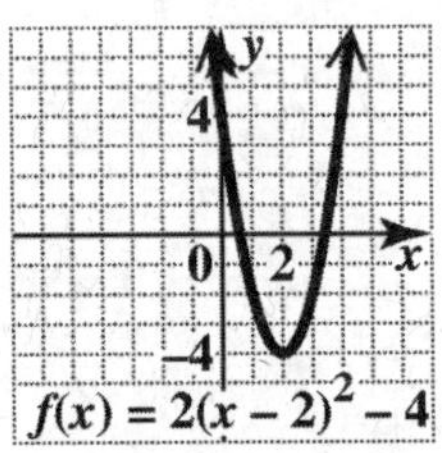

33. $f(x) = -\frac{1}{2}(x+1)^2 + 2$ written in the form
$f(x) = a(x-h)^2 + k$ is
$f(x) = -\frac{1}{2}[x-(-1)]^2 + 2$.

Since $h = -1$ and $k = 2$, the vertex (h, k) is $(-1, 2)$ and the axis is $x = -1$. Here, $a = -0.5 < 0$, so the graph opens down. Also, $|a| = |-0.5| = 0.5 < 1$, so the graph is wider than the graph of $f(x) = x^2$.

$f(x) = -\frac{1}{2}(x+1)^2 + 2$

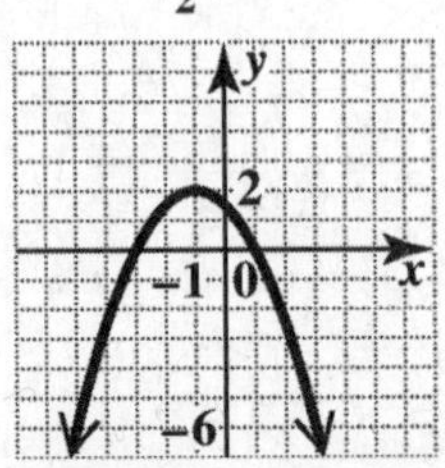

The points $(1, 0)$ and $(-3, 0)$ are on the graph. We can substitute any value for x, so the domain is $(-\infty, \infty)$. The value of y is less than or equal to 2, so the range is $(-\infty, 2]$.

35. $f(x) = 2(x-2)^2 - 3$ written in the form
$f(x) = a(x-h)^2 + k$ is
$f(x) = 2(x-2)^2 + (-3)$.

Here, $h = 2$ and $k = -3$, so the vertex (h, k) is $(2, -3)$ and the axis is $x = 2$. The graph opens up and is narrower than $f(x) = x^2$ because $|a| = |2| > 1$. Two other points on the graph are $(3, -1)$ and $(1, -1)$. We can substitute any value for x, so the domain is $(-\infty, \infty)$. The value of y is greater than or equal to -3, so the range is $[-3, \infty)$.

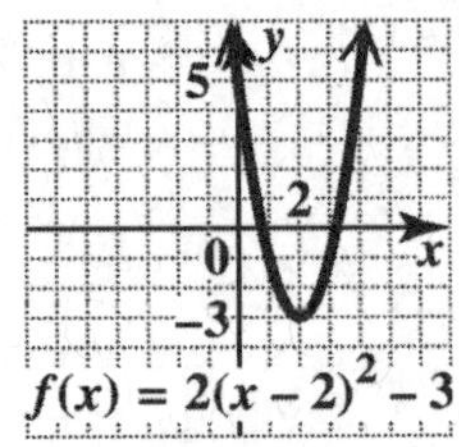

37. The points appear to lie on a line, so a *linear* function would be a more appropriate model. The line would rise, so it would have a *positive* slope.

39. The points appear to lie on a parabola, so a *quadratic* function would be a more appropriate model. The parabola would open up, so a would be *positive.*

41. The points appear to lie on a parabola, so a *quadratic* function would be a more appropriate model. The parabola would open down, so a would be *negative.*

43. **(a)**

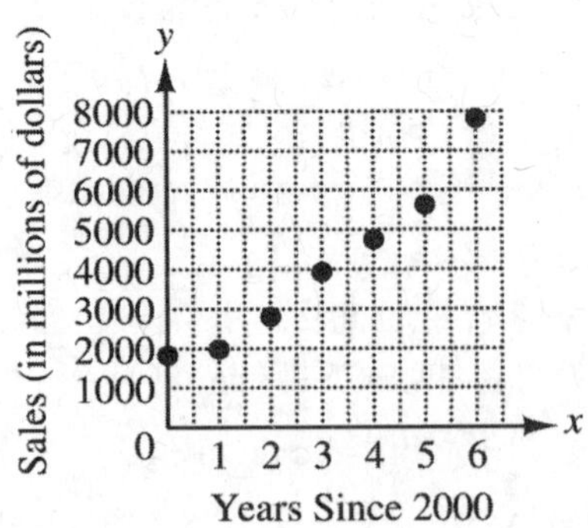

(b) Since the arrangement of the data points is approximately parabolic, a quadratic function would be the more appropriate model for the data set. The coefficient of x^2 should be positive, since the roughly parabolic shape of the graphed data set opens upward.

(c) Use $ax^2 + bx + c = y$ with $(0, 1825)$, $(3, 3921)$, and $(6, 7805)$.

$$\begin{aligned} 0a + 0b + c &= 1825 \quad (1) \\ 9a + 3b + c &= 3921 \quad (2) \\ 36a + 6b + c &= 7805 \quad (3) \end{aligned}$$

From (1), $c = 1825$, so the system becomes

$$\begin{aligned} 9a + 3b &= 2096 \quad (4) \\ 36a + 6b &= 5980 \quad (5) \end{aligned}$$

Now eliminate b.

$$\begin{aligned} -18a - 6b &= -4192 \qquad -2 \times (4) \\ 36a + 6b &= 5980 \qquad (5) \\ \hline 18a &= 1788 \\ a &= \tfrac{1788}{18} = 99\tfrac{1}{3} \approx 99.3 \end{aligned}$$

From (5) with $a = 99\frac{1}{3} = \frac{298}{3}$,

$$\begin{aligned} 36(\tfrac{298}{3}) + 6b &= 5980 \\ 3576 + 6b &= 5980 \\ 6b &= 2404 \\ b &= \tfrac{2404}{6} \approx 400.7 \end{aligned}$$

The quadratic function is approximately

$$y = f(x) = 99.3x^2 + 400.7x + 1825.$$

(d) $x = 2007 - 2000 = 7$ and $f(7) \approx 9496$. The sales of digital cameras in the United States in 2007 were about \$9496 million.

(e) No. The number of digital cameras sold in 2007 is far below the number approximated by the model. Rather than continuing to increase, sales of digital cameras fell in 2007.

45. **(a)** $y = f(x) = -69.15x^2 + 863.6x + 4973$
$x = 2006 - 1995 = 11$ and
$f(11) = 6105.45 \approx 6105$

(b) The approximation using the model is low.

47. $x^2 - x - 20 = 0$
From the screens, we see that the x-values of the x-intercepts are -4 and 5, so the solution set is $\{-4, 5\}$.

49. $-2x^2 + 6x = -2(x^2 - 3x)$

51.
$$\begin{aligned} x^2 + 3x - 4 &= 0 \\ (x + 4)(x - 1) &= 0 \end{aligned}$$

$$\begin{aligned} x + 4 &= 0 \quad \text{or} \quad x - 1 = 0 \\ x &= -4 \quad \text{or} \quad x = 1 \end{aligned}$$

Solution set: $\{-4, 1\}$

53.
$$\begin{aligned} x^2 + 6x - 3 &= 0 \\ x^2 + 6x &= 3 \\ x^2 + 6x + 9 &= 3 + 9 \\ (x + 3)^2 &= 12 \\ x + 3 &= \pm\sqrt{12} \\ x &= -3 \pm 2\sqrt{3} \end{aligned}$$

Solution set: $\left\{-3 \pm 2\sqrt{3}\right\}$

11.7 More About Parabolas and Their Applications

11.7 Now Try Exercises

N1. $f(x) = x^2 + 2x - 8$

To find the vertex, first complete the square on $x^2 + 2x$.

$$\left[\tfrac{1}{2}(2)\right]^2 = 1^2 = 1$$

Now add and subtract 1 on the right.

$$\begin{aligned} f(x) &= \left(x^2 + 2x + 1 - 1\right) - 8 \\ &= \left(x^2 + 2x + 1\right) - 1 - 8 \\ f(x) &= (x + 1)^2 - 9 \end{aligned}$$

The vertex of this parabola is $(-1, -9)$.

N2. $f(x) = -4x^2 + 16x - 10$

Factor out -4 from the first two terms.

$$f(x) = -4(x^2 - 4x) - 10$$

Complete the square on $x^2 - 4x$.

$$\left[\tfrac{1}{2}(-4)\right]^2 = (-2)^2 = 4$$

Add and subtract 4.

$$\begin{aligned} f(x) &= -4\left(x^2 - 4x + 4 - 4\right) - 10 \\ &= -4\left(x^2 - 4x + 4\right) - 4(-4) - 10 \\ &= -4\left(x^2 - 4x + 4\right) + 16 - 10 \\ f(x) &= -4(x - 2)^2 + 6 \end{aligned}$$

The vertex of this parabola is $(2, 6)$.

N3. For $f(x) = 3x^2 - 2x + 8$, $a = 3$, $b = -2$, and $c = 8$.

The vertex is $\left(\dfrac{-b}{2a}, f\left(\dfrac{-b}{2a}\right)\right)$, so the x-coordinate is

$$\frac{-b}{2a} = \frac{-(-2)}{2(3)} = \frac{2}{6} = \frac{1}{3},$$

and the y-coordinate is

$$\begin{aligned} f(\tfrac{1}{3}) &= 3(\tfrac{1}{3})^2 - 2(\tfrac{1}{3}) + 8 \\ &= 3(\tfrac{1}{9}) - \tfrac{2}{3} + 8 \\ &= \tfrac{1}{3} - \tfrac{2}{3} + \tfrac{24}{3} \\ &= \tfrac{23}{3}. \end{aligned}$$

The vertex is $\left(\frac{1}{3}, \frac{23}{3}\right)$.

N4. $f(x) = x^2 + 2x - 3$

Step 1
The graph opens up because $a = 1 > 0$.

Step 2
To find the vertex, complete the square on $x^2 + 2x$.

$$\left[\tfrac{1}{2}(2)\right]^2 = (1)^2 = 1$$

Add and subtract 1.

$$f(x) = x^2 + 2x + 1 - 1 - 3$$
$$f(x) = (x+1)^2 - 4$$

The vertex is at $(-1, -4)$.

Step 3
Find any x-intercepts. Let $f(x) = 0$.

$$\begin{aligned} f(x) &= x^2 + 2x - 3 \\ 0 &= x^2 + 2x - 3 \\ 0 &= (x+3)(x-1) \end{aligned}$$

$$x + 3 = 0 \quad \text{or} \quad x - 1 = 0$$
$$x = -3 \quad \text{or} \quad x = 1$$

The x-intercepts are $(-3, 0)$ and $(1, 0)$.

Find the y-intercept. Let $x = 0$.

$$f(0) = 0^2 + 2(0) - 3 = -3$$

The y-intercept is $(0, -3)$.

Step 4
The axis is $x = -1$. By symmetry, another point on the graph is $(-2, -3)$. Since x can be any real number, the domain is $(-\infty, \infty)$. Since the lowest point of the parabola is $(-1, -4)$, the range is $[-4, \infty)$.

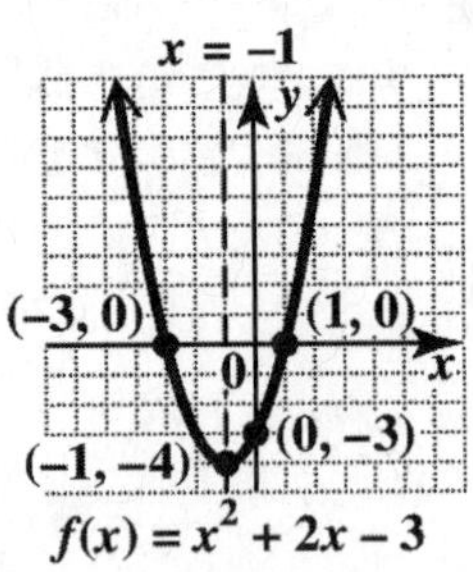

N5. **(a)** $f(x) = -2x^2 + 3x - 2$

Here $a = -2$, $b = 3$, and $c = -2$. The discriminant is

$$\begin{aligned} b^2 - 4ac &= 3^2 - 4(-2)(-2) \\ &= 9 - 16 \\ &= -7. \end{aligned}$$

Since the discriminant is negative, the graph has no x-intercepts.

(b) $f(x) = 3x^2 + 2x - 1$

Here $a = 3$, $b = 2$, and $c = -1$. The discriminant is

$$\begin{aligned} b^2 - 4ac &= 2^2 - 4(3)(-1) \\ &= 4 + 12 \\ &= 16. \end{aligned}$$

Since the discriminant is positive, the graph has two x-intercepts.

(c) $f(x) = 4x^2 - 12x + 9$

Here $a = 4$, $b = -12$, and $c = 9$. The discriminant is

$$\begin{aligned} b^2 - 4ac &= (-12)^2 - 4(4)(9) \\ &= 144 - 144 \\ &= 0. \end{aligned}$$

Since the discriminant is 0, the graph has only one x-intercept.

N6. In Example 6, replace 120 with 80, so

$$\mathcal{A}(x) = -2x^2 + 80x.$$

Here $a = -2$, $b = 80$, and $c = 0$, so the x-coordinate of the vertex is

$$x = \frac{-b}{2a} = \frac{-80}{2(-2)} = 20,$$

and the y-coordinate of the vertex, $\mathcal{A}(x)$, is

$$\begin{aligned} \mathcal{A}(20) &= -2(20)^2 + 80(20) \\ &= -2(400) + 1600 \\ &= -800 + 1600 \\ &= 800. \end{aligned}$$

The graph is a parabola that opens down with vertex at $(20, 800)$. The vertex of the graph shows that the maximum area will be 800 ft^2. This area will occur if the width, x, is 20 feet and the length is $80 - 2x$, or 40 feet.

N7. For $s(t) = -16t^2 + 48t$, $a = -16$, $b = 48$, and $c = 0$. Use the vertex formula.

$$t = \frac{-b}{2a} = \frac{-48}{2(-16)} = \frac{3}{2} = 1.5$$

This indicates that the maximum height is attained at 1.5 seconds. To find this maximum height, calculate $s(1.5)$.

$$\begin{aligned} s(t) &= -16t^2 + 48t \\ s(1.5) &= -16(1.5)^2 + 48(1.5) \\ &= -16(2.25) + 72 \\ &= -36 + 72 = 36 \end{aligned}$$

Therefore, the stomp rocket reaches a maximum height of 36 feet in 1.5 seconds.

N8. $x = (y+2)^2 - 1$

Since the roles of x and y are reversed, this graph has its vertex at $(-1, -2)$. It opens to the right, the positive x-direction, and has the same shape as $y = x^2$. The axis is $y = -2$, the domain is $[-1, \infty)$, and the range is $(-\infty, \infty)$. The points $(0, -1)$ and $(0, -3)$ are on the graph.

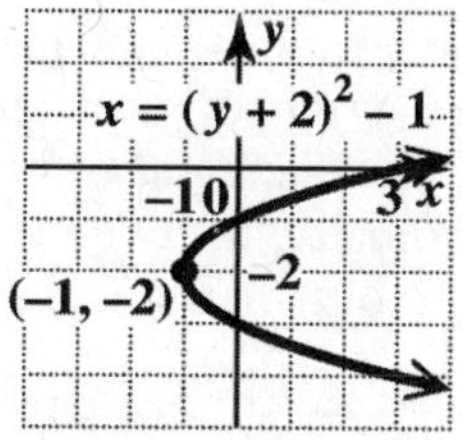

N9. $x = -3y^2 - 6y - 5$
$x = -3(y^2 + 2y) - 5$

Complete the square on $y^2 + 2y$.

$$\left[\tfrac{1}{2}(2)\right]^2 = (1)^2 = 1$$

Add and subtract 1.

$$\begin{aligned} x &= -3(y^2 + 2y + 1 - 1) - 5 \\ &= -3(y^2 + 2y + 1) - 3(-1) - 5 \\ x &= -3(y+1)^2 - 2 \end{aligned}$$

The vertex is $(-2, -1)$ and the axis is $y = -1$. Since $a = -3 < 0$, the graph opens to the left. The domain is $(-\infty, -2]$, and the range is $(-\infty, \infty)$.

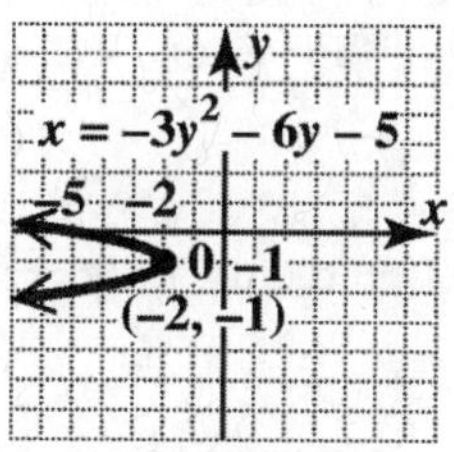

11.7 Section Exercises

1. If x is squared, the parabola has a vertical axis. If y is squared, the parabola has a horizontal axis.

3. Use the discriminant, $b^2 - 4ac$, of the function. If it is positive, there are two x-intercepts. If it is zero, there is one x-intercept (at the vertex), and if it is negative, there is no x-intercept.

5. As in Example 1, we'll complete the square to find the vertex.

$$\begin{aligned} f(x) &= x^2 + 8x + 10 \\ &= x^2 + 8x + \underline{16} + 10 - \underline{16} \qquad \left[\tfrac{1}{2}(8)\right]^2 = 16 \\ &= (x+4)^2 - 6 \end{aligned}$$

The vertex is $(-4, -6)$.

7. As in Example 2, we'll complete the square to find the vertex.

$$\begin{aligned} f(x) &= -2x^2 + 4x - 5 \\ &= -2(x^2 - 2x) - 5 \\ &= -2(x^2 - 2x + 1 - 1) - 5 \\ &= -2(x^2 - 2x + 1) + (-2)(-1) - 5 \\ &= -2(x-1)^2 - 3 \end{aligned}$$

The vertex is $(1, -3)$.

9. As in Example 3, we'll use the vertex formula to find the vertex.

$$f(x) = x^2 + x - 7$$

The x-coordinate of the vertex is

$$\frac{-b}{2a} = \frac{-1}{2(1)} = -\frac{1}{2}.$$

The y-coordinate of the vertex is

$$f(-\tfrac{1}{2}) = \tfrac{1}{4} - \tfrac{1}{2} - 7 = -\tfrac{29}{4}.$$

The vertex is $\left(-\tfrac{1}{2}, -\tfrac{29}{4}\right)$.

11. As in Example 2, we'll complete the square to find the vertex.

$$\begin{aligned} f(x) &= 2x^2 + 4x + 5 \\ &= 2(x^2 + 2x) + 5 \\ &= 2(x^2 + 2x + 1 - 1) + 5 \\ &= 2(x^2 + 2x + 1) + 2(-1) + 5 \\ &= 2(x+1)^2 - 2 + 5 \\ f(x) &= 2(x+1)^2 + 3 \end{aligned}$$

The vertex is $(-1, 3)$.
Because $a = 2 > 1$, the graph opens up and is narrower than the graph of $y = x^2$.

For $y = f(x) = 2x^2 + 4x + 5$, $a = 2$, $b = 4$, and $c = 5$. The discriminant is

$$\begin{aligned} b^2 - 4ac &= 4^2 - 4(2)(5) \\ &= 16 - 40 = -24. \end{aligned}$$

The discriminant is negative, so the parabola has no x-intercepts.

13. $f(x) = -x^2 + 5x + 3$

Use the vertex formula with $a = -1$ and $b = 5$.

The x-coordinate of the vertex is

$$\frac{-b}{2a} = \frac{-5}{2(-1)} = \frac{5}{2}.$$

The y-coordinate of the vertex is

$$\begin{aligned} f\left(\frac{-b}{2a}\right) &= f\left(\frac{5}{2}\right) \\ &= -\left(\tfrac{5}{2}\right)^2 + 5\left(\tfrac{5}{2}\right) + 3 \\ &= -\tfrac{25}{4} + \tfrac{25}{2} + 3 \\ &= \tfrac{-25+50+12}{4} = \tfrac{37}{4}. \end{aligned}$$

The vertex is

$$\left(\frac{-b}{2a}, f\left(\frac{-b}{2a}\right)\right) = \left(\frac{5}{2}, \frac{37}{4}\right).$$

Because $a = -1$, the parabola opens down and has the same shape as the graph of $y = x^2$.

$$\begin{aligned} b^2 - 4ac &= 5^2 - 4(-1)(3) \\ &= 25 + 12 = 37 \end{aligned}$$

The discriminant is positive, so the parabola has two x-intercepts.

15. Complete the square on the y-terms to find the vertex.

$$\begin{aligned} x &= \tfrac{1}{3}y^2 + 6y + 24 \\ &= \tfrac{1}{3}(y^2 + 18y) + 24 \\ &= \tfrac{1}{3}(y^2 + 18y + 81 - 81) + 24 \\ &= \tfrac{1}{3}(y^2 + 18y + 81) + \tfrac{1}{3}(-81) + 24 \\ &= \tfrac{1}{3}(y + 9)^2 - 27 + 24 \\ x &= \tfrac{1}{3}(y + 9)^2 - 3 \end{aligned}$$

The vertex is $(-3, -9)$.
The graph is a horizontal parabola. The graph opens to the right since $a = 0.\overline{3} > 0$ and is wider than the graph of $y = x^2$ since $|a| = |0.\overline{3}| < 1$.

17. The graph of $y = 2x^2 + 4x - 3$ is a vertical parabola opening up, so choice F is correct. **(F)**

19. The graph of $y = -\frac{1}{2}x^2 - x + 1$ is a vertical parabola opening down, so choices A and C are possibilities. The graph in C is wider than the graph in A, so it must correspond to $a = -\frac{1}{2}$ while the graph in A must correspond to $a = -1$. **(C)**

21. The graph of $x = -y^2 - 2y + 4$ is a horizontal parabola opening to the left, so choice D is correct. **(D)**

23. $y = f(x) = x^2 + 8x + 10$

Step 1
Since $a = 1 > 0$, the graph opens up and is the same shape as the graph of $y = x^2$.

Step 2
From Exercise 5, the vertex is $(-4, -6)$. Since the graph opens up, the axis goes through the x-coordinate of the vertex—its equation is $x = -4$.

Step 3
To find the y-intercept, let $x = 0$.
$f(0) = 10$, so the y-intercept is $(0, 10)$.
To find the x-intercepts, let $y = 0$.

$$0 = x^2 + 8x + 10$$

Now use the quadratic formula.

$$\begin{aligned} x &= \frac{-8 \pm \sqrt{64 - 40}}{2} = \frac{-8 \pm \sqrt{24}}{2} \\ &= \frac{-8 \pm 2\sqrt{6}}{2} = -4 \pm \sqrt{6} \end{aligned}$$

The x-intercepts are approximately $(-6.45, 0)$ and $(-1.55, 0)$.

Step 4
For an additional point on the graph, let $x = -2$ (two units to the right of the axis) to get $f(-2) = -2$. So the point $(-2, -2)$ is on the graph. By symmetry, the point $(-6, -2)$ (two units to the left of the axis) is on the graph.

$f(x) = x^2 + 8x + 10$

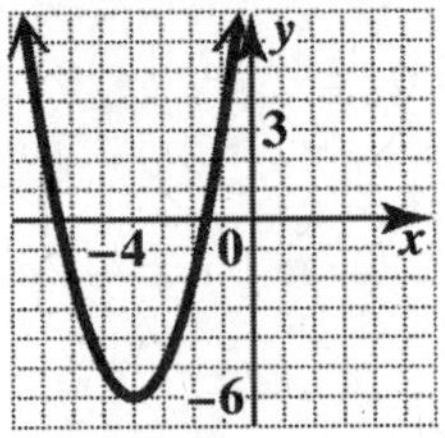

From the graph, we see that the domain is $(-\infty, \infty)$ and the range is $[-6, \infty)$.

25. $y = f(x) = -2x^2 + 4x - 5$

Step 1
Since $a = -2$, the graph opens down and is narrower than the graph of $y = x^2$.

Step 2
From Exercise 7, the vertex is $(1, -3)$. Since the graph opens down, the axis goes through the x-coordinate of the vertex—its equation is $x = 1$.

Step 3
If $x = 0$, $y = -5$, so the y-intercept is $(0, -5)$.
To find the x-intercepts, let $y = 0$.

$$\begin{aligned} 0 &= -2x^2 + 4x - 5 \\ x &= \frac{-4 \pm \sqrt{16 - 40}}{2(-2)} \end{aligned}$$

The discriminant is negative, so there are no x-intercepts.

Step 4
By symmetry, $(2, -5)$ is also on the graph.

$f(x) = -2x^2 + 4x - 5$

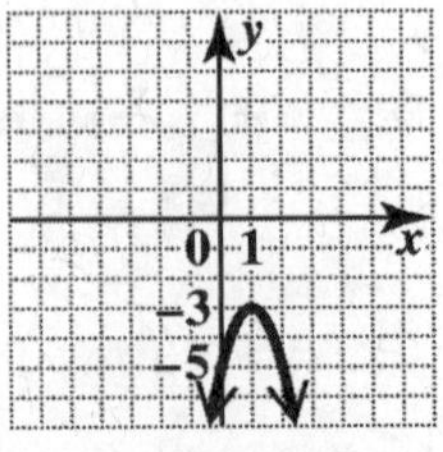

From the graph, we see that the domain is $(-\infty, \infty)$ and the range is $(-\infty, -3]$.

27. $x = (y+2)^2 + 1 = y^2 + 4y + 5$

The roles of x and y are reversed, so this is a horizontal parabola.

Step 1
The coefficient of y^2 is $1 > 0$, so the graph opens to the right.

Step 2
The vertex can be identified from the given form of the equation. When $y = -2$, $x = 1$, so the vertex is $(1, -2)$. Since the graph opens right, the axis goes through the y-coordinate of the vertex—its equation is $y = -2$.

Step 3
To find the x-intercept, let $y = 0$.
$x = 0^2 + 4(0) + 5 = 5$, so the x-intercept is $(5, 0)$.

To find the y-intercepts, let $x = 0$.

$$\begin{aligned} 0 &= (y+2)^2 + 1 \\ -1 &= (y+2)^2 \end{aligned}$$

Since $(y+2)^2$ cannot be negative, there are no y-intercepts.

Step 4
For an additional point on the graph, let $y = -4$ (two units below the axis) to get
$x = (-4+2)^2 + 1 = 5$.

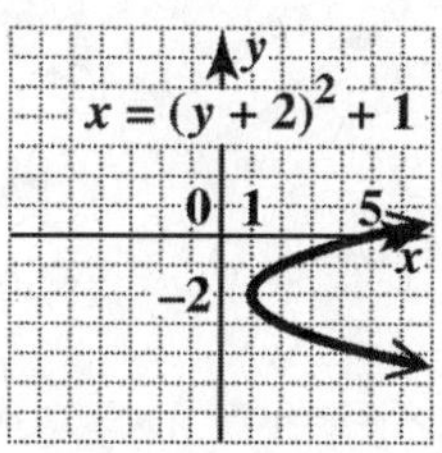

From the graph, we see the domain is $[1, \infty)$ and the range is $(-\infty, \infty)$.

29. $x = -\frac{1}{5}y^2 + 2y - 4$

The roles of x and y are reversed, so this is a horizontal parabola.

Step 1
Since $a = -\frac{1}{5} < 0$, the graph opens to the left and is wider than the graph of $y = x^2$.

Step 2
The y-coordinate of the vertex is

$$\frac{-b}{2a} = \frac{-2}{2\left(-\frac{1}{5}\right)} = \frac{-2}{-\frac{2}{5}} = 5.$$

The x-coordinate of the vertex is

$$-\tfrac{1}{5}(5)^2 + 2(5) - 4 = -5 + 10 - 4 = 1.$$

Thus, the vertex is $(1, 5)$. Since the graph opens left, the axis goes through the y-coordinate of the vertex—its equation is $y = 5$.

Step 3
To find the x-intercept, let $y = 0$.
If $y = 0$, $x = -4$, so the x-intercept is $(-4, 0)$.
To find the y-intercepts, let $x = 0$.

$$\begin{aligned} 0 &= -\tfrac{1}{5}y^2 + 2y - 4 \\ 0 &= y^2 - 10y + 20 \quad \textit{Multiply by } -5. \\ y &= \frac{10 \pm \sqrt{100-80}}{2} = \frac{10 \pm \sqrt{20}}{2} \\ &= \frac{10 \pm 2\sqrt{5}}{2} = 5 \pm \sqrt{5} \end{aligned}$$

The y-intercepts are approximately $(0, 7.2)$ and $(0, 2.8)$.

Step 4
For an additional point on the graph, let $y = 7$ (two units above the axis) to get $x = \frac{1}{5}$. So the point $\left(\frac{1}{5}, 7\right)$ is on the graph. By symmetry, the point $\left(\frac{1}{5}, 3\right)$ (two units below the axis) is on the graph.

$x = -\frac{1}{5}y^2 + 2y - 4$

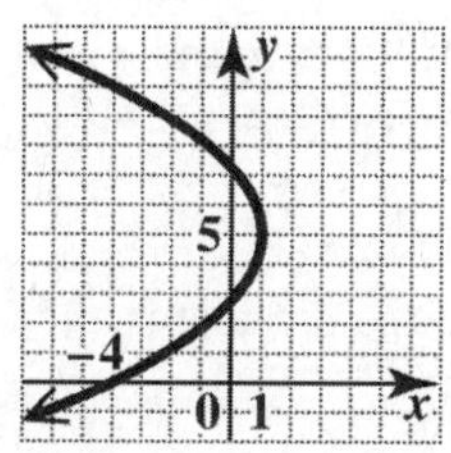

From the graph, we see that the domain is $(-\infty, 1]$ and the range is $(-\infty, \infty)$.

31. $x = 3y^2 + 12y + 5$

The roles of x and y are reversed, so this is a horizontal parabola.

Step 1
Since $a = 3 > 0$, the graph opens to the right and is narrower than the graph of $y = x^2$.

Step 2
Use the formula to find the y-value of the vertex.

$$\frac{-b}{2a} = \frac{-12}{2(3)} = -2$$

If $y = -2$, $x = -7$, so the vertex is $(-7, -2)$. Since the graph opens right, the axis goes through the y-coordinate of the vertex—its equation is $y = -2$.

Step 3
If $y = 0$, $x = 5$, so the x-intercept is $(5, 0)$.
To find the y-intercepts, let $x = 0$.

$$\begin{aligned} 0 &= 3y^2 + 12y + 5 \\ y &= \frac{-12 \pm \sqrt{144-60}}{6} = \frac{-12 \pm \sqrt{84}}{6} \\ &= \frac{-12 \pm 2\sqrt{21}}{6} = \frac{-6 \pm \sqrt{21}}{3} \end{aligned}$$

The y-intercepts are approximately $(0, -0.5)$ and $(0, -3.5)$.

Step 4
By symmetry, $(5, -4)$ is also on the graph.

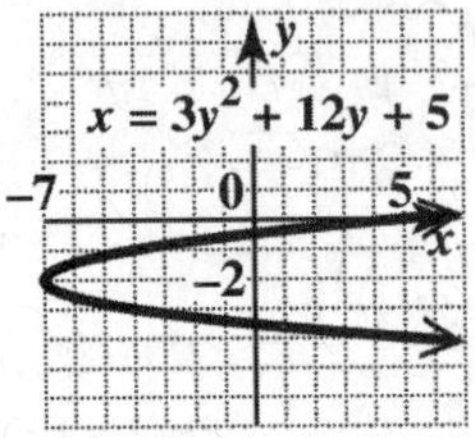

From the graph, we see that the domain is $[-7, \infty)$ and the range is $(-\infty, \infty)$.

33. Let $x =$ one number, $40 - x =$ the other number, and $P =$ the product.

$$\begin{aligned} P &= x(40 - x) \\ &= 40x - x^2 \quad \text{or} \quad -x^2 + 40x \end{aligned}$$

This parabola opens down so the maximum occurs at the vertex.

Here $a = -1$, $b = 40$, and $c = 0$.

$$\frac{-b}{2a} = \frac{-40}{2(-1)} = 20$$

$x = 20$ when the product is a maximum.

Since $x = 20$, $40 - x = 20$, and the two numbers are 20 and 20.

35. Let x represent the length of the two equal sides, and let $280 - 2x$ represent the length of the remaining side (the side parallel to the highway). Then substitute x for L and $280 - 2x$ for W in the formula for the area of a rectangle, $\mathcal{A} = LW$.

$$\begin{aligned} \mathcal{A} &= x(280 - 2x) \\ &= 280x - 2x^2 \text{ or } -2x^2 + 280x \end{aligned}$$

The maximum area will occur at the vertex.

$$x = \frac{-b}{2a} = \frac{-280}{2(-2)} = 70$$

When $x = 70$, $280 - 2x = 140$, and $\mathcal{A} = 9800$. Thus, the dimensions of the lot with maximum area are 140 feet by 70 feet and the maximum area is 9800 square feet.

37. $s(t) = -16t^2 + 32t$

Here, $a = -16 < 0$, so the parabola opens down. The time it takes to reach the maximum height and the maximum height are given by the vertex of the parabola. Use the vertex formula to find that

$$t = \frac{-b}{2a} = \frac{-32}{2(-16)} = \frac{-32}{-32} = 1,$$

and $$\begin{aligned} s(t) &= -16(1)^2 + 32(1) \\ &= -16 + 32 = 16. \end{aligned}$$

The vertex is $(1, 16)$, so the maximum height is 16 feet which occurs when the time is 1 second. The object hits the ground when $s = 0$.

$$\begin{aligned} 0 &= -16t^2 + 32t \\ 0 &= -16t(t - 2) \end{aligned}$$

$$\begin{aligned} -16t = 0 \quad &\text{or} \quad t - 2 = 0 \\ t = 0 \quad &\text{or} \quad t = 2 \end{aligned}$$

It takes 2 seconds for the object to hit the ground.

39. The graph of the height of the projectile,

$$s(t) = -16t^2 + 64t + 1,$$

is a parabola that opens down since $a = -16 < 0$. The time at which the cork reaches its maximum height and the maximum height are the t and s coordinates of the vertex.

$$t = \frac{-b}{2a} = \frac{-64}{2(-16)} = 2$$

$$\begin{aligned} s(2) &= -16(2)^2 + 64(2) + 1 \\ &= -64 + 128 + 1 = 65 \end{aligned}$$

The cork reaches a maximum height of 65 feet after 2 seconds.

41. $C(x) = x^2 - 40x + 610$

Find the vertex of the graph of C.

$$\frac{-b}{2a} = \frac{-(-40)}{2(1)} = 20$$

$$C(20) = 20^2 - 40(20) + 610 = 210$$

He should sell 20 tacos per day to minimize his costs of \$210.

43. $f(x) = 22.88x^2 - 141.3x + 1044$

(a) Since the graph opens up, the y-value of the vertex is a minimum.

(b) The x-value of the vertex is given by

$$x = \frac{-b}{2a} = \frac{-(-141.3)}{2(22.88)} \approx 3.088$$

The year was $2000 + 3 = 2003$.

$f(3.088) \approx \$825.8$ billion, which is the minimum amount of total receipts from individual taxes.

45. $f(x) = -20.57x^2 + 758.9x - 3140$

(a) The coefficient of x^2 is negative because a parabola that models the data must open down.

(b) Use the vertex formula.

$$x = \frac{-b}{2a} = \frac{-758.9}{2(-20.57)} \approx 18.45$$

$$f(18.45) \approx 3860$$

The vertex is approximately $(18.45, 3860)$.

(c) 18 corresponds to 2018, so in 2018 social security assets will reach their maximum value of \$3860 billion.

47. The number of people on the plane is $100 - x$ since x is the number of unsold seats. The price per seat is $200 + 4x$.

(a) The total revenue received for the flight is found by multiplying the number of seats by the price per seat. Thus, the revenue is

$$\begin{aligned} R(x) &= (100 - x)(200 + 4x) \\ &= 20{,}000 + 200x - 4x^2. \end{aligned}$$

(b) Use the formula for the vertex.

$$x = \frac{-b}{2a} = \frac{-200}{2(-4)} = 25$$

$R(25) = 22{,}500$, so the vertex is (25, 22,500). $R(0) = 20{,}000$, so the R-intercept is (0, 20,000). From the factored form for R, we see that the positive x-intercept is $(100, 0)$. (The factor $200 + 4x$ leads to a negative x-intercept, meaningless in this problem.)

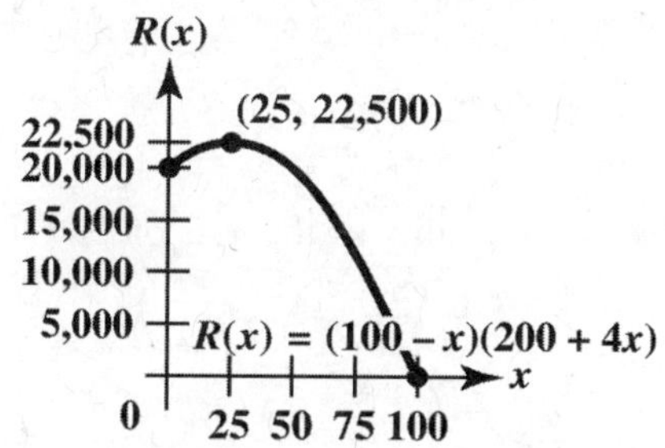

(c) The number of unsold seats x that produce the maximum revenue is 25, the x-value of the vertex.

(d) The maximum revenue is \$22,500, the y-value of the vertex.

49. $[1, 5] \Leftrightarrow 1 \leq x \leq 5$

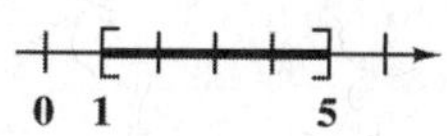

51. $(-\infty, 1] \cup [5, \infty) \Leftrightarrow x \leq 1$ or $x \geq 5$

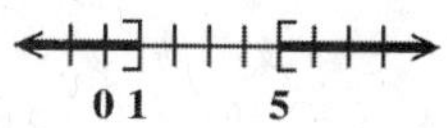

53.

$$\begin{aligned} -2x + 1 &< 4 \\ -2x &< 3 \\ x &> -\tfrac{3}{2} \end{aligned}$$

Solution set: $\left(-\frac{3}{2}, \infty\right)$

11.8 Polynomial and Rational Inequalities

11.8 Now Try Exercises

N1. **(a)** $x^2 - 3x - 4 > 0$

The x-intercepts are $(-1, 0)$ and $(4, 0)$.

Notice from the graph that x-values less than -1 or greater than 4 result in y-values *greater than* 0. Therefore, the solution set is

$$(-\infty, -1) \cup (4, \infty).$$

(b) $x^2 - 3x - 4 < 0$

Notice from the graph that x-values between -1 and 4 result in y-values *less than* 0. Therefore, the solution set is $(-1, 4)$.

N2. $x^2 + 2x - 8 > 0$

Use factoring to solve the quadratic equation

$x^2 + 2x - 8 = 0.$

$(x + 4)(x - 2) = 0$

$$\begin{aligned} x + 4 &= 0 \quad \text{or} \quad x - 2 = 0 \\ x &= -4 \quad \text{or} \quad x = 2 \end{aligned}$$

Locate the numbers -4 and 2 that divide the number line into three intervals A, B, and C.

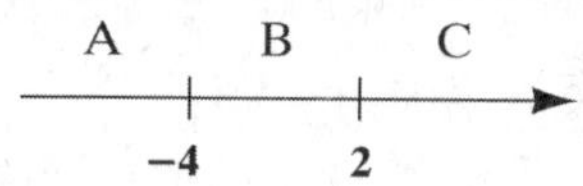

Choose a number from each interval to substitute in the inequality

$$x^2 + 2x - 8 > 0.$$

Interval A: Let $x = -5$.

$$\begin{aligned} (-5)^2 + 2(-5) - 8 &\overset{?}{>} 0 \\ 25 - 10 - 8 &\overset{?}{>} 0 \\ 7 &> 0 \qquad \textit{True} \end{aligned}$$

Interval B: Let $x = 0$.

$$\begin{aligned} (0)^2 + 2(0) - 8 &\overset{?}{>} 0 \\ -8 &> 0 \qquad \textit{False} \end{aligned}$$

Interval C: Let $x = 3$.

$$\begin{aligned} (3)^2 + 2(3) - 8 &\overset{?}{>} 0 \\ 9 + 6 - 8 &\overset{?}{>} 0 \\ 7 &> 0 \qquad \textit{True} \end{aligned}$$

The numbers in Intervals A and C are solutions. The numbers -4 and 2 are excluded because of $>$.

Solution set: $(-\infty, -4) \cup (2, \infty)$

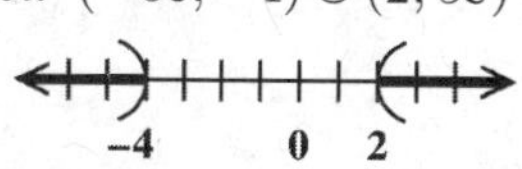

N3. **(a)** $(4x - 1)^2 > -3$

The square of any real number is always greater than or equal to 0, so any real number satisfies this inequality.

Solution set: $(-\infty, \infty)$

(b) $(4x - 1)^2 < -3$

The square of a real number is never negative.

Solution set: $\emptyset$

N4. $(x+4)(x-3)(2x+1) \le 0$

Solve the equation

$(x+4)(x-3)(2x+1) = 0.$

$x+4=0$ or $x-3=0$ or $2x+1=0$

$x=-4$ or $x=3$ or $x=-\frac{1}{2}$

Locate these numbers and the intervals A, B, C, and D on a number line.

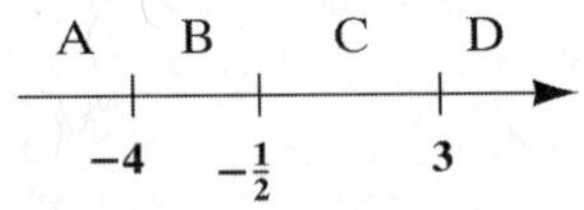

Test a number from each interval in the inequality

$$(x+4)(x-3)(2x+1) \le 0.$$

Interval A: Let $x=-5$.

$$(-1)(-8)(-9) \overset{?}{\le} 0$$
$$-72 \le 0 \qquad \textit{True}$$

Interval B: Let $x=-2$.

$$(2)(-5)(-3) \overset{?}{\le} 0$$
$$30 \le 0 \qquad \textit{False}$$

Interval C: Let $x=0$.

$$(4)(-3)(1) \overset{?}{\le} 0$$
$$-12 \le 0 \qquad \textit{True}$$

Interval D: Let $x=4$.

$$(8)(1)(9) \overset{?}{\le} 0$$
$$72 \le 0 \qquad \textit{False}$$

The numbers in Intervals A and C, including $-4, -\frac{1}{2}$, and 3 because of $\le$, are solutions.

Solution set: $(-\infty, -4] \cup [-\frac{1}{2}, 3]$

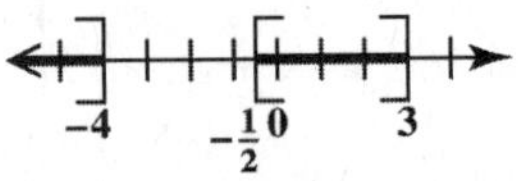

N5. $$\frac{3}{x+1} > 4$$

Write the inequality so that 0 is on one side.

$$\frac{3}{x+1} - 4 > 0$$
$$\frac{3}{x+1} - \frac{4(x+1)}{x+1} > 0$$
$$\frac{3-4x-4}{x+1} > 0$$
$$\frac{-4x-1}{x+1} > 0$$

The number $-\frac{1}{4}$ makes the numerator 0, and -1 makes the denominator 0. These two numbers determine three intervals.

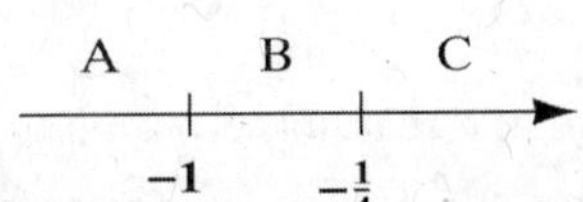

Test a number from each interval in the inequality

$$\frac{3}{x+1} > 4.$$

Interval A: Let $x=-2$.

$$\frac{3}{-2+1} \overset{?}{>} 4$$
$$-3 > 4 \qquad \textit{False}$$

Interval B: Let $x=-\frac{1}{2}$.

$$\frac{3}{-\frac{1}{2}+1} \overset{?}{>} 4$$
$$6 > 4 \qquad \textit{True}$$

Interval C: Let $x=0$.

$$\frac{3}{0+1} \overset{?}{>} 4$$
$$3 > 4 \qquad \textit{False}$$

The solution set includes numbers in Interval B, excluding endpoints.

Solution set: $(-1, -\frac{1}{4})$

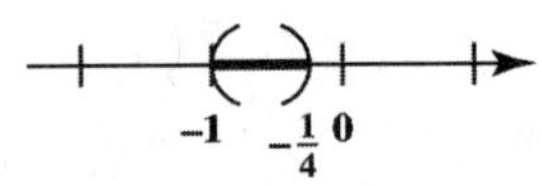

N6. $$\frac{x-3}{x+3} \le 2$$

Write the inequality so that 0 is on one side.

$$\frac{x-3}{x+3} - 2 \le 0$$
$$\frac{x-3}{x+3} - \frac{2(x+3)}{x+3} \le 0$$
$$\frac{x-3-2x-6}{x+3} \le 0$$
$$\frac{-x-9}{x+3} \le 0$$

The number -9 makes the numerator 0, and -3 makes the denominator 0. These two numbers determine three intervals.

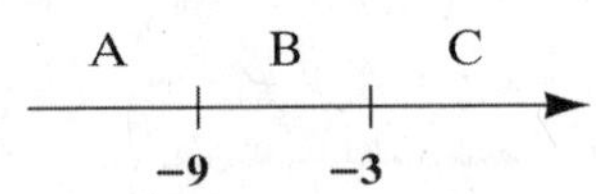

Test a number from each interval in the inequality

$$\frac{x-3}{x+3} \le 2.$$

Interval A: Let $x=-10$.

$$\frac{-10-3}{-10+3} \overset{?}{\le} 2$$
$$\frac{13}{7} \le 2 \qquad \textit{True}$$

Interval B: Let $x=-5$.

$$\frac{-5-3}{-5+3} \overset{?}{\le} 2$$
$$4 \le 2 \qquad \textit{False}$$

Interval C: Let $x = 0$.

$$\frac{0-3}{0+3} \overset{?}{\leq} 2$$

$$-1 \leq 2 \qquad \textit{True}$$

The numbers in Intervals A and C are solutions. -3 is not in the solution set (since it makes the denominator 0), but -9 is.

Solution set: $(-\infty, -9] \cup (-3, \infty)$

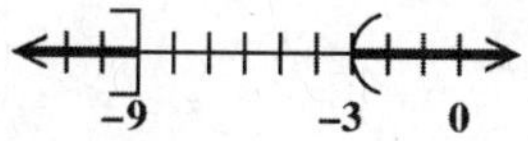

11.8 Section Exercises

1. **(a)** The x-intercepts determine the solutions of the equation $x^2 - 4x + 3 = 0$. From the graph, the solution set is $\{1, 3\}$.

(b) The x-values of the points on the graph that are *above* the x-axis form the solution set of the inequality $x^2 - 4x + 3 > 0$. From the graph, the solution set is $(-\infty, 1) \cup (3, \infty)$.

(c) The x-values of the points on the graph that are *below* the x-axis form the solution set of the inequality $x^2 - 4x + 3 < 0$. From the graph, the solution set is $(1, 3)$.

3. **(a)** The x-intercepts determine the solutions of the equation $-x^2 + 3x + 10 = 0$. From the graph, the solution set is $\{-2, 5\}$.

(b) The x-values of the points on the graph that are *above* the x-axis form the solution set of the inequality $-x^2 + 3x + 10 > 0$. From the graph, the solution set for $-x^2 + 3x + 10 \geq 0$ is $[-2, 5]$.

(c) The x-values of the points on the graph that are *below* the x-axis form the solution set of the inequality $-x^2 + 3x + 10 < 0$. From the graph, the solution set for $-x^2 + 3x + 10 \leq 0$ is $(-\infty, -2] \cup [5, \infty)$.

5. $(x + 1)(x - 5) > 0$
Solve the equation
$(x + 1)(x - 5) = 0$.

$$x + 1 = 0 \quad \text{or} \quad x - 5 = 0$$
$$x = -1 \quad \text{or} \quad x = 5$$

The numbers -1 and 5 divide a number line into three intervals: A, B, and C.

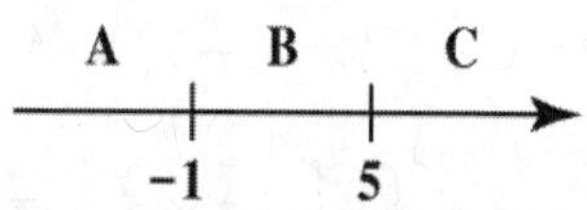

Test a number from each interval in the original inequality, $(x + 1)(x - 5) > 0$.

Interval A: Let $x = -2$.

$$(-2 + 1)(-2 - 5) \overset{?}{>} 0$$
$$-1(-7) \overset{?}{>} 0$$
$$7 > 0 \qquad \textit{True}$$

Interval B: Let $x = 0$.

$$(0 + 1)(0 - 5) \overset{?}{>} 0$$
$$-5 > 0 \qquad \textit{False}$$

Interval C: Let $x = 6$.

$$(6 + 1)(6 - 5) \overset{?}{>} 0$$
$$7 > 0 \qquad \textit{True}$$

The solution set includes the numbers in Intervals A and C, excluding -1 and 5 because of $>$.

Solution set: $(-\infty, -1) \cup (5, \infty)$

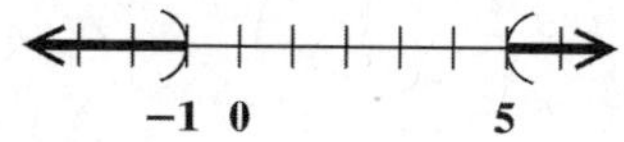

7. $(x + 4)(x - 6) < 0$
Solve the equation
$(x + 4)(x - 6) = 0$.

$$x + 4 = 0 \quad \text{or} \quad x - 6 = 0$$
$$x = -4 \quad \text{or} \quad x = 6$$

The numbers -4 and 6 divide a number line into three intervals: A, B, and C.

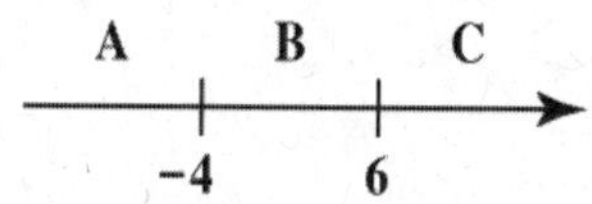

Test a number from each interval in the original inequality, $(x + 4)(x - 6) < 0$.

Interval A: Let $x = -5$.

$$(-5 + 4)(-5 - 6) \overset{?}{<} 0$$
$$-1(-11) \overset{?}{<} 0$$
$$11 < 0 \qquad \textit{False}$$

Interval B: Let $x = 0$.

$$4(-6) \overset{?}{<} 0$$
$$-24 < 0 \qquad \textit{True}$$

Interval C: Let $x = 7$.

$$(7 + 4)(7 - 6) \overset{?}{<} 0$$
$$11(1) \overset{?}{<} 0$$
$$11 < 0 \qquad \textit{False}$$

The solution set includes Interval B, where the expression is negative.

Solution set: $(-4, 6)$

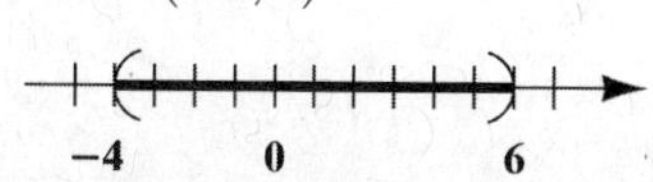

9. $x^2 - 4x + 3 \geq 0$

Solve the equation

$x^2 - 4x + 3 = 0.$

$(x - 1)(x - 3) = 0$

$x - 1 = 0$ or $x - 3 = 0$

$x = 1$ or $x = 3$

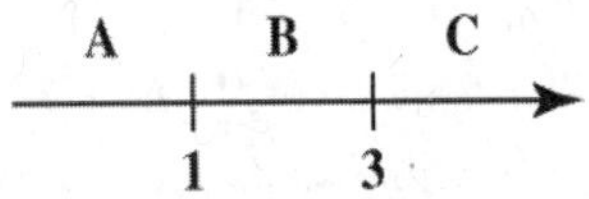

Test a number from each interval in the original inequality, $x^2 - 4x + 3 \geq 0$.

Interval A: Let $x = 0$.

$3 \geq 0$ *True*

Interval B: Let $x = 2$.

$2^2 - 4(2) + 3 \overset{?}{\geq} 0$

$-1 \geq 0$ *False*

Interval C: Let $x = 4$.

$4^2 - 4(4) + 3 \overset{?}{\geq} 0$

$3 \geq 0$ *True*

The solution set includes the numbers in Intervals A and C, including 1 and 3 because of $\geq$.

Solution set: $(-\infty, 1] \cup [3, \infty)$

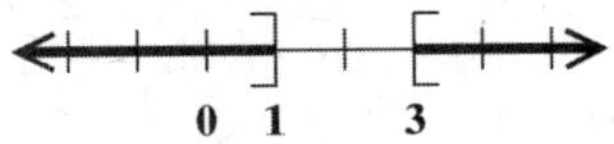

11. $10x^2 + 9x \geq 9$

$10x^2 + 9x - 9 \geq 0$

Solve the equation

$10x^2 + 9x - 9 = 0.$

$(2x + 3)(5x - 3) = 0$

$2x + 3 = 0$ or $5x - 3 = 0$

$x = -\frac{3}{2}$ or $x = \frac{3}{5}$

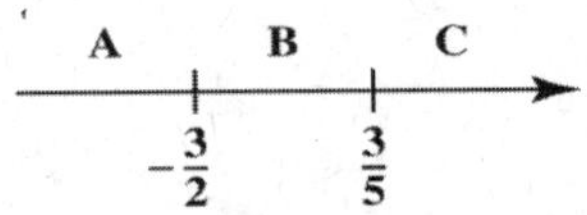

Test a number from each interval in the original inequality, $10x^2 + 9x \geq 9$.

Interval A: Let $x = -2$.

$10(-2)^2 + 9(-2) \overset{?}{\geq} 9$

$40 - 18 \overset{?}{\geq} 9$

$22 \geq 9$ *True*

Interval B: Let $x = 0$.

$0 \geq 9$ *False*

Interval C: Let $x = 1$.

$10(1)^2 + 9(1) \overset{?}{\geq} 9$

$10 + 9 \overset{?}{\geq} 9$

$19 \geq 9$ *True*

The solution set includes the numbers in Intervals A and C, including $-\frac{3}{2}$ and $\frac{3}{5}$ because of $\geq$.

Solution set: $(-\infty, -\frac{3}{2}] \cup [\frac{3}{5}, \infty)$

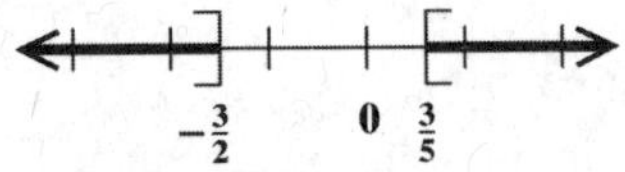

13. $4x^2 - 9 \leq 0$

Solve the equation

$4x^2 - 9 = 0.$

$(2x + 3)(2x - 3) = 0$

$2x + 3 = 0$ or $2x - 3 = 0$

$x = -\frac{3}{2}$ or $x = \frac{3}{2}$

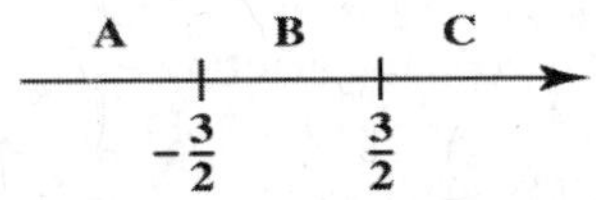

Test a number from each interval in the original inequality, $4x^2 - 9 \leq 0$.

Interval A: Let $x = -2$.

$4(-2)^2 - 9 \overset{?}{\leq} 0$

$7 \leq 0$ *False*

Interval B: Let $x = 0$.

$-9 \leq 0$ *True*

Interval C: Let $x = 2$.

$4(2)^2 - 9 \overset{?}{\leq} 0$

$7 \leq 0$ *False*

The solution set includes Interval B, including the endpoints.

Solution set: $[-\frac{3}{2}, \frac{3}{2}]$

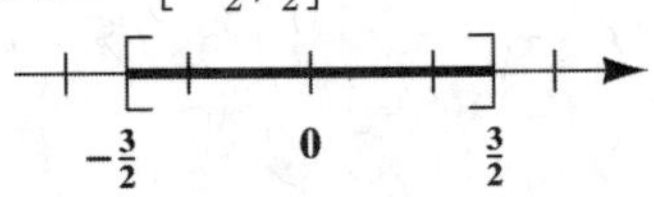

15. $6x^2 + x \geq 1$

$6x^2 + x - 1 \geq 0$

Solve the equation

$6x^2 + x - 1 = 0.$

$(2x + 1)(3x - 1) = 0$

$2x + 1 = 0$ or $3x - 1 = 0$

$x = -\frac{1}{2}$ or $x = \frac{1}{3}$

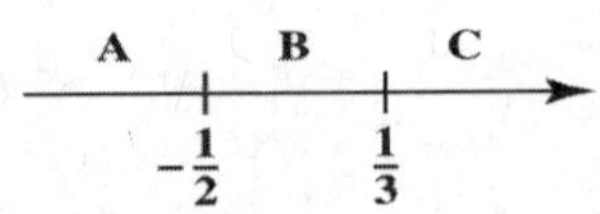

Test a number from each interval in the original inequality, $6x^2 + x \geq 1$.

Interval A: Let $x = -1$.

$6(-1)^2 + (-1) \overset{?}{\geq} 1$

$5 \geq 1$ *True*

Interval B: Let $x = 0$.

$0 \geq 1$ *False*

Interval C: Let $x = 1$.

$$6(1)^2 + 1 \overset{?}{\geq} 1$$
$$7 \geq 1 \quad \textit{True}$$

The solution set includes the numbers in Intervals A and C, including $-\frac{1}{2}$ and $\frac{1}{3}$ because of $\geq$.

Solution set: $\left(-\infty, -\frac{1}{2}\right] \cup \left[\frac{1}{3}, \infty\right)$

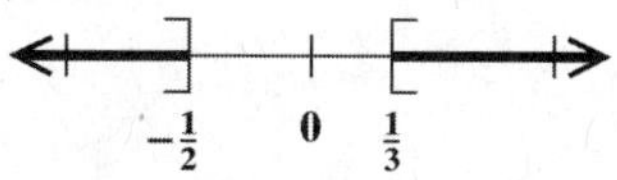

17. $z^2 - 4z \geq 0$
Solve the equation
$z^2 - 4z = 0$.
$z(z - 4) = 0$

$$z = 0 \quad \text{or} \quad z - 4 = 0$$
$$z = 4$$

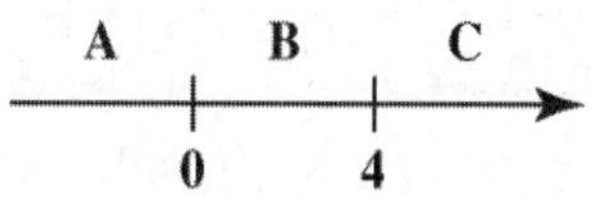

Test a number from each interval in the original inequality, $z^2 - 4z \geq 0$.

Interval A: Let $z = -1$.

$$(-1)^2 - 4(-1) \overset{?}{\geq} 0$$
$$5 \geq 0 \quad \textit{True}$$

Interval B: Let $z = 2$.

$$2^2 - 4(2) \overset{?}{\geq} 0$$
$$-4 \geq 0 \quad \textit{False}$$

Interval C: Let $z = 5$.

$$5^2 - 4(5) \overset{?}{\geq} 0$$
$$5 \geq 0 \quad \textit{True}$$

The solution set includes the numbers in Intervals A and C, including 0 and 4 because of $\geq$.

Solution set: $(-\infty, 0] \cup [4, \infty)$

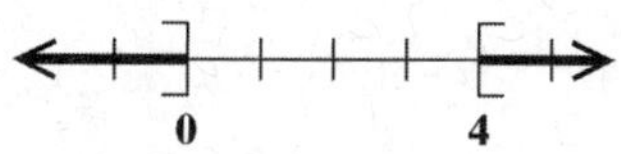

19. $3x^2 - 5x \leq 0$
Solve the equation
$3x^2 - 5x = 0$.
$x(3x - 5) = 0$

$$x = 0 \quad \text{or} \quad 3x - 5 = 0$$
$$x = \frac{5}{3}$$

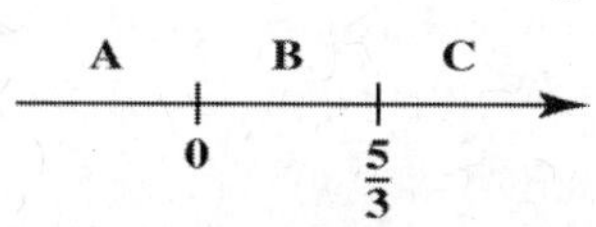

Test a number from each interval in the original inequality, $3x^2 - 5x \leq 0$.

Interval A: Let $x = -1$.

$$3(-1)^2 - 5(-1) \overset{?}{\leq} 0$$
$$8 \leq 0 \quad \textit{False}$$

Interval B: Let $x = 1$.

$$3(1)^2 - 5(1) \overset{?}{\leq} 0$$
$$-2 \leq 0 \quad \textit{True}$$

Interval C: Let $x = 2$.

$$3(2)^2 - 5(2) \overset{?}{\leq} 0$$
$$2 \leq 0 \quad \textit{False}$$

The solution set includes the numbers in Interval B, including the endpoints.

Solution set: $\left[0, \frac{5}{3}\right]$

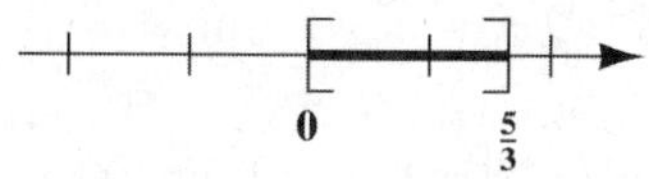

21. $x^2 - 6x + 6 \geq 0$
Solve the equation
$x^2 - 6x + 6 = 0$.

Since $x^2 - 6x + 6$ does not factor, let $a = 1$, $b = -6$, and $c = 6$ in the quadratic formula.

$$x = \frac{-(-6) \pm \sqrt{(-6)^2 - 4(1)(6)}}{2(1)}$$
$$= \frac{6 \pm \sqrt{12}}{2} = \frac{6 \pm 2\sqrt{3}}{2}$$
$$= \frac{2\left(3 \pm \sqrt{3}\right)}{2} = 3 \pm \sqrt{3}$$
$$x = 3 + \sqrt{3} \approx 4.7 \text{ or}$$
$$x = 3 - \sqrt{3} \approx 1.3$$

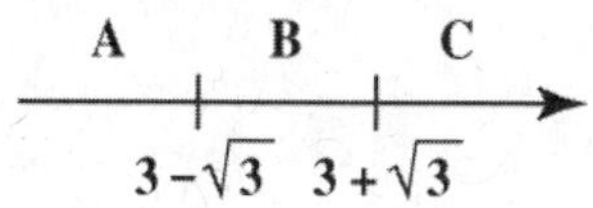

Test a number from each interval in the original inequality, $x^2 - 6x + 6 \geq 0$.

Interval A: Let $x = 0$.

$$6 \geq 0 \quad \textit{True}$$

Interval B: Let $x = 3$.

$$3^2 - 6(3) + 6 \overset{?}{\geq} 0$$
$$-3 \geq 0 \quad \textit{False}$$

Interval C: Let $x = 5$.

$$5^2 - 6(5) + 6 \overset{?}{\geq} 0$$
$$1 \geq 0 \quad \textit{True}$$

The solution set includes the numbers in Intervals A and C, including $3 - \sqrt{3}$ and $3 + \sqrt{3}$ because of $\geq$.

Solution set: $\left(-\infty, 3 - \sqrt{3}\right] \cup \left[3 + \sqrt{3}, \infty\right)$

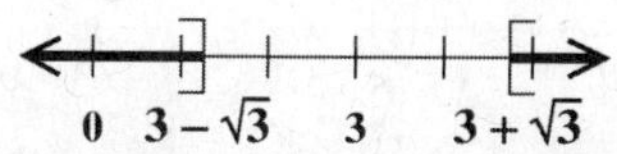

23. $(4-3x)^2 \geq -2$

Since $(4-3x)^2$ is either 0 or positive, $(4-3x)^2$ will always be greater than -2. Therefore, the solution set is $(-\infty, \infty)$.

25. $(3x+5)^2 \leq -4$

Since $(3x+5)^2$ is never negative, $(3x+5)^2$ will never be less than or equal to a negative number. Therefore, the solution set is $\emptyset$.

27. $(x-1)(x-2)(x-4) < 0$

The numbers 1, 2, and 4 are solutions of the cubic equation

$$(x-1)(x-2)(x-4) = 0.$$

These numbers divide a number line into four intervals.

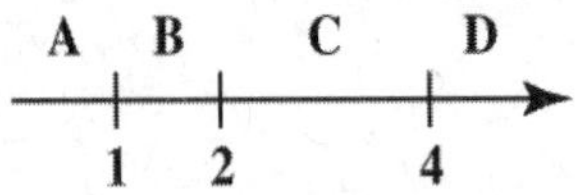

Test a number from each interval in the inequality

$$(x-1)(x-2)(x-4) < 0.$$

Interval A: Let $x = 0$.

$$-1(-2)(-4) \stackrel{?}{<} 0$$
$$-8 < 0 \quad \textit{True}$$

Interval B: Let $x = 1.5$.

$$(1.5-1)(1.5-2)(1.5-4) \stackrel{?}{<} 0$$
$$0.5(-0.5)(-2.5) \stackrel{?}{<} 0$$
$$0.625 < 0 \quad \textit{False}$$

Interval C: Let $x = 3$.

$$(3-1)(3-2)(3-4) \stackrel{?}{<} 0$$
$$2(1)(-1) \stackrel{?}{<} 0$$
$$-2 < 0 \quad \textit{True}$$

Interval D: Let $x = 5$.

$$(5-1)(5-2)(5-4) \stackrel{?}{<} 0$$
$$4(3)(1) \stackrel{?}{<} 0$$
$$12 < 0 \quad \textit{False}$$

The numbers in Intervals A and C, not including 1, 2, or 4, are solutions.

Solution set: $(-\infty, 1) \cup (2, 4)$

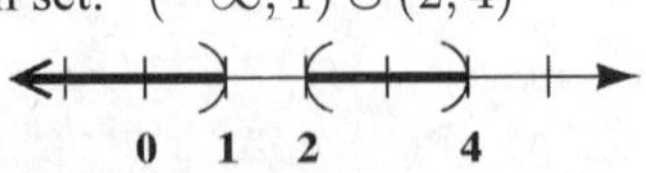

29. $(x-4)(2x+3)(3x-1) \geq 0$

The numbers 4, $-\frac{3}{2}$, and $\frac{1}{3}$ are solutions of the cubic equation

$$(x-4)(2x+3)(3x-1) = 0.$$

These numbers divide a number line into 4 intervals.

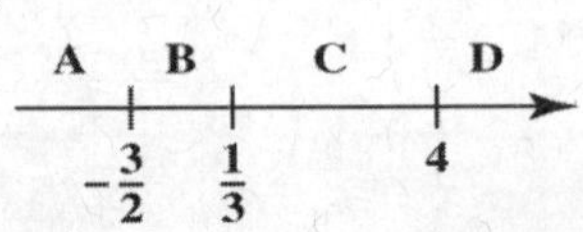

Interval A: Let $x = -2$.

$$-6(-1)(-7) \stackrel{?}{\geq} 0$$
$$-42 \geq 0 \quad \textit{False}$$

Interval B: Let $x = 0$.

$$-4(3)(-1) \stackrel{?}{\geq} 0$$
$$12 \geq 0 \quad \textit{True}$$

Interval C: Let $x = 1$.

$$-3(5)(2) \stackrel{?}{\geq} 0$$
$$-30 \geq 0 \quad \textit{False}$$

Interval D: Let $x = 5$.

$$1(13)(14) \stackrel{?}{\geq} 0$$
$$182 \geq 0 \quad \textit{True}$$

The solution set includes numbers in Intervals B and D, including the endpoints.

Solution set: $[-\frac{3}{2}, \frac{1}{3}] \cup [4, \infty)$

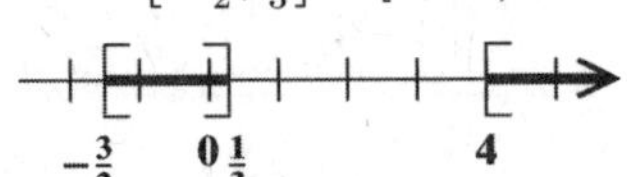

31. $\dfrac{x-1}{x-4} > 0$

The number 1 makes the numerator 0, and 4 makes the denominator 0. These two numbers determine three intervals.

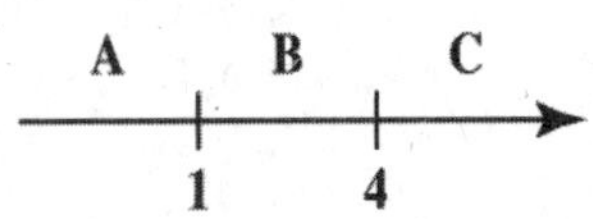

Test a number from each interval in the inequality

$$\frac{x-1}{x-4} > 0.$$

Interval A: Let $x = 0$.

$$\frac{0-1}{0-4} \stackrel{?}{>} 0$$
$$\frac{1}{4} > 0 \quad \textit{True}$$

Interval B: Let $x = 2$.

$$\frac{2-1}{2-4} \stackrel{?}{>} 0$$
$$\frac{1}{-2} > 0 \quad \textit{False}$$

Interval C: Let $x = 5$.

$$\frac{5-1}{5-4} \stackrel{?}{>} 0$$
$$4 > 0 \quad \textit{True}$$

The solution set includes numbers in Intervals A and C, excluding endpoints.

Solution set: $(-\infty, 1) \cup (4, \infty)$

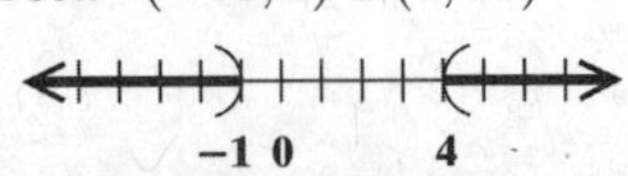

33. $\dfrac{2x+3}{x-5} \le 0$

The number $-\frac{3}{2}$ makes the numerator 0, and 5 makes the denominator 0. These two numbers determine three intervals.

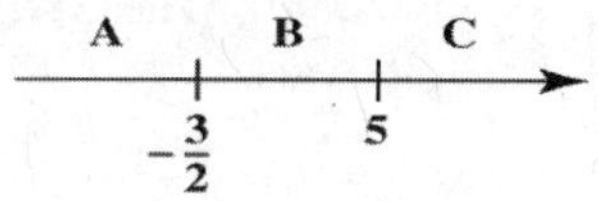

Test a number from each interval in the inequality

$$\frac{2x+3}{x-5} \le 0.$$

Interval A: Let $x = -2$.

$$\frac{2(-2)+3}{(-2)-5} \overset{?}{\le} 0$$
$$\tfrac{1}{7} \le 0 \qquad \textit{False}$$

Interval B: Let $x = 0$.

$$\frac{2(0)+3}{0-5} \overset{?}{\le} 0$$
$$-\tfrac{3}{5} \le 0 \qquad \textit{True}$$

Interval C: Let $x = 6$.

$$\frac{2(6)+3}{6-5} \overset{?}{\le} 0$$
$$15 \le 0 \qquad \textit{False}$$

The solution set includes the points in Interval B. The endpoint 5 is not included since it makes the left side undefined. The endpoint $-\frac{3}{2}$ is included because it makes the left side equal to 0.
Solution set: $\left[-\frac{3}{2}, 5\right)$

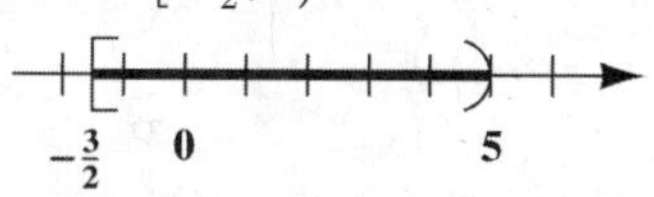

35. $\dfrac{8}{x-2} \ge 2$

Write the inequality so that 0 is on one side.

$$\frac{8}{x-2} - 2 \ge 0$$
$$\frac{8}{x-2} - \frac{2(x-2)}{x-2} \ge 0$$
$$\frac{8-2x+4}{x-2} \ge 0$$
$$\frac{-2x+12}{x-2} \ge 0$$

The number 6 makes the numerator 0, and 2 makes the denominator 0. These two numbers determine three intervals.

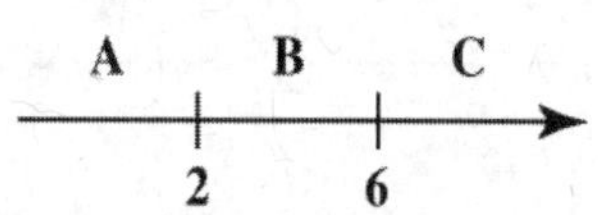

Test a number from each interval in the inequality

$$\frac{8}{x-2} \ge 2.$$

Interval A: Let $x = 0$.

$$\frac{8}{0-2} \overset{?}{\ge} 2$$
$$-4 \ge 2 \qquad \textit{False}$$

Interval B: Let $x = 3$.

$$\frac{8}{3-2} \overset{?}{\ge} 2$$
$$8 \ge 2 \qquad \textit{True}$$

Interval C: Let $x = 7$.

$$\frac{8}{7-2} \overset{?}{\ge} 2$$
$$\tfrac{8}{5} \ge 2 \qquad \textit{False}$$

The solution set includes numbers in Interval B, including 6 but excluding 2, which makes the fraction undefined.
Solution set: $(2, 6]$

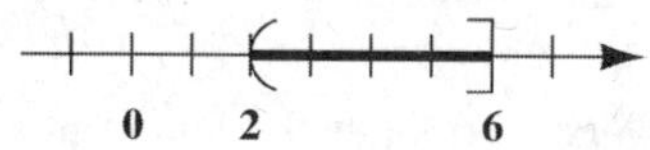

37. $\dfrac{3}{2x-1} < 2$

Write the inequality so that 0 is on one side.

$$\frac{3}{2x-1} - 2 < 0$$
$$\frac{3}{2x-1} - \frac{2(2x-1)}{2x-1} < 0$$
$$\frac{3-4x+2}{2x-1} < 0$$
$$\frac{-4x+5}{2x-1} < 0$$

The number $\frac{5}{4}$ makes the numerator 0, and $\frac{1}{2}$ makes the denominator 0. These two numbers determine three intervals.

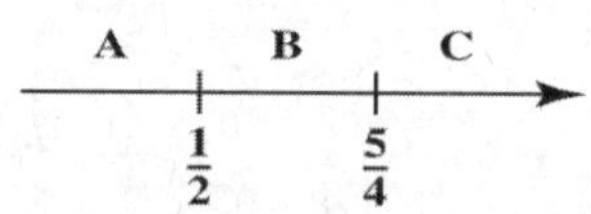

Test a number from each interval in the inequality

$$\frac{3}{2x-1} < 2.$$

Interval A: Let $x = 0$.

$$\frac{3}{2(0)-1} \overset{?}{<} 2$$
$$-3 < 2 \qquad \textit{True}$$

Interval B: Let $x = 1$.

$$\frac{3}{2(1)-1} \overset{?}{<} 2$$
$$3 < 2 \qquad \textit{False}$$

Interval C: Let $x = 2$.

$$\frac{3}{2(2)-1} \overset{?}{<} 2$$
$$1 < 2 \qquad \textit{True}$$

The solution set includes numbers in Intervals A and C, excluding endpoints.

Solution set: $\left(-\infty, \frac{1}{2}\right) \cup \left(\frac{5}{4}, \infty\right)$

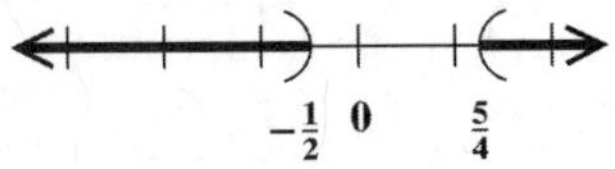

39. $$\frac{x-3}{x+2} \geq 2$$

Write the inequality so that 0 is on one side.

$$\frac{x-3}{x+2} - 2 \geq 0$$

$$\frac{x-3}{x+2} - \frac{2(x+2)}{x+2} \geq 0$$

$$\frac{x-3-2x-4}{x+2} \geq 0$$

$$\frac{-x-7}{x+2} \geq 0$$

The number -7 makes the numerator 0, and -2 makes the denominator 0. These two numbers determine three intervals.

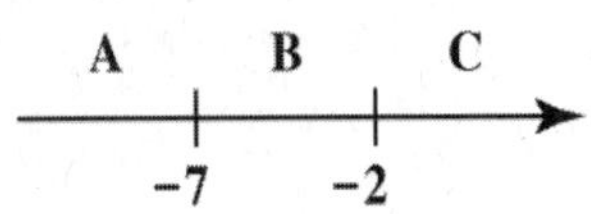

Test a number from each interval in the inequality

$$\frac{x-3}{x+2} \geq 2.$$

Interval A: Let $x = -8$.

$\frac{-11}{-6} \overset{?}{\geq} 2$

$\frac{11}{6} \geq 2$ *False*

Interval B: Let $x = -4$.

$\frac{-7}{-2} \overset{?}{\geq} 2$

$\frac{7}{2} \geq 2$ *True*

Interval C: Let $x = 0$.

$\frac{-3}{2} \geq 2$ *False*

The solution set includes numbers in Interval B, including -7 but excluding -2, which makes the fraction undefined.

Solution set: $[-7, -2)$

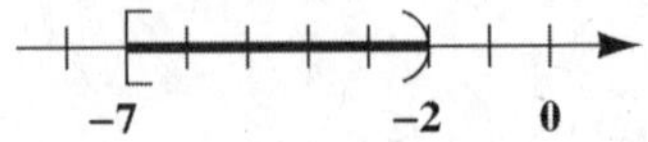

41. $$\frac{x-8}{x-4} < 3$$

Write the inequality so that 0 is on one side.

$$\frac{x-8}{x-4} - 3 < 0$$

$$\frac{x-8}{x-4} - \frac{3(x-4)}{x-4} < 0$$

$$\frac{x-8-3x+12}{x-4} < 0$$

$$\frac{-2x+4}{x-4} < 0$$

The number 2 makes the numerator 0, and 4 makes the denominator 0. These two numbers determine three intervals.

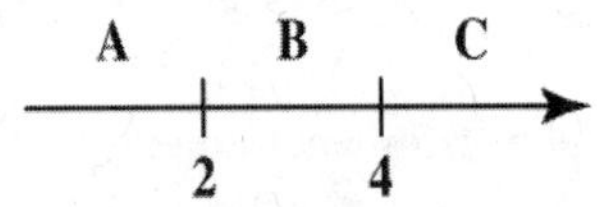

Test a number from each interval in the inequality

$$\frac{x-8}{x-4} < 3.$$

Interval A: Let $x = 0$.

$\frac{-8}{-4} \overset{?}{<} 3$

$2 < 3$ *True*

Interval B: Let $x = 3$.

$\frac{-5}{-1} \overset{?}{<} 3$

$5 < 3$ *False*

Interval C: Let $x = 5$.

$\frac{-3}{1} < 3$ *True*

The solution set includes numbers in Intervals A and C, excluding endpoints.

Solution set: $(-\infty, 2) \cup (4, \infty)$

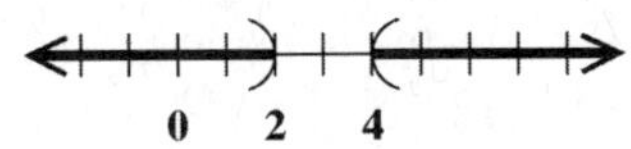

43. $$\frac{4k}{2k-1} < k$$

Write the inequality so that 0 is on one side.

$$\frac{4k}{2k-1} - k < 0$$

$$\frac{4k}{2k-1} - \frac{k(2k-1)}{2k-1} < 0$$

$$\frac{4k-2k^2+k}{2k-1} < 0$$

$$\frac{-2k^2+5k}{2k-1} < 0$$

$$\frac{k(-2k+5)}{2k-1} < 0$$

The numbers 0 and $\frac{5}{2}$ make the numerator 0, and $\frac{1}{2}$ makes the denominator 0. These three numbers determine four intervals.

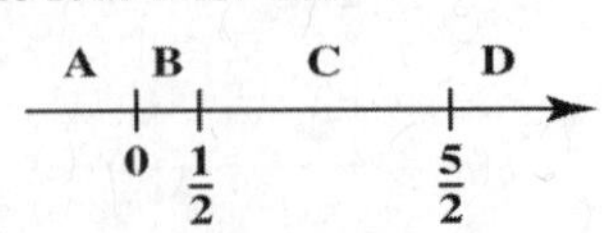

Test a number from each interval in the inequality

$$\frac{4k}{2k-1} < k.$$

Interval A: Let $k = -1$.

$$\frac{4(-1)}{2(-1)-1} \stackrel{?}{<} -1$$
$$\tfrac{4}{3} < -1 \qquad \textit{False}$$

Interval B: Let $k = \frac{1}{4}$.

$$\frac{4(\frac{1}{4})}{2(\frac{1}{4})-1} \stackrel{?}{<} \frac{1}{4}$$
$$-2 < \tfrac{1}{4} \qquad \textit{True}$$

Interval C: Let $k = 1$.

$$\frac{4(1)}{2(1)-1} \stackrel{?}{<} 1$$
$$4 < 1 \qquad \textit{False}$$

Interval D: Let $k = 3$.

$$\frac{4(3)}{2(3)-1} \stackrel{?}{<} 3$$
$$\tfrac{12}{5} < 3 \qquad \textit{True}$$

The solution set includes numbers in Intervals B and D. None of the endpoints are included.

Solution set: $(0, \frac{1}{2}) \cup (\frac{5}{2}, \infty)$

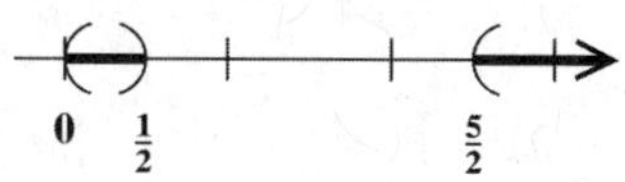

45. $\dfrac{2x-3}{x^2+1} \geq 0$

The denominator is positive for all real numbers x, so it has no effect on the solution set for the inequality.

$$2x - 3 \geq 0$$
$$2x \geq 3$$
$$x \geq \tfrac{3}{2}$$

Solution set: $[\frac{3}{2}, \infty)$

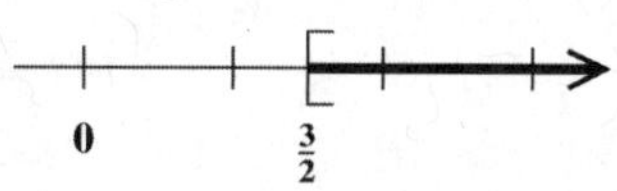

47. $\dfrac{(3x-5)^2}{x+2} > 0$

The numerator is positive for all real numbers x except $x = \frac{5}{3}$, which makes it equal to 0. If we solve the inequality $x + 2 > 0$, then we only have to be sure to exclude $\frac{5}{3}$ from that solution set to determine the solution set of the original inequality.

$$x + 2 > 0$$
$$x > -2$$

Solution set: $(-2, \frac{5}{3}) \cup (\frac{5}{3}, \infty)$

49. $\{(0,1), (1,2), (2,4), (3,8)\}$

The domain is the set of x-values: $\{0, 1, 2, 3\}$. The range is the set of y-values: $\{1, 2, 4, 8\}$.

51. Using the vertical line test, we find any vertical line will intersect the graph at most once. This indicates that the graph represents a function.

Chapter 11 Review Exercises

1. $t^2 = 121$

$$t = 11 \quad \text{or} \quad t = -11$$

Solution set: $\{\pm 11\}$

2. $p^2 = 3$

$$p = \sqrt{3} \quad \text{or} \quad p = -\sqrt{3}$$

Solution set: $\{\pm\sqrt{3}\}$

3. $(2x+5)^2 = 100$

$$2x + 5 = 10 \quad \text{or} \quad 2x + 5 = -10$$
$$2x = 5 \qquad\qquad 2x = -15$$
$$x = \tfrac{5}{2} \quad \text{or} \quad x = -\tfrac{15}{2}$$

Solution set: $\{-\frac{15}{2}, \frac{5}{2}\}$

4. $(3x-2)^2 = -25$

$$3x - 2 = \sqrt{-25} \quad \text{or} \quad 3x - 2 = -\sqrt{-25}$$
$$3x - 2 = 5i \qquad\qquad 3x - 2 = -5i$$
$$3x = 2 + 5i \qquad\qquad 3x = 2 - 5i$$
$$x = \frac{2+5i}{3} \qquad\qquad x = \frac{2-5i}{3}$$
$$x = \tfrac{2}{3} + \tfrac{5}{3}i \quad \text{or} \quad x = \tfrac{2}{3} - \tfrac{5}{3}i$$

Solution set: $\{\frac{2}{3} \pm \frac{5}{3}i\}$

5. By the square root property, the first step should be $x = \sqrt{12}$ or $x = -\sqrt{12}$. The solution set is $\{\pm 2\sqrt{3}\}$.

6. $4.9t^2 = d$

$$4.9t^2 = 165 \qquad \textit{Let } d = 165.$$
$$t^2 = \frac{165}{4.9}$$
$$t = \sqrt{\frac{165}{4.9}} \qquad t \geq 0$$
$$t \approx 5.8 \text{ seconds}$$

It would take about 5.8 seconds for the wallet to fall 165 meters.

7. $x^2 + 4x = 15$

Complete the square.

$\left(\frac{1}{2}\cdot 4\right)^2 = 2^2 = 4$

Add 4 to each side.

$x^2 + 4x + 4 = 15 + 4$

$(x+2)^2 = 19$

$x + 2 = \sqrt{19}$ or $x + 2 = -\sqrt{19}$

$x = -2 + \sqrt{19}$ or $x = -2 - \sqrt{19}$

Solution set: $\left\{-2 \pm \sqrt{19}\right\}$

8. $2x^2 - 3x = -1$

$x^2 - \frac{3}{2}x = -\frac{1}{2}$ *Divide by 2.*

Complete the square.

$\left[\frac{1}{2}\left(-\frac{3}{2}\right)\right]^2 = \left(-\frac{3}{4}\right)^2 = \frac{9}{16}$

Add $\frac{9}{16}$ to each side.

$x^2 - \frac{3}{2}x + \frac{9}{16} = -\frac{1}{2} + \frac{9}{16}$

$\left(x - \frac{3}{4}\right)^2 = -\frac{8}{16} + \frac{9}{16}$

$\left(x - \frac{3}{4}\right)^2 = \frac{1}{16}$

$x - \frac{3}{4} = \sqrt{\frac{1}{16}}$ or $x - \frac{3}{4} = -\sqrt{\frac{1}{16}}$

$x - \frac{3}{4} = \frac{1}{4}$ $\quad$ $x - \frac{3}{4} = -\frac{1}{4}$

$x = \frac{3}{4} + \frac{1}{4}$ $\quad$ $x = \frac{3}{4} - \frac{1}{4}$

$x = 1$ or $x = \frac{1}{2}$

Solution set: $\left\{\frac{1}{2}, 1\right\}$

9. $2z^2 + 8z - 3 = 0$

$z^2 + 4z - \frac{3}{2} = 0$ *Divide by 2.*

Rewrite the equation with the variable terms on one side and the constant on the other side.

$$z^2 + 4z = \frac{3}{2}$$

Take half the coefficient of z and square it.

$$\tfrac{1}{2}(4) = 2, \text{ and } 2^2 = 4.$$

Add 4 to both sides of the equation.

$z^2 + 4z + 4 = \frac{3}{2} + 4$

$(z+2)^2 = \frac{11}{2}$

$$z + 2 = \pm\sqrt{\frac{11}{2}}$$

$$z + 2 = \pm\frac{\sqrt{11}}{\sqrt{2}}\cdot\frac{\sqrt{2}}{\sqrt{2}} = \pm\frac{\sqrt{22}}{2}$$

$$z = -2 \pm \frac{\sqrt{22}}{2}$$

$$z = \frac{-4}{2} \pm \frac{\sqrt{22}}{2} = \frac{-4 \pm \sqrt{22}}{2}$$

Solution set: $\left\{\frac{-4\pm\sqrt{22}}{2}\right\}$

10. $4x^2 - 3x + 6 = 0$

$x^2 - \frac{3}{4}x + \frac{3}{2} = 0$ *Divide by 4.*

Rewrite the equation with the variable terms on one side and the constant on the other side.

$$x^2 - \tfrac{3}{4}x = -\tfrac{3}{2}$$

Take half the coefficient of x and square it.

$$\tfrac{1}{2}\left(-\tfrac{3}{4}\right) = -\tfrac{3}{8}, \text{ and } \left(-\tfrac{3}{8}\right)^2 = \tfrac{9}{64}.$$

Add $\frac{9}{64}$ to both sides of the equation.

$$x^2 - \tfrac{3}{4}x + \tfrac{9}{64} = -\tfrac{3}{2} + \tfrac{9}{64}$$

$$\left(x - \tfrac{3}{8}\right)^2 = -\tfrac{96}{64} + \tfrac{9}{64}$$

$$\left(x - \tfrac{3}{8}\right)^2 = -\tfrac{87}{64}$$

$$x - \tfrac{3}{8} = \pm\sqrt{-\tfrac{87}{64}}$$

$$x - \tfrac{3}{8} = \pm\tfrac{i\sqrt{87}}{8}$$

$$x = \tfrac{3}{8} \pm \tfrac{\sqrt{87}}{8}i$$

Solution set: $\left\{\frac{3}{8} \pm \frac{\sqrt{87}}{8}i\right\}$

11. $2x^2 + x - 21 = 0$

Here $a = 2$, $b = 1$, and $c = -21$.

$$x = \frac{-b \pm \sqrt{b^2 - 4ac}}{2a}$$

$$x = \frac{-1 \pm \sqrt{1^2 - 4(2)(-21)}}{2(2)}$$

$$= \frac{-1 \pm \sqrt{1 + 168}}{4}$$

$$= \frac{-1 \pm \sqrt{169}}{4} = \frac{-1 \pm 13}{4}$$

$$x = \frac{-1 + 13}{4} = \frac{12}{4} = 3 \text{ or}$$

$$x = \frac{-1 - 13}{4} = -\frac{14}{4} = -\frac{7}{2}$$

Solution set: $\left\{-\frac{7}{2}, 3\right\}$

12. $x^2 + 5x = 7$

$x^2 + 5x - 7 = 0$

Here $a = 1$, $b = 5$, and $c = -7$.

$$x = \frac{-b \pm \sqrt{b^2 - 4ac}}{2a}$$

$$x = \frac{-5 \pm \sqrt{5^2 - 4(1)(-7)}}{2(1)}$$

$$= \frac{-5 \pm \sqrt{25 + 28}}{2} = \frac{-5 \pm \sqrt{53}}{2}$$

Solution set: $\left\{\frac{-5\pm\sqrt{53}}{2}\right\}$

13. $(t+3)(t-4)=-2$
$t^2-t-12=-2$
$t^2-t-10=0$
Here $a=1$, $b=-1$, and $c=-10$.

$$t=\frac{-b\pm\sqrt{b^2-4ac}}{2a}$$
$$t=\frac{-(-1)\pm\sqrt{(-1)^2-4(1)(-10)}}{2(1)}$$
$$=\frac{1\pm\sqrt{1+40}}{2}=\frac{1\pm\sqrt{41}}{2}$$

Solution set: $\left\{\frac{1\pm\sqrt{41}}{2}\right\}$

14. $2x^2+3x+4=0$
Here $a=2$, $b=3$, and $c=4$.

$$x=\frac{-b\pm\sqrt{b^2-4ac}}{2a}$$
$$x=\frac{-3\pm\sqrt{3^2-4(2)(4)}}{2(2)}$$
$$=\frac{-3\pm\sqrt{9-32}}{4}=\frac{-3\pm\sqrt{-23}}{4}$$
$$=\frac{-3\pm i\sqrt{23}}{4}=-\frac{3}{4}\pm\frac{\sqrt{23}}{4}i$$

Solution set: $\left\{-\frac{3}{4}\pm\frac{\sqrt{23}}{4}i\right\}$

15. $3p^2=2(2p-1)$
$3p^2=4p-2$
$3p^2-4p+2=0$
Here $a=3$, $b=-4$, and $c=2$.

$$p=\frac{-b\pm\sqrt{b^2-4ac}}{2a}$$
$$p=\frac{-(-4)\pm\sqrt{(-4)^2-4(3)(2)}}{2(3)}$$
$$=\frac{4\pm\sqrt{16-24}}{6}=\frac{4\pm\sqrt{-8}}{6}$$
$$=\frac{4\pm 2i\sqrt{2}}{6}=\frac{2\left(2\pm i\sqrt{2}\right)}{6}$$
$$=\frac{2\pm i\sqrt{2}}{3}=\frac{2}{3}\pm\frac{\sqrt{2}}{3}i$$

Solution set: $\left\{\frac{2}{3}\pm\frac{\sqrt{2}}{3}i\right\}$

16. $x(2x-7)=3x^2+3$
$2x^2-7x=3x^2+3$
$0=x^2+7x+3$
Here $a=1$, $b=7$, and $c=3$.

$$x=\frac{-b\pm\sqrt{b^2-4ac}}{2a}$$
$$x=\frac{-7\pm\sqrt{7^2-4(1)(3)}}{2(1)}$$
$$=\frac{-7\pm\sqrt{49-12}}{2}$$
$$=\frac{-7\pm\sqrt{37}}{2}$$

Solution set: $\left\{\frac{-7\pm\sqrt{37}}{2}\right\}$

17. **(a)** $x^2+5x+2=0$
Here $a=1$, $b=5$, and $c=2$.
$b^2-4ac=5^2-4(1)(2)$
$=25-8=17$
Since the discriminant is positive, but not a perfect square, there are two distinct irrational number solutions. The answer is **C**.

(b) $4t^2=3-4t$
$4t^2+4t-3=0$
Here $a=4$, $b=4$, and $c=-3$.
$b^2-4ac=4^2-4(4)(-3)$
$=16+48$
$=64$ or 8^2
Since the discriminant is positive, and a perfect square, there are two distinct rational number solutions. The answer is **A**.

18. **(a)** $4x^2=6x-8$
$4x^2-6x+8=0$
Here $a=4$, $b=-6$, and $c=8$.
$b^2-4ac=(-6)^2-4(4)(8)$
$=36-128=-92$

Since the discriminant is negative, there are two distinct nonreal complex number solutions. The answer is **D**.

(b) $9z^2+30z+25=0$
Here $a=9$, $b=30$, and $c=25$.
$b^2-4ac=30^2-4(9)(25)$
$=900-900=0$
Since the discriminant is zero, there is exactly one rational number solution. The answer is **B**.

19. $\frac{15}{x}=2x-1$
$x\left(\frac{15}{x}\right)=x(2x-1)$ *Multiply by the LCD, x.*
$15=2x^2-x$
$0=2x^2-x-15$
$0=(2x+5)(x-3)$

$2x+5=0$ or $x-3=0$
$x=-\frac{5}{2}$ or $x=3$

Check $x=-\frac{5}{2}$: $-6=-5-1$ *True*
Check $x=3$: $5=6-1$ *True*
Solution set: $\left\{-\frac{5}{2},3\right\}$

20. $\frac{1}{n} + \frac{2}{n+1} = 2$

$n(n+1)\left(\frac{1}{n} + \frac{2}{n+1}\right)$ *Multiply by the LCD, n(n+1).*

$= n(n+1) \cdot 2$

$(n+1) + 2n = 2n^2 + 2n$

$0 = 2n^2 - n - 1$

$0 = (2n+1)(n-1)$

$2n + 1 = 0$ or $n - 1 = 0$

$n = -\frac{1}{2}$ or $n = 1$

Check $n = -\frac{1}{2}$: $-2 + 4 = 2$ *True*

Check $n = 1$: $1 + 1 = 2$ *True*

Solution set: $\{-\frac{1}{2}, 1\}$

21. $-2r = \sqrt{\frac{48 - 20r}{2}}$

Square each side.

$(-2r)^2 = \left(\sqrt{\frac{48 - 20r}{2}}\right)^2$

$4r^2 = \frac{48 - 20r}{2}$

$4r^2 = 24 - 10r$

$4r^2 + 10r - 24 = 0$

$2r^2 + 5r - 12 = 0$

$(r+4)(2r-3) = 0$

$r + 4 = 0$ or $2r - 3 = 0$

$r = -4$ or $r = \frac{3}{2}$

Check $r = -4$: $8 = \sqrt{64}$ *True*

Check $r = \frac{3}{2}$: $-3 = \sqrt{9}$ *False*

Solution set: $\{-4\}$

22. $8(3x+5)^2 + 2(3x+5) - 1 = 0$

Let $u = 3x + 5$, so $u^2 = (3x+5)^2$.

$8u^2 + 2u - 1 = 0$

$(2u+1)(4u-1) = 0$

$2u + 1 = 0$ or $4u - 1 = 0$

$u = -\frac{1}{2}$ or $u = \frac{1}{4}$

To find x, substitute $3x + 5$ for u.

$3x + 5 = -\frac{1}{2}$ or $3x + 5 = \frac{1}{4}$

$3x = -\frac{11}{2}$ $\qquad$ $3x = -\frac{19}{4}$

$x = -\frac{11}{6}$ or $x = -\frac{19}{12}$

Check $x = -\frac{11}{6}$: $2 - 1 - 1 = 0$ *True*

Check $x = -\frac{19}{12}$: $0.5 + 0.5 - 1 = 0$ *True*

Solution set: $\{-\frac{11}{6}, -\frac{19}{12}\}$

23. $2x^{2/3} - x^{1/3} - 28 = 0$

Let $u = x^{1/3}$, so $u^2 = (x^{1/3})^2 = x^{2/3}$.

$2u^2 - u - 28 = 0$

$(2u+7)(u-4) = 0$

$2u + 7 = 0$ or $u - 4 = 0$

$u = -\frac{7}{2}$ or $u = 4$

To find x, substitute $x^{1/3}$ for u.

$x^{1/3} = -\frac{7}{2}$ or $x^{1/3} = 4$

$(x^{1/3})^3 = \left(-\frac{7}{2}\right)^3$ $\qquad$ $(x^{1/3})^3 = 4^3$

$x = -\frac{343}{8}$ or $x = 64$

Check $x = -\frac{343}{8}$: $24.5 + 3.5 - 28 = 0$ *True*

Check $x = 64$: $32 - 4 - 28 = 0$ *True*

Solution set: $\{-\frac{343}{8}, 64\}$

24. $p^4 - 10p^2 + 9 = 0$

Let $x = p^2$, so $x^2 = p^4$.

$x^2 - 10x + 9 = 0$

$(x-1)(x-9) = 0$

$x - 1 = 0$ or $x - 9 = 0$

$x = 1$ or $x = 9$

To find p, substitute p^2 for x.

$p^2 = 1$ or $p^2 = 9$

$p = \pm\sqrt{1}$ $\qquad$ $p = \pm\sqrt{9}$

$p = \pm 1$ or $p = \pm 3$

Check $p = \pm 1$: $1 - 10 + 9 = 0$ *True*

Check $p = \pm 3$: $81 - 90 + 9 = 0$ *True*

Solution set: $\{-3, -1, 1, 3\}$

25. Let x = Bahaa's rate.

Make a table. Use $d = rt$, or $t = \frac{d}{r}$.

	Distance	Rate	Time
Upstream	20	$x - 3$	$\frac{20}{x-3}$
Downstream	20	$x + 3$	$\frac{20}{x+3}$

Time upstream plus time downstream equals 7 hr.

$\frac{20}{x-3} + \frac{20}{x+3} = 7$

Multiply each side by the LCD, $(x+3)(x-3)$.

$$(x+3)(x-3)\left(\frac{20}{x-3}+\frac{20}{x+3}\right)=(x+3)(x-3)(7)$$
$$20(x+3)+20(x-3)=7(x^2-9)$$
$$20x+60+20x-60=7x^2-63$$
$$0=7x^2-40x-63$$
$$0=(x-7)(7x+9)$$

$$x-7=0 \quad \text{or} \quad 7x+9=0$$
$$x=7 \quad \text{or} \quad x=-\tfrac{9}{7}$$

Reject $-\frac{9}{7}$. Bahaa's rate was 7 mph.

26. Let $x =$ Carol-Ann's rate on the trip to pick up her friend. Make a chart. Use $d = rt$, or $t = \frac{d}{r}$.

	Distance	Rate	Time
To Her Friend	8	x	$\frac{8}{x}$
To the Mall	11	$x+15$	$\frac{11}{x+15}$

Time to pick up her friend plus time to mall equals 24 min (or 0.4 hr).

$$\frac{8}{x} + \frac{11}{x+15} = 0.4$$

Multiply each side by the LCD, $x(x+15)$.

$$x(x+15)\left(\frac{8}{x}+\frac{11}{x+15}\right)=x(x+15)(0.4)$$
$$8(x+15)+11x=0.4x(x+15)$$
$$8x+120+11x=0.4x^2+6x$$
$$0=0.4x^2-13x-120$$

Multiply by 5 to clear the decimal.

$$0=2x^2-65x-600$$
$$0=(x-40)(2x+15)$$

$$x-40=0 \quad \text{or} \quad 2x+15=0$$
$$x=40 \quad \text{or} \quad x=-\tfrac{15}{2}$$

Reject $-\frac{15}{2}$. Carol-Ann's rate on the trip to pick up her friend was 40 mph.

27. Let $x =$ the amount of time for the old machine alone and
$x-1 =$ the amount of time for the new machine alone.

Make a chart.

Machine	Rate	Time Together	Fractional Part of the Job Done
Old	$\frac{1}{x}$	2	$\frac{2}{x}$
New	$\frac{1}{x-1}$	2	$\frac{2}{x-1}$

Part done by old machine plus Part done by new machine equals 1 whole job.

$$\frac{2}{x} + \frac{2}{x-1} = 1$$

Multiply each term by the LCD, $x(x-1)$.

$$x(x-1)\left(\frac{2}{x}+\frac{2}{x-1}\right)=x(x-1)\cdot 1$$
$$2(x-1)+2x=x^2-x$$
$$2x-2+2x=x^2-x$$
$$0=x^2-5x+2$$

Use the quadratic formula.

$$x=\frac{-b\pm\sqrt{b^2-4ac}}{2a}$$
$$x=\frac{-(-5)\pm\sqrt{(-5)^2-4(1)(2)}}{2(1)}$$
$$=\frac{5\pm\sqrt{25-8}}{2}=\frac{5\pm\sqrt{17}}{2}$$
$$x=\frac{5+\sqrt{17}}{2}\approx 4.6 \text{ or}$$
$$x=\frac{5-\sqrt{17}}{2}\approx 0.4$$

Reject 0.4 as the time for the old machine, because that would yield a negative time for the new machine. Thus, the old machine takes about 4.6 hours.

28. Let $x =$ the time for Claude alone and
$x-1 =$ the time for Zoran alone.

Worker	Rate	Time Together	Fractional Part of the Job Done
Claude	$\frac{1}{x}$	1.5	$\frac{1.5}{x}$
Zoran	$\frac{1}{x-1}$	1.5	$\frac{1.5}{x-1}$

Part by Claude plus part by Zoran equals 1 whole job.

$$\frac{1.5}{x} + \frac{1.5}{x-1} = 1$$

Multiply each term by the LCD, $x(x-1)$.

$$x(x-1)\left(\frac{1.5}{x}+\frac{1.5}{x-1}\right)=x(x-1)\cdot 1$$
$$1.5(x-1)+1.5x=x^2-x$$
$$1.5x-1.5+1.5x=x^2-x$$
$$0=x^2-4x+1.5$$

Use the quadratic formula.

$$x = \frac{-b \pm \sqrt{b^2 - 4ac}}{2a}$$
$$x = \frac{-(-4) \pm \sqrt{(-4)^2 - 4(1)(1.5)}}{2(1)}$$
$$= \frac{4 \pm \sqrt{16 - 6}}{2} = \frac{4 \pm \sqrt{10}}{2}$$
$$x = \frac{4 + \sqrt{10}}{2} \approx 3.6 \text{ or}$$
$$x = \frac{4 - \sqrt{10}}{2} \approx 0.4$$

Reject 0.4 as Claude's time, because that would make Zoran's time negative. Thus, Claude's time alone is about 3.6 hours and Zoran's time alone is about $x - 1 = 2.6$ hours.

29. Solve $k = \dfrac{rF}{wv^2}$ for v.

Multiply each side by v^2, then divide by k.

$$v^2 = \frac{rF}{kw}$$
$$v = \pm\sqrt{\frac{rF}{kw}} = \frac{\pm\sqrt{rF}}{\sqrt{kw}}$$
$$= \frac{\pm\sqrt{rF}}{\sqrt{kw}} \cdot \frac{\sqrt{kw}}{\sqrt{kw}} = \frac{\pm\sqrt{rFkw}}{kw}$$

30. Solve $p = \sqrt{\dfrac{yz}{6}}$ for y.

$$p^2 = \left(\sqrt{\frac{yz}{6}}\right)^2 \qquad \textit{Square each side.}$$
$$p^2 = \frac{yz}{6}$$
$$\frac{6p^2}{z} = y, \text{ or } y = \frac{6p^2}{z}$$

31. Solve $mt^2 = 3mt + 6$ for t.

$$mt^2 - 3mt - 6 = 0$$

Use the quadratic formula with $a = m$, $b = -3m$, and $c = -6$.

$$t = \frac{-b \pm \sqrt{b^2 - 4ac}}{2a}$$
$$t = \frac{3m \pm \sqrt{(-3m)^2 - 4(m)(-6)}}{2m}$$
$$= \frac{3m \pm \sqrt{9m^2 + 24m}}{2m}$$

32. Let $x =$ the length of the longer leg;
$\frac{3}{4}x =$ the length of the shorter leg;
$2x - 9 =$ the length of the hypotenuse.

Use the Pythagorean theorem.

$$c^2 = a^2 + b^2$$
$$(2x - 9)^2 = x^2 + \left(\tfrac{3}{4}x\right)^2$$
$$4x^2 - 36x + 81 = x^2 + \tfrac{9}{16}x^2$$
$$16\left(4x^2 - 36x + 81\right) = 16\left(x^2 + \tfrac{9}{16}x^2\right)$$
$$64x^2 - 576x + 1296 = 16x^2 + 9x^2$$
$$39x^2 - 576x + 1296 = 0$$
$$13x^2 - 192x + 432 = 0 \qquad \textit{Divide by 3.}$$
$$(13x - 36)(x - 12) = 0$$

$$13x - 36 = 0 \quad \text{or} \quad x - 12 = 0$$
$$x = \tfrac{36}{13} \quad \text{or} \quad x = 12$$

Reject $x = \frac{36}{13}$ since $2\left(\frac{36}{13}\right) - 9$ is negative.
If $x = 12$, then

$$\tfrac{3}{4}x = \tfrac{3}{4}(12) = 9$$

and

$$2x - 9 = 2(12) - 9 = 15.$$

The lengths of the three sides are 9 feet, 12 feet, and 15 feet.

33. Let $x =$ the amount removed from one dimension.

The area of the square is 256 cm^2, so the length of one side is $\sqrt{256}$ or 16 cm. The dimensions of the new rectangle are $16 + x$ and $16 - x$ cm. The area of the new rectangle is 16 cm^2 less than the area of the square.

$$(16 + x)(16 - x) = 256 - 16$$
$$256 - x^2 = 240$$
$$-x^2 = -16$$
$$x^2 = 16$$
$$x = \pm\sqrt{16} = \pm 4$$

Length cannot be negative, so reject -4. If $x = 4$, then $16 + x = 20$, and $16 - x = 12$. The dimensions are 20 cm by 12 cm.

34. Let $x =$ the width of the border.

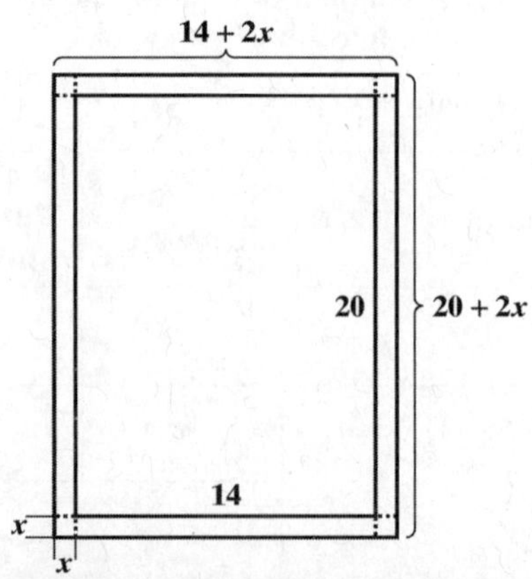

Area of mat = length • width

$$352 = (2x + 20)(2x + 14)$$
$$352 = 4x^2 + 68x + 280$$
$$0 = 4x^2 + 68x - 72$$
$$0 = x^2 + 17x - 18$$
$$0 = (x + 18)(x - 1)$$

$x + 18 = 0$ or $x - 1 = 0$
$x = -18$ or $x = 1$

Reject the negative answer for length. The mat is 1 inch wide.

35. $V = 3(x - 6)^2$

$432 = 3(x - 6)^2$ *Let V = 432.*

$144 = (x - 6)^2$ *Divide by 3.*

Use the square root property.

$\pm\sqrt{144} = x - 6$

$6 \pm 12 = x$

$x = 6 + 12 = 18$

or $x = 6 - 12 = -6$

Discard the negative solution since x represents the length. The original length is 18 inches.

36. $f(t) = -16t^2 + 45t + 400$

$200 = -16t^2 + 45t + 400$ *Let f(t) = 200.*

$0 = -16t^2 + 45t + 200$

$0 = 16t^2 - 45t - 200$

Here $a = 16$, $b = -45$, and $c = -200$.

$$t = \frac{-b \pm \sqrt{b^2 - 4ac}}{2a}$$
$$t = \frac{-(-45) \pm \sqrt{(-45)^2 - 4(16)(-200)}}{2(16)}$$
$$= \frac{45 \pm \sqrt{2025 + 12{,}800}}{32}$$
$$= \frac{45 \pm \sqrt{14{,}825}}{32}$$
$$t = \frac{45 + \sqrt{14{,}825}}{32} \approx 5.2 \quad \text{or}$$
$$t = \frac{45 - \sqrt{14{,}825}}{32} \approx -2.4$$

Reject the negative solution since time cannot be negative. The ball will reach a height of 200 ft above the ground after about 5.2 seconds.

37. $f(t) = 100t^2 - 300t$ is the distance of the light from the starting point at t minutes. When the light returns to the starting point, the value of $f(t)$ will be 0.

$0 = 100t^2 - 300t$
$0 = t^2 - 3t$ *Divide by 100.*
$0 = t(t - 3)$

$t = 0$ or $t - 3 = 0$
$t = 3$

Since $t = 0$ represents the starting time, the light will return to the starting point in 3 minutes.

38. **(a)** Use $f(x) = 230.5x^2 - 252.9x + 5987$ with $x = 7$.

$$f(7) = 230.5(7)^2 - 252.9(7) + 5987$$
$$= 15{,}511.2 \approx 15{,}511$$

The value of \$15,511 million is close to the number suggested by the graph.

(b) Let $f(x) = 14{,}000$.

$$14{,}000 = 230.5x^2 - 252.9x + 5987$$
$$0 = 230.5x^2 - 252.9x - 8013$$

Here $a = 230.5$, $b = -252.9$, and $c = -8013$.

$$x = \frac{-b \pm \sqrt{b^2 - 4ac}}{2a}$$
$$x = \frac{-(-252.9) \pm \sqrt{(-252.9)^2 - 4(230.5)(-8013)}}{2(230.5)}$$
$$= \frac{252.9 \pm \sqrt{7{,}451{,}944.41}}{461}$$
$$x = \frac{252.9 + \sqrt{7{,}451{,}944.41}}{461} \approx 6.47 \quad \text{or}$$
$$x = \frac{252.9 - \sqrt{7{,}451{,}944.41}}{461} \approx -5.37$$

Since time cannot be negative, the correct answer is $x \approx 6$, which represents 2006. Based on the graph, the revenue in 2006 was closer to \$13,000 million than \$14,000 million.

39. $f(x) = -(x - 1)^2$
Write in $y = a(x - h)^2 + k$ form as $y = -1(x - 1)^2 + 0$. The vertex (h, k) is $(1, 0)$.

40. $f(x) = (x - 3)^2 + 7$
The equation is in the form $y = a(x - h)^2 + k$, so the vertex (h, k) is $(3, 7)$.

41. $x = (y - 3)^2 - 4$

When $y = 3$, $x = -4$, so the vertex is $(-4, 3)$.

42. $y = f(x) = -3x^2 + 4x - 2$

Use the vertex formula with $a = -3$ and $b = 4$.

The x-coordinate of the vertex is

$$\frac{-b}{2a} = \frac{-4}{2(-3)} = \frac{2}{3}.$$

The y-coordinate of the vertex is

$$\begin{aligned} f\left(\frac{-b}{2a}\right) &= f\left(\frac{2}{3}\right) \\ &= -3\left(\tfrac{2}{3}\right)^2 + 4\left(\tfrac{2}{3}\right) - 2 \\ &= -\tfrac{4}{3} + \tfrac{8}{3} - 2 \\ &= -\tfrac{2}{3}. \end{aligned}$$

The vertex is $\left(\frac{2}{3}, -\frac{2}{3}\right)$.

43. $y = 2(x-2)^2 - 3$

The graph opens up since $a = 2 > 0$.

The vertex is $(2, -3)$ and the axis is $x = 2$. The domain is $(-\infty, \infty)$. The smallest y-value is -3, so the range is $[-3, \infty)$.

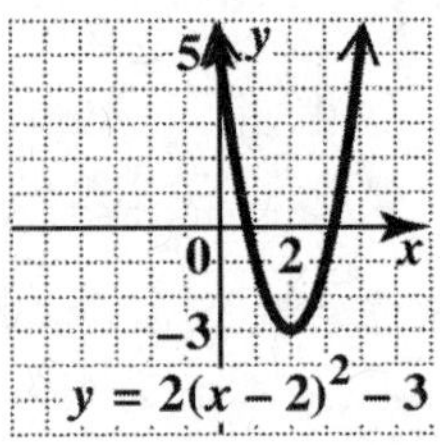

44. $f(x) = -2x^2 + 8x - 5$

Complete the square to find the vertex.

$$\begin{aligned} f(x) &= -2(x^2 - 4x) - 5 \\ &= -2(x^2 - 4x + 4 - 4) - 5 \\ &= -2(x^2 - 4x + 4) - 2(-4) - 5 \\ &= -2(x-2)^2 + 8 - 5 \\ &= -2(x-2)^2 + 3 \end{aligned}$$

The equation is in the form $y = a(x-h)^2 + k$, so the vertex (h, k) is $(2, 3)$ and the axis is $x = 2$. Here, $a = -2 < 0$, so the parabola opens down.

Also, $|a| = |-2| = 2 > 1$, so the graph is narrower than the graph of $y = x^2$. The points $(0, -5)$, $(1, 1)$, and $(3, 1)$ are on the graph.

$f(x) = -2x^2 + 8x - 5$

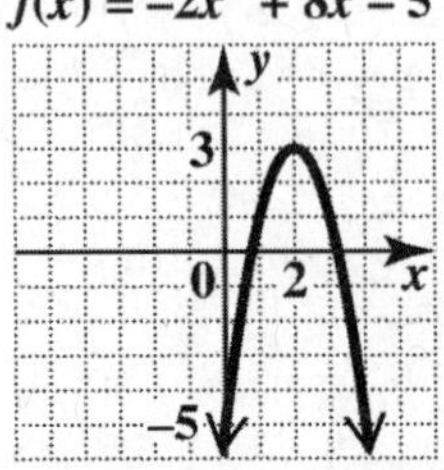

The domain is $(-\infty, \infty)$. The largest y-value is 3, so the range is $(-\infty, 3]$.

45. $x = 2(y+3)^2 - 4$

Since the roles of x and y are reversed, this is a horizontal parabola.

$x = 2[y - (-3)]^2 + (-4)$

The equation is in the form

$$x = a(y-k)^2 + h,$$

so the vertex (h, k) is $(-4, -3)$ and the axis is $y = -3$. Here, $a = 2 > 0$, so the parabola opens to the right and is narrower than the graph of $y = x^2$.

Two other points on the graph are $(4, -1)$ and $(4, -5)$.

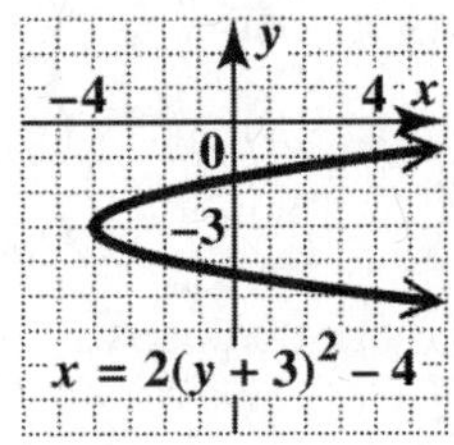

The smallest x-value is -4, so the domain is $[-4, \infty)$. The range is $(-\infty, \infty)$.

46. $x = -\frac{1}{2}y^2 + 6y - 14$

Since the roles of x and y are reversed, this is a horizontal parabola. Complete the square to find the vertex.

$$\begin{aligned} x &= -\tfrac{1}{2}y^2 + 6y - 14 \\ &= -\tfrac{1}{2}(y^2 - 12y) - 14 \\ &= -\tfrac{1}{2}(y^2 - 12y + 36 - 36) - 14 \\ &= -\tfrac{1}{2}(y^2 - 12y + 36) - \tfrac{1}{2}(-36) - 14 \\ &= -\tfrac{1}{2}(y-6)^2 + 18 - 14 \\ x &= -\tfrac{1}{2}(y-6)^2 + 4 \end{aligned}$$

The equation is in the form $x = a(y-k)^2 + h$, so the vertex (h, k) is $(4, 6)$ and the axis is $y = 6$. Here, $a = -\frac{1}{2} < 0$, so the parabola opens to the left.

Also, $|a| = \left|-\frac{1}{2}\right| = \frac{1}{2} < 1$, so the graph is wider than the graph of $y = x^2$. The points $(-14, 0)$, $(2, 4)$, and $(2, 8)$ are on the graph.

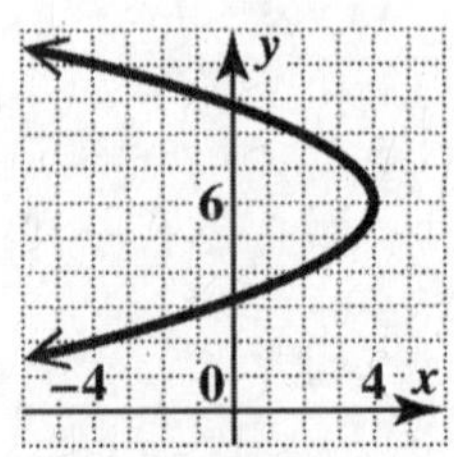

$x = -\frac{1}{2}y^2 + 6y - 14$

The largest x-value is 4, so the domain is $(-\infty, 4]$. The range is $(-\infty, \infty)$.

47. **(a)** Use $ax^2 + bx + c = y$ with $(0, 2.9)$, $(10, 24.3)$, and $(20, 56.5)$.

$$\begin{aligned} c &= 2.9 & (1)\\ 100a + 10b + c &= 24.3 & (2)\\ 400a + 20b + c &= 56.5 & (3) \end{aligned}$$

(b) Rewrite equations (2) and (3) with $c = 2.9$.

$$\begin{aligned} 100a + 10b + 2.9 &= 24.3\\ 400a + 20b + 2.9 &= 56.5 \end{aligned}$$

$$\begin{aligned} 100a + 10b &= 21.4 & (4)\\ 400a + 20b &= 53.6 & (5) \end{aligned}$$

Now eliminate b.

$$\begin{array}{rcrcrl} -200a & - & 20b & = & -42.8 & -2 \times (4)\\ 400a & + & 20b & = & 53.6 & (5)\\ \hline 200a & & & = & 10.8 & \\ & & a & = & \frac{10.8}{200} = 0.054 & \end{array}$$

Use (4) to find b.

$$\begin{aligned} 100a + 10b &= 21.4 \quad (4)\\ 100(0.054) + 10b &= 21.4\\ 5.4 + 10b &= 21.4\\ 10b &= 16\\ b &= \tfrac{16}{10} = 1.6 \end{aligned}$$

Thus, we get the quadratic function

$$f(x) = 0.054x^2 + 1.6x + 2.9.$$

(c) Use $f(x) = 0.054x^2 + 1.6x + 2.9$ with $x = 21$ for the year 2006.

$$\begin{aligned} f(21) &= 0.054(21)^2 + 1.6(21) + 2.9\\ &= 60.314 \approx \$60.3 \text{ billion} \end{aligned}$$

The result using the model is close to the table value of \$61.4 billion, but slightly low.

48. $s(t) = -16t^2 + 160t$
The equation represents a parabola. Since $a = -16 < 0$, the parabola opens down. The time and maximum height occur at the vertex of the parabola, given by

$$\left(\frac{-b}{2a}, s\left(\frac{-b}{2a}\right)\right).$$

Using the standard form of the equation, $a = -16$ and $b = 160$, so

$$t = \frac{-b}{2a} = \frac{-160}{2(-16)} = 5,$$

$$\begin{aligned} \text{and} \quad s(t) &= -16(5)^2 + 160(5)\\ &= -400 + 800 = 400. \end{aligned}$$

The vertex is $(5, 400)$. The time at which the maximum height is reached is 5 seconds. The maximum height is 400 feet.

49. Let L = the length of the rectangle and W = the width.

The perimeter of the rectangle is 200 m, so

$$\begin{aligned} 2L + 2W &= 200\\ 2W &= 200 - 2L\\ W &= 100 - L. \end{aligned}$$

Since the area is length times width, substitute $100 - L$ for W.

$$\begin{aligned} A &= LW\\ &= L(100 - L)\\ &= 100L - L^2 \quad \text{or} \quad -L^2 + 100L \end{aligned}$$

Use the vertex formula.

$$L = \frac{-b}{2a} = \frac{-100}{2(-1)} = 50$$

So $L = 50$ meters and
$W = 100 - L = 100 - 50 = 50$ meters.
The maximum area is

$$50 \cdot 50 = 2500 \text{ m}^2.$$

50. $(x - 4)(2x + 3) > 0$
Solve the equation
$(x - 4)(2x + 3) = 0$.

$$\begin{aligned} x - 4 = 0 \quad &\text{or} \quad 2x + 3 = 0\\ x = 4 \quad &\text{or} \quad x = -\tfrac{3}{2} \end{aligned}$$

The numbers $-\frac{3}{2}$ and 4 divide a number line into three intervals.

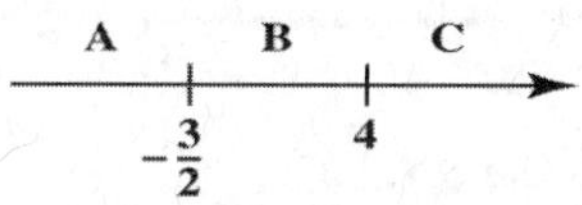

Test a number from each interval in the inequality

$$(x - 4)(2x + 3) > 0.$$

Interval A: Let $x = -2$.

$$-6(-1) \overset{?}{>} 0$$
$$6 > 0 \qquad \textit{True}$$

Interval B: Let $x = 0$.

$$-4(3) \overset{?}{>} 0$$
$$-12 > 0 \qquad \textit{False}$$

Interval C: Let $x = 5$.

$$1(13) \overset{?}{>} 0$$
$$13 > 0 \qquad \textit{True}$$

The solution set includes numbers in Intervals A and C, excluding endpoints.

Solution set: $\left(-\infty, -\frac{3}{2}\right) \cup (4, \infty)$

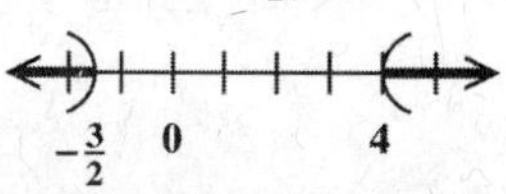

51. $x^2 + x \leq 12$

Solve the equation

$$x^2 + x = 12.$$
$$x^2 + x - 12 = 0$$
$$(x + 4)(x - 3) = 0$$

$$x + 4 = 0 \quad \text{or} \quad x - 3 = 0$$
$$x = -4 \quad \text{or} \quad x = 3$$

The numbers -4 and 3 divide a number line into three intervals.

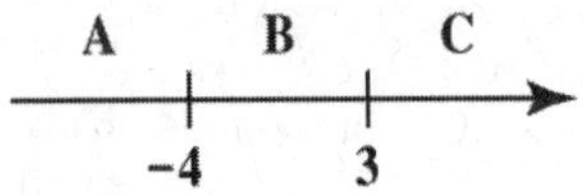

Test a number from each interval in the inequality

$$x^2 + x \leq 12.$$

Interval A: Let $x = -5$.
$$25 - 5 \overset{?}{\leq} 12$$
$$20 \leq 12 \quad \textit{False}$$

Interval B: Let $x = 0$.
$$0 \leq 12 \quad \textit{True}$$

Interval C: Let $x = 4$.
$$16 + 4 \overset{?}{\leq} 12$$
$$20 \leq 12 \quad \textit{False}$$

The numbers in Interval B, including -4 and 3, are solutions.

Solution set: $[-4, 3]$

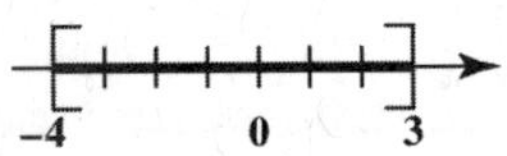

52. $(x + 2)(x - 3)(x + 5) \leq 0$

The numbers -2, 3, and -5 are solutions of the cubic equation

$$(x + 2)(x - 3)(x + 5) = 0.$$

These numbers divide a number line into four intervals.

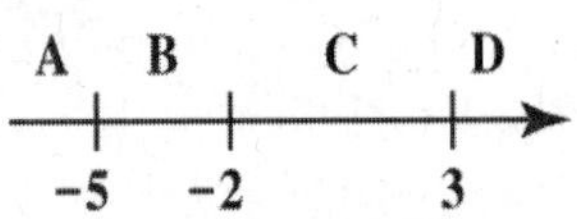

Test a number from each interval in the inequality

$$(x + 2)(x - 3)(x + 5) \leq 0.$$

Interval A: Let $x = -6$.
$$(-4)(-9)(-1) \leq 0 \quad \textit{True}$$

Interval B: Let $x = -3$.
$$(-5)(-6)(2) \leq 0 \quad \textit{False}$$

Interval C: Let $x = 0$.
$$(2)(-3)(5) \leq 0 \quad \textit{True}$$

Interval D: Let $x = 4$.
$$(6)(1)(9) \leq 0 \quad \textit{False}$$

The numbers in Intervals A and C, including -5, -2, and 3, are solutions.

Solution set: $(-\infty, -5] \cup [-2, 3]$

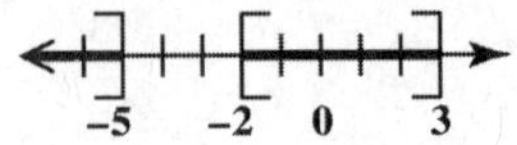

53. $(4x + 3)^2 \leq -4$

The square of a real number is never negative. So, the solution set of this inequality is $\emptyset$.

54. $\dfrac{6}{2z - 1} < 2$

Write the inequality so that 0 is on one side.

$$\frac{6}{2z - 1} - 2 < 0$$
$$\frac{6}{2z - 1} - \frac{2(2z - 1)}{2z - 1} < 0$$
$$\frac{6 - 4z + 2}{2z - 1} < 0$$
$$\frac{-4z + 8}{2z - 1} < 0$$

The number 2 makes the numerator 0, and $\frac{1}{2}$ makes the denominator 0. These two numbers determine three intervals.

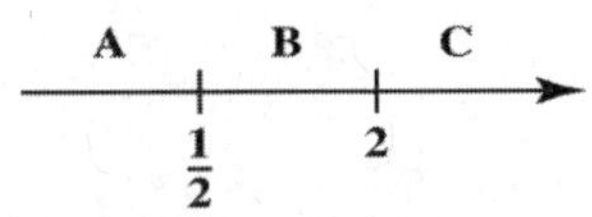

Test a number from each interval in the inequality

$$\frac{6}{2z - 1} < 2.$$

Interval A: Let $z = 0$.
$$-6 < 2 \quad \textit{True}$$

Interval B: Let $z = 1$.
$$6 < 2 \quad \textit{False}$$

Interval C: Let $z = 3$.
$$\frac{6}{5} < 2 \quad \textit{True}$$

The solution set includes numbers in Intervals A and C, excluding endpoints.

Solution set: $(-\infty, \frac{1}{2}) \cup (2, \infty)$

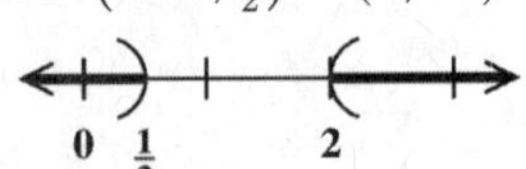

55. $\dfrac{3t + 4}{t - 2} \leq 1$

Write the inequality so that 0 is on one side.

$$\frac{3t + 4}{t - 2} - 1 \leq 0$$
$$\frac{3t + 4}{t - 2} - \frac{1(t - 2)}{t - 2} \leq 0$$
$$\frac{3t + 4 - t + 2}{t - 2} \leq 0$$
$$\frac{2t + 6}{t - 2} \leq 0$$

The number -3 makes the numerator 0, and 2 makes the denominator 0. These two numbers determine three intervals.

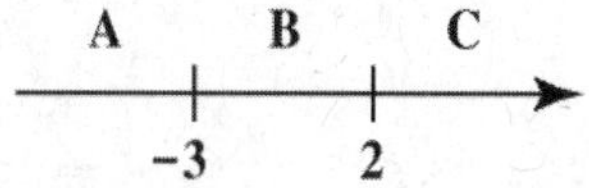

Test a number from each interval in the inequality

$$\frac{3t+4}{t-2} \leq 1.$$

Interval A: Let $t = -4$.

$$\tfrac{-8}{-6} \overset{?}{\leq} 1$$

$$\tfrac{4}{3} \leq 1 \qquad \textit{False}$$

Interval B: Let $t = 0$.

$$\tfrac{4}{-2} \overset{?}{\leq} 1$$

$$-2 \leq 1 \qquad \textit{True}$$

Interval C: Let $t = 3$.

$$\tfrac{13}{1} \overset{?}{\leq} 1$$

$$13 \leq 1 \qquad \textit{False}$$

The numbers in Interval B, including -3 but not 2, are solutions.

Solution set: $[-3, 2)$

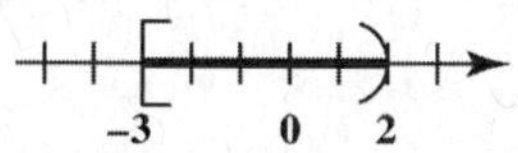

56. **[11.5]** Solve $V = r^2 + R^2h$ for R.

$$V - r^2 = R^2h$$

$$R^2h = V - r^2$$

$$R^2 = \frac{V - r^2}{h}$$

$$R = \pm\sqrt{\frac{V - r^2}{h}} = \frac{\pm\sqrt{V - r^2}}{\sqrt{h}}$$

$$= \frac{\pm\sqrt{V - r^2}}{\sqrt{h}} \cdot \frac{\sqrt{h}}{\sqrt{h}}$$

$$= \frac{\pm\sqrt{Vh - r^2h}}{h}$$

57. **[11.3]**

$$3t^2 - 6t = -4$$

$$3t^2 - 6t + 4 = 0$$

Use the quadratic formula.

$$t = \frac{-b \pm \sqrt{b^2 - 4ac}}{2a}$$

$$t = \frac{-(-6) \pm \sqrt{(-6)^2 - 4(3)(4)}}{2(3)}$$

$$= \frac{6 \pm \sqrt{-12}}{6} = \frac{6 \pm 2i\sqrt{3}}{6}$$

$$= \frac{2\left(3 \pm i\sqrt{3}\right)}{6} = \frac{3 \pm i\sqrt{3}}{3} = 1 \pm \frac{\sqrt{3}}{3}i$$

Solution set: $\left\{1 \pm \frac{\sqrt{3}}{3}i\right\}$

58. **[11.1]** $(3x + 11)^2 = 7$

$$3x + 11 = \sqrt{7} \quad \text{or} \quad 3x + 11 = -\sqrt{7}$$

$$3x = -11 + \sqrt{7} \qquad 3x = -11 - \sqrt{7}$$

$$x = \frac{-11 + \sqrt{7}}{3} \quad \text{or} \quad x = \frac{-11 - \sqrt{7}}{3}$$

Solution set: $\left\{\frac{-11 \pm \sqrt{7}}{3}\right\}$

59. **[11.5]** Solve $S = \dfrac{Id^2}{k}$ for d.

Multiply each side by k, then divide by I.

$$\frac{Sk}{I} = d^2$$

$$d = \pm\sqrt{\frac{Sk}{I}} = \frac{\pm\sqrt{Sk}}{\sqrt{I}}$$

$$= \frac{\pm\sqrt{Sk}}{\sqrt{I}} \cdot \frac{\sqrt{I}}{\sqrt{I}} = \frac{\pm\sqrt{SkI}}{I}$$

60. **[11.8]** $(8x - 7)^2 \geq -1$

The square of any real number is always greater than or equal to 0, so any real number satisfies this inequality.

Solution set: $(-\infty, \infty)$

61. **[11.4]**

$$2x - \sqrt{x} = 6$$

$$2x - \sqrt{x} - 6 = 0$$

Let $u = \sqrt{x}$, so $u^2 = x$.

$$2u^2 - u - 6 = 0$$

$$(2u + 3)(u - 2) = 0$$

$$2u + 3 = 0 \quad \text{or} \quad u - 2 = 0$$

$$u = -\tfrac{3}{2} \quad \text{or} \quad u = 2$$

$$\sqrt{x} = -\tfrac{3}{2} \quad \text{or} \quad \sqrt{x} = 2$$

Since $\sqrt{x} \geq 0$, we must have $\sqrt{x} = 2$, or $x = 4$.

Check $x = 4$: $8 - 2 = 6$ *True*

Solution set: $\{4\}$

62. **[11.4]**

$$x^4 - 8x^2 = -1$$

$$x^4 - 8x^2 + 1 = 0$$

Use the quadratic formula with $a = 1$, $b = -8$, and $c = 1$ to solve for x^2, not x.

$$x^2 = \frac{-b \pm \sqrt{b^2 - 4ac}}{2a}$$

$$x^2 = \frac{-(-8) \pm \sqrt{(-8)^2 - 4(1)(1)}}{2(1)}$$

$$= \frac{8 \pm \sqrt{64 - 4}}{2} = \frac{8 \pm \sqrt{60}}{2}$$

$$= \frac{8 \pm 2\sqrt{15}}{2} = 4 \pm \sqrt{15}$$

Both $4 + \sqrt{15}$ and $4 - \sqrt{15}$ are positive and can be set equal to x^2.

$$x^2 = 4 + \sqrt{15} \quad \text{or} \quad x^2 = 4 - \sqrt{15}$$
$$x = \pm\sqrt{4 + \sqrt{15}} \qquad x = \pm\sqrt{4 - \sqrt{15}}$$

The proposed solutions all check.

Solution set: $\left\{\pm\sqrt{4+\sqrt{15}}, \pm\sqrt{4-\sqrt{15}}\right\}$

63. [11.8] $\dfrac{-2}{x+5} \le -5$

Write the inequality so that 0 is on one side.

$$\frac{-2}{x+5} + 5 \le 0$$
$$\frac{-2}{x+5} + \frac{5(x+5)}{x+5} \le 0$$
$$\frac{-2 + 5x + 25}{x+5} \le 0$$
$$\frac{5x+23}{x+5} \le 0$$

The number $-\frac{23}{5}$ makes the numerator 0, and -5 makes the denominator 0. These two numbers determine three intervals.

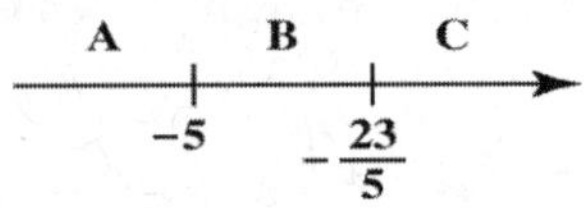

Test a number from each interval in the inequality

$$\frac{-2}{x+5} \le -5.$$

Interval A: Let $x = -6$.

$$\frac{-2}{-1} \overset{?}{\le} -5$$
$$2 \le -5 \qquad \textit{False}$$

Interval B: Let $x = -\frac{24}{5}$.

$$\frac{-2}{\frac{1}{5}} \overset{?}{\le} -5$$
$$-10 \le -5 \qquad \textit{True}$$

Interval C: Let $x = 0$.

$$\tfrac{-2}{5} \le -5 \qquad \textit{False}$$

The numbers in Interval B, including $-\frac{23}{5}$ but not -5, are solutions.

Solution set: $\left(-5, -\frac{23}{5}\right]$

64. [11.4] $6 + \dfrac{15}{s^2} = -\dfrac{19}{s}$

Multiply each term by the LCD, s^2.

$$s^2\left(6 + \frac{15}{s^2}\right) = s^2\left(-\frac{19}{s}\right)$$
$$6s^2 + 15 = -19s$$
$$6s^2 + 19s + 15 = 0$$
$$(3s+5)(2s+3) = 0$$
$$3s + 5 = 0 \quad \text{or} \quad 2s + 3 = 0$$
$$s = -\tfrac{5}{3} \quad \text{or} \quad s = -\tfrac{3}{2}$$

Check $s = -\frac{5}{3}$: $6 + \frac{27}{5} = \frac{57}{5}$ *True*

Check $s = -\frac{3}{2}$: $6 + \frac{20}{3} = \frac{38}{3}$ *True*

Solution set: $\left\{-\frac{5}{3}, -\frac{3}{2}\right\}$

65. [11.4] $(x^2 - 2x)^2 = 11(x^2 - 2x) - 24$

Let $u = x^2 - 2x$, so $u^2 = (x^2 - 2x)^2$.

$$u^2 = 11u - 24$$
$$u^2 - 11u + 24 = 0$$
$$(u-8)(u-3) = 0$$
$$u - 8 = 0 \quad \text{or} \quad u - 3 = 0$$
$$u = 8 \quad \text{or} \quad u = 3$$

To find x, substitute $x^2 - 2x$ for u.

$$x^2 - 2x = 8 \quad \text{or} \quad x^2 - 2x = 3$$
$$x^2 - 2x - 8 = 0 \qquad x^2 - 2x - 3 = 0$$
$$(x-4)(x+2) = 0 \qquad (x-3)(x+1) = 0$$
$$x - 4 = 0 \text{ or } x + 2 = 0 \qquad x - 3 = 0 \text{ or } x + 1 = 0$$
$$x = 4 \text{ or } x = -2 \text{ or } x = 3 \text{ or } x = -1$$

The proposed solutions all check.

Solution set: $\{-2, -1, 3, 4\}$

66. [11.8] $(r-1)(2r+3)(r+6) < 0$

Solve the equation

$(r-1)(2r+3)(r+6) = 0$.

$$r - 1 = 0 \quad \text{or} \quad 2r + 3 = 0 \quad \text{or} \quad r + 6 = 0$$
$$r = 1 \quad \text{or} \quad r = -\tfrac{3}{2} \quad \text{or} \quad r = -6$$

The numbers -6, $-\frac{3}{2}$, and 1 divide a number line into four intervals.

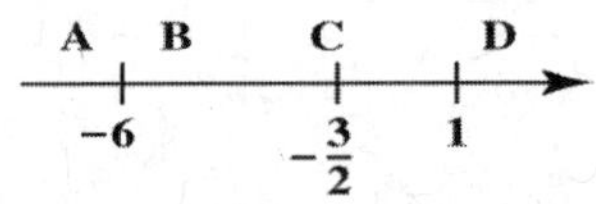

Test a number from each interval in the inequality

$$(r-1)(2r+3)(r+6) < 0.$$

Interval A: Let $r = -7$.

$$-8(-11)(-1) \overset{?}{<} 0$$
$$-88 < 0 \qquad \textit{True}$$

Interval B: Let $r = -2$.

$$-3(-1)(4) \overset{?}{<} 0$$
$$12 < 0 \qquad \textit{False}$$

Interval C: Let $r = 0$.

$$-1(3)(6) \overset{?}{<} 0$$
$$-18 < 0 \qquad \textit{True}$$

Interval D: Let $r = 2$.

$$1(7)(8) \overset{?}{<} 0$$
$$56 < 0 \qquad \textit{False}$$

The numbers in Intervals A and C, not including -6, $-\frac{3}{2}$, or 1, are solutions.

Solution set: $(-\infty, -6) \cup \left(-\frac{3}{2}, 1\right)$

67. **[11.6]** **(a)** $g(x) = x^2 - 5$ written in the form
$g(x) = a(x-h)^2 + k$ is
$g(x) = 1(x-0)^2 + (-5)$.

Here, $h = 0$ and $k = -5$, so the vertex (h, k) is $(0, -5)$. The graph is shifted 5 units down from the graph of $f(x) = x^2$. Since $a = 1 > 0$, the graph opens up. The correct figure is **F**.

(b) $h(x) = -x^2 + 4$
Because the coefficient of x^2 is negative, the parabola opens down. Because $k = 4$, the shift is upward 4 units. The figure that most closely resembles the graph of $h(x)$ is Choice **B**.

(c) $F(x) = (x-1)^2$ written in the form
$F(x) = a(x-h)^2 + k$ is
$F(x) = 1(x-1)^2 + 0$.

Here, $h = 1$ and $k = 0$, so the vertex (h, k) is $(1, 0)$. The graph is shifted 1 unit to the right of the graph of $f(x) = x^2$. Since $a = 1 > 0$, the graph opens up. The correct figure is **C**.

(d) $G(x) = (x+1)^2$
Because the coefficient of the x^2-term is positive, the graph opens up. Because $h = -1$, the shift is to the left 1 unit. The figure that most closely resembles the graph of $G(x)$ is Choice **A**.

(e) $H(x) = (x-1)^2 + 1$ is written in the form
$H(x) = a(x-h)^2 + k$.

Here, $h = 1$ and $k = 1$, so the vertex (h, k) is $(1, 1)$. The graph is shifted 1 unit to the right and 1 unit up from the graph of $f(x) = x^2$. Since $a = 1 > 0$, the graph opens up. The correct figure is **E**.

(f) $K(x) = (x+1)^2 + 1$
Because $h = -1$ and $k = 1$, the shift is to the left 1 unit and up 1 unit. The figure that most closely resembles the graph of $K(x)$ is choice **D**.

68. **[11.7]** $y = f(x) = 4x^2 + 4x - 2$
Complete the square to find the vertex.

$$\begin{aligned} y &= 4(x^2 + x) - 2 \\ &= 4(x^2 + x + \tfrac{1}{4} - \tfrac{1}{4}) - 2 \\ &= 4(x^2 + x + \tfrac{1}{4}) + 4(-\tfrac{1}{4}) - 2 \\ &= 4(x + \tfrac{1}{2})^2 - 1 - 2 \\ &= 4(x + \tfrac{1}{2})^2 - 3 \\ y &= 4(x + \tfrac{1}{2})^2 - 3 \end{aligned}$$

The equation is now in the form $y = a(x-h)^2 + k$, so the vertex (h, k) is $(-\frac{1}{2}, -3)$ and the axis is $x = -\frac{1}{2}$. Since $a = 4 > 0$, the parabola opens up. Also, $|a| = |4| = 4 > 1$, so the graph is narrower than the graph of $y = x^2$. The points $(-2, 6)$, $(0, -2)$, and $(1, 6)$ are on the graph.

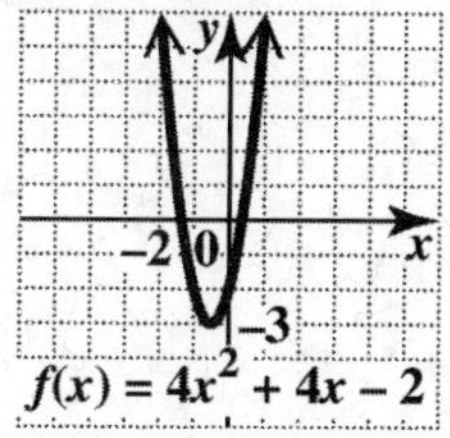

The domain is $(-\infty, \infty)$. The smallest y-value is -3, so the range is $[-3, \infty)$.

69. **[11.5]** Let x = the rate of the boat in still water. Use $t = d/r$ to make a table.

	Distance	Rate	Time
Upstream	15	$x - 5$	$\dfrac{15}{x-5}$
Downstream	15	$x + 5$	$\dfrac{15}{x+5}$

The total time is 4 hours.

$$\frac{15}{x-5} + \frac{15}{x+5} = 4$$

Multiply each term by the LCD, $(x+5)(x-5)$.

$$\begin{aligned} 15(x+5) + 15(x-5) &= 4(x+5)(x-5) \\ 15x + 75 + 15x - 75 &= 4x^2 - 100 \\ 0 &= 4x^2 - 30x - 100 \\ 0 &= 2x^2 - 15x - 50 \\ 0 &= (2x+5)(x-10) \end{aligned}$$

$$2x + 5 = 0 \quad \text{or} \quad x - 10 = 0$$
$$x = -\tfrac{5}{2} \quad \text{or} \quad x = 10$$

The rate must be positive, so check $x = 10$. The 15-mile trip upstream would take 3 hours at 5 mph. The trip downstream would take 1 hour at 15 mph. Thus, the total time would be 4 hours and we have the rate of the boat in still water equal to 10 mph.

70. **[11.5]** Let x = the width of the rectangle.
Then $2x - 1$ = the length of the rectangle.

Use the Pythagorean theorem.

$$\begin{aligned} x^2 + (2x-1)^2 &= (2.5)^2 \\ x^2 + 4x^2 - 4x + 1 &= 6.25 \\ 5x^2 - 4x - 5.25 &= 0 \end{aligned}$$

Multiply by 4.

$$\begin{aligned} 20x^2 - 16x - 21 &= 0 \\ (2x-3)(10x+7) &= 0 \end{aligned}$$

$$2x - 3 = 0 \quad \text{or} \quad 10x + 7 = 0$$
$$x = \tfrac{3}{2} \quad \text{or} \quad x = -\tfrac{7}{10}$$

Discard the negative solution.
If $x = 1.5$, then

$$2x - 1 = 2(1.5) - 1 = 2.$$

The width of the rectangle is 1.5 cm, and the length is 2 cm.

Chapter 11 Test

1. $t^2 = 54$

$$t = \sqrt{54} \quad \text{or} \quad t = -\sqrt{54}$$
$$t = 3\sqrt{6} \quad \text{or} \quad t = -3\sqrt{6}$$

Solution set: $\left\{\pm 3\sqrt{6}\right\}$

2. $(7x+3)^2 = 25$

$$7x+3 = 5 \quad \text{or} \quad 7x+3 = -5$$
$$7x = 2 \qquad\qquad 7x = -8$$
$$x = \tfrac{2}{7} \quad \text{or} \quad x = -\tfrac{8}{7}$$

Solution set: $\{-\frac{8}{7}, \frac{2}{7}\}$

3. $x^2 + 2x = 4$

$$x^2 + 2x + 1 = 4 + 1$$
$$(x+1)^2 = 5$$
$$x + 1 = \sqrt{5} \quad \text{or} \quad x + 1 = -\sqrt{5}$$
$$x = -1 + \sqrt{5} \quad \text{or} \quad x = -1 - \sqrt{5}$$

Solution set: $\left\{-1 \pm \sqrt{5}\right\}$

4. $2x^2 - 3x - 1 = 0$

Here $a = 2$, $b = -3$, and $c = -1$.

$$x = \frac{-b \pm \sqrt{b^2 - 4ac}}{2a}$$
$$x = \frac{-(-3) \pm \sqrt{(-3)^2 - 4(2)(-1)}}{2(2)}$$
$$= \frac{3 \pm \sqrt{17}}{4}$$

Solution set: $\left\{\frac{3 \pm \sqrt{17}}{4}\right\}$

5. $3t^2 - 4t = -5$

$$3t^2 - 4t + 5 = 0$$

Here $a = 3$, $b = -4$, and $c = 5$.

$$t = \frac{-b \pm \sqrt{b^2 - 4ac}}{2a}$$
$$t = \frac{-(-4) \pm \sqrt{(-4)^2 - 4(3)(5)}}{2(3)}$$
$$= \frac{4 \pm \sqrt{-44}}{6} = \frac{4 \pm 2i\sqrt{11}}{6}$$
$$= \frac{2\left(2 \pm i\sqrt{11}\right)}{6} = \frac{2 \pm i\sqrt{11}}{3}$$
$$= \frac{2}{3} \pm \frac{\sqrt{11}}{3}i$$

Solution set: $\left\{\frac{2}{3} \pm \frac{\sqrt{11}}{3}i\right\}$

6. If k is a negative number, then $4k$ is also negative, so the equation $x^2 = 4k$ will have two nonreal complex solutions. The answer is **A**.

7. $2x^2 - 8x - 3 = 0$

$$b^2 - 4ac = (-8)^2 - 4(2)(-3)$$
$$= 64 + 24 = 88$$

The discriminant, 88, is positive but not a perfect square, so there will be two distinct irrational number solutions.

8. $3x = \sqrt{\dfrac{9x+2}{2}}$

$$9x^2 = \frac{9x+2}{2} \qquad \textit{Square each side.}$$
$$18x^2 = 9x + 2$$
$$18x^2 - 9x - 2 = 0$$

Here $a = 18$, $b = -9$, and $c = -2$.

$$x = \frac{-b \pm \sqrt{b^2 - 4ac}}{2a}$$
$$x = \frac{-(-9) \pm \sqrt{(-9)^2 - 4(18)(-2)}}{2(18)}$$
$$= \frac{9 \pm \sqrt{225}}{36} = \frac{9 \pm 15}{36}$$
$$x = \frac{9+15}{36} = \frac{24}{36} = \frac{2}{3} \quad \text{or}$$
$$x = \frac{9-15}{36} = \frac{-6}{36} = -\frac{1}{6}$$

Check $x = \frac{2}{3}$: $\quad 2 = \sqrt{4} \quad$ *True*

Check $x = -\frac{1}{6}$: $\quad -\frac{1}{2} = \sqrt{\frac{1}{4}} \quad$ *False*

Solution set: $\{\frac{2}{3}\}$

9. $3 - \dfrac{16}{x} - \dfrac{12}{x^2} = 0$

Multiply each term by the LCD, x^2.

$$x^2\left(3 - \frac{16}{x} - \frac{12}{x^2}\right) = x^2 \cdot 0$$
$$3x^2 - 16x - 12 = 0$$
$$(3x+2)(x-6) = 0$$
$$3x + 2 = 0 \quad \text{or} \quad x - 6 = 0$$
$$x = -\tfrac{2}{3} \quad \text{or} \quad x = 6$$

Check $x = -\frac{2}{3}$: $\quad 3 + 24 - 27 = 0 \quad$ *True*

Check $x = 6$: $\quad 3 - \frac{8}{3} - \frac{1}{3} = 0 \quad$ *True*

Solution set: $\{-\frac{2}{3}, 6\}$

10. $4x^2 + 7x - 3 = 0$

Use the quadratic formula with $a = 4$, $b = 7$, and $c = -3$.

$$x = \frac{-b \pm \sqrt{b^2 - 4ac}}{2a}$$
$$x = \frac{-7 \pm \sqrt{7^2 - 4(4)(-3)}}{2(4)} = \frac{-7 \pm \sqrt{97}}{8}$$

Solution set: $\left\{\frac{-7 \pm \sqrt{97}}{8}\right\}$

11. $9x^4 + 4 = 37x^2$

$9x^4 - 37x^2 + 4 = 0$

Let $u = x^2$, so $u^2 = (x^2)^2 = x^4$.

$9u^2 - 37u + 4 = 0$

$(9u - 1)(u - 4) = 0$

$9u - 1 = 0$ or $u - 4 = 0$

$u = \frac{1}{9}$ or $u = 4$

To find x, substitute x^2 for u.

$x^2 = \frac{1}{9}$ or $x^2 = 4$

$x = \pm\sqrt{\frac{1}{9}}$ $\quad x = \pm\sqrt{4}$

$x = \pm\frac{1}{3}$ or $x = \pm 2$

Check $x = \pm\frac{1}{3}$: $\frac{1}{9} + 4 = \frac{37}{9}$ *True*

Check $x = \pm 2$: $144 + 4 = 37(4)$ *True*

Solution set: $\{\pm\frac{1}{3}, \pm 2\}$

12. $12 = (2n+1)^2 + (2n+1)$

Let $u = 2n + 1$, so $u^2 = (2n+1)^2$.

$12 = u^2 + u$

$0 = u^2 + u - 12$

$0 = (u+4)(u-3)$

$u + 4 = 0$ or $u - 3 = 0$

$u = -4$ or $u = 3$

To find n, substitute $2n + 1$ for u.

$2n + 1 = -4$ or $2n + 1 = 3$

$2n = -5$ $\quad 2n = 2$

$n = -\frac{5}{2}$ or $n = 1$

Check $n = -\frac{5}{2}$: $12 = 16 - 4$ *True*

Check $n = 1$: $12 = 9 + 3$ *True*

Solution set: $\{-\frac{5}{2}, 1\}$

13. Solve $S = 4\pi r^2$ for r.

$$\frac{S}{4\pi} = r^2$$

$$r = \pm\sqrt{\frac{S}{4\pi}} = \frac{\pm\sqrt{S}}{2\sqrt{\pi}}$$

$$= \frac{\pm\sqrt{S}}{2\sqrt{\pi}} \cdot \frac{\sqrt{\pi}}{\sqrt{\pi}}$$

$$r = \frac{\pm\sqrt{\pi S}}{2\pi}$$

14. Let x = Terry's time alone.
Then $x - 2$ = Callie's time alone. Make a table.

	Rate	Time Together	Fractional Part of the Job Done
Terry	$\frac{1}{x}$	5	$\frac{5}{x}$
Callie	$\frac{1}{x-2}$	5	$\frac{5}{x-2}$

Part done by Terry plus part done by Callie equals 1 whole job.

$$\frac{5}{x} + \frac{5}{x-2} = 1$$

Multiply each side by the LCD, $x(x-2)$.

$$x(x-2)\left(\frac{5}{x} + \frac{5}{x-2}\right) = x(x-2) \cdot 1$$

$$5x - 10 + 5x = x^2 - 2x$$

$$0 = x^2 - 12x + 10$$

Use the quadratic formula with $a = 1$, $b = -12$, and $c = 10$.

$$x = \frac{-b \pm \sqrt{b^2 - 4ac}}{2a}$$

$$x = \frac{-(-12) \pm \sqrt{(-12)^2 - 4(1)(10)}}{2(1)}$$

$$= \frac{12 \pm \sqrt{104}}{2} = \frac{12 \pm 2\sqrt{26}}{2}$$

$$= \frac{2\left(6 \pm \sqrt{26}\right)}{2} = 6 \pm \sqrt{26}$$

$x = 6 + \sqrt{26} \approx 11.1$ or

$x = 6 - \sqrt{26} \approx 0.9$

Reject 0.9 for Terry's time, because that would yield a negative time for Callie. Thus, Terry's time is about 11.1 hours and Callie's time is $x - 2 \approx 9.1$ hours.

15. Let x = Qihong's rate.
Make a table. Use $d = rt$, or $t = \frac{d}{r}$.

	Distance	Rate	Time
Upstream	10	$x - 3$	$\frac{10}{x-3}$
Downstream	10	$x + 3$	$\frac{10}{x+3}$

Time upstream plus Time downstream equals 3.5 hr.

$$\frac{10}{x-3} + \frac{10}{x+3} = \frac{7}{2}$$

Multiply each side by the LCD, $2(x+3)(x-3)$.

$$2(x+3)(x-3)\left(\frac{10}{x-3} + \frac{10}{x+3}\right) = 2(x+3)(x-3)\left(\frac{7}{2}\right)$$

$$20(x+3) + 20(x-3) = 7(x^2 - 9)$$

$$20x + 60 + 20x - 60 = 7x^2 - 63$$

$$0 = 7x^2 - 40x - 63$$

$$0 = (x-7)(7x+9)$$

$x - 7 = 0$ or $7x + 9 = 0$

$x = 7$ or $x = -\frac{9}{7}$

Reject $-\frac{9}{7}$ since the rate can't be negative.
Qihong's rate was 7 mph.

16. Let $x =$ the width of the walk.
The area of the walk is equal to the area of the outer figure minus the area of the pool.

$$\begin{aligned} 152 &= (10+2x)(24+2x)-(24)(10) \\ 152 &= 240+68x+4x^2-240 \\ 0 &= 4x^2+68x-152 \\ 0 &= x^2+17x-38 \\ 0 &= (x+19)(x-2) \end{aligned}$$

$$x+19=0 \quad \text{or} \quad x-2=0$$
$$x=-19 \quad \text{or} \quad x=2$$

Reject -19 since width can't be negative. The walk is 2 feet wide.

17. Let $x =$ the height of the tower. Then $2x+2 =$ the distance from the point to the top.

The distance from the base to the point is 30 m. These three segments form a right triangle, so the Pythagorean theorem applies.

$$\begin{aligned} a^2+b^2 &= c^2 \\ x^2+30^2 &= (2x+2)^2 \\ x^2+900 &= 4x^2+8x+4 \\ 0 &= 3x^2+8x-896 \\ 0 &= (x-16)(3x+56) \end{aligned}$$

$$x-16=0 \quad \text{or} \quad 3x+56=0$$
$$x=16 \quad \text{or} \quad x=-\tfrac{56}{3}$$

Reject $-\frac{56}{3}$ since height can't be negative. The tower is 16 m high.

18. $f(x)=a(x-h)^2+k$
Since $a<0$, the parabola opens down. Since $h>0$ and $k<0$, the x-coordinate is positive and the y-coordinate is negative. Therefore, the vertex is in quadrant IV. The correct graph is **A**.

19. $f(x)=\frac{1}{2}x^2-2$
$f(x)=\frac{1}{2}(x-0)^2-2$
The graph is a parabola in $f(x)=a(x-h)^2+k$ form with vertex (h,k) at $(0,-2)$. The axis is $x=0$. Since $a=\frac{1}{2}>0$, the parabola opens up. Also, $|a|=\left|\frac{1}{2}\right|=\frac{1}{2}<1$, so the graph of the parabola is wider than the graph of $f(x)=x^2$. The points $(2,0)$ and $(-2,0)$ are on the graph.

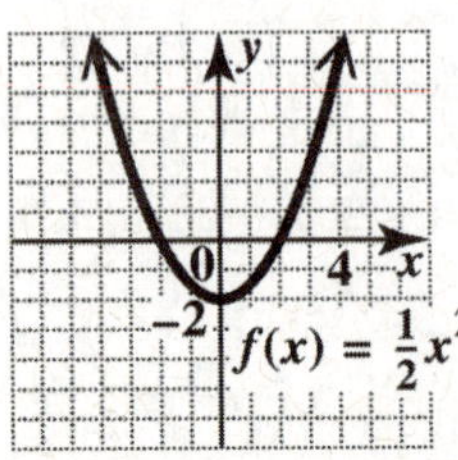

From the graph, we see that the x-values can be any real number, so the domain is $(-\infty,\infty)$. The y-values are greater than or equal to -2, so the range is $[-2,\infty)$.

20. $f(x)=-x^2+4x-1$

The x-coordinate of the vertex is

$$x=\frac{-b}{2a}=\frac{-4}{2(-1)}=2.$$

The y-coordinate of the vertex is

$$f(2)=-4+8-1=3.$$

The graph is a parabola with vertex (h,k) at $(2,3)$ and axis $x=2$. Since $a=-1<0$, the parabola opens down.
Also, $|a|=|-1|=1$, so the graph has the same shape as the graph of $f(x)=x^2$. The points $(0,-1)$ and $(4,-1)$ are on the graph.

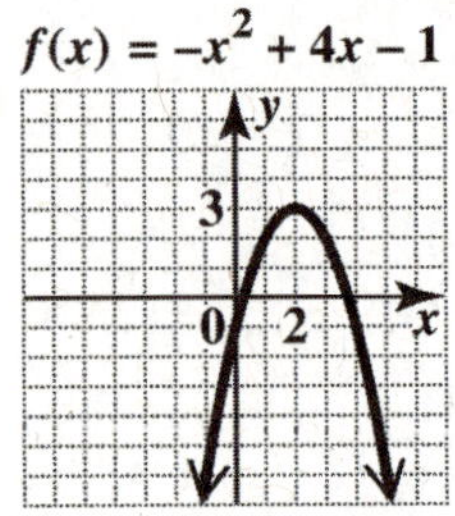

From the graph, we see that the x-values can be any real number, so the domain is $(-\infty,\infty)$.

The y-values are less than or equal to 3, so the range is $(-\infty,3]$.

21. $x=-(y-2)^2+2$
The equation is in $x=a(y-k)^2+h$ form. The graph is a horizontal parabola with vertex (h,k) at $(2,2)$ and axis $y=2$. Since $a=-1<0$, the graph opens to the left. Also, $|a|=|-1|=1$, so the graph has the same shape as the graph of $y=x^2$. The points $(-2,0)$ and $(-2,4)$ are on the same graph.

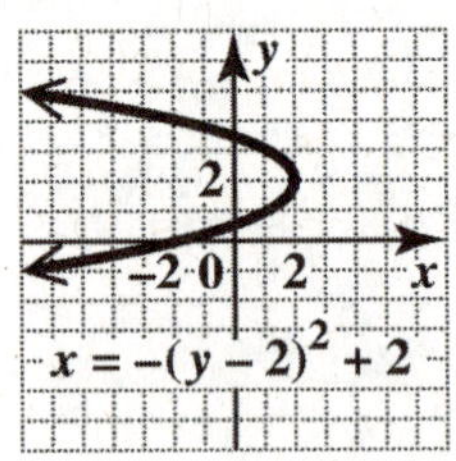

The largest value of x is 2, so the domain is $(-\infty,2]$.

The y-values can be any real number, so the range is $(-\infty,\infty)$.

22. **(a)** Use $f(x)=-0.529x^2+8.00x+115$ with $x=4$.

$$\begin{aligned} f(4) &= -0.529(4)^2+8.00(4)+115 \\ &= 138.536 \approx 139 \text{ million} \end{aligned}$$

In 2004, there were about 139 million civilians employed in the United States.

(b) Find the vertex.

$$\frac{-b}{2a} = \frac{-8.00}{2(-0.529)} \approx 7.6 \text{ (rounding)}$$

$$\begin{aligned} f(7.6) &= -0.529(7.6)^2 + 8.00(7.6) + 115 \\ &\approx 145 \end{aligned}$$

In 2007, there were about 145 million civilians employed in the United States.

23. Let $x =$ the width of the lot.
Then $640 - 2x =$ the length of the lot.

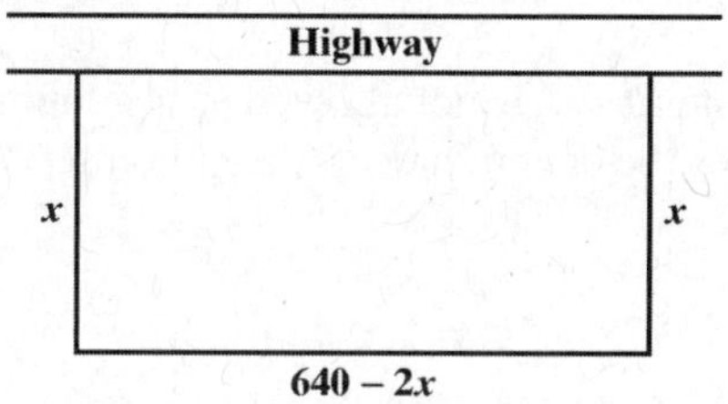

Area $\mathcal{A}$ is length times width.

$$\begin{aligned} \mathcal{A} &= x(640 - 2x) \\ \mathcal{A}(x) &= 640x - 2x^2 = -2x^2 + 640x \end{aligned}$$

Use the vertex formula.

$$x = \frac{-b}{2a} = \frac{-640}{2(-2)} = 160$$

$$\begin{aligned} \mathcal{A}(160) &= -2(160)^2 + 640(160) \\ &= -51{,}200 + 102{,}400 \\ &= 51{,}200 \end{aligned}$$

The graph is a parabola that opens down, so the maximum occurs at the vertex $(160, 51{,}200)$. The maximum area is $51{,}200 \text{ ft}^2$ if the width x is 160 feet and the length is $640 - 2x = 640 - 2(160) = 320$ feet.

24.
$$\begin{aligned} 2x^2 + 7x &> 15 \\ 2x^2 + 7x - 15 &> 0 \end{aligned}$$

Solve the equation

$$2x^2 + 7x - 15 = 0.$$
$$(2x - 3)(x + 5) = 0$$

$$2x - 3 = 0 \quad \text{or} \quad x + 5 = 0$$
$$x = \tfrac{3}{2} \quad \text{or} \quad x = -5$$

The numbers -5 and $\frac{3}{2}$ divide a number line into three intervals.

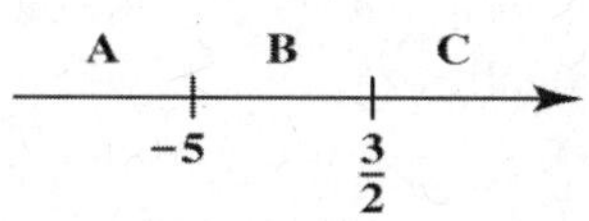

Test a number from each interval in the original inequality, $2x^2 + 7x > 15$.

Interval A: Let $x = -6$.
$$72 - 42 \overset{?}{>} 15$$
$$30 > 15 \qquad \textit{True}$$

Interval B: Let $x = 0$.
$$0 > 15 \qquad \textit{False}$$

Interval C: Let $x = 2$.
$$8 + 14 \overset{?}{>} 15$$
$$22 > 15 \qquad \textit{True}$$

The numbers in Intervals A and C, not including -5 and $\frac{3}{2}$, are solutions.

Solution set: $(-\infty, -5) \cup \left(\frac{3}{2}, \infty\right)$

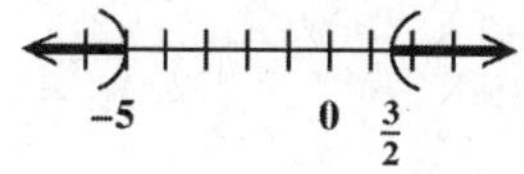

25. $\dfrac{5}{t-4} \leq 1$

Write the inequality so that 0 is on one side.

$$\frac{5}{t-4} - 1 \leq 0$$
$$\frac{5}{t-4} - \frac{1(t-4)}{t-4} \leq 0$$
$$\frac{5 - t + 4}{t-4} \leq 0$$
$$\frac{-t + 9}{t-4} \leq 0$$

The number 9 makes the numerator 0, and 4 makes the denominator 0. These two numbers determine three intervals.

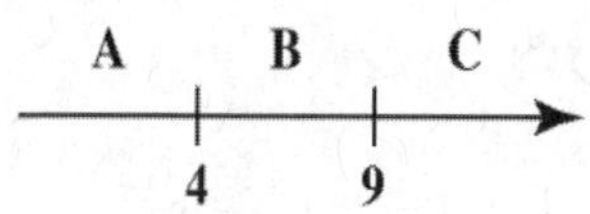

Test a number from each interval in the inequality

$$\frac{5}{t-4} \leq 1.$$

Interval A: Let $t = 0$.
$$\tfrac{5}{-4} \leq 1 \qquad \textit{True}$$

Interval B: Let $t = 7$.
$$\tfrac{5}{3} \leq 1 \qquad \textit{False}$$

Interval C: Let $t = 10$.
$$\tfrac{5}{6} \leq 1 \qquad \textit{True}$$

The numbers in Intervals A and C, including 9 but not 4, are solutions.

Solution set: $(-\infty, 4) \cup [9, \infty)$

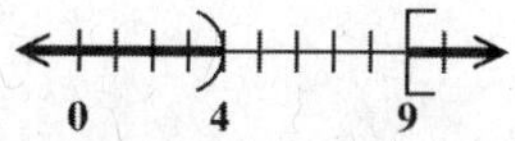

Cumulative Review Exercises (Chapters 1–11)

1. $S = \left\{-\frac{7}{3}, -2, -\sqrt{3}, 0, 0.7, \sqrt{12}, \sqrt{-8}, 7, \frac{32}{3}\right\}$

(a) The elements of S that are integers are -2, 0, and 7.

(b) The elements of S that are rational numbers are $-\frac{7}{3}$, -2, 0, 0.7, 7, and $\frac{32}{3}$.

(c) All the elements of S except $\sqrt{-8}$ are real numbers.

(d) All the elements of S are complex numbers.

2.
$$\begin{aligned} 7-(4+3t)+2t &= -6(t-2)-5 \\ 7-4-3t+2t &= -6t+12-5 \\ 3-t &= -6t+7 \\ 5t &= 4 \\ t &= \frac{4}{5} \end{aligned}$$

Check $t=\frac{4}{5}$: $\frac{11}{5}=\frac{11}{5}$ *True*

Solution set: $\left\{\frac{4}{5}\right\}$

3. $|6x-9| = |-4x+2|$

$$\begin{aligned} 6x-9 &= -4x+2 \quad \text{or} \quad & 6x-9 &= -(-4x+2) \\ 10x &= 11 & 6x-9 &= 4x-2 \\ & & 2x &= 7 \\ x &= \tfrac{11}{10} \quad \text{or} & x &= \tfrac{7}{2} \end{aligned}$$

Check $x=\frac{11}{10}$: $\left|-\frac{24}{10}\right| = \left|-\frac{24}{10}\right|$ *True*

Check $x=\frac{7}{2}$: $|12| = |-12|$ *True*

Solution set: $\left\{\frac{11}{10}, \frac{7}{2}\right\}$

4.
$$\begin{aligned} 2x &= \sqrt{\frac{5x+2}{3}} \\ (2x)^2 &= \left(\sqrt{\frac{5x+2}{3}}\right)^2 \quad \textit{Square.} \\ 4x^2 &= \frac{5x+2}{3} \\ 12x^2 &= 5x+2 \\ 12x^2-5x-2 &= 0 \\ (3x-2)(4x+1) &= 0 \end{aligned}$$

$$\begin{aligned} 3x-2 &= 0 \quad \text{or} \quad & 4x+1 &= 0 \\ x &= \tfrac{2}{3} \quad \text{or} & x &= -\tfrac{1}{4} \end{aligned}$$

Check $x=\frac{2}{3}$: $\frac{4}{3} = \sqrt{\frac{16}{9}}$ *True*

Check $x=-\frac{1}{4}$: $-\frac{1}{2} = \sqrt{\frac{1}{4}}$ *False*

Solution set: $\left\{\frac{2}{3}\right\}$

5.
$$\begin{aligned} \frac{3}{x-3} - \frac{2}{x-2} &= \frac{3}{x^2-5x+6} \\ \frac{3}{x-3} - \frac{2}{x-2} &= \frac{3}{(x-3)(x-2)} \end{aligned}$$

Multiply each term by the LCD, $(x-3)(x-2)$.

$$\begin{aligned} (x-3)(x-2)\left(\frac{3}{x-3} - \frac{2}{x-2}\right) &= (x-3)(x-2)\left[\frac{3}{(x-3)(x-2)}\right] \\ 3(x-2)-2(x-3) &= 3 \\ 3x-6-2x+6 &= 3 \\ x &= 3 \end{aligned}$$

The number 3 is not allowed as a solution since it makes the denominator 0. Solution set: $\emptyset$

6.
$$\begin{aligned} (r-5)(2r+3) &= 1 \\ 2r^2-7r-15 &= 1 \\ 2r^2-7r-16 &= 0 \end{aligned}$$

Here $a=2$, $b=-7$, and $c=-16$.

$$\begin{aligned} r &= \frac{-b \pm \sqrt{b^2-4ac}}{2a} \\ r &= \frac{-(-7) \pm \sqrt{(-7)^2-4(2)(-16)}}{2(2)} \\ &= \frac{7 \pm \sqrt{49+128}}{4} = \frac{7 \pm \sqrt{177}}{4} \end{aligned}$$

Solution set: $\left\{\frac{7 \pm \sqrt{177}}{4}\right\}$

7. $x^4 - 5x^2 + 4 = 0$

Let $u = x^2$, so $u^2 = (x^2)^2 = x^4$.

$$\begin{aligned} u^2-5u+4 &= 0 \\ (u-4)(u-1) &= 0 \end{aligned}$$

$$\begin{aligned} u-4 &= 0 \quad \text{or} \quad & u-1 &= 0 \\ u &= 4 \quad \text{or} & u &= 1 \end{aligned}$$

To find x, substitute x^2 for u.

$$\begin{aligned} x^2 &= 4 \quad \text{or} \quad & x^2 &= 1 \\ x = 2 \text{ or } x &= -2 \quad \text{or} & x = 1 \text{ or } x &= -1 \end{aligned}$$

The proposed solutions check.

Solution set: $\{\pm 1, \pm 2\}$

8.
$$\begin{aligned} -2x+4 &\le -x+3 \\ -x &\le -1 \end{aligned}$$

Multiply by -1, and reverse the direction of the inequality.

$$x \ge 1$$

Solution set: $[1, \infty)$

9. $|3x - 7| \leq 1$

$$-1 \leq 3x - 7 \leq 1$$
$$6 \leq 3x \leq 8 \quad \textit{Add 7.}$$
$$2 \leq x \leq \tfrac{8}{3} \quad \textit{Divide by 3.}$$

Solution set: $\left[2, \frac{8}{3}\right]$

10. $x^2 - 4x + 3 < 0$

Solve the equation

$$x^2 - 4x + 3 = 0.$$
$$(x - 3)(x - 1) = 0$$

$$x - 3 = 0 \quad \text{or} \quad x - 1 = 0$$
$$x = 3 \quad \text{or} \quad x = 1$$

The numbers 1 and 3 divide a number line into three intervals.

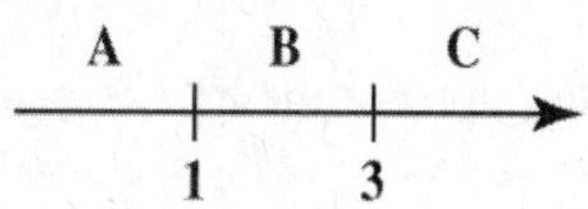

Test a number from each interval in the original inequality, $x^2 - 4x + 3 < 0$.

Interval A: Let $x = 0$.
$3 < 0$ *False*

Interval B: Let $x = 2$.
$4 - 8 + 3 \overset{?}{<} 0$
$-1 < 0$ *True*

Interval C: Let $x = 4$.
$16 - 16 + 3 \overset{?}{<} 0$
$3 < 0$ *False*

The numbers from Interval B, not including 1 or 3, are solutions.

Solution set: $(1, 3)$

11. $\dfrac{3}{p+2} > 1$

Write the inequality so that 0 is on one side.

$$\frac{3}{p+2} - 1 > 0$$
$$\frac{3}{p+2} - \frac{1(p+2)}{p+2} > 0$$
$$\frac{3 - p - 2}{p+2} > 0$$
$$\frac{-p+1}{p+2} > 0$$

The number 1 makes the numerator 0, and -2 makes the denominator 0. These two numbers determine three intervals.

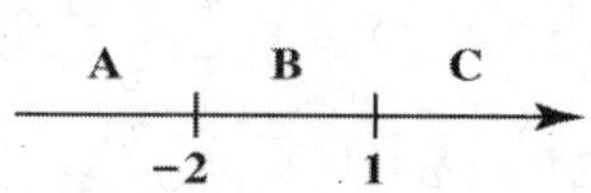

Test a number from each interval in the inequality

$$\frac{3}{p+2} > 1.$$

Interval A: Let $p = -4$.
$\frac{3}{-2} > 1$ *False*

Interval B: Let $p = 0$.
$\frac{3}{2} > 1$ *True*

Interval C: Let $p = 2$.
$\frac{3}{4} > 1$ *False*

The numbers from Interval B, not including -2 or 1, are solutions.

Solution set: $(-2, 1)$

12. $4x - 5y = 15$

Draw the line through its intercepts, $\left(\frac{15}{4}, 0\right)$ and $(0, -3)$. The graph passes the vertical line test, so the relation is a function. As with any line that is not horizontal or vertical, the domain and range are both $(-\infty, \infty)$.

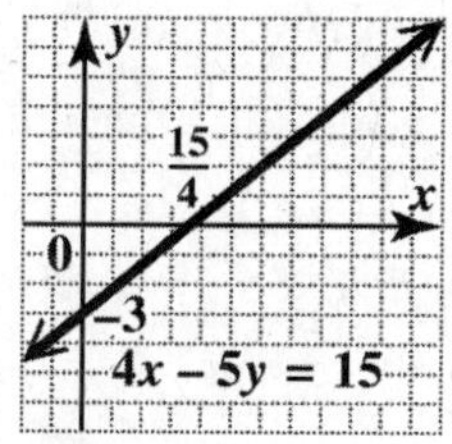

Solve the equation for y.

$$4x - 5y = 15$$
$$-5y = 15 - 4x$$
$$y = \frac{15 - 4x}{-5}, \quad \text{or} \quad \frac{4x - 15}{5}$$

Thus, $f(x) = \frac{4}{5}x - 3$.

13. $4x - 5y < 15$

Draw a dashed line through the points $\left(\frac{15}{4}, 0\right)$ and $(0, -3)$. Check the origin:

$$4(0) - 5(0) \overset{?}{<} 15$$
$$0 < 15 \quad \textit{True}$$

Shade the region that contains the origin.

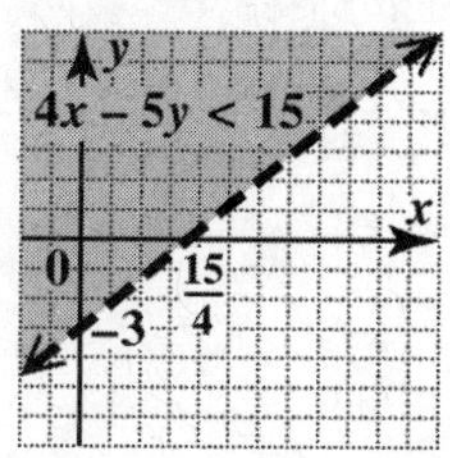

The relation is not a function since for any value of x, there is more than one value of y.

14. $y = -2(x-1)^2 + 3$

The equation is in $f(x) = a(x-h)^2 + k$ form, so the graph is a parabola with vertex (h, k) at $(1, 3)$. Since $a = -2 < 0$, the parabola opens down. Also $|a| = |-2| = 2 > 1$, so the graph is narrower than the graph of $f(x) = x^2$. The points $(0, 1)$ and $(2, 1)$ are on the graph.

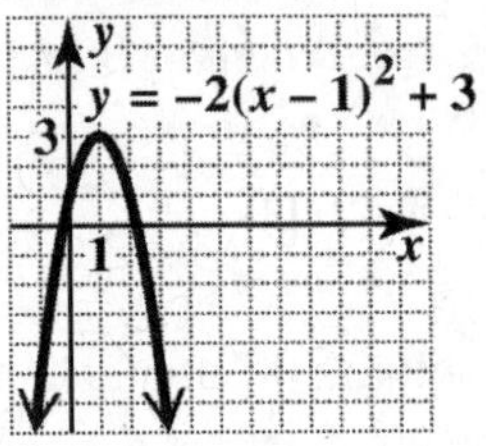

The relation is a function since it passes the vertical line test. The domain is $(-\infty, \infty)$. The largest value of y is 3, so the range is $(-\infty, 3]$. The equation is already solved for y, so using function notation we have $f(x) = -2(x-1)^2 + 3$.

15. $-2x + 7y = 16$

Solve the equation for y.

$$\begin{aligned} 7y &= 2x + 16 \\ y &= \tfrac{2}{7}x + \tfrac{16}{7} \end{aligned}$$

So the slope is $\frac{2}{7}$ and the y-intercept is $\left(0, \frac{16}{7}\right)$. Let $y = 0$ in $-2x + 7y = 16$ to find the x-intercept.

$$\begin{aligned} -2x + 7(0) &= 16 \\ -2x &= 16 \\ x &= -8 \end{aligned}$$

The x-intercept is $(-8, 0)$.

16. **(a)** Solve $5x + 2y = 6$ for y.

$$\begin{aligned} 2y &= -5x + 6 \\ y &= -\tfrac{5}{2}x + 3 \end{aligned}$$

So the slope of the given line and the desired line is $-\frac{5}{2}$. The required form is

$$\begin{aligned} y &= -\tfrac{5}{2}x + b. \\ -3 &= -\tfrac{5}{2}(2) + b \quad \textit{Let } x=2,\ y=-3. \\ -3 &= -5 + b \\ 2 &= b \end{aligned}$$

The equation is $y = -\frac{5}{2}x + 2$.

(b) The negative reciprocal of the slope in part (a) is

$$-\frac{1}{-\frac{5}{2}} = \frac{2}{5},$$

which is the slope of the line perpendicular to the given line.

$$\begin{aligned} \text{So } y &= \tfrac{2}{5}x + b. \\ 1 &= \tfrac{2}{5}(-4) + b \quad \textit{Let } x = -4,\ y = 1. \\ 1 &= -\tfrac{8}{5} + b \\ \tfrac{13}{5} &= b \end{aligned}$$

The equation is $y = \frac{2}{5}x + \frac{13}{5}$.

17.
$$\begin{aligned} \left(\frac{x^{-3}y^2}{x^5y^{-2}}\right)^{-1} &= \left(x^{-3-5}y^{2-(-2)}\right)^{-1} \\ &= \left(x^{-8}y^4\right)^{-1} \\ &= x^8y^{-4} \\ &= \frac{x^8}{y^4} \end{aligned}$$

18.
$$\begin{aligned} \frac{(4x^{-2})^2(2y^3)}{8x^{-3}y^5} &= \frac{16x^{-4}(2y^3)}{8x^{-3}y^5} \\ &= \frac{4x^{-4}y^3}{x^{-3}y^5} \\ &= 4x^{-4-(-3)}y^{3-5} \\ &= 4x^{-1}y^{-2} \\ &= \frac{4}{xy^2} \end{aligned}$$

19.
$$\begin{aligned} \left(\tfrac{2}{3}t + 9\right)^2 &= \left(\tfrac{2}{3}t\right)^2 + 2\left(\tfrac{2}{3}t\right)(9) + 9^2 \\ &= \tfrac{4}{9}t^2 + 12t + 81 \end{aligned}$$

20. Divide $4x^3 + 2x^2 - x + 26$ by $x + 2$.

$$\begin{array}{r}
4x^2 - 6x + 11 \\
x+2 \overline{\big) 4x^3 + 2x^2 - x + 26} \\
\underline{4x^3 + 8x^2} \qquad\qquad \\
-6x^2 - x \qquad \\
\underline{-6x^2 - 12x} \qquad \\
11x + 26 \\
\underline{11x + 22} \\
4
\end{array}$$

$(4x^3 + 2x^2 - x + 26)$ divided by $(x + 2)$ is

$$4x^2 - 6x + 11 + \frac{4}{x+2}.$$

21. $24m^2 + 2m - 15$

The two integers whose product is $24(-15) = -360$ and whose sum is 2 are 20 and -18.

$$\begin{aligned}
&24m^2 + 2m - 15 \\
&= 24m^2 + 20m - 18m - 15 \\
&= 4m(6m + 5) - 3(6m + 5) \\
&= (6m + 5)(4m - 3)
\end{aligned}$$

22. $8x^3 + 27y^3$
Use the sum of cubes formula,

$$a^3 + b^3 = (a + b)(a^2 - ab + b^2),$$

with $a = 2x$ and $b = 3y$.

$$\begin{aligned}
&8x^3 + 27y^3 \\
&= (2x + 3y)\left[(2x)^2 - (2x)(3y) + (3y)^2\right] \\
&= (2x + 3y)\left(4x^2 - 6xy + 9y^2\right)
\end{aligned}$$

23. $9x^2 - 30xy + 25y^2$
Use the perfect square formula,

$$a^2 - 2ab + b^2 = (a - b)^2,$$

with $a = 3x$ and $b = 5y$.

$$\begin{aligned}
&9x^2 - 30xy + 25y^2 \\
&= \left[(3x)^2 - 2(3x)(5y) + (5y)^2\right] \\
&= (3x - 5y)^2
\end{aligned}$$

24. $\dfrac{5x+2}{-6} \div \dfrac{15x+6}{5}$

Multiply by the reciprocal.

$$\begin{aligned}
&= \frac{5x+2}{-6} \cdot \frac{5}{15x+6} \\
&= \frac{5x+2}{-6} \cdot \frac{5}{3(5x+2)} \\
&= \frac{5}{-18}, \text{ or } -\frac{5}{18}
\end{aligned}$$

25. $\dfrac{3}{2-x} - \dfrac{5}{x} + \dfrac{6}{x^2 - 2x}$

$$\begin{aligned}
&= \frac{3}{2-x} - \frac{5}{x} + \frac{6}{x(x-2)} \\
&= \frac{-3}{x-2} - \frac{5}{x} + \frac{6}{x(x-2)}
\end{aligned}$$

The LCD is $x(x - 2)$.

$$\begin{aligned}
&= \frac{-3x}{(x-2)x} - \frac{5(x-2)}{x(x-2)} + \frac{6}{x(x-2)} \\
&= \frac{-3x - 5(x-2) + 6}{x(x-2)} \\
&= \frac{-3x - 5x + 10 + 6}{x(x-2)} \\
&= \frac{-8x + 16}{x(x-2)} \\
&= \frac{-8(x-2)}{x(x-2)} = -\frac{8}{x}
\end{aligned}$$

26. $\dfrac{\frac{r}{s} - \frac{s}{r}}{\frac{r}{s} + 1}$

Multiply the numerator and denominator by the LCD of all the fractions, rs.

$$\begin{aligned}
&= \frac{\left(\frac{r}{s} - \frac{s}{r}\right)rs}{\left(\frac{r}{s} + 1\right)rs} = \frac{r^2 - s^2}{r^2 + rs} \\
&= \frac{(r-s)(r+s)}{r(r+s)} = \frac{r-s}{r}
\end{aligned}$$

27.
$$\begin{aligned}
2x - 4y &= 10 \quad (1) \\
9x + 3y &= 3 \quad (2)
\end{aligned}$$

Simplify the equations.

$$\begin{aligned}
x - 2y &= 5 \quad (3) \quad \tfrac{1}{2} \times (1) \\
3x + y &= 1 \quad (4) \quad \tfrac{1}{3} \times (2)
\end{aligned}$$

To eliminate y, multiply (4) by 2 and add the result to (3).

$$\begin{aligned}
x - 2y &= 5 \quad (3) \\
6x + 2y &= 2 \quad 2 \times (4) \\
\hline
7x &= 7 \\
x &= 1
\end{aligned}$$

Substitute $x = 1$ into (4).

$$\begin{aligned}
3x + y &= 1 \quad (4) \\
3(1) + y &= 1 \\
y &= -2
\end{aligned}$$

Solution set: $\{(1, -2)\}$

28.
$$\begin{aligned} x + y + 2z &= 3 \quad (1)\\ -x + y + z &= -5 \quad (2)\\ 2x + 3y - z &= -8 \quad (3)\end{aligned}$$

Eliminate z by adding (2) and (3).

$$\begin{aligned} -x + y + z &= -5 \quad (2)\\ 2x + 3y - z &= -8 \quad (3)\\ \hline x + 4y \quad &= -13 \quad (4)\end{aligned}$$

To get another equation without z, multiply equation (3) by 2 and add the result to equation (1).

$$\begin{aligned} x + y + 2z &= 3 \quad (1)\\ 4x + 6y - 2z &= -16 \quad 2 \times (3)\\ \hline 5x + 7y \quad &= -13 \quad (5)\end{aligned}$$

To eliminate x, multiply (4) by -5 and add the result to (5).

$$\begin{aligned} -5x - 20y &= 65 \quad -5 \times (4)\\ 5x + 7y &= -13 \quad (5)\\ \hline -13y &= 52\\ y &= -4\end{aligned}$$

Use (4) to find x.

$$\begin{aligned} x + 4y &= -13 \quad (4)\\ x + 4(-4) &= -13\\ x - 16 &= -13\\ x &= 3\end{aligned}$$

Use (2) to find z.

$$\begin{aligned} -x + y + z &= -5 \quad (2)\\ -3 - 4 + z &= -5\\ -7 + z &= -5\\ z &= 2\end{aligned}$$

Solution set: $\{(3, -4, 2)\}$

29. Let $x =$ the amount of revenue for Microsoft and $y =$ the amount of revenue for Oracle.

The combined revenues were \$82.8 billion, so

$$x + y = 82.8. \quad (1)$$

Revenues for Microsoft were \$6.8 billion less than three times those of Oracle, so

$$x = 3y - 6.8. \quad (2)$$

Substitute $3y - 6.8$ for x in (1), then solve for y.

$$\begin{aligned} (3y - 6.8) + y &= 82.8\\ 4y &= 89.6\\ y &= 22.4\end{aligned}$$

From (2), $x = 3(22.4) - 6.8 = 60.4$. Thus, in 2009, Microsoft had \$60.4 billion in revenue and Oracle had \$22.4 billion in revenue.

30.
$$\begin{aligned}\sqrt[3]{\frac{27}{16}} &= \frac{\sqrt[3]{27}}{\sqrt[3]{16}} = \frac{\sqrt[3]{3^3}}{\sqrt[3]{8 \cdot 2}} = \frac{3}{2\sqrt[3]{2}}\\ &= \frac{3 \cdot \sqrt[3]{4}}{2\sqrt[3]{2} \cdot \sqrt[3]{4}} = \frac{3\sqrt[3]{4}}{2\sqrt[3]{8}}\\ &= \frac{3\sqrt[3]{4}}{2 \cdot 2} = \frac{3\sqrt[3]{4}}{4}\end{aligned}$$

31.
$$\begin{aligned}\frac{2}{\sqrt{7} - \sqrt{5}} &= \frac{2(\sqrt{7} + \sqrt{5})}{(\sqrt{7} - \sqrt{5})(\sqrt{7} + \sqrt{5})}\\ &= \frac{2(\sqrt{7} + \sqrt{5})}{7 - 5}\\ &= \frac{2(\sqrt{7} + \sqrt{5})}{2} = \sqrt{7} + \sqrt{5}\end{aligned}$$

32. Let $x =$ the distance traveled by the southbound car and
$2x - 38 =$ the distance traveled by the eastbound car.

Since the cars are traveling at right angles with one another, the Pythagorean theorem can be applied.

$$\begin{aligned} a^2 + b^2 &= c^2\\ x^2 + (2x - 38)^2 &= 95^2\\ x^2 + 4x^2 - 152x + 1444 &= 9025\\ 5x^2 - 152x - 7581 &= 0\end{aligned}$$

$$x = \frac{-b \pm \sqrt{b^2 - 4ac}}{2a}$$

$$\begin{aligned} x &= \frac{-(-152) \pm \sqrt{(-152)^2 - 4(5)(-7581)}}{2(5)}\\ &= \frac{152 \pm \sqrt{174{,}724}}{10} = \frac{152 \pm 418}{10}\end{aligned}$$

Thus, $x = \dfrac{152 + 418}{10} = 57$ (the other value is negative). The southbound car traveled 57 miles, and the eastbound car traveled $2x - 38 = 2(57) - 38 = 76$ miles.

CHAPTER 12 INVERSE, EXPONENTIAL, AND LOGARITHMIC FUNCTIONS

12.1 Inverse Functions

12.1 Now Try Exercises

N1. **(a)** $F = \{(-1,-2), (0,0), (1,-2), (2,-8)\}$

Since the y-value -2 corresponds to x-values -1 and 1, the function is not one-to-one. Therefore, it does not have an inverse.

(b) $G = \{(0,0), (1,1), (4,2), (9,3)\}$

Each x-value in the relation corresponds to only one y-value, so the relation is indeed a function. Furthermore, each y-value corresponds to only one x-value, so this is a one-to-one function.

The inverse function is found by interchanging the x- and y-values in each ordered pair.

The inverse is

$$G^{-1} = \{(0,0), (1,1), (2,4), (3,9)\}.$$

(c) Since the height 639 corresponds to the number of stories 31 and 41, the function is not one-to-one. Therefore, it does not have an inverse.

N2. **(a)** Since a horizontal line will intersect the graph in no more than one point, the function is one-to-one.

(b) Since a horizontal line could intersect the graph in two points, the function is not one-to-one.

N3. To find the equation of the inverse of a function,

(1) determine if the function is one-to-one;
(2) replace $f(x)$ with y;
(3) interchange x and y; *Step 1*
(4) solve for y; *Step 2*
(5) replace y with $f^{-1}(x)$. *Step 3*

(a) $f(x) = 5x - 7$

The graph of f is a line. By the horizontal line test, it is a one-to-one function. To find the inverse, first replace $f(x)$ with y.

Step 1 Interchange x and y.

$$y = 5x - 7$$
$$x = 5y - 7$$

Step 2 Solve for y.

$$x + 7 = 5y$$
$$\frac{x+7}{5} = y$$

Step 3 Replace y with $f^{-1}(x)$.

$$f^{-1}(x) = \frac{x+7}{5}, \text{ or } f^{-1}(x) = \frac{1}{5}x + \frac{7}{5}$$

(b) $f(x) = (x+1)^2$

This equation has a vertical parabola as its graph, so some horizontal lines will intersect the graph at two points. For example, both $x = -2$ and $x = 0$ correspond to $y = 1$. Thus, the function is not a one-to-one function and does not have an inverse.

(c) $f(x) = x^3 - 4$

Each value of x produces a different value of y, so this is a one-to-one function. To find the inverse, first replace $f(x)$ with y.

Step 1 Interchange x and y.

$$y = x^3 - 4$$
$$x = y^3 - 4$$

Step 2 Solve for y.

$$x + 4 = y^3$$

Take the cube root on each side.

$$\sqrt[3]{x+4} = y$$

Step 3 Replace y with $f^{-1}(x)$.

$$f^{-1}(x) = \sqrt[3]{x+4}$$

N4. The given graph goes through $(-3,-2)$, $(-2,-1)$, $(-1,2)$, and $(0,7)$. So, the inverse will go through $(-2,-3)$, $(-1,-2)$, $(2,-1)$, and $(7,0)$. Use these points and symmetry about $y = x$ to complete the graph of the inverse.

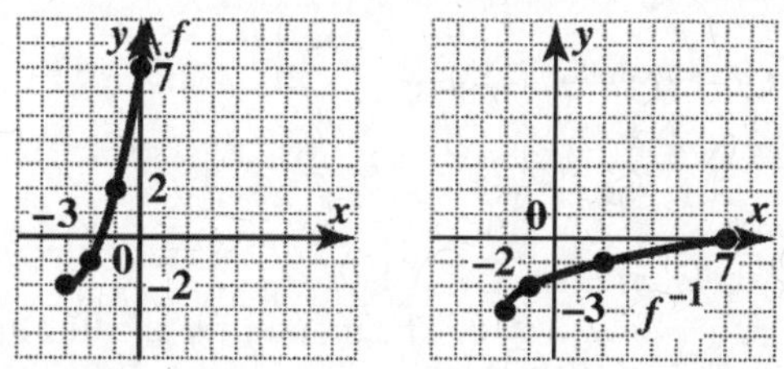

12.1 Section Exercises

1. This function is not one-to-one because both France and the United States are paired with the same trans fat percentage, 11.

3. The function in the table that pairs a city with a distance is a one-to-one function because for each city there is one distance and each distance has only one city to which it is paired.

If the distance from Indianapolis to Denver had 1 mile added to it, it would be $1058 + 1 = 1059$ mi, the same as the distance from Los Angeles to Denver. In this case, one distance would have two cities to which it is paired, and the function would not be one-to-one.

5. If a function is made up of ordered pairs in such a way that the same y-value appears in a correspondence with two different x-values, then the function is not one-to-one. Choice **B**

7. All of the graphs pass the vertical line test, so they all represent functions. The graph in choice **A** is the only one that passes the horizontal line test, so it is the one-to-one function.

9. $\{(3,6), (2,10), (5,12)\}$ is a one-to-one function, since each x-value corresponds to only one y-value and each y-value corresponds to only one x-value. To find the inverse, interchange x and y in each ordered pair. The inverse is

$$\{(6,3), (10,2), (12,5)\}.$$

11. $\{(-1,3), (2,7), (4,3), (5,8)\}$ is not a one-to-one function. The ordered pairs $(-1,3)$ and $(4,3)$ have the same y-value for two different x-values.

13. The graph of $f(x) = 2x + 4$ is a nonvertical, nonhorizontal line. By the horizontal line test, $f(x)$ is a one-to-one function. To find the inverse, replace $f(x)$ with y.

$$y = 2x + 4$$

Interchange x and y.

$$x = 2y + 4$$

Solve for y.

$$2y = x - 4$$

$$y = \frac{x-4}{2}$$

Replace y with $f^{-1}(x)$.

$$f^{-1}(x) = \frac{x-4}{2}, \text{ or } f^{-1}(x) = \frac{1}{2}x - 2$$

15. Write $g(x) = \sqrt{x-3}$ as $y = \sqrt{x-3}$. Since $x \geq 3$, $y \geq 0$. The graph of g is half of a horizontal parabola that opens to the right. The graph passes the horizontal line test, so g is one-to-one. To find the inverse, interchange x and y to get

$$x = \sqrt{y-3}.$$

Note that now $y \geq 3$, so $x \geq 0$.
Solve for y by squaring each side.

$$x^2 = y - 3$$

$$x^2 + 3 = y$$

Replace y with $g^{-1}(x)$.

$$g^{-1}(x) = x^2 + 3, x \geq 0$$

17. $f(x) = 3x^2 + 2$ is not a one-to-one function because two x-values, such as 1 and -1, both have the same y-value, in this case 5. The graph of this function is a vertical parabola which does not pass the horizontal line test.

19. The graph of $f(x) = x^3 - 4$ is the graph of $g(x) = x^3$ shifted down 4 units. (Recall that $g(x) = x^3$ is the elongated S-shaped curve.) The graph of f passes the horizontal line test, so f is one-to-one.

Replace $f(x)$ with y.

$$y = x^3 - 4$$

Interchange x and y.

$$x = y^3 - 4$$

Solve for y.

$$x + 4 = y^3$$

Take the cube root of each side.

$$\sqrt[3]{x+4} = y$$

Replace y with $f^{-1}(x)$.

$$f^{-1}(x) = \sqrt[3]{x+4}$$

In Exercises 21–24, $f(x) = 2^x$ is a one-to-one function.

21. **(a)** To find $f(3)$, substitute 3 for x.

$f(x) = 2^x$, so $f(3) = 2^3 = 8$.

(b) Since f is one-to-one and $f(3) = 8$, it follows that $f^{-1}(8) = 3$.

23. **(a)** To find $f(0)$, substitute 0 for x.

$f(x) = 2^x$, so $f(0) = 2^0 = 1$.

(b) Since f is one-to-one and $f(0) = 1$, it follows that $f^{-1}(1) = 0$.

25. **(a)** The function is one-to-one since any horizontal line intersects the graph at most once.

(b) In the graph, the two points marked on the line are $(-1,5)$ and $(2,-1)$. Interchange x and y in each ordered pair to get $(5,-1)$ and $(-1,2)$. Plot these points, then draw a dashed line through them to obtain the graph of the inverse function.

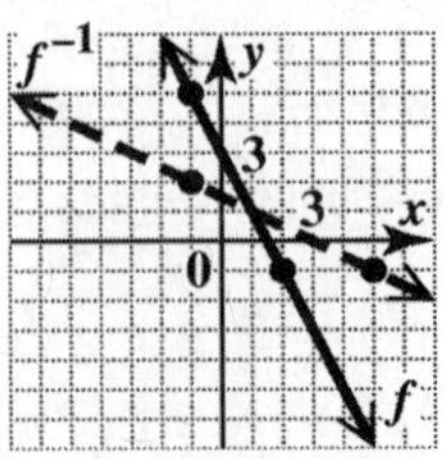

27. **(a)** The function is not one-to-one since there are horizontal lines that intersect the graph more than once. For example, the line $y = 1$ intersects the graph twice.

29. **(a)** The function is one-to-one since any horizontal line intersects the graph at most once.

(b) In the graph, the four points marked on the curve are $(-4, 2)$, $(-1, 1)$, $(1, -1)$, and $(4, -2)$. Interchange x and y in each ordered pair to get $(2, -4)$, $(1, -1)$, $(-1, 1)$, and $(-2, 4)$. Plot these points, then draw a dashed curve (symmetric to the original graph about the line $y = x$) through them to obtain the graph of the inverse.

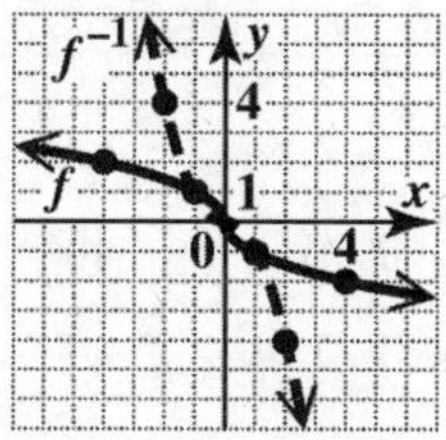

31. $f(x) = 2x - 1$ or $y = 2x - 1$
The graph is a line through $(-2, -5)$, $(0, -1)$, and $(3, 5)$. Plot these points and draw the solid line through them. Then the inverse will be a line through $(-5, -2)$, $(-1, 0)$, and $(5, 3)$. Plot these points and draw the dashed line through them.

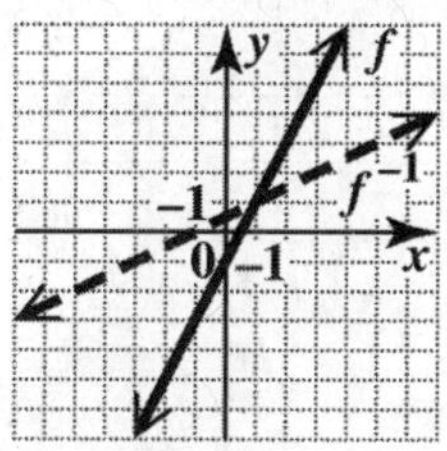

33. $g(x) = -4x$ or $y = -4x$
The graph is a line through $(0, 0)$ and $(1, -4)$. For the inverse, interchange x and y in each ordered pair to get the points $(0, 0)$ and $(-4, 1)$. Draw a dashed line through these points to obtain the graph of the inverse function.

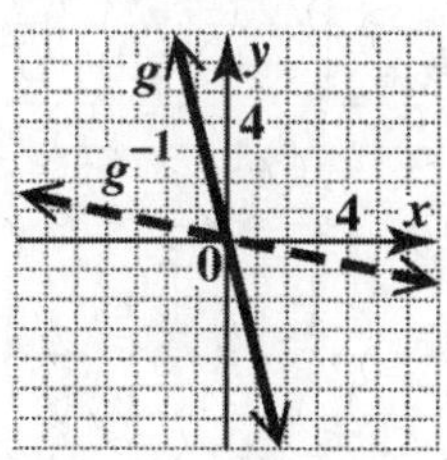

35. $f(x) = \sqrt{x}$, $x \geq 0$
Complete the table of values.

x	$f(x)$
0	0
1	1
4	2

Plot these points and connect them with a solid smooth curve.
Since $f(x)$ is one-to-one, make a table of values for $f^{-1}(x)$ by interchanging x and y.

x	$f^{-1}(x)$
0	0
1	1
2	4

Plot these points and connect them with a dashed smooth curve.

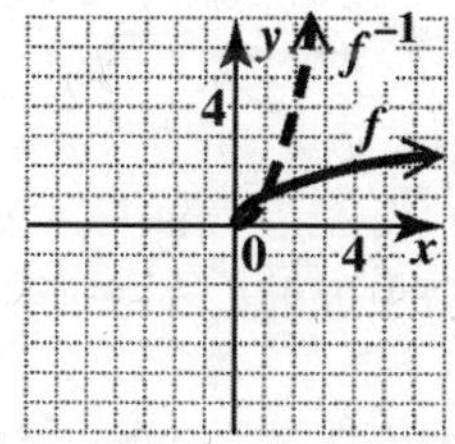

37. $f(x) = x^3 - 2$
Complete the table of values.

x	$f(x)$
−1	−3
0	−2
1	−1
2	6

Plot these points and connect them with a solid smooth curve.
Make a table of values for f^{-1}.

x	$f^{-1}(x)$
−3	−1
−2	0
−1	1
6	2

Plot these points and connect them with a dashed smooth curve.

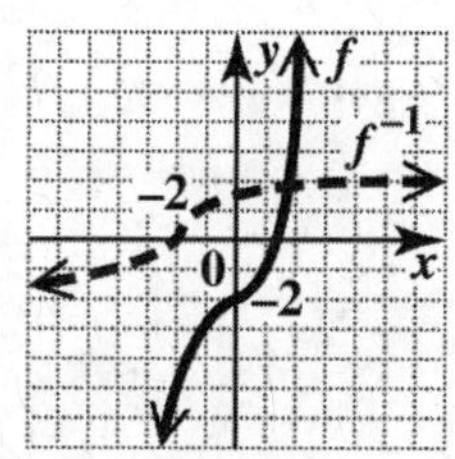

◆◆◆ Relating Concepts 39–42 ◆◆◆

39.
$$f(x) = 4x - 5$$
Replace $f(x)$ with y.
$$y = 4x - 5$$
Interchange x and y.
$$x = 4y - 5$$
Solve for y.
$$x + 5 = 4y$$
$$\frac{x + 5}{4} = y$$
Replace y with $f^{-1}(x)$.
$$\frac{x + 5}{4} = f^{-1}(x),$$
or $f^{-1}(x) = \frac{1}{4}x + \frac{5}{4}$

40. Insert each number in the inverse function found in Exercise 39,

$$f^{-1}(x) = \frac{x+5}{4}.$$

$$f^{-1}(47) = \frac{47+5}{4} = \frac{52}{4} = 13 = \text{M},$$

$$f^{-1}(95) = \frac{95+5}{4} = \frac{100}{4} = 25 = \text{Y},$$

and so on.

The decoded message is as follows:
My graphing calculator is the greatest thing since sliced bread.

41. A one-to-one code is essential to this process because if the code is not one-to-one, an encoded number would refer to two different letters.

42. Answers will vary according to the student's name. For example, Jane Doe is encoded as follows:

1004 5 2748 129 68 3379 129.

43. $Y_1 = f(x) = 2x - 7$
Replace $f(x)$ with y.

$$y = 2x - 7$$

Interchange x and y.

$$x = 2y - 7$$

Solve for y.

$$x + 7 = 2y$$

$$\frac{x+7}{2} = y$$

Replace y with $f^{-1}(x)$.

$$\frac{x+7}{2} = f^{-1}(x) = Y_2 \left(\text{or } \frac{1}{2}x + \frac{7}{2}\right)$$

Now graph Y_1 and Y_2.

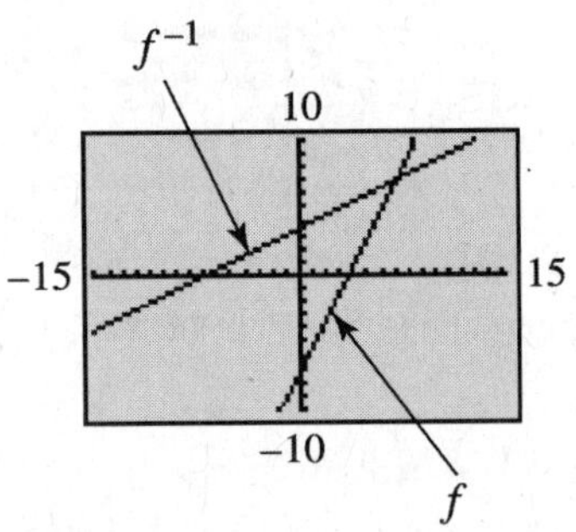

45. $Y_1 = f(x) = x^3 + 5$
Replace $f(x)$ with y.

$$y = x^3 + 5$$

Interchange x and y.

$$x = y^3 + 5$$

Solve for y.

$$x - 5 = y^3$$

Take the cube root of each side.

$$\sqrt[3]{x-5} = y$$

Replace y with $f^{-1}(x)$.

$$\sqrt[3]{x-5} = f^{-1}(x) = Y_2$$

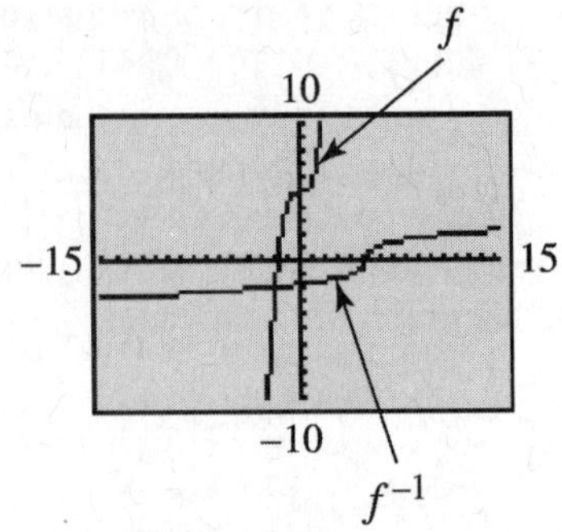

47. $f(x) = 4^x$, so $f(3) = 4^3 = 64$.

49. $f(x) = 4^x$, so $f\left(-\frac{1}{2}\right) = 4^{-1/2} = \frac{1}{4^{1/2}} = \frac{1}{2}$.

12.2 Exponential Functions

12.2 Now Try Exercises

In Now Try Exercises 1–3, choose values of x and find the corresponding values of y. Then plot the points, and draw a smooth curve through them.

N1. $f(x) = 4^x$

x	-2	-1	0	1	2
$f(x) = 4^x$	$\frac{1}{16}$	$\frac{1}{4}$	1	4	16

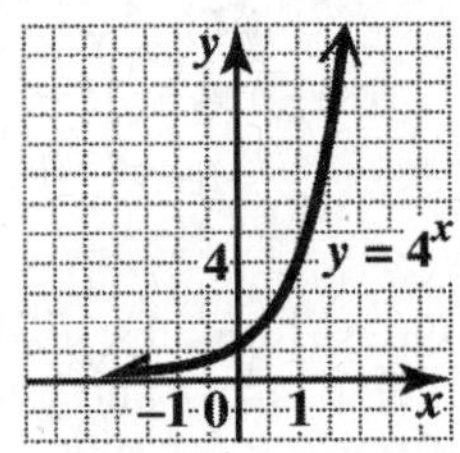

N2. $g(x) = \left(\frac{1}{10}\right)^x$

x	-2	-1	0	1	2
$g(x) = \left(\frac{1}{10}\right)^x$	100	10	1	$\frac{1}{10}$	$\frac{1}{100}$

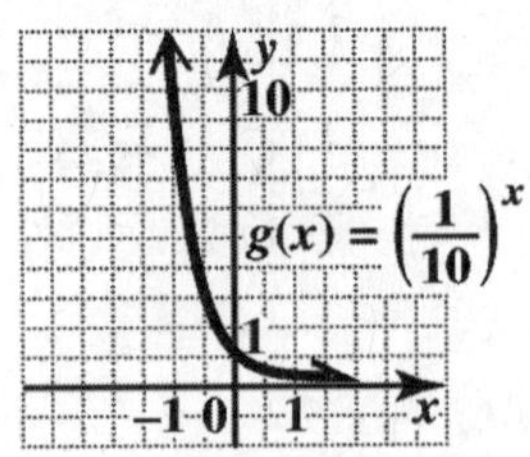

N3. $y = f(x) = 4^{2x-1}$

Make a table of values. It will help to find values for $2x - 1$ before you find y.

x	$2x - 1$	$y = 4^{2x-1}$
0	-1	$\frac{1}{4}$
$\frac{1}{2}$	0	1
1	1	4
-1	-3	$\frac{1}{64}$

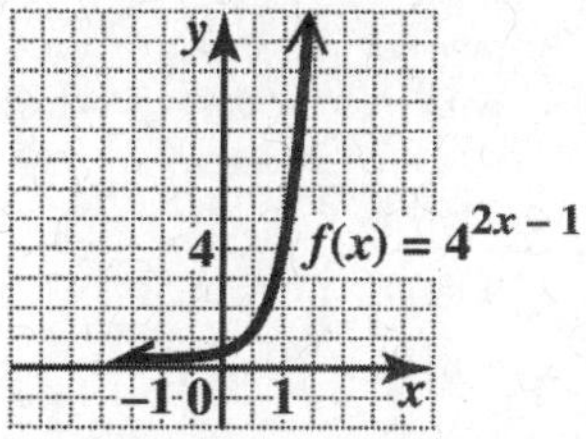

N4. $8^x = 16$

Step 1 Write each side with the base 2.

$(2^3)^x = 2^4$

Step 2 Simplify exponents.

$2^{3x} = 2^4$

Step 3 Set the exponents equal.

$3x = 4$

Step 4 Solve.

$x = \frac{4}{3}$

Check $x = \frac{4}{3}$ in the original equation.

$$8^{4/3} \stackrel{?}{=} 16$$
$$(8^{1/3})^4 \stackrel{?}{=} 16$$
$$2^4 \stackrel{?}{=} 16$$
$$16 = 16 \quad \textit{True}$$

The solution set is $\{\frac{4}{3}\}$.

N5. (a) $3^{2x-1} = 27^{x+4}$

$(3)^{2x-1} = (3^3)^{x+4}$ *Same base*

$3^{2x-1} = 3^{3x+12}$ *Simplify exponents.*

$2x - 1 = 3x + 12$ *Set exponents equal.*

$-13 = x$

Check $x = -13$:

$$3^{-26-1} \stackrel{?}{=} 27^{-13+4}$$
$$3^{-27} \stackrel{?}{=} (3^3)^{-9}$$
$$3^{-27} = 3^{-27} \quad \textit{True}$$

The solution set is $\{-13\}$.

(b) $5^x = \frac{1}{625}$

$5^x = \dfrac{1}{5^4}$

$5^x = 5^{-4}$ *Same base*

$x = -4$ *Set exponents equal.*

Check $x = -4$: $5^{-4} = \frac{1}{5^4} = \frac{1}{625}$

The solution set is $\{-4\}$.

(c) $(\frac{2}{7})^x = \frac{343}{8}$

$(\frac{2}{7})^x = (\frac{7}{2})^3$

$(\frac{2}{7})^x = (\frac{2}{7})^{-3}$

$x = -3$ *Set exponents equal.*

Check $x = -3$: $(\frac{2}{7})^{-3} = (\frac{7}{2})^3 = \frac{343}{8}$

The solution set is $\{-3\}$.

N6. $x = 2000 - 1750 = 250$

$$f(x) = 266(1.001)^x$$
$$f(250) = 266(1.001)^{250}$$
$$f(250) \approx 342 \text{ parts per million}$$

N7.
$$f(x) = 1038(1.000134)^{-x}$$
$$f(6000) = 1038(1.000134)^{-6000}$$
$$f(6000) \approx 465$$

The pressure is approximately 465 millibars.

12.2 Section Exercises

1. Since the graph of $F(x) = a^x$ always contains the point $(0, 1)$, the correct response is **C**.

3. Since the graph of $F(x) = a^x$ always approaches the x-axis, the correct response is **A**.

5. $f(x) = 3^x$

Make a table of values.

$f(-2) = 3^{-2} = \dfrac{1}{3^2} = \dfrac{1}{9}$,

$f(-1) = 3^{-1} = \dfrac{1}{3^1} = \dfrac{1}{3}$, and so on.

x	-2	-1	0	1	2
$f(x)$	$\frac{1}{9}$	$\frac{1}{3}$	1	3	9

Plot the points from the table and draw a smooth curve through them.

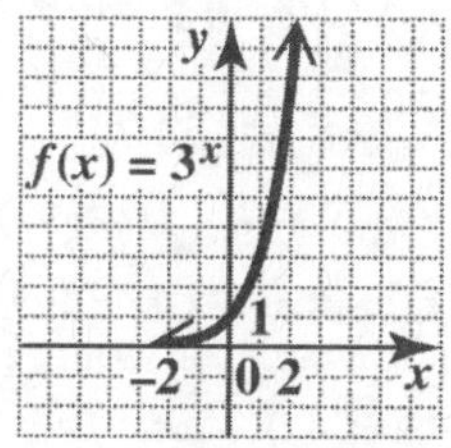

7. $g(x) = (\frac{1}{3})^x$

Make a table of values.

$g(-2) = (\frac{1}{3})^{-2} = (\frac{3}{1})^2 = 9$,

$g(-1) = (\frac{1}{3})^{-1} = (\frac{3}{1})^1 = 3$, and so on.

x	-2	-1	0	1	2
$g(x)$	9	3	1	$\frac{1}{3}$	$\frac{1}{9}$

Plot the points from the table and draw a smooth curve through them.

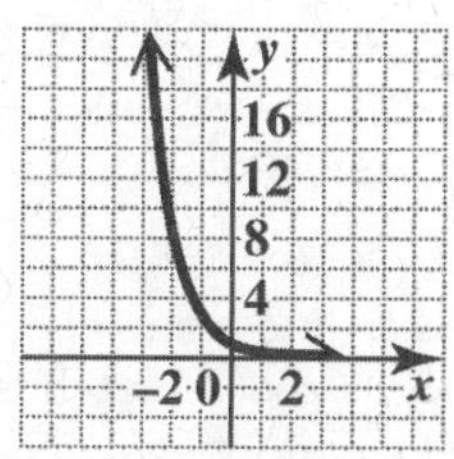

$g(x) = (\frac{1}{3})^x$

9. $y = 4^{-x}$
This equation can be rewritten as

$$y = (4^{-1})^x = \left(\tfrac{1}{4}\right)^x,$$

which shows that it is *falling* from left to right. Make a table of values.

x	-2	-1	0	1	2
y	16	4	1	$\frac{1}{4}$	$\frac{1}{16}$

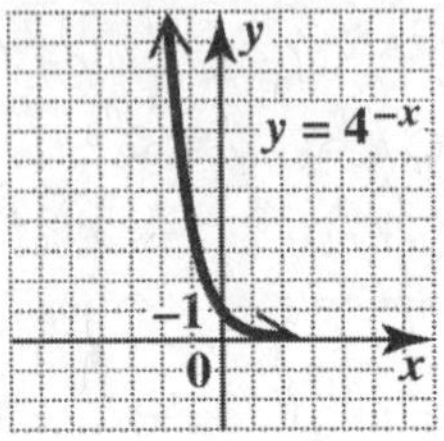

11. $y = 2^{2x-2}$
Make a table of values. It will help to find values for $2x - 2$ before you find y.

x	-2	-1	0	1	2	3
$2x-2$	-6	-4	-2	0	2	4
y	$\frac{1}{64}$	$\frac{1}{16}$	$\frac{1}{4}$	1	4	16

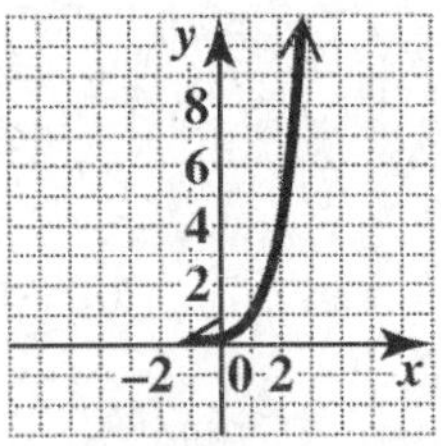

$y = 2^{2x-2}$

13. For an exponential function defined by $f(x) = a^x$, if $a > 1$, the graph rises from left to right. (See Example 1, $f(x) = 2^x$, in your text.) If $0 < a < 1$, the graph falls from left to right. (See Example 2, $g(x) = \left(\frac{1}{2}\right)^x = 2^{-x}$, in your text.)

15.
$$6^x = 36$$
Write each side as a power of 6.
$$6^x = 6^2$$
For $a > 0$ and $a \neq 1$, if $a^x = a^y$, then $x = y$. Set the exponents equal to each other.
$$x = 2$$

Check $x = 2$: $6^2 = 36$ *True*
The solution set is $\{2\}$.

17.
$$100^x = 1000$$
Write each side as a power of 10.
$$(10^2)^x = 10^3$$
$$10^{2x} = 10^3$$
For $a > 0$ and $a \neq 1$, if $a^x = a^y$, then $x = y$. Set the exponents equal to each other.
$$2x = 3$$
$$x = \tfrac{3}{2}$$

Check $x = \frac{3}{2}$: $100^{3/2} = 1000$ *True*
The solution set is $\left\{\frac{3}{2}\right\}$.

19. $16^{2x+1} = 64^{x+3}$
Write each side as a power of 4.
$(4^2)^{2x+1} = (4^3)^{x+3}$
$4^{4x+2} = 4^{3x+9}$
Set the exponents equal.
$4x + 2 = 3x + 9$
$x = 7$

Check $x = 7$: $16^{15} = 64^{10}$ *True*
The solution set is $\{7\}$.

21. $5^x = \frac{1}{125}$
$5^x = \left(\frac{1}{5}\right)^3$
Write each side as a power of 5.
$5^x = 5^{-3}$
Set the exponents equal.
$x = -3$

Check $x = -3$: $5^{-3} = \frac{1}{125}$ *True*
The solution set is $\{-3\}$.

23. $5^x = 0.2$
$5^x = \frac{2}{10} = \frac{1}{5}$
Write each side as a power of 5.
$5^x = 5^{-1}$
Set the exponents equal.
$x = -1$

Check $x = -1$: $5^{-1} = 0.2$ *True*
The solution set is $\{-1\}$.

25. $\left(\frac{3}{2}\right)^x = \frac{8}{27}$
$\left(\frac{3}{2}\right)^x = \left(\frac{2}{3}\right)^3$
Write each side as a power of $\frac{3}{2}$.
$\left(\frac{3}{2}\right)^x = \left(\frac{3}{2}\right)^{-3}$
Set the exponents equal.
$x = -3$

Check $x = -3$: $\left(\frac{3}{2}\right)^{-3} = \frac{8}{27}$ *True*
The solution set is $\{-3\}$.

27. $12^{2.6} \approx 639.545$

29. $0.5^{3.921} \approx 0.066$

31. $2.718^{2.5} \approx 12.179$

33. **(a)** $y = (1.046 \times 10^{-38})(1.0444^x)$
$= (1.046 \times 10^{-38})(1.0444^{2000})$
$\approx 0.57 \approx 0.6$

The increase for the exponential function in the year 2000 is about 0.6°C.

(b) $y = 0.009x - 17.67$
$= 0.009(2000) - 17.67$
$= 0.33 \approx 0.3$

The increase for the linear function in the year 2000 is about 0.3°C.

35. **(a)** $y = (1.046 \times 10^{-38})(1.0444^x)$
$= (1.046 \times 10^{-38})(1.0444^{2020})$
$\approx 1.35 \approx 1.4$

The increase for the exponential function in the year 2020 is about 1.4°C.

(b) $y = 0.009x - 17.67$
$= 0.009(2020) - 17.67$
$= 0.51 \approx 0.5$

The increase for the linear function in the year 2020 is about 0.5°C.

37. $f(x) = 4231(1.0174)^x$

(a) 1980 corresponds to $x = 10$.

$$f(10) = 4231(1.0174)^{10} \approx 5028$$

The answer has units in millions of metric tons.

(b) 1995 corresponds to $x = 25$.

$$f(25) = 4231(1.0174)^{25} \approx 6512$$

(c) 2000 corresponds to $x = 30$.

$$f(30) = 4231(1.0174)^{30} \approx 7099$$

The actual amount, 6735 millions of metric tons, is less than the 7099 millions of metric tons that the model provides.

39. $V(t) = 5000(2)^{-0.15t}$
(a) The original value is found when $t = 0$.

$$V(0) = 5000(2)^{-0.15(0)} = 5000(2)^0 = 5000(1) = 5000$$

The original value is \$5000.

(b) The value after 5 years is found when $t = 5$.

$$V(5) = 5000(2)^{-0.15(5)} = 5000(2)^{-0.75} \approx 2973.02$$

The value after 5 years is about \$2973.

(c) The value after 10 years is found when $t = 10$.

$$V(10) = 5000(2)^{-0.15(10)} = 5000(2)^{-1.5} \approx 1767.77$$

The value after 10 years is about \$1768.

(d) Use the results of parts (a) – (c) to make a table of values.

t	0	5	10
$V(t)$	5000	2973	1768

Plot the points from the table and draw a smooth curve through them.

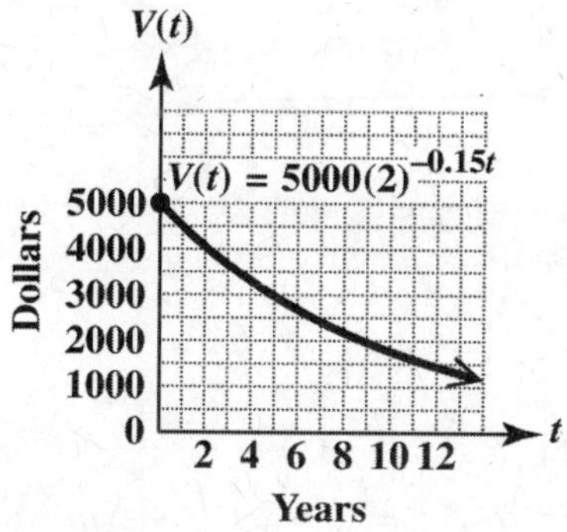

41. $V(t) = 5000(2)^{-0.15t}$

$2500 = 5000(2)^{-0.15t}$ *Let V(t) = 2500.*

$\frac{1}{2} = (2)^{-0.15t}$ *Divide by 5000.*

$2^{-1} = 2^{-0.15t}$

$-1 = -0.15t$ *Equate exponents.*

$t = \frac{-1}{-0.15} \approx 6.67$

The value of the machine will be \$2500 in approximately 6.67 years after it was purchased.

43. $16 = 2 \cdot 2 \cdot 2 \cdot 2 = 2^4$, so $\square = 4$.

45. $2^0 = 1$, so $\square = 0$.

12.3 Logarithmic Functions

12.3 Now Try Exercises

N1. **(a)** For the exponential form $6^3 = 216$, the logarithmic form is $\log_6 216 = 3$.

(b) For the logarithmic form $\log_{64} 4 = \frac{1}{3}$, the exponential form is $64^{1/3} = 4$.

N2. **(a)** $\log_2 x = -5$
Write in exponential form.

$$x = 2^{-5}$$
$$x = \frac{1}{2^5} = \frac{1}{32}$$

The argument (the input of the logarithm) must be a positive number, so $x = \frac{1}{32}$ is acceptable.

The solution set is $\{\frac{1}{32}\}$.

(b) $\log_{3/2}(2x - 1) = 3$

$2x - 1 = \left(\frac{3}{2}\right)^3$ *Exponential form*

$2x - 1 = \frac{27}{8}$ *Apply the exponent.*

$2x = \frac{35}{8}$ *Add 1.*

$x = \frac{35}{16}$ *Divide by 2.*

Substitute $\frac{35}{16}$ for x in the argument, $2x - 1$, $2\left(\frac{35}{16}\right) - 1 = \frac{35}{8} - 1 = \frac{27}{8} > 0$, so $\frac{35}{16}$ is acceptable.

The solution set is $\left\{\frac{35}{16}\right\}$.

(c) $\log_x 10 = 2$

$x^2 = 10$ *Exponential form*

$x = \pm\sqrt{10}$ *Take square roots.*

Reject $x = -\sqrt{10}$ since the base of a logarithm must be positive and not equal to 1.

The solution set is $\left\{\sqrt{10}\right\}$.

(d) $\log_{125} \sqrt[3]{5} = x$

$125^x = \sqrt[3]{5}$ *Exponential form*

$(5^3)^x = 5^{1/3}$ *Same base*

$5^{3x} = 5^{1/3}$ *Power rule*

$3x = \frac{1}{3}$ *Equate exponents.*

$x = \frac{1}{9}$ *Divide by 3.*

The solution set is $\left\{\frac{1}{9}\right\}$.

N3. For each expression, use the properties of logarithms,

$$\log_b b = 1 \text{ and } \log_b 1 = 0,$$

with $b > 0$ and $b \neq 1$.

(a) $\log_{10} 10 = 1$ **(b)** $\log_8 1 = 0$

(c) $\log_{0.1} 1 = 0$

N4. $y = f(x) = \log_6 x$ is equivalent to $x = 6^y$. Choose values for y and find x.

x	$\frac{1}{36}$	$\frac{1}{6}$	1	6	36
y	-2	-1	0	1	2

Plot the points, and draw a smooth curve through them.

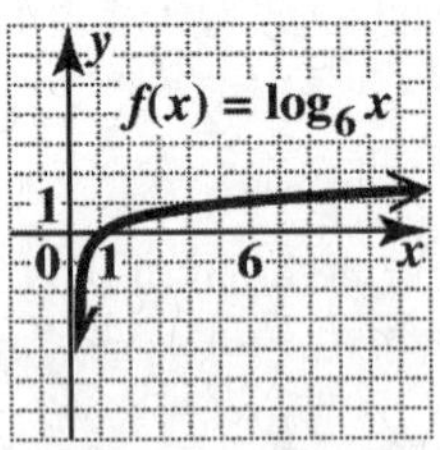

N5. $y = g(x) = \log_{1/4} x$ is equivalent to $x = \left(\frac{1}{4}\right)^y$.

Choose values for y and find x.

x	4	1	$\frac{1}{4}$	$\frac{1}{16}$
y	-1	0	1	2

Plot the points, and draw a smooth curve through them.

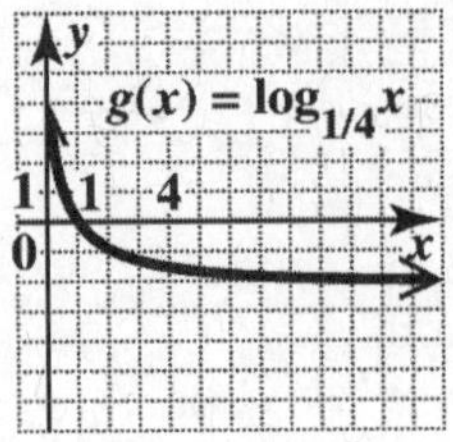

N6. $G(t) = 15.0 + 2.00 \log_{10} t$

(a) $G(1) = 15.0 + 2.00 \log_{10} 1$ *Let $t = 1$.*

$= 15.0 + 2.00(0)$ *$\log_{10} 1 = 0$*

$= 15.0$

The GNP in 2004 was \$15.0 million.

(b) $G(10) = 15.0 + 2.00 \log_{10} 10$ *Let $t = 10$.*

$= 15.0 + 2.00(1)$ *$\log_{10} 10 = 1$*

$= 17.0$

The GNP in 2013 is predicted to be \$17.0 million.

12.3 Section Exercises

1. **(a)** $\log_{1/3} 3 = -1$ is equivalent to $\left(\frac{1}{3}\right)^{-1} = 3$. **(B)**

(b) $\log_5 1 = 0$ is equivalent to $5^0 = 1$. **(E)**

(c) $\log_2 \sqrt{2} = \frac{1}{2}$ is equivalent to $2^{1/2} = \sqrt{2}$. **(D)**

(d) $\log_{10} 1000 = 3$ is equivalent to $10^3 = 1000$. **(F)**

(e) $\log_8 \sqrt[3]{8} = \frac{1}{3}$ is equivalent to $8^{1/3} = \sqrt[3]{8}$. **(A)**

(f) $\log_4 4 = 1$ is equivalent to $4^1 = 4$. **(C)**

3. The base is 4, the exponent (logarithm) is 5, and the number is 1024, so $4^5 = 1024$ becomes $\log_4 1024 = 5$ in logarithmic form.

5. $\frac{1}{2}$ is the base and -3 is the exponent, so $\left(\frac{1}{2}\right)^{-3} = 8$ becomes $\log_{1/2} 8 = -3$ in logarithmic form.

7. The base is 10, the exponent (logarithm) is -3, and the number is 0.001, so $10^{-3} = 0.001$ becomes $\log_{10} 0.001 = -3$ in logarithmic form.

9. $\sqrt[4]{625} = 625^{1/4} = 5$

The base is 625, the exponent (logarithm) is $\frac{1}{4}$, and the number is 5, so $\sqrt[4]{625} = 5$ becomes $\log_{625} 5 = \frac{1}{4}$ in logarithmic form.

11. $8^{-2/3} = \frac{1}{4}$ becomes $\log_8 \frac{1}{4} = -\frac{2}{3}$ in logarithmic form.

13. $5^0 = 1$ becomes $\log_5 1 = 0$ in logarithmic form.

15. In $\log_4 64 = 3$, 4 is the base and 3 is the logarithm (exponent), so $\log_4 64 = 3$ becomes $4^3 = 64$ in exponential form.

17. In $\log_{10} \frac{1}{10{,}000} = -4$, the base is 10, the logarithm (exponent) is -4, and the number is $\frac{1}{10{,}000}$, so $\log_{10} \frac{1}{10{,}000} = -4$ becomes $10^{-4} = \frac{1}{10{,}000}$ in exponential form.

19. In $\log_6 1 = 0$, 6 is the base and 0 is the logarithm (exponent), so $\log_6 1 = 0$ becomes $6^0 = 1$ in exponential form.

21. In $\log_9 3 = \frac{1}{2}$, the base is 9, the logarithm (exponent) is $\frac{1}{2}$, and the number is 3, so $\log_9 3 = \frac{1}{2}$ becomes $9^{1/2} = 3$ in exponential form.

23. $\log_{1/4} \frac{1}{2} = \frac{1}{2}$ becomes $(\frac{1}{4})^{1/2} = \frac{1}{2}$ in exponential form.

25. $\log_5 5^{-1} = -1$ becomes $5^{-1} = 5^{-1}$ in exponential form.

27. Use the properties of logarithms,

$$\log_b b = 1 \quad \text{and} \quad \log_b 1 = 0,$$

for $b > 0$, $b \neq 1$.

(a) $\log_8 8 = 1$ **(C)**

(b) $\log_{16} 1 = 0$ **(B)**

(c) $\log_{0.3} 1 = 0$ **(B)**

(d) $\log_{\sqrt{7}} \sqrt{7} = 1$ **(C)**

29.
$$x = \log_{27} 3$$
Write in exponential form.
$$27^x = 3$$
Write each side as a power of 3.
$$(3^3)^x = 3$$
$$3^{3x} = 3^1$$
Set the exponents equal.
$$3x = 1$$
$$x = \tfrac{1}{3}$$

Check $x = \frac{1}{3}$: $\frac{1}{3} = \log_{27} 3$ since $27^{1/3} = 3$.

The solution set is $\{\frac{1}{3}\}$.

31.
$$\log_x 9 = \tfrac{1}{2}$$
Change to exponential form.
$$x^{1/2} = 9$$
$$(x^{1/2})^2 = 9^2 \quad \textit{Square.}$$
$$x^1 = 81$$
$$x = 81$$

$x = 81$ is an acceptable base since it is a positive number (not equal to 1).

Check $x = 81$: $\log_{81} 9 = \frac{1}{2}$ since $81^{1/2} = 9$

The solution set is $\{81\}$.

33.
$$\log_x 125 = -3$$
Write in exponential form.
$$x^{-3} = 125$$
$$\frac{1}{x^3} = 125$$
$$1 = 125(x^3)$$
$$\frac{1}{125} = x^3$$
Take the cube root of each side.
$$\sqrt[3]{\frac{1}{125}} = \sqrt[3]{x^3}$$
$$x = \sqrt[3]{\frac{1}{5^3}} = \frac{1}{5}$$

$x = \frac{1}{5}$ is an acceptable base since it is a positive number (not equal to 1).

The solution set is $\{\frac{1}{5}\}$.

35. $\log_{12} x = 0$
$$12^0 = x \quad \textit{Write in exponential form.}$$
$$1 = x$$

The argument (the input of the logarithm) must be a positive number, so $x = 1$ is acceptable.

The solution set is $\{1\}$.

37. $\log_x x = 1$
$$x^1 = x \quad \textit{Write in exponential form.}$$

This equation is true for all the numbers x that are allowed as the base of a logarithm; that is, all positive numbers x, $x \neq 1$.

The solution set is $\{x \mid x > 0,\ x \neq 1\}$.

39. $\log_x \frac{1}{25} = -2$
$$x^{-2} = \frac{1}{25} \quad \textit{Write in exponential form.}$$
$$\frac{1}{x^2} = \frac{1}{25}$$
$$x^2 = 25 \quad \textit{Denominators must be equal.}$$
$$x = \pm 5$$

Reject $x = -5$ since the base of a logarithm must be positive and not equal to 1.

The solution set is $\{5\}$.

41. $\log_8 32 = x$
$$8^x = 32 \quad \textit{Exponential form}$$
Write each side as a power of 2.
$$(2^3)^x = 2^5$$
$$2^{3x} = 2^5$$
$$3x = 5 \quad \textit{Equate exponents.}$$
$$x = \tfrac{5}{3}$$

Check $x = \frac{5}{3}$: $\log_8 32 = \frac{5}{3}$ since $8^{5/3} = 2^5 = 32$.

The solution set is $\{\frac{5}{3}\}$.

43. $\log_\pi \pi^4 = x$

$\pi^x = \pi^4$ *Exponential form*

$x = 4$ *Equate exponents.*

Check $x = 4$: $\log_\pi \pi^4 = 4$ since $\pi^4 = \pi^4$.

The solution set is $\{4\}$.

45. $\log_6 \sqrt{216} = x$

$\log_6 216^{1/2} = x$ *Equivalent form*

$6^x = 216^{1/2}$ *Exponential form*

$6^x = (6^3)^{1/2}$ *Same base*

$6^x = 6^{3/2}$

$x = \frac{3}{2}$ *Equate exponents.*

Check $x = \frac{3}{2}$:

$\log_6 \sqrt{216} = \frac{3}{2}$ since $6^{3/2} = \sqrt{6^3} = \sqrt{216}$.

The solution set is $\{\frac{3}{2}\}$.

47. $\log_4 (2x + 4) = 3$

$2x + 4 = 4^3$ *Exponential form*

$2x = 64 - 4$

$2x = 60$

$x = 30$

Check $x = 30$: $\log_4 (2 \cdot 30 + 4) = \log_4 64 = 3$.

The solution set is $\{30\}$.

49. $y = \log_3 x$

$3^y = x$ *Exponential form*

Refer to Section 12.2, Exercise 5, for the graph of $f(x) = 3^x$. Since $y = \log_3 x$ (or $3^y = x$) is the inverse of $f(x) = y = 3^x$, its graph is symmetric about the line $y = x$ to the graph of $f(x) = 3^x$. The graph can be plotted by reversing the ordered pairs in the table of values belonging to $f(x) = 3^x$.

x	$\frac{1}{9}$	$\frac{1}{3}$	1	3	9
y	-2	-1	0	1	2

Plot the points, and draw a smooth curve through them.

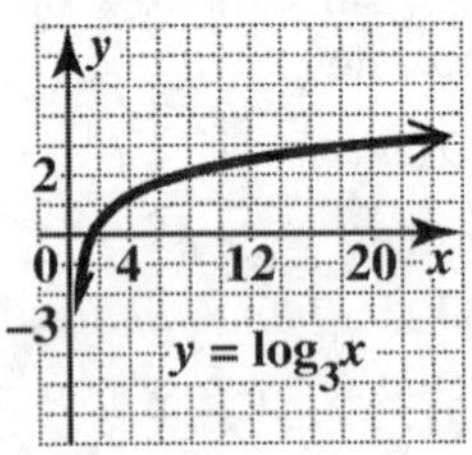

51. $y = \log_{1/3} x$

$\left(\frac{1}{3}\right)^y = x$ *Exponential form*

Refer to Section 12.2, Exercise 7, for the graph of $g(x) = \left(\frac{1}{3}\right)^x$. Since $y = \log_{1/3} x$ (or $\left(\frac{1}{3}\right)^y = x$) is the inverse of $y = \left(\frac{1}{3}\right)^x$, its graph is symmetric about the line $y = x$ to the graph of $y = \left(\frac{1}{3}\right)^x$. The graph can be plotted by reversing the ordered pairs in the table of values belonging to $g(x) = \left(\frac{1}{3}\right)^x$.

x	9	3	1	$\frac{1}{3}$	$\frac{1}{9}$
y	-2	-1	0	1	2

Plot the points, and draw a smooth curve through them.

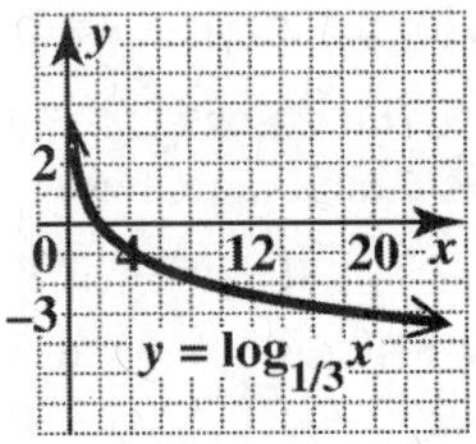

53. The number 1 is not used as a base for a logarithmic function since the function would look like $x = 1^y$ in exponential form. Then, for any real value of y, the statement $1 = 1$ would always be the result since every power of 1 is equal to 1.

55. The range of $f(x) = a^x$ is the domain of $g(x) = \log_a x$, that is, $\underline{(0, \infty)}$.
The domain of $f(x) = a^x$ is the range of $g(x) = \log_a x$, that is, $\underline{(-\infty, \infty)}$.

57. The values of t are on the horizontal axis, and the values of $f(t)$ are on the vertical axis. Read the value of $f(t)$ from the graph for the given value of t. At $t = 0$, $f(0) = 8$.

59. To find $f(60)$, find 60 on the t-axis, then go up to the graph and across to the $f(t)$ axis to read the value of $f(60)$. At $t = 60$, $f(60) = 24$.

61. $f(x) = 3800 + 585 \log_2 x$

(a) $x = 1982 - 1980 = 2$

$$\begin{aligned} f(2) &= 3800 + 585 \log_2 2 \\ &= 3800 + 585(1) \\ &= 4385 \end{aligned}$$

The model gives an approximate withdrawal of 4385 billion ft^3 of natural gas from crude oil wells in the United States for 1982.

(b) $x = 1988 - 1980 = 8$

$$\begin{aligned} f(8) &= 3800 + 585 \log_2 8 \\ &= 3800 + 585(3) \\ &= 5555 \end{aligned}$$

The model gives an approximate withdrawal of 5555 billion ft^3 of natural gas from crude oil wells in the United States for 1988.

(c) $x = 1996 - 1980 = 16$

$$\begin{aligned} f(16) &= 3800 + 585\log_2 16 \\ &= 3800 + 585(4) \\ &= 6140 \end{aligned}$$

The model gives an approximate withdrawal of 6140 billion ft^3 of natural gas from crude oil wells in the United States for 1996.

63. $S(t) = 100 + 30\log_3(2t+1)$

(a) $$\begin{aligned} S(1) &= 100 + 30\log_3(2 \cdot 1 + 1) \\ &= 100 + 30\log_3(3) \\ &= 100 + 30(1) = 130 \end{aligned}$$

After 1 year, the sales were 130 thousand units.

(b) $$\begin{aligned} S(13) &= 100 + 30\log_3(2 \cdot 13 + 1) \\ &= 100 + 30\log_3(27) \\ &= 100 + 30(3) = 190 \end{aligned}$$

After 13 years, the sales were 190 thousand units.

(c) Make a table of values, plot the points they represent, and draw a smooth curve through them.

To make the table, find values of t such that $2t + 1 = 3^k$, where $k = 0, 1, 2, 3, 4$.

k	0	1	2	3	4
3^k	1	3	9	27	81
$2t+1$	1	3	9	27	81
$2t$	0	2	8	26	80
t	0	1	4	13	40
$S(t)$	100	130	160	190	220

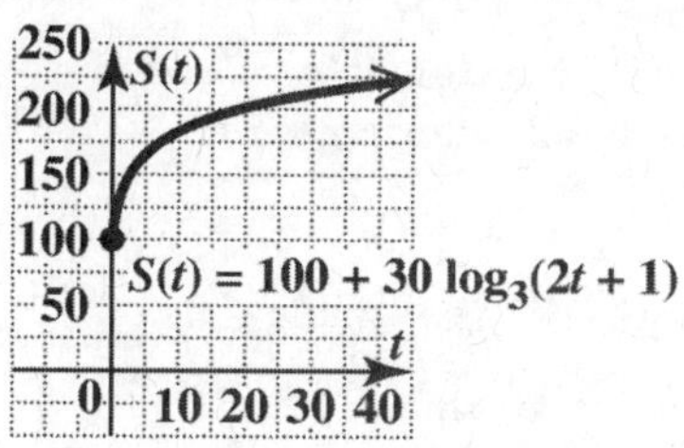

65. $R = \log_{10} \dfrac{x}{x_0}$

Change to exponential form.

$$10^R = \frac{x}{x_0}, \text{ so } x = x_0\,10^R.$$

Let $R = 6.7$ for the Northridge earthquake, with intensity x_1.

$$x_1 = x_0 10^{6.7}$$

Let $R = 7.3$ for the Landers earthquake, with intensity x_2.

$$x_2 = x_0 10^{7.3}$$

The ratio of x_2 to x_1 is

$$\frac{x_2}{x_1} = \frac{x_0 10^{7.3}}{x_0 10^{6.7}} = 10^{0.6} \approx 3.98.$$

The Landers earthquake was about 4 times more powerful than the Northridge earthquake.

67. $g(x) = \log_3 x$
On a TI-83/4, assign 3^x to Y_1. Then enter

DrawInv Y_1

on the home screen to obtain the figure that follows. DrawInv is choice 8 under the DRAW menu. Y_1 is choice 1 under VARS, Y-VARS, Function.

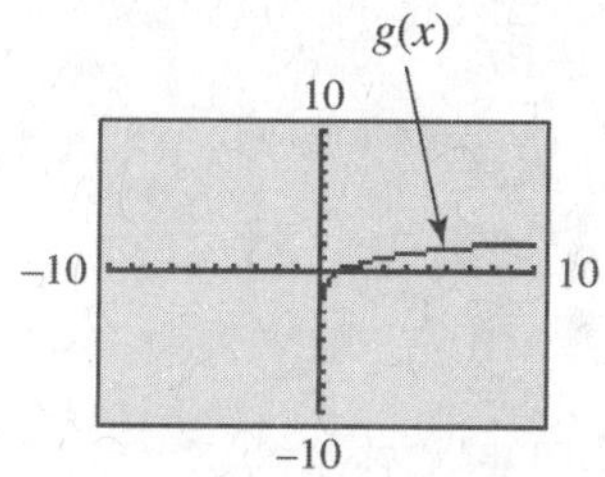

69. $g(x) = \log_{1/3} x$
Assign (1/3)^x to Y_1 and enter DrawInv Y_1.

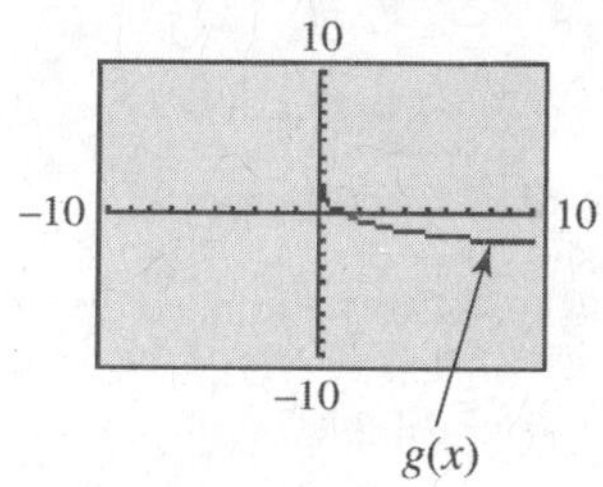

71. $4^7 \cdot 4^2 = 4^{7+2} = 4^9$

73. $\dfrac{7^8}{7^{-4}} = 7^{8-(-4)} = 7^{12}$

12.4 Properties of Logarithms

12.4 Now Try Exercises

N1. Use the product rule for logarithms.

$$\log_b xy = \log_b x + \log_b y$$

(a) $\log_{10}(7 \cdot 9) = \log_{10} 7 + \log_{10} 9$

(b) $$\begin{aligned} \log_5 11 + \log_5 8 &= \log_5(11 \cdot 8) \\ &= \log_5 88 \end{aligned}$$

(c) $$\begin{aligned} \log_5(5x) &= \log_5 5 + \log_5 x \\ &= 1 + \log_5 x \quad (x > 0) \end{aligned}$$

(d) $$\begin{aligned} \log_2 t^3 &= \log_2(t \cdot t \cdot t) \\ &= \log_2 t + \log_2 t + \log_2 t \\ &= 3\log_2 t \quad (t > 0) \end{aligned}$$

N2. Use the quotient rule for logarithms.

$$\log_b \frac{x}{y} = \log_b x - \log_b y$$

(a) $\log_{10} \frac{7}{9} = \log_{10} 7 - \log_{10} 9$

(b) $\log_4 x - \log_4 12 = \log_4 \frac{x}{12} \quad (x > 0)$

(c) $\log_5 \frac{25}{27} = \log_5 25 - \log_5 27$
$= 2 - \log_5 27$

N3. Use the power rule for logarithms.

$$\log_b x^r = r \log_b x$$

(a) $\log_7 5^3 = 3 \log_7 5$

(b) $\log_a \sqrt{10} = \log_a 10^{1/2}$
$= \frac{1}{2} \log_a 10 \quad (a > 0)$

(c) $\log_3 \sqrt[4]{x^3} = \log_3 x^{3/4}$
$= \frac{3}{4} \log_3 x \quad (x > 0)$

N4. Use the Special Properties,

$$b^{\log_b x} = x \quad \text{or} \quad \log_b b^x = x.$$

(a) By the second property,

$$\log_4 4^7 = 7.$$

(b) Using the second property,

$$\log_{10} 10{,}000 = \log_{10} 10^4 = 4.$$

(c) By the first property,

$$8^{\log_8 5} = 5.$$

N5. Use the properties of logarithms.

(a) $\log_3 9z^4$
$= \log_3 9 + \log_3 z^4$ *Product rule*
$= 2 + 4 \log_3 z$ *$\log_3 3^2 = 2$; Power rule*

(b) $\log_6 \sqrt{\frac{n}{3m}}$
$= \log_6 \left(\frac{n}{3m}\right)^{1/2}$
$= \frac{1}{2} \log_6 \left(\frac{n}{3m}\right)$ *Po. rule*
$= \frac{1}{2}(\log_6 n - \log_6 3m)$ *Qu. rule*
$= \frac{1}{2}[\log_6 n - (\log_6 3 + \log_6 m)]$
$= \frac{1}{2}(\log_6 n - \log_6 3 - \log_6 m)$

(c) $\log_2 x + 3 \log_2 y - \log_2 z$
$= \log_2 x + \log_2 y^3 - \log_2 z$ *Power rule*
$= \log_2 (xy^3) - \log_2 z$ *Product rule*
$= \log_2 \frac{xy^3}{z}$ *Quotient rule*

(d) $\log_5 (x + 10) + \log_5 (x - 10) - \frac{3}{5} \log_5 x \quad (x > 10)$
$= \log_5 [(x + 10)(x - 10)] - \frac{3}{5} \log_5 x$ *Product rule*
$= \log_5 (x^2 - 100) - \log_5 x^{3/5}$ *Power rule*
$= \log_5 \frac{x^2 - 100}{x^{3/5}}$ *Quotient rule*

(e) $\log_7 (49 + 2x)$ cannot be written as a sum of logarithms, since

$$\log_b (M + N) \neq \log_b M + \log_b N.$$

There is no property of logarithms to rewrite the logarithm of a sum.

N6. **(a)** $\log_2 70 = \log_2 (7 \cdot 10)$
$= \log_2 7 + \log_2 10$
$= 2.8074 + 3.3219$
$= 6.1293$

(b) $\log_2 0.7 = \log_2 \frac{7}{10}$
$= \log_2 7 - \log_2 10$
$= 2.8074 - 3.3219$
$= -0.5145$

(c) $\log_2 49 = \log_2 7^2$
$= 2 \log_2 7$
$= 2(2.8074)$
$= 5.6148$

N7. **(a)** $\log_2 16 + \log_2 16 = \log_2 32$

Evaluate each side.

LS $= \log_2 16 + \log_2 16$
$= \log_2 2^4 + \log_2 2^4 = 4 + 4 = 8$
RS $= \log_2 32 = \log_2 2^5 = 5$

The statement is *false* because $8 \neq 5$.

(b) $(\log_2 4)(\log_3 9) = \log_6 36$

Evaluate each side.

LS $= (\log_2 4)(\log_3 9)$
$= (\log_2 2^2)(\log_3 3^2) = (2)(2) = 4$
RS $= \log_6 36 = \log_6 6^2 = 2$

The statement is *false* because $4 \neq 2$.

12.4 Section Exercises

1. By the product rule,

$$\log_{10} (7 \cdot 8) = \log_{10} 7 + \log_{10} 8.$$

3. By a special property (see page 776 in the text),

$$3^{\log_3 4} = 4.$$

5. By a special property (see page 776 in the text),

$$\log_3 3^9 = 9.$$

7. Use the product rule for logarithms.

$$\log_7 (4 \cdot 5) = \log_7 4 + \log_7 5$$

9. Use the quotient rule for logarithms.

$$\log_5 \tfrac{8}{3} = \log_5 8 - \log_5 3$$

11. Use the power rule for logarithms.

$$\log_4 6^2 = 2 \log_4 6$$

13. $\log_3 \dfrac{\sqrt[3]{4}}{x^2 y} = \log_3 \dfrac{4^{1/3}}{x^2 y}$

Use the quotient rule for logarithms.

$= \log_3 4^{1/3} - \log_3 (x^2 y)$

Use the product rule for logarithms.

$= \log_3 4^{1/3} - \left(\log_3 x^2 + \log_3 y\right)$

$= \log_3 4^{1/3} - \log_3 x^2 - \log_3 y$

Use the power rule for logarithms.

$= \frac{1}{3} \log_3 4 - 2 \log_3 x - \log_3 y$

15. $\log_3 \sqrt{\dfrac{xy}{5}}$

$= \log_3 \left(\dfrac{xy}{5}\right)^{1/2}$

$= \frac{1}{2} \log_3 \left(\frac{xy}{5}\right)$ *Power rule*

$= \frac{1}{2} [\log_3 (xy) - \log_3 5]$ *Quotient rule*

$= \frac{1}{2} (\log_3 x + \log_3 y - \log_3 5)$ *Product rule*

$= \frac{1}{2} \log_3 x + \frac{1}{2} \log_3 y - \frac{1}{2} \log_3 5$

17. $\log_2 \dfrac{\sqrt[3]{x} \cdot \sqrt[5]{y}}{r^2}$

$= \log_2 \dfrac{x^{1/3} y^{1/5}}{r^2}$

$= \log_2 \left(x^{1/3} y^{1/5}\right) - \log_2 r^2$ *Quotient rule*

$= \log_2 x^{1/3} + \log_2 y^{1/5} - \log_2 r^2$ *Product rule*

$= \frac{1}{3} \log_2 x + \frac{1}{5} \log_2 y - 2 \log_2 r$ *Power rule*

19. In the notation $\log_a (x + y)$, the parentheses do not indicate multiplication. They indicate that $x + y$ is the result of raising a to some power.

21. By the product rule for logarithms,

$$\log_b x + \log_b y = \log_b xy.$$

23. By the quotient rule for logarithms,

$$\log_a m - \log_a n = \log_a \tfrac{m}{n}.$$

25. $(\log_a r - \log_a s) + 3 \log_a t$

$= \log_a \dfrac{r}{s} + \log_a t^3$ *Quotient rule; Power rule*

$= \log_a \dfrac{rt^3}{s}$ *Product rule*

27. $3 \log_a 5 - 4 \log_a 3$

$= \log_a 5^3 - \log_a 3^4$ *Power rule*

$= \log_a \dfrac{5^3}{3^4}$ *Quotient rule*

$= \log_a \frac{125}{81}$

29. $\log_{10} (x + 3) + \log_{10} (x - 3)$

$= \log_{10} [(x + 3)(x - 3)]$ *Product rule*

$= \log_{10} (x^2 - 9)$

31. By the power rule for logarithms,

$3 \log_p x + \frac{1}{2} \log_p y - \frac{3}{2} \log_p z - 3 \log_p a$

$= \log_p x^3 + \log_p y^{1/2} - \log_p z^{3/2} - \log_p a^3.$

Group the terms into sums.

$= (\log_p x^3 + \log_p y^{1/2}) - (\log_p z^{3/2} + \log_p a^3)$

$= \log_p \left(x^3 y^{1/2}\right) - \log_p \left(z^{3/2} a^3\right)$ *Product rule*

$= \log_p \dfrac{x^3 y^{1/2}}{z^{3/2} a^3}$ *Quotient rule*

In Exercises 33–44,
$\log_{10} 2 \approx 0.3010$ and $\log_{10} 9 \approx 0.9542$.

33. $\log_{10} 18 = \log_{10} (2 \cdot 9)$

$= \log_{10} 2 + \log_{10} 9$ *Product rule*

$\approx 0.3010 + 0.9542$

$= 1.2552$

35. $\log_{10} \frac{2}{9} = \log_{10} 2 - \log_{10} 9$ *Quotient rule*

$\approx 0.3010 - 0.9542$

$= -0.6532$

37. Use the product and power rules for logarithms.

$\log_{10} 36 = \log_{10} \left(2^2 \cdot 9\right)$

$= 2 \log_{10} 2 + \log_{10} 9$

$\approx 2(0.3010) + 0.9542$

$= 1.5562.$

39. $\log_{10} \sqrt[4]{9} = \log_{10} 9^{1/4}$

$= \frac{1}{4} \log_{10} 9$ *Power rule*

$\approx \frac{1}{4}(0.9542)$

$= 0.23855 \approx 0.2386$

41. $\log_{10} 3 = \log_{10} 9^{1/2}$ *Rename 3*

$= \frac{1}{2} \log_{10} 9$ *Power rule*

$\approx \frac{1}{2}(0.9542)$

$= 0.4771$

43. $\log_{10} 9^5 = 5 \log_{10} 9$ *Power rule*

$\approx 5(0.9542)$

$= 4.7710$

45. $\text{LS} = \log_2 (8 + 32) = \log_2 40$

$\text{RS} = \log_2 8 + \log_2 32 = \log_2 (8 \cdot 32)$

$= \log_2 256$

$\text{LS} \neq \text{RS}$, so the given statement is *false*.

47. $\log_3 7 + \log_3 7^{-1} = \log_3 7 + (-1)\log_3 7$
$= 0$

The given statement is *true.*

49. $\log_6 60 - \log_6 10 = \log_6 \frac{60}{10}$
$= \log_6 6 = 1$

The given statement is *true.*

51. $\dfrac{\log_{10} 7}{\log_{10} 14} \stackrel{?}{=} \dfrac{1}{2}$

$2\log_{10} 7 \stackrel{?}{=} 1\log_{10}(7 \cdot 2)$ *Cross products are equal.*

$2\log_{10} 7 \stackrel{?}{=} \log_{10} 7 + \log_{10} 2$

$\log_{10} 7 \neq \log_{10} 2$ *Subtract $\log_{10} 7$.*

The given statement is *false.*

53. The exponent of a quotient is the difference between the exponent of the numerator and the exponent of the denominator.

55. We can't determine the logarithm of 0 because no number allowed as a logarithmic base can be raised to a power with a result of 0.

57. $10^4 = 10{,}000$ becomes $\log_{10} 10{,}000 = 4$.

59. $10^{-2} = 0.01$ becomes $\log_{10} 0.01 = -2$.

61. $\log_{10} 1 = 0$ becomes $10^0 = 1$.

12.5 Common and Natural Logarithms

12.5 Now Try Exercises

N1. Evaluate each logarithm using a calculator.

(a) $\log 115 \approx 2.0607$

(b) $\log 0.25 \approx -0.6021$

N2.
$$\begin{aligned} \text{pH} &= -\log[H_3O^+] \\ &= -\log(3.4 \times 10^{-5}) \\ &= -(\log 3.4 + \log 10^{-5}) \\ &= -(\log 3.4 - 5\log 10) \\ &\approx -[0.5315 - 5(1)] \\ &= -0.5315 + 5 \\ &= 4.4685 \approx 4.5 \end{aligned}$$

Since the pH is between 4.0 and 6.0, the wetland is a poor fen.

N3.
$$\begin{aligned} \text{pH} &= -\log[H_3O^+] \\ 2.6 &= -\log[H_3O^+] \quad \textit{Let pH = 2.6.} \\ -2.6 &= \log[H_3O^+] \\ -2.6 &= \log_{10}[H_3O^+] \\ 10^{-2.6} &= [H_3O^+] \quad \textit{Exponential form} \\ [H_3O^+] &\approx 2.5 \times 10^{-3} \end{aligned}$$

N4.
$$\begin{aligned} D &= 10\log\left(\frac{I}{I_0}\right) \\ &= 10\log\left(\frac{6.312 \times 10^{13} I_0}{I_0}\right) \\ &= 10\log(6.312 \times 10^{13}) \\ &\approx 10(13.8) \approx 138 \end{aligned}$$

The level is about 138 dB.

N5. Evaluate each logarithm using a calculator.

(a) $\ln 0.26 \approx -1.3471$

(b) $\ln 12 \approx 2.4849$

(c) $\ln 150 \approx 5.0106$

N6.
$$\begin{aligned} f(x) &= 51{,}600 - 7457\ln x \\ f(600) &= 51{,}600 - 7457\ln 600 \\ &\approx 3898.1 \approx 3900 \end{aligned}$$

Atmospheric pressure is 600 millibars at approximately 3900 m.

N7. The change-of-base rule is

$$\log_a x = \frac{\log_b x}{\log_b a}.$$

Use common logarithms ($b = 10$).

$$\log_8 60 = \frac{\log_{10} 60}{\log_{10} 8} = \frac{\log 60}{\log 8} \approx 1.9690$$

N8. $f(x) = 2014 + 384.7\log_2 x$

2002 corresponds to $x = 2002 - 1989 = 13$.

$$\begin{aligned} f(13) &= 2014 + 384.7\log_2 13 \\ &= 2014 + 384.7\left(\frac{\log 13}{\log 2}\right) \\ &\approx 2014 + 384.7(3.7004) \\ &\approx 3438 \end{aligned}$$

The model gives about 3438 million barrels for 2002, which is greater than the actual amount of 3336 million barrels.

12.5 Section Exercises

1. Since $\log x = \log_{10} x$, the base is 10. The correct response is **C**.

3. $10^0 = 1$ and $10^1 = 10$, so $\log 1 = 0$ and $\log 10 = 1$. Thus, the value of $\log 6.3$ must lie between 0 and 1. The correct response is **C**.

5. By the special property, $\log_b b^x = x$,

$$\log 10^{31.6} = \log_{10} 10^{31.6} = 31.6.$$

7. To four decimal places,

$$\log 43 \approx 1.6335.$$

9. $\log 328.4 \approx 2.5164$

11. $\log 0.0326 \approx -1.4868$

13. $\log(4.76 \times 10^9) \approx 9.6776$
On a TI-83/4, enter

[LOG] 4.76 [2nd] [EE] 9).

15. $\ln 7.84 \approx 2.0592$

17. $\ln 0.0556 \approx -2.8896$

19. $\ln 388.1 \approx 5.9613$

21. $\ln(8.59 \times e^2) \approx 4.1506$
On a TI-83/4, enter

[LN] 8.59 [X] [2nd] [e^x] 2)).

23. $\ln 10 \approx 2.3026$

25. **(a)** $\log 356.8 \approx 2.552\,424\,846$

(b) $\log 35.68 \approx 1.552\,424\,846$

(c) $\log 3.568 \approx 0.552\,424\,846$

(d) The whole number part of the answers (2, 1, or 0) varies, whereas the decimal part (0.552 424 846) remains the same, indicating that the whole number part corresponds to the placement of the decimal point and the decimal part corresponds to the digits 3, 5, 6, and 8.

27. $$\begin{aligned}\text{pH} &= -\log[\text{H}_3\text{O}^+]\\ &= -\log\left(3.1 \times 10^{-5}\right) \approx 4.5\end{aligned}$$
Since the pH is between 3.0 and 6.0, the wetland is classified as a *poor fen*.

29. $$\begin{aligned}\text{pH} &= -\log[\text{H}_3\text{O}^+]\\ &= -\log\left(2.5 \times 10^{-2}\right) \approx 1.6\end{aligned}$$
Since the pH is less than 3.0, the wetland is classified as a *bog*.

31. $$\begin{aligned}\text{pH} &= -\log[\text{H}_3\text{O}^+]\\ &= -\log\left(2.7 \times 10^{-7}\right) \approx 6.6\end{aligned}$$
Since the pH is between 6.0 and 7.5, the wetland is classified as a *rich fen*.

33. Ammonia has a hydronium ion concentration of 2.5×10^{-12}.
$$\text{pH} = -\log[\text{H}_3\text{O}^+]$$
$$\text{pH} = -\log\left(2.5 \times 10^{-12}\right) \approx 11.6$$

35. Grapes have a hydronium ion concentration of 5.0×10^{-5}.
$$\text{pH} = -\log[\text{H}_3\text{O}^+]$$
$$\text{pH} = -\log\left(5.0 \times 10^{-5}\right) \approx 4.3$$

37. Human blood plasma has a pH of 7.4.
$$\begin{aligned}\text{pH} &= -\log[\text{H}_3\text{O}^+]\\ 7.4 &= -\log[\text{H}_3\text{O}^+]\\ \log[\text{H}_3\text{O}^+] &= -7.4\\ [\text{H}_3\text{O}^+] &= 10^{-7.4} \approx 4.0 \times 10^{-8}\end{aligned}$$

39. Spinach has a pH value of 5.4.
$$\begin{aligned}\text{pH} &= -\log[\text{H}_3\text{O}^+]\\ 5.4 &= -\log[\text{H}_3\text{O}^+]\\ \log[\text{H}_3\text{O}^+] &= -5.4\\ [\text{H}_3\text{O}^+] &= 10^{-5.4} \approx 4.0 \times 10^{-6}\end{aligned}$$

41. $D = 10\log\left(\dfrac{I}{I_0}\right)$

(a) $$\begin{aligned}D &= 10\log\left(\frac{5.012 \times 10^{10} I_0}{I_0}\right)\\ &= 10\log\left(5.012 \times 10^{10}\right) \approx 107\end{aligned}$$

The average decibel level for *Avatar* is about 107 dB.

(b) $$\begin{aligned}D &= 10\log\left(\frac{10^{10} I_0}{I_0}\right)\\ &= 10\log 10^{10} = 10 \cdot 10 = 100\end{aligned}$$

The average decibel level for *Iron Man 2* is 100 dB.

(c) $$\begin{aligned}D &= 10\log\left(\frac{6{,}310{,}000{,}000\, I_0}{I_0}\right)\\ &= 10\log(6{,}310{,}000{,}000) \approx 98\end{aligned}$$

The average decibel level for *Clash of the Titans* is about 98 dB.

43. $N(r) = -5000\ln r$

(a) 85% (or 0.85)

$N(0.85) = -5000\ln 0.85 \approx 813 \approx 800$

The number of years since the split for 85% of common words is about 800 years.

(b) 35% (or 0.35)

$N(0.35) = -5000\ln 0.35 \approx 5249 \approx 5200$

The number of years since the split for 35% of common words is about 5200 years.

(c) 10% (or 0.10)

$N(0.10) = -5000\ln 0.10 \approx 11{,}513 \approx 11{,}500$

The number of years since the split for 10% of common words is about 11,500 years.

45. $f(x) = -1317 + 304\ln x$

(a) $x = 1998 - 1900 = 98$

$f(98) = -1317 + 304\ln(98) \approx 77$

The prediction for 1998 is 77%.

(b) $f(x) = 50$

$$\begin{aligned} 50 &= -1317 + 304 \ln x \\ 1367 &= 304 \ln x \\ \ln x &= \tfrac{1367}{304} \\ e^{\ln x} &= e^{1367/304} \\ x &\approx 89.72 \end{aligned}$$

The outpatient surgeries reached 50% in 1989.

47. $T = -0.642 - 189 \ln (1 - p)$

(a) $$\begin{aligned} T &= -0.642 - 189 \ln (1 - 0.25) \\ &= -0.642 - 189 \ln (0.75) \approx 53.73 \end{aligned}$$

About \$54 per ton will reduce emissions by 25%.

(b) If $p = 0$, then $\ln (1 - p) = \ln 1 = 0$, so T would be negative. If $p = 1$, then $\ln (1 - p) = \ln 0$, but the domain of $\ln x$ is $(0, \infty)$.

49. The change-of-base rule is

$$\log_a x = \frac{\log_b x}{\log_b a}.$$

Use common logarithms ($b = 10$).

$$\log_3 12 = \frac{\log_{10} 12}{\log_{10} 3} = \frac{\log 12}{\log 3} \approx 2.2619$$

51. Use natural logarithms ($b = e$).

$$\log_5 3 = \frac{\log_e 3}{\log_e 5} = \frac{\ln 3}{\ln 5} \approx 0.6826$$

53. $\log_3 \sqrt{2} = \dfrac{\ln \sqrt{2}}{\ln 3} \approx 0.3155$

55. $\log_\pi e = \dfrac{\ln e}{\ln \pi} = \dfrac{1}{\ln \pi} \approx 0.8736$

57. $\log_e 12 = \dfrac{\ln 12}{\ln e} = \dfrac{\ln 12}{1} \approx 2.4849$

59. Let m = the number of letters in your first name and n = the number of letters in your last name.

Answers will vary, but suppose the name is Paul Bunyan, with $m = 4$ and $n = 6$.

(a) $\log_m n = \log_4 6$ is the exponent to which 4 must be raised in order to obtain 6.

(b) Use the change-of-base rule.

$$\begin{aligned} \log_4 6 &= \frac{\log 6}{\log 4} \\ &\approx 1.292\,481\,25 \end{aligned}$$

(c) Here, $m = 4$. Use the power key (y^x, x^y, ^) on your calculator.

$$4^{1.292\,481\,25} \approx 6$$

The result is 6, the value of n.

61. $f(x) = 3800 + 585 \log_2 x$

$x = 2003 - 1980 = 23$

$$\begin{aligned} f(23) &= 3800 + 585 \log_2 23 \\ &= 3800 + 585 \left(\frac{\log 23}{\log 2} \right) \\ &\approx 6446 \end{aligned}$$

The model gives an approximate withdrawal of 6446 billion ft^3 of natural gas from crude oil wells in the United States for 2003.

63. $$\begin{aligned} 4^{2x} &= 8^{3x+1} \\ (2^2)^{2x} &= (2^3)^{3x+1} \\ 2^{4x} &= 2^{9x+3} \\ 4x &= 9x + 3 \\ -3 &= 5x \\ x &= -\tfrac{3}{5} \end{aligned}$$

The solution set is $\{-\frac{3}{5}\}$.

65. $$\begin{aligned} \log_3 (x + 4) &= 2 \\ x + 4 &= 3^2 \\ x + 4 &= 9 \\ x &= 5 \end{aligned}$$

Substitute 5 for x in the argument, $x + 4$, $5 + 4 = 9 > 0$, so 5 is acceptable.

The solution set is $\{5\}$.

67. $$\begin{aligned} \log_{1/2} 8 &= x \\ \left(\tfrac{1}{2}\right)^x &= 8 \\ (2^{-1})^x &= 2^3 \\ 2^{-x} &= 2^3 \\ -x &= 3 \\ x &= -3 \end{aligned}$$

The solution set is $\{-3\}$.

69. $$\begin{aligned} &\log (x + 2) + \log (x + 3) \\ &= \log [(x + 2)(x + 3)], \quad \text{or} \quad \log (x^2 + 5x + 6) \end{aligned}$$

12.6 Exponential and Logarithmic Equations; Further Applications

12.6 Now Try Exercises

N1. $$\begin{aligned} 5^x &= 20 \\ \log 5^x &= \log 20 && \textit{Property 3} \\ x \log 5 &= \log 20 && \textit{Power rule} \\ x &= \frac{\log 20}{\log 5} && \textit{Divide.} \\ x &\approx 1.861 \end{aligned}$$

Check $x = 1.861$: $5^{1.861} \approx 20$

The solution set is $\{1.861\}$.

N2. $e^{0.12x} = 10$

Take base e logarithms on each side.

$$\begin{aligned} \ln e^{0.12x} &= \ln 10 & \\ 0.12x \ln e &= \ln 10 & \textit{Power rule} \\ 0.12x &= \ln 10 & \textit{ln e = 1} \\ x &= \frac{\ln 10}{0.12} \approx 19.188 & \end{aligned}$$

Check $x = 19.188$: $e^{0.12(19.188)} \approx 10$

The solution set is $\{19.188\}$.

N3. $\log_5 (x-1)^3 = 2$

$$\begin{aligned} (x-1)^3 &= 5^2 & \textit{Exponential form} \\ (x-1)^3 &= 25 & \\ x - 1 &= \sqrt[3]{25} & \textit{Cube root} \\ x &= 1 + \sqrt[3]{25} & \textit{Add 1.} \end{aligned}$$

Check $x = 1 + \sqrt[3]{25}$:

$$\begin{aligned} \log_5 \left(1 + \sqrt[3]{25} - 1\right)^3 &= \log_5 \left(\sqrt[3]{25}\right)^3 \\ &= \log_5 25 = \log_5 5^2 = 2 \end{aligned}$$

The solution set is $\left\{1 + \sqrt[3]{25}\right\}$.

N4.

$$\begin{aligned} \log_4 (2x+13) - \log_4 (x+1) &= \log_4 10 & \textit{Given equation} \\ \log_4 \frac{2x+13}{x+1} &= \log_4 10 & \textit{Quotient rule} \\ \frac{2x+13}{x+1} &= 10 & \textit{Property 4} \\ 2x + 13 &= 10(x+1) & \textit{Multiply by x + 1.} \\ 2x + 13 &= 10x + 10 & \textit{Distributive prop.} \\ 3 &= 8x & \textit{Subtract 2x; 10.} \\ \tfrac{3}{8} &= x & \textit{Divide by 8.} \end{aligned}$$

Check $x = \frac{3}{8}$:

$$\log_4 \tfrac{55}{4} - \log_4 \tfrac{11}{8} = \log_4 \frac{\frac{55}{4}}{\frac{11}{8}} = \log_4 10$$

The solution set is $\left\{\frac{3}{8}\right\}$.

N5. $\log_4 (x+2) + \log_4 2x = 2$

$$\begin{aligned} \log_4 [2x(x+2)] &= 2 & \textit{Product rule} \\ 2x(x+2) &= 4^2 & \textit{Exponential form} \\ 2x^2 + 4x &= 16 & \\ 2x^2 + 4x - 16 &= 0 & \\ x^2 + 2x - 8 &= 0 & \\ (x+4)(x-2) &= 0 & \end{aligned}$$

$$\begin{aligned} x + 4 &= 0 \quad \text{or} \quad x - 2 = 0 \\ x &= -4 \quad \text{or} \quad x = 2 \end{aligned}$$

Reject $x = -4$ since it yields an equation in which the logarithm of a negative number must be found.

Check $x = 2$: $\log_4 4 + \log_4 4 = 1 + 1 = 2$

The solution set is $\{2\}$.

N6. Use $A = P\left(1 + \frac{r}{n}\right)^{nt}$ with $P = 10{,}000$, $r = 2.5\% = 0.025$, $n = 12$ (compounded monthly), and $t = 10$.

$$\begin{aligned} A &= 10{,}000\left(1 + \frac{0.025}{12}\right)^{12(10)} \\ &= 10{,}000\left(1.00208\overline{3}\right)^{120} \approx 12{,}836.92 \end{aligned}$$

There will be \$12,836.92 in the account.

N7.

$$\begin{aligned} A &= P\left(1 + \frac{r}{n}\right)^{nt} & \\ 2P &= P\left(1 + \frac{0.04}{4}\right)^{4 \cdot t} & \textit{Double, so A = 2P; Quarterly, n = 4} \\ 2 &= (1.01)^{4t} & \textit{Divide by P.} \\ \log 2 &= \log (1.01)^{4t} & \textit{Property 3} \\ \log 2 &= 4t \log 1.01 & \textit{Power rule} \\ 4t &= \frac{\log 2}{\log 1.01} & \textit{Divide by log 1.01.} \\ t &= \frac{\log 2}{4 \log 1.01} & \textit{Divide by 4.} \\ t &\approx 17.42 & \end{aligned}$$

It will take about 17.42 years for any amount of money in an account paying 4% interest compounded quarterly to double.

N8. **(a)**

$$\begin{aligned} A &= Pe^{rt} \\ A &= 4000e^{0.03(2)} \\ A &\approx 4247.35 \end{aligned}$$

It will grow to \$4247.35.

(b)

$$\begin{aligned} A &= Pe^{rt} & \\ 2(4000) &= 4000e^{0.03t} & \\ 2 &= e^{0.03t} & \textit{Divide by 4000.} \\ \ln 2 &= \ln e^{0.03t} & \textit{Take natural logs.} \\ \ln 2 &= 0.03t & \ln e^k = k \\ t &= \frac{\ln 2}{0.03} & \textit{Divide by 0.03.} \\ t &\approx 23.1 & \end{aligned}$$

It would take about 23.1 years for the initial investment to double.

N9. $y = y_0 e^{-0.00043t}$

(a)

$$\begin{aligned} y &= 4.5e^{-0.00043(150)} & \textit{Let t = 150.} \\ y &\approx 4.22 & \end{aligned}$$

After 150 years, there will be about 4.2 grams.

(b)

$$\begin{aligned} \tfrac{1}{2}(4.5) &= 4.5e^{-0.00043t} \\ \tfrac{1}{2} &= e^{-0.00043t} \\ \ln \tfrac{1}{2} &= -0.00043t \\ t &= \frac{\ln \frac{1}{2}}{-0.00043} \approx 1612 \end{aligned}$$

The half-life is about 1612 years.

12.6 Section Exercises

1. $7^x = 5$

Take the logarithm of each side.

$\log 7^x = \log 5$

Use the power rule for logarithms.

$x \log 7 = \log 5$

$$x = \frac{\log 5}{\log 7} \approx 0.827$$

The solution set is $\{0.827\}$.

3.
$$\begin{aligned} 9^{-x+2} &= 13 \\ \log 9^{-x+2} &= \log 13 \\ (-x+2)\log 9 &= \log 13\,(*) \\ -x\log 9 + 2\log 9 &= \log 13 \\ -x \log 9 &= \log 13 - 2\log 9 \\ x\log 9 &= 2\log 9 - \log 13 \\ x &= \frac{2\log 9 - \log 13}{\log 9} \\ &\approx 0.833 \end{aligned}$$

$(*)$ Alternative solution steps:

$$\begin{aligned} (-x+2)\log 9 &= \log 13 \\ -x+2 &= \frac{\log 13}{\log 9} \\ 2 - \frac{\log 13}{\log 9} &= x \end{aligned}$$

The solution set is $\{0.833\}$.

5.
$$\begin{aligned} 3^{2x} &= 14 \\ \log 3^{2x} &= \log 14 \\ 2x\log 3 &= \log 14 \\ x &= \frac{\log 14}{2\log 3} \approx 1.201 \end{aligned}$$

The solution set is $\{1.201\}$.

7.
$$\begin{aligned} 2^{x+3} &= 5^x \\ \log 2^{x+3} &= \log 5^x \\ (x+3)\log 2 &= x\log 5 \\ x\log 2 + 3\log 2 &= x \log 5 \end{aligned}$$

Get x-terms on one side.

$x\log 2 - x\log 5 = -3\log 2$

$x(\log 2 - \log 5) = -3\log 2$ *Factor out x.*

$$\begin{aligned} x &= \frac{-3\log 2}{\log 2 - \log 5} \\ &\approx 2.269 \end{aligned}$$

The solution set is $\{2.269\}$.

9.
$$\begin{aligned} 2^{x+3} &= 3^{x-4} \\ \log 2^{x+3} &= \log 3^{x-4} \\ (x+3)\log 2 &= (x-4)\log 3 \\ x\log 2 + 3\log 2 &= x\log 3 - 4\log 3 \end{aligned}$$

Distributive property

Get x-terms on one side.

$x\log 2 - x\log 3 = -3\log 2 - 4\log 3$

Factor out x.

$$\begin{aligned} x(\log 2 - \log 3) &= -3\log 2 - 4\log 3 \\ x &= \frac{-3\log 2 - 4\log 3}{\log 2 - \log 3} \\ &\approx 15.967 \end{aligned}$$

The solution set is $\{15.967\}$.

11.
$$\begin{aligned} 4^{2x+3} &= 6^{x-1} \\ \log 4^{2x+3} &= \log 6^{x-1} \\ (2x+3)\log 4 &= (x-1)\log 6 \\ 2x\log 4 + 3\log 4 &= x\log 6 - 1\log 6 \end{aligned}$$

Distributive property

Get x-terms on one side.

$2x\log 4 - x\log 6 = -3\log 4 - \log 6$

Factor out x.

$$\begin{aligned} x(2\log 4 - \log 6) &= -3\log 4 - \log 6 \\ x &= \frac{-3\log 4 - \log 6}{2\log 4 - \log 6} \\ &\approx -6.067 \end{aligned}$$

The solution set is $\{-6.067\}$.

13. $e^{0.012x} = 23$

Take base e logarithms on each side.

$$\begin{aligned} \ln e^{0.012x} &= \ln 23 \\ 0.012x(\ln e) &= \ln 23 \\ 0.012x &= \ln 23 \quad \textit{ln e = 1} \\ x &= \frac{\ln 23}{0.012} \approx 261.291 \end{aligned}$$

The solution set is $\{261.291\}$.

15. $e^{-0.205x} = 9$

Take base e logarithms on each side.

$$\begin{aligned} \ln e^{-0.205x} &= \ln 9 \\ -0.205x(\ln e) &= \ln 9 \\ -0.205x &= \ln 9 \quad \textit{ln e = 1} \\ x &= \frac{\ln 9}{-0.205} \approx -10.718 \end{aligned}$$

The solution set is $\{-10.718\}$.

17.
$$\begin{aligned} \ln e^{3x} &= 9 \\ 3x &= 9 \quad \textit{Special property} \\ x &= 3 \end{aligned}$$

The solution set is $\{3\}$.

19.
$$\begin{aligned} \ln e^{0.45x} &= \sqrt{7} \\ 0.45x &= \sqrt{7} \\ x &= \frac{\sqrt{7}}{0.45} \approx 5.879 \end{aligned}$$

The solution set is $\{5.879\}$.

21. $\ln e^{-x} = \pi$

$$-x = \pi$$
$$x = -\pi \approx -3.142$$

The solution set is $\{-\pi\}$, or $\{-3.142\}$.

23. $e^{\ln 2x} = e^{\ln(x+1)}$

$$2x = x + 1 \quad \textit{Special property}$$
$$x = 1$$

The solution set is $\{1\}$.

25. Let's try Exercise 14.

$$e^{0.006x} = 30$$
$$\log e^{0.006x} = \log 30$$
$$0.006x\,(\log e) = \log 30$$
$$x = \frac{\log 30}{0.006 \log e} \approx 566.866$$

The natural logarithm is a better choice because $\ln e = 1$, whereas $\log e$ needs to be calculated.

27. $\log_3 (6x + 5) = 2$

$$6x + 5 = 3^2 \quad \textit{Exponential form}$$
$$6x + 5 = 9$$
$$6x = 4$$
$$x = \tfrac{4}{6} = \tfrac{2}{3}$$

Check $x = \frac{2}{3}$: $\log_3 9 = \log_3 3^2 = 2$

The solution set is $\{\frac{2}{3}\}$.

29. $\log_2 (2x - 1) = 5$

$$2x - 1 = 2^5 \quad \textit{Exponential form}$$
$$2x - 1 = 32$$
$$2x = 33$$
$$x = \tfrac{33}{2}$$

Check $x = \frac{33}{2}$: $\log_2 32 = \log_2 2^5 = 5$

The solution set is $\{\frac{33}{2}\}$.

31. $\log_7 (x + 1)^3 = 2$

$$(x + 1)^3 = 7^2 \quad \textit{Exponential form}$$
$$x + 1 = \sqrt[3]{49} \quad \textit{Cube root}$$
$$x = -1 + \sqrt[3]{49}$$

Check $x = -1 + \sqrt[3]{49}$: $\log_7 49 = \log_7 7^2 = 2$

The solution set is $\left\{-1 + \sqrt[3]{49}\right\}$.

33. 2 cannot be a solution because $\log(2 - 3) = \log(-1)$, and -1 is not in the domain of $\log x$.

35. $\log(6x + 1) = \log 3$

$$6x + 1 = 3 \quad \textit{Property 4}$$
$$6x = 2$$
$$x = \tfrac{2}{6} = \tfrac{1}{3}$$

Check $x = \frac{1}{3}$: $\log(2 + 1) = \log 3$ *True*

The solution set is $\{\frac{1}{3}\}$.

37. $\log_5 (3t + 2) - \log_5 t = \log_5 4$

$$\log_5 \frac{3t + 2}{t} = \log_5 4$$
$$\frac{3t + 2}{t} = 4$$
$$3t + 2 = 4t$$
$$2 = t$$

Check $t = 2$: $\log_5 8 - \log_5 2 = \log_5 \frac{8}{2} = \log_5 4$

The solution set is $\{2\}$.

39. $\log 4x - \log(x - 3) = \log 2$

$$\log \frac{4x}{x - 3} = \log 2$$
$$\frac{4x}{x - 3} = 2$$
$$4x = 2(x - 3)$$
$$4x = 2x - 6$$
$$2x = -6$$
$$x = -3$$

Reject $x = -3$, because $4x = -12$, which yields an equation in which the logarithm of a negative number must be found.

The solution set is $\emptyset$.

41. $\log_2 x + \log_2 (x - 7) = 3$

$$\log_2 [x(x - 7)] = 3$$
$$x(x - 7) = 2^3 \quad \textit{Exponential form}$$
$$x^2 - 7x = 8$$
$$x^2 - 7x - 8 = 0$$
$$(x - 8)(x + 1) = 0$$

$$x - 8 = 0 \quad \text{or} \quad x + 1 = 0$$
$$x = 8 \quad \text{or} \quad x = -1$$

Reject $x = -1$, because it yields an equation in which the logarithm of a negative number must be found.

Check $x = 8$: $\log_2 8 + \log_2 1 = \log_2 2^3 + 0 = 3$

The solution set is $\{8\}$.

43. $\log 5x - \log(2x - 1) = \log 4$

$$\log \frac{5x}{2x - 1} = \log 4$$
$$\frac{5x}{2x - 1} = 4$$
$$5x = 8x - 4$$
$$4 = 3x$$
$$\tfrac{4}{3} = x$$

Check $x = \frac{4}{3}$: $\log \frac{20}{3} - \log \frac{5}{3} = \log \frac{20/3}{5/3} = \log 4$

The solution set is $\{\frac{4}{3}\}$.

45. $\log_2 x + \log_2 (x - 6) = 4$

$$\log_2 [x(x-6)] = 4$$
$$x(x-6) = 2^4 \quad \textit{Exponential form}$$
$$x^2 - 6x = 16$$
$$x^2 - 6x - 16 = 0$$
$$(x-8)(x+2) = 0$$
$$x - 8 = 0 \quad \text{or} \quad x + 2 = 0$$
$$x = 8 \quad \text{or} \quad x = -2$$

Reject $x = -2$, because it yields an equation in which the logarithm of a negative number must be found.

Check $x = 8$:
$\log_2 8 + \log_2 2 = \log_2 16 = \log_2 2^4 = 4$

The solution set is $\{8\}$.

47. **(a)** Use the formula $A = P\left(1 + \frac{r}{n}\right)^{nt}$ with $P = 2000$, $r = 0.04$, $n = 4$, and $t = 6$.

$$A = 2000\left(1 + \tfrac{0.04}{4}\right)^{4 \cdot 6}$$
$$= 2000(1.01)^{24} \approx 2539.47$$

The account will contain \$2539.47.

(b)
$$3000 = 2000\left(1 + \tfrac{0.04}{4}\right)^{4t}$$
$$\frac{3000}{2000} = (1.01)^{4t}$$
$$\log\left(\tfrac{3}{2}\right) = \log(1.01)^{4t}$$
$$\log\left(\tfrac{3}{2}\right) = 4t \log(1.01)$$
$$t = \frac{\log\left(\frac{3}{2}\right)}{4\log(1.01)} \approx 10.2$$

It will take about 10.2 years for the account to grow to \$3000.

49. **(a)** Use the formula $A = Pe^{rt}$ with $P = 4000$, $r = 0.035$, and $t = 6$.

$$A = 4000e^{(0.035)(6)}$$
$$= 4000e^{0.21} \approx 4934.71$$

There will be \$4934.71 in the account.

(b) If the initial amount doubles, then $A = 2P$, or \$8000.

$$8000 = 4000e^{0.035t}$$
$$2 = e^{0.035t} \quad \textit{Divide by 4000.}$$
$$\ln 2 = \ln e^{0.035t}$$
$$\ln 2 = 0.035t$$
$$t = \frac{\ln 2}{0.035} \approx 19.8$$

The initial amount will double in about 19.8 years.

51. Use $A = P\left(1 + \frac{r}{n}\right)^{nt}$ with $P = 5000$, $r = 0.07$, and $t = 12$.

(a) If the interest is compounded annually, $n = 1$.

$$A = 5000\left(1 + \tfrac{0.07}{1}\right)^{1 \cdot 12}$$
$$= 5000(1.07)^{12} \approx 11{,}260.96$$

There will be \$11,260.96 in the account.

(b) If the interest is compounded semiannually, $n = 2$.

$$A = 5000\left(1 + \tfrac{0.07}{2}\right)^{2 \cdot 12}$$
$$= 5000(1.035)^{24} \approx 11{,}416.64$$

There will be \$11,416.64 in the account.

(c) If the interest is compounded quarterly, $n = 4$.

$$A = 5000\left(1 + \tfrac{0.07}{4}\right)^{4 \cdot 12}$$
$$= 5000(1.0175)^{48} \approx 11{,}497.99$$

There will be \$11,497.99 in the account.

(d) If the interest is compounded daily, $n = 365$.

$$A = 5000\left(1 + \tfrac{0.07}{365}\right)^{365 \cdot 12}$$
$$\approx 11{,}580.90$$

There will be \$11,580.90 in the account.

(e) Use the continuous compound interest formula.

$$A = Pe^{rt}$$
$$A = 5000e^{0.07(12)}$$
$$= 5000e^{0.84} \approx 11{,}581.83$$

There will be \$11,581.83 in the account.

53. In the continuous compound interest formula, let $A = 1850$, $r = 0.065$, and $t = 40$.

$$A = Pe^{rt}$$
$$1850 = Pe^{0.065(40)}$$
$$1850 = Pe^{2.6}$$
$$P = \frac{1850}{e^{2.6}} \approx 137.41$$

Deposit \$137.41 today.

55. $f(x) = 15.94e^{0.0656x}$

(a) 1980 corresponds to $x = 0$.

$$f(0) = 15.94e^{0.0656(0)}$$
$$= 15.94(1) = 15.94$$

The approximate volume of materials recovered from municipal solid waste collections in the United States in 1980 was 15.9 million tons.

(b) 1990 corresponds to $x = 10$.

$$f(10) \approx 30.7 \text{ million tons}$$

(c) 2000 corresponds to $x = 20$.

$$f(20) \approx 59.2 \text{ million tons}$$

(d) 2007 corresponds to $x = 27$.

$$f(27) \approx 93.7 \text{ million tons}$$

57. $S(x) = 112{,}047e^{0.0827x}$
2007 corresponds to $x = 3$.
$S(3) = 112{,}047e^{0.0827(3)}$
$\approx 143{,}598$
The approximate value of software-publisher revenues for 2007 was 143,598 million dollars.

59. $A(t) = 2.00e^{-0.053t}$

(a) Let $t = 4$.

$$A(4) = 2.00e^{-0.053(4)} = 2.00e^{-0.212} \approx 1.62$$

About 1.62 grams would be present.

(b) $A(10) = 2.00e^{-0.053(10)} \approx 1.18$

About 1.18 grams would be present.

(c) $A(20) = 2.00e^{-0.053(20)} \approx 0.69$

About 0.69 grams would be present.

(d) The initial amount is the amount $A(t)$ present at time $t = 0$.

$$A(0) = 2.00e^{-0.053(0)} = 2.00e^{0} = 2.00(1) = 2.00$$

Initially, 2.00 grams were present.

61. **(a)** Find $A(t) = 400e^{-0.032t}$ when $t = 25$.

$$A(25) = 400e^{-0.032(25)} = 400e^{-0.8} \approx 179.73$$

About 179.73 grams of lead will be left.

(b) Use $A(t) = 400e^{-0.032t}$, with $A(t) = \frac{1}{2}(400) = 200$.

$$200 = 400e^{-0.032t}$$
$$\frac{200}{400} = e^{-0.032t}$$
$$0.5 = e^{-0.032t}$$
$$\ln 0.5 = \ln e^{-0.032t}$$
$$\ln 0.5 = -0.032t(\ln e)$$
$$t = \frac{\ln 0.5}{-0.032} \approx 21.66$$

It would take about 21.66 years for the sample to decay to half its original amount.

63.
$$f(x) = 15.94e^{0.0656x}$$
$$130 = 15.94e^{0.0656x}$$
$$e^{0.0656x} = \frac{130}{15.94}$$
$$\ln e^{0.0656x} = \ln \frac{130}{15.94}$$
$$0.0656x\,(\ln e) = \ln \frac{130}{15.94}$$
$$0.0656x = \ln \frac{130}{15.94}$$
$$x = \frac{\ln \frac{130}{15.94}}{0.0656} \approx 32$$

Since $x = 0$ corresponds to 1980, $x = 32$ corresponds to 2012.

65. The display means that after 250 years, approximately 2.918 748 7 grams of the original sample of radium 226 remain.

67. $f(x) = 2x^2$

The graph of f is narrower than the graph of $y = x^2$.

x	0	± 1	± 2
y	0	2	8

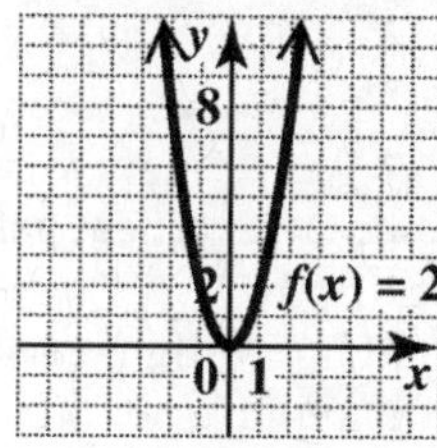

69. $f(x) = (x+1)^2$

The graph of f is the graph of $y = x^2$ shifted *left* 1 unit.

x	-3	-2	-1	0	1
y	4	1	0	1	4

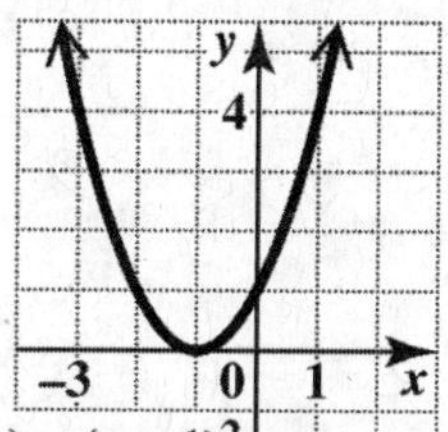

$f(x) = (x + 1)^2$

Chapter 12 Review Exercises

1. Since a horizontal line intersects the graph in two points, the function is not one-to-one.

2. Since every horizontal line intersects the graph in no more than one point, the function is one-to-one.

3. The function $f(x) = -3x + 7$ is a linear function. By the horizontal line test, it is a one-to-one function. To find the inverse, replace $f(x)$ with y.

$$y = -3x + 7$$

Interchange x and y.

$$x = -3y + 7$$

Solve for y.

$$x - 7 = -3y$$
$$\frac{x-7}{-3} = y, \text{ or } \frac{7-x}{3} = y$$

Replace y with $f^{-1}(x)$.

$$f^{-1}(x) = \frac{x-7}{-3}, \text{ or } f^{-1}(x) = -\frac{1}{3}x + \frac{7}{3}$$

4. $f(x) = \sqrt[3]{6x - 4}$
The function is one-to-one since each $f(x)$-value corresponds to exactly one x-value.
To find the inverse, replace $f(x)$ with y.

$y = \sqrt[3]{6x - 4}$
Interchange x and y.
$x = \sqrt[3]{6y - 4}$

Solve for y.

$x^3 = 6y - 4$ *Cube each side.*
$x^3 + 4 = 6y$
$\dfrac{x^3 + 4}{6} = y$

Replace y with $f^{-1}(x)$

$\dfrac{x^3 + 4}{6} = f^{-1}(x)$

5. $f(x) = -x^2 + 3$
This is an equation of a vertical parabola which opens down.
Since a horizontal line will intersect the graph in two points, the function is not one-to-one.

6. This function is not one-to-one because two sodas in the list have 41 mg of caffeine.

7. The graph is a linear function through $(0, 1)$ and $(3, 0)$. The graph of $f^{-1}(x)$ will include the points $(1, 0)$ and $(0, 3)$, found by interchanging x and y. Plot these points, and draw a straight line through them.

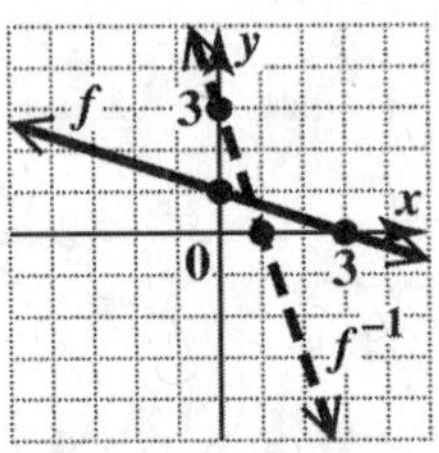

8. The graph is a curve through $(1, 2)$, $(0, 1)$, and $\left(-1, \frac{1}{2}\right)$. Interchange x and y to get $(2, 1)$, $(1, 0)$, and $\left(\frac{1}{2}, -1\right)$, which are on the graph of $f^{-1}(x)$. Plot these points, and draw a smooth curve through them.

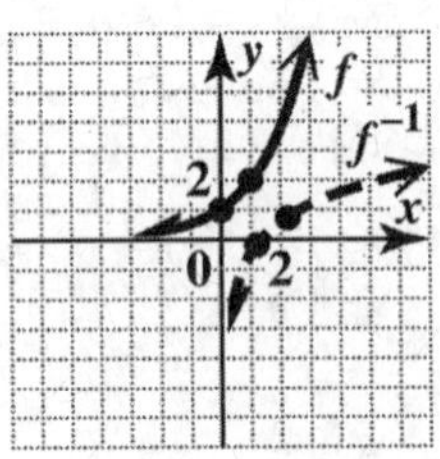

9. $f(x) = 3^x$
Make a table of values.

x	-2	-1	0	1	2
$f(x)$	$\frac{1}{9}$	$\frac{1}{3}$	1	3	9

Plot the points from the table and draw a smooth curve through them.

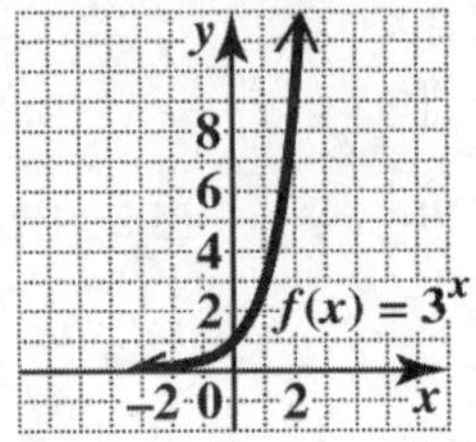

10. $f(x) = \left(\frac{1}{3}\right)^x$
Make a table of values.

x	-2	-1	0	1	2
$f(x)$	9	3	1	$\frac{1}{3}$	$\frac{1}{9}$

Plot the points from the table and draw a smooth curve through them.

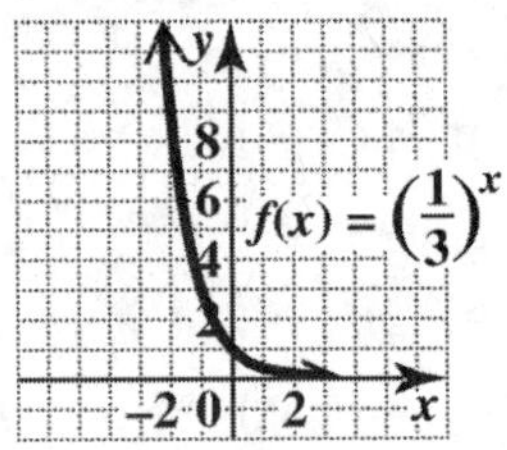

11. $y = 2^{2x+3}$
Make a table of values.

x	-2	$-\frac{3}{2}$	-1	0	$\frac{1}{2}$
y	$\frac{1}{2}$	1	2	8	16

Plot the points from the table and draw a smooth curve through them.

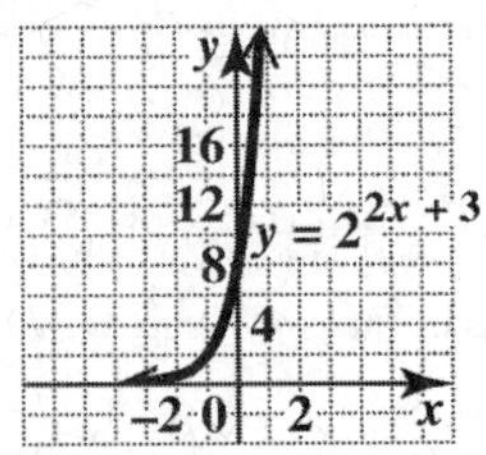

12. $5^{2x+1} = 25$
Write each side as a power of 5.
$(5)^{2x+1} = (5)^2$
$2x + 1 = 2$ *Equate exponents.*
$2x = 1$
$x = \frac{1}{2}$

Check $x = \frac{1}{2}$: $5^2 = 25$ *True*

The solution set is $\left\{\frac{1}{2}\right\}$.

13. $4^{3x} = 8^{x+4}$
Write each side as a power of 2.
$(2^2)^{3x} = (2^3)^{x+4}$
$2^{6x} = 2^{3x+12}$
$6x = 3x + 12$ *Equate exponents.*
$3x = 12$
$x = 4$

Check $x=4$: $4^{12}=8^8$ *True*

The solution set is $\{4\}$.

14.
$$\left(\tfrac{1}{27}\right)^{x-1}=9^{2x}$$
$$\left[\left(\tfrac{1}{3}\right)^3\right]^{x-1}=\left(3^2\right)^{2x}$$
Write each side as a power of 3.
$$\left(3^{-3}\right)^{x-1}=\left(3^2\right)^{2x}$$
$$3^{-3x+3}=3^{4x}$$
$$-3x+3=4x \quad \textit{Equate exponents.}$$
$$3=7x$$
$$\tfrac{3}{7}=x$$
Check $x=\frac{3}{7}$: $\left(\frac{1}{27}\right)^{-4/7}=9^{6/7}$ *True*

The solution set is $\left\{\frac{3}{7}\right\}$.

15. $S(x)=33.07(1.0241)^{-x}$

(a) $x=1975-1970=5$

$$S(5)=33.07(1.0241)^{-5}$$
$$\approx 29.4$$

The approximate amount of sulfur dioxide emissions in the United States in 1975 was 29.4 million tons.

(b) $x=1995-1970=25$

$S(25)\approx 18.2$ million tons

(c) $x=2005-1970=35$

$S(35)\approx 14.4$ million tons

16. $g(x)=\log_3 x$
Replace $g(x)$ with y, and write in exponential form.

$$y=\log_3 x$$
$$3^y=x$$

Make a table of values. Since $x=3^y$ is the inverse of $f(x)=y=3^x$ in Exercise 9, simply reverse the ordered pairs in the table of values belonging to $f(x)=3^x$.

x	$\frac{1}{9}$	$\frac{1}{3}$	1	3	9
y	-2	-1	0	1	2

Plot the points from the table and draw a smooth curve through them.

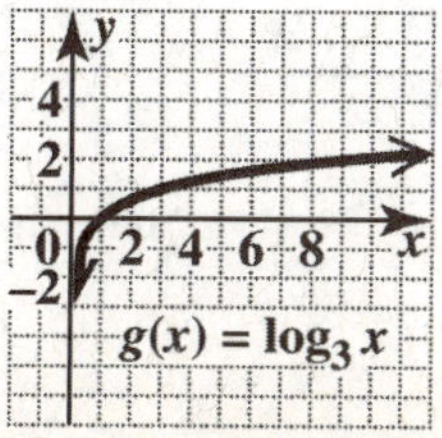

17. $g(x)=\log_{1/3} x$
Replace $g(x)$ with y, and write in exponential form.

$$y=\log_{1/3} x$$
$$\left(\tfrac{1}{3}\right)^y=x$$

Make a table of values. Since $x=\left(\frac{1}{3}\right)^y$ is the inverse of $f(x)=y=\left(\frac{1}{3}\right)^x$ in Exercise 10, simply reverse the ordered pairs in the table of values belonging to $f(x)=\left(\frac{1}{3}\right)^x$.

x	9	3	1	$\frac{1}{3}$	$\frac{1}{9}$
y	-2	-1	0	1	2

Plot the points from the table and draw a smooth curve through them.

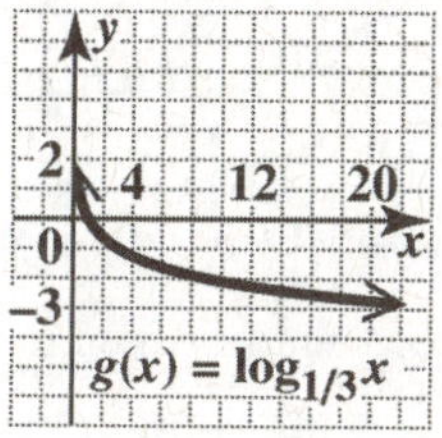

18. $\log_8 64=x$
$$8^x=64 \quad \textit{Exponential form}$$
Write each side as a power of 8.
$$8^x=8^2$$
$$x=2 \quad \textit{Equate exponents.}$$

The solution set is $\{2\}$.

19. $\log_2 \sqrt{8}=x$
$$2^x=\sqrt{8} \quad \textit{Exponential form}$$
$$2^x=8^{1/2}$$
Write each side as a power of 2.
$$2^x=\left(2^3\right)^{1/2}$$
$$2^x=2^{3/2}$$
$$x=\tfrac{3}{2} \quad \textit{Equate exponents.}$$

The solution set is $\left\{\frac{3}{2}\right\}$.

20. $\log_x\left(\frac{1}{49}\right)=-2$
$$x^{-2}=\frac{1}{49} \quad \textit{Exponential form}$$
$$\frac{1}{x^2}=\frac{1}{49}$$
$$x^2=49$$
$$x=\pm 7$$

Since x is the base, we cannot have a negative number.
The solution set is $\{7\}$.

21. $\log_4 x = \frac{3}{2}$

$x = 4^{3/2}$ *Exponential form*

$x = (\sqrt{4})^3 = 2^3 = 8$

The argument (the input of the logarithm) must be a positive number, so $x = 8$ is acceptable.

The solution set is $\{8\}$.

22. $\log_k 4 = 1$

$k^1 = 4$ *Exponential form*

$k = 4$

$k = 4$ is an acceptable base since it is a positive number (not equal to 1).

The solution set is $\{4\}$.

23. $\log_b b^2 = 2$

$b^2 = b^2$ *Exponential form*

This is an identity. Thus, b can be any real number, $b > 0$ and $b \neq 1$.

The solution set is $\{b \mid b > 0,\ b \neq 1\}$.

24. $\log_b a$ is the exponent to which b must be raised to obtain a.

25. From Exercise 24,

$$b^{\log_b a} = a.$$

26. $S(x) = 100 \log_2 (x + 2)$

(a) When $x = 6$,

$$S(6) = 100 \log_2 (6 + 2) = 100(3) = 300.$$

After 6 weeks the sales were 300 thousand dollars or \$300,000.

(b) To graph the function, make a table of values that includes the ordered pair from above.

x	0	2	6
$S(x)$	100	200	300

Plot the ordered pairs and draw the graph through them.

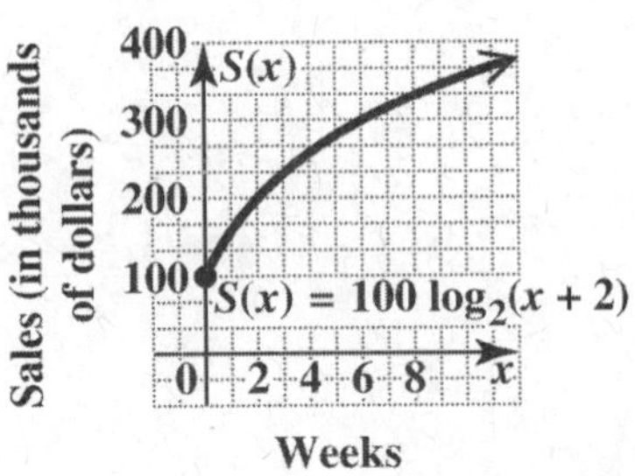

27. $\log_2 3xy^2$

$= \log_2 3 + \log_2 x + \log_2 y^2$ *Product rule*

$= \log_2 3 + \log_2 x + 2\log_2 y$ *Power rule*

28. $\log_4 \dfrac{\sqrt{x} \cdot w^2}{z}$

$= \log_4 (\sqrt{x} \cdot w^2) - \log_4 z$ *Quotient rule*

$= \log_4 x^{1/2} + \log_4 w^2 - \log_4 z$ *Product rule*

$= \frac{1}{2}\log_4 x + 2\log_4 w - \log_4 z$ *Power rule*

29. $\log_b 3 + \log_b x - 2\log_b y$

Use the product and power rules for logarithms.

$= \log_b (3 \cdot x) - \log_b y^2$

$= \log_b \dfrac{3x}{y^2}$ *Quotient rule*

30. $\log_3 (x + 7) - \log_3 (4x + 6)$

$= \log_3 \left(\dfrac{x + 7}{4x + 6}\right)$ *Quotient rule*

31. $\log 28.9 \approx 1.4609$

32. $\log 0.257 \approx -0.5901$

33. $\ln 28.9 \approx 3.3638$

34. $\ln 0.257 \approx -1.3587$

35. $\log_{16} 13 = \dfrac{\log 13}{\log 16} \approx 0.9251$

36. $\log_4 12 = \dfrac{\log 12}{\log 4} \approx 1.7925$

37. Milk has a hydronium ion concentration of 4.0×10^{-7}.

$pH = -\log [H_3O^+]$

$pH = -\log (4.0 \times 10^{-7}) \approx 6.4$

38. Crackers have a hydronium ion concentration of 3.8×10^{-9}.

$pH = -\log [H_3O^+]$

$pH = -\log (3.8 \times 10^{-9}) \approx 8.4$

39. Orange juice has a pH of 4.6.

$pH = -\log [H_3O^+]$

$4.6 = -\log [H_3O^+]$

$\log_{10} [H_3O^+] = -4.6$

$[H_3O^+] = 10^{-4.6} \approx 2.5 \times 10^{-5}$

40. Since we want to determine a ratio of intensities, we'll solve the given formula for the intensity I.

$$M = 6 - 2.5 \log \frac{I}{I_0}$$

$2.5 \log \dfrac{I}{I_0} = 6 - M$

$\log_{10} \dfrac{I}{I_0} = \dfrac{6 - M}{2.5}$ *Common log*

$\dfrac{I}{I_0} = 10^{(6-M)/2.5}$ *Exponential form*

$I = 10^{0.4(6-M)} I_0$ *1/2.5 = 0.4*

We now have I as a function of M. The ratio of intensities between stars of magnitude 1 and 3 is

$$\frac{I(1)}{I(3)} = \frac{10^{0.4(6-1)}I_0}{10^{0.4(6-3)}I_0} = \frac{10^2}{10^{1.2}} = 10^{0.8} \approx 6.3.$$

Thus, a star of magnitude 1 is about 6.3 times as intense as a star of magnitude 3.

41. $t(r) = \dfrac{\ln 2}{\ln(1+r)}$

(a) $4\% = 0.04;\ t(0.04) = \dfrac{\ln 2}{\ln(1+0.04)} \approx 18$

At 4%, it would take about 18 years.

(b) $6\% = 0.06;\ t(0.06) = \dfrac{\ln 2}{\ln(1+0.06)} \approx 12$

At 6%, it would take about 12 years.

(c) $10\% = 0.10;\ t(0.10) = \dfrac{\ln 2}{\ln(1+0.10)} \approx 7$

At 10%, it would take about 7 years.

(d) $12\% = 0.12;\ t(0.12) = \dfrac{\ln 2}{\ln(1+0.12)} \approx 6$

At 12%, it would take about 6 years.

(e) Each comparison shows approximately the same number. For example, in part (a) the doubling time is 18 yr (rounded) and $\frac{72}{4} = 18$. Thus, the formula $t = \dfrac{72}{100r}$ (called the *rule of 72*) is an excellent approximation of the doubling time formula.

42.
$$\begin{aligned} 3^x &= 9.42 \\ \log 3^x &= \log 9.42 \\ x \log 3 &= \log 9.42 \\ x &= \frac{\log 9.42}{\log 3} \approx 2.042 \end{aligned}$$

Check $x = 2.042$: $3^{2.042} \approx 9.425$

The solution set is $\{2.042\}$.

43.
$$\begin{aligned} 2^{x-1} &= 15 \\ \log 2^{x-1} &= \log 15 \\ (x-1)\log 2 &= \log 15 \\ x - 1 &= \frac{\log 15}{\log 2} \\ x &= \frac{\log 15}{\log 2} + 1 \approx 4.907 \end{aligned}$$

Check $x = 4.907$: $2^{3.907} \approx 15.0$

The solution set is $\{4.907\}$.

44. $e^{0.06x} = 3$

Take base e logarithms on each side.

$$\begin{aligned} \ln e^{0.06x} &= \ln 3 \\ 0.06x \ln e &= \ln 3 \\ 0.06x &= \ln 3 && \textit{ln e = 1} \\ x &= \frac{\ln 3}{0.06} \approx 18.310 \end{aligned}$$

Check $x = 18.310$: $e^{1.0986} \approx 3.0$

The solution set is $\{18.310\}$.

45. $\log_3(9x+8) = 2$

$$\begin{aligned} 9x + 8 &= 3^2 && \textit{Exponential form} \\ 9x + 8 &= 9 \\ 9x &= 1 \\ x &= \tfrac{1}{9} \end{aligned}$$

Check $x = \frac{1}{9}$: $\log_3 9 = \log_3 3^2 = 2$

The solution set is $\left\{\frac{1}{9}\right\}$.

46. $\log_5(x+6)^3 = 2$

Change to exponential form.

$$\begin{aligned} (x+6)^3 &= 5^2 \\ (x+6)^3 &= 25 \end{aligned}$$

Take the cube root of each side.

$$\begin{aligned} x + 6 &= \sqrt[3]{25} \\ x &= \sqrt[3]{25} - 6 \end{aligned}$$

Check $x = -6 + \sqrt[3]{25}$: $\log_5 25 = \log_5 5^2 = 2$

The solution set is $\left\{-6 + \sqrt[3]{25}\right\}$.

47. $\log_3(x+2) - \log_3 x = \log_3 2$

$$\begin{aligned} \log_3 \frac{x+2}{x} &= \log_3 2 && \textit{Quotient rule} \\ \frac{x+2}{x} &= 2 && \textit{Property 4} \\ x + 2 &= 2x \\ 2 &= x \end{aligned}$$

Check $x = 2$: $\log_3 4 - \log_3 2 = \log_3 \frac{4}{2} = \log_3 2$

The solution set is $\{2\}$.

48.
$$\begin{aligned} \log(2x+3) &= 1 + \log x \\ \log(2x+3) - \log x &= 1 \\ \log_{10} \frac{2x+3}{x} &= 1 && \textit{Quotient rule} \\ 10^1 &= \frac{2x+3}{x} && \textit{Exponential form} \\ 10x &= 2x + 3 \\ 8x &= 3 \\ x &= \tfrac{3}{8} \end{aligned}$$

Check $x = \frac{3}{8}$:

LS $= \log\left(\frac{3}{4} + 3\right) = \log \frac{15}{4}$

RS $= \log 10 + \log \frac{3}{8} = \log \frac{30}{8} = \log \frac{15}{4}$

The solution set is $\left\{\frac{3}{8}\right\}$.

49. $\log_4 x + \log_4 (8 - x) = 2$

$$\begin{aligned} \log_4 [x(8-x)] &= 2 && \textit{Product rule} \\ x(8-x) &= 4^2 && \textit{Exponential form} \\ 8x - x^2 &= 16 \\ x^2 - 8x + 16 &= 0 \\ (x-4)(x-4) &= 0 \\ x - 4 &= 0 \\ x &= 4 \end{aligned}$$

Check $x = 4$: $\log_4 4 + \log_4 4 = 1 + 1 = 2$

The solution set is $\{4\}$.

50. $\log_2 x + \log_2 (x + 15) = \log_2 16$

$$\begin{aligned} \log_2 [x(x+15)] &= \log_2 16 && \textit{Product rule} \\ x^2 + 15x &= 16 && \textit{Property 4} \\ x^2 + 15x - 16 &= 0 \\ (x+16)(x-1) &= 0 \end{aligned}$$

$x + 16 = 0$ or $x - 1 = 0$
$x = -16$ or $x = 1$

Reject $x = -16$, because it yields an equation in which the logarithm of a negative number must be found.

Check $x = 1$:
$\log_2 1 + \log_2 16 = 0 + \log_2 16 = \log_2 16$

The solution set is $\{1\}$.

51. When the power rule was applied in the second step, the domain was changed from $\{x \mid x \neq 0\}$ to $\{x \mid x > 0\}$. Instead of using the power rule for logarithms, we can change the original equation to the exponential form $x^2 = 10^2$ and get $x = \pm 10$. As you can see in the erroneous solution, the valid solution -10 was "lost." The solution set is $\{\pm 10\}$.

52. $A = P\left(1 + \frac{r}{n}\right)^{nt}$

Let $P = 20{,}000$, $r = 0.04$, and $t = 5$. For $n = 4$ (quarterly compounding),

$$A = 20{,}000\left(1 + \frac{0.04}{4}\right)^{4 \cdot 5} \approx 24{,}403.80.$$

There will be \$24,403.80 in the account after 5 years.

53. In the continuous compounding formula, let $P = 10{,}000$, $r = 0.0375$, and $t = 3$.

$$\begin{aligned} A &= Pe^{rt} \\ A &= 10{,}000e^{0.0375(3)} \approx 11{,}190.72 \end{aligned}$$

There will be \$11,190.72 in the account after 3 years.

54. Use $A = P\left(1 + \frac{r}{n}\right)^{nt}$.

Plan A:
Let $P = 1000$, $r = 0.04$, $n = 4$, and $t = 3$.

$$A = 1000\left(1 + \tfrac{0.04}{4}\right)^{4 \cdot 3} \approx 1126.83$$

Plan B:
Let $P = 1000$, $r = 0.039$, $n = 12$, and $t = 3$.

$$A = 1000\left(1 + \tfrac{0.039}{12}\right)^{12 \cdot 3} \approx 1123.91$$

Plan A is the better plan by \$2.92.

55. Let $Q(t) = \frac{1}{2}A_0$ to find the half-life of the radioactive substance.

$$\begin{aligned} Q(t) &= A_0 e^{-0.05t} \\ \tfrac{1}{2}A_0 &= A_0 e^{-0.05t} \\ 0.5 &= e^{-0.05t} \\ \ln 0.5 &= \ln e^{-0.05t} \\ \ln 0.5 &= -0.05t\,(\ln e) \\ -0.05t &= \ln 0.5 \\ t &= \frac{\ln 0.5}{-0.05} \approx 13.9 \end{aligned}$$

The half-life is about 13.9 days.

56. $S = C(1 - r)^n$

(a) Let $C = 30{,}000$, $r = 0.15$, and $n = 12$.

$$\begin{aligned} S &= 30{,}000(1 - 0.15)^{12} \\ &= 30{,}000(0.85)^{12} \approx 4267 \end{aligned}$$

The scrap value is about \$4267.

(b) Let $S = \frac{1}{2}C$ and $n = 6$.

$$\begin{aligned} S &= C(1-r)^n \\ \tfrac{1}{2}C &= C(1-r)^6 \\ 0.5 &= (1-r)^6 \quad (*) \\ \ln 0.5 &= \ln (1-r)^6 \\ \ln 0.5 &= 6 \ln (1-r) \\ \ln (1-r) &= \frac{\ln 0.5}{6} \\ \ln (1-r) &\approx -0.1155 \\ 1 - r &= e^{-0.1155} \\ 1 - r &\approx 0.89 \\ r &= 0.11 \end{aligned}$$

The rate is approximately 11%.

$(*)$ Alternative solution steps without logarithms:

$$\begin{aligned} 0.5 &= (1-r)^6 \\ \sqrt[6]{0.5} &= 1 - r \\ r &= 1 - \sqrt[6]{0.5} \approx 0.11 \end{aligned}$$

Note that $1 - r$ must be positive, so $\pm\sqrt[6]{0.5}$ is not needed.

57. $N(r) = -5000 \ln r$

Replace $N(r)$ with 2000, and solve for r.

$$\begin{aligned} 2000 &= -5000 \ln r \\ \ln r &= -\tfrac{2000}{5000} \\ \log_e r &= -0.4 \\ r &= e^{-0.4} && \textit{Exponential form} \\ r &\approx 0.67 && \textit{Approximate} \end{aligned}$$

About 67% of the words are common to both of the evolving languages.

58. **A.** Solve $7^x = 23$ by using the power rule with common logarithms.

$$\begin{aligned} 7^x &= 23 \\ \log 7^x &= \log 23 \\ x \log 7 &= \log 23 \\ x &= \frac{\log 23}{\log 7} \end{aligned}$$

B. Solve $7^x = 23$ by using the power rule with natural logarithms.

$$\begin{aligned} 7^x &= 23 \\ \ln 7^x &= \ln 23 \\ x \ln 7 &= \ln 23 \\ x &= \frac{\ln 23}{\ln 7} \end{aligned}$$

C. Use the change-of-base rule with the solution from **A.**

$$x = \frac{\log 23}{\log 7} = \log_7 23$$

D. $x = \dfrac{\log 23}{\log 7} \neq \log_{23} 7$

The answer is **D**.

59. **[12.4]** $\log_2 128 = \log_2 2^7 = 7$, by a special property of logarithms.

60. **[12.4]** By a special property of logarithms,

$$5^{\log_5 36} = 36.$$

61. **[12.5]** $e^{\ln 4} = 4$ since $e^{\ln x} = x$.

62. **[12.5]** $10^{\log e} = e$ since $10^{\log x} = x$.

63. **[12.4]** $\log_3 3^{-5} = -5$ since $\log_a a^x = x$.

64. **[12.5]** $\ln e^{5.4} = 5.4$ since $\ln e^x = x$.

65. **[12.6]** $\log_3 (x + 9) = 4$

$$\begin{aligned} x + 9 &= 3^4 && \textit{Exponential form} \\ x + 9 &= 81 \\ x &= 72 \end{aligned}$$

Check $x = 72$: $\log_3 81 = \log_3 3^4 = 4$

The solution set is $\{72\}$.

66. **[12.6]** $\ln e^x = 3$

$$x = 3$$

The solution set is $\{3\}$.

67. **[12.3]** $\log_x \frac{1}{81} = 2$

$$\begin{aligned} x^2 &= \tfrac{1}{81} && \textit{Exponential form} \\ x^2 &= \left(\tfrac{1}{9}\right)^2 \\ x &= \pm\tfrac{1}{9} && \textit{Square root property} \end{aligned}$$

The base x cannot be negative.

Check $x = \frac{1}{9}$: $\log_{1/9} \frac{1}{81} = 2$ since $\left(\frac{1}{9}\right)^2 = \frac{1}{81}$

The solution set is $\left\{\frac{1}{9}\right\}$.

68. **[12.2]** $27^x = 81$

Write each side as a power of 3.

$$\begin{aligned} (3^3)^x &= 3^4 \\ 3^{3x} &= 3^4 \\ 3x &= 4 && \textit{Equate exponents.} \\ x &= \tfrac{4}{3} \end{aligned}$$

Check $x = \frac{4}{3}$: $27^{4/3} = 3^4 = 81$

The solution set is $\left\{\frac{4}{3}\right\}$.

69. **[12.2]** $2^{2x-3} = 8$

Write each side as a power of 2.

$$\begin{aligned} 2^{2x-3} &= 2^3 \\ 2x - 3 &= 3 && \textit{Equate exponents.} \\ 2x &= 6 \\ x &= 3 \end{aligned}$$

Check $x = 3$: $2^3 = 8$ *True*

The solution set is $\{3\}$.

70. **[12.2]** $5^{x+2} = 25^{2x+1}$

$$\begin{aligned} 5^{x+2} &= (5^2)^{2x+1} && \textit{Equal bases} \\ 5^{x+2} &= 5^{4x+2} \\ x + 2 &= 4x + 2 && \textit{Equate exponents.} \\ 0 &= 3x \\ 0 &= x \end{aligned}$$

Check $x = 0$: $5^2 = 25^1$ *True*

The solution set is $\{0\}$.

71. **[12.6]**

$\log_3 (x + 1) - \log_3 x = 2$

$$\begin{aligned} \log_3 \frac{x+1}{x} &= 2 && \textit{Quotient rule} \\ \frac{x+1}{x} &= 3^2 && \textit{Exponential form} \\ 9x &= x + 1 \\ 8x &= 1 \\ x &= \tfrac{1}{8} \end{aligned}$$

Check $x = \frac{1}{8}$:

$\log_3 \frac{9}{8} - \log_3 \frac{1}{8} = \log_3 9 = \log_3 3^2 = 2$

The solution set is $\left\{\frac{1}{8}\right\}$.

72. **[12.6]** $\log(3x-1) = \log 10$
$$\begin{aligned} 3x - 1 &= 10 \\ 3x &= 11 \\ x &= \tfrac{11}{3} \end{aligned}$$

Check $x = \frac{11}{3}$: $\log(11-1) = \log 10$ *True*

The solution set is $\{\frac{11}{3}\}$.

73. **[12.6]** $\ln(x^2+3x+4) = \ln 2$
$$\begin{aligned} x^2+3x+4 &= 2 \\ x^2+3x+2 &= 0 \\ (x+2)(x+1) &= 0 \end{aligned}$$

$$x+2=0 \quad \text{or} \quad x+1=0$$
$$x=-2 \quad \text{or} \quad x=-1$$

Check $x=-2$: $\ln(4-6+4) = \ln 2$ *True*
Check $x=-1$: $\ln(1-3+4) = \ln 2$ *True*

The solution set is $\{-2,-1\}$.

74. **[12.6]** **(a)** $\log(2x+3) = \log x + 1$
$$\begin{aligned} \log(2x+3) - \log x &= 1 \\ \log_{10}\frac{2x+3}{x} &= 1 && \textit{Quotient rule} \\ \frac{2x+3}{x} &= 10^1 && \textit{Exponential form} \\ 2x+3 &= 10x \\ 3 &= 8x \\ \tfrac{3}{8} &= x \end{aligned}$$

Check $x = \frac{3}{8}$:
$LS = \log(\frac{3}{4}+3) = \log\frac{15}{4}$
$RS = \log\frac{3}{8} + \log 10 = \log\frac{30}{8} = \log\frac{15}{4}$

The solution set is $\{\frac{3}{8}\}$.

(b) From the graph, the x-value of the x-intercept is 0.375, the decimal equivalent of $\frac{3}{8}$. Note that the solutions of $Y_1 - Y_2 = 0$ are the same as the solutions of $Y_1 = Y_2$.

75. **[12.6]** For 2005, $x = 2005 - 1980 = 25$
$$\begin{aligned} R(x) &= 0.0997e^{0.0470x} \\ R(25) &= 0.0997e^{0.0470(25)} \\ &\approx 0.3228 \end{aligned}$$

In 2005, about 32.28% of municipal solid waste was recovered.

76. **[12.5]** **(a)** There are 90 of one species and 10 of another, so
$$p_1 = \tfrac{90}{100} = 0.9 \quad \text{and} \quad p_2 = \tfrac{10}{100} = 0.1.$$

Thus, the index of diversity is
$$\begin{aligned} &-(p_1 \ln p_1 + p_2 \ln p_2) \\ &= -(0.9\ln 0.9 + 0.1\ln 0.1) \\ &\approx 0.325. \end{aligned}$$

(b) There are 60 of one species and 40 of another, so
$$p_1 = \tfrac{60}{100} = 0.6 \quad \text{and} \quad p_2 = \tfrac{40}{100} = 0.4.$$

Thus, the index of diversity is
$$\begin{aligned} &-(p_1 \ln p_1 + p_2 \ln p_2) \\ &= -(0.6\ln 0.6 + 0.4\ln 0.4) \\ &\approx 0.673. \end{aligned}$$

Chapter 12 Test

1. **(a)** $f(x) = x^2 + 9$

This function is not one-to-one. The graph of $f(x)$ is a vertical parabola. A horizontal line will intersect the graph more than once.

(b) This function is one-to-one. A horizontal line will not intersect the graph in more than one point.

2. $f(x) = \sqrt[3]{x+7}$
Replace $f(x)$ with y.
$$y = \sqrt[3]{x+7}$$
Interchange x and y.
$$x = \sqrt[3]{y+7}$$
Solve for y.
$$\begin{aligned} x^3 &= y+7 \\ x^3 - 7 &= y \end{aligned}$$
Replace y with $f^{-1}(x)$.
$$f^{-1}(x) = x^3 - 7$$

3. By the horizontal line test, $f(x)$ is a one-to-one function and has an inverse. Choose some points on the graph of $f(x)$, such as $(4,0)$, $(3,-1)$, and $(0,-2)$. To graph the inverse, interchange the x- and y-values to get $(0,4)$, $(-1,3)$, and $(-2,0)$. Plot these points and draw a smooth curve through them.

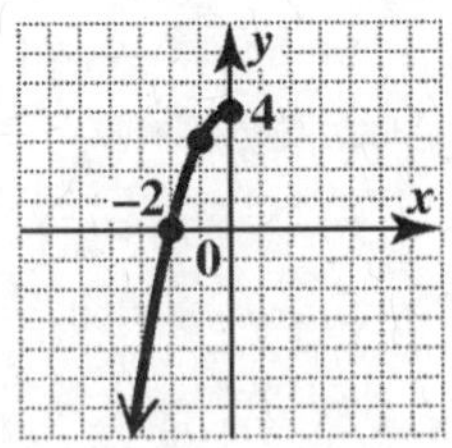

4. $f(x) = 6^x$
Make a table of values.

x	-2	-1	0	1
$f(x)$	$\frac{1}{36}$	$\frac{1}{6}$	1	6

Plot these points and draw a smooth exponential curve through them.

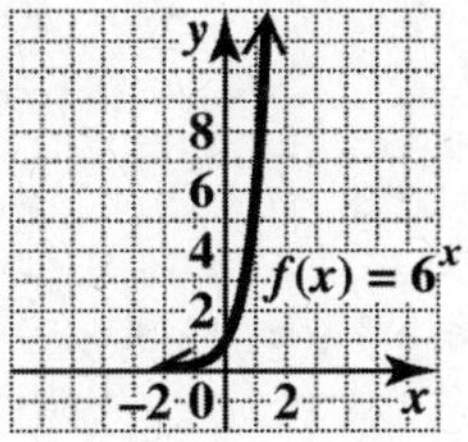

5. $g(x) = \log_6 x$
Make a table of values.

Powers of 6	6^{-2}	6^{-1}	6^0	6^1
x	$\frac{1}{36}$	$\frac{1}{6}$	1	6
$g(x)$	-2	-1	0	1

Plot these points and draw a smooth logarithmic curve through them.

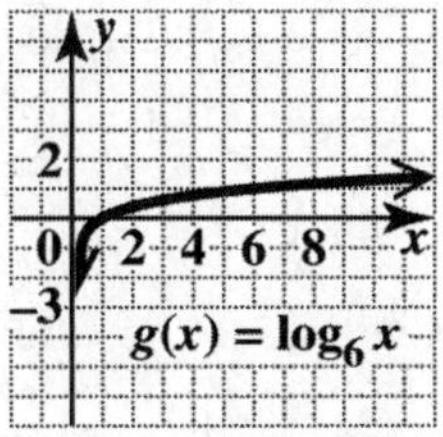

6. $y = 6^x$ and $y = \log_6 x$ are inverse functions. To use the graph from Exercise 4 to obtain the graph of the function in Exercise 5, interchange the x- and y-coordinates of the ordered pairs $\left(-2, \frac{1}{36}\right)$, $\left(-1, \frac{1}{6}\right)$, $(0, 1)$, and $(1, 6)$ to get $\left(\frac{1}{36}, -2\right)$, $\left(\frac{1}{6}, -1\right)$, $(1, 0)$, and $(6, 1)$.
Plot these points and draw a smooth logarithmic curve through them.

7.
$$5^x = \frac{1}{625}$$
$$5^x = \left(\frac{1}{5}\right)^4$$
Write each side as a power of 5.
$$5^x = 5^{-4}$$
$$x = -4 \qquad \textit{Equate exponents.}$$

The solution set is $\{-4\}$.

8.
$$2^{3x-7} = 8^{2x+2}$$
Write each side as a power of 2.
$$2^{3x-7} = \left(2^3\right)^{2x+2}$$
$$3x - 7 = 3(2x + 2) \qquad \textit{Equate exponents.}$$
$$3x - 7 = 6x + 6$$
$$-13 = 3x$$
$$-\frac{13}{3} = x$$

Check $x = -\frac{13}{3}$:
$$\text{LS} = 2^{-13-7} = 2^{-20}$$
$$\text{RS} = 8^{-20/3} = \left(2^3\right)^{-20/3} = 2^{-20}$$

The solution set is $\left\{-\frac{13}{3}\right\}$.

9. $f(x) = 46.9e^{0.0247x}$

(a) 2015: $x = 2015 - 2008 = 7$
$$f(7) = 46.9e^{0.0247(7)} \approx 55.8$$

The U.S. Hispanic population estimate for 2015 is 55.8 million.

(b) 2030: $x = 2030 - 2008 = 22$
$$f(22) = 46.9e^{0.0247(22)} \approx 80.8$$

The U.S. Hispanic population estimate for 2030 is 80.8 million.

10. The base is 4, the exponent (logarithm) is -2, and the number is 0.0625, so $4^{-2} = 0.0625$ becomes $\log_4 0.0625 = -2$ in logarithmic form.

11. The base is 7, the logarithm (exponent) is 2, and the number is 49, so $\log_7 49 = 2$ becomes $7^2 = 49$ in exponential form.

12. $\log_{1/2} x = -5$
$$x = \left(\tfrac{1}{2}\right)^{-5} \qquad \textit{Exponential form}$$
$$x = \left(\tfrac{2}{1}\right)^{5} = 32$$

The argument (the input of the logarithm) must be a positive number, so $x = 32$ is acceptable.

Check $x = 32$:
$\log_{1/2} 32 = -5$ since $\left(\frac{1}{2}\right)^{-5} = 2^5 = 32$

The solution set is $\{32\}$.

13.
$$x = \log_9 3$$
$$9^x = 3 \qquad \textit{Exponential form}$$
Write each side as a power of 3.
$$\left(3^2\right)^x = 3$$
$$3^{2x} = 3^1$$
$$2x = 1 \qquad \textit{Equate exponents.}$$
$$x = \frac{1}{2}$$

Check $x = \frac{1}{2}$:
$\frac{1}{2} = \log_9 3$ since $9^{1/2} = \sqrt{9} = 3$

The solution set is $\left\{\frac{1}{2}\right\}$.

14. $\log_x 16 = 4$
$$x^4 = 16 \qquad \textit{Exponential form}$$
$$x^2 = \pm 4 \qquad \textit{Square root property}$$
Reject -4 since $x^2 \geq 0$.
$$x^2 = 4$$
$$x = \pm 2 \qquad \textit{Square root property}$$

Reject -2 since the base cannot be negative.

Check $x = 2$: $\log_2 16 = 4$ since $2^4 = 16$

The solution set is $\{2\}$.

15. The value of $\log_2 32$ is $\underline{5}$. This means that if we raise $\underline{2}$ to the $\underline{5\text{th}}$ power, the result is $\underline{32}$.

16. $\log_3 x^2y$
$= \log_3 x^2 + \log_3 y$ *Product rule*
$= 2\log_3 x + \log_3 y$ *Power rule*

17. $\log_5\left(\frac{\sqrt{x}}{yz}\right)$
$= \log_5 \sqrt{x} - \log_5 yz$ *Quotient rule*
$= \log_5 x^{1/2} - (\log_5 y + \log_5 z)$ *Product rule*
$= \frac{1}{2}\log_5 x - \log_5 y - \log_5 z$ *Power rule*

18. $3\log_b s - \log_b t$
$= \log_b s^3 - \log_b t$ *Power rule*
$= \log_b \frac{s^3}{t}$ *Quotient rule*

19. $\frac{1}{4}\log_b r + 2\log_b s - \frac{2}{3}\log_b t$
Use the power rule for logarithms.
$= \log_b r^{1/4} + \log_b s^2 - \log_b t^{2/3}$
Use the product and quotient rules for logarithms.
$= \log_b \frac{r^{1/4}s^2}{t^{2/3}}$

20. **(a)** $\log 23.1 \approx 1.3636$

(b) $\ln 0.82 \approx -0.1985$

21. **(a)** $\log_3 19 = \frac{\log_{10} 19}{\log_{10} 3} = \frac{\log 19}{\log 3}$

(b) $\log_3 19 = \frac{\log_e 19}{\log_e 3} = \frac{\ln 19}{\ln 3}$

(c) The four-decimal-place approximation of either fraction is 2.6801.

22.
$$3^x = 78$$
$$\ln 3^x = \ln 78$$
$$x\ln 3 = \ln 78 \quad \textit{Power rule}$$
$$x = \frac{\ln 78}{\ln 3} \approx 3.966$$

Check $x = 3.966$: $3^{3.966} \approx 78.0$

The solution set is $\{3.966\}$.

23. $\log_8(x+5) + \log_8(x-2) = 1$
Use the product rule for logarithms.
$$\log_8[(x+5)(x-2)] = 1$$
$$(x+5)(x-2) = 8^1 \quad \textit{Exp. form}$$
$$x^2 + 3x - 10 = 8$$
$$x^2 + 3x - 18 = 0$$
$$(x+6)(x-3) = 0$$
$$x + 6 = 0 \quad \text{or} \quad x - 3 = 0$$
$$x = -6 \quad \text{or} \quad x = 3$$

Reject $x = -6$, because $x + 5 = -1$, which yields an equation in which the logarithm of a negative number must be found.

Check $x = 3$: $\log_8 8 + \log_8 1 = 1 + 0 = 1$

The solution set is $\{3\}$.

24. $A = P\left(1 + \frac{r}{n}\right)^{nt}$
$$A = 10{,}000\left(1 + \frac{0.045}{4}\right)^{4\cdot 5} \approx 12{,}507.51$$

\$10,000 invested at 4.5% annual interest, compounded quarterly, will increase to \$12,507.51 in 5 years.

25. $A = Pe^{rt}$

(a) $A = 15{,}000e^{0.05(5)} \approx 19{,}260.38$
There will be \$19,260.38 in the account.

(b) Let $A = 2(15{,}000)$ and solve for t.
$$2(15{,}000) = 15{,}000e^{0.05t}$$
$$2 = e^{0.05t}$$
$$\ln 2 = \ln e^{0.05t}$$
$$\ln 2 = 0.05t(\ln e)$$
$$0.05t = \ln 2$$
$$t = \frac{\ln 2}{0.05} \approx 13.9$$

The principal will double in about 13.9 years.

Cumulative Review Exercises (Chapters 1–12)

For Exercises 1–4,

$$S = \left\{-\tfrac{9}{4}, -2, -\sqrt{2}, 0, 0.6, \sqrt{11}, \sqrt{-8}, 6, \tfrac{30}{3}\right\}.$$

1. The integers are -2, 0, 6, and $\frac{30}{3}$ (or 10).

2. The rational numbers are $-\frac{9}{4}$, -2, 0, 0.6, 6, and $\frac{30}{3}$ (or 10). Each can be expressed as a quotient of two integers.

3. The irrational numbers are $-\sqrt{2}$ and $\sqrt{11}$.

4. $|-8| + 6 - |-2| - (-6+2)$
$= 8 + 6 - 2 - (-4)$
$= 14 - 2 + 4 = 16$

5. $2(-5) + (-8)(4) - (-3)$
$= -10 - 32 + 3 = -39$

6.
$$7 - (3 + 4x) + 2x = -5(x-1) - 3$$
$$7 - 3 - 4x + 2x = -5x + 5 - 3$$
$$4 - 2x = -5x + 2$$
$$3x = -2$$
$$x = -\tfrac{2}{3}$$

The solution set is $\left\{-\frac{2}{3}\right\}$.

7. $2x + 2 \le 5x - 1$
$-3x \le -3$
Divide by -3; reverse the inequality.
$x \ge 1$

The solution set is $[1, \infty)$.

8. $|2x - 5| = 9$

$$\begin{aligned} 2x - 5 &= 9 \quad \text{or} \quad 2x - 5 = -9 \\ 2x &= 14 \qquad\qquad 2x = -4 \\ x &= 7 \quad \text{or} \qquad x = -2 \end{aligned}$$

The solution set is $\{-2, 7\}$.

9. $|4x + 2| > 10$

$$\begin{aligned} 4x + 2 &> 10 \quad \text{or} \quad 4x + 2 < -10 \\ 4x &> 8 \qquad\qquad 4x < -12 \\ x &> 2 \quad \text{or} \qquad x < -3 \end{aligned}$$

The solution set is $(-\infty, -3) \cup (2, \infty)$.

10. $\sqrt{2x+1} - \sqrt{x} = 1$

$$\begin{aligned} \sqrt{2x+1} &= 1 + \sqrt{x} \\ \left(\sqrt{2x+1}\right)^2 &= \left(1 + \sqrt{x}\right)^2 \\ 2x + 1 &= 1 + 2\sqrt{x} + x \\ x &= 2\sqrt{x} \\ (x)^2 &= \left(2\sqrt{x}\right)^2 \\ x^2 &= 4x \\ x^2 - 4x &= 0 \\ x(x - 4) &= 0 \end{aligned}$$

$$x = 0 \quad \text{or} \quad x = 4$$

Check $x = 0$: $\sqrt{1} - \sqrt{0} = 1$ *True*

Check $x = 4$: $\sqrt{9} - \sqrt{4} = 1$ *True*

The solution set is $\{0, 4\}$.

11. $3x^2 - x - 1 = 0$
Here $a = 3$, $b = -1$, and $c = -1$.
Use the quadratic formula.

$$\begin{aligned} x &= \frac{-b \pm \sqrt{b^2 - 4ac}}{2a} \\ x &= \frac{-(-1) \pm \sqrt{(-1)^2 - 4(3)(-1)}}{2(3)} \\ &= \frac{1 \pm \sqrt{1 + 12}}{6} = \frac{1 \pm \sqrt{13}}{6} \end{aligned}$$

The solution set is $\left\{\frac{1 \pm \sqrt{13}}{6}\right\}$.

12. $x^2 + 2x - 8 > 0$
Solve the equation

$$\begin{aligned} x^2 + 2x - 8 &= 0. \\ (x + 4)(x - 2) &= 0 \end{aligned}$$

$$\begin{aligned} x + 4 &= 0 \quad \text{or} \quad x - 2 = 0 \\ x &= -4 \quad \text{or} \qquad x = 2 \end{aligned}$$

The numbers -4 and 2 divide a number line into three intervals.

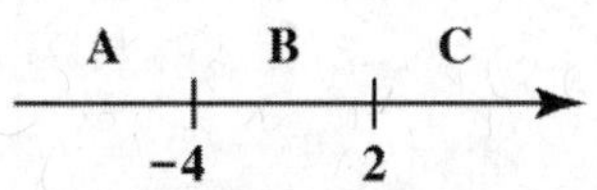

Test a number from each interval in the inequality

$$x^2 + 2x - 8 > 0.$$

Interval A: Let $x = -5$.

$$\begin{aligned} 25 - 10 - 8 &\overset{?}{>} 0 \\ 7 &> 0 \quad \textit{True} \end{aligned}$$

Interval B: Let $x = 0$.

$$-8 > 0 \quad \textit{False}$$

Interval C: Let $x = 3$.

$$\begin{aligned} 9 + 6 - 8 &\overset{?}{>} 0 \\ 7 &> 0 \quad \textit{True} \end{aligned}$$

The numbers in Intervals A and C, not including -4 or 2 because of $>$, are solutions. The solution set is $(-\infty, -4) \cup (2, \infty)$.

13. $x^4 - 5x^2 + 4 = 0$
Let $u = x^2$, so $u^2 = (x^2)^2 = x^4$.

$$\begin{aligned} u^2 - 5u + 4 &= 0 \\ (u - 1)(u - 4) &= 0 \end{aligned}$$

$$\begin{aligned} u - 1 &= 0 \quad \text{or} \quad u - 4 = 0 \\ u &= 1 \quad \text{or} \qquad u = 4 \end{aligned}$$

To find x, substitute x^2 for u.

$$\begin{aligned} x^2 &= 1 \quad \text{or} \quad x^2 = 4 \\ x &= \pm 1 \quad \text{or} \quad x = \pm 2 \end{aligned}$$

The solution set is $\{\pm 1, \pm 2\}$.

14.
$$\begin{aligned} 5^{x+3} &= \left(\tfrac{1}{25}\right)^{3x+2} \\ 5^{x+3} &= \left[\left(\tfrac{1}{5}\right)^2\right]^{3x+2} \end{aligned}$$

Write each side to the power of 5.

$$\begin{aligned} 5^{x+3} &= \left(5^{-2}\right)^{(3x+2)} \\ 5^{x+3} &= 5^{-2(3x+2)} \\ x + 3 &= -2(3x + 2) \\ x + 3 &= -6x - 4 \\ 7x &= -7 \\ x &= -1 \end{aligned}$$

Check $x = -1$: $5^2 = \left(\frac{1}{25}\right)^{-1}$ *True*
The solution set is $\{-1\}$.

15. $5x + 2y = 10$

Find the x- and y-intercepts. To find the x-intercept, let $y = 0$.

$$\begin{aligned} 5x + 2(0) &= 10 \\ 5x &= 10 \\ x &= 2 \end{aligned}$$

The x-intercept is $(2, 0)$.
To find the y-intercept, let $x = 0$.

$$5(0) + 2y = 10$$
$$2y = 10$$
$$y = 5$$

The y-intercept is $(0, 5)$.
Plot the intercepts and draw the line through them.

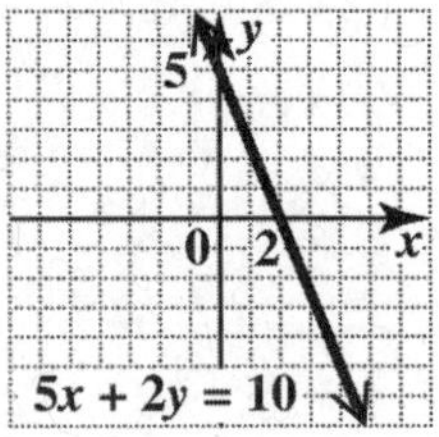

16. $-4x + y \leq 5$

Graph the line $-4x + y = 5$, which has intercepts $(0, 5)$ and $\left(-\frac{5}{4}, 0\right)$, as a solid line because the inequality involves $\leq$. Test $(0, 0)$, which yields $0 \leq 5$, a true statement. Shade the region that includes $(0, 0)$.

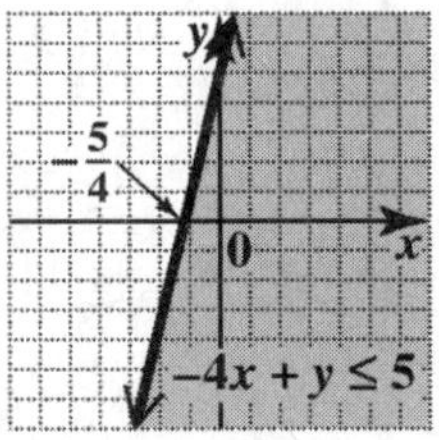

17. $f(x) = \frac{1}{3}(x-1)^2 + 2$ is in $f(x) = a(x-h)^2 + k$ form. The graph is a vertical parabola with vertex (h, k) at $(1, 2)$. Since $a = \frac{1}{3} > 0$, the graph opens up. Also, $|a| = \left|\frac{1}{3}\right| = \frac{1}{3} < 1$, so the graph is wider than the graph of $f(x) = x^2$. The points $\left(0, 2\frac{1}{3}\right)$, $(-2, 5)$, and $(4, 5)$ are also on the graph.

$f(x) = \frac{1}{3}(x-1)^2 + 2$

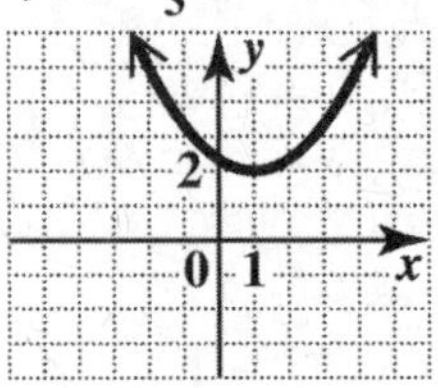

18. $f(x) = 2^x$
Make a table of values.

x	-2	-1	0	1	2
$f(x)$	$\frac{1}{4}$	$\frac{1}{2}$	1	2	4

Plot the ordered pairs from the table, and draw a smooth exponential curve through the points.

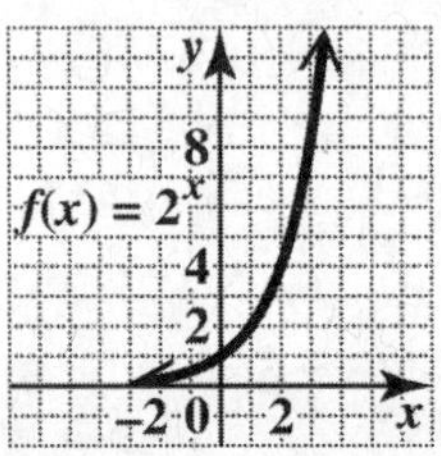

19. $f(x) = \log_3 x$
Make a table of values.

Powers of 3	3^{-2}	3^{-1}	3^0	3^1	3^2
x	$\frac{1}{9}$	$\frac{1}{3}$	1	3	9
y	-2	-1	0	1	2

Plot the ordered pairs and draw a smooth logarithmic curve through the points.

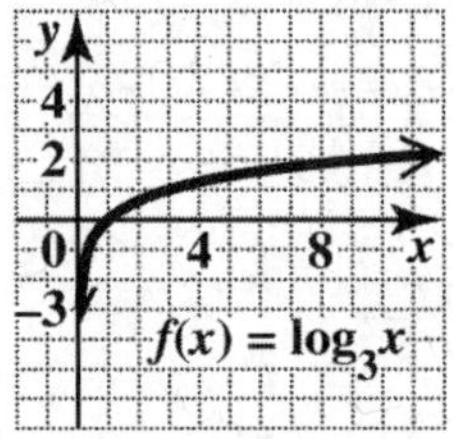

20. **(a)** Yes, this graph is the graph of a function because it passes the vertical line test.

(b) $(x_1, y_1) = (2003, 41{,}218)$ and $(x_2, y_2) = (2008, 57{,}949)$.

$$m = \frac{y_2 - y_1}{x_2 - x_1} = \frac{57{,}949 - 41{,}218}{2008 - 2003}$$
$$= \frac{16{,}731}{5} = 3346.2$$

The slope of the line in the graph is 3346.2 and can be interpreted as follows: The number of travelers increased by an average of 3346.2 thousand per year during the period 2003-2008.

21. Through $(5, -1)$; parallel to $3x - 4y = 12$
Find the slope of

$$3x - 4y = 12$$
$$-4y = -3x + 12$$
$$y = \tfrac{3}{4}x - 3.$$

The slope is $\frac{3}{4}$, so a line parallel to it also has slope $\frac{3}{4}$. Let $m = \frac{3}{4}$ and $(x_1, y_1) = (5, -1)$ in the point-slope form.

$$y - y_1 = m(x - x_1)$$
$$y - (-1) = \tfrac{3}{4}(x - 5)$$
$$y + 1 = \tfrac{3}{4}x - \tfrac{15}{4}$$
$$y = \tfrac{3}{4}x - \tfrac{19}{4}$$

22. $(2p + 3)(3p - 1) = 6p^2 - 2p + 9p - 3$
$= 6p^2 + 7p - 3$

23. $(4k - 3)^2 = (4k)^2 - 2(4k)(3) + 3^2$
$= 16k^2 - 24k + 9$

24. $(3m^3 + 2m^2 - 5m) - (8m^3 + 2m - 4)$
$= 3m^3 + 2m^2 - 5m - 8m^3 - 2m + 4$
$= 3m^3 - 8m^3 + 2m^2 - 5m - 2m + 4$
$= -5m^3 + 2m^2 - 7m + 4$

25.

$$
\begin{array}{r|rrrrr}
 & 2t^3 & +\,5t^2 & -\,3t & +\,4 & \\
\hline
3t+1 & 6t^4 & +\,17t^3 & -\,4t^2 & +\,9t & +\,4 \\
 & 6t^4 & +\,2t^3 & & & \\
\hline
 & & 15t^3 & -\,4t^2 & & \\
 & & 15t^3 & +\,5t^2 & & \\
\hline
 & & & -\,9t^2 & +\,9t & \\
 & & & -\,9t^2 & -\,3t & \\
\hline
 & & & & 12t & +\,4 \\
 & & & & 12t & +\,4 \\
\hline
 & & & & & 0
\end{array}
$$

The quotient is $2t^3 + 5t^2 - 3t + 4$.

26. $8x + x^3$
Factor out the GCF, x.

$$8x + x^3 = x(8 + x^2)$$

27. $24y^2 - 7y - 6$

Two integer factors whose product is $(24)(-6) = -144$ and whose sum is -7 are -16 and 9.
Rewrite the trinomial in a form that can be factored by grouping.

$$
\begin{aligned}
&24y^2 - 7y - 6 \\
&= 24y^2 - 16y + 9y - 6 \\
&= 8y(3y - 2) + 3(3y - 2) \\
&= (3y - 2)(8y + 3)
\end{aligned}
$$

Thus, $24y^2 - 7y - 6 = (3y - 2)(8y + 3)$.

28. $5z^3 - 19z^2 - 4z$
Factor out the GCF, z, and then factor by trial and error.

$$
\begin{aligned}
5z^3 - 19z^2 - 4z &= z(5z^2 - 19z - 4) \\
&= z(5z + 1)(z - 4)
\end{aligned}
$$

29. $16a^2 - 25b^4$

Use the difference of squares formula,

$$\boldsymbol{x^2 - y^2 = (x + y)(x - y)},$$

where $x = 4a$ and $y = 5b^2$.

$$16a^2 - 25b^4 = (4a + 5b^2)(4a - 5b^2)$$

30. $8c^3 + d^3$

Use the sum of cubes formula,

$$\boldsymbol{x^3 + y^3 = (x + y)(x^2 - xy + y^2)},$$

where $x = 2c$ and $y = d$.

$$8c^3 + d^3 = (2c + d)(4c^2 - 2cd + d^2)$$

31.
$$
\begin{aligned}
&16r^2 + 56rq + 49q^2 \\
&= (4r)^2 + 2(4r)(7q) + (7q)^2
\end{aligned}
$$

Use the perfect square formula,

$$\boldsymbol{x^2 + 2xy + y^2 = (x + y)^2},$$

where $x = 4r$ and $y = 7q$.

$$16r^2 + 56rq + 49q^2 = (4r + 7q)^2$$

32.
$$
\begin{aligned}
\frac{(5p^3)^4(-3p^7)}{2p^2(4p^4)} &= \frac{(5^4p^{12})(-3p^7)}{8p^6} \\
&= \frac{(625)(-3)p^{19}}{8p^6} \\
&= -\frac{1875p^{13}}{8}
\end{aligned}
$$

33. $\dfrac{x^2 - 9}{x^2 + 7x + 12} \div \dfrac{x - 3}{x + 5}$
Multiply by the reciprocal.

$$
\begin{aligned}
&= \frac{x^2 - 9}{x^2 + 7x + 12} \cdot \frac{x + 5}{x - 3} \\
&= \frac{(x + 3)(x - 3)}{(x + 3)(x + 4)} \cdot \frac{(x + 5)}{(x - 3)} \quad \textit{Factor.} \\
&= \frac{x + 5}{x + 4}
\end{aligned}
$$

34. $\dfrac{2}{k + 3} - \dfrac{5}{k - 2}$
The LCD is $(k + 3)(k - 2)$.

$$
\begin{aligned}
&= \frac{2(k - 2)}{(k + 3)(k - 2)} - \frac{5(k + 3)}{(k - 2)(k + 3)} \\
&= \frac{2k - 4 - 5k - 15}{(k + 3)(k - 2)} \\
&= \frac{-3k - 19}{(k + 3)(k - 2)}
\end{aligned}
$$

35.
$$
\begin{aligned}
5x - 3y &= 14 \quad (1) \\
2x + 5y &= 18 \quad (2)
\end{aligned}
$$

Multiply equation (1) by 5 and equation (2) by 3. Then add the results.

$$
\begin{array}{rcrcrl}
25x & - & 15y & = & 70 & \quad 5 \times (1) \\
6x & + & 15y & = & 54 & \quad 3 \times (2) \\
\hline
31x & & & = & 124 & \quad \textit{Add.} \\
 & & x & = & 4 &
\end{array}
$$

Substitute 4 for x in equation (1) to find y.

$$
\begin{aligned}
5x - 3y &= 14 \qquad (1) \\
5(4) - 3y &= 14 \\
20 - 3y &= 14 \\
-3y &= -6 \\
y &= 2
\end{aligned}
$$

The solution set is $\{(4, 2)\}$.

36.
$$\begin{aligned} x+2y+3z &= 11 \quad (1)\\ 3x-y+z &= 8 \quad (2)\\ 2x+2y-3z &= -12 \quad (3)\end{aligned}$$

To eliminate z, add equations (1) and (3).

$$\begin{array}{rcrcrcrl} x & + & 2y & + & 3z & = & 11 & (1)\\ 2x & + & 2y & - & 3z & = & -12 & (3)\\ \hline 3x & + & 4y & & & = & -1 & (4)\end{array}$$

To eliminate z again, multiply equation (2) by 3 and add the result to equation (3).

$$\begin{array}{rcrcrcrl} 9x & - & 3y & + & 3z & = & 24 & 3\times(2)\\ 2x & + & 2y & - & 3z & = & -12 & (3)\\ \hline 11x & - & y & & & = & 12 & (5)\end{array}$$

Multiply equation (5) by 4 and add the result to equation (4).

$$\begin{array}{rcrcrl} 44x & - & 4y & = & 48 & 4\times(5)\\ 3x & + & 4y & = & -1 & (4)\\ \hline 47x & & & = & 47 & \\ & & x & = & 1 & \end{array}$$

Substitute 1 for x in equation (5) to find y.

$$\begin{aligned} 11x-y &= 12 \quad (5)\\ 11(1)-y &= 12\\ 11-y &= 12\\ -y &= 1\\ y &= -1\end{aligned}$$

Substitute 1 for x and -1 for y in equation (2) to find z.

$$\begin{aligned} 3x-y+z &= 8 \quad (2)\\ 3(1)-(-1)+z &= 8\\ 3+1+z &= 8\\ 4+z &= 8\\ z &= 4\end{aligned}$$

The solution set is $\{(1,-1,4)\}$.

37. Let x = the amount of candy at \$1.00 per pound.

	Number of Pounds	Price per Pound	Value
First Candy	x	\$1.00	$1.00x$
Second Candy	10	\$1.96	$1.96(10)$
Mixture	$x+10$	\$1.60	$1.60(x+10)$

The sum of the values of each candy must equal the value of the mixture.

$$1.00x+1.96(10)=1.60(x+10)$$

Multiply by 10 to clear the decimals.

$$\begin{aligned} 10x+196 &= 16(x+10)\\ 10x+196 &= 16x+160\\ 36 &= 6x\\ 6 &= x\end{aligned}$$

Use 6 pounds of the \$1.00 candy.

38. $\sqrt{288}=\sqrt{144\cdot 2}=\sqrt{144}\sqrt{2}=12\sqrt{2}$

39.
$$\begin{aligned} 2\sqrt{32}-5\sqrt{98} &= 2\sqrt{16\cdot 2}-5\sqrt{49\cdot 2}\\ &= 2\cdot 4\sqrt{2}-5\cdot 7\sqrt{2}\\ &= 8\sqrt{2}-35\sqrt{2}\\ &= -27\sqrt{2}\end{aligned}$$

40.
$$\begin{aligned} (5+4i)(5-4i) &= 5^2-(4i)^2\\ &= 25-16i^2\\ &= 25-16(-1)\\ &= 25+16=41\end{aligned}$$

41. $\log \dfrac{x^3\sqrt{y}}{z}$

$$\begin{aligned} &= \log \frac{x^3y^{1/2}}{z}\\ &= \log\left(x^3y^{1/2}\right)-\log z && \textit{Quotient rule}\\ &= \log x^3+\log y^{1/2}-\log z && \textit{Product rule}\\ &= 3\log x+\tfrac{1}{2}\log y-\log z && \textit{Power rule}\end{aligned}$$

42. $B(t)=25{,}000e^{0.2t}$

(a) At noon, $t=0$.

$$\begin{aligned} B(0) &= 25{,}000e^{0.2(0)}\\ &= 25{,}000e^0=25{,}000(1)=25{,}000\end{aligned}$$

25,000 bacteria are present at noon.

(b) At 1 P.M., $t=1$.

$$B(1)=25{,}000e^{0.2(1)}\approx 30{,}535$$

About 30,500 bacteria are present at 1 P.M.

(c) At 2 P.M., $t=2$.

$$B(2)=25{,}000e^{0.2(2)}\approx 37{,}296$$

About 37,300 bacteria are present at 2 P.M.

(d) The population doubles when $B(t)=50{,}000$.

$$\begin{aligned} 50{,}000 &= 25{,}000e^{0.2t}\\ 2 &= e^{0.2t}\\ \ln 2 &= \ln e^{0.2t}\\ \ln 2 &= 0.2t\ln e\\ t &= \frac{\ln 2}{0.2}\approx 3.5\end{aligned}$$

The population will double in about 3.5 hours, or at about 3:30 P.M.

CHAPTER 13 NONLINEAR FUNCTIONS, CONIC SECTIONS, AND NONLINEAR SYSTEMS

13.1 Additional Graphs of Functions

13.1 Now Try Exercises

N1. $f(x) = \dfrac{1}{x+3}$

The graph of this function is found by shifting the graph of the reciprocal function, $f(x) = \frac{1}{x}$, 3 units left. Since $x \neq -3$ (or a denominator of 0 results), the line $x = -3$ is a vertical asymptote and the domain of this function is $(-\infty, -3) \cup (-3, \infty)$.

Since $\frac{1}{x+3} \neq 0$, the line $y = 0$ is a horizontal asymptote, and the range of this function is $(-\infty, 0) \cup (0, \infty)$.

x	y
-5	$-\frac{1}{2}$
-4	-1
-2	1
-1	$\frac{1}{2}$

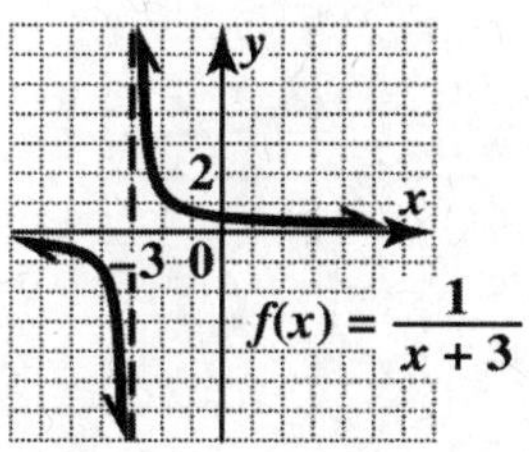

N2. $f(x) = \sqrt{x} + 2$

The graph of this function is found by shifting the graph of the square root function, $f(x) = \sqrt{x}$, 2 units up.

Since x must be at least 0 for f to be defined, the domain of this function is $[0, \infty)$. Since y is at least 2, the range of this function is $[2, \infty)$.

x	y
0	2
1	3
4	4

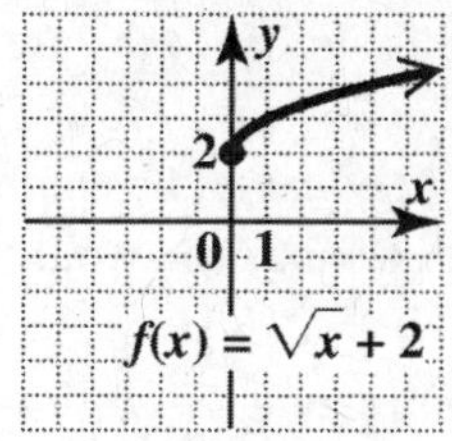

N3. $f(x) = |x + 1| - 3$

This is the graph of the absolute value function, $f(x) = |x|$, shifted 1 unit to the left (since $x + 1 = 0$ if $x = -1$) and 3 units downward (because of the -3).

The domain of this function is $(-\infty, \infty)$ and its range is $[-3, \infty)$.

x	y
-3	-1
-2	-2
-1	-3
0	-2
1	-1

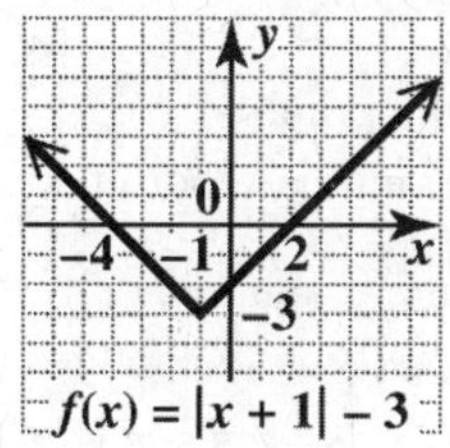

N4. **(a)** $[\![5]\!] = 5 \quad (5 \leq 5)$

(b) $[\![-6]\!] = -6 \quad (-6 \leq -6)$

(c) $[\![3.5]\!] = 3 \quad (3 \leq 3.5)$

(d) $[\![-4.1]\!] = -5 \quad (-5 \leq -4.1)$

N5. $f(x) = [\![x - 1]\!]$

The graph of f is the same as the graph of $y = [\![x]\!]$ shifted one unit to the right. Shifting the graph of $y = [\![x]\!]$ (right, left, up, down) does not change the domain and range. The domain of f is $(-\infty, \infty)$ and the range of f is the set of integers, that is, $\{\ldots, -2, -1, 0, 1, 2, \ldots\}$.

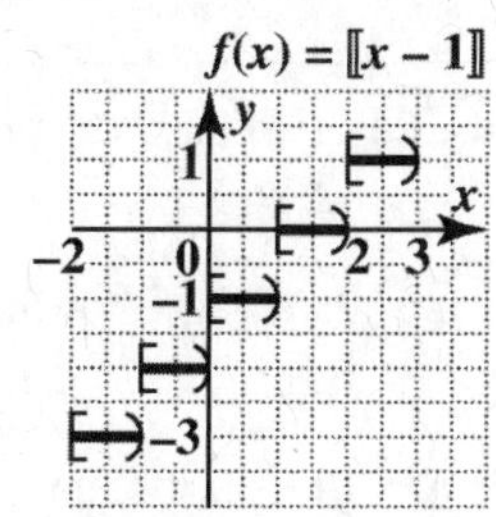

N6. This function is similar to the greatest integer function, but in this case, we use the integer that is *greater than* or equal to the number. For example, for $\frac{1}{2}$ hour of parking, you would be charged for 1 hour.

Interval	Hours Charged for	Cost
$(0, 1]$	1	\$4.00
$(1, 2]$	2	\$6.00
$(2, 3]$	3	\$8.00
$(3, 4]$	4	\$10.00

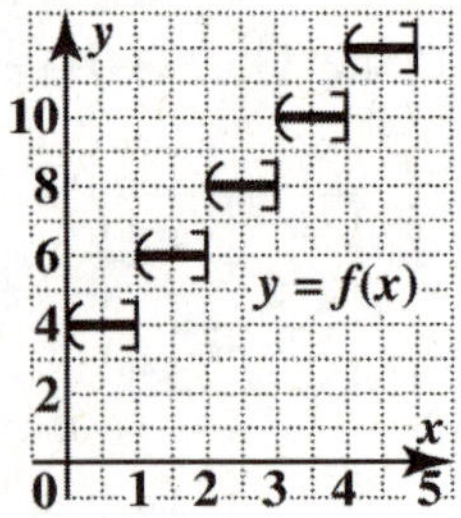

13.1 Section Exercises

1. Choice **E** is the graph of $f(x) = |x|$. The lowest point on its graph has coordinates $(\underline{\ 0\ }, \underline{\ 0\ })$.

3. Choice **A** is the graph of $f(x) = [\![x]\!]$. The domain of f is $(-\infty, \infty)$ and the range of f is the set of integers, that is, $\{\ldots, -2, -1, 0, 1, 2, \ldots\}$.

5. Choice **B** is not the graph of a function since it does not satisfy the conditions of the vertical line test.

7. $f(x) = |x - 2| + 2$

The graph of this function has its "vertex" at $(2, 2)$, so the correct graph is **B**.

9. $f(x) = |x - 2| - 2$

The graph of this function has its "vertex" at $(2, -2)$, so the correct graph is **A**.

11. $f(x) = |x + 1|$

Since x can be any real number, the domain is $(-\infty, \infty)$.

The value of y is always greater than or equal to 0, so the range is $[0, \infty)$.

The graph of $y = |x + 1|$ looks like the graph of the absolute value function $y = |x|$, but the graph is translated 1 unit to the left. The x-value of its "vertex" is obtained by setting $x + 1 = 0$ and solving for x:

$$\begin{aligned} x + 1 &= 0 \\ x &= -1. \end{aligned}$$

Since the corresponding y-value is 0, the "vertex" is $(-1, 0)$. Some additional points are $(-3, 2)$, $(-2, 1)$, $(0, 1)$, and $(1, 2)$.

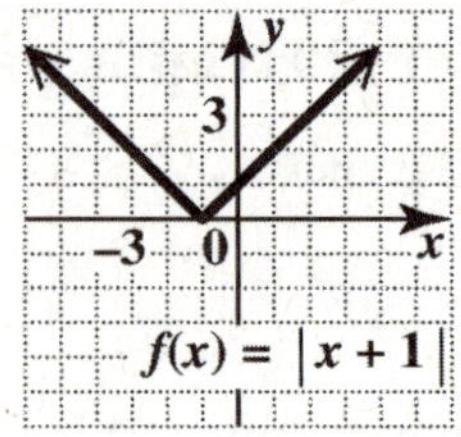

13. $f(x) = \dfrac{1}{x} + 1$

The graph of this function is similar to the graph of $g(x) = \frac{1}{x}$, except that each point is translated 1 unit upward. Just as with $g(x) = \frac{1}{x}$, $x = 0$ is the vertical asymptote, but this graph has $y = 1$ as its horizontal asymptote.

The domain is all real numbers except 0, that is, $(-\infty, 0) \cup (0, \infty)$.

The range is all real numbers except 1, that is, $(-\infty, 1) \cup (1, \infty)$.

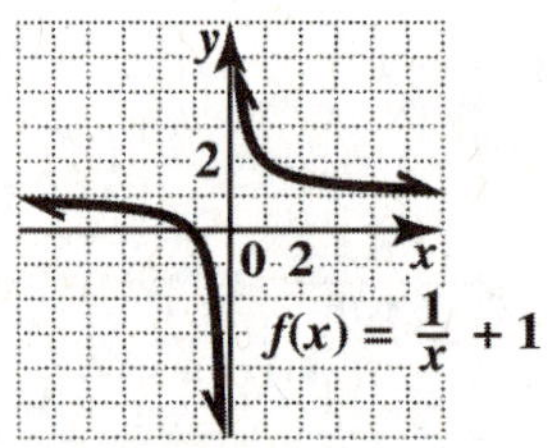

15. $f(x) = \sqrt{x - 2}$

The graph is found by shifting the graph of $y = \sqrt{x}$ two units to the right. The following table of ordered pairs gives some specific points the graph passes through.

x	2	3	6
y	0	1	2

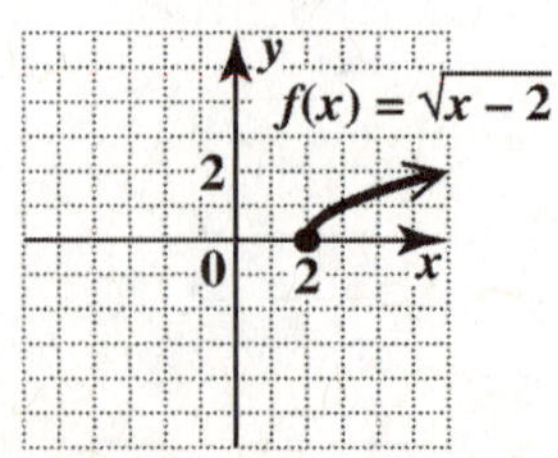

The domain of the function is $[2, \infty)$ and its range is $[0, \infty)$.

17. $f(x) = \dfrac{1}{x-2}$

This is the graph of the reciprocal function, $g(x) = \frac{1}{x}$, shifted 2 units to the right. Since $x \neq 2$ (or a denominator of 0 results), the line $x = 2$ is a vertical asymptote. Since $\dfrac{1}{x-2} \neq 0$, the line $y = 0$ is a horizontal asymptote.

x	-1	0	1	3	4	5
y	$-\frac{1}{3}$	$-\frac{1}{2}$	-1	1	$\frac{1}{2}$	$\frac{1}{3}$

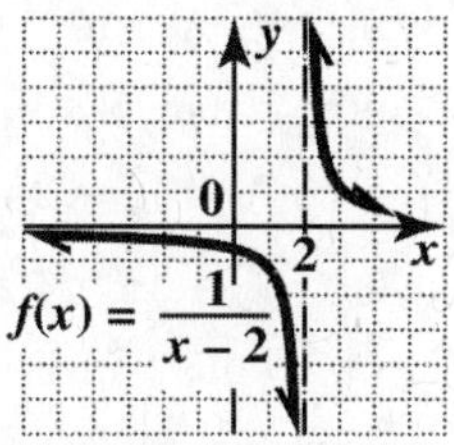

The domain of the function is $(-\infty, 2) \cup (2, \infty)$ and its range is $(-\infty, 0) \cup (0, \infty)$.

19. $f(x) = \sqrt{x+3} - 3$

The graph is found by shifting the graph of $y = \sqrt{x}$ three units to the left and three units down. The following table of ordered pairs gives some specific points the graph passes through.

x	-3	-2	1
y	-3	-2	-1

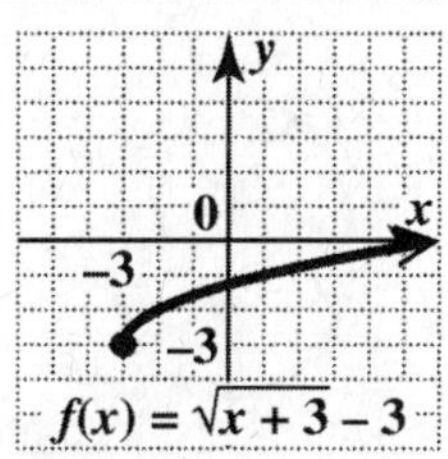

The domain of the function is $[-3, \infty)$ and its range is $[-3, \infty)$.

21. $f(x) = |x-3| + 1$

This is the graph of the absolute value function, $f(x) = |x|$, shifted 3 units to the right (since $x - 3 = 0$ if $x = 3$) and 1 unit upward (because of the $+1$).

The domain of this function is $(-\infty, \infty)$ and its range is $[1, \infty)$.

x	2	3	4
y	2	1	2

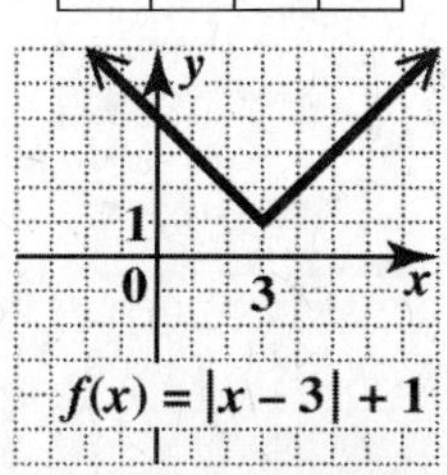

23. Shift the graph of $g(x) = \frac{1}{x}$ three units to the right and two units up to get the graph of
$$f(x) = \frac{1}{x-3} + 2.$$

In Exercises 25–34, the answer is the greatest integer that is less than or equal to the number in the greatest integer symbol.

25. $[\![3]\!] = 3 \quad (3 \leq 3)$

27. $[\![4.5]\!] = 4 \quad (4 \leq 4.5)$

29. $[\![\frac{1}{2}]\!] = 0 \quad (0 \leq \frac{1}{2})$

31. $[\![-14]\!] = -14 \quad (-14 \leq -14)$

33. $[\![-10.1]\!] = -11 \quad (-11 \leq -10.1)$

35. $f(x) = [\![x]\!] - 1$

This is the graph of the greatest integer function, $g(x) = [\![x]\!]$, shifted down 1 unit.

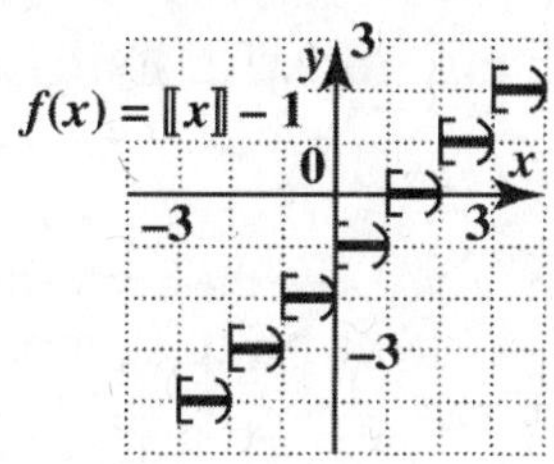

37. $f(x) = [\![x-3]\!]$

This is the graph of the greatest integer function, $g(x) = [\![x]\!]$, shifted 3 units to the right.

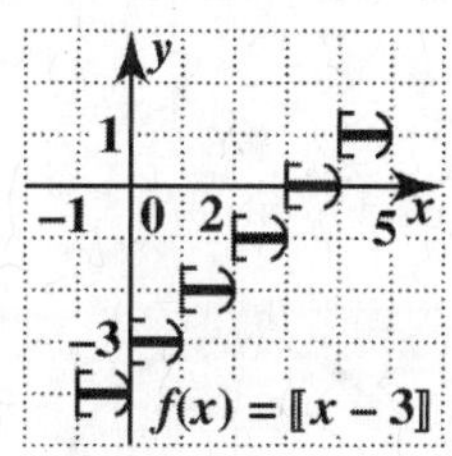

39. For any portion of the first ounce, the cost will be one 44¢ stamp. If the weight exceeds one ounce (up to two ounces), an additional 17¢ stamp is required. The following table summarizes the weight of a letter, x, and the number of stamps required, $p(x)$, on the interval $(0, 5]$.

x	$(0,1]$	$(1,2]$	$(2,3]$	$(3,4]$	$(4,5]$
$p(x)$	1	2	3	4	5

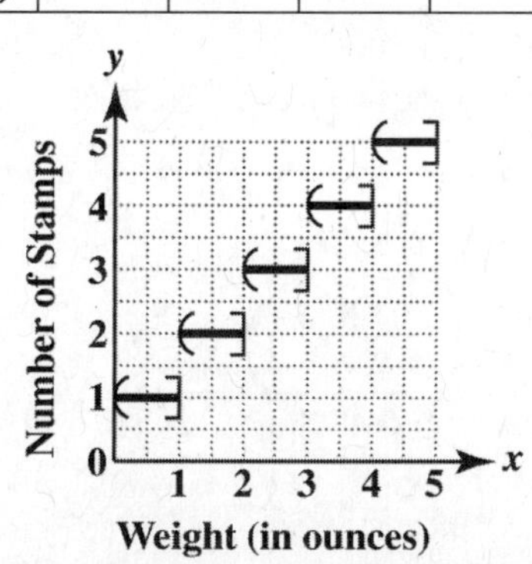

41.
$$\begin{aligned} f(x) &= 0.40[\![x]\!] + 0.75 \\ f(5.5) &= 0.40[\![5.5]\!] + 0.75 \\ &= 0.40(5) + 0.75 \\ &= 2 + 0.75 = 2.75 \end{aligned}$$

The cost of a 5.5-minute call is \$2.75.

In Exercises 43–45, use the distance formula

$$d = \sqrt{(x_2 - x_1)^2 + (y_2 - y_1)^2}.$$

43. $(2, -1)$ and $(4, 3)$

$$\begin{aligned} d &= \sqrt{(4-2)^2 + [3-(-1)]^2} \\ &= \sqrt{2^2 + 4^2} = \sqrt{4 + 16} \\ &= \sqrt{20} = \sqrt{4} \cdot \sqrt{5} = 2\sqrt{5} \end{aligned}$$

45. (x, y) and (h, k)

$$d = \sqrt{(x-h)^2 + (y-k)^2}$$

13.2 The Circle and the Ellipse

13.2 Now Try Exercises

N1. If the point (x, y) is on the circle, the distance from (x, y) to the center $(0, 0)$ is 6. Use the distance formula.

$$\begin{aligned} \sqrt{(x_2 - x_1)^2 + (y_2 - y_1)^2} &= d \\ \sqrt{(x-0)^2 + (y-0)^2} &= 6 \\ \sqrt{x^2 + y^2} &= 6 \\ x^2 + y^2 &= 36 \quad \textit{Square.} \end{aligned}$$

An equation of this circle is $x^2 + y^2 = 36$.

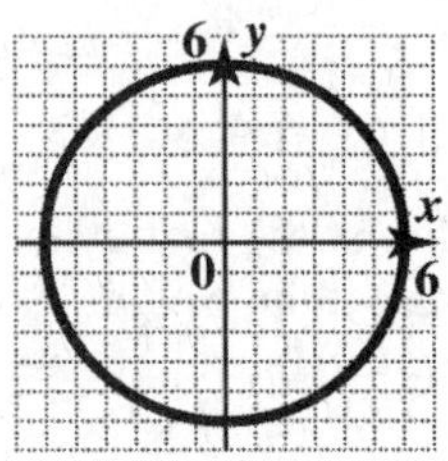

N2. Center at $(-2, 2)$; radius 3

$$\begin{aligned} \sqrt{(x_2 - x_1)^2 + (y_2 - y_1)^2} &= d \\ \sqrt{[x-(-2)]^2 + (y-2)^2} &= 3 \\ \sqrt{(x+2)^2 + (y-2)^2} &= 3 \\ (x+2)^2 + (y-2)^2 &= 9 \quad \textit{Square.} \end{aligned}$$

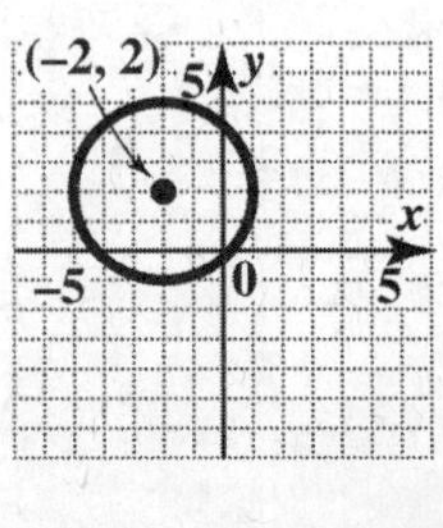

N3. Center at $(-5, 4)$; radius $\sqrt{6}$

Substitute $h = -5$, $k = 4$, and $r = \sqrt{6}$ in the center-radius form of the equation of a circle.

$$\begin{aligned} (x-h)^2 + (y-k)^2 &= r^2 \\ [x-(-5)]^2 + (y-4)^2 &= \left(\sqrt{6}\right)^2 \\ (x+5)^2 + (y-4)^2 &= 6 \end{aligned}$$

N4. To find the center and radius, add the appropriate constants to complete the squares on x and y.

$$\begin{aligned} x^2 + y^2 - 8x + 10y - 8 &= 0 \\ (x^2 - 8x) + (y^2 + 10y) &= 8 \\ (x^2 - 8x + 16) + (y^2 + 10y + 25) &= 8 + 16 + 25 \\ (x-4)^2 + (y+5)^2 &= 49 \\ (x-4)^2 + [y-(-5)]^2 &= 7^2 \end{aligned}$$

The circle has center at $(4, -5)$ and radius 7.

N5. $\dfrac{x^2}{16} + \dfrac{y^2}{25} = 1$

This ellipse has an equation of the form

$$\frac{x^2}{a^2} + \frac{y^2}{b^2} = 1.$$

Here, $a^2 = 16$, so $a = 4$ and the x-intercepts are $(4, 0)$ and $(-4, 0)$. Similarly, $b^2 = 25$, so $b = 5$ and the y-intercepts are $(0, 5)$ and $(0, -5)$. Plot the intercepts, and draw the ellipse through them.

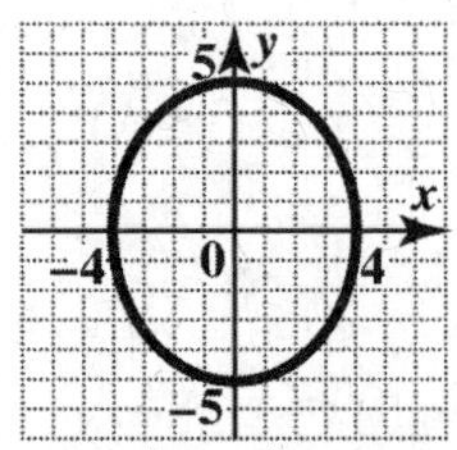

$$\frac{x^2}{16} + \frac{y^2}{25} = 1$$

N6. $\dfrac{(x-3)^2}{36} + \dfrac{(y-4)^2}{4} = 1$

The center is $(3, 4)$.

The ellipse passes through the four points:

$$\begin{aligned} (3-6, 4) &= (-3, 4) \\ (3+6, 4) &= (9, 4) \\ (3, 4+2) &= (3, 6) \\ (3, 4-2) &= (3, 2) \end{aligned}$$

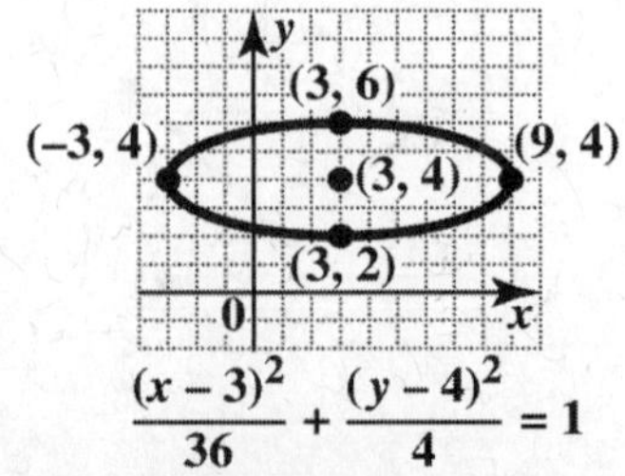

$$\frac{(x-3)^2}{36} + \frac{(y-4)^2}{4} = 1$$

13.2 Section Exercises

1. **(a)** $x^2 + y^2 = 25$ can be written in the center-radius form as

$$(x-0)^2 + (y-0)^2 = 5^2.$$

The center is the point $(0, 0)$.

(b) The radius is 5.

(c) The x-intercepts are $(5, 0)$ and $(-5, 0)$. The y-intercepts are $(0, 5)$ and $(0, -5)$.

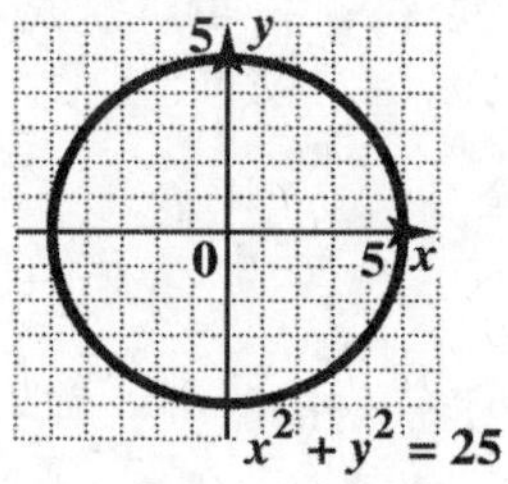

3. $(x-3)^2 + (y-2)^2 = 25$ is an equation of a circle with center $(3, 2)$ and radius 5, choice **B**.

5. $(x+3)^2 + (y-2)^2 = 25$ is an equation of a circle with center $(-3, 2)$ and radius 5, choice **D**.

7. Center: $(-4, 3)$; radius: 2

Substitute $h = -4$, $k = 3$, and $r = 2$ in the center-radius form of the equation of a circle.

$$(x-h)^2 + (y-k)^2 = r^2$$
$$[x-(-4)]^2 + (y-3)^2 = 2^2$$
$$(x+4)^2 + (y-3)^2 = 4$$

9. Center: $(-8, -5)$; radius: $\sqrt{5}$

Substitute $h = -8$, $k = -5$, and $r = \sqrt{5}$ in the center-radius form of the equation of a circle.

$$(x-h)^2 + (y-k)^2 = r^2$$
$$[x-(-8)]^2 + [y-(-5)]^2 = \left(\sqrt{5}\right)^2$$
$$(x+8)^2 + (y+5)^2 = 5$$

11. $$x^2 + y^2 + 4x + 6y + 9 = 0$$

Rewrite the equation keeping only the variable terms on the left and grouping the x-terms and y-terms.

$$x^2 + 4x + y^2 + 6y = -9$$

Complete both squares on the left, and add the same constants to the right.

$$(x^2 + 4x + \underline{4}) + (y^2 + 6y + \underline{9}) = -9 + \underline{4} + \underline{9}$$
$$(x+2)^2 + (y+3)^2 = 4$$

From the form $(x-h)^2 + (y-k)^2 = r^2$, we have $h = -2$, $k = -3$, and $r = 2$. The center is $(-2, -3)$, and the radius r is 2.

13. $$x^2 + y^2 + 10x - 14y - 7 = 0$$
$$(x^2 + 10x \qquad) + (y^2 - 14y \qquad) = 7$$
$$(x^2 + 10x + \underline{25}) + (y^2 - 14y + \underline{49}) = 7 + \underline{25} + \underline{49}$$
$$(x+5)^2 + (y-7)^2 = 81$$

The center is $(-5, 7)$, and the radius is $\sqrt{81} = 9$.

15. $$3x^2 + 3y^2 - 12x - 24y + 12 = 0$$
$$x^2 + y^2 - 4x - 8y + 4 = 0 \quad \textit{Divide by 3.}$$

Complete the square on x and the square on y.

$$(x^2 - 4x \qquad) + (y^2 - 8y \qquad) = -4$$
$$(x^2 - 4x + \underline{4}) + (y^2 - 8y + \underline{16}) = -4 + \underline{4} + \underline{16}$$
$$(x-2)^2 + (y-4)^2 = 16$$

The center is $(2, 4)$, and the radius is $\sqrt{16} = 4$.

17. $$x^2 + y^2 = 9$$
$$(x-0)^2 + (y-0)^2 = 3^2$$

Here, $h = 0$, $k = 0$, and $r = 3$, so the graph is a circle with center $(0, 0)$ and radius 3.

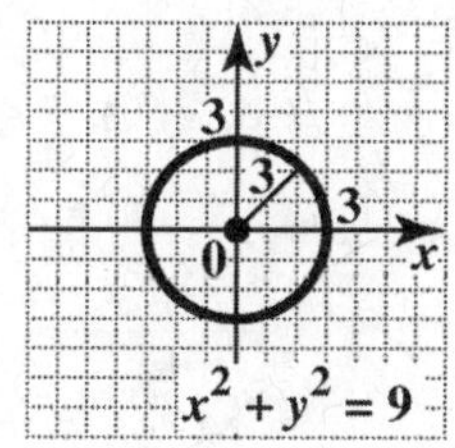

19. $$2y^2 = 10 - 2x^2$$
$$2x^2 + 2y^2 = 10$$
$$x^2 + y^2 = 5 \quad \textit{Divide by 2.}$$
$$(x-0)^2 + (y-0)^2 = \left(\sqrt{5}\right)^2$$

Here, $h = 0$, $k = 0$, and $r = \sqrt{5} \approx 2.2$, so the graph is a circle with center $(0, 0)$ and radius $\sqrt{5}$.

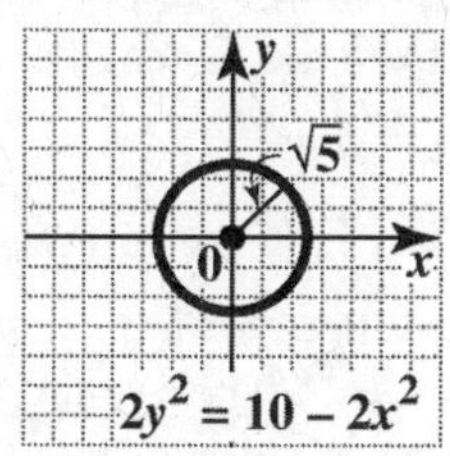

21. $(x+3)^2 + (y-2)^2 = 9$

Here, $h = -3$, $k = 2$, and $r = \sqrt{9} = 3$. The graph is a circle with center $(-3, 2)$ and radius 3.

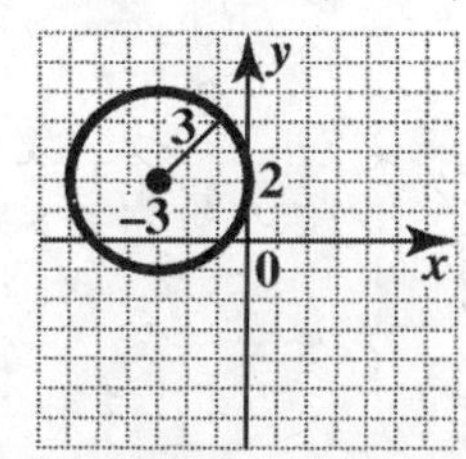

$(x+3)^2 + (y-2)^2 = 9$

23. $$x^2 + y^2 - 4x - 6y + 9 = 0$$
$$(x^2 - 4x \quad\;) + (y^2 - 6y \quad\;) = -9$$
Complete the square for each variable.
$$(x^2 - 4x + \underline{4}) + (y^2 - 6y + \underline{9}) = -9 + \underline{4} + \underline{9}$$
$$(x - 2)^2 + (y - 3)^2 = 4$$
Here, $h = 2$, $k = 3$, and $r = \sqrt{4} = 2$. The graph is a circle with center $(2, 3)$ and radius 2.

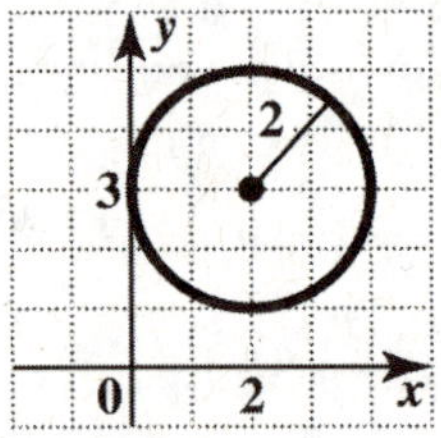

$x^2 + y^2 - 4x - 6y + 9 = 0$

25. $$x^2 + y^2 + 6x - 6y + 9 = 0$$
$$(x^2 + 6x \quad\;) + (y^2 - 6y \quad\;) = -9$$
Complete the square for each variable.
$$(x^2 + 6x + \underline{9}) + (y^2 - 6y + \underline{9}) = -9 + \underline{9} + \underline{9}$$
$$(x + 3)^2 + (y - 3)^2 = 9$$
The center is $(-3, 3)$, and the radius is $\sqrt{9} = 3$.

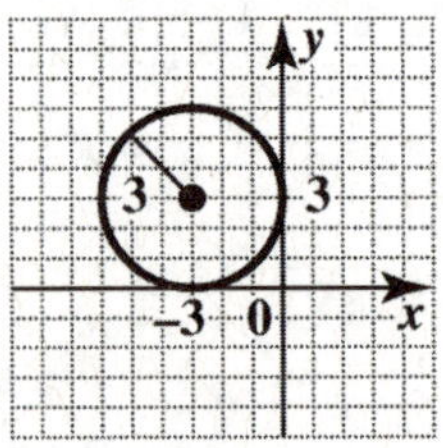

$x^2 + y^2 + 6x - 6y + 9 = 0$

27. This method works because the pencil is always the same distance from the fastened end. The fastened end works as the center, and the length of the string from the fastened end to the pencil is the radius.

29. $\dfrac{x^2}{9} + \dfrac{y^2}{25} = 1$ is in the form $\dfrac{x^2}{a^2} + \dfrac{y^2}{b^2} = 1$. The graph is an ellipse with $a^2 = 9$ and $b^2 = 25$, so $a = 3$ and $b = 5$. The x-intercepts are $(3, 0)$ and $(-3, 0)$. The y-intercepts are $(0, 5)$ and $(0, -5)$. Plot the intercepts, and draw the ellipse through them.

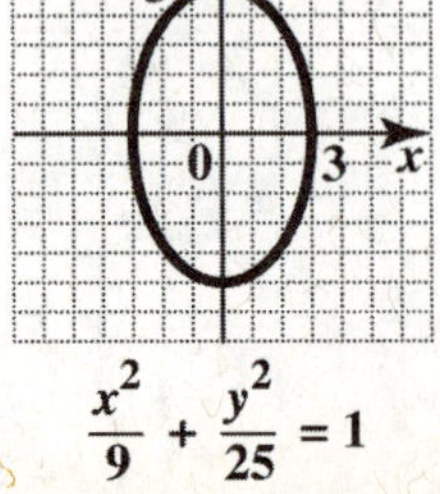

$\dfrac{x^2}{9} + \dfrac{y^2}{25} = 1$

31. $\dfrac{x^2}{36} + \dfrac{y^2}{16} = 1$ is in the form $\dfrac{x^2}{a^2} + \dfrac{y^2}{b^2} = 1$. The graph is an ellipse with $a^2 = 36$ and $b^2 = 16$, so $a = 6$ and $b = 4$. The x-intercepts are $(6, 0)$ and $(-6, 0)$. The y-intercepts are $(0, 4)$ and $(0, -4)$. Plot the intercepts, and draw the ellipse through them.

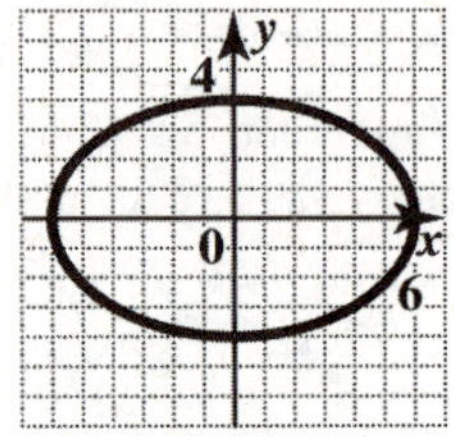

$\dfrac{x^2}{36} + \dfrac{y^2}{16} = 1$

33. $\dfrac{x^2}{16} + \dfrac{y^2}{4} = 1$ is in the form $\dfrac{x^2}{a^2} + \dfrac{y^2}{b^2} = 1$. The graph is an ellipse with $a^2 = 16$ and $b^2 = 4$, so $a = 4$ and $b = 2$. The x-intercepts are $(4, 0)$ and $(-4, 0)$. The y-intercepts are $(0, 2)$ and $(0, -2)$. Plot the intercepts, and draw the ellipse through them.

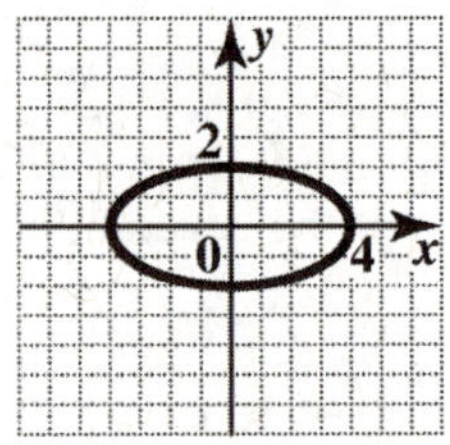

$\dfrac{x^2}{16} + \dfrac{y^2}{4} = 1$

35. $\dfrac{y^2}{25} = 1 - \dfrac{x^2}{49} \Leftrightarrow \dfrac{x^2}{49} + \dfrac{y^2}{25} = 1$ is in the form $\dfrac{x^2}{a^2} + \dfrac{y^2}{b^2} = 1$. The graph is an ellipse with $a^2 = 49$ and $b^2 = 25$, so $a = 7$ and $b = 5$. The x-intercepts are $(7, 0)$ and $(-7, 0)$. The y-intercepts are $(0, 5)$ and $(0, -5)$. Plot the intercepts, and draw the ellipse through them.

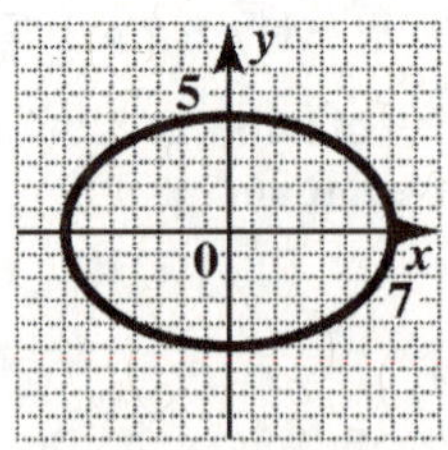

$\dfrac{y^2}{25} = 1 - \dfrac{x^2}{49}$

37. $\dfrac{(x + 1)^2}{64} + \dfrac{(y - 2)^2}{49} = 1$ is in the form $\dfrac{(x - h)^2}{a^2} + \dfrac{(y - k)^2}{b^2} = 1$, so the center of the ellipse is at $(-1, 2)$. Since $a^2 = 64$, $a = 8$. Since

$b^2 = 49, b = 7$. The graph includes the points $(-1 \pm 8, 2)$ and $(-1, 2 \pm 7)$, or equivalently, $(7, 2), (-9, 2), (-1, 9)$, and $(-1, -5)$. Plot the points, and draw the ellipse through them.

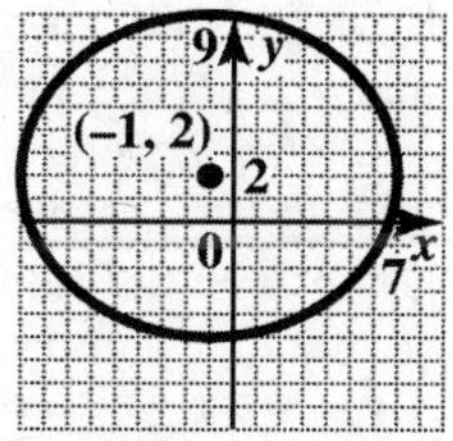

$$\frac{(x+1)^2}{64} + \frac{(y-2)^2}{49} = 1$$

39. $$\frac{(x-2)^2}{16} + \frac{(y-1)^2}{9} = 1$$

The center of the ellipse is at $(2, 1)$. Since $a^2 = 16, a = 4$. Since $b^2 = 9, b = 3$. Some points on the ellipse are $(2 \pm 4, 1)$ and $(2, 1 \pm 3)$, or equivalently, $(6, 1), (-2, 1), (2, 4)$, and $(2, -2)$. Plot the points, and draw the ellipse through them.

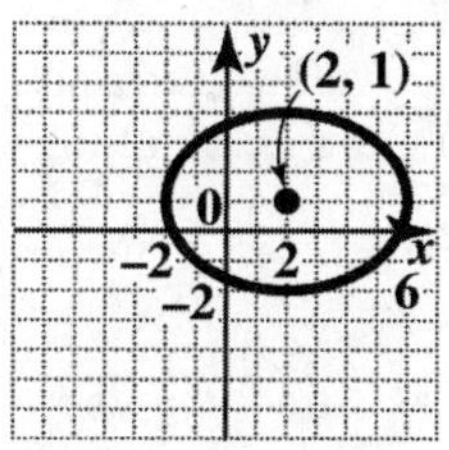

$$\frac{(x-2)^2}{16} + \frac{(y-1)^2}{9} = 1$$

41. By the vertical line test the set is not a function, because a vertical line may intersect the graph of an ellipse in two points.

43. $$(x+2)^2 + (y-4)^2 = 16$$
$$(y-4)^2 = 16 - (x+2)^2$$

Take the square root of each side.

$$y - 4 = \pm\sqrt{16 - (x+2)^2}$$
$$y = 4 \pm \sqrt{16 - (x+2)^2}$$

Therefore, the two functions used to obtain the graph were

$$y_1 = 4 + \sqrt{16 - (x+2)^2} \quad \text{and}$$
$$y_2 = 4 - \sqrt{16 - (x+2)^2}.$$

45. $$x^2 + y^2 = 36$$
$$y^2 = 36 - x^2$$

Take the square root of both sides.

$$y = \pm\sqrt{36 - x^2}$$

Therefore,

$$y_1 = \sqrt{36 - x^2} \quad \text{and} \quad y_2 = -\sqrt{36 - x^2}.$$

Use these two functions to obtain the graph.

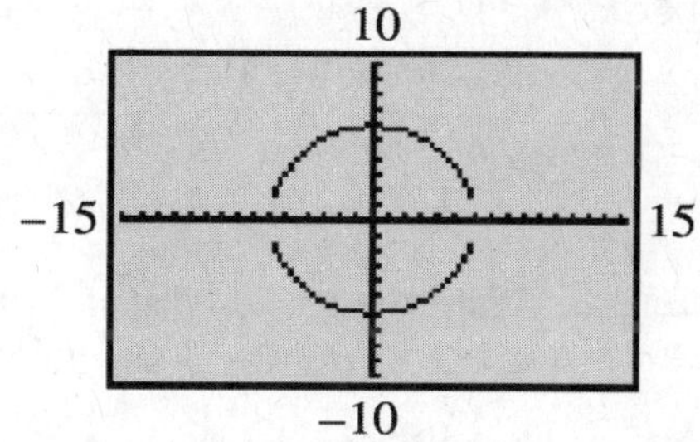

47. $$\frac{x^2}{16} + \frac{y^2}{4} = 1$$
$$\frac{y^2}{4} = 1 - \frac{x^2}{16}$$
$$y^2 = 4\left(1 - \frac{x^2}{16}\right)$$

Take the square root of both sides.

$$y = \pm 2\sqrt{1 - \frac{x^2}{16}}$$

Therefore,

$$y_1 = 2\sqrt{1 - \frac{x^2}{16}} \quad \text{and} \quad y_2 = -2\sqrt{1 - \frac{x^2}{16}}.$$

Use these two functions to obtain the graph.

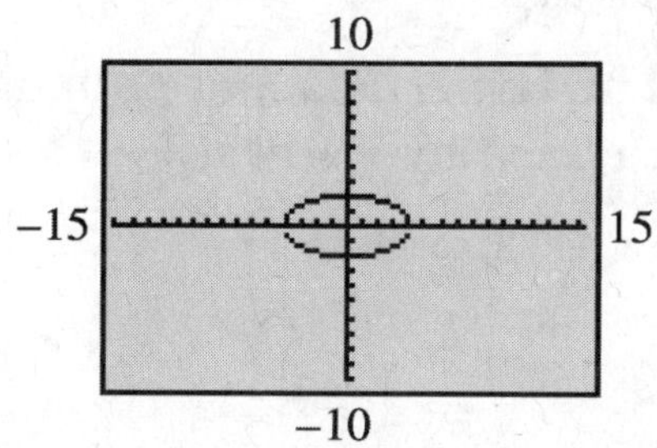

49. $\frac{x^2}{36} + \frac{y^2}{9} = 1$ is in the form $\frac{x^2}{a^2} + \frac{y^2}{b^2} = 1$.

$c^2 = a^2 - b^2 = 36 - 9 = 27$, so

$c = \sqrt{27} = 3\sqrt{3}$.

The kidney stone and the source of the beam must be placed $3\sqrt{3}$ units from the center of the ellipse.

51. **(a)** $100x^2 + 324y^2 = 32{,}400$

$$\frac{x^2}{324} + \frac{y^2}{100} = 1 \qquad \textit{Divide by 32,400.}$$
$$\frac{x^2}{18^2} + \frac{y^2}{10^2} = 1$$

The height in the center is the y-coordinate of the positive y-intercept. The height is 10 meters.

(b) The width of the ellipse is the distance between the x-intercepts, $(-18, 0)$ and $(18, 0)$. The width across the bottom of the arch is $18 + 18 = 36$ meters.

53. $\dfrac{x^2}{141.7^2} + \dfrac{y^2}{141.1^2} = 1$

(a) $c^2 = a^2 - b^2$, so

$$c = \sqrt{a^2 - b^2} = \sqrt{141.7^2 - 141.1^2}$$
$$= \sqrt{169.68} \approx 13.0$$

From the figure, the greatest distance (the **apogee**) is $a + c = 141.7 + 13.0 = 154.7$ million miles.

(b) The least distance (the **perigee**) is $a - c = 141.7 - 13.0 = 128.7$ million miles.

55. Plot the points $(3, 4)$, $(-3, 4)$, $(3, -4)$, and $(-3, -4)$.

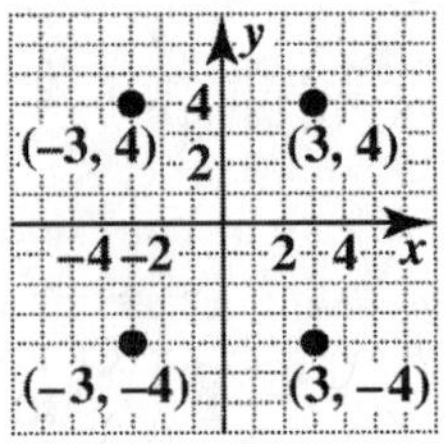

57. $4x + 3y = 12$

If $y = 0$, then $4x = 12$, so $x = 3$ and $(3, 0)$ is the x-intercept.
If $x = 0$, then $3y = 12$, so $y = 4$ and $(0, 4)$ is the y-intercept.

13.3 The Hyperbola and Functions Defined by Radicals

13.3 Now Try Exercises

N1. $\dfrac{x^2}{25} - \dfrac{y^2}{9} = 1$ or $\dfrac{x^2}{5^2} - \dfrac{y^2}{3^2} = 1$

Step 1
The graph is a hyperbola with $a = 5$ and $b = 3$. It has left and right branches with x-intercepts $(5, 0)$ and $(-5, 0)$.

Step 2
The four points $(5, 3)$, $(-5, 3)$, $(-5, -3)$, and $(5, -3)$ are the vertices of the fundamental rectangle.

Steps 3 and 4
The equations of the asymptotes are $y = \pm\frac{3}{5}x$, and the hyperbola approaches these lines as x and y get larger and larger in absolute value.

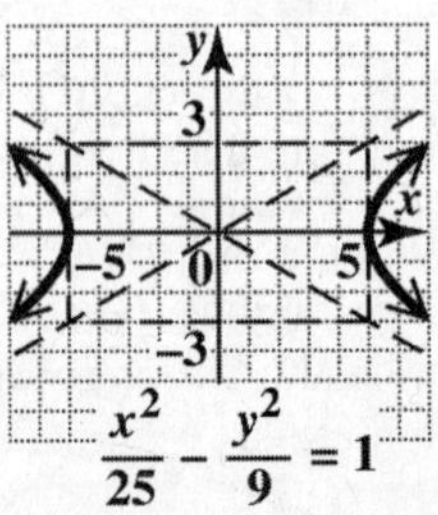

N2. $\dfrac{y^2}{9} - \dfrac{x^2}{16} = 1$ or $\dfrac{y^2}{3^2} - \dfrac{x^2}{4^2} = 1$

The graph is a hyperbola with $b = 3$ and $a = 4$. It has upper and lower branches with y-intercepts $(0, 3)$ and $(0, -3)$. The four points $(4, 3)$, $(-4, 3)$, $(-4, -3)$, and $(4, -3)$ are the vertices of the fundamental rectangle. Draw the asymptotes, $y = \pm\frac{3}{4}x$, through the corners of the fundamental rectangle. Then draw the branches of the hyperbola so that they approach the asymptotes.

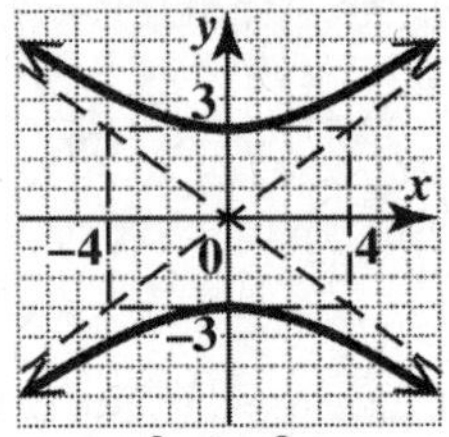

$\dfrac{y^2}{9} - \dfrac{x^2}{16} = 1$

N3. **(a)** $y^2 - 10 = -x^2$
$$x^2 + y^2 = 10$$

Since the x^2- and y^2-terms have equal positive coefficients, the graph of the equation is a **circle**.

(b) $y - 2x^2 = 8$
$$y = 2x^2 + 8$$

This is an equation of a vertical **parabola** since only one variable, x, is squared.

(c) $3x^2 + y^2 = 4$
$$\frac{3x^2}{4} + \frac{y^2}{4} = 1 \quad \textit{Divide by 4.}$$

Since the x^2- and y^2-terms have different positive coefficients, the graph of the equation is an **ellipse**.

N4.
$$f(x) = \sqrt{64 - x^2}$$
$$y = \sqrt{64 - x^2} \quad f(x) = y$$
$$y^2 = 64 - x^2 \quad \textit{Square both sides.}$$
$$x^2 + y^2 = 64 \quad \textit{Add } x^2.$$

The last equation is the graph of a circle with center $(0, 0)$ and radius 8. Since $f(x)$ represents a principal square root in the original equation, $f(x)$ is nonnegative and its graph is the upper half of the circle.

The domain of the function is $[-8, 8]$ and its range is $[0, 8]$.

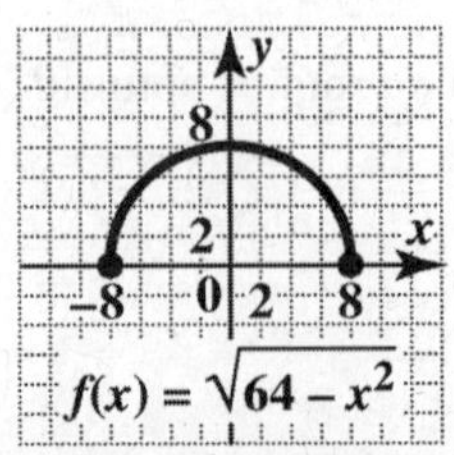

N5.

$$\frac{y}{4} = -\sqrt{1-\frac{x^2}{9}}$$

$$\frac{y^2}{16} = 1 - \frac{x^2}{9} \qquad \textit{Square both sides.}$$

$$\frac{x^2}{9} + \frac{y^2}{16} = 1 \qquad \textit{Add } \frac{x^2}{9}.$$

The last equation is the graph of an ellipse with intercepts $(3,0)$, $(-3,0)$, $(0,4)$, and $(0,-4)$. Since $\frac{y}{4}$ equals a negative square root in the original equation, y must be nonpositive, restricting the graph to the lower half of the ellipse.

The domain of the function is $[-3,3]$ and its range is $[-4,0]$.

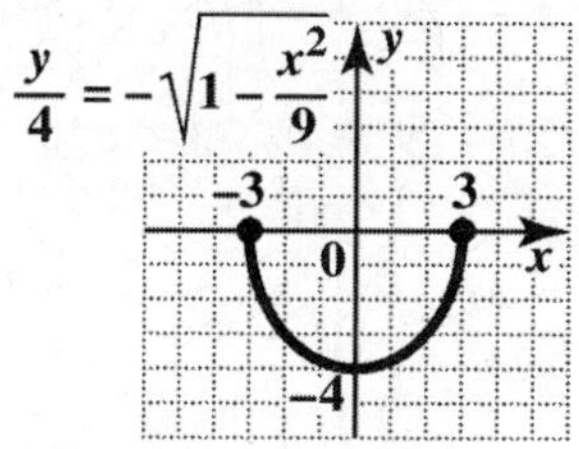

13.3 Section Exercises

1. $\dfrac{x^2}{25} + \dfrac{y^2}{9} = 1$

This is the standard form for the equation of an ellipse with x-intercepts $(5,0)$ and $(-5,0)$ and y-intercepts $(0,3)$ and $(0,-3)$. This is graph **C**.

3. $\dfrac{x^2}{9} - \dfrac{y^2}{25} = 1$

This is the standard form for the equation of a hyperbola that opens left and right. Its x-intercepts are $(3,0)$ and $(-3,0)$. This is graph **D**.

5. $\dfrac{x^2}{16} - \dfrac{y^2}{9} = 1$ is in the form $\dfrac{x^2}{a^2} - \dfrac{y^2}{b^2} = 1$. The graph is a hyperbola with $a = 4$ and $b = 3$. The x-intercepts are $(4,0)$ and $(-4,0)$. There are no y-intercepts. The vertices of the fundamental rectangle are $(4,3)$, $(4,-3)$, $(-4,-3)$, and $(-4,3)$. Extend the diagonals of the rectangle through these points to get the asymptotes. Graph a branch of the hyperbola through each intercept and approaching the asymptotes.

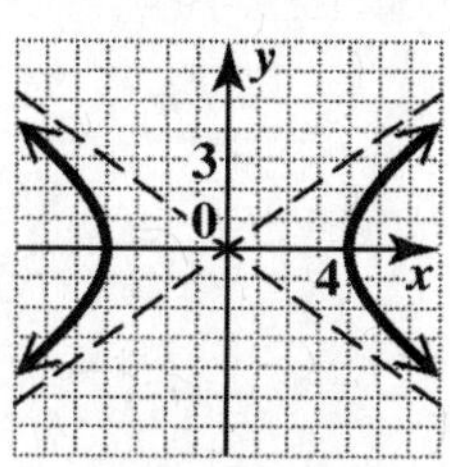

$$\frac{x^2}{16} - \frac{y^2}{9} = 1$$

7. $\dfrac{y^2}{4} - \dfrac{x^2}{25} = 1$ is in the form $\dfrac{y^2}{b^2} - \dfrac{x^2}{a^2} = 1$. The graph is a hyperbola with $a = 5$ and $b = 2$. The y-intercepts are $(0,2)$ and $(0,-2)$. There are no x-intercepts. The vertices of the fundamental rectangle are $(5,2)$, $(5,-2)$, $(-5,-2)$, and $(-5,2)$. Extend the diagonals of the rectangle through these points to get the asymptotes. Graph a branch of the hyperbola through each intercept and approaching the asymptotes.

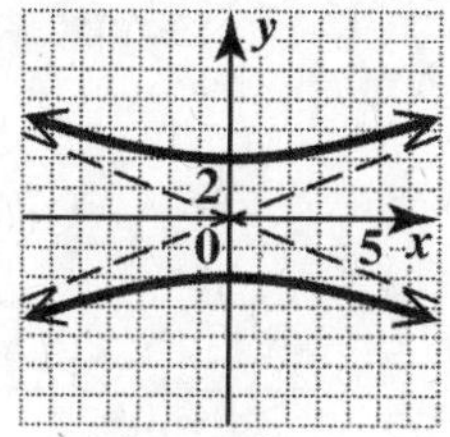

$$\frac{y^2}{4} - \frac{x^2}{25} = 1$$

9. $\dfrac{x^2}{25} - \dfrac{y^2}{36} = 1$ is in the form $\dfrac{x^2}{a^2} - \dfrac{y^2}{b^2} = 1$. The graph is a hyperbola with $a = 5$ and $b = 6$. The x-intercepts are $(5,0)$ and $(-5,0)$. There are no y-intercepts. The vertices of the fundamental rectangle are $(5,6)$, $(5,-6)$, $(-5,-6)$ and $(-5,6)$. Extend the diagonals of the rectangle through these points to get the asymptotes. Graph a branch of the hyperbola through each intercept and approaching the asymptotes.

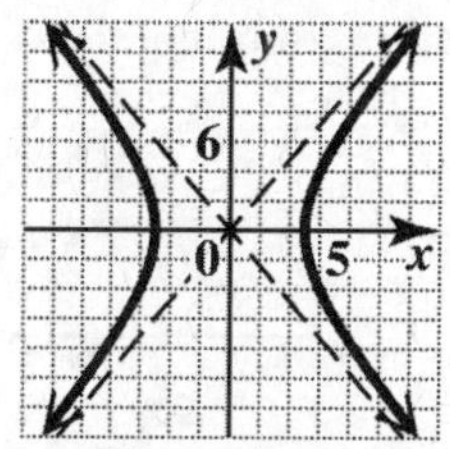

$$\frac{x^2}{25} - \frac{y^2}{36} = 1$$

11. $\dfrac{y^2}{16} - \dfrac{x^2}{16} = 1$ is in the form $\dfrac{y^2}{b^2} - \dfrac{x^2}{a^2} = 1$. The graph is a hyperbola with $a = 4$ and $b = 4$. The y-intercepts are $(0,4)$ and $(0,-4)$. There are no x-intercepts. The vertices of the fundamental rectangle are $(4,4)$, $(4,-4)$, $(-4,-4)$, and $(-4,4)$. Extend the diagonals of the rectangle through these points to get the asymptotes. Graph a branch of the hyperbola through each intercept and approaching the asymptotes.

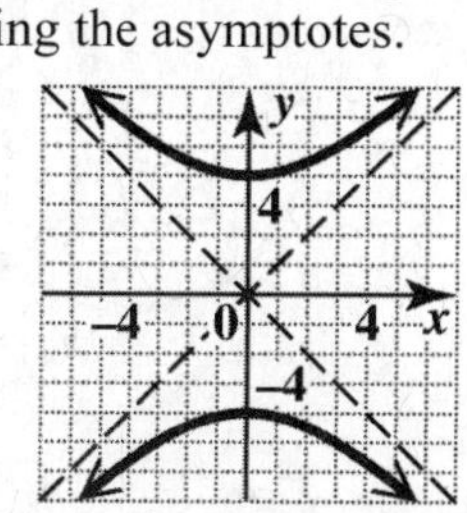

$$\frac{y^2}{16} - \frac{x^2}{16} = 1$$

13. $x^2 - y^2 = 16$

$\frac{x^2}{16} - \frac{y^2}{16} = 1$ *Divide by 16.*

This equation is in the form $\frac{x^2}{a^2} - \frac{y^2}{b^2} = 1$ with $a = 4$ and $b = 4$. The graph is a **hyperbola** with x-intercepts $(4, 0)$ and $(-4, 0)$ and no y-intercepts. One asymptote passes through $(4, 4)$ and $(-4, -4)$. The other asymptote passes through $(-4, 4)$ and $(4, -4)$. Sketch the graph through the intercepts and approaching the asymptotes.

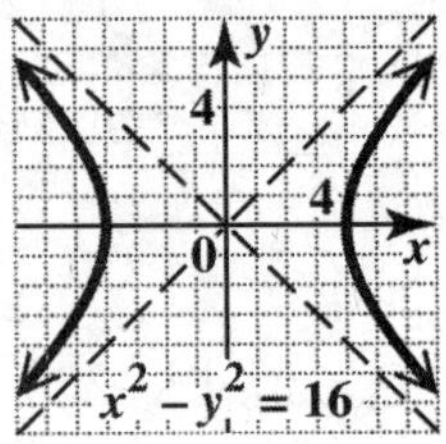

15. $4x^2 + y^2 = 16$

$\frac{x^2}{4} + \frac{y^2}{16} = 1$ *Divide by 16.*

This equation is in the form $\frac{x^2}{a^2} + \frac{y^2}{b^2} = 1$ with $a = 2$ and $b = 4$. The graph is an **ellipse**. The x-intercepts $(2, 0)$ and $(-2, 0)$. The y-intercepts are $(0, 4)$ and $(0, -4)$. Plot the intercepts and draw the ellipse through them.

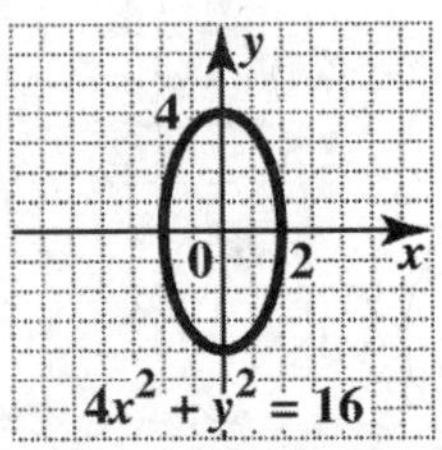

17. $y^2 = 36 - x^2$

$x^2 + y^2 = 36$

$(x - 0)^2 + (y - 0)^2 = 36$

The graph is a **circle** with center at $(0, 0)$ and radius $\sqrt{36} = 6$.

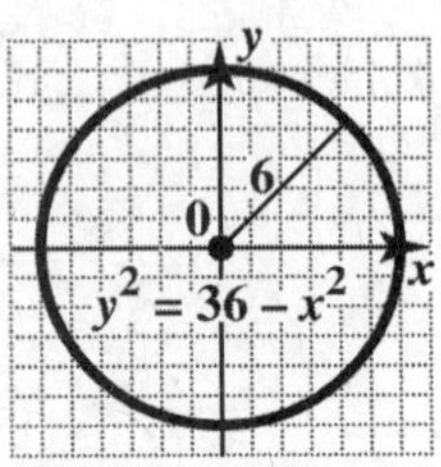

19. $x^2 - 2y = 0$

$x^2 = 2y$

$\frac{1}{2}x^2 = y$

The equation is in the form

$$f(x) = a(x - h)^2 + k,$$

with $a = \frac{1}{2}$, $h = 0$, and $k = 0$. The graph is a **parabola** that opens up and is wider than the graph of $f(x) = x^2$ because $|a| = \frac{1}{2} < 1$. The vertex is at $(0, 0)$.

x	1	2	3
y	$\frac{1}{2}$	2	4.5

Use symmetry about $x = 0$ to draw the graph.

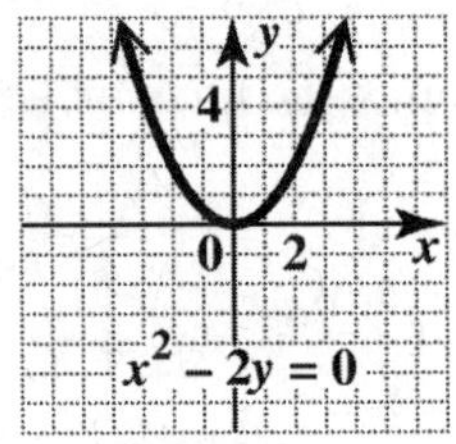

21. $y^2 = 4 + x^2$

$y^2 - x^2 = 4$

$\frac{y^2}{4} - \frac{x^2}{4} = 1$ *Divide by 4.*

The graph is a **hyperbola** with y-intercepts $(0, 2)$ and $(0, -2)$. One asymptote passes through $(2, 2)$ and $(-2, -2)$. The other asymptote passes through $(-2, 2)$ and $(2, -2)$.

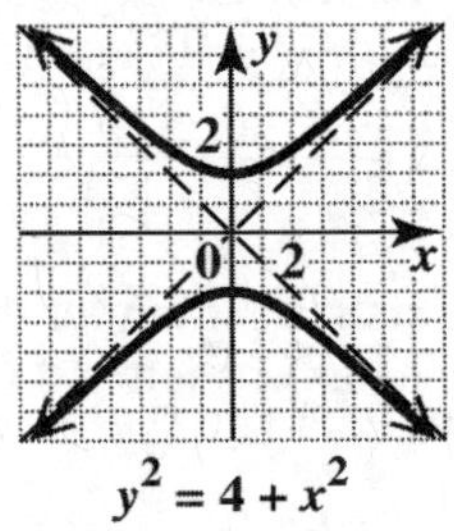

23. $f(x) = \sqrt{16 - x^2}$

Replace $f(x)$ with y and square both sides to get the equation

$$y^2 = 16 - x^2 \quad \text{or} \quad x^2 + y^2 = 16.$$

This is the graph of a circle with center $(0, 0)$ and radius 4. Since $f(x)$, or y, represents a principal square root in the original equation, $f(x)$ must be nonnegative. This restricts the graph to the upper half of the circle.

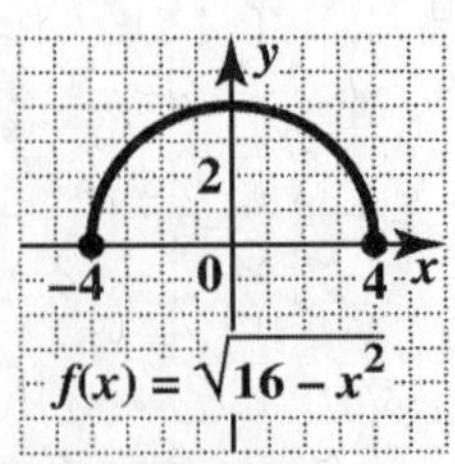

The domain is $[-4, 4]$, and the range is $[0, 4]$.

25. $f(x) = -\sqrt{36 - x^2}$

$y = -\sqrt{36 - x^2}$ *Replace f(x) with y.*

$y^2 = 36 - x^2$ *Square each side.*

$x^2 + y^2 = 36$

This is a circle centered at the origin with radius $\sqrt{36} = 6$. Since $f(x)$, or y, represents a nonpositive square root in the original equation, $f(x)$ must be nonpositive. This restricts the graph to the bottom half of the circle.

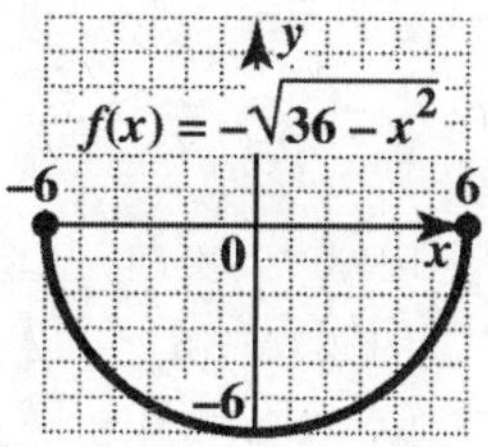

The domain is $[-6, 6]$, and the range is $[-6, 0]$.

27. $y = -2\sqrt{1 - \dfrac{x^2}{9}}$

$y^2 = 4\left(1 - \dfrac{x^2}{9}\right)$ *Square each side.*

$\dfrac{y^2}{4} = 1 - \dfrac{x^2}{9}$ *Divide by 4.*

$\dfrac{x^2}{9} + \dfrac{y^2}{4} = 1$

This is the equation of an ellipse with intercepts $(3, 0)$, $(-3, 0)$, $(0, 2)$, and $(0, -2)$. Since y equals a negative square root in the original equation, y must be nonpositive, restricting the graph to the lower half of the ellipse.

The domain of the function is $[-3, 3]$ and its range is $[-2, 0]$.

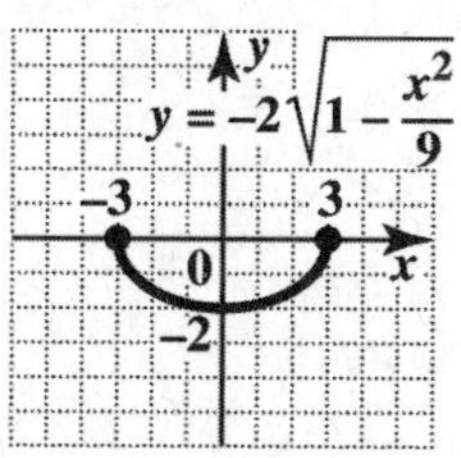

29. $\dfrac{y}{3} = \sqrt{1 + \dfrac{x^2}{9}}$

$\dfrac{y^2}{9} = 1 + \dfrac{x^2}{9}$ *Square each side.*

$\dfrac{y^2}{9} - \dfrac{x^2}{9} = 1$

This is a hyperbola opening up and down with y-intercepts $(0, 3)$ and $(0, -3)$. The four points $(3, 3)$, $(3, -3)$, $(-3, 3)$, and $(-3, -3)$ are the vertices of the rectangle that determine the asymptotes. Since $f(x)$, or y, represents a square root in the original equation, $f(x)$ must be nonnegative. This restricts the graph to the upper half of the hyperbola.

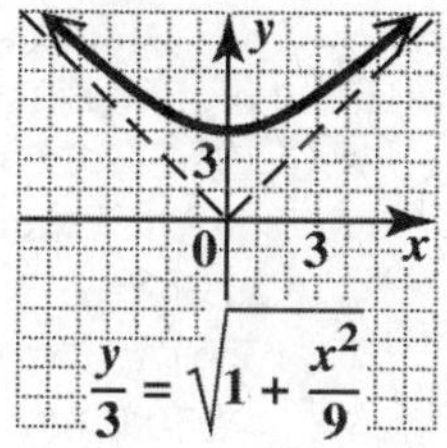

The domain is $(-\infty, \infty)$, and the range is $[3, \infty)$.

31. $\dfrac{(x-2)^2}{4} - \dfrac{(y+1)^2}{9} = 1$ is a hyperbola centered at $(2, -1)$, with $a = 2$ and $b = 3$. The hyperbola goes through the points $(2 \pm 2, -1)$, or $(4, -1)$ and $(0, -1)$. The asymptotes are the extended diagonals of the rectangle with vertices $(2, 3)$, $(2, -3)$, $(-2, -3)$ and $(-2, 3)$ shifted 2 units right and 1 unit down, or $(4, 2)$, $(4, -4)$, $(0, -4)$ and $(0, 2)$. Draw the hyperbola.

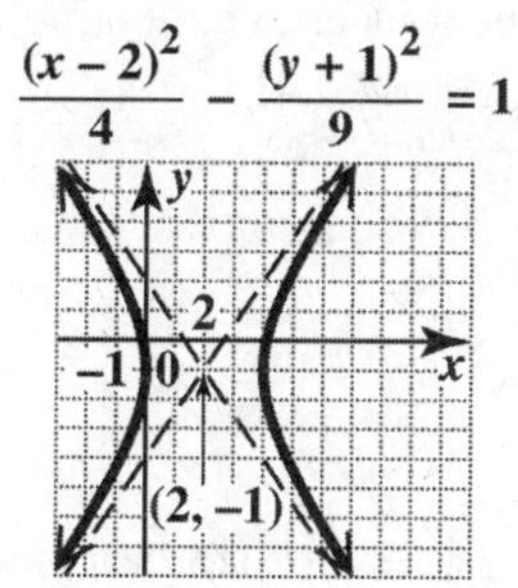

33. $\dfrac{y^2}{36} - \dfrac{(x-2)^2}{49} = 1$ is a hyperbola centered at $(2, 0)$ with $a = 7$, and $b = 6$. The lowest point on the upper branch is $(2, 6)$ and the highest point on the lower branch is $(2, -6)$. The asymptotes are the extended diagonals of the rectangle with vertices $(7, 6)$, $(7, -6)$, $(-7, -6)$, and $(-7, 6)$ shifted right 2 units, or $(9, 6)$, $(9, -6)$, $(-5, -6)$, and $(-5, 6)$. Draw the hyperbola.

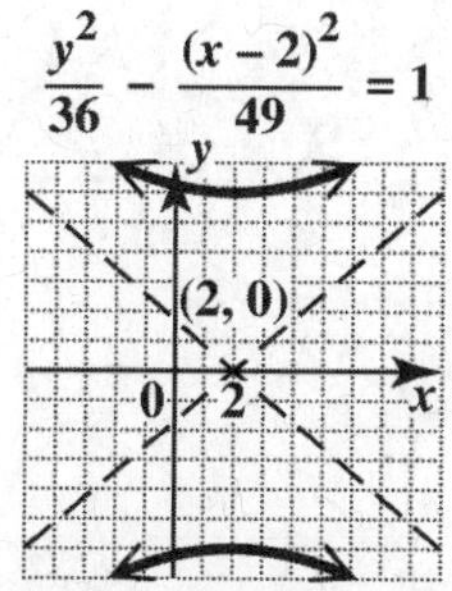

35. **(a)** $400x^2 - 625y^2 = 250{,}000$

$$\frac{x^2}{625} - \frac{y^2}{400} = 1 \quad \textit{Divide by 250,000.}$$

$$\frac{x^2}{25^2} - \frac{y^2}{20^2} = 1$$

The x-intercepts are $(25, 0)$ and $(-25, 0)$. The distance between the buildings is the distance between the x-intercepts. The buildings are $25 + 25 = 50$ meters apart at their closest point.

(b) At $x = 50$, $y = \frac{d}{2}$, so $d = 2y$.

$$400(50)^2 - 625y^2 = 250{,}000$$
$$1{,}000{,}000 - 625y^2 = 250{,}000$$
$$-625y^2 = -750{,}000$$
$$y^2 = 1200$$
$$y = \sqrt{1200}$$

The distance d is $2\sqrt{1200} \approx 69.3$ meters.

37. $$\frac{x^2}{9} - y^2 = 1$$

$$\frac{x^2}{9} - 1 = y^2 \quad \textit{Isolate } y^2.$$

$$\pm\sqrt{\frac{x^2}{9} - 1} = y \quad \textit{Take square roots.}$$

The two functions used to obtain the graph were

$$y_1 = \sqrt{\frac{x^2}{9} - 1} \text{ and } y_2 = -\sqrt{\frac{x^2}{9} - 1}.$$

39. $$\frac{x^2}{25} - \frac{y^2}{49} = 1$$

$$\frac{x^2}{25} - 1 = \frac{y^2}{49}$$

$$49\left(\frac{x^2}{25} - 1\right) = y^2 \quad \textit{Multiply by 49.}$$

$$\pm 7\sqrt{\frac{x^2}{25} - 1} = y \quad \textit{Take square roots.}$$

To obtain the graph, use the two functions

$$y_1 = 7\sqrt{\frac{x^2}{25} - 1} \text{ and } y_2 = -7\sqrt{\frac{x^2}{25} - 1}.$$

10

−15 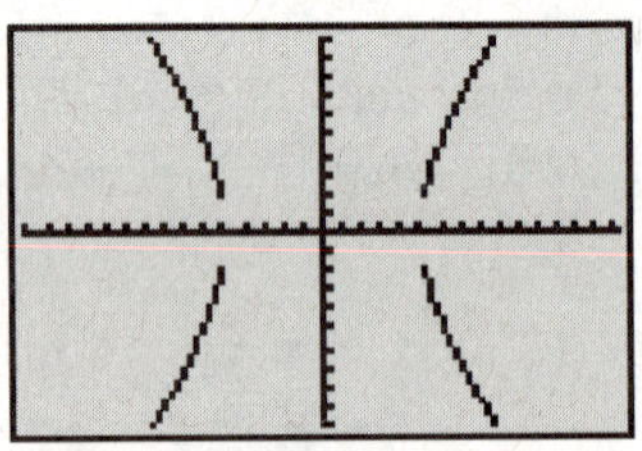 15

−10

41. $y^2 - 9x^2 = 9$

$$y^2 = 9 + 9x^2$$
$$y^2 = 9(1 + x^2) \quad \textit{Factor out 9.}$$
$$y = \pm 3\sqrt{1 + x^2} \quad \textit{Take square roots.}$$

To obtain the graphs, use the two functions

$$y_1 = 3\sqrt{1 + x^2} \text{ and } y_2 = -3\sqrt{1 + x^2}.$$

10

−15 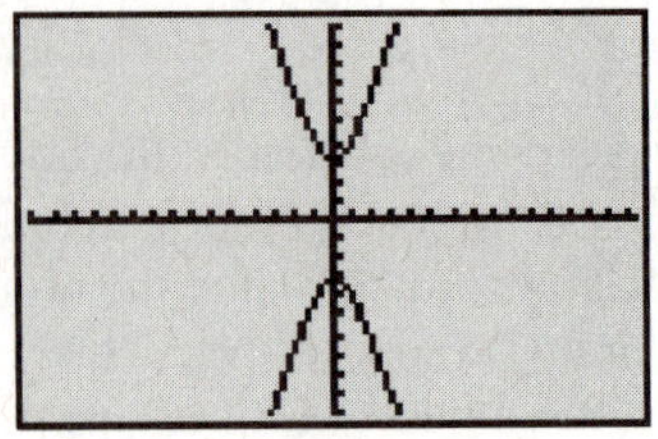 15

−10

43. $2x + y = 13 \quad (1)$

$y = 3x + 3 \quad (2)$

Substitute $3x + 3$ for y in (1).

$$2x + (3x + 3) = 13$$
$$5x + 3 = 13$$
$$5x = 10$$
$$x = 2$$

Substitute 2 for x in (2).

$$y = 3(2) + 3 = 9$$

The solution set is $\{(2, 9)\}$.

45. $4x - 3y = -10 \quad (1)$

$4x + 6y = 8 \quad (2)$

To eliminate x, multiply equation (1) by -1 and add the result to equation (2).

$-4x + 3y = 10$		$-1 \times (1)$
$4x + 6y = 8$	(2)	
$9y = 18$	*Add.*	
$y = 2$	*Solve for y.*	

To find x, substitute 2 for y in equation (1) or (2).

$$4x + 6y = 8 \quad (2)$$
$$4x + 6(2) = 8$$
$$4x + 12 = 8$$
$$4x = -4$$
$$x = -1$$

The solution set is $\{(-1, 2)\}$.

47. $2x^4 - 5x^2 - 3 = 0$

Let $u = x^2$, so that $u^2 = x^4$.

$$2u^2 - 5u - 3 = 0$$
$$(2u + 1)(u - 3) = 0$$

$2u + 1 = 0$ or $u - 3 = 0$

$u = -\frac{1}{2}$ or $u = 3$

$x^2 = -\frac{1}{2}$ or $x^2 = 3$

$x = \pm\sqrt{-\frac{2}{4}}$ or $x = \pm\sqrt{3}$

$x = \pm\frac{\sqrt{2}}{2}i$

The solution set is $\left\{-\sqrt{3}, \sqrt{3}, -\frac{\sqrt{2}}{2}i, \frac{\sqrt{2}}{2}i\right\}$.

13.4 Nonlinear Systems of Equations

13.4 Now Try Exercises

N1. $4x^2 + y^2 = 36 \quad (1)$
$x - y = 3 \quad (2)$

Solve equation (2) for y.

$$y = x - 3 \quad (3)$$

Substitute $x - 3$ for y in equation (1).

$$\begin{aligned} 4x^2 + y^2 &= 36 \quad (1) \\ 4x^2 + (x-3)^2 &= 36 \\ 4x^2 + (x^2 - 6x + 9) &= 36 \\ 5x^2 - 6x - 27 &= 0 \\ (x-3)(5x+9) &= 0 \end{aligned}$$

$$x - 3 = 0 \quad \text{or} \quad 5x + 9 = 0$$
$$x = 3 \quad \text{or} \quad x = -\tfrac{9}{5}$$

From (3) with $x = 3, y = 3 - 3 = 0$.
From (3) with $x = -\frac{9}{5}, y = -\frac{9}{5} - 3 = -\frac{24}{5}$.

Check $(3, 0)$: $36 + 0 = 36$; $3 - 0 = 3$
Check $\left(-\frac{9}{5}, -\frac{24}{5}\right)$: $\frac{324}{25} + \frac{576}{25} = 36$; $-\frac{9}{5} + \frac{24}{5} = 3$

The solution set is $\left\{(3, 0), \left(-\frac{9}{5}, -\frac{24}{5}\right)\right\}$.

N2. $xy = 2 \quad (1)$
$x - 3y = 1 \quad (2)$

Solve equation (1) for y.

$$y = \frac{2}{x} \quad (3)$$

Substitute $\frac{2}{x}$ for y in equation (2).

$$\begin{aligned} x - 3y &= 1 \quad (2) \\ x - 3\left(\frac{2}{x}\right) &= 1 \\ x - \frac{6}{x} &= 1 \\ x^2 - 6 &= x \quad \textit{Multiply by x.} \\ x^2 - x - 6 &= 0 \\ (x-3)(x+2) &= 0 \end{aligned}$$

$$x - 3 = 0 \quad \text{or} \quad x + 2 = 0$$
$$x = 3 \quad \text{or} \quad x = -2$$

From (3) with $x = 3$, $y = \frac{2}{3}$.

From (3) with $x = -2$, $y = \frac{2}{-2} = -1$.

Check $\left(3, \frac{2}{3}\right)$: $3\left(\frac{2}{3}\right) = 2$; $3 - 2 = 1$
Check $(-2, -1)$: $(-2)(-1) = 2$; $-2 + 3 = 1$

The solution set is $\left\{(-2, -1), \left(3, \frac{2}{3}\right)\right\}$.

N3. $x^2 + y^2 = 16 \quad (1)$
$4x^2 + 13y^2 = 100 \quad (2)$

Multiply equation (1) by -4 and add the result to (2).

$$\begin{aligned} -4x^2 - 4y^2 &= -64 \quad -4 \times (1) \\ 4x^2 + 13y^2 &= 100 \\ \hline 9y^2 &= 36 \quad \textit{Add.} \\ y^2 &= 4 \quad \textit{Divide by 9.} \\ y = 2 \quad \text{or} \quad y &= -2 \quad \textit{Square root property} \end{aligned}$$

Substitute 4 for y^2 in equation (1).

$$\begin{aligned} x^2 + 4 &= 16 \\ x^2 &= 12 \quad \textit{Add 5.} \\ x = 2\sqrt{3} \quad \text{or} \quad x &= -2\sqrt{3} \quad \textit{Square rt. prop.} \end{aligned}$$

Check: If $x = \pm 2\sqrt{3}$, $x^2 = 12$, and if $y = \pm 2$, $y^2 = 4$. Thus, in any case, we get $12 + 4 = 16$ in (1) and $48 + 52 = 100$ in (2). All *four* ordered pairs check.

The solution set is $\left\{\left(2\sqrt{3}, 2\right), \left(2\sqrt{3}, -2\right), \left(-2\sqrt{3}, 2\right), \left(-2\sqrt{3}, -2\right)\right\}$.

N4. $x^2 + 3xy - y^2 = 23 \quad (1)$
$x^2 - y^2 = 5 \quad (2)$

Multiply equation (2) by -1 and add the result to equation (1).

$$\begin{aligned} x^2 + 3xy - y^2 &= 23 \quad (1) \\ -x^2 + y^2 &= -5 \quad -1 \times (2) \\ \hline 3xy &= 18 \\ xy &= 6 \end{aligned}$$

Solving $xy = 6$ for y gives us $y = \frac{6}{x}$ (3).

Substitute $\frac{6}{x}$ for y in equation (2).

$$\begin{aligned} x^2 - y^2 &= 5 \quad (2) \\ x^2 - \frac{36}{x^2} &= 5 \\ x^4 - 36 &= 5x^2 \\ x^4 - 5x^2 - 36 &= 0 \\ (x^2 - 9)(x^2 + 4) &= 0 \end{aligned}$$

$$x^2 - 9 = 0 \quad \text{or} \quad x^2 + 4 = 0$$
$$x^2 = 9 \quad \text{or} \quad x^2 = -4$$
$$x = \pm 3 \quad \text{or} \quad x = \pm 2i$$

If $x = \pm 3$, then from (3), $y = \pm \frac{6}{3} = \pm 2$.
If $x = \pm 2i$, then from (3),
$y = \pm \frac{6}{2i} \cdot \frac{i}{i} = \pm \frac{3i}{-1} = \mp 3i$.

The solution set is
$\{(3, 2), (-3, -2), (2i, -3i), (-2i, 3i)\}$.

13.4 Section Exercises

1. The line intersects the ellipse in exactly one point, so there is *one* point in the solution set of the system.

3. The line does not intersect the hyperbola, so there are no points in the solution set of the system.

5. A line and a circle; no points ▪ Draw any circle, and then draw a line that does not cross the circle.

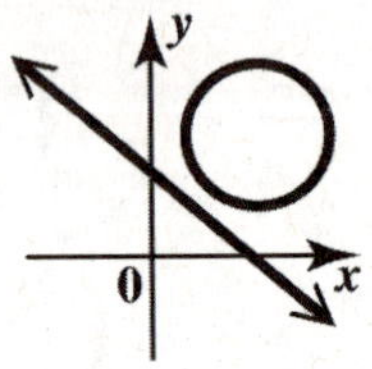

7. A line and a hyperbola; one point ▪ The line is tangent to the hyperbola.

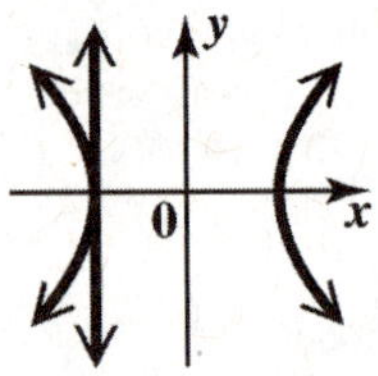

9. A circle and an ellipse; four points ▪ Draw any ellipse, and then draw a circle with the same center whose radius is just large enough so that there are four points of intersection. (If the radius of the circle is too large or too small, there may be fewer points of intersection.)

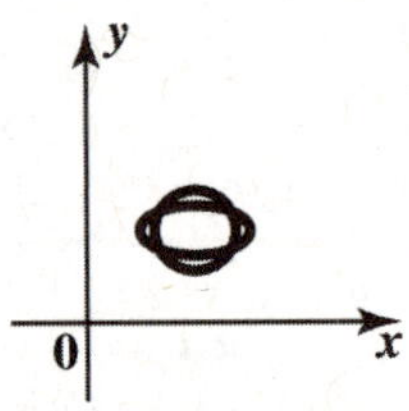

11. A parabola and an ellipse; four points ▪ Draw any parabola, and then draw an ellipse large enough so that there are four points of intersection. (If the ellipse is too large or too small, there may be fewer points of intersection.)

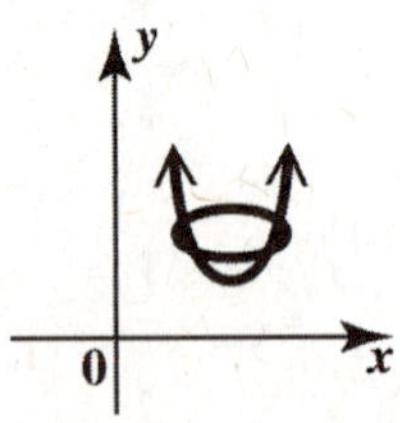

13. $y = 4x^2 - x \quad (1)$
$y = x \quad (2)$

Substitute x for y in equation (1).

$$y = 4x^2 - x \quad (1)$$
$$x = 4x^2 - x$$
$$0 = 4x^2 - 2x$$
$$0 = 2x(2x - 1)$$

$$2x = 0 \quad \text{or} \quad 2x - 1 = 0$$
$$x = 0 \quad \text{or} \quad x = \tfrac{1}{2}$$

Use equation (2) to find y for each x-value.

If $x = 0$, then $y = 0$.
If $x = \frac{1}{2}$, then $y = \frac{1}{2}$.

The solution set is $\left\{(0, 0), \left(\frac{1}{2}, \frac{1}{2}\right)\right\}$.

15. $y = x^2 + 6x + 9 \quad (1)$
$x + y = 3 \quad (2)$

Substitute $x^2 + 6x + 9$ for y in equation (2).

$$x + y = 3 \quad (2)$$
$$x + \left(x^2 + 6x + 9\right) = 3$$
$$x^2 + 7x + 9 = 3$$
$$x^2 + 7x + 6 = 0$$
$$(x + 6)(x + 1) = 0$$

$$x + 6 = 0 \quad \text{or} \quad x + 1 = 0$$
$$x = -6 \quad \text{or} \quad x = -1$$

Substitute these values for x in equation (2) and solve for y.

If $x = -6$, then $x + y = 3 \quad (2)$
$$-6 + y = 3$$
$$y = 9.$$

If $x = -1$, then $x + y = 3 \quad (2)$
$$-1 + y = 3$$
$$y = 4.$$

The solution set is $\{(-6, 9), (-1, 4)\}$.

17. $x^2 + y^2 = 2 \quad (1)$
$2x + y = 1 \quad (2)$

Solve equation (2) for y.

$$y = 1 - 2x \quad (3)$$

Substitute $1 - 2x$ for y in equation (1).

$$x^2 + y^2 = 2 \quad (1)$$
$$x^2 + (1 - 2x)^2 = 2$$
$$x^2 + 1 - 4x + 4x^2 = 2$$
$$5x^2 - 4x - 1 = 0$$
$$(5x + 1)(x - 1) = 0$$

$$5x + 1 = 0 \quad \text{or} \quad x - 1 = 0$$
$$x = -\tfrac{1}{5} \quad \text{or} \quad x = 1$$

Use equation (3) to find y for each x-value.

If $x = -\frac{1}{5}$, then
$$y = 1 - 2\left(-\tfrac{1}{5}\right) = 1 + \tfrac{2}{5} = \tfrac{7}{5}.$$

If $x = 1$, then
$$y = 1 - 2(1) = -1.$$

The solution set is $\left\{\left(-\frac{1}{5}, \frac{7}{5}\right), (1, -1)\right\}$.

19. $xy = 4 \quad (1)$
$3x + 2y = -10 \quad (2)$

Solve equation (1) for y to get $y = \frac{4}{x}$.

Substitute $\frac{4}{x}$ for y in equation (2) to find x.

$$3x + 2y = -10 \quad (2)$$
$$3x + 2\left(\frac{4}{x}\right) = -10$$

Multiply by the LCD, x.

$$3x^2 + 8 = -10x$$
$$3x^2 + 10x + 8 = 0$$
$$(3x+4)(x+2) = 0$$

$$3x + 4 = 0 \quad \text{or} \quad x + 2 = 0$$
$$x = -\tfrac{4}{3} \quad \text{or} \quad x = -2$$

Since $y = \frac{4}{x}$, if $x = -\frac{4}{3}$, then $y = \frac{4}{-\frac{4}{3}} = -3$.

If $x = -2$, then $y = \frac{4}{-2} = -2$.

The solution set is $\left\{(-2,-2), \left(-\frac{4}{3}, -3\right)\right\}$.

21. $xy = -3 \quad (1)$
$x + y = -2 \quad (2)$

Solve equation (2) for y.

$$y = -x - 2 \quad (3)$$

Substitute $-x - 2$ for y in equation (1).

$$xy = -3 \quad (1)$$
$$x(-x-2) = -3$$
$$-x^2 - 2x = -3$$
$$-x^2 - 2x + 3 = 0$$
$$x^2 + 2x - 3 = 0 \quad \textit{Multiply by } -1.$$
$$(x+3)(x-1) = 0$$

$$x + 3 = 0 \quad \text{or} \quad x - 1 = 0$$
$$x = -3 \quad \text{or} \quad x = 1$$

Use equation (3) to find y for each x-value.

If $x = -3$, then

$$y = -(-3) - 2 = 1.$$

If $x = 1$, then

$$y = -(1) - 2 = -3.$$

The solution set is $\{(-3,1), (1,-3)\}$.

23. $y = 3x^2 + 6x \quad (1)$
$y = x^2 - x - 6 \quad (2)$

Substitute $x^2 - x - 6$ for y in equation (1) to find x.

$$y = 3x^2 + 6x \quad (1)$$
$$x^2 - x - 6 = 3x^2 + 6x$$
$$0 = 2x^2 + 7x + 6$$
$$0 = (2x+3)(x+2)$$

$$2x + 3 = 0 \quad \text{or} \quad x + 2 = 0$$
$$x = -\tfrac{3}{2} \quad \text{or} \quad x = -2$$

Substitute $-\frac{3}{2}$ for x in equation (1) to find y.

$$y = 3x^2 + 6x \quad (1)$$
$$y = 3\left(-\tfrac{3}{2}\right)^2 + 6\left(-\tfrac{3}{2}\right)$$
$$= 3\left(\tfrac{9}{4}\right) + 6\left(-\tfrac{6}{4}\right)$$
$$= \tfrac{27}{4} - \tfrac{36}{4} = -\tfrac{9}{4}$$

Substitute -2 for x in equation (1) to find y.

$$y = 3x^2 + 6x \quad (1)$$
$$y = 3(-2)^2 + 6(-2)$$
$$= 12 - 12 = 0$$

The solution set is $\left\{\left(-\frac{3}{2}, -\frac{9}{4}\right), (-2,0)\right\}$.

25. $2x^2 - y^2 = 6 \quad (1)$
$y = x^2 - 3 \quad (2)$

Substitute $x^2 - 3$ for y in equation (1).

$$2x^2 - y^2 = 6 \quad (1)$$
$$2x^2 - (x^2-3)^2 = 6$$
$$2x^2 - (x^4 - 6x^2 + 9) = 6$$
$$-x^4 + 8x^2 - 9 = 6$$
$$-x^4 + 8x^2 - 15 = 0$$
$$x^4 - 8x^2 + 15 = 0 \quad \textit{Multiply by } -1.$$

Let $z = x^2$, so $z^2 = x^4$.

$$z^2 - 8z + 15 = 0$$
$$(z-3)(z-5) = 0$$

$$z - 3 = 0 \quad \text{or} \quad z - 5 = 0$$
$$z = 3 \quad \text{or} \quad z = 5$$

Since $z = x^2$,

$$x^2 = 3 \quad \text{or} \quad x^2 = 5.$$

Use the square root property.

$$x = \pm\sqrt{3} \quad \text{or} \quad x = \pm\sqrt{5}$$

Use equation (2) to find y for each x-value.

If $x = \sqrt{3}$ or $-\sqrt{3}$, then

$$y = \left(\pm\sqrt{3}\right)^2 - 3 = 0.$$

(Note: We could substitute 3 for x^2 in (2) to obtain the same result.)

If $x = \sqrt{5}$ or $-\sqrt{5}$, then

$$y = \left(\pm\sqrt{5}\right)^2 - 3 = 2.$$

The solution set is
$\left\{(-\sqrt{3},0), (\sqrt{3},0), (-\sqrt{5},2), (\sqrt{5},2)\right\}$.

27. $x^2 - xy + y^2 = 0 \quad (1)$
$x - 2y = 1 \quad (2)$

Solve equation (2) for x.

$$x = 2y + 1 \quad (3)$$

Substitute $2y + 1$ for x in equation (1).

$$\begin{aligned}(2y+1)^2 - (2y+1)y + y^2 &= 0\\ 4y^2 + 4y + 1 - 2y^2 - y + y^2 &= 0\\ 3y^2 + 3y + 1 &= 0\end{aligned}$$

Use the quadratic formula with $a = 3$, $b = 3$, and $c = 1$.

$$\begin{aligned}y &= \frac{-b \pm \sqrt{b^2 - 4ac}}{2a}\\ y &= \frac{-3 \pm \sqrt{9 - 12}}{6}\\ &= \frac{-3 \pm \sqrt{-3}}{6} = \frac{-3 \pm i\sqrt{3}}{6}\end{aligned}$$

Use equation (3) to solve for x.

$$\begin{aligned}x &= 2y + 1 \qquad (3)\\ &= 2\left(\frac{-3 \pm i\sqrt{3}}{6}\right) + 1\\ &= \frac{-3 \pm i\sqrt{3}}{3} + \frac{3}{3}\\ &= \pm\frac{i\sqrt{3}}{3}\end{aligned}$$

The solution set is
$\left\{\left(\frac{\sqrt{3}}{3}i, -\frac{1}{2} + \frac{\sqrt{3}}{6}i\right), \left(-\frac{\sqrt{3}}{3}i, -\frac{1}{2} - \frac{\sqrt{3}}{6}i\right)\right\}$.

29. $3x^2 + 2y^2 = 12 \quad (1)$
$x^2 + 2y^2 = 4 \quad (2)$

Multiply equation (2) by -1 and add the result to equation (1).

$$\begin{array}{rcll} 3x^2 + 2y^2 &=& 12 & (1)\\ -x^2 - 2y^2 &=& -4 & -1 \times (2)\\ \hline 2x^2 &=& 8 &\\ x^2 &=& 4 &\\ x &=& \pm 2 &\end{array}$$

Substitute ± 2 for x in equation (2) to find y.

$$\begin{aligned}x^2 + 2y^2 &= 4 \qquad (2)\\ (\pm 2)^2 + 2y^2 &= 4\\ 4 + 2y^2 &= 4\\ 2y^2 &= 0\\ y^2 &= 0\\ y &= 0\end{aligned}$$

The solution set is $\{(-2, 0), (2, 0)\}$.

31. $2x^2 + 3y^2 = 6 \quad (1)$
$x^2 + 3y^2 = 3 \quad (2)$

Multiply equation (2) by -1 and add the result to equation (1).

$$\begin{array}{rcll} 2x^2 + 3y^2 &=& 6 & (1)\\ -x^2 - 3y^2 &=& -3 & -1 \times (2)\\ \hline x^2 &=& 3 &\\ x &=& \pm\sqrt{3} &\end{array}$$

Substitute $\pm\sqrt{3}$ for x in equation (2).

$$\begin{aligned}x^2 + 3y^2 &= 3 \qquad (2)\\ \left(\pm\sqrt{3}\right)^2 + 3y^2 &= 3\\ 3 + 3y^2 &= 3\\ y^2 &= 0\\ y &= 0\end{aligned}$$

The solution set is $\{(\sqrt{3}, 0), (-\sqrt{3}, 0)\}$.

33. $2x^2 + y^2 = 28 \quad (1)$
$4x^2 - 5y^2 = 28 \quad (2)$

Multiply (1) by 5 and add the result to equation (2).

$$\begin{array}{rcll} 10x^2 + 5y^2 &=& 140 & 5 \times (1)\\ 4x^2 - 5y^2 &=& 28 & (2)\\ \hline 14x^2 &=& 168 &\\ x^2 &=& 12 &\end{array}$$

$$x = \pm\sqrt{12} = \pm 2\sqrt{3}$$

Let $x = \pm 2\sqrt{3}$ in (1).

$$\begin{aligned}2\left(\pm 2\sqrt{3}\right)^2 + y^2 &= 28\\ 2(12) + y^2 &= 28\\ y^2 &= 4\\ y &= \pm 2\end{aligned}$$

The solution set is $\{(2\sqrt{3}, 2), (2\sqrt{3}, -2), (-2\sqrt{3}, 2), (-2\sqrt{3}, -2)\}$.

35. $2x^2 = 8 - 2y^2 \quad (1)$
$3x^2 = 24 - 4y^2 \quad (2)$

Multiply equation (1) by -2 and add the result to equation (2).

$$\begin{array}{rcll} -4x^2 &=& -16 + 4y^2 & -2 \times (1)\\ 3x^2 &=& 24 - 4y^2 & (2)\\ \hline -x^2 &=& 8 &\\ x^2 &=& -8 &\\ x &=& \pm\sqrt{-8} = \pm 2i\sqrt{2} &\end{array}$$

Substitute $\pm 2i\sqrt{2}$ for x in equation (1).

$$\begin{aligned}2x^2 &= 8-2y^2 \quad (1)\\ 2\left(\pm 2i\sqrt{2}\right)^2 &= 8-2y^2\\ 2(-8) &= 8-2y^2\\ -16 &= 8-2y^2\\ 2y^2 &= 24\\ y^2 &= 12\\ y &= \pm\sqrt{12} = \pm 2\sqrt{3}\end{aligned}$$

Since $2i\sqrt{2}$ can be paired with either $2\sqrt{3}$ or $-2\sqrt{3}$ and $-2i\sqrt{2}$ can be paired with either $2\sqrt{3}$ or $-2\sqrt{3}$, there are four possible solutions.

The solution set is

$\left\{\left(-2i\sqrt{2},-2\sqrt{3}\right),\left(-2i\sqrt{2},2\sqrt{3}\right),\left(2i\sqrt{2},-2\sqrt{3}\right),\left(2i\sqrt{2},2\sqrt{3}\right)\right\}$.

37. $x^2+xy+y^2=15 \quad (1)$
$x^2+y^2=10 \quad (2)$

Multiply equation (2) by -1 and add the result to equation (1).

$$\begin{array}{rcll} x^2+xy+y^2 &=& 15 & (1)\\ -x^2 \qquad -y^2 &=& -10 & -1\times(2)\\ \hline xy &=& 5 &\\ y &=& \dfrac{5}{x} & \end{array}$$

Substitute $\frac{5}{x}$ for y in equation (2).

$$\begin{aligned}x^2+y^2 &= 10 \quad (2)\\ x^2+\left(\frac{5}{x}\right)^2 &= 10\\ x^2+\frac{25}{x^2} &= 10\\ x^4+25 &= 10x^2 \quad \textit{Multiply by } x^2.\\ x^4-10x^2+25 &= 0\end{aligned}$$

Let $z=x^2$, so $z^2=x^4$.

$$\begin{aligned}z^2-10z+25 &= 0\\ (z-5)^2 &= 0\\ z-5 &= 0\\ z &= 5\end{aligned}$$

Since $z=x^2$,

$$x^2=5, \text{ and } x=\pm\sqrt{5}.$$

Using the equation $y=\frac{5}{x}$, we get the following.

If $x=-\sqrt{5}$, then

$$y=\frac{5}{-\sqrt{5}}=\frac{5\cdot\sqrt{5}}{-\sqrt{5}\cdot\sqrt{5}}=\frac{5\sqrt{5}}{-5}=-\sqrt{5}.$$

Similarly, if $x=\sqrt{5}$, then $y=\sqrt{5}$.

The solution set is $\left\{\left(-\sqrt{5},-\sqrt{5}\right),\left(\sqrt{5},\sqrt{5}\right)\right\}$.

39. $3x^2+2xy-3y^2=5 \quad (1)$
$-x^2-3xy+y^2=3 \quad (2)$

Multiply equation (2) by 3 and add the result to equation (1).

$$\begin{array}{rcll} 3x^2+2xy-3y^2 &=& 5 & (1)\\ -3x^2-9xy+3y^2 &=& 9 & 3\times(2)\\ \hline -7xy &=& 14 &\\ x &=& \dfrac{14}{-7y}=-\dfrac{2}{y} & \end{array}$$

Substitute $-\frac{2}{y}$ for x in equation (2).

$$\begin{aligned}-x^2-3xy+y^2 &= 3 \quad (2)\\ -\left(-\frac{2}{y}\right)^2-3\left(-\frac{2}{y}\right)y+y^2 &= 3\\ -\left(\frac{4}{y^2}\right)+6+y^2 &= 3\\ y^2+3-\frac{4}{y^2} &= 0\\ y^4+3y^2-4 &= 0 \quad \textit{Multiply by } y^2.\\ (y^2+4)(y^2-1) &= 0\end{aligned}$$

$$\begin{aligned}y^2+4=0 \quad &\text{or} \quad y^2-1=0\\ y^2=-4 \quad & \qquad y^2=1\\ y=\pm 2i \quad &\text{or} \qquad y=\pm 1\end{aligned}$$

Since $x=-\frac{2}{y}$, substitute these values for y to find the values of x.

If $y=2i$, then

$$x=-\frac{2}{2i}=-\frac{1}{i}=-\frac{1}{i}\cdot\frac{i}{i}=\frac{-i}{-1}=i.$$

If $y=-2i$, then

$$x=-\frac{2}{-2i}=\frac{1}{i}=\frac{1}{i}\cdot\frac{i}{i}=\frac{i}{-1}=-i.$$

If $y=1$, then $x=-\frac{2}{1}=-2$.

If $y=-1$, then $x=-\frac{2}{-1}=2$.

The solution set is
$\{(i,2i),(-i,-2i),(2,-1),(-2,1)\}$.

41. $xy=-6 \quad (1)$
$x+y=-1 \quad (2)$

Solve both equations for y.

$$y=-\frac{6}{x} \quad \text{and} \quad y=-x-1$$

Graph

$$\text{Y}_1=-\frac{6}{\text{X}} \quad \text{and} \quad \text{Y}_2=-\text{X}-1$$

on a graphing calculator to obtain the solution set $\{(2,-3),(-3,2)\}$.

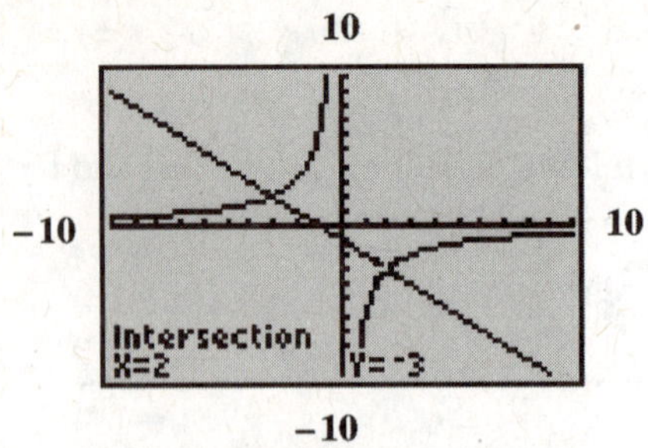

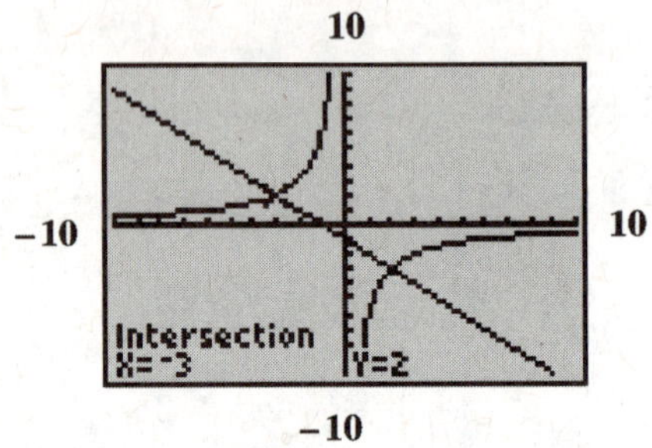

Now solve the system algebraically using substitution.

$$\begin{aligned} xy &= -6 \quad (1) \\ x + y &= -1 \quad (2) \end{aligned}$$

Solve equation (2) for y.

$$y = -1 - x \quad (3)$$

Substitute $-1 - x$ for y in equation (1).

$$\begin{aligned} x(-1 - x) &= -6 \\ -x - x^2 &= -6 \\ 0 &= x^2 + x - 6 \\ 0 &= (x+3)(x-2) \end{aligned}$$

$$x + 3 = 0 \quad \text{or} \quad x - 2 = 0$$
$$x = -3 \quad \text{or} \quad x = 2$$

Substitute these values for x in (3) to find y.

If $x = -3$, then $y = -1 - (-3) = 2$.

If $x = 2$, then $y = -1 - 2 = -3$.

The solution set, $\{(2, -3), (-3, 2)\}$, is the same as that obtained using a graphing calculator.

43. Let W = the width, and L = the length.

The formula for the area of a rectangle is $LW = \mathcal{A}$, so

$$LW = 84. \quad (1)$$

The perimeter of a rectangle is given by $2L + 2W = P$, so

$$2L + 2W = 38. \quad (2)$$

Solve equation (2) for L to get

$$L = 19 - W. \quad (3)$$

Substitute $19 - W$ for L in equation (1).

$$\begin{aligned} LW &= 84 \quad (1) \\ (19 - W)W &= 84 \\ 19W - W^2 &= 84 \\ -W^2 + 19W - 84 &= 0 \\ W^2 - 19W + 84 &= 0 \quad \textit{Multiply by } -1. \\ (W - 7)(W - 12) &= 0 \end{aligned}$$

$$W - 7 = 0 \quad \text{or} \quad W - 12 = 0$$
$$W = 7 \quad \text{or} \quad W = 12$$

Using equation (3), with $W = 7$,

$$L = 19 - 7 = 12.$$

If $W = 12$, then $L = 7$, which are the same two numbers. Length must be greater than width, so the length is 12 feet and the width is 7 feet.

45. $px = 16 \quad (1)$
$p = 10x + 12 \quad (2)$

Substitute $10x + 12$ for p in equation (1).

$$\begin{aligned} px &= 16 \quad (1) \\ (10x + 12)x &= 16 \\ 10x^2 + 12x &= 16 \\ 10x^2 + 12x - 16 &= 0 \\ 5x^2 + 6x - 8 &= 0 \\ (5x - 4)(x + 2) &= 0 \end{aligned}$$

$$5x - 4 = 0 \quad \text{or} \quad x + 2 = 0$$
$$x = \tfrac{4}{5} \quad \text{or} \quad x = -2$$

Since x cannot be negative, eliminate -2 as a value of x. Substitute $\frac{4}{5}$ for x in equation (2) to find p.

$$\begin{aligned} p &= 10x + 12 \quad (2) \\ p &= 10(\tfrac{4}{5}) + 12 \\ &= 8 + 12 = 20 \end{aligned}$$

Since the demand x is in thousands,

$$\text{demand} = \tfrac{4}{5}(1000) = 800.$$

The equilibrium price is \$20. The supply/demand at that price is 800 calculators.

47. $2x - y \leq 4$

Draw a solid line through the intercepts $(2, 0)$ and $(0, -4)$. Checking $(0, 0)$ gives us $0 \leq 4$, a true statement, so shade the side containing $(0, 0)$.

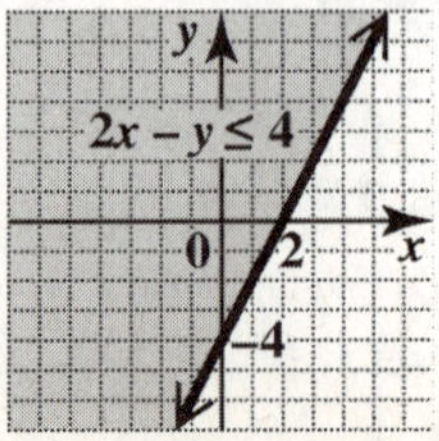

13.5 Second-Degree Inequalities and Systems of Inequalities

13.5 Now Try Exercises

N1. $x^2 + y^2 \geq 9$

The boundary, $x^2 + y^2 = 9$, is a circle with radius 3. It is drawn as a solid curve because of the $\geq$ symbol. Use the point $(0, 0)$ as a test point.

$$x^2 + y^2 \geq 9$$
$$0^2 + 0^2 \overset{?}{\geq} 9$$
$$0 \geq 9 \quad \textit{False}$$

Shade the region that does not contain $(0, 0)$. This is the region outside the circle.

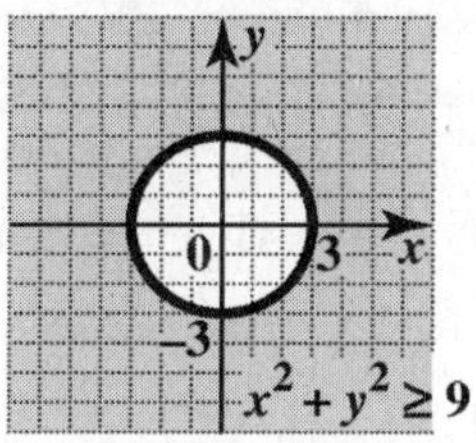

N2. $y \geq -(x + 2)^2 + 1$

The boundary, $y = -(x + 2)^2 + 1$ is a parabola opening downward with its vertex at $(-2, 1)$. It is drawn as a solid curve because of the $\geq$ symbol. Use the point $(0, 0)$ as a test point.

$$0 \overset{?}{\geq} -(0 + 2)^2 + 1$$
$$0 \overset{?}{\geq} -4 + 1$$
$$0 \geq -3 \quad \textit{True}$$

Shade the region that contains $(0, 0)$. This is the region outside the parabola.

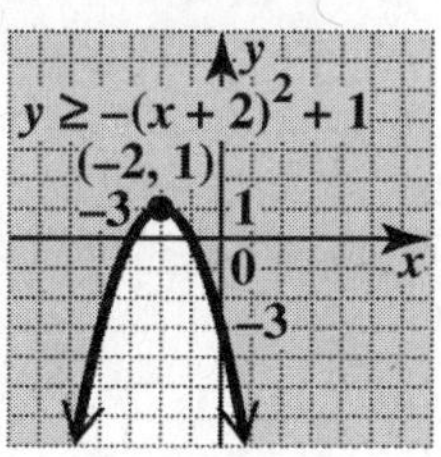

N3. $25x^2 - 16y^2 > 400$

$$\frac{x^2}{16} - \frac{y^2}{25} > 1 \quad \textit{Divide by 400.}$$

The boundary is the hyperbola

$$\frac{x^2}{16} - \frac{y^2}{25} = 1,$$

which has intercepts $(4, 0)$ and $(-4, 0)$. It is drawn as a dashed curve because of the $>$ symbol. Test $(0, 0)$.

$$25(0)^2 - 16(0)^2 \overset{?}{>} 400$$
$$0 > 400 \quad \textit{False}$$

Shade the region that does not contain $(0, 0)$. This is the region inside the branches of the hyperbola.

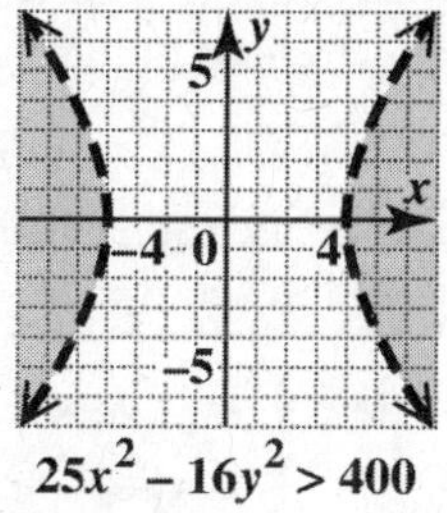

N4. $x^2 + y^2 > 9$
$y > x^2 - 1$

The boundary, $x^2 + y^2 = 9$, is a circle with radius 3. Draw this boundary as a dashed curve. Test $(0, 0)$.

$$0^2 + 0^2 \overset{?}{>} 9$$
$$0 > 9 \quad \textit{False}$$

Shade the region outside the circle.
The boundary $y = x^2 - 1$ is an upward opening parabola with vertical axis $x = 0$ and vertex $(0, -1)$. Draw this boundary as a dashed curve. Test $(0, 0)$.

$$0 \overset{?}{\geq} 0^2 - 1$$
$$0 \geq -1 \quad \textit{True}$$

Shade the region inside the parabola.
The graph of the solution set of the system includes all points common to the two graphs, that is, the intersection of the two shaded regions.

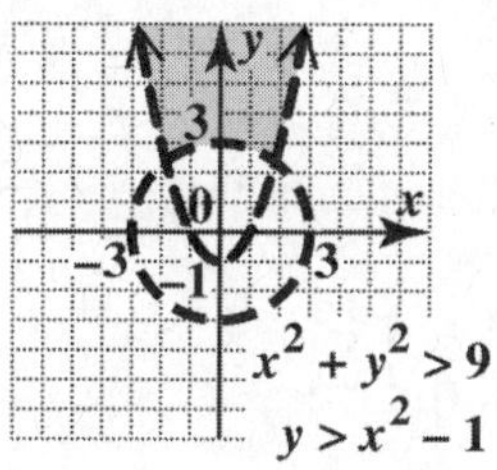

N5. $3x + 2y > 6$
$y \geq \frac{1}{2}x - 2$
$x \geq 0$

The boundary $3x + 2y = 6$ is a dashed line with intercepts $(2, 0)$ and $(0, 3)$. Test $(0, 0)$.

$$3(0) + 2(0) \overset{?}{>} 6$$
$$0 > 6 \quad \textit{False}$$

Shade the side of the line that does not contain $(0, 0)$.

The boundary $y = \frac{1}{2}x - 2$ is a solid line with intercepts $(4, 0)$ and $(0, -2)$. Test $(0, 0)$.

$$0 \overset{?}{\geq} \tfrac{1}{2}(0) - 2$$
$$0 \geq -2 \quad \textit{True}$$

Shade the side of the line that contains $(0,0)$.

The boundary $x = 0$ is a solid vertical line through $(0,0)$. Shade the region to the right of the line since $x \geq 0$.

The intersection of the three shaded regions is the graph of the solution set of the system.

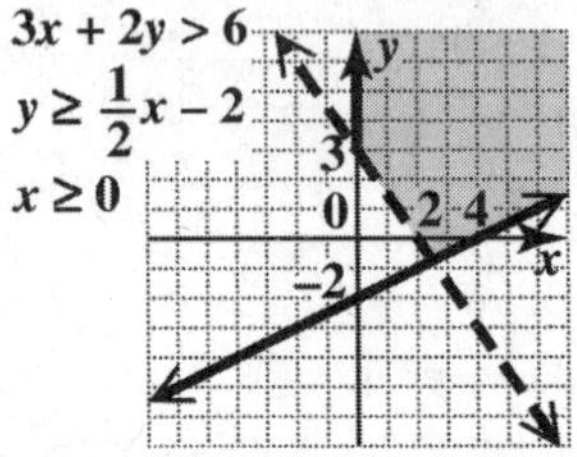

N6. $\frac{x^2}{4} + \frac{y^2}{16} \leq 1$
$y \leq x^2 - 2$
$y + 3 > 0$

The boundary $\frac{x^2}{4} + \frac{y^2}{16} = 1$ is an ellipse with intercepts $(2,0)$, $(-2,0)$, $(0,4)$, and $(0,-4)$. It is drawn as a solid curve because of the $\leq$ symbol. Test $(0,0)$.

$$\frac{0^2}{4} + \frac{0^2}{16} \stackrel{?}{\leq} 1$$
$$0 \leq 1 \quad \textit{True}$$

Shade the region that contains $(0,0)$. This is the region inside the ellipse.

The boundary $y = x^2 - 2$ is a parabola opening upward with vertex at $(0,-2)$. It is drawn as a solid curve because of the $\leq$ symbol. Test $(0,0)$.

$$0 \stackrel{?}{\leq} 0^2 - 2$$
$$0 \leq -2 \quad \textit{False}$$

Shade the region that does not contain $(0,0)$. This is the region outside the parabola.

The boundary $y + 3 = 0$, or $y = -3$, is a dashed horizontal line through $(0,-3)$. Shade the region above the line since $y + 3 > 0$.

The graph of the solution set of the system includes all points common to the three graphs, that is, the intersection of the three shaded regions.

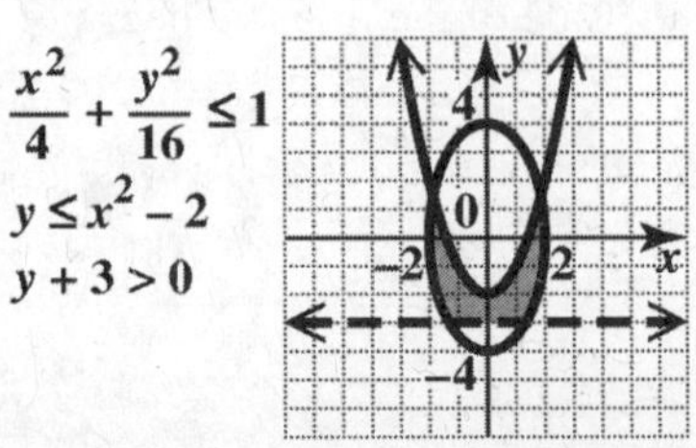

13.5 Section Exercises

1. $x^2 + y^2 < 25$
$y > -2$

The boundary, $x^2 + y^2 = 25$, is a circle (dashed because of the $<$ sign) with center $(0,0)$ and radius 5. When $(0,0)$ is tested, a true statement, $0 < 25$, results, so the inside of the circle is shaded. The graph of $y = -2$ is a horizontal line through $(0,-2)$ with shading above the dashed line, since $y > -2$. The correct answer is **C**.

3. $y \geq x^2 + 4$

This is an inequality whose boundary is a solid parabola, opening up, with vertex $(0,4)$. The inside of the parabola is shaded since $(0,0)$ makes the inequality false. This is graph **B**.

5. $y < x^2 + 4$

This is an inequality whose boundary is a dashed parabola, opening up, with vertex $(0,4)$. The outside of the parabola is shaded since $(0,0)$ makes the inequality true. This is graph **A**.

7.
$$y^2 > 4 + x^2$$
$$y^2 - x^2 > 4$$
$$\frac{y^2}{4} - \frac{x^2}{4} > 1$$

The boundary, $\frac{y^2}{4} - \frac{x^2}{4} = 1$, is a hyperbola with y-intercepts $(0,2)$ and $(0,-2)$ and asymptotes formed by the extended diagonals of the fundamental rectangle with vertices at $(2,2)$, $(2,-2)$, $(-2,-2)$, and $(-2,2)$. The hyperbola has dashed branches because of $>$. Test $(0,0)$.

$$0^2 \stackrel{?}{>} 4 + 0^2$$
$$0 > 4 \quad \textit{False}$$

Shade the sides of the hyperbola that do not contain $(0,0)$. These are the regions inside the branches of the hyperbola.

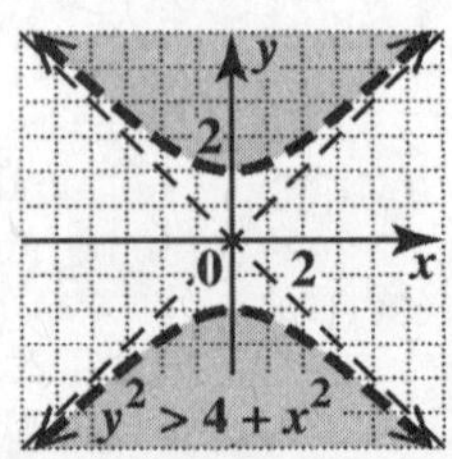

9. $y \geq x^2 - 2$

$y \geq (x-0)^2 - 2$

Graph the solid vertical parabola $y = x^2 - 2$ with vertex $(0, -2)$. Two other points on the parabola are $(2, 2)$ and $(-2, 2)$. Test a point not on the parabola, say $(0, 0)$, in $y \geq x^2 - 2$ to get $0 \geq -2$, a true statement. Shade that portion of the graph that contains the point $(0, 0)$. This is the region inside the parabola.

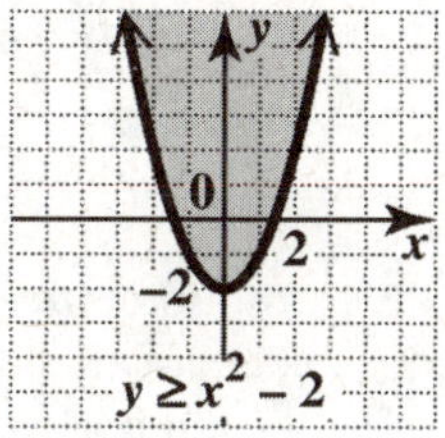

11. $2y^2 \geq 8 - x^2$

$x^2 + 2y^2 \geq 8$

$\frac{x^2}{8} + \frac{y^2}{4} \geq 1$

The boundary, $\frac{x^2}{8} + \frac{y^2}{4} = 1$, is the ellipse with intercepts $\left(2\sqrt{2}, 0\right)$, $\left(-2\sqrt{2}, 0\right)$, $(0, 2)$, and $(0, -2)$, drawn as a solid curve because of $\geq$. Test $(0, 0)$.

$$2(0)^2 \overset{?}{\geq} 8 - 0^2$$
$$0 \geq 8 \qquad \textit{False}$$

Shade the region of the ellipse that does not contain $(0, 0)$. This is the region outside the ellipse.

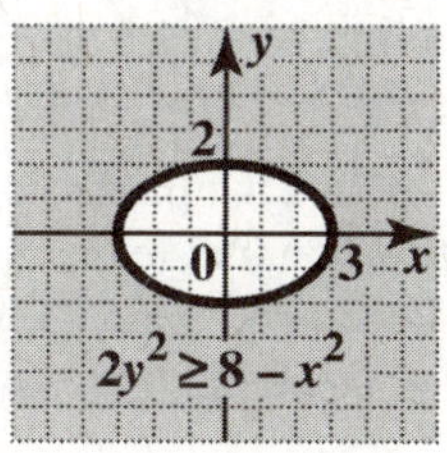

13. $y \leq x^2 + 4x + 2$

Graph the solid vertical parabola $y = x^2 + 4x + 2$. Use the vertex formula $x = \frac{-b}{2a}$ to obtain the vertex $(-2, -2)$. Two other points on the parabola are $(0, 2)$ and $(1, 7)$. Test a point not on the parabola, say $(0, 0)$, in $y \leq x^2 + 4x + 2$ to get $0 \leq 2$, a true statement. Shade outside the parabola, since this region contains $(0, 0)$.

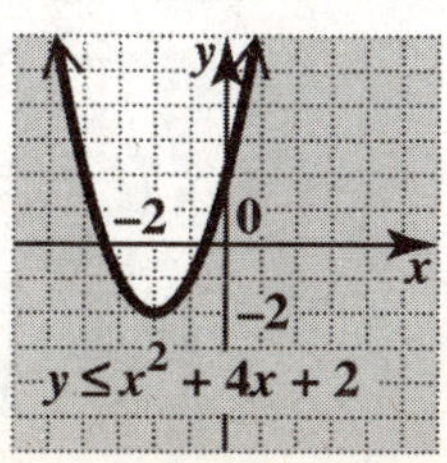

15. $9x^2 > 16y^2 + 144$

$9x^2 - 16y^2 > 144$

$\frac{x^2}{16} - \frac{y^2}{9} > 1$

The boundary, $\frac{x^2}{16} - \frac{y^2}{9} = 1$, is a hyperbola with x-intercepts $(4, 0)$ and $(-4, 0)$ and asymptotes formed by the extended diagonals of the fundamental rectangle with vertices at $(4, 3)$, $(4, -3)$, $(-4, -3)$, and $(-4, 3)$. The hyperbola has dashed branches because of $>$. Test $(0, 0)$.

$$9(0)^2 \overset{?}{>} 16(0)^2 + 144$$
$$0 > 144 \qquad \textit{False}$$

Shade the sides of the hyperbola that do not contain $(0, 0)$. These are the regions inside the branches of the hyperbola.

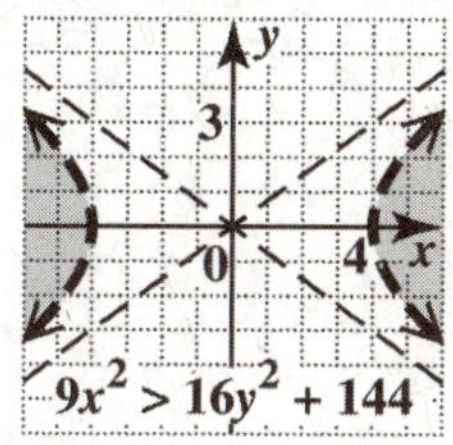

17. $x^2 - 4 \geq -4y^2$

$x^2 + 4y^2 \geq 4$

$\frac{x^2}{4} + \frac{y^2}{1} \geq 1$

Graph the solid ellipse $\frac{x^2}{4} + \frac{y^2}{1} = 1$ through the x-intercepts $(2, 0)$ and $(-2, 0)$ and y-intercepts $(0, 1)$ and $(0, -1)$. Test a point not on the ellipse, say $(0, 0)$, in $x^2 - 4 \geq -4y^2$ to get $-4 \geq 0$, a false statement. Shade outside the ellipse, since this region does *not* include $(0, 0)$.

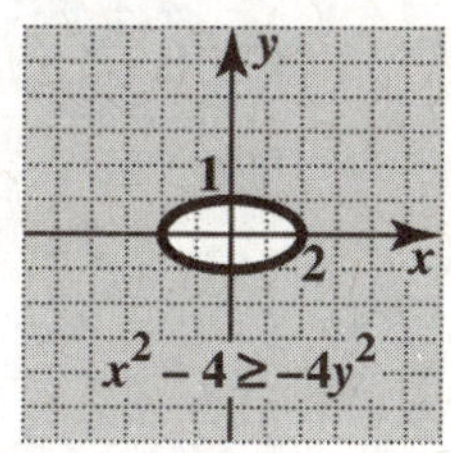

19. $x \leq -y^2 + 6y - 7$

Complete the square to find the vertex.

$$\begin{aligned} x &= -y^2 + 6y - 7 \\ &= -\left(y^2 - 6y\right) - 7 \\ &= -\left(y^2 - 6y + 9\right) + 9 - 7 \\ x &= -(y-3)^2 + 2 \end{aligned}$$

The boundary, $x = -(y-3)^2 + 2$, is a solid horizontal parabola with vertex at $(2, 3)$ that opens to the left. Test $(0, 0)$.

$$0 \overset{?}{\le} -0^2 + 6(0) - 7$$
$$0 \le -7 \qquad \textit{False}$$

Shade the region of the parabola that does not contain $(0, 0)$. This is the region inside the parabola.

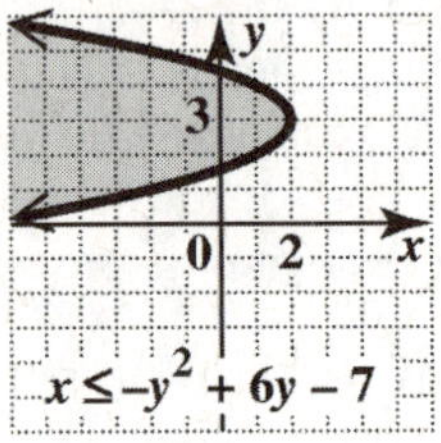

21. $2x + 5y < 10$
$x - 2y < 4$

Graph $2x + 5y = 10$ as a dashed line through $(5, 0)$ and $(0, 2)$. Test $(0, 0)$.

$$2x + 5y < 10$$
$$2(0) + 5(0) \overset{?}{<} 10$$
$$0 < 10 \qquad \textit{True}$$

Shade the region containing $(0, 0)$.
Graph $x - 2y = 4$ as a dashed line through $(4, 0)$ and $(0, -2)$. Test $(0, 0)$.

$$x - 2y < 4$$
$$0 - 2(0) \overset{?}{<} 4$$
$$0 < 4 \qquad \textit{True}$$

Shade the region containing $(0, 0)$. The graph of the system is the intersection of the two shaded regions.

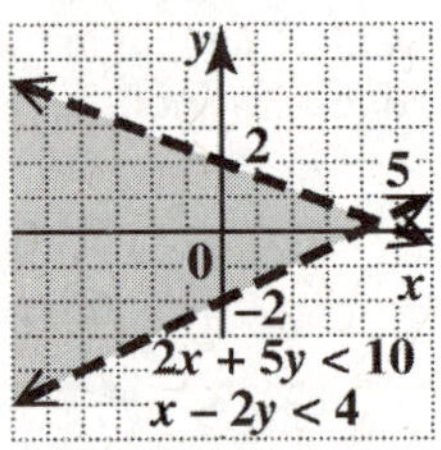

23. $5x - 3y \le 15$
$4x + y \ge 4$

The boundary, $5x - 3y = 15$, is a solid line with intercepts $(3, 0)$ and $(0, -5)$. Test $(0, 0)$.

$$5(0) - 3(0) \overset{?}{\le} 15$$
$$0 \le 15 \qquad \textit{True}$$

Shade the side of the line that contains $(0, 0)$.
The boundary, $4x + y = 4$, is a solid line with intercepts $(1, 0)$ and $(0, 4)$. Test $(0, 0)$.

$$4(0) + 0 \overset{?}{\ge} 4$$
$$0 \ge 4 \qquad \textit{False}$$

Shade the side of the line that does not contain $(0, 0)$.
The graph of the system is the intersection of the two shaded regions.

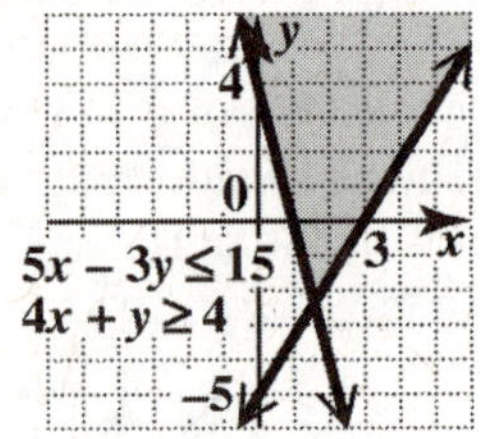

25. $x \le 5$
$y \le 4$

Graph $x = 5$ as a solid vertical line through $(5, 0)$. Since $x \le 5$, shade the left side of the line.
Graph $y = 4$ as a solid horizontal line through $(0, 4)$. Since $y \le 4$, shade below the line.
The graph of the system is the intersection of the two shaded regions.

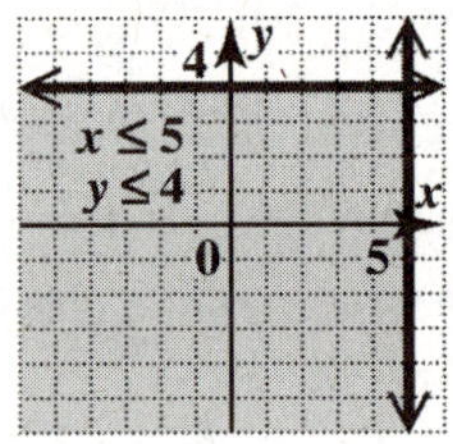

27. $y > x^2 - 4$
$y < -x^2 + 3$

The boundary, $y = x^2 - 4$, is a dashed parabola with vertex $(0, -4)$ that opens up. Test $(0, 0)$.

$$0 \overset{?}{>} 0^2 - 4$$
$$0 > -4 \qquad \textit{True}$$

Shade the side of the parabola that contains $(0, 0)$. This is the region inside the parabola.
The boundary, $y = -x^2 + 3$, is a dashed parabola with vertex $(0, 3)$ that opens down. Test $(0, 0)$.

$$0 \overset{?}{<} -0^2 + 3$$
$$0 < 3 \qquad \textit{True}$$

Shade the side of the parabola that contains $(0, 0)$. This is the region inside the parabola.
The graph of the system is the intersection of the two shaded regions.

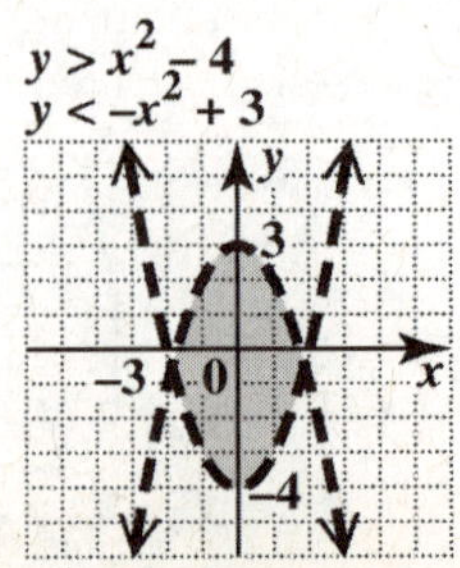

29. $x^2 + y^2 \geq 4$ (1)
$x + y \leq 5$ (2)
$x \geq 0$ (3)
$y \geq 0$ (4)

The graph of (1) is the region outside the solid circle boundary $x^2 + y^2 = 4$ with center $(0, 0)$ and radius 2.
The graph of (2), equivalent to $y \leq -x + 5$, is the region below the solid line $x + y = 5$ with intercepts $(5, 0)$ and $(0, 5)$.
The intersection of the graphs of (3) and (4) is quadrant I where both x and y are positive. The positive x- and y-axes are included.
The graph of the system is the intersection of the graphs of (1) and (2) in quadrant I.

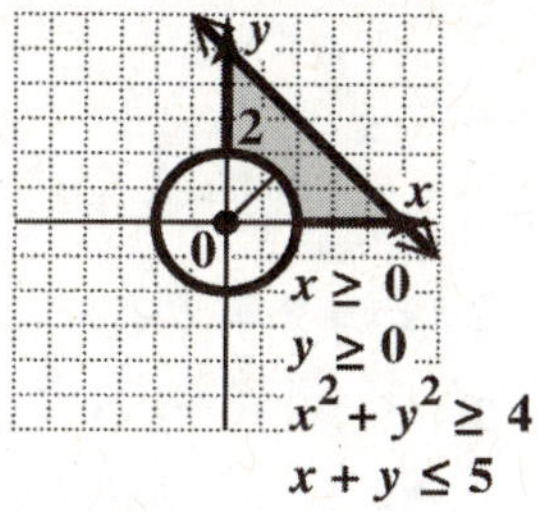

31. $y \leq -x^2$
$y \geq x - 3$
$y \leq -1$
$x < 1$

The boundary, $y = -x^2$, is a solid parabola with vertex at $(0, 0)$ that opens down. Test $(0, -1)$.

$$-1 \overset{?}{\leq} -(0)^2$$
$$-1 \leq 0 \qquad \textit{True}$$

Shade the side of the parabola that contains $(0, -1)$. This is the region inside the parabola.
The boundary, $y = x - 3$, is a solid line with intercepts $(3, 0)$ and $(0, -3)$. Test $(0, 0)$.

$$0 \overset{?}{\geq} 0 - 3$$
$$0 \geq -3 \qquad \textit{True}$$

Shade the side of the line that contains $(0, 0)$.
For $y \leq -1$, shade below the solid horizontal line $y = -1$.
For $x < 1$, shade to the left of the dashed vertical line $x = 1$.
The intersection of the four shaded regions is the graph of the system.

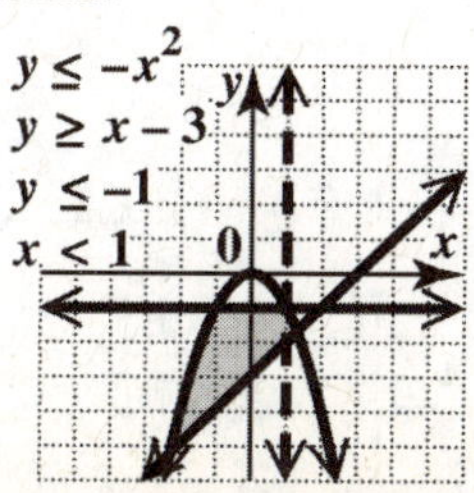

33. $x^2 + y^2 > 36, x \geq 0$

This is a circle of radius 6 centered at the origin. The graph is a dashed curve. Since $(0, 0)$ does not satisfy the inequality, the shading will be outside the circle. By including the restriction $x \geq 0$, we consider only the shading to the right of, and including, the y-axis.

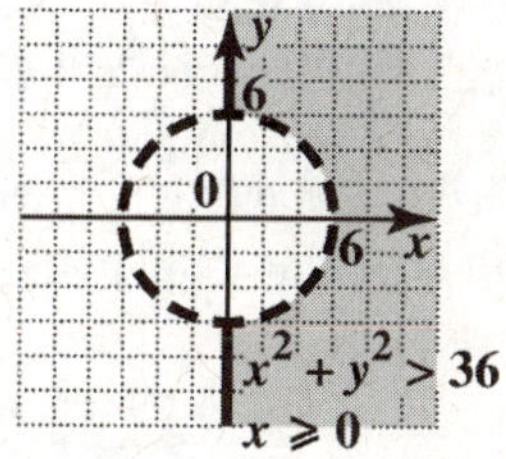

35. $x < y^2 - 3, x < 0$

Consider the equation.

$$x = y^2 - 3, \text{ or}$$
$$x = (y - 0)^2 - 3.$$

This is a parabola with vertex $(-3, 0)$, opening to the right having the same shape as $x = y^2$. The graph is a dashed curve. The shading will be outside the parabola, since $(0, 0)$ does not satisfy the inequality. By including the restriction $x < 0$, we consider only the shading to the left of, but not including, the y-axis. The y-axis is a dashed line.

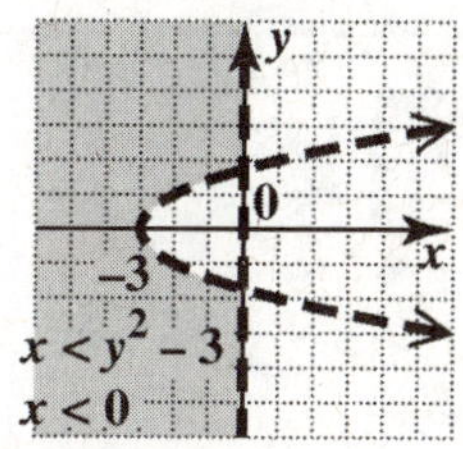

37. $4x^2 - y^2 > 16, x < 0$

Consider the equation

$$4x^2 - y^2 = 16, \text{ or}$$
$$\frac{x^2}{4} - \frac{y^2}{16} = 1.$$

This is a hyperbola with x-intercepts $(2, 0)$ and $(-2, 0)$. The graph is a dashed curve. The shading will be to the left of the left branch and to the right of the right branch of the hyperbola, since $(0, 0)$ does not satisfy the inequality. By including the restriction $x < 0$, we consider only the shading to the left of the y-axis.

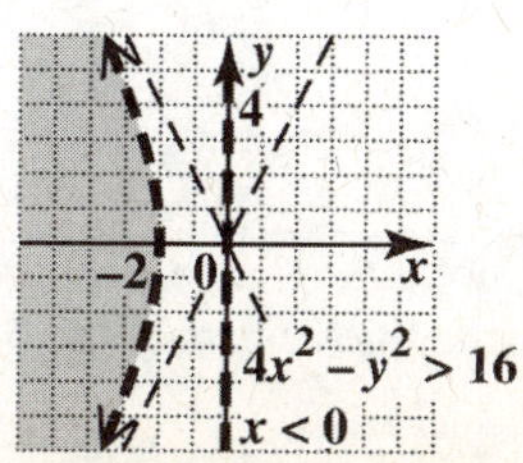

39. $x^2 + 4y^2 \geq 1, x \geq 0, y \geq 0$

Consider the equation

$$x^2 + 4y^2 = 1, \text{ or}$$
$$\frac{x^2}{1} + \frac{y^2}{\frac{1}{4}} = 1.$$

The graph is a solid ellipse with x-intercepts $(1, 0)$ and $(-1, 0)$ and y-intercepts $\left(0, \frac{1}{2}\right)$ and $\left(0, -\frac{1}{2}\right)$. The shading will be outside the ellipse, since $(0, 0)$ does not satisfy the inequality. By including the restrictions $x \geq 0$ and $y \geq 0$, we consider only the shading in quadrant I and portions of the x- and y-axis.

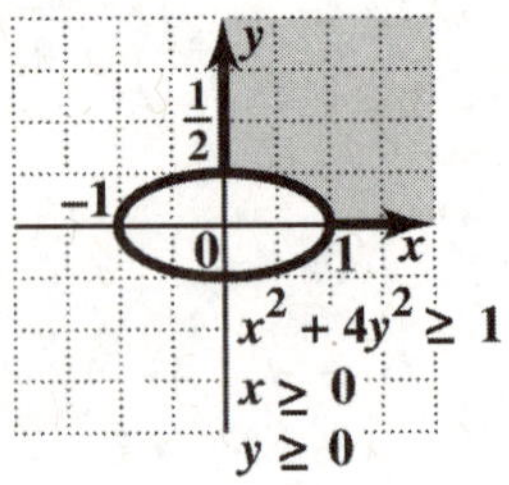

41. $y \geq x - 3$
$y \leq -x + 4$

The graphs of both inequalities include the points on the lines as part of the solution because of $\geq$ and $\leq$.
To produce the graph of the system on the TI-83/4, make the following Y-assignments:

To get the upper and lower darkened triangles to the left of Y_1 and Y_2, simply place the cursor in that spot and press ENTER to cycle through the graphing choices.

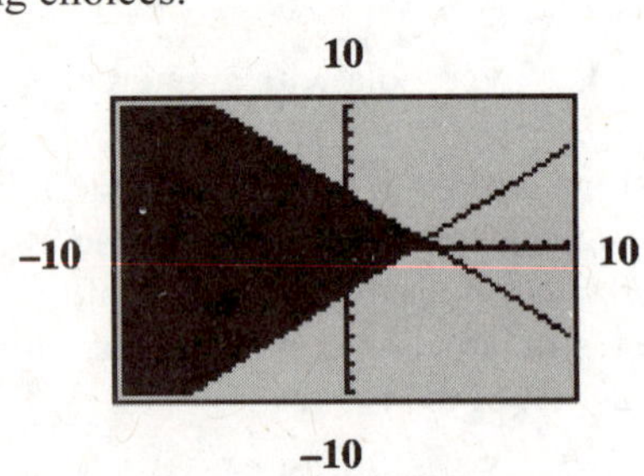

43. $y < x^2 + 4x + 4$
$y > -3$

The graphs do *not* include the points on the parabola or the line as part of the solution because of $<$ and $>$.

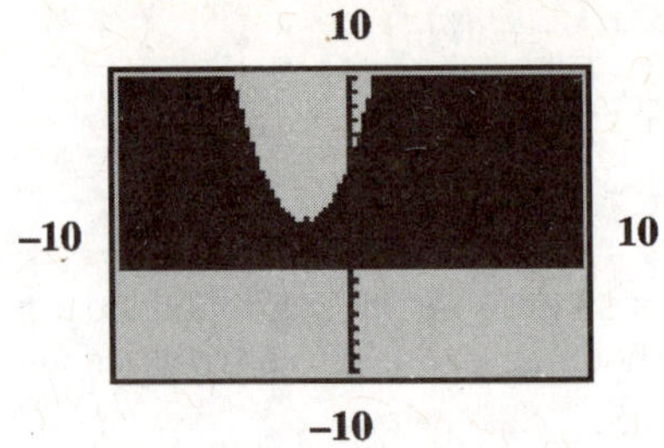

45. $\dfrac{n+5}{n}$

(a) $[n = 1]$ $\frac{1+5}{1} = \frac{6}{1} = 6$

(b) $[n = 2]$ $\frac{2+5}{2} = \frac{7}{2}$

(c) $[n = 3]$ $\frac{3+5}{3} = \frac{8}{3}$

(d) $[n = 4]$ $\frac{4+5}{4} = \frac{9}{4}$

47. $n^2 - n$

(a) $[n = 1]$ $1^2 - 1 = 1 - 1 = 0$

(b) $[n = 2]$ $2^2 - 2 = 4 - 2 = 2$

(c) $[n = 3]$ $3^2 - 3 = 9 - 3 = 6$

(d) $[n = 4]$ $4^2 - 4 = 16 - 4 = 12$

Chapter 13 Review Exercises

1. $f(x) = |x + 4|$

This is the graph of the absolute value function $g(x) = |x|$ shifted 4 units to the left (since $x + 4 = 0$ if $x = -4$).

x	y
-6	2
-5	1
-4	0
-3	1
-2	2

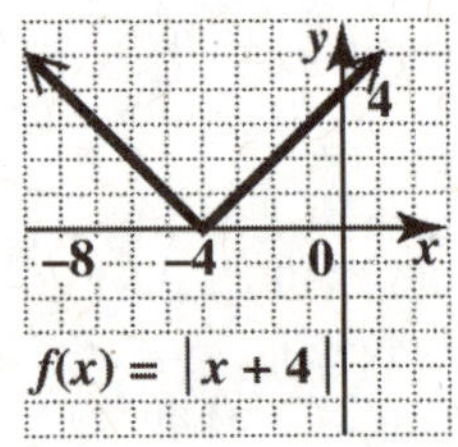

2. $f(x) = \dfrac{1}{x - 4}$

This is the graph of the reciprocal function, $g(x) = \frac{1}{x}$, shifted 4 units to the right. Since $x \neq 4$ (or a denominator of 0 results), the line $x = 4$ is a vertical asymptote. Since $\dfrac{1}{x-4} \neq 0$, the line $y = 0$ is a horizontal asymptote.

x	y
2	$-\frac{1}{2}$
3	-1
$3\frac{1}{2}$	-2
$4\frac{1}{2}$	2
5	1
6	$\frac{1}{2}$

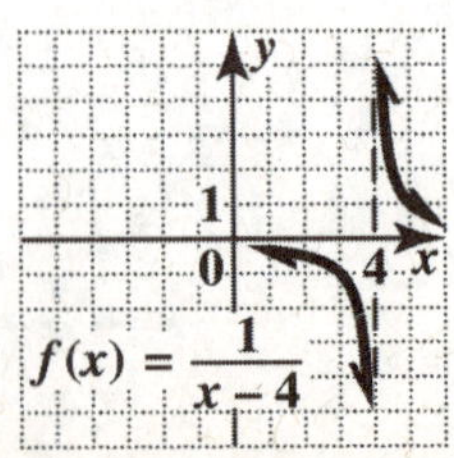

3. $f(x) = \sqrt{x} + 3$

This is the graph of the square root function, $g(x) = \sqrt{x}$, shifted 3 units upward.

x	y
0	3
1	4
4	5

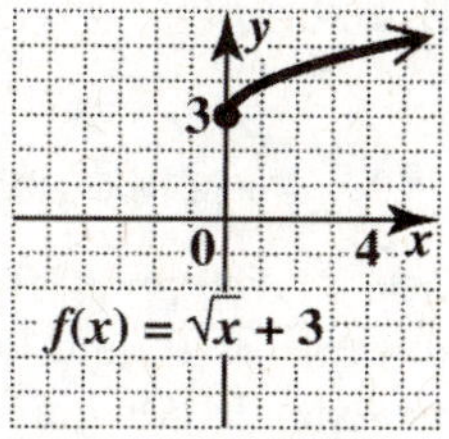

4. $f(x) = [\![x]\!] - 2$

This is the graph of the greatest integer function, $g(x) = [\![x]\!]$, shifted down 2 units.

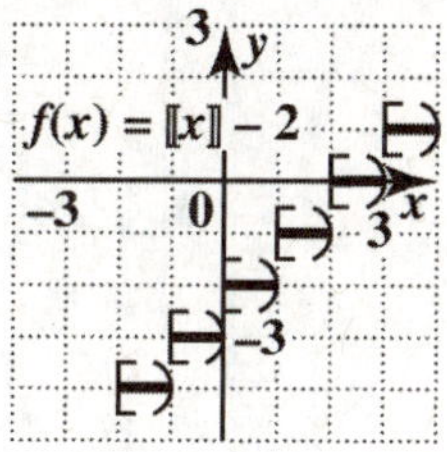

5. Center $(-2, 4)$, $r = 3$
Here $h = -2$, $k = 4$, and $r = 3$, so an equation of the circle is

$$(x-h)^2 + (y-k)^2 = r^2$$
$$[x-(-2)]^2 + (y-4)^2 = 3^2$$
$$(x+2)^2 + (y-4)^2 = 9.$$

6. Center $(-1, -3)$, $r = 5$
Here $h = -1$, $k = -3$, and $r = 5$, so an equation of the circle is

$$(x-h)^2 + (y-k)^2 = r^2$$
$$[x-(-1)]^2 + [y-(-3)]^2 = 5^2$$
$$(x+1)^2 + (y+3)^2 = 25.$$

7. Center $(4, 2)$, $r = 6$
Here $h = 4$, $k = 2$, and $r = 6$, so an equation of the circle is

$$(x-h)^2 + (y-k)^2 = r^2$$
$$(x-4)^2 + (y-2)^2 = 6^2$$
$$(x-4)^2 + (y-2)^2 = 36.$$

8. $x^2 + y^2 + 6x - 4y - 3 = 0$

Write the equation in center-radius form,

$$(x-h)^2 + (y-k)^2 = r^2,$$

by completing the squares on x and y.

$$(x^2 + 6x \qquad) + (y^2 - 4y \qquad) = 3$$
$$(x^2 + 6x + \underline{9}) + (y^2 - 4y + \underline{4})$$
$$= 3 + \underline{9} + \underline{4}$$
$$(x+3)^2 + (y-2)^2 = 16$$
$$[x-(-3)]^2 + (y-2)^2 = 16$$

The circle has center (h, k) at $(-3, 2)$ and radius $\sqrt{16} = 4$.

9. $x^2 + y^2 - 8x - 2y + 13 = 0$

Write the equation in center-radius form,

$$(x-h)^2 + (y-k)^2 = r^2,$$

by completing the squares on x and y.

$$(x^2 - 8x \qquad) + (y^2 - 2y \qquad) = -13$$
$$(x^2 - 8x + \underline{16}) + (y^2 - 2y + \underline{1})$$
$$= -13 + \underline{16} + \underline{1}$$
$$(x-4)^2 + (y-1)^2 = 4$$

The circle has center (h, k) at $(4, 1)$ and radius $\sqrt{4} = 2$.

10. $2x^2 + 2y^2 + 4x + 20y = -34$
$x^2 + y^2 + 2x + 10y = -17$

Write the equation in center-radius form,

$$(x-h)^2 + (y-k)^2 = r^2,$$

by completing the squares on x and y.

$$(x^2 + 2x \qquad) + (y^2 + 10y \qquad) = -17$$
$$(x^2 + 2x + \underline{1}) + (y^2 + 10y + \underline{25})$$
$$= -17 + \underline{1} + \underline{25}$$
$$(x+1)^2 + (y+5)^2 = 9$$
$$[x-(-1)]^2 + [y-(-5)]^2 = 9$$

The circle has center (h, k) at $(-1, -5)$ and radius $\sqrt{9} = 3$.

11. $4x^2 + 4y^2 - 24x + 16y = 48$
$x^2 + y^2 - 6x + 4y = 12$

Write the equation in center-radius form,

$$(x-h)^2 + (y-k)^2 = r^2,$$

by completing the squares on x and y.

$$(x^2 - 6x \qquad) + (y^2 + 4y \qquad) = 12$$
$$(x^2 - 6x + \underline{9}) + (y^2 + 4y + \underline{4})$$
$$= 12 + \underline{9} + \underline{4}$$
$$(x-3)^2 + (y+2)^2 = 25$$
$$(x-3)^2 + [y-(-2)]^2 = 25$$

The circle has center (h, k) at $(3, -2)$ and radius $\sqrt{25} = 5$.

12.
$$x^2 + y^2 = 16$$
$$(x-0)^2 + (y-0)^2 = 4^2$$

This is a circle with center $(0, 0)$ and radius 4.

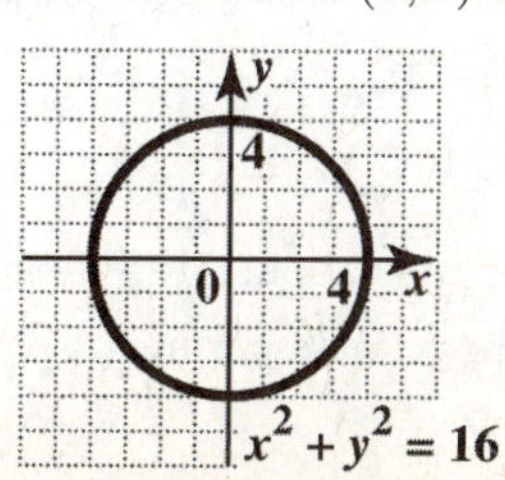

13. $\frac{x^2}{16} + \frac{y^2}{9} = 1$ is in $\frac{x^2}{a^2} + \frac{y^2}{b^2} = 1$ form with $a = 4$ and $b = 3$. The graph is an ellipse with x-intercepts $(4, 0)$ and $(-4, 0)$ and y-intercepts $(0, 3)$ and $(0, -3)$. Plot the intercepts, and draw the ellipse through them.

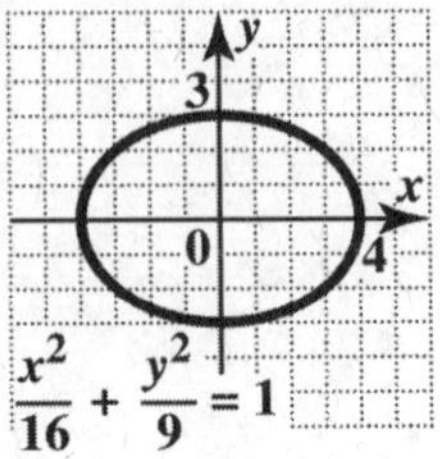

14. $\frac{x^2}{49} + \frac{y^2}{25} = 1$ is in $\frac{x^2}{a^2} + \frac{y^2}{b^2} = 1$ form with $a = 7$ and $b = 5$. The graph is an ellipse with x-intercepts $(7, 0)$ and $(-7, 0)$ and y-intercepts $(0, 5)$ and $(0, -5)$. Plot the intercepts, and draw the ellipse through them.

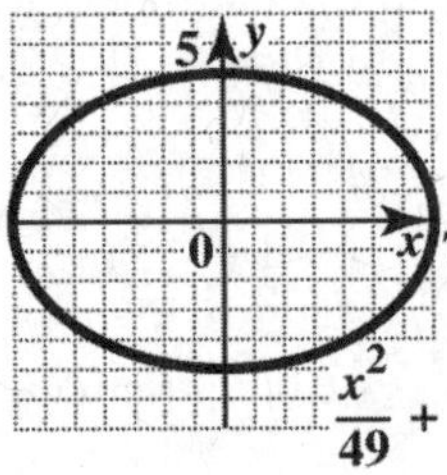

15. The total distance on the horizontal axis is $160 + 16{,}000 = 16{,}160$ km. This represents $2a$, so $a = \frac{1}{2}(16{,}160) = 8080$. The distance from Earth to the center of the ellipse is

$$8080 - 160 = 7920,$$

which is the value of c. From the hint, we know that $c^2 = a^2 - b^2$, so

$$b^2 = a^2 - c^2.$$
$$b^2 = 8080^2 - 7920^2$$
$$= 2{,}560{,}000$$

Thus, $b = \sqrt{2{,}560{,}000} = 1600$ and the equation is

$$\frac{x^2}{8080^2} + \frac{y^2}{1600^2} = 1$$

or $$\frac{x^2}{65{,}286{,}400} + \frac{y^2}{2{,}560{,}000} = 1.$$

16. **(a)** The distance between the foci is $2c$, where c can be found using the relationship

$$c = \sqrt{a^2 - b^2}.$$
$$= \sqrt{310^2 - (513/2)^2}$$
$$\approx 174.1 \text{ feet}$$

So the distance is about 348.2 feet.

(b) The approximate perimeter of the Roman Colosseum is

$$P \approx 2\pi\sqrt{\frac{a^2 + b^2}{2}}.$$
$$= 2\pi\sqrt{\frac{310^2 + (513/2)^2}{2}} \approx 1787.6 \text{ feet}$$

17. $\frac{x^2}{16} - \frac{y^2}{25} = 1$ is in $\frac{x^2}{a^2} - \frac{y^2}{b^2} = 1$ form with $a = 4$ and $b = 5$. The graph is a hyperbola with x-intercepts $(4, 0)$ and $(-4, 0)$ and asymptotes that are the extended diagonals of the rectangle with vertices $(4, 5)$, $(4, -5)$, $(-4, -5)$, and $(-4, 5)$. Graph a branch of the hyperbola through each intercept approaching the asymptotes.

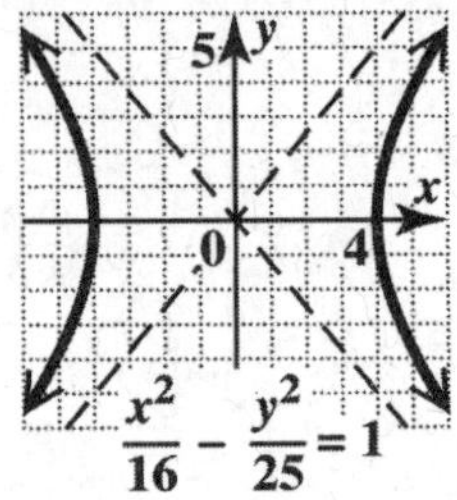

18. $\frac{y^2}{25} - \frac{x^2}{4} = 1$ is in $\frac{y^2}{b^2} - \frac{x^2}{a^2} = 1$ form with $a = 2$ and $b = 5$. The graph is a hyperbola with y-intercepts $(0, 5)$ and $(0, -5)$ and asymptotes that are the extended diagonals of the rectangle with vertices $(2, 5)$, $(2, -5)$, $(-2, -5)$, and $(-2, 5)$. Graph a branch of the hyperbola through each intercept approaching the asymptotes.

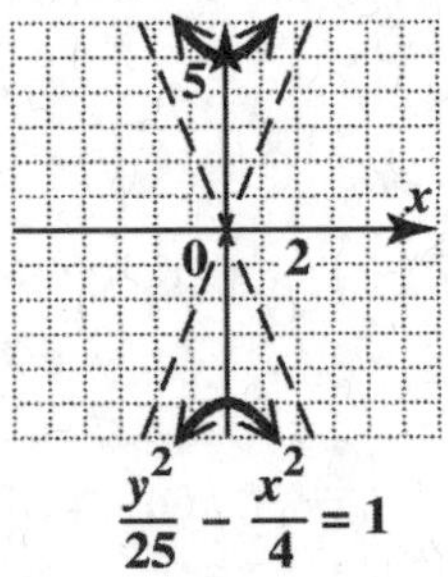

19. $$f(x) = -\sqrt{16 - x^2}$$

Replace $f(x)$ with y.

$$y = -\sqrt{16 - x^2}$$

Square both sides.

$$y^2 = 16 - x^2$$
$$x^2 + y^2 = 16$$

This equation is the graph of a circle with center $(0, 0)$ and radius 4. Since $f(x)$ represents a nonpositive square root, $f(x)$ is nonpositive and its graph is the lower half of the circle.

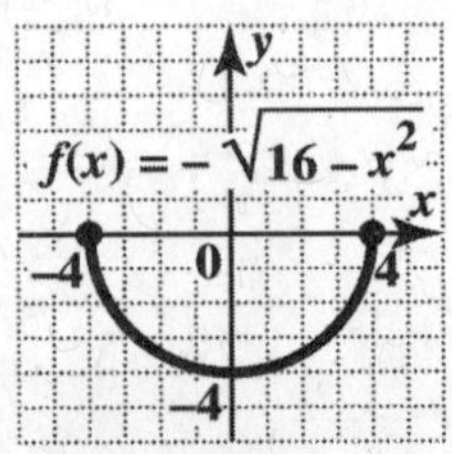

20.
$$x^2 + y^2 = 64$$
$$(x - 0)^2 + (y - 0)^2 = 8^2$$

The last equation is in $(x - h)^2 + (y - k)^2 = r^2$ form. The graph is a *circle*.

21. $y = 2x^2 - 3$
$$y = 2(x - 0)^2 - 3$$

The last equation is in $y = a(x - h)^2 + k$ form. The graph is a *parabola*.

22.
$$y^2 = 2x^2 - 8$$
$$2x^2 - y^2 = 8$$
$$\frac{x^2}{4} - \frac{y^2}{8} = 1$$

The last equation is in $\frac{x^2}{a^2} - \frac{y^2}{b^2} = 1$ form, so the graph is a *hyperbola*.

23.
$$y^2 = 8 - 2x^2$$
$$2x^2 + y^2 = 8$$
$$\frac{x^2}{4} + \frac{y^2}{8} = 1$$

The last equation is in $\frac{x^2}{a^2} + \frac{y^2}{b^2} = 1$ form, so the graph is an *ellipse*.

24. $x = y^2 + 4$
$$x = (y - 0)^2 + 4$$

The last equation is in $x = a(y - k)^2 + h$ form, so the graph is a *parabola*.

25. $x^2 - y^2 = 64$
$$\frac{x^2}{64} - \frac{y^2}{64} = 1$$

The last equation is in $\frac{x^2}{a^2} - \frac{y^2}{b^2} = 1$ form, so the graph is a *hyperbola*.

26. A hyperbola is defined as the set of all points in a plane such that the absolute value of the *difference* of the distances from two fixed points (called *foci*) is constant. The hyperbola shown in the text will have an equation of the form

$$\frac{x^2}{a^2} - \frac{y^2}{b^2} = 1. \quad (1)$$

The constant difference for this hyperbola is

$$|d_1 - d_2| = |80 - 30| = 50.$$

Let Q be the point on the right branch of the hyperbola that is on $\overline{MS}$. Let $w = \overline{MQ}$ and $z = \overline{QS}$. Since $\overline{MS} = 100$ and $\overline{MQ} - \overline{QS}$ must be 50 (since 50 is the constant difference for this hyperbola), we have the following system.

$$\begin{array}{rcll} w + z &=& 100 & (2) \\ w - z &=& 50 & \\ \hline 2w &=& 150 & \textit{Add.} \\ w &=& 75 & \end{array}$$

From (2), we see that $z = 25$. The distance from the center of the hyperbola, *C*, to *S*, is 50, so $a = \overline{CQ} = 50 - 25 = 25$.
To find b^2, we'll find a point on the hyperbola, substitute for a, x, and y, and then solve for b^2. Let *P* have coordinates (d, e). If we draw a perpendicular line from *P* to $\overline{MS}$, and call the point of intersection R, we see two right triangles, PRM and PRS. Using the Pythagorean formula, we get the following system of equations.

$$\begin{array}{rcll} (50 + d)^2 + e^2 &=& 80^2 & \\ (50 - d)^2 + e^2 &=& 30^2 & (3) \\ \hline \left(2500 + 100d + d^2\right) - \left(2500 - 100d + d^2\right) && & \\ &=& 6400 - 900 & \textit{Subtract.} \\ 200d &=& 5500 & \\ d &=& 27.5 & \end{array}$$

From (3) with $d = 27.5$, we get $e^2 = 30^2 - 22.5^2$, so $e = \sqrt{393.75}$. Now substitute 25 for a, 27.5 for x, and $\sqrt{393.75}$ for y in (1) to solve for b^2.

$$\frac{27.5^2}{25^2} - \frac{393.75}{b^2} = 1$$
$$\frac{756.25}{625} - \frac{625}{625} = \frac{393.75}{b^2}$$
$$\frac{131.25}{625} = \frac{393.75}{b^2}$$
$$\frac{625}{131.25} = \frac{b^2}{393.75}$$
$$b^2 = \frac{625}{131.25}(393.75)$$
$$b^2 = 1875$$

Thus, equation (1) becomes

$$\frac{x^2}{625} - \frac{y^2}{1875} = 1.$$

27.
$$\begin{array}{rll} 2y = 3x - x^2 & (1) \\ x + 2y = -12 & (2) \end{array}$$

Substitute $3x - x^2$ for $2y$ in equation (2).

$$x + 2y = -12 \quad (2)$$
$$x + \left(3x - x^2\right) = -12$$
$$-x^2 + 4x + 12 = 0$$
$$x^2 - 4x - 12 = 0$$
$$(x - 6)(x + 2) = 0$$
$$x - 6 = 0 \quad \text{or} \quad x + 2 = 0$$
$$x = 6 \quad \text{or} \quad x = -2$$

Substitute these values for x in equation (2) to find y.

If $x = 6$, then

$$\begin{aligned} x + 2y &= -12 \quad (2) \\ 6 + 2y &= -12 \\ 2y &= -18 \\ y &= -9. \end{aligned}$$

If $x = -2$, then

$$\begin{aligned} x + 2y &= -12 \quad (2) \\ -2 + 2y &= -12 \\ 2y &= -10 \\ y &= -5. \end{aligned}$$

The solution set is $\{(6, -9), (-2, -5)\}$.

28. $y + 1 = x^2 + 2x \quad (1)$
$y + 2x = 4 \quad (2)$

Solve equation (2) for y.

$$y = 4 - 2x \quad (3)$$

Substitute $4 - 2x$ for y in (1).

$$\begin{aligned} (4 - 2x) + 1 &= x^2 + 2x \\ 0 &= x^2 + 4x - 5 \\ 0 &= (x + 5)(x - 1) \end{aligned}$$

$$x + 5 = 0 \quad \text{or} \quad x - 1 = 0$$
$$x = -5 \quad \text{or} \quad x = 1$$

Substitute these values for x in equation (3) to find y.

If $x = -5$, then $y = 4 - 2(-5) = 14$.
If $x = 1$, then $y = 4 - 2(1) = 2$.

The solution set is $\{(1, 2), (-5, 14)\}$.

29. $x^2 + 3y^2 = 28 \quad (1)$
$y - x = -2 \quad (2)$

Solve equation (2) for y.

$$y = x - 2$$

Substitute $x - 2$ for y in equation (1).

$$\begin{aligned} x^2 + 3y^2 &= 28 \quad (1) \\ x^2 + 3(x - 2)^2 &= 28 \\ x^2 + 3(x^2 - 4x + 4) - 28 &= 0 \\ x^2 + 3x^2 - 12x + 12 - 28 &= 0 \\ 4x^2 - 12x - 16 &= 0 \\ 4(x^2 - 3x - 4) &= 0 \\ 4(x - 4)(x + 1) &= 0 \end{aligned}$$

$$x - 4 = 0 \quad \text{or} \quad x + 1 = 0$$
$$x = 4 \quad \text{or} \quad x = -1$$

Since $y = x - 2$, if $x = 4$, then $y = 4 - 2 = 2$.

If $x = -1$, then $y = -1 - 2 = -3$.

The solution set is $\{(4, 2), (-1, -3)\}$.

30. $xy = 8 \quad (1)$
$x - 2y = 6 \quad (2)$

Solve equation (2) for x.

$$x = 2y + 6 \quad (3)$$

Substitute $2y + 6$ for x in equation (1) to find y.

$$\begin{aligned} xy &= 8 \quad (1) \\ (2y + 6)y &= 8 \\ 2y^2 + 6y - 8 &= 0 \\ 2(y^2 + 3y - 4) &= 0 \\ 2(y + 4)(y - 1) &= 0 \end{aligned}$$

$$y + 4 = 0 \quad \text{or} \quad y - 1 = 0$$
$$y = -4 \quad \text{or} \quad y = 1$$

Substitute these values for y in equation (3) to find x.
If $y = -4$, then $x = 2(-4) + 6 = -2$.
If $y = 1$, then $x = 2(1) + 6 = 8$.

The solution set is $\{(-2, -4), (8, 1)\}$.

31. $x^2 + y^2 = 6 \quad (1)$
$x^2 - 2y^2 = -6 \quad (2)$

Multiply equation (2) by -1 and add the result to equation (1).

$$\begin{aligned} x^2 + y^2 &= 6 \quad (1) \\ -x^2 + 2y^2 &= 6 \quad -1 \times (2) \\ \hline 3y^2 &= 12 \\ y^2 &= 4 \end{aligned}$$

$$y = 2 \quad \text{or} \quad y = -2$$

Substitute these values for y in equation (1) to find x.
If $y = \pm 2$, then

$$\begin{aligned} x^2 + y^2 &= 6 \quad (1) \\ x^2 + (\pm 2)^2 &= 6 \\ x^2 + 4 &= 6 \\ x^2 &= 2. \end{aligned}$$

$$x = \sqrt{2} \quad \text{or} \quad x = -\sqrt{2}$$

Since each value of x can be paired with each value of y, there are four points and the solution set is
$\{(\sqrt{2}, 2), (-\sqrt{2}, 2), (\sqrt{2}, -2), (-\sqrt{2}, -2)\}$.

32. $3x^2 - 2y^2 = 12 \quad (1)$
$x^2 + 4y^2 = 18 \quad (2)$

Multiply equation (1) by 2 and add the result to equation (2).

$$\begin{aligned} 6x^2 - 4y^2 &= 24 \quad 2 \times (1) \\ x^2 + 4y^2 &= 18 \quad (2) \\ \hline 7x^2 &= 42 \\ x^2 &= 6 \end{aligned}$$

$$x = \sqrt{6} \quad \text{or} \quad x = -\sqrt{6}$$

Substitute these values for x in equation (2) to find y.
If $x = \pm\sqrt{6}$, then

$$\begin{aligned} x^2 + 4y^2 &= 18. \quad (2) \\ (\pm\sqrt{6})^2 + 4y^2 &= 18 \\ 6 + 4y^2 &= 18 \end{aligned}$$

$$\begin{aligned} 6 + 4y^2 &= 18 \\ 4y^2 &= 12 \\ y^2 &= 3 \end{aligned}$$

$$y = \sqrt{3} \quad \text{or} \quad y = -\sqrt{3}$$

The solution set is

$$\left\{(\sqrt{6}, \sqrt{3}), (\sqrt{6}, -\sqrt{3}), (-\sqrt{6}, \sqrt{3}), (-\sqrt{6}, -\sqrt{3})\right\}.$$

33. A circle and a line can intersect in zero, one, or two points, so zero, one, or two solutions are possible.

34. A parabola and a hyperbola can intersect in zero, one, two, three, or four points, so zero, one, two, three, or four solutions are possible.

35.
$$\begin{aligned} 9x^2 &\geq 16y^2 + 144 \\ 9x^2 - 16y^2 &\geq 144 \\ \frac{x^2}{16} - \frac{y^2}{9} &\geq 1 \end{aligned}$$

The boundary, $\frac{x^2}{16} - \frac{y^2}{9} = 1$, is a solid hyperbola with x-intercepts $(4, 0)$ and $(-4, 0)$. The asymptotes are the extended diagonals of the rectangle with vertices $(4, 3)$, $(4, -3)$, $(-4, -3)$, and $(-4, 3)$. Test $(0, 0)$.

$$\begin{aligned} 9(0)^2 &\stackrel{?}{\geq} 16(0)^2 + 144 \\ 0 &\geq 144 \qquad \textit{False} \end{aligned}$$

Shade the sides of the hyperbola that do not contain $(0, 0)$. These are the regions inside the branches of the hyperbola.

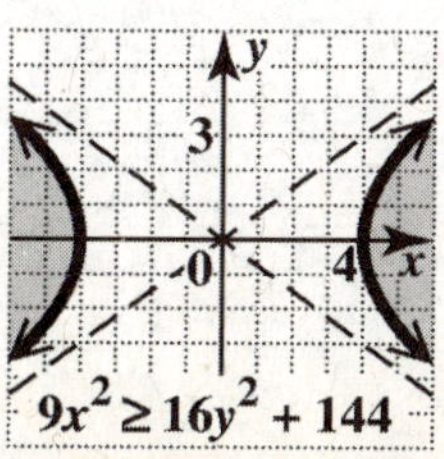

36. $4x^2 + y^2 \geq 16$

$$\frac{x^2}{4} + \frac{y^2}{16} \geq 1$$

The boundary, $\frac{x^2}{4} + \frac{y^2}{16} = 1$, is a solid ellipse with intercepts $(2, 0)$, $(-2, 0)$, $(0, 4)$, and $(0, -4)$. Test $(0, 0)$.

$$\begin{aligned} 4(0)^2 + 0^2 &\stackrel{?}{\geq} 16 \\ 0 &\geq 16 \qquad \textit{False} \end{aligned}$$

Shade the side of ellipse that does not contain $(0, 0)$. This is the region outside the ellipse.

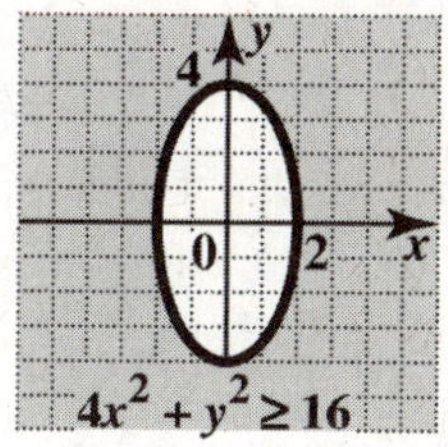

37. $y < -(x + 2)^2 + 1$

The boundary, $y = -(x + 2)^2 + 1$, is a dashed vertical parabola with vertex $(-2, 1)$. Since $a = -1 < 0$, the parabola opens down. Also, $|a| = |-1| = 1$, so the graph has the same shape as the graph of $y = x^2$. Test $(0, 0)$.

$$\begin{aligned} 0 &\stackrel{?}{<} -(0 + 2)^2 + 1 \\ 0 &\stackrel{?}{<} -(4) + 1 \\ 0 &< -3 \qquad \textit{False} \end{aligned}$$

Shade the side of the parabola that does not contain $(0, 0)$. This is the region inside the parabola.

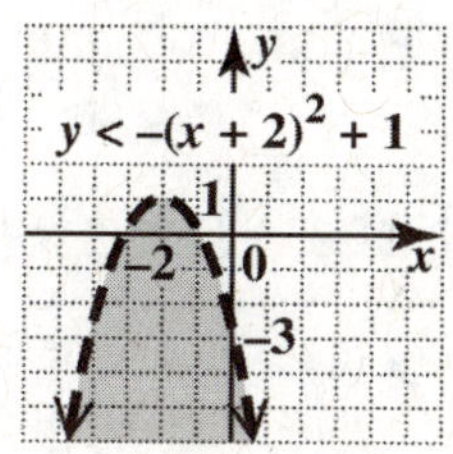

38.
$$\begin{aligned} 2x + 5y &\leq 10 \\ 3x - y &\leq 6 \end{aligned}$$

The boundary, $2x + 5y = 10$ is a solid line with intercepts $(5, 0)$ and $(0, 2)$. Test $(0, 0)$.

$$\begin{aligned} 2(0) + 5(0) &\stackrel{?}{\leq} 10 \\ 0 &\leq 10 \qquad \textit{True} \end{aligned}$$

Shade the side of the line that contains $(0, 0)$. The boundary, $3x - y = 6$, is a solid line with intercepts $(2, 0)$ and $(0, -6)$. Test $(0, 0)$.

$$\begin{aligned} 3(0) - 0 &\stackrel{?}{\leq} 6 \\ 0 &\leq 6 \qquad \textit{True} \end{aligned}$$

Shade the side of the line that contains $(0, 0)$.
The graph of the system is the intersection of the two shaded regions.

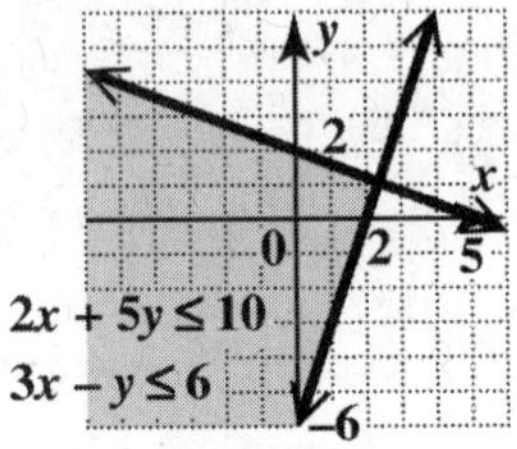

39.
$$|x| \le 2$$
$$|y| > 1$$
$$4x^2 + 9y^2 \le 36$$

The equation of the boundary, $|x| = 2$, can be written as

$$x = -2 \text{ or } x = 2.$$

The graph is these two solid vertical lines. Since $|0| \le 2$ is true, the region between the lines, containing $(0, 0)$, is shaded.
The boundary, $|y| = 1$, consists of the two dashed horizontal lines $y = 1$ and $y = -1$. Since $|0| > 1$ is false, the regions above $y = 1$ and below $y = -1$, not containing $(0, 0)$, are shaded.
The boundary given by

$$4x^2 + 9y^2 = 36$$
$$\text{or } \frac{x^2}{9} + \frac{y^2}{4} = 1$$

is graphed as a solid ellipse with intercepts $(3, 0)$, $(-3, 0)$, $(0, 2)$, and $(0, -2)$. Test $(0, 0)$.

$$4(0)^2 + 9(0)^2 \overset{?}{\le} 36$$
$$0 \le 36 \quad \textit{True}$$

The region inside the ellipse, containing $(0, 0)$, is shaded.
The graph of the system consists of the regions that include the common points of the three shaded regions.

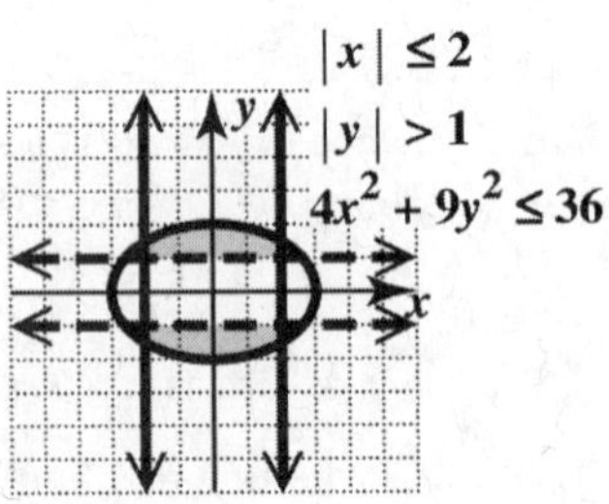

40.
$$9x^2 \le 4y^2 + 36$$
$$x^2 + y^2 \le 16$$

The equation of the first boundary is

$$9x^2 = 4y^2 + 36$$
$$9x^2 - 4y^2 = 36$$
$$\frac{x^2}{4} - \frac{y^2}{9} = 1.$$

The graph is a solid hyperbola with x-intercepts $(2, 0)$ and $(-2, 0)$. The asymptotes are the extended diagonals of the rectangle with vertices $(2, 3)$, $(2, -3)$, $(-2, -3)$, and $(-2, 3)$. Test $(0, 0)$.

$$9(0)^2 \overset{?}{\le} 4(0)^2 + 36$$
$$0 \le 36 \quad \textit{True}$$

Shade the region between the branches of the hyperbola that contains $(0, 0)$.
The equation of the second boundary is $x^2 + y^2 = 16$. This is a solid circle with center $(0, 0)$ and radius 4. Test $(0, 0)$.

$$0^2 + 0^2 \overset{?}{\le} 16$$
$$0 \le 16 \quad \textit{True}$$

Shade the region inside the circle.
The graph of the system is the intersection of the shaded regions which is between the two branches of the hyperbola and inside the circle.

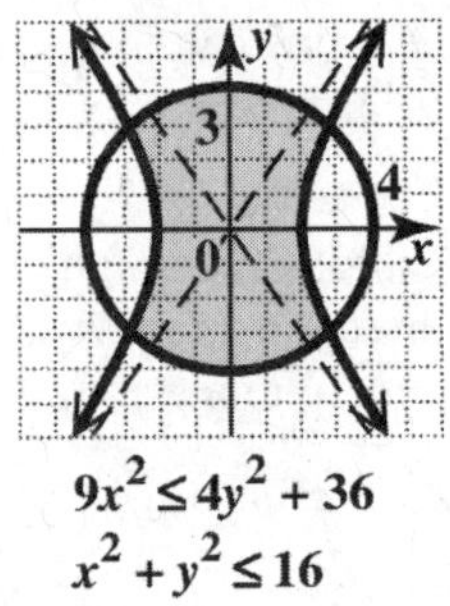

41. **[13.3]**
$$\frac{y^2}{4} - 1 = \frac{x^2}{9}$$
$$\frac{y^2}{4} - \frac{x^2}{9} = 1$$

The equation is in $\frac{y^2}{b^2} - \frac{x^2}{a^2} = 1$ form with $a = 3$ and $b = 2$. The graph is a hyperbola with y-intercepts $(0, 2)$ and $(0, -2)$ and asymptotes that are the extended diagonals of the rectangle with vertices $(3, 2)$, $(3, -2)$, $(-3, -2)$, and $(-3, 2)$. Draw a branch of the hyperbola through each intercept and approaching the asymptotes.

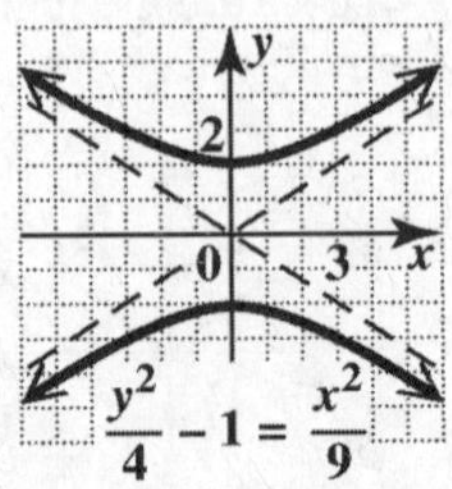

42. **[13.2]** $x^2 + y^2 = 25$ is in $(x-h)^2 + (y-k)^2 = r^2$ form. The graph is a circle with center at $(0,0)$ and radius 5.

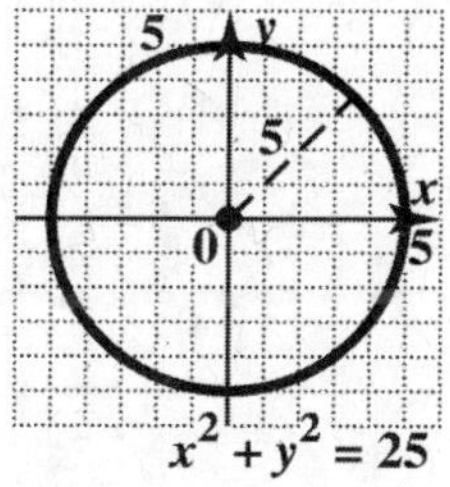

43. **[13.2]** $x^2 + 9y^2 = 9$ in $\frac{x^2}{a^2} + \frac{y^2}{b^2} = 1$ form is $\frac{x^2}{9} + \frac{y^2}{1} = 1$ with $a = 3$ and $b = 1$. The graph is an ellipse with x-intercepts $(3,0)$ and $(-3,0)$ and y-intercepts $(0,1)$ and $(0,-1)$. Plot the intercepts, and draw the ellipse through them.

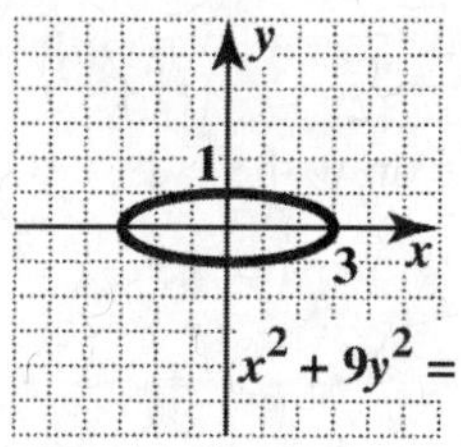

44. **[13.3]** $x^2 - 9y^2 = 9$

$$\frac{x^2}{9} - \frac{y^2}{1} = 1$$

The equation is in the $\frac{x^2}{a^2} - \frac{y^2}{b^2} = 1$ form with $a = 3$ and $b = 1$. The graph is a hyperbola with x-intercepts $(3,0)$ and $(-3,0)$ and asymptotes that are the extended diagonals of the rectangle with vertices $(3,1)$, $(3,-1)$, and $(-3,-1)$, and $(-3,1)$. Graph a branch of the hyperbola through each intercept and approaching the asymptotes.

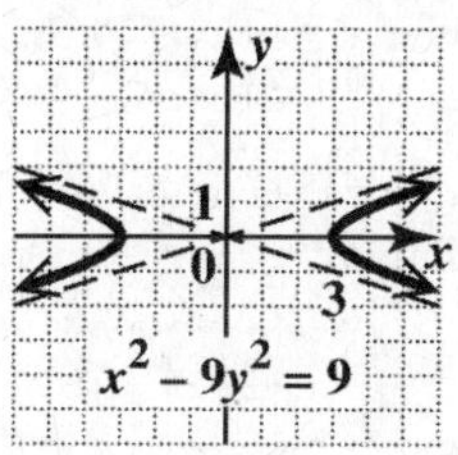

45. **[13.3]** $f(x) = \sqrt{4-x}$

Replace $f(x)$ with y.

$$y = \sqrt{4-x}$$

Square both sides.

$$y^2 = 4 - x$$
$$x = -y^2 + 4$$
$$x = -1(y-0)^2 + 4$$

This equation is the graph of a horizontal parabola with vertex $(4,0)$. Since $a = -1 < 0$, the graph opens to the left. Also, $|a| = |-1| = 1$, so the graph has the same shape as the graph of $x = y^2$. The points $(0,2)$ and $(3,1)$ are on the graph.

Since $f(x)$ represents a square root, $f(x)$ is nonnegative and its graph is the upper half of the parabola.

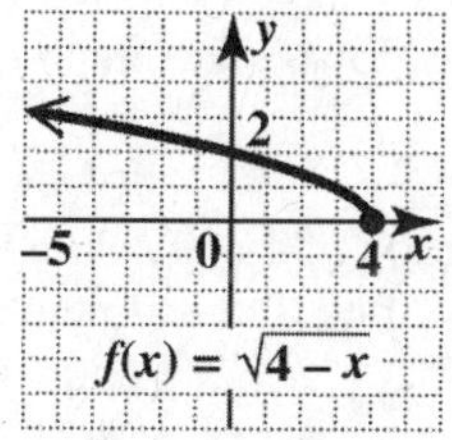

46. **[13.5]** $4y > 3x - 12$
$x^2 < 16 - y^2$

The boundary $4y = 3x - 12$ is a dashed line with intercepts $(0,-3)$ and $(4,0)$. Test $(0,0)$.

$$4(0) \overset{?}{>} 3(0) - 12$$
$$0 > -12 \qquad \textit{True}$$

Shade the side of the line that contains $(0,0)$. The boundary $x^2 = 16 - y^2$, or $x^2 + y^2 = 16$, is a dashed circle with center at $(0,0)$ and radius 4. Test $(0,0)$.

$$0^2 \overset{?}{<} 16 - 0^2$$
$$0 < 16 \qquad \textit{True}$$

Shade the region inside the circle.
The graph of the system is the intersection of the two shaded regions.

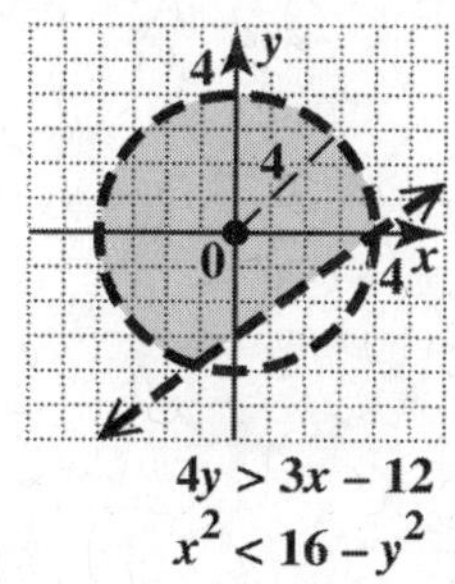

Chapter 13 Test

1. For the reciprocal function defined by $f(x) = 1/x$, $\underline{0}$ is the only real number not in the domain since division by 0 is undefined.

2. The range of the square root function, given by $f(x) = \sqrt{x}$, is $\underline{[0, \infty)}$ since the resulting values of the square root are positive or zero.

3. The range of $f(x) = [\![x]\!]$, the greatest integer function, is

$$\underline{\{\ldots, -2, -1, 0, 1, 2, \ldots\}}.$$

4. **(a)** $f(x) = \sqrt{x-2}$ is the graph of $g(x) = \sqrt{x}$ shifted 2 units right. The graph is **C**.

(b) $f(x) = \sqrt{x+2}$ is the graph of $g(x) = \sqrt{x}$ shifted 2 units left. The graph is **A**.

(c) $f(x) = \sqrt{x} + 2$ is the graph of $g(x) = \sqrt{x}$ shifted 2 units up. The graph is **D**.

(d) $f(x) = \sqrt{x} - 2$ is the graph of $g(x) = \sqrt{x}$ shifted 2 units down. The graph is **B**.

5. $f(x) = |x - 3| + 4$

This is the graph of the absolute value function, $g(x) = |x|$, shifted 3 units to the right (since $x - 3 = 0$ if $x = 3$) and 4 units upward (because of the $+4$). Its lowest point is at $(3, 4)$.

The function is defined for all values of x, so its domain is $(-\infty, \infty)$. The y-values are at least 4, so the range of f is $[4, \infty)$.

x	0	1	2	3	4	5	6
y	7	6	5	4	5	6	7

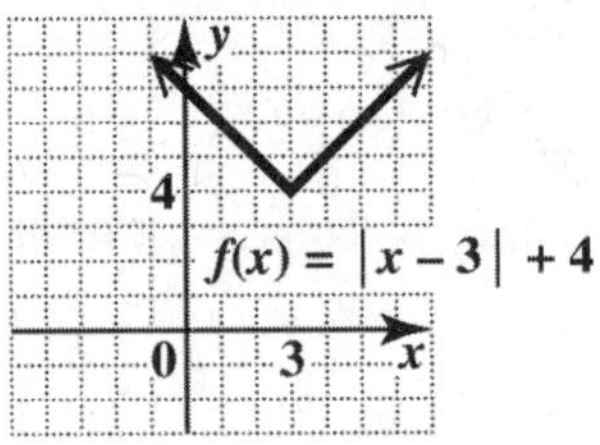

6.
$$(x - 2)^2 + (y + 3)^2 = 16$$
$$(x - 2)^2 + [y - (-3)]^2 = 4^2$$

The graph is a circle with center $(2, -3)$ and radius 4.

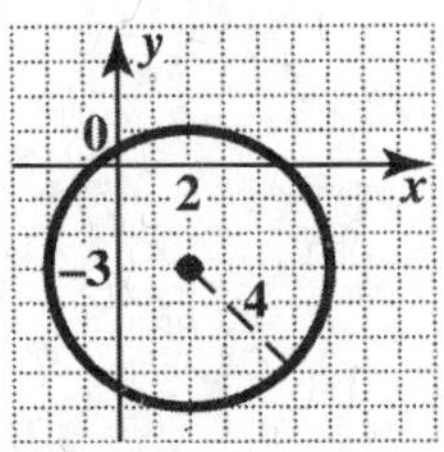

$(x - 2)^2 + (y + 3)^2 = 16$

7. $x^2 + y^2 + 8x - 2y = 8$

To find the center and radius, complete the squares on x and y.

$$(x^2 + 8x \qquad) + (y^2 - 2y \qquad) = 8$$
$$(x^2 + 8x + \underline{16}) + (y^2 - 2y + \underline{1}) = 8 + \underline{16} + \underline{1}$$
$$(x + 4)^2 + (y - 1)^2 = 25$$

The graph is a circle with center $(-4, 1)$ and radius $\sqrt{25} = 5$.

8.
$$f(x) = \sqrt{9 - x^2}$$

Replace $f(x)$ with y.

$$y = \sqrt{9 - x^2}$$

Square both sides.

$$y^2 = 9 - x^2$$
$$x^2 + y^2 = 9$$

The graph of $x^2 + y^2 = 9$ is a circle of radius $\sqrt{9} = 3$ centered at the origin. Since $f(x)$ is nonnegative, only the top half of the circle is graphed.

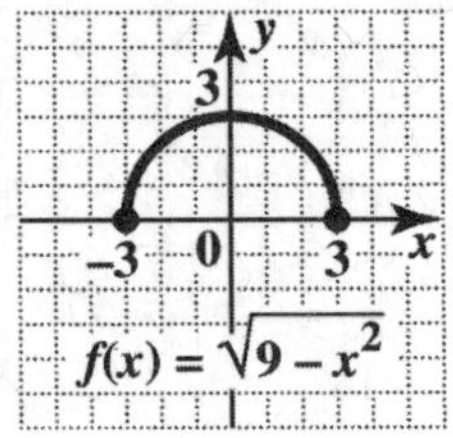

9. $4x^2 + 9y^2 = 36$

$$\frac{x^2}{9} + \frac{y^2}{4} = 1$$

The equation is in $\frac{x^2}{a^2} + \frac{y^2}{b^2} = 1$ form with $a = 3$ and $b = 2$. The graph is an ellipse with intercepts $(3, 0)$, $(-3, 0)$, $(0, 2)$, and $(0, -2)$. Plot these intercepts, and draw the ellipse through them.

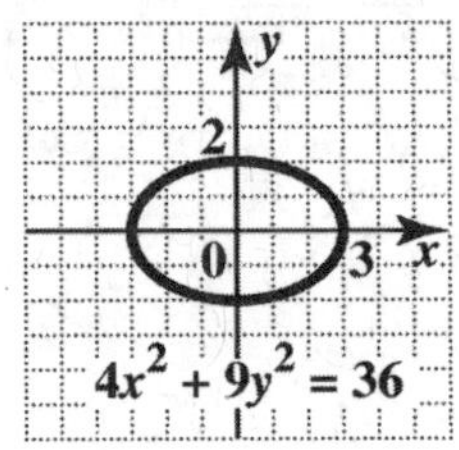

10. $16y^2 - 4x^2 = 64$

$$\frac{y^2}{4} - \frac{x^2}{16} = 1$$

The equation is in $\frac{y^2}{b^2} - \frac{x^2}{a^2} = 1$ form with $a = 4$ and $b = 2$. The graph is a hyperbola with y-intercepts $(0, 2)$ and $(0, -2)$ and asymptotes that are the extended diagonals of the rectangle with vertices $(4, 2)$, $(4, -2)$, $(-4, -2)$, and $(-4, 2)$.

Draw a branch of the hyperbola through each intercept and approaching the asymptotes.

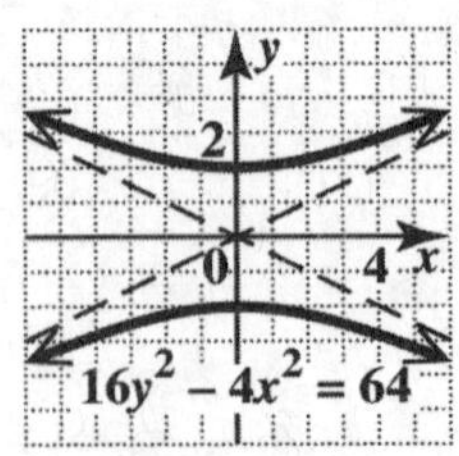

11.
$$\frac{y}{2} = -\sqrt{1 - \frac{x^2}{9}}$$

Square both sides.

$$\frac{y^2}{4} = 1 - \frac{x^2}{9}$$
$$\frac{x^2}{9} + \frac{y^2}{4} = 1$$

This is an ellipse with x-intercepts $(3, 0)$ and $(-3, 0)$ and y-intercepts $(0, 2)$ and $(0, -2)$. Since y represents a negative square root in the original equation, y must be nonpositive. This restricts the graph to the lower half of the ellipse.

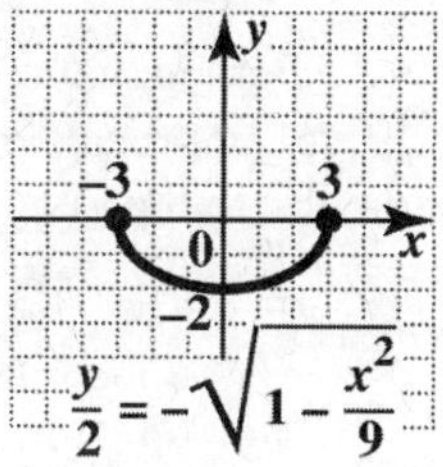

12. $6x^2 + 4y^2 = 12$

We have the *sum* of squares with different coefficients equal to a positive number, so this is an equation of an *ellipse*.

13.
$$16x^2 = 144 + 9y^2$$
$$16x^2 - 9y^2 = 144$$

We have the *difference* of squares equal to a positive number, so this is an equation of a *hyperbola*.

14.
$$y^2 = 20 - x^2$$
$$x^2 + y^2 = 20$$

We have the *sum* of squares with the same coefficients equal to a positive number, so this is an equation of an *circle*.

15.
$$4y^2 + 4x = 9$$
$$4x = -4y^2 + 9$$

We have an x-term and a y^2-term, so this is an equation of a horizontal *parabola*.

16.
$$2x - y = 9 \quad (1)$$
$$xy = 5 \quad (2)$$

Solve equation (1) for y.

$$y = 2x - 9 \quad (3)$$

Substitute $2x - 9$ for y in equation (2).

$$xy = 5 \quad (2)$$
$$x(2x - 9) = 5$$
$$2x^2 - 9x = 5$$
$$2x^2 - 9x - 5 = 0$$
$$(2x + 1)(x - 5) = 0$$

$$2x + 1 = 0 \quad \text{or} \quad x - 5 = 0$$
$$x = -\tfrac{1}{2} \quad \text{or} \quad x = 5$$

Substitute these values for x in equation (3) to find y.

If $x = -\frac{1}{2}$, then $y = 2\left(-\frac{1}{2}\right) - 9 = -10$.

If $x = 5$, then $y = 2(5) - 9 = 1$.

The solution set is $\left\{\left(-\frac{1}{2}, -10\right), (5, 1)\right\}$.

17.
$$x - 4 = 3y \quad (1)$$
$$x^2 + y^2 = 8 \quad (2)$$

Solve equation (1) for x.

$$x = 3y + 4$$

Substitute $3y + 4$ for x in equation (2).

$$x^2 + y^2 = 8 \quad (2)$$
$$(3y + 4)^2 + y^2 = 8$$
$$9y^2 + 24y + 16 + y^2 = 8$$
$$10y^2 + 24y + 8 = 0$$
$$2(5y^2 + 12y + 4) = 0$$
$$2(5y + 2)(y + 2) = 0$$

$$5y + 2 = 0 \quad \text{or} \quad y + 2 = 0$$
$$y = -\tfrac{2}{5} \quad \text{or} \quad y = -2$$

Since $x = 3y + 4$, substitute these values for y to find x.

If $y = -\frac{2}{5}$, then

$$x = 3\left(-\tfrac{2}{5}\right) + 4 = -\tfrac{6}{5} + 4 = \tfrac{14}{5}.$$

If $y = -2$, then $x = 3(-2) + 4 = -2$.

The solution set is $\left\{(-2, -2), \left(\frac{14}{5}, -\frac{2}{5}\right)\right\}$.

18.
$$x^2 + y^2 = 25 \quad (1)$$
$$x^2 - 2y^2 = 16 \quad (2)$$

Multiply equation (1) by 2 and add the result to equation (2).

$$\begin{array}{rcl} 2x^2 + 2y^2 &=& 50 \quad 2 \times (1) \\ x^2 - 2y^2 &=& 16 \quad (2) \\ \hline 3x^2 &=& 66 \\ x^2 &=& 22 \end{array}$$

$$x = \sqrt{22} \quad \text{or} \quad x = -\sqrt{22}$$

Substitute 22 for x^2 in equation (1).

$$x^2 + y^2 = 25 \quad (1)$$
$$22 + y^2 = 25$$
$$y^2 = 3$$

$$y = \sqrt{3} \quad \text{or} \quad y = -\sqrt{3}$$

The solution set is

$\left\{\left(\sqrt{22}, \sqrt{3}\right), \left(\sqrt{22}, -\sqrt{3}\right), \left(-\sqrt{22}, \sqrt{3}\right), \left(-\sqrt{22}, -\sqrt{3}\right)\right\}$.

19. $y < x^2 - 2$

The boundary, $y = x^2 - 2$, is a dashed parabola in $y = a(x-h)^2 + k$ form with vertex (h, k) at $(0, -2)$. Since $a = 1 > 0$, the parabola opens up. It also has the same shape as $y = x^2$. The points $(2, 2)$ and $(-2, 2)$ are on the graph. Test $(0, 0)$.

$$0 \overset{?}{\le} (0)^2 - 2$$
$$0 \le -2 \qquad \textit{False}$$

Shade the side of the parabola that does not contain $(0, 0)$. This is the region outside the parabola.

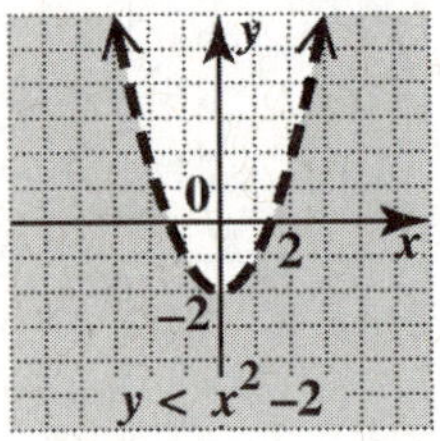

20. $x^2 + 25y^2 \le 25$
$x^2 + \quad y^2 \le 9$

The first boundary, $\frac{x^2}{25} + \frac{y^2}{1} = 1$, is a solid ellipse with intercepts $(5, 0)$, $(-5, 0)$, $(0, 1)$, and $(0, -1)$. Test $(0, 0)$.

$$0^2 + 25 \bullet 0^2 \overset{?}{\le} 25$$
$$0 \le 25 \qquad \textit{True}$$

Shade the region inside the ellipse.
The second boundary, $x^2 + y^2 = 9$, is a solid circle with center $(0, 0)$ and radius 3. Test $(0, 0)$.

$$0^2 + 0^2 \overset{?}{\le} 9$$
$$0 \le 9 \qquad \textit{True}$$

Shade the region inside the circle. The solution of the system is the intersection of the two shaded regions.

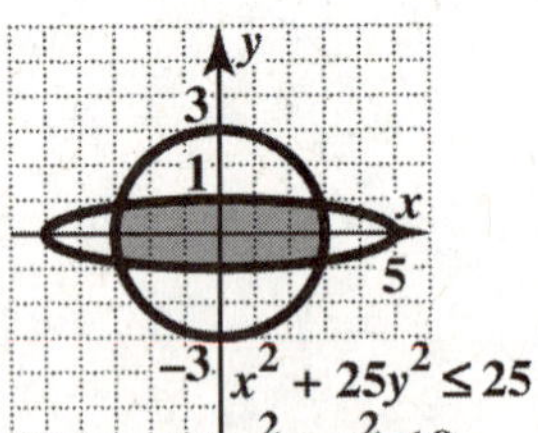

Cumulative Review Exercises (Chapters 1–13)

1. Let $(x_1, y_1) = (2, 5)$ and $(x_2, y_2) = (-4, 1)$.

$$m = \frac{y_2 - y_1}{x_2 - x_1} = \frac{1-5}{-4-2} = \frac{-4}{-6} = \frac{2}{3}$$

2. Through $(-3, -2)$; perpendicular to $2x - 3y = 7$

Write $2x - 3y = 7$ in slope-intercept form.

$$-3y = -2x + 7$$
$$y = \tfrac{2}{3}x - \tfrac{7}{3}$$

The slope is $\frac{2}{3}$. Perpendicular lines have slopes that are negative reciprocals of each other, so a line perpendicular to the given line will have slope $-\frac{3}{2}$. Let $m = -\frac{3}{2}$ and $(x_1, y_1) = (-3, -2)$ in the point-slope form.

$$y - y_1 = m(x - x_1)$$
$$y - (-2) = -\tfrac{3}{2}[x - (-3)]$$
$$y + 2 = -\tfrac{3}{2}(x + 3)$$

Multiply by 2 to clear the fraction.

$$2y + 4 = -3(x + 3)$$
$$2y + 4 = -3x - 9$$
$$3x + 2y = -13$$

3. $(5y - 3)^2 = (5y)^2 - 2(5y)3 + 3^2$
$= 25y^2 - 30y + 9$

4. $\dfrac{8x^4 - 4x^3 + 2x^2 + 13x + 8}{2x + 1}$

$$\begin{array}{r|l} & 4x^3 - 4x^2 + 3x + 5 \\ \hline 2x + 1 & 8x^4 - 4x^3 + 2x^2 + 13x + 8 \\ & 8x^4 + 4x^3 \\ & \quad -8x^3 + 2x^2 \\ & \quad -8x^3 - 4x^2 \\ & \qquad 6x^2 + 13x \\ & \qquad 6x^2 + 3x \\ & \qquad\quad 10x + 8 \\ & \qquad\quad 10x + 5 \\ & \qquad\qquad 3 \end{array}$$

The answer is

$$4x^3 - 4x^2 + 3x + 5 + \frac{3}{2x + 1}.$$

5. $12x^2 - 7x - 10$

Two integer factors whose product is $(12)(-10) = -120$ and whose sum is -7 are 8 and -15. Rewrite the trinomial in a form that can be factored by grouping.

$$12x^2 - 7x - 10$$
$$= 12x^2 + 8x - 15x - 10$$
$$= 4x(3x + 2) - 5(3x + 2)$$
$$= (3x + 2)(4x - 5)$$

6. $z^4 - 1 = (z^2 + 1)(z^2 - 1)$
$= (z^2 + 1)(z + 1)(z - 1)$

7. $a^3 - 27b^3 = a^3 - (3b)^3$
$= (a - 3b)(a^2 + 3ab + 9b^2)$

8. $\dfrac{y^2 - 4}{y^2 - y - 6} \div \dfrac{y^2 - 2y}{y - 1}$

Multiply by the reciprocal.

$= \dfrac{y^2 - 4}{y^2 - y - 6} \cdot \dfrac{y - 1}{y^2 - 2y}$

Factor and simplify.

$= \dfrac{(y+2)(y-2)}{(y-3)(y+2)} \cdot \dfrac{(y-1)}{y(y-2)}$

$= \dfrac{y-1}{y(y-3)}$

9. $\dfrac{5}{c+5} - \dfrac{2}{c+3}$

The LCD is $(c+5)(c+3)$.

$= \dfrac{5(c+3)}{(c+5)(c+3)} - \dfrac{2(c+5)}{(c+3)(c+5)}$

$= \dfrac{5c + 15 - 2c - 10}{(c+5)(c+3)}$

$= \dfrac{3c+5}{(c+5)(c+3)}$

10. $\dfrac{p}{p^2+p} + \dfrac{1}{p^2+p} = \dfrac{p+1}{p^2+p}$
$= \dfrac{p+1}{p(p+1)} = \dfrac{1}{p}$

11. Let $x =$ the time to do the job working together. Make a chart.

Worker	Rate	Time Together	Fractional Part of the Job Done
Henry	$\frac{1}{3}$	x	$\frac{x}{3}$
Lawrence	$\frac{1}{2}$	x	$\frac{x}{2}$

Part done by Henry	plus	part done by Lawrence	equals	1 whole job.
$\frac{x}{3}$	$+$	$\frac{x}{2}$	$=$	1

Multiply by the LCD, 6.

$$6\left(\frac{x}{3} + \frac{x}{2}\right) = 6 \cdot 1$$
$$2x + 3x = 6$$
$$5x = 6$$
$$x = \tfrac{6}{5} \text{ or } 1\tfrac{1}{5}$$

It takes $\frac{6}{5}$ or $1\frac{1}{5}$ hours to do the job together.

12.
$$\begin{aligned} 3x - y &= 12 \quad (1) \\ 2x + 3y &= -3 \quad (2) \end{aligned}$$

Multiply equation (1) by 3 and add the result to equation (2).

$$\begin{aligned} 9x - 3y &= 36 \quad 3 \times (1) \\ 2x + 3y &= -3 \quad (2) \\ \hline 11x &= 33 \\ x &= 3 \end{aligned}$$

Substitute 3 for x in equation (1) to find y.

$$\begin{aligned} 3x - y &= 12 \quad (1) \\ 3(3) - y &= 12 \\ 9 - y &= 12 \\ -y &= 3 \\ y &= -3 \end{aligned}$$

The solution set is $\{(3, -3)\}$.

13.
$$\begin{aligned} x + y - 2z &= 9 \quad (1) \\ 2x + y + z &= 7 \quad (2) \\ 3x - y - z &= 13 \quad (3) \end{aligned}$$

Add equation (2) and equation (3).

$$\begin{aligned} 2x + y + z &= 7 \quad (2) \\ 3x - y - z &= 13 \quad (3) \\ \hline 5x &= 20 \\ x &= 4 \end{aligned}$$

Multiply equation (1) by -1 and add the result to equation (2).

$$\begin{aligned} -x - y + 2z &= -9 \\ 2x + y + z &= 7 \quad (2) \\ \hline x + 3z &= -2 \quad (4) \end{aligned}$$

Substitute 4 for x in equation (4) to find z.

$$\begin{aligned} x + 3z &= -2 \quad (4) \\ 4 + 3z &= -2 \\ 3z &= -6 \\ z &= -2 \end{aligned}$$

Substitute 4 for x and -2 for z in equation (2) to find y.

$$\begin{aligned} 2x + y + z &= 7 \quad (2) \\ 2(4) + y - 2 &= 7 \\ y + 6 &= 7 \\ y &= 1 \end{aligned}$$

The solution set is $\{(4, 1, -2)\}$.

14.
$$\begin{aligned} xy &= -5 \quad (1) \\ 2x + y &= 3 \quad (2) \end{aligned}$$

Solve equation (2) for y.

$$y = -2x + 3 \quad (3)$$

Substitute $-2x+3$ for y in equation (1).

$$\begin{aligned} xy &= -5 \quad (1) \\ x(-2x+3) &= -5 \\ -2x^2 + 3x &= -5 \\ -2x^2 + 3x + 5 &= 0 \\ 2x^2 - 3x - 5 &= 0 \\ (2x-5)(x+1) &= 0 \end{aligned}$$

$$\begin{aligned} 2x - 5 = 0 \quad &\text{or} \quad x + 1 = 0 \\ x = \tfrac{5}{2} \quad &\text{or} \quad x = -1 \end{aligned}$$

Substitute these values for x in equation (3) to find y.
If $x = \frac{5}{2}$, then $y = -2\left(\frac{5}{2}\right) + 3 = -2$.

If $x = -1$, then $y = -2(-1) + 3 = 5$.

The solution set is $\left\{(-1, 5), \left(\frac{5}{2}, -2\right)\right\}$.

15. Let s = Al's rate, $2s$ = Bev's rate, t = Bev's time, and $t + \frac{1}{2}$ = Al's time.

	Distance	Rate	Time
Al	20	s	$t + \frac{1}{2}$
Bev	20	$2s$	t

Since $d = rt$, the system of equations is

$$\begin{aligned} s\left(t + \tfrac{1}{2}\right) &= 20 \quad (1) \\ 2st &= 20. \quad (2) \end{aligned}$$

Solve equation (2) for s.

$$\begin{aligned} 2st &= 20 \quad (2) \\ s &= \frac{20}{2t} = \frac{10}{t} \end{aligned}$$

Substitute $\frac{10}{t}$ for s in equation (1) to find t.

$$\begin{aligned} s\left(t + \tfrac{1}{2}\right) &= 20 \quad (1) \\ \frac{10}{t}\left(t + \frac{1}{2}\right) &= 20 \\ 10 + \frac{5}{t} &= 20 \end{aligned}$$

Multiply each term by the LCD, t.

$$\begin{aligned} 10t + 5 &= 20t \\ 5 &= 10t \\ t &= \tfrac{5}{10} = \tfrac{1}{2} \end{aligned}$$

Since $s = \frac{10}{t}$ and $t = \frac{1}{2}$,

$$s = \frac{10}{\frac{1}{2}} = 20.$$

So Al's rate was 20 mph and Bev's rate was $2 \cdot 20 = 40$ mph.

Another solution:

$$\begin{aligned} st + \tfrac{1}{2}s &= 20 \quad (1) \\ 2st &= 20 \quad (2) \end{aligned}$$

Multiply equation (1) by -2 and add the result to equation (2).

$$\begin{aligned} -2st - s &= -40 \quad -2 \times (1) \\ 2st &= 20 \quad (2) \\ \hline -s &= -20 \\ s &= 20 \end{aligned}$$

Substitute 20 for s in equation (2) to find $t = \frac{1}{2}$.

16.
$$\begin{aligned} \frac{(2a)^{-2}a^4}{a^{-3}} &= \frac{2^{-2}a^{-2}a^4}{a^{-3}} = \frac{2^{-2}a^2}{a^{-3}} \\ &= \frac{a^2a^3}{2^2} = \frac{a^5}{4} \end{aligned}$$

17.
$$\begin{aligned} 4\sqrt[3]{16} - 2\sqrt[3]{54} &= 4\sqrt[3]{8 \cdot 2} - 2\sqrt[3]{27 \cdot 2} \\ &= 4 \cdot 2\sqrt[3]{2} - 2 \cdot 3\sqrt[3]{2} \\ &= 8\sqrt[3]{2} - 6\sqrt[3]{2} = 2\sqrt[3]{2} \end{aligned}$$

18.
$$\begin{aligned} \frac{3\sqrt{5x}}{\sqrt{2x}} &= \frac{3\sqrt{5x} \cdot \sqrt{2x}}{\sqrt{2x} \cdot \sqrt{2x}} = \frac{3\sqrt{10x^2}}{2x} \\ &= \frac{3x\sqrt{10}}{2x} = \frac{3\sqrt{10}}{2} \end{aligned}$$

19. $\dfrac{5+3i}{2-i}$

Multiply the numerator and denominator by the conjugate of the denominator.

$$\begin{aligned} &= \frac{(5+3i)(2+i)}{(2-i)(2+i)} \\ &= \frac{10 + 5i + 6i + 3i^2}{4 - i^2} \\ &= \frac{10 + 11i + 3(-1)}{4 - (-1)} \\ &= \frac{7 + 11i}{5} = \frac{7}{5} + \frac{11}{5}i \end{aligned}$$

20.
$$\begin{aligned} 4 - (2x+3) + x &= 5x - 3 \\ 4 - 2x - 3 + x &= 5x - 3 \\ -x + 1 &= 5x - 3 \\ -6x &= -4 \\ x &= \tfrac{2}{3} \end{aligned}$$

The solution set is $\left\{\frac{2}{3}\right\}$.

21.
$$\begin{aligned} -4x + 7 &\geq 6x + 1 \\ -10x &\geq -6 \\ x &\leq \frac{-6}{-10} \quad \textit{Divide by } -10\textit{; reverse the inequality direction.} \\ x &\leq \tfrac{3}{5} \end{aligned}$$

The solution set is $\left(-\infty, \frac{3}{5}\right]$.

22. $|5x| - 6 = 14$

$$|5x| = 20$$

$$5x = 20 \quad \text{or} \quad 5x = -20$$

$$x = 4 \quad \text{or} \quad x = -4$$

The solution set is $\{-4, 4\}$.

23. $|2p - 5| > 15$

$$2p - 5 > 15 \quad \text{or} \quad 2p - 5 < -15$$

$$2p > 20 \qquad 2p < -10$$

$$p > 10 \quad \text{or} \quad p < -5$$

The solution set is $(-\infty, -5) \cup (10, \infty)$.

24. $2\sqrt{x} = \sqrt{5x + 3}$

$$4x = 5x + 3 \qquad \textit{Square each side.}$$

$$-x = 3$$

$$x = -3$$

Since x must be nonnegative so that $\sqrt{x}$ is a real number, -3 cannot be a solution. The solution set is $\emptyset$.

25. $10q^2 + 13q = 3$

$$10q^2 + 13q - 3 = 0$$

$$(5q - 1)(2q + 3) = 0$$

$$5q - 1 = 0 \quad \text{or} \quad 2q + 3 = 0$$

$$q = \tfrac{1}{5} \quad \text{or} \quad q = -\tfrac{3}{2}$$

The solution set is $\left\{\frac{1}{5}, -\frac{3}{2}\right\}$.

26. $3x^2 - 3x - 2 = 0$

Use the quadratic formula with $a = 3$, $b = -3$, and $c = -2$.

$$x = \frac{-b \pm \sqrt{b^2 - 4ac}}{2a}$$

$$x = \frac{-(-3) \pm \sqrt{(-3)^2 - 4(3)(-2)}}{2(3)}$$

$$= \frac{3 \pm \sqrt{9 + 24}}{6} = \frac{3 \pm \sqrt{33}}{6}$$

The solution set is $\left\{\frac{3 \pm \sqrt{33}}{6}\right\}$.

27. $2\left(x^2 - 3\right)^2 - 5\left(x^2 - 3\right) = 12$

Let $u = \left(x^2 - 3\right)$, so $u^2 = \left(x^2 - 3\right)^2$.

$$2u^2 - 5u = 12$$

$$2u^2 - 5u - 12 = 0$$

$$(2u + 3)(u - 4) = 0$$

$$2u + 3 = 0 \quad \text{or} \quad u - 4 = 0$$

$$u = -\tfrac{3}{2} \quad \text{or} \quad u = 4$$

Substitute $x^2 - 3$ for u to find x.

If $u = -\frac{3}{2}$, then

$$x^2 - 3 = -\tfrac{3}{2}$$

$$x^2 = \tfrac{3}{2}$$

$$x = \pm\sqrt{\frac{3}{2}} = \pm\frac{\sqrt{3}}{\sqrt{2}} \cdot \frac{\sqrt{2}}{\sqrt{2}} = \pm\frac{\sqrt{6}}{2}.$$

If $u = 4$, then

$$x^2 - 3 = 4$$

$$x^2 = 7$$

$$x = \pm\sqrt{7}.$$

The solution set is $\left\{\pm\frac{\sqrt{6}}{2}, \pm\sqrt{7}\right\}$.

28. $\log(x + 2) + \log(x - 1) = 1$

$$\log_{10}[(x + 2)(x - 1)] = 1$$

$$(x + 2)(x - 1) = 10^1$$

$$x^2 + x - 2 = 10$$

$$x^2 + x - 12 = 0$$

$$(x + 4)(x - 3) = 0$$

$$x + 4 = 0 \quad \text{or} \quad x - 3 = 0$$

$$x = -4 \quad \text{or} \quad x = 3$$

The original equation is undefined when $x = -4$.

The solution set is $\{3\}$.

29. Solve $F = \dfrac{kwv^2}{r}$ for v.

$$Fr = kwv^2$$

$$v^2 = \frac{Fr}{kw}$$

Take the square root of each side.

$$v = \pm\sqrt{\frac{Fr}{kw}} = \frac{\pm\sqrt{Fr}}{\sqrt{kw}} \cdot \frac{\sqrt{kw}}{\sqrt{kw}} = \frac{\pm\sqrt{Frkw}}{kw}$$

30. $f(x) = y = x^3 + 4$

Interchange x and y and solve for y.

$$x = y^3 + 4$$

$$y^3 = x - 4$$

$$y = \sqrt[3]{x - 4}$$

$$f^{-1}(x) = \sqrt[3]{x - 4}$$

31. **(a)** $a^{\log_a x} = x$, so $3^{\log_3 4} = 4$.

(b) $e^{\ln x} = x$, so $e^{\ln 7} = 7$.

32. $2\log(3x + 7) - \log 4 = \log(3x + 7)^2 - \log 4$

$$= \log\frac{(3x + 7)^2}{4}$$

33. **(a)** $y = 28.43(1.25)^x$

$$y = 28.43(1.25)^5 \qquad \textit{Let } x = 5.$$

$$\approx 86.8 \text{ billion dollars}$$

The sales are estimated to be \$86.8 billion in 2005.

(b) $y = 28.43(1.25)^8 \qquad \textit{Let } x = 8.$

$$\approx 169.5 \text{ billion dollars}$$

The sales are estimated to be \$169.5 billion in 2008.

34. $f(x) = |x - 3|$

The function is defined for any value of x, so the domain is $(-\infty, \infty)$. The result of the absolute value is nonnegative, so the range is $[0, \infty)$.

35. $f(x) = -3x + 5$

The equation is in slope-intercept form, so the y-intercept is $(0, 5)$ and $m = -3$ or $\frac{-3}{1}$.

Plot $(0, 5)$. From $(0, 5)$, move down 3 units and right 1 unit to the point $(1, 2)$. Draw the line through these two points.

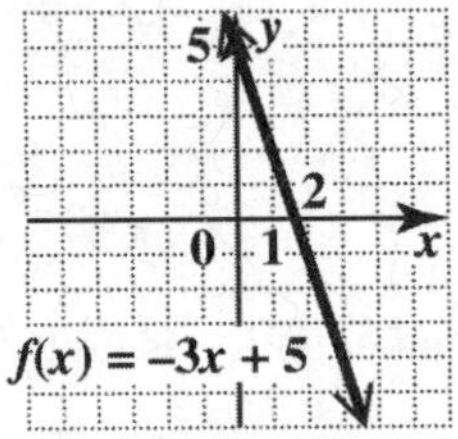

36. $f(x) = -2(x - 1)^2 + 3$

The graph is a parabola that has been shifted 1 unit to the right and 3 units upward from $(0, 0)$, so its vertex is at $(1, 3)$. Since $a = -2 < 0$, the parabola opens down. Also $|a| = |-2| = 2 > 1$, so the graph is narrower than the graph of $f(x) = x^2$. The points $(0, 1)$ and $(2, 1)$ are on the graph.

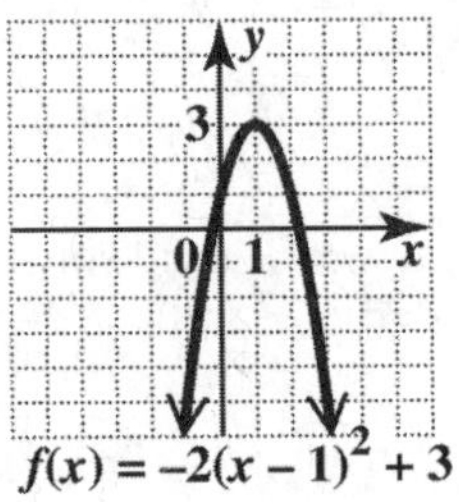

37. $\frac{x^2}{25} + \frac{y^2}{16} \le 1$

The boundary, $\frac{x^2}{25} + \frac{y^2}{16} = 1$, is a solid ellipse in $\frac{x^2}{a^2} + \frac{y^2}{b^2} = 1$ form with intercepts $(5, 0)$, $(-5, 0)$, $(0, 4)$, and $(0, -4)$. Test $(0, 0)$.

$$\frac{0^2}{25} + \frac{0^2}{16} \overset{?}{\le} 1$$
$$0 \le 1 \quad \textit{True}$$

Shade the region inside the ellipse.

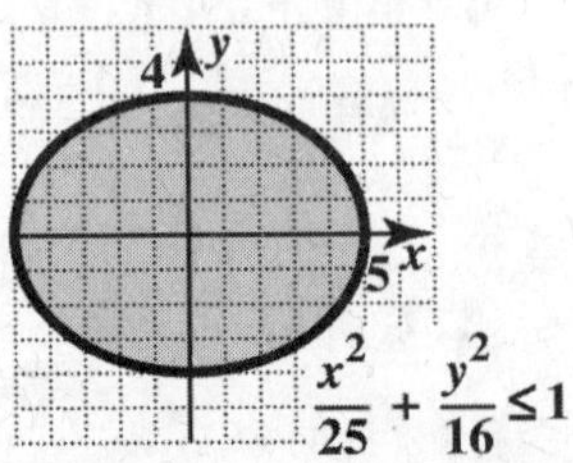

38. $f(x) = \sqrt{x - 2}$

This is the graph of $g(x) = \sqrt{x}$ shifted 2 units right.

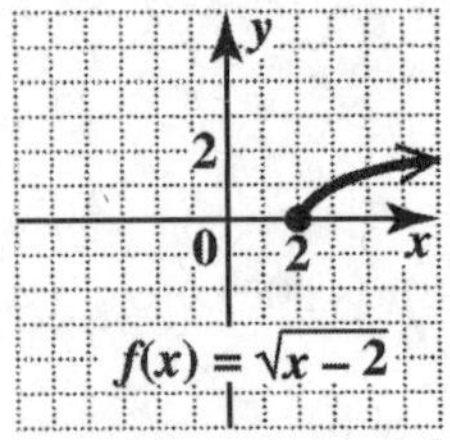

39. $\frac{x^2}{4} - \frac{y^2}{16} = 1$ is in $\frac{x^2}{a^2} - \frac{y^2}{b^2} = 1$ form.

The graph is a hyperbola with x-intercepts $(2, 0)$ and $(-2, 0)$ and asymptotes that are the extended diagonals of the rectangle with vertices $(2, 4)$, $(2, -4)$, $(-2, -4)$, and $(-2, 4)$. Draw a branch of the hyperbola through each intercept approaching the asymptotes.

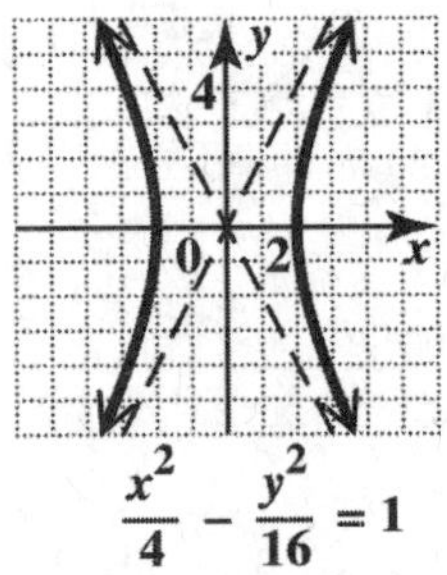

40. $f(x) = 3^x$

The graph of f is an increasing exponential.

x	-1	0	1	2
$f(x)$	$\frac{1}{3}$	1	3	9

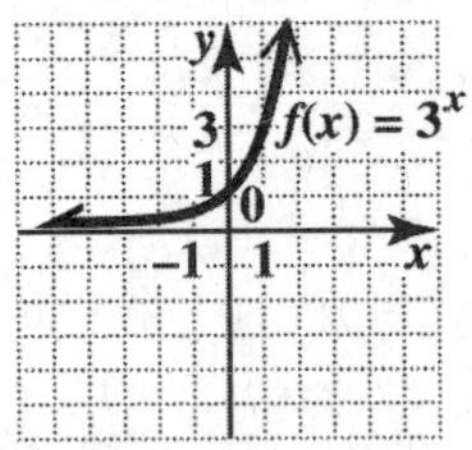

CHAPTER 14 SEQUENCES AND SERIES

14.1 Sequences and Series

14.1 Now Try Exercises

N1. $a_n = 5 - 3n$

To get a_3, the third term, replace n with 3.

$$a_3 = 5 - 3(3) = -4$$

N2. $-3, 9, -27, 81, \ldots$ can be written as

$(-3)^1, (-3)^2, (-3)^3, (-3)^4, \ldots$, so $a_n = (-3)^n$.

N3. Make a table as follows:

Month	Interest	Payment	Unpaid balance
0			8,000
1	$8000(0.02) = 160$	$400 + 160 = 560$	$8000 - 400 = 7600$
2	$7600(0.02) = 152$	$400 + 152 = 552$	$7600 - 400 = 7200$
3	$7200(0.02) = 144$	$400 + 144 = 544$	$7200 - 400 = 6800$
4	$6800(0.02) = 136$	$400 + 136 = 536$	$6800 - 400 = 6400$

The payments are \$560, \$552, \$544, and \$536; the unpaid balance is \$6400.

N4. $\sum_{i=1}^{5}(i^2 - 4)$

$$\begin{aligned}&= (1^2 - 4) + (2^2 - 4) + (3^2 - 4) \\ &\quad + (4^2 - 4) + (5^2 - 4) \\ &= (1 - 4) + (4 - 4) + (9 - 4) \\ &\quad + (16 - 4) + (25 - 4) \\ &= -3 + 0 + 5 + 12 + 21 \\ &= 35\end{aligned}$$

N5. (a) $3 + 5 + 7 + 9 + 11$

The terms increase by 2, so $2i$ is part of the general term. If $i = 1$, $2i = 2$, but since the first term is 3, we must add 1 to $2i$ to get 3. There are five terms, so

$$3 + 5 + 7 + 9 + 11 = \sum_{i=1}^{5}(2i + 1)$$

(b) $-1 - 4 - 9 - 16 - 25$

$$\begin{aligned}&= -1^2 - 2^2 - 3^2 - 4^2 - 5^2 \\ &= \sum_{i=1}^{5} -i^2\end{aligned}$$

N6. $\overline{x} = \dfrac{\sum_{i=1}^{n} x_i}{n} = \dfrac{\sum_{i=1}^{5} x_i}{5}$

$$= \frac{\begin{pmatrix}461{,}054 + 188{,}381 + 136{,}583 \\ + 107{,}630 + 93{,}958\end{pmatrix}}{5}$$

$$= \frac{987{,}606}{5} = 197{,}521.2$$

To the nearest whole number, the average number of Quarter Horses registered per state in the top five states is 197,521.

14.1 Section Exercises

1. $a_n = n + 1$

$a_1 = 1 + 1 = 2$ *Replace n with 1.*

$a_2 = 2 + 1 = 3$ *Replace n with 2.*

$a_3 = 3 + 1 = 4$ *Replace n with 3.*

$a_4 = 4 + 1 = 5$ *Replace n with 4.*

$a_5 = 5 + 1 = 6$ *Replace n with 5.*

Answer: 2, 3, 4, 5, 6

3. $a_n = \dfrac{n + 3}{n}$

To get a_i, replace n with i, where $i = 1, 2, 3, 4, 5$.

$a_1 = \dfrac{1 + 3}{1} = \dfrac{4}{1} = 4$ $a_2 = \dfrac{2 + 3}{2} = \dfrac{5}{2}$

$a_3 = \dfrac{3 + 3}{3} = \dfrac{6}{3} = 2$ $a_4 = \dfrac{4 + 3}{4} = \dfrac{7}{4}$

$a_5 = \dfrac{5 + 3}{5} = \dfrac{8}{5}$ **Answer:** $4, \dfrac{5}{2}, 2, \dfrac{7}{4}, \dfrac{8}{5}$

5. $a_n = 3^n$

To get a_i, replace n with i, where $i = 1, 2, 3, 4, 5$.

$a_1 = 3^1 = 3$ $a_2 = 3^2 = 9$

$a_3 = 3^3 = 27$ $a_4 = 3^4 = 81$

$a_5 = 3^5 = 243$ **Answer:** 3, 9, 27, 81, 243

7. $a_n = \dfrac{1}{n^2}$

To get a_i, replace n with i, where $i = 1, 2, 3, 4, 5$.

$a_1 = \dfrac{1}{1^2} = 1$ $a_2 = \dfrac{1}{2^2} = \dfrac{1}{4}$

$a_3 = \dfrac{1}{3^2} = \dfrac{1}{9}$ $a_4 = \dfrac{1}{4^2} = \dfrac{1}{16}$

$a_5 = \dfrac{1}{5^2} = \dfrac{1}{25}$ **Answer:** $1, \dfrac{1}{4}, \dfrac{1}{9}, \dfrac{1}{16}, \dfrac{1}{25}$

9. $a_n = 5(-1)^{n-1}$

$a_1 = 5(-1)^{1-1} = 5(-1)^0 = 5(1) = 5$

$a_2 = 5(-1)^{2-1} = 5(-1)^1 = 5(-1) = -5$

$a_3 = 5(-1)^{3-1} = 5(-1)^2 = 5(1) = 5$

$a_4 = 5(-1)^{4-1} = 5(-1)^3 = 5(-1) = -5$

$a_5 = 5(-1)^{5-1} = 5(-1)^4 = 5(1) = 5$

Answer: 5, −5, 5, −5, 5

11. $a_n = n - \frac{1}{n}$

$a_1 = 1 - \frac{1}{1} = 0 \qquad a_2 = 2 - \frac{1}{2} = \frac{3}{2}$

$a_3 = 3 - \frac{1}{3} = \frac{8}{3} \qquad a_4 = 4 - \frac{1}{4} = \frac{15}{4}$

$a_5 = 5 - \frac{1}{5} = \frac{24}{5}$ **Answer:** $0, \frac{3}{2}, \frac{8}{3}, \frac{15}{4}, \frac{24}{5}$

13. $a_n = -9n + 2$
To find a_8, replace n with 8.

$$a_8 = -9(8) + 2 = -72 + 2 = -70$$

15. $a_n = \frac{3n+7}{2n-5}$
To find a_{14}, replace n with 14.

$$a_{14} = \frac{3(14)+7}{2(14)-5} = \frac{42+7}{28-5} = \frac{49}{23}$$

17. $a_n = (n+1)(2n+3)$
To find a_8, replace n with 8.

$$\begin{aligned} a_8 &= (8+1)[2(8)+3] \\ &= 9(19) \\ &= 171 \end{aligned}$$

19. 4, 8, 12, 16, ... can be written as
$4 \cdot 1, 4 \cdot 2, 4 \cdot 3, 4 \cdot 4, \ldots$, so $a_n = 4n$.

21. $-8, -16, -24, -32, \ldots$ can be written as
$-8 \cdot 1, -8 \cdot 2, -8 \cdot 3, -8 \cdot 4, \ldots$, so $a_n = -8n$.

23. $\frac{1}{3}, \frac{1}{9}, \frac{1}{27}, \frac{1}{81}, \ldots$ can be written as
$\frac{1}{3^1}, \frac{1}{3^2}, \frac{1}{3^3}, \frac{1}{3^4}, \ldots$, so $a_n = \frac{1}{3^n}$.

25. $\frac{2}{5}, \frac{3}{6}, \frac{4}{7}, \frac{5}{8}, \ldots$ can be written as
$\frac{1+1}{1+4}, \frac{2+1}{2+4}, \frac{3+1}{3+4}, \frac{4+1}{4+4}, \ldots$, so $a_n = \frac{n+1}{n+4}$.

27. Make a table as follows:

Month	Interest	Payment	Unpaid balance
0			1000
1	1000(0.01) = 10	100 + 10 = 110	1000 − 100 = 900
2	900(0.01) = 9	100 + 9 = 109	900 − 100 = 800
3	800(0.01) = 8	100 + 8 = 108	800 − 100 = 700
4	700(0.01) = 7	100 + 7 = 107	700 − 100 = 600
5	600(0.01) = 6	100 + 6 = 106	600 − 100 = 500
6	500(0.01) = 5	100 + 5 = 105	500 − 100 = 400

The payments are \$110, \$109, \$108, \$107, \$106, and \$105; the unpaid balance is \$400.

29. When new, the car is worth \$20,000.
Let a_n = the value of the car after the nth year.
The car retains $\frac{4}{5}$ of its value each year.

$a_1 = \frac{4}{5}(20{,}000) = \$16{,}000$
(value after the first year)
$a_2 = \frac{4}{5}(16{,}000) = \$12{,}800$
$a_3 = \frac{4}{5}(12{,}800) = \$10{,}240$
$a_4 = \frac{4}{5}(10{,}240) = \$8192$
$a_5 = \frac{4}{5}(8192) = \$6553.60$

The value of the car after 5 years is about \$6554.

31. $\sum_{i=1}^{5}(i+3)$
$= (1+3) + (2+3) + (3+3) + (4+3) + (5+3)$
Let i = 1 to 5 and add terms.
$= 4 + 5 + 6 + 7 + 8$
$= 30$

33. $\sum_{i=1}^{3}(i^2+2)$
$= (1^2+2) + (2^2+2) + (3^2+2)$
Let i = 1 to 3 and add terms.
$= 3 + 6 + 11$
$= 20$

35. $\sum_{i=1}^{6}(-1)^i$
$= (-1)^1 + (-1)^2 + (-1)^3 + (-1)^4 + (-1)^5 + (-1)^6$
$= -1 + 1 - 1 + 1 - 1 + 1$
$= 0$

37. $\sum_{i=3}^{7}(i-3)(i+2)$
$= (3-3)(3+2) + (4-3)(4+2) + (5-3)(5+2) + (6-3)(6+2) + (7-3)(7+2)$
$= 0(5) + 1(6) + 2(7) + 3(8) + 4(9)$
$= 0 + 6 + 14 + 24 + 36$
$= 80$

39. $3 + 4 + 5 + 6 + 7$
$= (1+2) + (2+2) + (3+2) + (4+2) + (5+2)$
$= \sum_{i=1}^{5}(i+2)$

41. $-2 + 4 - 8 + 16 - 32$
$= (-1)^1 \cdot 2^1 + (-1)^2 \cdot 2^2 + (-1)^3 \cdot 2^3 + (-1)^4 \cdot 2^4 + (-1)^5 \cdot 2^5$
$= \sum_{i=1}^{5}(-1)^i \cdot 2^i$, or $\sum_{i=1}^{5}(-2)^i$

43. $1 + 4 + 9 + 16$
$= 1^2 + 2^2 + 3^2 + 4^2$
$= \sum_{i=1}^{4} i^2$

45. A sequence is a list of terms in a specific order, while a series is the indicated sum of the terms of a sequence.

47. $$\overline{x} = \frac{\sum_{i=1}^{n} x_i}{n} = \frac{\sum_{i=1}^{7} x_i}{7}$$
$$= \frac{8+11+14+9+7+6+8}{7} = \frac{63}{7} = 9$$

49. $$\overline{x} = \frac{5+9+8+2+4+7+3+2+0}{9} = \frac{40}{9}$$

51. $$\overline{x} = \frac{8041+7975+8117+8024+8022}{5}$$
$$= \frac{40{,}179}{5} = 8035.8$$

The average number of funds available for this five-year period was about 8036.

53. $$\begin{aligned} a + 3d &= 12 \quad (1) \\ a + 8d &= 22 \quad (2) \end{aligned}$$

Multiply (1) by -1 and add the result to (2).

$$\begin{aligned} -a - 3d &= -12 && -1 \times (1) \\ a + 8d &= 22 && (2) \\ \hline 5d &= 10 && \textit{Add.} \\ d &= 2 \end{aligned}$$

Substitute 2 for d in (1).

$$\begin{aligned} a + 3d &= 12 \quad (1) \\ a + 3(2) &= 12 \\ a + 6 &= 12 \\ a &= 6 \end{aligned}$$

Thus, $a = 6$ and $d = 2$.

55. Given $a = -2$, $n = 5$, and $d = 3$,

$$\begin{aligned} a + (n-1)d &= -2 + (5-1)3 \\ &= -2 + (4)3 \\ &= -2 + 12 \\ &= 10. \end{aligned}$$

14.2 Arithmetic Sequences

14.2 Now Try Exercises

N1. $-4, -13, -22, -31, -40, \ldots$

You should find the difference for all pairs of adjacent terms to determine if the sequence is arithmetic. In this case, we are *given* that the sequence is arithmetic, so d is the difference between any two adjacent terms. Choose the terms -13 and -4.

$$d = -13 - (-4) = -13 + 4 = -9$$

N2. Given $a_1 = 10$ and $d = -8$,

$$\begin{aligned} a_2 &= a_1 + d = 10 - 8 = 2 \\ a_3 &= a_2 + d = 2 - 8 = -6 \\ a_4 &= a_3 + d = -6 - 8 = -14 \\ a_5 &= a_4 + d = -14 - 8 = -22 \end{aligned}$$

The first five terms of the sequence are 10, 2, -6, -14, -22.

N3. $-5, 0, 5, 10, 15, \ldots$

To find d, subtract any two adjacent terms.

$$d = 5 - 0 = 5$$

The first term is $a_1 = -5$. Use it to find a_n.

$$\begin{aligned} a_n &= a_1 + (n-1)d \\ &= -5 + (n-1)(5) \\ &= -5 + 5n - 5 \\ &= 5n - 10 \end{aligned}$$

Now $a_{20} = 5(20) - 10 = 100 - 10 = 90$.

N4. After 1 month, the account will have

$$\$1000 + 1 \cdot \$120 = \$1120.$$

After 2 months, the account will have

$$\$1000 + 2 \cdot \$120 = \$1240.$$

In general, after n months, the account will have

$$\$1000 + n \cdot \$120.$$

Thus, after 96 months, the account will have

$$\$1000 + 96 \cdot \$120 = \$12{,}520.$$

N5. **(a)** Given $a_1 = 21$ and $d = -3$, find a_{22}.

$$\begin{aligned} a_n &= a_1 + (n-1)d \\ a_{22} &= a_1 + (22-1)d \\ &= 21 + 21(-3) \\ &= -42 \end{aligned}$$

(b) Given $a_7 = 25$ and $a_{12} = 40$, find a_{19}. Use $a_n = a_1 + (n-1)d$ to write a system of equations.

$$\begin{aligned} a_7 &= a_1 + (7-1)d \\ 25 &= a_1 + 6d && (1) \\ a_{12} &= a_1 + (12-1)d \\ 40 &= a_1 + 11d && (2) \end{aligned}$$

To eliminate a_1, multiply (1) by -1 and add the result to (2).

$$\begin{aligned} -25 &= -a_1 - 6d && -1 \times (1) \\ 40 &= a_1 + 11d && (2) \\ \hline 15 &= 5d && \textit{Add.} \\ 3 &= d \end{aligned}$$

From (1), $25 = a_1 + 6(3)$, so $a_1 = 7$. Now find a_{19}.

$$\begin{aligned} a_{19} &= a_1 + (19-1)d \\ &= 7 + 18(3) \\ &= 61 \end{aligned}$$

An alternative approach: There are 5 differences from a_7 to a_{12}, so

$$\begin{aligned} a_{12} - a_7 &= 5d. \\ 40 - 25 &= 5d \\ 15 &= 5d \\ 3 &= d \end{aligned}$$

There are 7 differences from a_{12} to a_{19}, so

$$\begin{aligned} a_{19} &= a_{12} + 7d. \\ &= 40 + 7(3) \\ &= 61 \end{aligned}$$

N6. $1, \frac{4}{3}, \frac{5}{3}, 2, \ldots, 11$

$a_n = a_1 + (n-1)d$	*Formula for a_n*
$11 = 1 + (n-1)(\frac{1}{3})$	$d = \frac{4}{3} - 1 = \frac{1}{3}$
$33 = 3 + (n-1)$	*Multiply by 3.*
$33 = n + 2$	*Combine terms.*
$31 = n$	*Subtract 2.*

The sequence has 31 terms.

N7. Since we want the sum of the first seven terms, we'll find a_1 and a_7 using $a_n = 5n - 7$.

$$\begin{aligned} a_1 &= 5(1) - 7 = -2 \\ a_7 &= 5(7) - 7 = 28 \end{aligned}$$

Now use the formula for the sum of the first n terms of an arithmetic sequence.

$S_n = \frac{n}{2}(a_1 + a_n)$	*Formula for S_n.*
$S_7 = \frac{7}{2}(a_1 + a_7)$	*Let $n = 7$.*
$S_7 = \frac{7}{2}(-2 + 28)$	*Substitute.*
$= \frac{7}{2}(26)$	*Add.*
$= 91$	*Multiply.*

N8. We are given $a_1 = -8$, $d = -5$, and $n = 9$. Use the second formula for the sum of an arithmetic sequence.

$$\begin{aligned} S_n &= \frac{n}{2}[2a_1 + (n-1)d] \\ S_9 &= \tfrac{9}{2}[2(-8) + (9-1)(-5)] \\ &= \tfrac{9}{2}[-16 + 8(-5)] \\ &= \tfrac{9}{2}[-16 - 40] \\ &= \tfrac{9}{2}(-56) = -252 \end{aligned}$$

N9. Evaluate $\sum_{i=1}^{11}(5i - 7)$.

To find the first and last (11th) terms, let $n = 1$ and $n = 11$ in $a_n = 5n - 7$.

$$\begin{aligned} a_1 &= 5(1) - 7 = -2 \\ a_{11} &= 5(11) - 7 = 48 \end{aligned}$$

Now find S_{11}.

$$\begin{aligned} S_n &= \frac{n}{2}(a_1 + a_n) \\ S_{11} &= \tfrac{11}{2}(-2 + 48) \\ &= \tfrac{11}{2}(46) = 253 \end{aligned}$$

14.2 Section Exercises

1. $1, 2, 3, 4, 5, \ldots$

d is the difference between any two adjacent terms. Choose the terms 3 and 2.

$$d = 3 - 2 = 1$$

The terms 2 and 1 would give

$$d = 2 - 1 = 1,$$

the same result. Therefore, the common difference is $d = 1$.

Note: You should find the difference for all pairs of adjacent terms to determine if the sequence is arithmetic.

3. $2, -4, 6, -8, 10, -12, \ldots$

The difference between the first two terms is $-4 - 2 = -6$, but the difference between the second and third terms is $6 - (-4) = 10$. The differences are not the same so the sequence is *not arithmetic*.

5. $10, 5, 0, -5, -10, \ldots$

Choose the terms 5 and 10, and find the difference.

$$d = 5 - 10 = -5$$

The terms -10 and -5 would give

$$d = -10 - (-5) = -10 + 5 = -5,$$

the same result. Therefore, the common difference is $d = -5$.

7. $a_1 = 5, d = 4$

$$\begin{aligned} a_2 &= a_1 + d = 5 + 4 = 9 \\ a_3 &= a_2 + d = 9 + 4 = 13 \\ a_4 &= a_3 + d = 13 + 4 = 17 \\ a_5 &= a_4 + d = 17 + 4 = 21 \end{aligned}$$

Answer: 5, 9, 13, 17, 21

9. $a_1 = -2, d = -4$

$$\begin{aligned} a_2 &= a_1 + d = -2 + (-4) = -6 \\ a_3 &= a_2 + d = -6 + (-4) = -10 \\ a_4 &= a_3 + d = -10 + (-4) = -14 \\ a_5 &= a_4 + d = -14 + (-4) = -18 \end{aligned}$$

Answer: $-2, -6, -10, -14, -18$

11. $a_1 = 2,\ d = 5$

$$\begin{aligned} a_n &= a_1 + (n-1)d \\ &= 2 + (n-1)5 \\ &= 2 + 5n - 5 \\ &= 5n - 3 \end{aligned}$$

13. $3, \frac{15}{4}, \frac{9}{2}, \frac{21}{4}, \ldots$

To find d, subtract any two adjacent terms.

$$d = \frac{15}{4} - 3 = \frac{15}{4} - \frac{12}{4} = \frac{3}{4}$$

The first term is $a_1 = 3$. Now find a_n.

$$\begin{aligned} a_n &= a_1 + (n-1)d \\ &= 3 + (n-1)\left(\tfrac{3}{4}\right) \\ &= 3 + \tfrac{3}{4}n - \tfrac{3}{4} \\ &= \tfrac{3}{4}n + \tfrac{9}{4} \end{aligned}$$

15. $-3, 0, 3, \ldots$

To find d, subtract any two adjacent terms.

$$d = 0 - (-3) = 3$$

The first term is $a_1 = -3$. Now find a_n.

$$\begin{aligned} a_n &= a_1 + (n-1)d \\ &= -3 + (n-1)3 \\ &= -3 + 3n - 3 \\ &= 3n - 6 \end{aligned}$$

17. Given $a_1 = 4$ and $d = 3$, find a_{25}.

$$\begin{aligned} a_n &= a_1 + (n-1)d \\ a_{25} &= 4 + (25-1)3 \\ &= 4 + 72 \\ &= 76 \end{aligned}$$

19. Given $2, 4, 6, \ldots,$ find a_{24}.

Here, $a_1 = 2$ and $d = 4 - 2 = 2$.

$$\begin{aligned} a_n &= a_1 + (n-1)d \\ a_{24} &= 2 + (24-1)2 \\ &= 2 + 46 \\ &= 48 \end{aligned}$$

21. Given $a_{12} = -45$ and $a_{10} = -37$, find a_1.

Use $a_n = a_1 + (n-1)d$ to write a system of equations.

$$\begin{aligned} a_{12} &= a_1 + (12-1)d \\ -45 &= a_1 + 11d \qquad (1) \\ a_{10} &= a_1 + (10-1)d \\ -37 &= a_1 + 9d \qquad (2) \end{aligned}$$

To eliminate d, multiply equation (1) by -9 and equation (2) by 11. Then add the results.

$$\begin{array}{rcrcrl} 405 &=& -9a_1 &-& 99d & \quad -9 \times (1) \\ -407 &=& 11a_1 &+& 99d & \quad 11 \times (2) \\ \hline -2 &=& 2a_1 & & & \\ -1 &=& a_1 & & & \end{array}$$

23. $3, 5, 7, \ldots, 33$

Let n represent the number of terms in the sequence. So, $a_n = 33,\ a_1 = 3$, and $d = 5 - 3 = 2$.

$$\begin{aligned} a_n &= a_1 + (n-1)d \\ 33 &= 3 + (n-1)2 \\ 33 &= 3 + 2n - 2 \\ 33 &= 2n + 1 \\ 32 &= 2n \\ n &= 16 \end{aligned}$$

The sequence has 16 terms.

25. $\frac{3}{4}, 3, \frac{21}{4}, \ldots, 12$

Let n represent the number of terms in the sequence. So, $a_n = 12,\ a_1 = \frac{3}{4}$, and $d = 3 - \frac{3}{4} = \frac{9}{4}$.

$$\begin{aligned} a_n &= a_1 + (n-1)d \\ 12 &= \tfrac{3}{4} + (n-1)\left(\tfrac{9}{4}\right) \\ \tfrac{45}{4} &= (n-1)\left(\tfrac{9}{4}\right) \\ 5 &= n - 1 \qquad \textit{Multiply by } \tfrac{4}{9}. \\ 6 &= n \end{aligned}$$

The sequence has 6 terms.

27. n represents the number of terms.

29. Find S_6 given $a_1 = 6,\ d = 3$, and $n = 6$.

$$\begin{aligned} S_n &= \tfrac{n}{2}\left[2a_1 + (n-1)d\right] \\ S_6 &= \tfrac{6}{2}\left[2 \cdot 6 + (6-1)3\right] \\ &= 3(12 + 5 \cdot 3) \\ &= 3(27) \\ &= 81 \end{aligned}$$

31. Find S_6 given $a_1 = 7,\ d = -3$, and $n = 6$.

$$\begin{aligned} S_n &= \tfrac{n}{2}\left[2a_1 + (n-1)d\right] \\ S_6 &= \tfrac{6}{2}\left[2 \cdot 7 + (6-1)(-3)\right] \\ &= 3\left[14 + 5(-3)\right] \\ &= 3(14 - 15) \\ &= 3(-1) \\ &= -3 \end{aligned}$$

33. Find S_6 given $a_n = 4 + 3n$.

Find the first and last terms.

$$\begin{aligned} a_1 &= 4 + 3(1) = 7 \\ a_6 &= 4 + 3(6) = 22 \end{aligned}$$

Now find the sum.

$$\begin{aligned} S_n &= \frac{n}{2}(a_1 + a_n) \\ S_6 &= \tfrac{6}{2}(7 + 22) \\ &= 3(29) \\ &= 87 \end{aligned}$$

35. $\sum_{i=1}^{10} (8i-5)$

$$a_n = 8n - 5$$
$$a_1 = 8(1) - 5 = 3$$
$$a_{10} = 8(10) - 5 = 75$$

Use $S_n = \frac{n}{2}(a_1 + a_n)$ with $n = 10$, $a_1 = 3$, and $a_{10} = 75$.

$$S_{10} = \tfrac{10}{2}(3 + 75) = 5(78) = 390$$

37. $\sum_{i=1}^{20} \left(\frac{3}{2}i + 4\right)$

$$a_n = \tfrac{3}{2}n + 4$$
$$a_1 = \tfrac{3}{2}(1) + 4 = \tfrac{11}{2}$$
$$a_{20} = \tfrac{3}{2}(20) + 4 = 34$$

Use $S_n = \frac{n}{2}(a_1 + a_n)$ with $n = 20$, $a_1 = \frac{11}{2}$, and $a_{20} = 34$.

$$S_{20} = \tfrac{20}{2}\left(\tfrac{11}{2} + 34\right) = \tfrac{20}{2}\left(\tfrac{79}{2}\right) = 395$$

39. $\sum_{i=1}^{250} i$ ▪ Here, $a_n = n$, $a_1 = 1$, and $a_{250} = 250$. Use $S_n = \frac{n}{2}(a_1 + a_n)$ with $n = 250$, $a_1 = 1$, and $a_{250} = 250$.

$$S_{250} = \tfrac{250}{2}(1 + 250) = 125(251) = 31{,}375$$

41. The sequence is 1, 2, 3, ..., 30.

$$S_n = \tfrac{n}{2}(a_1 + a_n)$$
$$S_{30} = \tfrac{30}{2}(1 + 30) = 15(31) = 465$$

The account will have \$465 deposited in it over the entire month.

43. Cherian's salary at six-month intervals form an arithmetic sequence with $a_1 = 1600$ and $d = 50$. Since Cherian's salary is increased every 6 months, or $2(5) = 10$ times, after 5 years it will equal the term a_{11}.

$$a_n = a_1 + (n-1)d$$
$$a_{11} = 1600 + (11-1)50 = 1600 + 500 = 2100$$

Cherian's salary will be \$2100/month.

45. Given the sequence 20, 22, 24, ..., for 25 terms, find a_{25}. Here, $a_1 = 20$, $d = 22 - 20 = 2$, and $n = 25$.

$$a_n = a_1 + (n-1)d$$
$$a_{25} = 20 + (25-1)2 = 20 + 48 = 68$$

There are 68 seats in the last row. Now find S_{25}.

$$S_n = \tfrac{n}{2}(a_1 + a_n)$$
$$S_{25} = \tfrac{25}{2}(20 + 68) = \tfrac{25}{2}(88) = 25(44) = 1100$$

There are 1100 seats in the section.

47. Given the sequence 35, 31, 27, ..., can the sequence end in 1? If not, find the last positive value. If the sequence ends in 1, we can find n, a whole number.

$$d = 31 - 35 = -4$$
$$a_n = a_1 + (n-1)d$$
$$1 = 35 + (n-1)(-4)$$
$$1 = 35 - 4n + 4$$
$$-38 = -4n$$
$$9.5 = n$$

Since n is not a whole number, the sequence cannot end in 1. The largest n possible is $n = 9$.

$$a_n = a_1 + (n-1)d$$
$$a_9 = 35 + (9-1)(-4) = 35 - 32 = 3$$

She can build 9 rows. There are 3 blocks in the last row.

49. $a = 2$, $r = 3$, $n = 2$

$$ar^n = 2(3)^2 = 2(9) = 18$$

51. $a = 4$, $r = \frac{1}{2}$, $n = 3$

$$ar^n = 4\left(\tfrac{1}{2}\right)^3 = 4\left(\tfrac{1}{8}\right) = \tfrac{1}{2}$$

14.3 Geometric Sequences

14.3 Now Try Exercises

N1. $\frac{1}{4}, -1, 4, -16, 64, \ldots$

You should find the ratio for all pairs of adjacent terms to determine if the sequence is geometric. In this case, we are *given* that the sequence is geometric, so to find r, choose any two adjacent terms and divide the second one by the first one. Choose the terms 4 and -16.

$$r = \tfrac{-16}{4} = -4$$

Notice that any other two adjacent terms could have been used with the same result. The common ratio is $r = -4$.

N2. $\frac{1}{4}, -1, 4, -16, 64, \ldots$

The common ratio is

$$r = \frac{-16}{4} = -4.$$

Substitute into the formula for a_n.

$$a_n = a_1 r^{n-1} = \tfrac{1}{4}(-4)^{n-1}$$

N3. **(a)** Given $a_1 = 3$ and $r = 2$, find a_8.

$$\begin{aligned} a_n &= a_1 r^{n-1} \\ a_8 &= 3(2)^{8-1} \\ &= 3(2)^7 = 3(128) = 384 \end{aligned}$$

(b) Given the geometric sequence $10, 2, \frac{2}{5}, \frac{2}{25}, \ldots$, find a_7.

The common ratio is

$$r = \tfrac{2}{10} = \tfrac{1}{5}.$$

Substitute into the formula for a_n.

$$\begin{aligned} a_n &= a_1 r^{n-1} \\ a_7 &= 10(\tfrac{1}{5})^{7-1} = \tfrac{10}{15{,}625} = \tfrac{2}{3125} \end{aligned}$$

N4.
$$\begin{aligned} a_1 &= 25 \\ a_2 &= a_1 r = 25(-\tfrac{1}{5}) = -5 \\ a_3 &= a_2 r = -5(-\tfrac{1}{5}) = 1 \\ a_4 &= a_3 r = 1(-\tfrac{1}{5}) = -\tfrac{1}{5} \\ a_5 &= a_4 r = -\tfrac{1}{5}(-\tfrac{1}{5}) = \tfrac{1}{25} \end{aligned}$$

N5. Given $a_1 = 4$ and $r = 2$, find S_6.

$$\begin{aligned} S_n &= \frac{a_1(1-r^n)}{1-r} \\ S_6 &= \frac{4(1-2^6)}{1-2} \\ &= \frac{4(1-64)}{-1} \\ &= 4(63) = 252 \end{aligned}$$

N6. For $\sum_{i=1}^{5} 8\left(\frac{1}{2}\right)^i$, $a_1 = 4$ and $r = \frac{1}{2}$. Find S_5.

$$\begin{aligned} S_n &= \frac{a_1(1-r^n)}{1-r} \\ S_5 &= \frac{4\left[1-\left(\frac{1}{2}\right)^5\right]}{1-\frac{1}{2}} \\ &= \frac{4\left(1-\frac{1}{32}\right)}{\frac{1}{2}} \\ &= 8\left(\tfrac{32}{32}-\tfrac{1}{32}\right) = 8\left(\tfrac{31}{32}\right) = \tfrac{31}{4} \end{aligned}$$

N7. **(a)** Use the formula for the future value of an ordinary annuity with $R = 600$, $i = 0.025$, and $n = 18$.

$$\begin{aligned} S &= R\left[\frac{(1+i)^n - 1}{i}\right] \\ S &= 600\left[\frac{(1+0.025)^{18} - 1}{0.025}\right] \\ &= 13{,}431.81 \end{aligned}$$

The future value of the annuity is \$13,431.81.

(b) Now use $r = \$100$, $i = \frac{0.03}{12} = 0.0025$, and $n = 12(18) = 216$.

$$\begin{aligned} S &= \$100\left[\frac{(1+0.0025)^{216} - 1}{0.0025}\right] \\ &= \$28{,}594.03 \end{aligned}$$

The future value of the annuity is \$28,594.03.

N8. Use the formula for the sum of the terms of an infinite geometric sequence with $a_1 = -4$ and $r = \frac{2}{3}$.

$$S = \frac{a_1}{1-r} = \frac{-4}{1-\frac{2}{3}} = \frac{-4}{\frac{1}{3}} = -12$$

N9. For $\sum_{i=1}^{\infty}\left(\frac{5}{8}\right)\left(\frac{3}{4}\right)^i$, $a_1 = \frac{15}{32}$ and $r = \frac{3}{4}$.

$$S = \frac{a_1}{1-r} = \frac{\frac{15}{32}}{1-\frac{3}{4}} = \frac{\frac{15}{32}}{\frac{1}{4}} = \frac{15}{8}$$

14.3 Section Exercises

1. $4, 8, 16, 32, \ldots$

To find r, choose any two adjacent terms and divide the second one by the first one.

$$r = \tfrac{8}{4} = 2$$

Notice that any two other adjacent terms could have been used with the same result. The common ratio is $r = 2$.

Note: You should find the ratio for all pairs of adjacent terms to determine if the sequence is geometric.

3. $\frac{1}{3}, \frac{2}{3}, \frac{3}{3}, \frac{4}{3}, \ldots$

Choose any two adjacent terms and divide the second by the first.

$$r = \frac{\frac{2}{3}}{\frac{1}{3}} = \frac{2}{3} \cdot 3 = 2$$

Confirm this result with any two other adjacent terms.

$$r = \frac{\frac{3}{3}}{\frac{2}{3}} = 1 \cdot \frac{3}{2} = \frac{3}{2}$$

Since $2 \neq \frac{3}{2}$, the ratios are not the same. The sequence is *not geometric*.

5. $1, -3, 9, -27, 81, \ldots$

[1st and 2nd terms] $r = \frac{-3}{1} = -3$
[2nd and 3rd terms] $r = \frac{9}{-3} = -3$

The common ratio is $r = -3$.

7. $1, -\frac{1}{2}, \frac{1}{4}, -\frac{1}{8}, \ldots$

$$\left[\frac{\text{2nd term}}{\text{1st term}}\right] \quad r = \frac{-\frac{1}{2}}{1} = -\frac{1}{2}$$

$$\left[\frac{\text{3rd term}}{\text{2nd term}}\right] \quad r = \frac{\frac{1}{4}}{-\frac{1}{2}} = \frac{1}{4}\left(-\frac{2}{1}\right) = -\frac{1}{2}$$

The common ratio is $r = -\frac{1}{2}$.

9. Find a general term for $-5, -10, -20, \ldots$.

First, find r. $r = \frac{-10}{-5} = 2$

Use $a_1 = -5$ and $r = 2$ to find a_n.

$$a_n = a_1 r^{n-1}$$
$$a_n = -5(2)^{n-1}$$

11. Find a general term for $-2, \frac{2}{3}, -\frac{2}{9}, \ldots$.

First, find r. $r = \frac{\frac{2}{3}}{-2} = \frac{2}{3}\left(-\frac{1}{2}\right) = -\frac{1}{3}$

Use $a_1 = -2$ and $r = -\frac{1}{3}$ to find a_n.

$$a_n = a_1 r^{n-1}$$
$$a_n = -2\left(-\tfrac{1}{3}\right)^{n-1}$$

13. Find a general term for $10, -2, \frac{2}{5}, \ldots$.

First, find r. $r = \frac{-2}{10} = -\frac{1}{5}$

Use $a_1 = 10$ and $r = -\frac{1}{5}$ to find a_n.

$$a_n = a_1 r^{n-1}$$
$$a_n = 10\left(-\tfrac{1}{5}\right)^{n-1}$$

15. Substitute $a_1 = 2$, $r = 5$, and $n = 10$ in the nth-term formula to find a_{10}.

$$a_n = a_1 r^{n-1}$$
$$a_{10} = a_1(r)^{10-1}$$
$$= 2(5)^9 = 3{,}906{,}250$$

17. Given $\frac{1}{2}, \frac{1}{6}, \frac{1}{18}, \ldots$, find a_{12}.

First, find r. $r = \frac{\frac{1}{6}}{\frac{1}{2}} = \frac{1}{6}\left(\frac{2}{1}\right) = \frac{1}{3}$

Substitute $a_1 = \frac{1}{2}$, $r = \frac{1}{3}$, and $n = 12$ in the nth-term formula.

$$a_n = a_1 r^{n-1}$$
$$a_{12} = a_1(r)^{12-1}$$
$$= \tfrac{1}{2}\left(\tfrac{1}{3}\right)^{11} \text{ or } \tfrac{1}{354{,}294}$$

19. Given $a_3 = \frac{1}{2}$ and $a_7 = \frac{1}{32}$, find a_{25}.

Find a_1 and r using the general term $a_n = a_1 r^{n-1}$.

$$a_3 = a_1 r^{3-1}$$
$$\tfrac{1}{2} = a_1 r^2 \qquad (1)$$
$$a_7 = a_1 r^{7-1}$$
$$\tfrac{1}{32} = a_1 r^6 \qquad (2)$$

Solve (1) for a_1.

$$a_1 = \frac{1}{2r^2}$$

Substitute $\frac{1}{2r^2}$ for a_1 in (2).

$$\frac{1}{32} = \frac{1}{2r^2} r^6$$
$$\tfrac{1}{16} = r^4$$
$$r^2 = \pm\tfrac{1}{4}$$

Since r^2 is positive,

$$r^2 = \tfrac{1}{4} \text{ and } r = \pm\tfrac{1}{2}.$$

Substitute $\frac{1}{4}$ for r^2 in (1).

$$\tfrac{1}{2} = a_1\left(\tfrac{1}{4}\right)$$
$$2 = a_1$$

Use $a_1 = 2$ and $r = \frac{1}{2}$ (or $-\frac{1}{2}$) to find a_{25}.

$$a_{25} = a_1(r)^{25-1}$$
$$= 2\left(\tfrac{1}{2}\right)^{24} = \frac{1}{2^{23}}$$

21. $a_1 = 2$, $r = 3$; use $a_n = a_1 r^{n-1}$.

$$a_2 = a_1 r^1 = 2(3) = 6$$
$$a_3 = a_1 r^2 = 2(3)^2 = 18$$
$$a_4 = a_1 r^3 = 2(3)^3 = 54$$
$$a_5 = a_1 r^4 = 2(3)^4 = 162$$

Answer: 2, 6, 18, 54, 162

23. $a_1 = 5$, $r = -\frac{1}{5}$; use $a_n = a_1 r^{n-1}$.

$$a_2 = a_1 r^1 = 5\left(-\tfrac{1}{5}\right) = -1$$
$$a_3 = a_1 r^2 = 5\left(-\tfrac{1}{5}\right)^2 = \tfrac{1}{5}$$
$$a_4 = a_1 r^3 = 5\left(-\tfrac{1}{5}\right)^3 = -\tfrac{1}{25}$$
$$a_5 = a_1 r^4 = 5\left(-\tfrac{1}{5}\right)^4 = \tfrac{1}{125}$$

Answer: $5, -1, \frac{1}{5}, -\frac{1}{25}, \frac{1}{125}$

25. $\frac{1}{3}, \frac{1}{9}, \frac{1}{27}, \frac{1}{81}, \frac{1}{243}$

$a_1 = \frac{1}{3}$, $n = 5$, and $r = \frac{\frac{1}{9}}{\frac{1}{3}} = \frac{1}{9} \cdot 3 = \frac{1}{3}$.

$$S_n = \frac{a_1(1 - r^n)}{1 - r}$$
$$S_5 = \frac{\frac{1}{3}\left[1 - \left(\frac{1}{3}\right)^5\right]}{1 - \frac{1}{3}}$$

$$= \frac{\frac{1}{3}\left(1 - \frac{1}{243}\right)}{\frac{2}{3}}$$

$$= \frac{\frac{1}{3}\left(\frac{242}{243}\right)}{\frac{2}{3}} = \frac{121}{243}$$

27. $-\frac{4}{3}, -\frac{4}{9}, -\frac{4}{27}, -\frac{4}{81}, -\frac{4}{243}, -\frac{4}{729}$

$a_1 = -\frac{4}{3}$, $n = 6$, and $r = \dfrac{-\frac{4}{9}}{-\frac{4}{3}} = -\dfrac{4}{9}\left(-\dfrac{3}{4}\right) = \dfrac{1}{3}$.

$$S_n = \frac{a_1(1 - r^n)}{1 - r}$$

$$S_6 = \frac{-\frac{4}{3}\left[1 - \left(\frac{1}{3}\right)^6\right]}{1 - \frac{1}{3}}$$

$$= \frac{-\frac{4}{3}\left(1 - \frac{1}{729}\right)}{\frac{2}{3}}$$

$$= \frac{-\frac{4}{3}\left(\frac{728}{729}\right)}{\frac{2}{3}} = -\frac{1456}{729} \approx -1.997$$

29. $\sum_{i=1}^{7} 4\left(\frac{2}{5}\right)^i$

Use $a_1 = 4\left(\frac{2}{5}\right) = \frac{8}{5}$, $n = 7$, and $r = \frac{2}{5}$.

$$S_n = \frac{a_1(1 - r^n)}{1 - r}$$

$$S_7 = \frac{\frac{8}{5}\left[1 - \left(\frac{2}{5}\right)^7\right]}{1 - \frac{2}{5}}$$

$$= \frac{\frac{8}{5}\left[1 - \left(\frac{2}{5}\right)^7\right]}{\frac{3}{5}}$$

$$= \frac{8}{3}\left[1 - \left(\frac{2}{5}\right)^7\right] \approx 2.662$$

31. $\sum_{i=1}^{10} (-2)\left(\frac{3}{5}\right)^i$

Use $a_1 = (-2)\left(\frac{3}{5}\right) = -\frac{6}{5}$, $n = 10$, and $r = \frac{3}{5}$.

$$S_n = \frac{a_1(1 - r^n)}{1 - r}$$

$$S_{10} = \frac{-\frac{6}{5}\left[1 - \left(\frac{3}{5}\right)^{10}\right]}{1 - \frac{3}{5}}$$

$$= \frac{-\frac{6}{5}\left[1 - \left(\frac{3}{5}\right)^{10}\right]}{\frac{2}{5}}$$

$$= -3\left[1 - \left(\frac{3}{5}\right)^{10}\right] \approx -2.982$$

33. There are 21 deposits, so $n = 21$.

$$S = R\left[\frac{(1 + i)^n - 1}{i}\right]$$

$$= 1000\left[\frac{(1 + 0.044)^{21} - 1}{0.044}\right]$$

$$= 33{,}410.84$$

There will be \$33,410.84 in the account.

35. Quarterly deposits for 10 years give us $n = 4 \cdot 10 = 40$. The interest rate per period is $i = \frac{0.05}{4} = 0.0125$.

$$S = R\left[\frac{(1 + i)^n - 1}{i}\right]$$

$$= 1200\left[\frac{(1 + 0.0125)^{40} - 1}{0.0125}\right]$$

$$= 61{,}787.47$$

We now use the compound interest formula to determine the value of this money after 5 more years.

$$A = P\left(1 + \frac{r}{n}\right)^{nt}$$

$$= 61{,}787.47\left(1 + \frac{0.06}{12}\right)^{12(5)}$$

$$= 83{,}342.04$$

The woman is also saving \$300 per month, so we use the annuity formula to determine that value.

$$S = R\left[\frac{(1 + i)^n - 1}{i}\right]$$

$$= 300\left[\frac{\left(1 + \frac{0.06}{12}\right)^{12(5)} - 1}{\frac{0.06}{12}}\right]$$

$$= 20{,}931.01$$

Adding \$20,931.01 to \$83,342.04 gives a total of \$104,273.05 in the account.

37. Find the sum if $a_1 = 6$ and $r = \frac{1}{3}$.
Since $|r| < 1$, the sum exists.

$$S = \frac{a_1}{1 - r} = \frac{6}{1 - \frac{1}{3}} = \frac{6}{\frac{2}{3}} = 6 \cdot \frac{3}{2} = 9$$

39. Find the sum if $a_1 = 1000$ and $r = -\frac{1}{10}$.
Since $|r| < 1$, the sum exists.

$$S = \frac{a_1}{1 - r} = \frac{1000}{1 - \left(-\frac{1}{10}\right)} = \frac{1000}{\frac{11}{10}}$$

$$= 1000 \cdot \frac{10}{11} = \frac{10{,}000}{11}$$

41. $\sum_{i=1}^{\infty} \frac{9}{8}\left(-\frac{2}{3}\right)^i$

$a_1 = \frac{9}{8}\left(-\frac{2}{3}\right)^1 = -\frac{3}{4}$ and $r = -\frac{2}{3}$.
Since $|r| < 1$, the sum exists.

$$S = \frac{a_1}{1 - r} = \frac{-\frac{3}{4}}{1 - \left(-\frac{2}{3}\right)} = \frac{-\frac{3}{4}}{\frac{5}{3}}$$

$$= -\frac{3}{4} \cdot \frac{3}{5} = -\frac{9}{20}$$

43. $\sum_{i=1}^{\infty} \frac{12}{5}\left(\frac{5}{4}\right)^i$

Since $|r| = \frac{5}{4} > 1$, the sum *does not exist.*

45. The ball is dropped from a height of 10 feet and will rebound $\frac{3}{5}$ of its original height.
Let $a_n =$ the ball's height on the nth rebound.

$$a_1 = 10 \quad \text{and} \quad r = \tfrac{3}{5}.$$

Since we must find the height after the fourth bounce, $n = 5$ (since a_1 is the starting point).
Use $a_n = a_1 r^{n-1}$.

$$a_5 = 10\left(\tfrac{3}{5}\right)^{5-1} = 10\left(\tfrac{3}{5}\right)^4 \approx 1.3$$

The ball will rebound approximately 1.3 feet after the fourth bounce.

47. This exercise can be modeled by a geometric sequence with $a_1 = 256$ and $r = \frac{1}{2}$. First we need to find n so that $a_n = 32$.

$$\begin{aligned} 32 &= a_1 r^{n-1} \\ 32 &= 256\left(\tfrac{1}{2}\right)^{n-1} \\ \tfrac{1}{8} &= \left(\tfrac{1}{2}\right)^{n-1} \\ \left(\tfrac{1}{2}\right)^3 &= \left(\tfrac{1}{2}\right)^{n-1} \\ 3 &= n-1 \\ 4 &= n \end{aligned}$$

Since n is 4, this means that 32 grams will be present on the day which corresponds to the 4th term of the sequence. That would be on day 3. To find what is left after the tenth day, we need to find a_{11} since we started with a_1.

$$a_{11} = a_1 r^{11-1} = 256\left(\tfrac{1}{2}\right)^{10} = \tfrac{256}{1024} = \tfrac{1}{4}$$

There will be $\frac{1}{4}$ gram of the substance after 10 days.

49. **(a)** Here, $a_1 = 1.1$ billion and $r = 106\% = 1.06$. Since we must find the consumption after 5 years, $n = 6$ (since a_1 is the starting point).

$$\begin{aligned} a_6 &= a_1 r^{n-1} \\ a_6 &= 1.1(1.06)^{6-1} \\ &= 1.1(1.06)^5 \approx 1.5 \end{aligned}$$

The community will use about 1.5 billion units 5 years from now.

(b) If consumption doubles, then the consumption would be $2a_1$.

$$\begin{aligned} 2a_1 &= a_1(1.06)^{n-1} \\ 2 &= (1.06)^{n-1} \\ \ln 2 &= \ln(1.06)^{n-1} \\ \ln 2 &= (n-1)\ln(1.06) \\ n-1 &= \frac{\ln 2}{\ln 1.06} \\ n-1 &\approx 12 \\ n &\approx 13 \end{aligned}$$

Since n is about 13, that would represent the 13th term of the sequence, which represents about 12 years after the start.

51. Since the machine depreciates by $\frac{1}{4}$ of its value, it retains $1 - \frac{1}{4} = \frac{3}{4}$ of its value. Since the cost of the machine new is \$50,000, $a_1 = 50{,}000$. We want the value after 8 years so since the original cost is a_1, we need to find a_9.

$$a_9 = a_1 r^{9-1} = 50{,}000\left(\tfrac{3}{4}\right)^8 \approx 5005.65$$

The machine's value after 8 years is about \$5005.65.

♦♦♦ Relating Concepts 53–58 ♦♦♦

53. $\frac{1}{3} = 0.33333\ldots$

54. $\frac{2}{3} = 0.66666\ldots$

55.
$$\begin{array}{r} 0.33333\ldots \\ +\,0.66666\ldots \\ \hline 0.99999\ldots \end{array}$$

56. $S = \dfrac{a_1}{1-r} = \dfrac{0.9}{1-0.1} = \dfrac{0.9}{0.9} = 1$

Therefore, $0.99999\ldots = 1$.

57. From Exercise 56, $0.99999\ldots = 1$, so choice **B** is correct.

58.
$$\begin{aligned} 0.49999\ldots &= 0.4 + 0.09999\ldots \\ &= \tfrac{4}{10} + \tfrac{1}{10}(0.9999\ldots) \\ &= \tfrac{4}{10} + \tfrac{1}{10}(1) \\ &= \tfrac{5}{10} = \tfrac{1}{2} \end{aligned}$$

59.
$$\begin{aligned} (3x+2y)^2 &= (3x)^2 + 2(3x)(2y) + (2y)^2 \\ &= 9x^2 + 12xy + 4y^2 \end{aligned}$$

61.
$$\begin{aligned} &(a-b)^3 \\ &= (a-b)^2(a-b)^1 \\ &= \left(a^2 - 2ab + b^2\right)(a-b) \\ &= a^3 - 2a^2b + ab^2 - a^2b + 2ab^2 - b^3 \\ &= a^3 - 3a^2b + 3ab^2 - b^3 \end{aligned}$$

14.4 The Binomial Theorem

14.4 Now Try Exercises

N1. $7! = 7 \cdot 6 \cdot 5 \cdot 4 \cdot 3 \cdot 2 \cdot 1 = 5040$

N2. **(a)**
$$\begin{aligned} \frac{8!}{6!\,2!} &= \frac{8 \cdot 7 \cdot 6 \cdot 5 \cdot 4 \cdot 3 \cdot 2 \cdot 1}{(6 \cdot 5 \cdot 4 \cdot 3 \cdot 2 \cdot 1)(2 \cdot 1)} \\ &= \frac{8 \cdot 7}{2 \cdot 1} = \frac{4 \cdot 7}{1} = 28 \end{aligned}$$

(b)
$$\begin{aligned} \frac{8!}{5!\,3!} &= \frac{8 \cdot 7 \cdot 6 \cdot 5 \cdot 4 \cdot 3 \cdot 2 \cdot 1}{(5 \cdot 4 \cdot 3 \cdot 2 \cdot 1)(3 \cdot 2 \cdot 1)} \\ &= \frac{8 \cdot 7 \cdot 6}{3 \cdot 2 \cdot 1} = \frac{8 \cdot 7}{1} = 56 \end{aligned}$$

(c) $\frac{6!}{6!\,0!} = \frac{1}{0!}$ [cancel 6!] $= \frac{1}{1} = 1$

(d) $\frac{6!}{5!\,1!} = \frac{6\cdot5\cdot4\cdot3\cdot2\cdot1}{(5\cdot4\cdot3\cdot2\cdot1)(1)} = \frac{6}{1} = 6$

N3. $_7C_2 = \frac{7!}{2!\,(7-2)!} = \frac{7!}{2!\,5!}$
$= \frac{7\cdot6\cdot5\cdot4\cdot3\cdot2\cdot1}{(2\cdot1)(5\cdot4\cdot3\cdot2\cdot1)} = 21$

N4. $(a+3b)^5$
$= (a)^5 + \frac{5!}{1!\,4!}(a)^4(3b)^1 + \frac{5!}{2!\,3!}(a)^3(3b)^2$
$+ \frac{5!}{3!\,2!}(a)^2(3b)^3 + \frac{5!}{4!\,1!}(a)^1(3b)^4 + (3b)^5$
$= a^5 + 5a^4(3b) + 10a^3(9b^2)$
$+ 10a^2(27b^3) + 5a(81b^4) + 243b^5$
$= a^5 + 15a^4b + 90a^3b^2$
$+ 270a^2b^3 + 405ab^4 + 243b^5$

N5. $\left(\frac{x}{3} - 2y\right)^4$
$= \left(\frac{x}{3}\right)^4 + \frac{4!}{1!\,3!}\left(\frac{x}{3}\right)^3(-2y)^1 + \frac{4!}{2!\,2!}\left(\frac{x}{3}\right)^2(-2y)^2$
$+ \frac{4!}{3!\,1!}\left(\frac{x}{3}\right)^1(-2y)^3 + (-2y)^4$
$= \frac{x^4}{3^4} + 4\cdot\frac{x^3}{3^3}(-2y) + 6\cdot\frac{x^2}{3^2}(4y^2)$
$+ 4\cdot\frac{x}{3}(-8y^3) + 16y^4$
$= \frac{1}{81}x^4 - \frac{8}{27}x^3y + \frac{8}{3}x^2y^2 - \frac{32}{3}xy^3 + 16y^4$

N6. In the sixth term of the expansion of $(2m - n^2)^8$, the exponent on $(-n^2)$ is $6 - 1 = 5$ and the exponent on $2m$ is $8 - 5 = 3$. The sixth term is

$$\frac{8!}{5!\,3!}(2m)^3(-n^2)^5 = \frac{8\cdot7\cdot6}{3\cdot2\cdot1}\cdot2^3m^3\cdot(-n^{10})$$
$$= -56\cdot8m^3n^{10}$$
$$= -448m^3n^{10}$$

14.4 Section Exercises

1. $6! = 6\cdot5\cdot4\cdot3\cdot2\cdot1 = 720$

3. $8! = 8\cdot7\cdot6\cdot5\cdot4\cdot3\cdot2\cdot1 = 40{,}320$

5. $\frac{6!}{4!\,2!} = \frac{6\cdot5\cdot4\cdot3\cdot2\cdot1}{(4\cdot3\cdot2\cdot1)(2\cdot1)} = \frac{6\cdot5}{2\cdot1} = 15$

7. $\frac{4!}{0!\,4!} = \frac{1}{0!}$ [cancel 4!] $= \frac{1}{1} = 1$

9. $4!\cdot5 = (4\cdot3\cdot2\cdot1)\cdot5$
$= 24\cdot5 = 120$

11. $_6C_2 = \frac{6!}{2!\,(6-2)!} = \frac{6!}{2!\,4!} = 15,$
by Exercise 5.

13. $_{13}C_{11} = \frac{13!}{11!\,(13-11)!} = \frac{13!}{11!\,2!}$
$= \frac{13\cdot12}{2\cdot1}$ *11! cancels*
$= 13\cdot6 = 78$

15. $(m+n)^4$
$= m^4 + \frac{4!}{1!\,3!}m^3n^1 + \frac{4!}{2!\,2!}m^2n^2$
$+ \frac{4!}{3!\,1!}m^1n^3 + n^4$
$= m^4 + 4m^3n + 6m^2n^2 + 4mn^3 + n^4$

17. $(a-b)^5 = [a + (-b)]^5$
$= a^5 + \frac{5!}{1!\,4!}a^4(-b)^1 + \frac{5!}{2!\,3!}a^3(-b)^2$
$+ \frac{5!}{3!\,2!}a^2(-b)^3 + \frac{5!}{4!\,1!}a^1(-b)^4 + (-b)^5$
$= a^5 - 5a^4b + 10a^3b^2 - 10a^2b^3 + 5ab^4 - b^5$

19. $(2x+3)^3$
$= (2x)^3 + \frac{3!}{1!\,2!}(2x)^2(3)^1 + \frac{3!}{2!\,1!}(2x)(3)^2$
$+ (3)^3$
$= 8x^3 + 3(4x^2)(3) + 3(2x)(9) + 27$
$= 8x^3 + 36x^2 + 54x + 27$

21. $\left(\frac{x}{2} - y\right)^4 = \left[\left(\frac{x}{2}\right) + (-y)\right]^4$
$= \left(\frac{x}{2}\right)^4 + \frac{4!}{1!\,3!}\left(\frac{x}{2}\right)^3(-y)^1 + \frac{4!}{2!\,2!}\left(\frac{x}{2}\right)^2(-y)^2$
$+ \frac{4!}{3!\,1!}\left(\frac{x}{2}\right)^1(-y)^3 + (-y)^4$
$= \frac{x^4}{16} - \frac{x^3y}{2} + \frac{3x^2y^2}{2} - 2xy^3 + y^4$

23. $(x^2+1)^4$
$= (x^2)^4 + \frac{4!}{1!\,3!}(x^2)^3(1)^1 + \frac{4!}{2!\,2!}(x^2)^2(1)^2$
$+ \frac{4!}{3!\,1!}(x^2)^1(1)^3 + (1)^4$
$= x^8 + 4x^6 + 6x^4 + 4x^2 + 1$

25. $(3x^2 - y^2)^3 = [3x^2 + (-y^2)]^3$
$= (3x^2)^3 + \frac{3!}{1!\,2!}(3x^2)^2(-y^2)^1$
$+ \frac{3!}{2!\,1!}(3x^2)^1(-y^2)^2 + (-y^2)^3$
$= 27x^6 - 3(9x^4)y^2 + 3(3x^2)y^4 - y^6$
$= 27x^6 - 27x^4y^2 + 9x^2y^4 - y^6$

27. $(r+2s)^{12} = r^{12} + \frac{12!}{1!\,11!}r^{11}(2s)^1$
$+ \frac{12!}{2!\,10!}r^{10}(2s)^2 + \frac{12!}{3!\,9!}r^9(2s)^3 + \cdots$

The first four terms of the expansion are

$$r^{12} + 24r^{11}s + 264r^{10}s^2 + 1760r^9s^3.$$

29. $(3x - y)^{14} = [3x + (-y)]^{14}$
$= (3x)^{14} + \frac{14!}{1!\,13!}(3x)^{13}(-y)^1$
$+ \frac{14!}{2!\,12!}(3x)^{12}(-y)^2$
$+ \frac{14!}{3!\,11!}(3x)^{11}(-y)^3 + \cdots$

The first four terms of the expansion are

$$3^{14}x^{14} - 14(3^{13})x^{13}y + 91(3^{12})x^{12}y^2 - 364(3^{11})x^{11}y^3.$$

31. $(t^2 + u^2)^{10}$
$= (t^2)^{10} + \frac{10!}{1!\,9!}(t^2)^9(u^2)^1$
$+ \frac{10!}{2!\,8!}(t^2)^8(u^2)^2$
$+ \frac{10!}{3!\,7!}(t^2)^7(u^2)^3 + \cdots$

The first four terms of the expansion are

$$t^{20} + 10t^{18}u^2 + 45t^{16}u^4 + 120t^{14}u^6.$$

33. The rth term of the expansion of $(x + y)^n$ is

$$\frac{n!}{(r-1)!\,[n-(r-1)]!}(x)^{n-(r-1)}(y)^{r-1}.$$

Start with the exponent on y, which is 1 less than the term number r. In this case, we are looking for the fourth term, so $r = 4$ and $r - 1 = 3$. Thus, the fourth term of $(2m + n)^{10}$ is

$$\frac{10!}{3!\,(10-3)!}(2m)^{10-3}(n)^3 = \frac{10!}{3!\,7!}2^7m^7n^3 = 120(2^7)m^7n^3.$$

35. The seventh term of $\left(x + \frac{y}{2}\right)^8$ is

$$\frac{8!}{6!\,(8-6)!}(x)^{8-6}\left(\frac{y}{2}\right)^6 = \frac{8!}{6!\,2!}x^2\frac{y^6}{2^6} = 28\,\frac{x^2y^6}{2^6} = \frac{7x^2y^6}{16}.$$

37. The third term of $(k - 1)^9$ is

$$\frac{9!}{2!\,(9-2)!}k^{9-2}(-1)^2 = \frac{9!}{2!\,7!}k^7 = 36k^7.$$

39. The expansion of $(x^2 + 2y)^6$ has seven terms, so the middle term is the fourth. The fourth term of $(x^2 + 2y)^6$ is

$$\frac{6!}{3!\,(6-3)!}(x^2)^{6-3}(2y)^3 = \frac{6!}{3!\,3!}(x^2)^3(8y^3) = 20x^6(8y^3) = 160x^6y^3.$$

41. The term of the expansion of $(3x^3 - 4y^2)^5$ with x^9y^4 in it is the term with $(3x^3)^3(-4y^2)^2$, since $(x^3)^3(y^2)^2 = x^9y^4$. The term is

$$\frac{5!}{2!\,3!}(3x^3)^3(-4y^2)^2 = 10(27x^9)(16y^4) = 4320x^9y^4.$$

Chapter 14 Review Exercises

1. $a_n = 2n - 3$
$a_1 = 2(1) - 3 = -1$ $\quad a_2 = 2(2) - 3 = 1$
$a_3 = 2(3) - 3 = 3$ $\quad a_4 = 2(4) - 3 = 5$

Answer: $-1, 1, 3, 5$

2. $a_n = \frac{n-1}{n}$
$a_1 = \frac{1-1}{1} = 0$ $\quad a_2 = \frac{2-1}{2} = \frac{1}{2}$
$a_3 = \frac{3-1}{3} = \frac{2}{3}$ $\quad a_4 = \frac{4-1}{4} = \frac{3}{4}$

Answer: $0, \frac{1}{2}, \frac{2}{3}, \frac{3}{4}$

3. $a_n = n^2$
$a_1 = (1)^2 = 1$ $\quad a_2 = (2)^2 = 4$
$a_3 = (3)^2 = 9$ $\quad a_4 = (4)^2 = 16$

Answer: 1, 4, 9, 16

4. $a_n = \left(\frac{1}{2}\right)^n$
$a_1 = \left(\frac{1}{2}\right)^1 = \frac{1}{2}$ $\quad a_2 = \left(\frac{1}{2}\right)^2 = \frac{1}{4}$
$a_3 = \left(\frac{1}{2}\right)^3 = \frac{1}{8}$ $\quad a_4 = \left(\frac{1}{2}\right)^4 = \frac{1}{16}$

Answer: $\frac{1}{2}, \frac{1}{4}, \frac{1}{8}, \frac{1}{16}$

5. $a_n = (n + 1)(n - 1)$
$a_1 = (1 + 1)(1 - 1) = 2(0) = 0$
$a_2 = (2 + 1)(2 - 1) = 3(1) = 3$
$a_3 = (3 + 1)(3 - 1) = 4(2) = 8$
$a_4 = (4 + 1)(4 - 1) = 5(3) = 15$

Answer: 0, 3, 8, 15

6. $a_n = n(-1)^{n-1}$
$a_1 = 1(-1)^{1-1} = 1(1) = 1$
$a_2 = 2(-1)^{2-1} = 2(-1) = -2$
$a_3 = 3(-1)^{3-1} = 3(1) = 3$
$a_4 = 4(-1)^{4-1} = 4(-1) = -4$

Answer: $1, -2, 3, -4$

7. $\sum_{i=1}^{5} i^2 = 1^2 + 2^2 + 3^2 + 4^2 + 5^2$

$= 1 + 4 + 9 + 16 + 25$

8. $\sum_{i=1}^{6} (i+1) = (1+1) + (2+1) + (3+1)$

$+ (4+1) + (5+1) + (6+1)$

$= 2 + 3 + 4 + 5 + 6 + 7$

9. $\sum_{i=3}^{6} (5i - 4)$

$= (15-4) + (20-4) + (25-4) + (30-4)$ *Let i = 3 to 6.*

$= 11 + 16 + 21 + 26$

10. $\sum_{i=1}^{4} (i+2)$

$= (1+2) + (2+2) + (3+2) + (4+2)$

$= 3 + 4 + 5 + 6$

$= 18$

11. $\sum_{i=1}^{6} 2^i$

$= 2^1 + 2^2 + 2^3 + 2^4 + 2^5 + 2^6$

$= 2 + 4 + 8 + 16 + 32 + 64$

$= 126$

12. $\sum_{i=4}^{7} \frac{i}{i+1}$

$= \frac{4}{4+1} + \frac{5}{5+1} + \frac{6}{6+1} + \frac{7}{7+1}$

$= \frac{4}{5} + \frac{5}{6} + \frac{6}{7} + \frac{7}{8}$ *LCD=$2^3 \cdot 3 \cdot 5 \cdot 7$=840*

$= \frac{672}{840} + \frac{700}{840} + \frac{720}{840} + \frac{735}{840} = \frac{2827}{840}$

13. $\bar{x} = \frac{13{,}778 + 14{,}862 + 16{,}680 + 17{,}916 + 13{,}985}{5}$

$\bar{x} = \frac{77{,}221}{5} = 15{,}444.2$

The average total of retirement assets of Americans for the years 2004–2008 was \$15,444 billion.

14. 2, 5, 8, 11, ... is an *arithmetic* sequence with

$d = 5 - 2 = 3.$

Note: You should find the difference for all pairs of adjacent terms to determine if the sequence is arithmetic.

15. $-6, -2, 2, 6, 10, \ldots$ is an *arithmetic* sequence with

$d = -2 - (-6) = 4.$

16. $\frac{2}{3}, -\frac{1}{3}, \frac{1}{6}, -\frac{1}{12}, \ldots$ is a *geometric* sequence with

$r = \frac{-\frac{1}{3}}{\frac{2}{3}} = -\frac{1}{3} \cdot \frac{3}{2} = -\frac{1}{2}.$

Note: You should find the ratio for all pairs of adjacent terms to determine if the sequence is geometric.

17. $-1, 1, -1, 1, -1, \ldots$ is a *geometric* sequence with

$r = \frac{1}{-1} = -1.$

18. $64, 32, 8, \frac{1}{2}, \ldots$

Find two differences:
$32 - 64 = -32$ and $8 - 32 = -24$, so the sequence is not arithmetic.

Find two ratios:
$\frac{32}{64} \neq \frac{8}{32}$, so the sequence is not geometric.

Therefore, the sequence is *neither*.

19. 64, 32, 16, 8, ... is a *geometric* sequence with

$r = \frac{32}{64} = \frac{1}{2}.$

20. We are given the first six terms: 1, 1, 2, 3, 5, 8

7th term: $5 + 8 = 13$ 8th term: $8 + 13 = 21$
9th term: $13 + 21 = 34$ 10th term: $21 + 34 = 55$
11th term: $34 + 55 = 89$

21. Given $a_1 = -2$ and $d = 5$, find a_{16}.

$$\begin{aligned} a_n &= a_1 + (n-1)d \\ a_{16} &= -2 + (16-1)5 \\ &= -2 + 15(5) \\ &= -2 + 75 = 73 \end{aligned}$$

22. Given $a_6 = 12$ and $a_8 = 18$, find a_{25}.

$a_n = a_1 + (n-1)d$

$a_6 = a_1 + 5d$, so $12 = a_1 + 5d$ (1)

$a_8 = a_1 + 7d$, so $18 = a_1 + 7d$ (2)

Multiply equation (1) by -1 and add the result to equation (2).

$$\begin{aligned} -12 &= -a_1 - 5d & -1 \times (1) \\ 18 &= a_1 + 7d & (2) \\ \hline 6 &= 2d \\ 3 &= d \end{aligned}$$

To find a_1, substitute $d = 3$ in equation (1).

$$\begin{aligned} 12 &= a_1 + 5d \quad (1) \\ 12 &= a_1 + 5(3) \\ 12 &= a_1 + 15 \\ -3 &= a_1 \end{aligned}$$

Use $a_1 = -3$ and $d = 3$ to find a_{25}.

$$\begin{aligned} a_n &= a_1 + (n-1)d \\ a_{25} &= -3 + (25-1)3 \\ &= -3 + 24(3) \\ &= -3 + 72 = 69 \end{aligned}$$

23. $a_1 = -4,\ d = -5$

$$\begin{aligned} a_n &= a_1 + (n-1)d \\ a_n &= -4 + (n-1)(-5) \\ &= -4 - 5n + 5 \\ &= -5n + 1 \end{aligned}$$

24. $6, 3, 0, -3, \ldots$

To get the general term, a_n, first find d.
$d = 3 - 6 = -3$

$$\begin{aligned} a_n &= a_1 + (n-1)d \\ a_n &= 6 + (n-1)(-3) \\ &= 6 - 3n + 3 \\ &= -3n + 9 \end{aligned}$$

25. $7, 10, 13, \ldots, 49$

Here, $a_1 = 7$ and $d = 10 - 7 = 3$.
Now find n, the number of terms.

$$\begin{aligned} a_n &= a_1 + (n-1)d \\ 49 &= 7 + (n-1)(3) \\ 42 &= 3(n-1) \\ 14 &= n - 1 \qquad \textit{Divide by 3.} \\ 15 &= n \end{aligned}$$

There are 15 terms in this sequence.

26. $5, 1, -3, \ldots, -79$

Here, $a_1 = 5$ and $d = 1 - 5 = -4$.
Now find n, the number of terms.

$$\begin{aligned} a_n &= a_1 + (n-1)d \\ -79 &= 5 + (n-1)(-4) \\ -79 &= 9 - 4n \\ -88 &= -4n \\ n &= 22 \end{aligned}$$

There are 22 terms in this sequence.

27. Find S_8 if $a_1 = -2$ and $d = 6$. Find a_8 first.

$$\begin{aligned} a_8 &= a_1 + (8-1)d \\ &= -2 + 7(6) \\ &= -2 + 42 = 40 \end{aligned}$$

Now find the sum.

$$\begin{aligned} S_n &= \frac{n}{2}(a_1 + a_n) \\ S_8 &= \tfrac{8}{2}(-2 + 40) \\ &= 4(38) = 152 \end{aligned}$$

28. Find S_8 if $a_n = -2 + 5n$.
Find the first and last terms.

$$\begin{aligned} a_1 &= -2 + 5(1) = 3 \\ a_8 &= -2 + 5(8) = 38 \end{aligned}$$

Now find the sum.

$$\begin{aligned} S_n &= \frac{n}{2}(a_1 + a_n) \\ S_8 &= \tfrac{8}{2}(3 + 38) \\ &= 4(41) \\ &= 164 \end{aligned}$$

29. Find the general term for the geometric sequence $-1, -4, -16, \ldots$.

We determine that $a_1 = -1$ and $r = \frac{-4}{-1} = 4$.

Using the formula $a_n = a_1 r^{n-1}$, we get

$$a_n = -1(4)^{n-1}.$$

30. Find the general term for the geometric sequence $\frac{2}{3}, \frac{2}{15}, \frac{2}{75}, \ldots$.

We determine that $a_1 = \frac{2}{3}$ and
$r = \dfrac{\frac{2}{15}}{\frac{2}{3}} = \dfrac{2}{15} \cdot \dfrac{3}{2} = \dfrac{1}{5}$.

Using the formula $a_n = a_1 r^{n-1}$, we get

$$a_n = \tfrac{2}{3}\left(\tfrac{1}{5}\right)^{n-1}.$$

31. Find a_{11} for $2, -6, 18, \ldots$.

We determine that $a_1 = 2$ and $r = \frac{-6}{2} = -3$.

$$\begin{aligned} a_n &= a_1 r^{n-1} \\ a_{11} &= 2(-3)^{11-1} \\ &= 2(-3)^{10} = 118{,}098 \end{aligned}$$

32. Given $a_3 = 20$ and $a_5 = 80$, find a_{10}. Using the formula $a_n = a_1 r^{n-1}$, we get the following.

$$\begin{aligned} \textbf{For } \boldsymbol{a_3}\textbf{:} \quad a_3 &= a_1 r^{3-1} \\ 20 &= a_1 r^2 \\ \textbf{For } \boldsymbol{a_5}\textbf{:} \quad a_5 &= a_1 r^{5-1} \\ 80 &= a_1 r^4 \end{aligned}$$

The ratio of a_5 to a_3 is

$$\begin{aligned} \frac{80}{20} &= \frac{a_1 r^4}{a_1 r^2} \\ 4 &= r^2 \\ r &= \pm 2. \end{aligned}$$

Since $20 = a_1 r^2$ and $r^2 = 4$, we get $20 = a_1(4)$, and hence, $5 = a_1$. Now find a_{10}.

$$\begin{aligned} a_n &= a_1 r^{n-1} \\ a_{10} &= 5(\pm 2)^{10-1} \\ &= 5(\pm 2)^9 \end{aligned}$$

Two answers are possible for a_{10}:

$$5(2)^9 = 2560 \quad \textbf{or} \quad 5(-2)^9 = -2560.$$

33. $\sum_{i=1}^{5} \left(\frac{1}{4}\right)^i$ ■ $a_1 = \frac{1}{4}$, $r = \frac{1}{4}$, and $n = 5$.

$$S_n = \frac{a_1(1-r^n)}{1-r}$$
$$S_5 = \frac{\frac{1}{4}\left[1-\left(\frac{1}{4}\right)^5\right]}{1-\frac{1}{4}}$$
$$= \frac{\frac{1}{4}\left(1-\frac{1}{1024}\right)}{\frac{3}{4}}$$
$$= \frac{1}{3}\left(\frac{1023}{1024}\right) = \frac{341}{1024}$$

34. $\sum_{i=1}^{8} \frac{3}{4}(-1)^i$ ■ $a_1 = -\frac{3}{4}$, $r = -1$, and $n = 8$.

$$S_n = \frac{a_1(1-r^n)}{1-r}$$
$$S_8 = \frac{-\frac{3}{4}\left[1-(-1)^8\right]}{1-(-1)}$$
$$= \frac{-\frac{3}{4}(1-1)}{2}$$
$$= \frac{-\frac{3}{4}(0)}{2} = 0$$

35. $\sum_{i=1}^{\infty} 4\left(\frac{1}{5}\right)^i$

The terms are the terms of an infinite geometric sequence with $a_1 = \frac{4}{5}$ and $r = \frac{1}{5}$.

$$S = \frac{a_1}{1-r} = \frac{\frac{4}{5}}{1-\frac{1}{5}} = \frac{\frac{4}{5}}{\frac{4}{5}} = 1$$

36. $\sum_{i=1}^{\infty} 2(3)^i$

The terms are the terms of an infinite geometric sequence with $a_1 = 6$ and $r = 3$.

$$S = \frac{a_1}{1-r} \quad \text{if} \quad |r| < 1,$$

but $r = 3$, so S does not exist and hence the sum does not exist.

37. $(2p-q)^5 = [2p+(-q)]^5$

$$= (2p)^5 + \frac{5!}{1!\,4!}(2p)^4(-q)^1$$
$$+ \frac{5!}{2!\,3!}(2p)^3(-q)^2 + \frac{5!}{3!\,2!}(2p)^2(-q)^3$$
$$+ \frac{5!}{4!\,1!}(2p)^1(-q)^4 + (-q)^5$$
$$= 32p^5 + 5(16p^4)(-q) + 10(8p^3)q^2$$
$$+ 10(4p^2)(-q^3) + 5(2p)q^4 - q^5$$
$$= 32p^5 - 80p^4q + 80p^3q^2 - 40p^2q^3$$
$$+ 10pq^4 - q^5$$

38. $(x^2+3y)^4$

$$= (x^2)^4 + \frac{4!}{1!\,3!}(x^2)^3(3y)^1$$
$$+ \frac{4!}{2!\,2!}(x^2)^2(3y)^2$$
$$+ \frac{4!}{3!\,1!}(x^2)^1(3y)^3 + (3y)^4$$
$$= x^8 + 4(x^6)(3y) + 6(x^4)(9y^2)$$
$$+ 4(x^2)(27y^3) + 81y^4$$
$$= x^8 + 12x^6y + 54x^4y^2 + 108x^2y^3$$
$$+ 81y^4$$

39. $(3t^3-s^2)^4 = \left[3t^3+(-s^2)\right]^4$

$$= (3t^3)^4 + \frac{4!}{1!\,3!}(3t^3)^3(-s^2)^1$$
$$+ \frac{4!}{2!\,2!}(3t^3)^2(-s^2)^2$$
$$+ \frac{4!}{3!\,1!}(3t^3)^1(-s^2)^3 + (-s^2)^4$$
$$= 3^4t^{12} - 4(3^3t^9)(s^2) + 6(9t^6)(s^4)$$
$$- 4(3t^3)(s^6) + s^8$$
$$= 81t^{12} - 108t^9s^2 + 54t^6s^4 - 12t^3s^6 + s^8$$

40. The fourth term ($r = 4$, so $r - 1 = 3$) of $(3a+2b)^{19}$ is

$$\frac{19!}{3!\,16!}(3a)^{16}(2b)^3$$
$$= 969(3)^{16}(a^{16})(8)b^3$$
$$= 7752(3)^{16}a^{16}b^3.$$

41. **[14.3]** The geometric sequence $-3, 6, -12, \ldots$ has $a_1 = -3$ and $r = \frac{6}{-3} = -2$. First find a_{10}.

$$a_n = a_1r^{n-1}$$
$$a_{10} = -3(-2)^9$$
$$= -3(-512) = 1536$$

Now find the sum of the first ten terms.

$$S_n = \frac{a_1(1-r^n)}{1-r}$$
$$S_{10} = \frac{-3\left[1-(-2)^{10}\right]}{1-(-2)}$$
$$= \frac{-3(1-1024)}{3}$$
$$= -(-1023) = 1023$$

42. **[14.2]** The arithmetic sequence $1, 7, 13, \ldots$ has $a_1 = 1$ and $d = 7 - 1 = 6$. First find a_{40}.

$$a_n = a_1 + (n-1)d$$
$$a_{40} = 1 + (40-1)6$$
$$= 1 + (39)6$$
$$= 1 + 234 = 235$$

Now find the tenth term so that we can find the sum of the first ten terms.

$$\begin{aligned} a_{10} &= 1 + (10-1)6 \\ &= 1 + (9)6 = 55 \end{aligned}$$

Now find the sum of the first ten terms.

$$\begin{aligned} S_n &= \frac{n}{2}(a_1 + a_n) \\ S_{10} &= \tfrac{10}{2}(1+55) \\ &= 5(56) = 280 \end{aligned}$$

43. **[14.2]** The arithmetic sequence with $a_1 = -4$ and $d = 3$ is $-4, -1, 2, 5, \ldots$. First find a_{15}.

$$\begin{aligned} a_n &= a_1 + (n-1)d \\ a_{15} &= -4 + (15-1)3 \\ &= -4 + 42 = 38 \end{aligned}$$

Now find the sum of the first ten terms.

$$\begin{aligned} S_n &= \frac{n}{2}[2a_1 + (n-1)d] \\ S_{10} &= \tfrac{10}{2}[2(-4) + (10-1)3] \\ &= 5(-8+27) \\ &= 5(19) = 95 \end{aligned}$$

44. **[14.3]** The geometric sequence has $a_1 = 1$ and $r = -3$. First find a_9.

$$\begin{aligned} a_n &= a_1 r^{n-1} \\ a_9 &= 1(-3)^{9-1} \\ &= (-3)^8 = 6561 \end{aligned}$$

Now find the sum of the first ten terms.

$$\begin{aligned} S_n &= \frac{a_1(r^n - 1)}{r-1} \\ S_{10} &= \frac{1[(-3)^{10} - 1]}{-3-1} \\ &= -\tfrac{1}{4}(3^{10} - 1) \\ &= -\tfrac{1}{4}(59{,}049 - 1) \\ &= -14{,}762 \end{aligned}$$

45. **[14.3]** $2, 8, 32, \ldots$ is a geometric sequence with $a_1 = 2$ and $r = \frac{8}{2} = 4$.

$$\begin{aligned} a_n &= a_1 r^{n-1} \\ a_n &= 2(4)^{n-1} \end{aligned}$$

46. **[14.2]** $2, 7, 12, \ldots$ is an arithmetic sequence with $a_1 = 2$ and $d = 7 - 2 = 5$.

$$\begin{aligned} a_n &= a_1 + (n-1)d \\ a_n &= 2 + (n-1)5 \\ &= 2 + 5n - 5 \\ &= 5n - 3 \end{aligned}$$

47. **[14.2]** $12, 9, 6, \ldots$ is an arithmetic sequence with $a_1 = 12$ and $d = -3$.

$$\begin{aligned} a_n &= a_1 + (n-1)d \\ a_n &= 12 + (n-1)(-3) \\ &= 12 - 3n + 3 \\ &= -3n + 15 \end{aligned}$$

48. **[14.3]** $27, 9, 3, \ldots$ is a geometric sequence with $a_1 = 27$ and $r = \frac{9}{27} = \frac{1}{3}$.

$$\begin{aligned} a_n &= a_1 r^{n-1} \\ a_n &= 27\left(\tfrac{1}{3}\right)^{n-1} \end{aligned}$$

49. **[14.2]** The distances traveled in successive seconds are

$$3, 7, 11, 15, 19, \ldots.$$

This is an arithmetic sequence with $a_1 = 3$ and $d = 4$. Since we know a_1 and d, we'll use the second formula for the sum of an arithmetic sequence with $S_n = 210$.

$$\begin{aligned} S_n &= \frac{n}{2}[2a_1 + (n-1)d] \\ 210 &= \frac{n}{2}[2(3) + (n-1)4] \\ 420 &= n(6 + 4n - 4) \\ 420 &= 6n + 4n^2 - 4n \\ 0 &= 4n^2 + 2n - 420 \\ 0 &= 2n^2 + n - 210 \\ 0 &= (2n+21)(n-10) \end{aligned}$$

$$\begin{aligned} 2n + 21 &= 0 \quad \text{or} \quad n - 10 = 0 \\ n &= -\tfrac{21}{2} \quad \text{or} \quad n = 10 \end{aligned}$$

Discard $-\frac{21}{2}$ since time cannot be negative. It takes her 10 seconds.

50. **[14.3]** Use the formula for the future value of an ordinary annuity with $R = 672$, $i = \frac{0.045}{4} = 0.01125$, and $n = 7(4) = 28$.

$$\begin{aligned} S &= R\left[\frac{(1+i)^n - 1}{i}\right] \\ S &= 672\left[\frac{(1+0.01125)^{28} - 1}{0.01125}\right] \\ &= 21{,}973.00 \end{aligned}$$

The future value of the annuity is \$21,973.00.

51. **[14.1]** Since $100\% - 3\% = 97\% = 0.97$, the population after 1 year is $0.97(50{,}000)$, after 2 years is $0.97[0.97(50{,}000)]$ or $(0.97)^2(50{,}000)$, and after n years is $(0.97)^n(50{,}000)$. After 6 years, the population is

$$(0.97)^6(50{,}000) \approx 41{,}649 \approx 42{,}000.$$

52. **[14.1]** $\left(\frac{1}{2}\right)^n$ is left after n strokes. So $\left(\frac{1}{2}\right)^7 = \frac{1}{128} = 0.0078125$ is left after 7 strokes.

53. **[14.3]** **(a)** We can write the repeating decimal number $0.55555\ldots$ as an infinite geometric sequence as follows:

$$\tfrac{5}{10} + \tfrac{5}{10}\left(\tfrac{1}{10}\right) + \tfrac{5}{10}\left(\tfrac{1}{10}\right)^2 + \tfrac{5}{10}\left(\tfrac{1}{10}\right)^3 + \cdots$$

(b) The common ratio r is

$$\frac{\frac{5}{10}\left(\frac{1}{10}\right)}{\frac{5}{10}} = \frac{1}{10}.$$

(c) Since $|r| < 1$, the sum exists.

$$S = \frac{a_1}{1-r} = \frac{\frac{5}{10}}{1-\frac{1}{10}} = \frac{\frac{5}{10}}{\frac{9}{10}} = \frac{5}{9}$$

54. **[14.3]** No, the sum cannot be found, because $r = 2$. This value of r does not satisfy $|r| < 1$.

Chapter 14 Test

1.
$$\begin{aligned} a_n &= (-1)^n + 1 \\ a_1 &= (-1)^1 + 1 = 0 \\ a_2 &= (-1)^2 + 1 = 1 + 1 = 2 \\ a_3 &= (-1)^3 + 1 = -1 + 1 = 0 \\ a_4 &= (-1)^4 + 1 = 1 + 1 = 2 \\ a_5 &= (-1)^5 + 1 = -1 + 1 = 0 \end{aligned}$$

Answer: 0, 2, 0, 2, 0

2.
$$\begin{aligned} a_1 &= 4,\ d = 2 \\ a_2 &= a_1 + d = 4 + 2 = 6 \\ a_3 &= a_2 + d = 6 + 2 = 8 \\ a_4 &= a_3 + d = 8 + 2 = 10 \\ a_5 &= a_4 + d = 10 + 2 = 12 \end{aligned}$$

Answer: 4, 6, 8, 10, 12

3. $a_4 = 6,\ r = \frac{1}{2}$
First find a_1.

$$\begin{aligned} a_n &= a_1 r^{n-1} \\ a_4 &= a_1\left(\tfrac{1}{2}\right)^{4-1} \\ 6 &= a_1\left(\tfrac{1}{8}\right) \\ a_1 &= 48 \end{aligned}$$

Now find the remaining terms.

$$\begin{aligned} a_2 &= \tfrac{1}{2}a_1 = \tfrac{1}{2}(48) = 24 \\ a_3 &= \tfrac{1}{2}a_2 = \tfrac{1}{2}(24) = 12 \\ a_4 &= \tfrac{1}{2}a_3 = \tfrac{1}{2}(12) = 6 \\ a_5 &= \tfrac{1}{2}a_4 = \tfrac{1}{2}(6) = 3 \end{aligned}$$

Answer: 48, 24, 12, 6, 3

4. Given $a_1 = 6$ and $d = -2$, find a_4.

$$\begin{aligned} a_n &= a_1 + (n-1)d \\ a_4 &= a_1 + (4-1)d \\ &= 6 + (3)(-2) \\ &= 6 - 6 = 0 \end{aligned}$$

5. Given $a_5 = 16$ and $a_7 = 9$, find a_4. This is a geometric sequence, so

$$a_6 = a_5 r \quad \text{and} \quad a_7 = a_6 r.$$

From the last equation, $a_6 = \dfrac{a_7}{r}$, so by substitution,

$$a_5 r = \frac{a_7}{r}.$$

Multiply by r and divide by a_5 to get

$$\begin{aligned} r^2 &= \frac{a_7}{a_5} \\ r^2 &= \tfrac{9}{16} \\ r &= \pm\sqrt{\tfrac{9}{16}} = \pm\tfrac{3}{4} \end{aligned}$$

Since $a_5 = a_4 r$, we know that $a_4 = \dfrac{16}{r}$. Substituting each value of r in the last equation gives us two values of a_4.

$$a_4 = \frac{16}{\frac{3}{4}} = \frac{64}{3} \quad \text{or} \quad a_4 = \frac{16}{-\frac{3}{4}} = -\frac{64}{3}$$

6. Given the arithmetic sequence with $a_2 = 12$ and $a_3 = 15$, find S_5. First find d.

$$d = a_3 - a_2 = 15 - 12 = 3$$

Now find the first and fifth terms.

$$\begin{aligned} a_1 &= a_2 - 3 = 12 - 3 = 9 \\ a_5 &= a_4 + 3 = a_3 + 6 = 15 + 6 = 21 \end{aligned}$$

Now find the sum of the first five terms.

$$\begin{aligned} S_n &= \frac{n}{2}(a_1 + a_n) \\ S_5 &= \tfrac{5}{2}(9 + 21) \\ &= \tfrac{5}{2}(30) = 75 \end{aligned}$$

7. Given the geometric sequence with $a_5 = 4$ and $a_7 = 1$, find S_5.

$$r^2 = \frac{a_7}{a_5} = \frac{1}{4}, \text{ so } r = \frac{1}{2} \quad \text{or} \quad r = -\frac{1}{2}.$$

Use $a_7 = a_1 r^6$ to get $1 = a_1\left(\pm\frac{1}{2}\right)^6$, and so $a_1 = 64$. Use $S_n = \dfrac{a_1(1-r^n)}{1-r}$.

$$\begin{aligned} S_5 &= \frac{64\left[1-\left(\frac{1}{2}\right)^5\right]}{1-\frac{1}{2}} && \text{or } S_5 = \frac{64\left[1-\left(-\frac{1}{2}\right)^5\right]}{1-\left(-\frac{1}{2}\right)} \\ &= \frac{64}{\frac{1}{2}}\left(1-\frac{1}{32}\right) && = \frac{64}{\frac{3}{2}}\left(1+\frac{1}{32}\right) \\ &= 128\left(\tfrac{31}{32}\right) && = \tfrac{128}{3}\left(\tfrac{33}{32}\right) \\ &= 124 && = 44 \end{aligned}$$

8. $\overline{x} = \dfrac{\text{total}}{5}$, where total $=$ $78{,}473 + 80{,}967 + 83{,}860 + 86{,}150 + 97{,}103$.

Thus, $\overline{x} = \dfrac{426{,}553}{5} = 85{,}310.6 \approx 85{,}311.$

The average number of commercial bank offices for this five-year period was 85,311.

9. $S = R\left[\frac{(1+i)^n - 1}{i}\right]$

$= 4000\left[\frac{\left(1+\frac{0.06}{4}\right)^{4(7)} - 1}{\frac{0.06}{4}}\right]$

$= 4000\left[\frac{(1.015)^{28} - 1}{0.015}\right]$

$\approx 137{,}925.91$

The account will have \$137,925.91 at the end of this term.

10. An infinite geometric series has a sum if $|r| < 1$, where r is the common ratio.

11. $\sum_{i=1}^{5}(2i+8)$

$= [2(1)+8] + [2(2)+8] + [2(3)+8] + [2(4)+8] + [2(5)+8]$

$= 10 + 12 + 14 + 16 + 18$

$= 70$

12. $\sum_{i=1}^{6}(3i-5)$

Find the first and sixth terms.

$a_1 = 3(1) - 5 = -2$
$a_6 = 3(6) - 5 = 13$

Now find the sum of the first six terms.

$S_6 = \frac{n}{2}(a_1 + a_6)$
$S_6 = \frac{6}{2}(-2+13)$
$= 3(11) = 33$

13. $\sum_{i=1}^{500} i$

Use the formula $S_{500} = \frac{n}{2}(a_1 + a_{500})$ with $a_1 = 1$ and $a_{500} = 500$.

$S_{500} = \frac{500}{2}(1+500)$
$= 250(501) = 125{,}250$

14. $\sum_{i=1}^{3}\frac{1}{2}(4^i) = \frac{1}{2}\sum_{i=1}^{3}(4^i)$

$= \frac{1}{2}(4^1 + 4^2 + 4^3)$

$= \frac{1}{2}(4 + 16 + 64)$

$= \frac{1}{2}(84) = 42$

15. $\sum_{i=1}^{\infty}\left(\frac{1}{4}\right)^i$

This is an infinite geometric series with $a_1 = \frac{1}{4}$ and $r = \frac{1}{4}$. Since $|r| = \frac{1}{4} < 1$, the sum exists.

$$S = \frac{a_1}{1-r} = \frac{\frac{1}{4}}{1-\frac{1}{4}} = \frac{\frac{1}{4}}{\frac{3}{4}} = \frac{1}{4}\cdot\frac{4}{3} = \frac{1}{3}$$

16. $\sum_{i=1}^{\infty} 6\left(\frac{3}{2}\right)^i$

This is an infinite geometric series with $a_1 = 6\left(\frac{3}{2}\right)^1 = 9$ and $r = \frac{3}{2}$. Since $|r| = \frac{3}{2} > 1$, the sum does not exist.

17. $8! = 8\cdot7\cdot6\cdot5\cdot4\cdot3\cdot2\cdot1$
$= 40{,}320$

18. By definition, $0! = 1$.

19. $\frac{6!}{4!\,2!} = \frac{6\cdot5\cdot4\cdot3\cdot2\cdot1}{(4\cdot3\cdot2\cdot1)(2\cdot1)} = \frac{6\cdot5}{2\cdot1} = 15$

20. ${}_{12}C_{10} = \frac{12!}{10!\,(12-10)!}$

$= \frac{12\cdot11\cdot10!}{10!\,(2!)} = \frac{12\cdot11}{2} = 66$

21. $(3k-5)^4 = [3k + (-5)]^4$

$= (3k)^4 + \frac{4!}{1!\,3!}(3k)^3(-5)^1 + \frac{4!}{2!\,2!}(3k)^2(-5)^2 + \frac{4!}{3!\,1!}(3k)^1(-5)^3 + (-5)^4$

$= 81k^4 - 4(27k^3)(5) + 6(9k^2)(25) - 4(3k)(125) + 625$

$= 81k^4 - 540k^3 + 1350k^2 - 1500k + 625$

22. The fifth term ($r = 5$, so $r - 1 = 4$) of $\left(2x - \frac{y}{3}\right)^{12}$ is

$$\frac{12!}{4!\,(12-4)!}(2x)^{12-4}\left(-\frac{y}{3}\right)^4$$
$$= \frac{12!}{4!\,8!}(2x)^8\left(-\frac{y}{3}\right)^4$$
$$= \frac{14{,}080x^8y^4}{9}.$$

23. The amounts of unpaid balance during 15 months form an arithmetic sequence

300, 280, 260, ..., 40, 20,

which is the sequence with $n = 15$, $a_1 = 300$, and $a_{15} = 20$. Find the sum of these balances.

$S_n = \frac{n}{2}(a_1 + a_n)$
$S_{15} = \frac{15}{2}(300+20)$
$= \frac{15}{2}(320) = 2400$

Since 1% interest is paid on this total, the interest paid is 1% of \$2400 or \$24. The sewing machine cost \$300 (paid monthly at \$20), so the total cost is \$300 + \$24 = \$324.

24. The weekly populations form a geometric sequence with $a_1 = 20$ and $r = 3$ since the colony begins with 20 insects and triples each week. Find the general term of this geometric sequence.

$$a_n = a_1 r^{n-1}$$
$$a_n = 20(3)^{n-1}$$

We're assuming that from the beginning of July to the end of September is 12 weeks, so find a_{12}.

$$a_n = 20(3)^{n-1}$$
$$a_{12} = 20(3)^{11}$$

At the end of September, $20(3)^{11} = 3{,}542{,}940$ insects will be present in the colony.

Cumulative Review Exercises (Chapters 1–14)

1. $|-7| + 6 - |-10| - (-8+3)$
$= 7 + 6 - 10 - (-5)$
$= 13 - 10 + 5 = 8$

2. $4(-6) + (-8)(5) - (-9)$
$= -24 - 40 + 9 = -55$

In Exercises 3–4, use the set

$P = \left\{-\frac{8}{3}, 10, 0, \sqrt{13}, -\sqrt{3}, \frac{45}{15}, \sqrt{-7}, 0.82, -3\right\}$.

3. The rational numbers are

$-\frac{8}{3}$, 10, 0, $\frac{45}{15}$ (or 3), 0.82, and -3.

4. The irrational numbers are $\sqrt{13}$ and $-\sqrt{3}$.

5.
$$\begin{aligned} 9 - (5 + 3x) + 5x &= -4(x-3) - 7 \\ 9 - 5 - 3x + 5x &= -4x + 12 - 7 \\ 4 + 2x &= -4x + 5 \\ 6x &= 1 \\ x &= \tfrac{1}{6} \end{aligned}$$

The solution set is $\left\{\frac{1}{6}\right\}$.

6.
$$\begin{aligned} 7x + 18 &\le 9x - 2 \\ -2x &\le -20 \\ x &\ge 10 \end{aligned}$$
Divide by -2; reverse the inequality symbol.

The solution set is $[10, \infty)$.

7. $|4x - 3| = 21$

$4x - 3 = 21$ or $4x - 3 = -21$
$4x = 24$ $\qquad 4x = -18$
$x = 6$ or $x = -\frac{18}{4} = -\frac{9}{2}$

The solution set is $\left\{-\frac{9}{2}, 6\right\}$.

8. $\dfrac{x+3}{12} - \dfrac{x-3}{6} = 0$

Multiply by the LCD, 12.

$$\begin{aligned} 12\left(\frac{x+3}{12} - \frac{x-3}{6}\right) &= 12(0) \\ x + 3 - 2(x-3) &= 0 \\ x + 3 - 2x + 6 &= 0 \\ 9 - x &= 0 \\ 9 &= x \end{aligned}$$

Check $x = 9$: $\quad 1 - 1 = 0 \quad$ *True*

The solution set is $\{9\}$.

9. $2x > 8$ or $-3x > 9$
$x > 4$ or $x < -3$

The solution set is $(-\infty, -3) \cup (4, \infty)$.

10. $|2x - 5| \ge 11$

$2x - 5 \ge 11$ or $2x - 5 \le -11$
$2m \ge 16$ $\qquad 2m \le -6$
$m \ge 8$ or $m \le -3$

The solution set is $(-\infty, -3] \cup [8, \infty)$.

11.
$$\begin{aligned} 2x^2 + x &= 10 \\ 2x^2 + x - 10 &= 0 \\ (2x+5)(x-2) &= 0 \end{aligned}$$

$2x + 5 = 0$ or $x - 2 = 0$
$x = -\frac{5}{2}$ or $x = 2$

Check $x = -\frac{5}{2}$: $\quad \frac{25}{2} - \frac{5}{2} = 10 \quad$ *True*
Check $x = 2$: $\quad 8 + 2 = 10 \quad$ *True*

The solution set is $\left\{-\frac{5}{2}, 2\right\}$.

12. $x^2 - x - 6 \le 0$

Solve the equation

$$x^2 - x - 6 = 0.$$
$$(x-3)(x+2) = 0$$

$x - 3 = 0$ or $x + 2 = 0$
$x = 3$ or $x = -2$

The numbers -2 and 3 divide a number line into three intervals.

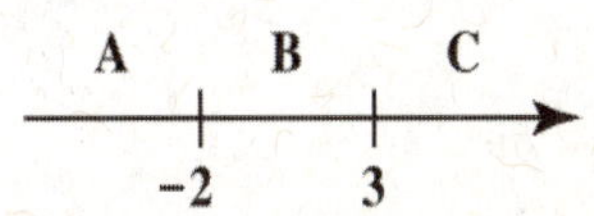

Test a number from each interval in the original inequality.

$$x^2 - x - 6 \le 0$$

Interval A: Let $x = -3$.

$$(-3)^2 - (-3) - 6 \overset{?}{\le} 0$$
$$6 \le 0 \quad \textit{False}$$

Interval B: Let $x = 0$.

$$0^2 - 0 - 6 \overset{?}{\le} 0$$
$$-6 \le 0 \quad \textit{True}$$

Interval C: Let $x = 4$.

$$4^2 - 4 - 6 \overset{?}{\le} 0$$
$$6 \le 0 \quad \textit{False}$$

The numbers in Interval B, including the endpoints -2 and 3 because of $\le$, are solutions.

The solution set is $[-2, 3]$.

13.
$$\frac{4}{x-3} - \frac{6}{x+3} = \frac{24}{x^2-9}$$
$$\frac{4}{x-3} - \frac{6}{x+3} = \frac{24}{(x+3)(x-3)}$$

Multiply by the LCD, $(x+3)(x-3)$. $(x \ne \pm 3)$

$$4(x+3) - 6(x-3) = 24$$
$$4x + 12 - 6x + 18 = 24$$
$$-2x + 30 = 24$$
$$-2x = -6$$
$$x = 3$$

Since x cannot equal 3, the solution set is $\emptyset$.

14.
$$6x^2 + 5x = 8$$
$$6x^2 + 5x - 8 = 0$$

Use the quadratic formula with $a = 6$, $b = 5$, and $c = -8$.

$$x = \frac{-b \pm \sqrt{b^2 - 4ac}}{2a}$$
$$x = \frac{-5 \pm \sqrt{5^2 - 4(6)(-8)}}{2(6)}$$
$$= \frac{-5 \pm \sqrt{25 + 192}}{12}$$
$$= \frac{-5 \pm \sqrt{217}}{12}$$

Solution set: $\left\{\frac{-5 \pm \sqrt{217}}{12}\right\}$

15.
$$3^{2x-1} = 81$$
$$3^{2x-1} = 3^4$$
$$2x - 1 = 4 \quad \textit{Equate exponents.}$$
$$2x = 5$$
$$x = \tfrac{5}{2}$$

Check $x = \frac{5}{2}$: $3^{5-1} = 3^4 = 81$

The solution set is $\left\{\frac{5}{2}\right\}$.

16.
$$\log_8 x + \log_8 (x+2) = 1$$
$$\log_8 x(x+2) = 1 \quad \textit{Product rule}$$
$$x(x+2) = 8^1 \quad \textit{Exponential form}$$
$$x^2 + 2x - 8 = 0$$
$$(x+4)(x-2) = 0$$

$$x + 4 = 0 \quad \text{or} \quad x - 2 = 0$$
$$x = -4 \quad \text{or} \quad x = 2$$

$x \ne -4$ because $\log_8 (-4)$ does not exist.

Check $x = 2$:
$$\log_8 2 + \log_8 4 = \log_8 (2 \cdot 4) = \log_8 8 = 1$$

The solution set is $\{2\}$.

17. $(4p + 2)(5p - 3)$

F O I L

$$= 20p^2 - 12p + 10p - 6$$
$$= 20p^2 - 2p - 6$$

18.
$$(3k - 7)^2 = (3k)^2 - 2(3k)(7) + 7^2$$
$$= 9k^2 - 42k + 49$$

19.
$$(2m^3 - 3m^2 + 8m) - (7m^3 + 5m - 8)$$
$$= 2m^3 - 7m^3 - 3m^2 + 8m - 5m + 8$$
$$= -5m^3 - 3m^2 + 3m + 8$$

20. Divide $6t^4 + 5t^3 - 18t^2 + 14t - 1$ by $3t - 2$.

$$\begin{array}{r|rrrrr}
 & 2t^3 & +\,3t^2 & -\,4t & +\,2 & \\
3t-2 & 6t^4 & +\,5t^3 & -\,18t^2 & +\,14t & -\,1 \\
 & 6t^4 & -\,4t^3 & & & \\ \hline
 & & 9t^3 & -\,18t^2 & & \\
 & & 9t^3 & -\,6t^2 & & \\ \hline
 & & & -\,12t^2 & +\,14t & \\
 & & & -\,12t^2 & +\,8t & \\ \hline
 & & & & 6t & -\,1 \\
 & & & & 6t & -\,4 \\ \hline
 & & & & & 3
\end{array}$$

Remainder

Answer: $2t^3 + 3t^2 - 4t + 2 + \dfrac{3}{3t - 2}$

21. $$6z^3+5z^2-4z = z(6z^2+5z-4) \\ = z(3z+4)(2z-1)$$

22. $$49a^4-9b^2 = (7a^2)^2-(3b)^2 \\ = (7a^2+3b)(7a^2-3b)$$

23. $$c^3+27d^3 = c^3+(3d)^3 \\ = (c+3d)(c^2-3cd+9d^2)$$

24. $$\left(\frac{2}{3}\right)^{-2} = \left(\frac{3}{2}\right)^2 = \frac{3}{2}\cdot\frac{3}{2} = \frac{9}{4}$$

25. $$\frac{(3p^2)^3(-2p^6)}{4p^3(5p^7)} = \frac{3^3p^6(-2)p^6}{20p^{10}} \\ = \frac{-54p^{12}}{20p^{10}} \\ = -\frac{27}{10}p^{12-10} \\ = -\frac{27p^2}{10}$$

26. $$\frac{x^2-16}{x^2+2x-8} \div \frac{x-4}{x+7} \\ = \frac{x^2-16}{x^2+2x-8}\cdot\frac{x+7}{x-4} \\ = \frac{(x+4)(x-4)(x+7)}{(x+4)(x-2)(x-4)} \\ = \frac{x+7}{x-2}$$

27. $$\frac{5}{p^2+3p} - \frac{2}{p^2-4p} \\ = \frac{5}{p(p+3)} - \frac{2}{p(p-4)}$$

The LCD is $p(p+3)(p-4)$.

$$= \frac{5(p-4)}{p(p+3)(p-4)} - \frac{2(p+3)}{p(p-4)(p+3)} \\ = \frac{5p-20-2p-6}{p(p+3)(p-4)} \\ = \frac{3p-26}{p(p+3)(p-4)}$$

28. $$5\sqrt{72}-4\sqrt{50} = 5\sqrt{36\cdot 2}-4\sqrt{25\cdot 2} \\ = 5\cdot 6\sqrt{2}-4\cdot 5\sqrt{2} \\ = 30\sqrt{2}-20\sqrt{2} \\ = 10\sqrt{2}$$

29. $$(8+3i)(8-3i) = 8^2-(3i)^2 \\ = 64-9i^2 \\ = 64-9(-1) \\ = 64+9 = 73$$

30. Let $(x_1,y_1)=(4,-5)$ and $(x_2,y_2)=(-12,-17)$. Then

$$m = \frac{y_2-y_1}{x_2-x_1} = \frac{-17-(-5)}{-12-4} = \frac{-12}{-16} = \frac{3}{4}.$$

The slope is $\frac{3}{4}$.

31. To find the equation of the line through $(-2,10)$ and parallel to $3x+y=7$, find the slope of

$$3x+y=7 \\ y=-3x+7.$$

The slope is -3, so a line parallel to it also has slope -3. Use $m=-3$ and $(x_1,y_1)=(-2,10)$ in the point-slope form.

$$y-y_1 = m(x-x_1) \\ y-10 = -3[x-(-2)] \\ y-10 = -3(x+2)$$

Write in standard form.

$$y-10 = -3x-6 \\ 3x+y=4$$

Alternative solution: The line must be of the form $3x+y=k$ since it is parallel to $3x+y=7$. Substitute -2 for x and 10 for y to find k.

$$3(-2)+10 = k \\ 4 = k$$

The equation is $3x+y=4$.

32. $\{(-3,2),(-2,6),(0,4),(1,2),(2,6)\}$

(a) The set of ordered pairs is a function since every first coordinate is paired with a unique second coordinate.

(b) The domain is the set of *first* coordinates, that is, $\{-3,-2,0,1,2\}$.

(c) The range is the set of *second* coordinates, that is, $\{2,6,4\}$.

33. $$y = 5x+3 \quad (1) \\ 2x+3y=-8 \quad (2)$$

From equation (1), substitute $5x+3$ for y in equation (2). Then solve for x.

$$2x+3y=-8 \quad (2) \\ 2x+3(5x+3)=-8 \\ 2x+15x+9=-8 \\ 17x=-17 \\ x=-1$$

From (1), $y=5(-1)+3=-2$.

The solution $(-1,-2)$ checks.

The solution set is $\{(-1,-2)\}$.

34.
$$\begin{aligned} x + 2y + z &= 8 \\ 2x - y + 3z &= 15 \\ -x + 3y - 3z &= -11 \end{aligned}$$

Write the augmented matrix.

$$\left[\begin{array}{ccc|c} 1 & 2 & 1 & 8 \\ 2 & -1 & 3 & 15 \\ -1 & 3 & -3 & -11 \end{array}\right]$$

$$\left[\begin{array}{ccc|c} 1 & 2 & 1 & 8 \\ 0 & -5 & 1 & -1 \\ 0 & 5 & -2 & -3 \end{array}\right] \quad \begin{array}{r} \\ -2R_1 + R_2 \\ R_1 + R_3 \end{array}$$

$$\left[\begin{array}{ccc|c} 1 & 2 & 1 & 8 \\ 0 & 1 & -\frac{1}{5} & \frac{1}{5} \\ 0 & 5 & -2 & -3 \end{array}\right] \quad -\tfrac{1}{5}R_2$$

$$\left[\begin{array}{ccc|c} 1 & 2 & 1 & 8 \\ 0 & 1 & -\frac{1}{5} & \frac{1}{5} \\ 0 & 0 & -1 & -4 \end{array}\right] \quad -5R_2 + R_3$$

$$\left[\begin{array}{ccc|c} 1 & 2 & 1 & 8 \\ 0 & 1 & -\frac{1}{5} & \frac{1}{5} \\ 0 & 0 & 1 & 4 \end{array}\right] \quad -R_3$$

This matrix gives the system

$$\begin{aligned} x + 2y + z &= 8 \\ y - \tfrac{1}{5}z &= \tfrac{1}{5} \\ z &= 4. \end{aligned}$$

Substitute $z = 4$ in the second equation.

$$\begin{aligned} y - \tfrac{1}{5}(4) &= \tfrac{1}{5} \\ y &= \tfrac{1}{5} + \tfrac{4}{5} = 1 \end{aligned}$$

Substitute $y = 1$ and $z = 4$ in the first equation.

$$\begin{aligned} x + 2(1) + 4 &= 8 \\ x + 6 &= 8 \\ x &= 2 \end{aligned}$$

The solution set is $\{(2, 1, 4)\}$.

35.
$$\begin{aligned} xy &= -5 \quad (1) \\ 2x + y &= 3 \quad (2) \end{aligned}$$

Solve equation (2) for y.

$$y = -2x + 3 \quad (3)$$

Substitute $-2x + 3$ for y in equation (1).

$$\begin{aligned} xy &= -5 \quad (1) \\ x(-2x + 3) &= -5 \\ -2x^2 + 3x &= -5 \\ -2x^2 + 3x + 5 &= 0 \\ 2x^2 - 3x - 5 &= 0 \\ (2x - 5)(x + 1) &= 0 \end{aligned}$$

$$\begin{aligned} 2x - 5 = 0 \quad &\text{or} \quad x + 1 = 0 \\ x = \tfrac{5}{2} \quad &\text{or} \quad x = -1 \end{aligned}$$

Substitute these values for x in equation (3) to find y.
If $x = \frac{5}{2}$, then $y = -2\left(\frac{5}{2}\right) + 3 = -2$.

If $x = -1$, then $y = -2(-1) + 3 = 5$.

The solution set is $\left\{(-1, 5), \left(\frac{5}{2}, -2\right)\right\}$.

36. Let $x =$ the number of pounds of \$3 per lb nuts.

	Number of Pounds	Price per Pound	Value
\$3/lb nuts	x	3	$3x$
\$4.25/lb nuts	8	4.25	4.25(8)
Mixture	$x + 8$	4	$4(x + 8)$

The last column gives the equation.

$$\begin{aligned} 3x + 4.25(8) &= 4(x + 8) \\ 3x + 34 &= 4x + 32 \\ 2 &= x \end{aligned}$$

Use 2 pounds of the \$3 per pound nuts.

37. $x - 3y = 6$

Find the x- and y-intercepts. To find the x-intercept, let $y = 0$.

$$\begin{aligned} x - 3(0) &= 6 \\ x &= 6 \end{aligned}$$

The x-intercept is $(6, 0)$.

To find the y-intercept, let $x = 0$.

$$\begin{aligned} 0 - 3y &= 6 \\ y &= -2 \end{aligned}$$

The y-intercept is $(0, -2)$.

Plot the intercepts and draw the line through them.

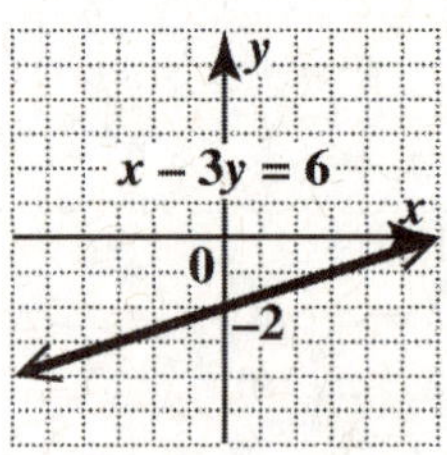

38. $4x - y < 4$

Graph the line $4x - y = 4$, which has intercepts $(0, -4)$ and $(1, 0)$, as a dashed line because the inequality involves $<$. Test $(0, 0)$, which yields $0 < 4$, a true statement. Shade the region on the side of the line that includes $(0, 0)$.

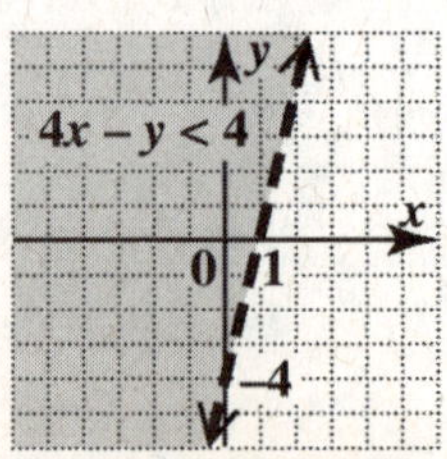

39. $f(x) = 2(x-2)^2 - 3$ is in
$f(x) = a(x-h)^2 + k$ form.

The graph is a vertical parabola with vertex (h, k) at $(2, -3)$. Since $a = 2 > 0$, the graph opens up. Also, $|a| = |2| = 2 > 1$, so the graph is narrower than the graph of $f(x) = x^2$. The points $(0, 5)$ and $(4, 5)$ are on the graph.

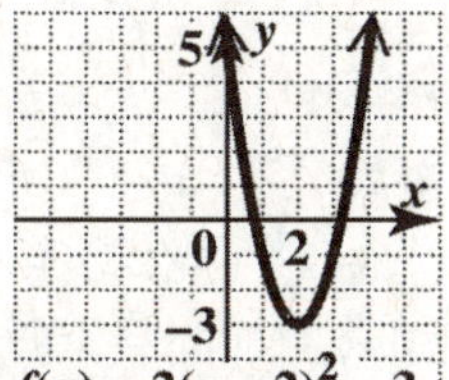

$f(x) = 2(x-2)^2 - 3$

40. $\dfrac{x^2}{9} + \dfrac{y^2}{25} = 1$ is in $\dfrac{x^2}{a^2} + \dfrac{y^2}{b^2} = 1$ form with $a = 3$ and $b = 5$. The graph is an ellipse centered at $(0, 0)$ with x-intercepts $(3, 0)$ and $(-3, 0)$ and y-intercepts $(0, 5)$ and $(0, -5)$. Plot the intercepts and draw the ellipse through them.

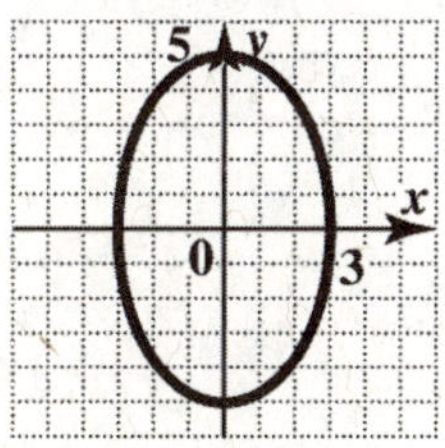

$\dfrac{x^2}{9} + \dfrac{y^2}{25} = 1$

41.
$$x^2 - y^2 = 9$$
$$\frac{x^2}{9} - \frac{y^2}{9} = 1$$

The graph is a hyperbola centered at $(0, 0)$ with x-intercepts $(3, 0)$ and $(-3, 0)$. The asymptotes are $y = \pm x$. Draw the right and left branches through the intercepts and approaching the asymptotes.

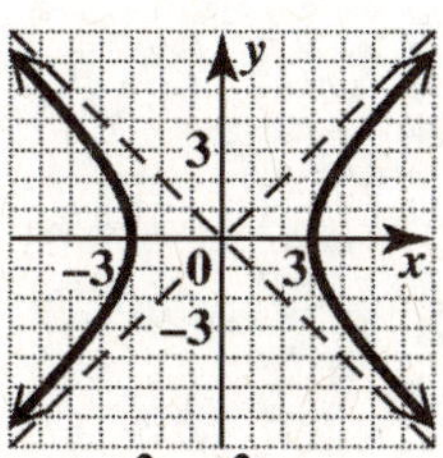

$x^2 - y^2 = 9$

42. Graph $g(x) = \left(\frac{1}{3}\right)^x$. Make a table of values.

x	-2	-1	0	1	2
$g(x)$	9	3	1	$\frac{1}{3}$	$\frac{1}{9}$

Plot these points, and draw a smooth decreasing exponential curve through them.

$g(x) = \left(\frac{1}{3}\right)^x$

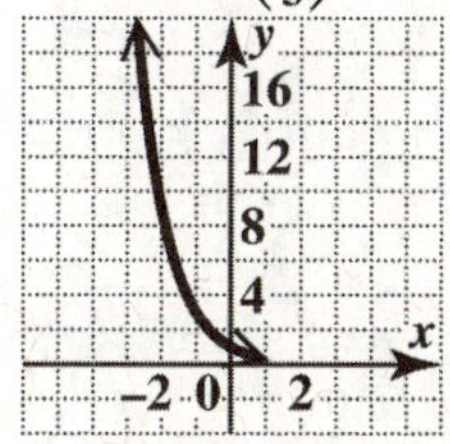

43. Graph $y = \log_{1/3} x$.
Change to exponential form.

$$\left(\tfrac{1}{3}\right)^y = x$$

This is the inverse of the graph of $g(x) = y = \left(\frac{1}{3}\right)^x$ in Exercise 42. To find points on the graph, interchange the x- and y-values in the table.

x	9	3	1	$\frac{1}{3}$	$\frac{1}{9}$
y	-2	-1	0	1	2

Plot these points, and draw a smooth decreasing logarithmic curve through them.

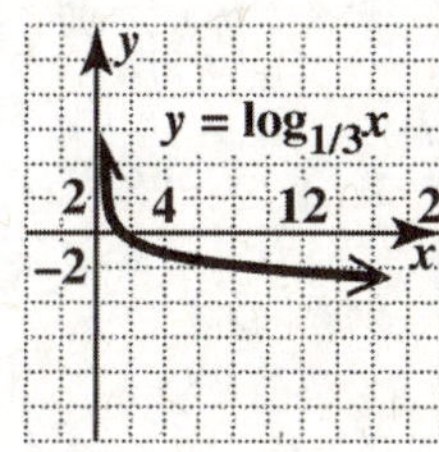

44. The graph of $f(x) = 9x + 5$ is a line. To find the inverse, replace $f(x)$ with y.

$$y = 9x + 5$$

Interchange x and y.

$$x = 9y + 5$$

Solve for y.

$$x - 5 = 9y$$
$$\frac{x-5}{9} = y$$

Replace y with $f^{-1}(x)$.

$$f^{-1}(x) = \frac{x-5}{9}, \quad \text{or} \quad f^{-1}(x) = \frac{1}{9}x - \frac{5}{9}$$

45. Center at $(-5, 12)$; radius 9

Use the equation of a circle with $h = -5$, $k = 12$, and $r = 9$.

$$(x-h)^2 + (y-k)^2 = r^2$$
$$[x-(-5)]^2 + (y-12)^2 = 9^2$$
$$(x+5)^2 + (y-12)^2 = 81$$

46.
$$a_n = 5n - 12$$
$$a_1 = 5(1) - 12 = 5 - 12 = -7$$
$$a_2 = 5(2) - 12 = 10 - 12 = -2$$
$$a_3 = 5(3) - 12 = 15 - 12 = 3$$
$$a_4 = 5(4) - 12 = 20 - 12 = 8$$
$$a_5 = 5(5) - 12 = 25 - 12 = 13$$

Answer: $-7, -2, 3, 8, 13$

47. **(a)** $a_1 = 8, d = 2$

$$S_n = \frac{n}{2}[2a_1 + (n-1)d]$$
$$S_6 = \frac{6}{2}[2(8) + (6-1)2]$$
$$= 3(16+10) = 3(26) = 78$$

(b) $15 - 6 + \frac{12}{5} - \frac{24}{25} + \cdots$

This is an infinite geometric series with $a_1 = 15$ and $r = \frac{-6}{15} = -\frac{2}{5}$. The sum is

$$S = \frac{a_1}{1-r} = \frac{15}{1-\left(-\frac{2}{5}\right)} = \frac{15}{\frac{7}{5}} = 15 \cdot \frac{5}{7} = \frac{75}{7}.$$

48.
$$\sum_{i=1}^{4} 3i = 3\sum_{i=1}^{4} i = 3(1+2+3+4)$$
$$= 3(10) = 30$$

49.
$$(2a-1)^5 = [2a + (-1)]^5$$
$$= (2a)^5 + \frac{5!}{1!\,4!}(2a)^4(-1)^1$$
$$+ \frac{5!}{2!\,3!}(2a)^3(-1)^2 + \frac{5!}{3!\,2!}(2a)^2(-1)^3$$
$$+ \frac{5!}{4!\,1!}(2a)^1(-1)^4 + (-1)^5$$
$$= 32a^5 + 5(16a^4)(-1) + 10(8a^3)(1)$$
$$+ 10(4a^2)(-1) + 5(2a)(1) + (-1)$$
$$= 32a^5 - 80a^4 + 80a^3 - 40a^2 + 10a - 1$$

50. The fourth term ($r = 4$, so $r - 1 = 3$) of $\left(3x^4 - \frac{1}{2}y^2\right)^5$ is

$$\frac{5!}{3!\,(5-3)!}\left(3x^4\right)^{5-3}\left(-\frac{1}{2}y^2\right)^3$$
$$= \frac{5!}{3!\,2!}\left(3x^4\right)^2\left(-\frac{1}{2}\right)^3\left(y^2\right)^3$$
$$= 10\left(9x^8\right)\left(-\tfrac{1}{8}\right)y^6$$
$$= -\frac{45x^8y^6}{4}.$$

APPENDIX A SETS

Appendix A Now Try Exercises

N1. The set of *odd* natural numbers less than 13 is

$$\{1, 3, 5, 7, 9, 11\}.$$

N2. **(a)** The set of negative integers
One way to list the elements is
$\{-1, -2, -3, -4, \dots\}$. The set is *infinite*.

(b) The set of even natural numbers between 11 and 19 is $\{12, 14, 16, 18\}$. The set is *finite*.

N3. **(a)** $B \subseteq A$ is *true* since all the elements of $B = \{1, 5, 7, 9\}$ are also elements of $A = \{1, 3, 5, 7, 9, 11\}$.

(b) $C \subseteq B$ is *false* since $C = \{1, 9, 11\}$ has an element, 11, which is not in B.

(c) $C \nsubseteq A$ (C is *not* a subset of A) is *false* since all the elements of C are also elements of A.

N4. $U = \{2, 4, 6, 8, 10, 12, 14\}$

$M = \{2, 10, 12, 14\}$, so $M' = \{4, 6, 8\}$, which is the set of elements that are in U but not in M.

N5. $M = \{1, 3, 5, 7, 9\}$ and $N = \{0, 3, 6, 9\}$, so $M \cup N = \{0, 1, 3, 5, 6, 7, 9\}$, the set of elements that are in *either M or N*.

N6. $M = \{1, 3, 5, 7, 9\}$ and $N = \{0, 3, 6, 9\}$, so $M \cap N = \{3, 9\}$, the set of elements that are in *both M and N*.

N7. $U = \{1, 2, 4, 5, 7, 8, 9, 10\}$, $A = \{1, 4, 7, 9, 10\}$, $B = \{2, 5, 8\}$, and $C = \{5\}$.

(a) $B \cup C = \{2, 5, 8\} = B$, the set of elements that are in either B or C.

(b) $A \cap B = \emptyset$ (the empty set), the set of elements that are in both A and B.

(c) $C' = \{1, 2, 4, 7, 8, 9, 10\}$, the set of elements that are in U but not in C.

Appendix A Section Exercises

1. The set of all natural numbers less than 8 is

$$\{1, 2, 3, 4, 5, 6, 7\}.$$

3. The set of seasons is

$$\{\text{winter, spring, summer, fall}\}.$$

The seasons may be written in any order within the braces.

5. To date, there have been no women presidents, so this set is the empty set, written $\emptyset$, or $\{\ \}$.

7. The set of letters of the alphabet between K and M is {L}.

9. The set of positive even integers is

$$\{2, 4, 6, 8, 10, \dots\}.$$

11. The sets in Exercises 9 and 10 are infinite, since each contains an unlimited number of elements.

13. $5 \in \{1, 2, 5, 8\}$ ▪ 5 is an element of the set, so the statement is true.

15. $2 \in \{1, 3, 5, 7, 9\}$ ▪ 2 is not an element of the set, so the statement is false.

17. $7 \notin \{2, 4, 6, 8\}$ ▪ 7 is not an element of the set, so the statement is true.

19. $\{2, 4, 9, 12, 13\} = \{13, 12, 9, 4, 2\}$ ▪ The two sets have exactly the same elements, so they are equal. The statement is true. (The order in which the elements are written does not matter.)

21. $A \subseteq U$ ▪

Since all the elements of $A = \{1, 3, 4, 5, 7, 8\}$ are elements of $U = \{1, 2, 3, 4, 5, 6, 7, 8, 9, 10\}$, the statement $A \subseteq U$ is true.

23. $\emptyset \subseteq A$ ▪ Since the empty set contains no elements, the empty set is a subset of every set. The statement $\emptyset \subseteq A$ is true.

25. $C \subseteq A$ ▪

Since all the elements of $C = \{1, 3, 5, 7\}$ are elements of $A = \{1, 3, 4, 5, 7, 8\}$, the statement $C \subseteq A$ is true.

27. $D \subseteq B$ ▪

Since 1 and 3 are elements of $D = \{1, 2, 3\}$ but are not elements of $B = \{2, 4, 6, 8\}$, the statement $D \subseteq B$ is false.

29. $D \nsubseteq E$ ▪

Since 1 and 2 are elements of $D = \{1, 2, 3\}$ and are not elements of $E = \{3, 7\}$, D is not a subset of E, so the statement $D \nsubseteq E$ is true.

31. There are exactly 4 subsets of E. ▪

Since $E = \{3, 7\}$ has 2 elements, the number of subsets is $2^2 = 4$. The statement is true.

33. There are exactly 12 subsets of C. ▪

Since $C = \{1, 3, 5, 7\}$ has 4 elements, the number of subsets is $2^4 = 16$. The statement is false.

35. $\{4, 6, 8, 12\} \cap \{6, 8, 14, 17\} = \{6, 8\}$ ▪

The symbol $\cap$ means the intersection of the two sets, which is the set of elements that belong to both sets. Since 6 and 8 are the only elements belonging to both sets, the statement is true.

37. $\{3, 1, 0\} \cap \{0, 2, 4\} = \{0\}$ ▪ Only 0 belongs to both sets, so the statement is true.

39. $\{3, 9, 12\} \cap \emptyset = \{3, 9, 12\}$ ▪

Since 3, 9, and 12 are not elements of the empty set, they are not in the intersection of the two sets. The intersection of any set with the empty set is the empty set. The statement is false.

41. $\{3, 5, 7, 9\} \cup \{4, 6, 8\} = \emptyset$ ▪

The union of the two sets is the set of all elements that belong to either one of the sets or to both sets. Thus, $\{3, 5, 7, 9\} \cup \{4, 6, 8\} = \{3, 4, 5, 6, 7, 8, 9\} \neq \emptyset$. The statement is false.

43. $\{4, 9, 11, 7, 3\} \cup \{1, 2, 3, 4, 5\}$

$$= \{1, 2, 3, 4, 5, 7, 9, 11\}$$ ▪

The union of the two sets is the set of all elements that belong to either one of the sets or to both sets. The statement is true.

45. A' ▪ $U = \{\text{a, b, c, d, e, f, g, h}\}$
$A = \{\text{a, b, c, d, e, f}\}$

A' contains all elements in U that are not in A, so

$$A' = \{\text{g, h}\}.$$

47. C' ▪ $U = \{\text{a, b, c, d, e, f, g, h}\}$
$C = \{\text{a, f}\}$

C' contains all elements in U that are not in C, so

$$C' = \{\text{b, c, d, e, g, h}\}.$$

49. $A \cap B$ ▪ $A = \{\text{a, b, c, d, e, f}\}$, $B = \{\text{a, c, e}\}$

The intersection of A and B is the set of all elements belonging to both A and B, so

$$A \cap B = \{\text{a, c, e}\} = B.$$

51. $A \cap D$ ▪ $A = \{\text{a, b, c, d, e, f}\}$, $D = \{\text{d}\}$

Since d is the only element in both A and D,

$$A \cap D = \{\text{d}\} = D.$$

53. $B \cap C$ ▪ $B = \{\text{a, c, e}\}$, $C = \{\text{a, f}\}$

Since a is the only element that belongs to both sets, the intersection is the set with a as its only element, so

$$B \cap C = \{\text{a}\}.$$

55. $B \cup D$ ▪ $B = \{\text{a, c, e}\}$, $D = \{\text{d}\}$

The union of B and D is the set of elements belonging to either B or D or both, so

$$B \cup D = \{\text{a, c, d, e}\}.$$

57. $C \cup B$ ▪ $C = \{\text{a, f}\}$, $B = \{\text{a, c, e}\}$

The union of C and B is the set of elements belonging to either C or B or both, so

$$C \cup B = \{\text{a, c, e, f}\}.$$

59. $A \cap \emptyset$ ▪ Since $\emptyset$ has no elements, there is no element that belongs to both A and $\emptyset$, so the intersection is the empty set. Thus,

$$A \cap \emptyset = \emptyset.$$

61. $A = \{\text{a, b, c, d, e, f}\}$ $\qquad C = \{\text{a, f}\}$
$B = \{\text{a, c, e}\}$ $\qquad D = \{\text{d}\}$

Disjoint sets are sets which have no elements in common.

B and D are disjoint since they have no elements in common. Also, C and D are disjoint since they have no elements in common.

APPENDIX B REVIEW OF EXPONENTS, POLYNOMIALS, AND FACTORING (Transition from Beginning to Intermediate Algebra)

Appendix B Now Try Exercises

N1. **(a)** $(m^{-8}n^4)(m^4n^{-3}) = m^{-8+4}n^{4+(-3)}$
$= m^{-4}n^1$
$= \frac{1}{m^4} \cdot n^1$
$= \frac{n}{m^4}$

(b) $-8^0 + 8^0 = -1 + 1 = 0$

(c) $\frac{(p^{-3}q)^4}{(p^2q^5)^2} = \frac{(p^{-3})^4(q)^4}{(p^2)^2(q^5)^2}$
$= \frac{p^{-12}q^4}{p^4q^{10}}$
$= \frac{1}{p^{16}q^6}$

(d) $\left(\frac{2x^{-2}y}{x^2y^{-4}}\right)^{-4} = \left(\frac{x^2y^{-4}}{2x^{-2}y}\right)^4$
$= \frac{(x^2y^{-4})^4}{(2x^{-2}y)^4}$
$= \frac{(x^2)^4(y^{-4})^4}{2^4(x^{-2})^4(y)^4}$
$= \frac{x^8y^{-16}}{16x^{-8}y^4}$
$= \frac{x^8x^8}{16y^{16}y^4}$
$= \frac{x^{16}}{16y^{20}}$

N2. **(a)** $(3x^3 + x^2 - 5x - 6) + (-6x^3 + 2x^2 + 4x - 1)$
$= (3-6)x^3 + (1+2)x^2 + (-5+4)x + (-6-1)$
$= -3x^3 + 3x^2 - x - 7$

(b) Subtract.
$$\begin{array}{r} 4x^2 + 7x - 5 \\ \underline{-\ 5x^2 - 2x + 3} \end{array}$$

Change the sign of each term in $-5x^2 - 2x + 3$, and add.

$$\begin{array}{r} 4x^2 + 7x - 5 \\ \underline{5x^2 + 2x - 3} \\ 9x^2 + 9x - 8 \end{array}$$

N3. **(a)** $(6x - 5)(2x - 3)$
F O I L
$= 6x(2x) + 6x(-3) - 5(2x) - 5(-3)$
$= 12x^2 - 18x - 10x + 15$
$= 12x^2 - 28x + 15$

(b) $(4m - 3n)(4m + 3n) = (4m)^2 - (3n)^2$
$= 16m^2 - 9n^2$

(c) $(7z + 1)^2 = (7z)^2 + 2(7z)(1) + 1^2$
$= 49z^2 + 14z + 1$

(d) $(r + 3)(r^2 - 3r + 9)$
Multiply vertically.

$$\begin{array}{r} r^2 - 3r + 9 \\ \underline{r + 3} \\ 3r^2 - 9r + 27 \\ \underline{r^3 - 3r^2 + 9r \quad\ } \\ r^3 \qquad\qquad + 27 \end{array}$$

The product is $r^3 + 27$.

N4. **(a)** $5x^2 - 20x - 60$

The greatest common factor is 5, so

$5x^2 - 20x - 60 = 5(x^2 - 4x - 12)$.

To factor $x^2 - 4x - 12$, we recognize that the two numbers that have a product of -12 and a sum of -4 are -6 and 2. Thus,

$x^2 - 4x - 12 = (x - 6)(x + 2)$.

The complete factorization is

$5x^2 - 20x - 60 = 5(x - 6)(x + 2)$.

(b) $10t^2 + 13t - 3$

To find the factors, find two terms that multiply to give $10t^2$ ($10t$ and t, $5t$ and $2t$) and two terms that multiply to give -3 ($+3$ and -1, -3 and $+1$). Make sure that the sum of the outer and inner products in the factored form is $+13t$.

$10t^2 + 13t - 3 = (5t - 1)(2t + 3)$

(c) $49x^2 + 42x + 9 = (7x)^2 + 2(7x)(3) + 3^2$
$= (7x + 3)^2$

(d) $mn - 2n + 5m - 10$
$= (mn + 5m) + (-2n - 10)$
$= m(n + 5) - 2(n + 5)$
$= (n + 5)(m - 2)$

(e) $27x^3 - 1000$
$= (3x)^3 - 10^3$ *Difference of cubes*
$= (3x - 10)[(3x)^2 + 3x(10) + 10^2]$
$= (3x - 10)(9x^2 + 30x + 100)$

Appendix B Section Exercises

1. $(a^4b^{-3})(a^{-6}b^2) = (a^{4+(-6)})(b^{-3+2})$
$= a^{-2}b^{-1}$
$= \dfrac{1}{a^2b}$

3. $(5x^{-2}y)^2(2xy^4)^2 = (5^2x^{-4}y^2)(2^2x^2y^8)$
$= 25 \bullet 4x^{-4+2}y^{2+8}$
$= 100x^{-2}y^{10}$
$= \dfrac{100y^{10}}{x^2}$

5. $-6^0 + (-6)^0 = -1 \bullet 6^0 + 1$
$= -1 \bullet 1 + 1$
$= -1 + 1 = 0$

7. $\dfrac{(2w^{-1}x^2y^{-1})^3}{(4w^5x^{-2}y)^2} = \dfrac{2^3w^{-3}x^6y^{-3}}{4^2w^{10}x^{-4}y^2}$
$= \dfrac{8x^6x^4}{16w^{10}y^2w^3y^3}$
$= \dfrac{x^{10}}{2w^{13}y^5}$

9. $\left(\dfrac{-4a^{-2}b^4}{a^3b^{-1}}\right)^{-3} = \left(\dfrac{a^3b^{-1}}{-4a^{-2}b^4}\right)^3$
$= \dfrac{a^9b^{-3}}{(-4)^3a^{-6}b^{12}}$
$= \dfrac{a^9a^6}{-64b^{12}b^3}$
$= \dfrac{a^{15}}{-64b^{15}}$

11. $(7x^{-4}y^2z^{-2})^{-2}(7x^4y^{-1}z^3)^2$
$= (7^{-2}x^8y^{-4}z^4)(7^2x^8y^{-2}z^6)$
$= \dfrac{7^2x^8z^4x^8z^6}{7^2y^4y^2}$
$= \dfrac{x^{16}z^{10}}{y^6}$

13. $(2a^4 + 3a^3 - 6a^2 + 5a - 12)$
$+ (-8a^4 + 8a^3 - 14a^2 + 21a - 3)$
$= (2-8)a^4 + (3+8)a^3 + (-6-14)a^2$
$+ (5+21)a + (-12-3)$
$= -6a^4 + 11a^3 - 20a^2 + 26a - 15$

15. $(6x^3 - 12x^2 + 3x - 4)$
$- (-2x^3 + 6x^2 - 3x + 12)$
$= 6x^3 - 12x^2 + 3x - 4 + 2x^3 - 6x^2 + 3x - 12$
$= 8x^3 - 18x^2 + 6x - 16$

17. Add.
$$\begin{array}{rrr} 5x^2y & +\ 2xy^2 & +\ y^3 \\ -4x^2y & -\ 3xy^2 & +\ 5y^3 \\ \hline x^2y & -\ xy^2 & +\ 6y^3 \end{array}$$

19. $3(5x^2 - 12x + 4) - 2(9x^2 + 13x - 10)$
$= 15x^2 - 36x + 12 - 18x^2 - 26x + 20$
$= -3x^2 - 62x + 32$

21. Subtract.
$$\begin{array}{rrrr} 6x^3 & -\ 2x^2 & +\ 3x & -\ 1 \\ \hline -4x^3 & +\ 2x^2 & -\ 6x & +\ 3 \end{array}$$

Change the sign of each term in $-4x^3 + 2x^2 - 6x + 3$, and add.
$$\begin{array}{rrrr} 6x^3 & -\ 2x^2 & +\ 3x & -\ 1 \\ 4x^3 & -\ 2x^2 & +\ 6x & -\ 3 \\ \hline 10x^3 & -\ 4x^2 & +\ 9x & -\ 4 \end{array}$$

23. $(3x + 1)(2x - 7)$
$= (3x)(2x) + (3x)(-7)$
$+ 1(2x) + 1(-7)$ *FOIL*
$= 6x^2 - 21x + 2x - 7$
$= 6x^2 - 19x - 7$

25. $(4x - 1)(x - 2)$
$= 4x^2 - 8x - x + 2$
$= 4x^2 - 9x + 2$

27. $(4t + 3)(4t - 3) = (4t)^2 - 3^2$
$= 16t^2 - 9$

29. $(2y^2 + 4)(2y^2 - 4) = (2y^2)^2 - 4^2$
$= 4y^4 - 16$

31. $(4x - 3)^2 = (4x)^2 - 2(4x)(3) + 3^2$
$= 16x^2 - 24x + 9$

33. $(6r + 5y)^2 = (6r)^2 + 2(6r)(5y) + (5y)^2$
$= 36r^2 + 60ry + 25y^2$

35. $(c + 2d)(c^2 - 2cd + 4d^2)$
Multiply vertically.
$$\begin{array}{rrrr} & c^2 & -\ 2cd & +\ 4d^2 \\ & & c & +\ 2d \\ \hline & 2c^2d & -\ 4cd^2 & +\ 8d^3 \\ c^3 & -\ 2c^2d & +\ 4cd^2 & \\ \hline c^3 & & & +\ 8d^3 \end{array}$$

37. $(4x - 1)(16x^2 + 4x + 1)$
Multiply vertically.
$$\begin{array}{rrrr} & 16x^2 & +\ 4x & +\ 1 \\ & & 4x & -\ 1 \\ \hline & -\ 16x^2 & -\ 4x & -\ 1 \\ 64x^3 & +\ 16x^2 & +\ 4x & \\ \hline 64x^3 & & & -\ 1 \end{array}$$

39. $(7t + 5s)(2t^2 + 5st - s^3)$
Multiply vertically.

$$\begin{array}{rrrr} & 2t^2 + & 5st - & s^2 \\ & & 7t + & 5s \\ \hline & 10st^2 + & 25s^2t - & 5s^3 \\ 14t^3 + & 35st^2 - & 7s^2t & \\ \hline 14t^3 + & 45st^2 + & 18s^2t - & 5s^3 \end{array}$$

41. $8x^3y^4 + 12x^2y^3 + 36xy^4$
The GCF is $4xy^3$.
$= 4xy^3(2x^2y + 3x + 9y)$

43. $x^2 - 2x - 15 = (x+3)(x-5)$

45. $2x^2 - 9x - 18 = (2x+3)(x-6)$

47. $36t^2 - 25 = (6t)^2 - 5^2$
$= (6t+5)(6t-5)$

49. $16t^2 + 24t + 9 = (4t)^2 + 2(4t)(3) + 3^2$
$= (4t+3)^2$

51. $4m^2p - 12mnp + 9n^2p$
$= p(4m^2 - 12mn + 9n^2)$
$= p[(2m)^2 - 2(2m)(3n) + (3n)^2]$
$= p(2m - 3n)^2$

53. $x^3 + 1$
$= (x)^3 + (1)^3$
$= (x+1)[(x)^2 - (x)(1) + (1)^2]$ *Sum of cubes*
$= (x+1)(x^2 - x + 1)$

55. $8t^3 + 125 = (2t)^3 + (5)^3$
$= (2t+5)[(2t)^2 - (2t)(5) + (5)^2]$
$= (2t+5)(4t^2 - 10t + 25)$

57. $t^6 - 125 = (t^2)^3 - (5)^3$
$= (t^2 - 5)[(t^2)^2 + (t^2)(5) + (5)^2]$
Difference of cubes
$= (t^2 - 5)(t^4 + 5t^2 + 25)$

59. $5xt + 15xr + 2yt + 6yr$
$= (5xt + 15xr) + (2yt + 6yr)$
$= 5x(t + 3r) + 2y(t + 3r)$
$= (t + 3r)(5x + 2y)$

61. $6ar + 12br - 5as - 10bs$
$= (6ar + 12br) + (-5as - 10bs)$
$= 6r(a + 2b) - 5s(a + 2b)$
$= (a + 2b)(6r - 5s)$

63. $t^4 - 1 = (t^2)^2 - 1^2$
$= (t^2 + 1)(t^2 - 1)$
$= (t^2 + 1)(t + 1)(t - 1)$

65. $4x^2 + 12xy + 9y^2 - 1$
The first three terms form a perfect square trinomial.
$= (4x^2 + 12xy + 9y^2) - 1$
$= [(2x)^2 + 2(2x)(3y) + (3y)^2] - 1$
$= (2x + 3y)^2 - 1^2$
Now factor the difference of squares.
$= [(2x + 3y) + 1][(2x + 3y) - 1]$
$= (2x + 3y + 1)(2x + 3y - 1)$

67. $4x^2 - 28x + 40$
$= 4(x^2 - 7x + 10)$
$= 4(x - 5)(x - 2)$

APPENDIX C SYNTHETIC DIVISION

Appendix C Now Try Exercises

N1. $\dfrac{4x^3+18x^2+19x+7}{x+3}$

$x+3=x-(-3)$, so use -3.

$$\begin{array}{r|rrrr} -3 & 4 & 18 & 19 & 7 \\ & & -12 & -18 & -3 \\ \hline & 4 & 6 & 1 & 4 \end{array} \leftarrow \textit{Remainder}$$

The quotient polynomial is $4x^2+6x+1$ and the remainder is 4.

Answer: $4x^2+6x+1+\dfrac{4}{x+3}$

N2. $\dfrac{-3x^4+13x^3-6x^2+31}{x-4}$

Insert a 0 for the missing x-term.

$$\begin{array}{r|rrrrr} 4 & -3 & 13 & -6 & 0 & 31 \\ & & -12 & 4 & -8 & -32 \\ \hline & -3 & 1 & -2 & -8 & -1 \end{array} \leftarrow \textit{Remainder}$$

Answer: $-3x^3+x^2-2x-8+\dfrac{-1}{x-4}$

N3. Evaluate $P(-2)$ for $P(x)=3x^3-2x^2+5x+30$.

Divide $P(x)$ by $x-(-2)=x+2$.

$$\begin{array}{r|rrrr} -2 & 3 & -2 & 5 & 30 \\ & & -6 & 16 & -42 \\ \hline & 3 & -8 & 21 & -12 \end{array} \leftarrow \textit{Remainder}$$

By the remainder theorem, $P(-2)=-12$.

N4. To decide whether -4 is a solution of $P(x)=0$, divide the polynomial by $x+4$. If the remainder is 0, then -4 is a solution. Otherwise, it is not.

$P(x)=5x^3+19x^2-2x+8$

$$\begin{array}{r|rrrr} -4 & 5 & 19 & -2 & 8 \\ & & -20 & 4 & -8 \\ \hline & 5 & -1 & 2 & 0 \end{array} \leftarrow \textit{Remainder}$$

The remainder is 0, so -4 is a solution of $P(x)=0$.

Appendix C Section Exercises

1. $\dfrac{x^2-6x+5}{x-1}$

$$\begin{array}{r|rrr} 1 & 1 & -6 & 5 \\ & & 1 & -5 \\ \hline & 1 & -5 & 0 \end{array}$$

$\leftarrow$ *Coefficients of numerator*

Write the answer from the bottom row.

↓ ↓

$x-5$

Answer: $x-5$

3. $\dfrac{4m^2+19m-5}{m+5}$

$m+5=m-(-5)$, so use -5.

$$\begin{array}{r|rrr} -5 & 4 & 19 & -5 \\ & & -20 & 5 \\ \hline & 4 & -1 & 0 \end{array}$$

Answer: $4m-1$

5. $\dfrac{2a^2+8a+13}{a+2}$

$a+2=a-(-2)$, so use -2.

$$\begin{array}{r|rrr} -2 & 2 & 8 & 13 \\ & & -4 & -8 \\ \hline & 2 & 4 & 5 \end{array} \leftarrow \textit{Remainder}$$

The quotient polynomial is $2a+4$ and the remainder is 5.

Answer: $2a+4+\dfrac{5}{a+2}$

7. $(p^2-3p+5)\div(p+1)$

$$\begin{array}{r|rrr} -1 & 1 & -3 & 5 \\ & & -1 & 4 \\ \hline & 1 & -4 & 9 \end{array}$$

Answer: $p-4+\dfrac{9}{p+1}$

9. $\dfrac{4a^3-3a^2+2a-3}{a-1}$

$$\begin{array}{r|rrrr} 1 & 4 & -3 & 2 & -3 \\ & & 4 & 1 & 3 \\ \hline & 4 & 1 & 3 & 0 \end{array}$$

Answer: $4a^2+a+3$

11. $(x^5 - 2x^3 + 3x^2 - 4x - 2) \div (x - 2)$

Insert 0 for the missing x^4-term.

$$\begin{array}{r|rrrrrr} 2 & 1 & 0 & -2 & 3 & -4 & -2 \\ & & 2 & 4 & 4 & 14 & 20 \\ \hline & 1 & 2 & 2 & 7 & 10 & 18 \end{array} \leftarrow \textit{Remainder}$$

Answer: $x^4 + 2x^3 + 2x^2 + 7x + 10 + \dfrac{18}{x - 2}$

13. $(-4r^6 - 3r^5 - 3r^4 + 5r^3 - 6r^2 + 3r + 3) \div (r - 1)$

$$\begin{array}{r|rrrrrrr} 1 & -4 & -3 & -3 & 5 & -6 & 3 & 3 \\ & & -4 & -7 & -10 & -5 & -11 & -8 \\ \hline & -4 & -7 & -10 & -5 & -11 & -8 & -5 \end{array} \leftarrow \textit{Remainder}$$

Answer:

$-4r^5 - 7r^4 - 10r^3 - 5r^2 - 11r - 8 + \dfrac{-5}{r - 1}$

15. $(-3y^5 + 2y^4 - 5y^3 - 6y^2 - 1) \div (y + 2)$

Insert 0 for the missing y-term.

$$\begin{array}{r|rrrrrr} -2 & -3 & 2 & -5 & -6 & 0 & -1 \\ & & 6 & -16 & 42 & -72 & 144 \\ \hline & -3 & 8 & -21 & 36 & -72 & 143 \end{array} \leftarrow \textit{Remainder}$$

Answer:

$-3y^4 + 8y^3 - 21y^2 + 36y - 72 + \dfrac{143}{y + 2}$

17. $P(x) = 2x^3 - 4x^2 + 5x - 3;\ k = 2$

To find $P(2)$, divide the polynomial by $x - 2$. $P(2)$ will be the remainder.

$$\begin{array}{r|rrrr} 2 & 2 & -4 & 5 & -3 \\ & & 4 & 0 & 10 \\ \hline & 2 & 0 & 5 & 7 \end{array} \leftarrow \textit{Remainder}$$

By the remainder theorem, $P(2) = 7$.

19. $P(x) = -x^3 - 5x^2 - 4x - 2;\ k = -4$

Divide by $x + 4$. The remainder is equal to $P(-4)$.

$$\begin{array}{r|rrrr} -4 & -1 & -5 & -4 & -2 \\ & & 4 & 4 & 0 \\ \hline & -1 & -1 & 0 & -2 \end{array} \leftarrow \textit{Remainder}$$

By the remainder theorem, $P(-4) = -2$.

21. $P(x) = 2x^3 - 4x^2 + 5x - 33;\ k = 3$

Divide by $x - 3$. The remainder is equal to $P(3)$.

$$\begin{array}{r|rrrr} 3 & 2 & -4 & 5 & -33 \\ & & 6 & 6 & 33 \\ \hline & 2 & 2 & 11 & 0 \end{array} \leftarrow \textit{Remainder}$$

By the remainder theorem, $P(3) = 0$.

23. By the remainder theorem, a zero remainder means that $P(k) = 0$. That is, k is a number that makes $P(x) = 0$.

25. Is $x = -2$ a solution of

$$x^3 - 2x^2 - 3x + 10 = 0?$$

To decide whether -2 is a solution to the given equation, divide the polynomial by $x + 2$.

$$\begin{array}{r|rrrr} -2 & 1 & -2 & -3 & 10 \\ & & -2 & 8 & -10 \\ \hline & 1 & -4 & 5 & 0 \end{array} \leftarrow \textit{Remainder}$$

Since the remainder is 0, -2 is a solution of the equation.

27. Is $x = -2$ a solution of

$$3x^3 + 2x^2 - 2x + 11 = 0?$$

To decide whether -2 is a solution to the given equation, divide the polynomial by $x + 2$.

$$\begin{array}{r|rrrr} -2 & 3 & 2 & -2 & 11 \\ & & -6 & 8 & -12 \\ \hline & 3 & -4 & 6 & -1 \end{array} \leftarrow \textit{Remainder}$$

Since the remainder is not 0, -2 is not a solution of the equation.

29. Is $x = -3$ a solution of

$$2x^3 - x^2 - 13x + 24 = 0?$$

To decide whether -3 is a solution to the given equation, divide the polynomial by $x + 3$.

$$\begin{array}{r|rrrr} -3 & 2 & -1 & -13 & 24 \\ & & -6 & 21 & -24 \\ \hline & 2 & -7 & 8 & 0 \end{array} \leftarrow \textit{Remainder}$$

Since the remainder is 0, -3 is a solution of the equation.

31. Is $x = -2$ a solution of

$$x^4 + 2x^3 - 3x^2 + 8x - 8 = 0?$$

To decide whether -2 is a solution to the given equation, divide the polynomial by $x + 2$.

$$\begin{array}{r|rrrrr} -2 & 1 & 2 & -3 & 8 & -8 \\ & & -2 & 0 & 6 & -28 \\ \hline & 1 & 0 & -3 & 14 & -36 \end{array} \leftarrow \textit{Remainder}$$

Since the remainder is not 0, -2 is not a solution of the equation.

♦♦♦ Relating Concepts 33–38 ♦♦♦

33. Factor $P(x) = 2x^2 + 5x - 12$. The two integers whose product is $2(-12) = -24$ and whose sum is 5 are 8 and -3.

$$\begin{aligned} &2x^2 + 5x - 12 \\ &= 2x^2 - 3x + 8x - 12 \\ &= x(2x-3) + 4(2x-3) \\ &= (2x-3)(x+4) \end{aligned}$$

34. Solve $P(x) = 2x^2 + 5x - 12 = 0$.

$$2x^2 + 5x - 12 = 0$$
$$(2x-3)(x+4) = 0$$

$$\begin{aligned} 2x - 3 &= 0 \quad \text{or} \quad x + 4 = 0 \\ 2x &= 3 \qquad\qquad x = -4 \\ x &= \tfrac{3}{2} \end{aligned}$$

The solution set is $\left\{-4, \frac{3}{2}\right\}$.

35. $$\begin{aligned} P(x) &= 2x^2 + 5x - 12 \\ P(-4) &= 2(-4)^2 + 5(-4) - 12 \\ &= 2(16) - 20 - 12 \\ &= 32 - 20 - 12 = 0 \end{aligned}$$

Another method: We can divide $P(x)$ by $x + 4$. The remainder is equal to $P(-4)$.

$$\begin{array}{r|rrr} -4 & 2 & 5 & -12 \\ & & -8 & 12 \\ \hline & 2 & -3 & 0 \end{array} \leftarrow \textit{Remainder}$$

By the remainder theorem, $P(-4) = 0$.

36. $$\begin{aligned} P(x) &= 2x^2 + 5x - 12 \\ P\left(\tfrac{3}{2}\right) &= 2\left(\tfrac{3}{2}\right)^2 + 5\left(\tfrac{3}{2}\right) - 12 \\ &= 2\left(\tfrac{9}{4}\right) + \tfrac{15}{2} - 12 \\ &= \tfrac{9}{2} + \tfrac{15}{2} - \tfrac{24}{2} = 0 \end{aligned}$$

37. If $P(a) = 0$, then $x - \underline{a}$ is a factor of $P(x)$.

38. $$\begin{aligned} Q(x) &= 3x^3 - 4x^2 - 17x + 6 \\ Q(3) &= 3(3)^3 - 4(3)^2 - 17(3) + 6 \\ &= 81 - 36 - 51 + 6 = 0 \end{aligned}$$

Since $Q(3) = 0$, $x - 3$ is a factor of $Q(x)$. To check, use synthetic division to see if 3 is a solution of the equation $Q(x) = 0$.

$$\begin{array}{r|rrrr} 3 & 3 & -4 & -17 & 6 \\ & & 9 & 15 & -6 \\ \hline & 3 & 5 & -2 & 0 \end{array}$$

Therefore, $x - 3$ is a factor of $Q(x)$ and

$$Q(x) = (x-3)(3x^2 + 5x - 2).$$

Now factor $3x^2 + 5x - 2$. The two integers whose product is $3(-2) = -6$ and whose sum is 5 are 6 and -1.

$$\begin{aligned} &3x^2 + 5x - 2 \\ &= 3x^2 - x + 6x - 2 \\ &= x(3x-1) + 2(3x-1) \\ &= (3x-1)(x+2) \end{aligned}$$

Thus, $Q(x)$ completely factored is

$$(x-3)(3x-1)(x+2).$$